Game Theory for Intelligent Network Control Paradigm

Sungwook Kim
Sogang University, South Korea

IGI Global
Scientific Publishing
Publishing Tomorrow's Research Today

Vice President of Editorial	Melissa Wagner
Director of Acquisitions	Mikaela Felty
Director of Book Development	Jocelynn Hessler
Production Manager	Mike Brehm
Cover Design	Jose Rosado

Published in the United States of America by
IGI Global Scientific Publishing
701 East Chocolate Avenue
Hershey, PA, 17033, USA
Tel: 717-533-8845
Fax: 717-533-7115
Website: https://www.igi-global.com E-mail: cust@igi-global.com

Library of Congress Cataloging-in-Publication Data

Names: Kim, Sungwook editor
Title: Game theory for intelligent network control paradigm / edited by: Sungwook Kim.
Description: Hershey PA : IGI Global Scientific Publishing, [2026] | Includes bibliographical references. | Summary: "The first part of my manuscript introduces the fundamental concepts of game theory while emphasizing the interconnection between game theory and intelligent learning algorithms in network control problems such as power control, access control, bandwidth adaptation, routing, and game theoretic solutions for the network management"-- Provided by publisher.
Identifiers: LCCN 2025001388 (print) | LCCN 2025001389 (ebook) | ISBN 9798337320137 hardcover | ISBN 9798337320144 paperback | ISBN 9798337320151 ebook
Subjects: LCSH: Intelligent control systems | Game theory
Classification: LCC TJ217.5 .G36 2025 (print) | LCC TJ217.5 (ebook) | DDC 004.6--dc23/eng/20250404
LC record available at https://lccn.loc.gov/2025001388
LC ebook record available at https://lccn.loc.gov/2025001389

British Cataloguing in Publication Data
A Cataloguing in Publication record for this book is available from the British Library.

Table of Contents

Preface

As the Teacher grew old and infirm, the disciples begged him not to die. Said the Teacher, "If I did not go, how would you ever see?"

"What is it we fail to see when you are with us?" they asked.

But the Teacher would not say.

When the moment of his death was near, they said, "What is it we will see when you are gone?"

With a twinkle in his eye, the Teacher said, "All I did was sit on the riverbank handing out river water. After I'm gone, I trust you will notice the river."

The Teacher was an advocate both of learning and of Wisdom.

"Learning," he said when asked, "is gotten by reading books or listening to lectures."

"And Wisdom?"

"By reading the book that is you."

He added as an afterthought: "Not an easy task at all, for every minute of the day brings a new edition of the book!"

"Laboratories and libraries, halls and porch and arch and learned lectures - all shall be of no avail if the wise heart and the seeing eye are absent."

"Because," said the Teacher solemnly, "people have learned to read printed books. They have forgotten the art of reading unprinted ones."

"Give us an example of an unprinted book."

- Anthony De Mello -

Figure 1.

The forthcoming 5G and beyond-5G (B5G) technologies are set to unlock vast potential, impacting not only the communications sector but society as a whole. These advanced networks are expected to be pivotal across diverse industries, including healthcare, manufacturing, agriculture, and entertainment. Additionally, the surge in user demand - driven by the rapid rollout and widespread adoption of various applications - has compelled network operators to upgrade their infrastructures. This evolving network environment calls for a complete reevaluation of network control and design strategies. Consequently, even though the first commercial 5G networks are already operational, numerous 5G standardization efforts are still underway. Looking ahead, it is likely that users will demand broader global coverage, faster data rates, and seamless access to both current and emerging Internet services and applications. To meet these expectations, current 5G technologies must continue to evolve in line with emerging 6G concepts.

With each new generation, network management becomes more challenging due to a growing user base, increasingly intricate infrastructures, and limited spectrum availability. To address these issues, 5G and 6G networks need to integrate smart management systems that automate critical operational processes. Consequently, the enhancement of network management is a primary driver behind the development of these next-generation networks, achievable by adopting advanced, state-of-the-art network control paradigms. From a network control paradigm standpoint, game theory offers an extensive suite of mathematical methods and models to analyze multilayered strategic decision-making, and it has been widely utilized in network management.

Classical game theory began to take shape in the 1920s. In 1928, Von Neumann introduced the minimax theorem in his work on social games, addressing the simplest case of a two-person contest. Later, in 1950, John Nash introduced the concept of Nash equilibrium to tackle decision convergence issues in multi-person non cooperative games, and his doctoral thesis not only proved the existence of such an equilibrium but also identified the conditions required for its solvability. These pioneering contributions laid the foundation for the further development of game theory. While theoretical advancements in classical game theory have largely plateaued since then, its applications have broadened, and many scholars continue to use game theory to address challenges in the network management. Traditionally, game theory assumes that all decision-makers are completely rational; however, researchers have identified limitations in this assumption due to the reality of bounded rationality. To address these challenges, learning algorithms have been in the spotlight recently.

Learning algorithm, a crucial branch of computer algorithms, is frequently employed to decipher the intricate environments of communication networks, thereby reducing management complexity. Typically, learning algorithms assume that the statistical properties of the environment remain constant over time. This assumption does not guarantee optimal learning outcomes, since in practice, the environment evolves dynamically through interactions among agents, rendering static strategies vulnerable in competitive settings. In contrast, game theory is viewed as a promising mathematical tool that is widely used to model and coordinate interactions among users. Usually, game theory applications presuppose complete knowledge of the environment; a condition rarely met in real-world communication scenarios. In reality, much valuable information is embedded in raw data and must be extracted, meaning that game theory often has to contend with imperfect information.

Although both learning algorithm and game theory models have individually addressed many challenges, each approach relies on ideal assumptions when used in isolation. Given that both learning algorithm and game theory are grounded in decision theory, combining these approaches appears both feasible and promising. Game theory can effectively model the interactions among multiple users and tackle the optimization challenges encountered by learning methods, while the data mining and sensing capabilities of learning algorithm can extract crucial information from raw data, thus alleviating the imperfect information problems faced by game theory models. Thus, integrating learning algorithm with game theory proves highly effective for implementing the intelligent network control paradigm, which deals with topical issues related mainly to congestion control, network routing, load balancing, multi-commodity flow, resource allocation, quality of service provisioning and multi-criteria optimization.

This book focuses on the latest research and applications of game theoretic techniques and learning methods, both of which play pivotal and unique roles in the network control paradigm. It includes numerous algorithm examples that enable readers to clearly see how game theoretic solutions can be developed for specific network control challenges. The content is designed to provide a solid foundation in the essential background, concepts, and principles needed to effectively apply both game theory and learning algorithm approaches. As a result, the book is an invaluable resource for undergraduate and graduate students in computer science and information technology, as well as for educators, research strategists, scientists, researchers, and engineers working in the field of intelligent network control mechanisms.

OUTLINE OF THE BOOK

Key challenge in this book is to provide the basic idea of game theory and fundamental understanding the interactive learning mechanism among network entities. The research presented in this book focuses on modeling, analyzing and solving the network control problems. This book also guides readers through a fascinating landscape, unveiling the diverse applications and implications that arise from merging game theory with learning algorithms. Each chapter highlights the transformative power of this synergy. Whether you are a seasoned researcher, an emerging practitioner, or simply a curious intellectual explorer, this book provides a roadmap for unraveling the complexities of strategic models and understanding their real-world impact.

To provide a didactic approach for studying game theory, this book is organized part I and part II. In part I, we will study the basic concepts and implementation areas. Generally, game theory's capacity to simulate the strategic interactions among rational agents has captivated both scholars and network engineers for years. In our digital era, where algorithms and autonomous agents are increasingly prevalent, the fusion of game theory with learning algorithms has unlocked a new realm of exploration. Merging these two fields paves the way for exciting new possibilities, as this work explores how artificial agents can engage in strategic thinking, adjust to ever-changing environments, and make decisions that align with both their objectives and the dynamics of their surroundings. Considering this situation, chapter I in part I covers the fundamental ideas of game theory and learning algorithm. And then, the basic principles of intelligent network control paradigm are explained. In Chapter II, advanced future network management topics are introduced, including cloud-based offloading services, network virtualization, IoT-related communications, heterogeneous network management, Device-to-Device (D2D) communications, vehicular ad hoc networks, non-terrestrial networks, wireless body area networks, new technologies for future communications, and emerging network control frameworks. In chapters I and II, I have gathered what I consider to be the essential material for anyone seeking to become a well-informed user of contemporary game theory.

In part II, I introduce network control models where strategic thinking and learning converge, paving a path for a future that enables intelligent network systems to make cognitive decisions. I have carefully assembled network control models, which grew out of my published papers between 2014 and 2024. These models aims to give readers a quick and accessible introduction to many of the most important control mechanisms with representative solutions. Therefore, this part deals with the case study of modeling, design, and analysis of game-theoretic protocols in future network areas. Part II consists of a total of 10 chapters (chapter III - chapter XII), and each chapter is also composed of various subchapters. Each subchapter has the same goal and is structured in the same format such as development motivation, contribution, control model, and final summary.

Over the past 30 years, telecommunications and game theory have engaged in a dynamic exchange, giving rise to a new discipline known as algorithmic game theory or, alternatively, the learning game model. Central network control challenges, such as resource allocation in expansive networks and power control in wireless communications, inherently involve interactions among numerous self-interested agents. Both learning mechanism and game theory contribute a variety of models and definitions that help analyze these challenges. Chapter III through chapter IV were written in accordance with this research trend. Chapter III covers the control field of cloud based offloading services, and provides the Two-tier Communications, Caching, and Computing (*TCCC*) scheme, the Edge Fog and Cloud Computing (*EFCC*) scheme, the Edge-assisted Federated Learning and Offloading (*EFLO*), the Two-phase Wireless Bandwidth Allocation

(*TWBA*) scheme, the Distributed Mobile Cloud Computing (*DMCC*) scheme, the Cooperative Federated Learning and Offloading (*CFLO*) scheme, the Bargaining Game based Resource Management (*BGRM*) scheme, and the Cloud Edge Offload Computing (*CEOC*) scheme. Chapter IV covers the control field of network virtualization, and provides the Adaptive Multi-service Resource Allocation (*AMRA*) scheme, the Dynamic Virtual Machine Migration (*DVMM*) scheme, the Hierarchical Multi-service Resource Allocation (*HMRA*) scheme, the Multicasting Device Communication Control (*MDCC*) scheme, and the Effective Sensor Cloud Control (*ESCC*) scheme.

Chapters V and VI begin by addressing coordination challenges and distributed problem solving. When facing a communication control problem, it is often necessary or beneficial to divide the task among multiple agents whose actions must be coordinated. As a result, these chapters have a strong algorithmic focus. Later, they expand to explore not only coordination but also competition. Although multi-agent systems are broader than game theory, mastering game theory is undoubtedly essential in this field. Chapter V covers the control field of IoT related communications, and provides the Behavioral Learning Game based Connection (*BLGC*) scheme, the Bargaining Game based Time Scheduling (*BGTS*) scheme, the Bargaining based Unlicensed Spectrum Sharing (*BUSS*) scheme, the Fog Computing Resource Allocation (*FCRA*) scheme, the Collaborative Coded Distributed Computing (*CCDC*) scheme, the Multi-agent Cooperative Spectrum Sharing (*MCSS*) scheme, the Learning based Service Quality Control (*LSQC*) scheme, and the Asymptotic Shapley Value based Allocation (*ASVA*) scheme. Chapter VI covers the control field of heterogeneous network management, and provides the Bargaining based Hierarchical Spectrum Allocation (*BHSA*) scheme, the Heterogeneous Network Bandwidth Control (*HNBC*) scheme, the Heterogeneous Network Spectrum Allocation (*HNSA*) scheme, the Cooperative Game based Resource Allocation (*CGRA*) scheme, the Fair Efficient Spectrum Allocation (*FESA*) scheme, the Coexistent Dynamic Spectrum Sharing (*CDSS*) scheme, the Two-level Game based Spectrum Allocation (*TGSA*) scheme, and the Interval Values based Spectrum Allocation (*IVSA*) scheme.

In an era marked by escalating complexity and groundbreaking innovation, the convergence of game theory and learning method has become a beacon for intellectual exploration. Chapters VII-IX illustrate the deep connection between these two influential fields - a union that is set to transform our understanding of decision-making, optimization, and the dynamics of artificial intelligence. Initially designed to decipher strategic interactions, game theory has been revitalized in the realm of learning algorithm, providing essential tools to model and analyze the intricate interactions among network agents. Meanwhile, learning algorithms have revolutionized machine intelligence by enabling agents to glean insights from data, forecast outcomes, and adapt to evolving environments. When these two methodologies intersect, they create a powerful synergy that unlocks a multitude of promising possibilities. To cover these topics, chapters VII-IX are organized. Chapter VII covers the control field of D2D communications, and provides the Small Cell Network Control (*SCNC*) scheme, the Cooperative Dual-level Game based Control (*CDGC*) scheme, the Heterogeneous Network Spectrum Allocation (*HNSA*) scheme, the Cellular Network Spectrum Allocation (*CNSA*) scheme, and the Spectrum and Energy Efficiency Control (*SEEC*) scheme. Chapter VIII covers the control field of vehicular ad hoc network, and provides the Joint Crowdsensing and Offloading Control (*JCOC*) scheme, the Cooperative Game based Traffic Control (*CGTC*) scheme, the Feedback Game based Service Control (*FGSC*) scheme, the Timed Bargaining-based Opportunistic Routing (*TBOR*) scheme, the Vehicular Sensing, Clouding and Routing (*VSCR*) scheme, the Radio Resource Management Control (*RRMC*) scheme, and the Effective Crowdsensing and Routing Control (*ECRC*) scheme. Chapter IX covers the control field of non-terrestrial networks, and provides the Hierarchical Aerial Offload Computing (*HAOC*) scheme, the Collaborative Game based Task Offloading (*CGTO*) scheme, the Bargaining based Computation Offloading Control (*BCOC*) scheme, the High-speed Railway Communication Control (*HRCC*) scheme, the Cognitive Satellite Spectrum Management (*CSSM*) scheme, and the Cooperative Multi-path Routing Control (*CMRC*) scheme.

As you explore chapters X-XII, you'll discover how game theory empowers learning algorithms to tackle challenges in adversarial decision-making, resource allocation, and strategic cooperation. This approach reveals new dimensions in network control mechanisms and promises to shape the future of integrated paradigms. These chapters are more than just a collection of ideas - they offer a gateway to a future where the fusion of game theory and learning algorithms leaves a lasting impact. Chapter X covers the control field of wireless body area network, and provides the Intra and Inter Network Control (*IINC*) scheme, the Learning and Game based Spectrum Allocation (*LGSA*) scheme, the Reverse Biform based Resource Sharing (*RBRS*) scheme, and the Bargaining Solution based Resource Allocation (*BSRA*) scheme. Chapter XI covers the control field of new technologies for future communications, and provides the Wireless Energy Harvesting Control (*WEHC*) scheme, the Meta-Bargaining based Tethering Control (*MBTC*) scheme, the Price and Service Distribution Control (*PSDC*) scheme, the Smart Home Traffic Control (*SHTC*) scheme, the Data Sponsoring Model Control (*DSMC*) scheme, the Hybrid Network Spectrum Allocation (*HNSA*) scheme, and the Rate Splitting Multiple Access (*RSMA*) scheme. Chapter XII covers the control field of emerging network control frameworks, and provides the Cooperative Network Traffic Control (*CNTC*) scheme, the Data Trading and Power Control (*DTPC*) scheme, the User Provided Network Control (*UPNC*) scheme, the Triple Bargaining based Energy Management (*TBEM*) scheme, the Underwater Network Distributed Routing (*UNDR*) scheme, and the Ultra-Dense Network Control (*UDNC*) scheme.

In summary, modern networks have become multifaceted, providing support for numerous bandwidth-hungry applications and a variety of devices, ranging from smart phones to tablet PCs to high-end servers. Also, modern users expect the network to be available all the time, from any device, and to let them securely collaborate with coworkers. Game theory and modeling techniques have greatly facilitated the implementation of mathematical programming theory into the practice of commercial network design. This book systematically introduces the application of game theory in network design, and surveys state-of-the-art game models for design, analysis, and management of networks, such as cellular, femtocell, mobile ad-hoc, overlay, sensor and wireless local area networks. It provides a comprehensive technical guide covering introductory concepts, fundamental techniques, recent advances, and open issues in related topics. The content of each chapter showcases the remarkable potential that arises when robust modeling, mathematics, and computational power intersect. I hope that readers will not only find inspiration and valuable insights within these chapters, but also be motivated to further explore the applications of game theory in learning.

Required Background

This book is rigorous and demands a mathematical mindset, though it only requires a basic level of background knowledge. For most of its content, a foundational understanding of computer science and elementary probability theory is assumed. In more technical sections, familiarity with Markov decision problems, mathematical programming, and classical logic is expected. Ultimately, however, the most essential prerequisite is the ability to think clearly.

Intended Audience

This book is primarily aimed at network engineers, architects, and technicians who have practical experience with network protocols and technologies. It also offers hands-on guidance for graduate students in computer science, mathematics, and electrical engineering on applying these concepts to internetwork design. Additionally, because it addresses both the engineering challenges of algorithm design and the associated mathematical foundations, the book is an excellent resource for technical professionals engaging in self-study. I always appreciate suggestions and corrections from readers - *Mone me si erro.*

Acknowledgement

Crafting a book is a heartfelt endeavor that usually relies on the contributions, support, and encouragement of many people. I am profoundly thankful to everyone who played a key role in bringing this work to life, and I wish to extend my sincere appreciation for their invaluable support.

First and foremost, I wish to express my sincere gratitude to my father, Sangjoon Kim, and my mother, Jungsun Shin, whose steadfast support and deep understanding have carried me through this endeavor. Their continuous encouragement and patience have served as my foundation during every challenge I've faced. For as long as I can remember, they have showered me with love, emotional support, and unwavering motivation. Furthermore, I also hold my grandparents close in my thoughts, imagining them watching over me with a gentle smile. No words can fully capture the depth of my appreciation for them - they will always be cherished in my heart, now and forever.

I am profoundly thankful to my academic advisor, Dr. Pramod K. Varshney, whose invaluable guidance, insightful criticism, steadfast faith, and unwavering encouragement - coupled with his remarkable thoughtfulness - have been essential in bringing this work to fruition. His consistent academic support has accompanied me at every step, and his wisdom and expertise have played a pivotal role in shaping the ideas within this book. His insights have truly served as a beacon on my intellectual journey.

I can say that I have enjoyed my life. This is because of all the wonderful people I've been surrounded by. There are too many angels in my life that I need to give my thanks to, and I could not possibly give thanks to them all. Therefore, I would like to give my deepest and most sincere thanks to God, who kindly sent these angels to me. Also, I am profoundly grateful to the many game theory researchers and learning algorithm experts who have so generously shared their insights and perspectives, thereby enriching the content of this book. Without their efforts and contributions, this work would not have been possible.

Last but not the least, I must extend my heartfelt thanks to my graduate students. They truly are the best. Through countless seminars and discussions, my students have provided me with invaluable academic inspiration and have been instrumental in bringing this book to completion. Their enthusiasm for the subject and their readiness to explore new ideas have made this writing journey exceptionally rewarding. I am deeply grateful for each one of them, but I cannot mention everyone by name, here. Therefore, on their behalf, I would like to sincerely offer special thanks to Dr. Youngjae Park, my first PhD student and now a cherished my friend.

Sungwook Kim

Section 1
Basic Concepts and Implementation Areas

Chapter 1
Fundamental Ideas of Game Theory and Learning Algorithms

ABSTRACT

Game theory is concerned with the decision making of utility-maximizing individuals in their interactions with one another and their environment. Nowadays, the field of machine learning is progressing at a faster pace than ever before. Many organizations leverage machine learning tools to extract useful information from a massive amount of data. From its earliest days of study, this chapter has recognized the important relationship between game theory and learning algorithms, and introduce a new solution concept, called the intelligent network control paradigm, which can provide an innovative guideline to solve different control problems for future networks.

1. INTRODUCTION

Recent advances in network communications and smart device technologies have promoted the proliferation of Internet of Everything (IoE) with ubiquitous sensing and computing capabilities to interconnect millions of physical objects to the Internet. The paradigm of IoE is advancing toward enriching people's lives by adding value to the Internet of things (IoT). While the IoT is a dynamic global network infrastructure concerned about physical devices, accessed through the Internet, IoE lays an upper foundation over IoT and is concerned with intelligent network connections and technologies. Therefore, the benefit of IoE is derived from the compound impact of connecting people, process, data, and things, and the value this increased connectedness creates as 'everything' comes online. The advancement of IoE propels the fast growth of next-generation, such as sixth-generation (6G) networks, leading to a new era of coverage, connectivity, and technological innovation, which calls for novel approaches to address rising energy consumption and maximize resource use. Nowadays, the IoE-based future network concept has received much attention from both academia and industry due to its great potential to deliver customer services in many aspects of modern life. This approach enables seamless communications and automatic management between heterogeneous smart devices without human intervention, which has the

DOI: 10.4018/979-8-3373-2013-7.ch001

potential to revolutionize industries and provide significant benefits to society through fully intelligent and automated control systems (Nguyen, 2021).

To pave the way for the development in IoE-based networks and beyond, many efforts have been put into research in different control mechanisms. The use of control mechanisms has been generalized over the years in many fields of industry due to its multiple advantages. Nevertheless, there are also many issues that need to be solved, and one of the principal challenges is the management of limited network resources, which are always inherent in network control problems. Control methods are critical components that can help the network resource management. Generally, there are two major control methods, namely, centralized and distributed control methods. Since the mid 1980s, network and systems management has steadily evolved from centralized methods, where the management application runs on a single management entity, to distributed methods, where it is distributed over many agents (Martin-Flatin, 1999).

Centralized methods are built around a single, centralized control server, which handles all major data processing and stores data and user information that other users can access. From there, client nodes can be connected to the main server and submit data requests instead of performing them directly. Typically, centralized methods are the most cost-effective options for small systems and require fewer resources to set up and maintain. This reduces the time and overhead necessary to keep a network up to date. Furthermore, it's easier to standardize interactions between the main server and client nodes. However, this approach has some drawbacks. Since centralized methods have a single point of failure, if the main server crashes, the entire system will likely shut down; it will likely result in service interruptions and consequent inconvenience and diminished reliability from a user perspective. Furthermore, if the network experiences traffic spikes beyond what the network was designed to handle, information bottlenecks may occur, and client nodes will experience the increased latency (Martin-Flatin, 1999).

In contrast to traditional centralized methods, distributed control methods do not have a central entity with access to all information or authority over all components. Distributed control involves the design of decision rules for systems of interconnected components to achieve a collective objective in a dynamic or uncertain environment. This distributed control approach is developed for distributed systems. The term 'distributed system' is sometimes used to describe a system, which is composed of equal, interconnected nodes, meaning that data ownership and computational resources are shared evenly across the entire system. Because a distributed system does not have a central server, the burden of data processing is crowdsourced across the system. Therefore, the decision-making process on distributed methods typically involves individual determinations, and the final behavior of the system changes in accordance with the aggregate results of the decisions each individual node. To achieve a common goal, the specific processes by which a distributed system makes decisions is contingent on the system's consensus mechanism. With distributed systems, a single node can fail independently without affecting the rest of the system, since the computational workload will simply be rebalanced among the remaining nodes. Furthermore, distributed methods are more scalable than centralized methods. Even though distributed methods exploit the flexibility and versatility of systems in a contested environment, there are several issues to consider. In the absence of a single server, there are no superior nodes overseeing the behavior of subordinate nodes, and consequently there is no way to regulate individual nodes on the system. Therefore, it can be difficult to make timely decisions. Furthermore, it is difficult for any individual node to gain a global view of the entire system. Therefore, it is harder for individual nodes to make well-informed decisions based on the state of other nodes in the system. In addition, since any meaningful change requires updating every individual node, distributed methods require more resources to maintain or reconfigure (Martin-Flatin, 1999).

With the rapid advancement and use of Internet, embedded systems, wireless communication technologies and novel control strategies, distributed control methods have been increasingly more paramount at the core of critical infrastructures and service systems. Current wired and wireless networks have, in turn, given rise to new network application domains, such as data-based networks, robotic networks, unmanned aerial vehicle systems, social and economic networks, smart power networks, and epidemic networks. Such applications often require distributed in-network control and optimization techniques to support various operations, including resource allocation, coordination, learning, and estimation. Therefore, it becomes clear that any progress made along this research line would then necessarily expand knowledge about broad theory and wide applications of distributed control methods (Ge, 2017; Nedić, 2018).

The analysis and synthesis of distributed methods require an integration and coordination of communication, computation and control. Especially, the problems arising in distributed methods stem mainly from two aspects: a lack of central authority or coordinator and an inherent dynamic of the system connectivity structure. Therefore, distributed control methods may pose a number of challenging issues that may not be suitably addressed by traditional control schemes. Over the last decade, system and control researchers have developed several system sciences and control methodologies, and computer scientists and networks engineers have made promising advancements such as in system software, high-speed computing and new networking techniques. However, some fundamental work is still needed to address real-world challenges integrating communication, computation and control in a uniform and systematic way (Ge, 2017; Nedić, 2018).

One approach to distributed control methods is to view the problem from the perspective of game theory. Game theory is the study of decision problems in which there are multiple decision makers and the quality of a decision maker's choice depends on both that choice and the choices of others. Originally, game theory has been employed as a modeling tool for describing and influencing behaviors in societal systems. Therefore, it is formally defined as 'the study of mathematical models of conflict and cooperation between intelligent rational decision-makers'. As a modeling paradigm in the mathematical control sciences, game theory can be applied in various fields such as economics, political science, psychology, computer science, biology, telecommunications, and even philosophy. Recently, game theory has emerged as a valuable tool for controlling or prescribing behaviors in distributed network systems (Marden, 2015; Marden, 2018).

1.1 Game Theory

As a branch of applied mathematics, game theory is the study of mathematical models of conflict and cooperation between intelligent rational decision-makers, and concerns the interactions of them. While game theory has been studied predominantly as a modeling paradigm in the mathematical social sciences, there is a strong connection to control systems in that a controller can be viewed as a decision-making entity. Accordingly, game theory is relevant in settings with multiple interacting controllers. Therefore, the appropriateness of game theory to distributed control is easily recognized. Nowadays, one approach to distributed control is to view the control problem from the perspective of game theory. Looked at objectively, this perspective is a departure from the traditional study of game theory, where the focus has been the development of models and methods for applications in economic and social sciences (Marden, 2015; Marden, 2018).

Many factors contribute to the appeal of game theory for distributed control. One of the features of game theory is that it provides a hierarchical decomposition between the distribution of the optimization problem, called game design, and the specific local decision rules, called distributed learning algorithms. For example, if the game is designed as a potential game model, there is an inherent robustness to decision making rules. Under a variety of informational dependencies, a wide class of distributed learning algorithms can achieve convergence to a pure Nash equilibrium. Therefore, one must specify the dynamics through which agents will arrive at an outcome or select from a set of possible outcomes. In other words, the central goal of game theory is to design control laws for the individual agents to ensure that the emergent global behavior is desirable with respect to a given objective. Ideally, a game designer seeks to satisfy this goal while conditioning each agent's control law on the least amount of information possible (Li, 2013).

Utilizing game theory for this purpose requires two steps. The first step in deriving a game-theoretic model is to identify the basic elements of the game, namely game players and their admissible strategies. This step is to model the agent as self-interested decision makers in a game theoretic environment, and to define a set of choices for each decision maker. With the basic elements such as players and strategies, the next step is to specify players' utility functions. Based on the utility functions, the game theory can be categorized as two agendas; the traditional role of game theory is the 'descriptive' agenda, and its application to distributed control is the 'engineering' agenda. Whereas in the descriptive agenda, utility functions are part of the modeling process and in the engineering agenda, utility functions constitute a design choice. Another important consideration in specifying the utility functions is the implication on the global objective. With utility functions in place, the game is fully specified. If one takes a solution concept such as Nash equilibrium to represent the outcome of the game, then these outcomes should be desirable as measured by the global objective. From this perspective, there is another important point that highlights a distinction between the descriptive and engineering agenda. In the descriptive agenda, learning algorithms in games seek to understand how a plausible adaptation disequilibrium process may converge, and thereby reinforce the role of Nash equilibrium as a predictive outcome. By contrast, in the engineering agenda, the role of such learning processes is to guide the agents towards a solution. Accordingly, the specification of the learning process also constitutes a design choice. There is some coupling between the two agendas and learning processes. In particular, it can be advantageous in designing utility functions to assure that the resulting game has an underlying structure so that one can exploit learning processes that converge for such games (Marden, 2015).

1.2 Reinforcement Learning Algorithms

As a branch of machine learning, reinforcement learning is concerned with how an agent can learn to take subsequent actions in a setting to maximize cumulative rewards. It has gained importance within the Artificial Intelligence (AI) community and computer science in general. In the reinforcement learning framework, an agent learns via trial and error, and seeks a strategy that maximizes the expected cumulative long-term benefit. Reinforcement learning was originally developed for Markov Decision Processes (MDPs). It allows a single agent to learn a policy that maximizes a possibly delayed reward signal in a stochastic stationary environment. However, when multiple agents apply reinforcement learning in a shared environment, this might be beyond the MDP model. In such systems, the optimal policy of an agent depends not only on the environment, but on the policies of the other agents as well. These situations arise naturally in the domains of telecommunications, distributed control, and wireless network systems, etc. At each time step, multiple agents interact with the environment discretely, observe the environment, follows their policies, and are quantitatively rewarded by the environment. Value functions help the agent estimate future rewards for specific states or state-action combinations. These value functions help the agent evaluate and refine the policy (Jain, 2024; Nowé, 2012).

Multi-agent system is a subfield of distributed AI that has experienced rapid growth because of the flexibility and the intelligence available solve distributed problems. Usually, multi-agent system can be defined as a group of autonomous, interacting entities sharing a common environment, which they perceive with sensors and upon which they act with actuators. Although the agents in a multi-agent system can be programmed with behaviors designed in advance, it is often necessary that they learn new behaviors online, such that the performance of the agent or of the whole multi-agent system gradually improves. This is usually because the complexity of the environment makes the a priori design of a good agent behavior difficult or even impossible. Therefore, individual agents collaboratively solve tasks yet they offer more flexibility due to their inherent ability to learn and make autonomous decisions. Agents use their interactions with neighboring agents or with the environment to learn new contexts and actions. Subsequently, agents use their knowledge to decide and perform an action on the environment to solve their task. Multi-agent systems have received tremendous attention from scholars in different disciplines, and they are rapidly finding applications in a variety of domains, including robotics, distributed control, computer science, civil engineering, telecommunications, and economics (Busoniu, 2008; Dorri, 2018).

The complexity of many tasks arising in multi-agent system domains makes them difficult to solve with hardwired behaviors. The agents must instead discover a solution on their own, using multi-agent learning. In the multi-agent setting, the assumptions that are needed to guarantee convergence are often violated. When agents' objectives are aligned and all agents try to maximize the same reward signal, coordination is still required to reach the global optimum. If agents have opposing goals, a clear optimal solution may no longer exist. In this case, an equilibrium between agent strategies is usually searched for. If a dynamic environment is adopted, the problem becomes even more complex and further complicated by the fact that agents typically have only limited information about the system. In the most challenging case, agents may not even be aware of the presence of other agents, making the environment seem non-stationary. However, despite the added learning complexity, a real need for multi-agent systems exists with decentralized systems (Nowé, 2012).

If game theory and reinforcement learning algorithms come together in the ever-changing world of decision-making, they create a fascinating combination that shows a new way to solve complex problems in many fields. Games may be solved using reinforcement learning by treating the game environment as a learning problem and training the model with strategies for playing the game. Research on the convergence of game theory and reinforcement learning has shown to be beneficial, providing essential insights into challenging decision-making issues in many different and various disciplines (Jain, 2024).

1.3 Intelligent Network Control Paradigm

The need for adaptive multi-agent systems, combined with the complexities of dealing with interacting learners has led to the development of a multi-agent reinforcement learning field, which is built on two basic pillars: the reinforcement learning within AI, and the interdisciplinary work on game theory. The field of game theory research has significantly benefited from reinforcement learning algorithms in designing intelligent agents with strong gaming skills. The commonality between game theory and reinforcement learning is their emphasis on decision-making in uncertain, complicated contexts. Reinforcement learning can be used to describe and solve games by treating them as learning problems, where the objective is to identify an optimal policy that maximizes the reward. In contrast, game theory provides a framework for evaluating strategic interactions between players and can assist in forecasting the equilibrium results of these interactions (Jain, 2024; Nowé, 2012).

Although reinforcement learning and game theory models have respectively solved many challenges to some extent, there are ideal assumptions for reinforcement learning or game theory individually. These assumptions have certain limitations. Analyzing from the perspective of learning algorithms, the applications always assume that the statistical distribution of the environment is fixed and will not change over time. In other words, the data in reinforcement learning algorithms are pre-existing, and its statistical properties cannot be changeable. However, practically, the real-world environment will dynamically change with the interaction among agents, and a fixed strategy is easy to be dominated when competing with others. The major goal of reinforcement learning is to analyze and learn the data from the environment, explore the potential distribution rules over the data, and use the rules to predict future events and make an optimal strategy and execute specific actions. Therefore, the assumption of reinforcement learning is unreasonable, and this approach cannot guarantee the optimal of the learning results. Another limitation of reinforcement learning is its heavy reliance on extensive interaction with an environment to learn effectively. It requires substantial data on the consequences of actions to refine their strategies, which can be resource-intensive and time-consuming to collect and can take a long time for the agent to learn optimal behaviors. Especially, the rate of data collection is limited by the dynamics of the environment especially when agents need to dynamically adjust their behavior. Therefore, the collected historical data will no longer be optimal when applying it to actual situations and will always lag behind the actual behavior. It is called as the 'dynamic interaction' problem of reinforcement learning algorithms (Shen, 2021).

Analyzing from the perspective of game theory, the applications always require complete and perfect information about the environment, and consider the agents as players in a specific model; the interaction between agents is direct. However, it always requires strong even unrealistic assumptions. In many real-world situations, information may be imperfect or asymmetrical, meaning that some players have more information than others. This can lead to situations where game theory fails to accurately predict behavior, as players may make decisions based on incomplete or inaccurate information. It is called as the 'imperfect information' problem of the applications of game theory. Another criticism of game theory is that it assumes that all players are rational, which is not always true in practice. In such case, game theory may not be the best tool for analyzing the complex situations (Shen, 2021).

In order to deal with the aforementioned two problems, we need technologies to guide reinforcement learning methods to interact with others and process the original data for game theory models. It inspires us to combine reinforcement learning and game theory, using the ability of reinforcement learning algorithms in data mining to overcome the problem of imperfect knowledge of game theory and using the game theory models to enhance the ability of reinforcement learning algorithms to deal with the dynamic interaction between agents.

Usually, an optimization problem can be established using the mathematical models of game theory (named as game modelling), and the equilibrium point is found through theoretical derivation. It is the reinforcement learning algorithm that achieves convergence of such equilibrium point by sensing the dynamic update of the external environment (named as strategy learning). Actually reinforcement learning and game theory are closely related because both approaches are based on decision theory, so we think combining reinforcement learning methods with game theory models is feasible and can achieve better performance gains. Especially, combining reinforcement learning and game theory has been used to create more intelligent agents capable of playing games at a high level, and can increase the understanding of strategic decision-making in complex and uncertain environments while creating more intelligent agents in various contexts. On the one hand, game theory can characterize the interactions of multiple agents and solve the optimization problems of reinforcement learning methods. From a game-theoretic viewpoint, the outcome of a multi-agent learning process is jointly determined by every agent's sequential moves. Based on the planning future strategic actions, each agent needs to take into account others' decision-making as well. On the other hand, reinforcement learning algorithms' ability of data mining and sensing can abstract useful information from the original data and solve the imperfect information problems for game theory models. For the dynamically changing situation, reinforcement learning is a powerful tool to practically implement the game theory model (Jain, 2024; Nowé, 2012; Shen, 2021).

The integration of game theory with multi-agent reinforcement learning can be considered a new control concept, called the *intelligent network control paradigm*, which explores and exploits the benefits of the two disciplines. The intelligent network control paradigm fundamentally combines the insights from strategic decision-making and autonomous learning process in multi-agent environments. Fundamentally, this approach explores how concepts from game theory influence the development and analysis of reinforcement learning algorithms for multiple agents. The introduction of intelligent network control paradigm shows us the huge gain in optimizing control strategies for the dynamics of network systems. As the basis for networks in the 21st century, this paradigm is a simple and powerful tool to emphasize how game-theoretic concepts enhance the development of learning algorithms.

1.4 Toward Beyond 5G Network Management

As future networks evolve to become more intricate and highly interconnected, managing the vast number of devices and systems they include has grown increasingly important. To address this, network management frameworks are playing a pivotal role in ensuring these complex infrastructures are properly maintained. The rising sophistication of telecommunication networks underscores the necessity for comprehensive management models. One well-established model is FCAPS - standing for Fault, Configuration, Accounting, Performance, and Security - which provides a structured approach to handle diverse management tasks. Fault management pertains to identifying and addressing network issues, while configuration management involves the oversight and maintenance of device settings. Accounting management is concerned with monitoring resource consumption, often for billing purposes. Performance management seeks to enhance overall network efficiency, and security management aims to defend the network from breaches and malicious activities. Originally introduced by the International Organization for Standardization (ISO) in 1977, FCAPS has become a foundational concept in network administration. Recently, approaches that incorporate machine learning into FCAPS have garnered considerable interest, reflecting the challenges posed by today's increasingly dynamic network environments. By leveraging the FCAPS model, network professionals can methodically tackle operational challenges and uphold the stability, performance, and security of their systems (Mekrache, 2024).

Fault management plays a critical role in maintaining a dependable and stable network infrastructure. It can be broadly divided into three main categories. The first, fault diagnosis, consists of two key processes: fault detection and prediction - which involve ongoing monitoring to identify or anticipate issues in real time - and fault isolation or localization, which focuses on accurately identifying the location of a fault within the network. The second category, fault analysis, entails fault correlation and pattern recognition across multiple incidents, as well as root cause analysis (RCA) to determine the fundamental reasons behind network failures. The third and final category is fault resolution, where appropriate corrective measures are taken based on the outcomes of RCA. These actions not only address the current problem but also help prevent similar issues in the future, thereby supporting uninterrupted network operations (Mekrache, 2024).

Configuration management in network operations involves several key functions designed to maintain an organized, efficient, and adaptable network environment. The first area, installation, refers to deploying network hardware and devices, ensuring they are correctly integrated and fully operational within the existing infrastructure. The second function, provisioning, is concerned with configuring network resources and services in accordance with organizational needs, guaranteeing that systems are prepared and accessible as required. The third component, service planning, involves designing and organizing service offerings to align with business goals and operational requirements. This includes establishing service parameters, technical specifications, and deployment protocols. Network planning, the fourth element, focuses on developing the network's architecture, topology, and layout to promote scalability, optimal performance, and robust connectivity. Lastly, status and control refers to the continuous monitoring and regulation of the network's condition, using real-time metrics to assess its functionality and maintain operational stability (Mekrache, 2024).

Accounting management refers to the process of recording and analyzing how network resources are utilized. It involves collecting data on user interactions, resource allocation, and overall system activity. This information plays a vital role in functions such as usage-based billing, forecasting future capacity needs, and detecting inefficiencies or improper use of resources. Performance management, in the networking domain, is a structured process aimed at maximizing the effectiveness and dependability of network systems and services. It comprises several integral components. Performance quality assurance ensures that network devices and services consistently meet established benchmarks, maintaining a high level of service reliability. Performance monitoring focuses on the continuous observation of network operations to promptly identify irregularities and address issues as they arise. Performance analysis involves evaluating performance-related data to recognize trends and extract actionable insights, which support decisions on network improvements and resource distribution. Together, these elements foster a well-optimized network environment characterized by reliability, consistent monitoring, and evidence-based enhancements (Mekrache, 2024).

Security management in digital environments is organized into several key areas that work together to protect data and maintain system integrity. The first area, detection, focuses on actively monitoring system activity and network traffic to identify potential threats before they cause harm. The second area, prevention, involves implementing protective mechanisms - such as firewalls, access control systems, and routine software updates - to block unauthorized access and vulnerabilities. The third area, administration, includes the development and enforcement of security policies, user authentication procedures, security audits, and staff training to ensure comprehensive management of security measures. Combined, these components form a robust security strategy that defends against cyberattacks and maintains the reliability of digital systems (Mekrache, 2024).

REFERENCES

Busoniu, L., Babuska, R., & De Schutter, B. (2008). A comprehensive survey of multiagent reinforcement learning. *IEEE Transactions on Systems, Man, and Cybernetics. Part C, Applications and Reviews*, *38*(2), 156–172. DOI: 10.1109/TSMCC.2007.913919

Dorri, A., Kanhere, S. S., & Jurdak, R. (2018). Multi-agent systems: A survey. *IEEE Access : Practical Innovations, Open Solutions*, *6*, 28573–28593. DOI: 10.1109/ACCESS.2018.2831228

Ge, X., Yang, F., & Han, Q. L. (2017). Distributed networked control systems: A brief overview. *Information Sciences*, *380*, 117–131. DOI: 10.1016/j.ins.2015.07.047

Jain, G., Kumar, A., & Bhat, S. A. (2024). Recent developments of game theory and reinforcement learning approaches: A systematic review. *IEEE Access : Practical Innovations, Open Solutions*, *12*, 9999–10011. DOI: 10.1109/ACCESS.2024.3352749

Li, N., & Marden, J. R. (2013). Designing games for distributed optimization. *IEEE Journal of Selected Topics in Signal Processing*, *7*(2), 230–242. DOI: 10.1109/JSTSP.2013.2246511

Marden, J. R., & Shamma, J. S. (2015). Game theory and distributed control. In *Handbook of game theory with economic applications* (Vol. 4, pp. 861–899). Elsevier.

Marden, J. R., & Shamma, J. S. (2018). Game theory and control. *Annual Review of Control, Robotics, and Autonomous Systems*, *1*(1), 105–134. DOI: 10.1146/annurev-control-060117-105102

Martin-Flatin, J. P., Znaty, S., & Hubaux, J. P. (1999). A survey of distributed enterprise network and systems management paradigms. *Journal of Network and Systems Management*, *7*(1), 9–26. DOI: 10.1023/A:1018761615354

Mekrache, A., Ksentini, A., & Verikoukis, C. (2024). Machine Learning in FCAPS: Toward Enhanced Beyond 5G Network Management. *IEEE Communications Surveys and Tutorials*, *26*(4), 2769–2797. DOI: 10.1109/COMST.2024.3395414

Nedić, A., & Liu, J. (2018). Distributed optimization for control. *Annual Review of Control, Robotics, and Autonomous Systems*, *1*(1), 77–103. DOI: 10.1146/annurev-control-060117-105131

Nguyen, D. C., Ding, M., Pathirana, P. N., Seneviratne, A., Li, J., Niyato, D., Dobre, O., & Poor, H. V. (2021). 6G Internet of Things: A comprehensive survey. *IEEE Internet of Things Journal*, *9*(1), 359–383. DOI: 10.1109/JIOT.2021.3103320

Nowé, A., Vrancx, P., & De Hauwere, Y. M. (2012). Game theory and multi-agent reinforcement learning. In *Reinforcement learning: State-of-the-art* (pp. 441–470). Springer Berlin Heidelberg. DOI: 10.1007/978-3-642-27645-3_14

Shen, J., Yang, C., Li, T., Wang, X., Song, Y., & Guizani, M. (2021). Interactive artificial intelligence meets game theory in next-generation communication networks. *IEEE Wireless Communications*, *28*(2), 128–135. DOI: 10.1109/MWC.001.1800554

Chapter 2
Application Fields of Intelligent Control Paradigm

ABSTRACT

As the deployment of 5G networks gains momentum across the globe, the research community is already planning the 6G. This paper highlights the application fields for future networks, and discusses some of the key technologies expected to support the move towards 6G communications. Specifically, the hot research issues and methodologies are surveyed for the areas of cloud services, network virtualization, IoT, heterogeneous network, D2D communications, vehicular networks, non-terrestrial networks, wireless body area networks while discussing the salient characteristics of the beyond 5G and 6G networks.

2. OVERVIEW OF FUTURE NETWORKS AND TRENDS

In the past decade, we witnessed and experienced the development of the fifth-generation (5G) technologies, standards, trails, and commercialization efforts. The development of 5G wireless technology has led to an explosion in the number of connected devices, enabling applications that were previously impossible. However, with the commercialization of the 5G network, it has exposed limitations that must be overcome for the realization of future applications. Therefore, it is time for both academia and industry to shift their attention towards the development of 6G technologies while expecting to address these limitations. From mobile Internet to the IoE, and then to the intelligent connection of all things, 6G will realize the transition from serving people and things to supporting the efficient connection of intelligent agents. Through the intelligent interconnection of people, machines, and things and collaborative integration, future network systems will offer major improvements in data transfer speed, latency, network capacity, and reliability to enable various applications. At this crossroad, an overview of the current state of the art and a vision of future communications are definitely of interest (Wang, 2022).

The future network technologies will realize the digitization and intelligence of physical entities in the real environment and penetrate all aspects of production, life, health, telecommunications, entertainment and social governance. Currently, research groups intend to study the capabilities of networks, when it is expected to support novel forward-looking scenarios, such as cloud based offloading services, network virtualization, IoT related communications, data center, heterogeneous network management, device-to-device (D2D) communications, ubiquitous intelligence, vehicular ad hoc network (VANET), non-terrestrial networks, wireless body area network (WBAN), etc. Based on breakthroughs in the infor-

DOI: 10.4018/979-8-3373-2013-7.ch002

mation technologies, these hottest research topics will further accelerate the fusion of communications with sensing, computing, and control for our daily lives, industrial production, and green development. Therefore, it is not surprising that nowadays there is urgent need to increase the efficiency of these network systems due to the growing demand of different services.

2.1 Cloud Based Offloading Services

Cloud computing is more user-friendly as compared to traditional computing because network agent can have access to data anytime anywhere using internet. It delivers services over the Internet using remote servers, offering scalability and accessibility. Traditional computing relies on local servers, requiring upfront hardware investments and maintenance. The key distinction lies in where and how services are hosted, impacting flexibility, cost-effectiveness, and ease of access. Cloud computing is a coalesce of many computing fields and it has gained much popularity in the recent years. As a service provisioning model, cloud computing provides computing, storage, services, and applications over the Internet. Moreover, cloud computing facilitates to reduce capital cost, decouple services from the underlying technology, and provides flexibility in terms of resource provisioning. The development of cloud computing technique enables resource-limited network agents to overcome the severity of scarce resource constrained by allowing them to offload computation and migrate several computation parts of an application to powerful cloud servers. In addition to the conventional fashion that mobile clients access cloud services as in the client-server model, a new rapidly growing work has proposed to explore cloud functionalities in another perspective (Othman, 2013).

With the increasing popularity of smart mobile devices such as tablets, smartphones, smart watches and notebooks, our lifestyles have been altered dramatically. However, they have limited resources in computational capacity, battery lifetime and network connectivity, which prevent them from running very complex applications. Recently, mobile applications are becoming increasingly intensive and sophisticated that require increasing amounts of computational capabilities. Especially for real-time and user-interactive applications, they have to wait a long time to obtain the results due to the limited processing speed of the mobile devices. Mobile cloud offloading is becoming a promising method to solve a number of concerns affecting mobile computing. It is defined as an integration of cloud computing technology with mobile devices to make the mobile devices resource-full in terms of computational power, memory, storage, energy, and context awareness. Therefore, this approach can take advantage of abundant resources hosted by clouds to offload part of the mobile codes to the cloud for remote execution in order to optimize the application performance and energy efficiency of the mobile device (Othman, 2013; Wu, 2018).

Although the mobile cloud computing is useful for computing and storage, the traditional computation offloading techniques cannot be used for the smart mobile devices directly, because these techniques are generally energy-unaware and bandwidth hungry. In the viewpoint of mobile devices, they also require specific applications that support computation offloading and should be optimized for mobile cloud environment in terms of heterogeneity, context awareness, application partitioning overhead, network data cost, bandwidth, and energy consumption. In addition, mobile network environments usually have a huge impact on the performance of offloading systems since mobile devices are easily subject to dynamically changing network conditions due to their mobility. Due to this reason, mobile devices still suffer from a number of shortcomings. Sometimes, the extra cost involved in data transfer via a wireless link may be greater than the cost savings from offloading in certain circumstances. Therefore, it will always be difficult to make offloading decision i) which portion of an application should be offloaded and which not, and ii) where to place the execution, i.e., either locally or remotely (Othman, 2013; Wu, 2018).

2.2 Network Virtualization

Network virtualization is the process of combining hardware and software network resources and network functionality into a single, software-based administrative entity, a virtual network. Inspired by the success of various virtualization technologies, virtualization has become an important research topic in communication networks. Generally, network virtualization involves platform virtualization, often combined with resource virtualization. It first abstracts the underlying physical network and then creates separate virtual networks through specific abstraction and isolation functional blocks. Recently, the network virtualization concept has appeared as a new alternative to help on the achievement of efficiency goal, reducing capital and operational costs. It is considered as a pioneer to achieve effective resource utilization with decreased operating expenses and capital expenses by decoupling the networks functionalities of coexisting virtual networks (Kitindi, 2017; Shen, 2021).

The concept and techniques of network virtualization have been evolving over more than three decades. In the point view of network operators, network virtualization has a very broad scope ranging from spectrum sharing, infrastructure virtualization, to air interface virtualization. It brings efficient resource utilization, reduces operational and investment cost, raise revenue, encourages innovations and allows new players in the business to enter into the markets at the lower cost. Early research on network virtualization includes virtual local area networks motivated by facilitating different types of services in distributed systems. It allows the flexible sharing of physical networking resources by multiple tenants. Each tenant runs its own applications over its virtual network, i.e., its slice of the actual physical network. Network slicing emphasizes a service-oriented perspective in network management by creating multiple end-to-end slices for different services on top of shared physical network infrastructure. Another example of network virtualization is virtual private networks, which establish efficient and secure communication links to connect geographically dispersed end users. It is a technology that creates a safe and encrypted connection over a less secure network, such as the Internet (Kitindi, 2017; Shen, 2021).

Regardless the vision of the network virtualization, wireless virtualization is challenging. Due to the inherent broadcast nature of the wireless communication and stochastic nature of the wireless channel, radio spectrum resource abstraction and isolation are not a straightforward. This is because wireless resources have to be shared and distributed in fairly manner; it is very difficult. Even though relatively little virtualization has occurred in infrastructure based wireless networks, the idea of virtualizing wireless access is gaining attention and has become one of the leading trends in future networks. This approach has the potential to improve spectrum utilization and perhaps create new services. As a result, it requires a comprehensive research effort to provide reliable and sustainable solutions for future wireless virtualization; many technical and managerial research challenges have to be addressed (Kitindi, 2017).

2.3 IoT Related Communications

The next wave of wireless technologies is proliferating in connecting things among themselves as well as to humans. It is rapidly becoming an integral part of our life and also multiple industries. In the era of the IoT, billions of sensors, machines, vehicles, drones, and robots will be connected, making the world around us smarter. It is expected to see the number of IoT connected devices explosively grows and will reach hundreds of billions during the next few years. Currently, the IoT technology is shaping the current and next generation of the Internet. The vision of IoT is to embed communication capabilities with a highly distributed, ubiquitous and dense heterogeneous devices network. In academics and industry, IoT-based systems have proliferated in the last few years, providing multiple new applications such as smart homes, intelligent transportation, smart grids, smart retail, smart hospitals, and smart cities. With the generated data from each device, the data will not remain raw but will be customized to the devices based on their needs and even converge with other data. However, the concept of IoT is simple, even though its capability is unlimited and its usage can change the entire paradigm of legacy technology (Lee, 2017).

The IoT phenomenon has rapidly emerged into a necessary ecosystem in which data, processes, humans, things and the Internet are associated with each other. It includes the adaptation of secure mobile networks, anytime, anywhere, by anyone or anything with new intelligent applications and services. The IoT domain is growing rapidly and it is finding applications in industries. Nowadays, manufacturing has seen the significant influence of IoT, as it underpins Industry 4.0, which is the trend that incorporates data-driven practices in manufacturing. In manufacturing, the IoT is used to create smart factories that are more efficient and productive with sensors and devices to gather data, optimize processes, and identify and offset potential problems. Recently, Industry 4.0 and IoT can provide a multitude of contemporary solutions, applications and services. Another important domain of IoT applications is smart cities including the use of IoT for monitoring water contamination, air quality, weather forecasting, and keeping a check on noise pollution. In addition, IoT has also made significant contributions to the healthcare domain in terms of remote monitoring of patients, biomedical wearables, intelligent health cloud recommendations for drugs, and e-health services (Fizza, 2022).

Along with the enormous growth of IoT, there are many issues and challenges to be addressed while developing IoT applications. First, IoT must be operated on large number of heterogeneous devices. Therefore, a unifying control framework is needed to implement IoT systems. Second, interoperability is another challenge where different devices must be able to communicate. To create a complete IoT environment, some protocols have been developed for different IoT applications. However, a common protocol structure is still required. Third, security issues must be considered during communications between IoT devices. Forth, scalability is another challenge to be addressed. When a lot of devices are connected simultaneously, scalability must be considered while designing control protocols and data access mechanisms. Finally, energy efficiency must be also taken care, since IoT devices have limited resources and battery power. Besides technological challenges, there are other aspects of challenges, such as government regulations. These challenges made IoT a little more critical. However, even though there are some limitations in the current IoT research and development, new and novel technologies to address these challenges can enable service providers and application programmers to implement their services efficiently (Sobin, 2020).

2.4 Heterogeneous Network Management

Nowadays, the proliferation of multimedia infotainment applications and high-end smart devices exacerbates the demand for high data rate services. Researchers in the field of telecommunications and Internet have reached a consensus that incremental improvements fail to meet the escalating data demands of the foreseeable future. Especially, the current homogeneous network is no longer capable of providing enough services to the ever-increasing user population and meet their demands. Therefore, a paradigm shift is required for the future networks, and the conventional networks are soon expected to be replaced by the heterogeneous network (HetNet). The HetNet is a network connecting computers and other devices where the operating systems and protocols have significant differences, and merges different types of networks into an integrated network system, which has become a hot research area in recent years towards next generation communication networks. The major goal of HetNets is to effectively exploit network resources and provide seamless connectivity for heterogeneous objects. Unlike other networks, HetNets hold such characteristics as heterogeneity, openness, distribution, and multi-domain involvement (Wang, 2022).

As a new evolution network structure, HetNets can increase the opportunity in the spatial resource reuse and improve service quality by developing small cells into the coverage of macro-cells. From the aspect of seamless services in small cells, efficient handoff schemes for HetNets are also important to support the excellent service quality and flawless mobility from one network to another network. Therefore, mobility management is a key technique in performing handoff in multi-tier networks. Another important research topic is a large-scale HetNet. Such a network comprises different types of networks, e.g., satellite networks, air networks, marine networks and terrestrial networks, intending to support anywhere and anytime networking. This approach enables HetNets to support good quality of service while serving diverse users in different areas (Wang, 2022).

Recently, researchers are beginning to discuss 6G, spectrum resource management for HetNets has emerged as a critical area due to the increased density of small-cell networks. It plays a pivotal role in enhancing spectrum utilization, load balancing, and network energy efficiency. Therefore, recent researches have mostly concentrated on resource management, including spectrum utilization and interference mitigation. In addition, the new HetNet is expected to combine multiple technologies. With these mixed technologies in HetNets, future mobile devices will be equipped with multiple radio interfaces to enable users to take advantage of these radio access technologies and seamlessly switch between technologies. However, there are cautions concerning implementation; it is critical to handle resources efficiently. Additionally, the interference management of HetNets will become more complicated. In addition to spectrum control and interference management concerns, there are several other equally important issues such as power allocation, user association, fairness, allocated capacity, and complexity (Manap, 2020).

2.5 D2D Communications

Mobile wireless communication demands for higher data rates, lower energy consumption, low latency, coverage area, as well as efficient spectrum sharing are rapidly increasing day by day. The ever-increasing requirements and demands of smart devices in mobile wireless networks are the main drivers for further enhancement of the network capacity. Nowadays, many techniques and approaches have been implemented to cope with the increasing demand for network capacity. One of the most promising approaches is the D2D communication technology, which is believed to be able to significantly enhance the current network capacity, coverage, peak rates, throughput, latency, user experience, etc. D2D communications can be seen as conceptually similar to cognitive radio principles. In both approaches, primary and secondary users exist in locality, and they either access the spectrum through cognitive sensing or reuse radio resources by means of D2D communication. Nevertheless, D2D communication is mostly managed by the network operators whereas cognitive radio is fully autonomous and uses cognitive sensing (Mach, 2015).

D2D communications can be categorized into the out-band D2D (O-D2D) and in-band D2D (I-D2D) methods.

In the O-D2D method, D2D communications exploit the unlicensed spectrum adopted by other wireless technologies supporting direct communication such as WiFi direct or Bluetooth. In general, the O-D2D method may result in poor service quality, because of uncontrolled interference. In addition, the discovery process and setup of connection in the O-D2D method is quite complicated; it needs the special intervention to establish the connection between the two devices. In the I-D2D method, D2D communications take place in a licensed spectrum allocated to the network operators. Therefore, D2D devices can access the licensed spectrum either in a dedicated mode or a shared mode. In contrast to the O-D2D method, the I-D2D method can handle the above mentioned problems in a more efficient way. The reason is that the network operator dynamically controls all D2D communication and adopts sophisticated allocation and interference mitigation techniques to provide high service quality (Mach, 2015).

Another advantage of D2D communications is to help alleviate load on backhaul network and improve overall network capacity while providing better reuse of available spectrum. Moreover, direct transmission between devices can be achieved with lower transmission power, resulting in improved energy efficiency. Furthermore, D2D communication can provide the benefits of fairness, congestion control and enhancing cell coverage. Although, the D2D communication technology has many advantages, there are still many open challenges to successfully implement this technique. In particular, D2D communications will require efficient spectrum allocation mechanisms, intelligent D2D mode selection algorithms, complex interference control techniques, mobility management procedures and robust security protocols (Jameel, 2018).

2.6 Vehicular ad hoc networks

The growing interest toward the possible applications of wireless technologies to a vehicular environment has recently led consortia and standardization bodies to develop technologies and protocols for data transmission between vehicles and between vehicles and road infrastructures. During the last couple of decades, VANETs gain a lot of attention from industry and academia to improve driving safety, efficiently manage traffic conditions, and enhance user experience during driving. It is a proposed type of mobile ad hoc network (MANET) involving road vehicles. VANETs are distributed, self-organizing communication networks built up from moving vehicles, and are thus characterized by very high speeds and limited degrees of freedom in nodes movement patterns. Such particular features often make standard networking protocols inefficient or unusable in VANETs. Currently, the major challenge of protocols developed for VANET is to benefit from the messages exchanged between vehicles to alter traffic, either for safety issues with advanced safety messages, or for traffic management in order to avoid traffic jams (Harri, 2009).

Message exchanging between vehicles occurs through i) vehicle-to-vehicle (V2V) links built dynamically, where any vehicle can be used as next hop, so to form an end-to-end path toward a final destination, and ii) vehicle-to-infrastructure (V2I) links, assuming ubiquitous deployment of fixed roadside units. V2V communication technology is a smart technology that enables vehicle data to exchange from one vehicle to another. It will have the greatest impact on vehicle safety applications once it is widely available, such as advancing crash avoidance. V2I communication technology captures data such as traffic congestion, weather advisories, bridge clearance levels, traffic light status, and then wirelessly transmits them to inform drivers of conditions they need to be aware of which aids in safety. Smart traffic signals powered by V2I help drivers understand traffic conditions better, helping to estimate accurate arrival times which can improve communication between drivers and customers (Vegni, 2015).

Particularly, VANETs are susceptible to the hidden node problem; it refers to a wireless issue in networks where some nodes are unable to detect signals from other nodes, leading to collisions and message corruption. In addition, VANETs must contend with limited spectral bandwidth and a highly variable channel influenced by both stationary and mobile obstructions and interference sources. In such an environment, infrastructure based network infrastructures hold a significant advantage over ad hoc networks. With computing and cache capabilities, the roadside unit is an important component, which mainly plays a role in collecting and analyzing traffic messages given from smart vehicles, or even providing drivers and passengers with Internet access. Furthermore, roadside unit allows optimal scheduling of channel access and distribution of network resources in a relatively simple manner, at the cost of needing to deploy a large number of roadside units throughout the intended coverage area. Advances in VANET, smart transportation systems, and intelligent roadside units have recently opened up vast possibilities for traffic safety, convenience, and effectiveness (Cooper, 2016).

2.7 Non-terrestrial Networks

The evolution of telecommunication technologies, the ever-increasing demand for new services, and the exponential growth of smart devices fuel the development of non-terrestrial network (NTN) systems as an effective solution to complement terrestrial networks in providing services over uncovered or under-served geographical areas. NTNs are wireless communication systems that operate above the earth's surface, involving satellites, high-altitude platforms and drones. Such components have the capability to offer wide area coverage by providing connectivity over the regions that are expensive or difficult to cover with terrestrial networks. Therefore, they are essential to realizing seamless coverage, bringing coverage even to remote areas such as rural areas, vessels, airplanes, ships, etc. Where the customer needs are changing radically, the NTN represents a coverage extension for the terrestrial network in a world market (Rinaldi, 2020).

Due to limited coverage areas and geographic constraints, it is challenging to guarantee ubiquitous coverage everywhere. Moreover, in times of natural disasters, connectivity outages are widespread, which can be detrimental to critical actions necessary for saving lives and properties. To address these challenges, NTNs can provide uninterrupted communications and high data transfer rates in remote, disaster-stricken, and rural areas where terrestrial networks with existing network infrastructures are not available. Recent studies have shown that NTNs can offer high availability and low latency, making them an ideal candidate for 6G communication systems. In addition, the use of NTNs in conjunction with the existing terrestrial infrastructure could provide a feasible and cost-effective solution for continuous and ubiquitous wireless coverage and thereby enable network scalability. In 6G communications, the integration of NTNs into existing terrestrial networks introduces a range of novel challenges, including task offloading, network routing, network slicing, and many more. In this approach, NTNs act as access agents to augment the performance of existing terrestrial networks while addressing the shortfalls of terrestrial infrastructure for future wireless applications (Iqbal, 2023).

Although NTN presents numerous potential benefits for the development of future 6G networks, it also entails several challenges that need to be addressed. Due to the long distances between the satellites and the ground devices, the propagation delay is usually higher in NTN environments. Another control issue is how to deploy and manage efficiently, including the physical and ground-based infrastructure, such as antennas, base stations, and backhaul networks. Additionally, high-speed air or space-borne platforms necessitate modifications to existing handover and paging protocols. The large path loss also increases the minimum power requirement for reliable transmission, initiating the need for novel beam and resource allocation strategies. Furthermore, the dynamic nature of NTNs, with platforms moving at high speeds and in different directions, introduces additional challenges related to signal propagation, interference, and handover management. Finally, spectrum sharing in the same frequency band with existing terrestrial or other services also requires further study in order to avoid interference between terrestrial and non-terrestrial devices (Mahboob, 2024).

2.8 Wireless Body Area Network

World population growth is facing three major challenges; demographic peak of baby boomers, increase of life expectancy leading to aging population and rise in health care costs. On the other hand, millions of people die from various fatal diseases every year. The common problem with all current diseases is that many people experience the symptoms and have disease diagnosed when it is too late. Usually, most diseases can be prevented if they are detected in their early stages. 5G technology is one of the essential technologies for societal digital transformation, and it is also a prerequisite for the interconnection of everything in smart healthcare. Promoting and implementing 5G smart healthcare can reduce inconsistencies in allocating medical resources and expedite medical advancements. Therefore, future health care systems should provide proactive wellness management and concentrate on early detection and prevention of diseases. One key solution to more affordable and proactive health care systems is through wearable monitoring systems capable of early detection of abnormal conditions resulting in major improvements in the quality of life (Movassaghi, 2014).

Wireless body area network (WBAN) connects low-power, miniaturized, invasive or non-invasive, lightweight devices that are situated in the clothes, on the body or under the skin of a person. The network typically expands over the whole human body and the devices are connected through a wireless communication channel. These devices, such as sensors and actuators, can monitor the human body functions and characteristics from the surrounding environment. The sensors are used to measure certain parameters of the human body, either externally or internally. On the other hand, the actuators take some specific actions according to the data they receive from the sensors or through interaction with network agents. Generally, WBANs enable new applications and thus new possible markets with respect to wireless sensor networks. Whereas, their design is affected by several issues that call for new paradigms and protocols. The diversity of envisioned applications in the medical field, which span from vital signs monitoring to automatic drug delivery, and ambient intelligence areas creates a set of technical requirements with a wide variation in terms of expected performance metrics, as throughput or delay. Therefore, flexible architectures and protocols are needed (Cavallari, 2014).

WBAN technology is improving the efficiency of doctor-patient activities such as remote patient monitoring, timely health status, notification, and emergency calling, anytime at anywhere. Therefore, the WBAN can offer automatic medical service through remote monitoring of a patient's vital organs. All the information can be monitored and stored from the control unit or remotely. In medical and healthcare service, another exciting application of WBAN is assisted living. Wearable medical sensors can be used at home to measure physiological data from patient's body and transmit them into particular medical center server in a regular interval. It helps the patient to stay at home and get continuous support instead of staying at hospital. Telemedicine is another fascinating application field of WBAN. It provides health care services over a distance with the help of information and communication technology. Nowadays, the WBAN technology can be integrated in the sector of telemedicine like online video consultation with doctors, transmission of medical reports and images, remote medical diagnosis etc. Interestingly, it is different from other communication technologies, several non-technical factors would also play crucial roles in the success of the WBAN technology development such as affordability, legal, regulatory and ethical issues, and user friendliness, comfort and acceptance (Arefin, 2017).

2.9 New Technologies for Future Communications

As we navigate through the rapidly changing terrain of technology, the domain of communication has become a critical area of focus, one that is perpetually molded by remarkable innovations and progressive thinking. Looking forward to the horizons of technological growth, we find ourselves at an exciting crossroads where different novel technologies intersect. Wi-Fi tethering is a function that allows smart devices to share their Internet connection with other devices. When this technology is enabled, each device acts as an acting router, capable of granting access to other devices that are associated with the router. As a traditional and widely-adopted incentive protocol, data sponsoring allows content providers to cooperate with the service providers of networks and provide 'Free Traffic' plan for mobile users, improving their retention ratio and achieving higher revenue; mobile users can enjoy free network traffic. This scheme achieves a win-win situation for content providers and mobile users (Zhang, 2020).

Simultaneous wireless information and power transfer (SWIPT) technology integrates wireless information transfer (WIT) and wireless power transfer (WPT) methods. This technology meets low power consumption and communication needs of the IoT devices and reduces communication system power consumption for green communication. As a pivotal component of SWIPT systems, the SWIPT receiver plays a crucial role in processing both energy and information. As a revolutionary paradigm for a connected home environment, a smart home can be defined as a residence developed with artificial IoT technologies to enable the provision of useful user-centric services to its occupants. Smart home network is a vital element of modern home automation systems, enabling the interconnectivity and control of different smart devices. This network allows homeowners to remotely control lighting, security, temperature, and entertainment systems via voice commands or smartphones. It offers energy efficiency, convenience, and improved security by permitting residents to monitor and modify their living surroundings (Alruwaili, 2024; Wu, 2024).

Rate-splitting multiple access (RSMA) provides a flexible and promising non-orthogonal multiple access paradigm. It is based on splitting the message into certain common and private parts while utilizing successive interference cancellation at the receivers. The common parts are jointly encoded into a common stream, and different encoders can be used for the private parts. Through information and communication theoretic analysis, the RSMA has been shown to be optimal in several transmission scenarios. With the centralized technology, cellular topologies do not seem to support the user-centric cell-free concept because each cell can only support a certain number of smart devices. In order to serve numerous devices and make distributed signal processing simpler, cell-free networks with decentralized technologies have been suggested. Cell-free networks have shown to outperform traditional networks under various performance metrics. Visible light communication (VLC) has attracted a lot of research interests, since this technology can provide higher data rate and higher security over visible light bands, which have advantages of low electromagnetic interferences, low costs, high bandwidth and high security, than traditional communication systems (Ammar, 2021; Can, 2023; Lin, 2019).

2.10 Emerging Network Paradigms

High-altitude balloons (HABs) are attracting increasing attention for future wireless communication networks, owing to their low deployment expense and large coverage range. In particular, they can be used for various services, including broadband Internet access, digital service request, and emergency response. To provide these services to ground users, HABs, acting as flying wireless base stations. Recently, there has been considerable interest in the development of unmanned aerial vehicles (UAVs). They have drawn attention for applications such as parcel delivery, aerial mapping, agriculture, and surveillance. The UAVs provide a new potential for business in civil and non-civil applications such as parcel delivery, aerial mapping, agriculture, wildlife conservation, and surveillance. Thanks to their advantages, UAVs can be exploited as a promising communication platform. However, UAV-enabled communications and networking may encounter several challenges against their maturity. Since the UAVs usually have stringent size, weight, and power constraints, they pose critical limits on their endurance and communication capabilities (Wang, 2021; Wei, 2017).

With the continuous expansion of the railway infrastructure network, the railway network has become more and more complex both in time and space dimensions. In recent decades, the complexity of railway transportation organizations has dramatically increased. Now the railway operators are seeking ways to accurately identify the critical stations in the railway operation network to formulate management measures more pertinently. Nowadays, the railway communication systems come with the challenges of deploying new communication technologies. Another challenge is to integrate into the rail communication network architecture the cutting-edge enabling techniques. Integrating some or all of these technologies could offer an advantage in terms of performance, flexible maintenance and management, cost reduction, flexibility, and scalability of railway communication networks. Satellite networks enable a whole new class of missions for navigation, communications, remote sensing, and scientific research for both civilian and military purposes. The proliferation of satellites will enable a better understanding of the near-Earth environment and provide an efficient and economical means to access the space through the use of multi-satellite solution. The satellite networks need to configure and maintain dynamic routes, manage intermediate satellites, and reconfigure themselves to achieve mission objectives. Hence, inter-satellite communication is a key aspect when satellites fly in formation (Narouwa, 2024; Radhakrishnan, 2016).

The recent emergence of communication technologies in mission-critical applications in the maritime industry has led to the introduction of the ship network paradigm. It is a novel application domain of 6G communications that refer to the network of smart interconnected maritime objects, which can be any physical device or infrastructure associated with a ship, a port, or the transportation itself to improve the safety, efficiency, and environmental sustainability of ship networks. Overall, future networks provide new motivation for researchers and scientists to explore new applications and to develop novel solutions for intelligent ship networks. Currently, mobile ships on the ocean mainly rely on maritime satellites to acquire services. With maritime satellites, shore and island-based base station are also used to extend the coverage of terrestrial 5G networks for maritime activities. To provide a practically affordable solution for broadband maritime communications, an efficient hybrid satellite-terrestrial maritime communication networks is urgently required to combine the advantage of satellites' wide coverage with shore-based systems' high capacity (Aslam, 2020).

REFERENCES

Alruwaili, F. F., Alohali, M. A., Aljaffan, N., Alhashmi, A. A., Mahmud, A., & Assiri, M. (2024). A decentralized approach to smart home security: Blockchain with red-tailed hawk-enabled deep learning. *IEEE Access : Practical Innovations, Open Solutions*, *12*, 14146–14156. DOI: 10.1109/ACCESS.2024.3352502

Ammar, H. A., Adve, R., Shahbazpanahi, S., Boudreau, G., & Srinivas, K. V. (2021). User-centric cell-free massive MIMO networks: A survey of opportunities, challenges and solutions. *IEEE Communications Surveys and Tutorials*, *24*(1), 611–652. DOI: 10.1109/COMST.2021.3135119

Arefin, M. T., Ali, M. H., & Haque, A. F. (2017). Wireless body area network: An overview and various applications. *Journal of Computer and Communications*, *5*(07), 53–64. DOI: 10.4236/jcc.2017.57006

Aslam, S., Michaelides, M. P., & Herodotou, H. (2020). Internet of ships: A survey on architectures, emerging applications, and challenges. *IEEE Internet of Things Journal*, *7*(10), 9714–9727. DOI: 10.1109/JIOT.2020.2993411

Can, M., Ilter, M. C., & Altunbas, I. (2023). Data-oriented downlink RSMA systems. *IEEE Communications Letters*, *27*(10), 2812–2816. DOI: 10.1109/LCOMM.2023.3309909

Cavallari, R., Martelli, F., Rosini, R., Buratti, C., & Verdone, R. (2014). A survey on wireless body area networks: Technologies and design challenges. *IEEE Communications Surveys and Tutorials*, *16*(3), 1635–1657. DOI: 10.1109/SURV.2014.012214.00007

Cooper, C., Franklin, D., Ros, M., Safaei, F., & Abolhasan, M. (2016). A comparative survey of VANET clustering techniques. *IEEE Communications Surveys and Tutorials*, *19*(1), 657–681. DOI: 10.1109/COMST.2016.2611524

Fizza, K., Banerjee, A., Jayaraman, P. P., Auluck, N., Ranjan, R., Mitra, K., & Georgakopoulos, D. (2022). A survey on evaluating the quality of autonomic internet of things applications. *IEEE Communications Surveys and Tutorials*, *25*(1), 567–590. DOI: 10.1109/COMST.2022.3205377

Harri, J., Filali, F., & Bonnet, C. (2009). Mobility models for vehicular ad hoc networks: A survey and taxonomy. *IEEE Communications Surveys and Tutorials*, *11*(4), 19–41. DOI: 10.1109/SURV.2009.090403

Iqbal, A., Tham, M. L., Wong, Y. J., Wainer, G., Zhu, Y. X., & Dagiuklas, T. (2023). Empowering non-terrestrial networks with artificial intelligence: A survey. *IEEE Access : Practical Innovations, Open Solutions*, *11*, 100986–101006. DOI: 10.1109/ACCESS.2023.3314732

Jameel, F., Hamid, Z., Jabeen, F., Zeadally, S., & Javed, M. A. (2018). A survey of device-to-device communications: Research issues and challenges. *IEEE Communications Surveys and Tutorials*, *20*(3), 2133–2168. DOI: 10.1109/COMST.2018.2828120

Kitindi, E. J., Fu, S., Jia, Y., Kabir, A., & Wang, Y. (2017). Wireless network virtualization with SDN and C-RAN for 5G networks: Requirements, opportunities, and challenges. *IEEE Access : Practical Innovations, Open Solutions*, *5*, 19099–19115. DOI: 10.1109/ACCESS.2017.2744672

Lee, S. K., Bae, M., & Kim, H. (2017). Future of IoT networks: A survey. *Applied Sciences (Basel, Switzerland)*, *7*(10), 1072. DOI: 10.3390/app7101072

Lin, X., & Zhang, L. (2019). Intelligent and practical deep learning aided positioning design for visible light communication receivers. *IEEE Communications Letters*, *24*(3), 577–580. DOI: 10.1109/LCOMM.2019.2958629

Mach, P., Becvar, Z., & Vanek, T. (2015). In-band device-to-device communication in OFDMA cellular networks: A survey and challenges. *IEEE Communications Surveys and Tutorials*, *17*(4), 1885–1922. DOI: 10.1109/COMST.2015.2447036

Mahboob, S., & Liu, L. (2024). Revolutionizing future connectivity: A contemporary survey on AI-empowered satellite-based non-terrestrial networks in 6G. *IEEE Communications Surveys and Tutorials*, *26*(2), 1279–1321. DOI: 10.1109/COMST.2023.3347145

Manap, S., Dimyati, K., Hindia, M. N., Talip, M. S. A., & Tafazolli, R. (2020). Survey of radio resource management in 5G heterogeneous networks. *IEEE Access : Practical Innovations, Open Solutions*, *8*, 131202–131223. DOI: 10.1109/ACCESS.2020.3002252

Movassaghi, S., Abolhasan, M., Lipman, J., Smith, D., & Jamalipour, A. (2014). Wireless body area networks: A survey. *IEEE Communications Surveys and Tutorials*, *16*(3), 1658–1686. DOI: 10.1109/SURV.2013.121313.00064

Narouwa, M., Mendiboure, L., Badis, H., Maaloul, S., Molla, D. M., Berbineau, M., & Langar, R. (2024). Enabling network technologies for flexible railway connectivity. *IEEE Access : Practical Innovations, Open Solutions*, *12*, 151532–151553. DOI: 10.1109/ACCESS.2024.3479879

Othman, M., Madani, S. A., & Khan, S. U. (2013). A survey of mobile cloud computing application models. *IEEE Communications Surveys and Tutorials*, *16*(1), 393–413.

Radhakrishnan, R., Edmonson, W. W., Afghah, F., Rodriguez-Osorio, R. M., Pinto, F., & Burleigh, S. C. (2016). Survey of inter-satellite communication for small satellite systems: Physical layer to network layer view. *IEEE Communications Surveys and Tutorials*, *18*(4), 2442–2473. DOI: 10.1109/COMST.2016.2564990

Rinaldi, F., Maattanen, H. L., Torsner, J., Pizzi, S., Andreev, S., Iera, A., Koucheryavy, Y., & Araniti, G. (2020). Non-terrestrial networks in 5G & beyond: A survey. *IEEE Access : Practical Innovations, Open Solutions*, *8*, 165178–165200. DOI: 10.1109/ACCESS.2020.3022981

Shen, X., Gao, J., Wu, W., Li, M., Zhou, C., & Zhuang, W. (2021). Holistic network virtualization and pervasive network intelligence for 6G. *IEEE Communications Surveys and Tutorials*, *24*(1), 1–30. DOI: 10.1109/COMST.2021.3135829

Sobin, C. C. (2020). A survey on architecture, protocols and challenges in IoT. *Wireless Personal Communications*, *112*(3), 1383–1429. DOI: 10.1007/s11277-020-07108-5

Vegni, A. M., & Loscri, V. (2015). A survey on vehicular social networks. *IEEE Communications Surveys and Tutorials*, *17*(4), 2397–2419. DOI: 10.1109/COMST.2015.2453481

Wang, J., Yan, Z., Wang, H., Li, T., & Pedrycz, W. (2022). A survey on trust models in heterogeneous networks. *IEEE Communications Surveys and Tutorials*, *24*(4), 2127–2162. DOI: 10.1109/COMST.2022.3192978

Wang, S., Chen, M., Yin, C., Saad, W., Hong, C. S., Cui, S., & Poor, H. V. (2021). Federated learning for task and resource allocation in wireless high-altitude balloon networks. *IEEE Internet of Things Journal*, *8*(24), 17460–17475. DOI: 10.1109/JIOT.2021.3080078

Wang, Z., Du, Y., Wei, K., Han, K., Xu, X., Wei, G., Tong, W., Zhu, P., Ma, J., Wang, J., Wang, G., Yan, X., Xiang, J., Huang, H., Li, R., Wang, X., Wang, Y., Sun, S., Suo, S., & Su, X. (2022). Vision, application scenarios, and key technology trends for 6G mobile communications. *Science China. Information Sciences*, *65*(5), 151301. DOI: 10.1007/s11432-021-3351-5

Wei, Z., Wu, H., Huang, S., & Feng, Z. (2017). Scaling laws of unmanned aerial vehicle network with mobility pattern information. *IEEE Communications Letters*, *21*(6), 1389–1392. DOI: 10.1109/LCOMM.2017.2671861

Wu, H. (2018). Multi-objective decision-making for mobile cloud offloading: A survey. *IEEE Access : Practical Innovations, Open Solutions*, *6*, 3962–3976. DOI: 10.1109/ACCESS.2018.2791504

Wu, S. H., Xu, X., Yue, Z., Yang, Y. L., Li, S., Lin, X. Q., Yu, Y., & Liu, Y. (2024). Integrated Rectifying-Mixer Circuit for Simultaneous Wireless Information and Power Transfer (SWIPT). *IEEE Microwave and Wireless Technology Letters*, *34*(7), 959–962. DOI: 10.1109/LMWT.2024.3403153

Zhang, C., Pang, H., Gao, L., Ding, Q., Yang, Y., & Sun, L. (2020). Incentivizing mobile video users with data sponsoring and edge caching. *IEEE Access : Practical Innovations, Open Solutions*, *8*, 9640–9654. DOI: 10.1109/ACCESS.2019.2963623

Section 2

Advanced Future Network Management Topics: Case Studies in Intelligent Control Paradigm

Chapter 3
Cloud-Based Offloading Services

ABSTRACT

Nowadays, cloud computing is one of the most promising ways to address the resource insufficiency of smart devices. Based on the cloud computing infrastructure, task offloading is an important challenge and it must practically decide the place and the time for performing each task. In this chapter, the potential protocols of cloud-based computation offloading services have investigated and discussed from a multidimensional view based on application goal, realizing approach, offloading direction, etc., and the main challenges of cloud computing services are presented.

3. OVERVIEW

With significant advances in communication and computing, modern day mobile devices are becoming increasingly intelligent. Therefore, they are now capable of supporting a wide range of applications, which has increased the computational requirement and energy consumption. Especially, the widespread use and increasing capabilities of smart mobiles devices are making them a viable platform for offering mobile services. Unfortunately, the advances in mobile device hardware and battery life have been slow to respond to the computational demands of current applications. Therefore, many applications are still unsuitable for mobile devices due to constraints, such as the limited computation power, memory, storage, and energy. This inadequacy can prevent mobile devices from exploiting their full potential as reliable service providers. Cloud computing is an essential solution to this problem because it provides powerful resources to efficiently process various applications. The elastic resource provisioning of cloud computing promises to bridge the gap between the limited resources of mobile devices and the growing resource demands of mobile services through offloading resource-intensive applications. Nowadays, we are entering an era of cloud computing and leverage cloud services from diverse aspects and enjoy various benefits of cloud computing. A key reason leading to an increasing commercial adoption of cloud computing is its unprecedented advantages over the conventional computing paradigm. Such advantages include reduced cost, easier maintenance, and automatic scaling.

DOI: 10.4018/979-8-3373-2013-7.ch003

3.1 Two-tier Communications, Caching, and Computing (*TCCC*) scheme

In recent years, demand for diverse multimedia services over 5G networks has escalated rapidly, calling for higher capacity, faster data rates, lower latency, and more reliable quality of service (QoS). To address these needs, various cutting-edge technologies have been integrated into cellular network architectures. In this context, S. Kim introduces the *Two-tier Communications, Caching, and Computing* (*TCCC*) scheme (Kim, 2018a), featuring hybrid control algorithms in smart base stations (SBSs) alongside novel communications, caching, and computing techniques. Within the *TCCC* scheme, SBSs are equipped with both computing capabilities and local storage, allowing them to offload computation tasks from user equipment (UE) and cache data from cloud servers. By uniting communications, caching, and computing algorithms under one framework, the *TCCC* scheme leverages game theory to capture both competitive and cooperative interactions among network entities. The principal benefit of this fully integrated approach lies in revealing the powerful synergy that arises from combined functionality, all while maintaining adaptability and flexibility to meet diverse performance requirements (Kim, 2018a).

3.1.1 Development Motivation

Over the past few years, 5G networks have been under intense development to cope with the rapid surge in global multimedia service demands, which necessitate greater capacity, higher data rates, reduced latency, and improved QoS. As such, efficient strategies have begun to reshape cellular network architectures. In this context, S. Kim proposes the *TCCC* scheme (Kim, 2018a). The *TCCC* scheme employs hybrid control algorithms in smart base stations (SBSs), underpinned by innovative communications, caching, and computing methods. Given that SBSs now possess both computational and storage capacities, they can offload the computational tasks of mobile user equipments (UEs) while caching relevant data from the cloud. By integrating communications, caching, and computing techniques into a cohesive control structure, the *TCCC* scheme leverages game theory to capture and manage both competitive and cooperative dynamics. Its core value lies in demonstrating the synergy that emerges from a fully integrated approach, while maintaining flexibility to fulfill diverse performance demands.

The future 5G standard aspires to substantially bolster network capacity, addressing the exponential rise of mobile traffic services. Advances in mobile communication technologies have made 5G the ideal choice for managing escalating wireless device counts and ubiquitous network connectivity (Wang, 2017). Traditional cellular network architectures, however, fall short due to radio link, front-haul, and back-haul constraints not keeping pace with explosive traffic growth (Li, 2017). In 4G networks, resource deployment has focused primarily on system performance maximization, but this conventional mindset struggles to adapt to the dynamic, repeated QoS requirements demanded by multiple users (Kim, 2010b, Kim, 2018a; Li, 2017).

As UEs become more capable, an array of computation-intensive and delay-sensitive applications - such as augmented reality and recognition assistance - have surfaced (Ma, 2017). Yet, because UEs generally have limited computing resources, running these applications locally can be difficult. Cloud computing offers a solution by expanding mobile devices' resource pool, effectively acting as a remote or distributed proxy for these constrained devices. This shift toward offloading high-complexity tasks to the cloud has fueled the fusion of communications and computing, a trend sometimes described as '*computing for communications*' (Andreev, 2016; Kim, 2018a; Lee, 2015). Between UEs and cloud servers, intercommunication serves as a crucial link, enabling collaborative execution and data sharing.

Typically, UEs acquire necessary data from the cloud, a process accompanied by unavoidable transmission delay. To mitigate these delays, data caching has been introduced, significantly diminishing redundant data transfers and alleviating front-haul bottlenecks (Bulti, 2015; Fan, 2016a). Coupling caching with computing allows content distribution to align with popularity, so that frequently requested data can be stored closer to UEs, enhancing content retrieval efficiency (Andreev, 2016; Chen, 2012; Kim, 2018a).

To support communications, caching, and computing in 5G, SBSs have become pivotal infrastructures. They handle computation offloading, data caching, and mobile communication collectively. Offloading computational workloads onto SBSs and caching data locally can significantly reduce delays inherent in UE-cloud data exchanges (Fan, 2016a). For optimal performance, all controlling factors should be examined in tandem, unlocking the maximum synergy among communications, caching, and computing (Andreev, 2016; Kim, 2018a).

Designing a modern SBS control protocol for 5G calls for a new control paradigm. In complex environments where rational agents have competing objectives, game theory is often employed. By examining strategic interactions among these agents, who each seek to maximize personal gains, game theory has found significant application in wireless communications (Kim, 2014). Thus, the *TCCC* scheme employs a game-theoretic perspective to formulate practical SBS control algorithms, reducing the computational burdens associated with purely centralized solutions (Kim, 2018a).

The *TCCC* scheme tackles computation offloading, cache splitting, and bandwidth allocation within a single, integrated SBS control mechanism. It aims to exploit synergies between SBS computation and communication with caching. The foundation of this integrated structure is a two-tier game model, orchestrated through an interactive process. In this model, SBSs and UEs make decisions in a stepwise manner, learning over time to improve outcomes. As 5G network conditions fluctuate, the *TCCC* scheme yields significant mutual benefits, demonstrating how competitive entities can also cooperate to enhance their advantages. This design, characterized by adaptability and flexibility, addresses a spectrum of 5G control challenges while proficiently matching system conditions (Kim, 2018a). By building on game-based learning, the *TCCC* scheme focuses on three central decisions; i) UE computation offloading - identifying the tasks to be offloaded, ii) cache splitting - determining how to allocate radio resources for caching content and data, and iii) bandwidth allocation - assigning channel resources to each SBS to ensure robust communications. To manage these decisions effectively, the *TCCC* scheme stresses feasibility, self-interactivity, and an overarching integration of dissimilar control algorithms. This harmonized methodology addresses conflicting performance goals across various 5G settings. While existing literature provides scattered algorithms for communications, caching, and computing, few if any integrate them within a unified framework. The *TCCC* scheme stands out as a pioneering solution that holistically weaves offloading, caching, and bandwidth management into a single game-driven 5G network control mechanism (Kim, 2018a).

3.1.2 Contribution

The *TCCC* scheme integrates communications, caching, and computing algorithms, comprehensively addressing all relevant practical factors to harness their combined synergy. To model the interactions among SBSs and UEs, the *TCCC* scheme introduces a two-tier hierarchical game. In the first-tier, SBSs serve as game players, sharing the total bandwidth through a dynamic bargaining process. In the second-tier, each SBS and its respective UEs form a Stackelberg game. Acting as a leader, the SBS allocates cache capacity for data and content caching and sets pricing for communication and computation services. In response, UEs (the followers) observe their leader's decisions and choose the strategies that maximize their individual payoffs. By adopting this two-tier game mechanism, the *TCCC* scheme achieves a fair, balanced outcome across various 5G network scenarios (Kim, 2018a).

3.1.3 Tow-tier Hierarchical Game Model

During the 5G network operation, SBSs and UEs make control decisions individually while taking considering their mutual relationship. The *TCCC* scheme considers a scenario formed by n SBSs $\mathbf{B} = \{\mathscr{B}_1 \ldots \mathscr{B}_n\}$ and m UEs $\mathbf{E} = \{\mathscr{E}_1 \ldots \mathscr{E}_m\}$. Each SBS $\mathscr{B}_i \in \mathfrak{B}$ is connected to Internet via a limited front-haul link with capacity $\mathscr{C}_{\mathscr{B}_i}$, whereas each UE $\mathscr{E}_j \in \mathscr{E}$ is connected to its serving SBS via a wireless link. Cache at each SBS has become an important means of offloading the traffic and tackling the front-haul bottleneck in order to reduce the latency and cost of services. There are two kind of caching for 5G network services; data caching for computation offloading and contents caching for mobile communications. Therefore, the capacity of $\mathscr{B}_i$'s cache storage $\mathscr{Q}_{\mathscr{B}_i}$ is divided. With the computation offloading and caching services, wireless bandwidth is assigned to each SBS to support the communications between UE and its corresponding SBS. Usually, the bandwidth needs for each SBS would vary temporally and spatially. To dynamically adapt the traffic fluctuations, the principle of allocating bandwidth to SBSs is a fair-efficient bargaining based on the self-monitoring (Kim, 2018a).

To combine computation offloading, caching and bandwidth allocation algorithms, the *TCCC* scheme develops a new two-tier hierarchical game model ($\mathbb{G}$) assuming dynamic 5G network situations. Formally, the *TCCC* scheme defines the $\mathrm{G} = \{\mathbb{G}^{first}, \mathbb{G}^{second}_{1 \le i \le n}\}$ where $\mathbb{G}^{first}$ is the first-tier bargaining game to formulate interactions among SBSs, and $\mathbb{G}^{second}_{1 \le i \le n}$ are multiple second-tier games to formulate interactions between the $\mathscr{B}_{1 \le i \le n}$ and its corresponding UEs; they work in parallel and independently. Firstly, the $\mathbb{G}^{first}$ can be defined as $\mathbb{G}^{first} = \left\{\mathfrak{B}, \mathbb{C}^{\mathscr{B}}, \mathbb{S}_{\mathscr{B}_1} \ldots \mathbb{S}_{\mathscr{B}_n}, \mathbb{U}_{\mathscr{B}_i \in \mathfrak{B}}, T\right\}$ at each time period t of gameplay (Kim, 2018a).

- $\mathbf{B} = \{\mathscr{B}_1 \ldots \mathscr{B}_n\}$ represents a set of SBSs; they are the first-tier game players.
- The bandwidth capacity of 5G network system is$\mathbb{C}$; it is divided up into SBSs.
- $\mathbb{S}_{\mathscr{B}_1} \ldots \mathbb{S}_{\mathscr{B}_n}$ is a strategy vector, which corresponds to the assigned bandwidth amount for each SBSs where $\mathbb{S}_{\mathscr{B}_i}$ is the allocation strategy set of $\mathscr{B}_i$.
- $\mathbb{U}_{\mathscr{B}_i}$ is the payoff received by the $\mathscr{B}_i$ during the bandwidth allocation operation.
- $T = \{\mathscr{H}_1, \ldots, \mathscr{H}_t, \mathscr{H}_{t+1}, \ldots\}$ denotes time, which is represented by a sequence of time steps with imperfect information for the $\mathbb{G}^{first}$ game process.

Secondly, the $\mathbb{G}_i^{second}$ is the ith second-tier game, and it can be defined as $\mathbb{G}_i^{second} = \left\{\{\mathscr{B}_i \in \mathfrak{B}, \mathscr{E}_{\mathscr{B}_i}\}, \left\{\{S^{\mathscr{P}}_{\mathscr{B}_i}, S^{\mathscr{Q}}_{\mathscr{B}_i}\}, S_{\mathscr{E}^j_{\mathscr{B}_i} \in \mathscr{E}_{\mathscr{B}_i}}\right\}, \left\{U_{\mathscr{B}_i}, \mathscr{U}^{\mathscr{B}_i}_{\mathscr{E}^j_{\mathscr{B}_i} \in \mathscr{E}_{\mathscr{B}_i}}\right\}, T\right\}$ at each time period t of gameplay. There are total n second-tier games $\left(\mathbb{G}^{second}_{1 \le i \le n}\right)$ in the two-tier game structure (Kim, 2018a).

- $\mathscr{E}_{\mathscr{B}_i} = \{\mathscr{E}^1_{\mathscr{B}_i} \ldots \mathscr{E}^r_{\mathscr{B}_i}\}$ is the set of UEs, who are in the $\mathscr{B}_i$'s coverage area. $\mathscr{B}_i$ and $\mathscr{E}_{\mathscr{B}_i}$ are game players for the second-tier $\mathbb{G}_i^{second}$ game, which is formulated as a Stackelberg model where $\mathscr{B}_i$ is a leader and $\mathscr{E}_{\mathscr{B}_i}$ are followers.
- $\left\{S^{\mathscr{P}}_{\mathscr{B}_i}, S^{\mathscr{Q}}_{\mathscr{B}_i}\right\}$ is the strategy sets of $\mathscr{B}_i$ where $S^{\mathscr{P}}_{\mathscr{B}_i} = \left\{\prod_{min \le k \le max} \mathfrak{P}^{\mathscr{B}_i}_k \middle| \mathfrak{P}^{\mathscr{B}_i}_k \in [\mathfrak{P}^{\mathscr{B}_i}_{min} \ldots \mathfrak{P}^{\mathscr{B}_i}_k \ldots \mathfrak{P}^{\mathscr{B}_i}_{max}]\right\}$ and $\mathfrak{P}^{\mathscr{B}_i}_k$ means the kth service price strategy. $S^{\mathscr{Q}}_{\mathscr{B}_i}$ is the $\mathscr{B}_i$'s strategy for cache splitting where $0 \le S^{\mathscr{Q}}_{\mathscr{B}_i} \le 1$ is the ratio for data caching. Therefore, the $\left[S^{\mathscr{Q}}_{\mathscr{B}_i} \times \mathscr{Q}_{\mathscr{B}_i}\right]$ cache capacity is assigned

to data caching $\left(\mathbb{Q}^{\mathbb{D}}_{\mathcal{B}_i}\right)$ for computation offloading services and the remaining cache capacity $\left[\left(1 - S^{\mathbb{Q}}_{\mathcal{B}_i}\right) \times \mathbb{Q}_{\mathcal{B}_i}\right]$ is assigned to contents caching $\left(\mathbb{Q}^{\mathbb{M}}_{\mathcal{B}_i}\right)$ for mobile communication services.

- $S_{\mathcal{E}^j_{\mathcal{B}_i} \in \mathcal{E}_{\mathcal{B}_i}}$ is the strategy set of $\mathcal{E}^j_{\mathcal{B}_i}$ where $S_{\mathcal{E}^j_{\mathcal{B}_i}} = \{(0,0),(0,1),(1,0),(1,1)\}$ is the decision combination for computation offloading and communication services, respectively; 0 represents no service and 1 represents service activation.
- In $\left\{U_{\mathcal{B}_i}, \mathcal{U}^{\mathcal{B}_i}_{\mathcal{E}^j_{\mathcal{B}_i} \in \mathcal{E}_{\mathcal{B}_i}}\right\}$, $U_{\mathcal{B}_i}$ is the $\mathcal{B}_i$'s payoff and $\mathcal{U}^{\mathcal{B}_i}_{\mathcal{E}^j_{\mathcal{B}_i}}$ is the $\mathcal{E}^j_{\mathcal{B}_i}$ payoff received from the $\mathbb{G}^{second}_i$ game process, respectively.
- The T is a time period. The $\mathbb{G}^{second}_i$ is repeated $t \in T < \infty$ time periods with imperfect information.

During the hierarchical two-tier game (G) operations, $\mathbb{G}^{first}$ and $\mathbb{G}^{second}_{1 \le i \le n}$ games work together in a coordinated manner. In the $\mathbb{G}^{first}$ game process, SBSs bargain with each other to fairly distribute the $\mathbb{C}^{\mathcal{B}}$. According to their contributions, SBSs make joint-agreements that give mutual advantage. In the multiple $\mathbb{G}^{second}_{1 \le i \le n}$ game processes, all individual game players select their strategies selfishly to maximize their payoffs. At the end of each game iteration, players examine their payoffs periodically and dynamically adapt their decisions in an entirely distributed fashion. During the step-by-step iteration, this feedback process is repeated until the best solution has been found (Kim, 2018a).

3.1.4. First-tier Game Model for Bandwidth Allocation Algorithm

Accommodating the surge in data volume and performance demands is among the most significant hurdles for 5G networks. SBSs have emerged as a viable solution for expanding cellular network capacity, yet ensuring better user services and maximizing resource usage also requires coordination protocols for bandwidth allocation. Recognizing the parallels between bargaining game decisions and bandwidth allocation processes, the *TCCC* scheme introduces a first-tier bargaining game model to guide bandwidth distribution among SBSs (Kim, 2018a).

Within the bandwidth allocation game model, SBSs serve as game players, and their strategies revolve around choosing specific bandwidth allocations. To gauge each player's satisfaction regarding the final outcome, the *TCCC* scheme defines a utility function for every participating SBS. By relying on the reciprocal interaction between resource utilization and user's QoS, the utility function for the $\mathcal{B}_i$ ($\mathbb{U}_{\mathcal{B}_i}$) can be derived from (Kim, 2018a; Niyato, 2008);

$$\mathbb{U}_{\mathcal{B}_i} = \frac{1}{1-\omega_{\mathcal{B}_i}} \times \max\left(0, \left[1 - \left(\omega_{\mathcal{B}_i} \times \exp\left(\mathfrak{N}_{\mathcal{B}_i}\right)\right)\right]\right) \tag{3.1.1}$$

$$\text{s.t., } \mathfrak{N}_{\mathcal{B}_i} = \frac{E^m}{m!} \Big/ \sum_{c=0}^{m} \frac{E^c}{c!} \text{ and } 0 \le \omega_{\mathcal{B}_i} \le 1$$

where $\mathfrak{N}_{\mathcal{B}_i}$ is the service blocking probability, and $\omega_{\mathcal{B}_i}$ is a control parameter of the $\mathcal{B}_i$'s utility function. Based on the traffic information such as bandwidth capacity, channel size, call duration time, $\mathfrak{N}_{\mathcal{B}_i}$ can be estimated according to the Erlang's formula (Niyato, 2008); E is the total amount of offered traffic in the $\mathcal{B}_i$ and m is the number of wireless channels for the $\mathcal{B}_i$. Due to the current traffic characteristics, $\omega_{\mathcal{B}_i}$ value is defined as the cache hitting ratio of $\mathcal{B}_i$; the larger the value of $\omega_{\mathcal{B}_i}$, the more sensitive is the utility to the $\mathfrak{N}_{\mathcal{B}_i}$.

Using a bargaining framework, SBSs in the first-tier game adopt a collaborative perspective, working jointly toward a collective decision. Drawing from the idea of *relative utilitarianism* (Dhillon, 1999; Sobel, 2001), the *TCCC* scheme introduces the *relative utilitarian bargaining solution* (*RUBS*) and devises an adaptive feedback bargaining model to iteratively refine the bandwidth allocation. This iterative process, designed for dynamically shifting 5G conditions, remains flexible in its pursuit of a fair and efficient network performance (Kim, 2018a).

In the bargaining solution, the solution set ($\mathbb{W}$) is normally interpreted as the set of feasible utility payoffs to game players. A point $\mathbf{P} = \left\{\mathbb{U}_{\mathcal{B}_1} \ldots \mathbb{U}_{\mathcal{B}_m}\right\} \in \mathbb{W}$ can be achieved if all SBSs agree to it. In case of agreement, game player $\mathcal{B}_i$ receives $\mathbb{U}_{\mathcal{B}_i}$. According to the *relative utilitarianism*, the desirable best bargaining solution for an utility vector $\left\{\mathbb{U}^*_{\mathcal{B}_1} \ldots \mathbb{U}^*_{\mathcal{B}_m}\right\}$ is obtained as follows (Kim, 2018a).

$$\left\{\mathbb{U}^* = \left\{\mathbb{U}^*_{\mathcal{B}_1} \ldots \mathbb{U}^*_{\mathcal{B}_m}\right\} | \mathbb{U}^* \text{ solves } \max_{\mathbb{S}_{\mathcal{B}_1} \ldots \mathbb{S}_{\mathcal{B}_m}} \sum_{\mathcal{B}_i \in \mathfrak{B}} \left(\frac{\mathbb{U}_{\mathcal{B}_i}}{\mathbb{U}^{max}_{\mathcal{B}_i} - \mathbb{U}^{min}_{\mathcal{B}_i}}\right)^{1-\rho_{\mathcal{B}_i}}\right\} \tag{3.1.2}$$

$$\text{s.t., } 0 \leq \frac{\mathbb{U}_{\mathcal{B}}}{\mathbb{U}^{max}_{\mathcal{B}} - \mathbb{U}^{min}_{\mathcal{B}}} \leq 1 \text{ and } \rho_{\mathcal{B}_i} = \frac{\mathfrak{P}^{\mathcal{B}_i}}{\sum_{\mathcal{B}_j \in \mathfrak{B}} \mathfrak{P}^{\mathcal{B}_j}}$$

where $\rho_{\mathcal{B}_i}$ is the $\mathcal{B}_i$'s service price strategy. It is the relative ability to exert influence over other SBSs. By considering the system efficiency of SBSs, the *TCCC* scheme can give appropriate incentives or punishments to each individual SBS; the SBS with a higher service price obtains a more bandwidth resource than other SBSs (Kim, 2018a).

3.1.5 Second-tier Game Model for Caching Algorithms

In recent years, mobile UEs are expected to manage increasingly complex applications such as multimedia processing, online gaming, virtual reality, and sensing. Given their constrained computation capabilities and power resources, cloud computing has emerged as an advanced solution. Within 5G network infrastructures, UEs can transfer computation-heavy tasks to resource-rich SBSs, thus allowing UEs to flexibly utilize resources in an on-demand manner (Andreev, 2016). In most cloud computing scenarios, application execution and related data access are strongly interconnected (Fan, 2016b), so communication among UEs, SBSs, and remote cloud servers becomes a primary contributor to delay latency. Integrating cloud computing more extensively into wireless environments introduces additional challenges to system performance (Kim, 2018a).

To optimize performance, the *TCCC* scheme employs a second-tier game modeled as a Stackelberg game. By definition, one participant is the leader and the remaining participants are followers (Kim, 2014). In this tier, each SBS plays the leader role, with its assigned UEs serving as followers. The leader's aim is to steer the followers' self-serving objectives toward outcomes beneficial to the entire network. From the follower perspective, UEs work to maximize their individual utility in a distributed online setting, with their strategic decisions serving as inputs to the SBS's control adjustments. This iterative, feedback-oriented exchange continues until players converge on an acceptable result. The SBS leader thus orchestrates caching and pricing decisions, while UE followers choose offloading and communication strategies (Kim, 2018a).

Recently, SBS caching has been introduced to store popular data and content at SBSs, and it is gaining traction as an effective method to reduce service latency in 5G networks. SBS caching offers multiple advantages. First, it reduces the distance between desired data or content and the UE, potentially lowering end-to-end latency. Second, it alleviates front-haul congestion by substituting cache capacity at local SBSs for front-haul transmission requirements. Without caching, UEs would otherwise download these files through the front-haul link. Third, caching at SBSs can reduce the request load on remote origin servers by directly supplying requested data or content (Andreev, 2016; Sheng, 2016; Kim, 2018a).

In the *TCCC* scheme, each SBS's caching infrastructure is split into two sections: one designated for popular content intended for communication services, and the other for popular data supporting computation offloading. By partitioning cache resources in this manner, system performance can be enhanced while maintaining a balance between communication and computation tasks. Since each SBS has a finite cache capacity, two key decisions must be made: how to split the cache for data vs. content, and how to determine a cache placement policy based on the observed request frequency. These decisions are integral to achieving optimal performance in a 5G network environment (Kim, 2018a).

In the second-tier game, each individual SBS adaptively split the total cache capacity ($\mathcal{Q}_{\mathcal{B}}$) for data file caching ($\mathcal{Q}_{\mathcal{B}}^{\mathrm{D}}$) and content file caching ($\mathcal{Q}_{\mathcal{B}}^{\mathrm{M}}$) where $\mathcal{Q}_{\mathcal{B}} = \mathcal{Q}_{\mathcal{B}}^{\mathrm{D}} + \mathcal{Q}_{\mathcal{B}}^{\mathrm{M}}$. To find the best solution of cache splitting problem, the *TCCC* scheme adopts the concepts of *Kalai-Smorodinsky bargaining solution* (KSBS) and *egalitarian bargaining solutions* (EBS). The main feature of KSBS is that the increasing of bargaining set size in a direction favorable to a specific player always benefits that player. Simply, the *TCCC* scheme can think that the KSBS is the maximal point which maintains the ratios of gains. Unlike KSBS, the EBS enjoys even stronger monotonicity requirements while satisfying independence conditions (Kim, 2014). Therefore, KSBS equalizes the ratios between players' payoffs and their ideal payoffs while the EBS equalizes players' payoffs. To design the cache splitting algorithm, data caching $\left(\mathcal{Q}_{\mathcal{B}}^{\mathrm{D}}\right)$ and contents caching $\left(\mathcal{Q}_{\mathcal{B}}^{\mathrm{M}}\right)$ are assumed as game players, and the *TCCC* scheme splits the $\mathcal{Q}_{\mathcal{B}_i}$ of $\mathcal{B}_i$ as follows (Kim, 2018a);

$$\frac{\delta_{\mathcal{H}_t}\left(\mathcal{Q}_{\mathcal{B}_i}^{\mathrm{D}}\right)}{\Omega_{\mathcal{H}_t}\left(\mathcal{Q}_{\mathcal{B}_i}^{\mathrm{M}}\right)} = \left(\frac{\delta_{\mathcal{H}_t}\left(\mathbb{O}_{\mathcal{B}_i}^{\mathrm{D}}\right) = \left\{\mathbb{O}_{\mathcal{B}_i}^{\mathrm{D}} = \mathcal{X}|\mathbf{optimal}\left(\delta_{\mathcal{H}_t}(\mathcal{X})\right)\right\}}{\Omega_{\mathcal{H}_t}\left(\mathbb{O}_{\mathcal{B}_i}^{\mathrm{M}}\right) = \left\{\mathbb{O}_{\mathcal{B}_i}^{\mathrm{M}} = \mathcal{Y}|\mathbf{optimal}\left(\Omega_{\mathcal{H}_t}(\mathcal{Y})\right)\right\}}\right)^{\gamma} \tag{3.1.3}$$

where $\delta_{\mathcal{H}_t}(\cdot)$ and $\Omega_{\mathcal{H}_t}(\cdot)$ are the cache hit ratio per cache size of $\mathcal{Q}_{\mathcal{B}_i}^{\mathrm{D}}$ and $\mathcal{Q}_{\mathcal{B}_i}^{\mathrm{M}}$, respectively, at time $\mathcal{H}_t$. $\mathbb{O}_{\mathcal{B}_i}^{\mathrm{D}}$ and $\mathbb{O}_{\mathcal{B}_i}^{\mathrm{M}}$ guarantee the ideal gains of $\delta_{\mathcal{H}_t}(\cdot)$ and $\Omega_{\mathcal{H}_t}(\cdot)$; they are the maximal possible ratios. The KSBS and EBS correspond to $\gamma = 1$ and $\gamma = 0$, respectively where $0 \leq \gamma \leq 1$ (Kim, 2018a).

Under the dynamic 5G network environment, a fixed value of γ cannot effectively adapt to the changing conditions. In the *TCCC* scheme, it is treated as an on-line decision problem and adaptively modify γ value. To fine-tune the system performance, it is a suitable approach. When the difference of $\delta\left(\mathbb{O}_{\mathcal{B}_i}^{\mathrm{D}}\right)$ and $\Omega\left(\mathbb{O}_{\mathcal{B}_i}^{\mathrm{M}}\right)$ is high, the *TCCC* scheme can put more emphasis on the axiom of *individual monotonicity*. In this case, KSBS is more suitable. When the difference of $\delta\left(\mathbb{O}_{\mathcal{B}_i}^{\mathrm{D}}\right)$ and $\Omega\left(\mathbb{O}_{\mathcal{B}_i}^{\mathrm{M}}\right)$ is low, the *TCCC* scheme strongly depends on the on the axiom of *strong equality*. In this case, EBS is more suitable. In the *TCCC* scheme, the value of γ is dynamically adjusted based on the current values of $\delta\left(\mathbb{O}_{\mathcal{B}_i}^{\mathrm{D}}\right)$ and $\Omega\left(\mathbb{O}_{\mathcal{B}_i}^{\mathrm{M}}\right)$. Therefore, the system can be more responsive to current 5G network conditions by the real-time network monitoring. The value of γ at time $\mathcal{H}_t$ is given by (Kim, 2018a);

$$\gamma = \left|\delta_{\mathcal{H}_i}(\mathbb{O}^{D}_{\mathcal{B}_i}) - \Omega_{\mathcal{H}_i}(\mathbb{O}^{M}_{\mathcal{B}_i})\right| \Big/ \max\left(\delta_{\mathcal{H}_i}(\mathbb{O}^{D}_{\mathcal{B}_i}), \Omega_{\mathcal{H}_i}(\mathbb{O}^{M}_{\mathcal{B}_i})\right) \tag{3.1.4}$$

Due to the limited cache capacity, it is impossible for caching all requested data and contents. Therefore, popular data and contents are properly cached in each SBS to reduce delay and system overhead. To develop a cache placement algorithm, the *TCCC* scheme assumes that a computation data file set D and a communication contents file set M, and files in D and M can be possibly cached in each SBS. The popularity distributions among D and M are represented by vectors, which are frequently requested by UEs. Generally, the vectors can be modeled by a Zipf distribution, which is a discrete probability distribution commonly used in the modelling of rare events (Li, 2016). As most of the cache placing work assumes perfect content popularity, the *TCCC* scheme also adopts the Zipf distribution to model the file popularity, and implement the cache placement algorithm according to this distribution (Kim, 2018a).

3.1.6. Second-tier Game Model for 5G Services

Within the second-tier game model, both SBSs and UEs act as self-interested participants, each aiming to maximize its own payoff. Because decisions are made iteratively, the choices of one player can influence the choices of others, creating a cascading effect within the control process. Guided by their utility functions, all players select the most advantageous strategies, taking into account the outcomes and responses of other game participants (Kim, 2018a).

Within the *TCCC* scheme, a follower's utility function is defined with two components reflecting cooperative and non-cooperative tendencies. The cooperative aspect employs a Taguchi loss function, where the follower's dissatisfaction grows in proportion to any deviation from the desired communication bandwidth (Konstantakopoulos, 2016). This mechanism encourages otherwise self-serving UEs to align with broader social objectives. On the other hand, the non-cooperative component models a user's own payoff - namely, the difference between the benefits received and the costs incurred. Based on this assumption, the utility function of $\mathcal{E}^k_{\mathcal{B}_i}$ $\left(\mathcal{U}^{\mathcal{B}_i}_{\mathcal{E}^k_{\mathcal{B}_i}}\right)$ is defined as follows (Kim, 2018a);

$$\begin{aligned}
&\mathcal{U}^{\mathcal{B}_i}_{\mathcal{E}^k_{\mathcal{B}_i}}\left(\mathcal{P}_{\mathcal{B}_i}, \Gamma(\mathcal{E}^k_{\mathcal{B}_i}), \Theta(\mathcal{E}^k_{\mathcal{B}_i})\right) = \left(\mu_{\mathcal{E}^k_{\mathcal{B}_i}} \times \log\left(\Gamma(\mathcal{E}^k_{\mathcal{B}_i})\right)\right) + \left(\psi_{\mathcal{E}^k_{\mathcal{B}_i}} \times \log\left(\Theta(\mathcal{E}^k_{\mathcal{B}_i})\right)\right) \\
&- \frac{\tau_{\mathcal{E}^k_{\mathcal{B}_i}}}{\left|\mathcal{E}_{\mathcal{B}_i}\right| - 1}\left(\sum_{\substack{\mathcal{E}^l_{\mathcal{B}_i} \in \mathcal{E}_{\mathcal{B}_i} \\ \mathcal{E}^l_{\mathcal{B}_i} \neq \mathcal{E}^k_{\mathcal{B}_i}}} \max\left\{\left(\Gamma(\mathcal{E}^l_{\mathcal{B}_i}) - \Gamma(\mathcal{E}^k_{\mathcal{B}_i}), 0\right)\right\}\right) - \mathcal{C}\left(\mathcal{P}_{\mathcal{B}_i}, \Gamma(\mathcal{E}^k_{\mathcal{B}_i})\right) \\
&- \frac{\phi_{\mathcal{E}^k_{\mathcal{B}_i}}}{\left|\mathcal{E}_{\mathcal{B}_i}\right| - 1}\left(\sum_{\substack{\mathcal{E}^l_{\mathcal{B}_i} \in \mathcal{E}_{\mathcal{B}_i} \\ \mathcal{E}^l_{\mathcal{B}_i} \neq \mathcal{E}^k_{\mathcal{B}_i}}} \max\left\{\left(\Gamma(\mathcal{E}^k_{\mathcal{B}_i}) - \Gamma(\mathcal{E}^l_{\mathcal{B}_i}), 0\right)\right\}\right) - \mathcal{D}\left(\mathcal{P}_{\mathcal{B}_i}, \Theta(\mathcal{E}^k_{\mathcal{B}_i})\right)
\end{aligned} \tag{3.1.5}$$

$$\text{s.t.,}\begin{cases} D\left(\mathcal{P}_{\mathcal{B}_i},\Theta\left(\mathscr{E}^k_{\mathcal{B}_i}\right)\right) = \begin{cases}\frac{1}{\zeta}\times\left[\mathfrak{X}\times\log\left(\Theta\left(\mathscr{E}^k_{\mathcal{B}_i}\right)\right)\right]\times\mathcal{P}_{\mathcal{B}_i}, ifdataiscacheed\\ \left[\mathfrak{X}\times\log\left(\Theta\left(\mathscr{E}^k_{\mathcal{B}_i}\right)\right)\right]\times\mathcal{P}_{\mathcal{B}_i}, otherwise\end{cases}\\ C\left(\mathcal{P}_{\mathcal{B}_i},\Gamma\left(\mathscr{E}^k_{\mathcal{B}_i}\right)\right) = \begin{cases}\frac{1}{\delta}\times\left[\mathfrak{Y}\times\log\left(\Gamma\left(\mathscr{E}^k_{\mathcal{B}_i}\right)\right)\right]\times\mathcal{P}_{\mathcal{B}_i}, ifcontentiscacheed\\ \left[\mathfrak{Y}\times\log\left(\Gamma\left(\mathscr{E}^k_{\mathcal{B}_i}\right)\right)\right]\times\mathcal{P}_{\mathcal{B}_i}, otherwise\end{cases}\end{cases}$$

where $\Gamma\left(\mathscr{E}^k_{\mathcal{B}_i}\right),\Theta\left(\mathscr{E}^k_{\mathcal{B}_i}\right)$ are $\mathscr{E}^k_{\mathcal{B}_i}$'s requested communication bandwidth and offloading computation amounts, respectively. $\mu_{\mathscr{E}^k_{\mathcal{B}_i}}$ and $\psi_{\mathscr{E}^k_{\mathcal{B}_i}}$ denote the $\mathscr{E}^k_{\mathcal{B}_i}$'s satisfaction factors for communication and computation services. $\mathscr{D}(\cdot),\mathscr{C}(\cdot)$ are offloading computation and communication cost functions of $\mathfrak{I}^h_k$, respectively. Let $\tau_{\mathscr{E}^k_{\mathcal{B}_i}}$ and $\phi_{\mathscr{E}^k_{\mathcal{B}_i}}$ be the $\mathscr{E}^k_{\mathcal{B}_i}$'s degree of envy and degree of guilty. Based on the concept of inequality aversion, social welfare preference is implemented in the follower's utility function (Kim, 2018a).

As a leader, each individual SBS considers its own payoff and social welfare. To deal with this multi-objective control problem, it is necessary to find the best compromise solution while keeping a good balance. Based on the reciprocal relationship of two objectives, the utility function of $\mathcal{B}_i$ at time $\mathscr{H}_t\left(U^{\mathscr{H}_t}_{\mathcal{B}_i}(\cdot)\right)$ is defined as follows (Kim, 2018a);

$$U^{\mathscr{H}_t}_{\mathcal{B}_i}\left(\mathcal{P}_{\mathcal{B}_i},\mathscr{E}_{\mathcal{B}_i},S^{\mathbb{Q}}_{\mathcal{B}_i}\right) = \left((1-\alpha)\times\sum\nolimits_{\mathscr{E}^l_{\mathcal{B}_i}\in\mathscr{E}_{\mathcal{B}_i}}\left(\mathscr{D}\left(\mathcal{P}_{\mathcal{B}_i},\Theta\left(\mathscr{E}^l_{\mathcal{B}_i}\right)\right)+\mathscr{C}\left(\mathcal{P}_{\mathcal{B}_i},\Gamma\left(\mathscr{E}^l_{\mathcal{B}_i}\right)\right)\right)\right)$$
$$+\left(\alpha\times\sum_{\mathscr{E}^l_{\mathcal{B}_i}\in\mathscr{E}_{\mathcal{B}_i}}\mathscr{U}^{\mathcal{B}_i}_{\mathscr{E}^l_{\mathcal{B}_i}}\left(\mathcal{P}_{\mathcal{B}_i},\Gamma\left(\mathscr{E}^l_{\mathcal{B}_i}\right),\Theta\left(\mathscr{E}^l_{\mathcal{B}_i}\right)\right)\right) \tag{3.1.6}$$

$$\text{s.t.,}\ \alpha = \frac{\left(\sum_{\mathscr{E}^l_{\mathcal{B}_i}\in\mathscr{E}_{\mathcal{B}_i}}\mathscr{U}^{\mathcal{B}_i}_{\mathscr{E}^l_{\mathcal{B}_i}}\left(\mathcal{P}_{\mathcal{B}_i},\Gamma\left(\mathscr{E}^l_{\mathcal{B}_i}\right),\Theta\left(\mathscr{E}^l_{\mathcal{B}_i}\right)\right)\right)^2}{\left|\mathscr{E}_{\mathcal{B}_i}\right|\times\sum_{\mathscr{E}^l_{\mathcal{B}_i}\in\mathscr{E}_{\mathcal{B}_i}}\left(\mathscr{U}^{\mathcal{B}_i}_{\mathscr{E}^l_{\mathcal{B}_i}}\left(\mathcal{P}_{\mathcal{B}_i},\Gamma\left(\mathscr{E}^l_{\mathcal{B}_i}\right),\Theta\left(\mathscr{E}^l_{\mathcal{B}_i}\right)\right)\right)^2}$$

To learn the traffic situation in 5G networks, the $\mathcal{B}_i$ periodically monitors the current local and global information; local information is its own payoff and global information represents the situation of neighboring SBSs. If the strategy $\mathfrak{P}^{\mathcal{B}_i}_k$ is selected at time $\mathscr{H}_{t-1}$ by the $\mathcal{B}_i$, the $\mathcal{B}_i$ updates the strategy $\mathfrak{P}^{\mathcal{B}_i}_k$'s learning value $\left(\mathbb{L}_{\mathscr{H}_t}\left(\mathfrak{P}^{\mathcal{B}_i}_k,\mathcal{B}_i\right)\right)$ for the next time $\mathscr{H}_t$ as follows (Kim, 2018a);

$$\mathbb{L}_{\mathscr{H}_t}\left(\mathfrak{P}^{\mathcal{B}_i}_k,\mathcal{B}_i\right) = \left((1-\chi)\times\mathbb{L}_{\mathscr{H}_{t-1}}\left(\mathfrak{P}^{\mathcal{B}_i}_k,\mathcal{B}_i\right)\right) + (\chi\times[\mathbb{H}+\mathbb{Y}]) \tag{3.1.7}$$

$$\text{s.t.,}\mathbb{H} = \xi\times U^{\mathscr{H}_{t-1}}_{\mathcal{B}_i}\left(\mathcal{P}_{\mathcal{B}_i},\mathscr{E}_{\mathcal{B}_i},S^{\mathbb{Q}}_{\mathcal{B}_i}\right),\mathbb{Y} = \left\{\frac{(1-\xi)}{|\mathfrak{B}|-1}\times\sum\nolimits_{\mathcal{B}_j\in\mathfrak{B},\mathcal{B}_j\neq\mathcal{B}_i}\mathbb{L}_{\mathscr{H}_{t-1}}\left(\mathfrak{P}^{\mathcal{B}_i}_k,\mathcal{B}_j\right)\right\}$$

$$\text{and}\ \xi = \min\left(\frac{1}{2},\left(\frac{1}{|\mathfrak{B}|-1}\times\sum\nolimits_{\mathcal{B}_j\in\mathfrak{B},\mathcal{B}_j\neq\mathcal{B}_i}\frac{\left|\mathfrak{M}_{\mathscr{H}_{t-1}}\left(\mathcal{B}_j\right)-\mathfrak{M}_{\mathscr{H}_{t-1}}\left(\mathcal{B}_i\right)\right|}{\mathfrak{M}_{\mathscr{H}_{t-1}}\left(\mathcal{B}_i\right)}\right)\right)$$

where χ is a learning rate that models how the L-values are updated. In the equation (3.1.7), H and Y represent local and global learning values, and ξ is a control factor for the weighted average between different learning approaches. $\mathfrak{M}_{\mathscr{H}_{t-1}}(\mathscr{B}_j)$ is the total traffic amount of $\mathscr{B}_j$ at time $\mathscr{H}_{t-1}$ (Kim, 2018a).

Based on the $\mathbb{L}(\cdot)$ values, a strategy selection distribution (P) for each SBS is defined. To respond the current 5G network situation at time $\mathscr{H}_t$, the *TCCC* scheme determines $\mathbb{P}_{\mathscr{H}_t}^{\mathscr{B}_i} = \left\{\mathscr{P}_{\mathscr{H}_t}(\mathfrak{P}_{min}^{\mathscr{B}_i}) \ldots \mathscr{P}_{\mathscr{H}_t}(\mathfrak{P}_{k}^{\mathscr{B}_i}) \ldots \mathscr{P}_{\mathscr{H}_t}(\mathfrak{P}_{max}^{\mathscr{B}_i})\right\}$ as the probability distribution of $\mathfrak{P}^{\mathscr{B}_i}$'s strategy selection; it is sequentially modified over time. From the $\mathbb{P}_{\mathscr{H}_t}^{\mathscr{B}_i}$, the $\mathfrak{P}_k^{\mathscr{B}_i}$ strategy selection probability by the $\mathscr{B}_i$ at time $\mathscr{H}_t$ $\left(\mathscr{P}_{\mathscr{H}_t}(\mathfrak{P}_k^{\mathscr{B}_i})\right)$ is defined as (Kim, 2018a):

$$\mathscr{P}_{\mathscr{H}_t}(\mathfrak{P}_k^{\mathscr{B}_i}) = \frac{\mathrm{EXP}\left(\mathbb{L}_{\mathscr{H}_t}(\mathfrak{P}_k^{\mathscr{B}_i}, \mathscr{B}_i)\right)}{\sum_{e=min}^{max} \mathrm{EXP}\left(\mathbb{L}_{\mathscr{H}_t}(\mathfrak{P}_e^{\mathscr{B}_i}, \mathscr{B}_i)\right)} \tag{3.1.8}$$

At each game rounds, the learning process will be taken sequentially according to the equation (3.1.5)-(3.1.8), and the $\mathscr{B}_i$ stochastically selects the $\mathfrak{P}^{\mathscr{B}_i}$ strategy using his strategy selection distribution ($\mathbb{P}^{\mathscr{B}_i}$). The *TCCC* scheme effectively implements the second-tier game model by adopting the dynamic learning based Stackelberg model. At each game period, leaders and followers attempt to maximize their payoffs by modifying their strategies (Kim, 2018a).

$$\max_{S_{\mathscr{E}_{\mathscr{B}_i}}}\left(\mathscr{U}_{\mathscr{E}_{\mathscr{B}_i}^k}^{\mathscr{B}_i}\left(\mathscr{P}_{\mathscr{B}_i}, \Gamma\left(\mathscr{E}_{\mathscr{B}_i}^k\right), \Theta\left(\mathscr{E}_{\mathscr{B}_i}^k\right)\right)\right) \text{ and } \max_{\mathscr{P}_{\mathscr{B}_i} \in S_{\mathscr{B}_i}^{\mathscr{P}}, S_{\mathscr{B}_i}^{\mathbb{Q}}}\left(U_{\mathscr{B}_i}^{\mathscr{H}_i}\left(\mathscr{P}_{\mathscr{B}_i}, \mathscr{E}_{\mathscr{B}_i}, S_{\mathscr{B}_i}^{\mathbb{Q}}\right)\right) \tag{3.1.9}$$

3.1.7 Summary

To accommodate the soaring demands of next-generation traffic services, a synergistic combination of communications, caching, and computing has emerged as a promising strategy for 5G networks. In response, the *TCCC* scheme proposes a holistic control approach that integrates bandwidth allocation, cache splitting, and computation offloading mechanisms. Anchored by a two-tier game model, the *TCCC* scheme addresses key questions about optimal control decisions to achieve balanced network performance. From an SBS perspective, both individual payoffs and overall social welfare are prioritized. For UEs, the *TCCC* scheme captures a balance between cooperative and self-interested (non-cooperative) inclinations. Through a step-by-step interactive process, SBSs and UEs sequentially refine their strategies to maximize expected gains. This repeated feedback method is particularly practical under conditions of incomplete information. Leveraging dynamic bargaining and Stackelberg game procedures, the *TCCC* scheme offers synergistic and complementary capabilities, adapting effectively to the changing demands of 5G network environments (Kim, 2018a).

3.2 Edge Fog and Cloud Computing (*EFCC*) Scheme

In recent years, the rapid escalation of computation-intensive applications has led to widespread use of fog and cloud computing paradigms on smart-chip-enabled mobile devices. These devices can partially offload processing tasks to either fog or cloud resources. Recognizing the challenges posed by future edge, fog, and cloud computing frameworks, S. Kim introduces the *Edge Fog and Cloud Computing* (*EFCC*) scheme (Kim, 2020a). The *EFCC* scheme seeks to address the task offloading problem through two cooperative bargaining game solutions - specifically, the *Tempered Aspirations Bargaining Solution* (*TABS*) and the *Gupta-Livne Bargaining Solution* (*GLBS*). By selecting an appropriate bargaining solution, the *EFCC* scheme aims to maximize application service quality. Within the *EFCC* scheme, the *TABS* is employed for time-sensitive offloading scenarios, while the *GLBS* is utilized to guarantee effective computation-centric offloading. A key benefit of this bargaining-based approach is its axiom-driven strategic framework, enabling task offloading decisions that adapt to changing network conditions (Kim, 2020a).

3.2.1 Development Motivation

The IoT connects billions of smart devices worldwide using standardized protocols and wireless communications, offering innovative 5G service applications. Concurrently, 5G networks are accelerating IoT development by providing broader coverage, higher throughput, lower latency, and dense connections for massive bandwidth. However, integrating diverse control criteria remains a significant challenge for legacy network infrastructures that rely on traditional computing paradigms (Hassan, 2018; Kim, 2020a; Wang, 2018). Despite improved capabilities in smart devices, they still trail behind server-grade hardware in terms of battery life, storage, compute power, and wireless bandwidth availability. These constraints hinder devices from seamlessly running content-rich or computation-intensive applications (e.g., real-time image processing, augmented reality, location-based services). Cloud computing has been proposed to bridge this gap by offloading computing, data storage, and mass information processing tasks to the cloud while ensuring reliable, high-availability services. This structure, called the Cloud of Things (CoT), helps lay the foundation for next-generation network architectures (Kim, 2020a; Li, 2015; Son, 2018).

In CoT scenarios, however, communicating with a remote cloud can introduce latency and limit real-time responsiveness. Because mobile devices cannot wait indefinitely for cloud responses, especially under stringent delay requirements, attention has turned to fog computing - a highly distributed approach that places cloudlets or fog nodes near the network edge. These nodes, which function like small-scale cloud data centers, offer local computation offloading with minimal latency and form the Fog-of-Things (FoT) paradigm (Funk, 2018; Kim, 2020a; Oteafy, 2018). Recently, improved embedded systems-on-a-chip have also produced more intelligent IoT devices capable of running partial computation-intensive tasks. This development paves the way for a new model known as the Edge-of-Things (EoT) (Hassan, 2018). Under the CoT or FoT paradigm, network dynamics can still yield overhead in response time and traffic management. EoT addresses this by enabling devices themselves to handle computing and communication tasks - often reducing latency, bandwidth load, and security concerns. However, edge devices remain constrained by cost and power limitations, restricting their processing capabilities (El, 2017; Kim, 2020a; Yu, 2018).

While FoT and EoT share the goal of moving processing closer to data sources, they differ in how and where intelligence is placed. In FoT, fog nodes or IoT gateways at the local network level carry out data processing and management - an intermediary between physical devices and the cloud. In EoT, processing and communications capacity are shifted directly onto edge devices. This local-first approach benefits latency, bandwidth, and security, but edge devices often remain limited in resources (El, 2017; Kim, 2020a; Yu, 2018). Notably, CoT, FoT, and EoT are not mutually exclusive. Instead, they can complement one another's limitations, prompting a joint framework to improve future network performance.

The *EFCC* scheme proposes a task offloading algorithm that embraces all three paradigms - CoT, FoT, and EoT - to produce a combined design with synergistic benefits. Yet, implementing a holistic solution demands innovative techniques (Kim, 2020a).

The *EFCC* scheme seeks to build a task offloading mechanism for a hierarchical next-generation network, harnessing CoT, FoT, and EoT paradigms simultaneously. It coordinates mobile devices, cloudlets, and cloud servers, ensuring these components co-exist and reinforce each other's capabilities. To analyze strategic interactions among these technologies, the *EFCC* scheme formulates cooperative games linking devices, cloudlets, and cloud servers. Two cooperative bargaining solutions - *Tempered Aspirations Bargaining Solution* (*TABS*) and *Gupta-Livne Bargaining Solution* (*GLBS*) - guide negotiations and allocation decisions. Each mobile device, together with its cloudlet and remote server, cooperates to resolve conflicting interests, guaranteeing both fairness and efficiency (Kim, 2020a).

3.2.2 Contribution

A major challenge in game-based task offloading is ensuring enough generality to accommodate future network developments. In all likelihood, new computing paradigms will emerge, and the existing three-layer hierarchical network model could evolve into more intricate configurations. Consequently, while CoT, FoT, and EoT serve as the current paradigms, they may eventually give way to fresh approaches. To address this uncertainty, the *EFCC* scheme maintains a dynamic, flexible framework capable of adaptively responding to emerging network infrastructures - an advantage not found in traditional task offloading algorithms (Kim, 2020a).

To deliver the potential benefits of three-layer hierarchical network architectures, various technical and operational hurdles must be overcome. In this context, the *EFCC* scheme employs *TABS* and *GLBS* to solve the task offloading problem. A new cooperative bargaining model is devised to represent interactions among mobile devices, cloudlets, and the cloud. By using two bargaining solutions, the *EFCC* scheme efficiently distributes hierarchical network resources in a fair and efficient way. With its adaptability and real-time responsiveness, the *EFCC* scheme aims to yield a well-balanced outcome and maximize the synergistic effect of CoT, FoT, and EoT paradigms (Kim, 2020a).

3.2.3. Hierarchical Network Architecture for Task Offloading Services

The *EFCC* scheme envisions a future network setup featuring a hierarchical computing structure, examining how different computing paradigms function based on their underlying physical resources. This hierarchical design primarily aims to enhance QoE for end users. In practice, edge devices can either process tasks locally or offload them to computing servers, such as the nearby cloudlet or the remote cloud. To handle task offloading, the *EFCC* scheme applies cooperative bargaining principles, incorporating cooperation, coordination, and collaboration among devices, cloudlets, and cloud servers (Kim, 2020a).

As shown in Figure 1, the *EFCC* scheme assumes a three-layer hierarchical network system comprised of multiple IoT devices, such as smart phones, surveillance cameras, personal digital assistants, laptops, on-board units and so on, denoted as the set of EoT devices $\mathbb{D} = \{\mathscr{D}_1, \mathscr{D}_2, \ldots, \mathscr{D}_n\}$. The $\mathscr{D}_{1 \le i \le n}$ generates different application service requests $\mathbb{A}^{\mathscr{D}_i} = \{\mathscr{A}_1^{\mathscr{D}_i}, \mathscr{A}_2^{\mathscr{D}_i}, \mathscr{A}_3^{\mathscr{D}_i}, \ldots\}$ and may offload certain amounts of computing tasks to the fog nodes, denoted as the set of cloudlets $\mathbb{F} = \{\mathscr{C}\mathscr{L}_1, \mathscr{C}\mathscr{L}_2, \ldots, \mathscr{C}\mathscr{L}_m\}$, and one cloud server ($\mathbb{C}$). The $\mathscr{D}_{1 \le i \le n}$, $\mathscr{C}\mathscr{L}_{1 \le j \le m}$ and C have their computation power capacities, i.e., $\mathfrak{P}^{\mathscr{D}_i}$, $\mathfrak{P}^{\mathscr{C}}$

$^{\mathscr{L}_j}$, and $\mathfrak{P}^{C}$, respectively, which can be consumed by a monotonic increasing function of the computation amount. Whereas in reality, the $\mathfrak{P}^{\mathscr{D}_i}$, $\mathfrak{P}^{\mathscr{CL}_j}$, and $\mathfrak{P}^{C}$ resources are limited and raced. When a lot of computation-intensive applications are executed, these resources will become exhausted rapidly. Due to the resource scarcity, it is impossible to guarantee all applications' needs. To maximize the overall system performance, it is necessary to effectively utilize these computation resources for different application requests (Kim, 2020a).

Figure 1. The infrastructure of the three-layer hierarchical network (Kim, 2020a)

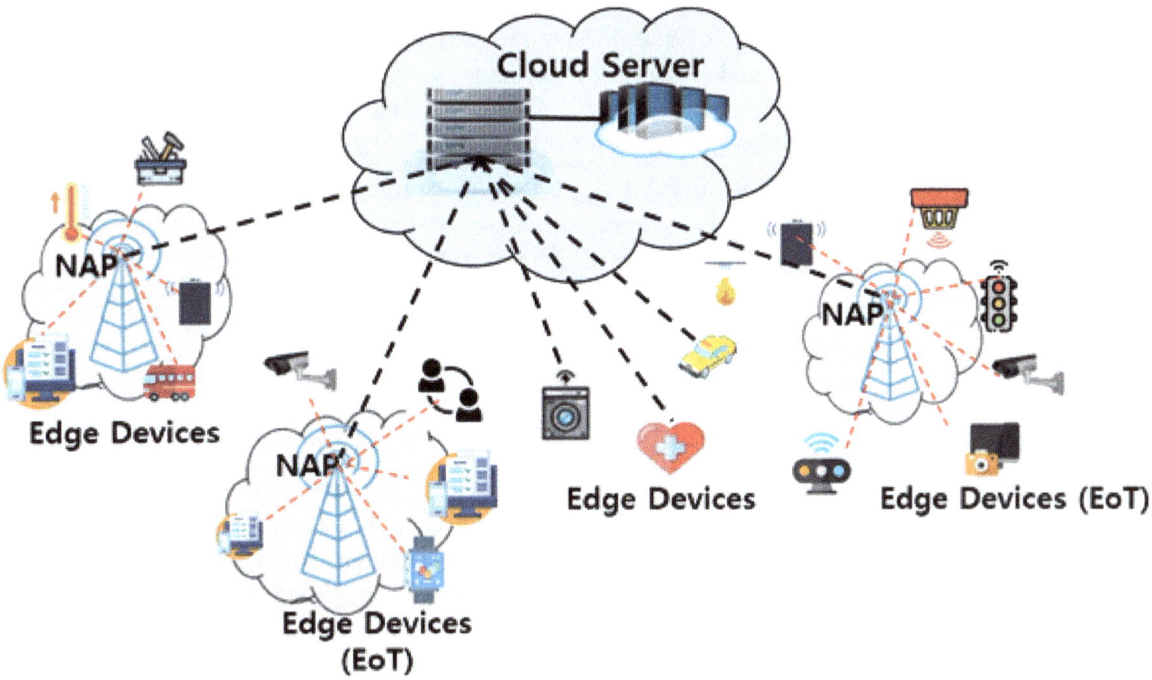

While offloading services offer clear benefits for processing IoT applications, the future network environment can still experience reduced QoE due to service latency. Different application services exhibit distinct computation intensities and delay sensitivities. Since the network may span an extensive range - from the edge device (D) to the central cloud (C) - communication delay must be factored into the decision-making. Based on QoE requirements, these applications generally fall into two categories: computation-intensive and delay-sensitive. Consequently, the *EFCC* scheme must consider QoE demands when determining offloading strategies, making resource management a pivotal element for improving system performance while also respecting service constraints (Kim, 2020a).

To tackle the future network task offloading problem, the *EFCC* scheme adopts two cooperative bargaining solutions; *Tempered Aspirations Bargaining Solution* (*TABS*) and *Gupta-Livne Bargaining Solution* (*GLBS*) (Balakrishnan, 2011). Each individual mobile device offloads its application task ($\mathscr{A}$) while partitioning the computation amount ($\Gamma^{\mathscr{A}}$) into three parts, i.e., $\mathfrak{P}^{\mathscr{D}}_{\mathscr{A}}$, $\mathfrak{P}^{\mathscr{CL}}_{\mathscr{A}}$ and $\mathfrak{P}^{C}_{\mathscr{A}}$; they are assigned to its own device D, the corresponding $\mathscr{CL}$ and the C, respectively. To adaptively partition its $\Gamma^{\mathscr{A}}$, the main ideas of *TABS* and *GLBS* are applied. Based on two bargaining solutions, the *EFCC* scheme can take various benefits in a fair-efficient way (Kim, 2020a).

3.2.4 Tempered Aspirations and Gupta-Livne Bargaining Solutions

Let *N* be the set of potential bargainers, and R, $\mathbb{R}_{+}$ and $\mathbb{R}_{++}$ are denoted as the sets of all, non-negative and positive real numbers, respectively. $\mathbb{R}^n$ is the *n*-fold Cartesian product of R. We use conventional notation for comparison of vectors: $\boldsymbol{x} \geq \boldsymbol{y}$ means that $x_{1\leq i\leq n} \geq y_{1\leq i\leq n}$, $\boldsymbol{x} > \boldsymbol{y}$ indicates that $\boldsymbol{x} \geq \boldsymbol{y}$ and $\boldsymbol{x} \neq \boldsymbol{y}$ and $\boldsymbol{x} \gg \boldsymbol{y}$ means $x_{1\leq i\leq n} > y_{1\leq i\leq n}$. Let *co*(*A*) denote the convex hull of the set *A* in $\mathbb{R}^n$; it is mathematically expressed as $co(A) = \{\mathcal{X} \in \mathbb{R}^n | \mathcal{X} = (\alpha \times x) + ((1\text{-}\alpha) \times y), x, y \in \text{A and } \alpha \in [0,1]\}$. Let *cch*(*A*) denote the convex and comprehensive hull of *A*, $cch(A) = \{y \in \mathbb{R}^n | y \leq \mathcal{X}, \mathcal{X} \in co(A)\}$. If *N* has more than one member, for every $x \in \mathbb{R}^n$ and every $i \in N$, define $x_{-i} = x_{N\setminus\{i\}}$ (Balakrishnan, 2011; Hinojosa, 2008). A disagreement point (d) is a vector $d = (d_1,...,d_n)$ that is expected to be the result if bargainers cannot reach an agreement. A bargaining problem for *N* is a pair (*S*, *d*) such that *S* is a bargaining set for *N*, *d* *S*, and there exists an $\boldsymbol{x} \in S$ satisfying $\boldsymbol{x}$ *d*. Let the aspirations vector $a(S, \boldsymbol{x})$ be defined by (Balakrishnan, 2011; Kim, 2020a);

$$a_i(S, \mathbf{x}) = max\{t \in \mathbb{R} | (t, x_{-i}) \in \text{S}\} \text{for every } i \in N \tag{3.2.1}$$

The ideal point of the problem (*S*, *d*) represents bargainers' expectations before bargaining negotiation and it is defined by *a*(*S*, *d*). Denote the family of all bargaining problems for *N* by Σ^N. The reference point $r \in S$ satisfies $r \geq d$. A solution concept on Σ^N is a function ϕ that associates with each triple $(S, d, r) \in \Sigma^N$, and a unique outcome of ϕ is denoted as $\phi(S, d, r) \in S$ (Balakrishnan, 2011; Kim, 2020a).

In 2011, P.V. Balakrishnan et al proposed a new bargaining solution, called *Tempered Aspirations Bargaining Solution* (*TABS*). With the reference point (r), the *TABS* is defined for every $(S, d, r) \in \Sigma^N$ as (Balakrishnan, 2011; Kim, 2020a);

$$TABS(S, d, r) = [\lambda^* \times \text{a(S, r)}] + [(1\text{-}\lambda^*) \times \text{d}] \tag{3.2.2}$$

$$\text{s.t., } \lambda^* = \mathbf{max}\{\lambda \in [0,1] | ([\lambda \times \text{a(S, r)}] + [(1\text{-}\lambda) \times \text{d}]) \in \text{S}\}$$

If a bargaining problem is translated so that the disagreement point is at the origin, the *TABS* is the only point along the frontier of *S* proportional to the aspirations vector a(*S*, *r*).

In 1988, Gupta and Livne proposed another bargaining solution, called *Gupta-Livne Bargaining Solution* (*GLBS*). The solution is 'dual' to the *TABS* in the sense that it exchanges the roles played by the reference and disagreement points. In the Gupta-Livne approach, the disagreement point *d* has no role to play as a threat in the bargain. It serves only to form the aspirations of players through the construction of the ideal aspiration point. For every $(S, d, r) \in \Sigma^N$, the *GLBS* is defined as follows (Balakrishnan, 2011; Kim, 2020a);

GLBS (S, d, r)

$= [\lambda^* \times a(S, d)] + [(1-\lambda^*) \times r]$

$$(3.2.3)$$

s.t., $\lambda^* = \max \{\lambda \in [0,1] | ([\lambda \times a(S, d)] + [(1-\lambda) \times r]) \in S\}$

The main difference of *TABS* and *GLBS* is the role of disagreement point *d*. In the *TABS*, the *d* is used as a reference vector from which proportional payoffs are measured. However, in the *GLBS*, it is used to set the ideal aspiration point. Both solution concepts are illustrated in Figure 2 (Balakrishnan, 2011; Kim, 2020a).

(a) TABS (b) GLBS

Figure 2. Tempered aspirations and Gupta-Livne bargaining solutions (Kim, 2020a)

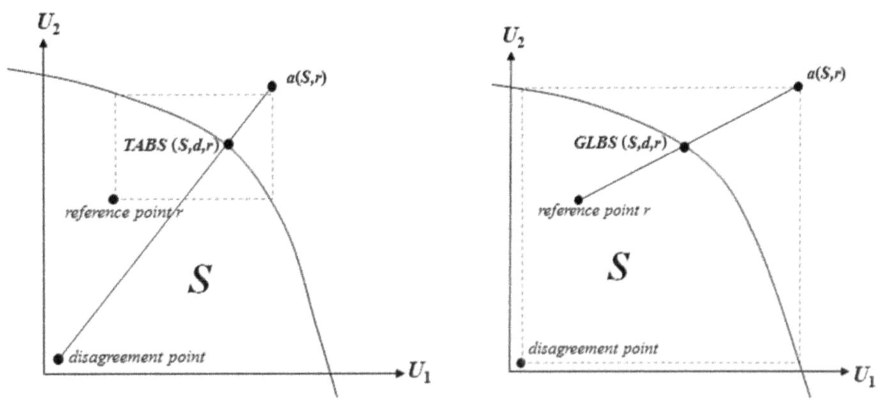

3.2.5 The Application Task Offloading Algorithms

The *EFCC* scheme designs two bargaining games for task offloading services. First, the idea of *TABS* is adopted to implement the time-sensitive application offloading algorithm. To fair-efficiently offload the task computation, the offload decision process for the $\mathcal{A}_k^{\mathcal{D}_i}$ is formulated as a cooperative bargaining game

$$\mathbb{G}_{TABS} = \left\{ \{\mathcal{D}_i, \mathcal{CL}_j, \mathbb{C}\}, \{\mathfrak{P}^{\mathcal{D}_i}, \mathfrak{P}^{\mathcal{CL}_j}, \mathfrak{P}^{\mathbb{C}}\}, \mathcal{A}_k^{\mathcal{D}_i}, \Gamma^{\mathcal{A}_k}, \{\mathbb{S}_{\mathcal{D}_i}, \mathbb{S}_{\mathcal{CL}_j}, \mathbb{S}_{\mathbb{C}}\}, \{U_{\mathcal{D}_i}, U_{\mathcal{CL}_j}, U_{\mathbb{C}}\}, r_{\mathcal{D}, \mathcal{CL}, \mathbb{C}}(S, x), a_{\mathcal{D}, \mathcal{CL}, \mathbb{C}}(S, x) \right\}$$

(Kim, 2020a):

- Players: in the $\mathbb{G}_{TABS}$, a smart device $\mathcal{D}_i \in \mathbb{D}$, the corresponding cloudlet $\mathcal{CL}_j \in \mathbb{F}$, and the cloud server C are assumed as game players $\{\mathcal{D}_i, \mathcal{CL}_j, \mathbb{C}\}$ to process the task offloading service.

- Computation powers of players: $\mathscr{D}_i$, $\mathscr{CL}_j$ and C have $\mathfrak{P}^{\mathscr{D}_i}$, $\mathfrak{P}^{\mathscr{CL}_j}$ and $\mathfrak{P}^{\mathrm{C}}$ computation powers, respectively; they are assumed as total CPU capacities of game players.
- Application task and computation amount: the application $\mathscr{A}_k^{\mathscr{D}_i}$ is generated from the $\mathscr{D}_i$, and total computation amount of $\mathscr{A}_k^{\mathscr{D}_i}$ is $\Gamma^{\mathscr{A}_k}$.
- Strategies: each player has a finite computation capacity. The set of strategies for each player consists of its discrete computation power levels. Let $\mathbb{S}_{\mathscr{D}_i} = \left\{\mathfrak{C}^{\mathscr{D}_i}_{1\le k\le r} \le \mathfrak{P}^{\mathscr{D}_i} | \left\{\mathfrak{C}^{\mathscr{D}_i}_1, \ldots, \mathfrak{C}^{\mathscr{D}_i}_r\right\}\right\}$ be the $\mathscr{D}_i$'s strategy set, $\mathbb{S}_{\mathscr{CL}} = \left\{\mathfrak{C}^{\mathscr{CL}_j}_{1\le k\le e} \le \mathfrak{P}^{\mathscr{CL}_j} | \left\{\mathfrak{C}^{\mathscr{CL}_j}_1, \ldots, \mathfrak{C}^{\mathscr{CL}_j}_e\right\}\right\}$ be the $\mathscr{CL}_j$'s strategy set, and $\mathbb{S}_{\mathrm{C}} = \left\{\mathfrak{C}^{\mathrm{C}}_{1\le k\le l} \le \mathfrak{P}^{\mathrm{C}} | \left\{\mathfrak{C}^{\mathrm{C}}_1, \ldots, \mathfrak{C}^{\mathrm{C}}_l\right\}\right\}$ be the C's strategy set.
- Utility functions: $\mathscr{D}_i$, $\mathscr{CL}_j$ and C players have their own utility functions $U^{\mathscr{A}_k}_{\mathscr{D}_i}$, $U^{\mathscr{A}_k}_{\mathscr{CL}_j}$ and $U^{\mathscr{A}_k}_{\mathrm{C}}$, respectively, to process the offload service of task $\mathscr{A}_k^{\mathscr{D}_i}$. Each utility function maps player's satisfaction to a real number, which represents the resulting payoff in the game $\mathbb{G}_{TABS}$.
- Reference point: the reference point of $\mathscr{D}_i$, $\mathscr{CL}_j$ and C is denoted as $r_{\mathscr{D}_i,\mathscr{CL}_j,\mathrm{C}}(S, x)$; it satisfies two features, i) $r_{\mathscr{D}_i,\mathscr{CL}_j,\mathrm{C}} \in S \setminus WPO(S)$, and ii) $r_{\mathscr{D}_i,\mathscr{CL}_j,\mathrm{C}} > d$.
- Aspirations point: the aspirations point of $\mathscr{D}_i$, $\mathscr{CL}_j$ and C is denoted as $a_{\mathscr{D}_i,\mathscr{CL}_j,\mathrm{C}}(S, x)$; it is defined based on the reference point.

To quantify service satisfaction, the utility functions of players in the *TABS* can be derived as follows (Kim, 2020a).

$$\begin{cases} U^{\mathscr{A}_k}_{\mathscr{D}_i}\left(\chi^{\mathscr{A}_k}_{\mathscr{D}_i}\right) = \frac{\psi_{\mathscr{D}_i}}{\eta_{\mathscr{D}_i}} \times \log\left(1 + \left(\eta_{\mathscr{D}_i} \times \frac{\chi^{\mathscr{A}_k}_{\mathscr{D}_i}}{\Gamma^{\mathscr{A}_k}}\right)\right) \\ U^{\mathscr{A}_k}_{\mathscr{CL}_j}\left(\chi^{\mathscr{A}_k}_{\mathscr{CL}_j}\right) = \frac{\psi_{\mathscr{CL}_j}}{\eta_{\mathscr{CL}_j}} \times \log\left(1 + \left(\eta_{\mathscr{CL}_j} \times \frac{\chi^{\mathscr{A}_k}_{\mathscr{CL}_j}}{\Gamma^{\mathscr{A}_k}}\right)\right) \\ U^{\mathscr{A}_k}_{\mathrm{C}}\left(\chi^{\mathscr{A}_k}_{\mathrm{C}}\right) = \frac{\psi_{\mathrm{C}}}{\eta_{\mathrm{C}}} \times \log\left(1 + \left(\eta_{\mathrm{C}} \times \frac{\chi^{\mathscr{A}_k}_{\mathrm{C}}}{\Gamma^{\mathscr{A}_k}}\right)\right) \end{cases} \tag{3.2.4}$$

$$\text{s.t., } \eta_{\mathscr{D}_i} = \frac{\beta^{\mathscr{D}_i}_{\mathrm{C}}}{\mathfrak{P}^{\mathscr{D}_i}},\ \eta_{\mathscr{CL}_j} = \frac{\beta^{\mathscr{CL}_j}_{\mathrm{C}}}{\mathfrak{P}^{\mathscr{CL}_j}},\ \eta_{\mathrm{C}} = \frac{\beta^{\mathrm{C}}_{\mathrm{C}}}{\mathfrak{P}^{\mathrm{C}}} \text{ and } \Gamma^{\mathscr{A}_k} = \left(\chi^{\mathscr{A}_k}_{\mathscr{D}_i} + \chi^{\mathscr{A}_k}_{\mathscr{CL}_j} + \chi^{\mathscr{A}_k}_{\mathrm{C}}\right)$$

where $\chi^{\mathscr{A}_k}_{\mathscr{D}_i}$, $\chi^{\mathscr{A}_k}_{\mathscr{CL}_j}$ and $\chi^{\mathscr{A}_k}_{\mathrm{C}}$ are assigned computation amounts to $\mathscr{D}_i$, $\mathscr{CL}_j$ and C, respectively. $\psi_{\mathscr{D}_i}$, $\psi_{\mathscr{CL}_j}$ and ψ_{C} are coefficient parameters to represent the QoE of $\mathscr{D}_i$, $\mathscr{CL}_j$ and C computation services, respectively. $\beta^{\mathscr{D}_i}_{\mathrm{C}}$, $\beta^{\mathscr{CL}_j}_{\mathrm{C}}$ and $\beta^{\mathrm{C}}_{\mathrm{C}}$ are the current computation loads of $\mathscr{D}_i$, $\mathscr{CL}_j$ and C, respectively. In the developed bargaining game, each player is a member of a team willing to compromise with other players. According to their utility functions and expected payoffs, team players make a collective decision to gain a total optimal solution. In the $\mathbb{G}_{TABS}$, the reference point, i.e., $r_{\mathscr{D}_i,\mathscr{CL}_j,\mathrm{C}}(S, x)$, is defined as follows (Kim, 2020a);

$$r_{\mathscr{D}_i,\mathscr{CL}_j,\mathrm{C}}(S, x) = \left(U^r_{\mathscr{D}_i}, U^r_{\mathscr{CL}_j}, U^r_{\mathrm{C}}\right) \tag{3.2.5}$$

$$\text{s.t., } U^r_{\mathscr{D}_i} = \frac{U^{\mathscr{A}_k}_{\mathscr{D}_i}(m^{\mathscr{A}_k})}{\varphi_{\mathscr{D}}},\ U^r_{\mathscr{CL}_j} = \frac{U^{\mathscr{A}_k}_{\mathscr{CL}_j}(m^{\mathscr{A}_k})}{\varphi_{\mathscr{CL}}},\ U^r_{\mathrm{C}}\left(\chi^{\mathscr{A}_k}_{\mathrm{C}}\right) = \frac{U^{\mathscr{A}_k}_{\mathrm{C}}(m^{\mathscr{A}_k})}{\varphi_{\mathrm{C}}}$$

where $\varphi_{\mathscr{D}}$, $\varphi_{\mathscr{CL}}$ and φ_{C} are control factors to decide the reference point values of $\mathscr{D}_i$, $\mathscr{CL}_j$ and C, respectively. $m^{\mathscr{A}_k}$ is the minimum computation capacity for the $\mathscr{A}_k$'s task offloading service. In the $\mathbb{G}_{TABS}$, the aspirations point of *TABS*, i.e., $a_{\mathscr{D}_i,\mathscr{CL}_j,\mathrm{C}}(S, x)$, is defined as follows (Kim, 2020a);

$$a_{\mathscr{D}_i,\mathscr{CL}_j,\mathrm{C}}(S, x) = \left(a_{\mathscr{D}_i}(S, x), a_{\mathscr{CL}_j}(S, x), a_{\mathrm{C}}(S, x)\right)$$

(3.2.6)

$$\text{s.t.}, \begin{cases} a_{\mathscr{D}_i}(S, x) = \max\left\{t \in \mathbb{R} \mid \left(t, U^r_{\mathscr{CL}_j}, U^r_{\mathrm{C}}\right) \in S\right\} \\ a_{\mathscr{CL}_j}(S, x) = \max\left\{t \in \mathbb{R} \mid \left(t, U^r_{\mathscr{D}_i}, U^r_{\mathrm{C}}\right) \in S\right\} \\ a_{\mathrm{C}}(S, x) = max\left\{t \in \mathbb{R} \mid \left(t, U^r_{\mathscr{D}_i}, U^r_{\mathscr{CL}_j}\right) \in S\right\} \end{cases}$$

Based on the disagree point d as a starting point, the line (***L***) forward to the aspirations point $a_{\mathscr{D}_i,\mathscr{CL}_j,\mathrm{C}}(S, x)$ is defined as follows (Kim, 2020a).

$$L = \left\{U \,\middle|\, \frac{U^{\mathscr{A}_k}_{\mathscr{D}_i}\left(\chi^{\mathscr{A}_k}_{\mathscr{D}_i}\right)}{a_{\mathscr{D}_i}(S, x)} = \frac{U^{\mathscr{A}_k}_{\mathscr{CL}_j}\left(\chi^{\mathscr{A}_k}_{\mathscr{CL}_j}\right)}{a_{\mathscr{CL}_j}(S, x)} = \frac{U^{\mathscr{A}_k}_{\mathrm{C}}\left(\chi^{\mathscr{A}_k}_{\mathrm{C}}\right)}{a_{\mathrm{C}}(S, x)}\right\}$$

(3.2.7)

Simply, the *EFCC* scheme can think that the *TABS* is a weak Pareto-optimal solution located in the S as well as in the line *L* in (3.2.7). Geometrically, the *TABS* is the intersection point $\left(U^{\mathscr{A}_k}_{\mathscr{D}_i}(\chi_{\mathscr{D}_i}), U^{\mathscr{A}_k}_{\mathscr{CL}_j}(\chi_{\mathscr{CL}_j}), U^{\mathscr{A}_k}_{\mathrm{C}}(\chi_{\mathrm{C}})\right)$ between the bargaining set S and the line *L*. Therefore, the *TABS* must satisfy (Kim, 2020a);

$$\left(U^{\mathscr{A}_k}_{\mathscr{D}_i}(\chi_{\mathscr{D}_i}) \middle/ a_{\mathscr{D}_i}(S, x)\right) = \left(U^{\mathscr{A}_k}_{\mathscr{CL}_j}(\chi_{\mathscr{CL}_j}) \middle/ a_{\mathscr{CL}_j}(S, x)\right) = \left(U^{\mathscr{A}_k}_{\mathrm{C}}(\chi_{\mathrm{C}}) \middle/ a_{\mathrm{C}}(S, x)\right)$$

(3.2.8)

Second, the idea of *GLBS* is adopted to develop the computation-intensive application offloading algorithm. To adaptively offload the delay tolerant task computation, the offload decision process for the $\mathscr{A}^{\mathscr{D}_i}_k$ is formulated as another cooperative game model

$$\mathbb{G}_{GLBS} = \left\{\{\mathscr{D}_i, \mathscr{CL}_j, \mathrm{C}\}, \{\mathfrak{P}^{\mathscr{D}_i}, \mathfrak{P}^{\mathscr{CL}_j}, \mathfrak{P}^{\mathrm{C}}\}, \mathscr{A}^{\mathscr{D}_i}_k, \Gamma^{\mathscr{A}_k}, \left\{\mathbb{S}_{\mathscr{D}_i}, \mathbb{S}_{\mathscr{CL}_j}, \mathbb{S}_{\mathrm{C}}\right\}, \left\{U_{\mathscr{D}_i}, U_{\mathscr{CL}_j}, U_{\mathrm{C}}\right\}, r_{\mathscr{D}_i\mathscr{CL}_j\mathrm{C}}(S, x), a_{\mathscr{D}_i\mathscr{CL}_j\mathrm{C}}(S, x)\right\}.$$

In the $\mathbb{G}_{GLBS}$ game, only utility functions and aspirations point are defined differently, and the other game elements are the same as the $\mathbb{G}_{TABS}$. In the $\mathbb{G}_{GLBS}$, the $a_{\mathscr{D}_i\mathscr{CL}_j\mathrm{C}}(S, x)$ is dynamically calculated according to (3.2.1), and the $\mathscr{D}_i$, $\mathscr{CL}_j$ and C's utility functions for the task $\mathscr{A}^{\mathscr{D}_i}_k$ can be derived as follows (Kim, 2020a);

$$U^{\mathscr{A}_k}_{\mathscr{D}_i}\left(\chi^{\mathscr{A}_k}_{\mathscr{D}_i}\right) = \frac{\psi_{\mathscr{D}_i}}{\eta_{\mathscr{D}_i}} \times \log\left(1 + \left(\eta_{\mathscr{D}_i} \times \frac{\chi^{\mathscr{A}_k}_{\mathscr{D}_i}}{\Gamma^{\mathscr{A}_k}}\right)\right) \times \mathscr{F}^{\mathscr{A}_k}_{\mathscr{D}_i}\left(\chi^{\mathscr{A}_k}_{\mathscr{D}_i}\right)$$

(3.2.9)

$$
\text{s.t., } \mathscr{F}^{\mathscr{A}_k}_{\mathscr{D}_i}\left(\chi^{\mathscr{A}_k}_{\mathscr{D}_i}\right) = \begin{cases} \dfrac{\chi^{\mathscr{A}_k}_{\mathscr{C}\mathscr{L}_j}}{\left(1+\dfrac{\omega^{\mathscr{A}_k}_{\mathscr{D}_i}}{\sigma}\right)\times\Gamma^{\mathscr{A}_k}}, if\left(\omega^{\mathscr{A}_k}_{\mathscr{D}_i}\times\dfrac{\chi^{\mathscr{A}_k}_{\mathscr{D}_i}}{m^{\mathscr{A}_k}}\right) \le \left(\dfrac{\chi^{\mathscr{A}_k}_{\mathscr{D}_i}}{\Gamma^{\mathscr{A}_k}}\times \mathrm{T}^{\mathscr{A}_k}\right) \\ 0, otherwise \end{cases}
$$

$$
U^{\mathscr{A}_k}_{\mathscr{C}\mathscr{L}_j}\left(\chi^{\mathscr{A}_k}_{\mathscr{C}\mathscr{L}_j}\right) = \frac{\psi_{\mathscr{C}\mathscr{L}_j}}{\eta_{\mathscr{C}\mathscr{L}_j}}\times\log\left(1+\left(\eta_{\mathscr{C}\mathscr{L}_j}\times\frac{\chi^{\mathscr{A}_k}_{\mathscr{C}\mathscr{L}_j}}{\Gamma^{\mathscr{A}_k}}\right)\right)\times\mathscr{F}^{\mathscr{A}_k}_{\mathscr{C}\mathscr{L}_j}\left(\chi^{\mathscr{A}_k}_{\mathscr{C}\mathscr{L}_j}\right)
$$

(3.2.10)

$$
\text{s.t., } \mathscr{F}^{\mathscr{A}_k}_{\mathscr{C}\mathscr{L}_j}\left(\chi^{\mathscr{A}_k}_{\mathscr{C}\mathscr{L}_j}\right) = \begin{cases} \dfrac{\chi^{\mathscr{A}_k}_{\mathscr{C}\mathscr{L}_j}}{\left(1+\dfrac{\left(\xi^{\mathscr{D}_i}_{\mathscr{C}\mathscr{L}_j}+\omega^{\mathscr{A}_k}_{\mathscr{C}\mathscr{L}_j}\right)+\omega^{\mathscr{A}_k}_{\mathscr{D}_i}}{\sigma}\right)\times\Gamma^{\mathscr{A}_k}}, \\ if\left(\left(\left(\xi^{\mathscr{D}_i}_{\mathscr{C}\mathscr{L}_j}+\omega^{\mathscr{A}_k}_{\mathscr{C}\mathscr{L}_j}\right)+\omega^{\mathscr{A}_k}_{\mathscr{D}_i}\right)\times\dfrac{\chi^{\mathscr{A}_k}_{\mathscr{C}\mathscr{L}_j}}{m^{\mathscr{A}_k}}\right) \le \left(\dfrac{\chi^{\mathscr{A}_k}_{\mathscr{C}\mathscr{L}_j}}{\Gamma^{\mathscr{A}_k}}\times \mathrm{T}^{\mathscr{A}_k}\right) \\ 0, otherwise \end{cases}
$$

$$
U^{\mathscr{A}_k}_{\mathrm{C}}\left(\chi^{\mathscr{A}_k}_{\mathrm{C}}\right) = \frac{\psi_{\mathrm{C}}}{\eta_{\mathrm{C}}}\times\log\left(1+\left(\eta_{\mathrm{C}}\times\frac{\chi^{\mathscr{A}_k}_{\mathrm{C}}}{\Gamma^{\mathscr{A}_k}}\right)\right)\times\mathscr{F}^{\mathscr{A}_k}_{\mathrm{C}}\left(\chi^{\mathscr{A}_k}_{\mathrm{C}}\right)
$$

(3.2.11)

$$
\text{s.t., } \mathscr{F}^{\mathscr{A}_k}_{\mathrm{C}}\left(\chi^{\mathscr{A}_k}_{\mathrm{C}}\right) = \begin{cases} \dfrac{\chi^{\mathscr{A}_k}_{\mathrm{C}}}{\left(1+\dfrac{\left(\xi^{\mathscr{D}_i}_{\mathrm{C}}+\omega^{\mathscr{A}_k}_{\mathrm{C}}\right)+\omega^{\mathscr{A}_k}_{\mathscr{D}_i}}{\sigma}\right)\times\Gamma^{\mathscr{A}_k}}, \\ if\left(\left(\left(\xi^{\mathscr{D}_i}_{\mathrm{C}}+\omega^{\mathscr{A}_k}_{\mathrm{C}}\right)+\omega^{\mathscr{A}_k}_{\mathscr{D}_i}\right)\times\dfrac{\chi^{\mathscr{A}_k}_{\mathrm{C}}}{m^{\mathscr{A}_k}}\right) \le \left(\dfrac{\chi^{\mathscr{A}_k}_{\mathrm{C}}}{\Gamma^{\mathscr{A}_k}}\times \mathrm{T}^{\mathscr{A}_k}\right) \\ 0, otherwise \end{cases}
$$

where $\omega^{\mathscr{A}_k}_{\mathscr{D}_i}$, $\omega^{\mathscr{A}_k}_{\mathscr{C}\mathscr{L}_j}$ and$\omega^{\mathscr{A}_k}_{\mathrm{C}}$ are computation delay factors of the $\mathscr{D}_i$,$\mathscr{C}\mathscr{L}_j$ and C, respectively. $\xi^{\mathscr{D}_i}_{\mathscr{C}\mathscr{L}_j}$ and $\xi^{\mathscr{D}_i}_{\mathrm{C}}$ are communication delay factors of the $\mathscr{C}\mathscr{L}_j$ and C, respectively. σ is the system's basic time-unit for the task offloading service. $\mathrm{T}^{\mathscr{A}_k}$ is the time deadline of the $\mathscr{A}_k$. Based on the reference point $r_{\mathscr{D}_i\mathscr{C}\mathscr{L}_j\mathrm{C}}(S, x)$ as a starting point, the line (L) forward to the aspirations point $a_{\mathscr{D}_i\mathscr{C}\mathscr{L}_j\mathrm{C}}(S, x)$ is defined as follows (Kim, 2020a).

$$
L = \left\{ U \middle| \frac{U^{\mathscr{A}_k}_{\mathscr{D}_i}\left(\chi^{\mathscr{A}_k}_{\mathscr{D}_i}\right) - U^{r}_{\mathscr{D}_i}\left(\chi^{\mathscr{A}_k}_{\mathscr{D}_i}\right)}{a_{\mathscr{D}_i}(S, x) - U^{r}_{\mathscr{D}_i}\left(\chi^{\mathscr{A}_k}_{\mathscr{D}_i}\right)} = \frac{U^{\mathscr{A}_k}_{\mathscr{C}\mathscr{L}_j}\left(\chi^{\mathscr{A}_k}_{\mathscr{C}\mathscr{L}_j}\right) - U^{r}_{\mathscr{C}\mathscr{L}_j}\left(\chi^{\mathscr{A}_k}_{\mathscr{C}\mathscr{L}_j}\right)}{a_{\mathscr{C}\mathscr{L}_j}(S, x) - U^{r}_{\mathscr{C}\mathscr{L}_j}\left(\chi^{\mathscr{A}_k}_{\mathscr{C}\mathscr{L}_j}\right)} = \frac{U^{\mathscr{A}_k}_{\mathrm{C}}\left(\chi^{\mathscr{A}_k}_{\mathrm{C}}\right) - U^{r}_{\mathrm{C}}\left(\chi^{\mathscr{A}_k}_{\mathrm{C}}\right)}{a_{\mathrm{C}}(S, x) - U^{r}_{\mathrm{C}}\left(\chi^{\mathscr{A}_k}_{\mathrm{C}}\right)} \right\}
$$

(3.2.12)

Simply, the *EFCC* scheme can think that the *GLBS* is a weak Pareto-optimal solution located in the S as well as in the line *L* in (3.2.12). Geometrically, the *GLBS* is the intersection point $\left(U_{\mathcal{D}_i}^{\mathcal{A}_k}(\chi_{\mathcal{D}_i}), U_{\mathcal{CL}_j}^{\mathcal{A}_k}(\chi_{\mathcal{CL}_j}), U_{\mathbb{C}}^{\mathcal{A}_k}(\chi_{\mathbb{C}})\right)$ between the bargaining set S and the line *L*. Therefore, the *GLBS* must satisfy (Kim, 2020a);

$$\left(U_{\mathcal{D}_i}^{\mathcal{A}_k}(\chi_{\mathcal{D}_i}) - U_{\mathcal{D}_i}^{r}(\chi_{\mathcal{D}_i}^{\mathcal{A}_k})\right) / \left(a_{\mathcal{D}_i}(S, x) - U_{\mathcal{D}_i}^{r}(\chi_{\mathcal{D}_i}^{\mathcal{A}_k})\right)$$

$$= \left(U_{\mathcal{CL}_j}^{\mathcal{A}_k}(\chi_{\mathcal{CL}_j}) - U_{\mathcal{CL}_j}^{r}(\chi_{\mathcal{CL}_j}^{\mathcal{A}_k})\right) / \left(a_{\mathcal{CL}_j}(S, x) - U_{\mathcal{CL}_j}^{r}(\chi_{\mathcal{CL}_j}^{\mathcal{A}_k})\right)$$

$$= \left(U_{\mathbb{C}}^{\mathcal{A}_k}(\chi_{\mathbb{C}}) - U_{\mathbb{C}}^{r}(\chi_{\mathbb{C}}^{\mathcal{A}_k})\right) / \left(a_{\mathbb{C}}(S, x) - U_{\mathbb{C}}^{r}(\chi_{\mathbb{C}}^{\mathcal{A}_k})\right)$$

(3.2.13)

3.2.6 Summary

The *EFCC* scheme examines application task offloading within edge, fog, and cloud computing architectures. Leveraging a three-tier network model consisting of mobile devices, cloudlets, and cloud servers, it employs cooperative bargaining principles to solve the offloading problem. Specifically, the *TABS* and the *GLBS* are applied to determine optimal allocation of computation for each task. By jointly evaluating computation intensity and delay sensitivity, the *EFCC* scheme intelligently selects the most fitting bargaining strategy. This bargaining game framework offers a compelling, real-world approach for managing the evolving needs of future network application services (Kim, 2020a).

3.3 Edge-assisted Federated Learning and Offloading (*EFLO*) Scheme

Unlike traditional learning methods, federated learning (FL) is an emerging distributed deep learning framework where smart devices conduct local computations on their own data, then upload the necessary updates to refine a global model. Recognizing this potential, S. Kim introduces the *Edge-assisted Federated Learning and Offloading* (*EFLO*) scheme (Kim, 2022b), which integrates computation offloading concepts. Specifically, edge computing platforms allow devices to partially offload compute-intensive tasks to edge clouds, thereby boosting the limited computation capabilities of resource-constrained devices. This approach utilizes idle edge server capacity to assist individual devices with model training. To manage the FL process, the *EFLO* scheme leverages bargaining solutions drawn from cooperative game theory for dynamic offloading. By doing so, the *EFLO* scheme seeks to maximize the combined benefits for devices and edge clouds, resulting in greater efficiency in federated learning. Consequently, the *EFLO* scheme can achieve a fair and efficient consensus under edge-assisted FL conditions, delivering mutual advantages and a full synergy among participating agents (Kim, 2022b).

3.3.1 Development Motivation

As IoT devices become more pervasive - in areas ranging from industrial control and cyber-physical systems to medical devices and environmental monitoring - they have rapidly evolved into critical enablers of future network services. The continued expansion of IoT, alongside the escalating computational and connectivity requirements of 5G and beyond-5G systems, has fostered the emergence of machine learning (ML) and artificial intelligence (AI) techniques for communication, networking, and context-aware services (Han, 2019; Kim, 2022b; Savazzi, 2020).

In 2017, McMahan et al. introduced federated learning (FL), a distributed machine learning model that processes locally stored data across a large number of IoT devices and aggregates learned parameters via a central server. FL's overarching aim is to establish a single, global model without collecting raw data centrally. Recently, the advent of mobile edge computing (MEC) has enhanced FL by embedding computation and storage capabilities within edge clouds, enabling large-scale distributed tasks that blend local processing with efficient remote coordination (Kim, 2022b; Savazzi, 2020; Wang, 2019).

As IoT devices generate exponential amounts of data, they also demand low latency and power efficiency. By offloading computational tasks to edge clouds, these devices can conserve energy while maintaining the necessary QoS. Nonetheless, a trade-off often emerges between efficiency and privacy - transferring sensitive data to an external server poses potential privacy risks. Such complexities demand careful consideration of both performance and data protection in FL-based MEC environments (Han, 2019; Kim, 2022b; Wang, 2019; Wu, 2020a).

The *EFLO* scheme aims to develop a novel mechanism for federated learning augmented by edge computing. While this is a prominent research topic, it has yet to be exhaustively explored. The key challenge is to balance the objectives of diverse network agents while guaranteeing global performance in a dynamically evolving MEC-FL environment. Consequently, the *EFLO* scheme calls for a fresh intelligent control paradigm grounded in innovative solution concepts.

In the *EFLO* scheme, autonomous, distributed, and intelligent IoT devices make rational, strategic decisions for running deep learning algorithms. This setting aligns well with cooperative game theory, which describes how rational players interact to share collective gains. A central issue in such cooperative games is the distribution of the joint surplus. Over time, a range of solution concepts has been introduced, each reflecting distinct fairness or efficiency criteria. Such solutions have broad application in economics, computing, wireless communications, and politics (Kim, 2022b; Yang, 2019).

Recent bargaining research increasingly highlights risk aversion, noting that in realistic scenarios, players prioritize avoiding losses more than achieving gains. Thus, risk aversion heavily shapes players' negotiation behaviors. Several studies now probe how risk influences bargaining outcomes (Driesen, 2011; Shalev, 2002). In 2002, J. Shalev investigated the link between risk aversion and the Nash solution, introducing the *loss aversive Nash bargaining solution* (*LNBS*). The *LNBS* incorporates reference points internally to represent loss aversion; heightened risk aversion typically worsens a player's outcome in bargaining. Psychological elements thus have a central role in determining a unique solution (Shalev, 2002).

In 2011, B. Driesen et al. proposed a loss aversion twist on the *Kalai-Smorodinsky bargaining solution* (*KSBS*), known as the *LKSBS*. Although the original *KSBS* is found by intersecting the bargaining set with a line that connects the disagreement point and the utopia point, the *LKSBS* introduces Pareto-optimal results considering players' respective loss aversion coefficients. This approach preserves the spirit of cooperative fairness while acknowledging the heightened focus on minimizing losses (Driesen, 2011; Kim, 2022b).

3.3.2 Contributions

Building on the fundamental principles of *LNBS* and *LKSBS*, the *EFLO* scheme presents a new edge cloud-based computing approach for IoT devices within the FL environment. Each IoT device makes adaptive decisions on how much of its computational workload to offload, guided by utilitarian considerations. To facilitate fair and efficient resource sharing, the edge cloud allocates its computation and communication resources among diverse devices in a dynamic, cooperative manner. By uniting the ideas behind *LNBS* and *LKSBS*, the *EFLO* scheme aims to balance overall system performance, ensuring that every participant benefits while adhering to favorable global outcomes (Kim, 2022b).

3.3.3 Edge Cloud System Infrastructure for Federated Learning

Within the *EFLO* scheme, a system model is established for an IoT environment integrated with edge clouds. In the *EFLO* scheme, IoT devices possess the necessary computational capabilities to participate in the FL process. Nevertheless, portable IoT devices often face limited computational resources and small onboard batteries, constraining their ability to handle intensive tasks independently. To mitigate these computational limitations, such devices can partially offload their processing tasks to nearby edge clouds. In the context of MEC infrastructure, an edge cloud functions as a small-scale datacenter that enhances traditional cloud computing by positioning it closer to the network's edge. The primary role of an edge cloud is to support computation-intensive tasks interactively by offering offloading services. This architectural innovation significantly expands the scope of current cloud computing services, tailoring them to meet the demands of the 5G IoT environment (Kim, 2022b).

Based on the edge cloud system model, $\mathscr{E} = \{E_1, \ldots, E_n\}$ is the set of edge clouds, and $\mathscr{D} = \{D_1, \ldots, D_m\}$ is the set of IoT devices. One edge cloud out of **E** can be chosen by nearby devices for offloading computation tasks. The $E_{1 \le k \le n} \in \mathscr{E}$ has its computation power Γ_{E_k} with wireless spectrum bandwidth $\mathscr{W}_{E_k}$. One IoT device $D_{1 \le i \le m} \in \mathscr{D}$ has its own battery cell and computation power. The computation power of D_i is Υ_{D_i}, and the battery cell stores electrical energy; the full-energy and remaining-energy of D_i are represented by $\mathfrak{E}_{D_i}$ and $e^c_{D_i}$, respectively. $\mathscr{T}_{D_i} = \left[\mathfrak{N}_{D_i}, \kappa_{D_i}, \mathscr{M}^{\mathscr{T}}_{D_i}, \mathscr{L}_{D_i}\right]$ is the learning task of D_i where $\mathfrak{N}_{D_i}$ is the data bit size of $\mathscr{T}_{D_i}$, and κ_{D_i} represents the number of CPU cycles required to process one-bit data input. $\mathscr{M}^{\mathscr{T}}_{D_i}$ is the maximum delay-latency to complete the $\mathscr{T}_{D_i}$, and $\mathscr{L}_{D_i}$ is the ratio of data offloading for the $\mathfrak{N}_{D_i}$. During the computation offloading process, time horizon is discretized into time epochs indexed by t_c with equivalent duration in seconds (Han, 2019; Kim, 2022b).

From the viewpoint of devices, how to leverage the trade-offs between local and offloading computing is a critical issue. Therefore, the D_i decides the value of $\mathscr{L}_{D_i}$, which can be viewed as a control parameter between 0 and 1. To effectively complete the learning process, the $\mathscr{L}_{D_i}$ value should be dynamically adjusted while coordinating its offloading and local computation amounts. In the point view of edge clouds, they are responsible for offloading services. Therefore, they adaptively provide their communication ($\mathscr{W}$)and computation (Γ) resources to successfully complete the requested offloading devices. Based on the MEC infrastructure, cloud edges and devices make control decisions intelligently in pursuit of their individual objectives (Kim, 2022b).

In the *EFLO* scheme, the interaction process of cloud edges and devices is formulated as a joint bargaining game ($\mathbb{G}$) in a coordination manner; G is subdivided into $\mathbb{G}^{\mathscr{L}}_D$, $\mathbb{G}^{\Gamma}_E$, and $\mathbb{G}^{\mathscr{W}}_E$ to solve different bargaining problems. The edge computing system operates in a slotted time structure and task offloading services are implemented at each time period. Formally, the *EFLO* scheme defines game entities, i.e.,

$$\mathbb{G} = \{\mathbb{G}^{\mathscr{L}}_D, \mathbb{G}^{\Gamma}_E, \mathbb{G}^{\mathscr{W}}_E\} = \{\{E_k \in \mathscr{E}, D_i \in \mathscr{D}\}, \{\mathbb{G}^{\mathscr{L}}_D | \mathscr{T}^O_D, \mathscr{T}^L_D, \mathscr{L}_D, U^L_D(\cdot), U^O_D(\cdot)\}, \{\mathbb{G}^{\Gamma}_E | \Gamma_E, \Gamma^D_E, U^{\Gamma}_D(\cdot)\}, \{\mathbb{G}^{\mathscr{W}}_E | \mathscr{W}_E, \mathscr{W}^D_E, U^{\mathscr{W}}_D(\cdot)\}, T\}$$

of gameplay (Kim, 2022b).

- $\mathbb{G}_D^{\mathcal{L}}$, $\mathbb{G}_E^{\Gamma}$ and $\mathbb{G}_E^{\mathcal{W}}$ are bargaining games for the L value decision by D, the Γ and W resource distributions by E, respectively. They are mutually and reciprocally interdependent in an interactive manner.
- $\mathcal{E},\mathcal{D}$ are the sets of cloud edges and devices; they are game entities for the $\mathbb{G}_D^{\mathcal{L}}$, $\mathbb{G}_E^{\Gamma}$ and $\mathbb{G}_E^{\mathcal{W}}$ bargaining games.
- In the $\mathbb{G}_D^{\mathcal{L}}$, $\mathcal{T}_D^O$ and $\mathcal{T}_D^L$ represent the D's computation offloading part and local computing part, respectively. They are game players, and $\mathcal{L}_D$and $(1-\mathcal{L}_D)$ are their strategies. $U_D^L(\cdot)$ and $U_D^O(\cdot)$ are utility functions of $\mathcal{T}_D^O$ and $\mathcal{T}_D^L$.
- In the $\mathbb{G}_E^{\Gamma}$, Γ_E is the computation power of$E$, and devices contacting to the E are game players. The Γ_E^D represents the assigned computing power for the D; it is the D's strategy and $U_D^{\Gamma}(\cdot)$ is the D's utility function for the $\Gamma_E$resource allocation problem.
- In the $\mathbb{G}_E^{\mathcal{W}}$, $\mathcal{W}_E$ is the wireless bandwidth of E, and devices contacting to the E are game players. The $\mathcal{W}_E^D$ represents the assigned bandwidth amount for the D; it is the D's strategy and $U_D^{\mathcal{W}}(\cdot)$ is the D's utility function for the $\mathcal{W}_E$ resource allocation problem.
- The limited Γ_E^D and $\mathcal{W}_E$ resources are rationally distributed according to the *LNBS* and *LKSBS*, respectively.
- $T=\{t_1,\ldots,t_c,t_{c+1},\ldots\}$ denotes time, which is represented by a sequence of time steps.

3.3.4 The Basic Concepts and Fundamental Ideas of *LNBS* and *LKSBS*

To characterize the basic concepts of bargaining solutions, we first define mathematical expressions. The set of bargaining game players is denoted by $N = \{1,\ldots,n\}$ with $n \geq 2$. For $x,y \in \mathbb{R}^N$, we write $x \geq y$ if $x_i \geq y_i$ for all $i \in N$, and $x > y$ if $x_i > y_i$ for all $i \in N$. The utility function of player $i \in N$ (u_i:$S \rightarrow \mathbb{R}$) can be represented as the degree of satisfaction received by player i; S is the set of feasible outcomes, which are translated into real numbers. Usually, bargaining games $(S,d) \in \mathfrak{B}^N$ consist of S and disagreement, d. If bargaining players fail to reach some other outcome $x = \{x_1,\ldots,x_n\} \in S$, d results; S contains a vector $x > d$ for all $x \in \mathfrak{B}^N$. An n-player bargaining problem is a set $S \subset \mathbb{R}^N$ with $S \subset d+\mathbb{R}_+^N$ where $\mathbb{R}_+^N = \{x \in \mathbb{R}^N | x \geq 0\}$. Players seek agreement on an outcome $x = \{x_1,\ldots,x_n\}$ in S, yielding an utility $x_{1\leq i\leq n}$ to player $i \in N$. The utility point of (S,d) is the highest possible utility payoff for all players; it is simply defined by the vector $\mathcal{I}(S,d) = (\mathcal{I}_1(S,d),\ldots,\mathcal{I}_n(S,d))$ where $\mathcal{I}_i(S,d) = \mathbf{max}\{x_i | x \in (S,d)\}$ (Driesen, 2011; Kim, 2022b; Shalev, 2002).

In the *LNBS* and *LKSBS*, the outcomes of players' utility functions are evaluated with respect to a reference point and loss aversion tendency. To capture the loss aversion aspect of a player's preference, its utility function has a constant level of loss aversion, called player's loss aversion coefficient (λ). A higher coefficient value indicates a higher level of loss aversion. Finally, player i's utility function with reference point Θ_i from an outcome x_i, i.e., $U_i(x_i,\Theta_i)$, is formally defined as follows (Kim, 2022b; Shalev, 2002);

$$U_i(x_i, \Theta_i) = \begin{cases} u_i(x_i), if\, u_i(x_i) \geq \Theta_i \\ u_i(x_i) - (\lambda_i \times (\Theta_i - u_i(x_i))), otherwise \end{cases} \tag{3.3.1}$$

s.t., $\Theta_i \in \mathbb{R}, x_i \in \mathfrak{B}^N$ and $\lambda_i \in \mathbb{R}_+$

where λ_i is the non-negative loss aversion coefficient of player i, and Θ_i is the expected utility to depict player i's reference level. When player i obtains a payoff x_i, which is below his reference outcome Θ_i, player i experiences a disutility that is equal to his loss $(\Theta_i - u_i(x_i))$, multiplied by the λ_i. If player i's payoff is above than the Θ_i, it is remains unchanged (Driesen, 2011; Kim, 2022b; Shalev, 2002).

To obtain the *LNBS*, the *EFLO* scheme can define a unique solution function $\varphi^{LNBS}(\cdot):\mathfrak{B}^N \rightarrow \mathbb{R}^N$ based on the $U_{1\leq i\leq n}(x_i, \Theta_i)$. It assigns to each bargaining problem $(S, d) \in \mathfrak{B}^N$ a single point $\varphi^{LNBS}(\cdot) \in S$ with $\lambda_{1\leq i\leq n}$ and $\Theta_{1\leq i\leq n}$ (Dietzenbacher, 2021; Kim, 2022b);

$$\varphi^{LNBS}(\mathfrak{B}^N, d, N, \Lambda, \mathscr{R}) = \max_{\substack{x\in\mathfrak{B}^N, \\ \lambda_i\in\Lambda, \\ \Theta_i\in\mathscr{R}}} \prod_{i\in N} \left(U_i(x_i, \Theta_i) - U_i(d_i, \Theta_i)\right) \tag{3.3.2}$$

s.t., $\Lambda = (\lambda_1, \ldots, \lambda_n) \in \mathbb{R}_+^N$ and $\mathscr{R} = (\Theta_1, \ldots, \Theta_n) \in \mathbb{R}^N$

The *LKSBS* is a map $\varphi^{LKSBS}(\cdot):\mathfrak{B}^N \rightarrow \mathbb{R}^N$ based on the $U_{1\leq i\leq n}(x_i)$, which also implies that a single point $\varphi^{LKSBS}(\cdot)$ is obtained in the bargaining problem $(S, d) \in \mathfrak{B}^N$ for all $\Lambda \in \mathbb{R}_+^N$ and $\mathscr{R} \in \mathbb{R}^N$. Finally, $\varphi^{LKSBS}(\cdot)$ is defined as follows (Driesen, 2011; Kim, 2022b);

$$\varphi^{LKSBS}(\mathscr{I}(S, d), N, \Lambda) =$$

$$\left\{x = \{x_1 \ldots x_n\} \in P(S) | \frac{(\mathscr{I}_n(S, d) - U_n(x_n))}{(U_n(x_n) - d_n)} = \left(\mathfrak{T}_k \times \frac{(\mathscr{I}_k(S, d) - U_k(x_k))}{(U_k(x_k) - d_k)}\right)\right\} \tag{3.3.3}$$

$$\text{s.t.,} \begin{cases} x \geq d, \widetilde{N} = N \backslash \{n\}, k \in \widetilde{N} and \Gamma \in \mathbb{R}_{++}^{\widetilde{N}} \\ \mathfrak{T}_k \in T = \left\{\frac{1+\lambda_n}{1+\lambda_1}, \frac{1+\lambda_n}{1+\lambda_2}, \ldots \frac{1+\lambda_n}{1+\lambda_{n-1}}\right\} \\ P(S) = \{x \in S \mid \text{for all } y \in \mathbb{R}^N, \text{ if } y \geq x \text{ and } x \neq y, \text{ then } y \notin S\} \end{cases}$$

3.3.5 The Cooperative Control Scheme for Federated Learning Process

In the FL of IoT devices, energy consumption rate and task's delay-constraint are major control factors to decide the $\mathscr{L}_D$ value. In addition, personal data in $\mathfrak{N}_D$ usually contain sensitive information with individual privacy. Therefore, a novel computation offloading mechanism is necessary to complete FL tasks while considering all current conditions; it is quite beneficial to maximize the learning system performance. For the task $\mathscr{T}_{D_i}$ in D_i, total computation work is $\left(\mathfrak{N}_{D_i} \times \kappa_{D_i}\right)$ CPU cycles, and it is partially offloaded to the edge cloud according to the $0 \leq \mathscr{L}_{D_i} \leq 1$; $\mathscr{T}^O_{D_i} = \left(\mathfrak{N}_{D_i} \times \kappa_{D_i} \times \mathscr{L}_{D_i}\right)$ indicates the computation offloading part, and $\mathscr{T}^L_{D_i} = \left(\mathfrak{N}_{D_i} \times \kappa_{D_i} \times \left(1 - \mathscr{L}_{D_i}\right)\right)$ is the local computing part. By considering the current situation of D_i and corresponding edge cloud, the $\mathscr{L}_{D_i}$ decision problem is formulated as a bargaining game model $\left(\mathbb{G}^{\mathscr{L}}_{D_i}\right)$ (Chen, 2015; Guo, 2018; Kim, 2022b; Lee, 2018).

In the game $\mathbb{G}^{\mathscr{L}}_{D_i}$, $\mathscr{T}^O_{D_i}$ and $\mathscr{T}^L_{D_i}$ are assumed as game players, and $\mathscr{L}_{D_i}$, $\left(1 - \mathscr{L}_{D_i}\right)$ are their strategies. To reduce computation complexity, strategies are specified in terms of basic control unit (Δ). $\mathscr{T}^O_{D_i}$ and $\mathscr{T}^L_{D_i}$ players have their own utility functions $U^O_{D_i}(\cdot)$ and $U^L_{D_i}(\cdot)$; each function maps player's satisfaction to a real number, which represents the resulting outcome in the game $\mathbb{G}^{\mathscr{L}}_{D_i}$. When the E_k is D_i's corresponding edge, $U^O_{D_i}(\cdot)$ and $U^L_{D_i}(\cdot)$ are defined as follows (Kim, 2022b);

$$\begin{cases} U^L_{D_i}\left(\mathscr{T}_{D_i}, \mathscr{L}_{D_i}\right) = \left(\frac{e^c_{D_i}}{\mathfrak{E}_{D_i}} \times \log\left(\alpha + \frac{\mathscr{T}^L_{D_i}}{\mathscr{X}_{D_i}}\right) \times \mathscr{G}^c_{D_i}\left(\Upsilon^c_{D_i}, \mathscr{T}_{D_i}, \mathscr{L}_{D_i}\right)\right) \\ U^O_{D_i}\left(\mathscr{T}_{D_i}, \mathscr{L}_{D_i}\right) = \left(\frac{\left(\Gamma_{E_k} - \Phi_{E_k}\right)}{\Gamma_{E_k}} \times \log\left(\psi + \frac{\mathscr{T}^O_{D_i}}{\mathscr{X}_{D_i}}\right) \times \mathscr{F}^P_{D_i}\left(\mathscr{T}_{D_i}, \mathscr{L}_{D_i}\right)\right) \\ \mathscr{F}^P_{D_i}\left(\mathscr{T}_{D_i}, \mathscr{L}_{D_i}\right) = \gamma - exp\left(\frac{\mathscr{T}^O_{D_i}}{\mathscr{X}_{D_i}}\right) \\ \mathscr{G}^c_{D_i}\left(\Upsilon^c_{D_i}, \mathscr{T}_{D_i}, \mathscr{L}_{D_i}\right) = \mu - exp\left(1 - \frac{\Upsilon^c_{D_i}}{\Upsilon_{D_i}}\right) \end{cases} \tag{3.3.4}$$

$$\text{s.t.,} \begin{cases} \mathscr{T}^L_{D_i} = \left(\mathfrak{N}_{D_i} \times \kappa_{D_i} \times \left(1 - \mathscr{L}_{D_i}\right)\right) \\ \mathscr{T}^O_{D_i} = \left(\mathfrak{N}_{D_i} \times \kappa_{D_i} \times \mathscr{L}_{D_i}\right) and\, \mathscr{X}_{D_i} = \left(\mathfrak{N}_{D_i} \times \kappa_{D_i}\right) \end{cases}$$

where $\alpha, \psi, \gamma, \mu$ are control factors for $U^O_{D_i}(\cdot)$ and $U^L_{D_i}(\cdot)$, and Φ_{E_k} is the currently used E_k's computing power. $\mathscr{G}^c_{D_i}(\cdot)$ is a function to represent the D_i's computation overhead, and $\mathscr{F}^P_{D_i}(\cdot)$ is a function to indicate the D_i's privacy sensitivity. $\mathbb{H}_{E_k}$ is the set of devices contacting to the E_k, and $\Upsilon^c_{D_i}$, $e^c_{D_i}$ are the currently available Υ_{D_i} and $\mathfrak{E}_{D_i}$ in the D_i. In the game $\mathbb{G}^{\mathscr{L}}_{D_i}$, game players, i.e., $\mathscr{T}^O_{D_i}$ and $\mathscr{T}^L_{D_i}$, work together for the D_i, itself. Therefore, bargaining solution for the $\mathbb{G}^{\mathscr{L}}_{D_i}$ may not consider the fairness issue among game players; only the sum of players' payoffs should be maximized in the viewpoint of efficiency. In this case, the utilitarian bargaining solution for the D_i $\left(UBS_{D_i}\right)$ is most appropriate (Garroppo, 2009). This solution can be formulated as the following (Kim, 2022b):

$$UBS_{D_i} = \arg \max_{0 \leq \mathscr{L}_{D_i} \leq 1} \sum \left(U^L_{D_i}\left(\mathscr{T}_{D_i}, \mathscr{L}_{D_i}\right) + U^O_{D_i}\left(\mathscr{T}_{D_i}, \mathscr{L}_{D_i}\right)\right) \tag{3.3.5}$$

According to (3.3.5), the D_i decides the $\mathcal{L}_{D_i}$ value to maximize the sum of $U^{O}_{D_i}(\cdot)$ and $U^{L}_{D_i}(\cdot)$. And then, the D_i's corresponding edge $E_k \in \mathcal{E}$ distributes its computation $\left(\Gamma_{E_k}\right)$ and communication $\left(\mathcal{W}_{E_k}\right)$ resources to individual devices in the $\mathbb{H}_{E_k}$. In the *EFLO* scheme, the Γ_{E_k} and $\mathcal{W}_{E_k}$ resource distribution problems are formulated as bargaining games, i.e., $\mathbb{G}^{\Gamma}_{E_k}$ and $\mathbb{G}^{\mathcal{W}}_{E_k}$. In these games, devices in the set $\mathbb{H}_{E_k}$ are game player and the allocated Γ_{E_k} and $\mathcal{W}_{E_k}$ resources for the D_i are strategies in the $\mathbb{G}^{\Gamma}_{E_k}$ and $\mathbb{G}^{\mathcal{W}}_{E_k}$. In the $\mathbb{G}^{\Gamma}_{E_k}$, the utility function for the D_i, i.e., $U^{\Gamma}_{D_i}(\cdot)$, is defined based on some factors; E_k's assigned computation power for the D_i, i.e., $\Gamma^{D_i}_{E_k}$, the currently used E_k's computing power, i.e., Φ_{E_k}, and D_i's loss aversion propensity, i.e., λ_{Γ,D_i}. Under the heavy computing overhead of E_k, the loss aversion tendency of the D_i is alleviated. However, if it was vice versa, i.e., the Φ_{E_k} is light, it means that the E_k has enough available computing power to support offloading services. In this case, the tendency toward loss aversion increases. Therefore, based on the current E_k conditions, the *EFLO* scheme dynamically modifies the loss aversion propensity λ_{Γ,D_i}. According to the $\Gamma^{D_i}_{E_k}$, Φ_{E_k} and Θ_{Γ,D_i} values, the D_i's utility function, i.e., $U^{\Gamma}_{D_i}(\cdot)$, can be derived as follows (Kim, 2022b).

$$U^{\Gamma}_{D_i}\left(\Gamma_{E_k},\Gamma^{D_i}_{E_k},\mathcal{T}^{O}_{D_i},\Phi_{E_k},\Theta_{\Gamma,D_i}\right) =$$

$$\begin{cases} u^{\Gamma}_{D_i}\left(\Gamma^{D_i}_{E_k}\right) = \left(\dfrac{T_{D_i}}{\eta + exp\left(-\dfrac{\Gamma^{D_i}_{E_k}}{\Gamma_{E_k}}\right)}\right), if\, u^{\Gamma}_{D_i}\left(\Gamma^{D_i}_{E_k}\right) \geq \Theta_{\Gamma,D_i} \\ u^{\Gamma}_{D_i}\left(\Gamma^{D_i}_{E_k}\right) - \left(\lambda_{\Gamma,D_i} \times \left(\Theta_{\Gamma,D_i} - u^{\Gamma}_{D_i}\left(\Gamma^{D_i}_{E_k}\right)\right)\right), otherwise \end{cases} \quad (3.3.6)$$

$$\text{s.t.,} \begin{cases} T_{D_i} = \left(\zeta + \dfrac{\left(\Omega_O \times \mathcal{T}^{O}_{D_i}\right) + \left(\Omega_L \times \mathcal{T}^{L}_{D_i}\right)}{\mathcal{M}^{\mathcal{T}}_{D_i}}\right) and\, D_i \in \mathbb{H}_{E_k} \\ \Theta_{\Gamma,D_i} = \left(u^{\Gamma}_{D_i}\left(\mathcal{T}^{O}_{D_i}\right) \times \rho_{D_i}\right) and\ \lambda_{\Gamma,D_i} = \left(\beta \times \dfrac{\left(\Gamma_{E_k} - \Phi_{E_k}\right)}{\Gamma_{E_k}}\right) \end{cases}$$

where η, ζ are control parameters for $U^{\Gamma}_{D}(\cdot)$, and Ω_O, Ω_L are time delay parameters for offloading and local computing processes, respectively. ρ_{D_i},β are the system adjustment factors for the loss aversion tendency. Especially, during the operation of offloading computation services, unexpected growth of offloading requests may develop in a specific edge cloud. To ensure a relative fairness, the idea of *LNBS* is applied for the game $\mathbb{G}^{\Gamma}_{E_k}$. It is given by (Kim, 2022b):

$$\varphi^{LNBS}\left(D_i \in \mathbb{H}_{E_k} \mid U^{\Gamma}_{D_i}(\cdot), \left(\Lambda_{E_k},\Theta_{\Gamma,D_i}\right)\right) = \max_{\Gamma^{D_i}_{E_k}} \prod_{D_i\in\mathbb{H}_{E_k}} \left(U^{\Gamma}_{D_i}(\cdot) - U^{d}_{D_i}(\cdot)\right) \quad (3.3.7)$$

$$\text{s.t., } \Lambda_{E_k} = \left(\ldots\ \lambda_{\Gamma,D_i}\ldots\right) \in \mathbb{R}^{|\mathbb{H}_{E_k}|}_{+}, \left(\ldots\Theta_{\Gamma,D_i}\ldots\right) \in \mathbb{R}^{|\mathbb{H}_{E_k}|} \text{ and } \sum\nolimits_{D_i\in\mathbb{H}_{E_k}} \Gamma^{D_i}_{E_k} \leq \Gamma_{E_k}$$

where $U^{d}_{D_i}(\cdot)$ is the disagreement point of D_i. According to (3.3.7), the $\Gamma^{D_i}_{E_k}$ value is adaptively decided in the $\mathbb{G}^{\Gamma}_{E_k}$ game. As the same manner as the $U^{\Gamma}_{D_i}(\cdot)$, the D_i's utility function in the spectrum allocation process, i.e., $U^{\mathscr{W}}_{D_i}(\cdot)$, is decided based on the $\mathscr{W}^{D_i}_{E_k}$ and $\Theta_{\mathscr{W},D_i}$; it can be derived as follows (Kim, 2022b).

$$U^{\mathscr{W}}_{D_i}\left(\mathscr{W}_{E_k}, \mathscr{W}^{D_i}_{E_k}, \mathfrak{R}_{D_i}, \mathscr{L}_{D_i}, \Theta_{\mathscr{W},D_i}, \varpi_{E_k}\right) =$$

$$\begin{cases} u^{\mathscr{W}}_{D_i}\left(\mathscr{W}^{D_i}_{E_k}\right) = \left(T_{D_i} \times \log\left(\sigma + \frac{\mathscr{W}^{D_i}_{E_k}}{\mathscr{W}_{E_k}}\right)\right), if\, u^{\mathscr{W}}_{D_i} \geq \Theta_{\mathscr{W},D_i} \\ u^{\mathscr{W}}_{D_i}\left(\mathscr{W}^{D_i}_{E_k}\right) - \left(\lambda_{\mathscr{W},D_i} \times \left(\Theta_{\mathscr{W},D_i} - u^{\mathscr{W}}_{D_i}\left(\mathscr{W}^{D_i}_{E_k}\right)\right)\right), otherwise \end{cases} \tag{3.3.8}$$

$$\text{s.t., } \Theta_{\mathscr{W},D_i} = \left(u^{\mathscr{W}}_{D_i}\left(\mathfrak{R}_{D_i} \times \mathscr{L}_{D_i}\right) \times \rho_{D_i}\right) \text{ and } \lambda_{\mathscr{W},D_i} = \left(\beta \times \frac{\left(\mathscr{W}_{E_k} - \varpi_{E_k}\right)}{\mathscr{W}_{E_k}}\right)$$

where σ is a control parameter for $U^{\mathscr{W}}_{D_i}(\cdot)$ and $\left(\mathfrak{R}_{D_i} \times \mathscr{L}_{D_i}\right)$ is the requested bandwidth of D_i, and ϖ_{E_k} is the currently used E_k's bandwidth amount. The concept of *LKSBS* is applied to distribute the $\mathscr{W}_{E_k}$ resource to devices in the set $\mathbb{H}_{E_k}$. It is given for the $\mathbb{G}^{\mathscr{W}}_{E_k}$ game, and the $\mathscr{W}^{D_i}_{E_k}$ value is adaptively decided according to (3.3.9) (Kim, 2022b).

$$\varphi^{LKSBS}\left(D_i, D_j \in \mathbb{H}_{E_k} | U^{\mathscr{W}}_{D_i}(\cdot), \left(\Lambda_{E_k}, \Theta_{\mathscr{W},D_i}\right)\right) =$$

$$\left\{\left\{\ldots \mathscr{W}^{D_i}_{E_k} \ldots\right\} | \frac{\left(\mathscr{I}_{D_j}(\cdot) - U^{\mathscr{W}}_{D_j}(\cdot)\right)}{\left(U^{\mathscr{W}}_{D_j}(\cdot) - U^{d}_{D_j}(\cdot)\right)} = \left(\mathfrak{T}_{D_i} \times \frac{\left(\mathscr{I}_{D_i}(\cdot) - U^{\mathscr{W}}_{D_i}(\cdot)\right)}{\left(U^{\mathscr{W}}_{D_i}(\cdot) - U^{d}_{D_i}(\cdot)\right)}\right)\right\} \tag{3.3.9}$$

$$\text{s.t.,} \Lambda_{E_k} = \left(\ldots \lambda_{\mathscr{W},D_i} \ldots\right) \in \mathbb{R}^{|\mathbb{H}_{E_k}|}_{+}, \mathfrak{T}_{D_i} = \frac{1 + \lambda_{\mathscr{W},D_j}}{1 + \lambda_{\mathscr{W},D_i}} \text{ and } \sum\nolimits_{D_i \in \mathbb{H}_{E_k}} \mathscr{W}^{D_i}_{E_k} \leq \mathscr{W}_{E_k}$$

3.3.6 Summary

The FL paradigm diverges from traditional centralized machine learning methods, which require all data samples to be uploaded to a single server. By adopting the concept of computation offloading from smart devices to edge clouds, the MEC has been introduced to support resource-constrained IoT devices within FL frameworks. Nonetheless, managing computational offloading introduces complex control challenges that must be addressed through effective cooperative strategies tailored to dynamic edge system environments. To tackle these challenges, the *EFLO* scheme proposes an innovative FL-based task offloading mechanism grounded in a joint cooperative game model. Utilizing the principles of utilitarianism, each device independently determines the extent of its computational offloading. Additionally, the *EFLO* scheme incorporates the *LNBS* and *LKSBS* to manage the allocation of communication and computational resources for each edge cloud. These three solutions are seamlessly integrated and interdependent within the *EFLO* scheme. By considering the current state of the FL system, various bargaining solutions work together cooperatively and synergistically in a real-time, online manner. Consequently, the *EFLO* scheme ensures that task offloading services are efficiently managed within the dynamically evolving MEC system environment (Kim, 2022b).

3.4 Two-phase Wireless Bandwidth Allocation (*TWBA*) scheme

Significant progress in embedded systems-on-a-chip has markedly increased the number of smart devices capable of executing applications both interactively and autonomously. A range of technologies pivotal to the proliferation of smart mobile devices has surfaced, including Edge of Things (EoT), social networks, cloud computing services, and 5G wireless systems. Despite the deployment of high-speed infrastructure, wireless bandwidth remains a notably scarce resource, underscoring the necessity for strategic resource allocation methodologies. Recently, S. Kim introduces the *Two-phase Wireless Bandwidth Allocation* (*TWBA*) scheme aimed at maximizing the performance of EoT systems efficiently (Kim,2022c). To ensure effective system operations, the *TWBA* scheme leverages two cooperative game-theoretic solutions: the *weighted Kalai-Smorodinsky bargaining solution* and the *interval Shapley Value*. These methodologies are instrumental in achieving mutually beneficial outcomes that strike a balance between efficiency and fairness. The primary advantage of adopting a cooperative game approach lies in its ability to deliver a fair and efficient control solution for the bandwidth allocation challenge, while dynamically adapting to changing system environments.

3.4.1 Development motivation

The IoT is rapidly advancing, with the potential to connect billions of mobile devices globally. According to forecasts by Gartner and Research Nester, there were 6.4 billion IoT devices in use worldwide in 2016, marking a 30 percent increase from the previous year, and this number is projected to reach 20.8 billion by 2020. The global IoT market is anticipated to grow to USD 724.2 billion by 2023. The vision for IoT emphasizes the critical role of smart devices in enabling various real-world applications such as smart cities, smart homes, and smart medical systems. A significant driver of this growth is the vast amount of data generated by big data-enabled IoT devices, with predictions suggesting that by 2020, IoT-generated data will total 600 Zettabytes annually, which is 275 times the expected Internet traffic (Georgakopoulos, 2016; Kim,2022c; Linthicum, 2017; Yang, 2017).

In pursuit of this IoT vision, cloud computing has emerged as a key technology, offering robust elastic computation and data management capabilities essential for IoT applications. The centralized cloud computing model has proven highly effective for large-scale, distributed applications, providing on-demand, pay-as-you-go services to end users. Cloud computing delivers flexible and scalable resources in terms of computing, storage, and networking, making it ideal for applications with adaptable and scalable needs. However, this centralized approach faces significant challenges within the IoT framework. Future IoT applications will generate such vast volumes and high velocities of data that transferring all data from edge smart devices to remote cloud servers becomes impractical. As the number of edge devices continues to grow exponentially, traditional cloud computing paradigms are expected to encounter increased difficulties related to traffic dynamics and network overheads (Kim,2022c; Pan, 2017; Zhu, 2016).

Currently, mobile smart devices are viewed as edge terminals equipped with substantial capabilities to serve both end users and their surrounding environments. With powerful edge devices like smartphones, on-board units, and tablets, we are witnessing a technological revolution. Edge computing has become a vital paradigm for numerous mission-critical applications due to its ability to provide real-time data processing, reduced latency, and enhanced QoS. This advancement has expanded the potential of IoT, leading to the development of the Edge-of-Things (EoT) paradigm. To ensure reliable QoS, the EoT paradigm offers an effective solution when edge smart devices possess sufficient resources to support ongoing processes (El-Sayed, 2017; Kim, 2010b; Kim, 2022c; Pande, 2013).

Building on existing technologies like EoT and cloud computing, the Cloud-supported Edge-of-Things (CEoT) has emerged as the next-generation network paradigm. CEoT facilitates the tight integration of EoT and cloud computing, allowing both EoT and cloud services to be coordinated in a dynamic and flexible manner. However, the CEoT infrastructure introduces several challenges; i) cloud assistance: cloud computing must assist resource-constrained edge smart devices, ii) task offloading: edge devices need to offload computational tasks to cloud servers when their own capabilities are insufficient, and iii) latency control: communication latency must be effectively managed, surpassing the limitations of traditional cloud computing models. Addressing these challenges requires fostering innovations within the CEoT framework (Kim, 2022c; Pan, 2017).

In the context of future 5G networks, it is envisioned that multiple EoT communities will form by integrating socially-aware smart devices. Within an EoT community, edge devices can share application outcomes with one another, enhancing the efficiency of the CEoT system. This cooperative approach necessitates a deep understanding of the social dynamics and the anticipated benefits among interacting smart devices (Du, 2018). Fortunately, cloud computing, EoT, and socially-aware interactions are inherently compatible and can complement each other's strengths and weaknesses. This convergence inspires the pursuit of a joint solution aimed at maximizing CEoT system performance by considering mutually dependent factors. However, integrating CoT, FoT, and EoT paradigms into a cohesive protocol is highly complex and challenging, highlighting the need for innovative solution concepts to address future 5G network requirements effectively (Kim, 2022c).

3.4.2 Contribution

The advancement of emerging applications and services is increasingly hindered by the limited computational power of edge devices. To address this limitation, the CEoT system facilitates computational offloading to cloud computing services. Ensuring the necessary QoS for computation-intensive applications requires not only sophisticated offloading techniques but also a robust communication infrastructure. The *TWBA* scheme aims to introduce a novel mechanism for wireless bandwidth allocation within the CEoT framework. To effectively resolve the resource allocation challenges while incorporating social features, the *TWBA* scheme integrates the fundamental principles of the *WKSBS* and the *ISV*. To explore the strategic interactions between edge devices and the Network Access Point (NAP), the *TWBA* scheme develops an innovative two-phase cooperative game model. This model is specifically designed to efficiently distribute the limited wireless bandwidth among participants. Following the guidelines of cooperative game solutions, individual edge devices collaborate to ensure both fairness and efficiency while managing their conflicting interests during negotiations (Kim, 2022c).

3.4.3 CEoT architecture for computation task offloading services

The *TWBA* scheme operates within a cloud-assisted EoT infrastructure, examining the functional capabilities associated with computational task offloading. The primary aim of computation offloading services is to enhance the QoS for end users. Edge devices have the option to either execute their application tasks locally or offload them to a remote cloud server to reduce computational burdens. These decision-making processes can be collaboratively managed by the devices themselves. Consequently, game theory is employed to model this collaborative scenario. The *TWBA* scheme addresses the challenge of limited communication resource allocation by utilizing the principles of the *WKSBS* and *ISV* (Kim, 2022c).

The *TWBA* scheme assumes a CEoT system consist of n smart edge devices, which is denoted as the set of EoT devices $\mathbb{D} = \{\mathscr{D}_1, \mathscr{D}_2, \ldots, \mathscr{D}_n\}$. The *TWBA* scheme also denotes the time period and the time period index set by t and T = {1,...,t,...∞}. The $\mathscr{D}_{1 \le i \le n}$ generates different application service requests $\mathbb{A}^{\mathscr{D}_i} = \{\mathscr{A}_1^{\mathscr{D}_i}, \mathscr{A}_2^{\mathscr{D}_i}, \mathscr{A}_3^{\mathscr{D}_i}, \ldots\}$ and may offload certain amounts of computing tasks to the cloud server ($\mathbb{C}$). The service requests generated from the $\mathscr{D}_i$ follow a Poisson process with an average arrival rate at each time period t. The requests are assumed to be delay sensitive and mutually independent. If the $\mathscr{D}_i$ does not get sufficient communication resource from the corresponding NAP, the requested cloud service in the $\mathscr{A}^{\mathscr{D}_i}$ is dropped out (Kim, 2022c; Lee, 2015; Liu, 2018).

Smart edge devices connect to the C via the NAP. Each NAP is accountable for the devices within its coverage area. However, in practice, the communication resources at each NAP, specifically wireless bandwidth, are limited and contested. When numerous offloaded tasks are processed simultaneously, these resources can be quickly depleted, making it impractical to meet the demands of all applications. Despite the clear benefits of offloading services, the CEoT system experiences reduced QoS due to scarcity of communication resources. To optimize CEoT system performance, implementing an effective communication resource management strategy is crucial. This strategy must enhance system performance while adhering to the service constraints posed by various task requests. However, developing such a strategy is complex and challenging within a dynamically changing CEoT environment. To address the wireless bandwidth allocation issue for task offloading services, the *TWBA* scheme incorporates two cooperative game-theoretic solutions: the *WKSBS* and the *ISV*. These solutions are employed to ensure a fair and efficient distribution of the limited wireless bandwidth, thereby maximizing the overall performance of the CEoT system while maintaining service quality for diverse application needs (Kim, 2022c).

3.4.4 Interval Shapley value

The *Shapley value* stands out as one of the most prevalent single-valued solution concepts within cooperative game theory. Recently, significant research efforts have been directed towards expanding the traditional *Shapley value* to accommodate scenarios involving interval data. The *TWBA* scheme specifically emphasizes the *ISV*, a concept introduced by Alparslan Gök et al. in 2009. The *ISV* assigns a payoff vector to each cooperative interval game, where the vector's components are compact intervals of real numbers. In recent years, the *ISV* has garnered considerable attention as it has been integrated into new game-theoretic models and widely applied to resource allocation problems across various fields such as economics, sociology, and computer science. Particularly in socially aware networks, situations involving social influence are often subject to interval uncertainty. In these contexts, representing influence factors as intervals of real numbers proves highly beneficial, allowing for a more nuanced and flexible analysis of social dynamics (Alparslan Gök, 2010; Branzei, 2010; Kim, 2022c).

Usually, a cooperative game in coalitional form is an ordered pair N, v, where $N = \{1, 2,..., n\}$ is the set of players, and v: $2^N \rightarrow \mathbb{R}$ is a map, assigning to each coalition $S \in 2^N$ a real number, such that $v(\emptyset) = 0$. Formally, a cooperative interval game in coalitional form is an ordered pair N, ω where ω: $2^N \rightarrow I(\mathbb{R})^N$ is the characteristic function such that $\omega(\emptyset) = [0, 0]$, where $I(\mathbb{R})^N$ is the set of all n-dimensional vectors with elements in $I(\mathbb{R})$, which is the set of all nonempty, compact intervals in R. Simply, the payoff vectors $\mathbf{x} = (x_1, x_2, ..., x_n) \in \mathbb{R}^N$ from the classical cooperative game theory are replaced by n-dimensional vectors $(I_1, I_2, ..., I_n) \in I(\mathbb{R})^N$. For all S 2^N, the worth interval $\omega(S)$ of the coalition S in the interval game N, ω is of the form $[\underline{\omega}(S), \overline{\omega}(S)]$, where $\underline{\omega}(S)$ is the coalition S's minimal payoff, and $\overline{\omega}(S)$ is the coalition S's maximal payoff. Note that if all the payoff intervals are degenerate intervals, i.e., $\underline{\omega}(S) = \overline{\omega}(S)$ for each $S \in 2^N$, then the interval game N, ω corresponds in a natural way to the classical cooperative game N, v where $v(S) = \underline{\omega}(S)$ (Alparslan Gök, 2010; Branzei, 2010; Kim, 2022c).

Let $I(\mathbb{R})$ be the set of all closed and bounded intervals in $\mathbb{R}$, and let $I, J \in I(\mathbb{R})$ with $I = [\underline{I}, \overline{I}]$, $J = [\underline{J}, \overline{J}]$, $|I| = \underline{I} - \overline{I}$ and $\alpha \in \mathbb{R}_+$. Then, $I + J = [\underline{I} + \underline{J}, \overline{I} + \overline{J}]$ and $\alpha I = [\alpha\underline{I}, \alpha\overline{I}]$. A partial subtraction operator has played a central role. We define $I - J$, only if $|I| \geq |J|$, by $I - J = [\underline{I} - \underline{J}, \overline{I} - \overline{J}]$. We say that I is weakly better than J, denoted by $I \succcurlyeq J$, if and only if, $\underline{I} \geq \underline{J}$ and $\overline{I} \geq \overline{J}$. IG^N is denoted as the family of all *Interval Games* with player set N. An interval solution concept F on a class of $IG^N$is a map assigning to each interval game ω in this class a set of n-dimensional vectors whose components belong to $I(\mathbb{R})$. Therefore, F in IG^N is a set of payoff vectors $x = (x_1, ..., x_n) \in \mathbb{R}^N$. Some classical games associated with an interval game $\omega \in IG^N$ have played a key role, namely the border games $N, \underline{\omega}, N, \overline{\omega}$, and the length game $N, |\omega|$, where $|\omega|(S) = \overline{\omega}(S) - \underline{\omega}(S)$ for each $S \in 2^N$. Note that $\overline{\omega} = \underline{\omega} + |\omega|$. For a game $\omega \in IG^N$ and a coalition $S \in 2^N \setminus \{\varnothing\}$, the interval subgame with player set $T \in 2^N \setminus \{\varnothing\}$ is the subgame ω_T defined by $\omega_T(S) = \omega(S)$ for all $S \in 2^T$, i.e. ω_T is the restriction of ω to the set 2^T. Such subgames are denoted as T, ω (Alparslan Gök, 2010; Branzei, 2010; Kim, 2022c).

Let a game N, ω be *size monotonic* if $N, |\omega|$ is monotonic, i.e., $|\omega|(S) \leq |\omega|(T)$ for all $S, T \in 2^N$ with $S \subset T$. For further use, we denote by $SMIG^N$ the class of *size monotonic interval games* with player set N. Denote by $\Pi(N)$ the set of permutations $\sigma: N \to N$ and $\omega \in SMIG^N$. We introduce the notions of interval marginal operator corresponding to σ, denoted by $m^\sigma: SMIG^N \to I(\mathbb{R})^N$, and of interval marginal vector of ω with respect to σ, denoted by $m^\sigma(\omega)$. The marginal vector $m^\sigma(\omega)$ corresponds to a situation, where players enter a room one by one in the order $\sigma(1), \sigma(2), \ldots, \sigma(n)$ and each player is given the marginal contribution he/she creates by entering. $P_\sigma(i)$ represents predecessors of i in σ and $\sigma^{-1}(i)$ denotes the entrance number of player i. Finally, the marginal vector of player i with ω, i.e., $m_i^\sigma(\omega)$, is defined as follows (Alparslan Gök, 2010; Branzei, 2010; Kim, 2022c).

$$m_i^\sigma(\omega) = \omega\left(P_\sigma(i) \cup \{i\}\right) - \omega\left(P_\sigma(i)\right) \tag{3.4.1}$$

s.t., $P_\sigma(i) = \{r \in N | \sigma^{-1}(r) \sigma^{-1}(i)\}$ and $i \in N$

The monotonicity of $|\omega|$ implies $\overline{\omega}(S \cup \{i\}) - \underline{\omega}(S \cup \{i\}) \geq \overline{\omega}(S) - \underline{\omega}(S)$, which can be rewritten as $\overline{\omega}(S \cup \{i\}) - \overline{\omega}(S) \geq \underline{\omega}(S \cup \{i\}) - \underline{\omega}(S)$. So, $\omega(S \cup \{i\}) - \omega(S)$ is defined for each $S \subset T$ and $i \notin N$. Therefore, all the interval marginal vectors of a *size monotonic* game are efficient interval payoff vectors. For each $\omega \in SMIG^N$, the *ISV* $\Phi: SMIG^N \to I(\mathbb{R})^N$ is defined as follows (Alparslan Gök, 2010; Kim, 2022c).

$$\Phi(\omega) = \frac{1}{n!} \sum_{\sigma \in \Pi(N)} m^\sigma(\omega) = \frac{1}{n!} \sum_{\sigma \in \Pi(N)} \left(\omega\left(P_\sigma(i) \cup \{i\}\right) - \omega\left(P_\sigma(i)\right)\right) \tag{3.4.2}$$

Note that there are exactly $|S|!(n-1-|S|)!$ orderings for which one has $P_{\sigma}(\{i\}) = S$. The first factor, i.e., $|S|!$, corresponds to the number of orderings of S and the second factor, i.e., $(n-1-|S|)!$, is just the number of orderings of $N\ (S \cup \{i\})$. If S is a subset of N not containing i, we can rewrite the *ISV* value $\Phi(\omega)$ as (Alparslan Gök, 2010; Kim, 2022c).

$$\Phi(\omega) = \sum_{S:i \notin S} \left(\frac{|S|!(n-1-|S|)!}{n!} \times (\omega(S \cup \{i\}) - \omega(S)) \right) \tag{3.4.3}$$

$$\text{s.t.,} \sum_{S:i \notin S} \frac{|S|!(n-1-|S|)!}{n!} = 1$$

Example: Let $\langle N, \omega \rangle$ be a cooperative interval game with $N = \{1, 2, 3\}$ and $\omega(1) = \omega(13) = [7, 7]$, $\omega(12) = [12, 17]$, $\omega(123) = [24, 29]$ and $\omega(S) = [0, 0]$ otherwise. The interval marginal vectors are given in the following Table 3.4.1, where $\sigma: N \to N$ is identified with $(\sigma(1), \sigma(2), \sigma(3))$. Firstly, for $\sigma_1 = (1, 2, 3)$, we can calculate the interval marginal vectors as follows (Kim, 2022c; Osman, 2015).

$$\begin{cases} m_1^{\sigma 1}(\omega) = \omega(1) = [7, 7] \\ m_2^{\sigma 1}(\omega) = \omega(12) - \omega(1) = [5, 10] \\ m_3^{\sigma 1}(\omega) = \omega(123) - \omega(12) = [12, 12] \end{cases} \tag{3.4.4}$$

Table 1. Interval marginal vectors

σ	$m_1^{\sigma}(\omega)$	$m_2^{\sigma}(\omega)$	$m_3^{\sigma}(\omega)$
$\sigma_1 = (1,2,3)$	$[7,7]$	$[5,10]$	$[12,12]$
$\sigma_2 = (1,3,2)$	$[7,7]$	$[17,22]$	$[0,0]$
$\sigma_3 = (2,1,3)$	$[12,17]$	$[0,0]$	$[12,12]$
$\sigma_4 = (2,3,1)$	$[24,29]$	$[0,0]$	$[0,0]$
$\sigma_5 = (3,1,2)$	$[7,7]$	$[17,22]$	$[0,0]$
$\sigma_6 = (3,2,1)$	$[24,29]$	$[0,0]$	$[0,0]$

The subsequent calculations follow a similar approach and are detailed in Table 1. This table presents the interval marginal vectors for the cooperative interval game discussed in the example. The *ISV* for this game is determined by averaging these six interval marginal vectors, as illustrated in (Kim, 2022c):

$$\Phi(\omega) = \left(\left[\frac{27}{2}, 16\right], \left[\frac{13}{2}, 9\right], [4, 4] \right)$$

3.4.5 Weighted *Kalai-Smorodinsky* Bargaining Solution

Bargaining scenarios are prevalent in today's world, especially within network management contexts. These situations can be interpreted through the lens of consensus theory, as it is generally presumed that a final decision is achievable only when every group member agrees to it. To formally articulate this concept, it is essential to introduce specific terminology. Υ is the set of all utility profiles that a bargaining process could possibly yield, and let d be the disagreement point, which is a point that specifies their utilities in case they do not reach a unanimous agreement on some point of Υ. Simply, a two person bargaining problem is a pair (Υ, d) where $\Upsilon \subseteq \mathbb{R}^2$ and $d \in \Upsilon$. For Υ, any function chooses an element of Υ is called a solution; usually denoted by F. For all $x, y \in \mathbb{R}^2$, $x \geq y$ if and only if $x_i \geq y_i$ for $i = 1, 2$. $x > y$ if and only if $x_i \geq y_i$ for $i = 1, 2$ and $x_j > y_j$ for some j. Finally, $x \gg y$ if and only if $x_i > y_i$ for $i = 1, 2$ (Dubra, 2001; Kim, 2022c).

In 1975, Kalai and Smorodinsky introduced a groundbreaking bargaining solution known as the *Kalai-Smorodinsky Bargaining Solution* (*KSBS*). A prominent feature of the *KSBS* is the *Symmetry* axiom. However, this axiom has faced direct criticism. Enforcing *Symmetry* implies that both parties possess equal bargaining skills, which is often unrealistic in situations where each participant has varying levels of bargaining power. For example, *Symmetry* becomes problematic when the bargaining scenario is seen as a complex non-cooperative game. To address these concerns, an important advancement in the *KSBS* involved relaxing the *Symmetry* axiom, allowing for asymmetric solutions and thereby making the theory more adaptable and flexible (Dubra, 2001; Kim, 2022c).

In order to provide simple and identifiable asymmetric *KSBS*, it is necessary to generalize the classical *KSBS* in a natural way. In 1994, the concept of *Weighted KSBS* (*WKSBS*) was introduced for two-player bargaining game. For notational simplicity, let $\{x \in \mathbb{R}_+^2 : x_2 = \infty x_1\}$ denote the y axis throughout. Then, the *WKSBS* is defined as follows (Dubra, 2001; Kim, 2022c);

$$\mathrm{D}_\lambda(\Upsilon) \equiv \left\{x \in \mathbb{R}_+^2 : \frac{x_2}{\mathrm{a}_2(\Upsilon)} = \lambda \times \frac{x_1}{\mathrm{a}_1(\Upsilon)}\right\} \cap P(\Upsilon) \tag{3.4.5}$$

$$s.t. \begin{cases} a_i(\Upsilon) \equiv \max\{x_i : (x_1, x_2) \in \Upsilon\}, i = 1, 2 \\ \lambda \in \mathbb{R}_+ \cup \{\infty\} \\ P(\Upsilon) \equiv \{x \in \Upsilon : y > x \Rightarrow y \notin \Upsilon\} \end{cases}$$

For any problem Υ, $a(\Upsilon)$ is the utopia point, the utility profile in which each individual achieves the greatest utility level when bargainer can get from the bargaining problem. If $\lambda = 1$, then $\mathrm{D}_1(\Upsilon)$ is the *KSBS* (Kim, 2022c).

3.4.6 Socially-aware EoT Communities and Interrelationship

Currently, a significant area of research focuses on designing and evaluating local IoT communities that integrate socialization features into IoT communications within EoT environments. This trend underscores the urgent need to comprehend the fundamental social characteristics and their effects on the EoT paradigm (Du, 2018). Typically, the activities of smart devices are inherently connected to those of their neighboring devices. Within EoT ecosystems, social relationships between devices can be defined by factors such as *Social Interest Similarity* (*SIS*) and *Interaction Trust* (*IT*) (Ai, 2018). By leveraging these online social relationships, the benefits derived from cloud services can be efficiently distributed among nearby smart devices. As a result, the intelligent management of socially-aware EoT communications not only boosts the efficiency of the CEoT system but also reduces the overhead associated with cloud services (Kim, 2022c).

Each individual device has the capability to communicate directly with one another. As a result, application outcomes can be shared among neighboring devices through device-to-device (D2D) communications. To anticipate future outcome sharing, the *Outcome Sharing Possibility* (*OSP*) between two devices can be assessed based on the *SIS* and *IT*. The *SIS* serves as a metric that quantifies the similarity of interests between two devices, derived from their social characteristics. On the other hand, the *IT* represents the proportion of outcomes shared to date and is dynamically estimated in a distributed online manner. The *OSP* of $\mathcal{D}_i$ and $\mathcal{D}_j$ devices, i.e., $OSP_{\mathcal{D}_j}^{\mathcal{D}_i}$, is defined as follows (Kim, 2022c);

$$OSP_{\mathcal{D}_j}^{\mathcal{D}_i} = SIS_{\mathcal{D}_j}^{\mathcal{D}_i} + IT_{\mathcal{D}_j}^{\mathcal{D}_i}\text{ , s.t., } SIS_{\mathcal{D}_j}^{\mathcal{D}_i} \in [0..1] \text{ and } IT_{\mathcal{D}_j}^{\mathcal{D}_i} = \frac{\mathcal{S}_{\mathcal{D}_i\mathcal{D}_j}}{\mathcal{C}_{\mathcal{D}_i} + \mathcal{C}_{\mathcal{D}_j}} \tag{3.4.6}$$

where the $SIS_{\mathcal{D}_j}^{\mathcal{D}_i}$ is a fixed given value and the $IT_{\mathcal{D}_j}^{\mathcal{D}_i}$ value is adaptively adjusted; $\mathcal{C}_{\mathcal{D}_i}$ is the total number of $\mathcal{D}_i$'s cloud assisted services, and $\mathcal{S}_{\mathcal{D}_i\mathcal{D}_j}$ is the outcome sharing number between $\mathcal{D}_i$ and $\mathcal{D}_j$. In the *TWBA* scheme, smart devices are self-organized into disjoint social communities. According to the proximity, edge devices in a neighborhood form a social group. Among the smart devices in a social group, a device with the highest social influence is selected as a *Group Leader* (*GL*), which is given by (Kim, 2022c);

$$GL = \max_{\mathcal{D}_i \in \mathbb{S}} \left\{ \mathfrak{X}(\mathcal{D}_i) | \mathfrak{X}(\mathcal{D}_i) = \sum_{\mathcal{D}_j \in \mathbb{S}} OSP_{\mathcal{D}_j}^{\mathcal{D}_i} \right\} \tag{3.4.7}$$

where S is the device set of a social group. As a representative of community, each *GL* has its own social influence power interval $\left[\underline{\Theta_{GL}}, \overline{\Theta_{GL}}\right]$ in $\mathbb{R}$; $\underline{\Theta_{GL}}$ and $\overline{\Theta_{GL}}$ are lower and upper bounds of social influence. Based on the *SIS* and *IT* values of the *GL*'s neighboring devices, $\underline{\Theta_{GL}}$ and $\overline{\Theta_{GL}}$ are estimated as follows (Kim, 2022c);

$$\underline{\Theta_{GL}} = \sum_{\mathcal{D}_i \in \mathbb{S}} IT_{GL}^{\mathcal{D}_i} \text{ and } \overline{\Theta_{GL}} = \sum_{\mathcal{D}_i \in \mathbb{S}} \left(SIS_{GL}^{\mathcal{D}_i} + IT_{GL}^{\mathcal{D}_i}\right) \tag{3.4.8}$$

To quantify the $\mathcal{D}_i$'s satisfaction with $\mathcal{A}_k$ application service, the utility function of $\mathcal{D}_i$ can be derived as follows (Kim, 2022c).

$$U_{\mathscr{D}_i}^{\mathscr{A}_k}\left(\chi_{\mathscr{D}_i}^{\mathscr{A}_k}\right) = \left(1+\sum_{\mathscr{D}_j \in \mathbb{S}} IT_{\mathscr{D}_j}^{\mathscr{D}_i}\right)^{\mathbb{I}_{\mathscr{D}_i}} \times \left(\chi_{\mathscr{D}_i}^{\mathscr{A}_k} \times \log\left(1+\frac{\chi_{\mathscr{D}_i}^{\mathscr{A}_k}}{\Gamma_{\mathscr{D}_i}^{\mathscr{A}_k}}\right)\right)$$

(3.4.9)

$$\text{s.t.,}\begin{cases} \mathbb{I}_{\mathscr{D}_i} = 1, if \mathscr{D}_i is GL \\ \mathbb{I}_{\mathscr{D}_i} = 0, otherwise \end{cases}$$

where $\Gamma_{\mathscr{D}_i}^{\mathscr{A}_k}$ and $\chi_{\mathscr{D}_i}^{\mathscr{A}_k}$ are the requested and assigned wireless bandwidth amounts of $\mathscr{D}_i$ for the application task $\mathscr{A}_k$, respectively (Kim, 2022c).

3.4.7 The two-phase Bandwidth Allocation Algorithms

The *TWBA* scheme is designed for the CEoT system comprising a number of NAPs. Each NAP communicates with multiple edge devices through wireless links. Generally, individual NAPs are responsible for the wireless bandwidth allocation strategy in an entirely distributed fashion. In the *TWBA* scheme, NAP allocate its bandwidth amount by employing a two-phase game process. In the first phase, NAP partitions available bandwidth into two parts: i) bandwidth for individual devices, i.e., $\mathfrak{B}_{\mathscr{D}}^{AP}$, and ii) bandwidth for *GLs*, i.e., $\mathfrak{B}_{GL}^{AP}$. This partition is decided according to the *WKSBS* approach. In the second phase, the $\mathfrak{B}_{GL}^{AP}$ is distributed among *GLs* based on the *ISV* (Kim, 2022c).

To divide the NAP's total bandwidth into $\mathfrak{B}_{\mathscr{D}}^{AP}$ and $\mathfrak{B}_{GL}^{AP}$, the *TWBA* scheme adopts the concept of two-player *WKSBS*. The set of individual devices $(\mathbb{I}_{\mathscr{D}})$ and the set of *GLs* $(\mathbb{G}_{GL})$ in the NAP covering area are assumed as game players, i.e., $\mathscr{P}_{\mathscr{D}}$, and $\mathscr{P}_{GL}$, respectively. Based on the allocated $\mathfrak{B}_{\mathscr{D}}^{AP}$ and $\mathfrak{B}_{GL}^{AP}$, the utility functions of $\mathscr{P}_{\mathscr{D}}$ and $\mathscr{P}_{GL}$, i.e., $\mathbb{U}_{\mathscr{D}}$ and $\mathbb{U}_{GL}$, are defined as follows (Kim, 2022c);

$$\begin{cases} \mathbb{U}_{\mathscr{D}}\left(\mathfrak{B}_{\mathscr{D}}^{AP}\right) = \sum_{\substack{\mathscr{D}_i \in \mathbb{I}_{\mathscr{D}}; \\ \chi_{\mathscr{D}_i}^{\mathscr{A}_k} > 0}} U_{\mathscr{D}_i}^{\mathscr{A}_k}\left(\chi_{\mathscr{D}_i}^{\mathscr{A}_k}\right), s.t., \sum_{\substack{\mathscr{D}_i \in \mathbb{I}_{\mathscr{D}}; \\ \chi_{\mathscr{D}_i}^{\mathscr{A}_k} > 0}} \chi_{\mathscr{D}_i}^{\mathscr{A}_k} = \mathfrak{B}_{\mathscr{D}}^{AP} \\ \mathbb{U}_{GL}\left(\mathfrak{B}_{GL}^{AP}\right) = \sum_{\substack{\mathscr{D}_j \in \mathbb{G}_{GL}; \\ \chi_{\mathscr{D}_j}^{\mathscr{A}_h} > 0}} U_{\mathscr{D}_j}^{\mathscr{A}_h}\left(\chi_{\mathscr{D}_j}^{\mathscr{A}_h}\right), s.t., \sum_{\substack{\mathscr{D}_j \in \mathbb{G}_{GL}; \\ \chi_{\mathscr{D}_j}^{\mathscr{A}_h} > 0}} \chi_{\mathscr{D}_j}^{\mathscr{A}_h} = \mathfrak{B}_{GL}^{AP} \end{cases}$$

(3.4.10)

In the bandwidth partitioning algorithm, each individual NAP partitions the available bandwidth based on the *WKSBS* approach. Therefore, the values of $\mathfrak{B}_{\mathscr{D}}^{AP}$ and $\mathfrak{B}_{GL}^{AP}$ are decided while satisfying the below mathematical equation (Dubra, 2001; Kim, 2022c).

$\text{WKSBS}\left(\mathfrak{B}_{\mathscr{D}}^{AP}, \mathfrak{B}_{GL}^{AP}, \lambda\right) =$

$$\left\{ \frac{\mathbb{U}_{\mathscr{D}}(\mathfrak{B}^{AP}_{\mathscr{D}})}{\sum_{\mathscr{D}_i \in \mathrm{I}_{\mathscr{D}}} U^{\mathscr{A}_k}_{\mathscr{D}_i}(\Gamma^{\mathscr{A}_k}_{\mathscr{D}_i})} = \lambda \times \frac{\mathbb{U}_{GL}(\mathfrak{B}^{AP}_{GL})}{\sum_{\mathscr{D}_j \in \mathrm{G}_{GL}} U^{\mathscr{A}_k}_{\mathscr{D}_j}(\Gamma^{\mathscr{A}_k}_{\mathscr{D}_j})} \right\} \cap P(S = \{\mathscr{P}_{\mathscr{D}}, \mathscr{P}_{GL}\}) \tag{3.4.11}$$

$$\text{s.t., } \lambda = 1/\left(1 + \sum_{\mathscr{D}_j \in \mathrm{G}_{GL}} IT^{GL}_{\mathscr{D}_j}\right)^{\xi}$$

where ξ is a control factor to adjust the weight of $\mathscr{P}_{GL}$. After the bandwidth partitioning, the NAP distributes its $\mathfrak{B}^{AP}_{\mathscr{D}}$ and $\mathfrak{B}^{AP}_{GL}$ to the corresponding $\mathscr{D}$*s* and *GLs*. Simply, the $\mathfrak{B}^{AP}_{\mathscr{D}}$ is distributed to maximize the $\mathbb{U}_{\mathscr{D}}(\mathfrak{B}^{AP}_{\mathscr{D}})$ according to (3.4.10). However, for the distribution of $\mathfrak{B}^{AP}_{GL}$, the *TWBA* scheme would consider the social features of *GLs*. To effectively solve the $\mathfrak{B}^{AP}_{GL}$ distribution problem, the *TWBA* scheme designs the bandwidth distribution algorithm by using the concept of *ISV* (Kim, 2022c).

In general, the resource distribution problem leads to the similar conflict situation as in the money division problem for n claimants. Therefore, the *TWBA* scheme adopts the bankruptcy game model to define the characteristic function ω for the NAPs' coalitions ($\omega(\cdot)$). Simply, the payoff of a coalition *S*, which is consists of *GLs*, is calculated as the remaining payoff while paying idealistic payoff for other players, who are not included in the *S* (Kim, 2018b; Niyato, 2006). Based on the *GL*'s interval $\left[\underline{\Theta_{GL}}, \overline{\Theta_{GL}}\right]$, the *TWBA* scheme can simply estimate the distribution amount of kth *GL* $\left(\mathscr{A}_{GL_k}\right)$ (Kim, 2022c).

$$\mathscr{A}_{GL_k} = \frac{\mathfrak{R}_{GL_k}}{\sum_{l=1}^{e} \mathfrak{R}_{GL_l}} \times \mathfrak{B}^{AP}_{GL}, \text{ s.t., } \mathfrak{R}_{GL_k} = \mathscr{T}_{GL_k} \times \frac{\left(\underline{\Theta_{GL_k}} + \overline{\Theta_{GL_k}}\right)/2}{\sum_{l=1}^{e}\left(\underline{\Theta_{GL_l}} + \overline{\Theta_{GL_l}}\right)/2} \tag{3.4.12}$$

where e is the total number of *GLs* in the NAP. The *TWBA* scheme establishes a two-phase bandwidth allocation method by leveraging the fundamental principles of the *WKSBS* and the *ISV*. In the *TWBA* scheme, each NAP first segments the available bandwidth using *WKSBS* and then allocates the divided bandwidth based on the *ISV*. Through the sophisticated integration of these two cooperative game-theoretic solutions, the *TWBA* scheme is able to achieve a well-balanced performance that effectively manages conflicting service requirements (Kim, 2022c).

3.4.8 Summary

In the EoT era, the integration of both centralized cloud computing and the emerging mobile-edge computing with existing network infrastructures has become a widely adopted platform for 5G computation-intensive applications. Nevertheless, this integrated approach faces challenges related to insufficient communication resources due to limited wireless bandwidth. The *TWBA* scheme explores the bandwidth allocation issue within the CEoT system. Drawing on the core principles of the *WKSBS* and *ISV*, the *TWBA* scheme is architected as an innovative two-phase cooperative game process. By leveraging the complementary strengths of these two cooperative game-theoretic solutions, they interact synergistically to achieve a 'win-win' outcome even in environments with dynamic application loads. This integrated methodology effectively encourages edge devices to collaborate in a coordinated manner, resulting in mutual benefits such as enhanced system throughput, improved bandwidth utilization, and reduced service failure probability (Kim,2022b).

3.5 Distributed Mobile Cloud Computing (*DMCC*) scheme

Within the cloud service framework, service providers strive to deliver computation-intensive applications to IoT devices in a cost-effective manner. Taking into account various IoT applications and a distributed cloud infrastructure, S. Kim has developed the *Distributed Mobile Cloud Computing* (*DMCC*) scheme aimed at facilitating data offloading services (Kim, 2021a). Utilizing principles from cooperative game theory, the *DMCC* scheme incorporates five distinct value solutions, each tailored to specific cooperative scenarios. Depending on the service characteristics, the *DMCC* scheme investigates the mutual advantages derived from the interactions between local and global mobile cloud computing (MCC) providers, ensuring the efficient distribution of cloud resources. To remain responsive to the dynamic conditions of the distributed MCC system, the *DMCC* scheme offers strategic guidance that enhances both the efficiency of cloud resource utilization and the quality of services provided. The primary innovation of the *DMCC* scheme lies in its capacity to foster a reciprocal consensus among multiple service providers, accommodating diverse control perspectives. This collaborative approach enables the *DMCC* scheme to effectively share distributed cloud resources while simultaneously improving service quality and resource efficiency.

3.5.1 Development Motivation

With the significant increase in the number of IoT devices, 6G communication systems are striving to achieve high spectral and energy efficiency, low latency, and massive connectivity. It is projected that the number of IoT devices will reach 25 billion by 2025, presenting a substantial challenge for existing multi-process techniques to manage such a vast number of devices. To efficiently handle the enormous data generated by these devices, cloud computing will become a crucial component of 6G networks. This integration allows computing power and software applications to be delivered and accessed over the Internet as versatile commodities, facilitating rapid service access (Akhtar, 2020; Kim, 2021a; Wen, 2019).

As a novel paradigm supporting mobile users, cloud computing technology has been integrated into mobile environments, giving rise to mobile cloud computing (MCC). The MCC involves delivering resources and computing power over the Internet, providing IoT devices with on-demand access to shared hardware, software, databases, and information. By offloading data processing tasks from IoT devices to MCC servers, this paradigm reduces the execution time of mobile applications and decreases the energy consumption of IoT devices. The ability of MCC to offer on-demand, always-on computing utilities has attracted numerous industry players due to its favorable cost-benefit ratios, leading to rapid growth in the cloud computing market. However, the diverse demands of IoT devices necessitate a restructuring of the MCC business model to enhance its dynamic resource scaling capabilities (Kaewpuang, 2013; Kim, 2021a; Mashayekhy, 2014).

Future MCC systems will rely on highly heterogeneous infrastructures. Currently, multiple MCC service providers have been deployed to support a wide array of innovative cloud services. While the centralized infrastructure of traditional MCC platforms has been extensively studied, it presents several drawbacks, including limited resource sharing, high communication bandwidth requirements, and potentially high service delays - issues that are particularly problematic for time-sensitive applications. It is anticipated that the MCC infrastructure will evolve into distributed infrastructures, allowing Service Providers (SPs) to leverage geo-diversity to enhance revenue and system performance. This evolution has led to the development of a Distributed Mobile Cloud Computing (DMCC) infrastructure, where cloud resources are geographically dispersed across a wide-area network. In scenarios where an SP lacks sufficient resources to meet various service demands, other SPs with idle capacities can join a cooperative coalition, pooling their DMCC resources to achieve greater scalability and performance. By forming a cloud coalition with other SPs, the DMCC system can dynamically scale up its resource capabilities. Despite its potential, the DMCC system remains in its early stages, and optimal sharing of limited cloud resources continues to pose significant challenges (Darzanos, 2019; Kaewpuang, 2013; Kim, 2021a; Mashayekhy, 2014; Yu, 2015).

Today's wide range of IoT-based applications exhibit significant temporal variations in their workloads and demand high QoS from DMCC services. To meet these increasingly stringent performance requirements, the *DMCC* scheme emphasizes cooperation among SPs through resource sharing. The *DMCC* scheme categorizes SP cooperation into two-phase mechanisms: local and global resource sharing cooperation. In the local phase cooperation, a local coalition of different SPs is formed, enabling them to collaboratively share cloud resources within the same local area. In the global phase cooperation, resource sharing occurs among geographically distributed cloud resources owned by specific SPs. These local and global cooperation mechanisms enhance resource utilization despite uncertainties within the DMCC system. A key challenge in implementing this two-phase cooperation scheme is understanding the behaviors of self-interested SPs within the DMCC platform. To address this, the *DMCC* scheme employs cooperative game theory to mediate and reconcile conflicting viewpoints and preferences among SPs (Kim, 2021a).

In cooperative games, a primary challenge is determining how to share the gains from cooperation. Often, players form coalitions to share payoffs, which is represented by a coalition structure - a predefined partitioning of players into groups. The evaluation of players' expectations in such games is based on a coalitional value. A foundational study by L. Shapley in 1953 introduced the *Shapley value*, a solution concept that measures the average marginal contribution of a player to every possible coalition, satisfying properties such as *efficiency*, *symmetry*, *dummy*, and *additivity*. Despite its central role, the *Shapley value* has faced critiques. Consequently, the *DMCC* scheme introduces and discusses new coalition game methods to address these limitations (Ju, 2007; Kim, 2014, 2021a).

The *DMCC* scheme adopts the *egalitarian non-individual contribution* (*ENIC*) axiom to distribute the limited DMCC resources. This principle ensures an equal division of the total surplus profits, assuming each player has already received some form of individual contribution. Based on the *ENIC* characteristic, the concept of individual contribution can be defined in five distinct ways. The *Center of the Imputation Set Value* (*CIS Value*) defines the individual worth based on the center of the imputation set. The *α-CIS Value* combines the CIS value with the equal division concept through a convex combination. The *Egalitarian Non-Separable Contribution Value* (*ENSC Value*) utilizes a separable contribution approach. The *Egalitarian Non-Pairwise Averaged Contribution Value* (*ENPAC Value*) applies pairwise-averaged contributions. The *Egalitarian Non-Banzhaf Contribution Value* (*ENBC Value*) incorporates the Banzhaf contribution concept. These five variants - the *CIS value*, *α-CIS value*, *ENSC value*, *ENPAC value*, and *ENBC value* - are developed by selecting different notions of individual contribution, ensuring a fair and efficient distribution of resources within the *DMCC* scheme (Driessen, 1997; Kim, 2021a; Namekata, 1999).

3.5.2 Contributions

Utilizing coalitional value solutions, the *DMCC* scheme successfully addresses resource allocation challenges within the DMCC platform. By taking into account various IoT applications and a distributed cloud infrastructure, the *DMCC* scheme develops an innovative two-phase cooperative game model. This model is not only defined by its underlying axioms but is also practically constructed based on the DMCC infrastructure. In the local phase, resources from different SPs within each BS are shared following the principles of the *CIS value*, *α-CIS value*, and *ENSC value*. These methodologies ensure that local resources are allocated fairly and efficiently among the participating SPs. During the global phase, a service provider's geographically dispersed cloud resources are allocated based on the *ENPAC value* and *ENBC value concepts*. This approach facilitates the optimal distribution of cloud resources across a wide-area network, enhancing overall system performance. Furthermore, the *DMCC* scheme categorizes heterogeneous IoT applications into two distinct types, managing each category differently based on their specific service characteristics. This tailored approach allows the *DMCC* scheme to address the unique requirements of each application type effectively. Through continuous cooperative interactions, the *DMCC* scheme adeptly manages the increasing demand for computation offloading requests. This iterative process ensures that the system performance remains optimized, even as the volume and complexity of tasks grow (Kim, 2021a).

3.5.3 DMCC Infrastructure and two-phase Cooperative Game Model

A real-world public DMCC system consists of multiple SPs where their resources are geographically distributed in each region. The *DMCC* scheme considers a service region in the DMCC environment; it is covered by wireless BSs. Let there be n number of BSs, which are represented as a set $\mathfrak{B} = \left\{\mathscr{B}_1, \ldots \mathscr{B}_n\right\}$. The set of SPs is denoted by P = $\{\mathscr{P}_1, \ldots \mathscr{P}_m\}$, and the set of IoT devices is denoted by D = $\{\mathscr{D}_1, \ldots \mathscr{D}_k\}$. IoT devices are distributed across in the DMCC service area, and directly contact with their corresponding BSs. A cloudlet ($\mathscr{C}$) is a mobility-enhanced small-scale cloud datacenter that is located at each BS. In each BS, all SPs install their cloudlets $\left(\mathscr{C}_{\mathscr{P}}^{\mathscr{B}}\right)$. Therefore, each BS has m cloudlets, and each SP install its cloudlets in all BSs where $\mathscr{C}_{\mathscr{P}_j}^{\mathscr{B}_i}$ is the $\mathscr{P}_j$'s cloudlet in the $\mathscr{B}_i$. Let $\mathscr{C}_{\mathscr{B}_i} = \left\{\mathscr{C}_{\mathscr{P}_1}^{\mathscr{B}_i}, \ldots \mathscr{C}_{\mathscr{P}_m}^{\mathscr{B}_i}\right\}$ be the set of locally clustered cloudlets in the $\mathscr{B}_i$, and $\mathscr{C}_{\mathscr{P}_j} = \left\{\mathscr{C}_{\mathscr{P}_j}^{\mathscr{B}_1}, \ldots \mathscr{C}_{\mathscr{P}_j}^{\mathscr{B}_n}\right\}$ is the set of globally distributed cloudlets of the $\mathscr{P}_j$; cloudlets in the $\mathscr{C}_{\mathscr{P}}$ are connected through the wired backbone Internet links. Let $\mathfrak{M}_{\mathscr{P}_j}^{\mathscr{B}_i}$ be the computation power of $\mathscr{C}_{\mathscr{P}_j}^{\mathscr{B}_i}$, and the total computation power of $\mathscr{P}_j$ is $\sum_{\mathscr{B}_i \in \mathfrak{B}} \mathfrak{M}_{\mathscr{P}_j}^{\mathscr{B}_i}$ (Darzanos, 2019; Kaewpuang, 2013; Kim, 2021a; Yu, 2015).

The *DMCC* scheme assumes that each individual $\mathscr{C}_{\mathscr{P}}$ serves a large number of computation offloading tasks where each of them may be generated from the $\mathscr{C}_{\mathscr{P}}$'s corresponding IoT devices. Usually, the cloud offloading mechanism for IoT applications combines two separate domains; hard real-time (*Hr*) and soft real-time (*Sr*) computing services. *Hr* service requires a guarantee that all processing is completed within a given time constraint. Therefore, timeliness is a primary measure of correctness. *Sr* service has a less rigorous notion of time constraint and time delay is not catastrophic. Therefore, it strives for good average-case performance while tolerating slow response times (Goddard, 2003; Kim, 2004). According to *Hr* and *Sr* service requirements, it is necessary to consider one key question: How to integrate different data services in a unified way to facilitate the DMCC processing while striking an appropriate system performance? (Kim, 2021a).

The two-phase cooperative game model ($\mathbb{G}$) consists of local and global processes to share the DMCC system computation resources. In the local cooperative games $\left(\mathscr{G}_{L,\mathscr{P}}^{\mathscr{B}}, \mathbb{G}_{L}^{\mathscr{B}}\right)$, each cloudlet partitions its computation resource for the *Hr* and *Sr* services, and this resource is shared with other cloudlets in the local B. Therefore, the received offloading computation tasks in the B are adaptively distributed its locally clustered cloudlets. In the global cooperative game $\left(\mathbb{G}_{G}^{\mathscr{P}}\right)$, the geographically distributed cloudlets of a SP work together to support the allocated computation tasks. In a parallel and distributed manner, $\mathscr{G}_{L,\mathscr{P}}^{\mathscr{B}}$, $\mathbb{G}_{L}^{\mathscr{B}}$ and $\mathbb{G}_{G}^{\mathscr{P}}$ act independently and make their decisions toward an appropriate DMCC system performance. Formally, the *DMCC* scheme defines the two-phase cooperative game entities, i.e.,

$$\mathbb{G} = \left\{\mathscr{G}_{L,\mathscr{P}}^{\mathscr{B}}, \mathbb{G}_{L}^{\mathscr{B}}, \mathbb{G}_{G}^{\mathscr{P}}\right\}$$
$$= \left\{\left\{\mathscr{G}_{L,\mathscr{P}_j \in \mathbf{P}}^{\mathscr{B}_i \in \mathbf{B}} | Hr_{\mathscr{B}_i \mathscr{P}_j}, Sr_{\mathscr{B}_i \mathscr{P}_j}, \mathscr{U}_{\mathscr{B}_i \mathscr{P}_j}^{Hr}, \mathscr{U}_{\mathscr{B}_i \mathscr{P}_j}^{Sr}, \mathscr{S}_{\mathscr{B}_i \mathscr{P}_j}, \mathfrak{M}_{\mathscr{P}_j}^{\mathscr{B}_i}\right\}, \left\{\mathbb{G}_{L}^{\mathscr{B}_i \in \mathbf{B}} | \mathscr{S}_{\mathscr{B}_i \mathscr{P}_j}^{Hr}, \mathscr{S}_{\mathscr{B}_i \mathscr{P}_j}^{Sr}, \mathscr{C}_{\mathscr{B}_i}, \mathfrak{P}_{\mathscr{B}_i \mathscr{P}_j}^{Hr}, \mathfrak{P}_{\mathscr{B}_i \mathscr{P}_j}^{Sr}, U_{L,\mathscr{B}_i}^{Hr,\mathscr{P}_j}, U_{L,\mathscr{B}_i}^{Sr,\mathscr{P}_j}\right\}, \left\{\mathbb{G}_{G}^{\mathscr{P}_j \in \mathbf{P}} | \mathscr{C}_{\mathscr{P}_j}, \mathfrak{U}_{G,\mathscr{P}_j}^{Hr,\mathscr{B}_i}, \mathfrak{U}_{G,\mathscr{P}_j}^{Sr,\mathscr{B}_i}\right\}, T\right\}$$

(Kim, 2021a):

- In the game G, $\mathscr{G}_{L,\mathscr{P}}^{\mathscr{B}}$ and $\mathbb{G}_{L}^{\mathscr{B}}$ are the local cooperative games in the B; the $\mathscr{G}_{L,\mathscr{P}}^{\mathscr{B}}$ is for service types and the $\mathbb{G}_{L}^{\mathscr{B}}$ is for local cloudlets. $\mathbb{G}_{G}^{\mathscr{P}}$ is the global cooperative game for SPs. $\mathscr{G}_{L,\mathscr{P}}^{\mathscr{B}}$, $\mathbb{G}_{L}^{\mathscr{B}}$ and $\mathbb{G}_{G}^{\mathscr{P}}$ games are mutual and reciprocal interdependent in an interactive manner.
- In the $\mathscr{G}_{L,\mathscr{P}_j}^{\mathscr{B}_i}$, $Hr_{\mathscr{B}_i \mathscr{P}_j}$ and $Sr_{\mathscr{B}_i \mathscr{P}_j}$ represent the offloading *Hr* and *Sr* services for the $\mathscr{C}_{\mathscr{P}_j}^{\mathscr{B}_i}$. $Hr_{\mathscr{B}_i \mathscr{P}_j}$ and $Sr_{\mathscr{B}_i \mathscr{P}_j}$ are game players, and $\mathscr{U}_{\mathscr{B}_i \mathscr{P}_j}^{Hr}$, $\mathscr{U}_{\mathscr{B}_i \mathscr{P}_j}^{Sr}$ are their utility functions, respectively.
- The $\mathscr{S}_{\mathscr{B}_i \mathscr{P}_j}$ is the total amount of computation offload for the $\mathscr{C}_{\mathscr{P}_j}^{\mathscr{B}_i}$ where $\mathscr{S}_{\mathscr{B}_i \mathscr{P}_j} = Hr_{\mathscr{B}_i \mathscr{P}_j} + Sr_{\mathscr{B}_i \mathscr{P}_j}$. $\mathfrak{M}_{\mathscr{P}_j}^{\mathscr{B}_i}$ is the computation power of $\mathscr{C}_{\mathscr{P}_j}^{\mathscr{B}_i}$
- In the $\mathbb{G}_{L}^{\mathscr{B}_i}$, $\mathscr{S}_{\mathscr{B}_i \mathscr{P}_j}^{Hr}$ and $\mathscr{S}_{\mathscr{B}_i \mathscr{P}_j}^{Sr}$ are the currently allocated cloud resources for the $Hr_{\mathscr{B}_i \mathscr{P}_j}$ and $Sr_{\mathscr{B}_i \mathscr{P}_j}$ in the $\mathscr{C}_{\mathscr{P}_j}^{\mathscr{B}_i}$; they are decided according to the $\mathscr{G}_{L}^{\mathscr{B}_i}$.
- $\mathscr{C}_{\mathscr{B}_i}$ is the set of locally clustered cloudlets in the $\mathscr{B}_i$, and $\mathfrak{P}_{\mathscr{B}_i \mathscr{P}_j}^{Hr}$, $\mathfrak{P}_{\mathscr{B}_i \mathscr{P}_j}^{Sr}$ are the $\mathscr{C}_{\mathscr{P}_j}^{\mathscr{B}_i}$'s currently available computation capacities for $Hr_{\mathscr{B}_i \mathscr{P}_j}$ and $Sr_{\mathscr{B}_i \mathscr{P}_j}$, respectively.
- In the $\mathbb{G}_{L}^{\mathscr{B}_i}$, $\mathscr{C}_{\mathscr{P}_j}^{\mathscr{B}_i} \in \mathscr{C}_{\mathscr{B}_i}$ is a game player, and $U_{L,\mathscr{B}_i}^{Hr,\mathscr{P}_j}$ and $U_{L,\mathscr{B}_i}^{Sr,\mathscr{P}_j}$ are the utility functions of $\mathscr{C}_{\mathscr{P}_j}^{\mathscr{B}_i}$ for $Hr_{\mathscr{B}_i \mathscr{P}_j}$ and $Sr_{\mathscr{B}_i \mathscr{P}_j}$, respectively.
- In the $\mathbb{G}_{G}^{\mathscr{P}_j}$, $\mathscr{C}_{\mathscr{P}_j}$ is the set of globally distributed cloudlets of $\mathscr{P}_j$, and $\mathscr{C}_{\mathscr{P}_j}^{\mathscr{B}_i} \in \mathscr{C}_{\mathscr{P}_j}$ is a game player. $\mathfrak{U}_{G,\mathscr{P}_j}^{Hr,\mathscr{B}_i}$, $\mathfrak{U}_{G,\mathscr{P}_j}^{Sr,\mathscr{B}_i}$ are utility functions of $\mathscr{C}_{\mathscr{P}_j}^{\mathscr{B}_i}$ for the $Hr_{\mathscr{B}_i \mathscr{P}_j}$ and $Sr_{\mathscr{B}_i \mathscr{P}_j}$, respectively.
- $T = \left\{t_1, \ldots, t_c, t_{c+1}, \ldots\right\}$ denotes time period, which is represented by a sequence of time steps.

3.5.4 The Basic Idea and Fundamental Concept of Value Solutions

To characterize the basic concepts of value solutions, we assume a n-player game. Let (N, v) be a cooperative n-player game with characteristic function where $N = \{1, \ldots, n\}$ is a finite set of players and $v : 2^N \rightarrow \mathbb{R}$ is a characteristic function. Coalitions are defined on the power set 2^N of all subsets of N. With every coalition $S \in 2^N$, there is associated the real number *v*(*S*), which represents the joint profit that players in S can achieve due to their cooperative behaviors. For coalition games, the one-point solution concept is called *value*, and its major goal is how to divide the overall profits *v*(*N*) among all players of (N, v). Any value can be interpreted as an allocation rule of the total profit among players in a game. Mathematically, it is a function Φ that assigns a payoff vector $\Phi(N, v) = \left(\Phi_i(N, v)\right)_{i \in N}$ to every game (N, v) where $\sum_{i \in N} \Phi_i(N, v) = v(N)$ to meet the efficiency principle. Finally, $\Phi_i(N, v)$ is the value of player i in the game (N, v); it represents player i's gain from participating in the coalitional game (Driessen, 1997; Kim, 2021a; Namekata, 1999).

The *CIS value* assigns to every player its singleton worth, and then distributes the remaining worth of the grand coalition equally among all players. It concerns only about the worth of individuals and the grand coalition since the grand coalition sometimes forms directly from singletons, without other intermediate coalitions forming. To reconcile the major economic allocation ideas, i.e., egalitarianism and marginalism, the *CIS value* $\left(\Phi_{i\in N}^{CIS}(N, v)\right)$ is formally defined as follows (Kim, 2021a; Namekata, 1999; Xu, 2015);

$$\Phi_i^{CIS}(N, v) := v(\{i\}) + \left(\tfrac{1}{n} \times \left[v(N) - \sum_{j\in N} v(\{j\})\right]\right), \text{ s.t., } i \in N \tag{3.5.1}$$

The α-*CIS value* considers the convex combinations of the *CIS value* and the equal division solution. Therefore, the α-*CIS value* can be represented as a payoff first assigning to every player only a fraction α of his individual worth, and then distributing the remaining worth of the grand coalition equally among all players. The α-*CIS value*, i.e., $\Phi_{i\in N}^{\alpha-CIS}(N, v)$, is given by (Kim, 2021a; Xu, 2015).

$$\Phi_i^{\alpha-CIS}(N, v) := \alpha \cdot v(\{i\}) + \left(\tfrac{1}{n} \times \left[v(N) - \alpha \cdot \sum_{j\in N} v(\{j\})\right]\right), \text{ s.t., } \alpha \in [0, 1] \tag{3.5.2}$$

The *ENSC value* is the dual of the *CIS value*, and has several dual properties. Therefore, it assigns to every player in a game its marginal contribution to the grand coalition and distributes the positive or negative remainder equally among players. Mathematically, the *ENSC value*, i.e., $\Phi_{i\in N}^{ENSC}(N, v)$, is defined as follows (Kim, 2021a; Namekata, 1999; van den Brink, 2016);

$$\Phi_{i\in N}^{ENSC}(N, v) := \psi_i(N, v) + \left(\tfrac{1}{n} \times \left[v(N) - \sum_{j\in N} \psi_j(N, v)\right]\right) \tag{3.5.3}$$

$$\text{s.t., } \psi_i(N, v) = (v(N) - v(N - \{i\})) \text{ and } i \in N$$

The *ENPAC value* is basically composed of pairwise contributions with respect to the formation of the grand coalition. According to pairs of players, it involves two stages. First, the pairwise-average contribution of a player is estimated. For example, player i's pairwise-average contribution is obtained by averaging over all other player $j \in N \setminus \{i\}$. Second, the resulting allocation to player i is obtained by the egalitarian division of the surplus of the overall profits after each player is conceded to get his pairwise-average contribution. That is the remaining amount of the overall profits is equally distributed to all players of the game. The *ENPAC value*, i.e., $\Phi_{i\in N}^{ENPAC}(N, v)$, is given by (Driessen, 1997; Kim, 2021a; Namekata, 1999).

$$\Phi_{i\in N}^{ENPAC}(N,v) := \Upsilon_i(N,v) + \left(\frac{1}{n} \times \left[v(N) - \sum_{j\in N} \Upsilon_j(N,v)\right]\right)$$

(3.5.4)

$$\text{s.t.}\ ,\ \Upsilon_i(N,v) = \left(v(N) - \left(\frac{1}{n-2} \times \sum_{j\in N\setminus\{i\}} v(N - \{i,j\})\right)\right)$$

Usually, the Banzhaf contribution (Γ) of player is interpretable as the average of all players' marginal contributions to participate in the game. By assuming that an individual contribution of any player is the Banzhaf contribution, the *ENBC value*, i.e., $\Phi_{i\in N}^{ENBC}(N,v)$, is defined as follows (Driessen, 1997; Kim, 2021a);

$$\Phi_{i\in N}^{ENBC}(N,v) := \Gamma_i(N,v) + \left(\frac{1}{n} \times \left[v(N) - \sum_{j\in N} \Gamma_j(N,v)\right]\right)$$

(3.5.5)

$$\text{s.t.}\ ,\ \Gamma_i(N,v) = \frac{1}{2^{n-1}} \times \left(\sum_{S\subseteq N\setminus\{i\}} (v(S\cup\{i\}) - v(S))\right)$$

3.5.5 The Local and Global Cooperative Games in the DMCC Platform

At each time $t_c \in T$, IoT devices in D offload independently their computation tasks to the corresponding SP. Their tasks can be categorized into *Hr* and *Sr* services. In each $\mathscr{B} \in \mathfrak{B}$, multiple SPs coexist, and the *DMCC* scheme mediates inter and intra SP interactions to support DMCC computation offloading services. Usually, resource sharing problem is analogous to the bankruptcy problem. A bankruptcy problem is a pair (E, c); E represents the total value of the estate and $\mathbf{c} = (c_1, \ldots, c_n)$ is the vector of the creditors' claims where $0 < E < \sum_{i=1}^{n} c_i$. Traditionally, to estimate value solutions, the characteristic function $v(S)$ for the coalition S is defined based on the bankruptcy model (Kim, 2010a, 2021a).

$$v(S) = \max\left(0\ , E - \sum_{i\in S} c_i\right)$$

(3.5.6)

The *DMCC* scheme designs a new two-phase DMCC resource sharing mechanism. At first, in the local level, the total computation amount of $\mathscr{C}_{\mathscr{P}}^{\mathscr{B}}$ is divided two parts; one part is for the *Hr* service, and the other part is for the *Sr* service. For the $\mathscr{G}_{L,\mathscr{P}_j}^{\mathscr{B}_i}$, $Hr_{\mathscr{B}_i\mathscr{P}_j}$ and $Sr_{\mathscr{B}_i\mathscr{P}_j}$ are game players and their utility functions, i.e., $\mathscr{U}_{\mathscr{B}_i\mathscr{P}_j}^{Hr}(\cdot)$ and $\mathscr{U}_{\mathscr{B}_i\mathscr{P}_j}^{Sr}(\cdot)$, are defined as follows (Kim, 2021a);

$$
\begin{cases}
\mathcal{U}^{Hr}_{\mathcal{B}_i\mathcal{P}_j}\left(R^{Hr}_{\mathcal{B}_i\mathcal{P}_j}, \mathcal{S}^{Hr}_{\mathcal{B}_i\mathcal{P}_j}, \mathfrak{M}^{\mathcal{B}_i}_{\mathcal{P}_j}\right) = \dfrac{\theta}{\eta + exp\left(\xi \times \dfrac{\min\left(R^{Hr}_{\mathcal{B}_i\mathcal{P}_j}, \mathcal{S}^{Hr}_{\mathcal{B}_i\mathcal{P}_j}\right)}{\mathfrak{M}^{\mathcal{B}_i}_{\mathcal{P}_j}}\right)} \\
\mathcal{U}^{Sr}_{\mathcal{B}_i\mathcal{P}_j}\left(R^{Sr}_{\mathcal{B}_i\mathcal{P}_j}, \mathcal{S}^{Sr}_{\mathcal{B}_i\mathcal{P}_j}, \mathfrak{M}^{\mathcal{B}_i}_{\mathcal{P}_j}\right) = \dfrac{\theta}{\eta + \left(\varepsilon \times exp\left(\dfrac{\min\left(R^{Sr}_{\mathcal{B}_i\mathcal{P}_j}, \mathcal{S}^{Sr}_{\mathcal{B}_i\mathcal{P}_j}\right)}{\mathfrak{M}^{\mathcal{B}_i}_{\mathcal{P}_j}}\right)\right)}
\end{cases}
\tag{3.5.7}
$$

where $R^{Hr}_{\mathcal{B}_i\mathcal{P}_j}$ and $R^{Sr}_{\mathcal{B}_i\mathcal{P}_j}$ $\left(or\ \mathcal{S}^{Hr}_{\mathcal{B}_i\mathcal{P}_j}\text{ and }\mathcal{S}^{Sr}_{\mathcal{B}_i\mathcal{P}_j}\right)$ are the requested amounts (or currently allocating cloud computing resource) for $Hr_{\mathcal{B}_i\mathcal{P}_j}$ and $Sr_{\mathcal{B}_i\mathcal{P}_j}$, respectively. η, θ, ξ and ε are the adjustment parameters for the $\mathcal{U}_{\mathcal{B}_i\mathcal{P}_j}(\cdot)$. To decide the $\mathcal{S}^{Hr}_{\mathcal{B}_i\mathcal{P}_j}$ and $\mathcal{S}^{Sr}_{\mathcal{B}_i\mathcal{P}_j}$ values for the $\mathcal{C}^{\mathcal{B}_i}_{\mathcal{P}_j}$, the DMCC scheme adopts the idea of α-CIS value. Based on the bankruptcy problem, the DMCC scheme can get the $v(S)$ value where $c_{Hr} = \mathcal{U}^{Hr}_{\mathcal{B}_i\mathcal{P}_j}\left(\mathcal{S}^{Hr}_{\mathcal{B}_i\mathcal{P}_j} = R^{Hr}_{\mathcal{B}_i\mathcal{P}_j}\right)$, $c_{Sr} = \mathcal{U}^{Sr}_{\mathcal{B}_i\mathcal{P}_j}\left(\mathcal{S}^{Sr}_{\mathcal{B}_i\mathcal{P}_j} = R^{Sr}_{\mathcal{B}_i\mathcal{P}_j}\right)$ and $E = \vartheta \times (c_{Hr} + c_{Sr})$; ϑ is an adjustment factor to estimate the grand coalition value. According to (3.5.2), the $\mathcal{S}^{Hr}_{\mathcal{B}_i\mathcal{P}_j}$ and $\mathcal{S}^{Sr}_{\mathcal{B}_i\mathcal{P}_j}$ are decided such as $\mathcal{S}^{Hr}_{\mathcal{B}_i\mathcal{P}_j} = \Phi^{\alpha-CIS}_{Hr_{\mathcal{B}_i\mathcal{P}_j}}$ and $\mathcal{S}^{Sr}_{\mathcal{B}_i\mathcal{P}_j} = \Phi^{\alpha-CIS}_{Sr_{\mathcal{B}_i\mathcal{P}_j}}$ (Kim, 2021a).

$$
\Phi^{\alpha-CIS}_{i\in\mathcal{N}}(\mathcal{N}, v) := \alpha \cdot v(\{i\}) + \left(\frac{1}{2} \times \left[v(\mathcal{N}) - \alpha \cdot \sum\nolimits_{j\in\mathcal{N}, i\neq j} v(\{j\})\right]\right)
\tag{3.5.8}
$$

$$
\text{s.t., } \mathcal{N} = \left\{Hr_{\mathcal{B}_i\mathcal{P}_j}, Sr_{\mathcal{B}_i\mathcal{P}_j}\right\} \text{ and } \alpha = \max\left(\frac{R^{Hr}_{\mathcal{B}_i\mathcal{P}_j}}{R^{Hr}_{\mathcal{B}_i\mathcal{P}_j} + R^{Sr}_{\mathcal{B}_i\mathcal{P}_j}}, \sigma\right)
$$

where $0 \leq \alpha \leq 1$ and σ is the priority factor. For the $\mathbb{G}^{\mathcal{B}_i}_L$ game, each cloudlet in the $\mathcal{B}_i$, i.e., $\mathcal{C}^{\mathcal{B}_i}_{\mathcal{P}_j} \in \mathcal{C}_{\mathcal{B}_i}$, is a game player, and the $\mathbb{G}^{\mathcal{B}_i}_L$ game consists of two subgames; one is for the Hr service and the other is for the Sr service. In the $\mathbb{G}^{\mathcal{B}_i}_L$, $\mathcal{C}^{\mathcal{B}_i}_{\mathcal{P}_j}$'s utility function for $Hr_{\mathcal{B}_i\mathcal{P}_j}$, i.e., $U^{Hr,\mathcal{P}_j}_{L,\mathcal{B}_i}(\cdot)$, is defined as follows (Kim, 2021a);

$$
U^{Hr,\mathcal{P}_j}_{L,\mathcal{B}_i}\left(\mathfrak{P}^{Hr}_{\mathcal{B}_i\mathcal{P}_j}, \mathcal{S}^{Hr}_{\mathcal{B}_i\mathcal{P}_j}, A^{Hr}_{\mathcal{B}_i\mathcal{P}_j}\right) = \left(\left(\gamma_{Hr} + \frac{\mathfrak{P}^{Hr}_{\mathcal{B}_i\mathcal{P}_j}}{\mathcal{S}^{Hr}_{\mathcal{B}_i\mathcal{P}_j}}\right) \times \log\left(\frac{A^{Hr}_{\mathcal{B}_i\mathcal{P}_j}}{\mathcal{S}^{Hr}_{\mathcal{B}_i\mathcal{P}_j}} + \omega_{Hr}\right)\right)
\tag{3.5.9}
$$

where $A^{Hr}_{\mathcal{B}_i\mathcal{P}_j}$ is the currently allocating $\mathcal{C}^{\mathcal{B}_i}_{\mathcal{P}_j}$'s cloud resource for the $Hr_{\mathcal{B}_i\mathcal{P}_j}$ and γ_{Hr}, ω_{Hr} are the adjustment parameters for the $U^{Hr,\mathcal{P}_j}_{L,\mathcal{B}_i}(\cdot)$. For the Sr service in the $\mathbb{G}^{\mathcal{B}_i}_L$ game, the $\mathcal{C}^{\mathcal{B}_i}_{\mathcal{P}_j}$'s utility function for the $Sr_{\mathcal{B}_i\mathcal{P}_j}$, i.e., $U^{Sr,\mathcal{P}_j}_{L,\mathcal{B}_i}(\cdot)$, is defined as the same manner as $U^{Hr,\mathcal{P}_j}_{L,\mathcal{B}_i}(\cdot)$; only $\mathcal{S}^{Hr}_{\mathcal{B}_i\mathcal{P}_j}$, $A^{Hr}_{\mathcal{B}_i\mathcal{P}_j}$, γ_{Hr} and ω_{Hr} are replaced by $\mathcal{S}^{Sr}_{\mathcal{B}_i\mathcal{P}_j}$, $A^{Sr}_{\mathcal{B}_i\mathcal{P}_j}$, γ_{Sr} and ω_{Sr}, respectively. The $A^{Sr}_{\mathcal{B}_i\mathcal{P}_j}$ is the currently allocating $\mathcal{C}^{\mathcal{B}_i}_{\mathcal{P}_j}$'s cloud resource for the $Sr_{\mathcal{B}_i\mathcal{P}_j}$, and γ_{Sr}, ω_{Sr} are the adjustment parameters for the $U^{Sr,\mathcal{P}_j}_{L,\mathcal{B}_i}(\cdot)$ (Kim, 2021a).

To decide the $A^{Hr}_{\mathcal{B}_i\mathcal{P}_j}$ value, the DMCC scheme adopts the idea of CIS value, and can get the $v(S)$ value where $c_{\mathcal{C}^{\mathcal{B}_i}_{\mathcal{P}_j}} = U^{Hr,\mathcal{P}_j}_{L,\mathcal{B}_i}\left(A^{Hr}_{\mathcal{B}_i\mathcal{P}_j} = \mathcal{S}^{Hr}_{\mathcal{B}_i\mathcal{P}_j}\right)$ and $E = \vartheta \times \left(\sum_{\mathcal{C}^{\mathcal{B}_i}_{\mathcal{P}_k}\in\mathcal{C}_{\mathcal{B}_i}} U^{Hr,\mathcal{P}_k}_{L,\mathcal{B}_i}\left(A^{Hr}_{\mathcal{B}_i\mathcal{P}_k} = \mathcal{S}^{Hr}_{\mathcal{B}_i\mathcal{P}_k}\right)\right)$. According to (3.5.1), the $A^{Hr}_{\mathcal{B}_i\mathcal{P}_j}$ value is decided such as $A^{Hr}_{\mathcal{B}_i\mathcal{P}_j} = \Phi^{CIS}_{\mathcal{C}^{\mathcal{B}_i}_{\mathcal{P}_j}}\left(\mathcal{C}_{\mathcal{B}_i}, v\right)$ (Kim, 2021a).

$$\Phi^{CIS}_{\mathcal{C}^{\mathcal{B}_i}_{\mathcal{P}_j}}\left(\mathcal{C}_{\mathcal{B}_i}, v\right) := v\left(\left\{\mathcal{C}^{\mathcal{B}_i}_{\mathcal{P}_j}\right\}\right) + \left(\frac{1}{\mathcal{C}_{\mathcal{B}_i}} \times \left[v\left(\mathcal{C}_{\mathcal{B}_i}\right) - \sum_{\mathcal{C}^{\mathcal{B}_i}_{\mathcal{P}_k}\in\mathcal{C}_{\mathcal{B}_i}} v\left(\left\{\mathcal{C}^{\mathcal{B}_i}_{\mathcal{P}_k}\right\}\right)\right]\right) \tag{3.5.10}$$

where $\mathcal{C}_{\mathcal{B}_i}$ is the cardinality of $\mathcal{C}_{\mathcal{B}_i}$. To decide the $A^{Sr}_{\mathcal{B}_i\mathcal{P}_j}$ value, the *DMCC* scheme adopts the idea of *ENSC value*, and can get the $v(S)$ value as the same manner as the *CIS value*. According to (3.5.3), the $A^{Sr}_{\mathcal{B}_i\mathcal{P}_j}$ value is decided such as $A^{Sr}_{\mathcal{B}_i\mathcal{P}_j} = \Phi^{ENSC}_{\mathcal{C}^{\mathcal{B}_i}_{\mathcal{P}_j}}\left(\mathcal{C}_{\mathcal{B}_i}, v\right)$ (Kim, 2021a).

$$\Phi^{ENSC}_{\mathcal{C}^{\mathcal{B}_i}_{\mathcal{P}_j}}\left(\mathcal{C}_{\mathcal{B}_i}, v\right) := \psi_{\mathcal{C}^{\mathcal{B}_i}_{\mathcal{P}_j}}\left(\mathcal{C}_{\mathcal{B}_i}, v\right) + \left(\frac{1}{\mathcal{C}_{\mathcal{B}_i}} \times \left[v\left(\mathcal{C}_{\mathcal{B}_i}\right) - \sum_{\mathcal{C}^{\mathcal{B}_i}_{\mathcal{P}_k}\in\mathcal{C}_{\mathcal{B}_i}} \psi_{\mathcal{C}^{\mathcal{B}_i}_{\mathcal{P}_k}}\left(\mathcal{C}_{\mathcal{B}_i}, v\right)\right]\right) \tag{3.5.11}$$

$$\text{s.t., } \psi_{\mathcal{C}^{\mathcal{B}_i}_{\mathcal{P}_j}}\left(\mathcal{C}_{\mathcal{B}_i}, v\right) = \left(v\left(\mathcal{C}_{\mathcal{B}_i}\right) - v\left(\mathcal{C}_{\mathcal{B}_i} - \left\{\mathcal{C}^{\mathcal{B}_i}_{\mathcal{P}_j}\right\}\right)\right)$$

For the $\mathbb{G}^{\mathcal{P}_j}_{G}$ game, globally distributed cloudlets of $\mathcal{C}^{\mathcal{B}_i\in\mathfrak{B}}_{\mathcal{P}_j}$ in $\mathcal{C}_{\mathcal{P}_j}$ are game players. The $\mathcal{C}^{\mathcal{B}_i}_{\mathcal{P}_j}$'s utility functions for the *Hr* and *Sr* services, i.e., $\mathfrak{U}^{Hr,\mathcal{B}_i}_{G,\mathcal{P}_j}(\cdot)$ and $\mathfrak{U}^{Sr,\mathcal{B}_i}_{G,\mathcal{P}_j}(\cdot)$, are defined as follows (Kim, 2021a);

$$\begin{cases} \mathfrak{U}^{Hr,\mathcal{B}_i}_{G,\mathcal{P}_j}\left(\mathfrak{P}^{Hr}_{\mathcal{B}_i\mathcal{P}_j}, \mathfrak{P}^{Sr}_{\mathcal{B}_i\mathcal{P}_j}, A^{Hr}_{\mathcal{B}_i\mathcal{P}_j}, a^{Hr}_{\mathcal{B}_i\mathcal{P}_j}\right) = \exp\left(\dfrac{\min\left(\left(\mathfrak{P}^{Hr}_{\mathcal{B}_i\mathcal{P}_j} + \mathfrak{P}^{Sr}_{\mathcal{B}_i\mathcal{P}_j}\right), a^{Hr}_{\mathcal{B}_i\mathcal{P}_j}\right)}{A^{Hr}_{\mathcal{B}_i\mathcal{P}_j}}\right) - \zeta \\ \\ \mathfrak{U}^{Sr,\mathcal{B}_i}_{G,\mathcal{P}_j}\left(\mathfrak{P}^{Sr}_{\mathcal{B}_i\mathcal{P}_j}, A^{Sr}_{\mathcal{B}_i\mathcal{P}_j}, a^{Sr}_{\mathcal{B}_i\mathcal{P}_j}\right) = \mu^{\left(\frac{\min\left(\mathfrak{P}^{Sr}_{\mathcal{B}_i\mathcal{P}_j}, a^{Sr}_{\mathcal{B}_i\mathcal{P}_j}\right)}{A^{Sr}_{\mathcal{B}_i\mathcal{P}_j}}\right)} - \zeta \end{cases} \tag{3.5.12}$$

where $a^{Hr}_{\mathcal{B}_i\mathcal{P}_j}$ and $a^{Sr}_{\mathcal{B}_i\mathcal{P}_j}$ are the currently allocating $\mathcal{C}^{\mathcal{B}_i}_{\mathcal{P}_j}$'s cloud resources for the *Hr* and *Sr* services, respectively. ζ, μ are the adjustment parameters for the $\mathfrak{U}_{G,\mathcal{P}_j}(\cdot)$. To decide the $a^{Hr}_{\mathcal{B}_i\mathcal{P}_j}$ value, the *DMCC* scheme adopts the idea of *ENPAC value*, and the $v(S)$ value is estimated as $c_{\mathcal{C}^{\mathcal{B}_i}_{\mathcal{P}_j}} = \mathfrak{U}^{Hr,\mathcal{B}_i}_{G,\mathcal{P}_j}\left(a^{Hr}_{\mathcal{B}_i\mathcal{P}_j} = \left(\mathfrak{P}^{Hr}_{\mathcal{B}_i\mathcal{P}_j} + \mathfrak{P}^{Sr}_{\mathcal{B}_i\mathcal{P}_j}\right)\right)$ and $E = \vartheta \times \left(\sum_{\mathcal{C}^{\mathcal{B}_k}_{\mathcal{P}_j}\in\mathcal{C}_{\mathcal{P}_j}} \mathfrak{U}^{Hr,\mathcal{B}_k}_{G,\mathcal{P}_j}\left(a^{Hr}_{\mathcal{B}_k\mathcal{P}_j} = \left(\mathfrak{P}^{Hr}_{\mathcal{B}_k\mathcal{P}_j} + \mathfrak{P}^{Sr}_{\mathcal{B}_k\mathcal{P}_j}\right)\right)\right)$. According to (3.5.4), the $a^{Hr}_{\mathcal{B}_i\mathcal{P}_j}$ value is decided as $a^{Hr}_{\mathcal{B}_i\mathcal{P}_j} = \Phi^{ENPAC}_{\mathcal{C}^{\mathcal{B}_i}_{\mathcal{P}_j}}\left(\mathcal{C}_{\mathcal{P}_j}, v\right)$ (Kim, 2021a).

$$\Phi^{ENPAC}_{\mathcal{C}^{\mathcal{B}_i}_{\mathcal{P}_j}}\left(\mathcal{C}_{\mathcal{P}_j}, v\right) :=$$

$$\Upsilon_{\mathcal{C}_{\mathcal{P}_j}^{\mathcal{B}_i}}\left(\mathcal{C}_{\mathcal{P}_j},v\right)+\left(\frac{1}{\mathcal{C}_{\mathcal{P}_j}}\times\left[v\left(\mathcal{C}_{\mathcal{P}_j}\right)-\sum\nolimits_{\mathcal{C}_{\mathcal{P}_j}^{\mathcal{B}_i}\in\mathcal{C}_{\mathcal{P}_j}}\Upsilon_{\mathcal{C}_{\mathcal{P}_j}^{\mathcal{B}_i}}\left(\mathcal{C}_{\mathcal{P}_j},v\right)\right]\right) \tag{3.5.13}$$

$$\text{s.t. , } \Upsilon_{\mathcal{C}_{\mathcal{P}_j}^{\mathcal{B}_i}}\left(\mathcal{C}_{\mathcal{P}_j},v\right)=\left(v\left(\mathcal{C}_{\mathcal{P}_j}\right)-\left(\frac{1}{\mathcal{C}_{\mathcal{P}_j}-2}\times\sum\nolimits_{\mathcal{C}_{\mathcal{P}_j}^{\mathcal{B}_i}\in N\setminus\left\{\mathcal{C}_{\mathcal{P}_j}^{\mathcal{B}_i}\right\}}v\left(\mathcal{C}_{\mathcal{P}_j}-\left\{\mathcal{C}_{\mathcal{P}_j}^{\mathcal{B}_i},\mathcal{C}_{\mathcal{P}_j}^{\mathcal{B}_k}\right\}\right)\right)\right)$$

To decide the $a^{Sr}_{\mathcal{B}_i,\mathcal{P}_j}$ value of $\mathcal{C}^{\mathcal{B}_i}_{\mathcal{P}_j}$, the DMCC scheme adopts the idea of ENBC value, and can get the $v(S)$ value where $c_{\mathcal{C}^{\mathcal{B}_i}_{\mathcal{P}_j}} = \mathfrak{U}^{Sr,\mathcal{B}_i}_{G,\mathcal{P}_j}\left(a^{Sr}_{\mathcal{B}_i,\mathcal{P}_j} = \mathfrak{P}^{Sr}_{\mathcal{B}_i,\mathcal{P}_j}\right)$ and $E = \vartheta \times \left(\sum\nolimits_{\mathcal{C}^{\mathcal{B}_k}_{\mathcal{P}_j}\in\mathcal{C}_{\mathcal{P}_j}} \mathfrak{U}^{Sr,\mathcal{B}_k}_{G,\mathcal{P}_j}\left(a^{Sr}_{\mathcal{B}_k,\mathcal{P}_j} = \mathfrak{P}^{Sr}_{\mathcal{B}_k,\mathcal{P}_j}\right)\right)$. According to (3.5.5), the $a^{Sr}_{\mathcal{B}_i,\mathcal{P}_j}$ value is decided as $a^{Sr}_{\mathcal{B}_i,\mathcal{P}_j} = \Phi^{ENBC}_{\mathcal{C}^{\mathcal{B}_i}_{\mathcal{P}_j}}\left(\mathcal{C}_{\mathcal{P}_j},v\right)$ (Kim, 2021a).

$$\Phi^{ENBC}_{\mathcal{C}_{\mathcal{P}_j}^{\mathcal{B}_i}}\left(\mathcal{C}_{\mathcal{P}_j},v\right):=\Gamma_{\mathcal{C}_{\mathcal{P}_j}^{\mathcal{B}_i}}\left(\mathcal{C}_{\mathcal{P}_j},v\right)+\left(\frac{1}{\mathcal{C}_{\mathcal{P}_j}}\times\left[v\left(\mathcal{C}_{\mathcal{P}_j}\right)-\sum\nolimits_{\mathcal{C}_{\mathcal{P}_j}^{\mathcal{B}_i}\in\mathcal{C}_{\mathcal{P}_j}}\Gamma_{\mathcal{C}_{\mathcal{P}_j}^{\mathcal{B}_i}}\left(\mathcal{C}_{\mathcal{P}_j},v\right)\right]\right) \tag{3.5.14}$$

$$\text{s.t. , } \Gamma_{\mathcal{C}_{\mathcal{P}_j}^{\mathcal{B}_i}}(N,v)=\frac{1}{2^{\left(\mathcal{C}_{\mathcal{P}_j}-1\right)}}\times\left(\sum\nolimits_{S\subseteq\mathcal{C}_{\mathcal{P}_j}\setminus\left\{\mathcal{C}_{\mathcal{P}_j}^{\mathcal{B}_i}\right\}}\left(v\left(S\cup\left\{\mathcal{C}_{\mathcal{P}_j}^{\mathcal{B}_i}\right\}\right)-v(S)\right)\right)$$

3.5.6 Summary

The cooperative DMCC paradigm is becoming a highly effective solution for future cloud infrastructures. To overcome the interoperability challenges faced by multiple SPs within the DMCC platform, the *DMCC* scheme utilizes coalition-based cooperative game models to support various types of cloud services. Leveraging the principles of *CIS*, *α-CIS*, *ENSC*, *ENPAC*, and *ENBC values*, the *DMCC* scheme incorporates both local and global resource sharing processes. At the local level, cloud resources within each BS are allocated based on the *CIS*, *α-CIS*, and *ENSC values*. This method ensures that resources are distributed equitably among local SPs, reflecting their individual contributions and needs. Following the local allocation, resources from geographically dispersed SPs are effectively shared using the *ENPAC* and *ENBC value* concepts. This global sharing mechanism ensures optimal utilization of resources across different regions, enhancing the overall performance of the DMCC system. These local and global resource sharing algorithms function in a parallel and distributed manner, interacting and coordinating seamlessly to achieve cohesive resource management. This integrated approach allows the *DMCC* scheme to efficiently manage and distribute resources in response to current computation offloading requests. By addressing the dynamic and fluctuating conditions of the DMCC system, the *DMCC* scheme provides mutual benefits and enhanced system performance. This adaptability ensures that the *DMCC* scheme can effectively handle increasing and variable demands, optimizing resource allocation and maintaining high service quality under changing circumstances (Kim, 2021a).

3.6 Cooperative Federated Learning and Offloading (*CFLO*) Scheme

Unlike traditional networks, tactical edge networks operate under dynamic and unpredictable scenarios and field conditions, making them challenging to model accurately. A wide range of control issues arises on tactical edge platforms due to the combination of individual IoT devices' behaviors, the edge cloud's strategies, and the timeliness of tactical tasks. Recently, S. Kim introduces the *Cooperative Federated Learning and Offloading* (*CFLO*) scheme tailored for tactical edge network platforms, aiming to manage computation-intensive tasks collaboratively and efficiently (Kim, 2021b). Driven by the incentives of Federated Learning (FL) model training, each device independently decides whether to offload its tactical tasks within resource-constrained network environments. Leveraging two distinct bargaining solutions - the *weighted average solution* and the *constant elasticity substitution solution* - the edge server and IoT devices collaborate in a coordinated manner to achieve a balanced system performance. As a result, network agents can reach agreements by identifying mutual benefits, fostering a reciprocal consensus among differing viewpoints.

3.6.1 Development Motivation

In today's conflicts, information at the tactical level is becoming increasingly crucial. In areas like battlefields and disaster zones, known as tactical edge environments, situational awareness is vital for successful operations. Effective tactical missions rely on the efficient sharing of information among all involved network agents. Additionally, timeliness poses a significant challenge in tactical networks, as missions must be completed promptly even when network resources are limited. To meet these demands, recent studies have proposed the integration of Mobile Edge Computing (MEC) technology. MEC is emerging as a compelling computing paradigm by bringing cloud computing capabilities closer to end agents, thereby supporting a variety of computation-intensive and latency-sensitive service applications. In these challenging and hostile tactical edge environments, MEC is essential for ensuring seamless operational tasks, even when the tactical edge is isolated (Kim, 2021b; Li, Y2019; Sun, 2020).

In recent years, mobile IoT devices have been equipped with increasingly sophisticated sensing, communication, and computing capabilities, enabling them to perform more complex tasks beyond their basic functionalities. Combined with advancements in 5G network technologies, IoT devices are now capable of being deployed for various tactical tasks. These tasks involve collaborative sensing, data submission, data analysis, and model training, all of which are based on large volumes of data. Moreover, the growing emphasis on privacy and security of personal information complicates data sharing and integration. Motivated by these challenges, the introduction of Federated Learning (FL) has partially addressed issues in tactical network operations. In FL, individual IoT devices maintain their private data and update model parameters through local training processes (Kim, 2021b; Lim, 2020; Ye, 2020).

As a distributed collaborative AI approach, the FL is revolutionizing edge intelligence architectures. The FL enables the training of high-quality AI models by averaging local updates from multiple IoT devices, eliminating the need for direct access to local data. Instead, IoT devices collaborate with the MEC server to train neural networks, exchanging only model parameters without sharing raw data. This method addresses the trend of massive data generation across numerous IoT devices and the increasing concerns over transmitting sensitive information to centralized servers. By allowing data storage and model training to occur locally on each device, the FL leverages the growing storage and computational power of IoT devices while safeguarding privacy. However, the varying computational capabilities of different IoT devices introduce challenges related to resource rationing in FL implementations (Kim, 2021b; Nguyen, 2021; Shen, 2021).

Within a tactical edge computing environment, the resource allocation process is critical for enabling cooperative computational offloading. Typically, there is a trade-off among various performance parameters, necessitating a new control paradigm for making these decisions across heterogeneous platforms.

Specifically, the MEC-assisted FL platform must account for multiple internal and external constraints, such as network dynamics, awareness of network limitations, large data volumes, and tactical task coordination. To implement an optimal computing control process, previous efforts have employed AI techniques, machine learning, heuristic methods, and optimization algorithms to enhance computational efficiency in constrained environments. However, this remains a highly complex and challenging task, making the development of a new control paradigm a top priority for designing future FL-based tactical edge computing algorithms (Dasari, 2019; Kim, 2021b).

Despite its significant advantages, the FL faces a critical challenge: most traditional FL studies assume that all IoT devices will contribute their computing resources unconditionally, without any incentives to encourage participation. This assumption is often impractical in real-world scenarios due to the resource costs associated with model training. From the perspective of the MEC server, providing incentives is necessary to motivate IoT devices to participate actively in the FL process. Conversely, IoT devices are primarily concerned with completing their tactical tasks while considering the task offloading mechanisms. Given these differing priorities, the MEC server and IoT devices adopt distinct control strategies that must work together to achieve optimal network performance. Therefore, the *CFLO* scheme should utilize a cooperative bargaining game paradigm to design a precise control mechanism for the MEC-assisted FL platform (Kim, 2020b, 2021b).

Nearly seventy years ago, J. Nash published a seminal paper that laid the foundation for the axiomatic theory of bargaining, becoming one of the most influential paradigms in cooperative game theory. The Nash solution is complemented by two other well-known bargaining solutions: the egalitarian solution and the utilitarian solution. The egalitarian solution selects a weakly efficient agreement where both players receive identical payoffs, while the utilitarian solution opts for an agreement that maximizes the sum of players' utilities. For any given bargaining problem, the Nash solution resides on the Pareto boundary of the utility space, positioned between the egalitarian and utilitarian solutions. To mediate between these two approaches, the average bargaining solution was introduced, which maximizes a convex combination of utilitarian and egalitarian objectives, particularly in the context of intergenerational equity. A generalized form of this is the *weighted average bargaining* (*WAB*) solution, which places augmented weights on one of the bargaining objectives to bridge the gap between utilitarianism and egalitarianism (Kim, 2021b; Rachmilevitch, 2015).

Another method to unify various bargaining solutions involves designing the *Constant Elasticity Substitution* (*CES*) function, leading to the development of the *CES* solution. This solution is derived by maximizing the *CES* function and encompasses the Nash and egalitarian solutions as specific instances. Originally, the *CES* function is a neoclassical production function that exhibits constant elasticity of substitution, meaning that the proportion of factors can be adjusted at a constant rate in response to changes in their marginal rate of technical substitution. The *CES* function is defined by its efficiency, distribution, and substitution parameters. In the *CES* solution, the elasticity of substitution determines how two factors replace each other when there is a change in their price ratio. This solution possesses numerous technical properties that are advantageous for analyzing utility functions (Haake, 2018; Kim, 2021b).

3.6.2 Contributions

Leveraging the principles of the *WAB* and *CES* solutions, the *CFLO* scheme effectively manages both FL and tactical task operations. Within the MEC infrastructure, which accommodates various IoT devices, the *CFLO* scheme introduces an innovative dual-interactive bargaining game model. This model plays a crucial role in determining overall system performance. In a distributed online manner, individual IoT devices autonomously decide their contributions to FL based on the *WAB* solution framework. Taking into account the current system information, the MEC server allocates incentives to participating IoT devices for their offloading services. This incentive distribution is formulated using the *CES* solution concept. Throughout the interactive and iterative process, control decisions become interdependent, enabling the system to approximate an optimized performance (Kim, 2021b).

3.6.3 MEC Assisted FL Process Platform and Dual Game Model

In the *CFLO* scheme, a system model in the FL environment with MEC server is taken for analysis. There are several kinds of IoT devices, i.e., $\mathbb{I} = \{\mathcal{D}_1,\ldots,\mathcal{D}_n\}$; they are smartphones, laptops, surveillance cameras, smartwatches, etc., and their computational abilities are limited. The MEC server ($\mathcal{E}$) is mainly deployed at the edge of backbone network, and it is composed of large amounts of computation resource as well as stable power supply. The E provides a distributed computing environment by running applications and performing related processing tasks closer to the IoT devices. Individual devices can simultaneously transfer their computation-heavy tasks to nearby MEC servers. This offloading not only accelerates task execution but also extends the battery life of mobile IoT devices. To manage these offloading services efficiently, the MEC resource allocation algorithm is essential within the edge computing framework. For the purpose of quantitatively analyzing computing resource allocation, the time horizon is divided into discrete epochs, each with an equal duration in seconds (Han, 2019; Ji, 2021; Kim, 2021b; Ye, 2020).

The FL is a collaborative machine learning framework that allows participating devices to periodically update their model parameters using their local datasets. Instead of sharing their local training data, IoT devices upload only their updated model parameters to the MEC server. The server then aggregates these parameters to create a global model, which it broadcasts back to the IoT devices for subsequent local updates. Before this aggregation step, each local device may perform one or multiple epochs of model training during the local update phase. In scenarios where the MEC assists FL, the operations of local IoT devices are split into two main components: i) FL-based distributed model training, and ii) tactical task computation. Through the local computing process, individual devices can dedicate more resources to task offloading to the MEC server. This fine-grained task offloading strategy allows IoT devices to become more lightweight and efficient, enhancing their practicality in real-world applications. These observations inspired S. Kim to propose an innovative MEC-assisted FL control scheme, which effectively balances the tradeoff between task offloading services and the FL process (Ji, 2021; Kim, 2021b; Ye, 2020).

Developing interactive mechanisms within MEC-assisted FL control algorithms has become both a crucial and complex area of focus. The *CFLO* scheme operates under the assumption that each individual IoT device manages multiple tactical tasks alongside FL processes. Within this framework, the *CFLO* scheme must strategically allocate computational resources between these two distinct operations to achieve balanced and optimal system performance. To solve this problem, the *CFLO* scheme has developed a new dual-interactive bargaining game model ($\mathbb{G}$), which consists of the MEC's bargaining game $\left(\mathbb{G}_{\mathcal{E}}\right)$ and the $\mathcal{D}_i$'s bargaining game $\left(\mathbb{G}_{\mathcal{D}_i}\right)$. In the $\mathbb{G}_{\mathcal{E}}$ and $\mathbb{G}_{\mathcal{D}_i}$ games, the computation resources of E and $\mathcal{D}_i$ are distributed and divided, respectively. In an interactive and sequential manner, $\mathbb{G}_{\mathcal{E}}$ and $\mathbb{G}_{\mathcal{D}_i}$ work together and act cooperatively with each other to enhance conflicting performance criteria. Formally, the *CFLO* scheme defines the dual-interactive bargaining game entities, i.e.,

$$\mathbb{G} = \left\{\mathbb{G}_{\mathscr{E}}, \mathbb{G}_{\mathscr{D}_{1\le i\le n}}\right\} = \left\{\left\{\mathbb{G}_{\mathscr{E}} | \mathfrak{M}_{\mathscr{E}}, \mathscr{U}_{\mathscr{D}_{1\le i\le n}}(\cdot), \mathscr{M}_{\mathscr{D}_i}, \mathbb{V}_{\mathscr{E}}\right\}, \left\{\mathbb{G}_{\mathscr{D}_i} | \mathfrak{N}_{\mathscr{D}_i}, \left(\mathscr{R}^T_{\mathscr{D}_i}, \mathscr{R}^{FL}_{\mathscr{D}_i}\right), \left(\mathfrak{U}^T_{\mathscr{D}_i}\left(\mathscr{R}^T_{\mathscr{D}_i}\right), \mathfrak{U}^{FL}_{\mathscr{D}_i}\left(\mathscr{R}^{FL}_{\mathscr{D}_i}\right)\right)\right\}, T\right\}$$

(Kim, 2021b):

- The dual bargaining game ($\mathbb{G}$) consists of $\mathbb{G}_{\mathscr{E}}$ and $\mathbb{G}_{\mathscr{D}_{1\le i\le n}}$; they are mutual and reciprocal interdependent in an interactive manner.
- In the $\mathbb{G}_{\mathscr{E}}$, $\mathfrak{M}_{\mathscr{E}}$ is the total computation capacity of E; it is distributed into multiple $\mathscr{D}_{1\le i\le n}$. $\mathscr{U}_{\mathscr{D}_i}(\cdot)$ is the utility function of $\mathscr{D}_i$, and $\mathscr{M}_{\mathscr{D}_i}$ is the assigned computation resource for the $\mathscr{D}_i$'s offloading service where $\sum_{\mathscr{D}_i \in \mathbb{I}} \mathscr{M}_{\mathscr{D}_i} \le \mathfrak{M}_{\mathscr{E}}$.
- In the $\mathbb{G}_{\mathscr{E}}$, $\mathscr{M}_{\mathscr{D}_{1\le i\le n}}$ is the strategy of E, and $\mathbb{V}_{\mathscr{E}} = \left[\mathscr{M}_{\mathscr{D}_1}, \ldots, \mathscr{M}_{\mathscr{D}_n}\right]$ is the vector of strategy profiles.
- In the $\mathbb{G}_{\mathscr{D}_i}$, $\mathfrak{N}_{\mathscr{D}_i}$ is the total computation capacity of $\mathscr{D}_i$, and $\mathscr{R}^T_{\mathscr{D}_i}$, $\mathscr{R}^{FL}_{\mathscr{D}_i}$ are the assigned computation resources for the tactical task operation and FL process, respectively.
- In the $\mathbb{G}_{\mathscr{D}_i}$, $\mathscr{R}^T_{\mathscr{D}_i}$ and $\mathscr{R}^{FL}_{\mathscr{D}_i}$ are the $\mathscr{D}_i$'s strategies where $\mathscr{R}^T_{\mathscr{D}_i} + \mathscr{R}^{FL}_{\mathscr{D}_i} \le \mathfrak{N}_{\mathscr{D}_i}$, and $\mathfrak{U}^T_{\mathscr{D}_i}\left(\mathscr{R}^T_{\mathscr{D}_i}\right)$ and $\mathfrak{U}^{FL}_{\mathscr{D}_i}\left(\mathscr{R}^{FL}_{\mathscr{D}_i}\right)$ are their utility functions, respectively.
- $T = \{t_1, \ldots, t_c, t_{c+1}, \ldots\}$ denotes time period, which is represented by a sequence of time steps.

3.6.4 The Basic Idea and Concept of *WAB* and *CES* Solutions

To characterize the basic ideas of *WAB* and *CES* solutions, we first start with some definitions. Let $N = \{1 \ldots i \ldots n\}$ denote the set of players and let $\mathbb{R}^n$ denote the *n*-dimensional Euclidean space. Given vectors $x, y \in \mathbb{R}^n$, $x \ge y$ and $x \gg y$ if $x_i \ge y_i$ and $x_i > y_i$ for all i, respectively. We use $x * y = ((x_1 \cdot y_1), (x_2 \cdot y_2), \cdots, (x_n \cdot y_n))$ to denote the *Hadamard product* of x and y. For $\lambda \in \mathbb{R}^n$ and $S \subseteq \mathbb{R}^n$ where S is the choice set in payoff space, $\lambda * S = \{\lambda * x | x \in S\}$. Set $\mathbb{R}^n_+ = \{x \in \mathbb{R}^n | x \ge 0\}$, $\mathbb{R}^n_{++} = \{x \in \mathbb{R}^n | x \gg 0\}$, $\Delta^n = \{x \in \mathbb{R}^n_+ | x_1 + \cdots + x_n = 1\}$, and $\Delta^n_{++} = \Delta^n \cap \mathbb{R}^n_{++}$. Elements in S are payoff allocations that players can achieve with agreements, whereas a disagreement point d specifies payoffs players end up getting in case of disagreement where $d \in S$. By a hyperplane problem, a bargaining problem is given by (Haake, 2018; Kim, 2021b);

$$S = \left\{x \in \mathbb{R}^n_+ | [(p_1 \cdot x_1) + \cdots + (p_n \cdot x_n)] \le \mathbb{B}\right\} \tag{3.6.1}$$

for some vector $p \in \mathbb{R}^n_{++}$ and number $\mathbb{B} > 0$. Denote by γ the elasticity of substitution, which measures the percentage change in the ratio of any two variables with respect to change in the marginal rate of substitution between these two variables. With the distribution parameter $\delta \in \Delta^n_{++}$ and $\gamma \in (0,1) \cup (1,\infty)$, the n-variable *CES* function over $\mathbb{R}^n_+$ is given by (Haake, 2018; Kim, 2021b);

$$H(x|\delta,\gamma) = \left[\left(\delta_1 \cdot x_1^{\frac{\gamma-1}{\gamma}}\right) + \left(\delta_2 \cdot x_2^{\frac{\gamma-1}{\gamma}}\right) + \ldots + \left(\delta_n \cdot x_n^{\frac{\gamma-1}{\gamma}}\right)\right]^{\frac{\gamma}{\gamma-1}} \tag{3.6.2}$$

$$
s.t.,\ \lim_{\gamma \to \bar{\gamma}} H(x|\delta, \varepsilon) = \begin{cases} min\{x_1, x_2, \cdots, x_n\}, if \bar{\gamma} = 0 \\ \left(x_1^{\delta_1} \cdot x_2^{\delta_2} \cdots x_n^{\delta_n}\right), if \bar{\gamma} = 1 \\ \delta_1 \cdot x_1 + \delta_2 \cdot x_2 + \ldots + \delta_n \cdot x_n, if \bar{\gamma} = \infty \end{cases}
$$

Based on the class of bargaining problems, a single-valued function $\mathscr{C}(S; \delta, \gamma)$ is defined as the *CES* solution. Note that the *CES* solution coincides with the Nash bargaining solution when $\gamma = 1$ and with the egalitarian solution when $\gamma = 0$ (Haake, 2018; Kim, 2021b);

$$
\mathscr{C}(S; \delta, \gamma) = \arg \max_{x \in S:\, x \geq 0} H(x|\delta, \gamma) \tag{3.6.3}
$$

For $x \in S$, the egalitarian solution is the unique maximizer of $\min_{i \in N}\{x_i\}$, and the utilitarian solution is the unique maximizer of utility sum, i.e., $\sum_{i \in N} x_i$. For any bargaining problem S, the Nash bargaining solution lies on S's Pareto boundary in between the egalitarian and utilitarian solutions. Research presented in paper (Rachmilevitch, 2015) demonstrates that when the Nash bargaining solution aligns with the egalitarian solution, this shared outcome also qualifies as the utilitarian solution. Consequently, the Nash bargaining solution tends to exhibit utilitarian characteristics more frequently than egalitarian ones. The Nash solution can be extended by incorporating players' weights, resulting in the weighted Nash bargaining solution, which resides between the weighted utilitarian and weighted egalitarian solutions. To bridge the gap between weighted utilitarianism and egalitarianism, the *WAB* solution is introduced, as detailed in papers (Kim, 2021b; Rachmilevitch, 2015).

$$
WAB(S; \alpha) = \max_{x \in S:\, x \geq 0} \left(\left[\alpha \times \sum_{i \in N} x_i\right] + \left[(1 - \alpha) \times \min_{i \in N} \{x_i\}\right] \right) \tag{3.6.4}
$$

$s.t., \alpha \in (0,1)$

3.6.5 The Dual-Interactive Bargaining Game in Tactical Networks

Within the *CFLO* scheme, each individual device, denoted as $\mathscr{D}_i$, manages its own tactical tasks alongside the FL process. To effectively synchronize the FL training across multiple IoT devices, the MEC server allocates its offloading computation capacity as incentives. This strategy encourages self-interested devices to collaborate, steering them towards achieving a socially optimal outcome. From the MEC server's viewpoint, the $\mathscr{D}_i$'s utility function, i.e., $\mathscr{U}_{\mathscr{D}_i}(\cdot)$, is defined based on the $\mathscr{D}_i$'s FL contribution and its incentive. Commonsensically, the FL contribution and incentive of $\mathscr{D}_i$ are in proportion to the $\mathscr{R}^{FL}_{\mathscr{D}_i}$ and $\mathscr{M}_{\mathscr{D}_i}$, but, they are properly adjusted according to the respective characteristics. Therefore, the $\mathscr{U}_{\mathscr{D}_i}(\cdot)$ is given by (Kim, 2021b);

$$\mathcal{U}_{\mathcal{D}_i}\left(\mathcal{R}^{FL}_{\mathcal{D}_i}, \mathcal{M}_{\mathcal{D}_i}, \vartheta_{\mathcal{D}_i}\right) = \left(\eta + \sqrt{\left(\frac{\mathcal{R}^{FL}_{\mathcal{D}_i}}{\mathfrak{N}_{\mathcal{D}_i}}\right)^{\vartheta_{\mathcal{D}_i}}}\right) \times \left(\frac{1}{1 + exp\left(-\frac{\mathcal{M}_{\mathcal{D}_i}}{\mathfrak{M}_{\mathcal{E}}}\right)} - \varrho\right) \tag{3.6.5}$$

where η and ϱ are control parameter for $\mathcal{U}_{\mathcal{D}_i}(\cdot)$, and $\vartheta_{\mathcal{D}_i}$ is the $\mathcal{D}_i$'s characteristic value of FL contribution. In the $\mathbb{G}_{\mathcal{E}}$, multiple IoT devices, i.e., $\mathcal{D}_{1 \le i \le n}$, are assumed as game players, and the E distribute its $\mathfrak{M}_{\mathcal{E}}$ into game players in a fair-efficient way. According to the $\mathcal{R}^{FL}_{\mathcal{D}}$ values, the bargaining approach can solve the $\mathfrak{M}_{\mathcal{E}}$ distribution problem. If the difference of $\mathcal{R}^{FL}_{\mathcal{D}}$ values is larger, the axiom of *scale invariance* is important. In this case, the Nash solution is suitable. Otherwise, the egalitarian solution is appropriate. Therefore, the *CFLO* scheme adopts the concept of *CES* solution for the $\mathbb{G}_{\mathcal{E}}$. According to (3.6.2) and (3.6.5), the *CES* solution is obtained as follows (Kim, 2021b);

$$CES_{\mathcal{E}}(\mathbb{V}_{\mathcal{E}} | \delta, \gamma) = \left[\sum_{\mathcal{D}_i \in \mathbb{I}} \left(\delta_{\mathcal{D}_i} \times \mathcal{U}_{\mathcal{D}_i}\left(\mathcal{R}^{FL}_{\mathcal{D}_i}, \mathcal{M}_{\mathcal{D}_i}, \vartheta_{\mathcal{D}_i}\right)\right)^{\frac{\gamma-1}{\gamma}}\right]^{\frac{\gamma}{\gamma-1}} \tag{3.6.6}$$

$$\text{s.t., } \gamma = \min\left(\max\left(\frac{\max_{\mathcal{D}_j \in \mathbb{I}}\left\{\mathcal{R}^{FL}_{\mathcal{D}_j}\right\} - \min_{\mathcal{D}_j \in \mathbb{I}}\left\{\mathcal{R}^{FL}_{\mathcal{D}_j}\right\}}{\max_{\mathcal{D}_j \in \mathbb{I}}\left\{\mathcal{R}^{FL}_{\mathcal{D}_j}\right\}}, \phi\right), \beta\right) \text{ and } \delta_{\mathcal{D}_i} = \frac{1}{|\mathbb{I}|}$$

where ϕ, β are the lower and upper bounds of γ. $|\mathbb{I}|$ is the cardinality of I and the distribution parameter δ is set to be evenly distributed. In each individual device, the $\mathbb{G}_{\mathcal{D}_{1 \le i \le n}}$ games are operated in a dispersive and parallel manner. In the $\mathbb{G}_{\mathcal{D}_i}$, the tactical task operation and FL process are assumed as game players, and the decisions of $\mathcal{R}^{T}_{\mathcal{D}_i}$ and $\mathcal{R}^{FL}_{\mathcal{D}_i}$ values are their strategies. Therefore, the $\mathfrak{N}_{\mathcal{D}_i}$ is divided to operate the tactical tasks and FL process. For these two different works, $\mathfrak{U}^{T}_{\mathcal{D}_i}(\cdot)$ and $\mathfrak{U}^{FL}_{\mathcal{D}_i}(\cdot)$ are defined as follows (Kim, 2021b);

$$\begin{cases} \mathfrak{U}^{T}_{\mathcal{D}_i}\left(\mathcal{R}^{T}_{\mathcal{D}_i}\right) = \dfrac{\mathcal{R}^{T}_{\mathcal{D}_i}/\mathfrak{N}_{\mathcal{D}_i}}{\mu + exp\left(-\left(\mathcal{R}^{T}_{\mathcal{D}_i}/\mathfrak{N}_{\mathcal{D}_i}\right)\right)} \\[2em] \mathfrak{U}^{FL}_{\mathcal{D}_i}\left(\mathcal{R}^{FL}_{\mathcal{D}_i}\right) = \dfrac{\left(\Gamma\left(\frac{\mathcal{P}(\mathcal{M}_{\mathcal{D}_i}) - \mathcal{P}(\mathcal{R}^{FL}_{\mathcal{D}_i})}{\mathcal{P}(\mathcal{R}^{FL}_{\mathcal{D}_i})}\right) \times \mathcal{R}^{FL}_{\mathcal{D}_i}\right)/\mathfrak{N}_{\mathcal{D}_i}}{\psi + exp\left(-\left(\left(\Gamma\left(\frac{\mathcal{P}(\mathcal{M}_{\mathcal{D}_i}) - \mathcal{P}(\mathcal{R}^{FL}_{\mathcal{D}_i})}{\mathcal{P}(\mathcal{R}^{FL}_{\mathcal{D}_i})}\right) \times \mathcal{R}^{FL}_{\mathcal{D}_i}\right)/\mathfrak{N}_{\mathcal{D}_i}\right)\right)} \end{cases} \tag{3.6.7}$$

where μ and ψ are control parameter for $\mathfrak{U}_{\mathcal{D}_i}(\cdot)$, and Γ is the modification factor for the offloading service. $\mathcal{P}\left(\mathcal{M}_{\mathcal{D}_i}\right)$, $\mathcal{P}\left(\mathcal{R}^{FL}_{\mathcal{D}_i}\right)$ are the $\mathcal{M}_{\mathcal{D}_i}$ and $\mathcal{R}^{FL}_{\mathcal{D}_i}$ values at previous time epoch, respectively. In the $\mathcal{D}_i$'s point of view, utilitarian and egalitarian objectives should be considered fairly. Therefore, the idea of *WAB* solution is chosen for the $\mathbb{G}_{\mathcal{D}_i}$. According to (3.6.4) and (3.6.7), the *WAB* solution is obtained as follows (Kim, 2021b);

$$WAB\left(\left(\mathscr{R}_{\mathscr{D}_i}^{T}, \mathscr{R}_{\mathscr{D}_i}^{FL}\right)|\alpha\right) = \max_{\left(\mathscr{R}_{\mathscr{D}_i}^{T}, \mathscr{R}_{\mathscr{D}_i}^{FL}\right)}\left(\left[\alpha \times \mathbb{H}_U\right] + \left[(1-\alpha) \times \mathbb{H}_E\right]\right) \tag{3.6.8}$$

$$\text{s.t.,}\begin{cases} \mathbb{H}_U = \left(\mathfrak{U}_{\mathscr{D}_i}^{T}\left(\mathscr{R}_{\mathscr{D}_i}^{T}\right) + \mathfrak{U}_{\mathscr{D}_i}^{FL}\left(\mathscr{R}_{\mathscr{D}_i}^{FL}\right)\right) \\ \mathbb{H}_E = \mathbf{min}\left\{\mathfrak{U}_{\mathscr{D}_i}^{T}\left(\mathscr{R}_{\mathscr{D}_i}^{T}\right), \mathfrak{U}_{\mathscr{D}_i}^{FL}\left(\mathscr{R}_{\mathscr{D}_i}^{FL}\right)\right\} \\ \mathscr{R}_{\mathscr{D}_i}^{T} + \mathscr{R}_{\mathscr{D}_i}^{FL} \leq \mathfrak{N}_{\mathscr{D}_i}; \alpha = \max\left\{\dfrac{\min\left\{\mathscr{P}\left(\mathscr{M}_{\mathscr{D}_i}\right), \mathscr{P}\left(\mathscr{R}_{\mathscr{D}_i}^{FL}\right)\right\}}{\max\left\{\mathscr{P}\left(\mathscr{M}_{\mathscr{D}_i}\right), \mathscr{P}\left(\mathscr{R}_{\mathscr{D}_i}^{FL}\right)\right\}}, \chi\right\} \end{cases}$$

where χ is the lower bound of α. In the $\mathbb{G}_{\mathscr{D}_i}$, the parameter α controls the relative weights given to utilitarian and egalitarian objectives. Under diverse system environments, a fixed α value cannot effectively adapt to the changing platform conditions. Therefore, the *CFLO* scheme dynamically modifies the α value at each time epoch. In the MEC-assisted FL processing, the difference between $\mathscr{P}\left(\mathscr{M}_{\mathscr{D}_i}\right)$ and $\mathscr{P}\left(\mathscr{R}_{\mathscr{D}_i}^{FL}\right)$ can be interpreted as an index of relative fairness for the $\mathfrak{N}_{\mathscr{D}_i}$ division problem. By considering the lower bound, the α value is adaptively adjusted in an online manner (Kim, 2021b).

3.6.6 Summary

The *CFLO* scheme delves into a fine-grained tactical task offloading mechanism integrated with the FL process. Initially, the *CFLO* scheme outlines the motivation for employing a dual-interactive bargaining game model, which is designed to dynamically manage the resource-constrained MEC-assisted FL platform. Subsequently, the *CFLO* scheme elucidates the fundamental principles of the *WAB* and *CES* solutions. These bargaining approaches are pivotal for understanding how their combination can further enhance system performance. Following this, the *CFLO* scheme introduces the $\mathbb{G}_{\mathscr{E}}$ and $\mathbb{G}_{\mathscr{D}_i}$ games, which are responsible for the distribution and allocation of $\mathfrak{M}_{\mathscr{E}}$ and $\mathfrak{N}_{\mathscr{D}_i}$ resources, respectively. To achieve optimal system performance, these games operate in a cooperative, interactive, and sequential manner, ensuring the maintenance of robust global properties. In the context of dynamically changing tactical network environments, the dual-interactive bargaining approach enables the establishment of a reciprocal consensus. This consensus effectively addresses the intricate control challenges associated with both the FL process and tactical task operations, thereby optimizing overall system performance (Kim, 2021b).

3.7 Bargaining Game based Resource Management (*BGRM*) Scheme

The rapid expansion of IoT devices is driving the growth of virtual reality applications, which can be achieved through service offloading supported by the pervasive edge computing (PEC) platform. However, the PEC system's limited computational and communication resources necessitate the creation of an innovative resource management algorithm. Recently, S. Kim develops the *Bargaining Game-based Resource Management* (*BGRM*) scheme (Kim, 2022a). Utilizing the principles of a unification bargaining solution, the *BGRM* scheme integrates various bargaining strategies to deliver a fair and efficient solution. By coordinating network agents, the *BGRM* scheme fosters a mutual consensus, enabling the system to achieve a balanced performance despite conflicting requirements. This approach is essential for addressing the trade-offs between efficiency and fairness. The key innovation of the *BGRM* scheme is its adaptability and flexibility, allowing it to effectively manage and share PEC resources in dynamic PEC system environments.

3.7.1 Development Motivation

The explosive expansion of IoT devices is fueling the advancement of virtual reality (VR) applications, which can be realized through service offloading facilitated by the pervasive edge computing (PEC) platform. However, given the limited computational and communication resources of PEC systems, there is a pressing need to develop innovative resource management algorithms. Recently, S. Kim introduces the *BGRM* scheme (Kim, 2022a). The *BGRM* scheme leverages the concept of a unification bargaining solution, which combines various bargaining strategies to deliver a fair and efficient solution. By coordinating network agents, the *BGRM* scheme fosters a mutual consensus, enabling the system to achieve balanced performance despite conflicting requirements. This approach is crucial for navigating the trade-offs between efficiency and fairness. A key innovation of the *BGRM* scheme lies in its adaptability and flexibility, allowing it to effectively respond to dynamic PEC system environments (Kim, 2022a).

With the swift progress of 5G network technology, the landscape of communication and computation is undergoing significant transformations. Innovations such as IoT, cloud and edge computing, tactile Internet, terahertz technologies, and blockchain are creating new business models and markets. These technologies offer numerous advantages across various sectors, leading to the emergence of novel concepts. Consequently, IP-based networking devices are becoming increasingly efficient and complex. Over the past decade, mobile and IoT devices have become integral to daily activities. As the IoT continues to evolve, the number of connected smart devices has surged and is projected to grow exponentially by 2025. In this context, heterogeneous devices possess diverse computational capabilities and are expected to run a wide array of applications with varying constraints and requirements (Apostolopoulos, 2020; Kim, 2022a; Li, 2020).

In IoT networks, applications that are both computation-intensive and latency-sensitive, such as VR, present significant challenges for resource-constrained IoT devices. Wireless VR is anticipated to become a flagship application in 5G and beyond, offering immersive experiences that transform communication methods. It is well-established that rendering is a major performance bottleneck in wireless VR systems, particularly for VR games, which require extensive foreground interactions and detailed background environment rendering. Consequently, these applications demand substantial computational resources from mobile IoT devices to ensure high Quality of Experience (QoE). Despite recent advancements in hardware, current smart devices are still inadequate for efficiently supporting computation-intensive VR applications, as their local computing power and energy resources remain insufficient. Technically, this means that local rendering on today's mobile devices cannot meet the QoE requirements of untethered VR users, such as achieving 14 ms latency and 60 frames per second. As a result, even with increasing market penetration, high-end VR systems today still require tethered setups (Guo, 2020; Kim, 2022a).

For future 5G and beyond networks, delivering VR videos with high QoE is a primary challenge. To overcome the limitations of current IoT devices, the PEC emerges as a new paradigm within information and communication technologies. The PEC can promptly support the processing of urgent and complex tasks by offloading a portion of rendering tasks to edge servers, thereby enhancing wireless VR performance. Unlike traditional edge computing methods, the PEC offers several advantages; i) Decentralization: no central authority is required, enabling feasible and diverse applications, ii) Infrastructure Independence: deploying and maintaining a dedicated cloud backend is not necessary, iii) Reduced Transmission Delay: data processing near end-users minimizes transmission delays, and iv) Enhanced Privacy and Security: privacy and security concerns are effectively addressed. Thus, the PEC paradigm is better suited for future network scenarios, facilitating decentralized decision-making, ensuring information security, achieving high accuracy, and maintaining high data transmission rates for resource-constrained IoT devices. In recent years, PEC applications have expanded beyond entertainment to include industries such as live gaming, multi-angle video viewing, and cooperative driving (Apostolopoulos, 2020; Chen, 2020; Khan, 2020; Kim, 2022a; Wang, 2020).

As a pivotal technology, the PEC has garnered significant attention in both academia and industry. However, despite its promising potential, several challenges remain. Given economic deployment constraints, individual edge servers have restricted computing power, making it impractical for a single server to handle all rendering tasks. An effective computation control strategy is necessary to enhance system efficiency. Wireless bandwidth can become a bottleneck for wireless VR, as immersive VR videos typically require data rates exceeding 25 Mbps. Meeting this demand is particularly challenging in time-varying network environments and under conditions of data traffic congestion. In a multi-device interactive environment, each IoT device operates as an autonomous, distributed, and intelligent network agent, aiming to maximize its own payoff in alignment with its preferences across different outcomes. This scenario has led researchers to explore methods for ensuring fairness among devices in a fully decentralized environment while maintaining the required QoE. However, this remains a complex and difficult task within dynamic PEC environments. Consequently, there is a need for a new intelligent control paradigm with novel solution concepts to guide multiple IoT devices in making rational and strategic decisions that lead to a fair and efficient consensus. In this context, cooperative game theory is a promising area of investigation for designing innovative PEC system control algorithms (Guo, 2020; ; Kim, 2014, 2022a; Lin, 2021).

The *Nash Bargaining Solution* (*NBS*) was the first cooperative game solution to be defined using a set of axioms. It is capable of predicting the outcome of a bargaining process based solely on each participant's preferences. The *NBS* is constructed using an expected utility function that evaluates all possible agreements and the potential outcome if the parties do not reach an agreement. Alternative solutions to bargaining problems include the *Egalitarian Bargaining Solution* (*EBS*) and the *Utilitarian Bargaining Solution* (*UBS*). The key difference among these three solutions lies in their respective approaches to balancing egalitarianism and utilitarianism. When addressing multi-criteria problems, the *NBS*, *EBS*, and *UBS* each negotiate based on these opposing axiomatic principles to find appropriate solutions. In the case of the *NBS*, if a bargaining solution derived from a larger set is applicable within a smaller domain, the solution remains unaffected by the size of the domain. In contrast, the *EBS* strives to ensure equal gains for all players, effectively identifying the point that maximizes the minimum payoff among participants. Meanwhile, the *UBS* focuses on maximizing the total sum of players' payoffs, considering their respective utility functions (Kim, 2014, 2022a; Pivato, 2009).

The *Unification Bargaining Solution* (*UniBS*) was introduced by C. Haakea and C. Qin to address scenarios where players in a cooperative game hold differing perspectives on the ideal solution. Notably, the *UniBS* encompasses the *NBS*, *EBS*, and *UBS* as specific instances, thereby providing a comprehensive framework for solving multi-criteria bargaining problems. A fundamental aspect of the *UniBS* is its unifying axiom, which dictates how the ratios of players' payoff gains in solutions derived from transformed problems - through affine transformations - respond to the ratios of players' payoff scales resulting from these transformations. This axiom ensures consistency and fairness across different bargaining scenarios by aligning the solution ratios with the underlying payoff scales. Consequently, the *UniBS* is defined by a control parameter that quantifies the balance between individual and collective orientations of players. This parameter allows the *UniBS* to adjust the degree of emphasis placed on individual versus collective benefits, providing a flexible approach to achieving fair and efficient outcomes in diverse bargaining contexts (Haake, 2018; Kim, 2022a).

3.7.2 Contributions

The *BGRM* scheme leverages the PEC platform alongside computation offloading technology. In an environment featuring heterogeneous IoT devices and multiple edge servers, each IoT device has the ability to partially offload its computational tasks to a designated PEC server for remote execution. The primary objective in developing a new computation offloading algorithm within the PEC infrastructure is to efficiently allocate the limited PEC resources amidst dynamically changing PEC environments. To accomplish this, multiple IoT devices and their corresponding servers collaborate and act cooperatively to balance and enhance conflicting performance criteria (Kim, 2022a).

Within the *BGRM* scheme, the scarce computational and communication resources of the PEC platform are adaptively distributed among individual IoT devices. To address these resource-sharing challenges, the *BGRM* scheme adopts the concept of the *UniBS* and implements a unified bargaining process. This process involves a sequential negotiation between each device-edge pair. While each device focuses on strategies to maximize its own payoff, the device-edge pair also shares a common goal and makes binding commitments through the exchange of current information. As a result, the strategies of the devices and edges become interconnected. In the unification bargaining game model, this interdependency plays a significant role in determining overall system performance. To the best of my knowledge, this is the first study where different bargaining concepts are selectively integrated into the design of a PEC resource allocation algorithm (Kim, 2022a).

3.7.3 IoT based PEC Platform and Unified Bargaining Game

The *BGRM* scheme assumes a PEC system platform consisting of multiple PEC servers, $\mathbb{E} = \{\mathscr{E}_1, \ldots, \mathscr{E}_n\}$, which provide computation services for a set of IoT devices, $\mathbb{D} = \{\mathscr{D}_1, \ldots, \mathscr{D}_m\}$. Individual PEC servers are connected via high speed fiber communications, and they are endowed with edge computing capabilities. Each device can offload its computation tasks to the corresponding PEC server via wireless communications. The computing capability of $\mathscr{E}_{1\leq i\leq n}$ is characterized by its computation service rate, that is CPU frequency. Each $\mathscr{E}_i$ serves a dedicated set of IoT devices in its serving area, denoted by $\mathbb{D}_{\mathscr{E}_i} \subseteq \mathbb{D}$. IoT devices are authorized to access the communication and computing service of deployed PEC server. The totally generated computation workload for the $\mathscr{D}_{1\leq j\leq m}$ is $\Upsilon_{\mathscr{D}_j}$ where $\Upsilon_{\mathscr{D}_j} = \Upsilon_{\mathscr{D}_j}^{\mathscr{E}_i} + \Upsilon_{\mathscr{D}_j}^{L}$ where $\Upsilon_{\mathscr{D}_j}^{\mathscr{E}_i}$ and $\Upsilon_{\mathscr{D}_j}^{L}$ are the $\mathscr{D}_j$'s offloading and local computation amounts, respectively. In the bargaining game model, the E-D association is a game entity to negotiate with each other in a cooperative manner (Kim, 2022a).

The operational timeline is segmented into discrete time slots to facilitate offloading decisions, which correspond to the timing of task arrivals. Within each time slot, computation tasks generated by each IoT device follow a Poisson process, a commonly used assumption for task arrivals in PEC systems. IoT devices may request various types of tasks that differ in input data size and required CPU cycles. Due to the often uneven distribution of task workload arrivals among IoT devices, tasks can be partially offloaded for processing. For computation offloading services, data transmission occurs via the wireless link between the device and its respective edge server. However, the computing power and wireless bandwidth of PEC servers are limited. Therefore, offloading strategies must take into account the current system conditions to enhance resource efficiency (Chen, 2018; Kim, 2022a).

To address the PEC resource sharing problem, the *BGRM* scheme formulates three bargaining games. At a time period, each individual $\mathscr{D}_{1 \le j \le m}$ and $\mathscr{E}_{1 \le i \le n}$ process their bargaining games. In the $\mathscr{D}_j$, the $\Upsilon_{\mathscr{D}_j}$ is partially offloaded to increase the $\mathscr{D}_j$'s profit, and the bargaining game $\left(\mathbb{G}_{\mathscr{D}_j}^{\mathscr{E}_i}\right)$ is designed to divide the $\Upsilon_{\mathscr{D}_j}$. In the $\mathscr{E}_i$, two bargaining games, i.e., $\mathbb{G}_{\mathscr{E}_i}^{\Gamma}$ and $\mathbb{G}_{\mathscr{E}_i}^{\Theta}$, are formulated to share the $\mathscr{E}_i$'s computation $\left(\Gamma_{\mathscr{E}_i}\right)$ and communication $\left(\Theta_{\mathscr{E}_i}\right)$ resources, respectively. Based on the concept of the *UniBS*, the $\mathbb{G}_{\mathscr{D}_j}^{\mathscr{E}_i}$, $\mathbb{G}_{\mathscr{E}_i}^{\Gamma}$ and $\mathbb{G}_{\mathscr{E}_i}^{\Theta}$ are operated in a cooperative manner, and they are repeated in a step-by-step interactive fashion at each time period. Formally, the *BGRM* scheme defines the $\mathbb{G}_{\mathscr{D}_j}^{\mathscr{E}_i}$, $\mathbb{G}_{\mathscr{E}_i}^{\Gamma}$ and $\mathbb{G}_{\mathscr{E}_i}^{\Theta}$ game entities, i.e., $\mathbb{G} = \left\{\mathbb{G}_{\mathscr{D}_{1 \le j \le m}}^{\mathscr{E}_i}, \mathbb{G}_{\mathscr{E}_{1 \le i \le n}}^{\Gamma}, \mathbb{G}_{\mathscr{E}_{1 \le i \le n}}^{\Theta}\right\} = \left\{\mathbb{E}, \mathbb{D}, \left\{\mathbb{G}_{\mathscr{D}_j}^{\mathscr{E}_i} | \mathfrak{M}_{\mathscr{D}_j}, \left(\Upsilon_{\mathscr{D}_j}^{\mathscr{E}_i}, \Upsilon_{\mathscr{D}_j}^{L}\right), U_{\mathscr{D}_j}^{O}(\cdot), U_{\mathscr{D}_j}^{L}(\cdot)\right\}, \left\{\mathbb{G}_{\mathscr{E}_i}^{\Gamma} | \Gamma_{\mathscr{E}_i}, \mathbb{D}_{\mathscr{E}_i}, \Gamma_{\mathscr{E}_i}^{\mathscr{D}_j \in \mathbb{D}_{\mathscr{E}_i}}, \mathscr{U}_{\mathscr{D}_j}(\cdot)\right\}, \left\{\mathbb{G}_{\mathscr{E}_i}^{\Theta} | \Theta_{\mathscr{E}_i}, \mathbb{D}_{\mathscr{E}_i}, \Theta_{\mathscr{E}_i}^{\mathscr{D}_j \in \mathbb{D}_{\mathscr{E}_i}}, \mathfrak{U}_{\mathscr{D}_j}(\cdot)\right\}, T\right\}$

of gameplay (Kim, 2022a).

- The $\mathbb{G}_{\mathscr{D}_j}^{\mathscr{E}_i}$, $\mathbb{G}_{\mathscr{E}_i}^{\Gamma}$ and $\mathbb{G}_{\mathscr{E}_i}^{\Theta}$ are mutually and reciprocally interdependent in an interactive manner, and they work together to share the PEC resources.
- E is the set of edge servers, and D is the set of IoT devices in the PEC system platform.
- In the $\mathbb{G}_{\mathscr{D}_j}^{\mathscr{E}_i}$, $\mathfrak{M}_{\mathscr{D}_j}$ is the total computing capacity of $\mathscr{D}_j$, and offloading computing and local computing are game players. $\Upsilon_{\mathscr{D}_j}^{\mathscr{E}_i}$ and $\Upsilon_{\mathscr{D}_j}^{L}$ are their strategies, and $U_{\mathscr{D}_j}^{O}(\cdot)$ and $U_{\mathscr{D}_j}^{L}(\cdot)$ are their utility functions.
- In the $\mathbb{G}_{\mathscr{E}_i}^{\Gamma}$, $\Gamma_{\mathscr{E}_i}$ is the $\mathscr{E}_i$'s total computation capacity, and IoT devices in the $\mathbb{D}_{\mathscr{E}_i}$ are game players. $\Gamma_{\mathscr{E}_i}^{\mathscr{D}_j \in \mathbb{D}_{\mathscr{E}_i}}$ is the $\mathscr{D}_j$'s strategy and $\mathscr{U}_{\mathscr{D}_j}(\cdot)$ is its utility function.
- In the $\mathbb{G}_{\mathscr{E}_i}^{\Theta}$, $\Theta_{\mathscr{E}_i}$ is the $\mathscr{E}_i$'s total communication capacity, and IoT devices in the $\mathbb{D}_{\mathscr{E}_i}$ are game players. $\Theta_{\mathscr{E}_i}^{\mathscr{D}_j \in \mathbb{D}_{\mathscr{E}_i}}$ is the $\mathscr{D}_j$'s strategy and $\mathfrak{U}_{\mathscr{D}_j}(\cdot)$ is its utility function.
- Discrete time model $T \in \{t_1, \ldots, t_c, t_{c+1}, \ldots\}$ is represented by a sequence of time steps. The length of t_c matches the event time-scale of $\mathbb{G}_{\mathscr{D}_j}^{\mathscr{E}_i}$, $\mathbb{G}_{\mathscr{E}_i}^{\Gamma}$ and $\mathbb{G}_{\mathscr{E}_i}^{\Theta}$.

3.7.4 The Fundamental Idea of the *UniBS*

To characterize the basic concepts of bargaining solutions, we assume a n-player bargaining problem. Let $N = \{1 \ldots i \ldots n\}$ be the set of players and let $\mathbb{R}^n$ denote the n-dimensional Euclidean space. Given vectors $x, y \in \mathbb{R}^n$, $x \ge y$ and $x \gg y$ if $x_i \ge y_i$ and $x_i > y_i$ for all i, respectively. Set $\mathbb{R}_+^n = \{x \in \mathbb{R}^n | x \ge 0\}$ and let S and d be the set of feasible outcomes and disagreement point where $S \subset \mathbb{R}^n$ and $d \in S$. Let $\sum$ be the set of all subsets of $\mathbb{R}_+^n$, and elements in $\sum$ are interpreted as bargaining problems. Mathematically, the *NBS* can be defined as follows (Haake, 2018; Kim, 2022a):

$$NBS(S) = \max_{i \in N, S \in \sum} \left\{u_i \in S | (u_1 - d_1) \times \ldots \times (u_i - d_i) \times \ldots \times (u_n - d_n)\right\} \tag{3.7.1}$$

The *EBS* can be computed via the following maximization problem (Haake, 2018; Kim, 2022a):

$$EBS(S) = \max_{i \in N} \left\{ u_i \in S | \min_{S \in \Sigma} \left((u_1 - d_1), \ldots, (u_i - d_i), \ldots, (u_n - d_n) \right) \right\}$$

(3.7.2)

The *UBS* can be defined as follows (Haake, 2018; Kim, 2022a):

$$UBS(S) = \max_{i \in N, S \in \Sigma} \left\{ u_i \in S | \left((u_1 - d_1) + \ldots + (u_i - d_i) + \ldots + (u_n - d_n) \right) \right\}$$

(3.7.3)

To unify different bargaining solutions, *unification function* (*UF*) can be defined with a control parameter ε where $\varepsilon \in (0,1) \cup (1,\infty)$. A new bargaining solution, called the *UniBS*, is obtained by maximizing the *UF*: this solution includes the *NBS*, *EBS* and *UBS* as special cases (Haake, 2018; Kim, 2022a).

$$UF(u_i \in S|\varepsilon) = \left[(u_1 - d_1)^{\frac{\varepsilon-1}{\varepsilon}} + \ldots + (u_i - d_i)^{\frac{\varepsilon-1}{\varepsilon}} + \ldots + (u_n - d_n)^{\frac{\varepsilon-1}{\varepsilon}} \right]^{\frac{\varepsilon}{\varepsilon-1}}$$

(3.7.4)

$$\text{s.t., } \lim_{\varepsilon \to \bar{\varepsilon}} UF(u_i \in S|\varepsilon) = \begin{cases} EBS, if \bar{\varepsilon} = 0 \\ NBS, if \bar{\varepsilon} = 1 \\ UBS, if \bar{\varepsilon} = \infty \end{cases}$$

3.7.5 The Unified Bargaining Game in the PEC Platform

To develop new unified bargaining game models, the *BGRM* scheme constructs the $\mathbb{G}_{\mathcal{D}_i}^{\mathcal{E}_i}$, $\mathbb{G}_{\mathcal{E}_i}^{\Gamma}$ and $\mathbb{G}_{\mathcal{E}_i}^{\Theta}$ games. They are interacting with one another during a sequence of time steps. At each time period, the $\mathbb{G}_{\mathcal{D}_j}^{\mathcal{E}_i}$ is designed for the $\mathcal{E}_i$-$\mathcal{D}_j$ pair; the $\mathcal{D}_j$ partially offloads its computation task to the $\mathcal{E}_i$. This game decides the computation offloading size, that is $\Upsilon_{\mathcal{D}_j}^{\mathcal{E}_i}$, to maximize the $\mathcal{D}_j$'s payoff. As game players, offloading and local computing services select their strategies, i.e., $\Upsilon_{\mathcal{D}_j}^{\mathcal{E}_i}$ and $\Upsilon_{\mathcal{D}_j}^{L}$, by considering the currently available $\mathcal{D}_j$'s computing power and the $\mathcal{E}_i$'s offloading cost. At the time t_c, the utility functions of offloading computing service $\left(U_{\mathcal{D}_j}^{O}(\cdot) \right)$ and the local computing service $\left(U_{\mathcal{D}_j}^{L}(\cdot) \right)$ are defined as follows (Kim, 2022a):

$$\begin{cases} U_{\mathcal{D}_j}^{O}\left(\Upsilon_{\mathcal{D}_j}, \Upsilon_{\mathcal{D}_j}^{\mathcal{E}_i}, c_{t_c}^{\mathcal{E}_i}\right) = exp\left(-\frac{\min\left(\Upsilon_{\mathcal{D}_j}, \Upsilon_{\mathcal{D}_j}^{\mathcal{E}_i}\right)}{\Upsilon_{\mathcal{D}_j}} \times c_{t_c}^{\mathcal{E}_i} \right) \times \Upsilon_{\mathcal{D}_j}^{\mathcal{E}_i} \\ U_{\mathcal{D}_j}^{L}\left(\Upsilon_{\mathcal{D}_j}, \Upsilon_{\mathcal{D}_j}^{L}, \rho_{t_c}^{\mathcal{D}_j}\right) = \left[\eta - \log\left(\frac{\min\left(\Upsilon_{\mathcal{D}_j}, \Upsilon_{\mathcal{D}_j}^{L}\right)}{\Upsilon_{\mathcal{D}_j}} \times \rho_{t_c}^{\mathcal{D}_j} \right) \right] \times \Upsilon_{\mathcal{D}_j}^{L} \end{cases} \quad (3.7.5)$$

$$\text{s.t., } \Upsilon_{\mathcal{D}_j} = \Upsilon_{\mathcal{D}_j}^{\mathcal{E}_i} + \Upsilon_{\mathcal{D}_j}^{L}, \quad c_{t_c}^{\mathcal{E}_i} = \frac{\mathcal{M}_{\mathcal{E}_i}^{t_c}}{\Gamma_{\mathcal{E}_i}} \text{ and } \rho_{t_c}^{\mathcal{D}_j} = exp\left(\frac{\Upsilon_{\mathcal{D}_j}}{\mathfrak{M}_{\mathcal{D}_j}} \right)$$

where $c_{t_c}^{\mathscr{E}_i}$ is the $\mathscr{E}_i$'s offloading computation cost, $\rho_{t_c}^{\mathscr{D}_j}$ is the $\mathscr{D}_j$'s working-load status, and $\mathscr{M}_{\mathscr{E}_i}^{t_c}$ is the using computing power of $\mathscr{E}_i$, at time t_c. Therefore, $c_{t_c}^{\mathscr{E}_i}$ and $\rho_{t_c}^{\mathscr{D}_j}$ may go up in direct proportion to the current computation overhead, and η is a control parameter for $U_{\mathscr{D}_j}^{L}(\cdot)$. Since all game players in the $\mathbb{G}_{\mathscr{D}_i}^{\mathscr{E}_i}$ are components of $\mathscr{D}_i$, the solution concept of $\mathbb{G}_{\mathscr{D}_i}^{\mathscr{E}_i}$ should be strongly concerned the egalitarianism for game players. In this case, the *EBS* is preferred for the $\mathscr{D}_i$. It is given by (Kim, 2022a):

$$\lim_{\varepsilon\to 0} UF\left(U_{\mathscr{D}_j}^{O}(\cdot), U_{\mathscr{D}_j}^{L}(\cdot), \Upsilon_{\mathscr{D}_j}, \left(d_{\mathscr{D}_j}^{O}, d_{\mathscr{D}_j}^{L}\right)|\varepsilon\right) \tag{3.7.6}$$

$$= \left[\left(U_{\mathscr{D}_j}^{O}(\cdot) - d_{\mathscr{D}_j}^{O}\right)^{\frac{\varepsilon-1}{\varepsilon}} + \left(U_{\mathscr{D}_j}^{L}(\cdot) - d_{\mathscr{D}_j}^{L}\right)^{\frac{\varepsilon-1}{\varepsilon}}\right]^{\frac{\varepsilon}{\varepsilon-1}}$$

$$\cong \max\left[\min\left\{\left(U_{\mathscr{D}_j}^{O}(\cdot) - d_{\mathscr{D}_j}^{O}\right), \left(U_{\mathscr{D}_j}^{L}(\cdot) - d_{\mathscr{D}_j}^{L}\right)\right\}\right]$$

where $d_{\mathscr{D}_j}^{O}$ and $d_{\mathscr{D}_j}^{L}$ are disagree points for offloading and local computing services. According to (3.7.6), the solution of $\mathbb{G}_{\mathscr{D}_i}^{\mathscr{E}_i}$ is obtained. Contrast to the $\mathbb{G}_{\mathscr{D}_i}^{\mathscr{E}_i}$, the $\mathbb{G}_{\mathscr{E}_i}^{\Gamma}$ and $\mathbb{G}_{\mathscr{E}_i}^{\Theta}$ games are operated in the $\mathscr{E}_i$. Through these two games, the $\mathscr{E}_i$ distribute its $\Gamma_{\mathscr{E}_i}$ and $\Theta_{\mathscr{E}_i}$ for the $\mathscr{D}_j \in \mathbb{D}_{\mathscr{E}_i}$. First, the $\mathbb{G}_{\mathscr{E}_i}^{\Gamma}$ game decides the $\Gamma_{\mathscr{E}_i}^{\mathscr{D}_j}$ value by considering each individual device's condition. As a game player, the utility function of $\mathscr{D}_j\left(\mathcal{U}_{\mathscr{D}_j}(\cdot)\right)$ is defined as follows (Kim, 2022a):

$$\mathcal{U}_{\mathscr{D}_j}\left(\Gamma_{\mathscr{E}_i}, \mathbb{D}_{\mathscr{E}_i}, \Upsilon_{\mathscr{D}_j}^{\mathscr{E}_i}, \Gamma_{\mathscr{E}_i}^{\mathscr{D}_j}\right) = \left[\frac{exp\left(\frac{\min\left(\Upsilon_{\mathscr{D}_j}^{\mathscr{E}_i}, \Gamma_{\mathscr{E}_i}^{\mathscr{D}_j}\right)}{\Upsilon_{\mathscr{D}_j}^{\mathscr{E}_i}}\right) - exp\left(-\frac{\min\left(\Upsilon_{\mathscr{D}_j}^{\mathscr{E}_i}, \Gamma_{\mathscr{E}_i}^{\mathscr{D}_j}\right)}{\Upsilon_{\mathscr{D}_j}^{\mathscr{E}_i}}\right)}{exp\left(\frac{\min\left(\Upsilon_{\mathscr{D}_j}^{\mathscr{E}_i}, \Gamma_{\mathscr{E}_i}^{\mathscr{D}_j}\right)}{\Upsilon_{\mathscr{D}_j}^{\mathscr{E}_i}}\right) + exp\left(-\frac{\min\left(\Upsilon_{\mathscr{D}_j}^{\mathscr{E}_i}, \Gamma_{\mathscr{E}_i}^{\mathscr{D}_j}\right)}{\Upsilon_{\mathscr{D}_j}^{\mathscr{E}_i}}\right)}\right] \times \Gamma_{\mathscr{E}_i}^{\mathscr{D}_j} \tag{3.7.7}$$

$$\text{s.t., } \mathscr{D}_j \in \mathbb{D}_{\mathscr{E}_i} \text{ and } \Gamma_{\mathscr{E}_i} \geq \sum\nolimits_{\mathscr{D}_j \in \mathbb{D}_{\mathscr{E}_i}} \Gamma_{\mathscr{E}_i}^{\mathscr{D}_j}$$

where $\Gamma_{\mathscr{E}_i}^{\mathscr{D}_j}$ is the allocated computation power for the $\mathscr{D}_j$. In the $\mathbb{G}_{\mathscr{E}_i}^{\Gamma}$, all game players are devices in $\mathbb{D}_{\mathscr{E}_i}$, and they share the limited $\Gamma_{\mathscr{E}_i}$ resource. In the viewpoint of game players, decreasing the difference of players' requirements for the $\Gamma_{\mathscr{E}_i}$ needs independence of alternatives other than the disagreement point. It says that the solution of a bargaining problem does not change as the set of feasible outcomes is reduced, so long as the disagreement point remains unchanged, and the solution originally selected remain feasible. This condition says that the selection of a feasible solution in an outcome set does not depend on any point except possibly the disagreement point (Roth, 1977). In this case, the *NBS* is preferred. Otherwise, the *UBS* is suitable. As a decision control parameter, the ε value is dynamically adjusted based on the current difference ratio of $\Upsilon_{\mathscr{D}_j \in \mathbb{D}_{\mathscr{E}_i}}^{\mathscr{E}_i}$ and the solution of $\mathbb{G}_{\mathscr{E}_i}^{\Gamma}$ is given by (Kim, 2022a):

$$UF\left(\mathbb{D}_{\mathscr{E}_i}, \mathcal{U}_{\mathscr{D}_j}(\cdot), d_{\mathscr{D}_j}^{\Gamma}|\varepsilon\right) = \left[\sum_{\mathscr{D}_j \in \mathbb{D}_{\mathscr{E}_i}} \left(\mathcal{U}_{\mathscr{D}_j}(\cdot) - d_{\mathscr{D}_j}^{\Gamma}\right)^{\frac{\varepsilon-1}{\varepsilon}}\right]^{\frac{\varepsilon}{\varepsilon-1}}$$

(3.7.8)

$$\text{s.t.}, \varepsilon = \frac{\min_{\mathcal{D}_k,\mathcal{D}_h \in \mathbb{D}_{\mathcal{E}_i}} \left(\left| \Upsilon_{\mathcal{D}_k}^{\mathcal{E}_i} - \Upsilon_{\mathcal{D}_h}^{\mathcal{E}_i} \right| \right)}{\max_{\mathcal{D}_k,\mathcal{D}_h \in \mathbb{D}_{\mathcal{E}_i}} \left(\left| \Upsilon_{\mathcal{D}_k}^{\mathcal{E}_i} - \Upsilon_{\mathcal{D}_h}^{\mathcal{E}_i} \right| \right)}$$

where $d_{\mathcal{D}_j}^{\Gamma}$ is the disagree point of $\mathcal{D}_j$ in the $\mathbb{G}_{\mathcal{E}_i}^{\Gamma}$. According to (3.7.8), the solution of $\mathbb{G}_{\mathcal{E}_i}^{\Gamma}$ is given. Second, the $\mathbb{G}_{\mathcal{E}_i}^{\Theta}$ game determines the $\Theta_{\mathcal{E}_i}^{\mathcal{D}_j}$ value to distribute the limited $\Theta_{\mathcal{E}_i}$ resource. Such as the $\mathbb{G}_{\mathcal{E}_i}^{\Gamma}$, the $\mathcal{D}_j \in \mathbb{D}_{\mathcal{E}_i}$ is a game player, and its utility function $\left(\mathfrak{U}_{\mathcal{D}_j}(\cdot) \right)$ is defined as follows (Kim, 2022a):

$$\mathfrak{U}_{\mathcal{D}_j}\left(\mathbb{D}_{\mathcal{E}_i}, \Theta_{\mathcal{E}_i}, \Theta_{\mathcal{E}_i}^{\mathcal{D}_j}, \Theta_{M}^{\mathcal{D}_j} \right) = \frac{\log\left(\psi + \frac{\min\left(\Theta_{\mathcal{E}_i}^{\mathcal{D}_j}, \Theta_{M}^{\mathcal{D}_j}\right)}{\Theta_{M}^{\mathcal{D}_j}} \right)}{\log\left(\psi + \frac{\Delta}{\Theta_{M}^{\mathcal{D}_j}} \right)} \times \Theta_{\mathcal{E}_i}^{\mathcal{D}_j}$$

(3.7.9)

$$\text{s.t.}, \mathcal{D}_j \in \mathbb{D}_{\mathcal{E}_i} \text{ and } \Theta_{\mathcal{E}_i} \geq \sum\nolimits_{\mathcal{D}_j \in \mathbb{D}_{\mathcal{E}_i}} \Theta_{\mathcal{E}_i}^{\mathcal{D}_j}$$

where ψ and Δ are adjustment factors for $\mathfrak{U}_{\mathcal{D}_j}(\cdot)$, and $\Theta_{M}^{\mathcal{D}_j}$ is the $\mathcal{D}_j$'s maximum requirement for the $\Theta_{\mathcal{E}_i}$.In the $\mathbb{G}_{\mathcal{E}_i}^{\Theta}$, all game players are devices in $\mathbb{D}_{\mathcal{E}_i}$, and they share the limited $\Theta_{\mathcal{E}_i}$ resource for wireless communications. The solution of $\mathbb{G}_{\mathcal{E}_i}^{\Theta}$ is obtained as the same idea as the solution of $\mathbb{G}_{\mathcal{E}_i}^{\Gamma}$. Therefore, it is given by (Kim, 2022a):

$$UF\left(\mathbb{D}_{\mathcal{E}_i}, \mathfrak{U}_{\mathcal{D}_j}(\cdot), d_{\mathcal{D}_j}^{\Theta} | \varepsilon \right) = \left[\sum_{\mathcal{D}_j \in \mathbb{D}_{\mathcal{E}_i}} \left(\mathfrak{U}_{\mathcal{D}_j}(\cdot) - d_{\mathcal{D}_j}^{\Theta} \right)^{\frac{\varepsilon-1}{\varepsilon}} \right]^{\frac{\varepsilon}{\varepsilon-1}}$$

(3.7.10)

$$\text{s.t.}, \varepsilon = \frac{\min_{\mathcal{D}_k,\mathcal{D}_h \in \mathbb{D}_{\mathcal{E}_i}} \left(\left| \Theta_{M}^{\mathcal{D}_k} - \Theta_{M}^{\mathcal{D}_h} \right| \right)}{\max_{\mathcal{D}_k,\mathcal{D}_h \in \mathbb{D}_{\mathcal{E}_i}} \left(\left| \Theta_{M}^{\mathcal{D}_k} - \Theta_{M}^{\mathcal{D}_h} \right| \right)}$$

where $d_{\mathcal{D}_j}^{\Theta}$ is the disagree point of $\mathcal{D}_j$ in the $\mathbb{G}_{\mathcal{E}_i}^{\Gamma}$. Using (3.7.10), the solution of $\mathbb{G}_{\mathcal{E}_i}^{\Theta}$ is determined (Kim, 2022a).

3.7.6 Summary

The *BGRM* scheme addresses the challenges associated with computation and communication resource allocation within the PEC platform. To effectively manage these issues, the *BGRM* scheme explores an appropriate unification solution concept by incorporating various bargaining strategies. This approach adapts seamlessly to the dynamic conditions of the PEC system, ensuring efficient resource management in ever-changing environments. Based on the ides of *UniBS*, the $\mathbb{G}_{\mathcal{D}}^{\mathcal{E}}$, $\mathbb{G}_{\mathcal{E}}^{\Gamma}$ and $\mathbb{G}_{\mathcal{E}}^{\Theta}$ games are formulated to achieve the goal of maximizing the system performance. Then, these games work together interactively in a step-by-step cooperative manner to select their optimal strategies in an online fashion (Kim, 2022a).

3.8 Cloud Edge Offload Computing (*CEOC*) Scheme

Looking ahead, the swift expansion of mobile communications and cloud systems has significantly advanced the edge computing paradigm. While edge computing technology has garnered considerable attention, most existing research tends to focus on specific applications, often neglecting the control perspective of cloud providers that offer general-purpose edge services. Recently, S. Kim introduces the *Cloud Edge Offload Computing* (*CEOC*) scheme, which integrates various resource sharing challenges to optimize the benefits of offloading computing services (Kim, 2022d). In particular, the *CEOC* scheme emphasizes two cooperative game-theoretic solutions: the *Sim Bargaining Solution* and the *Interval Banzhaf Value*. These solutions are employed to efficiently allocate computing, communication, and cache resources within the cloud-edge system platform. The *CEOC* scheme formalizes two separate cooperative games that collaborate to address and reconcile conflicting requirements. Across a wide range of service request scenarios, the *CEOC* scheme strives to enhance resource orchestration by implementing dynamic workload balancing and optimizing cloud resource utilization. The main advantage of the *CEOC* scheme lies in its ability to determine optimal control decisions, thereby ensuring appropriate system operations and maximizing the overall efficiency of the cloud-edge infrastructure (Kim, 2022d).

3.8.1 Development Motivation

Thanks to recent advancements in hardware and software technologies, the proliferation of intelligent IoT devices has surged dramatically. According to Cisco, over 30 billion active IoT devices generate roughly 2.5 exabytes (EB) of data daily. With the increasing capabilities of smart IoT devices—such as smartphones, wearables, tablets, and unmanned aerial vehicles - there has been a significant rise in real-time, latency-sensitive, and interactive mobile applications. This expanding array of innovative and diverse applications has become a fundamental component of mobile users' entertainment and social interactions, thereby elevating expectations for immersive QoS. To fulfill the QoS demands of applications like virtual reality (VR), e-Health, intelligent identification, autonomous driving, and interactive gaming, ample computational capacity is essential. However, mobile IoT devices typically face resource constraints in terms of computational power and battery life (Ma, 2019; Kim, 2022d; Taleb, 2017; Wu, 2020b).

In response to the explosive growth of computation-intensive applications, there has been a notable evolution in computing paradigms. The most prominent and established among these is Cloud Computing (CC), a relatively recent paradigm recognized for its potential to significantly enhance computational performance. Generally, a cloud server serves as a pooled and centralized resource, making its capabilities accessible to end users via the Internet. Unlike traditional dedicated servers, cloud computing operates on a pay-as-you-go pricing model, offering scalable processing power and storage. This flexibility makes CC services particularly advantageous for workloads that are temporary or infrequently used, allowing resources to be scaled up or down based on demand. With its substantial potential for next-generation networks, cloud computing has become a highly researched topic aimed at accommodating the overwhelming number of smart IoT devices and appliances (Guim, 2021; Kim, 2022d; Ren, 2019; Zhang, 2020).

Despite the benefits offered by cloud servers, the centralized cloud platform is becoming less viable as new applications increasingly demand real-time processing. The traditional cloud approach requires a significant amount of network resources to facilitate data communications with centralized cloud servers, which can be a major limitation. To overcome this challenge, Mobile Edge Computing

(MEC) has been proposed as an alternative network architecture. The fundamental concept of MEC is to bring cloud computing capabilities closer to the edge of cellular networks, effectively migrating relevant processing tasks to network edge nodes - also known as MEC nodes. This strategy enhances the computational capabilities of IoT devices by eliminating the constraints of a centralized architecture, thereby positioning computing resources nearer to end users. Consequently, mobile users can access sufficient computational resources without experiencing significant Internet delays (Guim, 2021; Kim, 2022d; Ren, 2019; Zhang, 2020).

While Mobile Edge Computing (MEC) offers significant potential, its primary limitation lies in its restricted computational capacity compared to traditional Cloud Computing (CC) servers. Typically, the MEC infrastructure incurs high construction and operational costs due to its widespread deployment requirements. Additionally, the profitability of MEC systems remains low because computing demands are time-varying and fluctuating, resulting in underutilized computing resources. In contrast, centralized cloud servers benefit from lower average construction and operational costs thanks to their centralized nature and economies of scale. However, as previously mentioned, the rapid growth of computational applications on traditional CC servers poses significant challenges in meeting Quality of Service (QoS) requirements. To address this conflict, it has become increasingly clear that relying on a single computing paradigm is insufficient for the evolving needs of future networks (Kim, 2022d; Ren, 2019; Wu, 2020b; Zhang, 2020).

In recent years, extensive research has focused on the integrated design of cloud servers and MEC nodes. Leveraging their heterogeneous characteristics, a hybrid computing infrastructure can be established, enabling application tasks to be opportunistically processed by both MEC nodes and cloud servers. Through the use of core networks, MEC nodes and CC servers can be interconnected to share computing resources with minimal communication delays. This synergy allows them to complement each other, thereby enhancing both profitability and QoS. As a result, the coexistence of cloud and MEC environments has spurred considerable research interest in areas such as system architecture, data transmission, computing resource optimization, and operational efficiency. Depending on the type of service and the current network conditions, CC and MEC resources should be dynamically shared to ensure guaranteed service quality for bursty offloading computing tasks (Kim, 2022d; Ren, 2019; Wu, 2020b; Zhang, 2020).

The *CEOC* scheme is focused on establishing a robust Cloud-assisted Mobile Edge Computing (C-MEC) platform and developing an innovative C-MEC control mechanism to efficiently handle dynamic requests while maintaining cost-effectiveness. Given that the computational capacity of MEC nodes may be inadequate, directly impacting both system costs and QoS performance, the *CEOC* scheme seeks to address two critical questions for achieving cost-efficiency; i) what is the best solution for managing the limited resources within the C-MEC framework? and ii) how can the inherent selfish motivations of MEC nodes be aligned with desirable outcomes? The *CEOC* scheme tackles these challenges by exploring diverse QoS requirements. To dynamically manage task offloading requests, the *CEOC* scheme emphasizes the integration of compatible CC servers and MEC nodes, and it designs a hierarchical task processing protocol aimed at enhancing the overall efficiency of the C-MEC platform. Despite these advancements, guiding self-interested system agents towards a socially optimal outcome remains a significant and complex challenge to address (Kim, 2022d).

The cooperative game approach effectively encourages participants to collaborate, thereby maximizing social welfare through joint efforts. There are two primary methods to motivate players to cooperate: value solutions and bargaining solutions. Value-type solutions are extensively utilized in cooperative games as they adhere to principles such as *efficiency*, *symmetry*, the *null player property*, and *additivity*. These solutions focus on the equitable distribution of both gains and costs among coalition members, especially in scenarios where players contribute unequally. Despite differing contributions, each player collaborates to achieve a collective gain or payoff. A novel extension of the *Banzhaf value*, known as the *Interval Banzhaf Value* (*IBV*), has been introduced for interval cooperative games. The *IBV* is not only consistently defined across all possible interval games but also remains median-indifferent within the category of size monotonic interval games. Utilizing the interval subtraction operator, the *IBV* can be precisely calculated by considering only the lower and upper bounds of the coalitions' interval payoffs (Kim, 2014, 2022d; Liang, 2019; Tan, 2020).

Conversely, bargaining-based solutions concentrate on the distribution characteristics of cooperative outcomes, specifically focusing on how the benefits from negotiations are divided among players. These solutions aim to optimize social welfare while ensuring that surplus payoffs are distributed fairly. The foundational concepts of bargaining solutions were introduced by J. Nash in his seminal papers from 1950 and 1953. Nash's ideas were further expanded through asymmetric bargaining models, which account for the varying preferences and bargaining powers of different players. These models are instrumental in maximizing payoffs and allocating limited resources to players who require them the most, thereby influencing the final bargaining outcome. Building on Nash's framework, the *Sim Bargaining Solution* (*SBS*) was developed to address bargaining scenarios involving multiple buyer-seller pairs. The *SBS* emphasizes the distinct characteristics and individual contributions of each player, ensuring that each receives a reasonable share. In cooperative settings, the *SBS* effectively distinguishes the attributes of game players by structuring their bargaining powers, thereby quantifying their preferences and contributions (Chen, 2021; Kim, 2014, 2022d; Sim, 2022).

3.8.2 Contribution

The primary objective of the *CEOC* scheme is to introduce a novel mechanism for allocating computing, communication, and cache resources within the C-MEC platform. To efficiently distribute these scarce resources, the *CEOC* scheme integrates the fundamental principles of *IBV* and *SBS*, thereby formulating two distinct cooperative games. These games are designed to address fluctuations in task offloading, whether they are temporal or spatial in nature, by examining the strategic interactions between the CC server and the MEC nodes. Through this approach, the *CEOC* scheme is able to significantly enhance the overall performance of the C-MEC system, ensuring improved system throughput, greater fairness among MEC nodes, and increased payoffs for end devices (Kim, 2022d).

3.8.3 C-MEC Platform Architecture and Cooperative Game Models

As illustrated in Figure 3, the *CEOC* scheme considers a cloud-edge collaboration platform with one centralized CC server ($\mathfrak{P}$) and n single antenna base stations (BSs), denoted by a set $\mathbb{B} = \{\mathcal{B}_1, \mathcal{B}_2, \ldots, \mathcal{B}_n\}$. Each BS is equipped with an MEC server, which has limited resource for data processing. Let $\mathbb{M} = \left\{\mathcal{M}_{\mathcal{B}_1}, \mathcal{M}_{\mathcal{B}_2}, \ldots, \mathcal{M}_{\mathcal{B}_n}\right\}$ denote the set of MEC nodes, and $\mathcal{M}_{\mathcal{B}_i}$ is embedded in the $\mathcal{B}_i$. Within the coverage area of $\mathcal{B}_{1 \le i \le n}$, there are multiple IoT devices, denoted by a set $\mathbb{D}_{\mathcal{B}_i} = \left\{\ldots \mathcal{D}^k_{\mathcal{B}_i} \ldots\right\}$, which generate different multimedia application tasks to be processed. Each individual $\mathcal{D}^k_{\mathcal{B}_i}$ is connected to its corresponding $\mathcal{M}_{\mathcal{B}_i}$ through wireless channels, and can offload its computing task to the $\mathcal{M}_{\mathcal{B}_i}$. The computation capacity of the $\mathcal{M}_{\mathcal{B}_i}$ is defined as $\mathcal{C}^{\mathcal{M}_{\mathcal{B}_i}}$, which are measured by the number of CPU cycles per second. To denote the $\mathcal{D}^k_{\mathcal{B}_i}$'s offloading task, the *CEOC* scheme uses a three-field notation $\mathcal{T}^k_{\mathcal{B}_i} = \left\{ m^k_{\mathcal{B}_i}, \mathfrak{M}^k_{\mathcal{B}_i}, c^k_{\mathcal{B}_i} \right\}$ where $m^k_{\mathcal{B}_i}$ and $\mathfrak{M}^k_{\mathcal{B}_i}$ represent the minimum and maximum bandwidth amounts for processing the $\mathcal{T}^k_{\mathcal{B}_i}$ task, and $c^k_{\mathcal{B}_i}$ denotes the number of CPU cycles that are required to compute this task (Kim, 2022d; Ren, 2019).

Figure 3. The infrastructure of the cloud assisted MEC (Kim, 2022d)

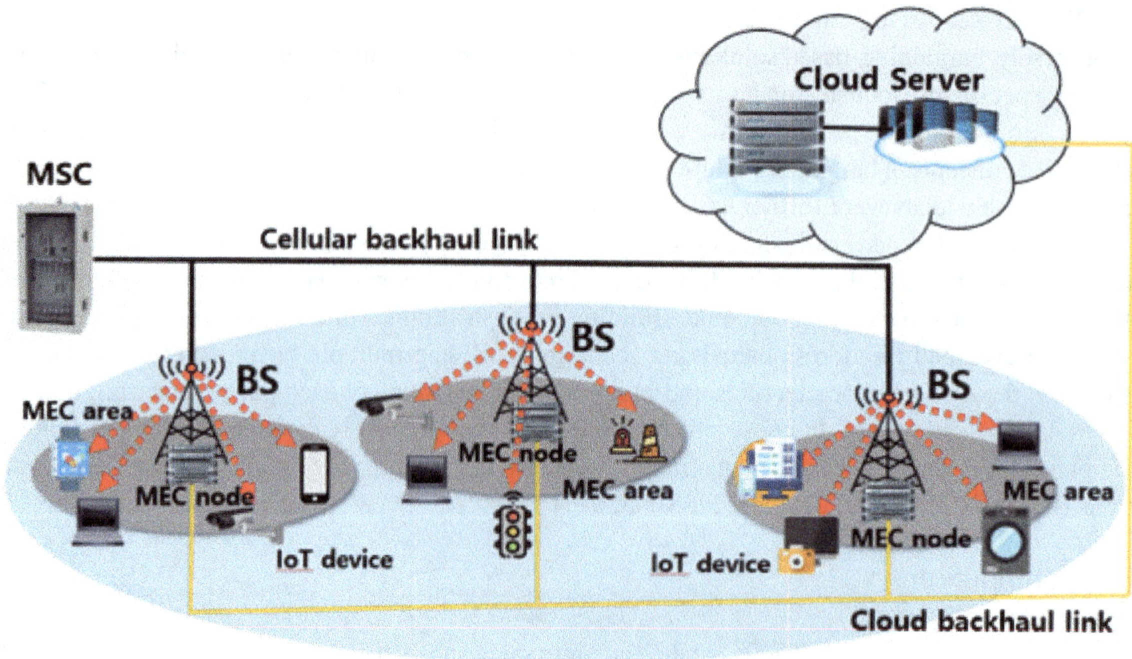

Typically, individual IoT devices are unable to handle their tasks directly due to their limited computational capabilities. As a result, these computational tasks are initially offloaded to their respective MEC nodes. Within the C-MEC platform, the *CEOC* scheme operates under the assumption that P is connected to multiple MEC nodes via cloud backhaul links. This setup allows each computation task to be processed concurrently at both the MEC node and the CC server. To ensure effective computation offloading services, MEC nodes have the capability to partially transmit tasks to the cloud server. Consequently, each MEC node must decide whether to handle the tasks independently or to collaborate with the cloud server for processing. If the assistance of the cloud service is required, individual MEC nodes determine the proportion of tasks that should be processed by the CC server. At this time, each MEC node attempts to share adaptively the P's computation ($\mathcal{C}^{\mathfrak{P}}$), communication ($\mathfrak{R}^{\mathfrak{P}}$) and cache ($\mathcal{Q}^{\mathfrak{P}}$) resources to ensure contradictory requirements, and the $\mathcal{C}^{\mathfrak{P}}$ and $\mathcal{C}^{\mathcal{M}}$ are dynamically allocated to the mobile devices by the virtual machine technique (Kim, 2022d; Ren, 2019).

The *CEOC* scheme discusses the functional capabilities of C-MEC system to provide a better QoS for end users. To effectively share the $\mathcal{C}^{\mathfrak{P}}$, $\mathcal{Q}^{\mathfrak{P}}$ and $\mathfrak{R}^{\mathfrak{P}}$ resources, the *CEOC* scheme employs the cooperative game theory. Under dynamically changing C-MEC system environments, the *CEOC* scheme approach is designed as a joint bargaining game model; control decisions are made by collaboration of system agents according to the ideas of *IBV* and *SBS*. Mathematically, the *CEOC* scheme formulates two bargaining games, i.e., $\mathbb{G}_{\mathscr{C}}$ and $\mathbb{G}_{\mathfrak{R}}^{\mathcal{Q}}$, at each time period. In the $\mathbb{G}_{\mathscr{C}}$, the MEC nodes bargain with the P to share the $\mathscr{C}^{\mathfrak{P}}$ resource according to the idea of *SBS*. In the $\mathbb{G}_{\mathfrak{R}}^{\mathcal{Q}}$, the MEC nodes bargain with each other to allocate the $\mathcal{Q}^{\mathfrak{P}}$ and $\mathfrak{R}^{\mathfrak{P}}$ resources based on the concept of *IBV*. The $\mathbb{G}_{\mathscr{C}}$ and $\mathbb{G}_{\mathfrak{R}}^{\mathcal{Q}}$ are operated sequentially in a step-by-step interactive fashion. It is noteworthy that the *CEOC* scheme formulates the P-M association in a cooperative manner. Formally, the *CEOC* scheme defines the $\mathbb{G}_{\mathscr{C}}$ and $\mathbb{G}_{\mathfrak{R}}^{\mathcal{Q}}$ game entities, i.e.,

$$\mathbb{G} = \left\{\mathbb{G}_{\mathscr{C}},\ \mathbb{G}_{\mathfrak{R}}^{\mathcal{Q}}\right\} = \left\{\mathfrak{P}, \mathbb{B}, \mathbb{M}, \mathbb{D}_{\mathscr{B}_{1\le i\le n}}, \left\{\mathbb{G}_{\mathscr{C}} | \mathscr{C}^{\mathfrak{P}}, \mathbb{M}, p_{\mathscr{M}_{\mathscr{B}_i}}^{\mathfrak{P}}, \left(x_{\mathscr{M}_{\mathscr{B}_i}}, x_{\mathfrak{P}}\right), U_{\mathscr{B}_i}^{\mathscr{M}}(\cdot), U_{\mathscr{B}_i}^{\mathfrak{P}}(\cdot), \mathfrak{U}_{\mathscr{M}_{\mathscr{B}_i}}(\cdot), \mathfrak{U}_{\mathfrak{P}}(\cdot)\right\}, \left\{\mathbb{G}_{\mathfrak{R}}^{\mathcal{Q}} | \mathcal{Q}^{\mathfrak{P}}, \mathfrak{R}^{\mathfrak{P}}, \mathbb{M}, \mathcal{Q}_{\mathscr{B}_i}, \mathfrak{R}_{\mathscr{B}_i}, \Theta_{\mathscr{B}_i}^{\mathscr{M}}\right\}, T\right\}$$

of gameplay (Kim, 2022d).

- P, B and M represent the cloud server, base stations, and MEC nodes, respectively. They are mutually and reciprocally interdependent in a coordinated manner, and they work together to share the limited C-MEC system resources.
- In the $\mathbb{G}_{\mathscr{C}}$, all MEC nodes in the M participate in the $\mathbb{G}_{\mathscr{C}}$, sequentially, and each $\mathscr{M}_{\mathscr{B}_{1\le i\le n}}, \mathfrak{P}$ pair is game players to share the $\mathscr{C}^{\mathfrak{P}}$. $p_{\mathscr{M}_{\mathscr{B}_i}}^{\mathfrak{P}}$ is the CC service price, and $x_{\mathscr{M}_{\mathscr{B}_i}}$ and $x_{\mathfrak{P}}$ are allocated computation offloading capacities to the $\mathscr{M}_{\mathscr{B}_i}$ and P, respectively.
- In the $\mathbb{G}_{\mathscr{C}}$, $U_{\mathscr{B}_i}^{\mathscr{M}}(\cdot)$ and $U_{\mathscr{B}_i}^{\mathfrak{P}}(\cdot)$ are utility functions of $\mathscr{M}_{\mathscr{B}_i}$ and P to decide the $p_{\mathscr{M}_{\mathscr{B}_i}}^{\mathfrak{P}}$. $\mathfrak{U}_{\mathscr{M}_{\mathscr{B}_i}}(\cdot)$ and $\mathfrak{U}_{\mathfrak{P}}(\cdot)$ are utility functions of $\mathscr{M}_{\mathscr{B}_i}$ and P to decide their offloading computation amount.
- In the $\mathbb{G}_{\mathfrak{R}}^{\mathcal{Q}}$, all MEC nodes in the M are game players, and $\mathcal{Q}_{\mathscr{B}_i}$ and $\mathfrak{R}_{\mathscr{B}_i}$ are the $\mathscr{M}_{\mathscr{B}_i}$'s strategies to share the $\mathcal{Q}^{\mathfrak{P}}$ and $\mathfrak{R}^{\mathfrak{P}}$, respectively. $\Theta_{\mathscr{B}_i}^{\mathscr{M}}$ is the $\mathscr{M}_{\mathscr{B}_i}$'s interval value, and it is the $\mathscr{M}_{\mathscr{B}_i}$'s payoff.
- Discrete time model $T \in \{t_1, \ldots, t_c, t_{c+1}, \ldots\}$ is represented by a sequence of time steps. The length of t_c matches the event time-scale of $\mathbb{G}_{\mathscr{C}}$ and $\mathbb{G}_{\mathfrak{R}}^{\mathcal{Q}}$.

3.8.4 Fundamental concepts of *IBV* and *SBS*

Usually, a cooperative game in coalitional form is an ordered pair N, ν where $N = \{1, 2, \ldots, n\}$ is the set of players, and $v: 2^N \rightarrow \mathrm{R}$ is a map where R is the set of real numbers. The v assigns each coalition $S \in 2^N$ to a real number, such that $v(\varnothing) = 0$. For cooperative interval games, an interval is expressed as $\overline{\alpha} = [\alpha_L, \alpha_U]$ where $\alpha_L, \alpha_U \in \mathbb{R}$ are the lower and upper bounds of the interval $\overline{\alpha}$, respectively. Some interval arithmetic are summarized as follows. Let: $\overline{\alpha} = [\alpha_L, \alpha_U]$ and $\overline{\beta} = [\beta_L, \beta_U]$ be two intervals: i) $\overline{\alpha} = \overline{\beta}$ if and only if $\alpha_L = \beta_L$ and $\alpha_U = \beta_U$, ii) $\overline{\alpha} + \overline{\beta} = [\alpha_L + \beta_L,\ \alpha_U + \beta_U]$, iii) if $\gamma \geq 0$, then $\gamma \cdot \overline{\alpha} = [\gamma \cdot \alpha_L, \gamma \cdot \alpha_U]$. A n-person interval cooperative game is expressed with an ordered-pair $N, \overline{\nu}$, where $\overline{\nu}$ is the interval characteristic function of players' coalitions and $\overline{\nu}(\varnothing) = [0,0]$. Generally, for any coalition $S \subseteq N$, $\overline{\nu}(S)$ is denoted by the interval $[\nu_L(S), \nu_U(S)]$, where $\nu_L(S) \leq \nu_U(S)$ (Kim, 2022d; Liang, 2019; Osman, 2015; Tan, 2020).

In the following, $N, \overline{\nu}$ is simply called the interval cooperative game $\overline{\nu}$, and the set of n-person interval cooperative games are denoted by $\overline{G}^n$. For any coalition $S \subseteq N$ and $\overline{\upsilon}, \overline{\nu} \in \overline{G}^n$, the interval characteristic function of $\overline{\upsilon} + \overline{\nu}$ is $(\overline{\upsilon} + \overline{\nu})(S) = \overline{\upsilon}(S) + \overline{\nu}(S) = [\upsilon_L(S) + \nu_L(S), \upsilon_U(S) + \nu_U(S)]$. For any interval cooperative game $\overline{\upsilon} \in \overline{G}^n$, the *CEOC* scheme can explicitly obtain its *IBV*, i.e., $IBV^N(\overline{\upsilon}) = \left(IBV_1^N(\overline{\upsilon}), \ldots, IBV_n^N(\overline{\upsilon})\right)$, whose components are given as follows (Kim, 2022d; Liang, 2019; Osman, 2015; Tan, 2020):

$$IB\,V^N_{1\le i\le n}(\overline{v}) = \frac{1}{2^{n-1}}\sum_{\substack{i\notin S\subseteq N\\ i\in N}} (\overline{v}(S\cup i) - \overline{v}(S))$$

(3.8.1)

$$= \left[\frac{1}{2^{n-1}}\sum\nolimits_{S\subseteq N\setminus i}\left(v_L(S\cup i) - v_L(S)\right), \frac{1}{2^{n-1}}\sum\nolimits_{S\subseteq N\setminus i}\left(v_U(S\cup i) - v_U(S)\right)\right]$$

For example, let $\langle N,\overline{v}\rangle$ be a cooperative interval game with $N = \{1, 2, 3\}$ and $\overline{v}(1) = \overline{v}(13) = [7, 7]$,$\overline{v}(12) = [12, 17]$,$\overline{v}(123) = [24, 29]$ and $\overline{v}(S) = [0,0]$ otherwise. Then, $IB\,V^N_1(\overline{v})$ is calculated as follows (Kim, 2022d; Osman, 2015):

$$IB\,V^N_1(\overline{v}) = \frac{1}{2^2}\sum_{1\in S}(\overline{v}(S) - \overline{v}(S\setminus 1))$$

(3.8.2)

$$= \frac{1}{2^2}(\overline{v}(1) + \overline{v}(12) - \overline{v}(2) + \overline{v}(13) - \overline{v}(3) + \overline{v}(123) - \overline{v}(23))$$

$$= \frac{1}{2^2}([7,\ 7] + [12,\ 17] - [0,0] + [7,\ 7] - [0,0] + [24,29] -)[0,0]$$

$$= \frac{1}{4}([50,60]) = \left[12\tfrac{1}{2}, 15\right]$$

The others can be calculated similarly: $IB\,V^N_2(\overline{v}) = \left[5\frac{1}{2}, 8\right]$, and $IB\,V^N_3(\overline{v}) = [3,3]$. Finally, the *IBV* can be shown as (Kim, 2022d; Osman, 2015).

$$IB\,V^N(\overline{v}) = \left(IB\,V^N_1(\overline{v}), IB\,V^N_2(\overline{v}), IB\,V^N_3(\overline{v})\right) = \left(\left[12\tfrac{1}{2},\ 15\right], \left[5\tfrac{1}{2},\ 8\right], [3,\ 3]\right)$$

(3.8.3)

To characterize the fundamental idea of *SBS* for a two-player bargaining problem, let $S \subset \mathbb{R}^2$ be a nonempty and finite set where $\mathbb{R}^2$ is the two-dimensional Euclidean space. A pair (S, d) is a bargaining game with $d \in S \subset \mathbb{R}^2$ where $d \le u \in S.u = \left(U_1(x_1), U_2(x_2)\right)$ is players' payoff pair utility where $U_1(x_1)$ is the level of satisfaction of player 1 for obtaining x_1. The $d = (d_1, d_2)$ is a disagreement pair, which is expected to be the result if two players cannot reach an agreement. It represents the worst outcome for each player in the cooperative game. Among the set of possible payoff pairs obtainable through agreement, the *SBS* maximizes the weighted geometric average of the payoffs of all players with $\alpha \in (0,1)$ (Kim, 2022d).

$$SBS(S,\ d) = \underset{(d_1,d_2)<\left(U_1(x_1),U_2(x_2)\right)}{\mathbf{argmax}} \left[\left(U_1(x_1) - d_1\right)^{\alpha} \times \left(U_2(x_2) - d_2\right)^{1-\alpha}\right] \quad (3.8.4)$$

$$\text{s.t.,}\begin{cases} \varphi = x_1 + x_2 \\ U_1(x_1) \ : = f_1^{(\varphi)}(x_1) = f_1^{(\varphi)}(\varphi - x_2) \\ U_2(x_2) \ : = f_2^{(\varphi)}(x_2) = f_2^{(\varphi)}(\varphi - x_1) \end{cases}$$

3.8.5 The C-MEC resource allocation algorithm

In the C-MEC system infrastructure, each individual IoT device generates its computation application $\left(\mathcal{T}_{\mathcal{B}}^{k}\right)$, and offloads them to its corresponding MEC node. Usually, each $\mathcal{T}_{\mathcal{B}}^{k}$ requests computing $\left(c_{\mathcal{B}}^{k}\right)$ and communication $\left(m_{\mathcal{B}}^{k}, \mathfrak{M}_{\mathcal{B}}^{k}\right)$resources for offloading services. By collecting these tasks, each MEC node estimates the total requested computation amount $\left(\varphi_{\mathcal{M}_{\mathcal{B}_i}}^{\mathfrak{P}}\right)$, and lower $\left(\theta_{\mathcal{M}_{\mathcal{B}_i}}^{L}\right)$ and upper $\left(\theta_{\mathcal{M}_{\mathcal{B}_i}}^{U}\right)$ bandwidth requests. In the $\mathcal{M}_{\mathcal{B}_i}$, they are calculated as follows (Kim, 2022d):

$$\varphi_{\mathcal{M}_{\mathcal{B}_i}}^{\mathfrak{P}} = \sum_{\mathcal{T}_{\mathcal{B}_i}^{k} \in \mathbb{D}_{\mathcal{B}_i}} c_{\mathcal{B}_i}^{k},\ \theta_{\mathcal{M}_{\mathcal{B}_i}}^{L} = \sum_{\mathcal{T}_{\mathcal{B}_i}^{k} \in \mathbb{D}_{\mathcal{B}_i}} m_{\mathcal{B}_i}^{k} \text{ and } \theta_{\mathcal{M}_{\mathcal{B}_i}}^{U} = \sum_{\mathcal{T}_{\mathcal{B}_i}^{k} \in \mathbb{D}_{\mathcal{B}_i}} \mathfrak{M}_{\mathcal{B}_i}^{k} \tag{3.8.5}$$

When the $\mathcal{M}_{\mathcal{B}_i}$ offloads its computing task to the P, the $\mathfrak{R}^{\mathfrak{P}}$ resource is shared among MEC nodes based on the requested interval information, i.e., $\overline{\Theta_{\mathcal{M}_{\mathcal{B}_i}}} = \left(\theta_{\mathcal{M}_{\mathcal{B}_i}}^{L}, \theta_{\mathcal{M}_{\mathcal{B}_i}}^{U}\right)$. For each $\mathcal{M}_{\mathcal{B}_i} \in \mathbb{M}$, the division problem on $\mathfrak{R}^{\mathfrak{P}}$ with sharp claim $\overline{\Theta_{\mathcal{M}_{\mathcal{B}_i}}}$ is formulated as a bankruptcy problem where $0 < \mathfrak{R}^{\mathfrak{P}} \leq \sum_{\mathcal{M}_{\mathcal{B}_i} \in \mathbb{M}} \theta_{\mathcal{M}_{\mathcal{B}_i}}^{L} \leq \sum_{\mathcal{M}_{\mathcal{B}_i} \in \mathbb{M}} \theta_{\mathcal{M}_{\mathcal{B}_i}}^{U}$. Let $\Theta = \left(\ldots \overline{\Theta_{\mathcal{M}_{\mathcal{B}_i}}} \ldots\right)$ is a vector of all claimants' interval, and all bankruptcy problems are denoted by the form $(\mathfrak{R}^{\mathfrak{P}}, \Theta)$. According to the traditional division problem with point claims, the *CEOC* scheme can consider an associated division game $\mathbb{M}, \upsilon_{\mathfrak{R}^{\mathfrak{P}},\Theta}$ based on the $(\mathfrak{R}^{\mathfrak{P}}, \Theta)$. Let F be the family of closed intervals in $\mathbb{R}_{+}$, and the characteristic function $\upsilon_{\mathfrak{R}^{\mathfrak{P}},\Theta}(S)$: $2^{\mathbb{M}} \rightarrow \mathbb{F}$ for each $S \in 2^{\mathbb{M}}$ is defined the below equation (Branzei, 2003; Kim, 2022d);

$$\upsilon_{\mathfrak{R}^{\mathfrak{P}},\Theta}(S) = \left[\left(\mathfrak{R}^{\mathfrak{P}} - \sum_{\mathcal{M}_{\mathcal{B}_i} \in \mathbb{M} \setminus S} \theta_{\mathcal{M}_{\mathcal{B}_i}}^{U}\right)_{+}, \left(\mathfrak{R}^{\mathfrak{P}} - \sum_{\mathcal{M}_{\mathcal{B}_i} \in \mathbb{M} \setminus S} \theta_{\mathcal{M}_{\mathcal{B}_i}}^{L}\right)_{+}\right] \tag{3.8.6}$$

$$\text{s.t., } (x)_{+} = \max(x, 0)$$

In the $\mathbb{M}, \upsilon_{\mathfrak{R}^{\mathfrak{P}},\Theta}$, each coalition S can receive an interval value which is between the amounts of $\mathfrak{R}^{\mathfrak{P}}$ that are left if all MEC nodes outside coalition S receive their upper and lower claims. According to (3.8.6), the $IBV^{\mathbb{M}}\left(\upsilon_{\mathfrak{R}^{\mathfrak{P}},\Theta}\right)$ of a game $\mathbb{M}, \upsilon_{\mathfrak{R}^{\mathfrak{P}},\Theta}$ is obtained as follows (Branzei, 2003; Kim, 2022d);

$$IBV^{\mathbb{M}}\left(\upsilon_{\mathfrak{R}^{\mathfrak{P}},\Theta}\right) = \left(IBV_{\mathcal{M}_{\mathcal{B}_i} \in \mathbb{M}} = \left(\Gamma_{\mathcal{M}_{\mathcal{B}_i}}^{L}, \Gamma_{\mathcal{M}_{\mathcal{B}_i}}^{U}\right), \ldots, IBV_{\mathcal{M}_{\mathcal{B}_i} \in \mathbb{M}} = \left(\Gamma_{\mathcal{M}_{\mathcal{B}_i}}^{L}, \Gamma_{\mathcal{M}_{\mathcal{B}_i}}^{U}\right)\right) \tag{3.8.7}$$

$$= \frac{1}{2^{|\mathbb{M}|-1}} \times \sum_{\mathcal{M}_{\mathcal{B}_i} \in S} \left(\upsilon_{\mathfrak{R}^{\mathfrak{P}},\Theta}(S) - \upsilon_{\mathfrak{R}^{\mathfrak{P}},\Theta}\left(S \setminus \left\{\mathcal{M}_{\mathcal{B}_i}\right\}\right)\right)$$

To effectively share the $\mathfrak{R}^{\mathfrak{P}}$ and $\mathcal{Q}^{\mathfrak{P}}$ resources in the P, the *CEOC* scheme designs the resource sharing algorithm by using the concept of *IBV*. According to (3.8.6) and (3.8.7), the *IBV* of each $\mathcal{M}_{\mathcal{B}_i}$ is obtained, and the distribution amounts of $\mathfrak{R}^{\mathfrak{P}}$ and $\mathcal{Q}^{\mathfrak{P}}$ for the $\mathcal{M}_{\mathcal{B}_i}$, i.e, $\chi_{\mathcal{M}_{\mathcal{B}_i}}^{\mathfrak{R}}$ and $\chi_{\mathcal{M}_{\mathcal{B}_i}}^{\mathcal{Q}}$, are estimated as follows (Kim, 2022d):

$$\chi^{\mathfrak{R}}_{\mathcal{M}_{\mathcal{B}_i}} = \frac{\Gamma^{L}_{\mathcal{M}_{\mathcal{B}_i}}}{\sum_{\mathcal{M}_{\mathcal{B}_i}\in \mathrm{M}} \Gamma^{L}_{\mathcal{M}_{\mathcal{B}_i}}} \times \mathfrak{R}^{\mathfrak{P}} \text{ and } \chi^{\mathcal{Q}}_{\mathcal{M}_{\mathcal{B}_i}} = \frac{\Gamma^{U}_{\mathcal{M}_{\mathcal{B}_i}}}{\sum_{\mathcal{M}_{\mathcal{B}_i}\in \mathrm{M}} \Gamma^{U}_{\mathcal{M}_{\mathcal{B}_i}}} \times \mathcal{Q}^{\mathfrak{P}} \tag{3.8.8}$$

In the C-MEC system infrastructure, each MEC node can partially offload its computation tasks to the CC server. To offload computation tasks, the *CEOC* scheme formulates a two-player bargaining game $\mathbb{G}_{\mathscr{C}}$, and adopt the *SBS* concept. In the $\mathbb{G}_{\mathscr{C}}$, each pair of $\mathcal{M}_{\mathcal{B}_{1\le i\le n}}$ in M and P are game players, and the *CEOC* scheme first decides the CC service price $\left(p^{\mathfrak{P}}_{\mathcal{M}_{\mathcal{B}_i}}\right)$ to effectively share the $\mathscr{C}^{\mathfrak{P}}$ resource. For this decision, the utility functions of $\mathcal{M}_{\mathcal{B}_i}$ and P, i.e., $U^{\mathcal{M}}_{\mathcal{B}_i}$ and $U^{\mathfrak{P}}_{\mathcal{B}_i}$, are defined as follows (Kim, 2022d):

$$\begin{cases} U^{\mathcal{M}}_{\mathcal{B}_i}\left(\varphi^{\mathfrak{P}}_{\mathcal{M}_{\mathcal{B}_i}}, p^{\mathfrak{P}}_{\mathcal{M}_{\mathcal{B}_i}}, \mathscr{C}^{\mathfrak{P}}_{t_c}\right) = \varphi^{\mathfrak{P}}_{\mathcal{M}_{\mathcal{B}_i}} \times exp\left(-\sigma \times p^{\mathfrak{P}}_{\mathcal{M}_{\mathcal{B}_i}}\right) \\ U^{\mathfrak{P}}_{\mathcal{B}_i}\left(\varphi^{\mathfrak{P}}_{\mathcal{M}_{\mathcal{B}_i}}, p^{\mathfrak{P}}_{\mathcal{M}_{\mathcal{B}_i}}, \mathscr{C}^{\mathfrak{P}}_{t_c}\right) = \varphi^{\mathfrak{P}}_{\mathcal{M}_{\mathcal{B}_i}} \times \left(\omega \times \left(\frac{\mathscr{C}^{\mathfrak{P}}_{t_c}}{\mathscr{C}^{\mathfrak{P}}}\right) \times \log\left(p^{\mathfrak{P}}_{\mathcal{M}_{\mathcal{B}_i}} + \gamma\right)\right) \end{cases} \tag{3.8.9}$$

where $\varphi^{\mathfrak{P}}_{\mathcal{M}_{\mathcal{B}_i}}$ is the total offloading computation amount in the $\mathcal{M}_{\mathcal{B}_i}$, and $\mathscr{C}^{\mathfrak{P}}_{t_c}$ is the P's current amount for offloading services. σ is a control parameter for the $U^{\mathcal{M}}_{\mathcal{B}_i}(\cdot)$, and ω, γ are control parameters for the $U^{\mathfrak{P}}_{\mathcal{B}_i}(\cdot)$. In the *CEOC* scheme, each $\mathcal{M}_{\mathcal{B}_i}, \mathfrak{P}$ pair negotiates over how to adjusts the service price $\left(p^{\mathfrak{P}}_{\mathcal{M}_{\mathcal{B}_i}}\right)$ by employing the concept of *SBS* (Kim, 2022d).

$$p^{\mathfrak{P}}_{\mathcal{M}_{\mathcal{B}_i}} = \underset{p^m \le p^{\mathfrak{P}}_{\mathcal{M}_{\mathcal{B}_i}} \le p^M}{\mathbf{argmax}} \left[\left(U^{\mathcal{M}}_{\mathcal{B}_i}(\cdot) - d^{p}_{\mathcal{M}_{\mathcal{B}_i}}\right) \times \left(U^{\mathfrak{P}}_{\mathcal{B}_i}(\cdot) - d^{p}_{\mathfrak{P}}\right)\right] \tag{3.8.10}$$

where p^m, p^M are the minimum and maximum price, respectively. According to (3.8.10), the price $p^{\mathfrak{P}}_{\mathcal{M}_{\mathcal{B}_i}}$ is dynamically decided. Based on the $p^{\mathfrak{P}}_{\mathcal{M}_{\mathcal{B}_i}}$, the $\varphi^{\mathfrak{P}}_{\mathcal{M}_{\mathcal{B}_i}}$ is split into two parts $x_{\mathcal{M}_{\mathcal{B}_i}}$ and $x_{\mathfrak{P}}$, which are finally assigned to the $\mathcal{M}_{\mathcal{B}_i}$ and P, respectively. This decision is also made based on the *SBS*. Finally, the *SBS* for the offloading services between $\mathcal{M}_{\mathcal{B}_i}$ and P, i.e., $SBS^{\mathfrak{P}}_{\mathcal{M}_{\mathcal{B}_i}}$, is given by (Kim, 2022d):

$$SBS^{\mathfrak{P}}_{\mathcal{M}_{\mathcal{B}_i}} = \underset{\left(x_{\mathcal{M}_{\mathcal{B}_i}}, x_{\mathfrak{P}}\right)}{\mathbf{argmax}} \left[\left(\mathfrak{U}_{\mathcal{M}_{\mathcal{B}_i}}(\cdot) - d^{x}_{\mathcal{M}_{\mathcal{B}_i}}\right) \times \left(\mathfrak{U}_{\mathfrak{P}}(\cdot) - d^{x}_{\mathfrak{P}}\right)\right] \tag{3.8.11}$$

$$
\text{s.t.,}\begin{cases}
\varphi^{\mathfrak{P}}_{\mathcal{M}_{\mathcal{B}_i}} = x_{\mathcal{M}_{\mathcal{B}_i}} + x_{\mathfrak{P}} \\
\mathfrak{U}_{\mathcal{M}_{\mathcal{B}_i}}(\cdot) = \mathfrak{U}_{\mathcal{M}_{\mathcal{B}_i}}\left(x_{\mathcal{M}_{\mathcal{B}_i}}, p^{\mathfrak{P}}_{\mathcal{M}_{\mathcal{B}_i}}\right) \; := x_{\mathcal{M}_{\mathcal{B}_i}} \times \left(\Gamma - \exp\left(-\left(p^{\mathfrak{P}}_{\mathcal{M}_{\mathcal{B}_i}} \times \frac{x_{\mathcal{M}_{\mathcal{B}_i}}}{\varphi^{\mathfrak{P}}_{\mathcal{M}_{\mathcal{B}_i}}}\right)\right)\right) \\
\mathfrak{U}_{\mathfrak{P}}(\cdot) = \mathfrak{U}_{\mathfrak{P}}\left(x_{\mathfrak{P}}, p^{\mathfrak{P}}_{\mathcal{M}_{\mathcal{B}_i}}\right) \; := x_{\mathfrak{P}} \times \left(\frac{\eta}{\eta + \exp\left(-\left(\frac{\theta}{p^{\mathfrak{P}}_{\mathcal{M}_{\mathcal{B}_i}}} \times \frac{x_{\mathfrak{P}}}{\varphi^{\mathfrak{P}}_{\mathcal{M}_{\mathcal{B}_i}}}\right)\right)} - \varrho\right)
\end{cases}
$$

where Γ is an adjustment factor for the $\mathfrak{U}_{\mathcal{M}_{\mathcal{B}_i}}(\cdot)$, and η, θ, ϱ are adjustment factors for the $\mathfrak{U}_{\mathfrak{P}}(\cdot)$. According to (3.8. 11), each individual MEC nodes sequentially bargains with the P, and split its $\varphi^{\mathfrak{P}}_{\mathcal{M}_{\mathcal{B}}}$ based on the current C-MEC system condition (Kim, 2022d).

3.8.6 Summary

Today, IoT devices are ubiquitous, being integrated into a wide array of applications and environments. Within this landscape, the cloud-assisted edge computing approach stands out as the most sophisticated computing paradigm. The *CEOC* scheme tackles the primary challenges associated with task offloading services in the C-MEC environments. In these settings, the CC server and MEC nodes collaborate to optimize overall system performance. To determine the optimal distribution of communication, cache, and computational resources within the C-MEC infrastructure, the *CEOC* scheme frames the resource allocation issues as cooperative games involving both the CC server and the MEC nodes. For the allocation of CC computing resources, the *CEOC* scheme utilizes the *SBS* concept. In this process, the CC server and each individual MEC node engage in sequential negotiations to arrive at a mutually beneficial solution. Additionally, to efficiently allocate the communication and cache resources of the CC server among multiple MEC nodes, the *CEOC* scheme employs the *IBV* concept. When addressing various offloading service requests, the two developed cooperative game models interact in a coordinated manner, facilitating a 'win-win' solution for all parties involved (Kim, 2022d).

REFERENCES

Ai, Y., Wang, L., Han, Z., Zhang, P., & Hanzo, L. (2018). Social networking and caching aided collaborative computing for the Internet of Things. *IEEE Communications Magazine*, *56*(12), 149–155. DOI: 10.1109/MCOM.2018.1701089

Akhtar, M. W., Hassan, S. A., Ghaffar, R., Jung, H., Garg, S., & Hossain, M. S. (2020). The shift to 6G communications: Vision and requirements. *Human-centric Computing and Information Sciences*, *10*(1), 1–27. DOI: 10.1186/s13673-020-00258-2

Alparslan Gök, S. Z., Branzei, R., & Tijs, S. (2010). The interval Shapley value: An axiomatization. *Central European Journal of Operations Research*, *18*(2), 131–140. DOI: 10.1007/s10100-009-0096-0

Andreev, S., Galinina, O., Pyattaev, A., Hosek, J., Masek, P., Yanikomeroglu, H., & Koucheryavy, Y. (2016). Exploring synergy between communications, caching, and computing in 5G-grade deployments. *IEEE Communications Magazine*, *54*(8), 60–69. DOI: 10.1109/MCOM.2016.7537178

Apostolopoulos, P. A., Tsiropoulou, E. E., & Papavassiliou, S. (2020). Risk-aware data offloading in multi-server multi-access edge computing environment. *IEEE/ACM Transactions on Networking*, *28*(3), 1405–1418. DOI: 10.1109/TNET.2020.2983119

Balakrishnan, P. S., Gómez, J. C., & Vohra, R. V. (2011). The tempered aspirations solution for bargaining problems with a reference point. *Mathematical Social Sciences*, *62*(3), 144–150. DOI: 10.1016/j.mathsocsci.2011.09.003

Branzei, R., Branzei, O., Alparslan Gök, S. Z., & Tijs, S. (2010). Cooperative interval games: A survey. *Central European Journal of Operations Research*, *18*(3), 397–411. DOI: 10.1007/s10100-009-0116-0

Branzei, R., Dimitrov, D., & Tijs, S. (2003). Shapley-like values for interval bankruptcy games. *Economic Bulletin*, *3*(9), 1–8.

Bulti, D. A., & Raimond, K. (2015). Optimizing caching in a patch streaming multimedia-on-demand system. *Journal of Computing Science and Engineering : JCSE*, *9*(3), 134–141. DOI: 10.5626/JCSE.2015.9.3.134

Chen, K. C. (2012). Machine-to-machine communications for healthcare. *Journal of Computing Science and Engineering : JCSE*, *6*(2), 119–126. DOI: 10.5626/JCSE.2012.6.2.119

Chen, L., Zhou, S., & Xu, J. (2018). Computation peer offloading for energy-constrained mobile edge computing in small-cell networks. *IEEE/ACM Transactions on Networking*, *26*(4), 1619–1632. DOI: 10.1109/TNET.2018.2841758

Chen, Q., Xu, X., Jiang, H., & Liu, X. (2020). An energy-aware approach for industrial internet of things in 5g pervasive edge computing environment. *IEEE Transactions on Industrial Informatics*, *17*(7), 5087–5097. DOI: 10.1109/TII.2020.3007973

Chen, W., Qiu, J., Zhao, J., Chai, Q., & Dong, Z. Y. (2021). Bargaining game-based profit allocation of virtual power plant in frequency regulation market considering battery cycle life. *IEEE Transactions on Smart Grid*, *12*(4), 2913–2928. DOI: 10.1109/TSG.2021.3053000

Chen, X., Jiao, L., Li, W., & Fu, X. (2015). Efficient multi-user computation offloading for mobile-edge cloud computing. *IEEE/ACM Transactions on Networking*, *24*(5), 2795–2808. DOI: 10.1109/TNET.2015.2487344

Darzanos, G., Koutsopoulos, I., & Stamoulis, G. D. (2019). Cloud federations: Economics, games and benefits. *IEEE/ACM Transactions on Networking*, *27*(5), 2111–2124. DOI: 10.1109/TNET.2019.2943810

Dasari, V. R., Geerhart, B. E., Alexander, D. M., & Shires, D. R. (2019, April). Distributed computation offloading framework for the tactical edge. In *IEEE INFOCOM 2019-IEEE Conference on Computer Communications Workshops (INFOCOM WKSHPS)* (pp. 1-6). IEEE. DOI: 10.1109/INFOCOMWKSHPS47286.2019.9093790

Dhillon, A., & Mertens, J. F. (1999). Relative utilitarianism. *Econometrica*, *67*(3), 471–498. DOI: 10.1111/1468-0262.00033

Dietzenbacher, B., Estévez-Fernández, A., Borm, P., & Hendrickx, R. (2021). Proportionality, equality, and duality in bankruptcy problems with nontransferable utility. *Annals of Operations Research*, *301*(1-2), 65–80. DOI: 10.1007/s10479-020-03643-3

Driesen, B., Perea, A., & Peters, H. (2011). The Kalai–Smorodinsky bargaining solution with loss aversion. *Mathematical Social Sciences*, *61*(1), 58–64. DOI: 10.1016/j.mathsocsci.2010.10.003

Driessen, T. S. H., & Funaki, Y. (1997). Reduced game properties of egalitarian division rules for TU-games. *Game theoretical applications to economics and operations research*, 85-103.

Du, Q., Song, H., & Zhu, X. (2018). Social-feature enabled communications among devices toward the smart IoT community. *IEEE Communications Magazine*, *57*(1), 130–137. DOI: 10.1109/MCOM.2018.1700563

Dubra, J. (2001). An asymmetric Kalai–Smorodinsky solution. *Economics Letters*, *73*(2), 131–136. DOI: 10.1016/S0165-1765(01)00426-8

El-Sayed, H., Sankar, S., Prasad, M., Puthal, D., Gupta, A., Mohanty, M., & Lin, C. T. (2017). Edge of things: The big picture on the integration of edge, IoT and the cloud in a distributed computing environment. *ieee access, 6*, 1706-1717.

Fan, W., Liu, Y. A., Tang, B., Wu, F., & Zhang, H. (2016a). Terminalbooster: Collaborative computation offloading and data caching via smart basestations. *IEEE Wireless Communications Letters*, *5*(6), 612–615. DOI: 10.1109/LWC.2016.2605694

Fan, W., Liu, Y. A., Tang, B., Wu, F., & Zhang, H. (2016b, December). Exploiting joint computation offloading and data caching to enhance mobile terminal performance. In *2016 IEEE Globecom Workshops (GC Wkshps)* (pp. 1-6). IEEE. DOI: 10.1109/GLOCOMW.2016.7848902

Funk, M. (2018, January). Designing the fog: towards an intranet of things. In *6th Workshop on Interacting with Smart Objects (SmartObjects 2018)* (pp. 31-38). CEUR-WS. org.

Garroppo, R. G., Giordano, S., & Iacono, D. (2009, November). Radio-aware scheduler for WiMAX systems based on time-utility function and game theory. In *GLOBECOM 2009-2009 IEEE Global Telecommunications Conference* (pp. 1-6). IEEE. DOI: 10.1109/GLOCOM.2009.5425581

Georgakopoulos, D., Jayaraman, P. P., Fazia, M., Villari, M., & Ranjan, R. (2016). Internet of Things and edge cloud computing roadmap for manufacturing. *IEEE Cloud Computing*, *3*(4), 66–73. DOI: 10.1109/MCC.2016.91

Goddard, S. (2003). *Variable rate resource allocation for mixed real-time/non-real-time systems* (Research project paper). University of Nebraska. http://cse.unl.edu/~goddard/ResearchProjects/VariableRateResourceAllocation.html

Guim, F., Metsch, T., Moustafa, H., Verrall, T., Carrera, D., Cadenelli, N., Chen, J., Doria, D., Ghadie, C., & Prats, R. G. (2021). Autonomous lifecycle management for resource-efficient workload orchestration for green edge computing. *IEEE Transactions on Green Communications and Networking*, *6*(1), 571–582. DOI: 10.1109/TGCN.2021.3127531

Guo, F., Yu, F. R., Zhang, H., Ji, H., Leung, V. C., & Li, X. (2020). An adaptive wireless virtual reality framework in future wireless networks: A distributed learning approach. *IEEE Transactions on Vehicular Technology*, *69*(8), 8514–8528. DOI: 10.1109/TVT.2020.2995877

Guo, F., Zhang, H., Ji, H., Li, X., & Leung, V. C. (2018). An efficient computation offloading management scheme in the densely deployed small cell networks with mobile edge computing. *IEEE/ACM Transactions on Networking*, *26*(6), 2651–2664. DOI: 10.1109/TNET.2018.2873002

Haake, C. J., & Qin, C. Z. (2018). On unification of solutions to the bargaining problem. *Available at SSRN* 3184007. DOI: 10.2139/ssrn.3184007

Han, Y., Li, D., Qi, H., Ren, J., & Wang, X. (2019, May). Federated learning-based computation offloading optimization in edge computing-supported internet of things. In *Proceedings of the ACM Turing Celebration Conference*-China (pp. 1-5). DOI: 10.1145/3321408.3321586

Hassan, N., Gillani, S., Ahmed, E., Yaqoob, I., & Imran, M. (2018). The role of edge computing in internet of things. *IEEE Communications Magazine*, *56*(11), 110–115. DOI: 10.1109/MCOM.2018.1700906

Hinojosa, M. A., Mármol, A. M., & Zarzuelo, J. M. (2008). Inequality averse multi-utilitarian bargaining solutions. *International Journal of Game Theory*, *37*(4), 597–618. DOI: 10.1007/s00182-008-0137-2

Ji, Z., Chen, L., Zhao, N., Chen, Y., Wei, G., & Yu, F. R. (2021). Computation offloading for edge-assisted federated learning. *IEEE Transactions on Vehicular Technology*, *70*(9), 9330–9344. DOI: 10.1109/TVT.2021.3098022

Ju, Y., Borm, P., & Ruys, P. (2007). The consensus value: A new solution concept for cooperative games. *Social Choice and Welfare*, *28*(4), 685–703. DOI: 10.1007/s00355-006-0192-1

Kaewpuang, R., Niyato, D., Wang, P., & Hossain, E. (2013). A framework for cooperative resource management in mobile cloud computing. *IEEE Journal on Selected Areas in Communications*, *31*(12), 2685–2700. DOI: 10.1109/JSAC.2013.131209

Khan, F., Jan, M. A., Rehman, A., Mastorakis, S., Alazab, M., & Watters, P. (2020). A secured and intelligent communication scheme for IIoT-enabled pervasive edge computing. *IEEE Transactions on Industrial Informatics*, *17*(7), 5128–5137. DOI: 10.1109/TII.2020.3037872 PMID: 33994885

Kim, K. S., Uno, S., & Kim, M. W. (2010b). Adaptive QoS mechanism for wireless mobile network. *Journal of Computing Science and Engineering : JCSE, 4*(2), 153–172. DOI: 10.5626/JCSE.2010.4.2.153

Kim, S. (2010a). Cooperative game theoretic online routing scheme for wireless network managements. *IET Communications, 4*(17), 2074–2083. DOI: 10.1049/iet-com.2009.0686

Kim, S. (Ed.). (2014). *Game theory applications in network design.* IGI Global. DOI: 10.4018/978-1-4666-6050-2

Kim, S. (2018a). 5G Network Communication, Caching, and Computing Algorithms Based on the Two-Tier Game Model. *ETRI Journal, 40*(1), 61–71. DOI: 10.4218/etrij.2017-0023

Kim, S. (2018b). A New Differential Privacy Crowdsensing Scheme Based on the Multilevel Interactive Game. *Wireless Communications and Mobile Computing, 2018*(1), 9867061. DOI: 10.1155/2018/9867061

Kim, S. (2020a). New application task offloading algorithms for edge, fog, and cloud computing paradigms. *Wireless Communications and Mobile Computing, 2020*(1), 8888074. DOI: 10.1155/2020/8888074

Kim, S. (2020b). Incentive design and differential privacy based federated learning: A mechanism design perspective. *IEEE Access : Practical Innovations, Open Solutions, 8*, 187317–187325. DOI: 10.1109/ACCESS.2020.3030888

Kim, S. (2021a). A novel local and global cooperative approach for distributed mobile cloud computing. *IEEE Access : Practical Innovations, Open Solutions, 9*, 117813–117822. DOI: 10.1109/ACCESS.2021.3105060

Kim, S. (2021b). Cooperative federated learning-based task offloading scheme for tactical edge networks. *IEEE Access : Practical Innovations, Open Solutions, 9*, 145739–145747. DOI: 10.1109/ACCESS.2021.3123313

Kim, S. (2022a). Bargaining game-based resource management for pervasive edge computing infrastructure. *IEEE Access : Practical Innovations, Open Solutions, 10*, 4072–4080. DOI: 10.1109/ACCESS.2022.3140663

Kim, S. (2022b). Bargaining game based offloading service algorithm for edge-assisted distributed computing model. *IEEE Access : Practical Innovations, Open Solutions, 10*, 63648–63657. DOI: 10.1109/ACCESS.2022.3183237

Kim, S. (2022c). Two-phase wireless bandwidth allocation scheme based on cooperative game solutions. *Wireless Personal Communications, 127*(4), 3061–3077. DOI: 10.1007/s11277-022-09910-9

Kim, S. (2022d). Collaborative resource sharing game based cloud-edge offload computing orchestration scheme. *IEEE Access : Practical Innovations, Open Solutions, 10*, 74523–74532. DOI: 10.1109/ACCESS.2022.3190857

Kim, S., & Varshney, P. K. (2004). An integrated adaptive bandwidth-management framework for QoS-sensitive multimedia cellular networks. *IEEE Transactions on Vehicular Technology, 53*(3), 835–846. DOI: 10.1109/TVT.2004.825704

Konstantakopoulos, I. C., Ratliff, L. J., Jin, M., Spanos, C., & Sastry, S. S. (2016, April). Smart building energy efficiency via social game: a robust utility learning framework for closing–the–loop. In *2016 1st International Workshop on Science of Smart City Operations and Platforms Engineering (SCOPE) in partnership with Global City Teams Challenge (GCTC)(SCOPE-GCTC)* (pp. 1-6). IEEE. DOI: 10.1109/SCOPE.2016.7515054

Lee, H. S., & Lee, J. W. (2018). Task offloading in heterogeneous mobile cloud computing: Modeling, analysis, and cloudlet deployment. *IEEE Access : Practical Innovations, Open Solutions*, *6*, 14908–14925. DOI: 10.1109/ACCESS.2018.2812144

Lee, K., & Shin, I. (2015). User mobility model based computation offloading decision for mobile cloud. *Journal of Computing Science and Engineering : JCSE*, *9*(3), 155–162. DOI: 10.5626/JCSE.2015.9.3.155

Li, H., Xu, H., Zhou, C., Lü, X., & Han, Z. (2020). Joint optimization strategy of computation offloading and resource allocation in multi-access edge computing environment. *IEEE Transactions on Vehicular Technology*, *69*(9), 10214–10226. DOI: 10.1109/TVT.2020.3003898

Li, J., Sun, J., Qian, Y., Shu, F., Xiao, M., & Xiang, W. (2016). A commercial video-caching system for small-cell cellular networks using game theory. *IEEE Access : Practical Innovations, Open Solutions*, *4*, 7519–7531. DOI: 10.1109/ACCESS.2016.2582836

Li, W., Zhao, Y., Lu, S., & Chen, D. (2015). Mechanisms and challenges on mobility-augmented service provisioning for mobile cloud computing. *IEEE Communications Magazine*, *53*(3), 89–97. DOI: 10.1109/MCOM.2015.7060487

Li, X., Wang, X., Li, K., & Leung, V. C. (2017). CaaS: Caching as a service for 5G networks. *IEEE Access : Practical Innovations, Open Solutions*, *5*, 5982–5993. DOI: 10.1109/ACCESS.2017.2689678

Li, Y., Wang, X., Gan, X., Jin, H., Fu, L., & Wang, X. (2019). Learning-aided computation offloading for trusted collaborative mobile edge computing. *IEEE Transactions on Mobile Computing*, *19*(12), 2833–2849. DOI: 10.1109/TMC.2019.2934103

Liang, K., & Li, D. (2019). A direct method of interval Banzhaf values of interval cooperative games. *Journal of Systems Science and Systems Engineering*, *28*(3), 382–391. DOI: 10.1007/s11518-019-5411-2

Lim, W. Y. B., Luong, N. C., Hoang, D. T., Jiao, Y., Liang, Y. C., Yang, Q., Niyato, D., & Miao, C. (2020). Federated learning in mobile edge networks: A comprehensive survey. *IEEE Communications Surveys and Tutorials*, *22*(3), 2031–2063. DOI: 10.1109/COMST.2020.2986024

Lin, P., Song, Q., Wang, D., Yu, F. R., Guo, L., & Leung, V. C. (2021). Resource management for pervasive-edge-computing-assisted wireless VR streaming in industrial Internet of Things. *IEEE Transactions on Industrial Informatics*, *17*(11), 7607–7617. DOI: 10.1109/TII.2021.3061579

Linthicum, D. S. (2017). Connecting fog and cloud computing. *IEEE Cloud Computing*, *4*(2), 18–20. DOI: 10.1109/MCC.2017.37

Liu, L., Chang, Z., & Guo, X. (2018). Socially aware dynamic computation offloading scheme for fog computing system with energy harvesting devices. *IEEE Internet of Things Journal*, *5*(3), 1869–1879. DOI: 10.1109/JIOT.2018.2816682

Ma, X., Wang, S., Zhang, S., Yang, P., Lin, C., & Shen, X. (2019). Cost-efficient resource provisioning for dynamic requests in cloud assisted mobile edge computing. *IEEE Transactions on Cloud Computing*, *9*(3), 968–980. DOI: 10.1109/TCC.2019.2903240

Ma, X., Zhang, S., Li, W., Zhang, P., Lin, C., & Shen, X. (2017, June). Cost-efficient workload scheduling in cloud assisted mobile edge computing. In *2017 IEEE/ACM 25th International Symposium on Quality of Service (IWQoS)* (pp. 1-10). IEEE. DOI: 10.1109/IWQoS.2017.7969148

Mashayekhy, L., Nejad, M. M., & Grosu, D. (2014). Cloud federations in the sky: Formation game and mechanism. *IEEE Transactions on Cloud Computing*, *3*(1), 14–27. DOI: 10.1109/TCC.2014.2338323

Namekata, T., & Driessen, T. S. (1999). The egalitarian non-k-averaged contribution (. *International Game Theory Review*, *1*(01), 45–61. DOI: 10.1142/S0219198999000050

Nguyen, D. C., Ding, M., Pathirana, P. N., Seneviratne, A., Li, J., Niyato, D., Dobre, O., & Poor, H. V. (2021). 6G Internet of Things: A comprehensive survey. *IEEE Internet of Things Journal*, *9*(1), 359–383. DOI: 10.1109/JIOT.2021.3103320

Niyato, D., & Hossain, E. (2006, June). A cooperative game framework for bandwidth allocation in 4G heterogeneous wireless networks. In *2006* [). IEEE.]. *IEEE International Conference on Communications : [proceedings]. IEEE International Conference on Communications*, *9*, 4357–4362.

Niyato, D., & Hossain, E. (2008). A noncooperative game-theoretic framework for radio resource management in 4G heterogeneous wireless access networks. *IEEE Transactions on Mobile Computing*, *7*(3), 332–345. DOI: 10.1109/TMC.2007.70727

Osman, P., Zeynep, A. G. S., & Gerhard-Wilhelm, W. (2015). An axiomatization of the interval Shapley value and on some interval solution concepts. *Contributions to Game Theory and Management*, *8*, 243–251.

Oteafy, S. M., & Hassanein, H. S. (2018). IoT in the fog: A roadmap for data-centric IoT development. *IEEE Communications Magazine*, *56*(3), 157–163. DOI: 10.1109/MCOM.2018.1700299

Pan, J., & McElhannon, J. (2017). Future edge cloud and edge computing for internet of things applications. *IEEE Internet of Things Journal*, *5*(1), 439–449. DOI: 10.1109/JIOT.2017.2767608

Pande, A., Ramamurthi, V., & Mohapatra, P. (2013). Quality-oriented video delivery over LTE. *Journal of Computing Science and Engineering : JCSE*, *7*(3), 168–176. DOI: 10.5626/JCSE.2013.7.3.168

Pivato, M. (2009). Twofold optimality of the relative utilitarian bargaining solution. *Social Choice and Welfare*, *32*(1), 79–92. DOI: 10.1007/s00355-008-0313-0

Rachmilevitch, S. (2015). The Nash solution is more utilitarian than egalitarian. *Theory and Decision*, *79*(3), 463–478. DOI: 10.1007/s11238-014-9477-5

Ren, J., Yu, G., He, Y., & Li, G. Y. (2019). Collaborative cloud and edge computing for latency minimization. *IEEE Transactions on Vehicular Technology*, *68*(5), 5031–5044. DOI: 10.1109/TVT.2019.2904244

Roth, A. E. (1977). Independence of irrelevant alternatives, and solutions to Nash's bargaining problem. *Journal of Economic Theory*, *16*(2), 247–251. DOI: 10.1016/0022-0531(77)90008-4

Savazzi, S., Nicoli, M., & Rampa, V. (2020). Federated learning with cooperating devices: A consensus approach for massive IoT networks. *IEEE Internet of Things Journal, 7*(5), 4641–4654. DOI: 10.1109/JIOT.2020.2964162

Shalev, J. (2002). Loss aversion and bargaining. *Theory and Decision, 52*(3), 201–232. DOI: 10.1023/A:1019674323804

Shen, C., Xu, J., Zheng, S., & Chen, X. (2021). Resource rationing for wireless federated learning: Concept, benefits, and challenges. *IEEE Communications Magazine, 59*(5), 82–87. DOI: 10.1109/MCOM.001.2000744

Sheng, M., Xu, C., Liu, J., Song, J., Ma, X., & Li, J. (2016). Enhancement for content delivery with proximity communications in caching enabled wireless networks: Architecture and challenges. *IEEE Communications Magazine, 54*(8), 70–76. DOI: 10.1109/MCOM.2016.7537179

Sim, K. M. (2022). Cooperative and parallel fog discovery and pareto optimal fog commerce bargaining. *IEEE Transactions on Computational Social Systems, 9*(4), 1112–1121. DOI: 10.1109/TCSS.2021.3129253

Sobel, J. (2001). Manipulation of preferences and relative utilitarianism. *Games and Economic Behavior, 37*(1), 196–215. DOI: 10.1006/game.2000.0839

Son, Y. H., & Lee, K. C. (2018, January). Cloud of things based on linked data. In *2018 International Conference on Information Networking (ICOIN)* (pp. 447-449). IEEE. DOI: 10.1109/ICOIN.2018.8343157

Sun, K., & Kim, Y. (2020, October). LISP-based hierarchical service mobility management for the tactical edge computing. In *2020 International Conference on Information and Communication Technology Convergence (ICTC)* (pp. 520-522). IEEE. DOI: 10.1109/ICTC49870.2020.9289539

Taleb, T., Samdanis, K., Mada, B., Flinck, H., Dutta, S., & Sabella, D. (2017). On multi-access edge computing: A survey of the emerging 5G network edge cloud architecture and orchestration. *IEEE Communications Surveys and Tutorials, 19*(3), 1657–1681. DOI: 10.1109/COMST.2017.2705720

Tan, C., Feng, W., & Han, W. (2020). On the banzhaf-like value for cooperative games with interval payoffs. *Mathematics, 8*(3), 372. DOI: 10.3390/math8030372

van den Brink, R., Chun, Y., Funaki, Y., & Park, B. (2016). Consistency, population solidarity, and egalitarian solutions for TU-games. *Theory and Decision, 81*(3), 427–447. DOI: 10.1007/s11238-016-9538-z

Wang, D., Chen, D., Song, B., Guizani, N., Yu, X., & Du, X. (2018). From IoT to 5G I-IoT: The next generation IoT-based intelligent algorithms and 5G technologies. *IEEE Communications Magazine, 56*(10), 114–120. DOI: 10.1109/MCOM.2018.1701310

Wang, Q., Chen, D., Zhang, N., Qin, Z., & Qin, Z. (2017). LACS: A lightweight label-based access control scheme in IoT-based 5G caching context. *IEEE Access : Practical Innovations, Open Solutions, 5*, 4018–4027. DOI: 10.1109/ACCESS.2017.2678510

Wang, S., Tuor, T., Salonidis, T., Leung, K. K., Makaya, C., He, T., & Chan, K. (2019). Adaptive federated learning in resource constrained edge computing systems. *IEEE Journal on Selected Areas in Communications, 37*(6), 1205–1221. DOI: 10.1109/JSAC.2019.2904348

Wang, X., Ning, Z., & Guo, S. (2020). Multi-agent imitation learning for pervasive edge computing: A decentralized computation offloading algorithm. *IEEE Transactions on Parallel and Distributed Systems*, *32*(2), 411–425. DOI: 10.1109/TPDS.2020.3023936

Wen, S., Yang, J., Chen, G., Tao, J., Yu, X., & Liu, A. (2019). Enhancing service composition by discovering cloud services community. *IEEE Access : Practical Innovations, Open Solutions*, *7*, 32472–32481. DOI: 10.1109/ACCESS.2019.2903825

Wu, H., Wolter, K., Jiao, P., Deng, Y., Zhao, Y., & Xu, M. (2020b). EEDTO: An energy-efficient dynamic task offloading algorithm for blockchain-enabled IoT-edge-cloud orchestrated computing. *IEEE Internet of Things Journal*, *8*(4), 2163–2176. DOI: 10.1109/JIOT.2020.3033521

Wu, Q., He, K., & Chen, X. (2020a). Personalized federated learning for intelligent IoT applications: A cloud-edge based framework. *IEEE Open Journal of the Computer Society*, *1*, 35–44. DOI: 10.1109/OJCS.2020.2993259 PMID: 32396074

Xu, G., Dai, H., & Shi, H. (2015). Axiomatizations and a noncooperative interpretation of the α-CIS value. *Asia-Pacific Journal of Operational Research*, *32*(05), 1550031. DOI: 10.1142/S0217595915500311

Yang, H., Wang, W., & Ding, Z. (2019). The weighted surplus division value for cooperative games. *Symmetry*, *11*(9), 1169. DOI: 10.3390/sym11091169

Yang, H., & Zhang, L. (2017). Response time prediction of IoT service based on time similarity. *Journal of Computing Science and Engineering : JCSE*, *11*(3), 100–108. DOI: 10.5626/JCSE.2017.11.3.100

Ye, Y., Li, S., Liu, F., Tang, Y., & Hu, W. (2020). EdgeFed: Optimized federated learning based on edge computing. *IEEE Access : Practical Innovations, Open Solutions*, *8*, 209191–209198. DOI: 10.1109/ACCESS.2020.3038287

Yu, R., Ding, J., Maharjan, S., Gjessing, S., Zhang, Y., & Tsang, D. H. (2015). Decentralized and optimal resource cooperation in geo-distributed mobile cloud computing. *IEEE Transactions on Emerging Topics in Computing*, *6*(1), 72–84. DOI: 10.1109/TETC.2015.2479093

Yu, R., Xue, G., Kilari, V. T., & Zhang, X. (2018). The fog of things paradigm: Road toward on-demand Internet of Things. *IEEE Communications Magazine*, *56*(9), 48–54. DOI: 10.1109/MCOM.2018.1701140

Zhang, Y., Lan, X., Ren, J., & Cai, L. (2020). Efficient computing resource sharing for mobile edge-cloud computing networks. *IEEE/ACM Transactions on Networking*, *28*(3), 1227–1240. DOI: 10.1109/TNET.2020.2979807

Zhu, C. (2016). Data-compression-based resource management in cloud computing for biology and medicine. *Journal of Computing Science and Engineering : JCSE*, *10*(1), 21–31. DOI: 10.5626/JCSE.2016.10.1.21

Chapter 4
Network Virtualization

ABSTRACT

Network virtualization involves separating network services and functions from the underlying physical infrastructure. Much like other virtualization techniques, its core objective is to insert an abstraction layer between the hardware itself and the applications or services that rely on it. In recent years, proponents have presented network virtualization as a long-term remedy for the Internet's increasing rigidity, suggesting it as a cornerstone in shaping next-generation networking paradigms. This chapter explores both the benefits and challenges associated with network virtualization in the context of future network design.

4. OVERVIEW

Network virtualization is the process of replacing hardware-based network functions with software-based ones in order to better optimize the network while increasing the efficiency, flexibility, and capacity of the infrastructure. Therefore, the underlying physical hardware remains, but network virtualization allows us to use software to build or layer virtual networks on top of that physical networks. This technology concerns a network that allows multiple service providers to form multiple separate and isolated virtual networks by sharing physical resources provided by one or more different physical network infrastructure providers. Especially, the virtualization of resources, functions, and networks enables resource sharing, and thereby increases resource utilization while reducing the cost for deploying and operating networks. Therefore, network virtualization separates the services a physical network provides from the hardware. It allows administrators to allocate those services virtually to create a new network or several networks.

Network virtualization already plays an important role in the system architecture. Analogously, virtual networks have recently emerged to flexibly allow for new network services without worrying about the specific details of the specific underlying networking hardware. Through virtualization, multiple virtual networks can flexibly operate over the same physical network infrastructure. In the context of virtual networking, the virtual networks are monitored and the networking resources are allocated to the individual virtual networks, which are the slices of overall network. By slicing, a physical network into several logical networks, each one can provide tailored services for a distinct application scenario. Especially, operating multiple virtual networks over a given network infrastructure with judicious resource allocation may improve the utilization of the networking hardware. A given slice provides a setting for examining novel networking paradigms, independent of the constraints imposed by the presently dominant Internet structures and protocols.

DOI: 10.4018/979-8-3373-2013-7.ch004

4.1 Adaptive Multi-service Resource Allocation (*AMRA*) Scheme

Wireless network virtualization (WNV) has emerged as a promising paradigm for delivering high-speed communications and is considered a key enabler for the realization of 5G networks. By using network slicing, physical networks can be segmented into multiple virtual networks, each tailored to meet distinct service requirements. Recently, S. Kim introduces the *Adaptive Multi-service Resource Allocation* (*AMRA*) scheme for WNV (Kim, 2022), which employs a two-tier resource allocation mechanism within multi-service wireless environments. To enhance flexibility and autonomy among different network slices, the *AMRA* approach is broken down into two separate bargaining games - one focused on intra-slice resource allocation and the other on inter-slice allocation. Both are designed to satisfy multiple traffic requirements while ensuring efficient and fair usage of available resources. By leveraging four distinct bargaining solutions, the *AMRA* scheme enables effective collaboration among network agents, resulting in a well-balanced performance across the system (Kim, 2022).

4.1.1 Development Motivation

With the rapid increase in demand for diverse communication services, wireless traffic has expanded dramatically in recent years. Market forecasts suggest that mobile data traffic will continue to grow at an annual rate of approximately 47% through 2030. In addition, subscriber growth of 5-15% each year is predicted to persist into the next decade, while user expectations regarding service quality keep rising, irrespective of their location or network load. In response to this surge in traffic, numerous technical studies focus on network densification to overcome capacity constraints. By intensifying the placement of low-power small base stations - known as ultra-dense small cell networks (UDNs) - the *AMRA* scheme aims to improve network capacity. Consequently, upcoming 5G networks are expected to employ UDN infrastructure to aggressively reuse spectrum (Ding, 2018; Kim, 2022; Wang, 2018).

Throughout the standardization of 5G networks, network densification has been regarded as a primary driver for cellular technology, leading to the complete integration of small cell methods into official specifications and the commercial release of small cell products. However, these new UDNs differ substantially from traditional cellular networks and cannot be deployed or operated as they have been in the past 25 years. As a result, UDNs introduce significant challenges related to network planning and optimization, including spectrum scarcity, uneven network load, and higher operating and capital expenses due to the widespread deployment of small base stations (Ding, 2018; González González, 2017; Kim, 2022; Wang, 2018).

Given these issues, network sharing mechanisms have emerged as promising solutions for future network environments. They enable infrastructure and wireless spectrum resources to be shared among end users. Furthermore, decoupling services and infrastructures is critical to fulfill diverse service needs, transitioning from physical to virtual entities and cutting costs. Without such an approach, introducing new technologies or modifying existing ones can be difficult because of the complex makeup of infrastructure providers (InPs). Hence, virtualization for UDNs has become a pivotal technology in the 5G era. Through virtualization, the physical infrastructure and resources of UDNs can be completely abstracted, pooled, and transformed into multiple virtual resources, which can then be effectively shared to enhance network capacity (Wang, 2018; Kim, 2022).

As a promising framework for network sharing, wireless network virtualization (WNV) allows different services and applications to coexist on a shared network infrastructure. Simply put, WNV may be explained as both decoupling and resource sharing: hardware, software resources, and network functionalities are combined to enable resource sharing while separating infrastructure from the services it provides. Via WNV, physical components - such as infrastructure, wireless spectrum, backhaul, and fronthaul - belonging to a base station (BS) operated by an InP are transformed into distinct virtual resources called slices. Mobile virtual network operators (MVNOs), which do not own the underlying network infrastructure or resources, can then transparently use these slices and thereby virtually access all BS resources. This technology offers several advantages; i) improved resource utilization through statistical multiplexing, ii) reduced deployment and operation costs through infrastructure sharing, and iii) enhanced service portfolio for service providers through virtualization (Jiang, 2018; Kim, 2022; Zhu, 2015). Despite the many benefits of WNV as a fresh control framework, designing an effective resource allocation strategy remains a major hurdle. In WNV, resource allocation addresses how physical resources can be partitioned into virtual networks to satisfy varying demand from multiple MVNOs. This problem becomes more complex when network agents act in their own self-interest, making it difficult to achieve ideal social efficiency. The main objectives of the *AMRA* scheme in this regard are; i) boost network capacity, ii) optimize spectrum resource utilization, iii) enhance quality of service, and iv) strike a balanced performance. Reaching these targets is challenging, and necessitates a radical change in the control paradigm (Ding, 2018; Kim, 2022; Zhu, 2015).

The *AMRA* scheme devises a novel spectrum control mechanism founded on WNV technology within UDNs. Under the highly dynamic environment of 5G networks, the *AMRA* scheme must facilitate inter- and intra-slice customization via efficient resource allocation. The primary challenge for the *AMRA* scheme is managing interactions among different MVNOs in a way that maintains favorable system-wide performance. To the best of my knowledge, although this is a key area of interest, it has not been extensively investigated. In the *AMRA* scheme, autonomous, decentralized, and intelligent MVNOs make rational, strategic decisions, guided by new solution concepts - an approach that naturally aligns with cooperative game theory. This field provides a robust model for cooperation among rational players, with the core issue being how to distribute the surplus outcome. Various solutions have been proposed to address resource distribution under different criteria (Kim, 2022; Yang, 2019).

Often, MVNOs care about both the proportion of their claims fulfilled and the total amount allocated to them. Proportional bargaining solutions are frequently cited as standard answers to this dilemma because they allow participants to compare how each party is treated in terms of the fraction of its claim that is met. From the standpoint of relative fairness, every participant receives the same fraction of its original claim, in line with egalitarian principles. In the case of conflicting claims, the Nash solution differs, since each participant's reference point influences its portion of the loss (Giménez-Gómez, 2015; Kim, 2022).

Starting in the 2000s, numerous studies have applied game-theoretic methods to bankruptcy -related issues, looking into broadly generalized properties. These works often cast the bankruptcy setting as a bargaining problem in the Nash tradition, representing a consensus-based idea: a final choice is made only if every group member endorses it. Expanding on proportional concepts in bargaining solutions, researchers have proposed multiple proportional solutions like the *Weighted Kalai-Smorodinsky Bargaining Solution* (*WKSBS*), *Claims-based Proportional Solution* (*CPS*), *Truncated Proportional Solution* (*TPS*), and *Restricted Truncated Proportional Solution* (*RTPS*). These solutions introduce new axiomatic characterizations in response to changes in claims and utilize a normalization procedure where zero utility corresponds to no allocation. Consequently, the zero vector replaces an external disagreement point, as often seen in bargaining models. *WKSBS*, *CPS*, *TPS*, and *RTPS* uphold numerous favorable properties under cooperative game theory (Dietzenbacher, 2022; Dubra, 2001; Kim, 2022).

4.1.2 Contribution

Building on the concept of proportional bargaining solutions, the *AMRA* scheme introduces a novel spectrum control mechanism tailored to the WNV technology within UDN environments. Under this scheme, each MVNO shares the limited BS spectrum resources in a fair and efficient manner, taking into account prevailing traffic conditions. For the inter-MVNO slice allocation, the *AMRA* scheme employs the *WKSBS* to manage inter-slice customization. Meanwhile, the intra-MVNO slice allocation relies on the *CPS*, *TPS*, and *RTPS* to handle the intra-slice customization. Together, these solutions promote a balanced system performance by enabling an interactive joint game approach that integrates various bargaining methods and ensures favorable global properties (Kim, 2022).

4.1.3 The UDN System Infrastructure with WNV

We consider a UDN infrastructure with an InP including multiple BSs and wireless resources. $\mathbb{B} = \left\{\mathcal{B}_1, \mathcal{B}_2, \ldots, \mathcal{B}_n\right\}$ is the set of BSs, and $\mathcal{B}_i \in \mathbb{B}$ has its own spectrum resource, i.e., $\mathfrak{M}_{\mathcal{B}_i}$. The $\mathcal{B}_i$ provides its service to a set $\mathbb{N}_{\mathcal{B}_i}$ of IoT devices, which are covered by the $\mathcal{B}_i$. Then, $\mathbb{N} = \bigcup_{\mathcal{B}_i \in \mathbb{B}} \mathbb{N}_{\mathcal{B}_i}$ represents the total number of IoT devices in the multi-cell UDN system. The *AMRA* scheme assumes that individual IoT devices generate three different type services such as primary services (PSs), secondary services (SSs) and tertiary services (TSs). The PSs are latency-sensitive services like as self-driving and drone control, which need low-latency QoS requirements. The SSs are characterized by high data rate like as high definition videos and virtual reality with high transmission data rates. For connection density like as smart city and smart home, the TSs need a large number of local and temporary transmissions with no strict QoS requirements in capacity, latency, and reliability. The InP virtualizes each BS's spectrum resource into three slices, i.e., $\mathcal{S}_P$, $\mathcal{S}_S$ and $\mathcal{S}_T$, and assigns them for PSs, SSs, and TSs, respectively. In each BS, three MVNOs, i.e., $\mathcal{M}_P$, $\mathcal{M}_S$ and $\mathcal{M}_T$, are installed to share and operate the virtualized $\mathcal{S}_P$, $\mathcal{S}_S$ and $\mathcal{S}_T$, respectively (Han, 2020; Jiang, 2018; Kim, 2022).

The *AMRA* scheme designs a two-tier spectrum control process for UDNs. Without a loss of generality, the *AMRA* scheme considers the $\mathcal{B}_i \in \mathbb{B}$ as an example. The $\mathcal{B}_i$ virtualizes its $\mathfrak{M}_{\mathcal{B}_i}$ into $\mathcal{S}_P^{\mathcal{B}_i}$, $\mathcal{S}_S^{\mathcal{B}_i}$ and $\mathcal{S}_T^{\mathcal{B}_i}$ where $\mathfrak{M}_{\mathcal{B}_i} \geq \left(\mathcal{S}_P^{\mathcal{B}_i} + \mathcal{S}_S^{\mathcal{B}_i} + \mathcal{S}_T^{\mathcal{B}_i}\right)$, and assigns these slices for its corresponding traffic services. For the $\mathcal{B}_i$, the *AMRA* scheme consists of two tiers. At the first tier, the inter-slice spectrum allocation problem is formulated as the game $\mathbb{G}_{\mathcal{B}_i}$. In this game, the $\mathcal{M}_P$, $\mathcal{M}_S$ and $\mathcal{M}_T$ work together to solve the matching problem of slice demands and physical spectrum resource $\left(\mathfrak{M}_{\mathcal{B}_i}\right)$ in the $\mathcal{B}_i$. At the second tier, the intra-slice scheduling problem for each type service is modeled as three games such as $\mathbb{G}_{\mathcal{B}_i}^{\mathcal{M}_P}$, $\mathbb{G}_{\mathcal{B}_i}^{\mathcal{M}_S}$, and $\mathbb{G}_{\mathcal{B}_i}^{\mathcal{M}_T}$. For these games, the $\mathcal{M}_P$, $\mathcal{M}_S$ and $\mathcal{M}_T$ work independently in a distributed fashion. Based on service types, they schedule their assigned $\mathcal{S}_P$, $\mathcal{S}_S$ and $\mathcal{S}_T$ slices to their corresponding PSs, SSs, and TSs, respectively. Under dynamically changing traffic environments, a fixed spectrum allocation and scheduling approach cannot effectively adapt to current UDN conditions. Therefore, the $\mathcal{M}_P$, $\mathcal{M}_S$ and $\mathcal{M}_T$ must be dynamically adjustable. To satisfy this goal, the *AMRA* scheme designs a joint interactive bargaining game ($\mathbb{G}$) in a coordination manner; G is subdivided into two-tier subgames; $\mathbb{G}_{\mathcal{B}_i \in \mathbb{B}}$, $\mathbb{G}_{\mathcal{B}_i}^{\mathcal{M}_P}$, $\mathbb{G}_{\mathcal{B}_i}^{\mathcal{M}_S}$, and $\mathbb{G}_{\mathcal{B}_i}^{\mathcal{M}_T}$. And, different bargaining solutions are adopted to effectively solve each bargaining games. Formally, the *AMRA* scheme defines game entities, i.e.,

$$\mathbb{G} = \left\{\mathcal{B}_i \in \mathbb{B} | \mathbb{G}_{\mathcal{B}_i}, \mathbb{G}_{\mathcal{B}_i}^{\mathcal{M}_P}, \mathbb{G}_{\mathcal{B}_i}^{\mathcal{M}_S}, \mathbb{G}_{\mathcal{B}_i}^{\mathcal{M}_T}\right\} =$$

$$\left\{\mathbb{B}, \mathbb{N}, \left\{\mathbb{G}_{\mathcal{B}_i} | \mathfrak{M}_{\mathcal{B}_i}, \mathbb{N}_{\mathcal{B}_i}, (\mathcal{M}_P, \mathcal{M}_S, \mathcal{M}_T), (\mathcal{S}_P^{\mathcal{B}_i}, \mathcal{S}_S^{\mathcal{B}_i}, \mathcal{S}_T^{\mathcal{B}_i}), (U_{\mathcal{M}_P}(\cdot), U_{\mathcal{M}_S}(\cdot), U_{\mathcal{M}_T}(\cdot))\right\},\right.$$

$$\left\{\mathbb{G}_{\mathcal{B}_i}^{\mathcal{M}_P} | \mathcal{M}_P, D_j^{\mathcal{B}_i,P} \in \mathbb{N}_{\mathcal{B}_i}^P, \mathcal{S}_j^{\mathcal{B}_i,P}, \mathfrak{U}_j^{\mathcal{B}_i,P}(\cdot)\right\}, \left\{\mathbb{G}_{\mathcal{B}_i}^{\mathcal{M}_S} | \mathcal{M}_S, D_k^{\mathcal{B}_i,S} \in \mathbb{N}_{\mathcal{B}_i}^S, \mathcal{S}_k^{\mathcal{B}_i,S}, \mathfrak{U}_k^{\mathcal{B}_i,S}(\cdot)\right\},$$

$$\left.\left\{\mathbb{G}_{\mathcal{B}_i}^{\mathcal{M}_T} | \mathcal{M}_T, D_l^{\mathcal{B}_i,T} \in \mathbb{N}_{\mathcal{B}_i}^T, \mathcal{S}_l^{\mathcal{B}_i,T}, \mathfrak{U}_l^{\mathcal{B}_i,T}(\cdot)\right\}, \mathbb{T}\right\}$$

of gameplay (Kim, 2022).

- $\mathbb{G}_{\mathcal{B}_i}$, $\mathbb{G}_{\mathcal{B}_i}^{\mathcal{M}_P}$, $\mathbb{G}_{\mathcal{B}_i}^{\mathcal{M}_S}$, and $\mathbb{G}_{\mathcal{B}_i}^{\mathcal{M}_T}$ are bargaining games for the $\mathcal{B}_i$, and they are mutually and reciprocally interdependent in an interactive manner.

- B and N are the sets of all BSs and IoT devices. For the $\mathcal{B}_i \in \mathbb{B}$, $\mathfrak{M}_{\mathcal{B}_i}$ is the $\mathcal{B}_i$'s physical spectrum resource, and $\mathbb{N}_{\mathcal{B}_i} \subset \mathbb{N}$ is the set of IoT devices covered by the $\mathcal{B}_i$.
- In the $\mathbb{G}_{\mathcal{B}_i}$, $\mathcal{M}_P$, $\mathcal{M}_S$ and $\mathcal{M}_T$ are game players, and the amounts of $\mathcal{S}_P^{\mathcal{B}_i}$, $\mathcal{S}_S^{\mathcal{B}_i}$ and $\mathcal{S}_T^{\mathcal{B}_i}$ are their strategies. $U_{\mathcal{M}_P}(\cdot)$, $U_{\mathcal{M}_S}(\cdot)$ and $U_{\mathcal{M}_T}(\cdot)$ are utility functions for the $\mathcal{M}_P$, $\mathcal{M}_S$ and $\mathcal{M}_T$, respectively.
- In the $\mathbb{G}_{\mathcal{B}_i}^{\mathcal{M}_P}$, $\mathbb{N}_{\mathcal{B}_i}^P$ is the set of devices, which run PSs in the $\mathcal{B}_i$. Each device $D_j^{\mathcal{B}_i,P} \in \mathbb{N}_{\mathcal{B}_i}^P$ is a game player, and $\mathcal{S}_j^{\mathcal{B}_i,P}$ is the assigned spectrum amount for the $D_j^{\mathcal{B}_i,P}$; it is the strategy of $D_j^{\mathcal{B}_i,P}$. $\mathfrak{U}_j^{\mathcal{B}_i,P}(\cdot)$ is the $D_j^{\mathcal{B}_i,P}$'s utility function where $\mathcal{S}_P^{\mathcal{B}_i} \geq \sum_{D_j^{\mathcal{B}_i,P} \in \mathbb{N}_{\mathcal{B}_i}^P} \mathcal{S}_j^{\mathcal{B}_i,P}$.
- In the $\mathbb{G}_{\mathcal{B}_i}^{\mathcal{M}_S}$, $\mathbb{N}_{\mathcal{B}_i}^S$ is the set of devices, which run SSs in the $\mathcal{B}_i$. Each device $D_k^{\mathcal{B}_i,S} \in \mathbb{N}_{\mathcal{B}_i}^S$ is a game player, and $\mathcal{S}_k^{\mathcal{B}_i,S}$ is the assigned spectrum amount for the $D_k^{\mathcal{B}_i,S}$; it is the strategy of $D_k^{\mathcal{B}_i,S}$. $\mathfrak{U}_j^{\mathcal{B}_i,S}(\cdot)$ is the $D_k^{\mathcal{B}_i,S}$'s utility function where $\mathcal{S}_S^{\mathcal{B}_i} \geq \sum_{D_k^{\mathcal{B}_i,S} \in \mathbb{N}_{\mathcal{B}_i}^S} \mathcal{S}_k^{\mathcal{B}_i,S}$.
- In the $\mathbb{G}_{\mathcal{B}_i}^{\mathcal{M}_T}$, $\mathbb{N}_{\mathcal{B}_i}^T$ is the set of devices, which run TSs in the $\mathcal{B}_i$. Each device $D_l^{\mathcal{B}_i,T} \in \mathbb{N}_{\mathcal{B}_i}^T$ is a game player, and $\mathcal{S}_l^{\mathcal{B}_i,T}$ is the assigned spectrum amount for the $D_l^{\mathcal{B}_i,T}$; it is the strategy of $D_l^{\mathcal{B}_i,T}$. $\mathfrak{U}_l^{\mathcal{B}_i,T}(\cdot)$ is the $D_l^{\mathcal{B}_i,T}$'s utility function where $\mathcal{S}_T^{\mathcal{B}_i} \geq \sum_{D_l^{\mathcal{B}_i,T} \in \mathbb{N}_{\mathcal{B}_i}^P} \mathcal{S}_l^{\mathcal{B}_i,T}$.
- $\mathrm{T} = \{t_1, \ldots, t_c, t_{c+1}, \ldots\}$ denotes time, which is represented by a sequence of time steps.

4.1.4 Bargaining Solutions for Cooperative Game Models

Cooperative game theory provides a highly suitable framework for examining bargaining with claims. In 1950, J. Nash introduced the Nash bargaining solution, which is grounded in axioms concerning shifts in the feasible set. Within the conventional Nash model, the outcome of a negotiation depends solely on the bargaining set and the disagreement point. However, numerous studies have highlighted the influence of additional environmental factors on the final bargaining result. Recently, attention has turned toward the proportional bargaining solution, a method that can effectively adapt the core principles of the Kalai-Smorodinsky solution; the utility gains at the compromise are proportional to what they would be at the ideal point, the point whose ith coordinate is the maximal utility player i could obtain in the part of bargaining set (Chun, 1992; Kim, 2022).

To characterize the basic concepts of proportional bargaining solutions, we preliminarily define some mathematical expressions. $E \subseteq \mathbb{R}_+^N$ is the set of attainable utility allocations where $E = \{x \in \mathbb{R}_+^N | \exists_{y \in E} : y \geq x\}$. Elements of E are assumed to be normalized such that allocating nothing to a player corresponds to zero utility. Let c be the vector of claims of N on E where $c \in \mathbb{R}_+^N$

with $c \nless x$ for all $x \in E$. The claim vector c represents the individual utility claims on the resource. With the limited resource amount $M \in \mathbb{R}_{+}$, a bargaining problem is denoted by (N, E, c) where $E = \left\{x \in \mathbb{R}_{+}^{N} | \sum_{i \in N} x_i \leq M\right\}$, $c \notin E$ and $E \neq \left\{0_N\right\}$. Let $(N, E, c) \in \Sigma^n$ and a solution is a function $f{:}\Sigma^n \rightarrow \mathbb{R}^n$ that associates a unique point of E, $f(N, E, c)$, called the solution outcome of (N, E, c). For any $t \in \mathbb{R}_{++}$, the set $(t \cdot E) \subseteq \mathbb{R}_{+}^{N}$ is defined by $(t \cdot E) = \{t \cdot x | \, x \in E\}$. Note that $u^{t \cdot E} = t \cdot u^E$ for all $t \in \mathbb{R}_{++}$. For any set of payoff allocations $E \subseteq \mathbb{R}_{++}^{n}$, u^E is the vector of utopia values; it is given by $u^E = \left(\mathbf{max}\{x_i | x \in E\}\right)_{i \in N}$ where $u^E \in \mathbb{R}_{++}^{N}$. $\hat{c}^E$ is the vector of truncated claims; it is given by $\hat{c}^E = \left(\mathbf{min}\{c_i, u_i^E\}\right)_{i \in N}$ where $\hat{c}^E \in \mathbb{R}_{++}^{N}$. The following solution definitions are *WKSBS*, *CPS*, *TPS* and *RTPS* (Dietzenbacher, 2018, Dietzenbacher, 2021; Kim, 2022).

Definition 1: The *WKSBS* for all $(N, E, c) \in \Sigma^n$ is $WKSBS(N, E, (\mathcal{W} \cdot u^E))$, which is the maximal point of E on the segment connecting $d = \left\{0_N\right\}$ and weighted utopian point. To express it as a mathematical formula type, $WKSBS(N, E, (\mathcal{W} \cdot u^E)) = (\lambda^{WKSBS} \cdot (\mathcal{W} \cdot u^E))$ where $\lambda^{WKSBS} = \mathbf{max}\{t \in (0, 1] | (t \cdot u^E) \in E\}$ and W is a weight vector.

Definition 2: The *CPS* for all $(N, E, c) \in \Sigma^n$ is $CPS(N, E, c)$, which is the maximal point of E on the segment connecting $d = \left\{0_N\right\}$ and c. To express it as a mathematical formula type, $CPS(N, E, c) = (\lambda^{CPS} \cdot c)$ where $\lambda^{CPS} = \mathbf{max}\{t \in (0, 1] | (t \cdot c) \in E\}$.

Definition 3: The TPS for all $(N, E, c) \in \Sigma^n$ is $TPS(N, E, \hat{c}^E)$, which is the maximal point of E on the segment connecting $d = \left\{0_N\right\}$ and $\hat{c}^E$. Therefore, $TPS(N, E, \hat{c}^E) = (\lambda^{CPS} \cdot \hat{c}^E)$.

Definition 4:: If $c < u^E$, the *RTPS* for all $(N, E, c) \in \Sigma^n$ is the maximal point of E on the segment connecting $d = \left\{0_N\right\}$ and c. Therefore, $RTPS(N, E, c) = (\lambda^{CPS} \cdot c) = CPS(N, E, c)$. Otherwise, if $c \geq u^E$, $RTPS(N, E, c)$ is $\left((t \cdot u_S^E), 0_{N \setminus S}\right)$ where $S = \left\{i \in N | c_i \geq u_i^E\right\}$ and $t \in (0, 1]$.

4.1.5 The Spectrum Control Scheme in the UDNs

In the UDN infrastructure, the InP virtualizes all base stations' spectrum resources with WNV technology. Each individual base station operate its two-tier game ($\mathbb{G}$) in a distributed manner. In the viewpoint of $\mathcal{B}_i \in \mathbb{B}$, the *AMRA* scheme explains the procedure of G, which consists of $\mathbb{G}_{\mathcal{B}_i}$, $\mathbb{G}_{\mathcal{B}_i}^{\mathcal{M}_P}$, $\mathbb{G}_{\mathcal{B}_i}^{\mathcal{M}_S}$, and $\mathbb{G}_{\mathcal{B}_i}^{\mathcal{M}_T}$. In the $\mathbb{G}_{\mathcal{B}_i}$, three MVNOs, i.e., $\mathcal{M}_P$, $\mathcal{M}_S$ and $\mathcal{M}_T$, are game players, who need $\mathcal{S}_P$, $\mathcal{S}_S$ and $\mathcal{S}_T$ slices to ensure their services. The utility payoff obtained by each MVNO depends on the current traffic flow and the amount of allocated slice. According to the satisfaction from MVNOs, the utility functions of $\mathcal{M}_P$, $\mathcal{M}_S$, $\mathcal{M}_T$, that is, $U_{\mathcal{M}_P}(\cdot), U_{\mathcal{M}_S}(\cdot), U_{\mathcal{M}_T}(\cdot)$, are derived as follows (Kim, 2022):

$$\begin{cases} U_{\mathcal{M}_P}\left(\mathcal{S}_P^{\mathcal{B}_i}, R_P^{\mathcal{B}_i}\right) = \chi \times \left(\sigma - exp\left(-\eta \times \dfrac{\min\left(\mathcal{S}_P^{\mathcal{B}_i}, R_P^{\mathcal{B}_i}\right)}{R_P^{\mathcal{B}_i}}\right)\right). \\ U_{\mathcal{M}_S}\left(\mathcal{S}_S^{\mathcal{B}_i}, R_S^{\mathcal{B}_i}\right) = \dfrac{\varepsilon}{\xi + exp\left(-\dfrac{\min\left(\mathcal{S}_S^{\mathcal{B}_i}, R_S^{\mathcal{B}_i}\right)}{R_S^{\mathcal{B}_i}}\right)} - \zeta \\ U_{\mathcal{M}_T}\left(\mathcal{S}_T^{\mathcal{B}_i}, R_T^{\mathcal{B}_i}\right) = \gamma \times \log\left(\dfrac{\min\left(\mathcal{S}_T^{\mathcal{B}_i}, R_T^{\mathcal{B}_i}\right)}{R_T^{\mathcal{B}_i}} + \Delta\right) \end{cases}$$

$$\text{s.t.,}\ \mathfrak{M}_{\mathcal{B}_i} \geq \left(\mathcal{S}_P^{\mathcal{B}_i} + \mathcal{S}_S^{\mathcal{B}_i} + \mathcal{S}_T^{\mathcal{B}_i}\right) \tag{4.1.1}$$

where χ, σ,η are control parameters for $U_{\mathcal{M}_P}(\cdot)$, ξ, ε, ζ are system adjustment factors for $U_{\mathcal{M}_S}(\cdot)$, and γ, Δ are control parameters for $U_{\mathcal{M}_T}(\cdot)$. $R_P^{\mathcal{B}_i}$, $R_S^{\mathcal{B}_i}$, $R_T^{\mathcal{B}_i}$ are the requested amount from the $\mathcal{M}_P$, $\mathcal{M}_S$, $\mathcal{M}_T$, respectively. To implement the spectrum allocation algorithm, the *AMRA* scheme concerns the different characteristics of PSs, SSs, and TSs. Therefore, the *AMRA* scheme should treat asymmetrically the PSs, SSs, and TSs. Usually, one main feature of traditional bargaining solutions is *Symmetry*. However, assuming *Symmetry* is unreasonable for the first-tier spectrum allocation problem; imposing *Symmetry* can mean, assuming equality of bargaining skill among the PSs, SSs, and TSs. In order to mediate the particularity of different traffic services, the *AMRA* scheme adopts the idea of *WKSBS* by relaxing *Symmetry*, and make the solution more flexible (Dubra, 2001). Therefore, the idea of *WKSBS* is adopted for the solution of $\mathbb{G}_{\mathcal{B}_i}$; it is given by (Kim, 2022):

$$WKSBS_{\mathbb{G}_{\mathcal{B}_i}}\left(\mathcal{M},\mathcal{S},\mathcal{W}_{\mathcal{M}},U_{\mathcal{M}}^{*}\right) = \left\{U_{\mathcal{M}}^{*} = \left(U_{\mathcal{M}_P}^{*},U_{\mathcal{M}_S}^{*},U_{\mathcal{M}_T}^{*}\right)|\left(\lambda^{WKSBS}\cdot\left(\mathcal{W}_{\mathcal{M}}\cdot U_{\mathcal{M}}^{*}\right)\right)\right\} \tag{4.1.2}$$

$$s.t.,\ \lambda^{WKSBS} = \mathbf{max}\left\{t \in (0,1]|\left(\mathcal{S}_P^{\mathcal{B}_i} + \mathcal{S}_S^{\mathcal{B}_i} + \mathcal{S}_T^{\mathcal{B}_i}\right) \le \mathfrak{M}_{\mathcal{B}_i}\right\}$$

where $\mathcal{W}_{\mathcal{M}}$ and $U_{\mathcal{M}}^{*}$ are the vectors of weights and utopia values for MVNOs, respectively. According to (2), the values of $\mathcal{S}_P^{\mathcal{B}_i}$, $\mathcal{S}_S^{\mathcal{B}_i}$ and $\mathcal{S}_T^{\mathcal{B}_i}$ are obtained. Based on the result of $\mathbb{G}_{\mathcal{B}_i}$, the second-tier spectrum scheduling algorithm is operated. Independently, individual MVNOs schedule their slices for their corresponding devices. Using the $\mathcal{S}_P^{\mathcal{B}_i}$, the $\mathbb{G}_{\mathcal{B}_i}^{\mathcal{M}_P}$ is designed to support multiple PSs. In the $\mathbb{G}_{\mathcal{B}_i}^{\mathcal{M}_P}$, devices in the $\mathbb{N}_{\mathcal{B}_i}^{P}$, i.e., $D_j^{\mathcal{B}_i,P} \in \mathbb{N}_{\mathcal{B}_i}^{P}$, are game players and the scheduled spectrum amount for the $D_j^{\mathcal{B}_i,P}$, i.e., $\mathcal{S}_j^{\mathcal{B}_i,P}$, is his strategy. The utility function of $D_j^{\mathcal{B}_i,P}$, i.e., $\mathfrak{U}_j^{\mathcal{B}_i,P}(\cdot)$, is defined as follows (Kim, 2022):

$$\mathfrak{U}_j^{\mathcal{B}_i,P}\left(\mathcal{S}_j^{\mathcal{B}_i,P},R_j^{\mathcal{B}_i,P},\mathbb{N}_{\mathcal{B}_i}^{P}\right) = \left(\delta - exp\left(\theta \times \frac{\min\left(\mathcal{S}_j^{\mathcal{B}_i,P},R_j^{\mathcal{B}_i,P}\right)}{R_j^{\mathcal{B}_i,P}}\right)\right) \tag{4.1.3}$$

where δ,θ are control parameters for $\mathfrak{U}_j^{\mathcal{B}_i,P}(\cdot)$. From the viewpoint of the $\mathcal{M}_P$, PSs should be operated monotonically due to the feature of latency sensitivity., Hence the *AMRA* scheme adopts the *CPS* for the solution of $\mathbb{G}_{\mathcal{B}_i}^{\mathcal{M}_P}$. According to each service request, devices in the $\mathbb{N}_{\mathcal{B}_i}^{P}$ have their claims, which simply maximize their payoffs. With their claims, the $\mathcal{S}_j^{\mathcal{B}_i,P}$ value is decided by using the concept of the *CPS*. For the $\mathbb{G}_{\mathcal{B}_i}^{\mathcal{M}_P}$, the *CPS* is given by (Kim, 2022):

$$CPS_{\mathbb{G}_{\mathcal{B}_i}^{\mathcal{M}_P}}\left(\mathbb{N}_{\mathcal{B}_i}^{P},\mathcal{S}_P^{\mathcal{B}_i},c_{\mathbb{N}_{\mathcal{B}_i}^{P}}\right) = \left\{c_{\mathbb{N}_{\mathcal{B}_i}^{P}} = \left(\cdots c_{D_j^{\mathcal{B}_i,P}\in\mathbb{N}_{\mathcal{B}_i}^{P}}\cdots\right)|\left(\lambda_P^{CPS}\cdot c_{\mathbb{N}_{\mathcal{B}_i}^{P}}\right)\right\} \tag{4.1.4}$$

$$\text{s.t.},\ \lambda_P^{CPS} = \mathbf{max}\left\{t \in (0,1]|\sum\nolimits_{D_j^{\mathcal{B}_i,P}\in\mathbb{N}_{\mathcal{B}_i}^{P}}\mathcal{S}_j^{\mathcal{B}_i,P} \le \mathcal{S}_P^{\mathcal{B}_i}\right\}$$

where $c_{D_j^{\mathcal{B}_i,P}}$ is the claim of $D_j^{\mathcal{B}_i,P}$; it is given by $\mathfrak{U}_j^{\mathcal{B}_i,P}(\cdot)$ with $\mathcal{S}_j^{\mathcal{B}_i,P}$ =

$R_j^{\mathcal{B}_i,P}$. According to (4.1.4), the $\mathcal{S}_j^{\mathcal{B}_i,P}$ is obtained for the PS of $D_j^{\mathcal{B}_i,P}$. For the $\mathcal{S}_S^{\mathcal{B}_i}$ scheduling process, the $\mathbb{G}_{\mathcal{B}_i}^{\mathcal{M}_S}$ is designed to support multiple SSs. In the $\mathbb{G}_{\mathcal{B}_i}^{\mathcal{M}_S}$, devices in the $\mathbb{N}_{\mathcal{B}_i}^S$, i.e., $D_k^{\mathcal{B}_i,S} \in \mathbb{N}_{\mathcal{B}_i}^S$, are game players and the scheduling spectrum amount for the $D_k^{\mathcal{B}_i,S}$, i.e., $\mathcal{S}_k^{\mathcal{B}_i,S}$, is his strategy. The utility function of $D_k^{\mathcal{B}_i,S}$, i.e., $\mathfrak{U}_k^{\mathcal{B}_i,S}(\cdot)$, is defined as follows (Kim, 2022):

$$\mathfrak{U}_k^{\mathcal{B}_i,S}\left(\mathcal{S}_k^{\mathcal{B}_i,S}, R_k^{\mathcal{B}_i,S}, \mathbb{N}_{\mathcal{B}_i}^S\right) = \frac{\phi}{\beta + exp\left(\rho \times \frac{\min\left(\mathcal{S}_k^{\mathcal{B}_i,S}, R_k^{\mathcal{B}_i,S}\right)}{R_k^{\mathcal{B}_i,S}}\right)} - \omega \tag{4.1.5}$$

where ϕ, β, ρ, ω are control parameters for $\mathfrak{U}_k^{\mathcal{B}_i,S}(\cdot)$. Owning to the high data rate characterization, the *AMRA* scheme accepts the concept of *TPS* for the $\mathbb{G}_{\mathcal{B}_i}^{\mathcal{M}_S}$, and it is obtained as follows (Kim, 2022):

$$TPS_{\mathbb{G}_{\mathcal{B}_i}^{\mathcal{M}_S}}\left(\mathbb{N}_{\mathcal{B}_i}^S, \mathcal{S}_S^{\mathcal{B}_i}, \hat{c}_{\mathbb{N}_{\mathcal{B}_i}^S}\right) = \left\{\hat{c}_{\mathbb{N}_{\mathcal{B}_i}^S} = (\ldots \hat{c}^{S,D_k^{\mathcal{B}_i,S} \in \mathbb{N}_{\mathcal{B}_i}^S} \ldots) | \left(\lambda_S^{TPS} \cdot \hat{c}_{\mathbb{N}_{\mathcal{B}_i}^S}\right)\right\} \tag{4.1.6}$$

$$s.t.,\ \lambda_S^{TPS} = max\left\{t \in (0,1] | \sum_{D_k^{\mathcal{B}_i,S} \in \mathbb{N}_{\mathcal{B}_i}^S} \mathcal{S}_k^{\mathcal{B}_i,S} \le \mathcal{S}_S^{\mathcal{B}_i}\right\}$$

where $\hat{c}^{S,D_k^{\mathcal{B}_i,S}}$ is the truncated claim of $D_k^{\mathcal{B}_i,S}$; it is given to ensure the device's SSs. For the $\mathcal{S}_T^{\mathcal{B}_i}$ scheduling process, the $\mathbb{G}_{\mathcal{B}_i}^{\mathcal{M}_T}$ is designed to support multiple TSs. In the $\mathbb{G}_{\mathcal{B}_i}^{\mathcal{M}_T}$, devices in the $\mathbb{N}_{\mathcal{B}_i}^T$, i.e., $D_l^{\mathcal{B}_i,T} \in \mathbb{N}_{\mathcal{B}_i}^T$, are game players and the scheduling spectrum amount for the $D_l^{\mathcal{B}_i,T}$, i.e., $\mathcal{S}_l^{\mathcal{B}_i,T}$, is his strategy. The utility function of $D_l^{\mathcal{B}_i,T}$, i.e., $\mathfrak{U}_l^{\mathcal{B}_i,T}(\cdot)$, is defined as follows (Kim, 2022):

$$\mathfrak{U}_l^{\mathcal{B}_i,T}\left(\mathcal{S}_l^{\mathcal{B}_i,T}, R_l^{\mathcal{B}_i,T}\right) = \mu \times \log\left(\frac{\min\left(\mathcal{S}_l^{\mathcal{B}_i,T}, R_l^{\mathcal{B}_i,T}\right)}{R_l^{\mathcal{B}_i,T}} + \tau\right) \tag{4.1.7}$$

where μ, τ are control parameters for $\mathfrak{U}_k^{\mathcal{B}_i,T}(\cdot)$. To support TSs, the *AMRA* scheme needs a large number of temporary transmissions with no strict QoS requirement. Therefore, the $\mathcal{M}_T$ strongly emphasizes the restricted truncated proportionality. Therefore, the *AMRA* scheme uses the approach of *RTPS* for the $\mathbb{G}_{\mathcal{B}_i}^{\mathcal{M}_T}$, and it is given by: Devices in the $\mathbb{N}_{\mathcal{B}_i}^T$ have also their truncated claims ($\hat{c}^{T,D_l}$) to ensure their TSs. With their truncated claims, the $\mathcal{S}_l^{\mathcal{B}_i,T}$ value is decided by using the concept of the *RTPS*. For the $\mathbb{G}_{\mathcal{B}_i}^{\mathcal{M}_T}$, the *RTPS* (Kim, 2022);

$$RTPS_{\mathbb{G}_{\mathcal{B}_i}^{\mathcal{M}_T}}\left(\mathbb{N}_{\mathcal{B}_i}^T, \mathcal{S}_T^{\mathcal{B}_i}, c_{\mathbb{N}_{\mathcal{B}_i}^T}\right) =$$

$$\begin{cases} \left\{c_{\mathbb{N}_{\mathcal{B}_i}^T} = \left(\cdots c_{D_l^{\mathcal{B}_i,T} \in \mathbb{N}_{\mathcal{B}_i}^T} \cdots\right) | \left(\lambda_T^{CPS} \cdot c_{\mathbb{N}_{\mathcal{B}_i}^T}\right)\right\}, if\, c_{\mathbb{N}_{\mathcal{B}_i}^T} < \mathfrak{U}^*_{\mathbb{N}_{\mathcal{B}_i}^T}. \\ \left\{\mathfrak{U}^*_{\mathbb{N}_{\mathcal{B}_i}^T} = \left(\ldots \mathfrak{U}^*_{D_l^{\mathcal{B}_i,T} \in \mathbb{N}_{\mathcal{B}_i}^T} \cdots\right) | \left(\lambda_T \cdot \mathfrak{U}^*_{\mathbb{N}_{\mathcal{B}_i}^T}\right)\right\}, otherwise. \end{cases} \tag{4.1.8}$$

$$
\text{s.t.,} \begin{cases} \lambda_T^{CPS} = max\left\{ t \in (0,1] | \sum_{D_l^{\mathcal{B}_i,T} \in \mathbb{N}_{\mathcal{B}_i}^T} \mathcal{S}_l^{\mathcal{B}_i,T} \leq \mathcal{S}_T^{\mathcal{B}_i} \right\} \\ \lambda_T = max\left\{ t \in (0,1] | \sum_{D_l^{\mathcal{B}_i,T} \in \mathbb{N}_{\mathcal{B}_i}^T} \mathcal{S}_l^{\mathcal{B}_i,T} \leq \mathcal{S}_T^{\mathcal{B}_i} \right\} \end{cases}
$$

where $c_{D_l^{\mathcal{B}_i,T}}$ is the claim of $D_l^{\mathcal{B}_i,T}$; it is given by $\mathfrak{U}_l^{\mathcal{B}_i,T}(\cdot)$ with $\mathcal{S}_l^{\mathcal{B}_i,T} = R_l^{\mathcal{B}_i,T}$. $\mathfrak{U}_{\mathbb{N}_{\mathcal{B}_i}^T}^{*}$ is the vector of utopia values for devices in the $\mathbb{N}_{\mathcal{B}_i}^T$, and $\mathfrak{U}_{D_l^{\mathcal{B}_i,T}}^E$ is the utopia value of $D_l^{\mathcal{B}_i,T}$ (Kim, 2022).

4.1.6 Summary

Confronted with the coexistence of various traffic types in future networks, wireless network virtualization can be employed to slice and share network resources among multiple service providers. The *AMRA* scheme introduces a new spectrum control algorithm for the UDNs, conceptualized as a two-tier bargaining game model. By jointly applying the *WKSBS*, *CPS*, *TPS*, and *RTPS* solutions, each MVNO collaborates at the first tier and operates independently at the second tier. Through this interactive bargaining approach - incorporating multiple bargaining solutions - the *AMRA* scheme achieves a well-balanced performance for the UDN system (Kim, 2022).

4.2 Dynamic Virtual Machine Migration (*DVMM*) scheme

As mobile networks advance toward 5G, Mobile Edge Computing (MEC) has emerged as a key paradigm designed to address the escalating computational demands of mobile applications. By positioning computing resources closer to end devices, MEC facilitates efficient mobile cloud computing services. Within this framework, Virtual Machine (VM) migration - the transfer of a virtual machine from one edge node to another - represents a critical challenge. To enhance both service quality and overall system efficiency, VM migration strategies must consider the computation and communication resources of the MEC platform. Recently, S. Kim models the VM migration challenge as a one-on-one contract game and proposes the *Dynamic Virtual Machine Migration* (*DVMM*) scheme to optimize resource utilization within MEC (Kim, 2018b). Through game theory and a learning-driven approach, the *DVMM* scheme adapts to the evolving MEC environment by continuously interacting with an unpredictable system context, effectively capturing system dynamics in real time.

4.2.1 Development Motivation

Next-generation 5G networks necessitate a fundamental shift in how they are planned and operated. The growing popularity of the IoT has heightened the significance of mobile devices, which now serve as crucial platforms for education, entertainment, social interaction, real-time updates, and business processes. As the appeal of smart mobile devices continues to soar, users increasingly expect to run computation-heavy applications anywhere and anytime. However, these devices often face hardware and energy limitations, making it challenging to support resource-intensive applications (Wang, 2016b). This shortfall can compromise user experience, as resource constraints become a bottleneck to maintaining satisfactory Quality of Experience (QoE) - potentially delaying the development of future networks (Kim, 2010, Kim, 2018b; Mao, 2016).

To bridge this gap and accommodate more sophisticated applications, external resources must supplement local computing on mobile devices. Mobile Cloud Computing (MCC) has emerged as a promising solution to enhance device capabilities by offloading compute-intensive tasks to resource-rich cloud servers, thus improving application performance and reducing energy consumption. Placing MCC servers closer to users (e.g., near Wi-Fi access points) can lower latency (Wu, 2016). Nonetheless, MCC has limitations. First, it may not guarantee ubiquitous service, given the restricted coverage of Wi-Fi networks. Second, MCC might be unable to maintain sufficient QoE for a significant number of users in cloudlet-based mobile computing, as resource constraints pose scalability challenges (Ahmed, 2016; Chen, 2015; Kim, 2018b; Lee, 2015; Liu, 2015; Zhu, 2014).

To address these drawbacks, the MEC has emerged as a powerful complement to MCC. In MEC, cloud computing platforms are placed at the network edge, enabling low-latency support for delay-sensitive, context-aware applications (Ahmed, 2016). By pushing computing resources closer to the end users, the MEC effectively accelerates data processing and streaming, which is critical for consumer-centric IoT services (Jaraweh, 2016). Consequently, MEC can deliver real-time operations and improved QoE, particularly for time-sensitive applications (Kim, 2018b).

Despite its potential, ensuring optimal resource utilization while maintaining the desired QoE in MEC systems is highly challenging. One successful approach involves using VMs for partitioning and sharing physical resources such as CPU, storage, and bandwidth. Because mobile users frequently change locations, MEC architectures must also account for VM migration, allowing ongoing applications to move smoothly from one physical node to another without service interruption (Becvar, 2014; Plachy, 2016). This raises a crucial question in MEC: how to migrate VMs to maximize overall system performance (Kim, 2018b).

A core consideration for designing an effective VM migration strategy is coordinating communication and computational resources. Recognizing the synergy between these resources has inspired the development of a combined resource pricing mechanism in MEC. Indeed, strategic pricing has a significant influence on user behavior and is one possible technique for steering the system toward globally desirable outcomes. However, because traffic loads in MEC environments can vary substantially over time, determining the optimal price is an ongoing challenge (Kim, 2018b).

Many researchers have investigated centralized control solutions, but these become problematic for distributed systems such as MEC due to the high overheads and limited scalability associated with a centralized approach. In contrast, distributed control shifts computational responsibilities to the network's edge nodes and is considered a more practical way to manage MEC systems. When autonomous or competing system agents coordinate their actions, the situation naturally lends itself to a game-theoretic framework (Kim, 2014, Kim, 2018b).

4.2.2 Contribution

The *DVMM* scheme introduces a one-on-one contract game model specifically designed for the MEC systems. In this scheme, a mobile user and a pair of edge nodes act as the primary players, each seeking to maximize their respective profits. Within the game, potential actions include; i) VM migration decision - choosing whether to migrate an actively running virtual machine, ii) contract formation - concluding agreements for VM migration, and iii) resource pricing - adjusting an edge node's resource price. Given the asymmetric information typical of MEC environments, these three control challenges are effectively managed through continual interactions in the one-on-one contract game. The *DVMM* scheme's key contribution lies in designing this game model to achieve optimal performance under given system conditions. Notably, research on VM migration in MEC remains sparse, with only a handful of related studies currently available in the literature (Kim, 2018b).

4.2.3 One-on-one Contract Game Model for the VM Migration

The MEC architecture is characterized by a distributed computing infrastructure consisting of multiple high-performance edge nodes (ENs) that maintain robust interconnections. This computing paradigm aims to bring cloud-like capabilities closer to mobile users (MUs), enabling ENs to directly receive workloads via wireless links. Additionally, ENs can exchange real-time information about running application workloads, promoting cooperative interactions. Whereas the MCC model extends mobile application capabilities by tapping into centralized cloud resources, the MEC approach pushes this concept further by distributing the bulk of computational tasks directly among the ENs in the local environment. To handle offloaded tasks, each EN allocates resources through the MEC system and spins up VMs for processing. A general depiction of the MEC system architecture is illustrated in Figure 1 (Kim, 2018b).

Figure 1. General F-RAN system structure (Kim, 2018b)

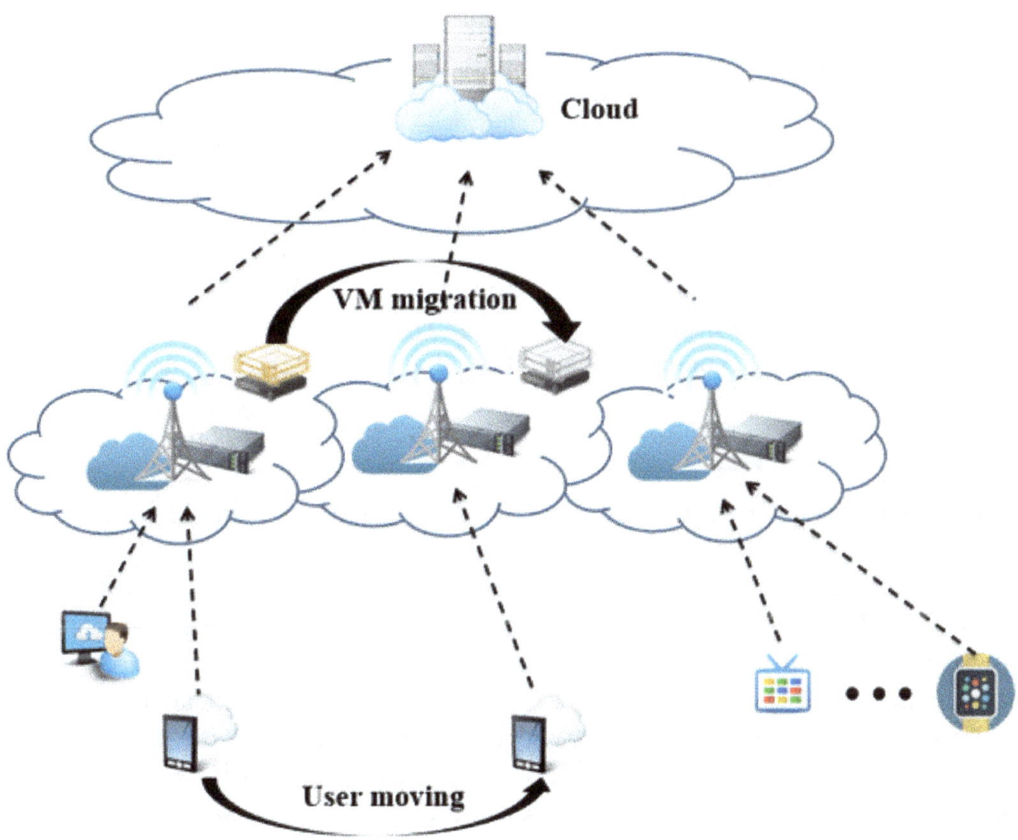

When MUs are on the move, ensuring consistent QoE in an MEC environment becomes more complex. During system operation, communication and computing resources must be jointly allocated to meet QoE requirements for offloaded applications. A widely used technique to maintain QoE in MEC scenarios is VM migration. When an MU transitions from one EN to another, the VM hosting that user's application is transferred from the previous EN to the new EN, thereby reducing communication delay by adapting to user mobility (Plachy, 2016). Generally, implementing VM migration involves multiple decision processes: determining whether to migrate, deciding how many MEC resources to allocate, and selecting the optimal price for those resources. Developing an adaptive VM migration algorithm is challenging due to the dynamic nature of MEC environments (Kim, 2018b).

In an MEC system, each agent - whether an EN or MU - makes decisions individually, but with awareness of their interdependencies. They aim to maximize their profits, either competitively or cooperatively. This setup naturally aligns with game theory. The *DVMM* scheme introduces a one-on-one contract game with three distinct phases. In the phase 1, an EN determines whether to migrate a specific VM, jointly considering communication and computation resources. In the phase 2, an EN-MU pair negotiates a contract based on the EN's offered price, deciding whether to utilize available resources. In the phase 3, the EN learns MUs' behaviors and dynamically adjusts its pricing strategy in real time. These three phases collectively facilitate an interactive decision-making environment for VM migration, enabling the system to adapt to evolving conditions and optimize resource usage. Formally, the *DVMM* scheme defines the one-on-one contract game model G= $\{\{\mathscr{N},\mathscr{M}\},\mathscr{V}, \{\mathscr{S}^{\mathscr{N}},\mathscr{S}^{\mathscr{M}}\}, \{\mathscr{U}^{\mathscr{N}},\mathscr{U}^{\mathscr{M}}\}, \mathrm{T}\}$ at each time period t of gameplay (Kim, 2018b).

- N and M are the finite sets of game players where $\mathscr{N} = \{EN_1 \ldots EN_n\}$ is the set of ENs and $\mathscr{M} = \{MU_1 \ldots MU_m\}$ is the set of MUs; ENs and MUs are game players.
- **V** is the set of VMs where $\mathscr{V}_j^{EN_i} \in \mathscr{V}$ represents the jth running VM in the EN_i. $\mathscr{V}_j^{EN_i}$ is defined as $(\mathfrak{B}_j^{EN_i}, \mathfrak{A}_j^{EN_i}, \tau_j^d(\Delta))$ where $\mathfrak{B}_j^{EN_i}$ is the bandwidth size for the $\mathscr{V}_j^{EN_i}$ migration, i.e., bits per second, $\mathfrak{A}_j^{EN_i}$ is the computation capacity to execute the $\mathscr{V}_j^{EN_i}$, i.e., number of CPU cycles per second, and $\tau_j^d(\Delta)$is the time deadline for the $\mathscr{V}_j^{EN_i}$ application during the time period Δ.
- $\mathscr{S}^{\mathscr{N}}$ and $\mathscr{S}^{\mathscr{M}}$ are the finite sets of game strategies where $\mathscr{S}^{\mathscr{N}} = \{\vartheta_1^{\mathfrak{P}}, \vartheta_2^{\mathfrak{P}}, \ldots, \vartheta_l^{\mathfrak{P}}\}$ is the set of ENs' strategies and $\mathscr{S}^{\mathscr{M}}$=\{*migrate*, *maintain*, *disconnect*\} is the set of MUs' strategies. In the game model, $\vartheta_{1 \le i \le l}^{\mathfrak{P}}$ represents a discrete price for the MEC resource unit, and Õ $\in \mathscr{S}^{\mathscr{M}}$ represents a MU's strategy for the EN's price offering, e.g., ϒ = *migrate* means the migration strategy of $\mathscr{V}_j^{EN_i}$, ϒ = *maintain* means the non-migration strategy of $\mathscr{V}_j^{EN_i}$, and ϒ = *disconnect* means the stop of application service.
- $\mathscr{U}^{\mathscr{N}}$ and $\mathscr{U}^{\mathscr{M}}$ are the finite sets of payoffs where $\mathscr{U}^{\mathscr{N}} = \{U_1^{\mathscr{N}} \ldots U_n^{\mathscr{N}}\}$ is the set of ENs' received payoffs and $\mathscr{U}^{\mathscr{M}} = \{U_1^{\mathscr{M}} \ldots U_m^{\mathscr{M}}\}$ is the set of MUs' received payoffs.
- The T is a time period. The G is repeated $t \in T < \infty$ time periods with imperfect information.

In the G game, the ENs and mobile users MUs act as rational players striving for a mutually beneficial (win-win) outcome. In a one-to-one contract scenario, each player makes strategic decisions to achieve their respective goals. From an EN's perspective, the primary aim is to maximize system performance while maintaining QoE standards for MUs. Meanwhile, each MU focuses on maximizing its own payoff. To address these sometimes conflicting objectives, the G game model is divided into three phases; i) VM migration decision - deciding whether or not to migrate a virtual machine, ii) contract conclusion - finalizing an agreement for MEC resource allocation, and iii) learning based price adjustment

- dynamically refining resource pricing through a learning process. By combining these three decision phases in a coordinated manner, the G game model achieves a balanced MEC performance, effectively accommodating both system-wide and user-centric requirements (Kim, 2018b).

4.2.4 VM Migration Algorithm in the MEC system

When a MU transitions from one EN coverage area to another, it informs the network of its updated position through a location update procedure (Sun, 2016). This mobility management process is pivotal in VM migration, as it enables the EN currently hosting a VM to determine the next potential proxy EN for that user. As previously noted, migrating a running VM between ENs can reduce application latency for the corresponding MU. However, not all situations call for a VM migration immediately upon an MU's movement; some migrations may fail to lower latency and could add unnecessary traffic overhead to the MEC system. Consequently, whenever an MU connects to a new EN, it becomes essential to estimate the expected profit of migrating its VM, ensuring that only beneficial migrations are performed (Kim, 2018b; Sun, 2016).

In order to implement the $\mathcal{V}_j^{EN_i}$ migration process, the *DVMM* scheme partitions the time-axis into intervals of length, i.e., $\Delta_j^{EN_i}$. The value of $\Delta_j^{EN_i}$ represents the time that the MU_j is covered by the EN_i. Therefore, the $\Delta_j^{EN_i}$ is chosen based on the MU_j's moving speed. In general, different speed MUs have different QoE propensities. Therefore, each EN should learn the propensities of different speed MUs and decides service prices differentially for individual Mus (Kim, 2018b).

At the first phase in the VM migration algorithm, a EN makes a decision whether to migrate a running VM or not. For the VM migration process, one of decision criteria is the delay latency. The *DVMM* scheme defines a latency gap ($\mathfrak{X}$). When the MU_j roams into a new EN, the latency gap of the $\mathcal{V}_j^{EN_i}$ migration $\left(\mathfrak{X}\left(\mathcal{V}_j^{EN_i}\right)\right)$ is estimated by considering the cases with and without the $\mathcal{V}_j^{EN_i}$migration. If the MU_j is moving from the EN_i area to the EN_sarea, the total traffic amount generated from the $\mathcal{V}_j^{EN_i}$ comprises two parts; $\mathcal{L}_j^{EN_i}$ $(or\ \mathcal{L}_j^{EN_s})$ is the traffic amount when the MU_j in the EN_i $(or\ EN_s)$ area. Finally, $\mathfrak{X}\left(\mathcal{V}_j^{EN_i}\right)$ can be defined as (Kim, 2018b);

$$\mathfrak{X}\left(\mathcal{V}_j^{EN_i}\right) = \mathcal{H}\left(\mathcal{V}_j^{EN_i}, 2 \times \Delta_j^{EN_i}\right) - \mathcal{M}\left(\mathcal{V}_j^{EN_i}, 2 \times \Delta_j^{EN_i}\right) \tag{4.2.1}$$

$$\text{s.t.,} \begin{cases} H\left(\mathcal{V}_j^{EN_i}, 2 \times \Delta_j^{EN_i}\right) = \left\{\left(\mathbf{t}_j^{EN_i} \times \mathcal{L}_j^{EN_i}\right) + \left(d_j^{EN_s} \times \mathbf{t}_j^{EN_s} \times \mathcal{L}_j^{EN_s}\right)\right\} \\ M\left(\mathcal{V}_j^{EN_i}, 2 \times \Delta_j^{EN_i}\right) = \left\{\left(\mathbf{t}_j^{EN_i} \times \mathcal{L}_j^{EN_i}\right) + \left(\mathbf{t}_{EN_i}^{EN_s} \times \Delta\left(\mathcal{V}_j^{EN_i}, \mathcal{V}_j^{EN_s}\right)\right) + \left(\mathbf{t}_j^{EN_s} \times \mathcal{L}_j^{EN_s}\right)\right\} \end{cases}$$

where $\mathcal{H}\left(\mathcal{V}_j^{EN_i}, 2 \times \Delta_j^{EN_i}\right)$ $\left(or\mathcal{M}\left(\mathcal{V}_j^{EN_i}, 2 \times \Delta_j^{EN_i}\right)\right)$is the total running time of MU_j's application without (*or* with) the $\mathcal{V}_j^{EN_i}$ migration during the comparison time $\left(2 \times \Delta_j^{EN_i}\right)$. $\mathbf{t}_j^{EN_i}$and $\mathbf{t}_j^{EN_s}$ are the communication delay from the EN_iandEN_s to the MU_j, respectively. $d_j^{EN_s}$ is the communication delay from the EN_i to the EN_s; it considers the extra latency between the EN_i and the EN_s. $\mathbf{t}_{EN_i}^{EN_s}$ and $m\left(\mathcal{V}_j^{EN_i}, \mathcal{V}_j^{EN_s}\right)$ is the delay and the traffic amount transferred from the EN_i to the EN_s during the $\mathcal{V}_j^{EN_i}$ migration process. With the latency gap ($\mathfrak{X}$), other constraints conditions for the $\mathcal{V}_j^{EN_i}$ migration are related to the current availability of MEC computation and communication resources; they can be interpreted as follows (Kim, 2018b):

$$\begin{cases} X\left(\mathcal{V}_j^{EN_i}\right) > å, s.t., å > 0 \\ \left(\mathcal{F}_c(EN_s) - \mathfrak{A}_j^{EN_i}\right) > 0 \\ \left(\mathcal{F}_b(EN_i, EN_s) - \mathfrak{B}_j^{EN_i}\right) > 0 \end{cases} \quad (4.2.2)$$

where the $\mathcal{F}_c(EN_s)$ is the available computation power in the EN_s, and $\mathcal{F}_b(EN_i, EN_s)$ is the available bandwidth from the EN_i to the EN_s for the $\mathcal{V}_j^{EN_i}$ migration. If all the conditions in (4.2.2) are satisfied, the EN_i can suggest the $\mathcal{V}_j^{EN_i}$ migration to the MU_j (Kim, 2018b).

At the second phase in the $\mathcal{V}_j^{EN_i}$ migration algorithm, the main step is a one-on-one contract between the EN_i and MU_j. During the contract process, each *EN* is a *proposer* and individual MU is a *responder*; they interact with each other for the VM migration. At this time, the *proposer* EN_i offers the price $\left(\mathfrak{P}_j^{EN_i}\right)$ for the $\mathcal{V}_j^{EN_i}$ migration to the MU_j. Based on the offered price, the MU_j estimates its payoff $U_j^{\mathcal{M}}(\cdot)$ as follows (Kim, 2018b);

$$U_j^{\mathcal{M}}\left(\mathcal{V}_j^{EN_i}, \mathfrak{P}_j^{EN_i}(t), \tilde{O}\right) = \mathbb{B}_j\left(\mathcal{V}_j^{EN_i}\right) - \mathbb{C}_j\left(\mathfrak{P}_j^{EN_i}(t), \mathfrak{A}_j^{EN_i}, \mathfrak{B}_j^{EN_i}\right) \quad (4.2.3)$$

$$s.t., \mathfrak{P}_j^{EN_i} \in \mathcal{S}^{\mathcal{N}} \text{and } \tilde{O} \in \mathcal{S}^{\mathcal{M}}$$

where $\mathbb{B}_j(\cdot)$ and $\mathbb{C}_j(\cdot)$ are the outcome and cost of the MU_j. $\mathfrak{P}_j^{EN_i}(t)$ is the EN_i's offered price at time t. If the strategy of the MU_j is the $\mathcal{V}_j^{EN_i}$ non-migration, i.e., $\tilde{O} = maintain$, $\mathbb{B}_j(\cdot)$ and $\mathbb{C}_j(\cdot)$ are defined as follows (Kim, 2018b);

$$\begin{cases} \mathbb{B}_j\left(\mathcal{V}_j^{EN_i}\right) = \dfrac{\omega^{\mathfrak{D}_s}}{\left(\left(\mathfrak{t}_j^{EN_i} \times \mathcal{L}_j^{EN_i}\right) + \left(d_j^{EN_i} \times \mathfrak{t}_j^{EN_i} \times \mathcal{L}_j^{EN_i}\right)\right)} \\ s.t., \mathfrak{D}_s = \max\left\{\left(\tau_j^d\left(2 \times \Delta_j^{EN_i}\right) - \mathcal{H}\left(\mathcal{V}_j^{EN_i}, 2 \times \Delta_j^{EN_i}\right)\right), 0\right\} \times \left(\mathcal{L}_j^{EN_i} + \mathcal{L}_j^{EN_s}\right) \\ \mathbb{C}_j\left(\mathfrak{P}_j^{EN_i}(t), \mathfrak{A}_j^{EN_i}, \mathfrak{B}_j^{EN_i}\right) = \dfrac{\left\{\left(\mathfrak{P}_j^{EN_i}(t) \times \mathcal{L}_j^{EN_i}\right) + \left(\mathfrak{P}_j^{EN_i}(t) \times \left(\varphi_{EN_i} \times \mathfrak{t}_{EN_i}^{EN_s}\right) \times \mathcal{L}_j^{EN_s}\right)\right\}}{\left(\left(\mathfrak{t}_j^{EN_i} \times \mathcal{L}_j^{EN_i}\right) + \left(d_j^{EN_i} \times \mathfrak{t}_j^{EN_i} \times \mathcal{L}_j^{EN_i}\right)\right)} \end{cases} \quad (4.2.4)$$

Otherwise, if the strategy of the MU_j is the $\mathcal{V}_j^{EN_i}$ migration, i.e., $\tilde{O} = migrate$, the outcome $\left(\mathbb{B}_j(\cdot)\right)$ and cost $\left(\mathbb{C}_j(\cdot)\right)$ of the MU_j are defined as follows (Kim, 2018b);

$$\begin{cases} \mathbb{B}_j\left(\mathcal{L}_j^{EN_i} + \mathcal{L}_j^{EN_s}\right) = \dfrac{\omega^{\mathfrak{D}_s}}{\left(\left(\mathfrak{t}_j^{EN_i} \times \mathcal{L}_j^{EN_i}\right) + \left(\mathfrak{t}_{EN_i}^{EN_s} \times m\left(\mathcal{V}_j^{EN_i}, \mathcal{V}_j^{EN_s}\right)\right) + \left(\mathfrak{t}_j^{EN_s} \times \mathcal{L}_j^{EN_s}\right)\right)} \\ s.t., \mathfrak{D}_s = \max\left\{\left(\tau_j^d\left(2 \times \Delta_j^{EN_i}\right) - \mathcal{M}\left(\mathcal{V}_j^{EN_i}, 2 \times \Delta_j^{EN_i}\right)\right), 0\right\} \times \left(\mathcal{L}_j^{EN_i} + \mathcal{L}_j^{EN_s}\right) \\ \mathbb{C}_j\left(\mathfrak{P}_j^{EN_i}(t), \mathfrak{A}_j^{EN_i}, \mathfrak{B}_j^{EN_i}\right) = \dfrac{\left\{\left(\mathfrak{P}_j^{EN_i}(t) \times \left(\mathfrak{A}_j^{EN_i} + \mathfrak{B}_j^{EN_i}\right)\right) + \left(\mathfrak{P}_j^{EN_i}(t) \times \left(\mathcal{L}_j^{EN_i} + \mathcal{L}_j^{EN_s}\right)\right)\right\}}{\left(\left(\mathfrak{t}_j^{EN_i} \times \mathcal{L}_j^{EN_i}\right) + \left(d_j^{EN_i} \times \mathfrak{t}_j^{EN_i} \times \mathcal{L}_j^{EN_i}\right)\right)} \end{cases} \quad (4.2.5)$$

If the MU_j stops its application service, i.e., $\tilde{O} = disconnect$, $\mathbb{B}_j(\cdot)$ and $\mathbb{C}_j(\cdot)$ are 0. Therefore, the MU_j has three different utility functions based on its strategy. In the point view of MU_j, utility maximization is the main concern. To this aim, the objective of MU_j can be presented as (Kim, 2018b):

$$\max_{\tilde{O} \in \mathcal{S}^{\mathcal{M}}} \left\{ U_j^{\mathcal{M}} \left(\mathcal{V}_j^{EN_i}, \mathfrak{P}_j^{EN_i}(t), \tilde{O} \right) \right\}$$

(4.2.6)

$$\text{s.t.}, \mathcal{S}^{\mathcal{M}} = \{migrate,\ maintain,\ disconnect\}$$

At the third phase in the VM migration algorithm, each EN adjusts its price policy distribution. In the *DVMM* scheme, the price decision problem is assumed as one of online learning problems; by observing MUs' responses, each EN can learn their propensity. Therefore, the result of current contract is the input back to the *EN*'s price decision process. The dynamics of this interactive contract feedback can help to decide the real-time price, effectively. For simplicity, the *DVMM* scheme assumes that the ranges of the actual price value (P) are discretely fixed, i.e., $\mathfrak{P}_j^{EN_i}(t) \in \mathcal{S}^{\mathcal{N}} =$

$\left\{ \vartheta_1^{\mathfrak{P}}, \vartheta_2^{\mathfrak{P}}, \ldots, \vartheta_l^{\mathfrak{P}} \right\}$ where l is the number of possible price strategies; $\vartheta_1^{\mathfrak{P}}$ and $\vartheta_l^{\mathfrak{P}}$ are the pre-defined minimum and maximum price for the MEC resource unit (Kim, 2018b).

In recent years, a variety of learning algorithms have been developed for use in non-deterministic environments. Among these, Q-learning stands out as one of the most widely recognized methods, having been successfully implemented in areas such as game learning, optimization, and financial applications. A key advantage of Q-learning, compared to many model-based approaches, is that it operates without requiring knowledge about the underlying model or environment (Sarabakha, 2016). Consequently, the *DVMM* scheme employs Q-learning to handle the time-varying conditions found in MEC systems. In Q-learning, the Q-value represents the expected reward obtained from following a particular strategy over future actions. The learner updates this Q-value according to the following process (Kim, 2018b):

$$Q(s,a) = ((1-\alpha) \times Q(s,a)) + \left(\alpha \times \left(\mathcal{R} + \left\{ \gamma \times \max_{a} Q(\acute{s},\acute{a}) \right\} \right) \right), \ \text{s.t.}, \alpha \in [0, 1]$$

(4.2.7)

where γ is a discount factor for the future value, and R is the current reward. The parameter α controls the relative weights given to recent and past reward histories. The *DVMM* scheme develops an online learning mechanism. Therefore, γ is set to 0. In the algorithm, Q-values represent the propensities for $\vartheta_{1 \leq k \leq l}^{\mathfrak{P}}$ price strategies. When the EN_i selects the $\vartheta_{k, 1 \leq k \leq l}^{\mathfrak{P}}$ price at time t, i.e., $\mathfrak{P}_j^{EN_i}(t) = \vartheta_k^{\mathfrak{P}}$, R of the $\mathcal{V}_j^{EN_i}$ migration decision is defined as (Kim, 2018b);

$$\mathcal{R} = \mathfrak{Z}(\tilde{O}) \times \left(U_j^{\mathcal{M}} \left(\mathcal{V}_j^{EN_i}, \mathfrak{P}_j^{EN_i}(t) = \vartheta_k^{\mathfrak{P}}, \tilde{O} \right) \Big/ \mathbb{E} \left(U_j^{\mathcal{M}} \left(\mathcal{V}_j^{EN_i}, \mathfrak{P}_j^{EN_i}(t) = \vartheta_l^{\mathfrak{P}}, \tilde{O} \right) \right) \right)$$

(4.2.8)

$$\text{s.t.,}\,\mathfrak{Z}(\tilde{O}) = \begin{cases} 1, if\tilde{O} = migrate \\ 0, if\tilde{O} = maintaint \\ -1, if\tilde{O} = disconnec \end{cases}$$

where $\mathbb{E}\left(U_j^{\mathcal{M}}\left(\mathcal{V}_j^{EN_i}, \mathfrak{P}_j^{EN_i}(t) = \vartheta_l^{\mathfrak{P}}, \tilde{O}\right)\right)$ is the expected payoff if the MU_j accepts the EN_i's $\vartheta_l^{\mathfrak{P}}$ offer. When the MEC resource utilization ($_{i,s}^{EN}$) of EN_i and EN_s is light, the system has sufficient resource to support application services. In this case, the lower P values are more suitable. But under the higher $_{i,s}^{EN}$ situation, the *DVMM* scheme can put more emphasis on the higher P values. For a MU who is moving at a speed of $\mathfrak{S}_c$, the *DVMM* scheme defines the weighted Q-value (*WQ*) of the $\vartheta_{k,1\leq k\leq l}^{\mathfrak{P}}$ at the time t like as (Kim, 2018b);

$$WQ_k^{\mathfrak{S}_c}(EN_i(t), \vartheta_k^{\mathfrak{P}}) = \left(\frac{\mathbb{I}\left(_{i,s}^{EN}\right) + 1}{\phi}\right) \times Q^{\mathfrak{S}_c}(EN_i(t), \vartheta_k^{\mathfrak{P}})$$

(4.2.9)

$$\text{s.t.,}\,\mathbb{I}\left(_{i,s}^{EN}\right) = \begin{cases} \left(_{i,s}^{EN} - \eta\right) \times (k/l), if\text{Skipping malformed tag: msubsup} \geq \eta \\ \left(\eta - _{i,s}^{EN}\right) \times ((l-k)/l), otherwise \end{cases}$$

where $Q^{\mathfrak{S}_c}(EN_i(t), \vartheta_k^{\mathfrak{P}})$ is the Q-value for the state of the EN_i at the time t with the $\mathfrak{P}_j^{EN_i}(t) = \vartheta_k^{\mathfrak{P}}$strategy when a corresponding MU is moving at the speed of $\mathfrak{S}_c$. ϕ is a control factor to characterize *WQ* values, and η is a threshold for the *WQ* value adjustment. Under diverse MEC environments, the *WQ* value can be dynamically modified; this approach can make the price decision process more responsive to the current MEC system conditions. In the *DVMM* scheme, EN_i's price strategy attractions are mapped into probability distribution. Therefore, the EN_i adaptively selects its price strategy in an online manner during the one-on-one contract process. When a MU is moving at a speed of $\mathfrak{S}_c$, the selection probability for the kth strategy $\vartheta_k^{\mathfrak{P}}$ $\left(Prob_{EN_i}^k(\vartheta_k^{\mathfrak{P}}, \mathfrak{S}_c, t)\right)$ at time t is defined as follows (Kim, 2018b);

$$Prob_{EN_i}^k(\vartheta_k^{\mathfrak{P}}, \mathfrak{S}_c, t) = \frac{\exp(\chi \times WQ_k^{\mathfrak{S}_c}(EN_i(t), \vartheta_k^{\mathfrak{P}}))}{\sum_{\vartheta_j^{\mathfrak{P}} \in \mathcal{S}^{\mathcal{M}}} \exp(\chi \times WQ_k^{\mathfrak{S}_c}(EN_i(t), \vartheta_j^{\mathfrak{P}}))}$$

(4.2.10)

where χ is the response sensitivity, and $WQ_k^{\mathfrak{S}_c}(EN_i(t), \vartheta_k^{\mathfrak{P}})$ is the EN_i's attraction to choose the strategy $\vartheta_k^{\mathfrak{P}}$ at time t.

4.2.5 Summary

In the quest to fulfill the QoE demands of next-generation IoT applications, the MEC has emerged as a key paradigm for 5G networks. Recognizing the importance of VM migration in providing high QoE, the *DVMM* scheme introduces a specialized algorithm to tackle this complex issue. To address the computational challenges associated with solving the VM migration problem, the *DVMM* scheme models interactions between ENs and MUs as a one-on-one contract game, executed through an iterative mechanism. This game-based process aims to maximize overall MEC performance and simultaneously meet QoE requirements. The *DVMM* scheme employs three VM migration phases to reconcile potentially conflicting objectives; i) VM migration decision, ii) formation of one-on-one contracts between ENs and Mus, and iii) learning based price adjustment for MEC resources. By combining these phases, the *DVMM* scheme ensures comprehensive control over both communication and computation resources in dynamic MEC environments (Kim, 2018b).

4.3 Hierarchical Multi-service Resource Allocation (*HMRA*) scheme

Future networks are expected to operate as multi-service infrastructures, supporting diverse communication service types over a single physical platform. As a transformative approach, wireless network virtualization (WNV) enables multiple network slices to coexist on shared hardware, offering a significant advantage in accommodating various traffic requirements in a multi-service environment. Recently, S. Kim introduces the *Hierarchical Multi-service Resource Allocation* (*HMRA*) scheme for next-generation wireless systems (Kim, 2023). This scheme adopts a joint control paradigm with upper and lower control stages. In the upper control stage, it merges concepts from the *Kalai-Smorodinsky bargaining solution* (*KSBS*) and the *Banzhaf value* (*BV*) to distribute resources among virtual networks. In the lower control stage, each virtual network operator uses a multi-criteria bargaining solution - combining the *Nash bargaining solution* (*NBS*) and the *KSBS* - to allocate the assigned resources to various mobile devices, optimizing performance under diverse service demands. By coordinating across upper and lower control mechanisms, the *HMRA* scheme enables the efficient sharing of finite physical resources. Its primary innovation lies in achieving mutual consensus among competing network agents, ensuring that resource allocation meets a range of network performance objectives (Kim, 2023).

4.3.1 Development motivation

With the rapid expansion of Internet-based applications, there has been an immense surge in data generation, offering users an abundance of information and knowledge. As mobile multimedia applications continue to proliferate, network capacity and service quality requirements are escalating. To address these demanding needs, 5G wireless networks have emerged as the essential backbone for delivering high-speed, low-latency, and extensive connection capabilities. Projections indicate that 5G systems will require up to five times more base stations (BSs) and 200 times more radio spectrum than 4G networks. Orchestrating these numerous 5G components to fulfill performance goals introduces significantly more complexity than in previous generations. Both the academic and industrial sectors have thus undertaken comprehensive research and standardization efforts to promote the development and commercialization of 5G (Cheng, 2018; Feng, 2017a; Kim, 2023).

The primary objective of forthcoming 5G wireless networks is to provide wireless multimedia services at any time, in any location, and under varied conditions. Nonetheless, these bandwidth-heavy and time-sensitive services must operate on limited wireless resources, introducing new challenges rarely encountered in 4G environments. To overcome this discrepancy between restricted resources and expanding wireless multimedia service demands, wireless network virtualization (WNV) has gained recognition as a viable solution. WNV partitions and shares network resources among multiple slices, thus creating virtual versions of underlying physical assets (Han, 2020; Kim, 2023; Ren, 2022; Zhang, 2018).

In a typical WNV setup, the role of a conventional network operator is conceptually split between the infrastructure provider (InP) and various virtual network operators (VNOs). The InP holds physical network elements - such as numerous BSs, antennas, backhaul links, and radio spectrum - while the VNOs lease these resources to deliver services to their own subscribers. The *HMRA* scheme leverages this virtualization paradigm to reduce both capital and operating expenses for the InP, and efficiently share physical spectrum among multiple VNOs. However, a pressing question remains: how best to distribute virtualized resources in a way that meets dynamic, heterogeneous service demands. This becomes increasingly challenging when individual VNOs behave self-interestedly, paying little attention to system-wide efficiency (Han, 2020; Kim, 2023; Tun, 2019).

Given the heterogeneous nature of wireless service types, the quality of service (QoS) demands are likewise varied. According to the ITU radio communication sector, three major usage categories dominate: Ultra-Reliable and Low Latency Communications (URLLC), enhanced Mobile Broadband (eMBB), and massive Machine-Type Communication (mMTC). URLLC prioritizes minimal latency for services such as autonomous vehicles and drone control, while eMBB targets high data-rate activities - like HD video streaming and virtual reality - and mMTC accommodates large-scale connectivity for smart cities and smart homes. An effective approach to WNV lies in forming a resource pool by combining InP assets and creating logical partitions (slices) that cater to distinct QoS needs. Essentially, each slice operates as a virtual network - akin to a partition of the underlying physical network. This structure ensures robust resource sharing among different service classes, maximizing the advantages of virtualization (Han, 2020; Kim, 2023; Li, 2018).

Most existing network virtualization research focuses on user-level resource partitioning, leaving service-level needs overlooked. In multi-service wireless environments, both service-level and user-level requirements must be incorporated into resource allocation strategies. Consequently, a two-level resource allocation method is necessary, splitting the process into inter-slice and intra-slice allocation. Because traffic patterns vary over time, these two allocations must be conducted dynamically and interactively (Han, 2020; Kim, 2023).

Implementing a two-level resource allocation hinges on a joint control paradigm that seeks to maximize efficiency in the spectrum sharing process. Traditionally, 'joint control' is defined as the consensual sharing of strategic oversight between multiple parties - i.e., any significant decision demands the unanimous approval of all involved. As an extension, cooperative optimization has gained recognition for its ability to tackle challenging optimization problems, drawing on principles of coordination. One prominent branch of cooperative optimization is cooperative game theory, initially introduced in 1950 by J. Nash through his *Nash Bargaining Solution* (*NBS*). Since then, numerous additional cooperative game solutions have been proposed, further refining Nash's seminal ideas. A second noteworthy solution for cooperative scenarios is the *Kalai-Smorodinsky Bargaining Solution* (*KSBS*), which uses a utopia point (the highest aspirations for every players) to define a reference line intersecting the bargaining set (Han, 2020; Kim 2014, Kim, 2023; Monroy, 2017).

The driving interest behind cooperative games is fair distribution of collectively obtained outcomes among participants in line with their contributions. The *HMRA* scheme aims to formulate a novel joint control mechanism that addresses resource allocation at both the inter-slice and intra-slice stages. From a cooperative standpoint, network agents exchange information to find a globally optimal solution in a 5G multi-service ecosystem. Recently, the concept of multi-criteria cooperative games has emerged for situations involving multiple evaluation criteria - each of which depends on every players' decisions. As an extension of classic cooperative games, multi-criteria models are especially helpful for ranking and selecting various alternatives, facilitating players' decision processes. However, studies focusing on multi-criteria bargaining solutions remain relatively sparse (Kim 2014, Kim 2023; Monroy 2017).

Another noteworthy solution concept in cooperative games is the *Banzhaf Value* (*BV*), designed to equitably distribute collective profits among coalition members. In the value-based approach, players can form any possible subset, implying that a solution is a function defined over all potential coalitions. Within the *HMRA* scheme, the *BV* can be viewed as a probabilistic measure, assuming that coalitions not containing a specific player can arise with equal likelihood. Well-known in the context of voting scenarios, the *BV* helps evaluate each participant's influence on a coalition's outcome. By incorporating *NBS*, *KSBS*, and *BV* into a joint control paradigm, the *HMRA* scheme aims to optimize resource allocation in virtualized multi-service wireless networks through interactive, cooperative decision-making (Kim, 2023).

4.3.2 Contribution

In the *HMRA* scheme, the upper and lower control stages are meticulously coordinated to achieve a socially optimal outcome. Designed for virtualized multi-service network platforms, the *HMRA* scheme adopts a hierarchical cooperative game model that incorporates the *NBS*, *KSBS*, and *BV*. To the best of my knowledge, this joint control paradigm - rooted in cooperative game theory - is a pioneering approach in the context of 5G network virtualization. For the upper-level resource allocation, the *HMRA* scheme implements inter-slice allocation, selecting optimal strategies by combining the *KSBS* with *BV*. In the lower-level allocation, the *HMRA* strategy employs intra-slice resource distribution in accordance with a multi-criteria cooperative game model. Under different criteria, the *NBS* and *KSBS* are merged to arrive at a compromise solution. Through this joint control paradigm, both upper and lower stages operate in tandem, effectively sharing limited spectrum resources within a virtualized network environment. By leveraging cooperative game solutions, the *HMRA* scheme captures the sequential interactions among network agents and realizes mutual advantages that significantly enhance overall system performance (Kim, 2023).

4.3.3 Network virtualization infrastructure and cooperative game models

In the network virtualization environment, infrastructures are decoupled from the services it provides. As a result, the network operator is decoupled into InP and VNO in wireless virtualization. Specifically, the InP owns the network infrastructure and spectrum resource, and the VNO realizes virtualization while i) allocating the physical resource to each virtual network, and ii) scheduling multiple application services. The *HMRA* scheme assumes a virtualized wireless network with a single InP that has a set of BSs denoted by $\mathbb{B} = \{\mathcal{B}_1,\ldots,\mathcal{B}_n\}$. In each BS, there are three VNOs $(\mathcal{V}_I, \mathcal{V}_{II}, \mathcal{V}_{III})$ for three slice types; they are Type-I slice $(\mathcal{S}_I)$ for the URLLC network, Type-II slice $(\mathcal{S}_{II})$ for the eMBB network, and Type-III slice $(\mathcal{S}_{III})$ for the mMTC network. Mobile devices $\mathbb{D} = \{\mathcal{D}_1,\ldots,\mathcal{D}_m\}$ are randomly distributed in the network area while connected to their corresponding BSs. Each individual BS carries out the spectrum resource management, scheduling, admit control, and more functionalities for their corresponding devices. The *HMRA* scheme assumes that each BS is virtualized and sliced into three isolated slices (Han, 2020; Kim, 2023; Li, 2018).

The *HMRA* scheme assumes that each BS and spectrum resource are owned and managed by a single InP which provides its network infrastructure as a service to VNOs. In each BS, the resource allocation is performed in the two-level stages. In the upper level stage, the BS estimates service demands from all slices and its resource is orthogonally divided into three parts for each type slice; it is called the inter-slice resource allocation where $\mathfrak{M}^I_{\mathcal{B}}$, $\mathfrak{M}^{II}_{\mathcal{B}}$ and $\mathfrak{M}^{III}_{\mathcal{B}}$ are the amounts of allocated resources for the

Type-I, Type-II and Type-III slices, respectively. In the lower level stage, the $\mathscr{V}_I$, $\mathscr{V}_{II}$, and $\mathscr{V}_{III}$ distribute the assigned $\mathfrak{M}_{\mathscr{B}}^{I}$, $\mathfrak{M}_{\mathscr{B}}^{II}$ and $\mathfrak{M}_{\mathscr{B}}^{III}$ resources into their corresponding devices; it is called the intra-slice resource allocation (Han, 2020; Kim, 2023; Li, 2018).

The *HMRA* scheme develops two cooperative games in the upper and lower level stages; they are hierarchically organized based on the joint control paradigm, and work together to achieve reciprocal advantages. The upper level cooperative game $\left(\mathbb{G}_{\mathscr{V}}^{\mathscr{B}}\right)$ allocates the spectrum resource for VNOs, and the lower level cooperative game $\left(\mathbb{G}_{\mathscr{D}}^{\mathscr{V}}\right)$ distributes the allocated resource into corresponding mobile devices. Through the $\mathbb{G}_{\mathscr{V}}^{\mathscr{B}}$ and $\mathbb{G}_{\mathscr{D}}^{\mathscr{V}}$ games, the InP, VNOs and devices are interactively associated with each other in an coordinated manner. Therefore, a mutually acceptable solution is obtained for different network agents. Formally, the *HMRA* scheme defines the tuple entities in the $\mathbb{G}_{\mathscr{V}}^{\mathscr{B}}$ and $\mathbb{G}_{\mathscr{D}}^{\mathscr{V}}$ game models, such as $\mathbb{G} = \left\{\mathbb{G}_{\mathscr{V}}^{\mathscr{B}},\ \mathbb{G}_{\mathscr{D}}^{\mathscr{V}}\right\} = \left\{\mathbb{B}, (\mathscr{V}_I, \mathscr{V}_{II}, \mathscr{V}_{III}), \mathbb{D}, \mathbb{G}_{\mathscr{V}}^{\mathscr{B}} = \left\{X \in \{I, II, III\} \mid \mathfrak{M}_{\mathscr{B}}, \mathscr{V}_X, \mathfrak{M}_{\mathscr{B}}^{X}, U_{\mathscr{V}_X}(\cdot)\right\}, \mathbb{G}_{\mathscr{D}}^{\mathscr{V}} = \left\{\mathbb{G}_{\mathscr{D}}^{\mathscr{V}_{X\in\{I,II,III\}}}, \mathfrak{M}_{\mathscr{B}}^{X} \mid \mathbb{D}_{\mathscr{V}_X}, \tilde{\mathbb{O}}_{\mathscr{D}}, \mathfrak{U}_{\mathscr{D}}(\cdot)\right\}, T\right\}$

(Kim, 2023).

- $\mathbb{B}$ and $\mathbb{D}$ represent the sets of BSs, and mobile devices, and the $\mathscr{V}_I$, $\mathscr{V}_{II}$ and $\mathscr{V}_{III}$ are the VNOs of the URLLC, eMBB, and mMTC networks, respectively.
- In the upper level game $\mathbb{G}_{\mathscr{V}}^{\mathscr{B}}$, VNOs are game players, which exist in each $\mathbb{B}$. $\mathfrak{M}_{\mathscr{B}}$ is the total spectrum amount of $\mathbb{B}$, and $\mathfrak{M}_{\mathscr{B}}^{X}$ is an allocated spectrum amount for the $\mathscr{V}_X$ where $\mathfrak{M}_{\mathscr{B}} = \bigcup_{X\in\{I,II,III\}} \mathfrak{M}_{\mathscr{B}}^{X}$.
- In the $\mathbb{G}_{\mathscr{V}}^{\mathscr{B}}$, the $\mathfrak{M}_{\mathscr{B}}^{X}$ and $U_{\mathscr{V}_X}(\cdot)$ are the strategy and utility function of $\mathscr{V}_X$, respectively.
- The lower game $\mathbb{G}_{\mathscr{D}}^{\mathscr{V}}$ consists of three sub-games, such as $\mathbb{G}_{\mathscr{D}}^{\mathscr{V}_I}$, $\mathbb{G}_{\mathscr{D}}^{\mathscr{V}_{II}}$, and $\mathbb{G}_{\mathscr{D}}^{\mathscr{V}_{III}}$, for three type slices. They are operated in a distributed parallel fashion.
- In the $\mathbb{G}_{\mathscr{D}}^{\mathscr{V}}$, the $\mathbb{D}_{\mathscr{V}_I}$, $\mathbb{D}_{\mathscr{V}_I}$ and $\mathbb{D}_{\mathscr{V}_{III}}$ are the corresponding device sets of $\mathscr{V}_I$, $\mathscr{V}_{II}$ and $\mathscr{V}_{III}$, respectively, where $\mathbb{D} = \mathbb{D}_{\mathscr{V}_I} \cup \mathbb{D}_{\mathscr{V}_I} \cup \mathbb{D}_{\mathscr{V}_{III}}$.
- In the $\mathbb{G}_{\mathscr{D}}^{\mathscr{V}_I}$, $\mathbb{G}_{\mathscr{D}}^{\mathscr{V}_{II}}$, and $\mathbb{G}_{\mathscr{D}}^{\mathscr{V}_{III}}$ games, mobile devices in the $\mathbb{D}_{\mathscr{V}_I}$, $\mathbb{D}_{\mathscr{V}_I}$, and $\mathbb{D}_{\mathscr{V}_{III}}$ are game players, and the $\tilde{\mathbb{O}}_{\mathscr{D}}$ is assigned spectrum resource of each device. For the $\mathscr{D}_k \in \mathbb{D}_{\mathscr{V}_X}$, the $\tilde{\mathbb{O}}_{\mathscr{D}_k}$ and $\mathfrak{U}_{\mathscr{D}_k}(\cdot)$ are its strategy and utility function, respectively, where $\mathfrak{M}_{\mathscr{B}}^{X} = \bigcup_{\mathscr{D}_k\in\mathbb{D}_{\mathscr{V}_X}} \mathfrak{M}_{\mathscr{D}_k}^{X}$.
- Discrete time model $T \in \{t_1, \ldots, t_c, t_{c+1}, \ldots\}$ is represented by a sequence of time steps. The length of t_c matches the event time-scale of $\mathbb{G}_{\mathscr{V}}^{\mathscr{B}}$ and $\mathbb{G}_{\mathscr{D}}^{\mathscr{V}}$ games.

4.3.4 The resource allocation scheme for multi-service networks

To develop the joint control mechanism for multi-service networks, the *HMRA* scheme constructs the upper $\left(\mathbb{G}_{\mathscr{V}}^{\mathscr{B}}\right)$ and lower $\left(\mathbb{G}_{\mathscr{D}}^{\mathscr{V}}\right)$ game models. In each BS, the $\mathbb{G}_{\mathscr{V}}^{\mathscr{B}}$ is operated in a distributed manner by the URLLC, eMBB, and mMTC VNOs. For example, the $\mathbb{G}_{\mathscr{V}}^{\mathscr{B}_i}$ is designed by the $\mathscr{B}_i \in \mathbb{B}$ to share the $\mathfrak{M}_{\mathscr{B}_i}$ for its $\mathscr{V}_I$, $\mathscr{V}_{II}$ and $\mathscr{V}_{III}$. The *HMRA* scheme adopts the *KSBS* for the solution of $\mathbb{G}_{\mathscr{V}}^{\mathscr{B}_i}$. Especially, the *KSBS* has the property of *individual monotonicity*, which means that the increasing of bargaining set (S) in a direction favorable to a specific player always benefits that player. Therefore, the idea of *KSBS* is suitable to share the $\mathfrak{M}_{\mathscr{B}_i}$ for different type services. To get the *KSBS* for the $\mathbb{G}_{\mathscr{V}}^{\mathscr{B}_i}$, the *HMRA* scheme first defines the disagree point in the $\mathbb{G}_{\mathscr{V}}^{\mathscr{B}_i}$. For this decision, the *HMRA* scheme employs the concept of *BV*; it is commonly used to measure the bargaining power. The *BV* of $\mathscr{V}_I$, i.e., $\Phi_I^{\mathrm{N}}(\upsilon)$, is calculated as follows (Kim, 2023);

$$\Phi_I^{\mathbb{N}}(\upsilon) = \sum_{T \subseteq \mathbb{N} \setminus \mathscr{V}_I} \left(\frac{1}{2^{|\mathbb{N}|-1}} \times \left(\upsilon(T \cup \mathscr{V}_I) - \upsilon(T) \right) \right), \text{ s.t., } \mathbb{N} = \{\mathscr{V}_I, \mathscr{V}_{II}, \mathscr{V}_{III}\}$$

(4.3.1)

where $\upsilon(\cdot)$ is estimated based on the bankruptcy allocation method; it is an economic concept where the available resource amount is not sufficient enough to fulfill the demands of all resource claiming agents. In the *HMRA* scheme, the bankruptcy allocation of $\mathscr{B}_i$ is formulated in a triple $\left(\mathbb{N}, \mathfrak{M}_{\mathscr{B}_i}, c\right)$ where $c = \left(c_{\mathscr{V}_I}, c_{\mathscr{V}_{II}}, c_{\mathscr{V}_{III}}\right)$ is a claim vector of VNOs and $0 \leq \mathfrak{M}_{\mathscr{B}_i} \leq \sum_{i \in \mathbb{N}} c_i$. The $\upsilon(\cdot) : 2^{\mathbb{N}} \rightarrow \mathbb{R}$ describes the worth of a coalition $T \subset \mathbb{N}$ can be written as follows (Degefu, 2016; Kim, 2023);

$$\upsilon(T) = \max\left\{ \left(\mathfrak{M}_{\mathscr{B}_i} - \sum_{i \in \mathbb{N} \setminus T} c_i \right), 0 \right\}, \text{ s.t., } T \subset \mathbb{N}$$

(4.3.2)

According to (4.3.1)-(4.3.2), the $\Phi_I^{\mathbb{N}}(\upsilon)$, $\Phi_{II}^{\mathbb{N}}(\upsilon)$, and $\Phi_{III}^{\mathbb{N}}(\upsilon)$ are obtained in the same manner, and the disagree point of $\mathscr{V}_I$, i.e., $d_{\mathscr{V}_I}$, is calculated as follows (Kim, 2023).

$$d_{\mathscr{V}_I} = \frac{\Phi_I^{\mathbb{N}}(\upsilon)}{\Phi_I^{\mathbb{N}}(\upsilon) + \Phi_{II}^{\mathbb{N}}(\upsilon) + \Phi_{III}^{\mathbb{N}}(\upsilon)} \times \mathfrak{R}_{\mathscr{B}_i}$$

(4.3.3)

where $\mathfrak{R}_{\mathscr{B}_i}$ is the equity base amount of spectrum resource. By using (4.3.3), $d_{\mathscr{V}_{II}}$ and $d_{\mathscr{V}_{III}}$ are given in the same manner. In the $\mathbb{G}_{\mathscr{V}}^{\mathscr{B}_i}$, the $\mathscr{V}_I$, $\mathscr{V}_{II}$ and $\mathscr{V}_{III}$ are game players, and their utility functions, i.e., $U_{\mathscr{V}_I}(\cdot)$, $U_{\mathscr{V}_I}(\cdot)$, and $U_{\mathscr{V}_I}(\cdot)$, are given as follows (Kim, 2023);

$$\begin{cases} U_{\mathscr{V}_I}\left(\mathscr{R}_{\mathscr{V}_I}, \mathfrak{M}_{\mathscr{B}_i}^I\right) = \left(\exp\left(\frac{\min\left(\mathscr{R}_{\mathscr{V}_I}, \mathfrak{M}_{\mathscr{B}_i}^I\right)}{\mathscr{R}_{\mathscr{V}_I}} \right) - \delta \right) \times \mathfrak{M}_{\mathscr{B}_i}^I \\ U_{\mathscr{V}_{II}}\left(\mathscr{R}_{\mathscr{V}_{II}}, \mathfrak{M}_{\mathscr{B}_i}^{II}\right) = \left(\mu \times \log\left(\eta - \frac{\min\left(\mathscr{R}_{\mathscr{V}_{II}}, \mathfrak{M}_{\mathscr{B}_i}^{II}\right)}{\mathscr{R}_{\mathscr{V}_{II}}} \right) \right) \times \mathfrak{M}_{\mathscr{B}_i}^{II} \\ U_{\mathscr{V}_{III}}\left(\mathscr{R}_{\mathscr{V}_{III}}, \mathfrak{M}_{\mathscr{B}_i}^{III}\right) = \left(\varrho^{\left(\frac{\min\left(\mathscr{R}_{\mathscr{V}_{II}}, \mathfrak{M}_{\mathscr{B}_i}^{II}\right)}{\mathscr{R}_{\mathscr{V}_{II}}} \right)} - \varepsilon \right) \times \mathfrak{M}_{\mathscr{B}_i}^{III} \end{cases}$$

(4.3.4)

where δ is a adjustment parameter for the $U_{\mathscr{V}_I}(\cdot)$, and μ, η are adjustment parameters for the $U_{\mathscr{V}_{II}}(\cdot)$. ϱ and ε are control factors for the $U_{\mathscr{V}_{III}}(\cdot)$. The $\mathscr{R}_{\mathscr{V}_I}$, $\mathscr{R}_{\mathscr{V}_{II}}$ and $\mathscr{R}_{\mathscr{V}_{III}}$ (*or* $\mathfrak{M}_{\mathscr{B}_i}^I$, $\mathfrak{M}_{\mathscr{B}_i}^{II}$ and $\mathfrak{M}_{\mathscr{B}_i}^{III}$) are the total resource request sums from the $\mathscr{V}_I$, $\mathscr{V}_{II}$ and $\mathscr{V}_{III}$, (*or* the allocated resources for the $\mathscr{V}_I$, $\mathscr{V}_{II}$ and $\mathscr{V}_{III}$), respectively. With the $d_{\mathscr{V}_I}$, $d_{\mathscr{V}_{II}}$ and $d_{\mathscr{V}_{III}}$ values, the *KSBS* for $\mathbb{G}_{\mathscr{V}}^{\mathscr{B}_i}$, i.e., $KSBS_{\mathscr{V}}^{\mathscr{B}_i}\left(\mathfrak{M}_{\mathscr{B}_i}\right)$, is obtained as follows (Kim, 2023);

$$KSBS_{\mathcal{V}}^{\mathcal{B}_i}\left(\mathfrak{M}_{\mathcal{B}_i}\right) = \max_{\left(\mathfrak{M}_{\mathcal{B}_i}^{I},\mathfrak{M}_{\mathcal{B}_i}^{II},\mathfrak{M}_{\mathcal{B}_i}^{III}\right)}\left(\frac{U_{\mathcal{V}_I}(\cdot) - d_{\mathcal{V}_I}}{U_{\mathcal{V}_I}^{*}(\cdot) - d_{\mathcal{V}_I}} = \frac{U_{\mathcal{V}_{II}}(\cdot) - d_{\mathcal{V}_{II}}}{U_{\mathcal{V}_{II}}^{*}(\cdot) - d_{\mathcal{V}_{II}}} = \frac{U_{\mathcal{V}_{III}}(\cdot) - d_{\mathcal{V}_{III}}}{U_{\mathcal{V}_{III}}^{*}(\cdot) - d_{\mathcal{V}_{III}}}\right) \tag{4.3.5}$$

$$\text{s.t.}, \sum_{X\in\{I,II,III\}} \mathfrak{M}_{\mathcal{B}_i}^{X} \leq \mathfrak{M}_{\mathcal{B}_i}$$

where $U_{\mathcal{V}_l}^{*}(\cdot)$ is the maximized payoff for the $\mathcal{V}_l$, and thus it is his ideal payoff. According to (4.3.5), the $\mathfrak{M}_{\mathcal{B}_i}^{I}$, $\mathfrak{M}_{\mathcal{B}_i}^{II}$, and $\mathfrak{M}_{\mathcal{B}_i}^{III}$ are obtained. Based on these values, the intra-slice allocation process is executed in the lower level stage. Individually, the $\mathcal{V}_I$, $\mathcal{V}_{II}$ and $\mathcal{V}_{III}$ operate their $\mathbb{G}_{\mathcal{D}}^{\mathcal{V}_I}$, $\mathbb{G}_{\mathcal{D}}^{\mathcal{V}_{II}}$, and $\mathbb{G}_{\mathcal{D}}^{\mathcal{V}_{III}}$ games, respectively, in a distributed parallel fashion. In the $\mathbb{G}_{\mathcal{D}}^{\mathcal{V}_l}$ game, the $\mathcal{V}_l$'s corresponding devices in the $\mathbb{D}_{\mathcal{V}_l}$ are game players. In the viewpoint of individual devices, there are two performance criteria. One is the throughput based measure, and the other is QoS based measure. Therefore, the utility function of $\mathcal{D}_j \in \mathbb{D}_{\mathcal{V}_l}$, i.e., $\mathfrak{U}_{\mathcal{D}_j}(\cdot)$, consists of two sub-functions where $\mathfrak{U}_{\mathcal{D}_j}(\cdot) = \left\{\mathfrak{U}_{\mathcal{D}_j}^{T}(\cdot),\right.$ $\left.\mathfrak{U}_{\mathcal{D}_j}^{Q}(\cdot)\right\}$. They are given as follows (Kim, 2023);

$$\mathfrak{U}_{\mathcal{D}_j}\left(\mathfrak{U}_{\mathcal{D}_j}^{T}, \mathfrak{U}_{\mathcal{D}_j}^{Q} \mid \mathfrak{M}_{\mathcal{B}_i}^{I}, \mathcal{R}_{\mathcal{D}_j}, \tilde{O}_{\mathcal{D}_j}\right) =$$

$$\begin{cases} \mathfrak{U}_{\mathcal{D}_j}^{T}\left(\mathfrak{M}_{\mathcal{B}_i}^{I}, \mathcal{R}_{\mathcal{D}_j}, \tilde{O}_{\mathcal{D}_j}\right) = \left[\beta \times \left(\dfrac{1}{1+\exp\left(-\dfrac{\min\left(\mathcal{R}_{\mathcal{D}_j},\tilde{O}_{\mathcal{D}_j}\right)}{\mathcal{R}_{\mathcal{D}_j}}\right)} - \theta\right)\right] \times \tilde{O}_{\mathcal{D}_j} \\ \mathfrak{U}_{\mathcal{D}_j}^{Q}\left(\mathfrak{M}_{\mathcal{B}_i}^{I}, \mathcal{R}_{\mathcal{D}_j}, \tilde{O}_{\mathcal{D}_j}\right) = \left[\xi - \exp\left(\kappa \times \dfrac{\min\left(\mathcal{R}_{\mathcal{D}_j},\tilde{O}_{\mathcal{D}_j}\right)}{\mathcal{R}_{\mathcal{D}_j}}\right)\right] \times \tilde{O}_{\mathcal{D}_j} \end{cases} \tag{4.3.6}$$

$$\text{s.t.}, \sum_{\mathcal{D}_k\in\mathbb{D}_{\mathcal{V}_l}} \tilde{O}_{\mathcal{D}_k} \leq \mathfrak{M}_{\mathcal{B}_i}^{I}$$

where β, θ are control parameters for the throughput based function $\mathfrak{U}_{\mathcal{D}_j}^{T}(\cdot)$, and ξ, κ are control parameters for the QoS based function $\mathfrak{U}_{\mathcal{D}_j}^{Q}(\cdot)$. $\mathcal{R}_{\mathcal{D}_j}$ is the requested resource amount from the $\mathcal{D}_j$, and $\tilde{O}_{\mathcal{D}_j}$ is the allocated resource for the $\mathcal{D}_j$. In the $\mathbb{G}_{\mathcal{D}}^{\mathcal{V}_I}$, the assigned $\mathfrak{M}_{\mathcal{B}_i}^{I}$ is distributed for the devices in the $\mathbb{D}_{\mathcal{V}_I}$. To implement this intra-slice allocation process, the *HMRA* scheme develops a new multi-criteria cooperative game based on the of *NBS* and *KSBS*. By considering two different criteria, the multi-criteria solution for the $\mathbb{G}_{\mathcal{D}}^{\mathcal{V}_I}$, i.e., $MCS(\mathbb{G}_{\mathcal{D}}^{\mathcal{V}_I})$, is obtained as follows (Kim, 2023);

$$MCS(\mathbb{G}_{\mathcal{D}}^{\mathcal{V}_I}) = \left(\mathcal{D}_j \in \mathbb{D}_{\mathcal{V}_I} \mid \tilde{O}_{\mathcal{D}_j}^{T}, \tilde{O}_{\mathcal{D}_j}^{Q}\right) = \max_{\left(\tilde{O}_{\mathcal{D}_j}^{T}, \tilde{O}_{\mathcal{D}_j}^{Q}\right)} \mathcal{S}^{T}, \mathcal{S}^{Q} \tag{4.3.7}$$

$$s.t. \begin{cases} \mathcal{S}^{T} = \prod_{\mathcal{D}_j\in\mathbb{D}_{\mathcal{V}_I}}\left(\mathfrak{U}_{\mathcal{D}_j}^{T}\left(\mathfrak{M}_{\mathcal{B}_i}^{I}, \mathcal{R}_{\mathcal{D}_j}, \tilde{O}_{\mathcal{D}_j}^{T}\right) - d_{\mathcal{D}_j}\right) \\ \mathcal{S}^{Q} = \left\{\dfrac{\mathfrak{U}_{\mathcal{D}_j}^{Q}\left(\mathfrak{M}_{\mathcal{B}_i}^{I}, \mathcal{R}_{\mathcal{D}_j}, \tilde{O}_{\mathcal{D}_j}^{Q}\right) - d_{\mathcal{D}_j}}{\mathfrak{U}_{\mathcal{D}_j}^{*}\left(\mathfrak{M}_{\mathcal{B}_i}^{I}, \mathcal{R}_{\mathcal{D}_j}, \tilde{O}_{\mathcal{D}_j}^{Q}\right) - d_{\mathcal{D}_j}} = \ldots = \dfrac{\mathfrak{U}_{\mathcal{D}_k}^{Q}\left(\mathfrak{M}_{\mathcal{B}_i}^{I}, \mathcal{R}_{\mathcal{D}_j}, \tilde{O}_{\mathcal{D}_j}^{Q}\right) - d_{\mathcal{D}_k}}{\mathfrak{U}_{\mathcal{D}_k}^{*}\left(\mathfrak{M}_{\mathcal{B}_i}^{I}, \mathcal{R}_{\mathcal{D}_j}, \tilde{O}_{\mathcal{D}_j}^{Q}\right) - d_{\mathcal{D}_k}}\right\} \\ \mathcal{D}_j, \mathcal{D}_k \in \mathbb{D}_{\mathcal{V}_I} \text{ and } \mathfrak{U}_{\mathcal{D}_j}^{T}(\cdot), \mathfrak{U}_{\mathcal{D}_j}^{Q}(\cdot) > d_{\mathcal{D}_k} \end{cases}$$

where $\mathfrak{U}^*_{\mathscr{D}_i}(\cdot)$ is the $\mathscr{D}_j$'s ideal outcome, and $d_{\mathscr{D}_j}$ is his disagreement point. The $\tilde{O}^T_{\mathscr{D}_j}$, $\tilde{O}^Q_{\mathscr{D}_j}$ are obtained by maximizing $\mathcal{S}^T$ and $\mathcal{S}^Q$, respectively. Based on the $\tilde{O}^T_{\mathscr{D}_j}$, $\tilde{O}^Q_{\mathscr{D}_j}$ values, the finally allocated resource for the $\mathscr{D}_j$, i.e., $\Gamma_{\mathscr{D}_j}$, is decided as follows (Kim, 2023);

$$\Gamma_{\mathscr{D}_j} = \frac{\tilde{O}^*_{\mathscr{D}_j}}{\sum_{\mathscr{D}_k \in \mathbb{D}_{\mathscr{V}_i}} \tilde{O}^*_{\mathscr{D}_k}} \times \mathfrak{M}^I_{\mathscr{B}_i} \text{ s.t., } \tilde{O}^*_{\mathscr{D}_j} = \frac{\tilde{O}^T_{\mathscr{D}_j} + \tilde{O}^Q_{\mathscr{D}_j}}{2} \quad (4.3.8)$$

The resource allocation processes in the $\mathbb{G}^{\mathscr{V}_{II}}_{\mathscr{D}}$ and $\mathbb{G}^{\mathscr{V}_{II}}_{\mathscr{D}}$ games are operated as the same manner as the $\mathbb{G}^{\mathscr{V}_I}_{\mathscr{D}}$ game according to (4.3.7)-(4.3.8). In the lower level stage, the $\mathbb{G}^{\mathscr{V}_I}_{\mathscr{D}}$, $\mathbb{G}^{\mathscr{V}_{II}}_{\mathscr{D}}$ and $\mathbb{G}^{\mathscr{V}_{II}}_{\mathscr{D}}$ games are independently executed in a distributed parallel fashion (Kim, 2023).

4.3.5 Summary

Due to the rapid surge in demand for diverse wireless communication services, spectrum scarcity has increasingly become a critical challenge for network operators. In response, the WNV has been identified as a promising approach to enhance multi-service network performance. Building on WNV, the *HMRA* scheme presents a hierarchical resource allocation mechanism for multi-service network systems, introducing a hybrid control framework. At the upper level, the *HMRA* scheme tackles inter-slice allocation by applying the *KSBS* principles. Here, the *BV* method is employed to determine the disagreement point among VNOs. In the lower level, the intra-slice allocation process is modeled as a multi-criteria resource allocation problem. By combining the *NBS* with *KSBS*, the *HMRA* scheme effectively negotiates different metrics and preferences. Facilitated by a VNO-device coordinated arrangement, this joint control method can reach mutually acceptable solutions despite the dynamic conditions characterizing today's virtualized networks (Kim, 2023).

4.4 Multicasting Device Communication Control (*MDCC*) scheme

Amid the surge in wireless services, mobile data traffic has grown at an unprecedented rate, placing a significant burden on traditional cellular networks. To address these escalating demands, network virtualization combined with multicasting device-to-device (MD2D) technology has been identified as a promising solution for next-generation networks. Recently, S. Kim introduces the *Multicasting Device Communication Control* (*MDCC*) scheme for virtualized cellular networks, enabling device clustering for local MD2D services aimed at optimizing system performance (Kim, 2019). By recognizing the dynamic and competitive nature of network environments, the *MDCC* scheme structures its control algorithms as a game model under imperfect system information. Drawing on incentive mechanisms and evolutionary decision-making, the *MDCC* scheme encourages self-interested mobile devices to adopt more cooperative, honest behaviors. Ultimately, MD2D services are delivered through an iterative feedback process that facilitates gradual, step-by-step improvements.

4.4.1 Development motivation

Over the past few decades, the widespread adoption of smart mobile devices and multimedia applications has spurred a substantial rise in wireless traffic. One of the main distinctions between 4G and future 5G networks is their ability to provide seamless connectivity for diverse devices and applications in various user environments (Jang, 2013; Kim, 2010, Kim, 2011b; Yu, 2017). To address these demands, network virtualization has been proposed as a promising strategy for 5G systems. It involves merging both hardware and software-based network resources and functionalities into a single software-driven entity, commonly referred to as a virtual network (Wang, 2016a). Essentially, network virtualization partitions existing physical network resources into logical segments in an efficient way. A core component of this process is resource slicing, where multiple virtual networks - each tailored to different needs - are deployed atop the same physical infrastructure (Richart, 2016; Kim, 2019).

Another emerging technology is device-to-device (D2D) communication, which establishes direct links between mobile devices without sending traffic through the core network. This approach boosts spectral efficiency in congested networks, as it offloads traffic from cellular infrastructures, extends network coverage, increases energy efficiency, and potentially achieves higher data rates due to stronger direct channels (Hieu, 2010; Kiwan, 2018). By deploying D2D under cellular networks, system throughput improves significantly, especially when supporting location-oriented applications and services (Kim, 2019).

An extension of traditional D2D is multicast D2D (MD2D), wherein devices form clusters to broadcast identical content to multiple receivers within confined geographic regions. MD2D capitalizes on the broadcast characteristics of wireless communication to serve many users simultaneously, which is particularly advantageous given the rapid growth in local content sharing. Nonetheless, incorporating MD2D into cellular systems introduces added complexity and technical challenges, especially when shifting from a sender-driven to a receiver-driven network paradigm and attempting to lessen base station (BS) traffic (Bhardwaj, 2017; Feng, 2018; Kim 2011a, Kim, 2019).

By merging MD2D with virtualized cellular infrastructures, overall application service quality can be further enhanced. Compared to traditional cellular networks, this integrated strategy can bolster system efficiency while addressing different service needs (Rizvi, 2015; Wang, 2016a). Within MD2D-enabled environments, wireless bandwidth is sliced into multiple virtual units. To enable MD2D services, mobile devices within proximity form clusters, each led by a designated cluster header. Several clusters may operate concurrently, each allocated specific bandwidth slices through its cluster header. In this manner, localized virtual MD2D networks are activated at the cluster level (Kim, 2019).

The key to maximizing performance in these systems is the adaptive selection of MD2D headers. However, any device functioning as a header invests additional energy to deliver MD2D services. The *MDCC* scheme thus incorporates incentive payments that encourage otherwise self-interested devices to act as headers. The primary objective of the *MDCC* scheme is to engineer a novel control mechanism for MD2D-enabled cellular systems. By integrating header selection with incentive distribution algorithms, the approach aims to enhance cellular network efficiency, although this is far from trivial and calls for a distinct control paradigm (Kim, 2019).

In recent years, game theory has emerged as a powerful method for examining and understanding interactions among independent network agents. Nevertheless, classical game-theoretic results - such as Nash equilibrium - often fail to guarantee uniqueness or optimality. This limitation has prompted the development of new game models designed to encourage socially desirable outcomes through institutional rules (Kim, 2014; Xiao, 2014). Two particularly relevant models are incentive mechanism design - which incentivizes players to act truthfully as their dominant strategy - and evolutionary game theory, which addresses information scarcity in uncertain environments and leads to adaptive, evolving strategies (Kim, 2019).

4.4.2 Contribution

The *MDCC* scheme introduces both a new incentive mechanism - designed to compensate MD2D headers - and an evolutionary game model for selecting those headers within MD2D clusters. By integrating these two components, the *MDCC* scheme achieves a synergistic outcome that enhances overall system performance. Its primary contributions include; i) leveraging a game-theoretic approach to deliver high-quality services in 5G networks, ii) striking a balance between optimality and practicality through a mutually beneficial framework, iii) modeling the evolving interactions among network agents, who may have varying perspectives, and iv) improving cellular network efficiency by harnessing MD2D technology (Kim, 2019).

4.4.3. Incentive mechanism design for MD2D headers

In MD2D-enabled cellular networks, there are two transmission manners: one is the direct communication with the BS, the other is the MD2D transmission assisted by MD2D headers. Therefore, there are multiple cellular users and MD2D users. The *MDCC* scheme denotes the set of cellular users as $\mathscr{C} = \{c_1 \ldots c_n\}$ and the set of MD2D users as $\mathscr{D} = \{d_1 \ldots d_m\}$. For the MD2D service, the cluster formation process categorizes groups of some proximal MD2D users into multiple clusters. For each cluster, cluster header should be selected. Cellular users communicate via the evolved node base station (eNB), while MD2D users directly communicate from the cluster header. The general cellular network architecture with MD2D technique is shown in Figure 2 (Kim, 2019).

Figure 2. MD2D assisted cellular network infrastructure (Kim, 2019)

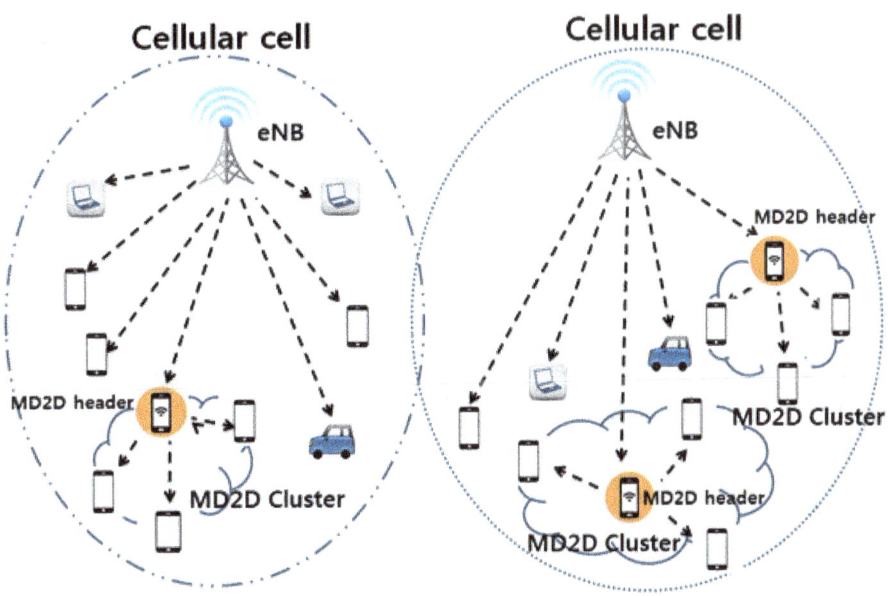

In cluster-based MD2D communications, selecting a cluster head is both critical and crucial. The *MDCC* scheme employs an auction-based selection procedure to identify the most suitable mobile devices for the role of MD2D head. As part of this decision process, the eNB requires private information from each MD2D-capable device, such as the number of neighboring devices eligible for MD2D and the cost associated with offering MD2D services. Because this information is not publicly accessible, an incentive mechanism is needed to encourage devices to truthfully report their private data. To address this need, the *MDCC* scheme adopts the foundational principles of incentive mechanism design, offering incentives that encourage honest disclosures from MD2D devices (Kim, 2019).

To improve the efficiency of MD2D services, MD2D headers cover as many as neighboring devices. In the *MDCC* scheme, the eNB has a predetermined threshold $(\tilde{\text{O}})$ to be MD2D header candidates; Υ is the number of reachable neighbor devices. The mobile devices, which can cover Υ neighbor devices, report their cost (ϕ) to serve MD2D services. From the eNB viewpoint, this information must be credible. However, the ϕ value is a privately-known information for each device. Therefore, the *MDCC* scheme should develop a procedure to induce header candidates to reveal truthfully their private information. The *MDCC* scheme designs a novel incentive mechanism to pay a reward for MD2D header devices (Kim, 2019).

Before explaining the incentive mechanism, the *MDCC* scheme first introduces the useful preliminaries. Assume that one mobile device in C has Υ neighboring devices. If this device has been selected as a header, i.e., $d_i \in \mathscr{D}$, the d_i's consuming energy cost is ϕ_{d_i} to provide MD2D services. As a rational game player, the d_i act self-interestedly. Therefore, the eNB should pay a reward for the d_i to compensate its cost. In the incentive mechanism, the function $\mathscr{F}_\chi\left(\phi_{d_i}\right): \mathbb{R} \to [0,1]$ is adopted to determine how much contribution to be procured from the d_i as a MD2D header. Based on the ϕ_{d_i} value, $\mathscr{F}_\chi\left(\phi_{d_i}\right)$ is defined as follows (Archer, 2001; Kim, 2019; Wu, 2017);

$$\mathscr{F}_\chi\left(\phi_{d_i}\right) = \ln\left(\mathrm{e} - \frac{\phi_{d_i}}{\chi}\right), \text{ s.t.}, 0 \le \frac{\phi_{d_i}}{\chi} \le (e-1) \tag{4.4.1}$$

where χ is a control parameter to adjust the range of $\mathscr{F}_\chi\left(\phi_{d_i}\right)$ function where $0 \le \mathscr{F}_\chi\left(\phi_{d_i}\right) \le 1$. According to (4.4.1), the incentive payment $\left(I_{d_i}\right)$ for the d_i can be calculated. Based on the d_i's declaration of his cost ϕ_{d_i}, the d_i's pro forma contribution is defined as $\phi_{d_i} \times \mathscr{F}_\chi\left(\phi_{d_i}\right)$. And then, $I_{d_i}\left(\phi_{d_i}\right)$ can be derived as follows (Archer, 2001; Kim, 2019; Wu, 2017);

$$I_{d_i}\left(\phi_{d_i}\right) = \left(\phi_{d_i} \times \mathscr{F}_\chi\left(\phi_{d_i}\right)\right) + \int_{\omega=\phi_{d_i}}^{\infty} \mathscr{F}_\chi\left(\phi_{d_i}\right) d\omega \tag{4.4.2}$$

Generally, utility function in game theory corresponds to player's received benefit minus the incurred cost. Therefore, the d_i's utility function with his reporting ϕ_{d_i} $\left(U_{d_i}\left(\phi_{d_i}\right)\right)$ is given by (Kim, 2019);

$$U_{d_i}\left(\phi_{d_i}\right) = I_{d_i}\left(\phi_{d_i}\right) - \left(\phi_{d_i} \times \mathscr{F}_\chi\left(\phi_{d_i}\right)\right) \tag{4.4.3}$$

To elicit the d_i's truthfulness, the payoff of $U_{d_i}\left(\phi_{d_i}\right)$ should be maximized when the d_i truthfully report his ϕ_{d_i} value (Kim, 2019).

Theorem 1. According to the incentive payment given by (4.4.1)-(4.4.3), truthful cost reporting is a dominant strategy for each MD2D device (Kim, 2019).

Proof: To prove the d_i's truthfulness, truthful reporting cost information, i.e., ϕ_{d_i}, by the d_i should guarantee a better payoff than any other falsely reported cost information, i.e., $\widetilde{\Phi}_{d_i}$. Therefore, we considers two case, i.e., $\phi_{d_i} < \widetilde{\Phi}_{d_i}$ and $\phi_{d_i} > \widetilde{\Phi}_{d_i}$, and confirm that $U_{d_i}\left(\phi_{d_i}\right)$ is always higher than $U_{d_i}\left(\widetilde{\Phi}_{d_i}\right)$.

Case I) When the d_i report a higher cost information $\widetilde{\Phi}_{d_i}$ than its truthful cost information $\left(\phi_{d_i} < \widetilde{\Phi}_{d_i}\right)$, we should verify $U_{d_i}\left(\phi_{d_i}\right) \geq U_{d_i}\left(\widetilde{\Phi}_{d_i}\right)$.

$$U_{d_i}\left(\phi_{d_i}\right) - U_{d_i}\left(\widetilde{\Phi}_{d_i}\right) = \left(I_{d_i}\left(\phi_{d_i}\right) - \left(\phi_{d_i} \times \mathcal{F}_{\chi}\left(\phi_{d_i}\right)\right)\right) - \left(I_{d_i}\left(\widetilde{\Phi}_{d_i}\right) - \left(\phi_{d_i} \times \mathcal{F}_{\chi}\left(\widetilde{\Phi}_{d_i}\right)\right)\right)$$

$$= \left(\left(\phi_{d_i} \times \mathcal{F}_{\chi}\left(\phi_{d_i}\right)\right) + \int_{\omega=\phi_{d_i}}^{\infty} \mathcal{F}_{\chi}(\omega)d\omega - \left(\phi_{d_i} \times \mathcal{F}_{\chi}\left(\phi_{d_i}\right)\right)\right) - \left(\left(\widetilde{\Phi}_{d_i} \times \mathcal{F}_{\chi}\left(\widetilde{\Phi}_{d_i}\right)\right) + \int_{\omega=\widetilde{\Phi}_{d_i}}^{\infty} \mathcal{F}_{\chi}(\omega)d\omega - \left(\phi_{d_i} \times \mathcal{F}_{\chi}\left(\widetilde{\Phi}_{d_i}\right)\right)\right)$$

$$= \left(\int_{\omega=\phi_{d_i}}^{\infty} \mathcal{F}_{\chi}(\omega)d\omega\right) - \left(\left(\widetilde{\Phi}_{d_i} \times \mathcal{F}_{\chi}\left(\widetilde{\Phi}_{d_i}\right)\right) + \int_{\omega=\widetilde{\Phi}_{d_i}}^{\infty} \mathcal{F}_{\chi}(\omega)d\omega - \left(\phi_{d_i} \times \mathcal{F}_{\chi}\left(\widetilde{\Phi}_{d_i}\right)\right)\right)$$

$$= \left(\left(\phi_{d_i} \times \mathcal{F}_{\chi}\left(\widetilde{\Phi}_{d_i}\right)\right) + \int_{\omega=\phi_{d_i}}^{\infty} \mathcal{F}_{\chi}(\omega)d\omega\right) - \left(\left(\widetilde{\Phi}_{d_i} \times \mathcal{F}_{\chi}\left(\widetilde{\Phi}_{d_i}\right)\right) + \int_{\omega=\widetilde{\Phi}_{d_i}}^{\infty} \mathcal{F}_{\chi}(\omega)d\omega\right)$$

$$= \left(\left(\phi_{d_i} \times \mathcal{F}_{\chi}\left(\widetilde{\Phi}_{d_i}\right)\right) - \left(\widetilde{\Phi}_{d_i} \times \mathcal{F}_{\chi}\left(\widetilde{\Phi}_{d_i}\right)\right)\right) + \int_{\omega=\phi_{d_i}}^{\widetilde{\Phi}_{d_i}} \mathcal{F}_{\chi}(\omega)d\omega$$

$$= \int_{\omega=\phi_{d_i}}^{\widetilde{\Phi}_{d_i}} \mathcal{F}_{\chi}(\omega)d\omega - \left(\left(\widetilde{\Phi}_{d_i} - \phi_{d_i}\right) \times \mathcal{F}_{\chi}\left(\widetilde{\Phi}_{d_i}\right)\right) \tag{4.4.4}$$

Figure 3. Payoff of d_i with $\phi_{d_i} < \widetilde{\Phi}_{d_i}$

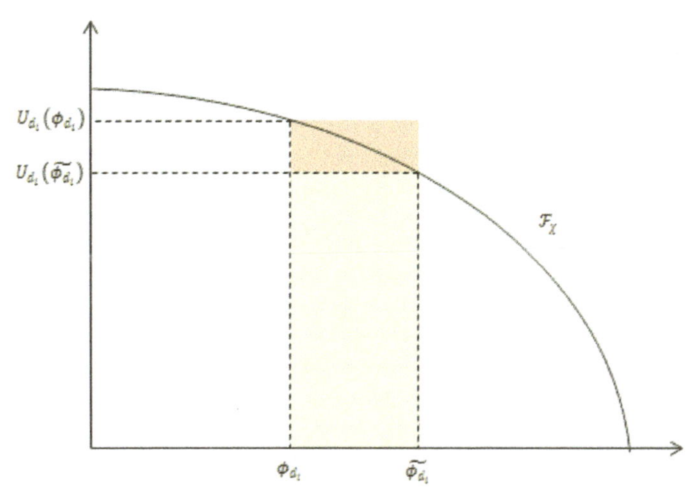

The $\mathcal{F}_{\chi}(\omega)$ function is a monotone decreasing function. Therefore, as seen in the Fig. 4.4.2,

$$\left(\left(\widetilde{\phi}_{d_i} - \phi_{d_i}\right) \times \mathcal{F}_{\chi}\left(\widetilde{\phi}_{d_i}\right)\right) \leq \int_{\omega=\phi_{d_i}}^{\widetilde{\phi}_{d_i}} \mathcal{F}_{\chi}(\omega)d\omega \leq \left(\left(\widetilde{\phi}_{d_i} - \phi_{d_i}\right) \times \mathcal{F}_{\chi}\left(\phi_{d_i}\right)\right) \tag{4.4.5}$$

Using (4.4.4) and (4.4.5) equations, it is obtained;

$$\left(\int_{\omega=\phi_{d_i}}^{\widetilde{\phi}_{d_i}} \mathcal{F}_{\chi}(\omega)d\omega - \left(\left(\widetilde{\phi}_{d_i} - \phi_{d_i}\right) \times \mathcal{F}_{\chi}\left(\widetilde{\phi}_{d_i}\right)\right)\right) \geq 0 \tag{4.4.6}$$

Therefore, we can verify that the case of $\phi_{d_i} < \widetilde{\phi}_{d_i}$, $U_{d_i}\left(\phi_{d_i}\right)$ is always higher than $U_{d_i}\left(\widetilde{\phi}_{d_i}\right)$, i.e., $U_{d_i}\left(\phi_{d_i}\right) \geq U_{d_i}\left(\widetilde{\phi}_{d_i}\right)$.

Case II) When the d_i report a lower cost information $\widetilde{\phi}_{d_i}$ than its truthful cost information $\left(\phi_{d_i} > \widetilde{\phi}_{d_i}\right)$, we should verify $U_{d_i}\left(\phi_{d_i}\right) \geq U_{d_i}\left(\widetilde{\phi}_{d_i}\right)$.

$$U_{d_i}\left(\phi_{d_i}\right) - U_{d_i}\left(\widetilde{\phi}_{d_i}\right) = \left(I_{d_i}\left(\phi_{d_i}\right) - \left(\phi_{d_i} \times \mathcal{F}_{\chi}\left(\phi_{d_i}\right)\right)\right) - \left(I_{d_i}\left(\widetilde{\phi}_{d_i}\right) - \left(\phi_{d_i} \times \mathcal{F}_{\chi}\left(\widetilde{\phi}_{d_i}\right)\right)\right)$$

$$= \left(\left(\phi_{d_i} \times \mathcal{F}_{\chi}\left(\phi_{d_i}\right)\right) + \int_{\omega=\phi_{d_i}}^{\infty} \mathcal{F}_{\chi}(\omega)d\omega - \left(\phi_{d_i} \times \mathcal{F}_{\chi}\left(\phi_{d_i}\right)\right)\right) - \left(\left(\widetilde{\phi}_{d_i} \times \mathcal{F}_{\chi}\left(\widetilde{\phi}_{d_i}\right)\right) + \int_{\omega=\widetilde{\phi}_{d_i}}^{\infty} \mathcal{F}_{\chi}(\omega)d\omega - \left(\phi_{d_i} \times \mathcal{F}_{\chi}\left(\widetilde{\phi}_{d_i}\right)\right)\right)$$

$$= \left(\int_{\omega=\phi_{d_i}}^{\infty} \mathcal{F}_{\chi}(\omega)d\omega\right) - \left(\left(\widetilde{\phi}_{d_i} \times \mathcal{F}_{\chi}\left(\widetilde{\phi}_{d_i}\right)\right) + \int_{\omega=\widetilde{\phi}_{d_i}}^{\infty} \mathcal{F}_{\chi}(\omega)d\omega - \left(\phi_{d_i} \times \mathcal{F}_{\chi}\left(\widetilde{\phi}_{d_i}\right)\right)\right)$$

$$= \left(\left(\phi_{d_i} \times \mathcal{F}_{\chi}\left(\widetilde{\phi}_{d_i}\right)\right) + \int_{\omega=\phi_{d_i}}^{\infty} \mathcal{F}_{\chi}(\omega)d\omega\right) - \left(\left(\widetilde{\phi}_{d_i} \times \mathcal{F}_{\chi}\left(\widetilde{\phi}_{d_i}\right)\right) + \int_{\omega=\widetilde{\phi}_{d_i}}^{\infty} \mathcal{F}_{\chi}(\omega)d\omega\right)$$

$$= \left(\left(\phi_{d_i} \times \mathcal{F}_{\chi}\left(\widetilde{\phi}_{d_i}\right)\right) - \left(\widetilde{\phi}_{d_i} \times \mathcal{F}_{\chi}\left(\widetilde{\phi}_{d_i}\right)\right)\right) - \int_{\omega=\widetilde{\phi}_{d_i}}^{\phi_{d_i}} \mathcal{F}_{\chi}(\omega)d\omega$$

$$= \left(\left(\phi_{d_i} - \widetilde{\phi}_{d_i}\right) \times \mathcal{F}_{\chi}\left(\widetilde{\phi}_{d_i}\right)\right) - \int_{\omega=\widetilde{\phi}_{d_i}}^{\phi_{d_i}} \mathcal{F}_{\chi}(\omega)d\omega \tag{4.4.7}$$

Figure 4. Payoff of d_i with $\phi_{d_i} > \widetilde{\phi}_{d_i}$

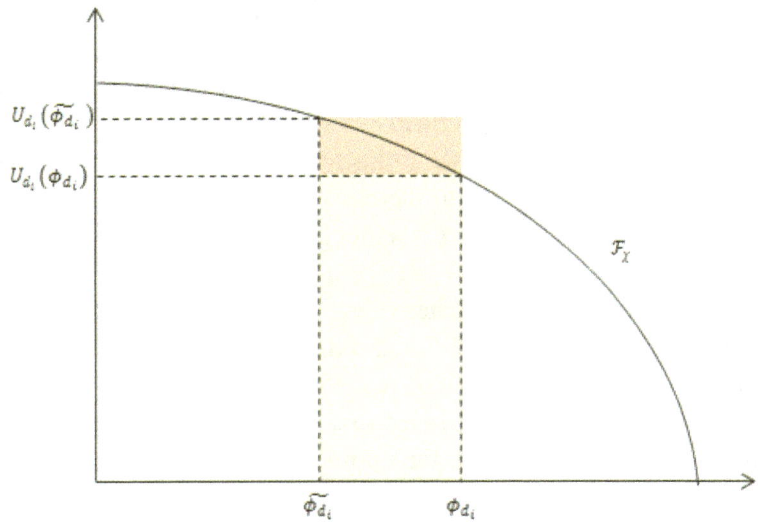

As seen in Figure 4.

$$\left(\left(\phi_{d_i} - \widetilde{\phi}_{d_i}\right) \times \mathcal{F}_\chi\left(\phi_{d_i}\right)\right) \leq \int_{\omega=\widetilde{\phi}_{d_i}}^{\phi_{d_i}} \mathcal{F}_\chi(\omega) d\omega \leq \left(\left(\phi_{d_i} - \widetilde{\phi}_{d_i}\right) \times \mathcal{F}_\chi\left(\widetilde{\phi}_{d_i}\right)\right) \tag{4.4.8}$$

Using (4.4.7) and (4.4.8) equations, it is obtained;

$$\left(\left(\left(\phi_{d_i} - \widetilde{\phi}_{d_i}\right) \times \mathcal{F}_\chi\left(\widetilde{\phi}_{d_i}\right)\right) - \int_{\omega=\widetilde{\phi}_{d_i}}^{\phi_{d_i}} \mathcal{F}_\chi(\omega) d\omega\right) \geq 0 \tag{4.4.9}$$

Therefore, we can verify that the case of $\phi_{d_i} > \widetilde{\phi}_{d_i}$, $U_{d_i}\left(\phi_{d_i}\right)$ is always higher than $U_{d_i}\left(\widetilde{\phi}_{d_i}\right)$, i.e., $U_{d_i}\left(\phi_{d_i}\right) \geq U_{d_i}\left(\widetilde{\phi}_{d_i}\right)$.

Based on the case I and case II), whether $\phi_{d_i} < \widetilde{\phi}_{d_i}$ or $\phi_{d_i} > \widetilde{\phi}_{d_i}$, we can always ensure $U_{d_i}\left(\phi_{d_i}\right) \geq U_{d_i}\left(\widetilde{\phi}_{d_i}\right)$. Therefore, truthful reporting is a dominant strategy for each MD2D device. ∎

4.4.4 Evolutionary game model in the MD2D header selection algorithm

Traditionally, game theory focuses on the decisions of individual players, with each player's payoff influenced by the collective choices of all participants. A key concern is predicting player behavior in any given game scenario. Classical game theory, however, assumes that each player makes perfectly rational decisions, based on comprehensive knowledge of available strategies - an assumption that often fails under real-world conditions, where limited information and bounded rationality come into play. To overcome this shortcoming, J. Smith proposed applying game theory within a framework of frequency-dependent selection, giving rise to evolutionary game theory (Kim, 2011c, Kim, 2019).

This approach views players' behavior as a dynamic, evolving process. During an evolutionary game, strategies that yield higher payoffs are favored and tend to prevail, while less successful strategies are gradually discarded. The main solution concept here is the evolutionarily stable strategy (ESS): in a repeated and randomized environment, players effectively converge on a stable strategy that no competing approach can replace. Replicator dynamics (RD) illustrate how different strategies shift in prevalence over time, where higher payoffs correspond to increased likelihood of the strategy being adopted. In other words, the fitness of a strategy determines whether it endures as the ESS (Feng 2017b; Kim 2011c, Kim 2019).

Building on these principles, the *MDCC* scheme proposes an evolutionary game model specifically for the eNB in MD2D scenarios. Unlike a traditional, symmetric evolutionary game, the version here is asymmetric, featuring two categories of players - namely, the eNB and MD2D headers - with distinct strategy sets. For the MD2D headers, strategies revolve around the cost of providing MD2D services, and under the scheme's incentive mechanism, the optimal choice for each header is truthfully disclosing its own cost. The eNB, in turn, decides how to select MD2D headers and then allocate virtualized bandwidth to each chosen header for local MD2D services. To optimize overall MD2D efficiency, the eNB dynamically updates its header selections, continually adjusting its strategic approach based on iterative feedback of system outcomes. Through these reported inputs and real-time performance checks, the incentive mechanism and the evolutionary game model interact in a mutually reinforcing manner (Kim, 2019).

To represent the *RD* for the MD2D header selection problem, let H be a number of MD2D header candidates and $\mathcal{P}_i$ is the propensity for the *i*th header candidate selection. H is the (H+1)-dimensional vector $[\mathcal{P}_1 \ldots \mathcal{P}_i \ldots \mathcal{P}_H, \mathcal{P}_{H+1}]$ where $\mathcal{P}_{H+1}$ represent a no-header selection strategy. $\overline{\mathcal{P}_j}$ stands for the RD of $\mathcal{P}_{1 \le j \le H+1}$. The RD of $\mathcal{P}_j$ $(\overline{\mathcal{P}_j})$ is defined as (Kim, 2019);

$$\overline{\mathcal{P}_j} = \left(\left(\mathcal{P}_j \times U_j^{eNB} \right) - \left(\frac{1}{(H+1)} \times \sum_{k=1}^{H+1} \left(\mathcal{P}_k \times U_k^{eNB} \right) \right) \right) \tag{4.4.10}$$

$$\text{s.t., } U_j^{eNB} = \delta \times \left(\frac{1}{\left(1 + exp\left(\frac{\mathcal{T}_j^{eNB} - \mathcal{T}_{-j}^{eNB}}{\mathcal{T}_{-j}^{eNB}} \right) \right)} - \gamma \right)$$

where $\mathcal{T}_j^{eNB}$, $\mathcal{T}_{-j}^{eNB}$ are the eNB's traffic load with the *j*th MD2D header candidate selection and without that candidate selection, respectively. δ and γ are adjusting parameters to measure the system payoff. In order to implement the RD process for MD2D services, the *MDCC* scheme partitions the time-axis into equal intervals, and the RD process is repeated at each time period. At the end of each iteration, the eNB evaluates periodically the current RD, and the RD of each eNB's strategy is used to update the H vector, which represents the strategy propensity to make a decision for the next iteration. If the *j*th MD2D header candidate is selected at time t period, the H vector for the next $t+1$ time period ($\mathbb{H}(t+1)$) is estimated as follows (Kim, 2019);

$$\mathbb{H}(t+1) = \mathbb{H}(t) + \left(\frac{\mathcal{P}_k}{\beta} \times \mathbb{E}(t)\right) - \left(\frac{\mathcal{P}_k}{\beta} \times [\mathbb{A}(t) - \mathbb{E}(t)]\right) \tag{4.4.11}$$

$$\text{s.t.,} \begin{cases} A(t) = [1_1, 1_2, \ldots 1_{H+1}] \\ E(t) = [e_1, e_2, \ldots, e_{H+1}] \, where \, e_{1 \leq k \leq H+1} = \begin{cases} 1, if \mathcal{P}_{k, k \neq j} > \mathcal{P}_j \\ 0, otherwise \end{cases} \end{cases}$$

where β is a control factor to adjust the convergence rate. A and E are also (H+1)-dimensional vector; all elements in A is 1, and elements in E, i.e., $e_{1 \leq k \leq H+1}$, are dynamically decided 0 or 1 based on the RD of each strategy. According to (4.4.11), H values are adjusted periodically at each time step for the next strategy selection. This interactive feedback process continues until the ESS is obtained while capturing how the eNB adapts his strategies to achieve a better payoff. If the eNB cannot benefit by changing his current selected strategies, a satisfactory solution is obtained over a game processing time. To evolve into the ESS, it is a practical and suitable approach in real world network operations (Kim, 2019).

4.4.5 Summary

The *MDCC* scheme presents an innovative control mechanism for MD2D-enabled cellular networks operating under uncertain conditions. Its primary objectives include selecting MD2D headers and determining incentive rewards for the mobile devices serving in these header roles. A detailed account of the MD2D header selection and incentive payment processes is provided, emphasizing the reciprocal interplay between them. By weaving these two essential algorithms into a jointly collaborative framework, the *MDCC* scheme leverages incentive mechanism design principles alongside evolutionary game theory. As network circumstances shift, the *MDCC* scheme follows a step-by-step interactive procedure, continually readjusting dynamic parameters to align with evolving conditions. Through this incremental approach, the *MDCC* scheme realizes significant mutual gains for both individual devices and the overall network (Kim, 2019).

4.5 Effective Sensor Cloud Control (*ESCC*) Scheme

With the goal of complementing ubiquitous sensor networks and cloud computing techniques, significant interest has recently shifted toward Sensor-Cloud (SC). SC builds on the principle of virtualizing physical sensor nodes, while introducing intermediate data processing layers between sensor nodes and end-users. In this context, S. Kim has proposed the *Effective Sensor Cloud Control* (*ESCC*) scheme, designed to offer on-demand sensing services for multiple applications (Kim, 2018a). Underpinned by game theory, the *ESCC* scheme adopts a two-stage game model that deftly integrates selection and incentive mechanisms. In this model, users can choose the most suitable data center to run their tasks, and each data center can offer incentives to sensors participating in the sensing operations. Notably, the two-stage game framework proves highly adaptable and flexible, making it particularly adept at satisfying diverse application requirements in practical SC control scenarios.

4.5.1 Development motivation

The swift progress in sensor electronics, digital communications, and miniaturized manufacturing has led to the development of compact, intelligent sensing devices. When combined with advanced 5G technology, these devices have significantly elevated living standards. In numerous practical scenarios - ranging from target tracking and battlefield monitoring to tele-monitoring and ubiquitous surveillance - they function autonomously to gather data. Moving forward, the concept of the Internet of Everything (IoE) envisions an interconnected future where such smart devices seamlessly integrate into global Internet infrastructures, thus further enhancing quality of life (Dinh, 2017; Kim, 2018a; Misra, 2014). Simultaneous to the rise of sophisticated sensor technology, Cloud Computing (CC) has become a prevalent paradigm. The term 'cloud' often denotes a vast collection of resources that can appear uniform from a distant perspective. In the information and communications technology (ICT) domain, cloud computing provides an Internet-based model for delivering on-demand data storage and processing services. Essentially, the CC introduces a distributed platform granting ubiquitous, convenient, on-demand network access to a shared pool of configurable computing resources. This capability allows for flexible and scalable service provision, especially when combined with the functionality of advanced smart sensing devices (Kim, 2018a; Lee, 2015; Liu, 2015; Pan, 2014; Zhu, 2014).

Amid these developments, Sensor-Cloud (SC) has drawn considerable interest from academia and industry alike. At its core, SC merges wireless sensor network (WSN) data collection with the extensive storage and computational capabilities of cloud servers. Conventional sensor networks tend to be single-user-centric and rarely share data beyond the owning organization. By contrast, SC seeks to provide multi-application services via shared sensor networks, thereby enhancing the cost-effectiveness and practical use of sensor data. Consequently, SC is seen as a potential solution for multi-organization sensor deployments that facilitate broader data access (Kim, 2018a; Misra, 2014; Zhu, 2017). According to MicroStrains (Misra, 2014), SC is defined as *a specialized platform for storing, visualizing, and remotely managing sensor data, leveraging cloud computing resources to achieve scalable data handling, rapid visualization, and user-programmable analytics*. Virtualization underpins this model by abstracting physical sensors into logical entities. This abstraction grants multiple independent users the ability to efficiently share the same physical hardware, allowing numerous applications to coexist on a single virtualized sensor network. The tight coupling between hardware deployments and services is thus significantly reduced (Guijarro, 2017; Kim, 2018a).

Although the SC has garnered widespread attention, various technical hurdles remain - particularly in areas such as sensor management, service quality regulation, and system efficiency optimization. While some research efforts have introduced SC control algorithms, a comprehensive framework addressing all these issues in tandem is still lacking. Moreover, fine-grained interactive control processes are needed to ensure high QoS for SC users (Kim, 2010, Kim, 2018a). Designing an interactive SC control scheme often requires handling conflicting objectives. Game theory has emerged as a robust tool for modeling and analyzing competition and cooperation among self-interested agents. Since it excels at managing multiple goal-driven players, game theory promises a scalable solution compared to centralized methods, which may become computationally burdensome. Drawing on these insights, the *ESCC* scheme uses a game-theoretic methodology to create a practical SC control mechanism (Guijarro, 2017; Kim, 2014, Kim, 2018a).

The *ESCC* scheme introduces a two-stage game model that captures both competitive and cooperative interactions among SC participants. At the upper level, the *ESCC* scheme addresses the problem of data center selection for each user; at the lower level, it defines an incentive mechanism for sensors. Together, these algorithms form a dynamic, adaptive approach that aligns with constantly shifting SC network conditions. In this setup, users, data centers, and sensors each choose optimal strategies to maximize their own utility or profit in a fully distributed manner. This decision process proceeds step by step, ultimately delivering reciprocal benefits and preserving flexibility under varying circumstances (Kim, 2018a).

Building on the two-stage game framework, the *ESCC* scheme addresses three critical questions; i) promoting truthful participation: ensuring SC participants achieve the best outcomes only by reporting information honestly, ii) optimal strategy selection: deciding on strategies that yield maximum payoff, and iii) efficient computation: designing a polynomial-time mechanism that remains feasible in real-world deployments. To tackle these challenges, the *ESCC* scheme emphasizes truthfulness, individual rationality, and computational feasibility. Although previous work on SC algorithms exists, a comprehensive study unifying these design principles has yet to be undertaken. As far as we know, the *ESCC* scheme stands out as the first solution to holistically combine feasibility, interactive decision-making, and conflicting control concerns within the SC ecosystem (Kim, 2018a).

4.5.2 Contribution

The *ESCC* scheme aims to dynamically manage SC operations for specific applications, seeking an effective solution that balances competing requirements. Thus far, this area remains largely unexplored. To capture the interactions among users, data centers, and sensors, the *ESCC* scheme introduces a two-stage game framework. In the upper-level game, users and data centers act as players, with users choosing the most appropriate data center to execute their tasks. While making these selections, application QoS considerations are also incorporated. In the lower-level game, each data center and its corresponding, geographically distributed physical sensors function as game players, and their incentive-based interactions are modeled as an incentive algorithm. Through this two-stage game mechanism, the *ESCC* scheme achieves a fair and balanced outcome across diverse SC network conditions (Kim, 2018a).

4.5.3 Two-stage game model for the SC system

The *ESCC* scheme views the SC infrastructure as a cloud-based sensor-management platform, operating as an interface between physical and cyber realms. Multiple user applications request diverse types of sensor data (e.g., temperature, humidity, GPS, wind speed) from different locations. Consequently, each user submits a service request to a specific Data Center (DC), specifying the desired data and the target region. To respond effectively to these application requests, the SC system provides *Sensors-as-a-Service* (*Se-aaS*), sharing a pool of sensor and cloud resources (EI Rachkidi, 2017). The scenario under consideration includes a set of user applications, multiple DCs, and Physical Sensors (PSs) for data collection. These PSs are dispersed across various geographic regions and can be grouped temporarily to meet a specific application's needs (Chatterjee, 2015; EI Rachkidi, 2017; Kim, 2018a).

Within the SC system, virtualization of PSs produces Virtual Sensors (VSs) - abstracted instances encapsulating the capabilities of physical devices and logical data sources. Often, VSs can span multiple, potentially overlapping regions and act as proxies between data sources and the *Se-aaS* of individual user applications. Geographically distributed DCs operate as service providers, where Virtual

Machines (VMs) are deployed to manage multiple VSs, contingent on the allocated resources. VS tasks for a given user application, originating from various regions, are stored and processed in a dedicated VM, which promptly forms a *Se-aaS* (Chatterjee, 2015; El Rachkidi, 2017). In the *ESCC* scheme, the primary objective is to design a dynamic SC control protocol that can deliver the requested data types from relevant VSs to the appropriate user applications. The main entities within the *ESCC* scheme are defined as follows (Kim, 2018a):

- **Physical Sensors (PSs)**: to form a physical wireless sensor network, each PS has its own attributes, i.e., ID and type where S is the set of PSs, i.e., $SN_i \in \mathbb{S} = \{SN_1 \ldots SN_n\}$ and I is the set of sensor types, i.e., $\tau_j \in \mathfrak{I} = \{\tau_1 \ldots \tau_m\}$; each *SN* has its own τ.
- **Virtual Sensors (VSs)**: VS is a software emulation of PS, and inherits all attributes from the corresponding PS. **V** is the set of VSs where $\mathscr{V}_i \in \mathscr{V}$ represents the ith VS. A mapping function to map a physical sensor SN_i to a virtual sensor $\mathscr{V}_i$ is defined as $\mathscr{F}_{p\rightarrow v}(SN_i) = \{\mathscr{V}_i, VM_A\}$ where VM_A is a running virtual machine including the $\mathscr{V}_i$.
- **Virtual Machines (VMs)**: VM consists of VSs which are software images of PSs and responsible to provide *Se-aaSs* to user applications. The VM receives the sensing data from the underlying VSs and provides the final information transparently to user applications. Let W be the set of VMs where $VM_{A_v} \in \mathfrak{W} = \{VM_{A_1} \ldots VM_{A_u}\}$.
- **Data Centers (DCs)**: the SC infrastructure is composed of several geographically distributed DCs. In each DC, a number of created VMs are running to handle user applications. Let $\mathscr{DC}$ be the set of DCs where $\mathfrak{A}_s \in \mathscr{DC} = \{\mathfrak{A}_1 \ldots \mathfrak{A}_h\}$.
- **Application**: $\mathbb{A} = \{A_1 \ldots A_v\}$ is the set of user applications, and the *ESCC* scheme characterizes a user application A with service type (τ), region of interest and QoS requirement.

The SC paradigm presents a compelling approach: when a user submits an application task to a designated DC, the user pays a fee to that DC, which in turn compensates the participating PSs through an incentive system. The DC sets up a VM to procure SC services and gather essential data from the VSs, ultimately completing the assigned task. Each VS corresponds to a PS, which acts as a self-interested agent - it bears a cost to deliver its service. To determine the most suitable DC for handling a task, it becomes crucial to consider both the user's payment obligations and the incentive design for the VSs providing the service (Kim, 2018a).

The *ESCC* scheme develops a new two-stage game model ($\mathbb{G}$) assuming dynamic SC situations. Formally, the *ESCC* scheme defines the G = $\{\mathbb{G}^{upper}, \mathbb{G}^{lower}_{VM_A \in \mathfrak{W}}\}$ where $\mathbb{G}^{upper}$ is the upper level game to formulate interactions among users and DCs to decide the corresponding DC, and $\mathbb{G}^{lower}_{VM_A \in \mathfrak{W}}$ is the lower level game to formulate interactions among the DC and its corresponding VSs to reward VSs' incentives. During the G game operations, $\mathbb{G}^{upper}$ and multiple $\mathbb{G}^{lower}_{VM_A \in \mathfrak{W}}$ are executed sequentially and repeatedly. These two-step iterations work together in a coordinated manner, and each application task is completed in an entirely distributed fashion (Kim, 2018a).

4.5.4 Upper level game model for the DC selection algorithm

In the $\mathbb{G}^{upper}$ game process, users and DCs negotiate with each other to decide the service price for each application task and the most adaptable DC is finally selected. From the viewpoint of users, they want to complete their tasks as low price as possible while meeting the requested QoS. To implement the DC selection algorithm, the $\mathbb{G}^{upper}$ can be defined as $\mathbb{G}^{upper} = \left\{ \{\mathbb{A}, \mathscr{DC}\}, \left\{ S_{A_v \in \mathbb{A}}, S^{\mathfrak{A}_s \in \mathscr{DC}}_{VM_{A_v} \in \mathfrak{B}} \right\}, T \right\}$ at each time period H of gameplay. Formally, the *ESCC* scheme defines the $\mathbb{G}^{upper}$ as follows (Kim, 2018a);

- $\{\mathbb{A}, \mathscr{DC}\}$ represents a set of $\mathbb{G}^{first}$'s game players.
- $S_{A_v \in \mathbb{A}} = \left\{ \prod_{min \le k \le max} \mathscr{P}^{A_v}_k \middle| \mathscr{P}^{A_v}_k \in [\mathscr{P}^{A_v}_{min} \ldots \mathscr{P}^{A_v}_k \ldots \mathscr{P}^{A_v}_{max}] \right\}$ is the A_v's strategy set where $\mathscr{P}^{A_v}_k$ is the A_v's announced price level to complete the requested task. For simplicity, S_{A_v} is discretely quantified.
- $S^{\mathfrak{A}_s \in \mathscr{DC}}_{VM_{A_v} \in \mathfrak{B}}$ is the VM_{A_v}'s strategy set and VM_{A_v} is in the $\mathfrak{A}_s$. $\mathfrak{P}^{\mathfrak{A}_s}_{VM_{A_v}} \in S^{\mathfrak{A}_s}_{VM_{A_v}}$ represents the VM_{A_v}'s QoS level. In the $\mathbb{G}^{first}$ game, the QoS level is estimated as the A_v' service delay; the range is varied from 0 to ∞.
- $T = \{\mathscr{H}_1, \ldots, \mathscr{H}_t, \mathscr{H}_{t+1}, \ldots\}$ denotes time, which is represented by a sequence of time steps with imperfect information for the $\mathbb{G}^{upper}$ game process.

To develop the DC selection algorithm, the $\mathbb{G}^{upper}$ is designed as a practical dynamic game model that is a suitable approach to study the interactions among users and DCs. The *ESCC* scheme considers that a user requests his application A_v with its type τ_m and he offers the price $\mathscr{P}^{A_v}_k$. According to the application type, QoS threshold (∂_A) is defined beforehand. Based on the $\mathscr{P}^{A_v}_k$, the $\mathfrak{A}_s \in \mathscr{DC}$ generates the VM_{A_v} and calculate the minimum cost to complete the A_v (Kim, 2018a).

$$\Gamma^{\mathfrak{A}_s}(A_v) = \mathscr{J}^{\mathfrak{A}_s}(A_v) + \mathscr{H}^{\mathfrak{A}_s}(A_v) + \sum_{\mathscr{V}_i \in \mathscr{V}_{VM_{A_v}}} \mathscr{R}^{VM_{A_v}}_{\mathscr{V}_i} \tag{4.5.1}$$

$$\text{s.t., } \mathscr{J}^{\mathfrak{A}_s}(\mathscr{P}^{A_v}_k) = \mathfrak{y}^{\mathfrak{A}_s} \times \mathfrak{C}^{A_v}_{\tau_m} \text{ and } \mathscr{H}^{\mathfrak{A}_s}(\mathscr{P}^{A_v}_k) = \mathfrak{p}^{\mathfrak{A}_s} \times \mathfrak{T}^{A_v}_{\tau_m}$$

where $\mathfrak{y}^{\mathfrak{A}_s}$, $\mathfrak{p}^{\mathfrak{A}_s}$ $\left(or\, \mathfrak{C}^{A_v}_{\tau_m}, \mathfrak{T}^{A_v}_{\tau_m}\right)$ are the unit costs (*or* needed capacities) for the CC and the user-DC communication to process the A_v, respectively. $\mathscr{R}^{VM_{A_v}}_{\mathscr{V}_i}$ is the payment to the $\mathscr{V}_i$ in VM_{A_v} and $\mathscr{V}_{VM_{A_v}}$ is the set of VSs for the VM_{A_v} where $\mathscr{V}_i \in \mathscr{V}_{VM_{A_v}}$. If $\Gamma^{\mathfrak{A}_s}(A_v)$ is less than the $\mathscr{P}^{A_v}_k$, i.e., $\Gamma^{\mathfrak{A}_s}(A_v) < \mathscr{P}^{A_v}_k$, the $\mathfrak{A}_s$ expects to be profitable in the A_v processing, and estimates the service QoS level $\mathfrak{P}^{\mathfrak{A}_s}_{VM_{A_v}}$ to complete the A_v by considering the current situation. To estimate the $\mathfrak{P}^{\mathfrak{A}_s}_{VM_{A_v}}$, the $\mathfrak{A}_s$ takes into account three main factors; information collection delay, CC delay and service transmission delay (Kim, 2018a).

$$\mathfrak{P}^{\mathfrak{A}_s}_{VM_{A_v}} = \max\left\{ \mathfrak{N}_{\mathscr{V}_j} | \mathscr{V}_j \text{ in } VM_{A_v} \right\} + \grave{\mathrm{E}}\left(A_v, \vartheta^{\mathfrak{A}_s}_{VM_{A_v}}\right) + \psi\left(A_v, \vartheta^{\mathfrak{A}_s}_{VM_{A_v}}\right) \tag{4.5.2}$$

$$s.t., \vartheta^{\mathfrak{A}_s}_{VM_{A_v}} = \mathscr{P}^{A_v}_k - \sum\nolimits_{\mathscr{V}_i \in \mathscr{V}_{VM_{A_v}}} \mathscr{R}^{VM_{A_v}}_{\mathscr{V}_i}$$

where $\mathfrak{N}^{SN_j}_{\mathscr{V}_j}$ is the data transmission delay from the SN_j to $\mathscr{V}_j$. $\grave{\mathrm{E}}\left(A_v, \vartheta^{\mathfrak{A}_s}_{VM_{A_v}}\right)$ and $\psi\left(A_v, \vartheta^{\mathfrak{A}_s}_{VM_{A_v}}\right)$ are defined as follows (Kim, 2018a);

$$\begin{cases} \grave{E}(A_v, \mathscr{P}_k^{A_v}) = \mathscr{D}^{CC}\left(\min\left\{ \left(\frac{\xi_{A_v} \times \vartheta_{VM_{A_v}}^{\mathfrak{A}_s}}{\grave{\mathrm{U}}_{\mathscr{H}_t}} \times \mathfrak{C}^{\mathfrak{A}_s} \right), \mathfrak{C}_{\mathscr{H}_t}^{\mathfrak{A}_s} \right\}, A_v \right) \\ \psi(A_v, \mathscr{P}_k^{A_v}) = \mathscr{D}^{Comm}\left(\min\left\{ \left(\frac{(1-\xi_{A_v}) \times \vartheta_{VM_{A_v}}^{\mathfrak{A}_s}}{\mu_{\mathscr{H}_t}} \times \mathfrak{T}^{\mathfrak{A}_s} \right), \mathfrak{T}_{\mathscr{H}_t}^{\mathfrak{A}_s} \right\}, A_v \right) \end{cases} \tag{4.5.3}$$

where ξ_{A_v} is a coefficient to assign service cost distributions between the CC and the user-DC communication. $\grave{\mathrm{U}}_{\mathscr{H}_t}$ and $\mu_{\mathscr{H}_t}$ are the costs for using the total CC capacity ($\mathfrak{C}^{\mathfrak{A}_s}$) and the total user-DC communication bandwidth ($\mathfrak{T}^{\mathfrak{A}_s}$). $\mathfrak{C}_{\mathscr{H}_t}^{\mathfrak{A}_s}$ and $\mathfrak{T}_{\mathscr{H}_t}^{\mathfrak{A}_s}$ are the available $\mathfrak{C}^{\mathfrak{A}_s}$ and $\mathfrak{T}^{\mathfrak{A}_s}$ capacities, at the time $\mathscr{H}_t$. $\mathscr{D}^{CC}(\cdot)$and $\mathscr{D}^{Comm}(\cdot)$ are the delay estimation functions of CC and user-DC communications to process the A_v, respectively. They Each individual DC estimate its QoS level using the equation (4.5.2). Finally, the user selects the best DC to complete the task A_v as follows (Kim, 2018a);

$$\min\left\{ \mathfrak{P}_{VM_{A_v}}^{\mathfrak{A}_s} \mid \mathfrak{A}_s \in \mathscr{DC} \right\}, \text{ s.t., } \mathfrak{P}_{VM_{A_v}}^{\mathfrak{A}_s} \leq \partial_{A_v} \text{ and } \Gamma^{\mathfrak{A}_s}(A_v) < \mathscr{P}_k^{A_v} \tag{4.5.4}$$

where ∂_{A_v} is the predefined minimum delay requirement. If all DCs cannot meet the ∂_{A_v} or the $\Gamma^{\mathfrak{A}}(A_v)$, the user offers a new $\mathscr{P}^{A_v}$ for the next $\mathbb{G}^{upper}$ game round. At this time, the new $\mathscr{P}^{A_v}$ is larger than the $\mathscr{P}_k^{A_v}$. Therefore, A_v and $\mathfrak{A}_s$ re-negotiate with each other based on the new proposing $\mathscr{P}^{A_v}$. To prevent infinite negotiation rounds, negotiation process must have completed within a negotiation constraint. The $\mathbb{G}^{upper}$ game process must have completed within a round constraint such as the cardinality of S_{A_v}. Therefore, less than the number of $|S_{A_v}|$ negotiation rounds, the A_v and a specific A finally reach an agreement or not (Kim, 2018a).

4.5.5 Lower level game model for the SC incentive algorithm

According to the result of $\mathbb{G}^{upper}$ game, the $\mathbb{G}_{VM}^{lower}$ game is triggered. In the selected DC, one VM is generating for the assigned task, and this VM distributes incentives to encourage VSs to participate SC services. For example, to complete the application task A_v, the selected DC generate the VM_{A_v}, which consists of VSs, i.e., $\mathscr{V}_i \in \mathscr{V}_{VM_{A_v}}$. Sensors are geographically distributed in a local area, and the VM allocates the sensing work to each V and pays the incentive $\tilde{\mathrm{O}}_{\mathscr{V}}^{VM_{A_v}}$. In the $\mathbb{G}_{VM}^{lower}$ game, the *ESCC* scheme employs a truthful incentive algorithm, which is able to ensure the properties including individual rationality and truthfulness of participation (Kim, 2018a).

To design the SC incentive algorithm, the lower level game is defined as $\mathbb{G}_{VM_{A_v} \in \mathfrak{B}}^{lower} = \left\{ \mathbb{S}_{VM_{A_v}}, \mathscr{P}_k^{A_v}, \mathcal{S}_{\mathscr{V}_j^{VM_{A_v}} \in \mathbb{S}_{VM_{A_v}}}, \mathcal{U}_{\mathscr{V}_j^{VM_{A_v}} \in \mathbb{S}_{VM_{A_v}}}, T \right\}$ at each time period $\mathscr{H} \in T$ of gameplay. The $\mathbb{G}_{VM_{A_v}}^{lower}$ game is operated based on the selected strategy $\mathscr{P}_k^{A_v}$ made in the $\mathbb{G}^{upper}$ game process. In the SC environment, there can be multiple VMs. Therefore, multiple $\mathbb{G}_{VM}^{lower}$ games work in parallel and independently in a distributed fashion. Formally, the ESCC scheme defines the $\mathbb{G}_{VM_{A_v}}^{lower}$ as follows (Kim, 2018a);

- $\mathbb{S}_{VM_{A_v}}$ is the set of VSs which form the VM_{A_v}, i.e., $\mathcal{V}_j^{VM_{A_v}} \in \mathbb{S}_{VM_{A_v}}$. They are game players of $\mathbb{G}_{VM_{A_v}}^{lower}$.
- $\mathcal{P}_k^{A_v}$ is the decided price to complete the A_v where $\mathcal{P}_k^{A_v} \in S_{A_v \in \mathbb{A}}$.
- $\mathcal{S}_{\mathcal{V}_j^{VM_{A_v}} \in \mathbb{S}_{VM_{A_v}}}$ is the $\mathcal{V}_j^{VM_{A_v}}$'s strategy set. In other word, $\mathcal{S}_{\mathcal{V}_j^{VM_{A_v}}}$ is the pronounced cost range of $\mathcal{V}_j^{VM_{A_v}}$ to process the requested task A_v.
- $\mathcal{U}_{\mathcal{V}_j^{VM_{A_v}} \in \mathbb{S}_{VM_{A_v}}}$ is the $\mathcal{V}_j^{VM_{A_v}}$'s received payoff from the $\mathbb{G}_{VM_{A_v}}^{lower}$ game process.
- The T is a time period. The $\mathbb{G}_{VM_{A_v}}^{lower}$ is repeated $\mathcal{H}_t \in T < \infty$ time periods with imperfect information.

In the lower game process, assume that the VM_{A_v} includes multiple VSs, i.e., $\{\mathcal{V}_1^{VM_{A_v}} \ldots \mathcal{V}_m^{VM_{A_v}}\}$ and each $\mathcal{V}^{VM_{A_v}}$ has a heterogeneous cost in provisioning the requested service. For example, when the sensing task is assigned to the $\mathcal{V}_{1 \le j \le m}^{VM_{A_v}}$, the $\mathcal{V}_j^{VM_{A_v}}$ incurs a cost $\delta_{\mathcal{V}_j^{VM_{A_v}}}$ to contribute its effort in the sensing process. To stimulate each $\mathcal{V}^{VM_{A_v}}$'s truthful participation, the incentive algorithm is formulated as follows. Each sensor is allowed to report its private cost information to perform the assigned sensing task. In order to increase its payoff, each sensor may untruthfully report its cost information to the corresponding VM (Wu, 2017). Let $\bar{\delta}_{\mathcal{V}_j^{VM_{A_v}}}$ be the cost bid of $\mathcal{V}_j^{VM_{A_v}}$ and $\delta_{\mathcal{V}_j^{VM_{A_v}}}$ be its true cost. After receiving the bids, the VM_{A_v} needs to decide how much reward, denoted as $\mathcal{R}_{\mathcal{V}_j}^{VM_{A_v}}$, should be paid to $\mathcal{V}_j^{VM_{A_v}}$. The payoff of $\mathcal{V}_j^{VM_{A_v}}$ is accordingly (Kim, 2018a);

$$\mathcal{U}_{\mathcal{V}_j^{VM_{A_v}}}\left(\bar{\delta}_{\mathcal{V}_j^{VM_{A_v}}}, \bar{\delta}_{-\mathcal{V}_j^{VM_{A_v}}}\right) = \mathcal{R}_{\mathcal{V}_j}^{VM_{A_v}} - \delta_{\mathcal{V}_j^{VM_{A_v}}} \tag{4.5.5}$$

$$\text{s.t.}, \bar{\delta}_{-\mathcal{V}_j^{VM_{A_v}}} = \left(\bar{\delta}_{\mathcal{V}_1^{VM_{A_v}}} \ldots \bar{\delta}_{\mathcal{V}_{j-1}^{VM_{A_v}}}, \bar{\delta}_{\mathcal{V}_{j+1}^{VM_{A_v}}} \ldots \bar{\delta}_{\mathcal{V}_m^{VM_{A_v}}}\right)$$

Based on the real contribution of $\mathcal{V}_j^{VM_{A_v}}$, the *ESCC* scheme needs to design an incentive algorithm satisfying the following desirable properties: i) *truthfulness*, ii) *individual rationality*, and iii) *computational feasibility* (Kim, 2018a; Wu, 2017).

i) **truthfulness**: truthful reporting is a dominant strategy for the $\mathcal{V}_j^{VM_{A_v}}$, i.e., $\mathcal{U}_{\mathcal{V}_j^{VM_{A_v}}}\left(\delta_{\mathcal{V}_j^{VM_{A_v}}}, \bar{\delta}_{-\mathcal{V}_j^{VM_{A_v}}}\right) \ge \mathcal{U}_{\mathcal{V}_j^{VM_{A_v}}}\left(\bar{\delta}_{\mathcal{V}_j^{VM_{A_v}}}, \bar{\delta}_{-\mathcal{V}_j^{VM_{A_v}}}\right)$, for any $\delta_{\mathcal{V}_j^{VM_{A_v}}} \ne \bar{\delta}_{\mathcal{V}_j^{VM_{A_v}}}$ and any $\bar{\delta}_{-\mathcal{V}_j^{VM_{A_v}}}$.
ii) **individual rationality**: the payoff of $\mathcal{V}_j^{VM_{A_v}}$ is non-negative, i.e., $\mathcal{U}_{\mathcal{V}_j^{VM_{A_v}}}\left(\delta_{\mathcal{V}_j^{VM_{A_v}}}, \bar{\delta}_{-\mathcal{V}_j^{VM_{A_v}}}\right) \ge 0$.
iii) ***computational feasibility***: the reward $\mathcal{R}(\cdot)$ should be computed in polynomial time.

To design a truthful incentive algorithm, the *ESCC* scheme first presents a payment mechanism determining how much to pay an incentive reward for the $\mathcal{V}_j^{VM_{A_v}}$. The main factors of the incentive algorithm are the cost per payment rate $\tilde{O}_{\mathcal{V}_j^{VM_{A_v}}}$ by the VM_{A_v}, and the contribution function $\mathbb{F}\left(\delta_{\mathcal{V}_j^{VM_{A_v}}}, \tilde{O}_{\mathcal{V}_j^{VM_{A_v}}}\right)$ from the $\mathcal{V}_j^{VM_{A_v}}$. $\mathbb{F}(\cdot)$ is inspired by the recent work devoted to the contribution maximization problem with limited budget while determining the payment as little as possible to meet the target contribution (Anari, 2014; Kim, 2018a; Myerson, 1981; Wu, 2017).

$$\mathcal{R}_{\mathcal{V}_j}^{VM_A} = \mathcal{R}_{\mathcal{V}_j}^{VM_A}\left(\bar{\delta}_{\mathcal{V}_j^{VM_A}}, \tilde{\mathrm{O}}_{\mathcal{V}_j^{VM_A}}\right) =$$

$$\left(\bar{\delta}_{\mathcal{V}_j^{VM_A}} \times \mathbb{F}\left(\bar{\delta}_{\mathcal{V}_j^{VM_A}}, \tilde{\mathrm{O}}_{\mathcal{V}_j^{VM_A}}\right)\right) + \int_{\bar{\delta}_{\mathcal{V}_j}}^{\infty} \mathbb{F}\left(\delta_{\mathcal{V}_j^{VM_A}}, \tilde{\mathrm{O}}_{\mathcal{V}_j^{VM_A}}\right) d\delta_{\mathcal{V}_j} \tag{4.5.6}$$

$$\text{s.t.,} \begin{cases} F\left(\bar{\delta}_{\mathcal{V}_j^{VM_A}}, \tilde{\mathrm{O}}_{\mathcal{V}_j^{VM_A}}\right) = 0, if\left(e - \dfrac{\bar{\delta}_{\mathcal{V}_j^{VM_A}}}{\tilde{\mathrm{O}}_{\mathcal{V}_j^{VM_A}}}\right) < 1 \\ F\left(\bar{\delta}_{\mathcal{V}_j^{VM_A}}, \tilde{\mathrm{O}}_{\mathcal{V}_j^{VM_A}}\right) = ln\left(e - \dfrac{\bar{\delta}_{\mathcal{V}_j^{VM_A}}}{\tilde{\mathrm{O}}_{\mathcal{V}_j^{VM_A}}}\right), otherwise \end{cases}$$

If $\left(e - \bar{\delta}_{\mathcal{V}_j^{VM_A}} / \tilde{\mathrm{O}}_{\mathcal{V}_j^{VM_A}}\right) < 1$, the $\mathcal{V}_j^{VM_A}$ does not contribute any effort. According to the Myerson's characterization, $\mathcal{R}(\cdot)$ function maps the cost for a sensing task into a payment for the service provisioning (Anari, 2014; Kim, 2018a; Wu, 2017).

4.5.6 Summary

In recent years, the integration of sensor networks and cloud computing has drawn significant attention from researchers. This SC technology makes network connectivity more advantageous and applicable by combining sensor and cloud resources to fulfill a wide range of application requirements. Within the *ESCC* scheme, the processes of selecting a DC and establishing an incentive mechanism are unified. Specifically, the *ESCC* scheme models the SC control challenge as a two-stage game that represents the relationships among users, DCs, and sensors in coordinating the completion of application tasks. Through a sequential interactive game procedure, the *ESCC* scheme addresses fundamental questions, such as how to achieve cost-effective task fulfillment and how to design an incentive mechanism that upholds desirable properties. By balancing cooperative and non-cooperative tendencies, every SC agent in ESCC interacts progressively, choosing optimal strategies to maximize their respective benefits (Kim, 2018a).

REFERENCES

Ahmed, A., & Ahmed, E. (2016). A survey on mobile edge computing. In *2016 10th International Conference on Intelligent Systems and Control (ISCO)* (pp. 1-8). IEEE. DOI: 10.1109/ISCO.2016.7727082

Anari, N., Goel, G., & Nikzad, A. (2014, October). Mechanism design for crowdsourcing: An optimal 1-1/e competitive budget-feasible mechanism for large markets. In *2014 IEEE 55th Annual Symposium on Foundations of Computer Science* (pp. 266-275). IEEE.

Archer, A., & Tardos, É. (2001, October). Truthful mechanisms for one-parameter agents. In *Proceedings 42nd IEEE Symposium on Foundations of Computer Science* (pp. 482-491). IEEE. DOI: 10.1109/SFCS.2001.959924

Becvar, Z., Plachy, J., & Mach, P. (2014, September). Path selection using handover in mobile networks with cloud-enabled small cells. In 2014 IEEE 25th Annual International Symposium on Personal, Indoor, and Mobile Radio Communication (PIMRC) (pp. 1480-1485). IEEE. DOI: 10.1109/PIMRC.2014.7136402

Bhardwaj, A., & Agnihotri, S. (2017, March). Interference-aware D2D-multicast session provisioning in LTE-A networks. In *2017 IEEE Wireless Communications and Networking Conference (WCNC)* (pp. 1-6). IEEE. DOI: 10.1109/WCNC.2017.7925491

Chatterjee, S., Ladia, R., & Misra, S. (2015). Dynamic optimal pricing for heterogeneous service-oriented architecture of sensor-cloud infrastructure. *IEEE Transactions on Services Computing*, *10*(2), 203–216. DOI: 10.1109/TSC.2015.2453958

Chen, X., Jiao, L., Li, W., & Fu, X. (2015). Efficient multi-user computation offloading for mobile-edge cloud computing. *IEEE/ACM Transactions on Networking*, *24*(5), 2795–2808. DOI: 10.1109/TNET.2015.2487344

Cheng, X., Wu, Y., Min, G., & Zomaya, A. Y. (2018). Network function virtualization in dynamic networks: A stochastic perspective. *IEEE Journal on Selected Areas in Communications*, *36*(10), 2218–2232. DOI: 10.1109/JSAC.2018.2869958

Chun, Y., & Thomson, W. (1992). Bargaining problems with claims. *Mathematical Social Sciences*, *24*(1), 19–33. DOI: 10.1016/0165-4896(92)90003-N

Degefu, D. M., & He, W. (2016, August). Power allocation among socio-economic sectors with overlapping demands during power shortage: A bankruptcy approach. In *2016 IEEE Smart Energy Grid Engineering (SEGE)* (pp. 6-10). IEEE. DOI: 10.1109/SEGE.2016.7589191

Dietzenbacher, B., Estévez-Fernández, A., Borm, P., & Hendrickx, R. (2021). Proportionality, equality, and duality in bankruptcy problems with nontransferable utility. *Annals of Operations Research*, *301*(1-2), 65–80. DOI: 10.1007/s10479-020-03643-3

Dietzenbacher, B., & Peters, H. (2018). *Characterizing NTU-bankruptcy rules using bargaining axioms* (Research Memorandum No. 006). Maastricht University, Graduate School of Business and Economics (GSBE).

Dietzenbacher, B., & Peters, H. (2022). Characterizing NTU-bankruptcy rules using bargaining axioms. *Annals of Operations Research, 318*(2), 871–888. DOI: 10.1007/s10479-020-03828-w

Ding, M., Lopez-Perez, D., Claussen, H., & Kaafar, M. A. (2018). On the fundamental characteristics of ultra-dense small cell networks. *IEEE Network, 32*(3), 92–100. DOI: 10.1109/MNET.2018.1700096

Dinh, T., & Kim, Y. (2017, May). An efficient sensor-cloud interactive model for on-demand latency requirement guarantee. In *2017 IEEE International Conference on Communications (ICC)* (pp. 1-6). IEEE. DOI: 10.1109/ICC.2017.7996817

Dubra, J. (2001). An asymmetric Kalai–Smorodinsky solution. *Economics Letters, 73*(2), 131–136. DOI: 10.1016/S0165-1765(01)00426-8

El Rachkidi, E., Agoulmine, N., Chendeb, N., & Belaid, D. (2017, May). Resources optimization and efficient distribution of shared virtual sensors in sensor-cloud. In *2017 IEEE international conference on communications (ICC)* (pp. 1-6). IEEE. DOI: 10.1109/ICC.2017.7996643

Feng, J., Zhang, Q., Dong, G., Cao, P., & Feng, Z. (2017a, March). An approach to 5G wireless network virtualization: Architecture and trial environment. In *2017 IEEE Wireless Communications and Networking Conference (WCNC)* (pp. 1-6). IEEE.

Feng, L., Yang, Q., & Kwak, K. S. (2017b). Incentive-compatible packet forwarding in mobile social networks via evolutionary game theory. *IEEE Access : Practical Innovations, Open Solutions, 5*, 13557–13569. DOI: 10.1109/ACCESS.2017.2689775

Feng, L., Zhao, P., Zhou, F., Yin, M., Yu, P., Li, W., & Qiu, X. (2018). Resource allocation for 5G D2D multicast content sharing in social-aware cellular networks. *IEEE Communications Magazine, 56*(3), 112–118. DOI: 10.1109/MCOM.2018.1700667

Giménez-Gómez, J. M., Osório, A., & Peris, J. E. (2015). From bargaining solutions to claims rules: A proportional approach. *Games, 6*(1), 32–38. DOI: 10.3390/g6010032

González González, D., Mutafungwa, E., Haile, B., Hämäläinen, J., & Poveda, H. (2017). A planning and optimization framework for ultra dense cellular deployments. *Mobile Information Systems, 2017*(1), 9242058. DOI: 10.1155/2017/9242058

Guijarro, L., Pla, V., Vidal, J. R., & Naldi, M. (2017). Game theoretical analysis of service provision for the Internet of Things based on sensor virtualization. *IEEE Journal on Selected Areas in Communications, 35*(3), 691–706. DOI: 10.1109/JSAC.2017.2672239

Han, Y., Tao, X., Zhang, X., & Jia, S. (2020). Hierarchical resource allocation in multi-service wireless networks with wireless network virtualization. *IEEE Transactions on Vehicular Technology, 69*(10), 11811–11827. DOI: 10.1109/TVT.2020.3019217

Hieu, C. T., & Hong, C. S. (2010). A Connection Entropy-based Multi-Rate Routing Protocol for Mobile Ad Hoc Networks. *Journal of Computing Science and Engineering : JCSE, 4*(3), 225–239. DOI: 10.5626/JCSE.2010.4.3.225

Jang, I., Pyeon, D., Kim, S., & Yoon, H. (2013). A survey on communication protocols for wireless sensor networks. *Journal of Computing Science and Engineering : JCSE, 7*(4), 231–241. DOI: 10.5626/JCSE.2013.7.4.231

Jararweh, Y., Doulat, A., Darabseh, A., Alsmirat, M., Al-Ayyoub, M., & Benkhelifa, E. (2016, April). SDMEC: Software defined system for mobile edge computing. In *2016 IEEE international conference on cloud engineering workshop (IC2EW)* (pp. 88-93). IEEE.

Jiang, H., Wang, T., & Wang, S. (2018). Multi-scale hierarchical resource management for wireless network virtualization. *IEEE Transactions on Cognitive Communications and Networking*, *4*(4), 919–928. DOI: 10.1109/TCCN.2018.2878028

Kim, D.-H., Lim, W.-S., & Suh, Y.-J. (2011a, December 30). Multicast Extension to Proxy Mobile IPv6 for Mobile Multicast Services. *Journal of Computing Science and Engineering*. Korean Institute of Information Scientists and Engineers.

Kim, J. H., Lee, K. J., Kim, T. H., & Yang, S. B. (2011b). Effective routing schemes for double-layered peer-to-peer systems in MANET. *Journal of Computing Science and Engineering : JCSE*, *5*(1), 19–31. DOI: 10.5626/JCSE.2011.5.1.019

Kim, K. S., Uno, S., & Kim, M. W. (2010). Adaptive QoS mechanism for wireless mobile network. *Journal of Computing Science and Engineering : JCSE*, *4*(2), 153–172. DOI: 10.5626/JCSE.2010.4.2.153

Kim, S. (2011c). Adaptive online power control scheme based on the evolutionary game theory. *IET Communications*, *5*(18), 2648–2655. DOI: 10.1049/iet-com.2011.0093

Kim, S. (Ed.). (2014). *Game theory applications in network design.* IGI Global. DOI: 10.4018/978-1-4666-6050-2

Kim, S. (2018a). An effective sensor cloud control scheme based on a two-stage game approach. *IEEE Access : Practical Innovations, Open Solutions*, *6*, 20430–20439. DOI: 10.1109/ACCESS.2018.2815578

Kim, S. (2018b). One-on-one contract game–based dynamic virtual machine migration scheme for Mobile Edge Computing. *Transactions on Emerging Telecommunications Technologies*, *29*(1), e3204. DOI: 10.1002/ett.3204

Kim, S. (2019). A New Multicasting Device-to-Device Communication Control Scheme for Virtualized Cellular Networks. *wireless communications and Mobile computing, 2019*(1), 3540674.

Kim, S. (2022). Adaptive multiservice resource allocation algorithm with wireless network virtualization. *IEEE Access : Practical Innovations, Open Solutions*, *10*, 51515–51524. DOI: 10.1109/ACCESS.2022.3174056

Kim, S. (2023). Hierarchical Multi-Service Resource Allocation Scheme for Future Wireless Network Virtualization. *IEEE Access : Practical Innovations, Open Solutions*, *11*, 87859–87868. DOI: 10.1109/ACCESS.2023.3304632

Kiwan, D., El Sherif, A., & ElBatt, T. (2018, April). Cooperative D2D communications in the uplink of cellular networks with time and power division. In *2018 Wireless Days (WD)* (pp. 80-85). IEEE. DOI: 10.1109/WD.2018.8361698

Lee, K., & Shin, I. (2015). User mobility model based computation offloading decision for mobile cloud. *Journal of Computing Science and Engineering : JCSE*, *9*(3), 155–162. DOI: 10.5626/JCSE.2015.9.3.155

Li, L., Deng, N., Ren, W., Kou, B., Zhou, W., & Yu, S. (2018). Multi-service resource allocation in future network with wireless virtualization. *IEEE Access : Practical Innovations, Open Solutions, 6*, 53854–53868. DOI: 10.1109/ACCESS.2018.2871506

Liu, Y., Sun, Y. L., Ryoo, J., & Vasilakos, A. V. (2015). A survey of security and privacy challenges in cloud computing: solutions and future directions.

Mao, Y., Zhang, J., & Letaief, K. B. (2016). Dynamic computation offloading for mobile-edge computing with energy harvesting devices. *IEEE Journal on Selected Areas in Communications, 34*(12), 3590–3605. DOI: 10.1109/JSAC.2016.2611964

Misra, S., Chatterjee, S., & Obaidat, M. S. (2014). On theoretical modeling of sensor cloud: A paradigm shift from wireless sensor network. *IEEE Systems Journal, 11*(2), 1084–1093. DOI: 10.1109/JSYST.2014.2362617

Monroy, L., Rubiales, V., & Mármol, A. M. (2017). The conservative Kalai–Smorodinsky solution for multiple scenario bargaining. *Annals of Operations Research, 251*(1-2), 285–299. DOI: 10.1007/s10479-015-1894-5

Myerson, R. B. (1981). Optimal auction design. *Mathematics of Operations Research, 6*(1), 58–73. DOI: 10.1287/moor.6.1.58

Pan, Y., & Hu, N. (2014, August). Research on dependability of cloud computing systems. In *2014 10th International Conference on Reliability, Maintainability and Safety (ICRMS)* (pp. 435-439). IEEE. DOI: 10.1109/ICRMS.2014.7107234

Plachy, J., Becvar, Z., & Strinati, E. C. (2016, September). Dynamic resource allocation exploiting mobility prediction in mobile edge computing. In *2016 IEEE 27th Annual International Symposium on Personal, Indoor, and Mobile Radio Communications (PIMRC)* (pp. 1-6). IEEE. DOI: 10.1109/PIMRC.2016.7794955

Ren, J., & Li, X. (2022). Wireless network virtualization resource sharing based on dynamic resource allocation algorithm. *Wireless Communications and Mobile Computing, 2022*, 1–11. DOI: 10.1155/2022/5654188

Richart, M., Baliosian, J., Serrat, J., & Gorricho, J. L. (2016). Resource slicing in virtual wireless networks: A survey. *IEEE Transactions on Network and Service Management, 13*(3), 462–476. DOI: 10.1109/TNSM.2016.2597295

Rizvi, S., Karpinski, K., & Razaque, A. (2015). Novel architecture of self-organized mobile wireless sensor networks. *Journal of Computing Science and Engineering : JCSE, 9*(4), 163–176. DOI: 10.5626/JCSE.2015.9.4.163

Sarabakha, A., & Kayacan, E. (2016, December). Y6 tricopter autonomous evacuation in an indoor environment using Q-learning algorithm. In *2016 IEEE 55th conference on decision and control (CDC)* (pp. 5992-5997). IEEE.

Sun, X., & Ansari, N. (2016). EdgeIoT: Mobile edge computing for the Internet of Things. *IEEE Communications Magazine, 54*(12), 22–29. DOI: 10.1109/MCOM.2016.1600492CM

Tun, Y. K., Pandey, S. R., Alsenwi, M., Zaw, C. W., & Hong, C. S. (2019, January). Weighted proportional allocation based power allocation in wireless network virtualization for future wireless networks. In *2019 International Conference on Information Networking (ICOIN)* (pp. 284-289). IEEE. DOI: 10.1109/ICOIN.2019.8718185

Wang, K., Li, H., Yu, F. R., & Wei, W. (2016a). Virtual resource allocation in software-defined information-centric cellular networks with device-to-device communications and imperfect CSI. *IEEE Transactions on Vehicular Technology*, *65*(12), 10011–10021. DOI: 10.1109/TVT.2016.2529660

Wang, L., Yang, C., Wang, X., Yu, F. R., & Leung, V. C. (2018). User oriented resource management with virtualization: A hierarchical game approach. *IEEE Access : Practical Innovations, Open Solutions*, *6*, 37070–37083. DOI: 10.1109/ACCESS.2018.2845913

Wang, Y., Sheng, M., Wang, X., Wang, L., & Li, J. (2016b). Mobile-edge computing: Partial computation offloading using dynamic voltage scaling. *IEEE Transactions on Communications*, *64*(10), 4268–4282. DOI: 10.1109/TCOMM.2016.2599530

Wu, W., Wang, W., Li, M., Wang, J., Fang, X., Jiang, Y., & Luo, J. (2017). Incentive mechanism design to meet task criteria in crowdsourcing: How to determine your budget. *IEEE Journal on Selected Areas in Communications*, *35*(2), 502–516. DOI: 10.1109/JSAC.2017.2659278

Wu, Z., Gui, L., Chen, J., Zhou, H., & Hou, F. (2016, October). Mobile cloudlet assisted computation offloading in heterogeneous mobile cloud. In *2016 8th International Conference on Wireless Communications & Signal Processing (WCSP)* (pp. 1-6). IEEE. DOI: 10.1109/WCSP.2016.7752655

Xiao, Y., Han, Z., Chen, K. C., & DaSilva, L. A. (2014). Bayesian hierarchical mechanism design for cognitive radio networks. *IEEE Journal on Selected Areas in Communications*, *33*(5), 986–1001. DOI: 10.1109/JSAC.2014.2361092

Yang, H., Wang, W., & Ding, Z. (2019). The weighted surplus division value for cooperative games. *Symmetry*, *11*(9), 1169. DOI: 10.3390/sym11091169

Yu, B., Yang, K., & Yang, G. (2017). A novel 28 GHz beam steering array for 5G mobile device with metallic casing application. *IEEE Transactions on Antennas and Propagation*, *66*(1), 462–466. DOI: 10.1109/TAP.2017.2772084

Zhang, X., & Zhu, Q. (2018). Scalable virtualization and offloading-based software-defined architecture for heterogeneous statistical QoS provisioning over 5G multimedia mobile wireless networks. *IEEE Journal on Selected Areas in Communications*, *36*(12), 2787–2804. DOI: 10.1109/JSAC.2018.2871327

Zhu, C., Leung, V. C., Wang, K., Yang, L. T., & Zhang, Y. (2017). Multi-method data delivery for green sensor-cloud. *IEEE Communications Magazine*, *55*(5), 176–182. DOI: 10.1109/MCOM.2017.1600822

Zhu, K., & Hossain, E. (2015). Virtualization of 5G cellular networks as a hierarchical combinatorial auction. *IEEE Transactions on Mobile Computing*, *15*(10), 2640–2654. DOI: 10.1109/TMC.2015.2506578

Zhu, W., & Lee, C. (2014). A new approach to web data mining based on cloud computing. *Journal of Computing Science and Engineering : JCSE*, *8*(4), 181–186. DOI: 10.5626/JCSE.2014.8.4.181

Chapter 5
IoT-Related Communications

ABSTRACT

Today, the Internet becomes more and more important for everyone in both personal life and professional life. The IoT refers to using the Internet to facilitate communication with 'things', such as smart phones, sensors, mobile computers, and more other smart objects. It is considered as a pillar of future networks and expected to enable intelligent operations and advanced communications. Indeed, it is a new revolution in communication technology which means that everything will be connected to other things and exchange information. In this chapter, different IoT scenarios are provided, enabling technologies are presented, control algorithms in the IoT paradigm are explained in detail.

5. OVERVIEW

We have been witnessing the growth of communication technology over the last decades. Evolution technology has shown many challenges in network designs and applications. Considering all these challenges in the existing network, future network has come up with a huge revolution in the wireless technology. During recent years, one of the most familiar names scaling new heights and creating a benchmark in the world is the Internet of Things (IoT). As an emerging paradigm, it is rapidly finding its path throughout our modern life, aiming to improve the quality of life by connecting many smart devices, technologies, and applications. The functional aspect of IoT is to unite every object of the world under one common infrastructure; the connectivity of each object in the environment with Internet has opened many avenues of research and development. It is the most promising area which penetrates the advantages of wireless sensor and actuator networks and pervasive computing domains. Overall, the IoT paradigm would allow for the automation of everything around us.

There are various domains and environments in which the IoT can play a remarkable role. Matching the smart services with different requirements, IoT communications have facilitated the interconnections of things between applications, users, and smart devices in order to gain extra advantage of the numerous services of the Internet. However, realizing the vision of the IoT is not an easy task due to the many challenges that need to be addressed. Examples of key challenges include availability, reliability, mobility, performance management, scalability, interoperability, and security and trust. For future networks, applications, algorithms, trust models, devices, all have evolved to accommodate the demands of user needs in the most optimal way possible. In the coming years, diverse technologies are expected the IoT

DOI: 10.4018/979-8-3373-2013-7.ch005

is expected to bridge the gap between research and practical aspects in support of intelligent decision making. It will open a door for new control methods and innovative operating frameworks.

5.1 Behavioral Learning Game based Connection (*BLGC*) scheme

Game theory is a groundbreaking framework for understanding human behaviors across various fields, including economics and political science. However, because game players often exhibit bounded rationality, game theory alone cannot entirely capture the complexities of human behavior. Therefore, it needs to be supplemented with other key concepts from behavioral sciences. Recently, S. Kim introduces the Behavioral Learning Game-based Connection (*BLGC*) scheme, which merges game theory with behavioral science approaches (Kim, 2016b). Firstly, the *BLGC* scheme develops a novel behavioral learning game model aimed at exploring how bounded rationality manifests within the cognitive capabilities of game players. Secondly, this developed game model is applied to manage and operate a socio-physical IoT system, demonstrating its practical utility. Finally, the *BLGC* scheme examines strategies for effective negotiation among players who, while primarily focused on their own welfare, are also inclined to consider the interests of other participants within the IoT system (Kim, 2016b).

5.1.1 Development motivation

The rapid advancement of IoT technology has enabled the connection of diverse smart objects via the Internet, offering enhanced methods for data interoperability tailored to various applications. Typically, IoT is envisioned to provide advanced connectivity among devices, systems, and services that surpass traditional machine-to-machine communications, encompassing a wide range of protocols, domains, and applications. The integration of these smart devices is anticipated to drive automation across nearly all sectors, significantly reducing the need for human intervention. The future outlook of IoT has been shaped by the convergence of multiple technologies, including wireless communication and micro-electromechanical systems (MEMS). This convergence implies that every traditional technology contributes to enabling the full potential of IoT (Kim, 2016b; Mazhelis, 2014; Singh, 2014; Vermesan, 2011).

Recently, the cascade of information within IoT systems has been shown to profoundly influence social behavior patterns. There is a growing recognition of the essential role played by coupled socio-physical networks as conduits for information dissemination. The socio-physical approach simultaneously considers the social interactions and physical interactions within IoT systems, resulting in significant qualitative and quantitative distinctions compared to approaches that analyze these aspects separately. During the operation of IoT systems, this collaborative methodology delivers services to both individuals and technical systems, thereby realizing the vision of future pervasive computing environments (Kim, 2016b; Nitti, 2014; Stai, 2014).

In IoT systems, individual devices typically make local control decisions aimed at maximizing their own profits, framing the situation as a game theory problem. Game theory serves as a decision-making framework where independent players engage in strategies to reach a joint decision that is acceptable to all participants. The standard application of game theory involves each player developing a utility function that quantifies the benefits they receive based on their own actions and those of other players. In traditional game theory, a solution concept is a rule that determines what constitutes an acceptable decision vector for all players, considering the conflict and cooperation dynamics of the environment (Archibald, 2006; Kim, 2016b).

Game theory typically operates under the assumption that players possess complete information about the game's dynamics throughout its duration. Consequently, many game models presuppose that players are perfectly rational and have the capability to act strictly according to their preferences. However, this assumption does not hold true in real-world scenarios; empirical studies have demonstrated that players do not always behave rationally or intelligently (Kim, 2011). Additionally, traditional game theory approaches focus on optimizing the individual profits of players. This often results in the collective interests of the group not being optimized or even adequately addressed, as each player pursues their own benefit. This individualistic optimization stems from the principle that each participant aims to maximize their own satisfaction without considering the well-being of others (Kim, 2016b; Stirling, 2005).

To address these shortcomings of traditional game theory, behavioral science has emerged as a novel research approach. Behavioral science posits that in decision-making processes, the rationality of individuals is constrained by the information they possess, the cognitive limitations of their minds, and the limited time available to make decisions. In real-world contexts, decision-makers often lack the capacity and resources to identify the optimal solution. As a result, they tend to settle for satisfactory solutions rather than seeking the best possible outcome. This perspective serves as an alternative to the conventional optimal decision-making framework, which views decision-making as a wholly rational endeavor aimed at finding the optimal choice based on available information. To better comprehend complex social behaviors, behavioral science has gained significant popularity and increasing interest within the social sciences (Camerer, 2003; Kim, 2016b).

The *BLGC* scheme introduces an innovative behavioral learning game model and presents a novel solution concept. Unlike traditional game models that assume fully rational decision-making, the *BLGC* scheme incorporates the bounded rationality of decision-makers during the solution-seeking process. Additionally, the *BLGC* scheme examines the tradeoff between selfish interests and social welfare, aiming to achieve a well-balanced system performance despite conflicting requirements. By leveraging a learning-based approach, the *BLGC* scheme explores the underlying motivations and probable behavioral patterns of players. The fundamental characteristic of this model is its real-world applicability, in contrast to conventional game theory models, which often assume perfectly optimal decision-making - an approach that is rarely feasible in practical scenarios (Kim, 2016b).

In the behavioral learning game, players adopt reasonable yet potentially suboptimal strategies. The game process involves analyzing and mapping decision-making strategies to enhance the effectiveness of the decision-making process. During gameplay, participants engage in iterative negotiations, continuously refining their strategy choices to reach a mutually satisfactory decision outcome. The *BLGC* scheme also introduces a new solution concept, referred to as *behavioral learning equilibrium* (*BLE*). This equilibrium represents a state where players' decision-making aligns with their behavioral learning process. Unlike traditional equilibrium concepts, the *BLE* accounts for the possibility that players may not fully grasp how different decisions would have altered their learning experience. As a result, they modify their behaviors when their real-world experiences contradict their prior beliefs. This dynamic learning-based equilibrium offers a fresh interpretation of decision-making, making it more adaptable to real-world scenarios (Kim, 2016b).

The IoT has been extensively studied across various domains, enabling seamless interconnection between uniquely identifiable smart objects or network agents within the existing internet framework. The rapid evolution of IoT technology has facilitated device-to-device communication, enhancing data interoperability for diverse applications (Wu, 2014). However, network optimization still has significant untapped potential, particularly in leveraging socially interactive relationships among agents. Exploring the continuum space of these relationships presents an exciting research opportunity (Chen, 2014; Jang, 2013; Raazi, 2010). Inspired by these insights, the *BLGC* scheme introduces a novel IoT communication control mechanism built upon the behavioral learning game model. By incorporating socio-physical relationships, the *BLGC* scheme accounts for both the social interactions and physical interdependencies among IoT agents, ensuring a more holistic and adaptive networking approach (Kim, 2016b).

5.1.2 Contribution

A fundamental insight of the *BLGC* scheme is that network agents are interconnected not only through physical interactions but also through social relationships, creating a dual-layer coupling within IoT environments. Given the dynamic and socially connected nature of IoT networks, the *BLGC* scheme models an agent's payoff as a combination of individual utility and social-group utility functions. The *BLGC* scheme allows agents to strategically optimize their decisions by balancing self-interest with collective welfare considerations. By leveraging autonomous agent behaviors, the *BLGC* scheme is structured as a behavioral learning game, wherein agents iteratively adjust their strategies to achieve an equilibrium state that reflects real-world decision-making patterns - referred to as the *behavioral learning equilibrium* (Kim, 2016b).

5.1.3 Behavioral Learning Game Model

To effectively model strategic interactions that incorporate learning dynamics, the *BLGC* scheme introduces a novel behavioral learning game. In this game framework, players aim to practically select their optimal strategies while dynamically adjusting their decisions. The model accounts for social relationships, ensuring that different players receive varying payoffs based on their level of interaction and influence within the network. The *BLGC* scheme formalizes this approach with the following definition (Kim, 2016b).

Definition 1. A behavioral learning game model constitutes a 5-tuple G = ($\mathbf{N}$, $S_{k,1\le k\le n}$, $U_{k,1\le k\le n}$, $\varphi_{k,1\le k\le n}$, $\mathbf{\Lambda}^{\varepsilon}$), where

(i) $\mathbf{N}$ is a set of game players $\mathbf{N} = \{p_1, \ldots, p_n\}$,
(ii) $S_k = \{s_1^k, s_2^k, \ldots, s_m^k\}$ is a nonempty finite set of all pure strategies of player *k* where *m* is the number of possible strategies,
(iii) $U_k = \{u_1^k, u_2^k, \ldots, u_m^k\}$ is the utility set of all payoffs of player *k*'s strategy; it is defined as a satisfaction levels of players. $\mathbf{U}$:$S_1 \times S_2 \ldots \times S_n \rightarrow \mathfrak{R}$ is the utility function where N represents the set of real numbers.
(iv) $\varphi_k = \{\varphi_{k1}, \ldots, \varphi_{kn}\}$ is the set of each player's social tie strength with other players. The *BLGC* scheme assumes that each player's social tie strength to itself is normalized as 1, i.e., $\varphi_{kn} = 1$ and $0 \le \varphi_{kj} \le 1$.
(v) $\mathbf{\Lambda}^{\varepsilon} = \{\varepsilon_1, \ldots, \varepsilon_n\}$ is the set of each player's cooperative degree for the total system performance. It is an indicator of non-cooperative actions for each player.

The behavioral learning game is built upon the fundamental principle of 'good enough' decisions rather than absolute optimization. In this framework, players search for an optimal solution but halt the search once they identify an option that meets or exceeds their individual aspiration levels. This means that instead of endlessly striving for profit maximization, players prioritize adequacy over strict selfish rationality, ensuring that their selected strategy yields a satisfactory outcome. By shifting the focus from profit maximization to sufficiency, the *BLGC* scheme presents a realistic decision-making model that aligns with practical constraints on information availability and computational resources in real-world scenarios. This approach enhances adaptability while maintaining efficiency in decision-making under dynamic IoT environments (Kim, 2016b).

The behavioral learning game uniquely integrates both cooperative and non-cooperative perspectives in multi-agent decision-making processes. According to social utility theory, each game player embodies two distinct personas; i) *non-cooperative persona* - this perspective focuses solely on maximizing individual payoff, where decisions are made purely based on self-interest, and ii) *cooperative persona* - this viewpoint prioritizes social relationships and public interests, ensuring decisions consider the well-being of other players in the network. By accounting for both selfish and cooperative behaviors, the behavioral learning game captures sophisticated decision-making dynamics, making it a realistic and systematic approach for modeling bounded rational behaviors. The formulation of utility functions within this model incorporates social utilities, which balance individual and collective interests, leading to a more adaptive and socially-aware IoT system (Kim, 2016b).

A practical example of behavioral learning games can be found in the electricity market model, where decision-making is influenced by uncertainties and varying levels of rationality among power suppliers. Usually, the market power released depends on suppliers' level of rationality, which is related to not only their information processing manner but also their utility function. The objective of power demander is to minimize the purchasing cost while facing the diversity of the market. Traditional hyper-rational game models assume perfect decision-making by market participants, which is often unrealistic in real-world markets with fluctuating conditions. These models struggle to accurately predict supplier behaviors and optimize decision-making under variable rationality levels. By integrating behavioral learning principles, the electricity market model can achieve more effective pricing strategies, optimized resource distribution, and improved market stability (Kim, 2016b; Shao, 2006).

5.1.4 Behavioral Learning Equilibrium

In non-cooperative game theory, the Nash equilibrium serves as a fundamental solution concept. However, one of its primary drawbacks lies in its unrealistic assumptions. Specifically, Nash equilibrium presumes that all players are perfectly rational, possess complete information, and operate within a well-defined and static environment. In practice, such conditions are rarely met (Kim, 2014). When players encounter unknown opponents or lack insight into others' individual preferences, achieving a Nash equilibrium becomes nearly impossible (Kim, 2016b).

Since the introduction of the Nash equilibrium, numerous alternative solution concepts have been developed by game theorists, including Pareto Equilibrium, Subgame Perfect Nash Equilibrium, Bayesian-Nash Equilibrium, ε-Equilibrium, Correlated Equilibrium, and Wardrop Equilibrium, among others (Kim, 2014). These refinements aim to address the limitations of Nash's original formulation. Nevertheless, despite these modifications, all equilibrium concepts fundamentally retain Nash's core idea: they examine decision-making by considering the strategic interactions among players. As a result, these equilibrium models still exhibit several inherent limitations: i) multiple equilibria can exist within a game, meaning there is often no guarantee of a unique solution; ii) in real-world scenarios, players do not always behave with perfect rationality, making this assumption impractical; iii) most equilibrium concepts have been formulated in static settings, thereby failing to account for players' ability to adapt their strategies dynamically over time; and iv) computing equilibria is computationally demanding, often requiring significant computational resources (Hemmati, 2010; Kim, 2014, 2016b).

The *BLGC* scheme introduces a new solution concept, called *Behavioral Learning Equilibrium* (*BLE*). Unlike traditional equilibrium concepts that rely on rational decision-making and static optimization, the *BLE* emerges as the result of a learning process over repeated interactions. Instead of assuming that players can instantly compute an equilibrium, the *BLE* suggests that they gradually adjust their strategies based on observed outcomes and feedback. Therefore, players try to maximize their satisfaction level through a repetitive learning process. In addition, the *BLE* solution includes the concept of mutual co-operation. If all players are satisfied while maintaining a social welfare, the *BLE* can be obtained. The *BLE* is formally defined as follows (Kim, 2016b).

Definition 2. *Behavioral Learning Equilibrium* (*BLE*) is a set of strategies that can be obtained by repeating a symmetric game with receiving feedbacks. When a set of strategies has chosen by all players and all the cooperative degrees of players are higher than a pre-defined minimum bound (Γ), this set of strategies and the corresponding payoffs constitute the *BLE*. That is formally formulated as

$$\mathbf{U}: U_1 \times U_2 \ldots \times U_n, \text{ s.t., } \left\{ u^i(s^i, \varepsilon_i) \in U_i, s^i \in S_i \text{ and } \varepsilon_i \in \mathbf{\Lambda}^{\varepsilon} \mid \min_{i, 1 \le i \le n} \{\varepsilon_i \mid \varepsilon_i \in \mathbf{\Lambda}^{\varepsilon}\} > \Gamma \right\} \tag{5.1.1}$$

s.t., $\Gamma < 1$ and $i \in N$

where U_i is the utility set of all consequent payoffs of player *i*'s strategy and $\ddot{E}^{\varepsilon}$ is the set of each player's cooperative degree. *n* is the number of game players.

BLE is the state is that all players' current cooperative degrees are above the pre-defined minimum bound (Γ). Therefore, the *BLE* is a strategy profile that approximately satisfies the condition of perfect equilibrium ($\Gamma = 1$). In the *BLE*, players have no incentives to deviate given their beliefs about the consequences of deviating. These beliefs are consistent with the information obtained from the actual equilibrium play of all players (Kim, 2016b).

The *BLGC* scheme does not focus on trying to get an optimal solution based on the traditional approach, but instead, an adaptive online interactive model is proposed. This approach can dramatically reduce the computational complexity and overheads. Even though, the *BLE* solution does not guarantee the performance optimization, the *BLE* concept can make this equilibrium possible in the real world operations. By replacing rigid optimization with adaptive behavioral learning, the *BLGC* scheme bridges the gap between theory and practice, offering a scalable and efficient solution for modern IoT-driven environments (Kim, 2016b).

5.1.5 Socio-Physical IoT Management Algorithm

The *BLGC* scheme introduces a novel IoT management algorithm based on the behavioral learning game model. The primary goal of the *BLGC* scheme is to strike a balance between optimality and 'good enough' performance, ensuring that IoT agents dynamically adapt to real-world system conditions. In the *BLGC* scheme, IoT agents are treated as game players, each having a set of packet transmission probabilities that serve as their strategies. Unlike traditional static decision-making models, players do not select a fixed best strategy. Instead, players dynamically adjust their decisions based on their preferences and system feedback. IoT agents are naturally self-interested, which may lead to non-cooperative behavior. To encourage collaboration, the *BLGC* scheme provides service cost compensation for players who act cooperatively. This incentivization mechanism ensures that even selfish players have a reason to participate in cooperative behaviors, thereby enhancing system efficiency (Kim, 2016b).

The coordinating entity (e.g., a system operator) monitors the IoT system state in real-time and intervenes adaptively to guide the system toward the *BLE*. Instead of relying on complete information, players use only local information and adjust their strategies through interactive learning techniques. This fully distributed implementation ensures that players' decisions evolve independently, avoiding the need for centralized control. Multi-player game models often require cooperative behavior, but in practice, the interests of all players rarely align perfectly. To maximize social welfare, players must have the ability to consider the well-being of others while still protecting their own interests. The *BLGC* scheme ensures that players can negotiate effectively, leading to a socially balanced decision-making process that fosters system-wide efficiency while maintaining fairness (Kim, 2016b).

5.1.6 Data Transmission Algorithm in IoT Systems

To implement the *BLGC* scheme, IoT agents are grouped as a cluster with a contention-based medium access communication mechanism. If there are n communication agents, they are game players; player $k(0 \leq k \leq n)$ contends for the opportunity of data transmission with probability q^k(i.e., $q^k \in S_k$) in a time slot where $0 \leq q^k \leq 1$. If a persistence mechanism is implemented, the data transmission probability is just the persistence probability. If multiple players contend in the same time slot, a collision occurs and no player can get the transmission opportunity. To define the utility function for each player, the *BLGC* scheme employs three different components. From the viewpoint of self-interest, the individual utility function of each player k ($\grave{\mathrm{E}}^k(\cdot)$) is defined as follows (Kim, 2016b).

$$\grave{\mathrm{E}}^k(q^k, \mathbf{q}^{-k}) = \log\left(\varrho_k \times q^k \times \prod_{\substack{l=1,\dots,n \\ l \neq k}} (1 - q^l)\right) - \mathscr{C}(k, q^k) \tag{5.1.2}$$

s.t., $\mathbf{q}^{-k} = (q^1, \dots, q^{k-1}, q^{k+1}, \dots q^N)$

where ϱ_k represents player k's action willingness (e.g., player k's efficiency of utilizing the transmission opportunity). Traditionally, the logarithmic function is widely used in literature for modeling utility of network users. $\mathscr{C}(k, q^k)$ is the cost function for player k with probability q^k. Thus, the individual payoff function ($\grave{\mathrm{E}}^k(\cdot)$) has a nice interpretation: the net gain of utility from channel access decreases by the network cost (Kim, 2016b).

In the *BLGC* scheme, the system operator monitors and coordinates explicitly by adjusting the cost. During game operations, the system operator periodically observes players' behaviors and cooperative degree of each player. Every time period (*unit_time*), the system operator dynamically adjusts the cost based on this measured information. Therefore, the system operator can induce players to cooperate each other as well as to take appropriate actions. In the *BLGC* scheme, cost is obtained based on the concept of fairness and cooperative degree. To characterize this fairness notion, the *BLGC* scheme follows the Jain's fairness index (F), which has been frequently used to measure the fairness of network resource allocations (Dianati, 2005). Based on the F and εinformation, player k's cost function $\mathscr{C}(k, q^k)$ is defined as follows (Kim, 2016b).

$$\mathscr{C}(k, q^k) = (1+\xi) \times \left(\frac{\gamma_k(q^k, \mathbf{q}^{-k})}{\sum_{i=1}^{n} \gamma_i(q^i, \mathbf{q}^{-i})} \right)^{\mathscr{F}} \times \log \left(\varrho_k \times q^k \times \prod_{\substack{l=1,\ldots,n \\ l \neq k}} (1 - q^l) \right)$$

(5.1.3)

$$\text{s.t., } \xi = \begin{cases} 0, if \varepsilon_k > \Gamma \\ \Gamma, if \varepsilon_k \leq \Gamma \end{cases}, \mathscr{F} = \frac{\left(\sum_{i=1}^{n} \gamma_i(q^i, \mathbf{q}^{-i}) \right)^2}{n \times \sum_{i=1}^{n} (\gamma_i(q^i, \mathbf{q}^{-i}))^2} \text{ and } i \in N$$

where$\gamma_k(q^k, \mathbf{q}^{-k})$ is the amount of actually transmitted data bits per *unit_time* for player k, and the range of F is varied from 0 to 1. Therefore, if a player has a higher fairness value (F) and cooperative degree (ε), he can reduce the $\mathscr{C}(\cdot)$ value (Kim, 2016b).

The *BLGC* scheme considers that two players are connected by a directed edge if one has a social tie towards the other. Let $\mathbb{G}_k$ denote the group of player k; $\mathbb{G}_k$ is the set of players connected by player k. From the perspective of group, the group utility function of $\mathbb{G}_k$ ($\mathscr{J}^k(\cdot)$) is defined based on the social relationships. The strength of the social tie from player k to player i is quantified by φ_{ki} where $0 \leq \varphi_{ki} \leq 1$. Each player k aims to maximize its $\mathscr{J}^k(\cdot)$, which is given by (Kim, 2016b);

$$\mathscr{J}^k(q^k, \mathbf{q}^{-k}) = \sum_{i=1, i \neq k}^{N} \left(\varphi_{ki} \times \grave{\mathrm{E}}^k(q^k, \mathbf{q}^{-k}) \right)$$

(5.1.4)

where $\mathbf{q}^{-k}$is the probability set of players without player k. From the viewpoint of social cooperation, the *BLGC* scheme should consider the total performance of participating players. Therefore, the social utility function of each player k ($\mathfrak{T}^k(\cdot)$) is defined as follows (Kim, 2016b).

$$\mathfrak{T}^k(q^k, \mathbf{q}^{-k}) = \varepsilon_k \times \sum_{\substack{l=1,\ldots,n \\ l \neq k}} \grave{\mathrm{E}}^l(q^l, \mathbf{q}^{-l})$$

(5.1.5)

By considering three different perspectives (i.e., individual utility (Θ), group utility (J) and social utility (T)), the *BLGC* scheme can make possible comparisons of the selfish option in contrast to the socially cooperative option. Finally, the total utility function (U^k) of player k is defined as follows (Kim, 2016b).

$$U^k = \max_{q^k \in S_k; \varepsilon_i \in \Lambda^{\varepsilon}} \left(\grave{\mathrm{E}}^k(q^k, \mathbf{q}^{-k}) + \mathscr{J}^k(q^k, \mathbf{q}^{-k}) + \mathfrak{T}^k(q^k, \mathbf{q}^{-k}) \right)$$

(5.1.6)

where ε_k is a weighted parameter for different objectives; it provides the evaluation of the relative degree of achieving a social payoff. Under the diverse preference and uncertainty, players dynamically adjust ε value to adapt the preference of social welfare (Kim, 2016b).

5.1.7 Summary

In the modern digital landscape, the IoT is considered the next technological and economic revolution, following the widespread adoption of the Internet. IoT enables intelligent interconnectivity by linking smart objects to the Internet, facilitating seamless data exchange and communication under agreed-upon protocols. The *BLGC* scheme takes a new direction to develop a socio-physical IoT connection algorithm and it represents a paradigm shift in IoT communication control, offering a scalable, intelligent, and adaptive solution for the next-generation connected world (Kim, 2016b).

5.2 Bargaining Game based Time Scheduling (*BGTS*) scheme

Backscatter communication has been recognized as a fundamental technology in IoT applications. Given that it relies on coordination among network agents, cooperative bargaining theory serves as an effective approach to achieving optimal system performance. Recently, S. Kim introduces the *Bargaining Game-based Time Scheduling* (*BGTS*) scheme for a backscatter-assisted radio-frequency (RF)-powered cognitive radio (CR) network (Kim, 2021b). In the *BGTS* scheme, multiple secondary transmitters can dynamically switch between backscatter communication, energy harvesting, and active data transmission modes. The primary goal is to enhance RF-CR system performance while simultaneously fostering mutual benefits and reaching a reciprocal agreement among various control parameters. By integrating two distinct bargaining solutions - the *modified Nash bargaining solution* and the *equitable Nash bargaining solution* - the *BGTS* scheme establishes a dual bargaining game model designed to efficiently allocate limited time resources. A key innovation of the *BGTS* scheme lies in its adaptability, flexibility, and responsiveness to the ever-changing conditions of the RF-CR system (Kim, 2021b).

5.2.1 Development Motivation

Over the past decades, wireless communication technologies have experienced significant advancements. The transition from first-generation (1G) to fourth-generation (4G) networks has posed numerous challenges in both physical design and practical applications. With the rapid evolution of modern communication systems, ultra-dense IoT is emerging as a key application paradigm for the upcoming fifth-generation (5G) and future wireless networks. According to Ericsson, approximately 22.3 billion IoT devices are projected to be deployed globally by 2024, with the number expected to exceed 80 billion by 2030. As a result, the proliferation of IoT within 5G networks is anticipated to be a transformative force in next-generation communication, accelerating the development of new technologies to accommodate diverse service demands. This evolution will pave the way for novel wireless architectures and intelligent services. To support these advanced services, it is crucial to ensure seamless connectivity among wireless IoT devices, enabling real-time data generation, monitoring, and processing (Chettri, 2020; Kim, 2021b; Ye, 2020).

The rapid increase in resource-intensive IoT applications has introduced several challenges in the large-scale deployment of IoT devices. One primary limitation is the constrained battery capacity of

IoT devices due to their compact size, which restricts their operational lifespan. Additionally, the rise of smart IoT devices has significantly heightened spectrum demand in contemporary wireless networks. Therefore, minimizing power consumption for communication data transmission is a critical concern. To address this issue, the *BGTS* scheme should incorporate innovative energy harvesting and spectrum-sharing strategies, offering a more practical and holistic solution compared to conventional energy and spectrum management techniques (Gao, 2020; Kim, 2021b; Zhang, 2019).

Cognitive Radio (CR) has been introduced as a solution to the impending spectrum scarcity issue. This technology enables secondary transmitters (STs) to dynamically access underutilized spectrum resources without interfering with primary transmitters (PTs). Consequently, CR has been recognized as a promising technique to accommodate the surging spectrum demand. Among various CR technologies, radio-frequency (RF)-powered CR networks have gained increasing attention due to advancements in RF energy harvesting. A major advantage of this approach is that IoT devices can harvest energy from consistently available RF sources, thereby extending their operational duration. When a PT is active, multiple STs collect energy from the emitted RF signals. Conversely, when a PT is idle, STs utilize the harvested energy to transmit data to the secondary gateway (SG). Consequently, RF-powered CR (RF-CR) systems enhance both spectrum efficiency and energy efficiency (Gao, 2019; Kim, 2021b; Wang, 2018).

In RF-CR systems, data transmission typically follows a harvest-then-transmit mechanism, where ST performance is largely influenced by PT activity. Ideally, PT activity should be maintained at a moderate level to ensure satisfactory ST performance. Otherwise, excessive PT activity may significantly reduce the total number of transmitted bits by STs. To mitigate this limitation, backscatter communication has recently emerged as a promising alternative. This technique enables IoT devices to transmit data by reflecting incident RF signals rather than generating their own. By leveraging backscatter technology, STs can communicate by reflecting PT signals even when PTs are actively transmitting. This approach has been recognized as a valuable enhancement to RF-CR systems, and integrating backscatter communication into these networks is gaining growing research interest (Gao, 2020; Kim, 2021b; Wang, 2018).

In backscatter-assisted RF-CR networks, multiple STs typically transmit their data to the SG. Notably, the SG is responsible for informing STs about the status of the primary channel. While STs do not require SG coordination during energy harvesting, they do rely on it for data transmission via backscatter communication. Each ST has the flexibility to alternate between harvest-then-transmit (HTT) and backscatter modes for data transmission. When a PT is utilizing the channel, STs cannot transmit data conventionally. However, at any given moment, an ST can either harvest energy from PT signals or use backscatter communication to transmit data at a lower bitrate. As a result, secondary transmissions are conducted through time-division multiple access (TDMA), where each time slot is allocated for either energy harvesting, backscattering, or ST data transmission (Gao, 2020; Kim, 2021b; Wang, 2018).

With the increasing number of IoT devices connected to RF-CR networks, meeting the transmission demands of all STs becomes increasingly challenging. In a TDMA-based system, time slots are highly valuable and limited resources, often failing to accommodate all ST transmission requests. Since individual STs have different service requirements, their need for time resources varies. Consequently, the SG is expected to dynamically allocate time slots to multiple STs in a manner that maximizes the overall welfare of the RF-CR system. Given their distinct characteristics, STs adopt different control strategies; however, they must collaborate to achieve optimal RF-CR system performance. To address this, the *BGTS* scheme employs a cooperative bargaining game framework to develop a refined time scheduling algorithm tailored for RF-assisted CR networks (Gao, 2019; Kim, 2021b).

Recently, A. Nazari et al. introduced the *modified Nash bargaining solution* (*MNBS*) to implement cooperative frameworks in a fair and equitable manner. Traditional *Nash bargaining solution* ensures equal surplus distribution among all participants. However, the *MNBS* refines this approach by allocating higher surpluses to participants with greater investment contributions, thereby ensuring fairness. Specifically, each participant assumes a proportional share of the shared investment costs, ensuring that all stakeholders benefit from the cooperation. In 2019, Y. Xu and N. Yoshihara further improved the original *Nash bargaining solution* by integrating considerations of equity and efficiency. This refinement, termed the equitable Nash bargaining solution (*ENBS*), combines the Nash solution with a variation of the *Kalai-Smorodinsky bargaining solution*. Under the principle of equity, the *ENBS* determines the optimal outcome by ensuring that the utility gains of all participants, relative to their maximum attainable utilities, are equalized (Kim, 2021b; Nazari, 2021; Xu, 2019).

5.2.2 Contribution

The *BGTS* scheme leverages a cooperative data transmission mechanism that integrates backscatter technology. Within the framework of a backscatter-assisted RF-powered CR system, its primary goal is to optimize system throughput in dynamically evolving environments. To achieve this, each ST is capable of adaptively switching between the HTT mode and the backscatter mode. To efficiently manage this switching process, the *BGTS* scheme incorporates the concept of the *ENBS*. When the primary channel becomes idle, multiple STs engage in active data transmission. Utilizing the principles of the *MNBS*, the available idle time is distributed among STs. By integrating both *ENBS* and *MNBS*, the *BGTS* scheme introduces a novel dual bargaining game model, which plays a crucial role in optimizing overall system performance. Through an interactive and iterative decision-making process, the allocation of limited time slots is dynamically adjusted, ensuring that time scheduling strategies are interdependent and collectively drive the system toward optimal performance. To the best of my knowledge, this is the first study to incorporate two distinct bargaining solutions into the control algorithm design for RF-CR systems (Kim, 2021b).

5.2.3 Backscatter-aided CR platform and dual bargaining game

In the *BGTS* scheme, a backscatter-aided RF-CR system model is taken for analysis. There are one PT $(\mathcal{P}_T)$, one SG $(\mathcal{S}_{\mathcal{G}})$, and n STs, that is, $\mathbb{D} = \{\mathcal{D}_1, \ldots, \mathcal{D}_n\}$, as shown in Figure 1. $\mathcal{P}_T$, $\mathcal{S}_{\mathcal{G}}$ and $\mathcal{D}_{1 \leq i \leq n}$ are equipped with omnidirectional antennas, and STs are located within a coverage area of the $\mathcal{S}_{\mathcal{G}}$. In the primary network, the channel state is divided into the busy (T^B) and the idle (T^I) periods depending on whether the PT emits signals or not. In the secondary network, the TDMA method is employed to avoid interference. Therefore, secondary transmission is executed in time slots. When the channel is occupied by the primary transmission, STs can either harvest energy from the PT's signal or backscatter the PT's signal for their own data transmission. If the channel is idle, STs can transmit their data to the SG actively (Gao, 2019, 2020; Kim, 2021b; Wang, 2018).

Figure 1. System model of a backscatter-aided RF-powered CR network (Kim, 2021b)

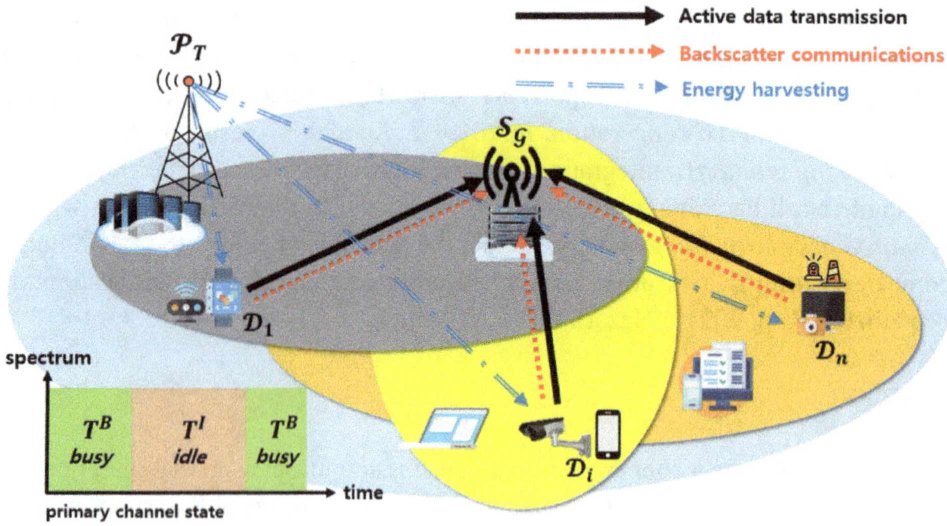

Each ST $(\mathcal{D}_{1\leq i\leq n})$ consists of a controller, a backscatter circuit, an energy harvesting module, a transmission data allocator, and a rechargeable battery. Note that the actions of ST are dependent on the channel state of the primary network. In particular, during the T^B period, the ST can backscatter ambient signals or harvest energy, which is selected by its controller. The length of T^B is evenly divided by multiple time slots, and each time slot is assigned to each ST; the time slot τ_i is allocated to the $\mathcal{D}_i$ where $T^B = \sum_{\mathcal{D}_i \in \mathrm{D}} \tau_i$. The τ_i is split by two time sub-slots, that is, $[(1-\rho_i)\cdot\tau_i]$ and $[\rho_i\cdot\tau_i]$; $[(1-\rho_i)\cdot\tau_i]$ is used to harvest energy and $[\rho_i\cdot\tau_i]$ is used to backscatter ambient signals to the $\mathcal{S}_\mathcal{G}$. To effectively adapt the current backscatter-aided communication environments, each ρ_i value should be dynamically adjusted. On the one hand, during the T^I period, the $\mathcal{S}_\mathcal{G}$ schedules the time slots to meet the data transmission demands of STs. To maximize the RF-CR system throughput, the time scheduling mechanism of $\mathcal{S}_\mathcal{G}$ must ensure a relevant tradeoff between efficiency and fairness. The channel frame structure for the T^B and T^I periods are illustrated in Fig. 5.2.2 (Kim, 2021b).

Figure 2. The channel frame structure in the RF-CR platform (Kim, 2021b)

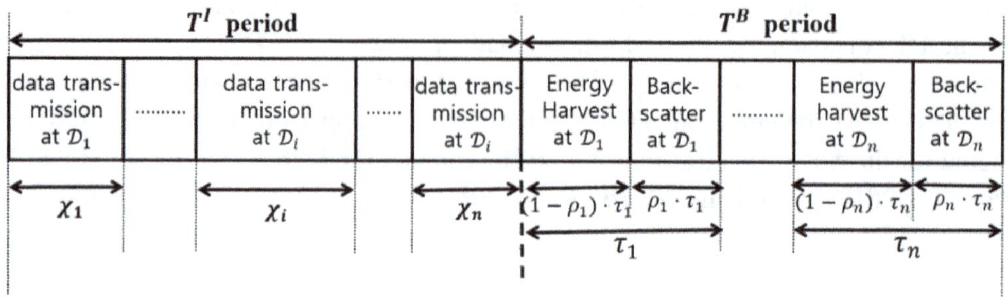

At each time period, the *BGTS* scheme formulates a bargaining game to address the time sharing problem. For the T^I period, the idle time bargaining game ($\mathbb{G}^I$) is designed. In the $\mathbb{G}^I$, individual STs in the D are game players, and the T^I period is shared through the concept of the *MNBS* to actively transmit the STs' data. When the primary channel is busy, the T^B period is evenly shared by STs. Each time slot, which is assigned for each ST, is divided for the energy harvest and backscatter modes. For the τ_i division problem, the busy time bargaining game $\left(\mathbb{G}_i^B\right)$ is formulated, and the τ_i is divided based on the idea of the *ENBS*. During the T^B period, multiple $\mathbb{G}^B_{1\leq i\leq n}$ games are operated in a parallel manner. Therefore, the $\mathbb{G}^I$ and $\mathbb{G}_i^B$ are repeated in a step-by-step manner, and the time sharing process is operated interactively at each time period. Formally, the *BGTS* scheme defines the $\mathbb{G}^I$ and $\mathbb{G}_i^B$ game entities, i.e,

$$\mathbb{G} = \left\{\mathbb{G}^I,\mathbb{G}^B_{1\leq i\leq n}\right\} = \left\{\mathcal{P}_T,\ \mathcal{S}_{\mathcal{G}},\mathbb{D},\left\{\mathbb{G}^I|\mathcal{D}_{1\leq i\leq n},\chi_{1\leq i\leq n},U_{\mathcal{D}_{1\leq i\leq n}}\right\},\left\{\mathbb{G}_i^B|\tau_i,\left(m_i^{EH},m_i^{BS}\right),\rho_i,\left(\mathfrak{U}_i^{EH},\mathfrak{U}_i^{BS}\right)\right\},T\right\}$$

of gameplay (Kim, 2021b).

- The $\mathbb{G}^I$ and $\mathbb{G}^B_{1\leq i\leq n}$ are mutually and reciprocally interdependent in an interactive manner, and they work together to share each limited time resource.
- $\mathcal{P}_T$ is the PT, $\mathcal{S}_{\mathcal{G}}$ is the SG, and D is the set of STs in the backscatter-aided RF-powered CR system platform.
- In the $\mathbb{G}^I$, $\mathcal{D}_{1\leq i\leq n} \in \mathbb{D}$ are the game players, $\chi_{1\leq i\leq n}$ are their strategies, and $U_{\mathcal{D}_{1\leq i\leq n}}$ are their utility functions.
- In the $\mathbb{G}_i^B$, τ_i is the time slot assigned for the $\mathcal{D}_i$; it is divided by the time splitting ratio ρ_i where $0 \leq \rho_i \leq 1$.
- m_i^{EH} is the energy harvest mode and m_i^{BS} is the backscatter mode. They are game players in the $\mathbb{G}_i^B$, and the ρ_i is the strategy. The $\mathfrak{U}_i^{EH}$ and $\mathfrak{U}_i^{BS}$ are utility functions of m_i^{EH} and m_i^{BS}, respectively.
- Discrete time model $T \in \left\{t_1,\ldots,t_c,t_{c+1},\ldots\right\}$ is represented by a sequence of time steps. The t_c can be the T^I or T^B, and the length of t_c matches the event time-scale of $\mathbb{G}^I$ or $\mathbb{G}^B_{1\leq i\leq n}$.

5.2.4 The basic ideas and concepts of the *ENBS* and *MNBS*

To characterize the basic ideas of the *ENBS* and *MNBS*, we first starts with some definitions. Let $\sum$ be the set of all subsets of $\mathbb{R}^n_+$, and elements in $\sum$ are interpreted as bargaining problems. For every bargaining problem $A \in \sum$, a bargaining solution F assigns a nonempty subset *F*(*A*) of A. For all $A \in \sum$ and all $i \in N$, let $\mathcal{M}_i(A) = \max\left\{a_i|\left(a_1,\cdots,a_i,\cdots,a_n\right) \in A\right\}$. $\mathcal{M}(A) \equiv \left(\mathcal{M}_i(A)\right)_{i\in N}$ is the ideal point of A. The classical Nash bargaining solution (*NBS*) over $\sum$ is mathematically defined as follows (Kim, 2021b; Xu, 2019):

$$NBS(A) = \max_{a\in A}\prod_{i\in N} a_i,\ \text{s.t., all } A \in \sum \tag{5.2.1}$$

All the points in the *NBS*(*A*) lie on the highest indifference surface given by $\min\left(\frac{a_1}{\mathcal{M}_1(A)},\ldots,\frac{a_n}{\mathcal{M}_n(A)}\right)$ attainable in A where $a \in A$. According to (5.2.1), the *NBS* can be rewritten as follows (Kim, 2021b; Xu, 2019):

$NBS(A) =$

$$\left\{a \in A \middle| \min\left(\frac{a_1}{\mathcal{M}_1(A)}, \dots, \frac{a_n}{\mathcal{M}_n(A)}\right) \geq \mathbf{min}\left(\frac{x_1}{\mathcal{M}_1(A)}, \dots, \frac{x_n}{\mathcal{M}_n(A)}\right), \forall x \in A\right\}$$

(5.2.2)

s.t., all$A \in \sum$

As a refinement of the classical *NBS*, the *ENBS* over $\sum$ is defined by considering the concept of egalitarianism. Formally, the *ENBS* is given as follows (Kim, 2021b; Xu, 2019):

$$ENBS(A) = \left\{a \in NBS(A) \middle| \min_{i\in N}\left\{\frac{a_i}{\mathcal{M}_i(A)}\right\} \geq \mathbf{min}_{i\in N}\left\{\frac{x_i}{\mathcal{M}_i(A)}\right\} \text{ for all } x \in NBS(A)\right\}$$

(5.2.3)

s.t., all$A \in \sum$

Another bargaining approach, known as the *MNBS*, is developed from a different perspective. While the traditional Nash solution ensures equal surplus distribution among all participants, the *MNBS* is introduced to establish a more equitable bargaining framework. In this model, players who contribute more to the investment receive proportionally higher surpluses, making participation more incentivizing. To achieve this, the original Nash solution is adjusted, and the *MNBS* is formally defined as follows (Kim, 2021b; Nazari, 2021):

$$MNBS = \left\{a \in A \middle| \left(\mathcal{H}_N^a \cdot \prod_{i=1}^{n} a_i\right) \geq \left(\mathcal{H}_N^x \cdot \prod_{i=1}^{n} x_i\right) \text{ for all } x \in A\right\}$$

(5.2.4)

s.t.,$\mathcal{H}_N^a = \prod_{i=1}^{n} f(a_i)$ and $\mathcal{H}_N^x = \prod_{i=1}^{n} f(x_i)$

where $f(\cdot)$ is the reward function for the i's investment. According to (5.2.4), the *MNBS* can be rewritten as follows (Kim, 2021b):

$$MNBS = \max_{x\in A}\left(\mathcal{H}_N^x \cdot \prod_{i=1}^{n} x_i\right), \text{ s.t.}, \mathcal{H}_N^x = \prod_{i=1}^{n} f(x_i)$$

(5.2.5)

5.2.5 The dual bargaining game in the RF-CR platform

To develop a new dual bargaining game model, the *BGTS* scheme constructs the $\mathbb{G}^I$ and $\mathbb{G}_i^B$ games by relying on the influence of reciprocal interaction. For the T^I period, the $\mathbb{G}^I$ is designed to share the limited time resource among the STs. As a game player, the $\mathcal{D}_i$ select his strategy χ_i by considering not only its generated data amount, but also the current energy level. At the time t_c, the $\mathcal{D}_i$'s utility function, that is, $U^{t_c}_{\mathcal{D}_i}(\cdot)$, is defined as follows (Kim, 2021b):

$$U^{t_c}_{\mathcal{D}_i}\left(\chi_i, \mathscr{C}_{T^l}, \mathcal{T}_{\mathcal{D}_i}, E_{\mathcal{D}_i}, \mathscr{E}^{t_c}_{\mathcal{D}_i}, \mathfrak{M}_{\mathbb{D}}\right) = \sqrt{\left(\frac{\min\left(\chi_i, \mathscr{C}_{T^l}\right)}{\mathscr{C}_{T^l}}\right)^{\beta-\vartheta_{\mathcal{D}_i}} \Big/ \left(\frac{E_{\mathcal{D}_i} - \mathscr{E}^{t_c}_{\mathcal{D}_i}}{E_{\mathcal{D}_i}}\right)} \tag{5.2.6}$$

$$\text{s.t., } \vartheta_{\mathcal{D}_i} = \left(\frac{1}{1 + exp\left(-\frac{\mathcal{T}_{\mathcal{D}_i}}{\mathfrak{M}_{\mathbb{D}}}\right)} - \varrho\right) \text{ and } \mathfrak{M}_{\mathbb{D}} = \sum\nolimits_{\mathcal{D}_j \in \mathbb{D}} \mathcal{T}_{\mathcal{D}_j}$$

where $\mathscr{C}_{T^l}$ is the total communication capacity of T^l period, and $\mathcal{T}_{\mathcal{D}_i}$ is the generated data amount of $\mathcal{D}_i$. $E_{\mathcal{D}_i}$ is the full energy capacity of $\mathcal{D}_i$ and $\mathscr{E}^{t_c}_{\mathcal{D}_i}$ is the $\mathcal{D}_i$'s energy status at time t_c. β and ϱ are control parameters for $U^{t_c}_{\mathcal{D}_i}(\cdot)$. In the $\mathbb{G}^l$, multiple ST devices, that is, $\mathcal{D}_{1\leq i\leq n}$, are assumed as game players, and they share the T^l period in a fair-efficient way. To implement this sharing process, the *BGTS* scheme should consider each player's contribution for the social welfare. Usually, spectrum sensing is the ground work for the realization of CR technology. In the *BGTS* scheme, the cooperative spectrum sensing method has been presented as an effective sensing way to improve the performance of sensing in the CR system. The main idea of cooperative sensing is to enhance the sensing performance if a group of SUs voluntarily contribute to sensing and share their local sensing information to get a better picture of the spectrum usage. For making a combined decision, which is more accurate than the individual decisions, spectrum sensing information are collected at the $\mathcal{S}_{\mathscr{g}}$. Then, the $\mathcal{S}_{\mathscr{g}}$ decides whether any primary user is active or not. However, STs are intuitively reluctant to cooperatively sense and would prefer to free-ride without contributing anything. It is the well-known free-ride problem (Kim, 2017b, 2021b).

To avoid the ST's free-ride problem, the *BGTS* scheme should consider the each ST's contribution in the $\mathbb{G}^l$ game process. Therefore, the *BGTS* scheme adopts the concept of *MNBS* for the $\mathbb{G}^l$ game. According to (4)-(6), the *MNBS* is obtained as follows (Kim, 2021b):

$$MNBS\left(\chi_{1\leq i\leq n} | \mathbb{D}\right) = \max_{\chi}\left(\mathscr{H}_{\mathbb{D}}(\chi) \cdot \prod_{\mathcal{D}_i \in \mathbb{D}} U^{t_c}_{\mathcal{D}_i}\left(\chi_i, \mathscr{C}_{T^l}, \mathcal{T}_{\mathcal{D}_i}, E_{\mathcal{D}_i}, \mathscr{E}^{t_c}_{\mathcal{D}_i}, \mathfrak{M}_{\mathbb{D}}\right)\right) \tag{5.2.7}$$

$$\text{s.t.,}\begin{cases} \mathscr{H}_{\mathbb{D}}(\chi) = \max_{\chi_i} \prod\nolimits_{i=1}^{n}\left(\left(\frac{\min\left(\chi_i, \mathscr{C}_{T^l}\right)}{\mathscr{C}_{T^l}} \times \log\left(\alpha + \frac{\tilde{O}^{t_c}_i}{\tilde{O}^{t_c}_i + \Phi^{t_c}_i}\right)\right) + \gamma\right) \\ \sum\nolimits_{i=1}^{n} \chi_i \leq T^l \\ \begin{cases} \tilde{O}^{t_c}_i = \tilde{O}^{t_{c-1}}_i + 1, \Phi^{t_c}_i = \Phi^{t_{c-1}}_i, if the\, \mathcal{D}_i\, cooperatively sensed \\ \tilde{O}^{t_c}_i = \tilde{O}^{t_{c-1}}_i, \Phi^{t_c}_i = \Phi^{t_{c-1}}_i + 1, otherwise \end{cases} \end{cases}$$

where α, γ are control parameters for the $\mathscr{H}_{D}(\cdot)$ function, which considers each ST's investment. For the T^B period, individual ST devices operate their $\mathbb{G}^{B}_{1\leq i\leq n}$ games in a dispersive and parallel manner. In the $\mathbb{G}^{B}_{i}$, the backscatter mode $\left(m_i^{BS}\right)$ and energy harvesting mode $\left(m_i^{EH}\right)$ are assumed as game players, and the decision of ρ_i value is their strategies. For the m_i^{BS} and m_i^{EH}, their utility functions, i.e., $\mathfrak{U}_i^{BS}(\cdot)$ and $\mathfrak{U}_i^{EH}(\cdot)$, are defined as follows (Kim, 2021b):

$$\begin{cases} \mathfrak{U}_i^{BS}\left(t_c,\rho_i,E_{\mathscr{D}_i},\mathscr{E}^{t_c}_{\mathscr{D}_i}\right) = \left(\eta + \left(\frac{\mathscr{E}^{t_c}_{\mathscr{D}_i}}{E_{\mathscr{D}_i}}\right)^{\theta}\right) \times \frac{(\rho_i)^{\mu}}{\eta + exp(-\rho_i)} \\ \mathfrak{U}_i^{EH}\left(t_c,\rho_i,E_{\mathscr{D}_i},\mathscr{E}^{t_c}_{\mathscr{D}_i}\right) = \left(\eta + \left(\frac{E_{\mathscr{D}_i} - \mathscr{E}^{t_c}_{\mathscr{D}_i}}{E_{\mathscr{D}_i}}\right)^{\theta}\right) \times \log\left(\left(\sigma \times (1-\rho_i)^{\zeta}\right) + \eta\right) \end{cases} \tag{5.2.8}$$

where η and θ are common control parameters for $\mathfrak{U}_i(\cdot)$. μ is a modification factor for $\mathfrak{U}_i^{BS}(\cdot)$ and σ, ζ are modification factors for $\mathfrak{U}_i^{EH}(\cdot)$. In the $\mathscr{D}_i$'s point of view, the equity principle should be considered for the m_i^{BS} and m_i^{EH}. Therefore, the idea of *ENBS* is chosen for the $\mathbb{G}_i^B$ game. According to (5.2.3) and (5.2.8), the *ENBS* is obtained as follows (Kim, 2021b):

$$ENBS\left(\left(\mathfrak{U}_i^{BS}(\cdot),\mathfrak{U}_i^{EH}(\cdot)\right)|\rho_i\right) =$$

$$\left\{\min\left\{\frac{\mathfrak{U}_i^{BS}(\cdot|\rho_i)}{\mathscr{M}_i^{BS}},\frac{\mathfrak{U}_i^{EH}(\cdot|1-\rho_i)}{\mathscr{M}_i^{EH}}\right\} \geq \mathbf{min}\left\{\frac{\mathfrak{U}_i^{BS}(\cdot|\rho_i')}{\mathscr{M}_i^{BS}},\frac{\mathfrak{U}_i^{EH}(\cdot|1-\rho_i')}{\mathscr{M}_i^{EH}}\right\}\right\} \tag{5.2.9}$$

$$\text{s.t.}, \left(\left(\mathfrak{U}_i^{BS}(\cdot),\mathfrak{U}_i^{EH}(\cdot)\right)|\rho_i\right), \left(\left(\mathfrak{U}_i^{BS}(\cdot),\mathfrak{U}_i^{EH}(\cdot)\right)|\rho_i'\right) \in NBS\left(\mathfrak{U}_i^{BS}(\cdot),\mathfrak{U}_i^{EH}(\cdot)\right)$$

5.2.6 Summary

The *BGTS* scheme addresses the issue of time allocation in a backscatter-assisted RF-CR platform by integrating principles from cooperative game theory. Initially, it establishes a comprehensive network framework featuring intelligent and rational system agents. Following this, the fundamental concepts of the *ENBS* and the *MNBS* are introduced to develop an innovative dual bargaining game model. These two bargaining approaches play a crucial role in making effective control decisions within dynamically evolving RF-CR system environments. To achieve optimal performance, the $\mathbb{G}^I$ and $\mathbb{G}_i^B$ games interact cooperatively, ensuring desirable global system properties. Through an iterative game-based process, the *BGTS* scheme continuously adapts to real-time system conditions, enabling an intelligent and responsive approach to time scheduling (Kim, 2021b).

5.3 Bargaining based Unlicensed Spectrum Sharing (*BUSS*) scheme

The rapid expansion of the IoT paradigm, combined with the limited availability of spectrum resources, has posed significant challenges to enhancing capacity in 5G wireless networks. To address this issue, it is essential to develop effective spectrum-sharing techniques that utilize unlicensed frequency bands. Recently, S. Kim introduces the *Bargaining-based Unlicensed Spectrum Sharing* (*BUSS*) scheme within the television white space (TVWS) framework (Kim, 2021a). Grounded in bargaining game theory, the *BUSS* scheme is designed to facilitate mutual benefits and establish a reciprocal consensus among various data application services. By employing a sequential and iterative game-based approach, the *BUSS* scheme dynamically determines control decisions for spectrum sharing in the TVWS environment. Leveraging an innovative bargaining solution along with established rationing principles, the *BUSS* scheme reconciles the conflicting interests of IoT devices and fosters cooperative agreements to optimize their service performance.

5.3.1 Development motivation

The rapid expansion of heterogeneous wireless-capable IoT devices, coupled with the continuous surge in data traffic within wireless networks, has made spectrum-sharing techniques indispensable for achieving high-throughput wireless Internet access. To ensure seamless coexistence among various network technologies, spectrum-sharing strategies must be carefully designed. In recent years, policymakers, researchers, and industry professionals have been engaged in discussions to develop more flexible spectrum access mechanisms aimed at enhancing spectrum utilization. To address this challenge, many scholars and experts have introduced the concept of *spectrum commons*, which grants equal spectrum access rights to all users, operators, and technologies. Advocates of *spectrum commons* argue that advanced spectrum-sharing technologies enable a virtually unlimited number of users to access the same spectrum resources without interfering with one another (Kim, 2021a; Voicu, 2016; Weiser, 2005).

From both regulatory and economic standpoints, spectrum allocation can generally be classified into licensed and unlicensed spectrum. Licensed spectrum typically grants exclusive access rights, making it highly limited and costly. Additionally, due to the priority given to primary license holders, spectrum-sharing mechanisms in licensed bands are subject to significant constraints. Conversely, unlicensed spectrum is a prime example of the *spectrum commons* concept, as it allows equal access to all users. One of the primary challenges in facilitating the coexistence of multiple IoT devices in dense network environments is the development of an innovative spectrum-sharing mechanism tailored for unlicensed spectrum. By enabling efficient utilization of spectrum resources without requiring regulatory approval, such mechanisms can help maximize the value of available spectrum (Kim, 2021a; Peha, 2005; Voicu, 2016).

In telecommunications, white spaces refer to spectrum bands allocated for broadcasting services but left unused in certain locations. While regulatory authorities assign white space bands for specific applications - often licensing them to designated entities - these bands remain underutilized at various times and locations. Despite advancements in spectrum-sharing technologies, licensed white spaces are often inefficiently managed. To address this inefficiency, the Federal Communications Commission (FCC) recently authorized opportunistic access to unused TV white space (TVWS), provided that unlicensed users ensure they do not interfere with incumbent license holders. This opportunistic access model enhances communication capacity while improving overall spectrum utilization. With ongoing standardization efforts in TVWS spectrum management, enabling the coexistence of diverse IoT devices in TVWS environments has emerged as a key research area (Bahrak, 2014; Elias, 2017; Kim, 2021a; Zaeri-Amirani, 2018).

TVWS technology faces a significant control challenge in implementing dynamic spectrum-sharing mechanisms to accommodate diverse requirements. Until recently, technical advancements in TVWS management have been driven by the use of spectrum databases, overseen by database operators (DOs) appointed by regulatory authorities to manage TVWS operations. DOs are responsible for distributing available TVWS spectrum among multiple IoT devices based on a hybrid control and service model. It is assumed that each IoT device is equipped with a single tunable antenna that can operate on a subset of TVWS bands. These devices dynamically query the spectrum database for temporary access to available spectrum for various services. Additionally, IoT devices can aggregate non-contiguous spectrum bands using spectrum bonding and aggregation techniques, enhancing their communication capabilities (Elias, 2017; Kelechi, 2018; Kim, 2021a; Zaeri-Amirani, 2018).

Wireless communications traditionally encompass two distinct application domains: real-time data applications (Class I) and non-real-time data applications (Class II). Class I applications require responses to external events within a strictly defined time frame and include applications such as medical monitoring, fire detection, and online signal processing. These applications demand guaranteed processing completion within specified time constraints, as delays can lead to severe consequences. On the other hand, Class II applications - such as standard best-effort Internet services - exhibit a more flexible notion of time sensitivity, where response times are generally subjective and less rigorously defined. While occasional latency may be tolerated, these applications aim to optimize average-case performance rather than strict real-time guarantees (Choi, 1998; Kim, 2014, 2021a).

To efficiently allocate TVWS spectrum among various service requests, the *BUSS* scheme must formulate a spectrum management strategy that accommodates highly diverse communication scenarios. One of the critical challenges in achieving this goal is understanding the behavior of self-interested IoT devices while mediating potential conflicts in spectrum access. However, addressing these conflicts in a dynamic environment is inherently complex. Consequently, the *BUSS* scheme requires a novel control paradigm. Game theory serves as a natural framework for resolving conflicts among self-interested players, offering robust mechanisms for strategic decision-making. In this context, individual IoT devices are modeled as rational agents, making autonomous decisions based on their local objectives and available information. To ensure that these self-organizing agents contribute to globally optimal system performance, game-theoretic models provide an effective means of guiding their interactions (Kim, 2014, 2021a).

Since the seminal work of J. Nash, it has become standard practice to define bargaining problems in terms of a disagreement point, which represents the outcome that prevails if players fail to reach an agreement. The disagreement point serves as a crucial determinant of bargaining power, influencing negotiation dynamics. While traditional models emphasize the disagreement point as the primary reference for bargaining outcomes, some researchers have argued that it should not be the sole factor shaping negotiations. In recent years, there has been growing interest in experimental studies exploring the role of reference point effects in bargaining scenarios. The reference point hypothesis suggests that decision-makers evaluate outcomes relative to a predefined reference rather than in absolute terms. Conventionally, this reference point in bargaining models is endogenously determined (Karagözoğlu, 2015; Kim, 2021a).

One of the pioneering studies that integrated reference points into bargaining models is the *reference integrative bargaining situation* (*RIBS*). This approach posits that a reference point can significantly influence the negotiated outcome, particularly when all participants perceive it as fair. If a reference point is widely accepted, it may serve as the starting point for negotiations. Importantly, however, the reference point does not replace the conflict outcome - each participant retains the ability to reject agreements that fail to meet their minimum acceptable threshold, thereby ensuring that conflict payoffs remain viable alternatives. Instead, the reference point acts as an additional determinant in shaping the final bargaining outcome (Gupta, 1990; Karagözoğlu, 2015; Kim, 2021a).

5.3.2 Contribution

Based on the principles of the *RIBS*, the *BUSS* scheme provides an effective solution for spectrum sharing in the TVWS platform. Taking into account the varying requirements of different application types, the *BUSS* scheme introduces an inter-bargaining game model to allocate the available unlicensed TVWS spectrum between Class I and Class II data services. Subsequently, two distinct intra-bargaining game models are developed to facilitate spectrum sharing among multiple IoT devices within their allocated spectrum portions. While the *RIBS* is applied across all bargaining stages, the determination of reference points is tailored according to the characteristics of each service. Through an iterative, interactive, and step-by-step bargaining process, the *BUSS* scheme ensures a well-balanced system performance in TVWS environments while enabling efficient sharing of unlicensed spectrum resources (Kim, 2021a).

5.3.3 TVWS platform and bargaining game model

The *BUSS* scheme considers a simple TVWS network cell consisting of one DO (Γ_{DO}) and n IoT devices $\mathbb{N} = \{\mathcal{D}_1, \ldots, \mathcal{D}_n\}$, which are existed within the transmission range of Γ_{DO}. This setting can be easily extended recursively. Therefore, in a parallel manner, the one Γ_{DO} TVWS platform could be expanded into the larger-magnitude TVWS platform with multiple DOs. Each IoT device is equipped with a single antenna to communicate with the Γ_{DO}. As a central coordinator, the Γ_{DO} is responsible for assigning the unlicensed spectrum resource to devices in N. To be specific, regulators such as FCC prescribe rules and mechanisms for protecting licensed devices operating in TV bands, but no such rules exist for unlicensed devices. To fill this void, the Γ_{DO} makes control decisions in order to solve the unlicensed spectrum sharing problem. The overview of TVWS system platform is shown in Figure 3. (Bahrak, 2014; Kelechi, 2018; Kim, 2021a).

Figure 3. The overview of TVWS system platform (Kim, 2021a)

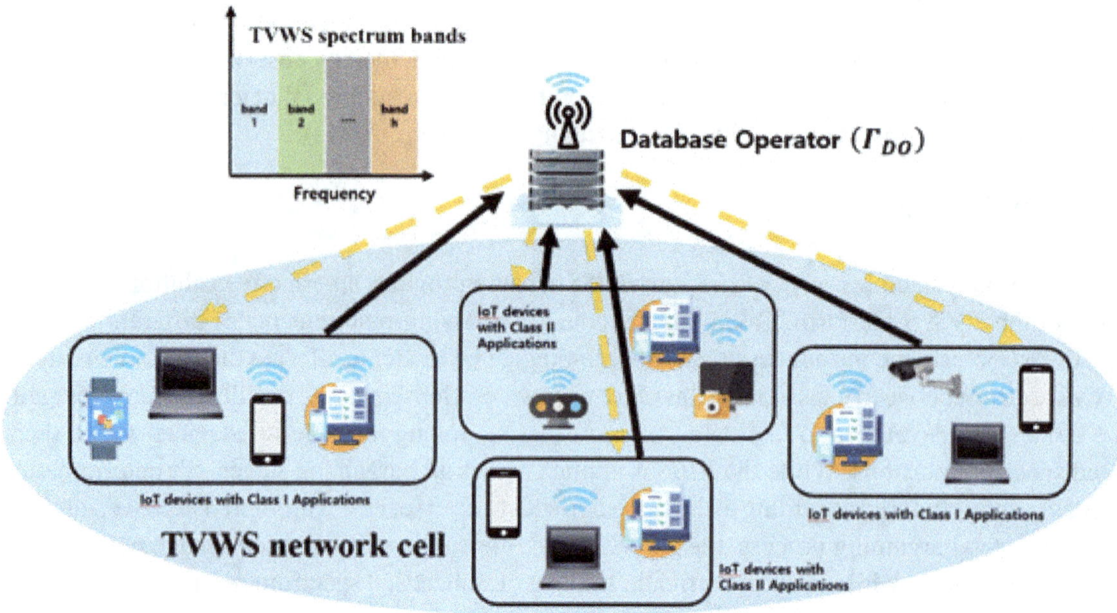

Let M denote the potentially available unlicensed spectrum; it is the set of idle TVWS spectrum bands. B is the spectrum band size, and it is same for all bands. Depending on the current spectrum conditions, individual frequency bands - whether contiguous or non-contiguous - can be aggregated. This capability leverages channel bonding technology to enhance spectrum utilization. Typically, the dynamic spectrum sharing problem is formulated by considering both spectrum demand and availability. However, it must also address several challenges related to the coexistence of heterogeneous IoT devices, taking into account various factors associated with different service types (Elias, 2017). Since the objective of all rational and autonomous devices is to establish a mutually acceptable agreement, this problem aligns with the principles of bargaining games. To tackle this challenge, the *BUSS* scheme integrates bargaining game theory to design an innovative TVWS spectrum-sharing mechanism. By leveraging this bargaining-based approach, the *BUSS* scheme effectively guides individual devices toward achieving an optimal global TVWS system performance (Kim, 2021a).

The *BUSS* scheme divides the operating time of the system into discrete time epochs, and in each epoch, the available TVWS spectrum is shared among IoT devices based on the bargaining game ($\mathbb{G}$). To support different type application services, the G consists of one inter-bargaining game ($\mathscr{G}$) and two intra-bargaining games $\left(G_I, G_{II}\right)$. In the G, class I and class II application types are game players, and the available TVWS spectrum amount at time epoch t $\left(\mathfrak{M}_t\right)$ is shared by them. As the interim results of the G, $\mathfrak{M}_t^I$ and $\mathfrak{M}_t^{II}$ are obtained; $\mathfrak{M}_t^I$ (*or* $\mathfrak{M}_t^{II}$) is allocated for class I (*or* class II) application services where $\mathfrak{M}_t = \mathfrak{M}_t^I + \mathfrak{M}_t^{II}$. In the G_I (*or* G_{II}), IoT devices with class I (*or* class II) applications are game players, and the $\mathfrak{M}_t^I$ (*or* $\mathfrak{M}_t^{II}$) is re-distributed to players. The inter and intra-bargaining games are operated interactively, and two intra-games, i.e.,G_I and G_{II}, are operated in a parallel manner. At each epoch, the G is repeated sequentially during the step-by-step iteration, and the M is adaptively shared based on the idea of bargaining solution. Formally, the *BUSS* scheme defines the G game entities, i.e,

$$\mathbb{G} = \left\{\mathscr{G}, \left(G_I, G_{II}\right)\right\} = \left\{\Gamma_{DO}, \mathbb{N}, \mathfrak{M}, \left\{\mathscr{G} | \left(\mathscr{P}_I, \mathscr{P}_{II}\right), \mathbb{S}_{\mathscr{G}}, \left(U_I, U_{II}\right)\right\}, \left\{G_I | \mathscr{D}_i \in \mathbb{N}_I, \mathbb{S}_{G_I}, \mathfrak{U}^I_{\mathscr{D}_i}\right\}, \left\{G_{II} | \mathscr{D}_j \in \mathbb{N}_{II}, \mathbb{S}_{G_{II}}, \mathfrak{U}^{II}_{\mathscr{D}_j}\right\}, T\right\}$$

of gameplay (Kim, 2021a).

- In the $\mathbb{G}$, $\mathscr{G}$ and G_I, G_{II} games are mutual and reciprocal interdependent in an interactive manner, and they work together to share the available TVWS spectrum ($\mathfrak{M}$).
- Γ_{DO} is the DO of TVWS platform, and it is a coordinator for the spectrum sharing process.
- $\mathbb{N}$ is the set of IoT devices, i.e., $\mathscr{D}_{1 \leq i \leq n}$, which are unlicensed secondary users to share the $\mathfrak{M}$.
- In the $\mathscr{G}$, the group of class I (*or* class II) applications are $\mathscr{P}_I$ (*or* $\mathscr{P}_{II}$), and they are game players. The utility function of $\mathscr{P}_I$ (*or* $\mathscr{P}_{II}$) is U_I (*or* U_{II}), and $\mathbb{S}_{\mathscr{G}}$ is their strategy set for the spectrum sharing problem.
- $\mathbb{N}_I$ is the set of IoT devices, which generate the class I application services. Devices in $\mathbb{N}_I$ are game players in the G_I. For each player, $\mathfrak{U}^I_{\mathscr{D}}$ is a utility function and $\mathbb{S}_{G_I}$ is his strategy set.
- $\mathbb{N}_{II}$ is the set of IoT devices, which generate the class II application services. Devices in $\mathbb{N}_{II}$ are game players in the G_{II}. For each player, $\mathfrak{U}^{II}_{\mathscr{D}}$ is a utility function and $\mathbb{S}_{G_{II}}$ is his strategy set.
- Discrete time model is represented by a sequence of time steps. Each time step, i.e., $t_c \in T = \left\{t_1, \ldots, t_c, t_{c+1}, \ldots\right\}$, matches the $\mathbb{G}$'s event time-scale.

5.3.4. The basic concepts of *RIBS*

To explain the idea of *RIBS*, the commonly used notations are adopted for reader friendliness. The feasible set $\left(S \subseteq \mathbb{R}^n_+\right)$ consists of all the utility vectors that can be achieved by game players. The disagreement point $d \in S$ represents the utility levels obtained by players if no agreement is reached. In the *RIBS*, another salient point, called the ideal point $\mathscr{I}(S,d) \notin S$, is employed. Originally, it was introduced by Kalai and Smorodinsky to represent an outcome in which each player is awarded the maximum utility he can hope to achieve, assuming that the other players will never accept outcomes which are worse for them than the conflict outcome. Usually, ideal point is infeasible in the bargaining problem. Let $\sum^n$ be the family of all n-player bargaining problems with the r. The *RIBS* on $\sum^n$ is a function that associates with each triple $(S, d, r) \in \sum^n$ a unique outcome $RIBS(S, d, r) \in S$. Mathematically, it can be formally defined as follows (Gupta, 1990; Karagözoğlu, 2015; Kim, 2021a);

$$RIBS(S, d, r) = ((1 - \lambda^*) \times r) + (\lambda^* \times \mathscr{I}(S, d)) \tag{5.3.1}$$

$$\text{s.t., } \lambda^* = \mathbf{max}\left\{\lambda \in [0, 1] | [((1 - \lambda) \times r) + (\lambda \times \mathscr{I}(S, d))] \in S\right\}$$

Technically, the *RIBS* provides the maximum point of the feasible set on the line segment connecting the $\mathscr{I}(S, d)$ point and the r point. Therefore, it is the intersection point of the *Pareto* frontier of S and that line segment. This is equivalent to saying that the *RIBS* chooses the maximum individually rational payoff profile at which each player's payoff gain from his reference point has the same proportion to the payoff difference between his ideal point and his reference point. Consequentially, the disagreement point d does not have a direct influence to get the *RIBS*. However, it has an indirect influence on the bargaining process as it is used to derive the ideal point. Figure 4 provides a graphical illustration of the *RIBS* for a two-player bargaining game (Gupta, 1990; Karagözoğlu, 2015; Kim, 2021a).

Figure 4. The idea of RIBS with two different references (Kim, 2021a)

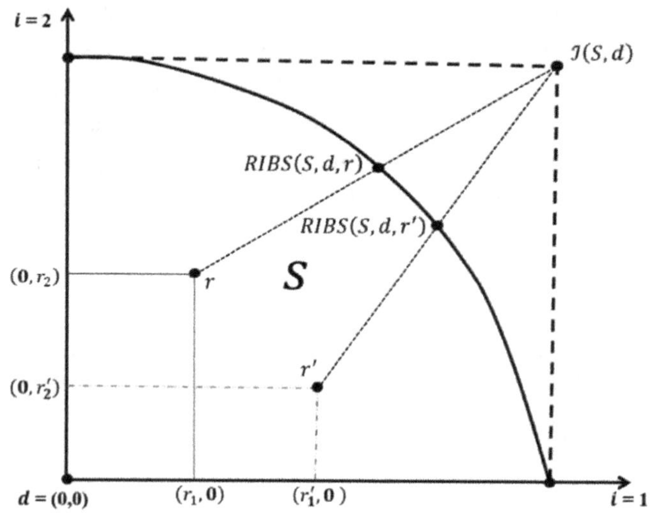

5.3.5. The Inter and intra bargaining games in the TVWS platform

At each time epoch $t_c \in T$, IoT devices generate independently their service requests. To support these requests, the Γ_{DO} adaptively allocates the currently available $\mathfrak{M}_{t_c}$ to devices while considering different QoS requirements. At the first step, the G game is started to divide the $\mathfrak{M}_{t_c}$ into two parts; $\mathfrak{M}^{I}_{t_c}$ is for the $\mathscr{P}_I$, and $\mathfrak{M}^{II}_{t_c}$ is for the $\mathscr{P}_{II}$. In the G, the utility functions of $\mathscr{P}_I$ and $\mathscr{P}_{II}$ at time t_c ($U_I(\cdot)$and $U_{II}(\cdot)$) are formally defined as follows (Kim, 2021a);

$$\begin{cases} U_I\left(t_c, \mathfrak{M}^{I}_{t_c}, \mathcal{Q}^{I}_{t_c}\right) = \left(\eta \times \left(\theta + \frac{\alpha}{\Gamma}\right)^{\Gamma}\right) - \phi \\ \\ U_{II}\left(t_c, \mathfrak{M}^{II}_{t_c}, \mathcal{Q}^{II}_{t_c}\right) = \gamma \times \log\left(\xi - \frac{\left(\min\left(\mathfrak{M}^{II}_{t_c}, \mathcal{Q}^{II}_{t_c}\right)/\mathcal{Q}^{II}_{t_c}\right)}{\psi}\right) \end{cases} \tag{5.3.2}$$

$$\text{s.t.}, \Gamma = \frac{\min\left(\mathfrak{M}^{I}_{t_c}, \mathcal{Q}^{I}_{t_c}\right)}{\mathcal{Q}^{I}_{t_c}} \text{ and } \left(\mathfrak{M}^{I}_{t_c} + \mathfrak{M}^{II}_{t_c}\right) \le \mathfrak{M}_{t_c}$$

where $\mathfrak{M}^{I}_{t_c}, \mathcal{Q}^{I}_{t_c}$ $\left(or\, \mathfrak{M}^{II}_{t_c}, \mathcal{Q}^{II}_{t_c}\right)$ are the allocated and requested spectrum resources for the $\mathscr{P}_I$ $(or\, \mathscr{P}_{II})$, respectively. η, θ,α and ϕ are adjustment parameters for the $U_I(\cdot)$, and γ, ξ and ψ are control factors for the $U_{II}(\cdot)$. To adaptively adjust the $\mathfrak{M}^{I}_{t_c}$ and $\mathfrak{M}^{II}_{t_c}$ values at time epoch t_c, the *BUSS* scheme adopts the idea of *RIBS*. Therefore, at time t_c, the *BUSS* scheme decides the reference point $\left(r_{t_c}^{\mathscr{G}}\right)$ for the G game. To adapt the current TVWS traffic conditions, the *BUSS* scheme assumes the $r_{t_c}^{\mathscr{G}}$ decision problem as a standard rationing problem. Traditionally, a rationing problem is formulated as follows. Let R be the resource and $\mathbb{C} = \{\ldots \mathbb{C}_i \ldots\} \in \mathbb{R}_+^{|\mathbb{C}|}$ be the claims of creditors where $|\mathbb{C}|$ is the cardinality of C and $0 \leq \mathscr{R} < \sum_{i=1}^{i=|\mathbb{C}|} \mathbb{C}_i$. There are two well-known rules defined for standard rationing problems: the *constrained equal awards* (*CEA*) rule, and the *constrained equal losses* (*CEL*) rule. Mathematically, they are defined as follows (Kim, 2021a; Timoner Lledó, 2015);

$$\begin{cases} CEA_i(\mathscr{R}, \mathbb{C}) = \min\{\mathbb{C}_i, \lambda\}, \\ s.t.,0 < \lambda \in R \text{ and } \sum_{i=1}^{i=|\mathbb{C}|} \min\{\mathbb{C}_i, \lambda\} = R \\ CEL_i(\mathscr{R}, \mathbb{C}) = \max\{0, (\mathbb{C}_i - \lambda)\} \\ s.t.,0 < \lambda \in R \text{ and } \sum_{i=1}^{i=|\mathbb{C}|} \max\{0, (\mathbb{C}_i - \lambda)\} = R \end{cases} \quad (5.3.3)$$

To estimate the $r_{t_c}^{\mathscr{G}}$, the *BUSS* scheme should consider the relative fairness between class I and class II application services. For this inter-bargaining process, it is necessary to adopt exogenous information, namely *ex-ante* condition ($\mathbb{E}$). In the rationing problem, the E reflects each creditor's initial condition of the corresponding resource. Hence, it reveals inequalities between creditors that might suggest payoff compensations in favor of some creditors and to the detriment of others. Mathematically, the *CEA* rule with E, i.e., *CEA*($\mathscr{R}$,$\mathbb{C}$,$\mathbb{E}$), is defined as follows (Kim, 2021a; Timoner Lledó, 2015);

$$CEA_i(\mathscr{R}, \mathbb{C}, \mathbb{E}) = \min\left\{\mathbb{C}_i, \left(\lambda - \mathbb{E}_i\right)_+\right\}$$

$$\text{s.t.},0 < \lambda \in \mathbb{R}, \mathbb{C}_i \in \mathbb{C}, \mathbb{E}_i \in \mathbb{E} \text{ and } \sum_{i=1}^{i=|\mathbb{C}|} CEA_i(\mathscr{R}, \mathbb{C}, \mathbb{E}) = \mathscr{R} \quad (5.3.4)$$

The *BUSS* scheme employs the *CEA*($\mathscr{R}$,$\mathbb{C}$,$\mathbb{E}$) to dynamically decide the $r_{t_c}^{\mathscr{G}}$ value. According to (5.3.4), the $r_{t_c}^{\mathscr{G}}$ is given by (Kim, 2021a);

$$r_{t_c}^{\mathscr{G}} = CEA_{\mathscr{L} \in \{\mathscr{P}_I, \mathscr{P}_{II}\}}(\mathscr{R}, \mathbb{C}, \mathbb{E}) = \min\left\{\mathbb{C}_{\mathscr{L}}, \left(\lambda - \mathbb{E}_{\mathscr{L}}\right)_+\right\} \quad (5.3.5)$$

$$\text{s.t.,}\begin{cases} R = \frac{\mathfrak{M}_{t_c}}{\delta}, C = \left(\mathbb{C}_{\mathcal{P}_I}, \mathbb{C}_{\mathcal{P}_{II}}\right) = \left(\frac{Q_{t_c}^{\mathcal{P}_I}}{\delta}, \frac{Q_{t_c}^{\mathcal{P}_{II}}}{\delta}\right) and 0 < \lambda \in R \\ E = \left[\nu_I \times \lambda, \nu_{II} \times \lambda\right] and \sum_{\mathcal{L} \in \{\mathcal{P}_I, \mathcal{P}_{II}\}} CEA_{\mathcal{L}}(\mathcal{R}, \mathbb{C}, \mathbb{E}) = R \end{cases}$$

where R is the cut-down spectrum resource and δ is the modification factor in the *CEA* rule. In the E, ν_I and ν_{II} are the compensating factors for the class I and class II data services, respectively. According to (5.3.5), the $r_{t_c}^{\mathcal{G}}$ is obtained, and the *RIBS* for the G, i.e., $RIBS_{\mathcal{G}}(\cdot)$, is given by (Kim, 2021a);

$$RIBS_{\mathcal{G}}\left(\left(U_I(\cdot), U_{II}(\cdot)\right), S_{\mathcal{G}}, r_{t_c}^{\mathcal{G}}, d_{\mathcal{G}}\right) =$$

(5.3.6)

$$\max_{\lambda \in [0,1]}\left\{\left(\mathcal{P}_I, \mathcal{P}_{II}\right) | \left[\left((1-\lambda) \times r_{t_c}^{\mathcal{G}}\right) + \left(\lambda \times \mathcal{J}\left(S_{\mathcal{G}}, d_{\mathcal{G}}\right)\right)\right] \in S_{\mathcal{G}}\right\}$$

$$\text{s.t.,}\ \mathcal{J}\left(S_{\mathcal{G}}, d_{\mathcal{G}}\right) = \max_{K \in \{I, II\}}\left\{U_K(\cdot) \in \mathbb{R} \middle| \left(U_K(\cdot), \left(d_{\mathcal{G}}\right)_{-K}\right) \in S_{\mathcal{G}}\right\}$$

where $d_{\mathcal{G}}$ is the disagree point in the G, and it is set to zeros. According to (5.3.6), the $\mathfrak{M}_{t_c}^I$ and $\mathfrak{M}_{t_c}^{II}$ values are estimated. And then, for the second step, the $\mathfrak{M}_{t_c}^I \left(or\, \mathfrak{M}_{t_c}^{II}\right)$ is distributed for the multiple IoT devices in $\mathbb{N}_I \left(or\, \mathbb{N}_{II}\right)$. At the G_I, each $\mathcal{D}_i \in \mathbb{N}_I$ is a game player, and the utility function of $\mathcal{D}_i$, i.e., $\mathfrak{U}_{\mathcal{D}_i}(\cdot)$, is mathematically given as follows (Kim, 2021a);

$$\mathfrak{U}_{\mathcal{D}_i \in \mathbb{N}_I}^I\left(t_c, \mathfrak{M}_{t_c}^{\mathcal{D}_i}, Q_{t_c}^{\mathcal{D}_i}\right) = \frac{exp(\Delta) - exp(-\Delta)}{exp(\Delta) + exp(-\Delta)}, \text{ s.t., } \Delta = \frac{\min\left(\mathfrak{M}_{t_c}^{\mathcal{D}_i}, Q_{t_c}^{\mathcal{D}_i}\right)}{Q_{t_c}^{\mathcal{D}_i}}$$

(5.3.7)

where $\mathfrak{M}_{t_c}^{\mathcal{D}_i}, Q_{t_c}^{\mathcal{D}_i}$ are the allocated and requested spectrum resources for the $\mathcal{D}_i \in \mathbb{N}_I$. To adaptively distribute the $\mathfrak{M}_{t_c}^I$ for devices in $\mathbb{N}_I$, the *BUSS* scheme also uses the idea of *RIBS*. Therefore, at time t_c, the *BUSS* scheme needs the reference point $\left(r_{t_c}^{G_I}\right)$ for the G_I game. In a similar fashion with the $r_{t_c}^{\mathcal{G}}$, the $r_{t_c}^{G_I}$ is decided based on the *CEA* rule. However, there is an equivalent priority in all game players. Therefore, the factor E is not necessary for this intra-bargaining process. Finally, the $r_{t_c}^{G_I}$ is given by (Kim, 2021a);

$$r_{t_c}^{G_I} = CEA_{\mathcal{D}_i \in \mathbb{N}_I}(\mathcal{R}^I, \mathbb{C}^I) = \min\left\{\mathbb{C}_{\mathcal{D}_i}^I, \lambda^I\right\}$$

(5.3.8)

$$\text{s.t.,}\begin{cases} \mathcal{R}^I = \frac{\mathfrak{M}_{t_c}^{\mathcal{P}_I}}{\varepsilon} and\, \mathbb{C}^I = \left\{\mathbb{C}_{\mathcal{D}_i \in \mathbb{N}_I}^I | \frac{Q_{t_c}^{\mathcal{D}_i}}{\varepsilon}\right\} \\ 0 < \lambda^I \in R and \sum_{\mathcal{D}_i \in \mathbb{N}_I} \min\left\{\mathbb{C}_{\mathcal{D}_i}^I, \lambda^I\right\} = \mathcal{R}^I \end{cases}$$

where ε is the modification factor in the *CEA* rule. According to (5.3.8), the $r_{t_c}^{G_I}$ is obtained, and the *RIBS* for the G_I, i.e., $RIBS_{G_I}(\cdot)$ is given by (Kim, 2021a);

$$RIBS_{G_I}\left(\mathfrak{U}^I_{\mathscr{D}_i \in \mathbb{N}_I}(\cdot), S_{G_I}, r_{t_c}^{G_I}, d_{G_I}\right) = \tag{5.3.9}$$

$$\max_{\lambda \in [0,1]}\left\{\mathscr{D}_i \in \mathbb{N}_I \middle| \left[\left((1-\lambda) \times r_{t_c}^{G_I}\right) + \left(\lambda \times \mathscr{I}\left(S_{G_I}, d_{G_I}\right)\right)\right] \in S_{G_I}\right\}$$

$$\text{s.t., } \mathscr{I}\left(S_{G_I}, d_{G_I}\right) = \max_{\mathscr{D}_i \in \mathbb{N}_I}\left\{\mathfrak{U}^I_{\mathscr{D}_i}(\cdot) \in \mathbb{R} \middle| \left(\mathfrak{U}^I_{\mathscr{D}_i}(\cdot), \left(d_{G_I}\right)_{-\mathscr{D}_i}\right) \in S_{G_I}\right\}$$

By using (5.3.9), the $\mathfrak{M}^I_{t_c}$ is adaptively distributed for the $\mathscr{D}_i \in \mathbb{N}_I$. With the G_I game, the other intra-bargaining game G_{II} is operated simultaneously and independently to allocate the $\mathfrak{M}^{II}_{t_c}$ to devices in $\mathbb{N}_{II}$. At the G_{II} game, each $\mathscr{D}_j \in \mathbb{N}_{II}$ is a game player, and the utility function of $\mathscr{D}_j$, i.e., $\mathfrak{U}^{II}_{\mathscr{D}_j}(\cdot)$, is mathematically given as follows (Kim, 2021a);

$$\mathfrak{U}^{II}_{\mathscr{D}_j}\left(t_c, \mathfrak{M}^{\mathscr{D}_j}_{t_c}, Q^{\mathscr{D}_j}_{t_c}\right) = \left(\frac{\mu \times \mathscr{H}}{\sigma + exp(\mathscr{H})}\right) \times exp(\log(\mathscr{H} + \beta)) \tag{5.3.10}$$

$$\text{s.t., } \mathscr{H} = \frac{\min\left(\mathfrak{M}^{\mathscr{D}_j}_{t_c}, Q^{\mathscr{D}_j}_{t_c}\right)}{Q^{\mathscr{D}_j}_{t_c}}$$

where $\mathfrak{M}^{\mathscr{D}_j}_{t_c}, Q^{\mathscr{D}_j}_{t_c}$ are the allocated and requested spectrum resources for the $\mathscr{D}_j \in \mathbb{N}_{II}$. μ, σ and β are adjustment parameters for the $\mathfrak{U}^{II}_{\mathscr{D}_j}(\cdot)$. Based on the characteristics of class II type data services, the *BUSS* scheme should decide the reference point for the G_{II} game $\left(r_{t_c}^{G_{II}}\right)$. In the *BUSS* scheme, the $r_{t_c}^{G_{II}}$ is given based on the *CEL* rule. Like as the G_I game model, each game player has an equivalent priority, and the factor E is also not necessary for this intra-bargaining process (Kim, 2021a; Timoner Lledó, 2015).

$$r_{t_c}^{G_{II}} = CEL_{\mathscr{D}_j \in \mathbb{N}_{II}}(\mathscr{R}^{II}, \mathbb{C}^{II}) = \max\left\{0, \left(\mathbb{C}^{II}_{\mathscr{D}_j} - \lambda^{II}\right)\right\} \tag{5.3.11}$$

$$\text{s.t., } \begin{cases} \mathscr{R}^{II} = \dfrac{\mathfrak{M}^{\mathscr{P}_{II}}_{t_c}}{\zeta} and\, \mathbb{C}^{II} = \left\{\mathbb{C}^{II}_{\mathscr{D}_j \in \mathbb{N}_{II}} \middle| \dfrac{Q^{\mathscr{D}_j}_{t_c}}{\zeta}\right\} \\ 0 < \lambda^{II} \in Rand \sum_{\mathscr{D}_j \in \mathbb{N}_{II}} \max\left\{0, \left(\mathbb{C}^{II}_{\mathscr{D}_j} - \lambda^{II}\right)\right\} = \mathscr{R}^{II} \end{cases}$$

where ε is the modification factor in the *CEL* rule. According to (5.3.11), the $r_{t_c}^{G_{II}}$ is given, and the *RIBS* for the G_{II}, i.e., $RIBS_{G_{II}}(\cdot)$ is given by (Kim, 2021a);

$$RIBS_{G_{II}}\left(\mathfrak{U}^{II}_{\mathcal{D}_j \in \mathbb{N}_{II}}(\cdot), S_{G_{II}}, r_{t_c}^{G_{II}}, d_{G_{II}}\right) =$$

$$\max_{\lambda \in [0, 1]}\left\{\mathcal{D}_j \in \mathbb{N}_{II} \mid \left[\left((1-\lambda) \times r_{t_c}^{G_{II}}\right) + \left(\lambda \times \mathcal{I}\left(S_{G_{II}}, d_{G_{II}}\right)\right)\right] \in S_{G_{II}}\right\} \tag{5.3.12}$$

$$\text{s.t.}, \mathcal{I}\left(S_{G_{II}}, d_{G_{II}}\right) = \max_{\mathcal{D}_j \in \mathbb{N}_{II}}\left\{\mathfrak{U}^{II}_{\mathcal{D}_j}(\cdot) \in \mathbb{R} \middle| \left(\mathfrak{U}^{II}_{\mathcal{D}_j}(\cdot), \left(d_{G_{II}}\right)_{-j}\right) \in S_{G_{II}}\right\}$$

5.3.6 Summary

The *BUSS* scheme introduces both inter-bargaining and intra-bargaining game models to address the TVWS spectrum-sharing challenge. Grounded in the fundamental principles of *RIBS*, multiple IoT devices collaborate and operate cooperatively to efficiently utilize the unlicensed TVWS spectrum. During the inter-bargaining phase, the available spectrum is divided between Class I and Class II application services based on the *RIBS* and the *CEA* rule with E. In the subsequent intra-bargaining processes, the allocated spectrum for each service type is dynamically distributed among the corresponding IoT devices. Within each intra-bargaining game, reference points are determined according to *CEA* and *CEA* rules, while the *RIBS* approach is also incorporated to optimize spectrum sharing. Throughout this interactive process, the primary objective of the *BUSS* scheme is to maximize system performance while maintaining an appropriate balance between efficiency and fairness (Kim, 2021a).

5.4 Fog Computing Resource Allocation (*FCRA*) scheme

The Social Internet of Things (SIoT) is a control paradigm that integrates social networking principles into IoT. Meanwhile, Fog Computing (FC) is an emerging technology designed to extend cloud computing capabilities to the access network. Recently, these two models have been combined by leveraging their complementary features, leading to the development of a new paradigm known as Social Fog IoT (SFIoT). In this context, S. Kim has introduced the *Fog Computing Resource Allocation* (*FCRA*) scheme for the SFIoT system (Kim, 2019). By considering social relationships and individual preferences, mobile devices within the SFIoT framework efficiently share the limited computation and communication resources provided by FC operators. To model the interaction between mobile devices and the FC operator, the *FCRA* scheme incorporates two fundamental game-theoretic approaches: voting games and bargaining games. The bicooperative voting mechanism is employed to facilitate control decisions regarding resource allocation, while the *Nash bargaining solution* (*NBS*) is utilized to optimally distribute computational resources across various application tasks. By implementing this two-phase game model, the *FCRA* scheme ensures a fair and efficient resource allocation process, offering multiple advantages in managing SFIoT system resources.

5.4.1 Development motivation

Since its inception in the early 1960s, the Internet has undergone remarkable advancements in research, development, and innovation within the realm of data communications. More recently, the Internet has evolved to establish connections among physical objects, leading to the emergence of IoT. This pervasive paradigm has the potential to significantly enhance the value of exchanged information by increasing the number of interconnected devices, thereby enabling a diverse range of intelligent services and applications that benefit both individuals and society. Consequently, IoT technologies facilitate the delivery of new services to end users while autonomously fostering social interactions among connected devices. Similar to human social networks, smart objects in the IoT ecosystem establish asymmetric connections based on preference similarity and shared interests (Kim, 2019; Li, 2016; Ortiz, 2014; Yang, 2017).

The integration of social networking concepts into IoT has given rise to an emerging research domain known as the Social Internet of Things (SIoT). Within the SIoT paradigm, smart objects can autonomously discover and access services through their social network connections in a decentralized manner. Here, the term social networking extends beyond online interactions to encompass natural social relationships within local environments. As a result, inter-object social networks can be established using short-range wireless connections via device-to-device communication. This convergence of social and communication networks has opened up vast business and technological opportunities within the SIoT ecosystem. However, while the SIoT paradigm effectively mitigates scalability challenges inherent in traditional IoT architecture, its reliance on social connectivity-based services introduces the risk of cascade effects. Therefore, to ensure the efficient and sustainable operation of SIoT systems, the primary objective is to maximize social welfare based on social relationships (Baccarelli, 2018; Kim, 2019; Li, 2016).

With the rise of big data-driven, computation-intensive applications, data generated by SIoT devices is typically processed in cloud infrastructures. However, transmitting large volumes of data from IoT devices to remote cloud servers can be inefficient, particularly for time-sensitive applications. To address this issue, Fog Computing (FC) has emerged as a viable alternative, offering low-latency computing services at the network edge, where data is generated. In FC architecture, edge agents evolve into edge clouds, known as cloudlets, equipped with computational capabilities. This architectural shift enables SIoT systems to handle the growing demand for computation-intensive applications. Compared to traditional cloud computing, the FC-enhanced SIoT system significantly improves efficiency by leveraging one-hop wireless connections to deliver powerful computing resources closer to IoT devices (Kim, 2019; Lee, 2015; Shah-Mansouri, 2018).

While SIoT and FC have traditionally been standalone technological paradigms, they have now been integrated within next-generation networks. The SIoT framework ensures the scalability of large-scale IoT ecosystems by enabling self-organized inter-device social relationships, while FC extends cloud computing capabilities to the network edge to support computation-intensive IoT applications. Recognizing their complementary features, researchers have introduced a new paradigm - the Social Fog IoT (SFIoT) - which merges the principles of both SIoT and FC technologies (Baccarelli, 2018; Kim, 2019).

Within the SFIoT framework, cloudlets function as small-scale virtualized data centers equipped with interconnected computational resources. These cloudlets, positioned at the edge of access networks, are responsible for handling thing-to-FC offloading tasks from SFIoT devices. Through the intra-FC resource pooling mechanism, each cloudlet partitions available resources among different application types (Baccarelli, 2018). In the SFIoT system, resources can be categorized into computation and communication resources. However, in practical scenarios, these resources are often scarce and highly contested. When a substantial number of computation-intensive applications are offloaded, cloudlet resources may become rapidly depleted, leading to a degradation in QoE (Kim, 2010; Liu, 2016; Pande, 2013). To ensure efficient resource allocation for SFIoT services, a robust strategy is required to optimize system performance while maintaining QoE provisioning. The *FCRA* scheme addresses this challenge by managing cloudlet resource allocation for SIoT devices, which are interconnected based on social ties (Kim, 2004, 2019).

Over the past decade, game theory has gained widespread attention as an essential tool for strategic decision-making in distributed optimization problems. As a subfield of applied mathematics, game theory provides a framework for analyzing interactions among self-interested entities. It is typically classified into non-cooperative games and cooperative games. In a cooperative game, players can form binding agreements, whereas in a non-cooperative game, such agreements are not enforceable. In communication networks, cooperative game theory has been extensively applied to analyze distributed optimization problems involving self-interested players (Kim, 2014, 2019).

In the SFIoT framework, individual IoT devices interact through a peer-to-peer communication model, autonomously forming inter-device social relationships based on shared preferences (Baccarelli, 2018). To optimize overall system performance, devices must collaborate while simultaneously pursuing individual benefits. The relationship among interdependent SFIoT devices can thus be modeled as a cooperative game. A key research question in this context is how to establish consensus-driven resource allocation that ensures fairness and efficiency. This challenge is the primary focus of the *FCRA* scheme (Kim, 2019).

To effectively resolve the resource allocation problem in SFIoT systems, the *FCRA* scheme leverages two distinct cooperative game-theoretic approaches. For computation resource allocation, the *FCRA* scheme adopts the *Nash Bargaining Solution* (*NBS*), which formulates expected utility functions over a set of feasible agreements based solely on individual preferences. Meanwhile, for communication resource allocation, the *FCRA* scheme utilizes the voting game model. However, traditional voting games are not directly applicable to the SFIoT resource allocation problem. To address this limitation, the *FCRA* scheme incorporates a specialized voting mechanism known as the *Bicooperative Voting Game* (*BVG*). This approach models a more realistic decision-making process, enabling adaptive communication resource allocation within the SFIoT system. By integrating the desired properties of *NBS* and *BVG*, the *FCRA* scheme seeks to achieve an optimal resource allocation outcome that balances efficiency, fairness, and technological adaptability (Kim, 2019).

5.4.2 Contribution

The *FCRA* scheme offers several key contributions; i) the capability to maximize SFIoT resource efficiency, ii) the ability to dynamically adapt to real-time SFIoT conditions through an interactive bargaining and voting mechanism, iii) QoE provisioning tailored for various types of application tasks, iv) the synergistic integration of SIoT and FC, leveraging their complementary features, and v) ensuring a mutually beneficial allocation strategy while enhancing the overall performance of the SFIoT system. The *FCRA* scheme is particularly distinguished by its adaptability, practicality, and efficiency in managing real-world SFIoT system operations, even under highly diverse and dynamic application task loads (Kim, 2019).

5.4.3 Fog controlled social IoT system infrastructure

Conventionally, the SFIoT system architecture comprises multiple cloudlets and numerous mobile devices. Each cloudlet, functioning as a small-scale cloud data center, is strategically positioned at the network edge. Its primary role is to support resource-intensive and interactive mobile applications by offering high-performance computing resources with minimal latency. Within the SFIoT framework, every mobile device simultaneously participates in both the underlying IoT infrastructure and an overlay social network. Usually, the IoT accounts for more conventional network services, and the social network reflects somewhat less conventional social ties induced by inter-thing social relationships, which are formally described by a social graph $\mathbb{G} = (\mathbb{N}, \mathbb{E})$; N is the set of mobile devices and $\mathbb{E} = \left\{ e_{m_i m_j} | 0 \leq e_{m_i m_j} \leq 1 \text{and} \left(m_i, m_j\right) \in \mathbb{N} \times \mathbb{N} \right\}$ is the set of the social ties. $e_{m_i m_j}$ is a real-valued nonnegative number to measure the strength of the social tie between the m_i and the m_j. Therefore, $\mathbb{S}_{m_i} = \left\{ m_j | e_{m_i m_j} > 0 \text{and} m_j \in \mathbb{N} \right\}$ formally defines the set of the social friends of m_i (Baccarelli, 2018; Kim, 2019).

To calculate each user's influential power, the *FCRA* scheme develops a social power function $\mathcal{F}(x)$, in line with some properties. Under diverse scenarios, the function $\mathcal{F}(x)$ should obey the following three properties to effectively estimate each individual user's social power (Jiang, 2017; Kim, 2019).

i) non-negative property: $\mathcal{F}(x) \geq 0$,
ii) non-decreasing property: $d\mathcal{F}(x)/dx \geq 0$,
iii) saturation property: $d^2\mathcal{F}(x)/dx^2 < 0$

The *non-negative* property means that the outcome should be larger than zero, and if no social relationship, it is zero. The *non-decreasing* property simply implies that the more users are socially connected, the more outcome can be obtained. The *saturation* property suggests that when the social relationship is sufficient, any further increase of the outcome remains marginal (Jiang, 2017). In order to construct a function while satisfying above three properties, the social power function of m_i, i.e., $\mathcal{F}_{m_i}(x)$, is defined as follows (Kim, 2019);

$$\mathcal{F}_{m_i}(x) = \tilde{A} - \frac{1}{exp(\mu \times x)}, \text{ s.t.}, x = \sum_{m_j \in \mathbb{S}_{m_i}} e_{m_i m_j} \tag{5.4.1}$$

where μ controls the speed of social power saturation, and Γ represents the maximum outcome of a user to normalize the social power (Jiang, 2017; Kim, 2019).

The *FCRA* scheme assumes that there are k cloudlets and n mobile devices where $k \ll n$. In the SFIoT system infrastructure, cloudlets are evolved to the Fog-computing based Access Point (F-AP) by being equipped with a certain computation power and communication capability. The design of F-AP platform has been introduced to deliver various devices requests, which have different characteristics and resource demands; that is, CPU capacity for computation and wireless bandwidth assignment for communication. However, due to the limited CPU capacity and bandwidth scarcity in the F-AP, it is impossible to guarantee all applications' needs (Kim, 2017a). Therefore, an efficient resource management strategy becomes a key factor in enhancing the SFIoT performance while ensuring the required QoE (Kim, 2019).

To address the SFIoT resource management challenge, the *FCRA* scheme employs a two-phase game-based resource allocation mechanism. At the first phase, each mobile device engages in two *BVGs* - one to determine the allocation strategy for communication resources and the other to partition computation resources based on the voting power of each device. Within each F-AP coverage area, voting power is assessed using the pseudo-Banzhaf value. At the second phase, the partitioned computation capacity is then distributed among homogeneous application tasks using the *NBS* approach. For practical implementation, each F-AP and its associated mobile devices independently execute the two-phase game in a fully distributed manner, ensuring efficient and scalable resource allocation. For example, the kth F-AP, i.e., $\mathcal{F}_k^{AP}$, and the set of corresponding mobile devices in the $\mathcal{F}_k^{AP}$'s covering area, i.e., $\mathbb{N}_{\mathcal{F}_k^{AP}}$, work together for their two-phase game $\mathbb{G}_k$. Therefore, multiple two-phase games $\mathbb{G}_{1\leq i\leq k}$ are operated in parallel. To formally define the game model, the *FCRA* scheme characterizes the $\mathbb{G}_k$ as

$$\left\{\left\{\mathcal{F}_k^{AP},\mathbb{N}_{\mathcal{F}_k^{AP}}\right\},\left\{\left(\mathbb{S}_{m_i},\mathcal{F}_{m_i}(x),\Psi_{m_i}^{\mathbb{Q}}\right)\middle| m_i\in\mathbb{N}_{\mathcal{F}_k^{AP}},x=\sum\nolimits_{m_j\in\mathbb{S}_{m_i}} e_{m_i,m_j}\right\},\mathbb{A}_{\mathcal{F}_k^{AP}},\alpha_{\mathcal{F}_k^{AP}},U_{\mathcal{F}_k^{AP}}\right\}$$

at each time period $t \in \mathcal{T}$ of gameplay (Kim, 2019);

- $\left\{\mathcal{F}_k^{AP},\mathbb{N}_{\mathcal{F}_k^{AP}}\right\}$ is the set of game players for the $\mathbb{G}_k$ game where $\mathbb{N}_{\mathcal{F}_k^{AP}} \cap \mathbb{N}_{\mathcal{F}_i^{AP}} = \varnothing$.
- $\left(\mathbb{S}_{m_i},\mathcal{F}_{m_i}(x),\Psi_{m_i}^{\mathbb{Q}}\right)$ is a three-pair for the $m_i \in \mathbb{N}_{\mathcal{F}_k^{AP}}$ where $\mathbb{S}_{m_i}$ is the set of m_i's social friends. $\mathcal{F}_{m_i}(x)$ and $\Psi_{m_i}^{\mathbb{Q}}$ represent the social power and the voting power of m_i, respectively.
- $\mathbb{A}_{\mathcal{F}_k^{AP}}$ is the set of generated application tasks, which are offloaded in the $\mathcal{F}_k^{AP}$ for FC services.
- $\alpha_{\mathcal{F}_k^{AP}}$ is the ratio of relative service weights given to different type applications in the $\mathcal{F}_k^{AP}$ where $0 \leq \alpha_{\mathcal{F}_k^{AP}} \leq 1$. Based on the $\alpha_{\mathcal{F}_k^{AP}}$ value, the $\mathcal{F}_k^{AP}$'s computation power is partitioned.
- $U_{\mathcal{F}_k^{AP}}$ is the utility function of $\mathcal{F}_k^{AP}$; it is optimized based on the NBS.

5.4.4 Bicooperative game and ternary voting model

Usually, cooperative game theory studies situations where a group of game players are associated to obtain a profit as a result of their cooperation. Thus, a cooperative game model is defined as a pair (N,V), where N is a finite set of players and V: $2^N \to$ R is a function verifying that $\mathcal{V}(\varnothing) = 0$. For each coalition $S \in 2^N$, $\mathcal{V}(S)$ can be interpreted as the coalition S's maximal gain that players in S can achieve themselves against the best offensive threat by the complementary coalition $N \setminus S$ (Bilbao, 2008a; Kim, 2019).

In 2008, Bilbao et al proposed a new idea, called bicooperative game for the multi-criteria decision-making process. In the bicooperative game, game players are allowed to choose one among three alternatives. To think about the voting situation by analogy, the *FCRA* scheme considers ordered pairs (S, T), with $S, T \subseteq N$ and $S \cap T = \varnothing$. Each pair (S, T) yields a partition of N in three groups. Players in S are in favor of the voting item, and players in T object to it. The remaining players in $N \setminus (S \cup T)$ are non-voters. Formally, the *FCRA* scheme can define the set $3^N = \{(S, T) | S, T \subseteq N, S \cap T = \varnothing\}$ and the function Q: $3^N \to$ R. For each $(S, T) \in 3^N$, the outcome of $\mathcal{Q}(S, T)$ can be interpreted as the gain that S can achieve when T is the opposite coalition and $N \setminus (S \cup T)$ is the neutral coalition (Bilbao, 2008a, 2008b; Kim, 2019).

The pair $(\varnothing,N)$ represents the situation if all players object to the voting item and $(N,\varnothing)$ represents the situation where all players are supporters of the voting item. More generally, a relation of 3^N is given by (Bilbao, 2008a, 2008b; Kim, 2019);

$$(A,B) \sqsubseteq (C,D) \Leftrightarrow A \subseteq C, B \supseteq D \tag{5.4.2}$$

The set $(3^N, \sqsubseteq)$ is a finite distributive lattice and a partially ordered set with the following properties (Bilbao, 2008a, 2008b; Kim, 2019):

i) $(\varnothing, N)$ is the first element: $(\varnothing, N) \sqsubseteq (A,B)$ for all $(A,B) \in 3^N$
ii) $(N,\varnothing)$ is the last element: $(A,B) \sqsubseteq (N,\varnothing)$ for all $(A,B) \in 3^N$
iii) Every fair of elements of 3^N has a join

$(A,B) \vee (C,D) = (A \cup C, B \cap D)$ and$(A,B) \wedge (C,D) = (A \cap C, B \cup D)$

The set of all bicooperative games with players N is denoted by $\mathscr{B}\mathscr{G}^N$, which is a real vector space (Bilbao, 2008a, 2008b; Kim, 2019).

$$\mathscr{B}\mathscr{G}^N = \{\mathcal{Q} | 3^N \rightarrow \mathbb{R} \text{and } \mathcal{Q}(\varnothing,\varnothing) = 0\} \tag{5.4.3}$$

A value on $\mathscr{B}\mathscr{G}^N$ is a mapping ϕ: $\mathscr{B}\mathscr{G}^N \rightarrow \mathbb{R}^n$ that associates with each game $\mathcal{Q} \in \mathscr{B}\mathscr{G}^N$. In a vector $(\phi_1(\mathcal{Q}), \ldots, \phi_n(\mathcal{Q})) \in \mathbb{R}^n$, $\phi_{1 \leq i \leq n}(\mathcal{Q})$ represents a real number, which is player i's value in a game Q on $\mathscr{B}\mathscr{G}^N$. Therefore, the mapping $\phi_{i \in N}(\mathcal{Q})$: $\mathscr{B}\mathscr{G}^N \rightarrow$ R represents player i's payoff from playing bicooperative games (Bilbao, 2008a, 2008b). There are three special collections of games in $\mathscr{B}\mathscr{G}^N$ taking values in $\{-1, 0, 1\}$: the *identity* games, the *superior unanimity* games and the *inferior unanimity* games which are defined for any (S,T) ∈ 3^N (Bilbao, 2008a, 2008b; Kim, 2019).

The *identity* game $\left\{\breve{\mathcal{Q}}_{(S,T)}(A,B) | 3^N \rightarrow \mathbb{R}\right\}$ is defined by

$$\breve{\mathcal{Q}}_{(S,T)}(A, B) = \begin{cases} 1, if(A, B) = (S, T) and (S, T) \neq (\varnothing,\varnothing) \\ 0, otherwise \end{cases} \tag{5.4.4}$$

The *superior unanimity* game $\left\{\overline{\mathcal{Q}}_{(S,T)}(A,B) | 3^N \rightarrow \mathbb{R}\right\}$ is defined by

$$\overline{\mathcal{Q}}_{(S,T)}(A, B) = \begin{cases} 1, if(S, T) \sqsubseteq (A, B) and (A, B) \neq (\varnothing,\varnothing) \\ 0, otherwise \end{cases} \tag{5.4.5}$$

The *inferior unanimity* game $\left\{\underline{\mathcal{Q}}_{(S,T)}(A,B) | 3^N \rightarrow \mathbb{R}\right\}$ is defined by

$$\underline{\mathcal{Q}}_{(S,T)}(A, B) = \begin{cases} -1, if(A, B) \sqsubseteq (S, T) and (A, B) \neq (\varnothing,\varnothing) \\ 0, otherwise \end{cases} \tag{5.4.6}$$

The voting game is a type of cooperative game model that integrates concepts from social choice theory. This model is utilized to examine scenarios in which voters can either support or oppose a given decision (Kim, 2014). A simple voting game with abstention can be represented as a bicooperative game, which has been extensively studied under the framework of the ternary voting game (Bilbao, 2010; Kim, 2019; Tsurumi, 2005).

Let N be a set of n voters that has to choose between a pair of two alternatives S and T. Every voter $i \in N$ has a preference over the two alternatives S and T. Thus, the payoff of voter i $\left(\mathcal{Q}_i\right)$ is defined by $\mathcal{Q}_i$: $3^N \rightarrow \{-1,0,1\}$ based on the majority rule. Formally, a ternary voting game $\mathcal{Q} \in \mathcal{BG}^N$ satisfies the following conditions (Bilbao, 2010; Kim, 2019).

I) For every bicoalition pair $(S,T) \in 3^N$, its worth $\mathcal{Q}(S, T) \in \{-1,0,1\}$.
II) If $(S, T), \left(\hat{S}, \hat{T}\right) \in 3^N$ with $(S, T) \sqsubseteq \left(\hat{S}, \hat{T}\right)$, then $\mathcal{Q}(S, T) \leq \mathcal{Q}\left(\hat{S}, \hat{T}\right)$

$$\text{III) } \mathcal{Q}_i = \begin{cases} 1, if(i \in S \text{ and } |S| > |T|)or(i \in T \text{ and } |T| > |S|) \\ -1, if(i \in S \text{ and } |S| < |T|)or(i \in T \text{ and } |T| < |S|) \\ 0, ifvoteriabstainsformvotingor|S| = |T| \end{cases}$$

where $|S|$ is the cardinality of coalition S. For voting games, the analysis of voting power is the main issue. In 1965, Banzhaf value was introduced by John F. Banzhaf III based on probabilistic analysis of the individual voters in a block voting system. It depends on the number of ways in which each voter can affect a swing in the game. In 2010, J. Bilbao et al introduced a pseudo-Banzhaf value, i.e., $\Psi_i^{\mathcal{Q}}$, for the ternary voting game Q; it can be interpreted as a voter's probabilistic power index (Bilbao, 2010; Kim, 2019; Tsurumi, 2005).

$$\Psi_i^{\mathcal{Q}} = \frac{1}{3^{n-1}} \times \sum_{(S,T) \in 3^{N \setminus i}} \left(\mathcal{Q}_i(S \cup \{i\}, T) - \mathcal{Q}_i(S, T \cup \{i\})\right) \tag{5.4.7}$$

As a special ternary voting game model, the weighted ternary voting game can formulate a decision-making process that not all voters have the same amount of influence over the decision. Therefore, votes of different voters are given different weight. This type of voting game model is used in shareholder meetings, where votes are weighted by the number of shares that each shareholder owns (Kim, 2014). The *FCRA* scheme develops a ternary voting model based on the weighted voting approach. In the game model, mobile devices are treated differently according to their social powers $\left(\mathcal{F}_m(x)\right)$, and give different amounts of influence to different members. Therefore, each coalition of players $S \subseteq N$ has the sum of voting weights of its components; that is $\mathcal{W}(S) = \sum_{m_i \in S} \mathcal{F}_{m_i}(x)$. If the value of $\mathcal{W}(S)$ exceeds a certain quota, players in S win the game (Bilbao, 2010; Kim, 2014, 2019).

For the weighted ternary voting game, we need two quotas; one is to block a decision, and the other is to approve it. Formally, it is represented by the voting body $\left[\mathfrak{P}, \vartheta; \mathcal{F}_{m_1}(x), \ldots \mathcal{F}_{m_n}(x)\right]$ where P is the quota for the decision-taking and ϑ is the quota for the decision-blocking; it can be seen as a particular case of n-player games with three choices. This situation can be formulated by a weighted ternary voting game ($\mathfrak{X}$), i.e., $\mathfrak{X}(S,T)$:$3^N \rightarrow \{-1,0,1\}$ (Bilbao, 2008a; Kim, 2019).

$$\mathfrak{X}(S, T) = \begin{cases} 1, if W(S) \geq P and W(T) < \vartheta \\ -1, if W(S) < P and W(T) \geq \vartheta \\ 0, otherwise \end{cases} \tag{5.4.8}$$

$$s.t., \mathfrak{P} = \alpha \times \left(\sum_{i=1}^{n} \mathscr{F}_{m_i}(x)\right) \text{ and } \vartheta = \beta \times \left(\sum_{i=1}^{n} \mathscr{F}_{m_i}(x)\right)$$

5.4.5 Communication and computation mechanisms

In the SFIoT system, wireless bandwidth is both a valuable and limited resource. Therefore, it must be allocated in a fair and efficient manner among mobile users. If bandwidth allocation strategies are not carefully considered during the design phase of allocation algorithms, disparities in bandwidth demands may lead to highly uneven distributions, ultimately causing system inefficiencies. Typically, bandwidth allocation methods are classified into three main categories: Proportional Allocation (PA), Constrained Allocation (CA), and Equal Loss Allocation (EA) (Kim, 2019).

I) Proportional Allocation (PA):

$$\begin{cases} A_i = R_i, if \sum_{i=1}^{n} R_i \leq B \\ A_i = \max\left\{\left(R_i \times \frac{\mathfrak{B}}{\sum_{i=1}^{n} R_i}\right), m_i\right\}, otherwise \end{cases} \tag{5.4.9}$$

II) Constrained Allocation (CA):

$$\begin{cases} A_i = R_i, if \sum_{i=1}^{n} R_i \leq B \\ A_i = \max\left(\min\left\{R_i, T^C\right\}, m_i\right), otherwise \\ s.t., \sum_{i=1}^{n} \min\left\{R_i, T^C\right\} = \mathfrak{B} \end{cases} \tag{5.4.10}$$

III) Equal Loss Allocation (EA):

$$\begin{cases} A_i = R_i, if \sum_{i=1}^{n} R_i \leq B \\ A_i = \max\left\{\left(R_i - \mathscr{L}^C\right), m_i\right\}, otherwise \\ s.t., \sum_{i=1}^{n} \max\left\{\left(R_i - \mathscr{L}^C\right), 0\right\} = \mathfrak{B} \end{cases} \tag{5.4.11}$$

where B is the total available bandwidth amount, and m_i is the pre-defined minimum bandwidth amount for the ith service. $R_{1\leq i\leq N}$ is the bandwidth request of mobile device m_i, and $A_{1\leq i\leq N}$ represents the decided bandwidth amount for the m_i. T^C and $\mathscr{L}^C$ are the constraint thresholds, which are chosen to satisfy each condition. If the available bandwidth is enough or the F-AP can secure the new request A_i from m_i through the allocated bandwidth re-adjustment, the new requested application is accepted. Otherwise, the new requested application is rejected (Kim, 2019).

In the bandwidth allocation algorithm, each F-AP decides its resource allocation way based on the weighted ternary voting game. In the $\mathscr{F}_k^{AP}$, consider the X game with $\mathbb{N}_{\mathscr{F}_k^{AP}}$. Individual $m_i \in \mathbb{N}_{\mathscr{F}_k^{AP}}$ participates in the voting procedure with their weight. Usually, the PA way is the most common method, and it can be seen as implicit consent. Therefore, for each voter, we can assume that there are three options, i.e., pro-CA, pro-EA and abstention. According to (5.4.8), the coalition S consists of pro-CA voters, and the coalition T consists of pro-EA voters. If the outcome of $\mathfrak{X}(S,T)$ is 1 or -1, $\mathscr{F}_k^{AP}$ simply adopts the CA way or the EA way, respectively, in the bandwidth allocation process. If the outcome of $\mathfrak{X}(S,T)$ is 0, the $\mathscr{F}_k^{AP}$ adopts the PA way. Based on the decisions of individual devices, Each F-AP adaptively decide its communication resource allocation way (Kim, 2019).

Similar to communication resources, the computation capacity of each F-AP is inherently limited. As a result, the computing power of an F-AP must be dynamically allocated to accommodate the demands of various application tasks. Typically, applications are classified into two categories based on their QoE requirements: Class I (real-time data applications) and Class II (non-real-time data applications). These two classes differ not only in their computation resource requirements but also in their QoE expectations. To develop an effective computation resource allocation algorithm, each F-AP must partition its computational capacity to support different types of applications. The *FCRA* scheme incorporates the weighted ternary voting game model for this purpose. However, unlike the voting game approach used in communication resource allocation, the computation allocation algorithm primarily leverages the pseudo-Banzhaf value as its core decision-making metric (Kim, 2019).

Consider another X game in the $\mathscr{F}_k^{AP}$ with $\mathbb{N}_{\mathscr{F}_k^{AP}}$. Individual $m_i \in \mathbb{N}_{\mathscr{F}_k^{AP}}$ participates in the second voting procedure. At this time, there are three options for each voter, i.e., pro-class I service, pro-class II service and abstention. According to (5.4.8), the coalition S consists of pro-class I service voters, and the coalition T consists of pro-class II service voters. Based on the outcome of $\mathfrak{X}(S,T)$, the F-AP estimates the preference ratio (α_I) of class I application services as follows (Kim, 2019);

$$\alpha_I = \sum_{m_i\in S}\left(\Psi_{m_i}^{\mathbb{Q}} \times \mathscr{F}_{m_i}(x)\right) / \left(\sum_{m_i\in S}\left(\Psi_{m_i}^{\mathbb{Q}} \times \mathscr{F}_{m_i}(x)\right) + \sum_{m_j\in T}\left(\Psi_{m_j}^{\mathbb{Q}} \times \mathscr{F}_{m_i}(x)\right)\right) \tag{5.4.12}$$

According to each user's pseudo-Banzhaf value and social power, the α_I value is decided, and each F-AP partitions its total computation capacity ($\mathscr{C}$). The division ratio for class I (*or* class II) type application services is α_I (*or* $1 - \alpha_I$). And then, the partitioned computation capacity is distributed for its corresponding type application services while maximizing the system performance. To implement this process, the *FCRA* scheme adopts the fundamental notion of the NBS. This is captured by the following optimization problem (Kim, 2019).

$$\max_{[\ldots A_{m_i}\ldots]} \prod_{m_i \in S} U^{I}_{m_i}\left(A_{m_i}\right) + \max_{[\ldots A_{m_j}\ldots]} \prod_{m_j \in T} U^{II}_{m_j}\left(A_{m_j}\right) \tag{5.4.13}$$

$$\text{s.t.,} \begin{cases} U^{I}_{m_i}\left(A_{m_i}\right) = \eta_I \times \log_2\left(1 + \frac{A_{m_i}}{R_{m_i}}\right) \\ U^{II}_{m_j}\left(A_{m_j}\right) = \eta_{II} \times \left(1 - \exp\left(-\frac{A_{m_i}}{R_{m_i}}\right)\right) \\ \sum_{m_i \in S} A_{m_i} \leq \left(\alpha_I \times \mathscr{C}_{\mathscr{F}^{AP}_k}\right) and \sum_{m_j \in T} A_{m_j} \leq \left(\left(1 - \alpha_I\right) \times \mathscr{C}_{\mathscr{F}^{AP}_k}\right) \end{cases}$$

where η_I $(or\ \eta_{II})$ is a control parameter to adjust the payoff of m_i $(or\ m_j)$, and $\mathscr{C}_{\mathscr{F}^{AP}_k}$ is the total computation capacity of $\mathscr{F}^{AP}_k$ (Kim, 2019).

5.4.6 Summary

Over the past decade, IoT has experienced rapid expansion, with a diverse range of smart mobile devices supporting numerous applications. Within the framework of future networks, FC and social networking principles have been integrated into the IoT paradigm to enhance distributed self-cooperation and self-management among mobile devices. The *FCRA* scheme introduces an innovative resource allocation mechanism for the SFIoT system. Based on the principles of BVGs and NBS, the *FCRA* scheme operates in two phases; i) mobile devices participate in a voting process to determine resource allocation strategies using the weighted ternary BVG model, and ii) the allocated computation capacity is then fairly and efficiently distributed across different applications utilizing the *NBS* approach. These two game-theoretic models operate iteratively in a step-by-step manner, dynamically adjusting to real-time SFIoT system conditions (Kim, 2019).

5.5 Collaborative Coded Distributed Computing (*CCDC*) scheme

As an advanced communication paradigm, the Artificial Intelligence-based Internet of Things (AIoT) offers a globally interactive platform that enhances the quality of networking services. Within this AIoT framework, Coded Distributed Computing (CDC) has recently gained attention as a promising approach to mitigate straggler effects in conventional distributed computing systems. Recently, S. Kim introduced the *Collaborative Coded Distributed Computing* (*CCDC*) scheme for the AIoT platform (Kim, 2023a). Rooted in cooperative game theory, the *CCDC* scheme primarily addresses two key challenges: i) determining the optimal *k* value for the CDC process, and ii) allocating edge node resources for task offloading. To tackle these challenges, the *CCDC* scheme adopts a two-phase game model based on coalition game theory and the *Weighted Nash Social Welfare Solution* (*WNSWS*), ensuring a mutually beneficial resource allocation strategy. At the first phase, a dynamic coalition formation process selects the most suitable edge nodes for offloading subtasks. At the second phase, the *WNSWS* is utilized to efficiently allocate computational resources among selected edge nodes. By jointly integrating these two cooperative game models, the *CCDC* scheme leverages synergistic effects to optimize the CDC process. In an edge-assisted distributed computing infrastructure, this reciprocal and combinative approach enables a fair and efficient resource allocation strategy, achieved through sequential interactions between edge nodes and AIoT devices (Kim, 2023a).

5.5.1 Development motivation

The evolution of the Internet is continuously shifting from the Internet of Computers (IoC) to IoT. The IoT paradigm has enabled a ubiquitously connected world, generating diverse heterogeneous data across a wide range of fields and applications. By 2025, the total number of installed IoT devices is projected to reach approximately 41.6 billion, while global data generation and consumption could reach 79.4 Zettabytes (ZBs). The primary goal of IoT technology is to create a smart and intelligent world, and to achieve this, AI is essential. In recent years, there has been a growing demand for the convergence of AI and IoT to address high-performance data processing challenges in IoT-driven engineering applications. The integration of AI with IoT (AIoT) has significantly accelerated the development of AIoT systems, enhancing existing IoT infrastructures to form autonomous future communication architectures that support intelligent data exchange among millions of interconnected devices (Chang, 2021; Ghosh, 2018; Kim, 2023a).

Due to limited computational resources, AIoT devices often struggle to handle computation-intensive tasks. One of the most critical AIoT operations is matrix multiplication, a fundamental process in machine learning. Computation offloading has been considered a potential solution to this problem. Traditionally, cloud computing has played a pivotal role in the IoT ecosystem, leveraging abundant cloud resources to provide on-demand computing and storage capabilities for IoT devices. However, in a cloud-centric AIoT system, substantial amounts of data must be transmitted from AIoT devices to centralized cloud data centers for further processing. While cloud data centers offer virtually unlimited computational capacity, they face significant challenges, such as network bandwidth congestion and latency limitations in cloud-based communication (Kim, 2023a).

To overcome these limitations, the emerging trend in AIoT is shifting AI capabilities from centralized cloud systems to edge nodes, bringing AI computation closer to end devices. This shift is driving the adoption of edge intelligence, a promising paradigm that plays a crucial role in realizing AIoT's full potential. Edge computing enables computational resources to be placed closer to the data source, significantly reducing network congestion and transmission delays. Given these advantages, edge intelligence has become an ideal solution for AIoT platforms, attracting significant interest from both academia and industry (Chang, 2021; Jia, 2022; Kim, 2023a).

With the increasing use of AIoT devices, distributed computing has emerged as a highly efficient approach for handling large-scale computations in wireless edge networks. Compa* red to centralized computing models, distributed computing offers greater fault tolerance and scalability. Due to these advantages, distributed computing has been widely adopted in edge-based AIoT systems. In distributed computing, complex tasks can be partitioned into smaller subtasks and executed in parallel across multiple edge nodes. However, despite its benefits, distributed edge computing faces a major challenge: unpredictable computing latency. The overall execution time of a distributed computing task is often determined by the slowest edge node, known as the straggler effect. Since computation tasks require results from all participating edge nodes, the presence of slow nodes can significantly degrade system performance, thereby reducing the efficiency of distributed computing (Kim, 2023a; Nguyen, 2022, in press; Van Huynh, 2022; Zhao, 2019).

To mitigate the straggler problem, Coded Distributed Computing (CDC) has recently emerged as a highly effective solution. The core principle of CDC is to introduce computational redundancy using advanced coding techniques. By injecting redundancy, CDC can compensate for unpredictable processing delays, thereby enhancing stability and latency performance in distributed computing. Unlike traditional distributed computing, CDC-based systems do not require all edge nodes to return computed results. Instead, the computation latency in a CDC-enabled system is determined by the fastest group of edge nodes. This feature makes CDC-based distributed computing a widely adopted solution in AIoT systems (Kim, 2023a; Nguyen, 2022, in press; Van Huynh, 2022; Zhao, 2019).

Among various CDC techniques, the Maximum Distance Separable (MDS) code is commonly used for matrix multiplication, a fundamental operation in machine learning algorithms. Using MDS coding, a computation task is divided into k equally sized subtasks, which are then encoded into n coded subtasks and distributed across n edge nodes, where $k \leq n$. Once k edge nodes complete their computations and return results, the master node can reconstruct the final output by decoding any k results from the available n subtasks. This approach effectively mitigates the straggler effect, improving the speed of distributed matrix multiplication by a factor of $\log n$. However, choosing optimal k and n values significantly impact the efficiency of the CDC process. A higher k value increases the likelihood of straggling nodes, leading to longer processing delays. Conversely, a lower k value can result in computational inefficiencies, as the remaining $n - k$ edge nodes become underutilized. Therefore, optimizing the trade-off between computation efficiency and system performance is crucial for AIoT applications (Kim, 2023a; Nguyen, 2022, in press; Van Huynh, 2022; Zhao, 2019).

As edge-assisted CDC systems are still in their early stages of development, they have not yet received widespread research attention. To address this gap, the *CCDC* scheme applies cooperative game theory to optimize MDS code operations. Game theory is a well-established mathematical framework used to model cooperative and competitive interactions among rational decision-makers. The fundamental principle of game theory is to model strategic decision-making as a game among multiple players. In cooperative games, all participants seek to maximize their payoffs by adopting optimal strategies, ensuring that no player has an incentive to unilaterally change its decision. Unlike traditional optimization methods, which focus on a single objective function, cooperative game models identify mutually acceptable solutions that benefit all participants (Kim, 2023a; Tinh, 2022).

Given the complex interactions between intelligent edge nodes and AIoT devices, cooperative game theory provides a compelling framework for designing an efficient MDS code management algorithm in edge-assisted CDC platforms (Kim, 2023a). Bargaining solutions are mathematical formulations that define unique outcomes in cooperative game settings. J. Nash was a pioneer in applying game theory to develop bargaining solutions. One of the most significant welfare criteria in cooperative games is the *Weighted Nash Social Welfare Solution* (*WNSWS*), which balances both efficiency and fairness. The *WNSWS* seeks to maximize Nash social welfare functions by maximizing the product of individual player payoffs. This concept is rooted in Nash's renowned solution for cooperative bargaining problems. Beyond serving as a fairness criterion, the *WNSWS* effectively converges to the Nash-optimal outcome in a distributed negotiation framework (Charkhgard, 2022; Kim, 2023a; Saghand, 2021).

5.5.2 Contribution

The *CCDC* scheme introduces an innovative CDC control mechanism based on a two-phase cooperative game framework, integrating coalition formation games and resource bargaining games. At the first phase, a coalition of k edge nodes are dynamically formed to implement the MDS code mechanism, utilizing a coalition formation game strategy. At the second phase, the computing resources of each participating edge node are dynamically allocated to process AIoT offloading tasks according to WNSWS. By leveraging this hybrid game-theoretic approach, the *CCDC* scheme effectively manages the trade-off between service latency and system efficiency, while mitigating the impact of real-world straggling effects among edge nodes. Furthermore, the *CCDC* scheme ensures fair resource distribution among individual edge nodes within the CDC process (Kim, 2023a).

5.5.3 CDC platform and a two-phase cooperative game model

As illustrated in Figure 5, the *CCDC* scheme considers an edge assisted CDC system platform consisting of one edge computing server ($\mathscr{C}$), a set of mobile edge nodes $\mathbb{E} = \{\mathscr{E}_1, \ldots, \mathscr{E}_n\}$, and a set of AIoT devices $\mathbb{D} = \{\mathscr{D}_1, \ldots, \mathscr{D}_m\}$ in a given geographical area. The edge node $\mathscr{E}_{1 \leq i \leq n} \in \mathbb{E}$ has his computing power $\left(\mathfrak{M}_{\mathscr{E}_i}\right)$, and communicates with the C through wired links. Each AIoT device $\mathscr{D}_{1 \leq j \leq m} \in \mathbb{D}$ generates its computational workload task $\left(\mathscr{W}_{\mathscr{D}_j}\right)$; $\mathscr{W}_{\mathscr{D}_j}$ is assumed to be delay sensitive and follows a Poisson process with an average generation rate at each time period. To reduce the computation load, the $\mathscr{D}_j$ can offload its task $\mathscr{W}_{\mathscr{D}_j}$ to the C. Using the (n, k) MDS code, the C divides the computing task $\left(\mathscr{W}_{\mathscr{D}_j}\right)$ into k subtasks with equal size. Then, these subtasks are encoded into n coded subtasks and sent them to n edge nodes. Edge nodes, i.e., $\mathscr{E} \in \mathbb{E}$, locally perform their corresponding subtasks, which are distributed from the C. While performing the distributed computation, each edge node periodically measures the remaining workload and reports this information to the C (Kim, 2023a; Nguyen, in press).

Figure 5. The infrastructure of edge assisted CDC system (Kim, 2023a)

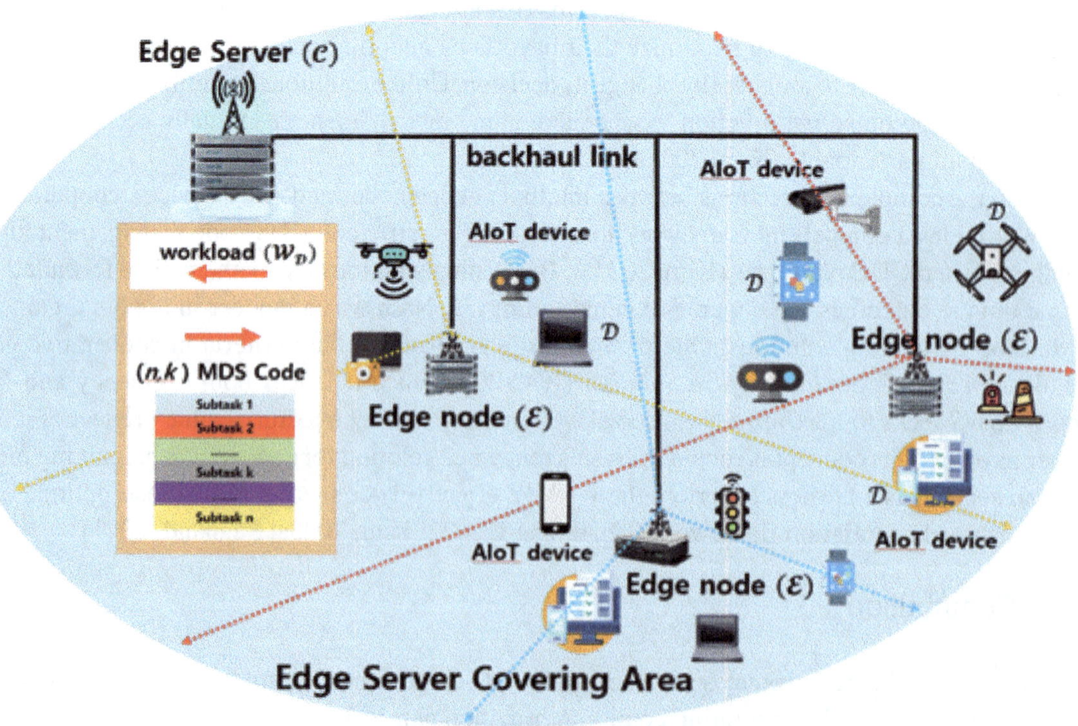

In the *CCDC* scheme, the (n, k) MDS code problem for the $\mathcal{D}_j$ and the $\mathcal{E}_i$'s resource sharing problem are formulated as cooperative games $\mathbb{G}_{\mathcal{D}_j}$ and $\mathbb{G}_{\mathcal{E}_i}$, respectively. As a consequence, server, edge nodes, and AIoT devices are sequentially interacted with each other; it is noteworthy that the *CCDC* scheme formulates the C-E-D association in a coordinated manner. Formally, the *CCDC* scheme defines the two-phase game G entities, i.e.,

$$\mathbb{G} = \left\{\mathbb{G}_{\mathcal{D}_j}, \mathbb{G}_{\mathcal{E}_i}\right\} = \left\{\mathscr{C}, \mathbb{E}, \mathbb{D}, \left\{\mathbb{G}_{\mathcal{D}_j} | \mathcal{D}_j \in \mathbb{D}, \mathbb{E}, \mathscr{W}_{\mathcal{D}_j}, v(\cdot)\right\}, \left\{\mathbb{G}_{\mathcal{E}_i} | \mathcal{E}_i \in \mathbb{E}, \mathfrak{M}_{\mathcal{E}_i}, \mathscr{W}^{\mathcal{E}_i}_{\mathcal{D}_j}, S^{\mathcal{E}_i}_{\mathcal{D}_j}, U^{\mathcal{E}_i}_{\mathcal{D}_j}(\cdot)\right\}, T\right\}$$

of gameplay (Kim, 2023a).

- C, E and D represent the edge computing server, the sets of edge nodes, and AIoT devices, respectively. They are mutually and reciprocally interdependent and work together in the edge assisted CDC platform.
- At the first phase, the $\mathbb{G}_{\mathcal{D}_j}$ is designed to form an edge structure to process the subtasks of $\mathscr{W}_{\mathcal{D}_j}$. In the $\mathbb{G}_{\mathcal{D}_j}$, edge nodes are game players, and $v(\cdot)$ is a characteristic function for players' coalition.
- At the second phase, the $\mathbb{G}_{\mathcal{E}_i}$ is developed to share the $\mathfrak{M}_{\mathcal{E}_i}$ for the subtasks of the AIoT devices. Each $\mathbb{G}_{\mathcal{E}_{1 \le i \le n}}$ game is operated in a distributed parallel fashion.
- In the $\mathbb{G}_{\mathcal{E}_i}$, $\mathscr{W}^{\mathcal{E}_i}_{\mathcal{D}_j}$ is the $\mathcal{D}_j$'s subtask assigned to the $\mathcal{E}_i$, and it is a game player. The $S^{\mathcal{E}_i}_{\mathcal{D}_j}$ and $U^{\mathcal{E}_i}_{\mathcal{D}_j}(\cdot)$ are player's strategy and utility function, respectively.
- Discrete time model $T \in \{t_1, \ldots, t_c, t_{c+1}, \ldots\}$ is represented by a sequence of time steps. The length of t_c matches the event's timescale of $\mathbb{G}_{\mathcal{D}_j}$ and $\mathbb{G}_{\mathcal{E}_i}$.

5.5.4 Fundamental ideas of coalition game model and *WNSWS*

Traditionally, forming effective coalitions is a major research challenge in cooperative games; a coalition of players can also do things more efficiently than individual players can do. Therefore, coalition formation entails finding a coalitional structure that maximizes the total payoff. In the coalition formation game, an external factor imposes a certain structure to form stable coalitions, and no player has an incentive to deviate. Formally, a coalition formation game is defined by (N, v) which $N = \{p_1, \ldots, p_n\}$ is the set of players and v is a real-valued characteristic function such that $v(S): 2^N \rightarrow \mathbb{R}$,s.t., $S \subset N$; a subset S is termed a coalition. Given a coalition game, a coalition structure $\mathscr{C} = \{S_1, \ldots, S_k\}$ is an exhaustive disjoint partition of the space of players into feasible coalitions and $v(\mathscr{C}) = \sum_{S \in \mathscr{C}} v(S)$. By considering game dynamics, the properties from resulting coalitions and its adaptability to environment variable or externalities are important research issues (Han, 2011; Kim, 2014, 2023a).

To implement a dynamic coalition formation algorithm, some definition and rules are necessary. First of all, the coalition composed of all players is referred as a grand coalition N, and a collection of coalitions in the N, denoted $\mathcal{S}$, is defined as the set $\mathcal{S} = \{S_1, \ldots, S_k\}$ of mutually disjoint coalitions $S_{1 \le i \le k} \subset \mathcal{N}$ and $\mathcal{N} = \cup_{j=1}^{k} S_j$. A *preference relation* $\triangleright$ is an order defined for comparing two coalition collections $\mathcal{S} = \{S_1, \ldots, S_l\}$ and $\widehat{\mathcal{S}} = \{\widehat{S}_1, \ldots, \widehat{S}_p\}$ that are partitions of the same subset $\mathcal{A} \subseteq \mathcal{N}$. Therefore, game players in $\mathcal{S}$ and $\widehat{\mathcal{S}}$ are same. In this case, $\mathcal{S} \triangleright \widehat{\mathcal{S}}$ implies that the way $\mathcal{S}$ partitions $\mathcal{A}$ is preferred to the way $\widehat{\mathcal{S}}$ partitions $\mathcal{A}$. Based on the concept of *preference relation*, merge and split rules can be defined as follows (Han, 2011; Kim, 2014, 2023a).

Merge Rule: Any set of coalitions $\{S_1,...,S_k\}$ may be merged whenever the merged form is preferred by players; i.e., where $\{\cup_{j=1}^{k}, S_j\} \triangleright \{S_1,...,S_k\}$, therefore, $\{S_1,...,S_k\} \rightarrow \{\cup_{j=1}^{k}, S_j\}$.

Split Rule: Any coalition $\cup_{j=1}^{k} S_j$ may be split whenever a split form is preferred by players; i.e., where $\{S_1,...,S_k\} \triangleright \{\cup_{j=1}^{k}, S_j\}$, thus, $\{\cup_{j=1}^{k}, S_j\} \rightarrow \{S_1,...,S_k\}$.

The basic idea behind the merger-and-split rules is that players enter into a binding agreement to form a coalition through the merger (*or* split) operation if all players are able to improve (*or* not decrease) their individual payoffs from the previous state (Han, 2011; Kim, 2014, 2023a).

The core concept of WNSWS originates from bargaining game theory. In this framework, all players collectively form a grand coalition to maximize their individual payoffs. To determine the optimal grand coalition, the *CCDC* scheme must optimize a super-criterion across all feasible actions. Within the bargaining problem literature, these super-criteria are commonly referred to as social welfare measures. As a result, the primary objective of *WNSWS* is to optimize the *Nash Social Welfare* (*NSW*) function, which serves as a super-criterion for feasible resource allocation, ensuring both fairness and efficiency simultaneously. To characterize the basic ideas of *WNSWS*, we first start with some definitions. Let $N = \{1...i...n\}$ be the set of imaginary players and let R be the set of all real numbers. $\mathbb{X} \subseteq \mathbb{R}^n$ represents the set of feasible solutions and it is assumed to be bounded where $\mathbb{R}^n$ is the n-fold Cartesian product of R. For all $i \in N, f(\mathbb{X}) := (x \in \mathbb{X} | f_1(x), \ldots, f_n(x))$ is a vector of payoff functions. Traditionally, a multi-objective optimization problem can be stated as (Charkhgard, 2022; Kim, 2023a; Saghand, 2021):

$$\begin{cases} \max \{i \in S^+, x \in \mathbb{X} | f_i(x)\} \\ \\ \min \{j \in S^-, x \in \mathbb{X} | f_j(x)\} \end{cases}, \text{ s.t., } S^+ \cup S^- = N \tag{5.5.1}$$

where S^+ and S^- are the index sets of payoff functions that need to be maximized and minimized, respectively. Denote $\mathbb{W} = (\omega_1, \ldots, \omega_n)$ is a vector of players' weights with $\omega_i > 0$ for all $i \in N$; they are players' corresponding degrees of importance. Based on the positive weights, the *CCDC* scheme can aggregate players' payoff functions to form a single-objective optimization problem (Charkhgard, 2022; Kim, 2023a).

$$\max \left\{ \sum_{i \in S^+} (\omega_i \times f_i(x)) - \sum_{j \in S^-} (\omega_j \times f_j(x)) \right\}, \text{s.t., } S^+ \cup S^- = N \text{ and } x \in \mathbb{X} \tag{5.5.2}$$

In the literature on multi-objective optimization problems, Equation (5.5.2) is commonly referred to as the weighted sum optimization approach. Based on the principles outlined in Equations (5.5.1)-(5.5.2), the *WNSWS* can be mathematically expressed as follows (Kim, 2023a):

$$WNSWS = \max_{x \in \mathbb{X}} \left(\prod_{i \in S^+} \left(f_i(x) - d_i \right)^{\omega_i} \times \prod_{j \in S^-} \left(d_j - f_j(x) \right)^{\omega_j} \right), \tag{5.5.3}$$

s.t., $f_i(x) \geq d_i, \forall i \in S^+$ and $f_j(x) \leq d_j, \forall j \in S^-$

where d_i is the i's reference point. In the literature of bargaining problems, the reference point is sometimes referred as the disagreement point, and it basically indicates the payoff of each player under no coalition. Note that since player $i \in S^+$ is interested in maximizing its payoff function, $(f_i(x) - d_i)^{\omega_i}$ captures the benefit that it will obtain as the result of creating a coalition. Similarly, player $j \in S^-$ is interested in minimizing their payoff function. Therefore, $\left(d_j - f_j(x)\right)^{\omega_j}$ captures the benefit that it obtains as the result of creating a coalition. Considering players' weights, the main goal of *WNSWS* is to maximize the product of benefits of players with respect to the reference point. The *WNSWS* in equation (5.5.3) can be transformed as a *local-benefit-scale-free WNSWS*, i.e., *lbsf-WNSWS*, by replacing the payoff function of the *WNSWS* with the arbitrary positive constants $\alpha_{1 \leq i \leq n}$ (Charkhgard, 2022; Kim, 2023a; Saghand, 2021).

$$lbsf - WNSWS = \tag{5.5.4}$$

$$\max_{x \in \mathbb{X}} \left(\prod_{i \in S^+} \left((\alpha_i \times f_i(x)) - (\alpha_i \times d_i) \right)^{\omega_i} \times \prod_{j \in S^-} \left((\alpha_j \times d_j) - (\alpha_j \times f_j(x)) \right)^{\omega_j} \right)$$

$$= \max_{x \in \mathbb{X}} \left(\prod_{k=1}^{n} (\alpha_k)^{\omega_k} \times \prod_{i \in S^+} \left(f_i(x) - d_i \right)^{\omega_i} \times \prod_{j \in S^-} \left(d_j - f_j(x) \right)^{\omega_j} \right)$$

$$\text{s.t.,} \begin{cases} \alpha_k = \dfrac{1}{\max_{x \in \mathbb{X}} \left(f_k(x) - d_k \right)}, \forall k \in S^+ \\ \\ \alpha_k = \dfrac{1}{\max_{x \in \mathbb{X}} \left(d_k - f_k(x) \right)}, \forall k \in S^- \end{cases}$$

Since $\prod_{k=1}^{n} (\alpha_k)^{\omega_k}$ is a positive constant, it has no impact on the optimization, and it can be dropped. Therefore, the maximum benefit that each player can obtain will be precisely one in the equivalent problem. This implies that an equivalent *WNSWS* with unit-maximum-benefit can be constructed from the original *WNSWS* formula (Charkhgard, 2022; Kim, 2023a; Saghand, 2021).

5.5.5 CDC control scheme for the AIoT paradigm

To develop the edge assisted CDC control mechanism for AIoT devices, the *CCDC* scheme formulates a new two-phase cooperative game model. At the first phase, the *CCDC* scheme defines the n and k values for the MDS code. In the distributed computing scenario, individual devices contact the C for their offloading services and have their corresponding edge nodes in the C's coverage area. Simply, the *CCDC* scheme sets the n value as the number of corresponding edge nodes. And then, the decision for k value is made based on the dynamic coalition formation game. For example, the $\mathcal{D}_j$ generates the $\mathcal{W}_{\mathcal{D}_j}$, and the $\mathbb{G}_{\mathcal{D}_j}$ is operated to form a coalition with efficiently workable edge nodes. The characteristic function of each individual edge node is defined as the worth of its contribution. For the $\mathcal{E}_i$, its characteristic function $v(\mathcal{E}_i)$ is given by (Kim, 2023a):

$$v\left(\mathscr{E}_i, \mathscr{W}_{\mathscr{D}_j}^{\mathscr{E}_i}\right) = \eta - \left(\mathscr{G}\left(C_{\mathscr{E}_i}\right) \times \log\left(\frac{\left(C_{\mathscr{E}_i} + f\left(\mathscr{W}_{\mathscr{D}_j}^{\mathscr{E}_i}\right)\right)}{\mathfrak{M}_{\mathscr{E}_i}} + \theta\right)\right) \tag{5.5.5}$$

$$\text{s.t., } \mathscr{G}\left(C_{\mathscr{E}_i}\right) = \begin{cases} \phi, iff\left(\mathscr{W}_{\mathscr{D}_j}^{\mathscr{E}_i}\right) \geq \left(\mathfrak{M}_{\mathscr{E}_i} - C_{\mathscr{E}_i}\right) \\ exp\left(\dfrac{f\left(\mathscr{W}_{\mathscr{D}_j}^{\mathscr{E}_i}\right)}{\left(\mathfrak{M}_{\mathscr{E}_i} - C_{\mathscr{E}_i}\right)}\right), otherwise \end{cases}$$

where η, θ, and ϕ are control parameters for the $v(\cdot)$. $C_{\mathscr{E}_i}$ is the currently used computing power of $\mathscr{E}_i$, and $f\left(\mathscr{W}_{\mathscr{D}_j}^{\mathscr{E}_i}\right)$ is the needed computing capacity for the $\mathscr{W}_{\mathscr{D}_j}^{\mathscr{E}_i}$ where $\mathscr{W}_{\mathscr{D}_j}^{\mathscr{E}_i}$ is the n-divided $\mathscr{W}_{\mathscr{D}_j}$ and assigned to the $\mathscr{E}_i$. $v(S)$ is the characteristic function that associates to each edge node coalition S (Kim, 2023a).

$$v(S) = \prod_{\mathscr{E}_i \in S} v(\mathscr{E}_i), \text{ s.t.,} S \subseteq \mathbb{E} \tag{5.5.6}$$

According to (5.5.6), the *CCDC* scheme can form the best coalition S, which consists of adaptable edge nodes to process the $\mathscr{W}_{\mathscr{D}_j}$. Finally, the k value is given by (Kim, 2023a):

$$k = \left|S_{\mathscr{D}_j}^{*}\right|, \text{ s.t.,} v\left(S_{\mathscr{D}_j}^{*}\right) = \max_{S_{\mathscr{D}_j}^{*} \subseteq \mathbb{E}}\left(\mathscr{E}_i \in \mathbb{E} \middle| \prod_{\mathscr{E}_i \in S_{\mathscr{D}_j}^{*}} v(\mathscr{E}_i)\right) \tag{5.5.7}$$

where $\left|S_{\mathscr{D}_j}^{*}\right|$ is the cardinality of $S_{\mathscr{D}_j}^{*}$. Based on the $S_{\mathscr{D}_j}^{*}$ formation, the k subtasks of $\mathscr{W}_{\mathscr{D}_j}$ are assigned to the edge nodes in the $S_{\mathscr{D}_j}^{*}$. In the $\mathbb{G}_{\mathscr{D}}$ for each AIoT device, the *CCDC* scheme can create its coalition $S_{\mathscr{D}}^{*}$. Based on this information, each edge node also creates two subtask coalitions, i.e., $S_{\mathscr{E}}^{+}$ and $S_{\mathscr{E}}^{-}$, where these subtasks are offloaded from its corresponding AIoT devices. For example, the $\mathscr{E}_i$ has two coalitions such as the $S_{\mathscr{E}_i}^{+}$ and $S_{\mathscr{E}_i}^{-}$. They are defined as follows (Kim, 2023a):

$$S_{\mathscr{E}_i}^{+} = \cup\left\{\mathscr{D}_j \in \mathbb{D} \middle| \mathscr{E}_i \in S_{\mathscr{D}_j}^{*}\right\} \text{ and } S_{\mathscr{E}_i}^{-} = \cup\left\{\mathscr{D}_k \in \mathbb{D} \middle| \mathscr{E}_i \notin S_{\mathscr{D}_k}^{*}\right\} \tag{5.5.8}$$

At the second stage, individual edge nodes allocate their computing powers $\left(\mathfrak{M}_{\mathscr{E}_{1 \leq i \leq n}}\right)$ for the assigned subtasks. For the $\mathscr{E}_i$, the $\mathbb{G}_{\mathscr{E}_i}$ game model is developed in a distributed manner, and the $\mathfrak{M}_{\mathscr{E}_i}$ is shared by the corresponding AIoT devices. If the $\mathscr{D}_j$ ask its offload task to the $\mathscr{E}_i$, his utility function, i.e., $U_{\mathscr{D}_j}^{\mathscr{E}_i}(\cdot)$, is given by (Kim, 2023a):

$$U_{\mathscr{D}_j}^{\mathscr{E}_i}\left(\mathscr{W}_{\mathscr{D}_j}^{\mathscr{E}_i}, \Delta_{\mathscr{D}_j}^{\mathscr{E}_i}, R_{\mathscr{D}_j}^{\mathscr{E}_i}, S_{\mathscr{D}_j}^{\mathscr{E}_i}\right) =$$

$$
U_{\mathcal{D}_j}^{\mathcal{E}_i}(\cdot) = \begin{cases} exp\left(\gamma \times \left(\frac{f\left(\mathcal{W}_{\mathcal{D}_j}^{\mathcal{E}_i}\right) - \Delta_{\mathcal{D}_j}^{\mathcal{E}_i}}{f\left(\mathcal{W}_{\mathcal{D}_j}^{\mathcal{E}_i}\right)}\right)\right) \times \frac{\min\left(S_{\mathcal{D}_j}^{\mathcal{E}_i}, R_{\mathcal{D}_j}^{\mathcal{E}_i}\right)}{R_{\mathcal{D}_j}^{\mathcal{E}_i}}, if \mathcal{D}_j \in S_{\mathcal{E}_i}^{+} \\ \left(\mathfrak{P}_{\mathcal{E}_i} \times \log\left(\left(\frac{f\left(\mathcal{W}_{\mathcal{D}_j}^{\mathcal{E}_i}\right) - \Delta_{\mathcal{D}_j}^{\mathcal{E}_i}}{f\left(\mathcal{W}_{\mathcal{D}_j}^{\mathcal{E}_i}\right)}\right) + \delta\right) + \Gamma\right) \times \frac{\min\left(S_{\mathcal{D}_j}^{\mathcal{E}_i}, R_{\mathcal{D}_j}^{\mathcal{E}_i}\right)}{R_{\mathcal{D}_j}^{\mathcal{E}_i}}, if \mathcal{D}_j \in S_{\mathcal{E}_i}^{-} \end{cases} \tag{5.5.9}
$$

where γ, δ and Γ are control parameters for the $U_{\mathcal{D}_j}^{\mathcal{E}_i}(\cdot)$, which is differently defined whether the $\mathcal{D}_j$ in $S_{\mathcal{E}_i}^{+}$ or $S_{\mathcal{E}_i}^{-}$. $\Delta_{\mathcal{D}_j}^{\mathcal{E}_i}$ is the remaining workload of $\mathcal{W}_{\mathcal{D}_j}^{\mathcal{E}_i}$. Therefore, the $\Delta_{\mathcal{D}_j}^{\mathcal{E}_i}$ is needed to be computed by the $\mathcal{E}_i$ for each offload task. To process the $f\left(\mathcal{W}_{\mathcal{D}_j}^{\mathcal{E}_i}\right)$, $R_{\mathcal{D}_j}^{\mathcal{E}_i}$ and $S_{\mathcal{D}_j}^{\mathcal{E}_i}$ are the requested and assigned computing powers at the next time period. Therefore, the $S_{\mathcal{D}_j}^{\mathcal{E}_i}$ is thought as the strategy for the $\mathcal{W}_{\mathcal{D}_j}^{\mathcal{E}_i}$. $\mathfrak{P}_{\mathcal{E}_i}$ represents the resource assignment ratio for services in $S_{\mathcal{E}_i}^{+}$. The *CCDC* scheme adopts the *WNSWS* to decide the $S_{\mathcal{D}_j}^{\mathcal{E}_i}$. According to (5.5.3)-(5.5.4), the *WNSWS* for the $\mathcal{E}_i$, i.e., $WNSWS_{\mathcal{E}_i}$, is obtained as follows (Kim, 2023a):

$$
WNSWS_{\mathcal{E}_i} = \max_{\ldots S_{\mathcal{D}_j}^{\mathcal{E}_i}, S_{\mathcal{D}_k}^{\mathcal{E}_i}, \ldots} \left(\prod_{\mathcal{D}_j \in S_{\mathcal{E}_i}^{+}} \left(U_{\mathcal{D}_j}^{\mathcal{E}_i}(\cdot) - d_{\mathcal{D}_j}^{+} \right)^{\omega_{\mathcal{D}_j}} \times \prod_{\mathcal{D}_k \in S_{\mathcal{E}_i}^{-}} \left(d_{\mathcal{D}_k}^{-} - U_{\mathcal{D}_k}^{\mathcal{E}_i}(\cdot) \right)^{\omega_{\mathcal{D}_k}} \right) \tag{5.5.10}
$$

$$
\text{s.t.,} \; \mathcal{D}_j, \mathcal{D}_k \in \mathbb{D}^{\mathcal{E}_i}, \; S_{\mathcal{E}_i}^{+} \cup S_{\mathcal{E}_i}^{-} = \mathbb{D}^{\mathcal{E}_i}, \text{ and } \sum_{\mathcal{D}_j \in \mathbb{D}^{\mathcal{E}_i}} S_{\mathcal{D}_j}^{\mathcal{E}_i} \leq \mathfrak{M}_{\mathcal{E}_i}
$$

where $d_{\mathcal{D}_j}^{+}$ is the $\mathcal{D}_j$'s reference point if $\mathcal{D}_j \in S_{\mathcal{E}_i}^{+}$, and $d_{\mathcal{D}_k}^{-}$ is the $\mathcal{D}_k$'s reference point if $\mathcal{D}_k \in S_{\mathcal{E}_i}^{-}$. $\omega_{\mathcal{D}_j}$ and $\omega_{\mathcal{D}_i}$ are the weights of $\mathcal{D}_j$ and $\mathcal{D}_k$, respectively. $\mathbb{D}^{\mathcal{E}_i}$ is the set of AIoT devices, which assign their offload workloads to the $\mathcal{E}_i$ (Kim, 2023a).

5.5.6 Summary

The *CCDC* scheme introduces a highly efficient approach to jointly optimize MDS code operations and edge node selection, thereby enhancing the performance of the CDC system within the AIoT framework. By leveraging cooperative game theory, the CDC control problem is formulated as a coalition formation game and a resource bargaining game between AIoT devices and edge nodes. Specifically, the coalition formation process is employed to identify the most suitable edge nodes, enabling them to efficiently execute the CDC process while mitigating straggler effects in the distributed computing network. To achieve this, the computing resources of selected edge nodes are dynamically allocated based on WNSWS. For a mutually beneficial outcome, AIoT devices and edge nodes collaboratively interact and coordinate their operations. By adopting a cooperative coordination paradigm, the *CCDC* scheme ensures an optimal performance balance, even under dynamic CDC system conditions. Consequently, the *CCDC* scheme achieves a 'win-win' solution by simultaneously optimizing;

i) system throughput maximization, ii) computation latency minimization, and iii) fairness in resource allocation among edge nodes (Kim, 2023a).

5.6 Multi-agent Cooperative Spectrum Sharing (*MCSS*) scheme

The shortage of spectrum resources has emerged as a significant technical bottleneck in the advancement of the Industrial Internet of Things (IIoT). To address this challenge, the Cognitive IIoT (CIIoT) paradigm, built on cognitive radio technology, enhances spectrum utilization by enabling opportunistic access to idle spectrum bands. Recently, S. Kim introduced the *Multi-agent Cooperative Spectrum Sharing* (*MCSS*) scheme for the CIIoT system platform (Kim, 2023b). The primary challenge of the *MCSS* scheme is to efficiently share limited spectrum resources through cooperative sensing and dynamic access techniques. To ensure a mutually beneficial solution for diverse CIIoT devices, the *MCSS* scheme employs a learning-based game model, integrating *multi-agent reinforcement learning* (*MARL*) with the *Negotiated Aspirations Bargaining Solution* (*NABS*). In the learning mechanism, individual CIIoT devices dynamically determine their cooperative sensing strategy based on the *MARL* model. In the bargaining mechanism, available spectrum resources are dynamically allocated through *NABS*, which is derived from the devices' chosen sensing strategies. By incorporating *MARL* principles into game theory, the *MCSS* scheme effectively guides intelligent CIIoT devices toward a socially optimal spectrum-sharing outcome (Kim, 2023b).

5.6.1 Development motivation

With the rapid advancement of technology, IoT has seen widespread adoption across various fields, including smart cities, network automation, smart manufacturing, and autonomous driving, among other cyber-physical systems. The industrial sector has particularly benefited from wireless network evolution, enabling industries to enhance efficiency, scalability, reliability, and QoS in production processes. Within an industrial environment, a diverse range of industrial devices - which generate access control data - can be connected to the Internet, forming the Industrial Internet of Things (IIoT). The primary challenge of IIoT lies in optimizing manufacturing processes through network interconnection, data interoperability, and seamless system integration of industrial resources (Aazam, 2018; Kim, 2023b; Liu, 2020, 2022).

First introduced by General Electric (GE) in 2012, the IIoT concept extends the IoT paradigm to industrial applications. While conventional IoT facilitates Internet connectivity for any 'thing', IIoT is specifically designed for industrial systems. The IIoT model aims to enhance system robustness, speed, and security, with a strong emphasis on the transmission and control of mission-critical information through machine-to-machine (M2M) communications. With the exponential growth of industrial data, IIoT devices require significantly larger spectrum resources to transmit massive amounts of data (Kim, 2023b).

Typically, IIoT networks rely on the unlicensed 2.4 GHz frequency band for wireless communication. However, this spectrum band is extensively shared by other wireless technologies such as ZigBee, Wi-Fi, and Bluetooth, making it highly congested. As a result, the scarcity of available spectrum resources has become a critical bottleneck in the advancement of IIoT technology (Aazam, 2018; Kim, 2023b; Liu, 2022, 2023).

To address the spectrum shortage problem, Cognitive Radio (CR) has emerged as an effective solution. CR technology enhances spectrum utilization by allowing unlicensed users, known as secondary users (SUs), to access licensed frequency bands without interfering with the primary users (PUs). By dynamically identifying underutilized spectrum bands, the CR system can significantly increase spectrum availability while preventing harmful interference with licensed users. By integrating CR technology into the IIoT paradigm, the Cognitive IIoT (CIIoT) model enables intelligent spectrum sharing, allowing industrial smart devices to dynamically access licensed spectrum bands. This approach enhances transmission capacity and improves overall system throughput, making it a viable solution to the IIoT spectrum scarcity challenge (Kim, 2023b; Liu, 2020, 2021, 2022; Shi, 2020).

As wireless communication technologies continue to evolve, multiple access techniques have shifted from Orthogonal Multiple Access (OMA) to Non-Orthogonal Multiple Access (NOMA), which is considered a core technology of 5G networks. NOMA enhances spectrum efficiency by enabling multiple devices to share the same frequency spectrum simultaneously, distinguishing them based on power levels. Given its high spectral efficiency, the combination of NOMA and CR has been widely recognized as a promising access technology for next-generation wireless communications. This integration is particularly relevant for industrial IoT networks, where high-capacity transmission is crucial (Kim, 2023b; Liu, 2020; Shi, 2020; Xu, 2020).

Introducing NOMA into the CIIoT framework allows for greater connectivity, enabling more IIoT devices to share the limited spectrum resource. However, this approach presents several challenges, particularly in spectrum sensing and interference management. A key aspect of the CIIoT platform is dynamic spectrum sensing, which is essential to prevent interference with PUs and ensure efficient spectrum utilization. If spectrum sensing is inaccurate, it can either lead to interference with PUs or result in wasted spectrum resources. Traditional spectrum sensing methods are typically non-cooperative, where each SU operates independently based on its local observations. However, non-cooperative sensing often struggles to detect weak signals in fading channels accurately. To address this issue, cooperative spectrum sensing (CSS) has gained attention, as it allows multiple SUs to collaborate, improving spectrum sensing accuracy by leveraging spatial diversity. Despite its benefits, CSS introduces additional complexity, requiring coordination mechanisms to manage interactions between multiple SUs (Huang, 2019; Kim, 2023b; Liu, 2020; Shi, 2020).

Recently, Multi-Agent Systems (MASs) have emerged as a promising approach for various applications, including robotic teams, distributed control, resource management, collaborative decision-making, and data mining. As a self-organizing framework, MAS enables intelligent agents to solve complex problems that would be difficult or impossible for a single agent to handle alone. Although MAS agents can be pre-programmed with specific behaviors, they must also learn new strategies in real time to continuously optimize system performance. Reinforcement learning is commonly used to enhance intelligence in MAS environments. Over the years, research on MAS design has drawn insights into computer science, psychology, operations research, and economics. Game theory, a fundamental concept in economics, provides a powerful framework for modeling strategic interactions in MAS environments. Both MAS and game theory focus on intelligent agents making rational decisions to maximize their individual and collective payoffs (Busoniu, 2008; Kim, 2023b; Pendharkar, 2012).

To maintain high-quality communication in the industrial IoT paradigm, an effective spectrum-sharing strategy is essential. Over the years, researchers and engineers have worked on developing spectrum-sharing solutions for the CIIoT ecosystem. The *MCSS* scheme introduces a novel spectrum-sharing algorithm for CIIoT networks. By leveraging insights from game theory and reinforcement learning, the *MCSS* scheme implements a cooperative spectrum sensing process while incorporating the NOMA technique. The primary objective of the *MCSS* scheme is to demonstrate how game theory can be utilized to design MASs in the CIIoT environment, ensuring an optimal balance between system performance and spectrum resource allocation among diverse CIIoT devices. The integration of cognitive radio, NOMA, and MAS-based cooperative spectrum sensing presents a new frontier in CIIoT optimization.

By adopting game-theoretic principles, the *MCSS* scheme enhances spectrum sharing efficiency, ensures fair resource distribution, and mitigates interference challenges in industrial IoT networks. As the demand for high-performance wireless communication in industrial environments continues to grow, solutions like the *MCSS* scheme are expected to play a critical role in shaping the future of intelligent IIoT platforms (Kim, 2023b).

The *negotiated aspirations bargaining solution* (*NABS*) is a new bargaining solution to deliver attainable allocations for any number of agents. The *NABS* fills a gap in the literature by providing a logical counterpart to the *KSBS*. The original *KSBS* idea considers the disagreement point (D^P) and utopian point (U^P) where D^P is the allocation which would result if negotiations broke down, and U^P is the allocation where each agent would be granted his maximal aspirations. Starting from D^P, increase the surplus allocated to every agent in the direction of U^P, increasing each agent's outcome in a proportional way. The *KSBS* selects the best allocation obtained through this procedure while maintaining feasibility. The *NABS* proposes the best allocation in the direction of utopia starting at an endogenous reference point which depends on both the U^P and bargaining power. Therefore, the *NABS* is the negotiation outcome, which takes into account feasibility, bargaining power, and the desire to approach the utopia point. Implicitly, the *NABS* can be seen as allocating gains from the endogenous reference point in the direction of utopia (Alós-Ferrer, 2018; Kim, 2023b).

In a multi-agent system, individual agents not only adapt and learn from their shared environment but also from the actions and learning behaviors of other agents. To achieve an effective solution, multi-agent learning relies on reinforcement learning techniques. Recently, the *MARL* has gained significant attention within the fields of machine learning, artificial intelligence, and game theory. As an interdisciplinary research area, the *MARL* is closely connected to game theory and MASs and can be viewed as a fusion of policy search techniques aimed at exploring both coordination and competition among multiple agents. By employing a game-theoretic approach, certain *MARL* models can be formulated as learning games, facilitating the discovery of fair and efficient solutions in dynamically evolving multi-agent environments (Hu, 2015; Kim, 2023b).

5.6.2 Contribution

The *MCSS* scheme is designed to enhance the performance of the CIIoT system by integrating *NABS* and *MARL*. By utilizing two control mechanisms - a learning mechanism and a bargaining mechanism - the *MCSS* scheme ensures that the limited spectrum resources are allocated in a fair and efficient manner. In the learning mechanism, each CIIoT device learns its optimal policy for cooperative spectrum sensing. In the bargaining mechanism, the *NABS* is employed to address the spectrum allocation challenge. For the effective operation of the CIIoT infrastructure, these two control mechanisms are seamlessly integrated into a new learning game model, enabling the system to achieve a socially optimal solution (Kim, 2023b).

5.6.3 CIIoT system infrastructure and a learning game model

We consider a CIIoT system platform, which comprises n control centers (CCs), i.e., $\mathbb{C} = \{\mathscr{C}_1, \ldots, \mathscr{C}_n\}$, and each CC has a coverage area and a fixed spectrum band $(\mathfrak{M}_C)$. The $\mathfrak{M}_C$ is licensed to the PU, and CCs have their secondary IIoT devices, i.e., $\mathbb{D} = \{\mathscr{D}_1, \ldots, \mathscr{D}_m\}$, which share the $\mathfrak{M}_C$ as SUs. IIoT devices are equipped with a single antenna to contact their corresponding CCs through wireless communications. In cognitive radio technology, it is important to perform spectrum sensing to find the idle spectrum, where the PU is not present temporarily. As an effective spectrum sensing method, cooperative spectrum sensing can improve detection performance by the collaborative sensing and decision of multiple SUs. Therefore, multiple IIoT devices first sense the spectrum band independently, and then send their sensing result to the CC. The CC makes a final decision on the status of the spectrum band, via combining the local sensing information from its corresponding IIoT devices. The general CIIoT network infrastructure is shown in Figure 6, (Kim, 2023b; Liu, 2023).

Figure 6. The industrial IoT system infrastructure for cognitive radio technology (Kim, 2023b)

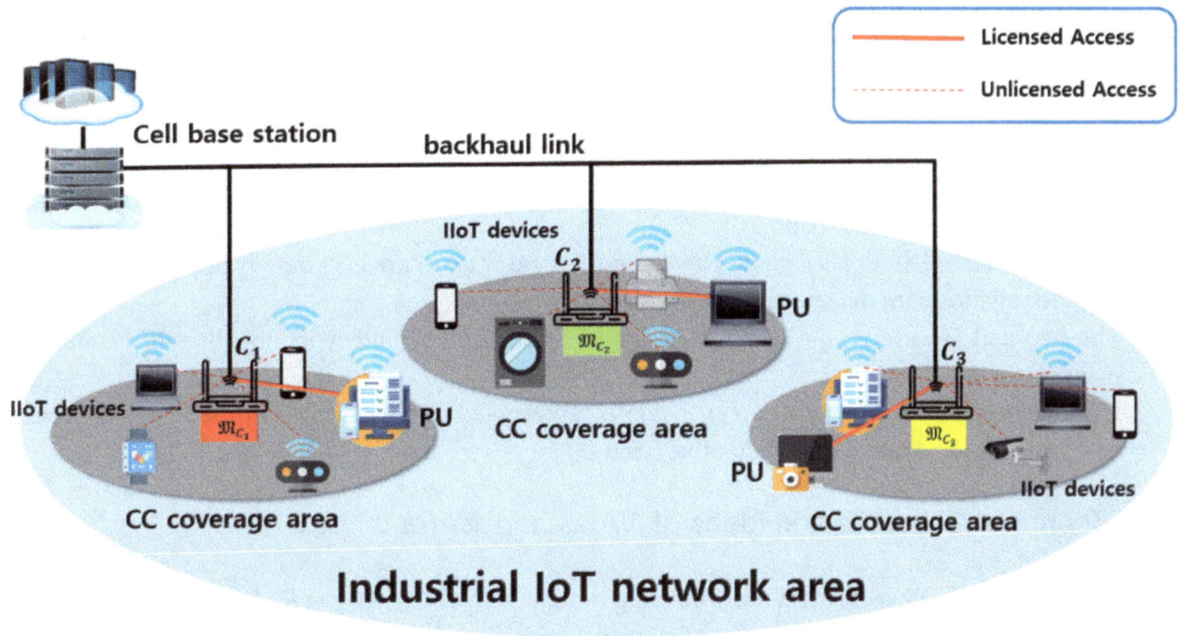

Without loss of generality, the *MCSS* scheme supposes that each spectrum band $(\mathfrak{M}_C)$ has two statuses, i.e., active state and idle state. In a cognitive radio network, the active state refers to a scenario where the PU is actively transmitting data, whereas the idle state indicates that the PU is not transmitting. To determine the current spectrum status, a cooperative spectrum sensing process takes place, where a group of SUs voluntarily participate in sensing and share their local sensing information to improve spectrum utilization. However, the sensing process requires energy and time, leading to a situation where self-interested SUs may attempt to behave as 'free riders', benefiting from the sensing efforts of others without actively contributing. This free-riding problem creates a risk where no SU is willing to sense the spectrum, ultimately hindering the cooperative sensing process. As a result, the key challenge in cooperative spectrum sensing is designing an effective incentive mechanism that encourages selfish SUs to actively collaborate with others. This strategic interaction can be effectively modeled using game theory (Kim, 2017b, 2023b; Shi, 2020).

The *MCSS* scheme adopts the learning game paradigm, which consists of learning model M and game model G. Through the M and G, multiple IIoT devices work as SUs, and they are sequentially interacted with each other to perfectly complete the cooperative spectrum sensing process. Formally, the *MCSS* scheme defines the tuple entities in the M and G, such as

$$\{\mathbb{M},\mathbb{G}\} = \left\{\mathbb{C},\mathbb{D},\left\{\mathscr{C}_{1\le i\le n}|\mathbb{M}_{C_i},\mathbb{G}_{C_i}\right\},\left\{\mathbb{M}_{C_i}|\mathscr{D}_j \in \mathbb{D}_{C_i},\mathfrak{M}_{C_i},a^{\mathscr{D}_j}_{1\le k\le l} \in \mathscr{L}_{\mathscr{D}_j},\mathscr{R}^{\mathscr{D}_j}_{a}\right\},\left\{\mathbb{G}_{C_i}|\mathscr{D}_j \in \mathbb{D}_{C_i},\mathfrak{M}_{C_i},S_{\mathscr{D}_j},\mathscr{U}_{\mathscr{D}_j}(\cdot)\right\},T\right\}$$

(Kim, 2023b).

- C and D represent the set of CCs and the set of IIoT devices, respectively.
- Each CC has its learning model $(\mathbb{M}_C)$ and game model $(\mathbb{G}_C)$.
- For the $C_i \in \mathbb{C}$, the $\mathbb{M}_{C_i}$ is developed as a *MARL* model for the $\mathscr{D}_j \in \mathbb{D}_{C_i}$ where $\mathbb{D}_{C_i}$ is the set of the C_i's corresponding IIoT devices. The $\mathbb{M}_{C_i}$ is operated in a distributed manner to learn the best action of $\mathscr{D}_j$.
- In the $\mathbb{M}_{C_i}$, $\mathfrak{M}_{C_i}$ is the spectrum band assigned to the PU. $\mathscr{L}_{\mathscr{D}_j}$ is the $\mathscr{D}_j$'s action set, which consists of total l actions $\left(a^{\mathscr{D}_j}_{1\le k\le l}\right)$. $\mathscr{R}^{\mathscr{D}_j}_{a}$ is the $\mathscr{D}_j$'s reward function with the joint action **a**.
- For the $C_i \in \mathbb{C}$, the $\mathbb{G}_{C_i}$ is designed as a cooperative game model for the devices in $\mathbb{D}_{C_i}$. The $\mathbb{G}_{C_i}$ is operated in a centralized manner.
- In the $\mathbb{G}_{C_i}$, the $\mathscr{D}_j \in \mathbb{D}_{C_i}$ is a game player, and $S_{\mathscr{D}_j}$ and $\mathscr{U}_{\mathscr{D}_j}(\cdot)$ are his strategy and utility function, respectively, to share the spectrum resource $\left(\mathfrak{M}_{C_i}\right)$.
- The $\mathbb{M}_{C_i}$ and $\mathbb{G}_{C_i}$ are reciprocally interdependent each other and work together in an iterative manner.
- Discrete time model $T \in \left\{t_1,\ldots,t_c,t_{c+1},\ldots\right\}$ is represented by a sequence of time steps. The length of t_c matches the event's timescale of $\mathbb{M}_{C_i}$ and $\mathbb{G}_{C_i}$.

5.6.4 Technical Concepts and Ideas of *NABS* and *MARL*

To characterize the idea of *NABS*, the following notations will be used. Let $\mathbb{R}^n_+$ be the n-fold Cartesian product of positive real numbers. A group of n agents, $i = 1,\ldots,n$, need to agree on a utility vector from a set of potential possibilities (called the bargaining set), $S \subseteq \mathbb{R}^n_+$. In case of disagreement, a disagreement point $d = [\ldots,0,\ldots] \in S$ will be implemented. Formally, a bargaining problem is a pair (S,d) where $d \in S$. Let $\sum^n$ be the set of all bargaining problems of the form (S,d). A bargaining solution is a function F: $\sum^n \rightarrow \mathbb{R}^n_+$ satisfying $\mathscr{F}(S,d) \in S$ for every $(S,d) \in \sum^n$; that is, given a bargaining problem (S,d), the solution F prescribes $\mathscr{F}(S,d)$. Simply, a bargaining problem $(S,d) \in \sum^n$ will be denoted by S from now on (Alós-Ferrer, 2018; Kim, 2023b).

For a given S, the utopia point $\mathscr{U}(S) \in \mathbb{R}^n_+$ is the point where each coordinate u_i contains the maximum conceivable outcome for agent i. Mathematically, the utopia points of i. i.e., $\mathscr{U}_i(S)$, is given by (Alós-Ferrer, 2018; Kim, 2023b);

$$\mathscr{U}_i(S) = \max\left\{u_i|\mathscr{U} \ge d \text{and} \mathscr{U} \in S\right\}, \text{ s.t.}, \mathscr{U} = \left[\ldots,u_i,\ldots\right] \tag{5.6.1}$$

To measure asymmetries in bargaining power, let $\mathscr{W} = \left[\mathscr{W}_1,\ldots,\mathscr{W}_n\right]$ be a vector of bargaining weights. The endogenous reference point, i.e., $\mathscr{U}^{\mathscr{W}}(S)$, is a key point for the *NABS*. In general, it might be in the interior of S, and respects the given bargaining weights. At the $\mathscr{U}^{\mathscr{W}}(S)$, each agent achieves a fraction of his utopian payoff which is proportional to the agent's bargaining power. The $\mathcal{U}^{\mathcal{W}}(S)$ is given by (Alós-Ferrer, 2018; Kim, 2023b);

$$\mathcal{U}^{\mathcal{W}}(S) = ((\mathcal{W}_1 \cdot \mathcal{U}_i(S)), \ldots, (\mathcal{W}_n \cdot \mathcal{U}_n(S))) \tag{5.6.2}$$

$$\text{s.t.}, \mathcal{W} = \left\{\mathcal{W} \in \mathbb{R}_+^n \mid \sum_{i=1}^{n} \mathcal{W}_i = 1\right\} \text{ and } \mathcal{U}^{\mathcal{W}}(S) \in S$$

Based on the $\mathcal{U}(S)$ and $\mathcal{U}^{\mathcal{W}}(S)$ points, the *NABS* can be obtained. As an anchor point, we start with the $\mathcal{U}(S)$ point and adjust it down to the $\mathcal{U}^{\mathcal{W}}(S)$ point while identifying the Pareto optimal point. Finally, the mathematical definition of *NABS*, i.e., $NABS(S, d, \mathcal{W})$, is given by (Alós-Ferrer, 2018; Kim, 2023b):

$$NABS(S,\ d, \mathcal{W}) = (\lambda^* \cdot \mathcal{U}(S)) + ((1 - \lambda^*) \cdot \mathcal{U}^{\mathcal{W}}(S)) \tag{5.6.3}$$

$$\text{s.t.}, \lambda^* = \max\{\lambda \in [0,1] \mid ((\lambda \cdot \mathcal{U}(S)) + ((1-\lambda) \cdot \mathcal{U}^{\mathcal{W}}(S))) \in S \in \textstyle\sum^n\}$$

Intuitively, the *NABS* minimizes the losses with respect to the utopia point, distributing it according to bargaining power. Therefore, we can think that the *NABS* is a weighted proportional losses solution, and it is the generalized *KSBS*; the original idea of *KSBS* corresponds to the particular case of the *NABS* (Kim, 2023b).

Traditionally, game theory has been closely associated with MASs. In contrast to single-agent reinforcement learning, the *MARL* is inherently linked to cooperative game theory, particularly in the study of multi-agent decision-making problems. A fundamental learning model for *MARL* is *Q*-learning, which has garnered significant attention over the past decade. In a multi-agent environment, each agent must not only consider the impact of its own actions but also account for the influence of other agents' decisions. From this perspective, agents are treated as joint-action learners, meaning they can observe all actions taken by any agent in the system. Consequently, *Q*-values are learned for every possible combination of actions among individual agents. To date, many *MARL* models have relied on precise state measurements and detailed tracking of other agents' actions. However, the general *Q*-learning formulation for the *MARL* processes is often overly complex for practical implementation. In real-world applications, this general approach is computationally infeasible due to the *Curse of Dimensionality*. In the multi-agent system designed for CIIoT control, a state representation is not required. Therefore, instead of employing a fully state-dependent *MARL* approach, the general multi-agent *Q*-learning model is simplified into a stateless version to enhance computational feasibility (Claus, 1998; Kim, 2023b; Morozs, 2016).

In the stateless *MARL* model M, we assume the *Q*-value of agent i, i.e., $Q_i\,(a_i, a_{-i})$, that provides an estimate of the value of performing joint action $a = (a_i, a_{-i})$. The sample $(a_i, a_{-i}), r$ is the experience obtained by the agent i where the joint action **a** was performed resulting in the reward r. Based on the $(a_i, a_{-i}), r$, the agent i updates its estimate $Q_i(a_i, a_{-i})$ as follows (Claus, 1998; Kim, 2023b):

$$Q_i(a_i, a_{-i}) = Q_i(a_i, a_{-i}) + \alpha \cdot (r - Q_i(a_i, a_{-i}))$$

(5.6.4)

where $\alpha \in [0,1]$ is the learning rate while governing to what extent the new sample replaces the current estimate. In the *MARL*, the primary objectives include ensuring the stability of learning dynamics and adapting to the changing behaviors of other agents. Stability, in this context, refers to the convergence of the learning process toward a coordinated equilibrium. To improve overall system performance during the *MARL* process, each agent is naturally inclined to favor better actions. While the theoretical understanding of this process remains limited, the Boltzmann strategy is one of the most widely used techniques for guiding agents toward a coordinated equilibrium. In this approach, each agent selects its next action probabilistically, with the likelihood of choosing a particular action being determined by its current estimate of that action's effectiveness (Busoniu, 2008; Claus, 1998; Kim, 2023b).

5.6.5 The spectrum sharing scheme for the cognitive IIoT paradigm

To develop the CIIoT spectrum sharing algorithms, the *MCSS* scheme constructs the learning ($\mathbb{M}$) and game ($\mathbb{G}$) models for each device. In the M, a single-state *MARL Q*-learning process is implemented in a distributed manner. From the viewpoint of each individual IIoT device, the Q value of the selected action is updated by receiving a reward, and what is the best action is gradually learned. In the *MCSS* scheme, the spectrum sensing activity of each device is discretely varied, and each activity level is defined as the device's action. Understandably, there is a trade-off for each action between sensing performance and sensing cost. Therefore, the main goal of IIoT devices is to maximize the spectrum sharing benefit while minimizing the sensing cost (Kim, 2023b).

In the stateless setting, the *MCSS* scheme assumes that the $\mathcal{D}_j$ device has his action set $\mathcal{L}_{\mathcal{D}} = \{a_1^{\mathcal{D}_j}, \ldots a_k^{\mathcal{D}_j} \ldots, a_l^{\mathcal{D}_j}\}$, which consists of its sensing participation levels. For example, the Q-value of $\mathcal{D}_j$'s k[th] action, i.e., $Q_{1\le k\le l}^{\mathcal{D}_j}\left(a_k^{\mathcal{D}_j}, a_{-\mathcal{D}_j}\right)$, provides an estimate of the value of performing the joint action $\mathbf{a} = \left(a_k^{\mathcal{D}_j}, a_{-\mathcal{D}_j}\right)$. The $\mathcal{D}_j$ updates its estimate $Q_k^{\mathcal{D}_j}(\cdot)$ value based on the experience sample a, $\mathcal{R}_a^{\mathcal{D}_j}$ where $\mathcal{R}_a^{\mathcal{D}_j}$ is the $\mathcal{D}_j$'s reward function of the joint action $\mathbf{a}$. The $\mathcal{R}_a^{\mathcal{D}_j}$ is defined based on the idea that the reward is assigned by considering its sensing contribution. In the *MCSS* scheme, the $\mathcal{R}_a^{\mathcal{D}_j}$ function is also used as the $\mathcal{D}_j$'s utility function, i.e., $\mathcal{U}_{\mathcal{D}_j}(\cdot)$, in the game model G. Therefore, the learning and game models are strongly connected to each other. Finally, the stateless *MARL Q*-function is defined as follows (Claus, 1998; Kim, 2023b).

$$Q_{1\le k\le l}^{\mathcal{D}_j}\left(a_k^{\mathcal{D}_j}, a_{-\mathcal{D}_j}\right) = Q_k^{\mathcal{D}_j}\left(a_k^{\mathcal{D}_j}, a_{-\mathcal{D}_j}\right) + \alpha \cdot \left(\mathcal{R}_a^{\mathcal{D}_j} - Q_k^{\mathcal{D}_j}\left(a_k^{\mathcal{D}_j}, a_{-\mathcal{D}_j}\right)\right)$$

(5.6.5)

$$\text{s.t., } a_{1\le k\le l}^{\mathcal{D}_j} \in \mathcal{L}_{\mathcal{D}_j} \text{ and } \mathcal{R}_a^{\mathcal{D}_j} = \mathcal{U}_{\mathcal{D}_j}(\cdot)$$

According to (5.6.5) and the Boltzmann strategy, the $\mathcal{D}_j$ can learn his best action for the cooperative spectrum sensing. Based on his action, the $\mathcal{D}_j$ can adaptively obtain the spectrum resource for its service. Usually, industrial equipments in the IIoT network are expected to support different application services. In the *MCSS* scheme, different application services over IIoT devices can be categorized into four data types according to their characteristics, i.e., type I, II, III and IV data. Based on the data type, multiple applications are grouped, and each CC assigns orthogonal spectrum sub-bands, i.e., $\mathfrak{M}^{I}_{C}$, $\mathfrak{M}^{II}_{C}$, $\mathfrak{M}^{III}_{C}$, and $\mathfrak{M}^{IV}_{C}$ for them. Then, each group is treated as a traffic unit via the NOMA technique. To achieve the strategic advantage within the CIIoT platform, each CC adaptively decides power levels for its corresponding IIoT devices within the same sub-band (Kim, 2020, 2023b).

To adaptively decide the devices' power levels, the *MCSS* scheme develops a cooperative game model G. In the $\mathbb{G}_{C_i}$, individual devices in the $\mathbb{D}_{C_i}$ share the $\mathfrak{M}_{C_i}$ in a centralized manner. In this game, the power level decision process is operated through the idea of *NABS*, and the *MCSS* scheme gets the $\mathbb{P}^{\mathbb{D}}_{C_i} = \mathscr{D}_j \in \mathbb{D}_{C_i}|\ldots,S_{\mathscr{D}_j},\ldots$; it is a $|\mathbb{P}^{\mathbb{D}}_{C_i}|$-dimensional power level vector for IIoT devices in the $\mathbb{D}_{C_i}$ where $S_{\mathscr{D}_j}$ is the power level of $\mathscr{D}_j$. Specifically, the $\mathbb{G}_{C_i}$ is divided into four sub games, i.e., $\mathbb{G}^{I}_{C_i}$, $\mathbb{G}^{II}_{C_i}$, $\mathbb{G}^{III}_{C_i}$, and $\mathbb{G}^{IV}_{C_i}$, based on the sub-band. Therefore, each sub game is operated separately. According to his data type $T \in \{I,II,III,IV\}$, the utility function for the $\mathscr{D}_j$, i.e., $\mathscr{U}^{T}_{\mathscr{D}_j}(\cdot)$, is defined as follows (Kim, 2023b):

$$\mathscr{U}^{T}_{\mathscr{D}_j}\left(\mathfrak{M}^{T}_{C_i},\mathbb{D}^{T}_{C_i},S_{\mathscr{D}_j},\mathbb{P}^{\mathbb{D}(T)}_{C_i},a^{\mathscr{D}_j}_{k},\theta^{T}_{\mathscr{D}_j},E^{t}_{\mathscr{D}_j}\right) = \mathscr{Q}_{\mathscr{D}_j} - \mathscr{C}_{\mathscr{D}_j} \tag{5.6.6}$$

$$\text{s.t.,}\begin{cases} \mathscr{Q}_{\mathscr{D}_j} = \left(\sigma \times \log\left(\mathscr{I}_{\mathscr{D}_j} + \kappa\right)\right) and \mathscr{C}_{\mathscr{D}_j} = \left(\exp\left(\zeta \times \left(\dfrac{\mathscr{E}^{\mathscr{D}_j} - E^{t}_{\mathscr{D}_j}}{\mathscr{E}^{\mathscr{D}_j}}\right)\right) - \phi\right) \\ \mathscr{I}_{\mathscr{D}_j} = \min\left[\left(\Upsilon \times \left(\dfrac{\mathfrak{M}^{T}_{C_i}}{\sum_{\mathscr{D}_k \in \mathbb{D}^{T}_{C_i}} \theta^{T}_{\mathscr{D}_k}}\right) \times \mathscr{H}\right), \eta\right] \\ H = S_{\mathscr{D}_j} + \left(S_{\mathscr{D}_j} - \left(\sum_{S_{\mathscr{D}_k} \in \mathbb{P}^{\mathbb{D}(T)}_{C_i}} S_{\mathscr{D}_k} / |\mathbb{P}^{\mathbb{D}(T)}_{C_i}|\right)\right) \end{cases}$$

where σ and κ are adjustment parameters for the $\mathscr{D}_j$'s outcome $\mathscr{Q}_{\mathscr{D}_j}$, and ζ and ϕ are adjustment parameters for the $\mathscr{D}_j$'s cost $\mathscr{L}_{\mathscr{D}_j}$. ϒ is an orthogonality factor for wireless communications, and η is a control factor. $\mathscr{E}^{\mathscr{D}_j}$ and $E^{t}_{\mathscr{D}_j}$ are the initial assigned energy amount and the currently remaining energy of $\mathscr{D}_j$, respectively. $\mathfrak{M}^{T}_{C_i}$, $\mathbb{D}^{T}_{C_i}$ and $\mathbb{P}^{\mathbb{D}(T)}_{C_i}$ are the sub-band, device set and power level vector, which are dedicated to the service type T devices in the C_i. $\theta^{T}_{\mathscr{D}_k}$ is the T type's service data generated from the $\mathscr{D}_k$ (Kim, 2023b).

According to (5.6.3) and (5.6.6), the idle spectrum sub-band $\left(\mathfrak{M}^{T}_{C_i}\right)$ is shared based on the idea of *NABS*. In the $\mathbb{G}^{T}_{C_i}$, the *MCSS* scheme defines the bargaining power of each device according to the spectrum sensing contribution, which is individually decided in the learning model M. For the $\mathbb{D}^{T}_{C_i}$, the bargaining power vector $\left(\mathbb{W}^{T}_{C_i}\right)$ of each devices is defined as follows (Kim, 2023b);

$$\mathbb{W}^{T}_{C_i} = \mathbb{W}^{T}_{C_i} \in \mathbb{R}^{|\mathbb{D}^{T}_{C_i}|}_{+}, \mathscr{D}_j \in \mathbb{D}^{T}_{C_i}|\left[\ldots,\mathscr{W}_{\mathscr{D}_j},\ldots\right] \tag{5.6.7}$$

$$\text{s.t., } \mathscr{W}_{\mathscr{D}_j} = \frac{a^{\mathscr{D}_j}}{\sum_{\mathscr{D}_k \in \mathbb{D}^{T}_{C_i}} a^{\mathscr{D}_k}} \text{ and } \sum_{\mathscr{D}_k \in \mathbb{D}^{T}_{C_i}} \mathscr{W}_{\mathscr{D}_k} = 1$$

where $a^{\mathscr{D}_k}$ is the $\mathscr{D}_k$'s selected action in the M. Finally, the allocated spectrum amount for the $\mathscr{D}_j$, i.e., $NABS_{\mathscr{D}_j}(\cdot)$, is given by (Kim, 2023b):

$$NABS_{\mathscr{D}_j}\left(\mathscr{D}_j \in \mathbb{D}_{C_i}^T | \mathbb{U}, d_{\mathbb{D}_{C_i}^T}, \mathbb{W}_{C_i}^T\right) = (\lambda^* \cdot \mathscr{U}^T(\mathbb{U})) + ((1-\lambda^*) \cdot \mathscr{U}_T^{\mathscr{W}}(\mathbb{U})) \tag{5.6.8}$$

$$\text{s.t., } \lambda^* = \begin{cases} U = \mathscr{D}_k \in \mathbb{D}_{C_i}^T | \ldots, \mathscr{U}_{\mathscr{D}_k}^T(\cdot), \ldots \\ \mathscr{U}_T^{\mathscr{W}}(\mathbb{U}) = \left(\ldots, \left(\mathscr{W}_{\mathscr{D}_k} \cdot \mathscr{U}_{\mathscr{D}_k}^T(\cdot)\right), \ldots\right) \in U \\ \lambda^* = \max\left\{\lambda \in [0,1] | \left((\lambda \cdot \mathscr{U}^T(\mathbb{U})) + ((1-\lambda) \cdot \mathscr{U}_T^{\mathscr{W}}(\mathbb{U}))\right) \in \mathbb{U} \in \sum^{|\mathbb{D}_{C_i}^T|}\right\} \end{cases}$$

where $\mathscr{U}_T^{\mathscr{W}}(\mathbb{U})$ is a reference point in the interior of U, and $d_{\mathbb{D}_{C_i}^T}$ is the disagreement points for devices in the $\mathbb{D}_{C_i}^T$.

5.6.6 Summary

The *MCSS* scheme addresses the key research challenge of efficiently sharing limited spectrum resources within a NOMA-based CIIoT system platform. By integrating spectrum sensing and access control algorithms, its primary objective is to maximize overall CIIoT system performance. To achieve this, the *MCSS* scheme employs a *MARL* process to optimize spectrum sensing strategies, while utilizing NABS to facilitate spectrum resource allocation. It is important to highlight the strong interconnection between MASs and game theory, as they are mutually dependent disciplines. In the *MCSS* scheme, the learning and bargaining algorithms are jointly integrated, working cooperatively to balance spectrum efficiency and service fairness. Additionally, the *MCSS* scheme dynamically blends both centralized and distributed control methods to reduce system complexity. From a practical standpoint, this hybrid approach is well-suited for real-world CIIoT system management, ensuring scalability, efficiency, and adaptability in dynamic industrial environments (Kim, 2023b).

5.7 Learning based Service Quality Control (*LSQC*) scheme

In the modern era, the Internet continues to evolve, increasing in significance and generating new value through its expansion and enhanced utilization. The IoT represents a new paradigm within the framework of the future Internet, gaining widespread adoption as a dynamic and globally interconnected network infrastructure. However, the IoT ecosystem presents challenges in meeting diverse QoS requirements, as well as in facilitating rapid service composition and deployment. To address these challenges, S. Kim recently introduces the *Learning-based Service Quality Control* (*LSQC*) scheme for IoT systems (Kim, 2015). Leveraging the Markov game model, the *LSQC* scheme effectively allocates IoT resources while optimizing overall system performance. In multi-agent environments, game theory serves as a powerful decision-making framework, offering an effective solution to resource allocation challenges in IoT networks.

5.7.1 Development motivation

Over the past four decades, the Internet has expanded into a vast network, now connecting approximately 1.8 billion users and achieving a global penetration rate of nearly 25%. Additionally, the convergence of telecommunications and the Internet has led to a highly integrated system for processing, storing, accessing, and distributing information, as well as content management. This technological integration is driven by the rapid advancements in digital technology and the widespread adoption of Internet-based systems. In recent years, significant progress has been made in the penetration of digital technologies, alongside the evolution of integrated telecommunications, information technology, and electronic media. These developments have fostered a vast and innovative technological landscape, facilitating advancements in the production, distribution, and consumption of information services (Ashton, 2009; Kim, 2015; Sallai, 2013).

In 1999, K. Ashton introduced the concept of the IoT (Ashton, 2009), marking the beginning of a technological revolution that has ushered in an era of ubiquitous connectivity, computing, and communication. The IoT vision extends the Internet into everyday life, creating a wireless network of uniquely identifiable smart objects. The continued development of IoT relies on dynamic technological advancements across multiple fields, including wireless sensors and nanotechnology (Wu, 2014). Additionally, the IoT service infrastructure is expected to assess QoS in real-time and deliver user-centric services by considering user preferences, device capabilities, and real-time network conditions. However, defining QoS in IoT environments remains a challenge due to limited research in this domain. To ensure effective and adaptive IoT system management, a new QoS control model is required - one that can balance network availability with information accuracy for optimal data delivery (Jang, 2013; Kim, 2012a, 2015; Li, 2014; Pujolle, 2013; Zhang, 2012).

A key challenge in QoS management is the judicious selection and allocation of scarce network resources to maximize system performance (Jang, 2013; Pujolle, 2013). To address this, game theory has been widely adopted for solving mission-critical network management problems. Game theory is commonly used to analyze strategic interactions where individual players select optimal actions to maximize their payoffs, while accounting for the decisions of others. Consequently, it serves as a powerful framework for modeling and analyzing complex interactions among intelligent, rational entities in conflict-driven environments (Kim, 2012a, 2015).

In conventional game-theoretic models, defining equilibrium strategies is a critical component of solving game scenarios. These equilibrium strategies assume that each player optimally reacts to the actions of others, given complete knowledge of payoffs and strategies. However, most equilibrium concepts require that players fully observe and anticipate the actions of others in advance - an assumption that rarely holds in real-world scenarios. In practice, players often possess only partial or no knowledge of their environment or competing agents, making traditional game-theoretic assumptions impractical for many real-world applications (Kim, 2012a). To overcome these limitations, J. van der Wal introduced the Markov game model (Van der Wal, 1977), which relaxes the strict assumptions of traditional game theory by incorporating learning algorithms. In this framework, players continuously learn and refine their strategies through repeated interactions, leveraging both current payoffs and historical observations of other players' behaviors. This adaptive learning mechanism enhances decision-making in dynamic and uncertain environments (Kim, 2015; Van der Wal, 1977; Vrancx, 2008).

5.7.2 Contribution

The primary objective of the *LSQC* scheme is to establish an effective QoS control mechanism for IoT systems. Utilizing the Markov game model, the *LSQC* scheme develops an intelligent decision-making framework to tackle critical QoS challenges in IoT environments. By incorporating a real-time learning feedback mechanism, the *LSQC* scheme dynamically adapts to the evolving requirements of IoT applications. Through online-oriented strategic decision-making, it aims to achieve a self-confirming equilibrium, a novel solution concept tailored for real-time network systems (Kim, 2015).

5.7.3 Markov Game Model for IoT Systems

Network services operate based on the OSI model, and the *LSQC* scheme implements a three-layered QoS architecture, comprising the application, network, and sensing layers. At the application layer, applications are selected to establish network connections, with decision-making handled by both the user and the QoS scheduling engine. Generally, the QoS module is responsible for allocating network resources to services that are selected at the application layer (Li, 2014). At the network layer, the QoS module manages the allocation of network resources to the selected services. The decision-making process at this layer may involve QoS attributes commonly used in traditional network QoS mechanisms (Li, 2014). At the sensing layer, decision-making at this level focuses on selecting the fundamental sensing infrastructure based on sensing capabilities and the QoS requirements of applications. The QoS module in the sensing layer is responsible for choosing appropriate sensing devices (Kim, 2015; Li, 2014).

The *LSQC* scheme explores learning algorithms that operate under uncertain, dynamic, and incomplete information, developing a novel adaptive QoS scheduling algorithm that employs an intelligent decision-making process for IoT systems. To facilitate interactive decision-making among IoT system agents, the *LSQC* scheme formulates a multi-agent decision-making process using a game-theoretic model, incorporating multi-agent learning techniques. This approach enables the *LSQC* scheme to significantly enhance QoS management within IoT environments (Kim, 2015).

Learning in this context refers to the ability to make intelligent decisions by self-adapting to environmental dynamics, leveraging past and present experiences, and optimizing long-term benefits. This concept can be likened to self-play, where either an individual agent or an entire population of agents evolves through repeated competitive interactions. During IoT system operations, learning processes are driven by the availability of information from each QoS scheduler (Imran, 2012). Traditional approaches demonstrate that complete information leads to improved system performance over partial observability. However, the associated control overhead often limits practical implementation. Consequently, a tradeoff must be carefully considered, particularly in self-organizing IoT systems, where the ability to make autonomous decisions is a highly desirable characteristic (Imran, 2012; Kim, 2015; Li, 2014).

The Markov decision process (MDP) is a well-established mathematical framework for solving sequential decision-making problems based on probabilistic modeling. It represents a decision-making system in which an action must be selected at each state, with probabilistic outcomes influencing state transitions. The primary goal of an MDP is to determine an optimal policy that prescribes the best action for each state. By employing a learning-based Markov game approach, the *LSQC* scheme enables distributed QoS schedulers to progressively learn an optimal strategy, refining their decisions step by step. Within each decision step, a repeated game strategy is applied to ensure cooperation among QoS schedulers. The Markov decision process can be naturally extended to multi-agent environments, resulting in the multi-player Markov game framework. In a Markov game, actions result from the joint action selection of all participating players, with payoffs and state transitions being determined by these joint decisions. As a result, payoffs are derived from the combined actions of multiple players, allowing them to learn within a joint action space. Using the obtained data, players can adapt to dynamic environments, enhance their decision-making capabilities, and deepen their understanding of fundamental system behaviors (Kim, 2015; Li, 2014; Van der Wal, 1977; Vrancx, 2008).

In the QoS control algorithm, the game model is defined as a tuple $S, N, \mathbb{A}_{k,1\leq k\leq N}, U_{k,1\leq k\leq N}, \mathcal{T}$ where S is the set of all possible states and N is the number of all players. In the model, each state means the resource allocation status in the IoT system. $\mathbb{A}_{k,1\leq k\leq N} = \{a_1, a_2, \ldots, a_m\}$ is the collection of strategies for player k where m is the number of possible strategies. Actions are the joint result of multiple players choosing a strategy individually. In the Markov game, QoS schedulers are assumed as game players and the collection of strategies for each player is the set of possible amounts of system resources. $U_{k,1\leq k\leq N}$: $S \times \mathbb{A}_1 \times \mathbb{A}_2 \times \ldots \times \mathbb{A}_N \rightarrow \mathfrak{R}$ is the utility function where N represents the set of real numbers. $T{:}S \times \mathbb{A}_1 \times \mathbb{A}_2 \times \ldots \times \mathbb{A}_N \rightarrow \Delta(S)$ is the state transition function where $\Delta(S)$is the set of discrete probability distributions over the set **S**. Therefore, $\mathcal{T}(s^t, a_1, a_2, \ldots, a_N, s^{t+1})$ gives the probability of arriving in state s^{t+1} when each agent takes an action $a_i$at state s^t, where s^t, $s^{t+1} \in \mathbf{S}$ (Kim, 2015; Li, 2014; Van der Wal, 1977; Vrancx, 2008).

In the developed game model, players seek to choose their strategy independently and self-interestedly to maximize their payoffs. Each strategy represents a large amount of system resources, and the utility function measures the outcome of this decision. Therefore, different players can receive different payoffs for the same state transition. By considering the allocated resource amount, delay and price, the utility function (U) of each player is defined as follows (Kim, 2015).

$$U(x) = \omega \times \exp\left(\left(\frac{\hat{O}(x)}{\mathbb{T}}\right)^{0.5-\epsilon} \times \left(\tau_M - \tau\right)^{0.5+\epsilon}\right) - c(x,\xi), s.t., \epsilon \in \{-0.5; 0.5\}$$

(5.7.1)

where ω represents player's willingness to pay for his perceived service worth. $\mathbb{T}$ is the system's average throughput and $\hat{O}(x)$ is player's current throughput with the allocated resource x. It is the rate of successful data delivery over a communication channel. $\tau_M$and τ are the maximum delay and the observed delay of application services, respectively. τ is measured from real network operations. Based on the real-time online manner, each QoS scheduler actually measures the $\mathbb{T}$, $\hat{O}(x)$ and τ. $c(x, \xi)$ is the cost function and ξ is the price for resource units. τ is obtained according to the processing and arrival service rates. In a distributed self-regarding fashion, each player (i.e., QoS scheduler) is independently interested in the sole goal of maximizing his utility function as follows (Kim, 2015).

$$\max_x U(x), where\, c(x,\xi) = \left(\frac{x \times \xi}{\overline{x}_A \times \xi}\right)^q$$

(5.7.2)

where$x$ is the allocated resource in its own QoS scheduler, and $\overline{x}_A$ is the average resource amount of all QoS schedulers and q is a cost parameter for the cost function $c(x, \xi)$ The cost function is defined as the ratio of its own obtained resource to the average resource amount of all QoS schedulers. Therefore, other players' decisions are the input back to each player. This iterative feedback procedure continues under IoT system dynamics. In the *LSQC* scheme, QoS schedulers can modify their actions in an effort to maximize their $U(x)$ in a distributed manner. This approach can significantly reduce computational complexity and control overheads. Therefore, it is practical and suitable for the real-world system implementation (Kim, 2015).

5.7.4 Markov Decision Process for QoS Control Problems

The *LSQC* scheme investigates how a player (i.e., QoS scheduler) in a dynamic IoT system learns the uncertain network situation and makes a control **decision** by taking into account the online feedback mechanism. With an iterative learning process, players' decision-making mechanism is developed as a Markov game model, which is an effective way for players' decision mechanism. Therefore, if players change their strategies, the system state may change. Based on the immediate payoff $(U(S_0, a_i(0)))$ of current state S_0 and action $a_i(0)$, players should take into account the future payoffs. With the current payoff, player i's long-term expected payoff $(V_i(S_0, a_i(0)))$ is given by (Kim, 2015; Li, 2014).

$$V_i(S_0, a_i(0)) = \max_{a_i(t),\, 0 \le t \le \infty} \left[U_i(S_0, a_i(0)) + \sum_{t=1}^{\infty} (\beta^t \times U_i(S_t, a_i(t))) \right] \tag{5.7.3}$$

$\text{s.t.}, a_i(t) \in \mathbb{A}_i$

where $a_i(t)$, $U_i(S_t, a_i(t))$ are player i's action and expected payoff at time t, respectively. β is a discount factor for the future state. During the game operations, each combination of starting state, action choice and next state has an associated transition probability. Based on the transition probability, the equation (5.7.3) can be re-written by a recursive Bellman equation form given in (Kim, 2015; Li, 2014).

$$V_i(S) = \max_{a_s} \left[U_i(S,\ a_s) + \gamma \times \sum_{S' \in S} \left\{ P^i(S'|S,\ a_s) \times V_i(S') \right\} \right],\ \text{s.t.}, a_s \in \mathbb{A}_i \tag{5.7.4}$$

where S' represents all possible next states of S and γ can be regarded as the probability that player keeps staying at the selected strategy. $P^i(S'|S, a_s)$ is the state transition probability from the state S to the state S' (Kim, 2015).

$P(S'|S, a_s)$ definition is a distributed multi-player probability decision problem. Through the multi-player learning algorithm, each player independently learns the current IoT system situation to dynamically decide the $P(S'|S, a_s)$. This approach can effectively control a Markov game process with unknown transition probabilities and payoffs. In the *LSQC* scheme, each player is assumed to be interconnected by letting them play in a repeated game with the same environment. Suppose there is a finite set of strategies $\mathbb{A}_{1 \le k \le N}(t) = \{a_1^k(t) \ldots a_m^k(t)\}$ chosen by player k at the game iteration t; m is the number of possible strategies. Correspondingly, the $U^k(t) = (u_1^k(t) \ldots u_m^k((t))$ is a vector of specifying payoffs for player k. If player k plays the action $a^k_{l, 1 \le l \le m}$, it earns a payoff $u^k_{l, 1 \le l \le m}$ with probability p_l^k. Over these actions, $P^k(t) = \{p_1^k(t), \ldots, p_m^k(t)\}$ is defined as player k's probability distribution (Kim, 2015).

Actions chosen by players are given as input to the environment, and the environmental response to these actions serves as an input to each player. Therefore, multiple players are connected in a feedback loop with its environment. When a player selects an action with his respective probability distribution $P(\cdot)$, the environment produces a payoff $U(\cdot)$ according to (1). Therefore, $P(\cdot)$ should be adjusted adaptively in order to cope with the payoff fluctuation. At every game round, all players update their probability distributions based on the online responses of the environment. If player k chooses a_l^k at the time t, this player updates $P^k(t+1)$ as follows (Kim, 2015).

$$p_j^k(t+1) = \begin{cases} f\left(p_j^k(t) + \psi \times \left[\frac{u_l^k(t) - u_l^k(t-1)}{u_l^k(t-1)}\right]\right), if j = l \\ \varphi \times p_j^k(t), if j \neq l \end{cases} \tag{5.7.5}$$

$$\text{s.t.}, \begin{cases} f(\chi) = 0, if \chi < 0 \\ f(\chi) = \chi, if 0 < \chi < 1 \\ f(\chi) = 1, if \chi > 1 \end{cases}$$

where φ is a discount factor and ψ is a parameter to control the learning size from $p(t)$ to $p(t+1)$. In general, small values of ψ correspond to slower rates of convergence, and vice versa. According to (5.7.5), $P^k(S'|S, a_S)$ is defined based on the Boltzmann distribution (Kim, 2015).

$$P^k(S'|S, a_S) = \frac{\exp\left(\left(\frac{1}{\lambda}\right) \times p_{a_s}^k(t)\right)}{\sum_{j \in \mathbb{A}_k} \exp\left(\left(\frac{1}{\lambda}\right) \times p_j^k(t)\right)} \tag{5.7.6}$$

$$\text{s.t.}, a_S \in \mathbb{A}_k(t) = \{a_1^k(t) \ldots a_m^k(t)\}$$

where λ is a control parameter. Strategies are chosen in proportion to their payoffs, but their relative probability is adjusted by λ. Value of λ close to zero allows very little randomization, and large value of λ results in complete randomization (Kim, 2015).

5.7.5 Summary

In today's world, IoT-based services and applications have become an integral part of everyday life. It is expected that IoT technology will play a crucial role in the future of the Internet, where interconnected 'things' will form a global wireless network, enabling dynamic services for both applications and users. By seamlessly integrating the virtual network with physical devices, IoT technology acts as a bridge between the digital and real worlds. The innovative adoption of IoT techniques not only enhances data accessibility for users but also introduces challenges related to retrieving heterogeneous application data, particularly in the dynamic environment of real-time IoT systems. The *LSQC* scheme is designed to address QoS control challenges in IoT environments. By leveraging a learning-based Markov game model, QoS schedulers iteratively observe system conditions, adjust their strategies, and optimize resource management. Through a step-by-step feedback process, the *LSQC* scheme incrementally refines its decisions, ensuring optimized system performance in a fully distributed manner. One of the key innovations of the *LSQC* scheme lies in its adaptability and responsiveness, allowing it to effectively adjust to real-time system conditions and maintain optimal QoS performance (Kim, 2015).

5.8 Asymptotic Shapley Value based Allocation (*ASVA*) scheme

Driven by advancements in modern communication technologies, the focus of society has shifted from basic network connectivity to pervasive information services. As a result, the IoT has experienced rapid growth. To enable efficient IoT services, spectrum resources must be dynamically allocated to meet service demands while ensuring optimal performance. Recently, S. Kim introduced the *Asymptotic Shapley Value-based Allocation* (*ASVA*) scheme, which addresses resource allocation challenges by leveraging cooperative game theory (Kim, 2016a). In this game-theoretic model, the *Shapley value* concept is extended and practically implemented to develop a novel bandwidth allocation algorithm that ensures fair and efficient resource distribution.

5.8.1 Development motivation

We are at the threshold of a new era characterized by true ubiquitous computing and communication, where society is increasingly adopting an 'always connected' paradigm. With the advent of emerging technologies, network coverage is expanding, and the number of smart objects capable of network connectivity is rapidly growing. This pervasive connectivity, referred to as the IoT, enhances the value of information generated by the large-scale interconnections between humans and smart devices. Consequently, IoT is expected to address various real-world challenges, enabling seamless connectivity between people and objects, anytime, anywhere, with anyone or anything (Kim, 2016a; Ortiz, 2014).

The next generation of IoT aims to deliver higher QoS across an increasingly diverse range of applications. However, wireless bandwidth remains a critical yet limited resource within the entire IoT ecosystem. Additionally, the demand for multimedia services - which include a variety of service classes with distinct traffic characteristics - has been rising. Each of these services requires different QoS guarantees, making bandwidth management even more complex. Given the unpredictable nature of bandwidth requirements, achieving efficient and dynamic bandwidth allocation for multimedia communications remains a significant challenge (Kim, 2010, 2016a; Luckshmi, 2011).

Recent research has explored the use of game theory in designing bandwidth allocation mechanisms. Game theory, which studies strategic interactions involving both conflict and cooperation, provides a mathematical framework to model multi-agent decision-making processes. It enables analysis of how participants interact and how their decisions impact overall system behavior. As a result, game theory offers a structured methodology for formulating, analyzing, and interpreting strategic resource allocation scenarios. Today, game-theoretic principles apply to a wide spectrum of behavioral models, encompassing both human decision-making and non-human agents, such as smart devices, computers, and network components (Kim, 2014, 2016a).

In network environments where multiple operations compete for resources, the cooperation of network agents is essential. Thus, cooperative game models present an effective approach to addressing the network resource allocation problem. In cooperative game theory, the primary objective is to fairly distribute shared resources based on each participant's contribution. To enforce fairness and adherence to cooperation, legal frameworks or predefined rules are necessary to ensure that all players honor their commitments. Among the most widely recognized cooperative game solutions are the *Shapley Value* (*SV*) and bargaining solutions, both of which provide equitable allocation mechanisms. These solutions ensure that collective benefits derived from collaborative interactions are fairly distributed among all participating entities (Kim, 2014, 2016a).

5.8.2 Contribution

The *ASVA* scheme explores the application of game theory to solve the bandwidth allocation challenge in IoT systems. By leveraging the *SV* concept, the *ASVA* scheme introduces a novel two-stage game model designed to ensure fair distribution of limited bandwidth resources. At first stage, bandwidth is allocated to each access point based on the fundamental principles of the asymptotic *SV*. At second stage, the allocated bandwidth is further distributed among various multimedia services, following the relative utilitarian value approach. In response to dynamic changes in IoT environments, this hierarchical game-theoretic framework enhances system performance, ensuring both efficiency and fair resource allocation (Kim, 2016a).

5.8.3 Game Model for the IoT Spectrum Resource Allocation

Smart IoT devices are widely utilized across various applications and are increasingly becoming an essential part of daily life. The IoT concept revolves around the seamless integration of numerous smart devices and machines with the Internet. To meet the demands of IoT applications, these devices must be capable of efficiently and reliably reporting various events and transmitting streaming data to a central server over an extended period. The *ASVA* scheme operates within a multi-tier hierarchical wireless network system. A fundamental aspect of multi-tier wireless networking is the integration of multiple heterogeneous networks - each with different coverage areas - into a single unified network, thereby offering optimal coverage. Typically, a geographic region is divided into two coverage zones; global coverage areas, served by a macro-cell network and local hotspot areas, covered by a micro-cell network. These heterogeneous networks coexist within a hierarchical system, featuring overlapping coverage zones to ensure seamless service provisioning (Kim, 2016a; Lohi, 1999; Ning, 2006).

The *ASVA* scheme assumes that there is one Base Station (*BS*) for a macro cell and multiple Access Points (*APs*) are deployed for each micro cell. The *BS* is responsible for the macro cell resource control, while distributing the bandwidth over multiple *APs*. Let there be K *APs* (M=$\{AP_1,AP_2,\dots,AP_K\}$) and L IoT devices. With the introduction of *APs*, the communications between the *BS* and an IoT device can be realized by either a direct link (*BS*-IoT device) or a two-hop link (*BS*-*AP*-IoT device) via an *AP*. $\mathfrak{B}_i^{AP}$ and $\mathfrak{B}^{BS}$ are the allocated bandwidth amount at the AP_i, where $1 \le i \le K$, and at the *BS*, respectively. B is the total amount of bandwidth resource, where $\mathscr{B} = \left\{\sum_{i=1}^{K}\left(\mathfrak{B}_i^{AP}\right) + \mathfrak{B}^{BS}\right\}$ (Kim, 2016a).

An *AP* covers a relatively small area, referred to as a micro cell. In each *AP*, multimedia data applications are serviced while it is ensured that QoS requirements are met. Usually, *APs* are situated around high traffic-density hot spots to improve communication capacity. When the requested bandwidth (∂_i) in the AP_i is less than the allocated bandwidth ($\mathscr{A}_i$), i.e., $\partial_i < \mathscr{A}_i$, the AP_i can sell this excess bandwidth to the system at the system unit price ($\mathscr{P}_u$). In this case, the value function ($v(AP_i)$) of the AP_i becomes $v(AP_i) = -\mathscr{P}_u \times (\mathscr{A}_i - \partial_i)$. Conversely, if $\partial_i > \mathscr{A}_i$, the deficient bandwidth amount $(\partial_i - \mathscr{A}_i)$ can be purchased from the system and the value function becomes $v(AP_i) = \mathscr{P}_u \times (\partial_i - \mathscr{A}_i)$. A pair (

$\mathcal{N} = \mathcal{M} \cup \{BS\}, v(\cdot))$ represents a cooperative game, where N is a finite set of players and $v(\cdot)$ is a real valued function defined on all subsets of N satisfying $v(\emptyset) = 0$. Therefore, in the game model, *APs* and the *BS* become game players, and a nonempty subset S of N is called a *coalition* (Kim, 2016a).

In 1953, L. Shapley introduced a solution concept for coalitional games, now widely recognized as the *Shapley Value* (*SV*). The *ASVA* scheme incorporates the core principles of SV to dynamically allocate bandwidth among *APs*. The *SV* determines a unique distribution of the total surplus generated through coalition formation among multiple players. A defining characteristic of *SV* is its ability to provide a single, fair solution for an n-player cooperative game. For instance, when a group of players forms a coalition, they collectively generate a total gain from cooperation. However, as some players contribute more than others, it becomes necessary to equitably distribute the resulting surplus among them. The *SV* serves as a mathematical solution for fair resource allocation in cooperative game scenarios (Han, 2011; Kamijo, 2009; Kim, 2014, 2016a).

To compute *SV*, let us define the value function (v) over the real line like as $v:2^{\mathcal{N}} \rightarrow \mathbb{R}$ with $v(\emptyset) = 0$, and characterize unique mapping $\phi = [\phi_1,\ldots,\phi_i,\ldots,\phi_n]$ that is the *SV*; ϕ_i is the payoff given to player *i* by the *SV* ϕ. Shapley showed that there exists a unique mapping, the *SV*, from the space of all coalitional games to$\mathbb{R}^{\mathcal{N}}$. Based on these properties, the *SV* can be obtained by considering the payoff depending on the order that player joins the coalition. In particular, the *SV* is the average payoff to a player if players enter into the coalition in a completely random order (Han, 2011; Kim, 2014). The *SV* $\phi = [\phi_1,\ldots,\phi_i,\ldots,\phi_n]$ can be computed as follows (Kim, 2016a):

$$\phi_i(\nu) = \sum_{\mathbb{S}\subset\mathbb{A}, i\in\mathbb{A}} \frac{(|\mathbb{S}| - 1)!(n - |\mathbb{S}|)!}{n!}(\nu(\mathbb{S}) - \nu(\mathbb{S} - \{i\})) \tag{5.8.1}$$

where $|\mathbb{S}|$ indicates the number of players in the set S. ν(S) is the minimum payoff which the coalition S can guarantee its members (i.e., players), and $\nu(\mathbb{S} - \{i\})$ is the payoff secured by a coalition with the same members in S except player *i*. In the *SV*, all the coalitions are regarded equal which means they have the same probability to appear. The first part of the formula can be interpreted as the probability of a coalition containing player *i* with the size of $|\mathbb{S}|$. The second part is the payoff difference between the coalitions with and without player *i*, which measures the contribution of player *i* to the coalition. The bigger the difference, the more player contributes to its coalition, then the more payoff he should earn (Han, 2011; Kim, 2014; Kim, 2016a).

In the *ASVA* scheme, a set of games with a finite number of players is denoted by Γ. Given a game $(\mathcal{N}, v(\cdot)) \in \tilde{A}$, let $\mathbb{C} = \{\mathscr{C}_1,\ldots,\mathscr{C}_j\}$ be a partition of N, that is, $\mathscr{C}_f \cap \mathscr{C}_h = \emptyset$ for $f \neq h$ and $\cup^j_{k=1} \mathscr{C}_k = \mathcal{N}$. Then C is called a *coalition structure* of N. Let θ be an order on N, that is, θ is a bijection on N. A set of all the orders on N is denoted by È($\mathcal{N}$) (Kamijo, 2009; Lee, 2014). A set of game players preceding to player i (i.e., $i \in \mathcal{N}$) at order θ is $\mathfrak{A}^\theta_i = \{j \in \mathcal{N}:\theta(j) < \theta(i)\}$. Therefore, $v(\mathfrak{A}^\theta_i)$ can be expressed as (Kim, 2016a);

$$v(\mathfrak{A}^\theta_i) = \mathscr{P}_u \times \left(\sum_{l\in\mathfrak{A}^\theta_i} \partial_l - \sum_{l\in\mathfrak{A}^\theta_i} \mathscr{A}_l\right) \tag{5.8.2}$$

A marginal contribution of player i at order θ in $(\mathcal{N}, v(\cdot))$ is defined by $\mathcal{S}_i^{\theta}(\mathcal{N}, v) = v(\mathfrak{A}_i^{\theta} \cup \{i\}) - v(\mathfrak{A}_i^{\theta})$. Then the *SV* of $(\mathcal{N}, v(\cdot))$ is defined as follows (Kamijo, 2009; Kim, 2016a):

$$SV_i(\mathcal{N}, v) = \frac{1}{|\grave{E}(\mathcal{N})|} \times \sum_{\theta \in \grave{E}(\mathcal{N})} \left(\mathcal{S}_i^{\theta}(\mathcal{N}, v)\right), \text{ for all } i \in \mathcal{N} \tag{5.8.3}$$

where $\cdots$ represents the cardinality of the set. Therefore, the *SV* is an average of marginal contribution vectors where each order $\theta \in \grave{E}(\mathcal{N})$ occurs in equal probability, that is, $1/|\grave{E}(\mathcal{N})|$ (Kim, 2016a).

Although the *SV* is quite an interesting concept and provides an optimal and fair solution for many applications, its main drawback is its computational complexity: the number of computations will increase prohibitively when the number of game players increases. Therefore, applications that utilize the *SV* remain scarce. In the *ASVA* scheme, if all possible orderings of *APs* ($\grave{E}(\mathcal{N})$) have to be taken into account in calculating equation (5.8.2), the computational complexity of calculating the *SV* can be very high and too heavy to be implemented in real network operations. To resolve this problem, the *ASVA* scheme adopts the new concept of *Asymptotic Shapley Value* (*A_SV*) approach, which is an approximation method for the *SV* under a large number of players (Kamijo, 2009; Kim, 2016a; Lee, 2014).

Consider the game players (N) who trade an amount of bandwidth $\lambda = (\mathcal{A} - \partial)$. According to the ratio of the average amount of bandwidth trading IoT systems, let the *A_SV* of player i be ϕ_i; it is given in (Kim, 2016a; Lee, 2014).

$$\phi_i = \begin{cases} \left(\mathcal{P}_u \times \int_0^1 \operatorname{erf}\left(\frac{\sqrt{\mathcal{P}_u} \times \tau}{\sqrt{2} \times \eta}\right) dp\right) \times \lambda_i, if \frac{\mu_S \times N_S}{\mu_B \times N_B} = 1 \\ \mathcal{P}_u \times \lambda_i, if \frac{\mu_S \times N_S}{\mu_B \times N_B} \neq 1 \end{cases} \tag{5.8.4}$$

$$\text{s.t.}, \operatorname{erf}(x) = \frac{1}{\sqrt{|\grave{E}(\mathcal{N})|}} \int_{-x}^{x} e^{-t^2} dt, \eta = \sqrt{\frac{\mu_B \times \sigma_S^2 + \mu_S \times \sigma_B^2}{\mu_B + \mu_S}}, \tau = \frac{\mu_S \times N_S - \mu_B \times N_B}{\sqrt{\mathcal{N}}}$$

where N_S and N_B are the number of players who have surplus bandwidth (i.e., $\mathcal{A} - \partial > 0$) and the number of players who need additional bandwidth (i.e., $\mathcal{A} - \partial < 0$), respectively. μ_S and μ_B (*or* σ_S^2 and σ_B^2) are the mean (*or* variance) of total surplus and requirable bandwidth amounts, respectively (Lee, 2014). The equation (5.8.1), (5.8.3) looks very simply. However, as the number of players increases, computation complexity also increases extremely. The equation (5.8.4) does seem very complicated, but all computation can be completed within a polynomial time (Kim, 2016a).

Under diverse IoT traffic environments, fixed bandwidth allocation methods cannot effectively adapt to changing network conditions. The *ASVA* scheme treats the bandwidth allocation for *APs* as an on-line decision problem. Therefore, according to ϕ values, B is dynamically reallocated over *APs*, periodically. In order to apply the time-driven implementation of the bandwidth reallocation, the *ASVA* scheme partitions the time-axis into equal intervals of length *unit_time*. At the end of each *unit_time* period, the reallocated bandwidth amount for the AP_i (Π_i^{AP}) is obtained as follows (Kim, 2016a).

$$Ð_i^{AP} = \mathscr{B} \times \frac{\phi_i + \left|\min_{j\in\mathcal{N}}\{\phi_j\}\right|}{\sum_{k\in\mathcal{N}}\left(\phi_k + \left|\min_{j\in\mathcal{N}}\{\phi_j\}\right|\right)}, \text{ s.t., } |r| \text{ is an absolute value of } r$$

(5.8.5)

5.8.4 Bandwidth Redistribution for Multimedia Applications

In IoT systems, *APs* are complex access points that handle wireless communications with multiple smart objects in micro cells. Therefore, *APs* are responsible for the bandwidth distribution for different multimedia traffic services. Usually, heterogeneous multimedia data can be categorized into different classes. By taking into account the characteristics among different multimedia traffic services, $\mathfrak{B}_i^{AP}$ in the AP_i, where $1 \leq i \leq K$, should be adaptively distributed to each traffic class in a distributed manner. It should guarantee QoS while ensuring bandwidth efficiency (Kim, 2016a).

Usually, in different kinds of multimedia applications, a group of applications cooperate, where data communications are required to be casually delivered. For this reason, some traffic services are tightly coupled, and they can be grouped for the purpose of payoff bargaining. This fact is incorporated into the game model by using a coalition structure, which is an exogenous partition of players into a set of groups or unions (Calvo, 2013; Kim, 2016a; Shojafar, 2013).

In the *ASVA* scheme, the assessment of a cooperative game within a coalition structure is determined by a coalitional value. This evaluation considers that interactions among players occur at two hierarchical levels. In the inter-coalition level, interaction takes place among different coalitions, treating each coalition as a single entity (player) in the game. In the intra-coalition level, within each coalition, individual players interact to distribute resources among themselves. By structuring interactions at these two levels, the *ASVA* scheme ensures a more structured and efficient allocation process within the cooperative game framework. Formally, the *ASVA* scheme categorizes service applications into six different traffic groups; *Class I, II, III, IV, V* and *VI* traffic services. In each *AP*, the intra-*AP* game model can be formulated where these traffic groups are assumed as players $\mathbb{P} = \{P_1,\ldots,P_6\}$, and coalition structure ($\mathcal{J}_{\mathrm{P}}$) over P is a partition of P where $\mathcal{J}_{\mathrm{P}} = \{\beta_1,\beta_2,\cdots,\beta_t\}$, $\beta_{f,1\leq f\leq t} \subseteq \mathrm{P}, t \leq 6, \beta_f \cap \beta_h = \varnothing$ for $f \neq h$ and $\cup_{f=1}^{t}\beta_f = \mathbb{P}$. According to the traffic service properties, this coalition structure ($\mathcal{J}_{\mathrm{P}}$) is static during IoT system operations (Kim, 2016a).

For the intra-*AP* cooperative games, the *ASVA* scheme develops utility functions ($\mathcal{U}(\cdot)$) for each player, i.e., traffic class. In general, real-time multimedia traffic services demand strict end-to-end performance guarantees and are designed to operate at a fixed bandwidth throughout their service duration. As a result, they follow step utility functions, where QoS requirements must be strictly met to ensure service continuity. Conversely, non-real-time multimedia traffic services are more tolerant to delays and can flexibly adjust their transmission rates. These services follow concave utility functions, which allow for multiple QoS levels depending on dynamically allocated bandwidth (Kim, 2012b). In the model, *Class I* and *II*, i.e., P_1 and P_2, are *real-time* traffic services and *Class III, IV, V*, and *VI*, i.e., P_3,P_4, P_5, and P_6, are *non real-time* traffic services. According to different data types, the utility function $\left(\mathcal{U}\left(\mathfrak{B}_j^{P_i}\right)\right)$ of each application *j* in player i (P_i) can be defined as follows (Kim, 2016a).

$$\mathcal{U}\left(\mathfrak{B}_j^{P_i}\right) = \begin{cases} \begin{cases} \rho_i \times 1, if\, \mathfrak{B}_j^{P_i} \geq m_j^{P_i} \\ 0, if\, \mathfrak{B}_j^{P_i} < m_j^{P_i} \end{cases}, where\, P_i \in classI, II \\ \\ \rho_i \times \dfrac{1 - exp\left(-\alpha_i \times \left(\mathfrak{B}_j^{P_i} - m_j^{P_i}\right)\right)}{1 - exp\left(-\alpha_i \times \left(M_j^{P_i} - m_j^{P_i}\right)\right)}, where\, P_i \in classIII, IV, V, VI \end{cases} \tag{5.8.6}$$

where $\mathfrak{B}_j^{P_i}$ is the currently allocated bandwidth of the application j in the P_i. ρ_i is the relative weight parameter for the P_i. $M_j^{P_i}$, $m_j^{P_i}$ are the maximum and minimum bandwidth requests of the application j, respectively. For the *non real-time* data services, the value α is selected based on the average slope of the linear utility function of the request. To reflect the difference in the quantitative QoS requirements, α can quantify the adaptability of an application (Kim, 2012b, 2016a).

For all $P_{i,1 \leq i \leq 6} \in \beta_f \subseteq \mathcal{J}_{\mathbb{P}}$, the *ASVA* scheme defines the coalition structure value ($\ddot{A}^{P_i}(\beta_f)$) of P_i within a coalition β_f. By using rescaled utility functions ($\mathcal{U}(\cdot)$), the value $\ddot{A}^{P_i}(\beta_f)$ is estimated based on the natural calibration mechanism while the utilitarian sum is maximized (Kim, 2016a):

$$\ddot{A}^{P_i}(\beta_f) = \max_{\mathfrak{B}^{P_i}} \sum_{P_i \in \beta_f} \left(\frac{\mathcal{U}(\mathfrak{B}^{P_i}) - \mathcal{U}(P_i)_{min}}{\mathcal{U}(P_i)_{max} - \mathcal{U}(P_i)_{min}} \right), \text{ s.t., } \mathfrak{B}^{P_i} = \sum_{j \in P_i} \mathfrak{B}_j^{P_i} \tag{5.8.7}$$

where $\mathcal{U}(P_i)_{max}$, $\mathcal{U}(P_i)_{min}$ are the maximum and minimum payoffs for player P_i such as $\mathcal{U}(P_i)_{max} = \max_{\mathfrak{B}^{P_i}} (\mathcal{U}(\mathfrak{B}^{P_i}))$ and $\mathcal{U}(P_i)_{min} = \mathbf{min}_{\mathfrak{B}^{P_i}}(\mathcal{U}(\mathfrak{B}^{P_i}))$. Based on the $\ddot{A}^{P_i}(\beta_f)$, the relative utilitarian value vector $\mathfrak{U}\mathcal{V}(\mathbb{P}, v^{\beta}) \in \mathbb{R}^{\mathbb{P}}$ of the cooperative game $(\mathbb{P}, v^{\beta})$ is defined by (Kim, 2016a);

$$\mathfrak{U}\mathcal{V}_{P_i}(\mathbb{P}, v^{\beta}) = \sum_{\beta_f \subseteq \mathbb{P}: P_i \in \beta_f} \frac{(|\mathbb{P}| - |\beta_f|)! \times (|\beta_f| - 1)!}{|\mathbb{P}|!} \times \ddot{A}^{P_i}(\beta_f) \tag{5.8.8}$$

for any $P_i \in \mathbb{P}$

The solution concept of $\mathfrak{U}\mathcal{V}(\mathbb{P}, v^{\beta})$ is considered to be an extension of the *SV* while a distributive analysis of a cooperative surplus among players is revisited when they already partition themselves into coalitions before realizing cooperation. To reflect the mutual aid of formed coalitions, players' marginal contributions are calculated as coalition structure values (Kim, 2016a).

According to $\mathfrak{U}\mathcal{V}(\mathbb{P}, v^{\beta})$, the allocated bandwidth amount in the *AP* ($Đ^{AP}$) is dynamically re-distributed over $\mathcal{J}_{\mathbb{P}}$. At the end of each *unit_time* period, the $Đ^{AP}$ in each *AP* is re-distributed to different traffic classes, i.e.,P_i, in a distributed manner. In the kth AP_k, the redistributed bandwidth amount to player P_i ($\tilde{O}_{P_i}^{AP_k}$) is obtained as (Kim, 2016a).

$$\tilde{O}_{P_i}^{AP_k} = Ð_k^{AP} \times \frac{\mathfrak{U}\mathscr{V}_{P_i}(\mathbb{P}, v^{\beta})}{\sum_{P_i \in \mathbb{P}} \left(\mathfrak{U}\mathscr{V}_{P_i}(\mathbb{P}, v^{\beta}) \right)} \tag{5.8.9}$$

5.8.5 Summary

Game theory provides a mathematical framework for analyzing complex interactions between cooperative or competing decision-makers, taking into account their preferences and requirements. In recent years, cooperative game theory has emerged as a powerful tool for addressing challenges in network resource allocation. The ASVA scheme employs a cooperative game-theoretic approach to develop an innovative bandwidth allocation mechanism for IoT systems. Its solution is structured as a two-step bandwidth allocation process. In the inter-AP cooperative game, available bandwidth at the BS is dynamically allocated among APs using the A_SV approach. In the intra-AP cooperative game, within each AP, the allocated bandwidth is redistributed among different traffic service groups, following the relative utilitarian value method. By integrating game-theoretic principles into practical network resource allocation, the ASVA scheme fosters further advancements in applied game theory and promotes collaborative research in real-world implementation techniques. Looking ahead, research in this domain is expected to witness continued growth in both theoretical advancements and practical applications, further accelerating progress in intelligent bandwidth management for IoT networks (Kim, 2016a).

REFERENCES

Aazam, M., Zeadally, S., & Harras, K. A. (2018). Deploying Fog Computing in Industrial Internet of Things and Industry 4.0. *IEEE Transactions on Industrial Informatics*, *14*(10), 4674–4682. DOI: 10.1109/TII.2018.2855198

Alós-Ferrer, C., García-Segarra, J., & Ginés-Vilar, M. (2018). Anchoring on Utopia: A generalization of the Kalai–Smorodinsky solution. *Economic Theory Bulletin*, *6*(2), 141–155. DOI: 10.1007/s40505-017-0130-7

Archibald, J. K., Hill, J. C., Johnson, F. R., & Stirling, W. C. (2006). Satisficing negotiations. *IEEE Transactions on Systems, Man, and Cybernetics. Part C, Applications and Reviews*, *36*(1), 4–18. DOI: 10.1109/TSMCC.2005.860571

Ashton, K. (2009). That 'Internet of Things' thing in the real world, things matter more than ideas. *RFID J.* Retrieved October 25, 2013, from https://www.rfidjournal.com/article/print/4986

Baccarelli, E., Scarpiniti, M., Naranjo, P. G. V., & Vaca-Cardenas, L. (2018). Fog of Social IoT: When the Fog Becomes Social. *IEEE Network*, *32*(4), 68–80. DOI: 10.1109/MNET.2018.1700031

Bahrak, B., & Park, J. J. (2014). Coexistence Decision Making for Spectrum Sharing Among Heterogeneous Wireless Systems. *IEEE Transactions on Wireless Communications*, *13*(3), 1298–1307. DOI: 10.1109/TWC.2014.012314.122009

Bilbao, J. M., Fernandez, J. R., Jimenez, N., & Lopez, J. J. (2008a). Biprobabilistic values for bicooperative games. *Discrete Applied Mathematics*, *156*(14), 2698–2711. DOI: 10.1016/j.dam.2007.11.007

Bilbao, J. M., Fernandez, J. R., Jimenez, N., & Lopez, J. J. (2008b). The Shapley value for bicooperative games. *Annals of Operations Research*, *158*(1), 99–115. DOI: 10.1007/s10479-007-0243-8

Bilbao, J. M., Fernández, J. R., Jiménez, N., & López, J. J. (2010). The Banzhaf power index for ternary bicooperative games. *Discrete Applied Mathematics*, *158*(9), 967–980. DOI: 10.1016/j.dam.2010.02.007

Busoniu, L., Babuska, R., & De Schutter, B. (2008). A Comprehensive Survey of Multiagent Reinforcement Learning. *IEEE Transactions on Systems, Man, and Cybernetics. Part C, Applications and Reviews*, *38*(2), 156–172. DOI: 10.1109/TSMCC.2007.913919

Calvo, E., & Gutierrez, E. (2013). The Shapley-solidarity value for games with a coalition structure. *International Game Theory Review*, *15*(1), 1–24. DOI: 10.1142/S0219198913500023

Camerer, C. F. (2003). *Behavioral Game Theory: Experiments in Strategic Interaction*. Princeton University Press.

Chang, Z., Liu, S., Xiong, X., Cai, Z., & Tu, G. (2021). A Survey of Recent Advances in Edge-Computing-Powered Artificial Intelligence of Things. *IEEE Internet of Things Journal*, *8*(18), 13849–13875. DOI: 10.1109/JIOT.2021.3088875

Charkhgard, H., Keshanian, K., Esmaeilbeigi, R., & Charkhgard, P. (2022). The Magic of Nash Social Welfare in Optimization: Do Not Sum, Just Multiply! *The ANZIAM Journal*, *64*(2), 119–134. DOI: 10.1017/S1446181122000074

Chen, X., Gong, X., Yang, L., & Zhang, J. (2014). A social group utility maximization framework with applications in database assisted spectrum access. *IEEE INFOCOM*, 1959–1967.

Chettri, L., & Bera, R. (2020). A comprehensive survey on Internet of Things (IoT) toward 5G wireless systems. *IEEE Internet of Things Journal*, *7*(1), 16–32. DOI: 10.1109/JIOT.2019.2948888

Choi, S., & Shin, K. G. (1998). A cellular wireless local area network with QoS guarantees for heterogeneous traffic. *Mobile Networks and Applications*, *3*(1), 89–100. DOI: 10.1023/A:1019152110549

Claus, C., & Boutilier, C. (1998). The dynamics of reinforcement learning in cooperative multiagent systems. *AAAI/IAAI'1998*, 746–752.

Dianati, M., Shen, X., & Naik, S. (2005). A New Fairness Index for Radio Resource Allocation in Wireless Networks. *IEEE WCNC*, 712–715.

Elias, J., Martignon, F., Chen, L., & Krunz, M. (2017). Distributed Spectrum Management in TV White Space Networks. *IEEE Transactions on Vehicular Technology*, *66*(5), 4161–4172.

Gao, X., Niyato, D., Wang, P., Yang, K., & An, J. (2020). Contract Design for Time Resource Assignment and Pricing in Backscatter-Assisted RF-Powered Networks. *IEEE Wireless Communications Letters*, *9*(1), 42–46. DOI: 10.1109/LWC.2019.2940942

Gao, X., Wang, P., Niyato, D., Yang, K., & An, J. (2019). Auction-Based Time Scheduling for Backscatter-Aided RF-Powered Cognitive Radio Networks. *IEEE Transactions on Wireless Communications*, *18*(3), 1684–1697. DOI: 10.1109/TWC.2019.2895340

Ghosh, A., Chakraborty, D., & Law, A. (2018). Artificial intelligence in Internet of things. *CAAI Transactions on Intelligence Technology*, *3*(4), 208–218. DOI: 10.1049/trit.2018.1008

Gupta, S., & Livne, Z. (1990). Testing the emergence and effect of the reference outcome in an integrative bargaining situation. *Marketing Letters*, *1*(2), 103–112. DOI: 10.1007/BF00435294

Han, Z., Niyato, D., Saad, W., Başar, T., & Hjørungnes, A. (2011). *Game Theory in Wireless and Communication Networks*. Cambridge University Press. DOI: 10.1017/CBO9780511895043

Hemmati, M., Sadati, N., & Nili, M. (2010). Towards a bounded-rationality model of multi-agent social learning in games. *IEEE Intelligent Systems Design and Applications (ISDA' 2010)*, 142–148.

Hu, Y., Gao, Y., & An, B. (2015). Multiagent Reinforcement Learning With Unshared Value Functions. *IEEE Transactions on Cybernetics*, *45*(4), 647–662. DOI: 10.1109/TCYB.2014.2332042 PMID: 25014990

Huang, Y., & Wang, J. (2019). Cooperative spectrum sensing in cognitive radio using Bayesian updating with multiple observations. *Journal of Electronic Science and Technology*, *17*(3), 252–259.

Imran, A., Bennis, M., & Giupponi, L. (2012). Use of learning, game theory and optimization as biomimetic approaches for Self-Organization in macro-femtocell coexistence. *IEEE WCNCW, 2012*, 103–108. DOI: 10.1109/WCNCW.2012.6215470

Jang, I., Pyeon, D., Kim, S., & Yoon, H. (2013). A Survey on Communication Protocols for Wireless Sensor Networks. *Journal of Computing Science and Engineering : JCSE*, *7*(4), 231–241. DOI: 10.5626/JCSE.2013.7.4.231

Jia, L., Zhou, Z., Xu, F., & Jin, H. (2022). Cost-Efficient Continuous Edge Learning for Artificial Intelligence of Things. *IEEE Internet of Things Journal*, *9*(10), 7325–7337. DOI: 10.1109/JIOT.2021.3104089

Jiang, C., Kuang, L., Han, Z., Ren, Y., & Hanzo, L. (2017). Information Credibility Modeling in Cooperative Networks: Equilibrium and Mechanism Design. *IEEE Journal on Selected Areas in Communications*, *35*(2), 432–448. DOI: 10.1109/JSAC.2017.2659078

Kamijo, Y. (2009). A two-step Shapley value in a cooperative game with a coalition structure. *International Game Theory Review*, *11*(2), 207–214. DOI: 10.1142/S0219198909002261

Karagözoğlu, E., & Keskin, K. (2015). A tale of two bargaining solutions. *Games*, *6*(2), 161–174. DOI: 10.3390/g6020161

Kelechi, A. H., Abdullah, N. F., Nordin, R., & Ismail, M. (2018). D-GRACE: Discounted Spectrum Price Game-Based resource allocation in a competitive environment for TVWS networks. *IEEE Access : Practical Innovations, Open Solutions*, *6*, 7579–7592. DOI: 10.1109/ACCESS.2017.2779598

Kim, K., Uno, S., & Kim, M. (2010). Adaptive QoS Mechanism for Wireless Mobile Network. *Journal of Computing Science and Engineering : JCSE*, *4*(2), 153–172. DOI: 10.5626/JCSE.2010.4.2.153

Kim, S. (2011). Adaptive online power control scheme based on the evolutionary game theory. *IET Communications*, *5*(18), 2648–2655. DOI: 10.1049/iet-com.2011.0093

Kim, S. (2012a). Adaptive Ad-hoc Network Routing Scheme by Using Incentive-based Model. *Ad-Hoc & Sensor Wireless Networks*, *15*(2), 107–125.

Kim, S. (2012b). Adaptive Call Admission Control Scheme for Heterogeneous Overlay Networks. *Journal of Communications and Networks (Seoul)*, *14*(4), 461–466. DOI: 10.1109/JCN.2012.6292253

Kim, S. (2014). *Game Theory Applications in Network Design*. IGI Global. DOI: 10.4018/978-1-4666-6050-2

Kim, S. (2015). Learning-Based QoS Control Algorithms for Next Generation Internet of Things. *Mobile Information Systems*, *2015*(1), 605357. DOI: 10.1155/2015/605357

Kim, S. (2016a). Asymptotic shapley value based resource allocation scheme for IoT services. *Computer Networks*, *100*, 55–63. DOI: 10.1016/j.comnet.2016.02.021

Kim, S. (2016b). Behavioral learning game for socio-physical IoT connections. *EURASIP Journal on Wireless Communications and Networking*, *2016*(1), 1–11. DOI: 10.1186/s13638-016-0521-8

Kim, S. (2017a). Fog radio access network system control scheme based on the embedded game model. *EURASIP Journal on Wireless Communications and Networking*, *2017*(113), 1–14. DOI: 10.1186/s13638-017-0900-9

Kim, S. (2017b). Inspection game based cooperative spectrum sensing and sharing scheme for cognitive radio IoT system. *Computer Communications*, *105*, 116–123. DOI: 10.1016/j.comcom.2017.01.015

Kim, S. (2019). Novel resource allocation algorithms for the social Internet of things based fog computing paradigm. *Wireless Communications and Mobile Computing*, *2019*(1), 3065438. DOI: 10.1155/2019/3065438

Kim, S. (2020). Heterogeneous Network Bandwidth Control Scheme for the Hybrid OMA-NOMA System Platform. *IEEE Access : Practical Innovations, Open Solutions*, *8*, 83414–83424. DOI: 10.1109/ACCESS.2020.2991780

Kim, S. (2021a). A new bargaining game-based unlicensed spectrum sharing scheme for TVWS platform. *IEEE Access : Practical Innovations, Open Solutions*, *9*, 95528–95537. DOI: 10.1109/ACCESS.2021.3094208

Kim, S. (2021b). Bargaining game based time scheduling scheme for ambient backscatter communications. *IEEE Access : Practical Innovations, Open Solutions*, *9*, 155526–155533. DOI: 10.1109/ACCESS.2021.3128207

Kim, S. (2023a). Collaborative Coded Distributed Computing Scheme: A Two-Phase Cooperative Game Approach. *IEEE Access : Practical Innovations, Open Solutions*, *11*, 42042–42051. DOI: 10.1109/ACCESS.2023.3252559

Kim, S. (2023b). Multi-Agent learning and bargaining scheme for cooperative spectrum sharing process. *IEEE Access : Practical Innovations, Open Solutions*, *11*, 47863–47872. DOI: 10.1109/ACCESS.2023.3268754

Kim, S., & Varshney, P. K. (2004). An integrated adaptive bandwidth-management framework for QoS-sensitive multimedia cellular networks. *IEEE Transactions on Vehicular Technology*, *53*(3), 835–846. DOI: 10.1109/TVT.2004.825704

Lee, K., & Shin, I. (2015). User Mobility Model Based Computation Offloading Decision for Mobile Cloud. *Journal of Computing Science and Engineering : JCSE*, *9*(3), 155–162. DOI: 10.5626/JCSE.2015.9.3.155

Lee, W., Xiang, L., Schober, R., & Wong, V. W. S. (2014). Direct Electricity Trading in Smart Grid: A Coalitional Game Analysis. *IEEE Journal on Selected Areas in Communications*, *32*(7), 1398–1411. DOI: 10.1109/JSAC.2014.2332112

Li, L., Li, S., & Zhao, S. (2014). QOS-Aware Scheduling of Services-Oriented Internet of Things. *IEEE Transactions on Industrial Informatics*, *10*(2), 1497–1505. DOI: 10.1109/TII.2014.2306782

Li, Z., Chen, R., Liu, L., & Min, G. (2016). Dynamic Resource Discovery Based on Preference and Movement Pattern Similarity for Large-Scale Social Internet of Things. *IEEE Internet of Things Journal*, *3*(4), 581–589. DOI: 10.1109/JIOT.2015.2451138

Liu, X., Hu, S., Li, M., & Lai, B. (2021). Energy-Efficient Resource Allocation for Cognitive Industrial Internet of Things With Wireless Energy Harvesting. *IEEE Transactions on Industrial Informatics*, *17*(8), 5668–5677. DOI: 10.1109/TII.2020.2997768

Liu, X., Jia, M., Zhou, M., Wang, B., & Durrani, T. S. (2023). Integrated Cooperative Spectrum Sensing and Access Control for Cognitive Industrial Internet of Things. *IEEE Internet of Things Journal*, *10*(3), 1887–1896. DOI: 10.1109/JIOT.2021.3137408

Liu, X., Sun, C., Yu, W., & Zhou, M. (2022). Reinforcement-Learning-Based Dynamic Spectrum Access for Software-Defined Cognitive Industrial Internet of Things. *IEEE Transactions on Industrial Informatics*, *18*(6), 4244–4253. DOI: 10.1109/TII.2021.3113949

Liu, X., & Zhang, X. (2020). NOMA-Based Resource Allocation for Cluster-Based Cognitive Industrial Internet of Things. *IEEE Transactions on Industrial Informatics*, *16*(8), 5379–5388. DOI: 10.1109/TII.2019.2947435

Liu, Y., Lee, M. J., & Zheng, Y. (2016). Adaptive Multi-Resource Allocation for Cloudlet-Based Mobile Cloud Computing System. *IEEE Transactions on Mobile Computing*, *15*(10), 2398–2410. DOI: 10.1109/TMC.2015.2504091

Lohi, M., Weerakoon, D., & Aghvami, A. H. (1999). Trends in multi-layer cellular system design and handover design. *IEEE Wireless Communications and Networking Conference*, 898–902. DOI: 10.1109/WCNC.1999.796801

Luckshmi, A. P. I., Visalakshi, P., & Karthikeyan, N. K. (2011). Intelligent Schemes for Bandwidth Allocation in Cellular Mobile Networks. *IEEE PACC*, *2011*, 1–6.

Mazhelis, O., & Tyrvainen, P. (2014). A framework for evaluating Internet-of-Things platforms: Application provider viewpoint. *IEEE World Forum on Internet of Things (WF-IoT' 2014)*, 147–152.

Morozs, N., Clarke, T., & Grace, D. (2016). Distributed Heuristically Accelerated Q-Learning for Robust Cognitive Spectrum Management in LTE Cellular Systems. *IEEE Transactions on Mobile Computing*, *15*(4), 817–825. DOI: 10.1109/TMC.2015.2442529

Nazari, A., Keypour, R., & Amjady, N. (2021). Joint investment of community energy storage systems in distribution networks using modified Nash bargaining theory. *Applied Energy*, *301*(117475), 1–16. DOI: 10.1016/j.apenergy.2021.117475

Nguyen, C. T., Nguyen, D. N., Hoang, D. T., Pham, H., & Dutkiewicz, E. (2022). Optimize Coding and Node Selection for Coded Distributed Computing over Wireless Edge Networks. *IEEE WCNC*, 1–6.

Nguyen, C. T., Nguyen, D. N., Hoang, D. T., Phan, K. T., Niyato, D., Pham, H., & Dutkiewicz, E. (2023, April). Elastic Resource Allocation for Coded Distributed Computing over Heterogeneous Wireless Edge Networks. *IEEE Transactions on Wireless Communications*, *22*(4), 2636–2649. DOI: 10.1109/TWC.2022.3213256

Ning, G., Zhu, G., Li, Q., & Wu, R. (2006). Dynamic Load Balancing Based on Sojourn Time in Multitier Cellular Systems. *IEEE Vehicular Technology Conference*, 111–116.

Nitti, M., Girau, R., & Atzori, L. (2014). Trustworthiness Management in the Social Internet of Things. *IEEE Transactions on Knowledge and Data Engineering*, *26*(5), 1253–1266. DOI: 10.1109/TKDE.2013.105

Ortiz, A. M., Hussein, D., Park, S., Han, S. N., & Crespi, N. (2014). The Cluster Between Internet of Things and Social Networks: Review and Research Challenges. *IEEE Internet of Things Journal*, *1*(3), 206–215. DOI: 10.1109/JIOT.2014.2318835

Pande, A., Ramamurthi, V., & Mohapatra, P. (2013). Quality-Oriented Video Delivery over LTE. *Journal for Corrosion Science and Engineering*, *7*(3), 168–176.

Peha, J. M. (2005). Approaches to spectrum sharing. *IEEE Communications Magazine*, *43*(2), 10–12. DOI: 10.1109/MCOM.2005.1391490

Pendharkar, P. C. (2012). Game theoretical applications for multi-agent systems. *Expert Systems with Applications, 39*(1), 273–279. DOI: 10.1016/j.eswa.2011.07.017

Pujolle, G. (2013). Metamorphic Networks. *Journal of Computing Science and Engineering : JCSE, 7*(3), 198–203. DOI: 10.5626/JCSE.2013.7.3.198

Raazi, S. M. K., & Lee, S. (2010). A Survey on Key Management Strategies for Different Applications of Wireless Sensor Networks. *Journal of Computing Science and Engineering : JCSE, 4*(1), 23–51. DOI: 10.5626/JCSE.2010.4.1.023

Saghand, P. G., & Charkhgard, H. (2021). A cooperative game solution approach for intensity modulated radiation therapy design: Nash Social Welfare optimization. *Physics in Medicine and Biology, 66*(7), 1–23. DOI: 10.1088/1361-6560/abed95 PMID: 33691291

Sallai, G. (2013). Chapters of Future Internet research. [CogInfoCom]. *IEEE Cognitive Infocommunications, 2013*, 161–166.

Shah-Mansouri, H., & Wong, V. W. S. (2018). Hierarchical Fog-Cloud Computing for IoT Systems: A Computation Offloading Game. *IEEE Internet of Things Journal, 5*(4), 3246–3257. DOI: 10.1109/JIOT.2018.2838022

Shao, K., Wang, R., & Zhou, X. Y. (2006). Behavioral game model and evolutionary analysis in electricity market. *IEEE Power Engineering Society General Meeting*, 1–7.

Shi, Z., Gao, W., Zhang, S., Liu, J., & Kato, N. (2020). Machine Learning-Enabled Cooperative Spectrum Sensing for Non-Orthogonal Multiple Access. *IEEE Transactions on Wireless Communications, 19*(9), 5692–5702. DOI: 10.1109/TWC.2020.2995594

Shojafar, M., Pooranian, Z., Abawajy, J. H., & Meybodi, M. R. (2013). An Efficient Scheduling Method for Grid Systems Based on a Hierarchical Stochastic Petri Net. *Journal of Computing Science and Engineering : JCSE, 7*(1), 44–52. DOI: 10.5626/JCSE.2013.7.1.44

Singh, D., Tripathi, G., & Jara, A. J. (2014). A survey of Internet-of-Things: Future vision, architecture, challenges and services. *IEEE World Forum on Internet of Things (WF-IoT' 2014)*, 287–292.

Stai, E., Karyotis, V., & Papavassiliou, S. (2014). Exploiting socio-physical network interactions via a utility-based framework for resource management in mobile social networks. *IEEE Wireless Communications, 21*(1), 10–17. DOI: 10.1109/MWC.2014.6757892

Stirling, W. C., & Frost, R. L. (2005). Social utility Functions-part II: Applications. *IEEE Transactions on Systems, Man, and Cybernetics. Part C, Applications and Reviews, 35*(4), 533–543. DOI: 10.1109/TSMCC.2004.843200

Timoner Lledó, P., & Izquierdo Aznar, J. M. (2015). *Generalized rationing problems and solutions* (Working Paper No. 329). University of Barcelona. https://www.ub.edu/ubeconomics/wp-content/uploads/2015/10/329WEB.pdf

Tinh, B. T., Nguyen, L. D., Kha, H. H., & Duong, T. Q. (2022). Practical Optimization and Game Theory for 6G Ultra-Dense Networks: Overview and Research Challenges. *IEEE Access : Practical Innovations, Open Solutions, 10*, 13311–13328. DOI: 10.1109/ACCESS.2022.3146335

Tsurumi, M., Inuiguchi, M., & Nishimura, A. (2005). Pseudo-Banzhaf Values in Bicooperative Games. *IEEE International Conference on Systems, Man and Cybernetics*, 1138–1143.

Van der Wal, J. (1977). Discounted Markov games: Successive approximation and stopping times. *International Journal of Game Theory*, *6*(1), 11–22. DOI: 10.1007/BF01770870

Van Huynh, N., Hoang, D. T., Nguyen, D. N., & Dutkiewicz, E. (2022). Joint Coding and Scheduling Optimization for Distributed Learning Over Wireless Edge Networks. *IEEE Journal on Selected Areas in Communications*, *40*(2), 484–498. DOI: 10.1109/JSAC.2021.3118432

Vermesan, O., & Friess, P. (2011). *Internet of Things - Global Technological and Societal Trends*. River Publishers.

Voicu, A. M., Simić, L., & Petrova, M. (2016). Inter-Technology Coexistence in a Spectrum Commons: A Case Study of Wi-Fi and LTE in the 5-GHz Unlicensed Band. *IEEE Journal on Selected Areas in Communications*, *34*(11), 3062–3077. DOI: 10.1109/JSAC.2016.2615246

Vrancx, P., Verbeeck, K., & Nowé, A. (2008). Decentralized Learning in Markov Games. *IEEE Transactions on Systems, Man, and Cybernetics. Part B, Cybernetics*, *38*(4), 976–981. DOI: 10.1109/TSMCB.2008.920998 PMID: 18632387

Wang, W., Hoang, D. T., Niyato, D., Wang, P., & Kim, D. I. (2018). Stackelberg Game for Distributed Time Scheduling in RF-Powered Backscatter Cognitive Radio Networks. *IEEE Transactions on Wireless Communications*, *17*(8), 5606–5622. DOI: 10.1109/TWC.2018.2847658

Weiser, P. J., & Hatfield, D. N. (2005). Policing the Spectrum Commons. *Fordham Law Review*, *74*(2), 663–694.

Wu, Q., Ding, G., Xu, Y., Feng, S., Du, Z., Wang, J., & Long, K. (2014). Cognitive Internet of Things: A New Paradigm Beyond Connection. *IEEE Internet of Things Journal*, *1*(2), 129–143. DOI: 10.1109/JIOT.2014.2311513

Xu, L., Yin, W., Zhang, X., & Yang, Y. (2020). Fairness-Aware Throughput Maximization Over Cognitive Heterogeneous NOMA Networks for Industrial Cognitive IoT. *IEEE Transactions on Communications*, *68*(8), 4723–4733. DOI: 10.1109/TCOMM.2020.2992720

Xu, Y., & Yoshihara, N. (2019). An equitable Nash solution to nonconvex bargaining problems. *International Journal of Game Theory*, *48*(3), 769–779. DOI: 10.1007/s00182-019-00658-4

Yang, H., & Zhang, L. (2017). Response Time Prediction of IoT Service Based on Time Similarity. *Journal of Computing Science and Engineering : JCSE*, *11*(3), 100–108. DOI: 10.5626/JCSE.2017.11.3.100

Ye, Y., Shi, L., Chu, X., & Lu, G. (2020). On the Outage Performance of Ambient Backscatter Communications. *IEEE Internet of Things Journal*, *7*(8), 7265–7278. DOI: 10.1109/JIOT.2020.2984449

Zaeri-Amirani, M., Afghah, F., & Zeadally, S. (2018). A Hierarchical Spectrum Access Scheme for TV White Space Coexistence in Heterogeneous Networks. *IEEE Access : Practical Innovations, Open Solutions*, *6*, 78992–79004. DOI: 10.1109/ACCESS.2018.2885732

Zhang, Q., & Peng, D. (2012). Intelligent Decision-Making Service Framework Based on QoS Model in the Internet of Things. *IEEE DCABES, 2012*, 103–107. DOI: 10.1109/DCABES.2012.26

Zhang, Y., Li, B., Gao, F., & Han, Z. (2019). A Robust Design for Ultra Reliable Ambient Backscatter Communication Systems. *IEEE Internet of Things Journal, 6*(5), 8989–8999. DOI: 10.1109/JIOT.2019.2925843

Zhao, S. (2019). A Node-Selection-Based Sub-Task Assignment Method for Coded Edge Computing. *IEEE Communications Letters, 23*(5), 797–801. DOI: 10.1109/LCOMM.2019.2909034

Chapter 6
Heterogeneous Network Management

ABSTRACT

Heterogeneous networks are one of the main features and the future trend of mobile communication technology. Nowadays, they have received a lot of attention in the wireless industry and in academic research communities. In this chapter, the key challenges and emerging heterogeneous network technologies are summarized and their significance is highlighted. And then, the concept of network interoperability is presented, and developing paradigms are identified and discussed. In addition, this chapter provides the state of the art in technology development, the main challenges and trade-offs, and the future research and development directions.

6. OVERVIEW

Due to the fast development of current Internet and wireless communication industry, network systems have been designed and improved with more complexity of infrastructure, higher diversity of associated devices and resources, and more dynamical formations of networks, and they are evolving towards the promising future networking paradigm, the heterogeneous networks (HetNets). The term 'HetNets' comes from the integration of different telecommunication technologies with the aim of using their additional characteristics. In HetNets, there are various types of networks with different working mechanisms. Therefore, the complicated network infrastructure is facing to troubles for effectively organizing, managing and optimizing the network resources. Another problem of HetNets is the skyrocketing mobile traffic demand; it is generating more pressure.

With the rapid increase in the number of mobile devices, wireless access technologies are evolving to provide mobile devices with high data rates and support new applications. HetNets are considered an important solution to deal with the increasing network capacity demands and provide high coverage to wireless devices in future wireless networks. To implement HetNets, the concept of sharing can be defined as a joint use of resources enabled by on-demand exchange. Sharing brings its beneficial nature in many domains including social and economic systems, as well as in nature. From the viewpoint of advantages and characteristics, the research on HetNets is vitally important to ensure heterogeneity, openness, distribution and multi-domain involvement including cross domain communications.

DOI: 10.4018/979-8-3373-2013-7.ch006

6.1 Bargaining based Hierarchical Spectrum Allocation (*BHSA*) scheme

To address the rising demand for application data traffic, which is expected to grow rapidly, 5G wireless networks have been proposed. One of the essential technologies for handling this increasing data challenge is the heterogeneous ultra-dense network (HetUDN). Recently, S. Kim has investigated the issue of spectrum allocation within the HetUDN framework and introduces the *Bargaining-based Hierarchical Spectrum Allocation* (*BHSA*) scheme to enhance both spectrum efficiency and QoS provisioning (Kim, 2019). The *BHSA* scheme employs *right-egalitarian, extended right-egalitarian,* and *sequential Raiffa bargaining solutions* to dynamically distribute the constrained spectrum resources. To minimize computational complexity, these bargaining solutions are applied in a hierarchical manner. By leveraging the characteristics of each solution, the *BHSA* scheme can flexibly choose the most suitable allocation policy, optimizing resource distribution while effectively managing cascading interactions (Kim, 2019).

6.1.1 Development motivation

One of the critical challenges in 5G networks is accommodating the exponentially increasing data demand. With numerous new applications extending beyond personal communications, the number of connected devices could potentially reach tens or even hundreds of billions. Over the span of a decade, the volume of IP data processed by wireless networks has experienced significant growth, increasing from below 3 exabytes in 2010 to over 190 exabytes by 2018, with projections suggesting it could surpass 500 exabytes by 2020. Consequently, the demand for application data traffic has been estimated to rise by a factor of 500 between 2010 and 2020. This surge is largely fueled by the proliferation of smartphones, tablets, and video streaming services. According to recent forecast reports, conventional network control methods will be insufficient to meet the projected demands of networks by 2020 (Andrews, 2014; Kim, 2019).

To handle the dramatic capacity expansion required by 5G communication systems, the heterogeneous ultra-dense network (HetUDN) has emerged as a promising solution due to its cost-effective approach to enhancing system coverage and capacity. In a HetUDN framework, macro base stations (MBSs) are deployed to provide extensive coverage and support high mobility, whereas small base stations (SBSs), including pico base stations, femto base stations, and WiFi access points, contribute to achieving higher data rates in localized hotspot areas. By integrating SBSs within local coverage zones, HetUDNs offer multiple advantages, such as enhanced coverage, cost efficiency, reduced power consumption, and improved QoS levels. To effectively address the engineering challenges associated with 5G communications, it is crucial to determine the specific requirements of HetUDN systems (Hajir, 2016; Kim, 2019; Niu, 2017).

Different applications impose varying performance requirements on network operations. For instance, video streaming services demand low latency but can tolerate lower reliability, whereas non-real-time applications prioritize reliability over high data rates. In the context of HetUDNs, efficient spectrum management among MBSs and SBSs is a crucial factor in supporting diverse applications. To facilitate effective spectrum sharing, the *BHSA* scheme must account for potential interference issues. One possible approach to mitigating interference is the split-spectrum policy, which assigns dedicated spectrum bands to MBSs and SBSs within the total available spectrum. This strategy enhances cooperative opportunities between MBSs and SBSs, thereby improving overall spectrum efficiency. Both academic and industry researchers have devoted significant efforts to developing innovative resource allocation frameworks that optimize spectrum utilization while maintaining globally desirable system characteristics (Andrews, 2014; Hajir, 2016; Kim, 2019).

To enable efficient spectrum management among various base stations, a novel hierarchical network architecture has been designed to streamline the complex control mechanisms of HetUDNs. Implementing a hierarchical and semi-distributed HetUDN system involves classifying system controllers into three levels: the centralized superior controller (CSC), the MBS, and SBS (Kim, 2019; Wang, 2017; Yu, 2018).

In conventional spectrum allocation models, traffic loads are typically managed at a localized level, simplifying network coverage regulation and spectrum allocation. However, due to the unique characteristics of HetUDNs, spectrum dynamics become more pronounced, making the network environment increasingly complex. As a result, spectrum allocation in HetUDNs presents several challenges. Specifically, as cell sizes shrink, traffic loads become increasingly unbalanced, and the dynamic nature of HetUDNs leads to frequent fluctuations in traffic distribution across spatial and temporal domains. Consequently, localized spectrum management fails to deliver optimal system performance. On the other hand, centralized spectrum management strategies, while effective in theory, face limitations in practical implementation due to their excessive computational complexity and control overhead. To strike a balance between implementation feasibility and system optimization, a novel control paradigm is required (Kim, 2019; Yu, 2018).

6.1.2 Contribution

Building on the previous discussion, the *BHSA* scheme is formulated using three bargaining approaches: *right-egalitarian*, *extended right-egalitarian*, and *sequential Raiffa bargaining solutions*. These solutions are strategically applied within the hierarchical HetUDN framework to ensure mutual benefits. By incorporating different bargaining strategies into a multi-tier game model, the *BHSA* scheme facilitates iterative feedback interactions until an optimal solution for the HetUDN system is reached. Throughout the bargaining game process, the participants - namely, the network controllers - make strategic control decisions aimed at achieving a globally optimal system performance while simultaneously maximizing their individual gains (Kim, 2019).

6.1.3 The hierarchical HetUDN architecture

The HetUDN architecture represents an emerging multi-tier cellular network framework comprising two types of base stations: MBSs and SBSs. Within a macro cell, the MBS is responsible for providing extensive coverage, while multiple SBSs are deployed to accommodate the rapidly growing network traffic by establishing small cells. This architecture divides a geographical region into smaller sub-areas, where numerous small cells are densely arranged within the larger cell's coverage. These heterogeneous cells coexist within a hierarchical network structure, often featuring overlapping coverage areas. In this irregular network topology, time is divided into discrete slots, with spectrum resource allocation remaining constant throughout each time slot. To mitigate both intra-tier and inter-tier interference among wireless spectrum links, spectrum resources are assigned to each base station using orthogonal frequency bands (Du, 2017; Haque, 2016; Kim, 2019; Niu, 2017; Yu, 2018).

Within the *BHSA* scheme, a three-tier HetUDN system is implemented. At the top tier, the CSC manages network operations from a global perspective. Using collected data, the CSC dynamically allocates spectrum resources to individual MBSs. In the second tier, each MBS distributes the assigned spectrum among its respective SBSs. Finally, at the third tier, each SBS further divides its allocated spectrum among the connected mobile devices. Since mobile devices generate diverse application service requests, a surge in application demand can lead to the rapid depletion of available spectrum resources. Given the inherent scarcity of these resources, it is impossible to fully accommodate the requirements of all applications. Therefore, to optimize overall system performance, it is crucial to implement an efficient spectrum-sharing strategy that accounts for the interdependent interactions within the system (Du, 2017; Haque, 2016; Kim, 2019; Niu, 2017; Yu, 2018).

The *BHSA* scheme assumes the HetUDN system comprised of one CSC, multiple MBSs $\mathbb{M} = \{\mathcal{M}_1, \mathcal{M}_2, \ldots, \mathcal{M}_n\}$, various SBSs $\mathbb{S} = \{\mathcal{S}_1, \mathcal{S}_2, \ldots, \mathcal{S}_m\}$ and numerous mobile devices $\mathbb{D} = \{\mathcal{D}_1, \mathcal{D}_2, \ldots, \mathcal{D}_k\}$ where $n > m \geqslant k$. This model is flexible to be applied expansively into a system with multiple CSCs. Each SBS $\mathcal{S}_{1 \le j \le m}$ is associated with a specific MBS and each mobile device $\mathcal{D}_{1 \le l \le k}$ is associated with a specific SBS; $\mathcal{S}_j^{\mathcal{M}_i}$ and $\mathcal{D}_l^{\mathcal{S}_j}$ represent the $\mathcal{S}_j$ and $\mathcal{D}_l$ associated with the $\mathcal{M}_i$ and $\mathcal{S}_j$, respectively. Let $\mathfrak{N} = \{\mathfrak{N}_{min} \ldots \mathfrak{N}_l \ldots \mathfrak{N}_{max}\}$ be the a set of spectrum allocation strategies where $\mathfrak{N}_l$ means the allocation amount level of spectrum resource; strategies $\mathfrak{N}_{min \le l \le max}$ are discretely specified in terms of basic spectrum units (BSUs). At the first tier, the CSC is a game planner and multiple $\mathcal{M}_{1 \le i \le n} \in \mathbb{M}$ are bargaining game players. In this game, the total spectrum resource (C) in the HetUDN system is allocated to each MBS while maximizing the payoff ($\mathbb{U}^{BS}$) received by the MBSs. At the second tier, each individual MBS M is a game planner and multiple $\mathcal{S}^{\mathcal{M}}_{1 \le j \le m} \in \mathbb{S}$ are bargaining game players. Independently and dispersively, second-tier games are operated in a parallel manner by multiple MBSs. At the third tier, each individual SBS is a game planner and $\mathcal{D}^{\mathcal{S}}_{1 \le l \le k} \in \mathbb{D}$ are bargaining game players. In the same manner as the second-tier games, third-tier games are operated independently and dispersively by various SBSs (Kim, 2019).

6.1.4 The main concepts of bargaining solutions

Bargaining involves coalitions formed by two or more participants working together with a shared objective, with the central challenge being the distribution of total profit among players. Since rationality and intelligence are fundamental principles in any bargaining process, participants must consider their individual payoff maximization when engaging in cooperative interactions. This research domain is commonly referred to as bargaining game theory. The concept of a bargaining solution was first introduced by J. Nash, who provided both a procedural framework and an axiomatic characterization. Around the same time, H. Raiffa proposed an alternative approach, known as arbitration mechanisms, to analyze bargaining solutions. For surplus-sharing problems, bargaining solutions can be extended in various ways. Despite its conceptual simplicity, there exists a wide range of bargaining solutions, leading to an extensive body of research focused on their properties. Generally, two primary generalizations of bargaining solutions can be distinguished based on their defining characteristics. The first is the proportional solution, which corresponds to the *Kalai-Smorodinsky solution* used in rationing problems. The second is the *Nash solution*, which aligns with the standard Nash bargaining approach for surplus allocation. While the proportional solution maintains the self-duality nature of the right-egalitarian solution, the Nash solution upholds the principle of egalitarian allocation (Herrero, 2010; Kim, 2019; Trockel, 2011).

To formalize cooperative bargaining ideas, let $\mathbb{R}$ denote the set of all (non-negative, positive) real numbers and $\mathbb{R}^n$ be the n-fold Cartesian product of R. A bargaining problem is a triple $[N, E, c]$ where $N = \{1, 2, \ldots, n\}$ represents the set of game players, $E \in \mathbb{R}$ is the total resource amount, and $c = \{c_1 \ldots c_n\} \in \mathbb{R}^n$ is the vector of each player's claim. Let Ω be the family of all problems. For any $\omega = [N, E, c] \in$ Ù,, let $C(\omega)$ be $\sum_{i \in N} c_i$ and let $H(\omega)$ be $\{z \in \mathbb{R}^n | \sum_{i \in N} z_i = E\}$. If $C(\omega) > E$, we are facing a problem of sharing losses from the aggregate entitlements, whereas if $C(\omega) < E$, the problem is one of surplus-sharing. In 1999, Herrero et al proposed the *rights-egalitarian* solution (F^{RE}) like as (Herrero, 2010; Kim, 2019);

$$F_i^{RE}(\omega) = c_i + \left[\frac{1}{n} \times (E - C(\omega))\right] \tag{6.1.1}$$

The *rights-egalitarian* solution assigns to each player his claim plus an equal share of the difference between the E and the total claims $C(\omega)$. When $E > C(\omega)$ (respectively, $E < C(\omega)$), this corresponds to a surplus sharing (respectively a rationing) problem that is solved by distributing equally the net proceeds among game players. That is, this solution assumes that players are owners of the surplus profit, if positive, but they are also fully responsible for the total losses, if negative. Therefore, the *rights-egalitarian* solution can be viewed as a combination of the equal-awards and equal-loss principles (Herrero, 2010; Kim, 2019).

In order to extend the concept of *rights-egalitarian* solution to more general environments, we consider a simple model when the utility possibility set is defined by a hyperplane. A problem is defined by means of an hyperplane $H(\omega)$ with normal (1, 1,..., 1) and a point c, both in $\mathbb{R}^n$. A solution is a function F:Ù $\rightarrow \mathbb{R}^n$, and it implies $F(\omega) = c$ for those problems ω with $c \in H(\omega)$. In general, $c \notin H(\omega)$ lies in one of the semispaces in which the hyperplane $H(\omega)$ divides $\mathbb{R}^n$. For the family of hyperplane problems $(\acute{O}_H)$, the allocation function $H(\Gamma, E)$ is defined as follows (Herrero, 2010; Kim, 2019);

$$H(\Gamma,E) = \left\{ s \in \mathbb{R}^n \;\middle|\; \sum\nolimits_{i\in N} (\gamma_i \times s_i) \le E \right\}, \tag{6.1.2}$$

s.t., $\Gamma = [\gamma_1 \dots \gamma_n]$ and $\gamma_{1\le i\le n} > 0$

Based on the function $H(\Gamma, E)$ and the mapping F: $\acute{O}_H \rightarrow \mathbb{R}^n$, the *extended rights-egalitarian* solution (F^{ERE}) is defined as follows (Herrero, 2010; Kim, 2019);

$$\gamma_i \times F_i^{ERE}(N, H(\Gamma,E), c) = (\gamma_i \times c_i) + \left|\frac{1}{n} \times \left(E - \sum_{\substack{j=1,\\ j\ne i}}^{n} (\gamma_j \times c_j)\right)\right| \tag{6.1.3}$$

s.t., $\sum_{i=1}^{n} (\gamma_i \times F_i^{ERE}(N, H(\Gamma,E), c)) = E$ and $i \in N$

The *extended rights-egalitarian* solution corresponds to the application of the same principle that defines the *rights egalitarian* solution, when units are different for different players. The axioms involved in the characterization of F^{RE} and F^{ERE} are stated below (Herrero, 2010; Kim, 2019).

- Symmetry: For all $\omega = [N, H(p, E), c]$, if $(p_i \times c_i) = (p_j \times c_j)$ for all $i, j \in N$, then $(p_i \times F_i(\omega)) = (p_j \times F_j(\omega))$ for all $i, j \in N$.
- Composition: For any $\omega = [N, H(p, E), c]$,and any $E_1, E_2 \in \mathbb{R}$ such that $E_1 + E_2 = E$, it follows that $F(\omega) = F[N, H(p, E_1), c]+F[N, H(p, E_2), c - F[N, E_1, c]]$.

The principle of *symmetry* dictates that when all players have equally weighted claims, they should receive equally weighted payoffs. Meanwhile, the *composition* property states that any bargaining problem can be resolved sequentially. Specifically, the solution to the original problem corresponds to the sum of the solutions derived from two sub-problems: one where a portion of the total payoff is allocated first, and another where the remaining payoff is distributed while adjusting the original claims based on what players have already received (Herrero, 2010; Kim, 2019).

In many scenarios, bargaining is an iterative process where players reach interim agreements step by step, with each agreement serving as a foundation for subsequent negotiations. Following this perspective, H. Raiffa proposed a bargaining solution for two-player negotiation problems. In this model, the negotiation progresses in discrete steps, with each interim agreement acting as a temporary settlement. Given a disagreement point, the most favorable outcome for a player is the one that maximizes their utility while ensuring that the other player remains at their disagreement utility level. The interim agreement is determined as the average of these two optimal points. By continuously repeating this process and using each interim agreement as the new disagreement point, the bargaining gradually converges to a Pareto-optimal solution within the bargaining set (Diskin, 2011; Kim, 2019).

To characterize the basic concept of Raiffa solution, we preliminarily define some mathematical expressions. As illustrated in Figure 1, a two-person bargaining problem consists of a compact, convex set $S \in \mathbb{R}^2$ of feasible payoff vectors (x_1, x_2) for player 1 and player 2. The disagreement point $d = (d_1, d_2) \in S$ is a threat point or status quo point. If players agree on some vector $x = (x_1, x_2) \in S$ they receive the payoffs x_1 and x_2, respectively. If they do not agree on any vector $x \in S$ they receive d_1 and d_2, respectively. The individually rational part of S, i.e. $S_d := S \cap (\{d\} + \mathbb{R}^2_+)$, is *comprehensive* in the sense that for any $x \in S_d$ also the line segment $[d, x] := co(x, d) \subset S_d$ where $co(x, d)$ is the convex hull in $\mathbb{R}^2$. This (S_d, d) is itself a bargaining problem. Let $proj_i(S_d, d_{3-i}) = \max\{x_i | (x_i, d_{3-i}) \in S_d\}, i = 1,2$. The *Pareto frontier* of S_d, denoted ∂S_d, is the graph of the two functions $proj_i(S_d, d_{3-i}), i = 1,2$ (Kim, 2019; Trockel, 2011).

Figure 1. Raiffa bargaining solution for two players (Kim, 2019)

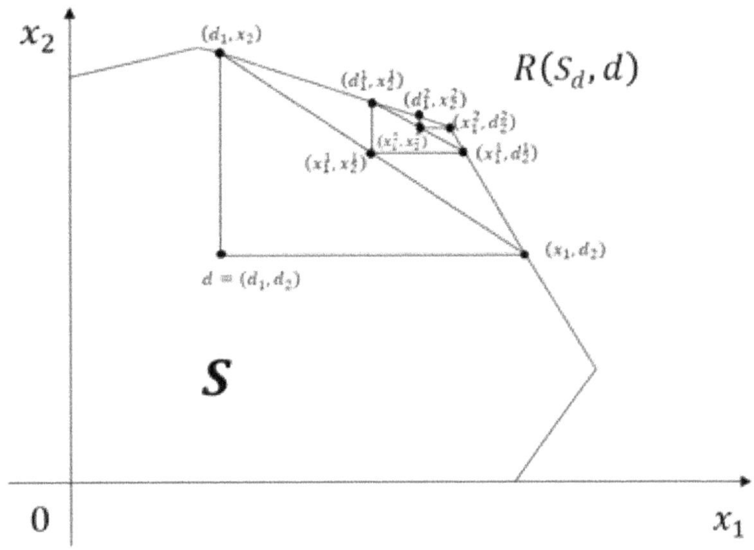

The Raiffa bargaining solution is determined through step-by-step negotiation. In the first step, the Raiffa procedure determines the midpoint of the line segment $\left[(d_1, proj_2(S_d, d_1)), (proj_1(S_d, d_2), d_2)\right]$. In the second step, the new bargaining problem (S, d^1) is considered. If $d^1 \in \partial S_d$ we are finished and d^1 is the vector of payoffs. Otherwise, the midpoint d^2 of the line segment $[(d_1^1, proj_2(S_d, d_1^1)), (proj_1(S_d, d_2^1), d_2^1)]$ is chosen, leading to the new bargaining problem (S, d^2). In the nth step, d^n is the midpoint of the line segment $[(d_1^{n-1}, proj_2(S_d, d_1^{n-1})), (proj_1(S_d, d_2^{n-1}), d_2^{n-1})]$. For any bargaining problem (S, d), the sequence of midpoints converges to a point $R(S,d)$, which is called the *sequential Raiffa* bargaining solution. To examine the axiomatic basis of *sequential Raiffa* bargaining solution, we let B be the set of all two person bargaining games. For any set $C \subseteq B$, a solution on $C$is a mapping $F: C \rightarrow \mathbb{R}^2 : (S, d) \mapsto f(S, d) \in S$ where the disagreement point $d := 0 \in \mathbb{R}^2$ (Kim, 2019; Trockel, 2011).

6.1.5 The bargaining based spectrum allocation algorithms

In the HetUDN system, the transmission data rate can be used to evaluate the utility payoff of each MBS, and it is related to the traffic load and the allocated spectrum usability. To quantify the service satisfaction, the utility function of $\mathcal{M}_i$, i.e., $\mathbb{U}^{\mathcal{M}_i}(\mathfrak{N}_k^{\mathcal{M}_i}, \mathbb{C})$, can be derived according to the spectrum allocation strategy $(\mathfrak{N}_k^{\mathcal{M}_i})$ and $\mathbb{C}$ (Kim, 2019).

$$\mathbb{U}^{\mathcal{M}_i}(\mathfrak{N}_k^{\mathcal{M}_i}, \mathbb{C}) = \log\left(exp\left(\alpha + \frac{\mathfrak{N}_k^{\mathcal{M}_i}}{\mathbb{C}} \right)^{\psi^{\mathcal{M}_i}} \right) \times \frac{\mathcal{F}(\mathbb{S}^{\mathcal{M}_i}, \mathfrak{N}_k^{\mathcal{M}_i})}{\mathfrak{N}_k^{\mathcal{M}_i}} \quad (6.1.4)$$

$$\text{s.t.,} \begin{cases} \psi^{\mathcal{M}_i} = \sum_{\mathcal{S}_j^{\mathcal{M}_i} \in \mathbb{S}^{\mathcal{M}_i}} \left(\beta + \frac{\psi^{\mathcal{S}_j^{\mathcal{M}_i}} \times R^{\mathcal{S}_j^{\mathcal{M}_i}}}{\psi_{\max} \times R^{\mathcal{S}_j^{\mathcal{M}_i}}} \right) \\ F\left(\mathbb{S}^{\mathcal{M}_i}, \mathfrak{N}_k^{\mathcal{M}_i}\right) = \begin{cases} \mathfrak{N}_k^{\mathcal{M}_i}, if \mathfrak{N}_k^{\mathcal{M}_i} \leq \sum_{\mathcal{S}_j^{\mathcal{M}_i} \in \mathbb{S}^{\mathcal{M}_i}} R^{\mathcal{S}_j^{\mathcal{M}_i}} \\ \sum_{\mathcal{S}_j^{\mathcal{M}_i} \in \mathbb{S}^{\mathcal{M}_i}} R^{\mathcal{S}_j^{\mathcal{M}_i}}, otherwise \end{cases} \end{cases}$$

where α, β are the control parameters for the $\mathbb{U}^{\mathcal{M}_i}\left(\mathfrak{N}_k^{\mathcal{M}_i}, \mathbb{C}\right)$ and $\psi^{\mathcal{S}_j^{\mathcal{M}_i}}$ is the weighting factor of the $\mathcal{S}_j^{\mathcal{M}_i}$; $\psi^{\mathcal{S}_j^{\mathcal{M}_i}}$ is estimated by the $\mathcal{S}_j^{\mathcal{M}_i}$. The $\psi_{\max}$ is the maximum weighting value and $R^{\mathcal{S}_j^{\mathcal{M}_i}}$ is the spectrum amount requested from the $\mathcal{S}_j^{\mathcal{M}_i}$. $\mathbb{S}^{\mathcal{M}_i}$ is the set of SBSs associated to the $\mathcal{M}_i$. To quantify the service satisfaction of the $\mathcal{S}_j^{\mathcal{M}_i}$, the utility function of $\mathcal{S}_j^{\mathcal{M}_i}$, i.e.,$\mathcal{U}^{\mathcal{S}_j^{\mathcal{M}_i}}\left(\mathcal{A}_{\mathcal{M}_i}, \mathfrak{N}_l^{\mathcal{S}_j}\right)$, can be derived according to the spectrum allocation strategy $\left(\mathfrak{N}_l^{\mathcal{S}_j}\right)$ and the allocated spectrum amount for the $\mathcal{M}_i$ $\left(\mathcal{A}_{\mathcal{M}_i}\right)$ (Kim, 2019).

$$\mathcal{U}^{\mathcal{S}_j^{\mathcal{M}_i}}\left(\mathcal{A}_{\mathcal{M}_i}, \mathfrak{N}_l^{\mathcal{S}_j}\right) = \log\left(exp\left(\gamma + \frac{\mathfrak{N}_l^{\mathcal{S}_j}}{\mathcal{A}_{\mathcal{M}_i}}\right)^{\psi^{\mathcal{S}_j}} \right) \times \frac{\mathcal{H}\left(\mathbb{S}^{\mathcal{S}_j}, \mathfrak{N}_l^{\mathcal{S}_j}\right)}{\mathfrak{N}_l^{\mathcal{S}_j}} \tag{6.1.5}$$

$$\text{s.t.}, \begin{cases} \psi^{\mathcal{S}_j} = \sum_{\mathcal{D}_x^{\mathcal{S}_j} \in \mathbb{S}^{\mathcal{S}_j}} \left(\delta + \frac{\psi^{\mathcal{D}_x^{\mathcal{S}_j}} \times R^{\mathcal{D}_x^{\mathcal{S}_j}}}{\psi_{\max} \times R^{\mathcal{D}_x^{\mathcal{S}_j}}} \right) \\ H\left(\mathbb{S}^{\mathcal{S}_j}, \mathfrak{N}_l^{\mathcal{S}_j}\right) = \begin{cases} \mathfrak{N}_l^{\mathcal{S}_j}, if \mathfrak{N}_l^{\mathcal{S}_j} \le \sum_{\mathcal{D}_x^{\mathcal{S}_j} \in \mathbb{S}^{\mathcal{S}_j}} R^{\mathcal{D}_x^{\mathcal{S}_j}} \\ \sum_{\mathcal{D}_x^{\mathcal{S}_j} \in \mathbb{S}^{\mathcal{S}_j}} R^{\mathcal{D}_x^{\mathcal{S}_j}}, otherwise \end{cases} \end{cases}$$

where γ, δ are the control parameters for the $\mathcal{U}^{\mathcal{S}_j^{\mathcal{M}_i}}\left(\mathcal{A}_{\mathcal{M}_i}, \mathfrak{N}_l^{\mathcal{S}_j}\right)$ and $\psi^{\mathcal{D}_x^{\mathcal{S}_j}}$is the weighting factor of the $\mathcal{D}_x^{\mathcal{S}_j}$; $\psi^{\mathcal{D}_x^{\mathcal{S}_j}}$ is reported by the $\mathcal{D}_x^{\mathcal{S}_j}$. $R^{\mathcal{D}_x^{\mathcal{S}_j}}$ is the spectrum amount requested from the $\mathcal{D}_x^{\mathcal{S}_j}$. $\mathbb{S}^{\mathcal{S}_j}$ is the set of mobile devices associated to the $\mathcal{S}_j$. To quantify the service satisfaction of the $\mathcal{D}_x^{\mathcal{S}_j}$, the utility function of $\mathcal{D}_x^{\mathcal{S}_j}$, i.e.,$U^{\mathcal{D}_x^{\mathcal{S}_j}}\left(\mathcal{A}_{\mathcal{S}_j}, \mathfrak{N}_y^{\mathcal{D}_x}\right)$, can be derived according to the spectrum allocation strategy $\left(\mathfrak{N}_y^{\mathcal{D}_x}\right)$ and the allocated spectrum amount for the $\mathcal{S}_j$ $\left(\mathcal{A}_{\mathcal{S}_j}\right)$ (Kim, 2019).

$$U^{\mathcal{D}_x^{\mathcal{S}_j}}\left(\mathcal{A}_{\mathcal{S}_j}, \mathfrak{N}_y^{\mathcal{D}_x}\right) = \log\left(exp\left(\eta + \frac{\mathfrak{N}_y^{\mathcal{D}_x}}{\mathcal{A}_{\mathcal{S}_j}}\right)^{\psi^{\mathcal{D}_x^{\mathcal{S}_j}}} \right) \times \frac{\mathscr{E}\left(R^{\mathcal{D}_x^{\mathcal{S}_j}}, \mathfrak{N}_y^{\mathcal{D}_x}\right)}{\mathfrak{N}_y^{\mathcal{D}_x}} \tag{6.1.6}$$

$$\text{s.t.}, \mathscr{E}\left(R^{\mathcal{D}_x^{\mathcal{S}_j}}, \mathfrak{N}_y^{\mathcal{D}_x}\right) = \begin{cases} \mathfrak{N}_y^{\mathcal{D}_x}, if \mathfrak{N}_y^{\mathcal{D}_x} \le R^{\mathcal{D}_x^{\mathcal{S}_j}} \\ R^{\mathcal{D}_x^{\mathcal{S}_j}}, otherwise \end{cases}$$

where η is the control parameter for the $U^{\mathcal{D}_x^{\mathcal{S}_j}}\left(\mathcal{A}_{\mathcal{S}_j}, \mathfrak{N}_y^{\mathcal{D}_x}\right)$ and $\psi^{\mathcal{D}_x^{\mathcal{S}_j}}$ is the weighting factor of the $\mathcal{D}_x^{\mathcal{S}_j}$ (Kim, 2019).

In the *BHSA* scheme, the spectrum resource of HetUDN system is sequentially allocated through a three-tier infrastructure. At the first tier, the CSC dynamically selects the spectrum allocation method to allocate the spectrum resource ($\mathbb{C}$) among the associated MBSs. When the available spectrum resource is enough, i.e., service blocking situation does not occur, the *right-egalitarian* method (F^{RE}) is selected. For the F^{RE}, $\omega = [N, E, c]$ is defined as $\left[\mathbb{M} = \left\{\mathcal{M}_1, \ldots \mathcal{M}_i, \ldots, \mathcal{M}_n\right\}, \mathbb{C}, \mathscr{C}^{\mathcal{M}} = \left\{\mathscr{C}^{\mathcal{M}_1}, \ldots \mathscr{C}^{\mathcal{M}_i}, \ldots, \mathscr{C}^{\mathcal{M}_n}\right\}\right]$ where $\mathscr{C}^{\mathcal{M}_i}$ is the requested spectrum amount from the $\mathcal{M}_i$. The F^{RE} for $\mathcal{M}_i$ $\left(F_{\mathcal{M}_i}^{RE}(\mathbb{M}, \mathbb{C}, \mathscr{C}^{\mathcal{M}})\right)$ is defined as follows (Kim, 2019);

$$F^{RE}_{\mathcal{M}_i}(\mathbb{M},\mathbb{C},\mathscr{C}^{\mathcal{M}}) = \mathscr{C}^{\mathcal{M}_i} + \left[\frac{1}{|\mathbb{M}|}\times\left(\mathbb{C} - \sum_{\mathcal{M}_i\in\mathbb{M}}\mathscr{C}^{\mathcal{M}_i}\right)\right] \tag{6.1.7}$$

When the available spectrum resource is not enough, i.e., a service blocking situation occurred, the *extended right-egalitarian* method (F^{ERE}) is selected. The F^{ERE} for $\mathcal{M}_i$ $\left(F^{ERE}_{\mathcal{M}_i}(\mathbb{M},H(\Gamma,\mathbb{C}),\mathscr{C}^{\mathcal{M}})\right)$ is defined as follows (Kim, 2019);

$$\gamma_{\mathcal{M}_i}\times F^{ERE}_{\mathcal{M}_i}(\mathbb{M},H(\Gamma,\mathbb{C}),\mathscr{C}^{\mathcal{M}}) =$$

$$\left(\gamma_{\mathcal{M}_i}\times\mathscr{C}^{\mathcal{M}_i}\right) + \left(\frac{1}{|\mathbb{M}|}\times\left(\mathbb{C} - \sum_{\substack{j=1,\\ j\neq i}}^{n}\left(\gamma_{\mathcal{M}_j}\times\mathscr{C}^{\mathcal{M}_j}\right)\right)\right) \tag{6.1.8}$$

$$\text{s.t.,}\begin{cases}\sum_{\mathcal{M}_i\in\mathbb{M}}\left(\gamma_{\mathcal{M}_i}\times F^{ERE}_{\mathcal{M}_i}(\mathbb{M},H(\Gamma,\mathbb{C}),\mathscr{C}^{\mathcal{M}})\right)\\ \Gamma = [\gamma_{\mathcal{M}_1}\cdots\gamma_{\mathcal{M}_i}\cdots\gamma_{\mathcal{M}_n}]\,and\,\gamma_{\mathcal{M}_i} \\ \quad = \psi^{\mathcal{M}_i}/\sum_{\mathcal{M}_j\in\mathbb{M}}\psi^{\mathcal{M}_j}\\ H(\Gamma,\mathbb{C}) = \{\, s = [s_1\ldots s_i\ldots s_n] \in \mathbb{R}^n \mid \\ \quad \sum_{\mathcal{M}_i\in\mathbb{M}}\left(\gamma_{\mathcal{M}_i}\times s_i\right) \leq C\}\end{cases}$$

At the second-tier, each $\mathcal{M}_{1\leq i\leq n}$ also dynamically selects the spectrum allocation method F^{RE} or F^{ERE} in the same manner as the CPC. Based on the selected method, the $\mathcal{M}_i$ distributes the allocated spectrum $\mathcal{A}_{\mathcal{M}_i}$ to the associated SBSs. For the F^{RE}, $\omega_{\mathcal{M}_i} = [N,E,c]$ is defined as $\left[\mathbb{S}^{\mathcal{M}_i} = \{\ldots\mathcal{S}^{\mathcal{M}_i}_j\ldots\},\mathcal{A}_{\mathcal{M}_i},\mathscr{C}^{\mathcal{S}} = \{\ldots\mathscr{C}^{\mathcal{S}^{\mathcal{M}_i}_j}\ldots\}\right]$ where $\mathscr{C}^{\mathcal{S}^{\mathcal{M}_i}_j}$ is the requested spectrum amount from the $\mathcal{S}^{\mathcal{M}_i}_j$. The F^{RE} for $\mathcal{S}^{\mathcal{M}_i}_j$ $\left(F^{RE}_{\mathcal{S}^{\mathcal{M}_i}_j}(\mathbb{S}^{\mathcal{M}_i},\mathcal{A}_{\mathcal{M}_i},\mathscr{C}^{\mathcal{S}})\right)$ is defined as follows (Kim, 2019);

$$F^{RE}_{\mathcal{S}^{\mathcal{M}_i}_j}\left(\mathbb{S}^{\mathcal{M}_i},\ \mathcal{A}_{\mathcal{M}_i},\mathscr{C}^{\mathcal{S}}\right) = \mathscr{C}^{\mathcal{S}^{\mathcal{M}_i}_j} + \left[\frac{1}{|\mathbb{S}^{\mathcal{M}_i}|}\times\left(\mathcal{A}_{\mathcal{M}_i} - \sum_{\mathcal{S}^{\mathcal{M}_i}_j\in\mathbb{S}^{\mathcal{M}_i}}\mathscr{C}^{\mathcal{S}^{\mathcal{M}_i}_j}\right)\right] \tag{6.1.9}$$

The F^{ERE} for $\mathcal{S}^{\mathcal{M}_i}_j$ $\left(F^{ERE}_{\mathcal{S}^{\mathcal{M}_i}_j}\left(\mathbb{S}^{\mathcal{M}_i},\ H\left(\Gamma,\mathcal{A}_{\mathcal{M}_i}\right),\mathscr{C}^{\mathcal{S}}\right)\right)$ is defined as follows;

$$\gamma_{\mathcal{S}^{\mathcal{M}_i}_j}\times F^{ERE}_{\mathcal{S}^{\mathcal{M}_i}_j}\left(F^{ERE}_{\mathcal{S}^{\mathcal{M}_i}_j}\left(\mathbb{S}^{\mathcal{M}_i},\ H\left(\Gamma,\mathcal{A}_{\mathcal{M}_i}\right),\mathscr{C}^{\mathcal{S}}\right)\right) =$$

$$\left(\gamma_{\mathcal{S}_j^{\mathcal{M}_i}} \times \mathcal{C}^{\mathcal{S}_j^{\mathcal{M}_i}}\right) + \left(\frac{1}{|\mathbb{S}^{\mathcal{M}_i}|} \times \left(\mathcal{A}_{\mathcal{M}_i} - \sum_{\mathcal{S}_j^{\mathcal{M}_i} \in \mathbb{S}^{\mathcal{M}_i}} \left(\gamma_{\mathcal{S}_j^{\mathcal{M}_i}} \times \mathcal{C}^{\mathcal{S}_j^{\mathcal{M}_i}}\right)\right)\right) \tag{6.1.10}$$

$$\text{s.t.,} \begin{cases} \sum_{\mathcal{S}_j^{\mathcal{M}_i} \in \mathbb{S}^{\mathcal{M}_i}} \left(\gamma_{\mathcal{S}_j^{\mathcal{M}_i}} \times F^{ERE}_{\gamma_{\mathcal{S}_j^{\mathcal{M}_i}}}\left(\mathbb{S}^{\mathcal{M}_i}, H\left(\Gamma, \mathcal{A}_{\mathcal{M}_i}\right), \mathcal{C}^{\mathcal{S}}\right)\right) \\ \qquad = \mathcal{A}_{\mathcal{M}_i} \\ \Gamma = \left[\dots \gamma_{\mathcal{S}_j^{\mathcal{M}_i}} \dots\right] and \gamma_{\mathcal{S}_j^{\mathcal{M}_i}} = \psi^{\mathcal{S}_j} / \sum_{\mathcal{S}_h^{\mathcal{M}_i} \in \mathbb{S}^{\mathcal{M}_i}} \psi^{\mathcal{S}_h} \\ H\left(\Gamma, \mathcal{A}_{\mathcal{M}_i}\right) = \left\{ s = \left[\dots s_{\mathcal{S}_j} \dots\right] \in \mathbb{R}^{|\mathbb{S}^{\mathcal{M}_i}|} \middle| \right. \\ \qquad \left. \sum_{\mathcal{S}_j^{\mathcal{M}_i} \in \mathbb{S}^{\mathcal{M}_i}} \left(\gamma_{\mathcal{S}_j^{\mathcal{M}_i}} \times s_{\mathcal{S}_j}\right) \leq \mathcal{A}_{\mathcal{M}_i} \right\} \end{cases}$$

At the third-tier, each individual SBS assigns the allocated spectrum to associated mobile devices $\mathcal{D}^{\mathcal{S}} \in \mathbb{D}$. For the adaptive spectrum sharing among mobile devices, service requests from devices are categorized into two different traffic classes according to the required QoS: delay sensitive *class I* (real-time) traffic services and delay tolerant *class II* (non real-time) traffic services. Based on different tolerance characteristics, *class I* data type applications have higher weighting factors than flexible *class II* data type applications. In the third-tier bargaining game, traffic classes are assumed as game players, i.e., P^{class_I} and P^{class_II}, and the basic concept of Raiffa solution is adopted to characterize the spectrum assignment for *class I* and *class II* traffic services. Let $\left(x_{class\_I}, x_{class\_II}\right) \in S \subset \mathbb{R}^2$ be feasible payoff vectors for the P^{class_I} and P^{class_II}. $\left(S_{d^1}, d^1\right)$ is an initial bargaining problem where $d^1 = \left(d^1_{class\_I}, d^1_{class\_II}\right) \in S$ is an initial disagreement point (Kim, 2019).

In the first step, the Raiffa procedure determines the midpoint of the line segment $\left[\left(d^1_{class\_I}, proj_{class\_II}\left(S_{d^1}, d^1_{class\_I}\right)\right), \left(proj_{class\_I}\left(S_{d^1}, d^1_{class\_II}\right), d^1_{class\_II}\right)\right]$ Raffia procedure determines in a first step the midpoint of the line segment. In the second step, the midpoint d^2 of the line segment $\left[\left(d^1_{class\_I}, proj_{class\_II}\left(S_{d^1}, d^1_{class\_I}\right)\right), \left(proj_{class\_I}\left(S_{d^1}, d^1_{class\_II}\right), d^1_{class\_II}\right)\right]$ is chosen, leading to the new bargaining problem $\left(S_{d^2}, d^2\right)$. In the next step, we can get the next midpoint sequentially. Finally, d^n, which is the midpoint of the line segment $\left[\left(d^{n-1}_{class\_I}, proj_{class\_II}\left(S_{d^{n-1}}, d^{n-1}_{class\_I}\right)\right), \left(proj_{class\_I}\left(S_{d^{n-1}}, d^{n-1}_{class\_II}\right), d^{n-1}_{class\_II}\right)\right]$ is obtained. Based on the sequential Raiffa bargaining process, the d^n can converge to the Raiffa point (Kim, 2019).

6.1.6 Summary

In recent years, the rapid expansion of wireless communication services has resulted in a surge in application data traffic. To efficiently accommodate this exponential increase, the HetUDN has emerged as a key technology, offering substantial improvements in network spectrum efficiency and overall system performance. The *BHSA* scheme introduces an innovative three-tier spectrum allocation framework, leveraging *right-egalitarian*, *extended right-egalitarian*, and *sequential Raiffa bargaining solutions.*

Through a novel hierarchical interactive control mechanism, bargaining game models dynamically interact to establish a mutually beneficial outcome within the ever-evolving HetUDN environment (Kim, 2019).

6.2 Heterogeneous Network Bandwidth Control (*HNBC*) scheme

Non-orthogonal multiple access (NOMA) has recently emerged as a promising technology to achieve the objectives of 5G cellular systems. With its exceptional spectral efficiency, the NOMA has garnered significant attention and is expected to play a crucial role in enhancing the capacity of future heterogeneous networks (HetNets). In this context, S. Kim recently introduces the *Heterogeneous Network Bandwidth Control* (*HNBC*) scheme, which integrates both NOMA and orthogonal multiple access (OMA) techniques into a unified framework (Kim, 2020b).

To facilitate the efficient coordination of different multiple access methods, the *HNBC* scheme incorporates principles from cooperative bargaining solution concepts. In the *HNBC* scheme, bandwidth allocation for base stations and channels is managed through the OMA mode, while mobile devices operating within the same channel are grouped using the NOMA mode. The OMA mode employs the traditional Nash bargaining solution to distribute bandwidth resources, whereas in the NOMA mode, individual device groups negotiate both internally and across groups based on the group bargaining solution. In response to dynamically evolving HetNet environments, this joint bargaining strategy optimizes bandwidth efficiency while ensuring rational resource allocation (Kim, 2020b).

6.2.1 Development motivation

With the rapid advancement of wireless communication technologies, the number of mobile devices (MDs) has surged at an unprecedented rate. These devices require seamless mobile internet connectivity with ultra-high data throughput, minimal latency, and enhanced energy efficiency, enabling a wide range of advanced applications. The demands for 5G communication systems have escalated, aiming to achieve a system capacity 1,000 times greater and spectral efficiency 10 times higher than that of 4G networks. Consequently, future cellular networks will encounter significant challenges, including spectrum scarcity and massive connectivity requirements. To optimize spectrum utilization and accommodate a growing number of MDs with ubiquitous services, a new network infrastructure paradigm is essential. In response to these challenges, the Heterogeneous Network (HetNet) platform has gained considerable attention in both 5G and beyond 5G (B5G) cellular networks (Kim, 2020b; Liu, 2018b, 2019b; Xu, 2019).

Recently, HetNets have emerged as an advanced networking architecture structured around a hierarchy of macro cells and small cells. Macro cells provide extensive coverage for large geographical areas, while small cells - which operate on both licensed and unlicensed spectrum - enhance network capacity with low-power radio access nodes. These two types of cells coexist while integrating various radio access technologies and are interconnected through high-speed, low-latency backhaul links, ensuring optimal network management flexibility. Within the HetNet infrastructure, multiple small cells are deployed within each macro cell region, functioning cooperatively to significantly improve network capacity and user connectivity. With recent technological advancements, individual handheld MDs are now capable of supporting multiple radio access technologies, a feature that is becoming increasingly common in 5G wireless communications (Gerasimenko, 2015; Kim, 2020b).

The coexistence of different types of base stations (BSs) presents considerable challenges in bandwidth management within HetNet systems. To maximize overall network capacity, intelligent bandwidth control mechanisms must be employed. In 4G networks, orthogonal multiple access (OMA) has been widely used as the air interface technique, demonstrating effectiveness in mitigating multi-path fading and enhancing system throughput. However, traditional OMA can only serve one MD per scheduling period to avoid interference, which limits spectral resource utilization by restricting each MD to a specific portion of the available bandwidth. Consequently, a more efficient radio access technology is needed to fully utilize the constrained bandwidth resources (Al-Abbasi, 2017; Kim, 2020b; Liu, 2019a).

Beyond OMA, non-orthogonal multiple access (NOMA) has emerged as a promising alternative, offering improved bandwidth utilization and higher connectivity density. Unlike OMA, NOMA enables multiple MDs to simultaneously share the same frequency bands while introducing only minimal inter-MD interference. This approach not only enhances the effective bandwidth available to each MD but also increases the number of devices that can be scheduled beyond the physical resource limitations. NOMA achieves this by superimposing multiple MDs' message signals in the power domain, leveraging their distinct channel gain differences. Recently, both industry and academic researchers have recognized the significance of integrating OMA and NOMA techniques, seeking innovative control paradigms to address the bandwidth management challenges in 5G HetNets (Ali, 2016; Kim, 2020b; Xu, 2019).

The Nash Bargaining Solution (NBS), introduced by J. Nash in 1950, serves as a foundational concept for fair and optimal resource allocation (Kim, 2014). As a key principle in cooperative game theory, NBS provides a strategic approach to achieving mutually beneficial solutions that strike a balance between efficiency and fairness. Building upon this concept, S. Chae and P. Heidhues proposed group bargaining solutions (GBSs) in 2004, extending the original NBS framework (Chae, 2004). Their model conceptualizes the bargaining process at two levels: intra-group bargaining, which occurs among individuals within a single group, and inter-group bargaining, which takes place between different groups. In this hierarchical structure, multiple individuals participate in decision-making groups that cooperate at both levels. According to traditional NBS principles, these group bargaining solutions adhere to the NBS framework, ensuring fairness in both intra-group and inter-group bargaining scenarios (Chae, 2004; Kim, 2020b).

6.2.2 Contribution

Building on the concepts of the NS and GBS, the *HNBC* scheme introduces an innovative bandwidth management mechanism tailored for the HetNet platform. In the *HNBC* scheme, the bandwidth available at MBSs is distributed among their associated SBSs, with each base station allocating its assigned bandwidth across multiple channels. Using the OMA mode, bandwidth resources are divided orthogonally among base stations and channels. MDs are grouped based on their service types and communicate with their corresponding base stations through designated channels. Within each group, MDs share a single channel according to the NOMA mode. By combining NOMA and OMA technologies, the *HNBC* scheme leverages two bargaining solutions - the NBS and GBS - within the HetNet infrastructure. This integrated control strategy maximizes the combined benefits of Nash and group bargaining approaches while effectively addressing complex group negotiation challenges (Kim, 2020b).

6.2.3 Two-tier HetNet system infrastructure with small cells

We consider a two-tier cellular network platform, which comprises two types of BSs. At the upper tier, MBSs $\mathbb{M} = \{\mathcal{M}_1,\ldots,\mathcal{M}_n\}$ exist, and they can provide cellular network coverage for a large area.

At the lower tier, SBSs $\mathbb{S} = \{\mathcal{S}_1,\dots,\mathcal{S}_m\}$ coexist, and they break up a macro-cell site into much smaller areas to increase the macro-cell's edge data capacity and overall network efficiency. One MBS (*or* one SBS) has a coverage area of radius $r_{\mathcal{M}}$ (*or* $r_{\mathcal{S}}$) and has a static portion of bandwidth resource where $\mathfrak{M}_{\mathcal{M}_i \in \mathbb{M}}$ for the $\mathcal{M}_i$ and $\mathfrak{N}_{\mathcal{S}_j \in \mathbb{S}}$ for the $\mathcal{S}_j$. By using $\mathfrak{M}_{\mathcal{M}}$ and $\mathfrak{N}_{\mathcal{S}}$, MBSs and SBSs provide the basic services to their corresponding MDs in their covering areas. In the cellular area, there are multiple MDs $\mathbb{D} = \{\mathcal{D}_1,\dots,\mathcal{D}_k\}$; they are assumed randomly distributed, and are equipped with a single antenna to contact their corresponding BSs through wireless communications (Wang, 2018; Zhao, 2017). The general two-tier HetNet infrastructure is shown in Figure 2 (Kim, 2020b).

Figure 2. Two-tier HetNet platform with OMA and NOMA modes (Kim, 2020b)

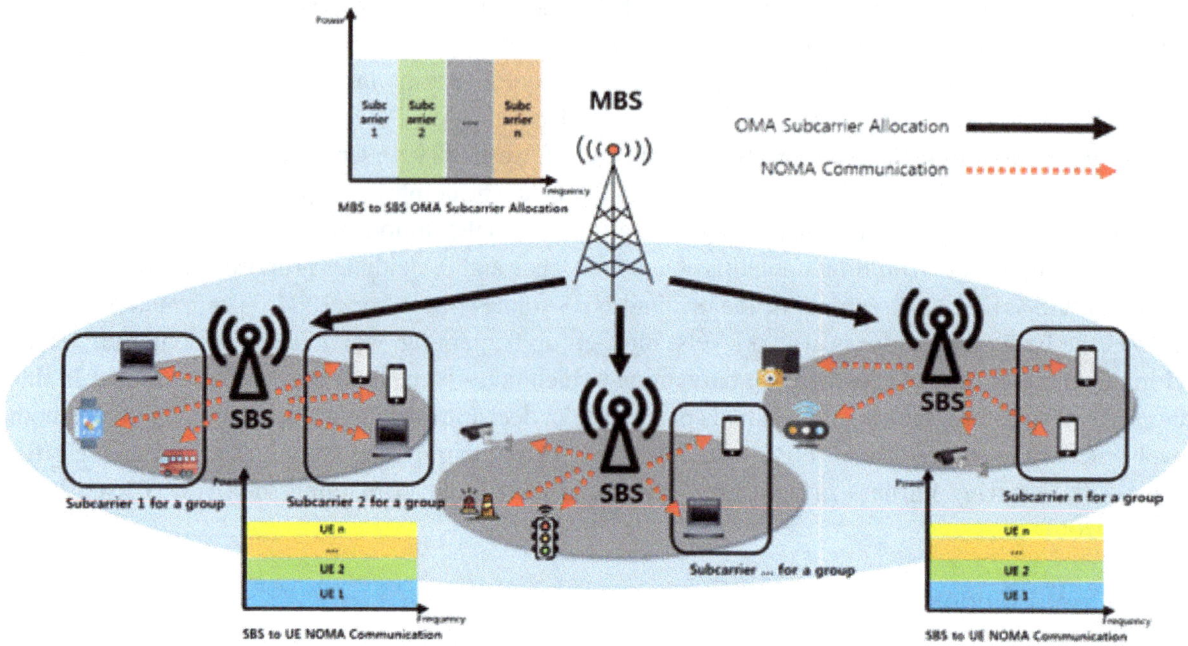

Each M (*or* S) divides its own available bandwidth into a set of channels, denoted by $\mathbb{L} = \{1,\dots,L\}$ (*or* $\mathcal{L} = \{1,\dots,l\}$). The *HNBC* scheme adopts the hybrid approach based on the combination NOMA and OMA technologies; it is realistic to effectively reduce the co-channel interference to perform the NOMA technique. Orthogonal bandwidth resources are allocated for different BSs and channels by using the OMA mode. In each BS, MDs in the same group share their corresponding channel. So, in sum, each group occupies one channel independently. In each group, MDs are operated by using the NOMA mode; the *HNBC* scheme assumes that the BS fully understands channel state information and decides the power levels of individual MDs (Kim, 2020b; Long, 2019).

In the *HNBC* scheme, the interactions between BSs and their corresponding MDs are formulated in a cooperative manner. In game theory, modeling situations are defined as a game to predict the outcome of complex interactions among game players. Formally, the *HNBC* scheme defines game entities, i.e.,

$\mathbb{G} = \left\{ \{\mathbb{M},\mathbb{S},\mathbb{D}\}, \left\{\mathfrak{M}_{\mathcal{M}_i},\mathfrak{N}^{\mathcal{M}_i}_{\mathcal{S}_j}\right\}, \{\mathbb{L},\mathscr{L}\}, \left\{\mathcal{T}^{\mathcal{M}_i}_{\mathcal{W}_{\mathcal{M}_i}},\mathcal{T}^{\mathcal{M}_i}_{\mathbb{L}},\mathcal{T}^{\mathcal{S}_j}_{\mathscr{L}}\right\}, \left\{\mathbb{P}^{\mathcal{M}_i}_{N_{\mathcal{M}_i}},\mathbb{P}^{\mathcal{S}_j}_{N_{\mathcal{S}_j}}\right\}, \left\{\mathfrak{X}_{\mathcal{M}_i},\mathfrak{X}_{\mathcal{S}_j}\right\}, \left\{U_{\mathcal{M}_i},U^{\mathcal{M}_i}_{\mathcal{S}_j},U^{\mathcal{M}_i}_{L},U^{\mathcal{S}_j}_{l},U^{l}_{\mathcal{D}_k}\right\},T \right.$ at each time period of gameplay (Kim, 2020b).

- $\{\mathbb{M},\mathbb{S},\mathbb{D}\}$ are the finite set of MBSs, SBSs and MDs; they are game players of the joint bargaining games.
- $\mathfrak{M}_{\mathcal{M}_i}$ is the total bandwidth amount for the $\mathcal{M}_i \in \mathbb{M}$, and $\mathfrak{N}^{\mathcal{M}_i}_{\mathcal{S}_j}$ is the assigned bandwidth amount for the $\mathcal{S}_j \in \mathcal{W}_{\mathcal{M}_i} \subset \mathbb{S}$ where $\mathcal{W}_{\mathcal{M}_i}$ is the set of SBSs within the $\mathcal{M}_i$'s area.
- L and L are the set of channels for MBS and SBS, respectively.
- $\mathcal{T}^{\mathcal{M}_i}_{\mathcal{W}_{\mathcal{M}_i}} = \mathcal{R}_{\mathcal{M}_i}, \ldots \mathcal{R}^{\mathcal{M}_i}_{\mathcal{S}_j \in \mathcal{W}_{\mathcal{M}_i}} \ldots$ is a $\left(1 + \left|\mathcal{W}_{\mathcal{M}_i}\right|\right)$-dimensional bandwidth allocation vector for the $\left\{\mathcal{M}_i \cup \mathcal{W}_{\mathcal{M}_i}\right\}$ where $\mathcal{R}_{\mathcal{M}_i}$ and $\mathcal{R}^{\mathcal{M}_i}_{\mathcal{S}_j}$ represent the bandwidth amount allocated for the $\mathcal{M}_i$ and $\mathcal{S}_j$, respectively.
- $\mathcal{T}^{\mathcal{M}_i}_{\mathbb{L}} = \mathfrak{C}^{\mathcal{M}_i}_{1}, \ldots, \mathfrak{C}^{\mathcal{M}_i}_{L}$ is a L-dimensional bandwidth allocation vector for the $\mathcal{M}_i$ where $\mathfrak{C}^{\mathcal{M}_i}_{L}$ represents the bandwidth amount allocated for the Lth channel in the $\mathcal{M}_i$.
- $\mathcal{T}^{\mathcal{S}_j}_{\mathscr{L}} = f^{\mathcal{S}_j}_1, \ldots, f^{\mathcal{S}_j}_l$ is a l-dimensional bandwidth allocation vector for the $\mathcal{S}_j$ where $f^{\mathcal{S}_j}_l$ represents the bandwidth amount allocated for the 1th channel in the $\mathcal{S}_j$.
- $\mathbb{P}^{\mathcal{M}_i}_{N_{\mathcal{M}_i}} = \ldots \mathcal{P}^{\mathcal{M}_i}_{\mathcal{D}_e \in N_{\mathcal{M}_i}} \ldots$ is a $\left|N_{\mathcal{M}_i}\right|$-dimensional power level vector for MDs where $N_{\mathcal{M}_i}$ is the set of MDs contacting to the $\mathcal{M}_i$, and $\mathcal{P}^{\mathcal{M}_i}_{\mathcal{D}_e}$ represents the power level of the $\mathcal{D}_e$ in the $\mathcal{M}_i$.
- $\mathbb{P}^{\mathcal{S}_j}_{N_{\mathcal{S}_j}} = \ldots P^{\mathcal{S}_j}_{\mathcal{D}_k \in N_{\mathcal{S}_j}} \ldots$ is a $\left|N_{\mathcal{S}_j}\right|$-dimensional power level vector for MDs where $N_{\mathcal{S}_j}$ is the set of MDs within the $\mathcal{S}_j$, and $P^{\mathcal{S}_j}_{\mathcal{D}_k}$ represents the power level of the $\mathcal{D}_k$ in the $\mathcal{S}_j$.
- $\mathfrak{X}_{\mathcal{M}_i} = \left\{\mathfrak{P}^{\mathcal{M}_i}_{min}, \ldots, \mathfrak{P}^{\mathcal{M}_i}_{max}\right\}$ is the set of available MBS power levels where $\mathcal{P}^{\mathcal{M}_i}_{\mathcal{D}_e} \in \mathfrak{X}_{\mathcal{M}_i}$, and $\mathfrak{X}_{\mathcal{S}_j} = \left\{\mathfrak{P}^{\mathcal{S}_j}_{min}, \ldots, \mathfrak{P}^{\mathcal{S}_j}_{max}\right\}$ is the set of available SBS power levels where $P^{\mathcal{S}_j}_{\mathcal{D}_k} \in \mathfrak{X}_{\mathcal{S}_j}$.
- $U_{\mathcal{M}_i}, U^{\mathcal{M}_i}_{\mathcal{S}_j}, U^{\mathcal{M}_i}_{L}, U^{\mathcal{S}_j}_{l}, U^{l}_{\mathcal{D}_k}$ are utility functions of $\mathcal{M}_i$, $\mathcal{S}_j$, L[th] channel in M, 1th channel in S, and $\mathcal{D}_k$, respectively.
- $T = \left\{t_1, \ldots, t_c, t_{c+1}, \ldots\right\}$ denotes time, which is represented by a sequence of time steps.

6.2.4 The basic concepts of cooperative bargaining solutions

The traditional NBS can be extended to encompass a broader class of group bargaining scenarios. In the GBS framework, NBS is applied both within individual groups and between different groups. At the inter-group level, groups negotiate with one another to establish feasible resource allocations, while at the intra-group level, members within each group engage in bargaining to determine their respective payoffs. Conceptually, bargaining occurs simultaneously at two levels - among groups and among individual members within each group - creating an interdependent negotiation process (Chae, 2004). When bargaining theory is utilized for OMA control, bandwidth resources are allocated fairly and efficiently using the NBS framework. In contrast, when applied to NOMA control, at least one bargaining entity typically consists of a group of MDs sharing the same frequency band. The GBS approach addresses the challenges of NOMA bandwidth management by incorporating a bargaining model where negotiations take place both within individual decision-making groups and between multiple groups (Chae, 2004). Under the GBS model, decision-making groups function as single bargaining agents, and negotiations are conducted using an extended version of the classical Nash bargaining model (Kim, 2020b).

The *HNBC* scheme introduces the notation and basic definitions of bargaining solution. Assuming that there are non-overlapping groups $G_1, \ldots, G_m$ of individuals. Denote the set of all individuals by $\mathbb{N} = \{1, \ldots, n\}$, and the group structure by $\mathbb{G} = \left\{G_1, \ldots, G_m\right\}$, which is formally a partition of the set N where

$n \leq m$. A payoff vector is an element of the payoff space $R^{\mathbb{N}}$, which is the n-dimensional Euclidean space indexed by the set of individuals. A feasible set S is a subset of the payoff space, and a disagreement point d is an element of the payoff space. The points in S represent the feasible utility levels that the individuals can reach if they agree. Otherwise, if the agreement is not reached, they obtain the utility levels given by the disagreement point (Chae, 2004; Kim, 2020b).

A bargaining problem is then specified as triple $(\mathbb{G}, S, d)$ consisting of a group structure, feasible set, and disagreement point. The traditional bargaining problem $(\mathbb{N}, S, d)$ can be regarded as a special case of the new bargaining problem where each individual forms one group, that is, $\mathbb{G} = \{\{1\}, ..., \{n\}\}$. A bargaining solution is a function F that associates to each problem $(\mathbb{G}, S, d)$ a feasible payoff vector $F(\mathbb{G}, S, d)$. Assume that the group G_j is replaced by a new group $\{j^*\}$ that consists of a single representative member $j^* \in G_j$. Denote the new set of individuals excluding all members of G_j but j^* by $\mathbb{N}^j \equiv (\mathbb{N} \backslash G_j) \cup \{j^*\}$, and denote the new group structure by $\mathbb{G}^j \equiv \{G_1, \ldots, G_{j-1}, \{j^*\}, G_{j+1}, \ldots, G_m\}$. To generate the reduced feasible set S^j from S, we set

$$S^j \equiv \{u \in R^{\mathbb{N}^j} : \text{there exists} v \in S \text{such that} v_{j^*} = u_i \text{for} i \in \mathbb{N} \backslash G_j \text{and} v_i = u_{j^*} \text{for all} i \in G_j\}.$$

In addition, generate the new disagreement point d^j from the original disagreement point d by replacing the payoff vector for the group $(d_i)_{i \in G_j}$ with the single payoff d_{j^*}. Thus, we have formally constructed the reduced bargaining problem $(\mathbb{G}^j, S^j, d^j)$ (Chae, 2004; Kim, 2020b).

For the general solution, let $r(c_j)$ represent the number of rights to talk that a group with c_j members maintain. As a homogenous group G_j was reduced to a singleton group $\{j^*\}$, we can replace group G_j with $r(c_j)$ identical singleton groups by replicating a representative member $j^* \in G_j$. Denote the new set of individuals by $\mathbb{N}^{j,r} \equiv (\mathbb{N} \backslash G_j) \cup \{j^1, ..., j^{r(c_j)}\}$, the new group structure by $\mathbb{G}^{j,r} \equiv \{G_1, ..., G_{j-1}, \{j^1\}, ..., \{j^{r(c_j)}\}, G_{j+1}, ..., G_m\}$, and the new feasible set by

$$S^{j,r} \equiv \{u \in R^{\mathbb{N}^{j,r}} : \text{there exists } v \in S \text{ such that } v_i = u_i \text{for } i \in \mathbb{N} \backslash G_j \text{ and } v_i = u_{j^k} \text{ for all } i \in G_j \text{ and } k = 1, ..., r(c_j)\}$$

(Chae, 2004; Kim, 2020b).

In the context of traditional bargaining problems, J. Nash demonstrated the existence of a unique solution that satisfies four fundamental properties, known as the NBS. This solution is derived by solving the following maximization problem (Chae, 2004; Kim, 2020b):

$$\max_{\mathbb{U} \in S, \mathbb{U} \geq d} \prod_{i=1}^{n} (u_i - d_i), \text{ s.t.}, \mathbb{U} = [u_1, \ldots, u_n] \text{ and } i \in \mathbb{N} \tag{6.2.1}$$

Building on the classical NBS, S. Chae and P. Heidhues conceptualized a group as a single bargaining entity and introduced a new GBS, referred to as NBS with the *RHG* property (NBS-*R*). This solution models a scenario where a decision-making group negotiates with external parties. As a result, the bargaining problem is structured into two hierarchical levels: inter-group bargaining, which occurs between different groups, and intra-group bargaining, where members within a group negotiate among themselves. This approach broadens the analytical framework for group bargaining problems. The NBS-*R* solution is derived by solving the following maximization problem (Chae, 2004; Kim, 2020b):

$$\max_{U \in S, U \geq d} \prod_{j=1}^{m} \left(\prod_{i \in G_j} (u_i - d_i)^{\mathcal{H}(G_j)} \right), \text{ s.t.}, \begin{cases} G_j \in G = \{G_1, \ldots, G_m\} \\ H(G_j) = \frac{1}{|G_j|} \end{cases} \tag{6.2.2}$$

where $|G_j|$ is the cardinality, i.e., the number of the members, of G_j. The NBS-R maximizes the weighted product of net utilities, where the weight for each individual is the reciprocal of the size of the group to which he belongs. The NBS-R can be generalized (NBS-**r**). For the general solution, let $r(G_j)$ represent the number of rights to talk that a group G_j maintains. Finally, the NBS-**r** is given by (Chae, 2004; Kim, 2020b);

$$\max_{U \in S, U \geq d} \prod_{j=1}^{m} \left(\prod_{i \in G_j} (u_i - d_i)^{\mathcal{H}(G_j)} \right)^{r(G_j)} \tag{6.2.3}$$

If $r(G_j) = 1$, the members' rights are reduced to a single right. If $r(G_j) = |G_j|$, the members of a group maintain their rights to talk. This is equivalent to the standard Nash solution ignoring the group structure. If $1 < r(G_j) < |G_j|$, the rights are reduced but not to a single right. If $r(G_j) > |G_j|$, the rights actually increase with the formation of a bargaining group. This class of solutions $(r(G_j) > |G_j|)$ can provide an independent reason why a group may form even in a pure bargaining situation (Kim, 2020b).

6.2.5 The bargaining control scheme for the OMA-NOMA system

The *HNBC* scheme concerns the appropriate bandwidth control algorithms for the hierarchical Het-Net platform. First, each MBS ($\mathcal{M}$) distributes its bandwidth resource $(\mathfrak{M}_{\mathcal{M}})$ to the set of SBSs $(\mathcal{W}_{\mathcal{M}})$ within the$\mathcal{M}$'s area. Therefore, the individual M partitions the $\mathfrak{M}_{\mathcal{M}}$ into disjoint portions, and allocates them for corresponding SBSs. This OMA based bandwidth allocation process can avoid complex interference management in the wireless network topology. To implement this process for the $\{\mathcal{M}_i \cup \mathcal{W}_{\mathcal{M}_i}\}$, the *HNBC* scheme focuses on the traditional NBS. As a fair-efficient solution, the $\mathcal{T}_{\mathcal{W}_{\mathcal{M}_i}}^{\mathcal{M}_i}$ is obtained as follows (Kim, 2020b);

$$\max_{\mathcal{T}_{\mathcal{W}_{\mathcal{M}_i}}^{\mathcal{M}_i} = \mathcal{R}_{\mathcal{M}_i}, \ldots \mathcal{R}_{\mathcal{S}_j \in \mathcal{W}_{\mathcal{M}_i}}^{\mathcal{M}_i}, \ldots} \left(\left(U_{\mathcal{M}_i}(\mathcal{R}_{\mathcal{M}_i}) - d_{\mathcal{M}_i} \right) \times \prod_{\mathcal{S}_j \in \mathcal{W}_{\mathcal{M}_i}} \left(U_{\mathcal{S}_j}^{\mathcal{M}_i}(\mathcal{R}_{\mathcal{S}_j}^{\mathcal{M}_i}) - d_{\mathcal{S}_j} \right) \right) \tag{6.2.4}$$

$$\text{s.t.,}\begin{cases} U_{\mathcal{M}_i}\left(\mathcal{R}_{\mathcal{M}_i}\right) = \Psi_{\mathcal{M}} \times \log\left(\frac{\min\left(\mathcal{R}_{\mathcal{M}_i}, r^{t_c}_{\mathcal{M}_i}\right)}{\mathfrak{M}_{\mathcal{M}_i}} + \varrho\right) \\ U^{\mathcal{M}_i}_{\mathcal{S}_j}\left(\mathcal{R}^{\mathcal{M}_i}_{\mathcal{S}_j}\right) = \Psi_{\mathcal{S}} \times \log\left(\frac{\min\left(\mathcal{R}^{\mathcal{M}_i}_{\mathcal{S}_j}, r^{t_c}_{\mathcal{S}_j}\right)}{\mathfrak{M}_{\mathcal{M}_i}} + \xi\right) \\ \left(\mathcal{R}_{\mathcal{M}_i} + \sum_{\mathcal{S}_j \in \mathcal{W}_{\mathcal{M}_i}} \mathcal{R}^{\mathcal{M}_i}_{\mathcal{S}_j}\right) \leq \mathfrak{M}_{\mathcal{M}_i} \end{cases}$$

where $\Psi_{\mathcal{M}}$ and $\Psi_{\mathcal{S}}$ are profit coefficient factors, and ϱ and ξ are control parameters for the $U(\cdot)$. At time t_c, $r^{t_c}_{\mathcal{M}_i}$ and $r^{t_c}_{\mathcal{S}_j}$ are the total requested bandwidth amounts in the $\mathcal{M}_i$ and $\mathcal{S}_j$, respectively. And then, each individual M splits its assigned bandwidth resource $\left(\mathcal{R}_{\mathcal{M}}\right)$ to different L channels. Based on the OMA mode, the bandwidth assignment for each channel is also implemented according to the NBS. Within the $\mathcal{M}_i$, $\mathcal{T}^{\mathcal{M}_i}_{\mathrm{L}}$ is obtained as follows (Kim, 2020b);

$$\max_{\mathcal{T}^{\mathcal{M}}_{\mathrm{L}} = \mathfrak{C}^{\mathcal{M}_i}_1, \ldots, \mathfrak{C}^{\mathcal{M}_i}_L} \prod_{L \in \mathrm{L} = \{1, \ldots, L\}} \left(U^{\mathcal{M}_i}_L\left(\mathfrak{C}^{\mathcal{M}_i}_L\right) - d^{\mathcal{M}_i}_L\right) \tag{6.2.5}$$

$$\text{s.t.,}\begin{cases} U^{\mathcal{M}_i}_L\left(\mathfrak{C}^{\mathcal{M}_i}_L\right) = \eta_L \times \log\left(\frac{\min\left(\mathfrak{C}^{\mathcal{M}_i}_L, r^{t_c}_L\right)}{\mathcal{R}_{\mathcal{M}_i}} + \alpha\right) \\ \sum_{b \in \mathrm{L}} \mathfrak{C}^{\mathcal{M}_i}_b \leq \mathcal{R}_{\mathcal{M}_i} \end{cases}$$

where η_L and α are control parameters for the $U^{\mathcal{M}}_L(\cdot)$, and $r^{t_c}_L$ is the total requested bandwidth amount in the channel L of $\mathcal{M}_i$ at time t_c. Recursively, each S in $\mathcal{W}_{\mathcal{M}_i}$ splits its assigned resource $\left(\mathcal{R}^{\mathcal{M}}_{\mathcal{S}}\right)$ to different l channels by using the OMA mode. Therefore, $\mathcal{T}^{\mathcal{S}_j}_{\mathcal{L}} = f^{\mathcal{S}_j}_1, \ldots f^{\mathcal{S}_j}_l$ is obtained as the same manner as the equation (6.2.5) (Kim, 2020b).

Finally, each BS decides power levels for its corresponding MDs by using the NOMA mode. Usually, 5G cellular networks are expected to support multimedia services. The *HNBC* scheme assumes that different multimedia services over MDs can be categorized into four types according to their characteristics, i.e., type I, II, III and IV applications. Based on the type, multiple applications are grouped, and each group is treated as a unit under diverse traffic environments. For each group, the *HNBC* scheme assigns an orthogonal bandwidth portion, i.e., channel, separately. In the *HNBC* scheme, each individual MBS and SBS adopt the NOMA technique to share the channel bandwidth for same group applications. Due to its high bandwidth efficiency nature, this approach is necessary to achieve the strategic advantage within a cell area. For the NOMA operation in the MBS, the *HNBC* scheme concerns the number of MDs in each channel. Therefore, the NBS-*R* is applied to decide each MD's power level in the MBS, and $\mathbb{P}^{\mathcal{M}_i}_{N_{\mathcal{M}_i}} = \ldots \mathcal{P}^{\mathcal{M}_i}_{\mathcal{D}_e \in N_{\mathcal{M}_i}} \ldots$ is obtained as follows (Kim, 2020b);

$$\max_{\mathbb{P}^{\mathcal{M}_i}_{N_{\mathcal{M}_i}} = \ldots \mathcal{P}^{\mathcal{M}_i}_{\mathcal{D}_e \in N_{\mathcal{M}_i}} \ldots} \prod_{L \in \mathrm{L}} \left(\prod_{\mathcal{D}_e \in \grave{\mathrm{E}}^L_{\mathcal{M}_i}} \left(U^L_{\mathcal{D}_e}\left(\mathcal{P}^{\mathcal{M}_i}_{\mathcal{D}_e}, \mathbb{P}^{\mathcal{M}_i}_{\grave{\mathrm{E}}^L_{\mathcal{M}_i}}\right) - d^L_{\mathcal{D}_e}\right)^{\frac{1}{\left|\grave{\mathrm{E}}^L_{\mathcal{M}_i}\right|}}\right)$$

(6.2.6)

$$\text{s.t.,}\begin{cases} U^L_{\mathcal{D}_e}\left(\mathcal{P}^{\mathcal{M}_i}_{\mathcal{D}_e},\mathbb{P}^{\mathcal{M}_i}_{\grave{E}_{\mathcal{M}_i}}\right) = \dfrac{\left(\omega_{\mathcal{D}_e}\times\mathcal{P}^{\mathcal{M}_i}_{\mathcal{D}_e}\right)}{\left(\left(\Upsilon\times\left(\left(\sum_{\mathcal{D}_d\in\grave{E}^L_{\mathcal{M}_i}}\mathcal{P}^{\mathcal{M}_i}_{\mathcal{D}_d}\right)-\mathcal{P}^{\mathcal{M}_i}_{\mathcal{D}_e}\right)\right)+\Gamma_{\mathcal{D}_e}\right)} \\ \omega_{\mathcal{D}_e} = r^{t_c}_{L},\ \Gamma_{\mathcal{D}_e} = \left|\grave{E}^L_{\mathcal{M}_i}\right| and \sum_{\mathcal{D}_e\in\grave{E}^L_{\mathcal{M}_i}}\mathcal{P}^{\mathcal{M}_i}_{\mathcal{D}_e} \le \mathscr{E}^{\mathcal{M}} \end{cases}$$

where Υ is an orthogonality factor for wireless communications. $\omega_{\mathcal{D}_e}, \Gamma_{\mathcal{D}_e}$ are a path gain from the BS, and a background noise to the $\mathcal{D}_e$, respectively. $\grave{E}^L_{\mathcal{M}_i}$ is the set of MDs on the channel L of $\mathcal{M}_i$. $\mathbb{P}^{\mathcal{M}_i}_{\grave{E}_{\mathcal{M}_i}}$ is a $\left|\grave{E}^L_{\mathcal{M}_i}\right|$-dimensional vector to represent power levels of MDs in $\grave{E}^L_{\mathcal{M}_i}$. $\mathscr{E}^{\mathcal{M}}$ is the maximum power limit of each MBS; it allocates power to each MD within the power limit. Therefore, the sum of powers allocated to all MDs in the channel L cannot exceed the power limit. Each MD has its own utility function that represents the degree of MD's satisfaction of the received communication service, and it is a function of the generic signal quality for a MD. According to (6.2.6), the $\mathcal{D}_e$'s utility function $U^L_{\mathcal{D}_e}(\cdot)$ depends not only on its own power level, but also on the power levels of all the other MDs in the same channel (Lee, 2005). In the case of SBS, the *HNBC* scheme considers the relative power of different channel groups. Therefore, the NOMA operation in the $\mathcal{S}_j$ is implemented by using the NBS-*r*, and $\mathbb{P}^{\mathcal{S}_j}_{N_{\mathcal{S}_j}} = \ldots P^{\mathcal{S}_j}_{\mathcal{D}_k\in N_{\mathcal{S}_j}}\ldots$ is obtained as follows (Kim, 2020b);

$$\max_{\mathbb{P}^{\mathcal{S}_j}_{N_{\mathcal{S}_j}}=\ldots P^{\mathcal{S}_j}_{\mathcal{D}_k\in N_{\mathcal{S}_j}}\ldots}\prod_{l\in\mathcal{L}}\left(\prod_{\mathcal{D}_k\in\grave{E}^l_{\mathcal{S}_j}}\left(U^l_{\mathcal{D}_k}\left(P^{\mathcal{S}_j}_{\mathcal{D}_k},\mathbb{P}^{\mathcal{S}_{ij}}_{\grave{E}_{\mathcal{S}_j}}\right)-d^l_{\mathcal{D}_k}\right)^{r\left(\grave{E}^l_{\mathcal{S}_j}\right)}\right)$$

(6.2.7)

$$\text{s.t.,}\begin{cases} U^l_{\mathcal{D}_k}\left(P^{\mathcal{S}_j}_{\mathcal{D}_k},\mathbb{P}^{\mathcal{S}_{ij}}_{\grave{E}_{\mathcal{S}_j}}\right) = \dfrac{\left(\omega_{\mathcal{D}_k}\times P^{\mathcal{S}_j}_{\mathcal{D}_k}\right)}{\left(\left(\psi\times\left(\left(\sum_{\mathcal{D}_d\in\grave{E}^l_{\mathcal{S}_j}}P^{\mathcal{S}_j}_{\mathcal{D}_d}\right)-P^{\mathcal{S}_j}_{\mathcal{D}_e}\right)\right)+\Gamma_{\mathcal{D}_e}\right)} \\ r\left(\grave{E}^l_{\mathcal{S}_j}\right) = \div(l),\ \omega_{\mathcal{D}_e} = r^{t_c}_{l},\ \Gamma_{\mathcal{D}_e} = \left|\grave{E}^l_{\mathcal{S}_j}\right| and \sum_{\mathcal{D}_k\in\grave{E}^l_{\mathcal{S}_j}}P^{\mathcal{S}_j}_{\mathcal{D}_k} \le \mathscr{E}^{\mathcal{S}} \end{cases}$$

where ψ is an orthogonality factor for wireless communications. $\grave{E}^l_{\mathcal{S}_j}$ is the set of MDs on the channel l of $\mathcal{S}_j$. $\mathbb{P}^{\mathcal{M}_i}_{\grave{E}_{\mathcal{S}_j}}$ is a $\left|\grave{E}^l_{\mathcal{S}_j}\right|$-dimensional vector to represent power levels of MDs in $\grave{E}^l_{\mathcal{S}_j}$. $r\left(\grave{E}^l_{\mathcal{S}_j}\right)$ is the function of application types; it is dependent on channels. $\mathscr{E}^{\mathcal{S}}$ is the maximum power limit of each SBS's channel. Each MD in SBSs has same characteristics as the MDs in MBSs. Therefore, $U^l_{\mathcal{D}_k}(\cdot)$ is designed as the same manner as the $U^L_{\mathcal{D}_e}(\cdot)$ (Kim, 2020b).

6.2.6 Summary

In wireless communication systems, limited bandwidth must be shared among multiple users, making efficient bandwidth management crucial for optimizing the performance of future wireless networks. The *HNBC* scheme addresses the 5G HetNet bandwidth control problem, accommodating diverse application requirements. This *HNBC* scheme integrates NBS and GBS concepts within a hybrid OMA-NOMA system framework. Within the OMA-enabled bandwidth control process, the NBS dynamically allocates orthogonal bandwidth resources across different SBSs and channels. In the NOMA-enabled bandwidth control process, two GBS models are employed to determine the power levels for MDs: NBS-*R* for MBSs and NBS-***r*** for SBSs. Under the dynamic HetNet environment, various bargaining solutions are seamlessly integrated within the *HNBC* scheme, working cooperatively to enhance network efficiency. At each time interval, the limited bandwidth resources are managed in a distributed manner, enabling the system to adapt to real-time network fluctuations while maintaining a well-balanced overall performance (Kim, 2020b).

6.3 Heterogeneous Network Spectrum Allocation (*HNSA*) scheme

The 5G network is designed to accommodate the surging demand for mobile data traffic while ensuring enhanced QoS. One promising approach to managing this traffic explosion is the Heterogeneous Network (HetNet) platform. However, the deployment of HetNets presents several technical challenges that must be addressed. To optimize the utilization of the limited HetNet spectrum resources, S. Kim recently proposes the *Heterogeneous Network Spectrum Allocation* (*HNSA*) scheme (Kim, 2020c). The *HNSA* scheme is built upon key principles from the *relative utilitarian*, *equality-averse*, and *inequality-averse bargaining solutions*. It follows a three-stage spectrum allocation process, aiming to strike a balance between efficiency and fairness. By leveraging the distinctive properties of each bargaining solution, the *HNSA* scheme applies these solutions in a hierarchical manner; i) spectrum assignment among base stations, ii) spectrum allocation among small-cell base stations, and iii) spectrum sharing among applications. Through the cooperative bargaining approach, the *HNSA* scheme effectively maximizes resource utilization while ensuring a rational and equitable distribution of spectrum resources (Kim, 2020c).

6.3.1 Development motivation

Since the emergence of cellular mobile communications in the 1980s, there has been remarkable progress in both research and technological advancements, leading to the widespread deployment of wireless communication networks. From the basic voice calls of 1G networks to the multimedia-rich applications in 5G, network operators have continually focused on addressing the ever-growing demand for bandwidth. With the rapid proliferation of smartphones, tablets, wearable devices, and laptops, the need for high data rate services has significantly increased. According to a Cisco report, global mobile data traffic was projected to grow nearly tenfold between 2014 and 2019, reaching 24.3 exabytes per month by 2019. Researchers in the 5G communications field widely agree that the conventional single-tier macro cellular network is insufficient to handle the soaring data demands expected in the future (Kim, 2020c; Liu, 2016).

To address this traffic explosion, the Heterogeneous Network (HetNet) architecture has recently been introduced as a potential solution. It represents a significant paradigm shift in cellular networks, introducing new research challenges. The HetNet model aims to improve coverage, enhance spatial spectrum reuse, and boost throughput, particularly in response to the growing congestion in 3G and 4G networks. Within a HetNet, low-power small base stations (SBSs) are deployed alongside traditional macro cells, forming a multi-tier network that offloads macro cell traffic. This architecture supports hotspot distributions and enhances spectral efficiency. However, the complex topology of HetNets presents challenges in spectrum resource allocation among SBSs. To fully exploit the advantages of HetNets, designing intelligent resource management protocols has become a key research focus for both academia and industry. Leading companies such as Alcatel-Lucent, Nokia Siemens, AT&T, and Qualcomm are actively working on HetNet management strategies (Kim, 2020c; Lin, 2015; Zhou, 2015).

Traditional single-tier macro cellular networks provide broad coverage for mobile devices. However, the HetNet platform introduces varied SBS types and small cells, creating new challenges. In macro cell areas, mobile devices located far from their base stations (BSs) often transmit data at higher power levels, causing strong interference to neighboring SBSs. One possible solution is to partition the total spectrum into disjoint segments, assigning each SBS a dedicated portion, thereby mitigating cross-tier interference in irregular network topologies. Given the scarcity and high cost of wireless spectrum, the design of efficient spectrum allocation protocols is crucial for optimizing spectrum utilization (Lin, 2015). Over the past few years, various spectrum allocation techniques have been proposed for multi-tier HetNet systems. However, research on HetNets is still in its early stages, and many technical challenges must be addressed to develop a practical and efficient spectrum allocation framework (Kim, 2020c).

To tackle these challenges, the *HNSA* scheme has been developed specifically for the HetNet platform. Given the complexity and scale of multi-tier network infrastructures, the *HNSA* scheme follows a three-phase spectrum distribution process; i) phase 1 - spectrum is allocated among BSs, ii) phase 2 - each BS further divides its allocated spectrum among its corresponding SBSs, and iii) phase 3 - each SBS assigns its spectrum to user devices for processing application tasks. By integrating these three phases, the resource allocation mechanism is designed to enhance self-adaptability and real-time efficiency. More importantly, the *HNSA* scheme leverages interdependence between the three allocation stages, ensuring an optimized and balanced approach. However, identifying the optimal solution remains a challenging task, necessitating the development of a new resource allocation model (Kim, 2020c).

Using game theory, many resource allocation problems can be effectively modeled by defining the set of possible agreements among network players. This model is known as a bargaining game, and the solution that assigns an optimal allocation to each bargaining scenario is referred to as a bargaining solution. Bargaining solutions have gained significant attention in resource allocation research due to their effectiveness in ensuring fair and optimal resource distribution. The *HNSA* scheme integrates three distinct bargaining solutions to optimize HetNet spectrum allocation at each phase. These solutions are interlinked and interdependent, working cooperatively to efficiently distribute the limited spectrum resources while maintaining a fair and balanced allocation (Kim, 2020c).

6.3.2 Contribution

To fully realize the potential benefits of HetNet platforms, various novel concepts and protocols have been introduced. While numerous methods have been proposed, the multi-tier spectrum allocation problem remains a challenge, as no existing approach has perfectly balanced the conflicting requirements to achieve an optimal system performance. The *HNSA* scheme addresses this issue by dividing the spectrum allocation process into three distinct phases. At the first phase, the *HNSA* scheme captures dynamic interactions among multiple BSs in a coordinated manner and allocates spectrum resources based on the *Relative Utilitarian Bargaining Solution* (*RUBS*). At the second phase, the *Equality Averse Bargaining Solution* (*EABS*) is employed to distribute the assigned BS spectrum among its corresponding

SBSs. At the third phase, the *Inequality Averse Bargaining Solution* (*IABS*) is utilized to allocate the SBS spectrum resources for application tasks (Kim, 2020c).

Each of these three bargaining solutions is rooted in the fundamental principle of rationality, ensuring that the utility space is maximized according to a well-defined ordering mechanism. By integrating these bargaining solutions within a hierarchical interactive framework, the *HNSA* scheme creates an interdependent allocation mechanism, enabling the system to effectively balance network performance. The primary challenge of the *HNSA* scheme is designing a novel spectrum resource allocation process that effectively harnesses the hierarchical interactions among the *RUBS*, *EABS*, and *IABS* models. The synergistic effect of this approach lies in the reciprocal integration of these three bargaining solutions, allowing them to complement and enhance one another. To maximize the benefits of this combined strategy, the *HNSA* scheme implements an integrated three-phase bargaining game model, which optimally leverages the synergistic and complementary properties of HetNet services (Kim, 2020c).

6.3.3 Multi-utilitarian bargaining solutions

Cooperative bargaining problems involve a group of game players and a set of feasible utility payoffs. Over time, various bargaining solutions have been introduced, each based on slightly different assumptions regarding the desired allocation properties. The *HNSA* scheme incorporates the core principles of multi-utilitarian bargaining solutions, which are rationalized through piecewise linear increasing utility functions. These functions define game players' utility gains based on a set of weighted utilitarian criteria and the decision rules used to aggregate them (Hinojosa, 2007, 2008; Kim, 2020c).

Multi-utilitarian bargaining solutions can be classified into different categories based on their defining characteristics. Using reported utility functions, a bargaining planner determines an allocation that maximizes the weighted sum of reported preferences. In a specific case of multi-utilitarian bargaining solutions, each player's weight is determined as the inverse of their maximum utility gain. This approach is referred to as the *Relative Utilitarian Bargaining Solution* (*RUBS*). If a conservative principle is applied when integrating different utilitarian criteria, then the outcome is determined by the minimum utility value among the options. This method is known as the *Inequality Averse Bargaining Solution* (*IABS*). Additionally, when feasible outcomes are evaluated based on the highest utility value, a distinct class of solutions emerges. This approach, which tends to favor extreme utilitarian payoffs, is called the *Equality Averse Bargaining Solution* (*EABS*). Like the other methods, the *EABS* relies on piecewise linear functions and weighed utilitarian criteria to define allocations (Hinojosa, 2007, 2008; Kim, 2020c).

An n-player bargaining problem can be described by a set of feasible utility vectors, $S \subseteq \mathbb{R}^n_+$, where the *HNSA* scheme assumes that S is convex, compact, and comprehensive and that there exists $x \in S \cap \mathbb{R}^n_{++}$. The set of n-player bargaining problems is denoted by $\sum$. Let $\ddot{A}^n$ denote the simplex in $\mathbb{R}^n_+$, $\ddot{A}^n = \{\lambda \in \mathbb{R}^n_+ | \lambda \cdot 1^n = 1\}$. A polyhedric subset $\Lambda \subseteq \ddot{A}^n$ is the intersection of $\ddot{A}^n$ and a finite set of half-spaces in $\mathbb{R}^n$. For each $S \in \sum$, the *RUBS* is simply defined as follows (Hinojosa, 2007, 2008; Kim, 2020c);

$$U_\lambda(S) = \left\{x \in S \middle| \sum_{i=1}^{n} (\lambda_i \times x_i) \geq \sum_{i=1}^{n} (\lambda_i \times y_i) \text{ for all } y \in S\right\} \text{ with} \in \mathbb{R}^n_{++} \tag{6.3.1}$$

The concept of the *IABS* is introduced as a method for analyzing cooperative bargaining problems. The rationale behind the *IABS* is a fusion of both the egalitarian and utilitarian principles. This approach seeks to strike a balance between these two principles by adopting a conservative stance, wherein the outcome is evaluated based on the minimum value derived from different utilitarian functions. The *IABS* framework is structured as follows; i) multiple bargaining planners employ different weighted utilitarian criteria to assess feasible outcomes, ii) the *IABS* selects an outcome where the minimum level of these criteria is maximized, and iii) the *IABS* solution is obtained by solving a max-min optimization problem, which can be formulated as a linear programming model if the set of feasible utilities forms a polyhedral structure. Given the polyhedron $\Lambda \subseteq \mathrm{int}(\Delta^n)$ with extreme points $\lambda_{1 \leq l \leq k}$, the *IABS* for each $S \in \sum$ is defined as follows (Hinojosa, 2007, 2008; Kim, 2020c);

$$IABS_{\Lambda}(S) = \arg \max_{x \in S} \mathfrak{W}_{\Lambda}(x), \text{ s.t., } \mathfrak{W}_{\Lambda}(x) = min \left\{ \lambda_1 \cdot x,\ \lambda_2 \cdot x, \ldots, \lambda_k \cdot x \right\}$$

(6.3.2)

For a vector $x \in \mathbb{R}^n_+$, as a consequence of the linearity of the scalar product, if $\mathfrak{W}_{\Lambda}(x)$ attains its minimum¸ then the real number $\mathfrak{W}_{\Lambda}(x)$ represents the minimum level attainable, across the axis determined by 1^n. In this sense, the *IABS* can be interpreted as a compromise solution between the egalitarian solution and the λ-utilitarian solutions for $\lambda \in \Lambda$. The *IABS* also can be computed by solving the below optimization problem (Hinojosa, 2007, 2008; Kim, 2020c).

$$IABS_{\Lambda}(S) = \arg \max_{x \in S} \mathcal{T}, \text{ s.t., } \lambda_{1 \leq i \leq n} \cdot x \geq \mathcal{T} \text{ and } \mathcal{T} \geq 0$$

(6.3.3)

Additionally, the *HNSA* scheme takes into account scenarios where feasible bargaining solutions are evaluated based on the criterion that yields the highest utility value. To address this, the *HNSA* scheme introduces a corresponding bargaining solution known as the *EABS*, which is derived from the same utility functions as the other bargaining models. EABS is also based on weighted utilitarian criteria, but in this case, it is not *inequality averse*; it is *inequality loving*. For each $S \in \sum$ with extreme points $\lambda_{1 \leq i \leq k}$, the *EABS* is defined as follows (Hinojosa, 2007, 2008; Kim, 2020c);

$$EABS_{\Lambda}(S) = \arg \max_{x \in S} \overline{\mathfrak{W}}_{\Lambda}(x), \text{ s.t., } \overline{\mathfrak{W}}_{\Lambda}(x) = max \left\{ \lambda_1 \cdot x,\ \lambda_2 \cdot x, \ldots, \lambda_n \cdot x \right\}$$

(6.3.4)

EABS is also obtained with the outcome produced by the egalitarian and the utilitarian approaches. In another manner, the *EABS* can be computed by solving *k* optimization problems. Based on the optimal value of the jth problem for $j = 1, \ldots, k$,the *EABS* can be denoted as follows (Hinojosa, 2007, 2008; Kim, 2020c);

$$EABS_{\Lambda}(S) = max_{1 \leq j \leq k} \left\{ \mathfrak{A}^*_j \right\}, \text{ s.t., } \mathfrak{A}^*_j = \arg \max_{x \in S} (\lambda^{1 \leq j \leq k} \cdot x)$$

(6.3.5)

Figure 3 graphically illustrates the *RUBS*, *IABS* and *EABS* for the simple two-person bargaining game (Hinojosa, 2007, 2008; Kim, 2020c).

Figure 3. The RUBS, IABS and EABS for the simple two-person game model (Kim, 2020c)

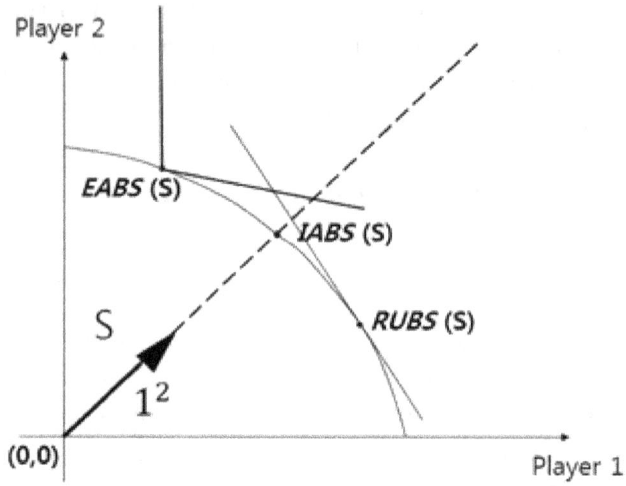

6.3.4 Three-phase spectrum allocation model for HetNet systems

The *HNSA* scheme is designed within a multi-tier HetNet framework, where the first tier consists of multiple macro cells, and the second tier comprises small cells deployed within the macro cell coverage area to enhance system performance. Each macro cell is managed by a BS, which functions as a gateway for Internet access. Within each macro cell, multiple small cells are randomly distributed, with SBSs handling communication with their respective mobile devices to deliver application services. Each SBS is assigned a dedicated licensed spectrum, ensuring exclusive access. Consequently, the *HNSA* scheme operates under the assumption that macro cells and small cells utilize separate spectrum bands. To maximize HetNet system efficiency, it is crucial to optimize the allocation of limited spectrum resources while accounting for the mutual interactions among network components. The general HetNet infrastructure is depicted in Figure 4, (Kim, 2020c).

Figure 4. The infrastructure of the HetNet (Kim, 2020c)

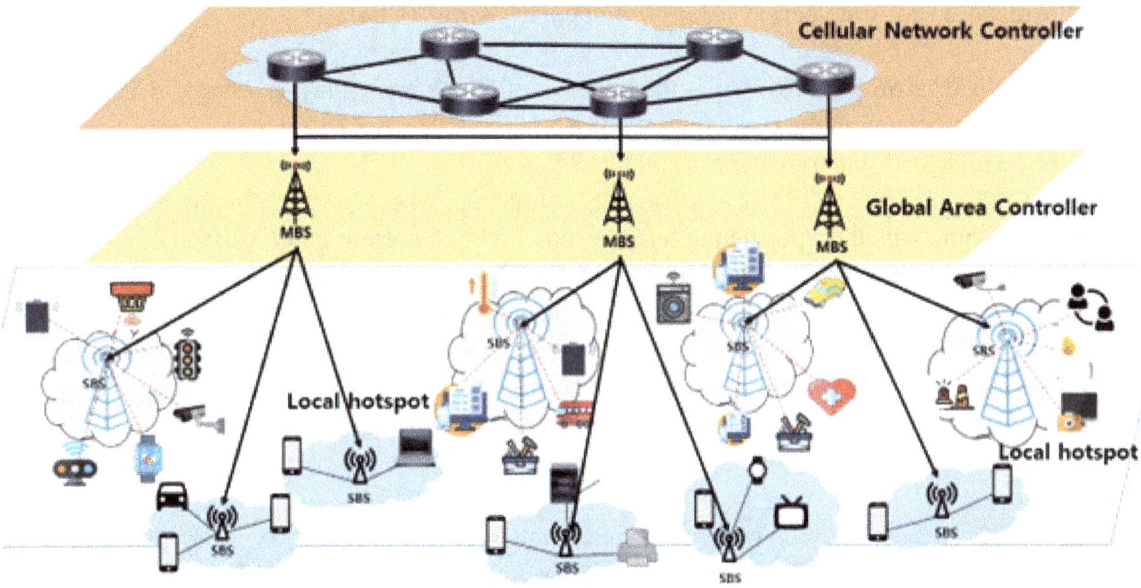

The three-phase bargaining game model is structured around three distinct spectrum resource allocation algorithms. In the first phase, the *HNSA* scheme models individual BSs as game players, where each BS aims to maximize its own payoff through the spectrum assignment process. In the second phase, the focus shifts to solving the spectrum allocation problem among the SBSs associated with each BS. In the third phase, the allocated spectrum resources are distributed among the respective application services. To achieve an optimal allocation of the limited resources within the HetNet system, the *HNSA* scheme incorporates multi-utilitarian bargaining solutions, ensuring an efficient and fair distribution of available spectrum. For the first phase bargaining game, the *HNSA* scheme formally defines the game model $\mathbb{G}^{First} = \{\mathbb{N}, \mathbb{C}, \mathbb{S}^{BS}, \mathscr{W}^{BS}, \mathbb{U}^{BS}, \Lambda^{BS}, T\}$ at each time period of gameplay (Kim, 2020c).

- $\mathbb{N} = \{BS_1 \ldots BS_i, \ldots BS_n\}$ is the finite set of game players where BS_i represents the ith BS in the coverage area of HetNet system.
- C is the total spectrum capacity in the HetNet system.
- $\mathbb{S}^{BS} = \{\mathfrak{P}^{BS}_{min}, \ldots \mathfrak{P}^{BS}_{k} \ldots \mathfrak{P}^{BS}_{max}\}$ is a set of BS's strategies where $\mathfrak{P}^{BS}_{k}$ means the *k*th level of allocated spectrum, which is specified in terms of basic BS spectrum unit (BBSSU); one BBSSU is the minimum amount of spectrum allocation.
- $\mathscr{W}^{BS}$ is the weighting factor of the BS during the $\mathbb{G}^{First}$ game process. Based on the $\mathscr{W}^{BS}$ value, the set of relative weight vector $\Lambda^{BS} = [\lambda^{BS}_1, \ldots, \lambda^{BS}_n]$ is defined where $\sum_{i=1}^{n} \lambda^{BS}_i = 1$.
- $\mathbb{U}^{BS}$ is the payoff received by the BS during the $\mathbb{G}^{First}$ game process.
- $T = \{\mathscr{H}_1, \ldots, \mathscr{H}_t, \mathscr{H}_{t+1}, \ldots\}$ denotes time, which is represented by a sequence of time steps for the $\mathbb{G}^{First}$ game process.

For the second phase, the *HNSA* scheme defines the game model $\mathbb{G}_{BS_i}^{Second} = \left\{\mathbb{P}_{BS_i}, \mathbb{A}_{BS_i}, \mathcal{S}^{SBS}, w^{SBS}, \mathcal{U}^{SBS}, \Lambda^{SBS}, T\right\}$ for the BS_i. Each individual BS operates its $\mathbb{G}_{BS}^{Second}$ in a parallel manner at each time period of $\mathbb{G}_{BS}^{Second}$ gameplay (Kim, 2020c).

- $\mathbb{P}_{BS_i} = \left\{SBS_1^{BS_i}\ldots SBS_j^{BS_i}\ldots SBS_m^{BS_i}\right\}$ is the finite set of $\mathbb{G}_{BS}^{Second}$ game players where $SBS_j^{BS_i}$ represents the jth SBS associated in the coverage area of BS_i.
- $\mathbb{A}_{BS_i}$ is the allocated spectrum capacity in the BS_i.
- $\mathcal{S}^{SBS} = \left\{\mathfrak{J}_{min}^{SBS}\ldots\mathfrak{J}_l^{SBS}\ldots\mathfrak{J}_{max}^{SBS}\right\}$ is a set of SBS's strategies where $\mathfrak{J}_l^{SBS}$ means the lth level of allocated spectrum, which is specified in terms of basic SBS spectrum unit (BSBSSU).
- w^{SBS} is the weighting factor of the SBS during the $\mathbb{G}_{BS}^{Second}$ game process.
- $\mathcal{U}^{SBS}$ is the payoff received by the SBS during the $\mathbb{G}_{BS}^{Second}$ game process.
- $\Lambda^{SBS} = \left\{\Lambda_1^{SBS},\ldots,\Lambda_z^{SBS}\right\}$ is the set of z points where each $\Lambda_{1\le i\le z}^{SBS} = \left[\lambda_1^{SBS},\ldots,\lambda_m^{SBS}\right]$ represents m element vector. The HNSA scheme assumes the basic m element vector $\Lambda_{basic}^{SBS} = \left[\frac{1}{m},\frac{2}{m},\frac{3}{m}\ldots,\frac{m-1}{m},\frac{m}{m}\right]$, and each Λ_i^{SBS} is defined as randomly ordered of Λ_{basic}^{SBS}, and the members of Λ^{SBS}, i.e., $\Lambda_{1\le i\le z}^{SBS}$, are differently defined.

For the third phase, the *HNSA* scheme defines the game model $\mathbb{G}_{SBS_j}^{Third} = \left\{\mathcal{P}_{SBS_j}, \mathfrak{C}_{SBS_j}, S^{\mathcal{A}}, \omega^{\mathcal{A}}, U^{\mathcal{A}}, \Lambda^{\mathcal{A}}, T\right\}$ for the SBS_j. Each individual SBS operates its $\mathbb{G}_{SBS}^{Third}$ in a parallel manner at each time period of $\mathbb{G}_{SBS}^{Third}$ gameplay (Kim, 2020c).

- $\mathcal{P}_{SBS_j} = \left\{\mathcal{A}_1^{SBS_j}\ldots\mathcal{A}_e^{SBS_j}\ldots\mathcal{A}_r^{SBS_j}\right\}$ is the finite set of $\mathbb{G}_{SBS}^{Third}$ game players where $\mathcal{A}_e^{SBS_j}$ represents the eth application service, which is associated to the SBS_j.
- $\mathfrak{C}_{SBS_j}$ is the allocated spectrum capacity in the SBS_j.
- $S^{\mathcal{A}} = \left\{\mathfrak{N}_{min}^{\mathcal{A}}\ldots\mathfrak{N}_f^{\mathcal{A}}\ldots\mathfrak{N}_{max}^{\mathcal{A}}\right\}$ is a set of application's strategies where $\mathfrak{N}_f^{\mathcal{A}}$ means the fth level of allocated spectrum level, which is specified in terms of basic application spectrum unit (BASU).
- $\omega^{\mathcal{A}}$ is the weighting factor of application A during the $\mathbb{G}_{SBS}^{Third}$ game process.
- $U^{\mathcal{A}}$ is the payoff received by the application during the $\mathbb{G}_{SBS}^{Third}$ game process.
- $\Lambda^{\mathcal{A}} = \left\{\Lambda_1^{\mathcal{A}},\ldots,\Lambda_z^{\mathcal{A}}\right\}$ is the set of z points where each $\Lambda_{1\le i\le z}^{\mathcal{A}} = \left[\lambda_1^{\mathcal{A}},\ldots,\lambda_r^{\mathcal{A}}\right]$ represents r element vector. The HNSA scheme assumes the basic r element vector $\Lambda_{basic}^{\mathcal{A}} = \left[\frac{1}{r},\frac{2}{r},\frac{3}{r}\ldots,\frac{r-1}{r},\frac{r}{r}\right]$, and each $\Lambda_i^{\mathcal{A}}$ is defined as randomly ordered of $\Lambda_{basic}^{\mathcal{A}}$ as the same manner as the Λ_i^{SBS} with Λ_{basic}^{SBS}.

Since the HetNet system generally serves more than one BS, BSs need to share the spectrum resource such as time and frequency slots. The utility payoff obtained by a BS depends on the traffic load and will be only a fraction of the allocated spectrum. Therefore, the BS's payoff is directly proportional to the allocated spectrum. At the first phase, the C of HetNet system is adaptively allocated to each BS according to the idea of *RUBS*. To quantify the BS service satisfaction, the payoff of BS_i with the strategy $\mathfrak{P}_k^{BS}$ and $\mathbb{C}$ $\left(\mathbb{U}^{BS_i}\left(\mathbb{C},\mathfrak{P}_k^{BS_i}\right)\right)$ can be derived as follows (Kim, 2020c).

$$\mathbb{U}^{BS_i}\left(\mathbb{C},\mathfrak{P}_k^{BS_i}\right) = \left(\frac{1}{1+exp\left(-\left(\mathcal{W}^{BS_i}\times\frac{\mathfrak{P}_k^{BS_i}}{\mathbb{C}}\right)\right)}\right)^{\mathcal{X}} \tag{6.3.6}$$

$$\text{s.t., } \mathscr{W}^{BS_i} = \left(\beta + \frac{\sum_{SBS_j^{BS_i} \in \mathbb{P}_{BS_i}} \left(w^{SBS_j^{BS_i}} \times \mathfrak{M}^{SBS_j^{BS_i}}\right)}{\sum_{SBS_j^{BS_i} \in \mathbb{P}_{BS_i}} \left(\mathbb{W}_{max} \times \mathfrak{M}^{SBS_j^{BS_i}}\right)}\right)$$

$$\text{and } \mathfrak{P}_k^{BS} \leq \sum_{SBS_j^{BS_i} \in \mathbb{P}_{BS_i}} \mathfrak{M}^{SBS_j^{BS_i}}$$

where X, β are the control parameters for the $\mathbb{U}^{BS_i}\left(\mathbb{C}, \mathfrak{P}_k^{BS_i}\right)$ and $w^{SBS_j^{BS_i}}$ is the weighting factor of the $SBS_j^{BS_i}$, which is the SBS_j associated with the BS_i; $w^{SBS_j^{BS_i}}$ is estimated in the game process. The $\mathbb{W}_{max}$ is the maximum weighting value and $\mathfrak{M}^{SBS_j^{BS_i}}$ is the spectrum amount requested from the $SBS_j^{BS_i}$. To quantify the SBS service satisfaction, the payoff of $SBS_j^{BS_i}$ with the strategy $\mathfrak{J}_l^{SBS}$ and $\mathbb{A}_{BS_i}\left(\mathscr{U}^{SBS_j}\left(\mathbb{A}_{BS_i}, \mathfrak{J}_l^{SBS_j}\right)\right)$ can be derived as follows (Kim, 2020c).

$$\mathscr{U}^{SBS_j}\left(\mathbb{A}_{BS_i}, \mathfrak{J}_l^{SBS_j}\right) = \left(\frac{1}{1 + exp\left(-\left(w^{SBS_j} \times \frac{\mathfrak{J}_l^{SBS_j}}{\mathbb{A}_{BS_i}}\right)\right)}\right)^{\xi} \tag{6.3.7}$$

$$\text{s.t., } w^{SBS_j} = \left(\gamma + \frac{\sum_{\mathscr{A}_e^{SBS_j} \in \mathscr{P}_{SBS_j}} \left(\omega^{\mathscr{A}_e^{SBS_j}} \times \mathfrak{M}^{\mathscr{A}_e^{SBS_j}}\right)}{\sum_{\mathscr{A}_e^{SBS_j} \in \mathscr{P}_{SBS_j}} \left(\mathbb{W}_{max} \times \mathfrak{M}^{\mathscr{A}_e^{SBS_j}}\right)}\right)$$

$$\text{and } \mathfrak{J}_l^{SBS_j} \leq \sum_{\mathscr{A}_e^{SBS_j} \in \mathscr{P}_{SBS_j}} \mathfrak{M}^{\mathscr{A}_e^{SBS_j}}$$

where ξ, γ are the control parameters for the $\mathscr{U}^{SBS_j}\left(\mathbb{A}_{BS_i}, \mathfrak{J}_l^{SBS_j}\right)$ and $\omega^{\mathscr{A}_e^{SBS_j}}$ is the weighting factor of the $\mathscr{A}_e^{SBS_j}$, which is the $\mathscr{A}_e$ associated with the SBS_j; $\omega^{\mathscr{A}_e^{SBS_j}}$ is estimated in the $\mathbb{G}_{SBS}^{Third}$ game process. $\mathfrak{M}^{\mathscr{A}_e^{SBS_j}}$ is the spectrum amount requested from the $\mathscr{A}_e^{SBS_j}$. To quantify the service satisfaction for the A, the payoff of $\mathscr{A}_e^{SBS_j}$ with the strategy $\mathfrak{N}_f^{\mathscr{A}_e}$ and $\mathfrak{C}_{SBS_j}\left(U^{\mathscr{A}_e}\left(\mathfrak{C}_{SBS_j}, \mathfrak{N}_f^{\mathscr{A}_e}\right)\right)$ can be derived as follows (Kim, 2020c).

$$U^{\mathscr{A}_e}\left(\mathfrak{C}_{SBS_j}, \mathfrak{N}_f^{\mathscr{A}}\right) = \left(\frac{1}{1 + exp\left(-\left(\omega^{\mathscr{A}_e^{SBS_j}} \times \frac{\mathfrak{N}_f^{\mathscr{A}_e}}{\mathfrak{C}_{SBS_j}}\right)\right)}\right)^{\psi} \tag{6.3.8}$$

$$\text{s.t., } \omega^{\mathscr{A}_e^{SBS_j}} = \left(\delta + \frac{\omega^{\mathscr{A}_e^{SBS_j}}}{\mathbb{W}_{max}}\right) \text{ and } \mathfrak{N}_f^{\mathscr{A}_e} \leq \mathfrak{M}^{\mathscr{A}_e^{SBS_j}}.$$

where ψ, δ are the control parameters for the $U^{\mathscr{A}_e}\left(\mathfrak{C}_{SBS_j}, \mathfrak{N}_f^{\mathscr{A}}\right)$. In the first phase, the *RUBS* is adopted to allocate the spectrum C to the associated BSs. With the relative weight vector Λ^{BS} and C, the *RUBS* for each $S \in \sum$ is defined as follows (Kim, 2020c);

$$RUBS_{\Lambda}(\mathbb{C}, S) = \arg\max_{x \in S}\left\{\sum_{i=1}^{n}(\lambda_i \times x_i)\right\} \tag{6.3.9}$$

$$= \arg \max_{\begin{bmatrix} \mathbb{U}^{BS_1}, \\ \dots \\ \mathbb{U}^{BS_n} \end{bmatrix} \in S} \left\{ \sum_{BS_i \in \mathbb{N}} (\lambda_i \times \mathbb{U}^{BS_i}(\mathbb{C}, \mathfrak{P}^{BS_i})) \right\}$$

$$\text{s.t., } \lambda_i = \mathscr{W}^{BS_i} / \sum_{BS_j \in \mathbb{N}} \mathscr{W}^{BS_j}$$

The *RUBS* is obtained through adjusting the BS's strategy $\mathfrak{P}^{BS}$. In the second phase, the *EABS* is used to allocate the assigned BS's spectrum into the associated SBSs. With the points $\Lambda^{SBS}_{1 \le i \le z}$, the *EABS* for each $S \in \sum$ is defined as follows (Kim, 2020c);

$$EABS_\Lambda(\mathbb{A}_{BS}, S) = \arg \max_{x \in S} \left\{ \max \left\{ (\Lambda_1^{SBS} \cdot x), (\Lambda_2^{SBS} \cdot x), \dots, (\Lambda_z^{SBS} \cdot x) \right\} \right\} \tag{6.3.10}$$

$$= \arg \max_{\begin{bmatrix} \mathcal{U}^{SBS_1}, \\ \dots \\ \mathcal{U}^{SBS_m} \end{bmatrix} \in S} \left\{ \max \left\{ \begin{matrix} \left(\Lambda_1^{SBS} \cdot \left[\mathcal{U}^{SBS_1}(\mathbb{A}_{BS}, \mathfrak{F}^{SBS_1}), \dots, \mathcal{U}^{SBS_m}(\mathbb{A}_{BS}, \mathfrak{F}^{SBS_m})\right]\right), \\ \dots\dots \\ \left(\Lambda_{1<i<Z}^{SBS} \cdot \left[\mathcal{U}^{SBS_1}(\mathbb{A}_{BS}, \mathfrak{F}^{SBS_1}), \dots, \mathcal{U}^{SBS_m}(\mathbb{A}_{BS}, \mathfrak{F}^{SBS_m})\right]\right), \\ \dots\dots \\ \left(\Lambda_z^{SBS} \cdot \left[\mathcal{U}^{SBS_1}(\mathbb{A}_{BS}, \mathfrak{F}^{SBS_1}), \dots, \mathcal{U}^{SBS_m}(\mathbb{A}_{BS}, \mathfrak{F}^{SBS_m})\right]\right) \end{matrix} \right\} \right\},$$

$$\text{s.t., } \mathbb{A}_{BS} = \sum_{SBS_j \in \mathbb{P}_{BS}} \mathfrak{F}^{SBS_j}$$

The *EABS* is obtained through adjusting the SBS's strategy $\mathfrak{F}^{SBS}$. In the third phase, the method of *IABS* is applied to share the allocated SBS's spectrum by the associated application services. With the points $\Lambda^{\mathscr{A}}_{1 \le i \le z}$, the *IABS* for each $S \in \sum$ is defined as follows; the *IABS* is obtained through adjusting the A's strategy $\mathfrak{N}^{\mathscr{A}}$ (Kim, 2020c).

$$IABS_\Lambda(\mathfrak{C}_{SBS}, S) = \arg \max_{x \in S} \left\{ \min \left\{ (\Lambda_1^{\mathscr{A}} \cdot x), (\Lambda_2^{\mathscr{A}} \cdot x), \dots, (\Lambda_z^{\mathscr{A}} \cdot x) \right\} \right\} \tag{6.3.11}$$

$$= \arg \max_{\begin{bmatrix} U^{\mathscr{A}_1}, \\ \dots \\ U^{\mathscr{A}_r} \end{bmatrix} \in S} \left\{ \min \left\{ \begin{matrix} \left(\Lambda_1^{\mathscr{A}} \cdot \left[U^{\mathscr{A}_1}(\mathfrak{C}_{SBS}, \mathfrak{N}^{\mathscr{A}_1}), \dots, U^{\mathscr{A}_r}(\mathfrak{C}_{SBS}, \mathfrak{N}^{\mathscr{A}_r})\right]\right), \\ \dots\dots \\ \left(\Lambda_{1<i<Z}^{\mathscr{A}} \cdot \left[U^{\mathscr{A}_1}(\mathfrak{C}_{SBS}, \mathfrak{N}^{\mathscr{A}_1}), \dots, U^{\mathscr{A}_r}(\mathfrak{C}_{SBS}, \mathfrak{N}^{\mathscr{A}_r})\right]\right), \\ \dots\dots \\ \left(\Lambda_z^{\mathscr{A}} \cdot \left[U^{\mathscr{A}_1}(\mathfrak{C}_{SBS}, \mathfrak{N}^{\mathscr{A}_1}), \dots, U^{\mathscr{A}_r}(\mathfrak{C}_{SBS}, \mathfrak{N}^{\mathscr{A}_r})\right]\right) \end{matrix} \right\} \right\},$$

$$\text{s.t., } \mathfrak{C}_{SBS} = \sum_{\mathfrak{N}^{\mathscr{A}_e} \in \mathscr{P}_{SBS}} \mathfrak{N}^{\mathscr{A}_e}.$$

6.3.5 Summary

The *HNSA* scheme is a spectrum allocation algorithm tailored for the hierarchical multi-tier HetNet platform. To enhance spectrum efficiency, it employs a novel three-phase bargaining game model, incorporating the fundamental principles of *RUBS*, *IABS*, and *EABS*. At each phase of the allocation process, BSs, SBSs, and application tasks serve as game players, bargaining models are formulated based on their respective utility functions, and the three bargaining processes are executed sequentially in a step-by-step manner. This interactive feedback mechanism generates cascade interactions between bargaining solutions, enabling the system to determine the most optimal and profitable allocation. Ultimately, spectrum resources are distributed to the corresponding game players at each network tier, ensuring a balanced integration of both egalitarian and utilitarian principles (Kim, 2020c).

6.4 Cooperative Game based Resource Allocation (*CGRA*) scheme

Heterogeneous Networks (HetNets) are widely regarded as a promising solution for enhancing communication capacity. However, managing HetNet control presents significant challenges due to inter-tier interference. To address this issue, enhanced *inter-cell interference coordination* (eICIC) technology has been introduced, providing several advantages, such as more balanced traffic distribution across macro cells and embedded small cells. To optimize resource allocation in eICIC-enabled HetNets, S. Kim proposes the *Cooperative Game-Based Resource Allocation* (*CGRA*) scheme (Kim, 2023a). The *CGRA* scheme is formulated as a joint cooperative game model to effectively manage conflicting network requirements. Especially, the *CGRA* scheme integrates principles from multiple bargaining solutions: *Kalai and Smorodinsky Solution* (*KSS*), *Multi-Criteria Kalai and Smorodinsky Solution* (*MCKSS*), and *Sequential Raiffa Solution* (*SRS*). These bargaining strategies are combined to create a hybrid control algorithm, enabling adaptive resources sharing among different base stations. To dynamically adjust eICIC fraction rates, the *CGRA* scheme applies the *MCKSS* and *SRS* interactively, ensuring efficient spectrum utilization. Meanwhile, for mobile devices in the HetNet, resources are allocated based on the *KSS* approach. The core innovation of the *CGRA* scheme lies in transforming a traditionally competitive resource allocation problem into a hierarchical cooperative framework, ensuring a socially optimal outcome. Its flexibility allows the system to achieve reciprocal consensus in the face of dynamic HetNet environments, making it a novel and adaptive solution for managing interference and optimizing resource allocation (Kim, 2023a).

6.4.1 Development motivation

With the commercial deployment of 5G networks progressing rapidly across various countries, both academic researchers and industry experts are focusing on improving network system performance. The exponential growth of mobile IoT devices has driven the development of 5G wireless communication systems, designed to provide a 1,000-fold increase in capacity compared to 4G networks. To achieve this, 5G networks integrate multiple communication technologies, including mobile edge computing, cloud computing, cloud radio access networks, and device-to-device communications, thereby enabling the IoT paradigm. However, the integration of these diverse technologies introduces several challenges, such as limited wireless spectrum resources, high infrastructure costs, and difficulties in maintaining service quality. To address these complex control challenges in 5G networks, researchers have been actively exploring innovative technologies (Kim, 2023a; Xu, 2021).

From a network architecture perspective, HetNets have gained significant interest due to their ability to enhance system performance. Deploying low-power small base stations (SBSs) within macro base station (MBS) coverage areas provides a cost-effective method for handling growing traffic demands. To accommodate increasing network loads, HetNets can support a large number of SBSs, allowing simultaneous data transmissions across multiple IoT devices. Unlike homogeneous networks (HomNets), HetNets improve spatial resource reuse by integrating small cells into macro cell coverage areas. By enabling coexistence between various types of small cells and macro cells, HetNets optimize spectrum efficiency while minimizing coverage gaps. Due to these advantages, wireless networks have evolved from HomNets to HetNets as a dominant architectural model (Kim, 2023a; Lee, 2023; Xu, 2021).

Despite its advantages, one of the primary challenges in HetNets is interference between macro and small cells. Since SBSs operate at lower transmission power, most IoT devices rely on the maximum *signal-to-interference-plus-noise ratio* (*SINR*) criterion to establish network connections. As a result, IoT devices may connect to distant MBSs due to their higher transmission power, rather than nearby SBSs, leading to suboptimal load distribution. This issue highlights the critical need for efficient resource management between high-power MBSs and low-power SBSs, ensuring balanced traffic distribution in HetNets (Kim, 2023a; Li, 2019; Marabissi, 2016; Vasudevan, 2013).

To address resource management challenges in HetNets, the 3rd Generation Partnership Project (3GPP) introduced *enhanced inter-cell interference coordination* (eICIC). As a time-domain interference mitigation solution, eICIC employs two key mechanisms. The *Cell Range Extension* (CRE) expands the coverage area of small cells by adding a bias value to the received power during user association. This technique improves small-cell coverage, cell-edge throughput, and overall network efficiency. The *Almost Blank Subframes* (*ABS*) reduces the MBS interference in the CRE region by temporarily muting MBS transmissions, except for essential control signals. This technique enhances the SINR for small-cell devices, allowing multiple IoT devices to communicate with SBSs with minimal interference (Kim, 2023a; Li, 2019; Marabissi, 2016; Vasudevan, 2013).

For effective eICIC deployment, joint resource allocation among MBSs and SBSs is essential to minimize inter-tier interference and enable spectrum sharing. Since the proportion of ABSs directly affects user association decisions and SBS resource allocation, these three factors must be jointly optimized to maximize HetNet performance. The *CGRA* scheme aims to design a novel control framework addressing three fundamental challenges; i) user association for load balancing - determining which devices should associate with MBSs or SBSs, ii) optimal ABS proportion and device allocation - deciding the ideal ABS ratio and determining which devices operate on ABS or non-ABS subframes in the time domain, and iii) spectrum resource allocation - efficiently distributing spectrum resources in the frequency domain. These three issues are tightly interconnected, necessitating a holistic optimization approach to fully leverage the potential of HetNets (Kim, 2023a; Li, 2014; Tang, 2014).

Cooperative optimization is inspired by coordination principles, wherein agents interact to maximize their payoffs based on defined objective functions. With a solid theoretical foundation and strong practical applications, cooperative optimization has gained significant attention across various research domains. A major branch of cooperative optimization is cooperative game theory, which focuses on game participants forming alliances to achieve shared objectives. A central challenge in cooperative games is how to fairly distribute the rewards generated by collective cooperation. Cooperative game models are widely applied across disciplines, including economics, biology, sociology, politics, computer science, and telecommunications. The *CGRA* scheme utilizes cooperative game theory principles to develop an advanced resource control mechanism for eICIC-based HetNets. By leveraging coordination strategies, MBSs, SBSs, and IoT devices interact and exchange information to obtain a globally optimal resource allocation, maximizing overall HetNet system performance (Kim, 2023a, 2023b; Wang, 2022).

Recently, multi-criteria cooperative games have gained interest due to their ability to account for multiple decision factors simultaneously. These models extend traditional cooperative games, allowing each player to evaluate multiple decision criteria rather than a single objective. Two major decision-making challenges arise in multi-criteria cooperative games: players' preferences - each participant considers individual criteria when evaluating possible decisions, and consensus solution selection - the challenge of finding a solution accepted by all rational players. Despite its potential, research on multi-criteria cooperation remains limited. In 2009, the *Multi-Criteria Kalai and Smorodinsky Solution* (*MCKSS*) was introduced to account for multi-dimensional payoffs. A key advantage of *MCKSS* is that feasible outcomes form a polyhedral set, allowing solutions to be computed via multi-criteria linear programming (Kim, 2023a; Monroy, 2009).

Another notable cooperative game solution is the *Sequential Raiffa Solution* (*SRS*), introduced by H. Raiffa for two-player cooperative problems. Given a disagreement point, the most preferred outcome for each player is the one that maximizes their utility while ensuring the other player remains at their disagreement payoff. The interim agreement is determined as the average of these two preferred points. By iteratively updating the disagreement point with each interim agreement, the SRS process converges to a Pareto-optimal solution within the cooperative game set. This approach effectively bridges the gap between cooperative and non-cooperative games, facilitating relative gains and concessions to achieve mutually beneficial outcomes (Diskin, 2011; Kim, 2023a; Trockel, 2015).

6.4.2 Contribution

The *CGRA* scheme is designed to enhance the performance of eICIC-enabled HetNets by employing a joint control paradigm based on the *KSS*, *MCKSS*, and *SRS*. The *CGRA* scheme operates in a step-by-step iterative manner, facilitating continuous interaction between MBSs and SBSs. Given the multi-criteria cooperation involved in these interactions, the MBS-SBS control problem is formulated as a multi-criteria cooperative game. From the perspective of an individual SBS, resource control challenges can be modeled as a two-player cooperative game. At the IoT device level, the allocated resources are managed according to a single-criterion cooperative game. In the *CGRA* scheme, these three cooperative game models are seamlessly integrated in a dynamically interactive framework, ensuring efficient collaboration throughout HetNet operations. By leveraging the synergies of cooperation among MBSs, SBSs, and IoT devices, the joint control approach enables the system to achieve a socially optimal resource allocation within the eICIC-based HetNet infrastructure (Kim, 2023a).

6.4.3 HetNet system infrastructure and cooperative game models

The *CGRA* scheme focuses on an orthogonal frequency-division multiple-access HetNet consisting of MBSs overlaid with multiple open access low power SBSs. Assume that in the two-tier HetNets, SBSs are randomly distributed within the coverage of each MBS, and the sets of MBSs and SBSs are denoted by $\mathbb{M} = \{\mathcal{M}_1,\dots,\mathcal{M}_n\}$ and $\mathbb{S} = \{\mathcal{S}_1,\dots,\mathcal{S}_m\}$, respectively. MBSs and SBSs are co-channel deployed and synchronous configured. The IoT devices $\mathbb{D} = \{\mathcal{D}_1,\dots,\mathcal{D}_l\}$, which are connected to MBSs or SBSs, are assigned to a number of sub-channels. It is assumed that the entire channel experiences flat fading, and each BS allocates equal power across all sub-channels. The IoT devices are randomly distributed within the cellular network area, following a uniform distribution model. Base stations are responsible for spectrum resource management, scheduling, admission control, and other key network functions for their associated devices. As illustrated in Figure 5, IoT devices have the capability to connect to different base stations, including MBSs and SBSs (Deb, 2014; Jia, 2016; Kim, 2023a; Shi, 2019).

Figure 5. A general two-tier heterogeneous network infrastructure (Kim, 2023a)

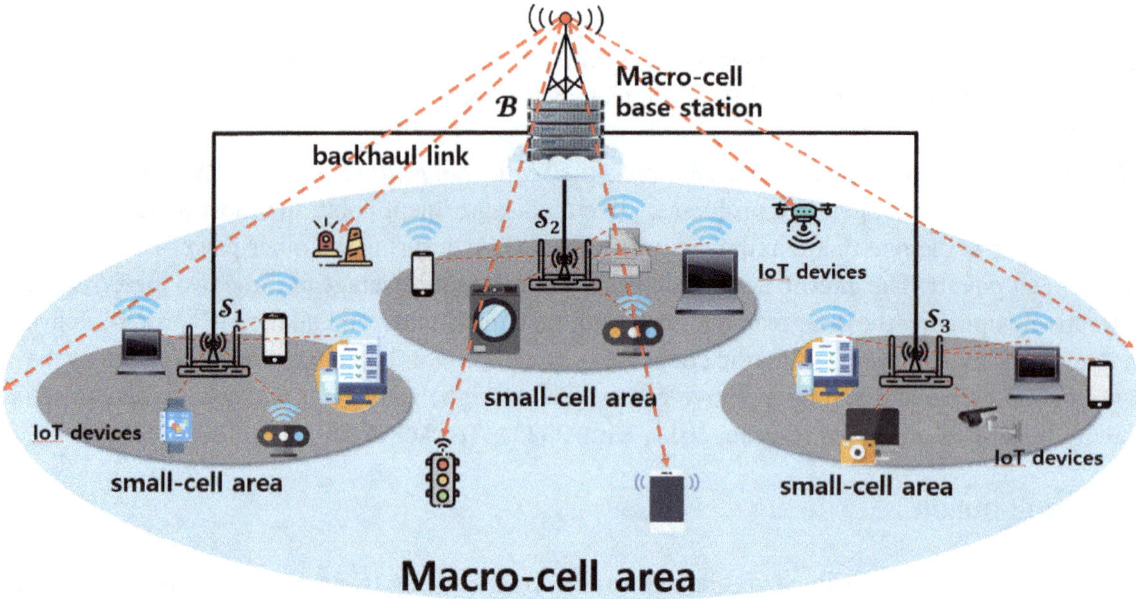

To alleviate the traffic load of MBSs in HetNets, the *CGRA* scheme employs time-domain eICIC mechanism. Assume a same ABS pattern that each MBS has the same ABS ratio α where α fraction is assigned to ABS and (1-α) fraction is assigned to non-ABS. In the ABS mechanism, the MBS remains silent during ABS periods but transmits at normal power levels during non-ABS periods. As a result, each MBS can only operate in non-ABS mode to serve its associated devices. However, SBSs have the flexibility to serve their corresponding devices in both ABS and non-ABS modes. Within the HetNet platform, the *CGRA* scheme categorizes IoT devices located within the nominal coverage of an SBS as SBS-centric devices. By applying a positive bias to the reference signal received power, the coverage area of the SBS can be expanded, and the devices operating in this extended region are referred to as SBS-extended devices. In non-ABS mode, SBS-extended devices experience significant interference from MBS transmissions, whereas SBS-centric devices are subject to relatively lower levels of interference (Deb, 2014; Jia, 2016; Kim, 2023a; Shi, 2019; Tang, 2014).

Without loss of generality, it is assumed that the SBS-centric and SBS-extended devices are served by non-ABS and ABS, respectively. However, some SBS-centric devices are seriously interfered with by the MBS. In order to solve this problem, the *CGRA* scheme configures a certain proportion of ABS fraction by using ratio factor β. Therefore, the resource scheduling policy of SBS is to serve the SBS-extended devices by using the β proportion of ABS fraction while (1-β) proportion of ABS fraction is assigned for the SBS-centric devices. Based on eICIC technology, the values of **α** and **β** are dynamically adjusted to share the limited spectrum resource to maximize the HetNet system. By combining the time domain and the frequency domain, the diagram of spectrum allocation for the MBS and SBS is shown in Figure 6, (Kim, 2023a).

Figure 6. Principle of allocation of spectrum resource in the eICIC technique (Kim, 2023a)

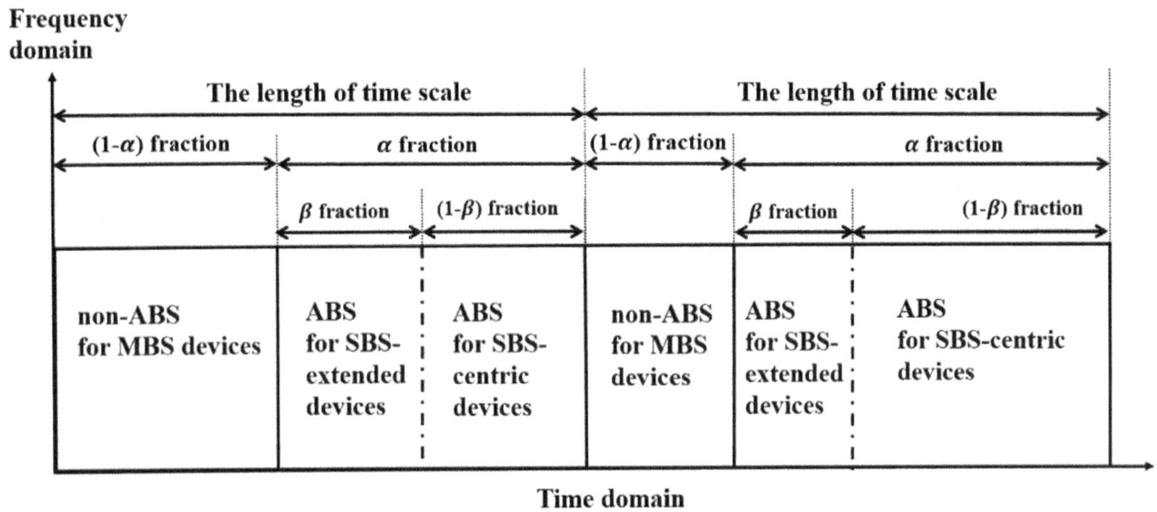

During the eICIC operations, each SBS has two types of devices. Devices are served on ABS mode (the set denoted by $\mathbb{D}^{A}_{\mathscr{S}}$), and the other devices are served on non-ABS mode (the set denoted by $\mathbb{D}^{nA}_{\mathscr{S}}$). Each MBS only uses the non-ABS mode to serve their corresponding devices, which are denoted by the set $\mathbb{D}_{\mathscr{M}}$. Therefore, the *CGRA* scheme has $|\mathbb{M}| + 2 \times |\mathbb{S}|$ virtual base stations in total. The *SINR* of $\mathscr{D}_j \in \mathbb{D}$ from virtual base station V is given by (Jia, 2016; Kim, 2023a):

$$SINR_{\mathscr{D}_j\mathscr{V}} = \frac{P_{\mathscr{V}} \times \left|h_{\mathscr{D}_j\mathscr{V}}\right|^2}{\sum_{\mathscr{Q} \in \mathscr{T}_{\mathscr{D}_j}} \left(P_{\mathscr{Q}} \times \left|h_{\mathscr{D}_j\mathscr{Q}}\right|^2\right) + \sigma^2} \tag{6.4.1}$$

where $P_{\mathscr{V}}$ is the transmission power of V, $\left|h_{\mathscr{D}_j\mathscr{V}}\right|^2$ is the channel gain from V to $\mathscr{D}_j$, and σ^2 is the noise power of each base station. $\mathscr{T}_{\mathscr{D}_j}$ denotes the interfering device set of $\mathscr{D}_j$. When $\mathscr{D}_j$ is served on ABS mode, $\mathscr{T}_{\mathscr{D}_j} = \mathbb{D}^{A}_{\mathscr{S}}/\{\mathscr{V}\}$; otherwise, $\mathscr{T}_{\mathscr{D}_j} = \mathbb{M} \cup \left(\mathbb{D}^{nA}_{\mathscr{S}}/\{\mathscr{V}\}\right)$. Based on the quasi-static fading channel, $SINR_{\mathscr{D}_j\mathscr{V}}$ is assumed to be constant during an allocation period. According to the Shannon capacity, the instantaneous communication rate from V to $\mathscr{D}_j$, i.e., $r_{\mathscr{D}_j\mathscr{V}}$, can be obtained, and the *CGRA* scheme finally gets the normalized link rate from V to $\mathscr{D}_j$, i.e., $R_{\mathscr{D}_j\mathscr{V}}$ as follows (Jia, 2016;Kim, 2023a; Li, 2019);

$$R_{\mathscr{D}_j\mathscr{V}} = \begin{cases} (1-\alpha) \times r_{\mathscr{D}_j\mathscr{V}}, if V \in M \\ \alpha \times \beta \times r_{\mathscr{D}_j\mathscr{V}}, if V \in \mathbb{D}^{A}_{\mathscr{S}} \\ (1-(\alpha \times \beta)) \times r_{\mathscr{D}_j\mathscr{V}}, if V \in \mathbb{D}^{nA}_{\mathscr{S}} \end{cases} \tag{6.4.2}$$

$$\text{s.t.}, \; r_{\mathscr{D}_j\mathscr{V}} = W \times \log\left(1 + SINR_{\mathscr{D}_j\mathscr{V}}\right)$$

where W is the channel bandwidth from V to $\mathscr{D}_j$. The *CGRA* scheme develops three cooperative games, which are hierarchically organized based on the joint control paradigm. The first cooperative game $\left(\mathbb{G}_{\mathscr{S}}^{\mathscr{M}}\right)$ decides the ABS ratio α, and the second cooperative game $\left(\mathbb{G}_{\mathscr{S}}\right)$ decides the proportion ratio factor β for each SBS. Finally, the third cooperative game $\left(\mathbb{G}_{\mathscr{D}}\right)$ distributes the assigned resource for corresponding IoT devices. Through the $\mathbb{G}_{\mathscr{S}}^{\mathscr{M}}$, $\mathbb{G}_{\mathscr{S}}$ and $\mathbb{G}_{\mathscr{D}}$ games, the MBS, SBSs and devices are sequentially interacted with each other to reach a mutually acceptable solution, which is called social optimal consensus. It is noteworthy that the *CGRA* scheme formulates the M-S-D association in an iterative coordinated manner. Formally, the *CGRA* scheme defines the tuple entities in the $\mathbb{G}_{\mathscr{S}}^{\mathscr{M}}$, $\mathbb{G}_{\mathscr{S}}$ and $\mathbb{G}_{\mathscr{D}}$ game models, such as

$$\mathbb{G} = \left\{\mathbb{G}_{\mathscr{S}}^{\mathscr{M}},\mathbb{G}_{\mathscr{S}},\mathbb{G}_{\mathscr{D}}\right\} = \left\{\mathbb{M},\mathbb{S},\mathbb{D},\mathbb{G}_{\mathscr{S}}^{\mathscr{M}_i} = \left\{\left(\mathscr{M}_i,\mathbb{S}_{\mathscr{M}_i}\right),\mathfrak{M}_{\mathscr{M}_i},\alpha_{\mathscr{M}_i},\left(U_{\mathscr{M}_i}(\cdot),U_{\mathbb{S}_{\mathscr{M}_i}}(\cdot)\right)\right\},\mathbb{G}_{\mathscr{S}_j} = \left\{\left(\mathbb{D}_e^{\mathscr{S}_j},\mathbb{D}_c^{\mathscr{S}_j}\right),\beta_{\mathscr{S}_j},\left(\mathscr{U}_e^{\mathscr{S}_j}(\cdot),\mathscr{U}_c^{\mathscr{S}_j}(\cdot)\right)\right\},\mathbb{G}_{\mathscr{D}} = \left\{\mathbb{G}_{\mathscr{D}}^{\mathscr{M}_i},\mathbb{G}_e^{\mathscr{S}_j},\mathbb{G}_c^{\mathscr{S}_j}|\mathbb{D}_{\mathscr{M}_i},\mathbb{D}_e^{\mathscr{S}_j},\mathbb{D}_c^{\mathscr{S}_j},\mathscr{D}_k,\mathfrak{R}_{\mathscr{D}_k},\mathfrak{U}_{\mathscr{D}_k}(\cdot)\right\},T\right\}$$

(Kim, 2023a).

- M, S and D represent the sets of MBSs, SBSs and IoT devices, respectively.
- In the first game for the $\mathscr{M}_i$, i.e., $\mathbb{G}_{\mathscr{S}}^{\mathscr{M}_i}$, the $\mathbb{S}_{\mathscr{M}_i}$ represents the set of SBSs, which exist in the $\mathscr{M}_i$'s coverage area, and the *CGRA* scheme assumes that the $\mathbb{S}_{\mathscr{M}_i}$ is an individual game player. The $\mathfrak{M}_{\mathscr{M}_i}$ is the $\mathscr{M}_i$'s spectrum resource amount.
- In the $\mathbb{G}_{\mathscr{S}}^{\mathscr{M}_i}$, the $\mathscr{M}_i$, $\mathbb{S}_{\mathscr{M}_i}$ are game players, and $(1-\alpha_{\mathscr{M}_i})$, $\alpha_{\mathscr{M}_i}$ are their strategies, and $U_{\mathscr{M}_i}(\cdot)$, $U_{\mathbb{S}_{\mathscr{M}_i}}(\cdot)$ are their utility functions, respectively.
- In the second game for the $\mathscr{S}_j$, i.e., $\mathbb{G}_{\mathscr{S}_j}$, the $\mathbb{D}_e^{\mathscr{S}_j}$ is the set of $\mathscr{S}_j$-extended devices, and the $\mathbb{D}_c^{\mathscr{S}_j}$ is the set of $\mathscr{S}_j$-centric devices. The *CGRA* scheme also assumes that the $\mathbb{D}_e^{\mathscr{S}_j}$ and $\mathbb{D}_c^{\mathscr{S}_j}$ are individual game players.
- In the $\mathbb{G}_{\mathscr{S}_j}$, the $\mathbb{D}_e^{\mathscr{S}_j}$, $\mathbb{D}_c^{\mathscr{S}_j}$ are game players, and $\beta_{\mathscr{S}_j}$, $(1-\beta_{\mathscr{S}_j})$ are their strategies, and $\mathscr{U}_e^{\mathscr{S}_j}(\cdot)$, $\mathscr{U}_c^{\mathscr{S}_j}(\cdot)$ are their utility functions, respectively.
- In the third game for IoT devices, i.e., $\mathbb{G}_{\mathscr{D}}$, there are three sub-games such as $\mathbb{G}_{\mathscr{D}}^{\mathscr{M}_i}$, $\mathbb{G}_e^{\mathscr{S}_j}$, and $\mathbb{G}_c^{\mathscr{S}_j}$. When the $\mathscr{S}_j$ is included in the $\mathscr{M}_i$'s covering area, the $\mathbb{D}_{\mathscr{M}_i}$ is the set of $\mathscr{M}_i$'s corresponding devices, $\mathbb{D}_e^{\mathscr{S}_j}$ is the set of $\mathscr{S}_j$-extended devices, and $\mathbb{D}_c^{\mathscr{S}_j}$ is the set of $\mathscr{S}_j$-centric devices.
- In the $\mathbb{G}_{\mathscr{D}}^{\mathscr{M}_i}$, $\mathbb{G}_e^{\mathscr{S}_j}$, and $\mathbb{G}_c^{\mathscr{S}_j}$, devices in the $\mathbb{D}_{\mathscr{M}_i}$, $\mathbb{D}_e^{\mathscr{S}_j}$, and $\mathbb{D}_c^{\mathscr{S}_j}$ are game players, respectively. For the $\mathscr{D}_k \in \left\{\mathbb{D}_{\mathscr{M}_i} \cup \mathbb{D}_e^{\mathscr{S}_j} \cup \mathbb{D}_c^{\mathscr{S}_j}\right\}$, the $\mathfrak{R}_{\mathscr{D}_k}$ and $\mathfrak{U}_{\mathscr{D}_k}(\cdot)$ are the $\mathscr{D}_k$'s strategy and utility function, respectively.
- Discrete time model $T \in \{t_1,\ldots,t_c,t_{c+1},\ldots\}$ is represented by a sequence of time steps. The length of t_c matches the event's time-scale of $\mathbb{G}_{\mathscr{S}}^{\mathscr{M}}$, $\mathbb{G}_{\mathscr{S}}$ and $\mathbb{G}_{\mathscr{D}}$.

6.4.4 Technical concepts and ideas of *KSS*, *MCKSS* and *SRS*.

To characterize the basic concept of *MCKSS*, we preliminarily define some mathematical expressions. Let $\mathbb{N} = \{1,2,\ldots,n\}$ be the set of players, and each player considers the same m criteria, $\mathscr{M} = \{1,2,\ldots,m\}$, to value the possible agreements. An n-person multi-criteria cooperative game is formally described by the pair (S,d) where $S \subseteq \mathbb{R}^{n\times m}$ is the set of all feasible outcomes and $d = \times_{i=1}^{n} d_i$ is the disagreement point or *status quo*. The outcomes in S are obtained as the result of a joint decision of all players. Therefore, there exists an agreement $X = \left(x_1,x_2,\ldots,x_n\right) \in S$, that gives player $i \in \mathbb{N}$ an outcome vector $x_i = \left(x_i^1,x_i^2,\ldots,x_i^m\right) \in \mathbb{R}^m$, where $x_i^{1\leq j\leq m}$ is the payoff of player i in criterion j. The d is the result obtainable if players fail to reach an agreement. In a multi-criteria cooperative game, individual rationality establishes that each player will only negotiate at or above those outcomes that improve upon the disagreement point d. Thus, the set of outcomes where players will negotiate is $d^{\geq} = \{X \in \mathbb{R}^{n \times m} | X \geq d^{\geq}\}$ (Kim, 2023a; Monroy, 2009).

Usually, a multi-criteria cooperative game solution is a procedure to incorporate partial information on the importance of the criteria in multi-criteria linear problems. Based on this idea, the *MCKSS* takes into account each player's maximum payoffs with respect to the criteria, which are represented by the ideal outcomes, they are denoted as $\mathcal{I} = \left(\mathcal{I}_{i \in \mathbb{N}}^{1 \leq j \leq m}\right)$. The *MCKSS* is obtained by replacing the payoff gains of players by the proportion with respect to their ideal outcomes. Therefore, for each feasible outcome $X = \left(x_1, x_2, \ldots, x_n\right) \in S$, the quotient $\left(x_i^j - d_i^j\right) / \left(\mathcal{I}_i^j - d_i^j\right)$ is a major decision factor where $i \in \mathbb{N}$ and $1 \leq j \leq m$. Originally, E. Kalai and M. Smorodinsky introduced the idea of *KSS* as a single criterion cooperative game solution. It provides a payoff that is proportional to the achievable maximum payoff while ensuring efficiency. Therefore, each player gets the same fraction of his maximum possible payoff in the *KSS*. Geometrically, the *KSS* is the intersection point between the cooperative game set S and the straight line between the disagreement point and the utopian point. By considering single criterion X and I, the *KSS*, i.e., $X^{KSS} = \left(x_1^*, x_2^*, \ldots, x_n^*\right)$, must satisfy (Kim, 2014, 2023a; Monroy, 2009);

$$X^{KSS} = \max_X \left\{ X | \frac{x_1^* - d_1}{\mathcal{I}_1 - d_1} = \frac{x_2^* - d_2}{\mathcal{I}_2 - d_2} = \ldots = \frac{x_n^* - d_n}{\mathcal{I}_n - d_n} \right\} \tag{6.4.3}$$

$$\text{s.t., } X^{KSS} = \left(x_1^*, x_2^*, \ldots, x_n^*\right) \in S$$

To define the *MCKSS*, let $\mathfrak{P}(X)$ denote the minimum proportional payoff vector whose components are $\mathfrak{P}^j(X) = \min_{1 \leq i \leq n} \left\{1 \leq j \leq m | \left(x_i^j - d_i^j\right) / \left(\mathcal{I}_i^j - d_i^j\right)\right\}$. For each feasible outcome $X \in S$, the minimum utility gains vector is $z(X) = (z^1(X), z^2(X), \ldots, z^m(X))$, where $z^{1 \leq j \leq m}(X) = \min_{1 \leq i \leq n} \left\{x_i^j - d_i^j\right\}$. Each component of $z(X)$ represents the guaranteed minimum utility gains of the set of players for the corresponding criterion. The *MCKSS* is based on the idea that players jointly agree on those outcomes whose minimum utility gain levels cannot be simultaneously improved with respect to all the criteria. Therefore, players will choose an outcome such that there is no other outcome whose minimum utility gain vector is better in terms of components. For the multi-criteria cooperative game (S, d), a feasible outcome $X \in S \cap d^{\geq}$ is the *MCKSS* if there does not exist $Y \in S \cap d^{\geq}$ such that $\mathfrak{P}(Y) \geq \mathfrak{P}(X)$. Mathematically, the *MCKSS*, i.e., $X^{MCKSS} = \left(\hat{x}_1, \hat{x}_2, \ldots, \hat{x}_n\right)$, is given by (Kim, 2023a; Monroy, 2009):

$$X^{MCKSS} = \left(\hat{x}_1, \hat{x}_2, \ldots, \hat{x}_n\right) = \max z^1, z^2, \ldots, z^m \tag{6.4.4}$$

$$
\text{s.t.,}\begin{cases}
\dfrac{x_i^1 - d_i^1}{\mathcal{I}_i^1 - d_i^1} \geq z^1, \forall i = 1,2,\ldots,n \\
\vdots \\
\vdots \\
\dfrac{x_i^m - d_i^m}{\mathcal{I}_i^m - d_i^m} \geq z^m, \forall i = 1,2,\ldots,n \\
X \in S \cap d^{\geq} and \hat{x}_{1 \leq i \leq n} = (x_i^1, x_i^2, \ldots, x_i^m)
\end{cases}
$$

Therefore, the *MCKSS* is Pareto optimal by criteria in $S \cap d^{\geq}$ and the set of efficient solutions for multi-criteria cooperative game (S, d) (Kim, 2023a).

A pair (S, d) is a cooperative game with $d \in S \subset \mathbb{R}^2$ where $d \leq y \leq x \in S \Rightarrow y \in S$. The set of all two-person cooperative games is denoted as B. For any non-empty subset C of B, a mapping $F{:}C \rightarrow \mathbb{R}^2 : (S,d) \mapsto F(S,d) \in S$ is called a cooperative solution on C. With $x \in \mathbb{R}^2$, let 1_S be indicator function of S where $1_S : \mathbb{R}^2 \rightarrow \mathbb{R} : x \mapsto (1, \text{if} x \in S | 0, \text{if} x \notin S)$. For any game $(S, d) \in \mathbb{B}$, we consider the mapping $f^{(S,d)} : \mathbb{R}^2 \rightarrow \mathbb{R}^2$; it defined as follows (Kim, 2023a; Trockel, 2015):

$$
f^{(S,d)}\left(x_{1,} x_2\right) := \left(f_1^{(S,d)}(x_2), f_2^{(S,d)}(x_1)\right) \tag{6.4.5}
$$

$$
\text{s.t.,}\begin{cases}
f_1^{(S,d)}{:}R \rightarrow R : x_2 \mapsto f_1^{(S,d)}(x_2) := \max\limits_{x_1 \in \mathbb{R}} (x_1 - d_1) \cdot 1_s(x_1, x_2) \\
f_2^{(S,d)}{:}R \rightarrow R : x_1 \mapsto f_2^{(S,d)}(x_1) := \max\limits_{x_2 \in \mathbb{R}} (x_2 - d_2) \cdot 1_s(x_1, x_2)
\end{cases}
$$

Now consider the sequence $\left(m_k^{(S,d)}\right)_{k \in \mathcal{N} = \{1,2,3,\ldots\}}$ defined by (Kim, 2023a; Trockel, 2015):

$$
m_{k \in \mathcal{N}}^{(S,d)} = \frac{1}{2} \times \left[\left(f_1^{(S,d)}\left(m_{k-1,2}^{(S,d)}\right), m_{k-1,2}^{(S,d)}\right) + \left(m_{k-1,1}^{(S,d)}, f_2^{(S,d)}\left(m_{k-1,1}^{(S,d)}\right)\right) \right] \tag{6.4.6}
$$

$$
\text{s.t.,}\ m_0^{(S,d)} := \left(m_{0,1}^{(S,d)}, m_{0,2}^{(S,d)}\right) := \left(d_1, d_2\right) = d
$$

According to (6.4.5) and (6.4.6), the *SRS*(S, d) is defined by (Kim, 2023a; Trockel, 2015):

$$
SRS(S,\ d) := \lim_{k \in (\mathcal{N} \cup \{0\})} m_k^{(S,d)},\ \text{s.t.,}\ \forall (S,d) \in \mathbb{B} \tag{6.4.7}
$$

6.4.5 The joint game model for the eICIC based HetNets

To develop the joint control mechanism for the eICIC based HetNet platform, the *CGRA* scheme constructs the $\mathbb{G}_{\mathcal{S}}^{\mathcal{M}}$, $\mathbb{G}_{\mathcal{S}}$ and $\mathbb{G}_{\mathcal{D}}$ game models. For the n MBSs in the M, total n $\mathbb{G}_{\mathcal{S}}^{\mathcal{M}}$ game models are operated in a parallel manner. The $\mathbb{G}_{\mathcal{S}}^{\mathcal{M}_i}$ is designed to decide the ABS ratio for the $\mathcal{M}_i$ and $\mathbb{S}_{\mathcal{M}_i}$. In the $\mathbb{G}_{\mathcal{S}}^{\mathcal{M}_i}$ game, the *CGRA* scheme concerns two decision criteria such as real-time sensitivity and service productivity for communication services. Therefore, the *CGRA* scheme adopts the idea of *MCKSS* for the $\mathbb{G}_{\mathcal{S}}^{\mathcal{M}_i}$'s solution concept. In the $\mathbb{G}_{\mathcal{S}}^{\mathcal{M}_i}$ game, the $\mathcal{M}_i$ and $\mathbb{S}_{\mathcal{M}_i}$ are game players. For game player $\mathcal{M}_i$, the *CGRA* scheme defines his utility function $U_{\mathcal{M}_i}(\cdot)$. However, the $U_{\mathcal{M}_i}(\cdot)$ consists of two sub-functions for the criteria of sensitivity $\left(U_{\mathcal{M}_i}^{S}(\cdot)\right)$ and productivity $\left(U_{\mathcal{M}_i}^{P}(\cdot)\right)$ where $U_{\mathcal{M}_i}(\cdot) = \left\{U_{\mathcal{M}_i}^{S}(\cdot), U_{\mathcal{M}_i}^{P}(\cdot)\right\}$. They are given as follows (Kim, 2023a);

$$U_{\mathcal{M}_i}\left(\left(\alpha_{\mathcal{M}_i}^{S}, \alpha_{\mathcal{M}_i}^{P}\right), \mathfrak{M}_{\mathcal{M}_i}, \mathfrak{R}_{\mathcal{M}_i}\right) = \begin{cases} U_{\mathcal{M}_i}^{S}(\cdot) = \eta \times \left(\exp\left(\delta \times \dfrac{\min\left(\mathfrak{R}_{\mathcal{M}_i}, \left(\left(1-\alpha_{\mathcal{M}_i}^{S}\right) \times \mathfrak{M}_{\mathcal{M}_i}\right)\right)}{\mathfrak{R}_{\mathcal{M}_i}}\right) - 1\right) \\ U_{\mathcal{M}_i}^{P}(\cdot) = \log\left(\zeta \times \dfrac{\min\left(\mathfrak{R}_{\mathcal{M}_i}, \left(\left(1-\alpha_{\mathcal{M}_i}^{P}\right) \times \mathfrak{M}_{\mathcal{M}_i}\right)\right)}{\mathfrak{R}_{\mathcal{M}_i}} + 1\right) \end{cases} \tag{6.4.8}$$

where η and δ are control parameters for the $U_{\mathcal{M}_i}^{S}(\cdot)$. $\mathfrak{R}_{\mathcal{M}_i}$ is the total resource request sum from the set $\left(\mathbb{D}_{\mathcal{M}_i}\right)$ of $\mathcal{M}_i$'s corresponding devices, and $\mathfrak{M}_{\mathcal{M}_i}$ is the $\mathcal{M}_i$'s spectrum resource amount. ζ is a control parameter for the $U_{\mathcal{M}_i}^{P}(\cdot)$. For another game player $\mathbb{S}_{\mathcal{M}_i}$, the *CGRA* scheme defines his utility function $U_{\mathbb{S}_{\mathcal{M}_i}}(\cdot)$ as the same manner as the $U_{\mathcal{M}_i}(\cdot)$ where $U_{\mathbb{S}_{\mathcal{M}_i}}(\cdot) = \left\{U_{\mathbb{S}_{\mathcal{M}_i}}^{S}(\cdot), U_{\mathbb{S}_{\mathcal{M}_i}}^{P}(\cdot)\right\}$. They are given as follows (Kim, 2023a);

$$U_{\mathbb{S}_{\mathcal{M}_i}}\left(\left(\alpha_{\mathcal{M}_i}^{S}, \alpha_{\mathcal{M}_i}^{P}\right), \mathfrak{M}_{\mathcal{M}_i}, \mathfrak{R}_{\mathcal{M}_i}\right) = \begin{cases} U_{\mathbb{S}_{\mathcal{M}_i}}^{S}(\cdot) = -\log\left(1 - \mu \times \dfrac{\min\left(\mathfrak{R}_{\mathbb{S}_{\mathcal{M}_i}}, \left(\alpha_{\mathcal{M}_i}^{S} \times \mathfrak{M}_{\mathcal{M}_i}\right)\right)}{\mathfrak{R}_{\mathbb{S}_{\mathcal{M}_i}}}\right) \\ U_{\mathbb{S}_{\mathcal{M}_i}}^{P}(\cdot) = \varrho \times \left(\dfrac{1}{1 + \exp\left(-\omega \times \dfrac{\min\left(\mathfrak{R}_{\mathbb{S}_{\mathcal{M}_i}}, \left(\alpha_{\mathcal{M}_i}^{P} \times \mathfrak{M}_{\mathcal{M}_i}\right)\right)}{\mathfrak{R}_{\mathbb{S}_{\mathcal{M}_i}}}\right)} - \psi\right) \end{cases} \tag{6.4.9}$$

where μ is a control parameters for the $U_{\mathbb{S}_{\mathcal{M}_i}}^{S}(\cdot)$. $\mathfrak{R}_{\mathbb{S}_{\mathcal{M}_i}}$ is the total resource request sum from the corresponding devices in $\mathbb{S}_{\mathcal{M}_i}$. ϱ, ω and ψ are control parameters for the $U_{\mathbb{S}_{\mathcal{M}_i}}^{P}(\cdot)$. For the multi-criteria $\mathbb{G}_{\mathcal{S}}^{\mathcal{M}_i}$ game, the *MCKSS*, i.e., $X^{MCKSS}\left(\mathbb{G}_{\mathcal{S}}^{\mathcal{M}_i}\right)$, is obtained as follows (Kim, 2023a);

$$X^{MCKSS}\left(\mathbb{G}_{\mathcal{S}}^{\mathcal{M}_i}\right) = \left(\widehat{U}_{\mathcal{M}_i}(\cdot), \widehat{U}_{\mathbb{S}_{\mathcal{M}_i}}(\cdot)\right) = \max z^S, z^P$$

(6.4.10)

$$\text{s.t.,}\begin{cases} \dfrac{U_i^S(\cdot) - d_i^S}{\mathcal{I}_i^S - d_i^S} \geq z^S, i = \left\{\mathcal{M}_i, \mathbb{S}_{\mathcal{M}_i}\right\} \\ \dfrac{U_i^P(\cdot) - d_i^P}{\mathcal{I}_i^P - d_i^P} \geq z^P, i = \left\{\mathcal{M}_i, \mathbb{S}_{\mathcal{M}_i}\right\} \\ \widehat{U}_{\mathcal{M}_i}(\cdot) = \left(U_{\mathcal{M}_i}^{S*}(\cdot), U_{\mathcal{M}_i}^{P*}(\cdot)\right) and\, \widehat{U}_{\mathbb{S}_{\mathcal{M}_i}}(\cdot) = \left(U_{\mathbb{S}_{\mathcal{M}_i}}^{S*}(\cdot), U_{\mathbb{S}_{\mathcal{M}_i}}^{P*}(\cdot)\right) \end{cases}$$

where $\mathcal{I}_i^S$ and $\mathcal{I}_i^P$ are the ideal outcomes, and d_i^S and d_i^P are the disagreement points of sensitivity and productivity criteria, respectively. $\widehat{U}_{\mathcal{M}_i}^S(\cdot), \widehat{U}_{\mathbb{S}_{\mathcal{M}_i}}^S(\cdot)\left(or\widehat{U}_{\mathcal{M}_i}^P(\cdot), \widehat{U}_{\mathbb{S}_{\mathcal{M}_i}}^P(\cdot)\right)$ are obtained by maximizing z^S, $(or\, z^P)$ and the $\alpha_{\mathcal{M}_i}^S\left(or\alpha_{\mathcal{M}_i}^P\right)$ value is given. Finally, the $\alpha_{\mathcal{M}_i}$ and $\alpha_{\mathbb{S}_{\mathcal{M}_i}}$ values for the $\mathbb{G}_{\mathcal{S}}^{\mathcal{M}_i}$ game are decided as follows (Kim, 2023a);

$$\alpha_{\mathcal{M}_i} = \frac{\alpha_{\mathcal{M}_i}^S + \alpha_{\mathcal{M}_i}^P}{2} \text{ and } \alpha_{\mathbb{S}_{\mathcal{M}_i}} = \left(1 - \alpha_{\mathcal{M}_i}\right)$$

(6.4.11)

To decide the $\beta_{\mathcal{S}_j}$ value in the $\mathcal{S}_j$, the *CGRA* scheme constructs the $\mathbb{G}_{\mathcal{S}_j}$ game. In this game, the $\mathbb{D}_e^{\mathcal{S}_j}$ and $\mathbb{D}_c^{\mathcal{S}_j}$ are game players, and they interact sequentially to reach a mutual consensus. In this case, the *SRS* is preferred for the solution concept. Through the interactive process, the utility functions of $\mathbb{D}_e^{\mathcal{S}_j}$ and $\mathbb{D}_c^{\mathcal{S}_j}$, i.e., $\mathcal{U}_e^{\mathcal{S}_j}(\cdot)$ and $\mathcal{U}_c^{\mathcal{S}_j}(\cdot)$, are defined as follows:

$$\begin{cases} \mathcal{U}_e^{\mathcal{S}_j}\left(\mathfrak{R}_{\mathbb{S}_{\mathcal{M}_i}}^e, \beta_{\mathcal{S}_j}\right) = \exp\left(\kappa \times \dfrac{\min\left(\mathfrak{R}_{\mathbb{S}_{\mathcal{M}_i}}^e, \left(\beta_{\mathcal{S}_j} \times \alpha_{\mathcal{M}_i} \times \mathfrak{M}_{\mathcal{M}_i}\right)\right)}{\mathfrak{R}_{\mathbb{S}_{\mathcal{M}_i}}^e}\right) - 1 \\ \mathcal{U}_c^{\mathcal{S}_j}\left(\mathfrak{R}_{\mathbb{S}_{\mathcal{M}_i}}^c, \beta_{\mathcal{S}_j}\right) = \varepsilon \times \log\left(1 - \left(\xi \times \dfrac{\min\left(\mathfrak{R}_{\mathbb{S}_{\mathcal{M}_i}}^c, \left(\left(1 - \beta_{\mathcal{S}_j}\right) \times \alpha_{\mathcal{M}_i} \times \mathfrak{M}_{\mathcal{M}_i}\right)\right)}{\mathfrak{R}_{\mathbb{S}_{\mathcal{M}_i}}^c}\right)\right) \end{cases} \quad (6.4.12)$$

where κ is a control parameter for the $\mathcal{U}_e^{\mathcal{S}_j}(\cdot)$, and ε, ξ are control parameters for the $\mathcal{U}_c^{\mathcal{S}_j}(\cdot)$. The $\mathfrak{R}_{\mathbb{S}_{\mathcal{M}_i}}^e$ and $\mathfrak{R}_{\mathbb{S}_{\mathcal{M}_i}}^c$ are total requested resource amounts from the $\mathbb{D}_e^{\mathcal{S}_j}$ and $\mathbb{D}_c^{\mathcal{S}_j}$, respectively. Based on the gradual negotiation, the *SRS* for the $\mathbb{G}_{\mathcal{S}_j}$ game, i.e., $m^{SRS}\left(\mathbb{G}_{\mathcal{S}_j}\right)$, converges to a fair-efficient solution while maintaining the viewpoints of $\mathbb{D}_e^{\mathcal{S}_j}$ and $\mathbb{D}_c^{\mathcal{S}_j}$. It is obtained as follows (Kim, 2023a);

$$m^{SRS}\left(\mathbb{G}_{\mathcal{S}_j}\right) = \left(\widehat{m}_e^{\mathcal{S}_j}, \widehat{m}_c^{\mathcal{S}_j}\right) := \lim_{k \in \mathcal{N}} m_k^{SRS}$$

(6.4.13)

$$= \left(\frac{1}{2} \times \left[\left(f_e^{\mathcal{S}_j}\left(m_{k-1,c}^{\mathcal{S}_j}\right), m_{k-1,c}^{\mathcal{S}_j}\right) + \left(m_{k-1,e}^{\mathcal{S}_j}, f_c^{\mathcal{S}_j}\left(m_{k-1,e}^{\mathcal{S}_j}\right)\right)\right]\right)$$

$$\text{s.t.}, \begin{cases} f_e^{\mathcal{S}_j}\left(m_c^{\mathcal{S}_j}\right) = \max_{m_e^{\mathcal{S}_j}}\left(m_e^{\mathcal{S}_j} - d_e\right) \cdot 1_s\left(m_e^{\mathcal{S}_j}, m_c^{\mathcal{S}_j}\right) \\ f_c^{\mathcal{S}_j}\left(m_e^{\mathcal{S}_j}\right) = \max_{m_c^{\mathcal{S}_j}}\left(m_c^{\mathcal{S}_j} - d_c\right) \cdot 1_s\left(m_e^{\mathcal{S}_j}, m_c^{\mathcal{S}_j}\right) \\ m_0^{\mathcal{S}_j} := \left(m_{0,e}^{\mathcal{S}_j}, m_{0,c}^{\mathcal{S}_j}\right) = \left(d_e, d_c\right) = d \end{cases}$$

where d_e and d_c are disagreement points for $\mathbb{D}_e^{\mathcal{S}_j}$ and $\mathbb{D}_c^{\mathcal{S}_j}$, respectively. When the change between $m_{k-1,e}$ and $m_{k,e}^{\mathcal{S}_j}$ is within a pre-defined minimum bound (φ), this change can be negligible. At this time, we can think that we converge a fair-efficient solution, and the negotiation process is terminated. Finally, the *CGRA* scheme can get $\beta_{\mathcal{S}_j}$ value, and the $\left(\alpha_{\mathcal{M}_i} \times \mathfrak{M}_{\mathcal{M}_i}\right)$ resource amount is proportionally distributed based on the $\beta_{\mathcal{S}_j}$ (Kim, 2023a).

For the devices in the D, the $\mathbb{G}_{\mathcal{D}}$ game is designed to distribute the assigned resource. In the $\mathcal{M}_i$, each individual device belongs to one of $\mathbb{D}_{\mathcal{M}_i}$, $\mathbb{D}_e^{\mathcal{S}_j}$ and $\mathbb{D}_c^{\mathcal{S}_j}$ sets where $\mathbb{D}_{\mathcal{M}_i} \cap \mathbb{D}_e^{\mathcal{S}_j} \cap \mathbb{D}_c^{\mathcal{S}_j} = \phi$. Devices in the $\mathbb{D}_{\mathcal{M}_i}$, $\mathbb{D}_e^{\mathcal{S}_j}$ and $\mathbb{D}_c^{\mathcal{S}_j}$ operate the $\mathbb{G}_{\mathcal{D}}^{\mathcal{M}_i}$, $\mathbb{G}_e^{\mathcal{S}_j}$, and $\mathbb{G}_c^{\mathcal{S}_j}$ games separately, in a parallel manner. For the $\mathcal{D}_k \in \left\{\mathbb{D}_{\mathcal{M}_i} \cup \mathbb{D}_e^{\mathcal{S}_j} \cup \mathbb{D}_c^{\mathcal{S}_j}\right\}$, its utility function is defined as follows (Kim, 2023a);

$$\mathfrak{U}_{\mathcal{D}_k}\left(\mathfrak{R}_{\mathcal{D}_k}, \Gamma_{\mathcal{D}_k}\right) = \frac{\exp\left(\frac{\min\left(\mathfrak{R}_{\mathcal{D}_k}, \Gamma_{\mathcal{D}_k}\right)}{\mathfrak{R}_{\mathcal{D}_k}}\right) - \exp\left(-\frac{\min\left(\mathfrak{R}_{\mathcal{D}_k}, \Gamma_{\mathcal{D}_k}\right)}{\mathfrak{R}_{\mathcal{D}_k}}\right)}{\exp\left(\frac{\min\left(\mathfrak{R}_{\mathcal{D}_k}, \Gamma_{\mathcal{D}_k}\right)}{\mathfrak{R}_{\mathcal{D}_k}}\right) + \exp\left(-\frac{\min\left(\mathfrak{R}_{\mathcal{D}_k}, \Gamma_{\mathcal{D}_k}\right)}{\mathfrak{R}_{\mathcal{D}_k}}\right)} \tag{6.4.14}$$

where $\mathfrak{R}_{\mathcal{D}_k}$ and $\Gamma_{\mathcal{D}_k}$ are the requested and allocated resource amounts for the $\mathcal{D}_k$, respectively. Based on the corresponding devices, the $\mathbb{G}_{\mathcal{D}}^{\mathcal{M}_i}$, $\mathbb{G}_e^{\mathcal{S}_j}$, and $\mathbb{G}_c^{\mathcal{S}_j}$ games are operated in a distributed fashion. For the $\mathbb{G}_{\mathcal{D}}^{\mathcal{M}_i}$, devices in the $\mathbb{D}_{\mathcal{M}_i}$ share the assigned resource, i.e., $(1-\alpha) \times \mathfrak{M}_{\mathcal{M}_i}$, according to the idea of *KSS*. Based on the utility function in (6.4.14), the *KSS*, i.e., $X^{KSS}\left(\mathbb{G}_{\mathcal{D}}^{\mathcal{M}_i}\right) = \left(\ldots, \mathfrak{U}_{\mathcal{D}_k}^*\left(\mathfrak{R}_{\mathcal{D}_k}, \Gamma_{\mathcal{D}_k}^*\right), \ldots\right)$, is obtained as follows (Kim, 2023a);

$$X^{KSS}\left(\mathbb{G}_{\mathcal{D}}^{\mathcal{M}_i}\right) = \max_{\mathfrak{U}}\left\{\mathcal{D}_k, \mathcal{D}_l \in \mathbb{D}_{\mathcal{M}_i} \mid \frac{\mathfrak{U}_{\mathcal{D}_k}^*(\cdot) - d_{\mathcal{D}_k}}{\mathcal{I}_{\mathcal{D}_k} - d_{\mathcal{D}_k}} = \ldots = \frac{\mathfrak{U}_{\mathcal{D}_l}^*(\cdot) - d_{\mathcal{D}_l}}{\mathcal{I}_{\mathcal{D}_l} - d_{\mathcal{D}_l}}\right\} \tag{6.4.15}$$

$$\text{s.t.}, \sum_{\mathcal{D}_k \in \mathbb{D}_{\mathcal{M}_i}} \Gamma_{\mathcal{D}_k}^* \leq \left(1 - \alpha_{\mathcal{M}_i}\right) \times \mathfrak{M}_{\mathcal{M}_i}$$

where $\mathcal{I}_{\mathcal{D}_k}$ is the ideal outcome for the $\mathcal{D}_k$. For the $\mathbb{G}_e^{\mathcal{S}_j}$ and $\mathbb{G}_c^{\mathcal{S}_j}$ games, the utility function for each game player is defined as the same by using (6.4.14). Finally, devices in the $\mathbb{D}_e^{\mathcal{S}_j}$ and $\mathbb{G}_c^{\mathcal{S}_j}$ share the assigned resources, i.e., $\left(\alpha \times \beta \times \mathfrak{M}_{\mathcal{M}_i}\right)$ and $\left[(1-(\alpha \times \beta)) \times \mathfrak{M}_{\mathcal{M}_i}\right]$, respectively, as the same manner as the $\mathbb{G}_{\mathcal{D}}^{\mathcal{M}_i}$ game (Kim, 2023a).

6.4.6 Summary

In cellular networks, traffic load fluctuates continuously due to user mobility and dynamic traffic patterns. The HetNet architecture has emerged as a promising solution to adaptively manage these variations, while also improving network coverage, spatial spectrum reuse, and system throughput. The *CGRA* scheme introduces a joint resource allocation mechanism specifically designed for eICIC-based HetNet systems. This approach offers several advantages, including a more balanced distribution of network traffic between MBSs and SBSs.

Previous studies on eICIC-based HetNets focused primarily on static optimization of eICIC parameters, failing to account for dynamic network conditions in HetNets. In contrast, the *CGRA* scheme is dynamic and enhances system performance by employing a hierarchical and iterative cooperative game model. A key feature of the *CGRA* scheme is its ability to optimize both: the non-ABS ratio for MBS devices, and the ABS fraction for SBS-extended devices. By dynamically adjusting these parameters, the *CGRA* scheme ensures adaptive resource allocation, leading to optimized network performance in response to real-time traffic fluctuations. These ratios are adaptively adjusted based on the concept of *MCKSS* and *SRS* through the and game models. And then, the assigned resources for MBS, SBS-centric, and SBS-extended devices are fair-efficiently distributed by using the idea of *KSS* through the game. By tightly coupling the , and games, the *CGRA* scheme can lead to consensus outcomes that may be accepted by all IoT devices of MBS and SBSs. The key innovation of the joint game-based approach lies in its ability to leverage reciprocal advantages through sequential interactions among MBSs, SBSs, and IoT devices. In dynamically evolving eICIC-based HetNet environments, the *CGRA* scheme continuously optimizes resource allocation by following a step-by-step iterative process, ultimately achieving a socially optimal solution (Kim, 2023a).

6.5 Fair Efficient Spectrum Allocation (*FESA*) scheme

Long-Term Evolution (LTE) carrier aggregation utilizing the unlicensed WiFi spectrum has been identified as a promising solution to accommodate the rapid surge in wireless data traffic. With both LTE and WiFi networks coexisting, the investigation of their mutual compatibility has become a high-priority research topic, drawing significant attention from both industry and academia. Recently, S. Kim introduced the *Fair Efficient Spectrum Allocation* (*FESA*) scheme to address LTE/WiFi coexistence challenges (Kim, 2022b). The *FESA* scheme employs a cooperative game-theoretic framework to design a novel control mechanism aimed at maximizing system throughput while ensuring relative fairness between the heterogeneous networks. By incorporating the *Proportional Nash Bargaining Solution* (*PNBS*) and *Zhang's Group Bargaining Solution* (*ZGBS*), the *FESA* scheme fosters a mutual consensus to achieve a fair-efficient spectrum sharing solution for LTE/WiFi coexistence. Using an online interactive bargaining approach, network agents collaborate to efficiently share LTE and WiFi spectrum resources. Given the dynamic nature of wireless networks, the *FESA* scheme plays a crucial role in balancing the trade-off between conflicting objectives, specifically fairness and efficiency in spectrum allocation (Kim, 2022b).

6.5.1 Development motivation

The global mobile traffic is projected to grow exponentially due to the rapid expansion of mobile devices and the emergence of new mobile applications. By 2023, nearly two-thirds of the world's population is expected to have Internet access, with the number of IoT devices surpassing three times the global population. The number of IoT devices per capita is estimated to increase from 2.4 in 2018 to 3.6 by 2023, resulting in a total of 29.3 billion IoT devices, up from 18.4 billion in 2018. However, existing cellular networks struggle to accommodate this growing demand due to limited dedicated spectrum resources. To mitigate spectrum scarcity and meet the rising demand for wireless data transmission, both academia and industry have begun exploring the use of unlicensed spectrum. In response, a variation of Long-Term Evolution (LTE) technology, known as Licensed-Assisted Access (LAA), has been proposed. This technology enhances LTE services by sharing the unlicensed 5 GHz band, which is predominantly utilized by WiFi networks (Cisco Systems, 2020; Kim, 2022b; Saleem, 2021; Tan, 2019).

Introduced in 3GPP Release 13 as part of LTE-Advanced, LAA technology facilitates the simultaneous operation of LTE and IEEE 802.11 wireless local area networks. It achieves this by employing carrier aggregation, which combines licensed LTE bands with unlicensed industrial, scientific, and medical bands, such as 2.4 GHz and 5.1 GHz. Recent performance evaluations highlight LAA technology as an appealing solution for network operators, offering higher data rates and more responsive user experience. However, LAA implementation does not explicitly standardize spectrum access procedures, leaving them open to vendor-specific designs. One of the key methods for LAA spectrum access is the Listen-Before-Talk Functionality (LBTF), which operates based on the ALOHA protocol. While several studies have examined this approach, LBTF primarily focuses on minimizing interference, neglecting essential LTE-specific features such as strict QoS guarantees. For instance, the LBTF method fails to incorporate non-queue-based network information and does not account for long-term network statistics, limiting its effectiveness in dynamic wireless environments (Kim, 2022b; Markova, 2019).

To effectively manage the shared spectrum, LAA-compatible LTE base stations (BSs) must ensure fair spectrum allocation while maintaining QoS requirements for LAA connections. A schedule-based LAA approach can enhance spectrum utilization, leading to higher system throughput and fairer resource distribution between WiFi and LTE networks. However, several challenges must be addressed when designing QoS-aware LAA connections; i) flexibility in LTE/WiFi coexistence protocols - the coexistence framework should be adaptable to meet QoS requirements through mutually acceptable agreements, ii) performance optimization for both LTE and WiFi networks - the LTE network's performance must be maximized while ensuring adequate protection for WiFi transmissions, iii) heterogeneous system integration - due to the fundamental differences between LTE and WiFi architectures, achieving a seamless and practical coexistence mechanism remains a challenge, and iv) coupling parameter optimization - effective solutions are required to fine-tune resource allocation parameters, ensuring a fair and efficient LTE/WiFi coexistence model (Kim, 2022b; Markova, 2019; Tan, 2019).

QoS policies play a crucial role in managing network capabilities and resource allocation to support IoT services. These policies enable service providers to gain clear visibility into their service performance, while ensuring optimal usability for end users. On the LTE/WiFi coexistence platform, QoS policies help regulate service quality between IoT service providers and users. Typically, heterogeneous IoT applications fall into two categories: the *best-effort services* (*BESs*) - these applications do not require strict QoS guarantees and generally prefer WiFi networks due to their cost-effectiveness, and the *QoS-preferred services* (*QPSs*) - these applications demand strict QoS guarantees and opt for LTE/WiFi hybrid systems for better service reliability. Given these differences, *QPSs* are assigned higher priority than BESs during service operations (Kim, 2022b; Singh, 2018; Tan, 2019).

Since LTE/WiFi networks fall within the category of heterogeneous networks, their performance is highly interdependent. Consequently, resource allocation in these networks must address both internetwork fairness (between LTE and WiFi) and intra-network fairness (between *BESs* and *QPSs*). Despite the importance of fair spectrum sharing, achieving consensus on LTE/WiFi coexistence protocols has proven difficult. Stakeholders are often divided into pro-*BESs* and pro-*QPSs* factions, with differing perspectives on resource allocation priorities. The *FESA* scheme provides a systematic approach to LTE/WiFi spectrum allocation, aiming to establish a balanced and efficient resource-sharing mechanism. By adopting this approach, the *FESA* scheme can maximize LTE/WiFi system performance while ensuring fair service distribution in diverse traffic environments (Gao, 2020; Kim, 2022b; Saleem, 2021).

A major challenge in LTE/WiFi coexistence is achieving a balance between fairness and efficiency in spectrum sharing. However, this is difficult, as each network agent operates selfishly to maximize its own resources. To address this issue, a new control paradigm is needed. In 2011, Y. Xu and N. Yoshihara introduced the *Proportional Nash Bargaining Solution* (*PNBS*), an extension of classical Nash bargaining theory. The *PNBS* model is designed to handle non-convex bargaining problems, reconciling efficiency and fairness by providing equitable recommendations for spectrum allocation. The *PNBS* approach introduces a step-by-step elimination process, enabling a structured negotiation framework for resolving conflicting criteria in bargaining scenarios (Kim, 2022b; Xu, 2011).

Traditional cooperative bargaining structures are often insufficient for addressing complex network interactions, as they fail to account for external cooperation between multiple groups. To overcome this limitation, X. Zhang proposed *Zhang's Group Bargaining Solution* (*ZGBS*), a model specifically designed for group-based bargaining scenarios. In the *ZGBS*, groups of network players negotiate collectively, determining optimal spectrum allocation through a cooperative framework. This approach defines each group's bargaining preferences and utilizes a structured bargaining process to ensure a fair distribution of resources. A similar concept was proposed by S. Chae, whose mathematical approach to group bargaining differs from the *ZGBS* but ultimately produces equivalent outcomes (Chae, 2004; Kim, 2022b; Zhang, 2009).

6.5.2 Contribution

The primary objective of the *FESA* scheme is to efficiently distribute the limited LTE and WiFi spectrum resources while optimizing social welfare through relative fairness. To achieve this, multiple system agents must collaborate in a coordinated manner to ensure an equitable and efficient allocation of spectrum resources. The *FESA* scheme leverages key principles from the *NBS*, *PNBS*, and *ZGBS* to tackle the LTE/WiFi spectrum allocation problem. To simplify the complexity of the problem, the *FESA* scheme decomposes it into three sub-problems. For the first sub-problem, the LTE and WiFi service providers engage in a cooperative negotiation based on the NBS model to fairly partition the available LTE spectrum. For the second sub-problem, the *FESA* scheme formulates a spectrum-sharing algorithm specifically for *QPSs*, using the *PNBS* approach, ensuring that high-priority services receive an optimal share of LTE resources. For the third sub-problem, the *BESs* and *QPSs* negotiate a mutual agreement using the *ZGBS* model, enabling a fair trade-off between service priorities. Given the dynamic nature of wireless networks, the *FESA* scheme operates as an adaptive, real-time online control framework. By continuously adjusting control decisions based on real-time system conditions, the *FESA* scheme ensures that the LTE/WiFi coexistence platform remains both efficient and practical for real-world deployment (Kim, 2022b).

6.5.3 LTE/WiFi platform and interactive bargaining game

In the *FESA* scheme, a heterogeneous network platform is considered to establish the joint LTE/WiFi platform. The *FESA* scheme considers a coexisting network of one LTE-LAA macro-cell and multiple WiFi micro-cells, which are overlaid on the macro-cell coverage area, and they are operated by WiFi access points (APs), i.e., $\mathbb{W} = \{\mathcal{W}_1, \ldots, \mathcal{W}_m\}$. Due to the small coverage of the APs, interference among them is neglected. LTE base station ($\mathfrak{B}$) operates on a licensed spectrum band $(\mathcal{S}_{\mathfrak{B}})$ while the $\mathcal{W}_{1 \leq i \leq m}$ has its unlicensed spectrum band $(\mathcal{S}_{\mathcal{W}_i})$. The B and $\mathcal{W}_i$ are connected via optical fiber networks. The *FESA* scheme assumes that there are a number of IoT devices, i.e., $\mathbb{D} = \{\mathcal{D}_1, \ldots, \mathcal{D}_n\}$, which are located in the coverage region of the LTE/WiFi network system. The traffic services of these devices are either served by the B or offloaded to the corresponding AP. It is assumed that the association among B, $\mathcal{W}_{1 \leq i \leq m}$ and $\mathcal{D}_{1 \leq j \leq n}$ has already been determined (Chung, 2017; Kim, 2022b).

Individual IoT devices have been equipped with technology to enable them to connect simultaneously with the LTE and WiFi networks. Therefore, the collaboration between B and W is realized by the dual connectivity capability, which enables IoT devices to aggregate the $\mathcal{S}_{\mathfrak{B}}$ and $\mathcal{S}_{\mathcal{W}}$ spectrum resources. According to a Poisson process, individual IoT devices in the D generate their communication tasks. In the coverage area of $\mathcal{W}_i$, the *FESA* scheme considers two different device sets. The set $\mathbb{B}_{\mathcal{W}_i}$ consists of devices, which generate *BESs* and the set of $\mathbb{Q}_{\mathcal{W}_i}$ consists of devices, which generate *QPSs*. All these devices can contact simultaneously to the B and their corresponding W to share the $\mathcal{S}_{\mathfrak{B}}$ and $\mathcal{S}_{\mathcal{W}_i}$. If the $\mathcal{D}_j$ is in the $\mathcal{W}_i$'s area, the totally allocated spectrum resource for the $\mathcal{D}_j$ is $\mathcal{A}_{\mathcal{D}_j} = \mathcal{A}^{\mathfrak{B}}_{\mathcal{D}_j} + \mathcal{A}^{\mathcal{W}_i}_{\mathcal{D}_j}$ where $\mathcal{A}^{\mathfrak{B}}_{\mathcal{D}_j}$ and $\mathcal{A}^{\mathcal{W}_i}_{\mathcal{D}_j}$ are the $\mathcal{D}_j$'s LTE and WiFi spectrum amounts, respectively. The operational timeline is discretized into time slots to make spectrum allocation; it is the same time scale of communication task arrivals, which is a common assumption in the LTE/WiFi system (Kim, 2014, 2022b).

In the LTE/WiFi system, the $\mathcal{S}_{\mathfrak{B}}$ and $\mathcal{S}_{\mathcal{W}_i}$ are limited resources. Therefore, effective spectrum allocation strategies should be considered to improve the system performance. To address the $\mathcal{S}_{\mathfrak{B}}$ and $\mathcal{S}_{\mathcal{W}_i}$ sharing problem, the *FESA* scheme formulates three bargaining games, i.e., $\mathbb{G}^{\mathfrak{B}}_{\mathcal{W}}$, $\mathbb{G}_{\mathfrak{B}}$ and $\mathbb{G}_{\mathcal{W}_i}$, at each time period. It is noteworthy that these three bargaining games formulate the B-W-D association in a cooperative manner. In the $\mathbb{G}^{\mathfrak{B}}_{\mathcal{W}}$, the $\mathcal{S}_{\mathfrak{B}}$ is divided for the B and $\mathcal{W}_{1 \leq i \leq m}$ using the idea of *NBS* where $\mathcal{R}_{\mathfrak{B}}$ for the B and $\mathcal{R}_{\mathcal{W}_i}$ for the $\mathcal{W}_i$. In the $\mathbb{G}_{\mathfrak{B}}$, all devices generated *QPSs* bargain with each other to share the $\mathcal{R}_{\mathfrak{B}}$ according to the *PNBS*. In the $\mathbb{G}_{\mathcal{W}_i}$, the devices in the $\mathbb{B}_{\mathcal{W}_i}$ and $\mathbb{Q}_{\mathcal{W}_i}$ bargain with each other to distribute the $\mathcal{R}_i$ and $\mathcal{S}_{\mathcal{W}_i}$ based on the *ZGBS*. The $\mathbb{G}^{\mathfrak{B}}_{\mathcal{W}}$, $\mathbb{G}_{\mathfrak{B}}$ and $\mathbb{G}_{\mathcal{W}_i}$ are operated sequentially in a step-by-step interactive fashion. Formally, the *FESA* scheme defines the game entities, i.e.,

$$\mathbb{G} = \left\{\mathbb{G}^{\mathfrak{B}}_{\mathcal{W}}, \mathbb{G}_{\mathfrak{B}}, \mathbb{G}_{\mathcal{W}_i}\right\} = \left\{\mathbb{W}, \mathbb{D}, \mathfrak{B}, \left\{\mathbb{G}^{\mathfrak{B}}_{\mathcal{W}} | \mathbb{Q}_{\mathfrak{B}}, \mathbb{B}_{\mathcal{W}_i}, \mathcal{W}_i \in \mathbb{W}, \mathcal{S}_{\mathfrak{B}}, \left(\mathcal{R}_{\mathfrak{B}}, \mathcal{R}_{\mathcal{W}}\right), \mathfrak{U}_{\mathfrak{B}}(\cdot), \mathfrak{U}_{\mathcal{W}_i}(\cdot)\right\}, \left\{\mathbb{G}_{\mathfrak{B}} | \mathbb{Q}_{\mathfrak{B}}, \mathcal{R}_{\mathfrak{B}}, \Upsilon_{\mathcal{D}_k \in \mathbb{Q}_{\mathfrak{B}}}, U_{\mathcal{D}_k}(\cdot)\right\}, \left\{\mathbb{G}_{\mathcal{W}_{1 \leq i \leq m}} | \mathcal{W}_i, \mathcal{D}_j \in \left(\mathbb{B}_{\mathcal{W}_i} \cup \mathbb{Q}_{\mathfrak{B}}\right), \mathcal{R}_{\mathcal{W}_i}, \mathcal{S}_{\mathcal{W}_i}, \mathcal{J}_{\mathcal{D}_j}, \mathcal{U}^{QPS}_{\mathcal{D}_j}(\cdot), \mathcal{U}^{BES}_{\mathcal{D}_j}(\cdot)\right\}, T\right\}$$

of gameplay (Kim, 2022b).

- The $\mathbb{G}^{\mathfrak{B}}_{\mathcal{W}}$, $\mathbb{G}_{\mathfrak{B}}$ and $\mathbb{G}_{\mathcal{W}_i}$ are mutually and reciprocally interdependent in an interactive manner, and they work together to share the $\mathcal{S}_{\mathfrak{B}}$ and $\mathcal{S}_{\mathcal{W}_i}$.
- $\mathbb{Q}_{\mathcal{W}_i}$ and $\mathbb{B}_{\mathcal{W}_i}$ are the sets of IoT devices, which are covered by the $\mathcal{W}_i$ and generate *QPSs* and *BESs*, respectively. $\mathbb{Q}_{\mathfrak{B}}$ is the set of all IoT devices, which generate *QPSs* where $\mathbb{Q}_{\mathfrak{B}} = \bigcup_{\mathcal{W}_i \in \mathbb{W}} \mathbb{Q}_{\mathcal{W}_i}$.
- In the $\mathbb{G}^{\mathfrak{B}}_{\mathcal{W}}$, the B and $\mathcal{W}_{1 \leq i \leq m}$ are game players, and $\mathcal{R}_{\mathfrak{B}}$, $\mathcal{R}_{\mathcal{W}_i}$ are their strategies. For the B and $\mathcal{W}_i$, the $\mathfrak{U}_{\mathfrak{B}}(\cdot), \mathfrak{U}_{\mathcal{W}_i}(\cdot)$ are their utility functions, respectively.

- In the $\mathbb{G}_{\mathfrak{B}}$, the device $\mathscr{D}_k \in \mathbb{Q}_{\mathfrak{B}}$ is a game player, and $\Upsilon_{\mathscr{D}_k}$ is its strategy to share the $\mathscr{R}_{\mathfrak{B}}$. $U_{\mathscr{D}_k}(\cdot)$ is the $\mathscr{D}_k$'s utility function.
- In the $\mathbb{G}_{\mathscr{W}_i}$, the $\mathscr{D}_j \in \left\{\mathbb{Q}_{\mathscr{W}_i} \cup \mathbb{B}_{\mathscr{W}_i}\right\}$ is a game player, and $\mathscr{J}_{\mathscr{D}_j}$ is its strategy. If $\mathscr{D}_j \in \mathbb{B}_{\mathscr{W}_i}$ $\left(or\ \mathscr{D}_j \in \mathbb{Q}_{\mathscr{W}_i}\right)$, $\mathscr{U}^{BES}_{\mathscr{D}_j}(\cdot)$ $\left(or\ \mathscr{U}^{QPS}_{\mathscr{D}_j}(\cdot)\right)$ is its utility function.
- Discrete time model $T \in \left\{t_1, \ldots, t_c, t_{c+1}, \ldots\right\}$ is represented by a sequence of time steps. The length of t_c matches the event's timescale of $\mathbb{G}^{\mathfrak{B}}_{\mathscr{W}}$, $\mathbb{G}_{\mathfrak{B}}$ and $\mathbb{G}_{\mathscr{W}_i}$.

6.5.4 The fundamental ideas of *PNBS* and *ZGBS*

To characterize the fundamental idea of *PNBS*, we assume an n-player bargaining problem. For any subset $A \subseteq \mathbb{R}^n_+$, A is said to be i) *non-trivial* if there exists $a \in A$ such that $a \gg 0$, and ii) *comprehensive* if for all $x, y \in \mathbb{R}^n_+$, $x > y$ and $x \in A$ implies $y \in A$. Let $\sum$ be the set of all *on-trivial*, *compact* and *comprehensive* subsets of $\mathbb{R}^n_+$. Elements in $\sum$ are interpreted as bargaining problems. Usually, a bargaining solution assigns a nonempty subset of A for every bargaining problem $A \in \sum$. For all $A \in \sum$ and all $i \in N$, let $m_i(A) = \max\left\{a_i | (a_1, \ldots a_i \ldots a_n) \in A\right\}$. Therefore, $m(A) \equiv \left(m_i(A)\right)_{i \in N}$ is the ideal point of A. Let $d_{i \in N}$ be a disagree point, which is expected to be the result of non-cooperative actions; it means a failure of the bargaining process. If for all $A \in \sum$, Nash bargaining solution, i.e., NBS(A), is defined by (Kim, 2022b; Xu, 2011):

$$NBS(A) = \prod_{i \in N} (a_i - d_i) \geq \prod_{i \in N} (x_i - d_i), \text{ s.t., } a \in A \text{ and } x \in A \tag{6.5.1}$$

Based on the *NBS*(A), the *PNBS*, i.e., *PNBS*(A), is mathematically formulated as follows (Xu, 2011):

$$PNBS(A) = \min_{i \in N}\left\{\frac{a_i - d_i}{m_i(A) - d_i}\right\} \geq \min_{i \in N}\left\{\frac{x_i - d_i}{m_i(A) - d_i}\right\}, \text{ s.t., } a, x \in NBS(A) \tag{6.5.2}$$

To define the concept of *ZGBS*, the bargaining problem is represented as a quadruple $\{\Gamma, S, u, d\}$, where Γ is a partition of a player set $N = \{1 \ldots i \ldots n\}$ and each element in Γ represents a formed coalition.S is a feasible payoff set in $\mathbb{R}^{|N|}$ where $\mathbb{R}^{|N|}$ is the n-dimensional Euclidean space, and $|N|$ denotes its size. The $d \in \mathbb{R}^{|N|}$ is a disagreement about the payoff vector when the coalitions fail to reach any joint contracts. There exists at least one element $u \in S$ where $u_i > d_i$ for any $i \in N$; u_i represents the i's payoff. When Γ is the finest partition of N, the pure bargaining problem with a coalition structure is the original Nash's pure bargaining problem. ZGBS is a direct extension to Nash's solution. To express the *ZGBS*, a function for two payoff vectors u and v with $A \in \Gamma$, i.e., $f(u, v, A)$, is defined as (Kim, 2022b; Zhang, 2009):

$$f(u, v, A) = \sum_{A \in \Gamma}\left(\frac{1}{|A|} \times \left(\sum_{i \in A} \frac{u_i - d_i}{v_i - d_i}\right)\right) \tag{6.5.3}$$

According to (6.5.3), the $ZGBS(\Gamma, S, u, d)$ is defined by (Zhang, 2009):

$$ZGBS(\Gamma,\ S, u)\ :\ =\ s^{*}\ =\ \max_{u \in S} f(u, s^{*}, A)\ =\ |\Gamma| \tag{6.5.4}$$

s.t., $u \in S,\ u_i > d_i$ and $\forall i \in N$

In 2004, Chae and Heidhues introduced a novel solution framework for the group bargaining problem. Their approach considers bargaining interactions occurring both within individual groups and across multiple groups. A key feature of their model is that it treats a bargaining group as a single bargaining entity, even when the group consists of heterogeneous individuals. Additionally, their study demonstrated that this solution leads to the joint-bargaining paradox, where a larger bargaining group tends to receive more favorable treatment. They further characterized a class of solutions in which larger groups gain an advantage in bargaining negotiations. The Chae and Heidhues solution are fundamentally based on the traditional Nash bargaining framework, applied both within individual groups and across multiple bargaining groups. This solution is mathematically defined as follows (Chae, 2004; Kim, 2022b):

$$\max_{u \in S} \prod_{j=1}^{k} \left(\prod_{i \in A_j} (u_i - d_i) \right),\ \text{s.t.},\ A_j \in \Gamma = \{A_1, \ldots, A_k\} \tag{6.5.5}$$

Although the *ZGBS* and Chae and Heidhues' group bargaining solution differs in their mathematical formulations, a study in (Zhang, 2009) demonstrated that applying a linear transformation to one of the problems results in their equivalence. As a result, both solutions are fundamentally identical in the context of group bargaining problems (Kim, 2022b).

6.5.5 The three-phase bargaining game in the LTE/WiFi platform

In the LTE/WiFi coexistent platform infrastructure, individual IoT devices generate their tasks. To effectively share the $\mathcal{S}_{\mathfrak{B}}$ and $\mathcal{S}_{\mathcal{W}}$ spectrum resources, three bargaining games $\mathbb{G}^{\mathfrak{B}}_{\mathcal{W}}$, $\mathbb{G}_{\mathfrak{B}}$ and $\mathbb{G}_{\mathcal{W}_i}$ are developed, and work together to allocate the $\mathcal{S}_{\mathfrak{B}}$ and $\mathcal{S}_{\mathcal{W}}$. At the first phase, the $\mathbb{G}^{\mathfrak{B}}_{\mathcal{W}}$ is designed to partition the $\mathcal{S}_{\mathfrak{B}}$ into $\mathcal{R}_{\mathfrak{B}}$ and $\mathcal{R}_{\mathcal{W}_{1 \le i \le m}}$. For this partitioning problem, the B and $\mathcal{W}_i$ bargain with each other based on the current information about *QPSs* and *BESs*. As game players, the utility functions for B and $\mathcal{W}_i$ at time t_c, i.e., $\mathfrak{U}^{t_c}_{\mathfrak{B}}(\cdot)$ and $\mathfrak{U}^{t_c}_{\mathcal{W}_i}(\cdot)$, are defined as follows (Kim, 2022b);

$$\begin{cases} \mathfrak{U}^{t_c}_{\mathfrak{B}}\left(\mathcal{S}_{\mathfrak{B}}, \mathcal{R}_{\mathfrak{B}}, \mathfrak{M}^{t_c}_{\mathfrak{B}}\right) = \log\left(\dfrac{\min\left(\mathcal{S}_{\mathfrak{B}}, \mathcal{R}_{\mathfrak{B}}\right)}{\mathfrak{M}^{t_c}_{\mathfrak{B}}} + \eta\right) - exp\left(-\dfrac{\min\left(\mathcal{S}_{\mathfrak{B}}, \mathcal{R}_{\mathfrak{B}}\right)}{\mathfrak{M}^{t_c}_{\mathfrak{B}}}\right) \\ \mathfrak{U}^{t_c}_{\mathcal{W}_i}\left(\mathcal{S}_{\mathfrak{B}}, \mathcal{S}_{\mathcal{W}_i}, \mathcal{R}_{\mathcal{W}_i}, \mathfrak{M}^{t_c}_{\mathcal{W}_i}\right) = \beta \times \left(\dfrac{\log\left(\dfrac{\min\left(\mathcal{S}_{\mathfrak{B}}, \mathcal{R}_{\mathcal{W}_i}\right) + \mathcal{S}_{\mathcal{W}_i}}{\mathfrak{M}^{t_c}_{\mathcal{W}_i}} + \kappa\right)}{\log\left(\dfrac{\min\left(\mathcal{S}_{\mathfrak{B}}, \mathcal{R}_{\mathcal{W}_i}\right) + \mathcal{S}_{\mathcal{W}_i}}{\mathfrak{M}^{t_c}_{\mathcal{W}_i}} + \omega\right)} \right) \end{cases} \tag{6.5.6}$$

s.t., $\mathfrak{M}^{t_c}_{\mathfrak{B}} = \sum_{\mathcal{D}_k \in \mathbb{Q}_{\mathfrak{B}}} \mathcal{T}_{\mathcal{D}_k}$, $\mathfrak{M}^{t_c}_{\mathcal{W}_i} = \sum_{\mathcal{D}_j \in \mathbb{B}_{\mathcal{W}_i}} \mathcal{T}_{\mathcal{D}_j}$ and $\mathcal{S}_{\mathfrak{B}} = \mathcal{R}_{\mathfrak{B}} + \sum_{\mathcal{W}_i \in \mathbb{W}} \mathcal{R}_{\mathcal{W}_i}$

where η is a control parameter for the $\mathfrak{U}_{\mathfrak{B}}^{t_c}(\cdot)$, and β, κ, ω are control parameters for the $\mathfrak{U}_{\mathcal{W}_i}^{t_c}(\cdot)$. $\mathcal{T}_{\mathcal{D}_k}$ is the $\mathcal{D}_k$'s maximum spectrum requirement. For the $\mathbb{G}_{\mathcal{W}}^{\mathfrak{B}}$, the features of *efficiency*, *anonymity* and *scale invariance* are important between *QPSs* and *BESs*. Therefore, the idea of classical *NBS* is adopted. In the *FESA* scheme, the utilities of disagreement points are assumed as zeros, and $\left[\mathcal{R}_{\mathfrak{B}}, \mathcal{R}_{\mathcal{W}_1}, \ldots, \mathcal{R}_{\mathcal{W}_m}\right]$ is obtained as the *NBS* for the $\mathbb{G}_{\mathcal{W}}^{\mathfrak{B}}$; it is given by (Kim, 2022b):

$$\underset{\left[\mathcal{R}_{\mathfrak{B}}, \mathcal{R}_{\mathcal{W}_1}, \ldots, \mathcal{R}_{\mathcal{W}_m}\right]}{NBS}\left(\mathfrak{U}_{\mathfrak{B}}^{t_c}(\cdot), \mathfrak{U}_{\mathcal{W}_{1 \le i \le m}}^{t_c}(\cdot)\right) =$$

$$\left(\mathfrak{U}_{\mathfrak{B}}^{t_c}(\mathcal{R}_{\mathfrak{B}}) \times \prod_{\mathcal{W}_i \in \mathbb{W}} \mathfrak{U}_{\mathcal{W}_i}^{t_c}(\mathcal{R}_{\mathcal{W}_i})\right) \ge \left(\mathfrak{U}_{\mathfrak{B}}^{t_c}(\mathcal{R}_{\mathfrak{B}}') \times \prod_{\mathcal{W}_i \in \mathbb{W}} \mathfrak{U}_{\mathcal{W}_i}^{t_c}(\mathcal{R}_{\mathcal{W}_i}')\right) \tag{6.5.7}$$

At the second phase, the $\mathbb{G}_{\mathfrak{B}}$ game is implemented to share the $\mathcal{R}_{\mathfrak{B}}$ among devices in the $\mathbb{Q}_{\mathfrak{B}}$. For the $\mathcal{D}_k \in \mathbb{Q}_{\mathfrak{B}}$, the utility function, i.e., $U_{\mathcal{D}_k}^{t_c}(\cdot)$, is defined as follows (Kim, 2022b):

$$U_{\mathcal{D}_k}^{t_c}\left(\Upsilon_{\mathcal{D}_k}, \mathcal{T}_{\mathcal{D}_k}, \mathcal{R}_{\mathfrak{B}}\right) = \left(\frac{exp\left(\frac{\Upsilon_{\mathcal{D}_k}}{\mathcal{T}_{\mathcal{D}_k}}\right) - exp\left(-\frac{\Upsilon_{\mathcal{D}_k}}{\mathcal{T}_{\mathcal{D}_k}}\right)}{exp\left(\frac{\Upsilon_{\mathcal{D}_k}}{\mathcal{T}_{\mathcal{D}_k}}\right) + exp\left(-\frac{\Upsilon_{\mathcal{D}_k}}{\mathcal{T}_{\mathcal{D}_k}}\right)}\right) + \frac{\log\left(\frac{\Upsilon_{\mathcal{D}_k}}{\mathcal{T}_{\mathcal{D}_k}} + \psi\right)}{\log\left(\frac{\Upsilon_{\mathcal{D}_k}}{\mathcal{T}_{\mathcal{D}_k}} + \varrho\right)} \tag{6.5.8}$$

$$\text{s.t., } \mathcal{R}_{\mathfrak{B}} \ge \sum\nolimits_{\mathcal{D}_k \in \mathbb{Q}_{\mathfrak{B}}} \Upsilon_{\mathcal{D}_k}$$

where ψ, ϱ are control parameters for the $U_{\mathcal{D}_k}^{t_c}(\cdot)$. To get a fair-efficient solution for *QPSs*, the *FESA* scheme concerns the features of *equity principle* and *weak contraction independence*. Therefore, PNBS is preferred for the solution concept of $\mathbb{G}_{\mathfrak{B}}$. As the *PNBS* for the $\mathbb{G}_{\mathfrak{B}}$, $\left[\ldots, \Upsilon_{\mathcal{D}_k \in \mathbb{Q}_{\mathfrak{B}}}, \ldots\right]$ is obtained by (Kim, 2022b):

$$\underset{\left[\ldots, \Upsilon_{\mathcal{D}_k \in \mathbb{Q}_{\mathfrak{B}}}, \ldots\right]}{PNBS}\left(U_{\mathcal{D}_k \in \mathbb{Q}_{\mathfrak{B}}}^{t_c}(\cdot)\right) = \min_{\mathcal{D}_k}\left\{\frac{U_{\mathcal{D}_k}^{t_c}(\Upsilon_{\mathcal{D}_k})}{m_{\mathcal{D}_k}(U_{\mathcal{D}_k}^{t_c}(\cdot))}\right\} \ge \min_{\mathcal{D}_k}\left\{\frac{U_{\mathcal{D}_k}^{t_c}(\Upsilon_{\mathcal{D}_k}')}{m_{\mathcal{D}_k}(U_{\mathcal{D}_k}^{t_c}(\cdot))}\right\} \tag{6.5.9}$$

$$\text{s.t., } \left[\ldots, \Upsilon_{\mathcal{D}_k \in \mathbb{Q}_{\mathfrak{B}}}, \ldots\right] \in \prod\nolimits_{\mathcal{D}_k \in \mathbb{Q}_{\mathfrak{B}}} U_{\mathcal{D}_k}^{t_c}\left(\Upsilon_{\mathcal{D}_k}\right) \ge \prod\nolimits_{\mathcal{D}_k \in \mathbb{Q}_{\mathfrak{B}}} U_{\mathcal{D}_k}^{t_c}\left(\Upsilon_{\mathcal{D}_k}'\right)$$

At the third phase, each $\mathcal{W}_i$ distributes its $\mathcal{R}_{\mathcal{W}_i}$ and $\mathcal{S}_{\mathcal{W}_i}$ for its corresponding device $\mathcal{D}_j \in \left\{\mathbb{B}_{\mathcal{W}_i} \cup \mathbb{Q}_{\mathcal{W}_i}\right\}$. During the $\mathbb{G}_{\mathcal{W}_i}$ game process, the $\mathcal{W}_i$ categorizes its corresponding devices into two groups, i.e., $\mathbb{B}_{\mathcal{W}_i}$ and $\mathbb{Q}_{\mathcal{W}_i}$, and the spectrum sharing problem is treated as the group bargaining problem. As a game player, the $\mathcal{D}_j$'s utility functions for *QPSs* and *BESs*, i.e., $\mathcal{U}_{\mathcal{D}_j}^{QPS}(\cdot)$ and $\mathcal{U}_{\mathcal{D}_j}^{BES}(\cdot)$, are defined as follows (Kim, 2022b):

$$\begin{cases} \mathcal{U}_{\mathcal{D}_j}^{QPS}\left(\mathcal{J}_{\mathcal{D}_j}, \mathcal{S}_{\mathcal{W}_i}, \mathcal{R}_{\mathcal{W}_i}, \mathcal{S}_{\mathcal{W}_i}, \mathbb{B}_{\mathcal{W}_i}, \mathbb{Q}_{\mathcal{W}_i}\right) = \\ \left(\dfrac{exp\left(\dfrac{\left(\mathcal{J}_{\mathcal{D}_j} + \Upsilon_{\mathcal{D}_j}\right)}{\mathcal{T}_{\mathcal{D}_j}}\right) - exp\left(-\dfrac{\left(\mathcal{J}_{\mathcal{D}_j} + \Upsilon_{\mathcal{D}_j}\right)}{\mathcal{T}_{\mathcal{D}_j}}\right)}{exp\left(\dfrac{\left(\mathcal{J}_{\mathcal{D}_j} + \Upsilon_{\mathcal{D}_j}\right)}{\mathcal{T}_{\mathcal{D}_j}}\right) + exp\left(-\dfrac{\left(\mathcal{J}_{\mathcal{D}_j} + \Upsilon_{\mathcal{D}_j}\right)}{\mathcal{T}_{\mathcal{D}_j}}\right)}\right) + \dfrac{\log\left(\dfrac{\left(\mathcal{J}_{\mathcal{D}_j} + \Upsilon_{\mathcal{D}_j}\right)}{\mathcal{T}_{\mathcal{D}_j}} + \psi\right)}{\log\left(\dfrac{\left(\mathcal{J}_{\mathcal{D}_j} + \Upsilon_{\mathcal{D}_j}\right)}{\mathcal{T}_{\mathcal{D}_j}} + \varrho\right)} \\ \mathcal{U}_{\mathcal{D}_j}^{BES}\left(\mathcal{J}_{\mathcal{D}_j}, \mathcal{S}_{\mathcal{W}_i}, \mathcal{R}_{\mathcal{W}_i}, \mathcal{S}_{\mathcal{W}_i}, \mathbb{B}_{\mathcal{W}_i}, \mathbb{Q}_{\mathcal{W}_i}\right) = \dfrac{\alpha}{\xi + \exp\left(-\dfrac{\mathcal{J}_{\mathcal{D}_j}}{\mathcal{T}_{\mathcal{D}_j}}\right)} - \tau \\ s.t., \left(\mathcal{R}_{\mathcal{W}_i} + \mathcal{S}_{\mathcal{W}_i}\right) \geq \sum_{\mathcal{D}_j \in \left\{\mathbb{B}_{\mathcal{W}_i} \cup \mathbb{Q}_{\mathcal{W}_i}\right\}} \mathcal{J}_{\mathcal{D}_j} \end{cases} \quad (6.5.10)$$

where ψ, ϱ are control parameters for the $\mathcal{U}_{\mathcal{D}_j}^{QPS}(\cdot)$; they are defined as the same in the $U_{\mathcal{D}_k}^{t_c}(\cdot)$. ξ, α, τ are control parameters for the $\mathcal{U}_{\mathcal{D}_j}^{BES}(\cdot)$. For the $\mathcal{D}_j$, $\mathcal{J}_{\mathcal{D}_j}$ is the allocated spectrum amount in the $\mathbb{G}_{\mathcal{W}_i}$ game. If $\mathcal{D}_j \in \mathbb{Q}_{\mathcal{W}_i}$, the $\mathcal{D}_j$ has already obtained the $\Upsilon_{\mathcal{D}_j}$ in the $\mathbb{G}_{\mathfrak{B}}$ game. In this case, the $\mathcal{J}_{\mathcal{D}_j}$ is additionally added, and the $\mathcal{D}_j$ has total $\left(\mathcal{J}_{\mathcal{D}_j} + \Upsilon_{\mathcal{D}_j}\right)$ spectrum amount for its *QPSs*. If $\mathcal{D}_j \in \mathbb{B}_{\mathcal{W}_i}$, the $\mathcal{D}_j$ gets the $\Upsilon_{\mathcal{D}_j}$ for its *BESs*. From the $\mathbb{G}_{\mathcal{W}_i}$, the *FESA* scheme can think the different type services, i.e., *BESs* and *QPSs*, are bargaining units. In this case, the *ZGBS* is preferred to reach a consensus in the $\mathbb{G}_{\mathcal{W}_i}$ game. It is given by (Kim, 2022b):

$$ZGBS\left(\mathbb{B}_{\mathcal{W}_i}, \mathbb{Q}_{\mathfrak{B}}, \mathcal{U}_{\mathcal{D}_j \in \left\{\mathbb{B}_{\mathcal{W}_i} \cup \mathbb{Q}_{\mathcal{W}_i}\right\}}(\cdot)\right) = \max_{\left[\ldots, \mathcal{J}_{\mathcal{D}_j}, \ldots\right]} \prod_{c=1}^{2}\left(\prod_{\mathcal{D}_j \in \Gamma_c} \mathcal{U}_{\mathcal{D}_j}(\cdot)\right) \quad (6.5.11)$$

$$s.t., \Gamma_1 = \mathbb{Q}_{\mathcal{W}_i},\ \Gamma_2 = \mathbb{B}_{\mathcal{W}_i} \text{ and } \mathcal{U}_{\mathcal{D}_j}(\cdot) = \begin{cases} \mathcal{U}_{\mathcal{D}_j}^{QPS}(\cdot), if \mathcal{D}_j \in \Gamma_1 \\ \mathcal{U}_{\mathcal{D}_j}^{BES}(\cdot), if \mathcal{D}_j \in \Gamma_2 \end{cases}$$

6.5.6 Summary

The LAA technology enables coexistence between LTE and WiFi networks, potentially enhancing network throughput. For efficient LTE/WiFi system operation, cooperation between LTE BSs and WiFi APs plays a crucial role in ensuring optimized spectrum utilization. The *FESA* scheme is designed to address the LTE/WiFi spectrum allocation problem, developing an effective spectrum allocation mechanism to achieve a balanced performance trade-off. To enhance efficiency, the *FESA* scheme integrates bargaining principles derived from the *NBS*, *PNBS*, and *ZGBS*. These bargaining approaches are applied in a step-by-step interactive manner to ensure fair and efficient distribution of spectrum. The *FESA* scheme employs a three-phase bargaining model, which facilitates mutual consensus to provide a fair-efficient spectrum allocation solution. In the *FESA* scheme, LTE BSs, multiple WiFi APs, and their associated IoT devices collaborate to effectively share limited spectrum resources. The iterative combination of spectrum partitioning and sharing strategies allows the *FESA* scheme to reach a reciprocal agreement, optimizing resource allocation. By ensuring fair QoS guarantees, the *FESA* scheme maintains service quality for multiple IoT devices while achieving an efficient LTE/WiFi coexistence model. By systematically integrating bargaining-based decision-making, the *FESA* scheme offers an adaptive and equitable approach to spectrum sharing in dynamic LTE/WiFi environments (Kim, 2022b).

6.6 Coexistent Dynamic Spectrum Sharing (*CDSS*) scheme

Recently, S. Kim introduced the *Coexistent Dynamic Spectrum Sharing* (*CDSS*) scheme, designed to accommodate diverse service requirements in the coexisting 4G/5G network platform (Kim, 2022a). The *CDSS* scheme is built on the fundamentals of cooperative game theory and incorporates various allocation rules to optimize spectrum sharing. The *CDSS* scheme employs a two-level dual bargaining model, formulated in a cooperative game framework. This hierarchical approach enables the efficient allocation of limited wireless spectrum resources, significantly reducing computational complexity and control overhead. To ensure mutual benefits for network agents, the *CDSS* scheme maximizes the synergy of the 4G/5G coexistence infrastructure, leading to enhanced system efficiency. By strategically balancing resource allocation, the *CDSS* scheme improves spectrum utilization while ensuring fair and effective network performance (Kim, 2022a).

6.6.1 Development motivation

As a next-generation mobile communication technology beyond 2020, the 5G network system introduces a new access mechanism, enabling seamless information retrieval anytime, anywhere, and for anyone or anything. However, the widespread adoption of the Internet of Everything (IoE) paradigm has significantly increased traffic loads, particularly with stringent low-latency requirements. Over the past decade, the volume of IP data handled by wireless networks has increased by more than 100 times, making it clear that emerging mobile services require new technologies capable of accommodating diverse user-demanded traffic patterns. Notably, 5G technology facilitates ultra-fast, highly reliable services across various industry sectors, driving innovation and transformation (Andrews, 2014; Dahlman, 2014; Kim, 2022a).

A major challenge in meeting future network demands is the limited availability of wireless spectrum resources, primarily due to the rigid fixed allocation policy currently in place. Addressing this spectrum scarcity requires the development of efficient spectrum-sharing technologies. One promising approach is Dynamic Spectrum Sharing (DSS), which allows for the gradual transition from 4G LTE to 5G New Radio (NR) by dynamically allocating spectrum resources based on the real-time demand of IoE devices. The DSS method is designed not only to enhance spectral efficiency but also to optimize spectrum utilization. The motivation behind DSS stems from the fact that a significant portion of wireless spectrum remains underutilized in both spatial and temporal domains. DSS technology seeks to bridge the paradox between spectrum scarcity and inefficient usage. Recent advancements in software defined networking, digital signal processing, and wideband transceiver technologies have made DSS a feasible and adaptive solution, capable of ensuring flexible spectrum management (Kim, 2022a; Sharma, 2018; Srinivasan, 2020).

Over the past decade, DSS technology has been a key research area in both academia and industry, and efforts to develop commercial system standards are currently underway. However, existing semi-static and hard spectrum partitioning methods fail to enable rapid wireless resource reallocation between devices and network operators. As a result, these traditional approaches lead to spectrum inefficiencies and underutilization in coexisting 4G/5G network environments. The *CDSS* scheme aims to address these inefficiencies by developing a novel DSS control algorithm that coordinates network agents while ensuring globally optimal system performance. Given the complexity of dynamically changing network conditions, the *CDSS* scheme requires an intelligent control paradigm capable of adapting to real-time spectrum demands. In the *CDSS* scheme, autonomous, distributed, and intelligent network agents interact strategically, making rational decisions to achieve a fair and efficient spectrum allocation consensus. This cooperative decision-making process aligns with cooperative game theory, providing a theoretically grounded approach to solving the 4G/5G spectrum-sharing problem (Jeon, 2019; Kim, 2014, 2022a).

In 1992, Y. Chun and W. Thomson introduced the *Chun and Thomson Bargaining Solution* (*CTBS*) to address bargaining problems where the reference point lies outside the traditional bargaining set. Their model interprets the estate of a bankruptcy problem as the feasible bargaining set, drawing from Nash's bargaining framework while incorporating a claims vector. In this context, the claims vector in bankruptcy problems functions similarly to the disagreement point in bargaining problems. The disagreement point in bargaining problems defines the minimum acceptable payoff for rational allocations. The claims vector in bankruptcy problems establishes an upper bound for feasible allocations, ensuring a structured distribution of resources (Chun, 1992; Dietzenbacher, 2018; Kim, 2022a).

The *CDSS* scheme integrates the *CTBS* model to optimize spectrum allocation in 4G/5G networks. Specifically, the *CDSS* scheme interprets the spectrum allocation vector as a feasible bargaining set and applies *CTBS* principles using three distinct allocation rules. From the perspective of cooperative game theory, these allocation rules satisfy key fairness and efficiency properties required for bargaining solutions. To construct an effective bargaining framework, the *CDSS* scheme develops specialized LTE and NR spectrum allocation algorithms, in which the spectrum bargaining problem is solved using the *CTBS* with different allocation rules. By leveraging classical cooperative game theory, the *CDSS* scheme characterizes its bargaining game models based on distinct theoretical properties (Kim, 2022a).

To minimize computational overhead, the LTE spectrum allocation algorithm in the *CDSS* scheme operates using a two-level hierarchical structure. Through a step-by-step interactive bargaining process, the limited LTE spectrum is dynamically distributed, ensuring that resources are allocated efficiently to maximize social benefits for end users (Kim, 2022a). Key contributions of the *CDSS* scheme are i) intelligent spectrum sharing: implements a novel DSS control algorithm to optimize dynamic spectrum allocation in 4G/5G networks, ii) cooperative game framework: adopts the *CTBS* principles to ensure fair and efficient spectrum allocation through cooperative bargaining, iii) adaptive and real-time decision making: employs an autonomous, distributed, and interactive control paradigm, enabling real-time spectrum management in a highly dynamic network environment, and iv) optimized LTE spectrum allocation: utilizes a two-level hierarchical model to reduce computational complexity while maximizing system efficiency. By integrating advanced cooperative game models with state-of-the-art spectrum-sharing techniques, the *CDSS* scheme represents a highly adaptable and scalable solution for optimizing spectrum utilization in coexisting 4G/5G networks (Kim, 2022a).

6.6.2 Contribution

Building on the *CTS* framework, the *CDSS* scheme introduces a novel DSS control algorithm tailored for the 4G/5G coexisting network platform. By employing a two-level dual bargaining model, each network agent autonomously and intelligently determines the optimal spectrum allocation in a distributed, online manner. To establish the disagreement point, the *CDSS* scheme incorporates three distinct allocation rules, ensuring adaptive spectrum distribution based on network condition; i) *proportional rule* (*PR*) - allocates spectrum proportionally based on predefined weightings, ii) *constrained relative equal awards rule* (*CREAR*) - distributes resources equitably while considering constraints on minimum allocations, and iii) *constrained relative equal loss rule* (*CRELR*) - ensures fair allocation by equalizing relative losses among network agents. Given the heterogeneous nature of network agents, the most suitable allocation rule is selected dynamically, ensuring efficient spectrum utilization. The *CTBS* model, in conjunction with its paired allocation rules, optimizes 4G/5G system performance, maintaining a well-balanced trade-off between efficiency and fairness. The interactive dual bargaining game approach enhances global network properties, making it highly adaptable across diverse and dynamic network environments (Kim, 2022a).

6.6.3 4G/5G coexisting network system infrastructure

The deployment of future wireless communication services should be approached as an evolutionary process, leveraging all available spectrum assets from both 4G LTE and 5G NR technologies. Ensuring a seamless migration from 4G to 5G is crucial to minimizing service disruptions while maintaining an optimal user experience. To enable the coexistence of LTE and NR spectrum resources within a shared network environment, the DSS mechanism plays a critical role in the transition of mobile networks from 4G to 5G. The DSS framework is built on three key design principles, which must be considered when developing a new DSS control algorithm in the *CDSS* scheme; i) preventing spectrum inefficiencies - ensuring that available spectrum resources are fully utilized to maximize network efficiency, ii) enhancing user experience via dual connectivity - providing users with the best possible experience through seamless LTE and 5G dual connectivity, and iii) maximizing operator benefits during 5G integration - allowing network operators to achieve optimal profitability by gradually introducing 5G NR into existing 4G network infrastructures. These fundamental principles serve as guidelines for the *CDSS* scheme, ensuring that its DSS control algorithm effectively supports the smooth transition from 4G to 5G while maintaining fair spectrum allocation and system efficiency (Kim, 2022a; Omana, 2019).

The coexistent platform consisting of 4G or 5G networks has a set of eNBs $\mathbb{E} = \left\{\mathscr{E}_1, \mathscr{E}_{2,} \ldots, \mathscr{E}_n\right\}$ for the LTE services, and a set of mobile network operators (MNOs) $\mathbb{M} = \left\{\mathscr{M}_1, \mathscr{M}_{2,} \ldots, \mathscr{M}_n\right\}$ for the NR services. Through E and M, multiple IoE devices, i.e., $\mathbb{D} = \left\{\mathscr{D}_1, \mathscr{D}_{2,} \ldots, \mathscr{D}_k\right\}$, can access the shared spectrum resources. Large numbers of devices can be scattered randomly over a large area, where they will automatically sense and contact their corresponding E and M. Each individual E has its own LTE spectrum resource $\left(\mathfrak{R}_{LTE}^{\mathscr{E}}\right)$ and covers a relatively large area. Each M has its own NR spectrum resource $\left(\mathfrak{R}_{NR}^{\mathscr{M}}\right)$ and covers a relatively small area. Therefore, multiple Ms' areas are overlapped by one E's covering area. Devices can share spectrum resources for asymmetrical traffic loads, as well as the shared spectrum is aggregated together, which is typically classified as the DSS technique. Simply, the *CDSS* scheme assumes that each D can be covered by one E and one M, simultaneously (Andrews, 2014; Dahlman, 2014; Jeon, 2019; Kim, 2022a; Sharma, 2018; Srinivasan, 2020).

In the *CDSS* scheme, a modeling situation for the E, M and D's interaction process is formulated as a dual bargaining game ($\mathbb{G}$) in a coordination manner; G is subdivided into $\mathbb{G}_{\mathscr{M},\mathscr{D}}^{LTE}$ and $\mathbb{G}_{\mathscr{D}}^{NR}$ to develop the LTE and NR resource allocation algorithms. The $\mathbb{G}_{\mathscr{M},\mathscr{D}}^{LTE}$ and $\mathbb{G}_{\mathscr{D}}^{NR}$ games are repeated sequentially in an interactive feedback manner. The 4G/5G network coexisting system operates in a slotted time structure and the spectrum allocation process is dynamically implemented at each time period. Formally, the *CDSS* scheme defines game entities, i.e.,

$$\mathbb{G} = \left\{\mathbb{G}_{\mathscr{D}}^{NR}, \mathbb{G}_{\mathscr{M},\mathscr{D}}^{LTE}\right\} = \left\{\{\mathbb{E}, \mathbb{M}, \mathbb{D}\}, \left\{\mathscr{M} \in \mathbb{M} | \mathfrak{R}_{NR}^{\mathscr{M}}, U_{NR}^{\mathscr{M}}, U_{LTE}^{\mathscr{M}}\right\}, \left\{\mathscr{E} \in \mathbb{E} | \mathfrak{R}_{LTE}^{\mathscr{E}}, U_{LTE}^{\mathscr{E}}\right\}, \left\{\mathscr{D} \in \mathbb{D} | U_{NR}^{\mathscr{D}}, U_{LTE}^{\mathscr{D}}\right\}, T\right\}$$

of gameplay (Kim, 2022a).

- G is a dual bargaining game consisting of $\mathbb{G}_{\mathcal{D}}^{NR}$ and $\mathbb{G}_{\mathcal{M},\mathcal{D}}^{LTE}$ games; they are related in a manner of mutual and reciprocal interdependency.
- $\mathbb{G}_{\mathcal{D}}^{NR}$ is a bargaining game model to solve the NR resource sharing problem. Each individual M distributes its $\mathfrak{R}_{NR}^{\mathcal{M}}$ to corresponding multiple Ds.
- $\mathbb{G}_{\mathcal{M},\mathcal{D}}^{LTE}$ is a bargaining game model to solve the LTE resource sharing problem. Each individual E distributes its $\mathfrak{R}_{LTE}^{\mathcal{E}}$ to corresponding multiple Ms. And then, each M re-distributes the allocated LTE resource to corresponding multiple Ds. Therefore, the $\mathbb{G}_{\mathcal{M},\mathcal{D}}^{LTE}$ game is designed as a two-level bargaining procedure.
- $\mathbb{E}, \mathbb{M}, \mathbb{D}$ are the sets of eNBs, MNOs and devices; they are game entities for the dual bargaining game.
- Initially, each E (*or* M) has its LTE resource $\mathfrak{R}_{LTE}^{\mathcal{E}}$ (*or* NR resource $\mathfrak{R}_{NR}^{\mathcal{M}}$); it is rationally distributed according to the concept of *CTBS*.
- $U_{LTE}^{\mathcal{E}}$ is the E's utility function from the $\mathfrak{R}_{LTE}^{\mathcal{E}}$ distribution.
- $U_{NR}^{\mathcal{M}}$ and $U_{LTE}^{\mathcal{M}}$ are the M's utility functions from the $\mathfrak{R}_{NR}^{\mathcal{M}}$ and its allocated LTE resource distributions, respectively.
- $U_{NR}^{\mathcal{D}}$ and $U_{LTE}^{\mathcal{D}}$ are the D's utility functions from the allocated NR and LTE resources, respectively.
- $T = \{t_1, \ldots, t_c, t_{c+1}, \ldots\}$ denotes time, which is represented by a sequence of time steps.

6.6.4 The basic concepts of proportional bargaining solutions

Cooperative game theory serves as a powerful framework for analyzing proportional bargaining solutions that incorporate claims-based allocations. Traditionally, axiomatic bargaining solutions are classified into three major categories; i) feasible set-based solutions - introduced by J. Nash (1950), this class focuses on how changes in the feasible set impact bargaining outcomes. The Nash bargaining solution is a primary example, ii) ideal point dependent solutions - since the 1980s, research has explored bargaining solutions that depend on ideal reference points, with the Kalai-Smorodinsky Solution being a prominent example, and iii) player number considerations - in this approach, the number of game participants is considered as a determining factor in the bargaining outcome. The egalitarian solution exemplifies this category. Recent studies have demonstrated that these three bargaining frameworks can be effectively integrated to derive a single optimal solution. In 1992, Y. Chun and W. Thomson introduced the *Chun and Thomson Bargaining Solution* (*CTBS*), a novel bargaining model incorporating claims-based allocations. Their approach was found to exhibit strong axiomatic properties, making it an attractive solution for various bargaining scenarios. The *CTBS* framework determines the maximum feasible allocation by selecting the optimal point along the line connecting the disagreement point (lower bound for rational payoffs) with the claims point (upper bound for feasible allocations). By following this principle, the *CTBS* provides a fair and structured approach to claims-based bargaining, ensuring proportional and efficient resource distribution (Chun, 1992; Kim, 2022a).

An n-player bargaining problem with claims, simply, (S, d, c) represents that players can achieve any point x of S if they unanimously agree on it; the coordinates of x are the utility payoff levels attained by the n players through the choice of some joint action. Therefore, x is a n-element vector. The d is also a n-element vector that is expected to be the result if players cannot reach an agreement; generally, it is a reference point to measure players' utility gains in order to evaluate a proposed compromise. The claim is a demand by a prospective co-operator for a particular co-operative surplus, made initially in the bargaining. Therefore, the claim point c is the point in utility space representing the outcome that would afford each player a utility equal to that of his claim. It is assumed that claims are made in good faith, and they should be taken into account in the determination of the final compromise (Chun, 1992; Kim, 2022a).

There exist $p \in \mathbb{R}^n_{++}$ and $r \in \mathbb{R}$ such that $\sum(p_i \cdot x_i) \leqq r$ for all $x \in S$. For all $x \in S$ and for all $y \in \mathbb{R}^n$, if $y \leq x$, then $y \in S$. There exists $x \in S$ with $x > d$. $c \notin S$, $c \geq d$ and $c \leqq a(S)$ where $a_i(S) \equiv \max\{x_j | x \in S\}$ for all player i if this maximum exists and $a_i(S) = \infty$, otherwise. A solution is a function $f: \acute{O}^n \rightarrow \mathbb{R}^n$ that associates with each $(S, d, c) \in \acute{O}^n$ a unique point of S, $f(S, d, c)$, called the solution outcome of (S, d, c) (Chun, 1992; Dietzenbacher, 2020; Kim, 2022a).

Definition 1: The *CTBS* for all $(S, d, c) \in \Sigma^n$ is $\mathfrak{P}(S, d, c)$, which has focused on the disagreement point and claims of game players. $\mathfrak{P}(S, d, c)$ is defined as the maximal point of S on the segment connecting d and c (Chun, 1992).

As a proportional bargaining solution, CTPS shares a conceptual relationship with the Kalai-Smorodinsky Solution. In general, it can be viewed as a normalized adaptation of the egalitarian solution, where bargaining outcomes are determined symmetrically among game players within the proportional framework. Given this relationship, the *CTPS* can be understood as a hybrid model that integrates key principles from both egalitarian and Kalai-Smorodinsky solutions, balancing fairness and proportionality in bargaining scenarios (Kim, 2022a).

With the idea of *CTPS*, there are some interesting principles for the limited resource allocation. Recently, B. Dietzenbacher et al. took an axiomatic approach to explore these principles, and clarify *proportional rules* (*PR*), *constrained relative equal awards rule* (*CREAR*), and *constrained relative equal loss rule* (*CRELR*). These allocation rules can be considered as an alternative interpretation of bargaining problems with claims. For any set of payoff allocations $E^{(\cdot)} \subseteq \mathbb{R}^n_+$ and $c^{(\cdot)}$, the *PR* allocates the payoffs as follows (Dietzenbacher, 2018, 2020; Kim, 2022a);

$$PR(E^{PR}, c^{PR}) = \varphi^{PR} \cdot c^{PR}, \text{ s.t.}, \varphi^{PR} = \max\{t \in (0,1] | (t \cdot c^{PR}) \in E^{PR}\} \tag{6.6.1}$$

For any $(E^{CREAR}, c^{CREAR}) \in \Sigma^n$, the *CREAR* allocates the payoffs as follows (Dietzenbacher, 2018, 2020; Kim, 2022a);

$$CREAR(E^{CREAR}, c^{CREAR}) = \left(\min\left\{c_i^{CREAR}, \left(\alpha^{AR} \cdot u_i^{E^{CREAR}}\right)\right\}\right)_{i \in \mathbb{N}} \tag{6.6.2}$$

$$\text{s.t.}, \begin{cases} \alpha^{AR} = \max\left\{t \in (0,1] | \left(\min\left\{c_i^{CREAR}, \left(t \cdot u_i^{E^{CREAR}}\right)\right\}\right)_{i \in \mathbb{N}} \in E^{CREAR}\right\} \\ u_i^{E^{CREAR}} = \left(\max\left\{x_i | x \in E^{CREAR}\right\}\right)_{i \in \mathbb{N}} \end{cases}$$

For any $(E^{CRELR}, c^{CRELR}) \in \Sigma^n$, the *CRELR* allocates the payoffs as follows (Dietzenbacher, 2018, 2020; Kim, 2022a);

$$CRELR(E^{CRELR}, c^{CRELR}) = \left(\max\left\{0,\ c_i^{CRELR} - \left(\beta^{LR} \cdot u_i^{E^{CRELR}}\right)\right\}\right)_{i \in \mathbb{N}} \tag{6.6.3}$$

$$\text{s.t.,} \begin{cases} \beta^{LR} = max\left\{\beta^{LR} \in \mathbb{R}_+ \mid \left(c_i^{CRELR} - \left(\beta^{LR} \cdot u_i^{E^{CRELR}}\right)\right)_{i \in \mathbb{N}} \in E^{CRELR}\right\} \\ u_i^{E^{CRELR}} = \left(\max\left\{x_i \mid x \in E^{CRELR}\right\}\right)_{i \in \mathbb{N}} \end{cases}$$

Figure 7. Bargaining solutions for two game players (Kim, 2022a)

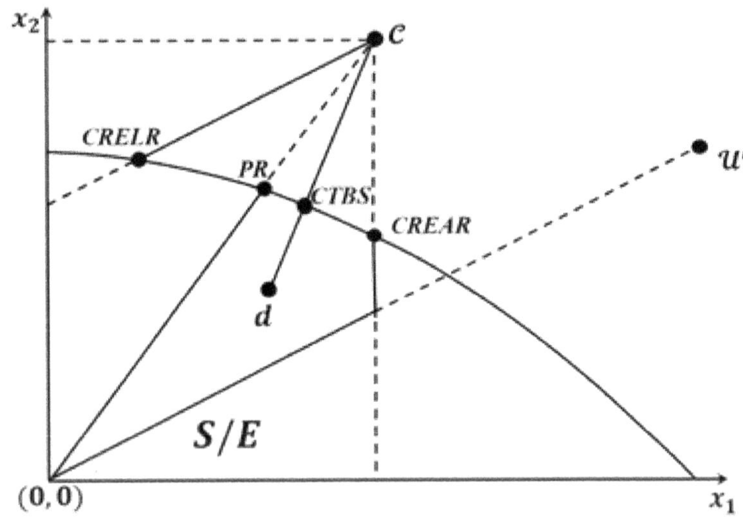

6.6.5 The two-level dual cooperative bargaining game process

The *CDSS* scheme operates within a two-tier 4G/5G coexisting network infrastructure. In the upper-tier, eNBs are deployed to support 4G LTE services. In the lower-tier, MNOs provide 5G NR services. In this coordinated architecture, MNOs and eNBs are interconnected through backhaul links, allowing them to work collaboratively to optimize network performance. Structurally, multiple MNOs are associated with their respective eNBs, ensuring efficient resource allocation and service continuity. Additionally, IoE devices are randomly distributed across the cell areas, with each device simultaneously communicating with its corresponding MNO and eNB. Through this dual connection, LTE and NR spectrum resources are effectively shared, enhancing network efficiency and service quality (Kim, 2022a).

The *CDSS* scheme consists of two bargaining game models for the LTE and NR resource allocations. In the NR resource bargaining game model, IoE devices are assumed as game players, and individual MNOs distribute their $\mathfrak{R}_{NR}^{\mathcal{M}}$ resources into game players through a fair-efficient spectrum allocation process. In the LTE resource bargaining game model, MNOs and IoE devices are assumed as game players in the two-level distribution mechanism. First, the $\mathfrak{N}^{\mathscr{E}}_{LTE}$ resource in each eNB is partitioned based on the requests of its corresponding MNOs. Second, based on the $\mathfrak{N}^{\mathscr{E}}_{LTE}$'s partitioned information, each MNO re-distributes the assigned LTE resource into its corresponding devices. Finally, IoE devices can share the $\mathfrak{N}^{\mathscr{E}}_{LTE}$ resource while maximizing the social benefit. To effectively allocate the limited LTE and NR resources, the LTE and NR spectrum allocation algorithms are designed based on the idea of *CTBS* (Kim, 2022a).

Implementing the NR bargaining game model presents a key challenge: maximizing the payoff for individual IoE devices by dynamically adjusting the resource allocation policy. To achieve this, the *CDSS* scheme must effectively leverage key system characteristics, including i) system capacity - ensuring that network resources are optimally utilized, ii) device utility - maximizing user satisfaction based on allocated spectrum and service quality, and iii) spectral efficiency - enhancing overall network performance by efficiently distributing spectrum resources. The *CDSS* scheme operates under the assumption that IoE devices generate application services, where each device is assigned a utility function based on two factors: allocated spectrum amount - the proportion of spectrum resources assigned to the device, and service price - the cost associated with accessing the allocated spectrum. By integrating these factors into the NR bargaining model, the *CDSS* scheme aims to optimize spectrum allocation, ensuring fair and efficient resource distribution across the 4G/5G coexisting infrastructure. The $\mathscr{D}_i$'s payoff from the NR resource allocation, i.e., $U^{\mathscr{D}_i}_{NR}(\cdot)$, can be derived as follows (Kim, 2022a);

$$U^{\mathscr{D}_i}_{NR}\left(\mathfrak{P}^{\mathscr{M}_j}_{\mathscr{D}_i}, \mathfrak{M}^{\mathscr{D}_i}\right) =$$

$$\left(\sigma \times exp\left(\log\left(\frac{\min\left(\mathfrak{P}^{\mathscr{M}_j}_{\mathscr{D}_i}, \mathfrak{M}^{\mathscr{D}_i}\right)}{\mathfrak{M}^{\mathscr{D}_i}} + \alpha\right)\right)\right) - \mathscr{C}^N\left(\mathfrak{P}^{\mathscr{M}_j}_{\mathscr{D}_i}, \mathfrak{M}^{\mathscr{D}_i}\right) \tag{6.6.4}$$

$$\text{s.t., } \mathscr{C}^N\left(\mathfrak{P}^{\mathscr{M}_j}_{\mathscr{D}_i}, \mathfrak{M}^{\mathscr{D}_i}\right) = \left((exp(\chi))^{\chi} - \psi\right) \text{ with } \chi = \mathfrak{P}^{\mathscr{M}_j}_{\mathscr{D}_i}/\mathfrak{M}^{\mathscr{D}_i}$$

where $\mathfrak{P}^{\mathscr{M}_j}_{\mathscr{D}_i}$ is the allocated NR resource for the $\mathscr{D}_i$, and $\mathfrak{M}^{\mathscr{D}_i}$ is the maximum request from the $\mathscr{D}_i$. σ and α are control parameters to evaluate the $U^{\mathscr{D}_i}_{NR}(\cdot)$ function, and ψ is a modification factor for the cost function $\mathscr{C}^N(\cdot)$. Based on the $U^{\mathscr{D}_i}_{NR}(\cdot)$, the $\mathscr{M}_j$'s utility function through the NR resource distribution, i.e., $U^{\mathscr{M}_j}_{NR}(\cdot)$, is defined as follows (Kim, 2022a);

$$U^{\mathscr{M}_j}_{NR}\left(\mathfrak{N}^{\mathscr{M}_j}_{NR}, \mathbb{S}^{\mathscr{M}_j}_{\mathscr{D}}\right) = \sum_{\mathscr{D}_i \in \mathbb{S}^{\mathscr{M}_j}_{\mathscr{D}}} U^{\mathscr{D}_i}_{NR}\left(\mathfrak{P}^{\mathscr{M}_j}_{\mathscr{D}_i}, \mathfrak{M}^{\mathscr{D}_i}\right), \text{ s.t., } \mathfrak{N}^{\mathscr{M}_j}_{NR} \le \sum_{\mathscr{D}_i \in \mathbb{S}^{\mathscr{M}_j}_{\mathscr{D}}} \mathfrak{P}^{\mathscr{M}_j}_{\mathscr{D}_i} \tag{6.6.5}$$

where $\mathbb{S}^{\mathscr{M}_j}_{\mathscr{D}}$ is the set of devices, which are covered by the $\mathscr{M}_j$. In the NR resource allocation algorithm, each device's claim for its NR spectrum is defined to maximize his utility function according to (6.6.4). When tasks are generated in D, each task has its pre-defined minimum $\left(\grave{E}^{\mathscr{D}}_m\right)$ and maximum $\left(\grave{E}^{\mathscr{D}}_M\right)$ spectrum requirements. Simply, the *CDSS* scheme can think the $\grave{E}^{\mathscr{D}}_M$ as an ideal point. To define the disagreement point d, the *CDSS* scheme needs an allocation rule with the $\grave{E}^{\mathscr{D}}_m$. The main idea of the *CDSS* scheme is that NR spectrum is primarily allocated for devices, and then the LTE spectrum can be supplementarily distributed. Therefore, the *PR* is adopted to define the d in the NR resource allocation algorithm while considering *TI, CFC* and *ILRC* properties. In the $PR(E^{PR}, c^{PR})$ of equation (6.6.1), c^{PR} is represented by the $\grave{E}_m$, and E is reduced by the diminution factor Γ_R to define the E^{PR} where $0 < \Gamma_R \ll 1$. Finally, the $\mathcal{M}_j$ can decide the $\mathfrak{P}^{\mathcal{M}_j}_{\mathcal{D}_i \in \mathbb{S}^{\mathcal{M}_j}_{\mathcal{D}}}$ for devices based on the *CTBS* with c and d (Kim, 2022a).

Under heavy traffic load situations, the limited NR resource is insufficient to satisfy all device's demands. Therefore, after the NR resource allocation process, some devices don't have enough spectrum resources. In the *CDSS* scheme, the LTE resource can be supplemented to maximize the device's payoff. To reduce computation complexity, the LTE spectrum allocation algorithm is designed as a two-level allocation mechanism. First, each M collects the total amount $\left(\mathfrak{M}^{\mathcal{M}_j}_{LTE}\right)$ of spectrum shortages from his covering devices and requests it for the corresponding eNB. The $\mathcal{M}_j$'s payoff from the LTE resource allocation, i.e., $U^{\mathcal{M}_j}_{LTE}(\cdot)$, can be derived as follows (Kim, 2022a);

$$U^{\mathcal{M}_j}_{LTE}\left(\mathfrak{P}^{\mathcal{E}_k}_{\mathcal{M}_j}, \mathfrak{M}^{\mathcal{M}_j}_{LTE}\right) =$$

$$\left(\zeta \times exp\left(\log\left(\frac{\min\left(\mathfrak{P}^{\mathcal{E}_k}_{\mathcal{M}_j}, \mathfrak{M}^{\mathcal{M}_j}_{LTE}\right)}{\mathfrak{M}^{\mathcal{M}_j}_{LTE}} + \beta\right)\right)\right) - \mathcal{C}^L\left(\mathfrak{P}^{\mathcal{E}_k}_{\mathcal{M}_j}, \mathfrak{M}^{\mathcal{M}_j}_{LTE}\right) \tag{6.6.6}$$

$$\text{s.t.,} \begin{cases} \mathfrak{M}^{\mathcal{M}_j}_{LTE} = \sum_{\mathcal{D}_i \in \mathbb{S}^{\mathcal{M}_j}_{\mathcal{D}}}\left(\max\left(\mathcal{J}_{\mathcal{D}_i}, 0\right)\right) with\, \mathcal{J}_{\mathcal{D}_i} = \left(\mathfrak{M}^{\mathcal{D}_i} - \mathfrak{P}^{\mathcal{M}_j}_{\mathcal{D}_i}\right) \\ \mathcal{C}^L\left(\mathfrak{P}^{\mathcal{E}_k}_{\mathcal{M}_j}, \mathfrak{M}^{\mathcal{M}_j}_{LTE}\right) = \left((exp(\chi))^{\chi} - \gamma\right) with \chi = \mathfrak{P}^{\mathcal{E}_k}_{\mathcal{M}_j} / \mathfrak{M}^{\mathcal{M}_j}_{LTE} \end{cases}$$

where $\mathfrak{P}^{\mathcal{E}_k}_{\mathcal{M}_j}$ is the allocated LTE resource for the $\mathcal{M}_j$. ζ and β are control parameters to evaluate the $U^{\mathcal{M}}_{LTE}(\cdot)$ function, and γ is a modification factor for the cost function $\mathcal{C}^L(\cdot)$. To share the LTE resource among MNOs, the concept of *CTBS* is also applied. In the upper-level LTE allocation algorithm, the c is defined to maximize the M's utility function according to (6.6.6), and the ideal point is defined in the same manner as c without considering the cost function $\mathcal{C}^L(\cdot)$. To define the d, the *CDSS* scheme adopts the idea of *CREAR*. In the $CREAR(E^{CREAR}, c^{CREAR})$ of equation (6.6.2), E^{CREAR} is obtained by using Γ_R, and c_j^{CREAR}, in this case, $c_{\mathcal{M}_j}^{CREAR}$, is given by (Kim, 2022a);

$$c^{CREAR}_{\mathcal{M}_j} = \mathfrak{M}^{\mathcal{M}_j}_{LTE} \times \left(\min\left(\max\left(f_{\mathcal{M}_j}, \kappa^L\right), \kappa^U\right)\right), \text{ s.t., } \mathcal{M}_j \in \mathbb{S}^{\mathcal{E}_k}_{\mathcal{M}} \tag{6.6.7}$$

where $f_{\mathcal{M}_j}$ is the current service failure probability of $\mathcal{M}_j$, and κ^L, κ^U are lower and upper boundary parameters. $\mathbb{S}^{\mathcal{E}_k}_{\mathcal{M}}$ is the set of MNOs, which are covered by the $\mathcal{E}_k$. Finally, the $\mathcal{E}_k$ can decide the $\mathfrak{P}^{\mathcal{E}_k}_{\mathcal{M} \in \mathbb{S}_{\mathcal{M}}}$ resource based on the *CTBS* with c and d. And then, the *CDSS* scheme implements the lower-level LTE allocation algorithm to re-distribute the $\mathfrak{P}^{\mathcal{E}_k}_{\mathcal{M}_j}$; the $\mathcal{M}_j$ allocates its assigned LTE resource $\left(\mathfrak{P}^{\mathcal{E}_k}_{\mathcal{M}}\right)$ into the $\mathcal{D}_i \in \mathbb{S}^{\mathcal{M}_j}_{\mathcal{D}}$. The $\mathcal{D}_i$'s second payoff from the LTE resource allocation, i.e., $U^{\mathcal{D}_i}_{LTE}(\cdot)$, can be derived as follows (Kim, 2022a);

$$U^{\mathcal{D}_i}_{LTE}\left(\mathfrak{X}^{\mathcal{E}_k}_{\mathcal{D}_i}, \mathcal{J}_{\mathcal{D}_i}\right) =$$

$$\left(\eta \times exp\left(\log\left(\frac{\min\left(\mathfrak{X}^{\mathcal{E}_k}_{\mathcal{D}_i}, \mathcal{J}_{\mathcal{D}_i}\right)}{\mathcal{J}_{\mathcal{D}_i}} + \varrho\right)\right)\right) - \mathscr{C}^T\left(\mathfrak{X}^{\mathcal{E}_k}_{\mathcal{D}_i}, \mathcal{J}_{\mathcal{D}_i}\right) \tag{6.6.8}$$

$$\text{s.t., } \mathscr{C}^T\left(\mathfrak{X}^{\mathcal{E}_k}_{\mathcal{D}_i}, \mathcal{J}_{\mathcal{D}_i}\right) = \left((exp(\chi))^{\chi} - \psi\right) \text{ with } \chi = \mathfrak{X}^{\mathcal{E}_k}_{\mathcal{D}_i}/\mathcal{J}_{\mathcal{D}_i}$$

where $\mathfrak{X}^{\mathcal{E}_k}_{\mathcal{D}_i}$ is the allocated LTE resource for the $\mathcal{D}_i$. η and ϱ are control parameters to evaluate the $U^{\mathcal{D}_i}_{LTE}(\cdot)$ function, and μ is a modification factor for the cost function $\mathscr{C}^T(\cdot)$. To share the LTE resource among $\mathcal{D}_i \in \mathbb{S}^{\mathcal{M}_j}_{\mathcal{D}}$, the idea of *CTBS* is used, too. In the algorithm, the c is defined to maximize the D's utility function according to (6.6.4), and the ideal point is defined in the same manner as c without considering the cost function $\mathscr{C}^T(\cdot)$. To define the d, the *CDSS* scheme adopts the idea of *CRELR*. In the *CRELR*(E^{CRELR}, c^{CRELR}) of equation (6.6.3), E^{CRELR} is obtained by using Γ_R, and c_i^{CRELR}, in this case, $c^{CREAR}_{\mathcal{D}_i}$, is given by (Kim, 2022a);

$$c_{\mathcal{D}_i} = \mathcal{J}_{\mathcal{D}_i} \times \left(\min\left(\max\left(\frac{\mathcal{J}_{\mathcal{D}_i}}{\mathfrak{M}^{\mathcal{D}_i}}, \text{Ä}^L\right), \text{Ä}^U\right)\right), \text{ s.t., } \mathcal{D}_i \in \mathbb{S}^{\mathcal{M}_j}_{\mathcal{D}} \tag{6.6.9}$$

where Ä^L, Ä^U are lower and upper boundary parameters for the $c_{\mathcal{D}_i}$. In the process, the $\mathcal{M}_j$ can assign the $\mathfrak{P}^{\mathcal{E}_k}_{\mathcal{M}_j}$ resource for the $\mathcal{D}_i \in \mathbb{S}^{\mathcal{M}_j}_{\mathcal{D}}$ based on the *CTBS* with c and d. Finally, the $\mathcal{D}_i$ gets its $\mathfrak{X}^{\mathcal{E}_k}_{\mathcal{D}_i}$, and the $\mathcal{D}_i$'s total payoff is obtained from the summation of $U^{\mathcal{D}_i}_{NR}(\cdot)$ and $U^{\mathcal{D}_i}_{LTE}(\cdot)$, and the $\mathcal{E}_k$'s utility function for the LTE resource, i.e., $U^{\mathcal{E}_k}_{LTE}(\cdot)$, is defined based on the $U^{\mathcal{D}_i}_{LTE}(\cdot)$ (Kim, 2022a);

$$U^{\mathcal{E}_k}_{LTE}\left(\mathfrak{N}^{\mathcal{E}_k}_{LTE}, \mathfrak{X}^{\mathcal{E}_k}_{\mathcal{D}_i}, \mathbb{S}^{\mathcal{M}_j}_{\mathcal{D}}, \mathbb{S}^{\mathcal{E}_k}_{\mathcal{M}}\right) = \sum_{\left\{\mathcal{D}_i \in \mathbb{S}^{\mathcal{M}_j}_{\mathcal{D}} \middle| \mathcal{M}_j \in \mathbb{S}^{\mathcal{E}_k}_{\mathcal{M}}\right\}} U^{\mathcal{D}_i}_{LTE}\left(\mathfrak{X}^{\mathcal{E}_k}_{\mathcal{D}_i}, \mathcal{J}_{\mathcal{D}_i}\right) \tag{6.6.10}$$

$$\text{s.t., } \mathfrak{N}^{\mathcal{E}_k}_{LTE} \leq \sum_{\left\{\mathcal{D}_i \in \mathbb{S}^{\mathcal{M}_j}_{\mathcal{D}} \middle| \mathcal{M}_j \in \mathbb{S}^{\mathcal{E}_k}_{\mathcal{M}}\right\}} \mathfrak{X}^{\mathcal{E}_k}_{\mathcal{D}_i}$$

6.6.6 Summary

The deployment of 5G NR wireless mobile communications promises higher data rates, lower latency, and enhanced system capacity. The initial implementation of 5G NR will leverage the existing 4G LTE infrastructure, ensuring a gradual and efficient transition. In the *CDSS* scheme, the DSS is introduced as an innovative spectrum utilization mechanism for the 4G/5G coexisting network infrastructure. This powerful and flexible spectrum control paradigm enables seamless coexistence between 4G LTE and 5G NR technologies, optimizing spectrum utilization without disrupting service continuity. To address diverse service requirements in future networks, the *CDSS* scheme incorporates a new DSS control algorithm, built upon the *CTBS*, *PR*, *CREAR*, and *CRELR*. The *CDSS* scheme comprises two resource allocation algorithms - the NR resource allocation algorithm and the LTE resource allocation algorithm. These two algorithms are interactively integrated, forming a mutually dependent system that dynamically adjusts spectrum allocation in response to real-time network conditions. The dual bargaining game model within the *CDSS* scheme ensures that each network agent actively participates in coordinated decision-making to optimize spectrum sharing. Given the dynamically evolving 4G/5G network environment, the two algorithms operate cooperatively and in real-time, continuously adapting to current network conditions to enhance system efficiency and service quality (Kim, 2022a).

6.7 Two-level Game based Spectrum Allocation (*TGSA*) scheme

With the rapid growth of mobile devices, the demand for spectrum resources has increased significantly, while available carrier resources remain limited. To address this challenge, Multi-Flow Carrier Aggregation (MCA) has been recently introduced as a promising technique to enhance capacity in 5G heterogeneous network (HetNet) systems. Recently, S. Kim proposed the *Two-Level Game-Based Spectrum Allocation* (*TGSA*) scheme for next-generation HetNets (Kim, 2020d). The *TGSA* scheme, designed for MCA-enabled HetNets, seeks to maximize system throughput while ensuring optimal benefits for individual mobile devices. To achieve a balanced trade-off between network operators and mobile users, the *TGSA* scheme employs a two-level game model. The upper-level game represents one type of game model, focusing on global spectrum allocation at the network operator level. The lower-level game adopts a different game-theoretical approach, addressing individual mobile device spectrum distribution.

These two levels interact dynamically, working together to reach a fair and efficient spectrum allocation solution. By leveraging a hierarchical game-theoretic approach, the *TGSA* scheme ensures reciprocal advantages between all stakeholders, providing insights into the optimal spectrum allocation strategy for MCA-enabled systems. This approach enhances overall network efficiency while maintaining fairness in resource distribution (Kim, 2020d).

6.7.1 Development motivation

In recent years, the demand for multimedia services has surged, driven by data-intensive applications such as video streaming, image uploads, and cloud-based computing. These applications impose significant demands on future mobile networks, requiring solutions that ensure high performance, security, and ease of use. However, wireless spectrum remains a scarce and valuable resource in 5G networks. Given the rapid expansion of data-intensive services and the limited availability of spectrum, one of the greatest challenges is ensuring efficient spectrum utilization. In both academic research and industry, various strategies have been explored to enhance network performance and optimize resource allocation (Chavarria-Reyes, 2016; Kim, 2020d; Yuan, 2017).

To address these challenges, Heterogeneous Networks (HetNets) have been proposed as a key architectural solution for 5G wireless communication systems. A HetNet infrastructure consists of macro-cells providing wide-area coverage, and overlaid small cells enhancing seamless connectivity and high-speed data transmission. These small cells operate across licensed and unlicensed spectrum and support hybrid mobile devices (MDs) that can access multiple radio access technologies (RATs). The HetNet system improves data communication rates by enabling spatial spectrum reuse, giving mobile devices the flexibility to select their preferred connectivity options. Given the availability of multiple service providers, MDs make dynamic decisions based on QoS metrics and service pricing. As a result, network operators

(COs) compete with one another, adjusting their pricing and QoS offerings to attract MDs (Gerasimenko, 2015; Kim, 2020d; Kumar, 2014).

In future 5G networks, MDs at the cell edge can aggregate spectrum bands from different COs to boost data transmission rates. This concept, known as Multi-Flow Carrier Aggregation (MCA), is an extension of Carrier Aggregation (CA), originally introduced in LTE-Advanced to merge fragmented spectrum bands into a single virtual carrier, enhancing throughput and spectrum efficiency. The MCA technique extends this capability to multi-tier HetNets, incorporating multi-band modeling and analysis to optimize spectrum allocation across multiple COs. By distributing MD traffic across multiple spectrum bands, MCA significantly improves network flexibility and utilization efficiency. However, to fully leverage MCA, it is crucial to implement effective spectrum allocation mechanisms, allowing MDs to seamlessly access resources from multiple COs simultaneously (Chen, 2016; Kim, 2020d; Wang, 2016a).

Despite initial efforts, MCA research remains in its early stages, with several unresolved technical challenges. To facilitate efficient multi-operator spectrum aggregation, a new control paradigm is required. The *TGSA* scheme employs game-theoretic principles to optimize spectrum sharing in future MCA-enabled HetNets. Game theory provides a structured framework for analyzing distributed optimization problems in strategic network environments. It enables independent and competing players to make rational, strategic decisions, ensuring optimal resource allocation in a limited-spectrum scenario (Kim, 2014, 2020d).

A Stackelberg game is a non-cooperative game model where players are hierarchically structured into leaders and followers. The leader makes the first move, determining key strategic decisions. The followers respond based on the leader's actions, aiming to maximize their individual satisfaction. In the MCA system, multiple COs and MDs interact dynamically, requiring an extended Stackelberg game that incorporates multiple leaders. COs act as leaders, setting their service prices while considering potential MD reactions. MDs act as followers, selecting spectrum resources for their multi-flow transmissions based on CO pricing strategies. Since control decisions are interdependent, each player's behavior influences the others' responses, creating a strategically coupled decision-making process. The *TGSA* scheme formulates this interaction as a Multiple-Leaders Multiple-Followers (MLMF) Stackelberg game (Kim, 2020d; Kulkarni, 2015).

In addition to Stackelberg games, the *TGSA* scheme incorporates the *Efficient Max-Min Bargaining Solution* (*EMBS*), a cooperative bargaining model based on the *Kalai-Smorodinsky solution*. Bargaining problems share similarities with decision-making models, as players negotiate to generate surplus resources that must be fairly distributed. The *EMBS* model allows players to determine fair payoffs, balancing efficiency and equity through joint agreements. The core principle of *EMBS* is to balance proportional fairness and Pareto optimality, making it a generalized extension of the *Kalai-Smorodinsky solution* to nonconvex, multi-player bargaining games. In the *TGSA* scheme, the *EMBS* governs the follower's reaction process, ensuring efficient spectrum allocation across multiple COs (Hougaard, 2003; Kim, 2020d).

The *TGSA* scheme introduces a novel spectrum allocation framework for the MCA system, structured as a two-level game model. In the upper-level, the MLMF Stackelberg game determines the service pricing strategy for each CO, influencing MDs' spectrum selection. In the lower-level, the *EMBS* based spectrum allocation optimizes multi-flow spectrum aggregation for MDs, ensuring fair resource allocation among COs. This hierarchical interconnection enables dynamic decision-making, where CO pricing decisions affect MD spectrum selection, and MD allocation choices provide feedback to COs, influencing future pricing strategies. Through an online interactive process, this fine-grained two-level game model iterates continuously, ultimately reaching a fair-efficient spectrum allocation equilibrium (Kim, 2020d).

6.7.2 Contribution

Based on the two-level game model, the *TGSA* scheme effectively integrates both non-cooperative and cooperative game strategies to address various control challenges holistically. The *TGSA* scheme incorporates MCA technology as a key component of future network systems. In designing the MCA system control framework, two primary concerns arise: (i) determining the spectrum pricing strategy for COs and (ii) allocating spectrum resources from multiple COs to accommodate each user's multiple data flows. To address these challenges, the *TGSA* scheme applies game theory principles and introduces an innovative two-level game model. These two levels employ distinct game models, which operate interactively to establish mutually beneficial agreements among players. At the upper level, the model is structured as an MLMF Stackelberg game, where COs function as leaders and MDs act as followers. Leaders set spectrum prices in a non-cooperative manner, yet their decisions must consider the actions of other leaders to achieve a balanced consensus. Meanwhile, the lower level is designed as a bargaining game model, where within each MD, traffic flows serve as individual game players. These traffic flows negotiate among themselves over the allocation of spectrum resources from various COs, leveraging the *EMBS*-based approach (Kim, 2020d).

6.7.3 The MCA system infrastructure and operating assumptions

We consider a multi-tier cellular network platform with $\mathscr{K} = \{1,\ldots,K\}$ denoting the set of K tiers which may include different coverage areas. The small cells in lower tier are regularly deployed within the large cell in higher tier. COs of different type cells may differ in terms of corresponding coverage size, spectrum resource, access technology and service price. Without loss of generality, one CO in the k-tier has a coverage area of radius r_k, and has a static portion of the spectrum resource to provide the basic services to its users in its area. There are n MDs $\mathbb{N} = \{\mathscr{D}_1,\ldots,\mathscr{D}_n\}$ and $\mathscr{D}_{1\leq i\leq n}$ is assumed to have the capability to aggregate multiple traffic flows over several COs. MDs are randomly distributed in the cellular area, and they are heterogeneous in their experienced carrier qualities, carrier aggregation capabilities, and QoS requirements. Therefore, each device's application service can be represented by multiple subcarriers (Kim, 2020d).

From the viewpoint of COs, the main goal is to maximize the system revenue while dynamically adjusting their price policy. In the point view of MDs, their major concern is to maximize the service payoff by adaptively operating the MCA technique. To satisfy the different goals of COs and MDs, the *TGSA* scheme develops a two-level game model. In the upper-level game, the COs' prices are effectively adjusted based on the Stackelberg game model. In the lower-level game, each MD divides its application into multiple flows and allocates the available COs' spectrum resources according to the cooperative bargaining game. The two-level game model defines the phenomenon that co-exists competition and cooperation face in the MCA system operation and adapts the current network environments while satisfying multi-objective goals. Formally, the *TGSA* scheme defines the game model $\mathbb{G}^{MCA} = \{\{\mathbb{O},\mathbb{F}^{\mathscr{D}_{1\leq i\leq n}}\},\{\mathbb{S}^{CO\in\mathbb{O}},\mathbb{S}^{\mathscr{D}\in\mathbb{N}}\},\mathbb{V}^{\mathfrak{P}^{co}},\{\mathbb{U}^{CO},\mathbb{U}^{\mathscr{D}}\},\mathfrak{M},T\}$ at each time period of gameplay (Kim, 2020d).

- $\{\mathbb{O},\mathbb{F}^{\mathscr{D}_i}\}$ are game players for the two-level game; $\mathbb{O} = \left\{\ldots,CO_j^{l\in\mathscr{K}},\ldots\right\}$ is the finite set of multiple COs where CO_j^l represents the jth CO in the lth tier, and $\mathbb{F}^{\mathscr{D}_i} = \left\{\ldots,f^{\mathscr{D}_i}_{CO\in\mathbb{C}_{\mathscr{D}_i}},\ldots\right\}$ is the set of multiple flows in the $\mathscr{D}_i$ in the MCA cellular area where $f^{\mathscr{D}_i}_{CO\in\mathbb{C}_{\mathscr{D}_i}}$ represents the $\mathscr{D}_i$'s flow connected to the CO in the $\mathbb{C}_{\mathscr{D}_i}$; $\mathbb{C}_{\mathscr{D}_i}$ is the set of COs, which are reachable by the $\mathscr{D}_i$.

- $\mathbb{S}^{CO} = \{\mathfrak{P}^{CO}_{min}\ldots\mathfrak{P}^{CO}_{h}\ldots\mathfrak{P}^{CO}_{max}\}$ is a set of CO's strategies where $\mathfrak{P}^{\mathscr{R}}_{h}$ means the price level to provide one basic spectrum unit (BSU). $\mathbb{S}^{\mathscr{D}} = \left\{\ldots r^{\mathscr{D}}_{CO\in\mathbb{C}_{\mathscr{D}}}\ldots\right\}$ is the set of D's available strategies where $r^{\mathscr{D}}_{CO\in\mathbb{C}_{\mathscr{D}}}$ represents the allocated spectrum amount from the CO for the D; $f^{\mathscr{D}}_{CO}$ is assigned for $f^{\mathscr{D}}_{CO}$.
- $\mathbb{V}^{\mathfrak{P}^{CO}} = [\eta^{\mathfrak{P}^{CO}_{min}}\ldots\eta^{\mathfrak{P}^{CO}_{h}}\ldots\eta^{\mathfrak{P}^{CO}_{max}}]$ is a vector of CO's strategy selection distribution where $\eta^{\mathfrak{P}^{CO}_{h}}$ represents the strategy $\mathfrak{P}^{CO}_{h}$'s selection probability.
- $\mathbb{U}^{CO}$ is the payoff received by the CO, and $\mathbb{U}^{\mathscr{D}}$ is the payoff received by the D during the MCA process.
- $\mathfrak{M} = \{\ldots,\mathfrak{M}^{CO^{i}_{j}},\ldots\}$ is a set of spectrum resource amounts for COs.
- $T = \{\mathscr{T}_1,\ldots,\mathscr{T}_t,\mathscr{T}_{t+1},\ldots\}$ denotes time, which is represented by a sequence of time steps.

6.7.4 Price control mechanism for MCA services

Usually, price control mechanisms are designed based on the elastic-demand paradigm; according to the service price (P_s), customers adapt their service requests. It is relevant under the MCA system operation. Figure 8 depicts graphically the elastic-demand paradigm by using demand curve $(\mathscr{C}_d)$ and performance curve $(\mathscr{C}_p)$. The $\mathscr{C}_d$ shows the aggregated spectrum request of users $(\mathscr{R}_s)$, and the $\mathscr{C}_p$ describes the relationship between the $\mathscr{R}_s$ and P_S. Traditionally, the $\mathscr{C}_d$ can be expected to be monotonically decreasing in the generalized P_s while the $\mathscr{C}_p$ is monotonically increasing in the generalized $\mathscr{R}_u$. Under the fixed spectrum capacity $(\mathscr{F}_c)$, the equilibrium point can be determined by combining the $\mathscr{C}_d$ and $\mathscr{C}_p$. In Fig. 6.7.1, let assume that the $\mathscr{C}^3_p$ and $\mathscr{C}_d$ are selected. When the current price is P^1_s (*or* P^2_s), the associated request amount is $\mathscr{R}^1_u$ (*or* $\mathscr{R}^2_u$). In order to toward the network equilibrium point $(\mathscr{E}_*)$, the current price should increase (*or* decrease) by ÄP^1_s (*or* ÄP^2_s). (Kim, 2011b, 2020d; Yang, 1997).

Figure 8. Demand and performance curves in the elastic-demand model (Kim, 2020d)

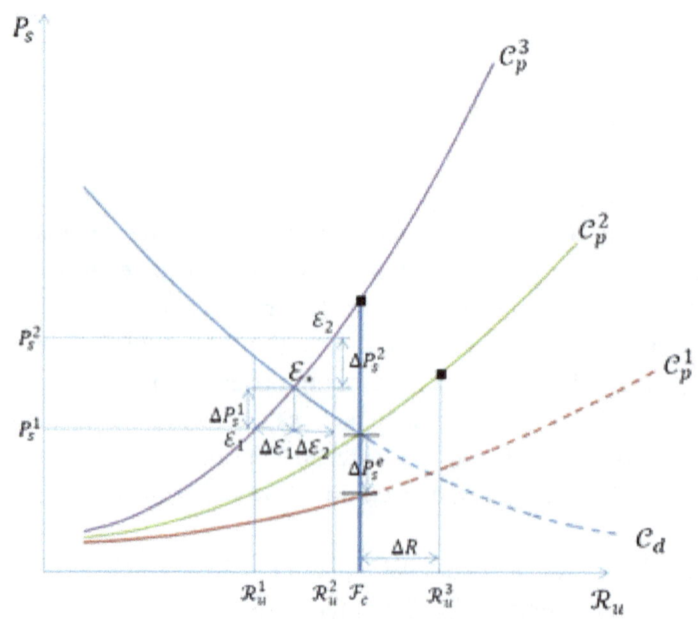

If the total spectrum amount is $\mathscr{R}_u^3$ under the $\mathscr{C}_p^2$, the available spectrum resource is not enough to support all requests; traffic congestion occurs. The over requests (ÄR) should be rejected to meet the $\mathscr{F}_c$. To deal with this traffic congestion problem, extra charge (ÄP_s^e) is used to reduce the potential spectrum demand. Finally, the demand-supply balance is obtained. Under the assume that the $\mathscr{C}_p^3$ and $\mathscr{C}_d$ are selected, $\mathscr{C}_p^3$, the collective user-satisfaction ($\mathscr{S}_{\mathscr{C}}$) can be maximized at the $\mathscr{E}_*$ state. Therefore, each CO attempts to reach the $\mathscr{E}_*$ by adjusting the P_s. To maximize the $\mathscr{S}_{\mathscr{C}}$, a mathematical programming formulation can be described as follows (Kim, 2011b, 2020d; Yang, 1997).

$$\max \mathscr{S}_{\mathscr{C}} \quad = \max_{i;P_s^i=(\mathscr{C}_d,\mathscr{C}_p)} \left(\int_0^{\mathscr{R}_u^A} \mathscr{C}_d(\mathscr{R}_u^i)\, d\mathscr{R}_u - \int_0^{\mathscr{R}_u^A} \mathscr{C}_p(\mathscr{R}_u^i)\, d\mathscr{R}_u \right) \tag{6.7.1}$$

where $\mathscr{R}_u^A$ is the total amount of all users requested spectrum resources. Based on the adjustment of price policy P_s^i, the $\mathscr{C}_d$ is dynamically decided by MDs. In Figure 8, the solution of Equation (6.7.1) can be viewed pictorially as a maximization of the area between the $\mathscr{C}_d$ and $\mathscr{C}_p\mathscr{C}_p$; the area under the $\mathscr{C}_d$ minus the area under the $\mathscr{C}_p$ within the $\mathscr{E}_*$ (Kim, 2020d; Yang, 1997).

Traditional non-cooperative game theory assumes that players possess perfect rationality and complete information about their environment. However, this idealized scenario rarely occurs in real-world situations and has been subject to criticism. Consequently, a key challenge is the development of learning algorithms that allow players to adaptively select optimal strategies to achieve their desired game outcomes. Through learning mechanisms, players can reassess their current strategies, evaluate the surrounding conditions, and adjust their actions dynamically to maximize their expected payoffs (Hofbauer, 2003; Kim, 2011a, 2020d).

In 1974, J. M. Smith introduced the foundational concept of the evolutionary learning game model, which was initially developed in the field of biological sciences to explain the evolution of genetically driven social behaviors. Within this framework, the rate at which players adjust their strategy selection is referred to as *Replicator Dynamics* (*RD*), which characterizes the evolutionary process by describing changes in the proportion of each strategy until an equilibrium is reached (Hofbauer, 2003; Kim, 2011a). The *TGSA* scheme incorporates the *RD* approach to designing a price control algorithm for COs. When a CO selects a pricing strategy, it influences the existing MCA system environment, prompting responses from other COs and MDs. As these adjustments continue, the probability of selecting a specific strategy evolves at a rate determined by the difference between its individual payoff and the average payoff. Throughout MCA operations, COs continuously refine their pricing strategies and engage in iterative interactions with other entities in the system. For example, if the payoff of ith price strategy $\left(\mathfrak{P}_i^{CO}\right)$ is higher compared to other strategies, the selection probability for i[th] strategy $(\eta^{\mathfrak{P}_i^{CO}})$ increases in proportion to the expected payoff increment. Therefore, the desirable strategy that will improve player's payoff is more likely to be selected. This interaction mechanism gradually leads the MCA system into a stable state; it means no individual CO can improve his payoff by unilaterally changing his strategy and can be immune from being changed. It is relevant to the Darwinian evolution mechanism (Kim, 2011a, 2020d).

To represent the *RD* for the price control problem, let L be a number of possible price strategies, i.e., $L = |\mathbb{S}^{CO}|$, and $\mathbb{V}^{\mathfrak{P}^{CO}}$ is the L-dimensional vector $(\eta^{\mathfrak{P}_1^{CO}} \ldots \eta^{\mathfrak{P}_i^{CO}} \ldots \eta^{\mathfrak{P}_L^{CO}})$. $\dot{\eta}_{\mathfrak{P}_i^{CO}}$ stands for the variation of $\eta^{\mathfrak{P}_i^{CO}}$; it is the *RD* for the $\mathfrak{P}_i^{CO}$. $E^p\left(\mathfrak{P}_i^{CO}, \mathfrak{P}_l^{CO}\right)$ is denoted by the expected payoff for a player using the $\mathfrak{P}_i^{CO}$ when it encounters a player with the $\mathfrak{P}_l^{CO}$, and $E^p\left(\mathfrak{P}_i^{CO}, \mathbb{V}^{\mathfrak{P}^{CO}}\right)$ is the payoff for a CO using the $\mathfrak{P}_i^{CO}$ when it encounters the rest of other COs whose strategies are distributed in $\mathbb{V}^{\mathfrak{P}^{co}}$, which can be expressed like as $\sum_{\mathfrak{P}_j^{co}}\left(E^p\left(\mathfrak{P}_i^{CO},\mathfrak{P}_j^{CO}\right)\times\eta^{\mathfrak{P}_j^{co}}\right)$ Finally, the *RD* is defined as (Kim, 2011a, 2020d);

$$\dot{\eta}_{\mathfrak{P}_i^{co}} = \eta^{\mathfrak{P}_i^{co}}\times\left(E^p\left(\mathfrak{P}_i^{CO},\ \mathbb{V}^{\mathfrak{P}^{co}}\right) - \sum_{\mathfrak{P}_j^{co}}\left(\eta^{\mathfrak{P}_j^{co}}\times E^p\left(\mathfrak{P}_l^{CO},\ \mathbb{V}^{\mathfrak{P}^{co}}\right)\right)\right)$$

$$=\eta^{\mathfrak{P}_i^{co}}\times\left(\sum_{\mathfrak{P}_j^{co}}\left(E^p\left(\mathfrak{P}_i^{CO},\ \mathfrak{P}_j^{CO}\right)\times\eta^{\mathfrak{P}_j^{co}}\right) - \sum_{\mathfrak{P}_j^{co}}\sum_{\mathfrak{P}_l^{co}}\left(E^p\left(\mathfrak{P}_j^{CO},\mathfrak{P}_l^{CO}\right)\times\eta^{\mathfrak{P}_j^{co}}\times\eta^{\mathfrak{P}_l^{co}}\right)\right)$$

(6.7.2)

6.7.5 The concept of efficient max-min bargaining solution

In 1950, J. Nash introduced a bargaining solution, later known as the Nash bargaining solution, which adheres to certain principles, including the *independence of irrelevant alternatives*. However, this particular axiom has faced significant criticism over time. In response, Kalai and Smorodinsky, in 1975, proposed an alternative approach by replacing the *independence* axiom with a *monotonicity* axiom, thereby formulating the Kalai-Smorodinsky bargaining solution. This solution ensures that utilities are maximized in proportion to the ideal outcome relative to the threat point. Since the *monotonicity* axiom is less controversial, it is often regarded as a more general and widely accepted principle for determining bargaining outcomes. Unfortunately, the Kalai-Smorodinsky solution does not extend directly to n-player bargaining problem. Recently, several EMBSs have been proposed for n-player bargaining problems. They can be seen as generalizations of the Kalai-Smorodinsky solution in the sense that no individual player gets a smaller payoff in a unique solution (Hougaard, 2003; Kim, 2020d).

Let U be the set of n-player bargaining problems, (a, S), where $a \in \mathrm{R}^n$ denotes the threat point, and S is a feasible set. If players fail to reach some other outcome $x = \{x_1,\ldots,x_n\} \in S$, this disagreement result is a threat point. $S \subset \mathbb{R}^n$ denotes the set of feasible payoffs, that satisfy i) S is compact and, ii) $S \cap \left(\{a\} + \mathbb{R}^n_{++}\right)$ is non-empty. For an n-player bargaining problem, (a, S), $M(S)$ is denoted the ideal point (Hougaard, 2003), and it is defined as follows (Kim, 2020d);

$$M(S) = \left(M_1(S),\ldots,M_{1\leq j\leq n}(S),\ldots M_n(S)\right),\ \text{s.t.},\ M_j(S) = \max_{x\in S}\ \mathcal{J}_j(x)$$

(6.7.3)

where $\mathcal{J}_j(x)$ is the projection on the jth coordinate of x, and $M_j(S)$ is the maximal value of the j'th coordinate. With $f(a, S) \in S$, let $f\!:\!\mathcal{U} \rightarrow \mathbb{R}^n$ be a solution to the bargaining problem (a, S) and let $P(S)$ be the set of strongly Pareto optimal payoffs in S. Moreover, let $D(a, S)$ be the diagonal of (a, S), and $\mathcal{M}^d$ (a, S) is the maximal payoff on the diagonal (Hougaard, 2003; Kim, 2020d);

$$\mathcal{M}^d(a, S) = P(S \cap D(a, S))$$

(6.7.4)

$$\text{s.t.},\ D(a, S) = \left\{x \in \mathbb{R}^n | x = ((1-\beta)\cdot a) + (\beta\cdot M(S))\ ;\ \beta \in \mathbb{R}\right\}$$

By using the above terminologies, the *TGSA* scheme can define the generalize the Kalai-Smorodinsky solution, i.e., $\mathscr{G}_{KS}(a, S)$, be as follows (Hougaard, 2003; Kim, 2020d);

$$\mathscr{G}^{KS}(a,\ S)\ =\ \left[a + (\beta \cdot (M(S) - a))\right]\ =\ ((1-\beta) \cdot a + (\beta \cdot M(S))) \tag{6.7.5}$$

s.t., $\beta \in [0,\ 1]$ and $\mathscr{G}^{KS}(a,\ S)\ \geq\ \mathscr{M}^{d}(a, S)\ \in\ (S \cap D(a, S))\ \neq\ \varnothing$

It can be interpreted as the maximal payoff in the intersection of the comprehensive hull of S and the diagonal $D(a, S)$. Clearly $\max_{x\in S}\min_{j\in J}\left(x_j - \mathscr{G}_j^{KS}(a,S)\right)\ =\ 0$ with $J\ =\ \left\{j\ \in\ N \middle| M_j(a,S)\right\rangle a_j\}$ (Hougaard, 2003). Therefore,

$$\{x\ \in\ S | x\ \geq\ \mathscr{G}^{KS}(a,\ S)\}\ =\ \left\{x\ \in\ S | \min_{j\in J}\left\{x_j - \mathscr{G}_j^{KS}(a,S)\right\}\ =\ 0\right\} \tag{6.7.6}$$

$$= \left\{x\ \in\ S | \min_{j\in J}\left\{\frac{x_j - a_j}{M_j(S) - a_j}\right\}\ =\ \beta\right\}$$

$$= \arg\max_{x\in S}\min_{j\in J}\left\{\frac{x_j - a_j}{M_j(S) - a_j}\right\}$$

Finally, a bargaining solution, $f_{Mm}^{e}:(a,S)\ \rightarrow\ \mathbb{R}^n$ with $f_{Mm}^{e}(a,S)\ \in\ S$, is an EMBS if and only if (Hougaard, 2003; Kim, 2020d);

$$f_{Mm}^{e}(a,\ S)\ \in\ \left\{x\ \in\ S\ \in\ \mathbb{R}^n | \min_{j\in J}\left\{\frac{x_j - a_j}{M_j(S) - a_j}\right\}\ =\ \beta\right\} \cap P(S) \tag{6.7.7}$$

Main property of EMBS is no worse than the Kalai-Smorodinsky solution in the sense that no coordinate is smaller (Kim, 2020d).

6.7.6 The MCA spectrum allocation scheme

To implement the MCA service, the main goal of COs is to maximize the system throughput by adjusting their price policy. The *TGSA* scheme adopts the MLMF Stackelberg game model to decide each CO's price strategy where MDs are followers and COs are leaders. As leaders, COs exploit their price strategies to induce MDs while increasing their payoffs. The CO's payoff $\left(U_{CO}^{\mathscr{T}_t}(\mathfrak{P}_i^{CO})\right)$ with the price strategy $\mathfrak{P}_i^{CO}$ at time $\mathscr{T}_t$ can be derived as follows (Kim, 2020d).

$$U_{CO}^{\mathscr{T}_t}(\mathfrak{P}_i^{CO})\ =\ \left(\frac{\sigma}{\left(1 + exp\left(-\eta \times \frac{\mathfrak{N}_{CO}^{\mathscr{T}_{t-1}}}{\mathfrak{M}^{CO}}\right)\right)} - \chi\right) \times \mathfrak{P}_i^{CO}$$

(6.7.8)

where σ, η and χ are control parameters to evaluate the $U_{CO}(\cdot)$ function, and $\mathfrak{N}_{CO}^{\mathcal{T}_{t-1}}$, $\mathfrak{M}^{CO}$ are the CO's allocated spectrum amount at time $\mathcal{T}_{t-1}$, and its total spectrum amount, respectively. According to the *RD*, each CO can get the variation of each price strategy in a step-by-step interactive feedback procedure. In conclusion, the variation of $\eta^{\mathfrak{P}_i^{co}}$ at time $\mathcal{T}_t$, i.e., $\dot{\eta}_{\mathfrak{P}_i^{co}}^{\mathcal{T}_t}$, is defined as follows (Kim, 2020d);

$$\dot{\eta}_{\mathfrak{P}_i^{co}}^{\mathcal{T}_t} = \eta^{\mathfrak{P}_i^{co}} \times \left(U_{CO}^{\mathcal{T}_t}(\mathfrak{P}_i^{CO}) - \left(\sum_{\eta^{\mathfrak{P}_j^{co}} \in \mathbb{L}} \left(\eta^{\mathfrak{P}_j^{co}} \times U_{CO}^{\mathcal{T}_{t-1}}(\mathfrak{P}_j^{CO})\right) \times \frac{1}{|\mathbb{S}^{CO}|} \right) \right)$$

(6.7.9)

where $|\mathbb{S}^{CO}|$ is the cardinality of $\mathbb{S}^{CO}$. Based on a sequence of time steps, the vector $\mathbb{V}^{\mathfrak{P}^{co}}$ is updated using the information of each strategy variation. Therefore, COs select the most profitable strategy while ensuring relevant adaptability (Kim, 2020d).

The major concern of MD is to optimize the user payoff by effectively allocating the spectrum resources from its contactable COs. To develop the resource allocation algorithm for MDs, the concept of EMBS is applied. As a follower, each individual MD is a game planner, and the multiple flows of its application task are cooperative game players. The MD $\mathcal{D}_i$'s payoff $\left(U_{\mathcal{D}_i}^{\mathcal{T}_t}(\mathbb{C}_{\mathcal{D}_i}, \mathcal{A}_{\mathcal{D}_i}) \right)$ with the set of its contactable COs $\left(\mathbb{C}_{\mathcal{D}_i}\right)$ and application task $\left(\mathcal{A}_{\mathcal{D}_i}\right)$ at time $\mathcal{T}_t$ can be derived as follows (Kim, 2020d).

$$U_{\mathcal{D}_i}^{\mathcal{T}_t}\left(\mathbb{C}_{\mathcal{D}_i}, \mathcal{A}_{\mathcal{D}_i}\right) = \sum_{CO_k \in \mathbb{C}_{\mathcal{D}_i}} \mathcal{PF}_{CO_k}^{\mathcal{D}_i}\left(r_{CO_k}^{\mathcal{D}_i}\right)$$

(6.7.10)

$$\text{s.t.,} \begin{cases} \mathcal{PF}_{CO_k}^{\mathcal{D}_i}\left(r_{CO_k}^{\mathcal{D}_i}\right) = \left(\omega \times \log\left(\frac{r_{CO_k}^{\mathcal{D}_i}}{\psi\left(\mathcal{A}_{\mathcal{D}_i}\right)}\right)\right) - \left(r_{CO_k}^{\mathcal{D}_i} \times \mathfrak{X}_{\mathcal{T}_t}(CO_k)\right) \\ \psi\left(\mathcal{A}_{\mathcal{D}_i}\right) = \sum_{CO_k \in \mathbb{C}_{\mathcal{D}_i}} r_{CO_k}^{\mathcal{D}_i} and \omega = \frac{\left(\grave{E} + \left(\tilde{A}^{CO_k} - d_{\mathcal{D}_i}^{CO_k}\right)\right)}{\tilde{A}^{CO_k}} \end{cases}$$

where $r_{CO_k}^{\mathcal{D}_i}$ is the spectrum amount, which is assigned to the $\mathcal{D}_i$ from the CO_k. $\psi\left(\mathcal{A}_{\mathcal{D}_i}\right)$ is the total spectrum request of the $\mathcal{A}_{\mathcal{D}_i}$. $\mathfrak{X}_{\mathcal{T}_t}(CO_k)$ is the CO_k's price strategy at time $\mathcal{T}_t$, and Θ is a control factor for each MD's payoff. $d_{\mathcal{D}_i}^{CO_k}$ is the distance from the CO_k to the $\mathcal{D}_i$, and $\tilde{A}^{CO_k}$ is the CO_k's radius of coverage area. To effectively decide the $r_{CO_k \in \mathbb{C}_{\mathcal{D}_i}}^{\mathcal{D}_i}$, the *TGSA* scheme adopts the concept of EMBS. According to (6.7.6) and (6.7.7), the $r_{CO_k}^{\mathcal{D}_i}$ can be adaptively adjusted, and the $\mathcal{D}_i$ effectively assigns the spectrum resource from the CO_k (Kim, 2020d).

$$\mathbb{S}^{\mathcal{D}_i} = \left[\ldots, r_{CO_k}^{\mathcal{D}_i}, \ldots\right]_{CO_k \in \mathbb{C}_{\mathcal{D}_i}} =$$

$$\max_{f_{CO_k}^{\mathcal{D}_i} \le \mathcal{A}_{\mathcal{D}_i}} \min_{CO_k \in \mathbb{C}_{\mathcal{D}_i}} \left\{ \frac{\mathcal{PF}_{CO_k}^{\mathcal{D}_i}\left(r_{CO_k}^{\mathcal{D}_i}\right) - d_{f_{CO_k}^{\mathcal{D}_i}}}{\mathcal{PF}_{CO_k}^{\mathcal{D}_i}\left(\psi\left(\mathcal{A}_{\mathcal{D}_i}\right)\right) - d_{f_{CO_k}^{\mathcal{D}_i}}} \right\} \cap \left(\psi\left(\mathcal{A}_{\mathcal{D}_i}\right) = \sum_{CO_k \in \mathbb{C}_{\mathcal{D}_i}} r_{CO_k}^{\mathcal{D}_i} \right)$$

(6.7.11)

To implement the minimax decision rule, the formulation of eq. (6.7.11) is a function for minimizing the possible loss for a worst case (maximum loss) scenario. When dealing with gains, it is referred to as maxmin - to maximize the minimum gain. In (6.7.11), $\mathbb{S}^{\mathscr{D}_i} = \left[\ldots, r^{\mathscr{D}_i}_{CO_k}, \ldots\right]$ is the set of $\mathscr{D}_i$'s available strategies where $r^{\mathscr{D}_i}_{CO_k \in \mathbb{C}_{\mathscr{D}_i}}$ represents the allocated spectrum amount from the CO_k for the $\mathscr{D}_i$, and $f^{\mathscr{D}_i}_{CO_k}$ represents the $\mathscr{D}_i$'s flow, which is connected to the CO_k in the $\mathbb{C}_{\mathscr{D}_i}$ (Kim, 2020d).

6.7.7 Summary

MCA technology has been introduced as a means to enhance spectrum efficiency for MDs within wireless HetNet systems. However, existing protocols primarily focus on optimizing spectrum utilization to maximize network capacity, often overlooking the complexities of diverse system conditions. Developing an effective MCA-enabled control mechanism is significantly more challenging, introducing new obstacles that must be addressed. To tackle these issues, the *TGSA* scheme proposes an innovative spectrum resource allocation mechanism tailored for MCA systems. Its implementation relies on a two-level game paradigm, designed for COs and MDs while considering the prevailing network conditions. At the upper level, the MLMF Stackelberg game model is employed to efficiently regulate COs' spectrum pricing strategies. Meanwhile, at the lower level, the *EMBS* approach is utilized to allocate spectrum resources among multiple COs for MCA services. These two game models are intricately integrated within the *TGSA* scheme, creating a mutually interdependent framework. As a result, COs and MDs continuously adapt their strategies based on the interactions of other participants. In the dynamic MCA system environment, this two-level game approach enables the achievement of a well-balanced consensus, effectively reconciling the conflicting interests of COs and MDs (Kim, 2020d).

6.8 Interval Values based Spectrum Allocation (*IVSA*) scheme

The ongoing evolution of 5G HetNets has resulted in unprecedented traffic demands and increasingly stringent QoS requirements for users. Addressing these challenges in next-generation communication systems is highly complex. Multi-flow carrier aggregation (MCA) has emerged as a promising technique aimed at enhancing the capacity of cellular networks. Recently, S. Kim introduced the *Interval Values-based Spectrum Allocation* (*IVSA*) scheme, designed for the HetNet platform under MCA technology (Kim, 2020a). To facilitate joint co-tier and cross-tier collaboration in HetNets, the *IVSA* scheme leverages interval Shapley and Banzhaf values as solution concepts. By incorporating these principles, the *IVSA* scheme ensures an optimal balance between performance and the dynamic distribution of limited spectrum resources. The *IVSA* scheme is formulated as a coalition game model, where interval values are strategically applied based on the requirements of different applications. Rooted in the key principles of cooperative game theory, the *IVSA* scheme effectively maximizes benefits in a rational and systematic manner (Kim, 2020a).

6.8.1 Development motivation

With the increasing demand for seamless mobile connectivity, 5G wireless networks face significant challenges in meeting the ever-growing network capacity requirements. These networks are designed to offer enhanced, highly adaptable connectivity that is faster, more cost-effective, reliable, secure, and flexible while maintaining energy efficiency. To successfully implement 5G communications, a combination of existing technological advancements and innovations is essential, as they serve as the foundation for new business models, applications, experiences, products, and services. However, 5G networks are not intended to replace previous generations but rather to coexist with them. As a result, 4G network devices, control mechanisms, and infrastructure will need to be customized and upgraded to ensure seamless integration with 5G networks (Akhtar, 2019; Hajir, 2016; Kim, 2020a).

To address the demands of 5G mobile data networks, Heterogeneous Networks (HetNets) are regarded as a promising solution. A HetNet consists of macro-cells and densely deployed small-cells, requiring the coexistence of macro-cell base stations (MBSs) and small-cell base stations (SBSs), with multiple SBSs deployed within each macro-cell coverage area. The extensive deployment of SBSs offers several advantages, including the spatial reuse of spectrum resources, leading to improved network throughput and efficiency. However, despite these benefits, HetNets introduced new research challenges. Among them, spectrum resource allocation in 5G HetNets is significantly more complex than in previous network generations due to evolving infrastructure and standardization requirements. Therefore, advanced and intelligent techniques are necessary to manage HetNet spectrum allocation effectively. This process is particularly intricate as it involves addressing multiple challenges such as load balancing, spectrum efficiency, fairness, QoS, and various other system constraints (Akhtar, 2019; Hajir, 2016; Kim, 2020a; Liu, 2018a).

One of the key distinguishing features of 5G HetNets is Multi-Flow Carrier Aggregation (MCA), which was originally designed to ensure backward compatibility with LTE-WiFi aggregation technologies. In 5G HetNets, MCA enhances wireless communication by increasing the data transmission rate for individual mobile devices (MDs), allowing them to be assigned to multiple spectrum blocks, referred to as component carriers. This approach enables each MD to aggregate multiple spectrum fragments from different base stations (BSs), leading to more efficient spectrum utilization. The MCA configuration is specific to each MD, providing a flexible and adaptable approach to per-device spectrum management. To optimize throughput and spectrum efficiency, individual MDs can selectively activate and aggregate spectrum carriers. However, implementing MCA in 5G HetNets also presents certain challenges, particularly in determining which and how many spectrum carriers should be activated for each MD. These issues are influenced by factors such as spectrum availability, coverage, efficiency, and MD-specific QoS requirements (Kim, 2020a; Wang, 2016a).

The MCA spectrum allocation process in 5G HetNets is typically categorized into two mechanisms: cross-tier (between MBS and SBSs) and co-tier (among SBSs) allocation. The cross-tier mechanism facilitates spectrum sharing between different BS tiers, whereas the co-tier mechanism distributes spectrum resources among SBSs within the same tier. Effective spectrum allocation policies are required for both mechanisms. A major challenge arises from fluctuations in mobile traffic demand, which can lead to significant load imbalances among BSs. To optimize system performance - including throughput, user fairness, resource utilization, and processing delays - it is crucial to implement control policies that dynamically redistribute traffic from heavily loaded BSs to their less congested counterparts. However, designing an optimal control policy remains a highly complex task (Hajir, 2016; Kim, 2020a; Wang, 2016b).

Coordination and cooperation play a vital role in developing an effective control policy for MCA operations, involving interactions across both cross-tier and co-tier BSs. Multiple BSs contribute significantly to the spectrum allocation process, making cooperative control approaches preferable to non-cooperative strategies in HetNet environments. From a control theory perspective, cooperative game theory offers a robust mathematical framework for addressing modern complex challenges. This approach models interactions as a game with distinct utility functions, ensuring that the resulting payoffs account for the costs associated with various actions. Recently, cooperative game models have gained traction as a valuable paradigm for designing networked multi-agent systems, effectively solving numerous challenges in 5G HetNets and beyond. A fundamental aspect of cooperative games is structuring agents' payoffs in a way that aligns their self-interest with desirable collective outcomes (Akhtar, 2019; Kim, 2014, 2020a).

To develop an innovative MCA spectrum allocation algorithm, the *IVSA* scheme integrates concepts from cooperative coalition games and employs two solution methodologies: the *interval Shapley value* (*ISV*) for cross-tier processes and the *interval Banzhaf value* (*IBV*) for co-tier processes. In cooperative interval games, the *ISV* and *IBV* solutions are determined by leveraging the lower and upper bounds of coalition interval payoffs, ensuring that they possess desirable theoretical properties (Alparslan Gök, 2010; Branzei, 2003; Osman, 2015; Ridaoui, 2018). Within the MCA service framework, BSs function as game players, exchanging information and engaging in negotiations to achieve shared objectives. As rational players, BSs tend to form a grand coalition to maximize overall profits. The coalition's collective profit is then distributed among participating players in a way that prevents any individual BS from having an incentive to leave the coalition. By incorporating the interval uncertainty of game players, the spectrum allocation process is modeled as a size-monotonic interval coalition game, with the *ISV* and *IBV* principles utilized to allocate spectrum resources among different BSs (Kim, 2020a).

6.8.2 Contribution

Utilizing the principles of *ISV* and *IBV*, the *IVSA* scheme introduces an innovative MCA spectrum allocation mechanism tailored for the multi-tier HetNet platform. For delay-sensitive application tasks, the cross-tier spectrum allocation process is more appropriate, incorporating the *ISV* concept to prioritize resource efficiency. Conversely, for delay-tolerant applications, the co-tier spectrum allocation process is preferable, applying the *IBV* approach to ensure relative fairness in resource distribution. By leveraging cooperative interval game theory, the dynamic spectrum allocation strategy effectively integrates the advantages of *ISV* and *IBV*, addressing a range of MCA control challenges in a comprehensive manner (Kim, 2020a).

6.8.3 Coalition game model for the HetNet system infrastructure

We consider a two-tier cellular network platform, which comprises two types of BSs; *n*MBSs $\mathbb{M} = \{\mathcal{M}_1,\ldots,\mathcal{M}_n\}$ and m SBSs $\mathbb{S} = \{\mathcal{S}_1,\ldots,\mathcal{S}_m\}$. They have different coverage areas, spectrum resource and access technology. One MBS (*or* one SBS) has a coverage area of radius $r_{\mathcal{M}}$ (*or* $r_{\mathcal{S}}$) and has a static portion of spectrum resource to provide the basic services to its corresponding users in its area. There are k MDs $\mathbb{N} = \{\mathcal{D}_1,\ldots,\mathcal{D}_k\}$ and they are randomly distributed in the cellular area. Based on the MCA technique, MDs' application services can be represented by multiple subcarriers, and they are heterogeneous in their QoS requirements. Therefore, $\mathcal{D}_{1\leq i\leq k}$ is assumed to have the capability to aggregate multiple traffic flows over several BSs for each application service. In the point view of MDs, their major concern is to maximize the service payoff by adaptively adjusting their subcarriers. To adapt the current HetNet environments, each individual MD allocates the available BSs' spectrum resources into multiple application flows. In the *IVSA* scheme, the interaction between an MD and its corresponding BSs is formulated as a coalitional game in a cooperative manner. Formally, the *IVSA* scheme defines the game model for the $\mathcal{D}_i$'s application $\mathcal{A}_j$, i.e., $\mathbb{G}_{\mathcal{A}_j}^{\mathcal{D}_i} = \left\{\mathbb{B}^{\mathcal{D}_i}, \mathcal{T}^{\mathcal{A}_j}, \mathbb{I}_{\mathbb{B}^{\mathcal{D}_i}}^{\mathcal{A}_j}, \mathbb{U}_{\mathcal{D}_i}^{\mathcal{A}_j}, \mathfrak{M}^{\mathcal{A}_j}, T\right\}$ at each time period of gameplay (Kim, 2020a).

- $\mathbb{B}^{\mathcal{D}_i}$ is the finite set of BSs, which are contactable by the $\mathcal{D}_i$. They are game players for the game $\mathbb{G}_{\mathcal{A}_j}^{\mathcal{D}_i}$.
- $\mathcal{T}^{\mathcal{A}_j} = \ldots, f_{\Psi \in \mathbb{B}^{\mathcal{D}_i}}^{\mathcal{A}_j}, \ldots$ is a $|\mathbb{B}^{\mathcal{D}_i}|$-dimensional vector where $f_{\Psi}^{\mathcal{A}_j}$ represents the spectrum resource amount allocated by one BS $\Psi \in \mathbb{B}^{\mathcal{D}_i}$ for the $\mathcal{A}_j$; Ψ can be a MBS or a SBS.
- $\mathbb{I}_{\mathbb{B}^{\mathcal{D}_i}}^{\mathcal{A}_j} = \ldots, I_{\Psi \in \mathbb{B}^{\mathcal{D}_i}}^{\mathcal{A}_j}, \ldots$ is the vector of interval values of $\Psi \in \mathbb{B}^{\mathcal{D}_i}$ where $I_{\Psi}^{\mathcal{A}_j}$ represents the low and high bounds of Ψ for the $\mathcal{A}_j$ service operation.
- $\mathbb{U}_{\mathcal{D}_i}^{\mathcal{A}_j}$ is the payoff received by the $\mathcal{D}_i$ for the $\mathcal{A}_j$ during the MCA process.
- $\mathfrak{M}^{\mathcal{A}_j}$ is the total requested spectrum amount for the $\mathcal{A}_j$.
- $T = \left\{\mathcal{T}_1, \ldots, \mathcal{T}_t, \mathcal{T}_{t+1}, \ldots\right\}$ denotes time, which is represented by a sequence of time steps.

6.8.4 The basic concepts of ISV and IBV solutions

Cooperative game theory describes a situation in which a finite set of n players can generate certain payoffs by cooperation. In cooperative game theory, a central problem is to define a value, that is, a payoff to be given to each player, taking into account his contribution to the game. As a solution concept, value is a function to calculate a n-dimensional real vector which represents a payoff distribution over players. To date, two well-known solution concepts have been introduced: the *Shapley value* and the *Banzhaf value*. In 1953, L. Shapley proposed the *Shapley value*, which provides a unique method for distributing the total surplus generated by a coalition among all participating players. This value is characterized by a set of desirable mathematical properties. Later, in 1965, J. Banzhaf introduced the *Banzhaf value* within the context of voting games. The *Banzhaf value* is derived as an expectation based on the probability distribution of the original partition function. Initially formulated in the context of voting power analysis, it quantifies the likelihood of a single vote altering an outcome, particularly when voting rights are not equally distributed among participants (Branzei, 2003; Kim, 2020a; Osman, 2015).

The *Shapley* and *Banzhaf values* have garnered significant attention due to their advantageous properties. Both values share a fundamental structure, as they represent a weighted average of a player's marginal contribution to coalitions. However, the key distinction lies in the weighting mechanisms. In the *Shapley value*, the marginal contributions are weighted according to coalition size to ensure efficiency, meaning the total payoff allocated to all players equals the total worth of the game. In contrast, the *Banzhaf value* assigns equal weights to all contributions, making it independent of coalition size (Kim, 2020a; Ridaoui, 2018).

Recent research has focused on extending the original concepts of the *Shapley* and *Banzhaf values* to accommodate interval data. The interval value model is particularly useful in scenarios where players engage in cooperation while only having precise knowledge of the lower and upper bounds of potential revenues or costs. The concept of interval values was initially introduced to address bankruptcy problems, where the total estate is known with certainty, but individual claims are defined within specified intervals. In this framework, an interval value is represented as a payoff vector, where each component consists of a compact interval of real numbers. To construct a cooperative game using interval values, both interval arithmetic and classical cooperative game theory play a crucial role. More recently, the *interval Shapley value* (*ISV*) and the *interval Banzhaf value* (*IBV*) have gained considerable attention as they have been extended into new game-theoretic models and widely applied in various domains, including resource allocation in economic systems, sociology, and computer science. In the dynamic HetNet environment, where the capacity of each base station can be represented as an interval of real numbers, these interval values provide a practical approach to modeling network system uncertainties (Alparslan Gök, 2010), (Kim, 2020a).

A bankruptcy game is a well-known example to explain the idea of value solutions. It is a distribution model where the amount of a perfectly divisible good is insufficient to satisfy all demands. Let $N = \{1, \ldots, n\}$ be the set of claimants among which a good E has to be divided. We denote by $I_i = [a_i, b_i]$ the claim interval of claimant $i \in N$ where a_i, b_i are the lower and upper bounds of the claim interval. R and $\mathbb{R}_+$ denote the sets of all and non-negative real numbers, respectively. The pair (E, I) will denote a bankruptcy problem with interval claims where $0 < E \leq \sum_{i \in N} a_i$ and $I = (I_1, \ldots, I_n)$ is a vector of all claimants' interval. We denote by $\mathfrak{S}^N$ the set of all bankruptcy problems of the form (E, I). All division problems on N with sharp claims w.r.t. the available amount E, of the form (E, d) with $d = (d_1, \ldots d_n)$ and $0 < E \leq \sum_{i \in N} a_i \leq \sum_{i \in N} d_i$ appear as particular cases of a problem $(E, I) \in \mathfrak{S}^N$ (Branzei, 2003; Kim, 2020a).

Note that if all claim intervals $I_i, i \in N$ are degenerated intervals, i.e. $I_i = [a_i, a_i]$, the problem (E, I) coincides with the classical bankruptcy problem (E, a) with $a = (a_1, \ldots, a_n)$ and $0 < E \leq \sum_{i \in N} a_i$. Moreover, In the following, we use the notation $\mathscr{D}^N$ to refer to the family of classical division problems related to a division problem with interval claims. Each problem in $\mathfrak{S}^N$ can be defined based on the idea to compromise the interval claims of each player $i \in N$ by weighting the upper bound of his claim interval with $t_i \in [0, 1]$ and the lower bound with $(1 - t_i)$ where t_i in $t = (t_1, t_2, \ldots, t_n)$. Given the t, we introduce the t-compromise claim $c^t = (c_1^{t_1}, c_2^{t_2}, \ldots, c_n^{t_n})$ by $c_i^{t_i} = (t_i \times b_i) + ((1 - t_i) \times a_i)$ for each $i \in N$. With the E and c^t, we can consider the t-compromise division problem (E, c^t) (Branzei, 2003; Kim, 2020a).

Let $I = (I_1, \ldots, I_n)$ be the vector of interval claims in the problem $(E, I) \in \mathfrak{S}^N$ and $t = (t_1, t_2, \ldots, t_n)$, where $t_i \in [0, 1]$ for each $i \in N$. The set of all division games N, υ_{E,c^t} is denoted by $\mathbb{F}^N$. According to the traditional division problem with point claims, the *IVSA* scheme can be considered an associated division game N, υ_{E,c^t} based on the (E, c^t). It has the characteristic function $\upsilon_{E,c^t}: 2^N \to \mathbb{R}$ as follows (Branzei, 2003; Kim, 2020a);

$$\upsilon_{E,c^t}(S) = \left(E - \sum_{i \in N \setminus S} c_i^{t_i}\right)_+ \text{ s.t., } x_+ = \max(x, 0) \text{ and } S \in 2^N \tag{6.8.1}$$

$\upsilon_{E,c^t}(S)$ can be interpreted as the amount of the estate that is left if all players outside coalition S receive their claim. Now, the *IVSA* scheme constructs the cooperative interval game $N, \upsilon_{E,I}$ with interval claims $(E, I) \in \mathfrak{S}^N$. Let F be the family of closed intervals in $\mathbb{R}_+$ and $\mathbb{F}^N$ be the set of all vectors of the form $I = (I_1, \ldots, I_n)$. The characteristic function $\upsilon_{E,I}: 2^N \to \mathbb{F}$ each $S \in 2^N$ is defined as $\upsilon_{E,I}(S) = \{\upsilon_{E,c^t}(S) | t \in [0,1]^N\}$. It is equivalent to the below equation (Branzei, 2003; Kim, 2020a);

$$\upsilon_{E,I}(S) = \left[\left(E - \sum_{i \in N \setminus S} b_i\right)_+, \left(E - \sum_{i \in N \setminus S} a_i\right)_+\right] = \left[\upsilon_{E,c^1}(S), \upsilon_{E,c^0}(S)\right] \tag{6.8.2}$$

s.t., $c^1 = (c_1^1, c_2^1, \ldots, c_n^1)$ and $c^0 = (c_1^0, c_2^0, \ldots, c_n^0)$

In the $N, v_{E,I}$, each coalition S can receive an amount which is between the amount of the estate that is left if all players outside coalition S receive their upper claim, and the amount of the estate that is left if all players outside coalition S receive their lower claim. The set of all cooperative interval games $N, v_{E,I}$ is denoted by $\mathcal{G}^N$. Based on the $N, v_{E,I} \in \mathcal{G}^N$, the marginal vectors $m_i^\sigma(v_{E,I})$ with permutations $\sigma: N \rightarrow N$ is given by (Branzei, 2003; Kim, 2020a);

$$m_i^\sigma(v_{E,I}) = v_{E,I}(P_\sigma(i) \cup \{i\}) - v_{E,I}(P_\sigma(i))$$

(6.8.3)

$$= \left[v_{E,c^1}(P_\sigma(i) \cup \{i\}),\ v_{E,c^0}(P_\sigma(i) \cup \{i\})\right] - \left[v_{E,c^1}(P_\sigma(i)),\ v_{E,c^0}(P_\sigma(i))\right]$$

$$\text{s.t., } P_\sigma(i) = \{r \in N | \sigma^{-1}(r) < \sigma^{-1}(i)\} \text{ and } i \in N$$

where $P_\sigma(i)$ the set of predecessors of i in σ. Finally, the ISV $\Phi_{ISV}(v_{E,I})$ of a game $N, v_{E,I} \in \mathcal{G}^N$, which is the average of the marginal vectors of the game, is defined as follows (Branzei, 2003; Kim, 2020a);

$$\Phi_{ISV}(v_{E,I}) = \frac{1}{n!} \times \sum_{\sigma \in \pi(N)} m^\sigma(v_{E,I}) = \left[\underline{\Phi_{ISV}(v_{E,I})},\ \overline{\Phi_{ISV}(v_{E,I})}\right]$$

(6.8.4)

$$\text{s.t.,} \begin{cases} \underline{\Phi_{ISV}(v_{E,I})} = \frac{1}{n!} \times \sum_{\sigma \in \pi(N)} \left(v_{E,c^1}(P_\sigma(i) \cup \{i\}) - v_{E,c^0}(P_\sigma(i))\right) \\ \overline{\Phi_{ISV}(v_{E,I})} = \frac{1}{n!} \times \sum_{\sigma \in \pi(N)} \left(v_{E,c^0}(P_\sigma(i) \cup \{i\}) - v_{E,c^1}(P_\sigma(i))\right) \end{cases}$$

where $\pi(N)$ is the set of all permutations π on N. To sum up, the *Shapley value* assumes that every player is equally likely to join any coalition of the same size and all coalitions with the same size are equally likely. However, the *Banzhaf value* considers that every player is equally likely to enter any coalition. To define the *IBV*, let $v_{E,I}$ be the characteristic function $v_{E,I}: 2^N \rightarrow I$ where $v_{E,I}(\emptyset) = [0,0]$. For each $S \in 2^N$, the worth interval of $v_{E,I}(S)$ is $\left[\underline{v_{E,I}(S)}, \overline{v_{E,I}(S)}\right]$ in the same manner as (6.8.1)-(6.8.2). Let $I_i, I_j \in$ I with $I_i = \left[\underline{I}_i, \overline{I}_i\right]$, $I_j = \left[\underline{I}_j, \overline{I}_j\right]$ and $|I_i| = \overline{I}_i - \underline{I}_i$. Then, $I_i + I_j = \left[\underline{I}_i + \underline{I}_j, \overline{I}_i + \overline{I}_j\right]$. A partial subtraction operator has played a central role. We define $I_i - I_j$, only if $|I_i| \geq |I_j|$, by $I_i - I_j = \left[\underline{I}_i - \underline{I}_j, \overline{I}_i - \overline{I}_j\right]$. We say that I_i is weakly better than I_j, denoted by $I_i \succcurlyeq I_j$, if and only if, $\underline{I}_i \geq \underline{I}_j$ and $\overline{I}_i \geq \overline{I}_j$. Finally, the *IBV* $\Phi_{IBV}(v_{E,I})$ of a game $N, v_{E,I} \in \mathcal{G}^N$ is defined as follows (Kim, 2020a; Osman, 2015);

$$\Phi_{IBV}(v_{E,I}) = \frac{1}{2^{|N|-1}} \times \sum_{i \in S} \left(v_{E,I}(\text{S}) - v_{E,I}(\text{S} \setminus \{i\})\right)$$

(6.8.5)

6.8.5 The MCA spectrum allocation scheme

For the implementation of MCA services, each BS communicates its service capacity to the corresponding MDs. Typically, a BS's service capacity is determined by two key factors: *spectrum availability* (*SA*) and *communication distance* (*CD*). To effectively represent these dual criteria, the interval value concept is well-suited. Within the MCA service framework, the *SA* serves as the primary determinant in evaluating the suitability of each BS for service provisioning. To decide the *SA* value, let $\mathfrak{N}_{\Psi}^{T}$ and $\mathfrak{N}_{\Psi}^{C}$ be the total and currently used spectrum amounts of BS Ψ, respectively. Based on the information of $\mathfrak{N}_{\Psi}^{T}$ and $\mathfrak{N}_{\Psi}^{C}$, Ψ's *SA* is defined as its relative spectrum capacity, and it is used as the upper bound of Ψ. From the viewpoint of each MD, the closer a BS, the more attractive for wireless communications. Therefore, the *CD* between MD and BS is a secondary factor for each BS's adaptability; *CD* is used to define the lower bound of Ψ. When the $\mathscr{D}_i$ attempts a MCA service for its $\mathscr{A}_j$ task, its corresponding BS's interval value $\left(I_{\Psi \in \mathbb{B}^{\mathscr{D}_i}}\right)$ is given by (Kim, 2020a);

$$I_{\Psi \in \mathbb{B}^{\mathscr{D}_i}} = \left[\underline{I_{\Psi}}, \overline{I_{\Psi}}\right] \tag{6.8.6}$$

$$\text{s.t.,} \begin{cases} \overline{I_{\Psi}} = \dfrac{\mathfrak{N}_{\Psi}^{T} - \mathfrak{N}_{\Psi}^{C}}{\mathfrak{N}_{\Psi}^{T}} \times \dfrac{\mathfrak{N}_{\Psi}^{T}}{\max\left\{\mathfrak{N}_{\Psi}^{T} | \Psi \in \mathbb{B}^{\mathscr{D}_i}\right\}} \, and \underline{I_{\Psi}} = \overline{I_{\Psi}} \times \log\left(\dfrac{\left(\tilde{A}^{\Psi} - d_{\mathscr{D}_i}^{\Psi}\right)}{\tilde{A}^{\Psi}} + \theta\right) \\ \quad , if \mathfrak{M}^{\mathscr{A}_j} \le \sum_{\Psi \in \mathbb{B}^{\mathscr{D}_i}} \left(\mathfrak{N}_{\Psi}^{T} - \mathfrak{N}_{\Psi}^{C}\right) \\ \left[\underline{I_{BS}}, \overline{I_{BS}}\right] = [0,0], otherwise \end{cases}$$

where $\tilde{A}^{\Psi}$ is the Ψ's radius of coverage area. If the Ψ is MBS (*or* SBS), $\tilde{A}^{\Psi}$ is $r_{\mathscr{M}}$ $(or\, r_{\mathscr{S}})$. $d_{\mathscr{D}_i}^{\Psi}$ is the distance from the Ψ to the $\mathscr{D}_i$. θ is a control parameter for the $\underline{I_{\Psi}}$. Based on the interval values of corresponding BSs in $\mathbb{B}^{\mathscr{D}_i}$, the $\mathscr{D}_i$ effectively allocates the $\mathbb{B}^{\mathscr{D}_i}$'s spectrum resources for its application's MCA service. As a game planner, the $\mathscr{D}_i$'s payoff for $\mathscr{A}_j$ $\left(U_{\mathscr{D}_i}\left(\mathbb{B}^{\mathscr{D}_i}, \mathscr{A}_j\right)\right)$ with the set $\mathbb{B}^{\mathscr{D}_i}$ can be derived as follows (Kim, 2020a);

$$U_{\mathscr{D}_i}\left(\mathbb{B}^{\mathscr{D}_i}, \mathscr{A}_j\right) = \sum_{\Psi \in \mathbb{B}^{\mathscr{D}_i}} \mathscr{P}_{\Psi}^{\mathscr{D}_i}\left(r_{\Psi}^{\mathscr{D}_i}\right) \tag{6.8.7}$$

$$\text{s.t.,} \begin{cases} \mathscr{P}_{\Psi}^{\mathscr{D}_i}\left(r_{\Psi}^{\mathscr{D}_i}\right) = \dfrac{\left(\tilde{A}^{\Psi} - d_{\mathscr{D}_i}^{\Psi}\right)}{\tilde{A}^{\Psi}} \times r_{\Psi}^{\mathscr{D}_i} \times \left(\dfrac{\alpha}{\xi + exp(X)} - \dfrac{\beta}{\varphi + exp(Y)}\right) \\ X = -\left(\dfrac{\mathfrak{N}_{\Psi}^{C} + r_{\Psi}^{\mathscr{D}_i}}{\mathfrak{N}_{\Psi}^{T}}\right) \times \gamma and Y = -\left(\dfrac{\mathfrak{N}_{\Psi}^{C}}{\mathfrak{N}_{\Psi}^{T}} \times \varrho\right) \\ \mathfrak{M}^{\mathscr{A}_j} = \sum_{\Psi \in \mathbb{B}^{\mathscr{D}_i}} r_{\Psi}^{\mathscr{D}_i} and \Psi \in MorS \end{cases}$$

where $r_{\Psi}^{\mathscr{D}_i}$ is the spectrum amount, which is assigned to the $\mathscr{D}_i$ from the Ψ.α, β, φ, ϱ and γ are control parameters to estimate the $\mathscr{P}_{\Psi}^{\mathscr{D}}(\cdot)$. In the $\mathscr{D}_i$, $r_{\Psi}^{\mathscr{D}_i}$ is assigned to a traffic flow of $\mathscr{A}_j$ for the MCA service (Kim, 2020a).

5G cellular networks are designed to support a wide range of multimedia services. These services, when deployed over HetNets, not only require varying amounts of spectrum but also have distinct QoS requirements. As a result, they can be broadly classified into two categories: Class I (real-time data) and Class II (non-real-time data), based on their respective QoS demands (Kim, 2004). Generally, Class I traffic is significantly more delay-sensitive than Class II traffic. Considering the overhead associated with handoffs, Class I traffic services tend to prefer connecting to MBSs rather than SBSs. Consequently, different MCA techniques are adaptively applied depending on the traffic environment. Within the *IVSA* scheme, Class I traffic services are prioritized for cross-tier MCA techniques, whereas Class II traffic services favor co-tier MCA techniques. In the cross-tier spectrum allocation process, resource efficiency is a critical concern, making the *ISV* approach a suitable choice. Conversely, in the co-tier spectrum allocation process, all game players are of the same type (i.e., SBSs). As a result, the *IVSA* scheme applies equal weighting to all possible coalitions, regardless of their size. Consequently, the *IBV* approach is employed to effectively address the MCA spectrum allocation problem (Kim, 2020a).

In the *IVSA* scheme, the MCA spectrum allocation problem leads to a similar conflict situation in the money division problem for claimants. Therefore, the *IVSA* scheme adopts the bankruptcy game model to define the characteristic function $v_{E,I}$ for each BS's coalitions $\left(v_{E,I}(\cdot)\right)$. When a new application task $\mathscr{A}_j$ is generated in the $\mathscr{D}_i$, contactable BSs by the $\mathscr{D}_i$ form the $\mathbb{B}^{\mathscr{D}_i}$ to create a coalition. If the $\mathscr{A}_j$ is a class, I (*or* class II). application, SBSs are preferred to form the $\mathbb{B}^{\mathscr{D}_i}$ where $\mathbb{B}^{\mathscr{D}_i} \cap \mathbb{M} \neq \varnothing$ (*or* $\mathbb{B}^{\mathscr{D}_i} \cap \mathbb{M} = \varnothing$), and their spectrum resources are assigned for the $\mathscr{A}_j$ through the MCA technique. To implement the characteristic function $v_{E,I}$, the E value for $\mathscr{A}_j$, i.e., $E_{\mathscr{A}_j}$, is defined based on the bankruptcy game concept (Kim, 2020a);

$$E_{\mathscr{A}_j} = \left(\sum_{\Psi \in \mathbb{B}^{\mathscr{D}_i}} \mathcal{T}_{\Psi}\right) \times \varepsilon \tag{6.8.8}$$

where ε is a modification factor. Simply, the payoff of a coalition *S*, which consists of BSs, is calculated as the remaining payoff while paying idealistic payoff for other players, who are not included in the *S*. For the Ψ, its ISV, i.e., $\Phi_{ISV}^{\Psi} = \left[\underline{I_{\Psi}^{ISV}}, \overline{I_{\Psi}^{ISV}}\right]$ and its IBV, i.e., $\Phi_{IBV}^{\Psi} = \left[\underline{I_{\Psi}^{IBV}}, \overline{I_{\Psi}^{IBV}}\right]$ are calculated according to (6.8.4)-(6.8.6). Finally, whether the ISV or the IBV is applied, the Ψ's spectrum allocation amount for the $\mathscr{A}_j$, i.e., $\mathfrak{R}_{\Psi}^{\mathscr{A}_j}$, is estimated as follows (Kim, 2020a);

$$\begin{cases} \mathfrak{R}_{\Psi}^{\mathcal{A}_j} = \mathfrak{M}^{\mathcal{A}_j} \times \frac{\left(\underline{I_{\Psi}^{ISV}} + \overline{I_{\Psi}^{ISV}}\right)/2}{\sum_{\eta \in \mathbb{B}^{\mathcal{D}_i}} \left(\underline{I_{\eta}^{ISV}} + \overline{I_{\eta}^{ISV}}\right)/2}, if the ISV is applied \\ \\ \mathfrak{R}_{\Psi}^{\mathcal{A}_j} = \mathfrak{M}^{\mathcal{A}_j} \times \frac{\left(\underline{I_{\Psi}^{IBV}} + \overline{I_{\Psi}^{IBV}}\right)/2}{\sum_{\eta \in \mathbb{B}^{\mathcal{D}_i}} \left(\underline{I_{\eta}^{IBV}} + \overline{I_{\eta}^{IBV}}\right)/2}, if the IBV is applied \end{cases} \quad (6.8.9)$$

6.8.6 Summary

In 5G wireless systems, the MCA has emerged as an innovative technique to expand the available spectrum for end users. However, most existing research has primarily focused on optimizing a single performance criterion - maximizing network capacity - while overlooking various critical factors in 5G HetNet environments. To address these limitations, the *IVSA* scheme introduces a novel spectrum allocation mechanism tailored for the MCA systems. The implementation of the *IVSA* scheme is based on interval value solutions, designed to accommodate different types of applications. For Class I application tasks, the *ISV* approach is employed, prioritizing the cross-tier spectrum allocation mechanism. In contrast, for Class II application tasks, the *IBV* approach is utilized, favoring the co-tier spectrum allocation mechanism. During each time interval, interval coalition games are executed in a distributed manner for individual applications, ensuring efficient allocation of spectrum resources among BSs. In the dynamically evolving HetNet environment, the cooperative game-based approach enables the system to achieve a well-balanced performance, effectively addressing the challenges of spectrum allocation (Kim, 2020a).

REFERENCES

Akhtar, T., Politis, I., Tselios, C., & Kotsopoulos, S. (2019). Cooperative Game Radio Resource Management Scheme for Small Cell Network. *IEEE 5G World Forum (5GWF)*, 69–73.

Al-Abbasi, Z. Q., & So, D. K. C. (2017). Resource Allocation in Non-Orthogonal and Hybrid Multiple Access System With Proportional Rate Constraint. *IEEE Transactions on Wireless Communications*, *16*(10), 6309–6320. DOI: 10.1109/TWC.2017.2721936

Ali, M. S., Tabassum, H., & Hossain, E. (2016). Dynamic User Clustering and Power Allocation for Uplink and Downlink Non-Orthogonal Multiple Access (NOMA) Systems. *IEEE Access : Practical Innovations, Open Solutions*, *4*, 6325–6343. DOI: 10.1109/ACCESS.2016.2604821

Alparslan Gök, S. Z., Branzei, R., & Tijs, S. (2010). The interval Shapley value: An axiomatization. *Central European Journal of Operations Research*, *18*(2), 131–140. DOI: 10.1007/s10100-009-0096-0

Andrews, J. G., Buzzi, S., Choi, W., Hanly, S. V., Lozano, A., Soong, A. C. K., & Zhang, J. C. (2014). What Will 5G Be? *IEEE Journal on Selected Areas in Communications*, *32*(6), 1065–1082. DOI: 10.1109/JSAC.2014.2328098

Branzei, R., Dimitrov, D., & Tijs, S. (2003). Shapley-like values for interval bankruptcy games. *Economic Bulletin*, *3*(8), 1–8.

Chae, S., & Heidhues, P. (2004). A group bargaining solution. *Mathematical Social Sciences*, *48*(1), 37–53. DOI: 10.1016/j.mathsocsci.2003.11.002

Chavarria-Reyes, E., Akyildiz, I. F., & Fadel, E. (2016). Energy-Efficient Multi-Stream Carrier Aggregation for Heterogeneous Networks in 5G Wireless Systems. *IEEE Transactions on Wireless Communications*, *15*(11), 7432–7443. DOI: 10.1109/TWC.2016.2602336

Chen, Q., Yu, G., Yin, R., & Li, G. Y. (2016). Energy-Efficient User Association and Resource Allocation for Multistream Carrier Aggregation. *IEEE Transactions on Vehicular Technology*, *65*(8), 6366–6376. DOI: 10.1109/TVT.2015.2472558

Chun, Y., & Thomson, W. (1992). Bargaining problems with claims. *Mathematical Social Sciences*, *24*(1), 19–33. DOI: 10.1016/0165-4896(92)90003-N

Chung, B. C., & Cho, D. (2017). Mobile Data Offloading With Almost Blank Subframe in LTE-LAA and Wi-Fi Coexisting Networks Based on Coalition Game. *IEEE Communications Letters*, *21*(3), 608–611. DOI: 10.1109/LCOMM.2016.2633341

Cisco Systems. (2020) *Cisco Annual Internet Report (2018–2023) White Paper* [White paper]. San Jose, CA: Cisco. https://www.cisco.com/c/en/us/solutions/collateral/executive-perspectives/annual-internet-report/white-paper-c11-741490.html

Dahlman, E., Mildh, G., Parkvall, S., Peisa, J., Sachs, J., Selén, Y., & Sköld, J. (2014). 5G wireless access: Requirements and realization. *IEEE Communications Magazine*, *52*(12), 42–47. DOI: 10.1109/MCOM.2014.6979985

Deb, S., Monogioudis, P., Miernik, J., & Seymour, J. P. (2014). Algorithms for Enhanced Inter-Cell Interference Coordination (eICIC) in LTE HetNets. *IEEE/ACM Transactions on Networking*, *22*(1), 137–150. DOI: 10.1109/TNET.2013.2246820

Dietzenbacher, B., Estévez-Fernández, A., Borm, P., & Hendrickx, R. (2020). Proportionality, equality, and duality in bankruptcy problems with nontransferable utility. *Annals of Operations Research*. Advance online publication. DOI: 10.1007/s10479-020-03643-3

Dietzenbacher, B., & Peters, H. (2018). *Characterizing NTU-Bankruptcy Rules Using Bargaining Axioms* (CentER Discussion Paper Series No. 2018-005). Graduate School of Business and Economics (GSBE), Maastricht University.

Diskin, A., Koppel, M., & Samet, D. (2011). Generalized Raiffa solutions. *Games and Economic Behavior*, *73*(2), 452–458. DOI: 10.1016/j.geb.2011.04.002

Du, J., Gelenbe, E., Jiang, C., Zhang, H., & Ren, Y. (2017). Contract Design for Traffic Offloading and Resource Allocation in Heterogeneous Ultra-Dense Networks. *IEEE Journal on Selected Areas in Communications*, *35*(11), 2457–2467. DOI: 10.1109/JSAC.2017.2760459

Gao, Y., & Roy, S. (2020). Achieving Proportional Fairness for LTE-LAA and Wi-Fi Coexistence in Unlicensed Spectrum. *IEEE Transactions on Wireless Communications*, *19*(5), 3390–3404. DOI: 10.1109/TWC.2020.2973145

Gerasimenko, M., Moltchanov, D., Florea, R., Andreev, S., Koucheryavy, Y., Himayat, N., Yeh, S., & Talwar, S. (2015). Cooperative Radio Resource Management in Heterogeneous Cloud Radio Access Networks. *IEEE Access : Practical Innovations, Open Solutions*, *3*, 397–406. DOI: 10.1109/ACCESS.2015.2422266

Hajir, M., Langar, R., & Gagnon, F. (2016). Coalitional Games for Joint Co-Tier and Cross-Tier Cooperative Spectrum Sharing in Dense Heterogeneous Networks. *IEEE Access : Practical Innovations, Open Solutions*, *4*, 2450–2464. DOI: 10.1109/ACCESS.2016.2562498

Haque, I. T., & Abu-Ghazaleh, N. (2016). Wireless Software Defined Networking: A Survey and Taxonomy. *IEEE Communications Surveys and Tutorials*, *18*(4), 2713–2737. DOI: 10.1109/COMST.2016.2571118

Herrero, C., & Villar, A. (2010). The rights egalitarian solution for NTU sharing problems. *International Journal of Game Theory*, *39*(1-2), 137–150. DOI: 10.1007/s00182-009-0177-2

Hinojosa, M. A., Mármol, A. M., & Zarzuelo, J. M. (2007). *Multi-Utilitarian Bargaining Solutions* (Working Papers No. 07.13). Universidad Pablo de Olavide, Department of Economics.

Hinojosa, M. A., Mármol, A. M., & Zarzuelo, J. M. (2008). Inequality averse multi-utilitarian bargaining solutions. *International Journal of Game Theory*, *37*(4), 597–618. DOI: 10.1007/s00182-008-0137-2

Hofbauer, J., & Sigmund, K. (2003). Evolutionary game dynamics. *Bulletin of the American Mathematical Society*, *40*(4), 479–519. DOI: 10.1090/S0273-0979-03-00988-1

Hougaard, J. L., & Tvede, M. (2003). Nonconvex n-person bargaining: Efficient maxmin solutions. *Economic Theory*, *21*(1), 81–95. DOI: 10.1007/s00199-001-0246-7

Jeon, J., Ford, R. D., Ratnam, V. V., Cho, J., & Zhang, J. (2019). Coordinated Dynamic Spectrum Sharing for 5G and Beyond Cellular Networks. *IEEE Access : Practical Innovations, Open Solutions*, *7*, 111592–111604. DOI: 10.1109/ACCESS.2019.2934385

Jia, Y., Zhao, M., & Zhou, W. (2016). Joint User Association and eICIC for Max-Min Fairness in HetNets. *IEEE Communications Letters*, *20*(3), 546–549. DOI: 10.1109/LCOMM.2016.2516002

Kim, S. (2011a). Adaptive online power control scheme based on the evolutionary game theory. *IET Communications*, *5*(18), 2648–2655. DOI: 10.1049/iet-com.2011.0093

Kim, S. (2011b). An Online Network Price Control Scheme by Using Stackelberg Game Model. *IEICE Transactions on Communications*, *94-B*(1), 322–325. DOI: 10.1587/transcom.E94.B.322

Kim, S. (2014). *Game Theory Applications in Network Design*. IGI Global. DOI: 10.4018/978-1-4666-6050-2

Kim, S. (2019). Cooperative bargaining based hierarchical spectrum allocation scheme for 5G heterogeneous networks. *IEEE Access : Practical Innovations, Open Solutions*, *7*, 102569–102579. DOI: 10.1109/ACCESS.2019.2930282

Kim, S. (2020a). Co-tier and cross-tier spectrum allocation scheme based on interval Shapley and Banzhaf values. *IEEE Access : Practical Innovations, Open Solutions*, *8*, 67766–67774. DOI: 10.1109/ACCESS.2020.2986218

Kim, S. (2020b). Heterogeneous network bandwidth control scheme for the hybrid OMA-NOMA system platform. *IEEE Access : Practical Innovations, Open Solutions*, *8*, 83414–83424. DOI: 10.1109/ACCESS.2020.2991780

Kim, S. (2020c). Heterogeneous network spectrum allocation scheme based on three-phase bargaining game. *Computer Networks*, *177*, 107301. DOI: 10.1016/j.comnet.2020.107301

Kim, S. (2020d). Two-level game based spectrum allocation scheme for multi-flow carrier aggregation technique. *IEEE Access : Practical Innovations, Open Solutions*, *8*, 89291–89299. DOI: 10.1109/ACCESS.2020.2993891

Kim, S. (2022a). 4G/5G coexistent dynamic spectrum sharing scheme based on dual bargaining game approach. *Computer Communications*, *181*, 215–223. DOI: 10.1016/j.comcom.2021.10.025

Kim, S. (2022b). A new fair-efficient spectrum allocation scheme for the LTE and WiFi coexistence platform. *EURASIP Journal on Wireless Communications and Networking*, *2022*(1), 80. DOI: 10.1186/s13638-022-02163-x

Kim, S. (2023a). Cooperative Game-Based Resource Allocation Scheme for Heterogeneous Networks with eICIC Technology. *Journal of Computer Networks and Communications*, *2023*, 5526441. DOI: 10.1155/2023/5526441

Kim, S. (2023b). Multi-Agent Learning and Bargaining Scheme for Cooperative Spectrum Sharing Process. *IEEE Access : Practical Innovations, Open Solutions*, *11*, 47863–47872. DOI: 10.1109/ACCESS.2023.3268754

Kim, S., & Varshney, P. K. (2004). An integrated adaptive bandwidth-management framework for QoS-sensitive multimedia cellular networks. *IEEE Transactions on Vehicular Technology*, *53*(3), 835–846. DOI: 10.1109/TVT.2004.825704

Kulkarni, A. A., & Shanbhag, U. V. (2015). An Existence Result for Hierarchical Stackelberg v/s Stackelberg Games. *IEEE Transactions on Automatic Control*, *60*(12), 3379–3384. DOI: 10.1109/TAC.2015.2423891

Kumar, A., Mallik, R. K., & Schober, R. (2014). A Probabilistic Approach to Modeling Users' Network Selection in the Presence of Heterogeneous Wireless Networks. *IEEE Transactions on Vehicular Technology*, *63*(7), 3331–3341. DOI: 10.1109/TVT.2013.2297437

Lee, H., Park, J., Lee, S. H., & Lee, I. (2023). Message-Passing Based User Association and Bandwidth Allocation in HetNets With Wireless Backhaul. *IEEE Transactions on Wireless Communications*, *22*(1), 704–717. DOI: 10.1109/TWC.2022.3197843

Lee, J., Mazumdar, R. R., & Shroff, N. B. (2005). Downlink power allocation for multi-class wireless systems. *IEEE/ACM Transactions on Networking*, *13*(4), 854–867. DOI: 10.1109/TNET.2005.852888

Li, A., & Qiu, L. (2014). Resource allocation in heterogeneous networks with eICIC. *IEEE WCSP*, *2014*, 1–6. DOI: 10.1109/WCSP.2014.6992105

Li, J., Wang, X., Li, Z., Wang, H., & Li, L. (2019). Energy Efficiency Optimization Based on eICIC for Wireless Heterogeneous Networks. *IEEE Internet of Things Journal*, *6*(6), 10166–10176. DOI: 10.1109/JIOT.2019.2936499

Lin, Y., Bao, W., Yu, W., & Liang, B. (2015). Optimizing User Association and Spectrum Allocation in HetNets: A Utility Perspective. *IEEE Journal on Selected Areas in Communications*, *33*(6), 1025–1039. DOI: 10.1109/JSAC.2015.2417011

Liu, D., Wang, L., Chen, Y., Elkashlan, M., Wong, K., Schober, R., & Hanzo, L. (2016). User Association in 5G Networks: A Survey and an Outlook. *IEEE Communications Surveys and Tutorials*, *18*(2), 1018–1044. DOI: 10.1109/COMST.2016.2516538

Liu, G., Wang, Z., Hu, J., Ding, Z., & Fan, P. (2019a). Cooperative NOMA Broadcasting/Multicasting for Low-Latency and High-Reliability 5G Cellular V2X Communications. *IEEE Internet of Things Journal*, *6*(5), 7828–7838. DOI: 10.1109/JIOT.2019.2908415

Liu, J., Wu, G., Xiao, S., Zhou, X., Li, G. Y., Guo, S., & Li, S. (2019b). Joint Power Allocation and User Scheduling for Device-to-Device-Enabled Heterogeneous Networks With Non-Orthogonal Multiple Access. *IEEE Access : Practical Innovations, Open Solutions*, *7*, 62657–62671. DOI: 10.1109/ACCESS.2019.2916921

Liu, L., Zhou, Y., Garcia, V., Tian, L., & Shi, J. (2018a). Load Aware Joint CoMP Clustering and Inter-Cell Resource Scheduling in Heterogeneous Ultra Dense Cellular Networks. *IEEE Transactions on Vehicular Technology*, *67*(3), 2741–2755. DOI: 10.1109/TVT.2017.2773640

Liu, Y., Yu, F. R., Li, X., Ji, H., & Leung, V. C. M. (2018b). Distributed Resource Allocation and Computation Offloading in Fog and Cloud Networks With Non-Orthogonal Multiple Access. *IEEE Transactions on Vehicular Technology*, *67*(12), 12137–12151. DOI: 10.1109/TVT.2018.2872912

Long, K., Wang, P., Li, W., & Chen, D. (2019). Spectrum Resource and Power Allocation With Adaptive Proportional Fair User Pairing for NOMA Systems. *IEEE Access : Practical Innovations, Open Solutions*, *7*, 80043–80057. DOI: 10.1109/ACCESS.2019.2908673

Marabissi, D., Bartoli, G., Fantacci, R., & Pucci, M. (2016). An Optimized CoMP Transmission for a Heterogeneous Network Using eICIC Approach. *IEEE Transactions on Vehicular Technology*, *65*(10), 8230–8239. DOI: 10.1109/TVT.2015.2510029

Markova, E., Moltchanov, D., Gudkova, I., Samouylov, K., & Koucharyavy, Y. (2019). Performance Assessment of QoS-Aware LTE Sessions Offloading Onto LAA/WiFi Systems. *IEEE Access : Practical Innovations, Open Solutions*, *7*, 36300–36311. DOI: 10.1109/ACCESS.2019.2905035

Monroy, L., Mármol, A. M., & Rubiales, V. (2009). A bargaining model for finite n-person multi-criteria games. *International Game Theory Review*, *11*(2), 121–139. DOI: 10.1142/S0219198909002212

Niu, C., Li, Y., Hu, R. Q., & Ye, F. (2017). Fast and Efficient Radio Resource Allocation in Dynamic Ultra-Dense Heterogeneous Networks. *IEEE Access : Practical Innovations, Open Solutions*, *5*, 1911–1924. DOI: 10.1109/ACCESS.2017.2653798

Omana, R. (2019). Using Spectrum Sharing to Deploy 4G/5G Capable Wireless Networks. *2019 Fall Technical Forum,* 1–9.

Osman, P., Alparslan G, , k, S. Z., & Weber, G. (2015). An Axiomatization of the Interval Shapley Value and on Some Interval Solution Concepts. *Contributions to Game Theory and Management*, *8*, 243–251.

Ridaoui, M., Grabisch, M., & Labreuche, C. (2018). *An axiomatisation of the Banzhaf value and interaction index for multichoice games* (Documents de travail du Centre d'Economie de la Sorbonne 18007). Université Panthéon-Sorbonne (Paris 1), Centre d'Economie de la Sorbonne.

Saleem, R., Alvi, S. A., & Durrani, S. (2021). Performance-Fairness Trade-off for Wi-Fi and LTE-LAA Coexistence. *IEEE Access : Practical Innovations, Open Solutions*, *9*, 62446–62459. DOI: 10.1109/ACCESS.2021.3075252

Sharma, S. K., Bogale, T. E., Le, L. B., Chatzinotas, S., Wang, X., & Ottersten, B. (2018). Dynamic Spectrum Sharing in 5G Wireless Networks With Full-Duplex Technology: Recent Advances and Research Challenges. *IEEE Communications Surveys and Tutorials*, *20*(1), 674–707. DOI: 10.1109/COMST.2017.2773628

Shi, Y., Qu, H., & Zhao, J. (2019). Dual-Connectivity Enabled Resource Allocation Approach With eICIC for Throughput Maximization in HetNets With Backhaul Constraint. *IEEE Wireless Communications Letters*, *8*(4), 1297–1300. DOI: 10.1109/LWC.2019.2917449

Singh, M., & Baranwal, G. (2018). Quality of Service (QoS) in Internet of Things. *IEEE IoT-SIU*, 1–6.

Srinivasan, M., & Murthy, C. S. R. (2020). Efficient Spectrum Slicing in 5G Networks: An Overlapping Coalition Formation Approach. *IEEE Transactions on Mobile Computing*, *19*(6), 1299–1316. DOI: 10.1109/TMC.2019.2908903

Tan, J., Xiao, S., Han, S., Liang, Y., & Leung, V. C. M. (2019). QoS-Aware User Association and Resource Allocation in LAA-LTE/WiFi Coexistence Systems. *IEEE Transactions on Wireless Communications, 18*(4), 2415–2430. DOI: 10.1109/TWC.2019.2904257

Tang, W., Zhang, R., Liu, Y., & Feng, S. (2014). Joint resource allocation for eICIC in heterogeneous networks. *IEEE Globecom Workshops, 2014*, 1–6. DOI: 10.1109/GLOCOM.2014.7037103

Trockel, W. (2011). *An Axiomatization of the Sequential Raiffa Solution* (Working Papers No. 425). Center for Mathematical Economics, Bielefeld University.

Trockel, W. (2015). Axiomatization of the discrete Raiffa solution. *Economic Theory Bulletin, 3*(1), 9–17. DOI: 10.1007/s40505-014-0046-4

Vasudevan, S., Pupala, R. N., & Sivanesan, K. (2013). Dynamic eICIC - A Proactive Strategy for Improving Spectral Efficiencies of Heterogeneous LTE Cellular Networks by Leveraging User Mobility and Traffic Dynamics. *IEEE Transactions on Wireless Communications, 12*(10), 4956–4969. DOI: 10.1109/TWC.2013.081413.121651

Wang, C., Wei, H., & Chen, W. (2016a). Resource Block Allocation with Carrier-Aggregation: A Strategy-Proof Auction Design. *IEEE Transactions on Mobile Computing, 15*(12), 3142–3155. DOI: 10.1109/TMC.2016.2524633

Wang, J., Hong, Y., Wang, J., Xu, J., Tang, Y., Han, Q., & Kurths, J. (2022). Cooperative and Competitive Multi-Agent Systems: From Optimization to Games. *IEEE/CAA Journal of Automatica Sinica, 9*(5), 763–783.

Wang, K., Wang, Y., Zeng, D., & Guo, S. (2017). An SDN-Based Architecture for Next-Generation Wireless Networks. *IEEE Wireless Communications, 24*(1), 25–31. DOI: 10.1109/MWC.2017.1600187WC

Wang, X., Zhang, H., Tian, Y., Zhu, C., & Leung, V. C. M. (2018). Optimal Distributed Interference Mitigation for Small Cell Networks With Non-Orthogonal Multiple Access: A Locally Cooperative Games. *IEEE Access : Practical Innovations, Open Solutions, 6*, 63107–63119. DOI: 10.1109/ACCESS.2018.2877181

Wang, Y., Xu, X., & Jin, Y. (2016b). QoS constraint optimal load balancing for Heterogeneous Ultra-dense Networks. *IEEE WPMC, 2016*, 1–7.

Xu, W., Qiu, R., & Jiang, X. (2019). Resource Allocation in Heterogeneous Cognitive Radio Network With Non-Orthogonal Multiple Access. *IEEE Access : Practical Innovations, Open Solutions, 7*, 57488–57499. DOI: 10.1109/ACCESS.2019.2914185

Xu, Y., Gui, G., Gacanin, H., & Adachi, F. (2021). A Survey on Resource Allocation for 5G Heterogeneous Networks: Current Research, Future Trends, and Challenges. *IEEE Communications Surveys and Tutorials, 23*(2), 668–695. DOI: 10.1109/COMST.2021.3059896

Xu, Y., & Yoshihara, N. (2011). *Proportional Nash solutions-A new and procedural analysis of nonconvex bargaining problems* (Discussion Paper Series No. 42). Institute of Economic Research, Hitotsubashi University.

Yang, H., & Bell, M. G. H. (1997). Traffic restraint, road pricing and network equilibrium. *Transportation Research Part B: Methodological, 31*(4), 303–314. DOI: 10.1016/S0191-2615(96)00030-6

Yu, G., Liu, R., Chen, Q., & Tang, Z. (2018). A Hierarchical SDN Architecture for Ultra-Dense Millimeter-Wave Cellular Networks. *IEEE Communications Magazine*, *56*(6), 79–85. DOI: 10.1109/MCOM.2018.1700490

Yuan, P., Xiao, Y., Bi, G., & Zhang, L. (2017). Toward Cooperation by Carrier Aggregation in Heterogeneous Networks: A Hierarchical Game Approach. *IEEE Transactions on Vehicular Technology*, *66*(2), 1670–1683. DOI: 10.1109/TVT.2016.2558318

Zhang, X. (2009). A note on the group bargaining solution. *Mathematical Social Sciences*, *57*(2), 155–160. DOI: 10.1016/j.mathsocsci.2008.09.001

Zhao, J., Liu, Y., Chai, K. K., Nallanathan, A., Chen, Y., & Han, Z. (2017). Spectrum Allocation and Power Control for Non-Orthogonal Multiple Access in HetNets. *IEEE Transactions on Wireless Communications*, *16*(9), 5825–5837. DOI: 10.1109/TWC.2017.2716921

Zhou, H., Ji, Y., Wang, X., & Zhao, B. (2015). Joint Resource Allocation and User Association for SVC Multicast Over Heterogeneous Cellular Networks. *IEEE Transactions on Wireless Communications*, *14*(7), 3673–3684. DOI: 10.1109/TWC.2015.2409834

Chapter 7
D2D Communications

ABSTRACT

D2D communications have emerged as a promising technology for the next generation mobile communication networks and wireless systems. This technology has shown great potential in improving communication capability, reducing communication delay and power dissipation, and fostering multifarious future network applications and services. In this chapter, the systematic overview of the D2D technical challenges and the prospective applications are presented to consolidate a background on D2D communications. In addition, this chapter discusses the fundamental motivations behind this new wave of technology development, followed by a brief discussion of the many definitions of D2D communications.

7. OVERVIEW

The exponential growth in the number of wireless communication devices and related applications in recent years has prompted dramatic rise in the wireless data traffic. Mobile wireless communication demands for higher data rates, lower energy consumption, low latency, coverage area, as well as efficient spectrum sharing are rapidly increasing. Therefore, the future wireless network systems must promise to improve the existing technology according to the future demands and provide a road map for reliable and resource-efficient solutions. In recent years, Device-to-Device (D2D) communication has attained significant attention in the research community and emerged as a promising technology for optimizing spectral efficiency in future networks. It proposes a new epitome in mobile networking to avail data exchange between physically proximate devices. The exploitation of D2D communication enables network operators to harvest short range communications for improving performance and corroborating proximity-based services.

The advantages of adopting D2D communications are the high data rates, low power consumption, short delays, enhanced spectral efficiency, and increased network capacity. Therefore, nowadays, the research community is actively investigating the D2D paradigm to realize its full potential and enable its smooth integration into the future network systems. Despite the advantages of D2D technology, there are also some limitations implied by the concept. One of the limiting factors is that D2D communication cannot be used very often because of distance restrictions. The other crucial problem regarding the D2D concept is its immaturity, resulting in many technical challenges. The most critical one is the interference management problem.

DOI: 10.4018/979-8-3373-2013-7.ch007

7.1 Small Cell Network Control (*SCNC*) scheme

In the context of current and future cellular networks, the integration of small cell architecture utilizing both licensed and unlicensed spectrum, content caching, and device-to-device (D2D) communication is considered essential. To address the numerous challenges faced by next-generation small cell networks, various control strategies have been proposed. Recently, S. Kim introduced the *Small Cell Network Control* (*SCNC*) scheme, which incorporates caching mechanisms and unlicensed D2D communication (Kim, 2017). Drawing on principles from game theory and learning algorithms, the *SCNC* scheme dynamically determines cached content and optimally allocates bandwidth between licensed and unlicensed channels. Given the dynamic nature of network environments, the *SCNC* scheme models network fluctuations through a novel dynamic Stackelberg game framework. By leveraging a hierarchical and feedback-driven control approach, small base stations and users can flexibly assume the roles of leaders or followers, ultimately enhancing the overall performance of 5G networks.

7.1.1 Development motivation

With the rapid increase in the number of mobile users and the growing demand for multimedia services, mobile data traffic is expanding at an unprecedented rate. The widespread use of ubiquitous devices such as smartphones has further intensified the need for data-heavy applications, including live video streaming, real-time social networking, and mobile gaming. As multimedia traffic continues to surge, wireless cellular networks require additional bandwidth to enhance system capacity. However, wireless bandwidth remains a highly valuable yet limited resource, with its availability nearly exhausted. Traditional methods of improving cellular network capacity face both physical and economic constraints. Consequently, current 5G network research is focused on developing innovative data dissemination techniques that move away from conventional network architectures (Afzal, 2016; Bulti, 2015; Jiang, 2015; Kim, 2017).

Small cells are widely recognized as crucial technology for 5G mobile wireless networks. By deploying low-power, cost-effective small cell base stations (SBSs) in a dense manner, network systems can enhance local coverage, increase bandwidth efficiency, boost throughput, and improve energy efficiency (Afzal, 2016; Cheng, 2017; Jiang, 2015). However, despite these advantages, small-cell networks do not resolve the issue of backhaul congestion. As multimedia data traffic grows, backhaul traffic overhead increases significantly, affecting the overall QoS throughout the network (Cheng, 2017). To mitigate this issue, caching mechanisms within SBSs have emerged as a promising approach, enabling higher transmission rates while alleviating backhaul load. In this caching-based framework, each SBS is equipped with a local cache that stores frequently requested content. If user requests match cached content, SBSs can directly transmit the data without relying on the backhaul, thereby minimizing traffic congestion and improving access latency (Cheng, 2017; Kim, 2010, Kim, 2017; Xu, 2017).

In next-generation cellular networks, device-to-device (D2D) communication has gained considerable attention from both academia and industry. This technique allows mobile users within proximity to communicate directly without routing data through the core network (Jiang, 2015). Initially introduced to enhance multi-hop network performance, D2D communication has evolved into a key enabler for achieving ultra-high data rates in 5G networks by offloading cellular traffic onto direct device connections. This approach not only alleviates backhaul congestion but also eliminates the need for additional network infrastructure. Despite its potential, D2D communication presents several technical challenges that must be addressed to ensure seamless integration into 5G networks (Atat, 2016; Chen, 2012; Kim, 2017).

To enable D2D communication, the bandwidth of each SBS must be divided into two sub-bands: licensed and unlicensed. The unlicensed band can be allocated to users for D2D communication, while the licensed band ensures QoS protection for primary users. However, due to fluctuations in service demands, a static allocation of licensed and unlicensed bandwidth is impractical. To achieve optimal 5G network performance, SBSs must dynamically adjust bandwidth allocation to maximize efficiency (Hajmohammad, 2013; Han, 2007; Kim, 2017).

Developing an advanced 5G network control scheme requires a novel control paradigm. Game theory, which analyzes strategic interactions among rational decision-makers with conflicting objectives, provides a suitable framework. It has been widely applied in wireless communications to resolve competition over network resources (Kim, 2014). In the context of D2D-enabled SBS operations, both SBSs and users act as rational agents. Inspired by the characteristics of 5G networks, the *SCNC* scheme employs a game-theoretic approach to optimize cache placement and bandwidth allocation. This method reduces the computational complexity of centralized optimization solutions, making them more practical for real-world implementation (Kim, 2017).

However, game theory has certain limitations. First, classical game theory assumes perfectly rational players with complete information, an assumption that is rarely valid in real-world network operations. Second, most game-theoretic models focus on static, one-sided stability and fail to account for the dynamic adaptation of players' strategies over time. Lastly, solving multiple high-order polynomial equations required in game-theoretic models can be computationally demanding, making real-time implementation challenging (Kim, 2014, Kim, 2017).

To address these limitations, the *SCNC* scheme introduces a dynamic Stackelberg game model tailored for D2D-enabled small cell networks. This model effectively captures the hierarchical relationship between SBSs and users, which aligns with the principles of the Stackelberg game. Traditionally, Stackelberg games involve a leader making strategic decisions, with followers reacting accordingly to optimize their own objectives (Kim, 2012, Kim, 2014). In contrast, the dynamic Stackelberg game allows roles to be interchangeable, enabling players to adapt their decision-making processes based on evolving network conditions. For cache placement, SBSs function as followers while their associated users act as leaders. Conversely, in bandwidth allocation, SBSs take on the leader role, while users act as followers. This dynamic and adaptive approach enhances decision-making efficiency in fluctuating 5G network environments (Kim, 2017).

To maintain network stability, the *SCNC* scheme incorporates learning and bargaining algorithms in a distributed manner. By integrating local, global, and social learning mechanisms along with a straightforward bargaining process, SBSs can intelligently manage caching and bandwidth allocation. Unlike existing approaches, the *SCNC* scheme prioritizes feasibility and adaptability to achieve an optimal balance between performance and practicality. Its primary innovation lies in its ability to effectively accommodate 5G network dynamics. While numerous D2D-enabled small cell network control protocols exist, few have successfully combined game theory with learning algorithms, making the *SCNC* scheme a significant advancement in this field (Kim, 2017).

7.1.2 Contribution

The *SCNC* scheme extends the cache placement and bandwidth allocation algorithms in several key-ways. To capture the interaction between each SBS and its users, it introduces a novel dynamic Stackelberg game model. In the *SCNC* scheme, the SBS acts as a follower when determining cache content by observing multiple leaders—its associated users. Conversely, as a leader, SBS dynamically allocates its bandwidth between licensed and unlicensed services. By leveraging learning and bargaining mechanisms, control decisions are made adaptively in real time. Ultimately, this approach ensures a well-balanced and fair solution across various D2D-enabled small cell network environments (Kim, 2017).

7.1.3 Dynamic Stackelberg Game Model

During the operation of D2D-enabled small cell networks, both SBSs and user equipment (UEs) independently make control decisions while considering their mutual interactions. This dynamic interplay makes game theory a suitable analytical tool. The *SCNC* scheme introduces a novel dynamic Stackelberg game model to govern interactions between each SBS and its associated UEs. This game process unfolds in two distinct phases. In the cache placement phase, each SBS monitors the file request frequency of its corresponding UEs and strategically determines which files to store within its limited cache capacity. In this context, the *SCNC* scheme models users as multiple leaders and the SBS as a single follower, who tracks the availability of cached content. This structure aligns well with a multiple-leaders, single-follower Stackelberg game model (Kim, 2017).

In the bandwidth allocation phase, each SBS analyzes the underlying D2D communications and allocates bandwidth between licensed and unlicensed communication services to enhance overall network capacity. Here, a single-leader, multiple-follower Stackelberg game model is more appropriate, where the SBS assumes the role of a leader, while the UEs act as followers. The UEs are categorized into two types within the SBS coverage area: S-UEs (Subscriber UEs), which connect directly to the SBS using licensed bandwidth, and D-UEs (Device UEs), which engage in direct communication without routing through the SBS via unlicensed bandwidth. As strategic game players, both SBSs and UEs adjust their strategies dynamically, seeking to optimize their respective payoffs based on an interactive feedback mechanism. At each time period of gameplay, the *SCNC* scheme formally defines the dynamic Stackelberg game model $\mathrm{G} = \left\{\{\mathfrak{B},\mathbb{E}\},\ \mathbb{C}^{\mathscr{B}},\ \left\{\left(\mathbb{S}_C^{\mathscr{B}},\mathbb{S}_S^{\mathscr{B}}\right),\mathbb{S}^{\mathscr{E}}\right\},\ \mathbb{L}(\mathfrak{P},\mathscr{B}),\ \{\mathbb{U}^{\mathscr{B}},\mathbb{U}^{\mathscr{E}}\},\ T\right\}$ as follows (Kim, 2017);

- In the $\{\mathfrak{B},\mathbb{E}\}$, $\mathrm{B} = \{\mathscr{B}_1 \ldots \mathscr{B}_n\}$ represents a set of SBSs and $\mathrm{E} = \{\mathscr{E}_1 \ldots \mathscr{E}_m\}$ is the set of UEs; they are game players.
- $\mathbb{E} = \{\mathscr{E}_1 \ldots \mathscr{E}_m\}$ can be divided into two subsets $\mathrm{E} = \mathfrak{I}^S \cup \mathfrak{I}^D$; $\mathfrak{I}^S$ is the subset of S-UEs, i.e., $\mathscr{E}^S_{1\leq k\leq m} \in \mathfrak{I}^S \subset \mathbb{E}$, and $\mathfrak{I}^D$ is the subset of D-UEs, i.e., $\mathscr{E}^D_{1\leq i\leq m} \in \mathfrak{I}^D \subset \mathbb{E}$.
- The bandwidth capacity of each SBS is $\mathbb{C}^{\mathscr{B}}$; it is divided up into licensed and unlicensed bandwidth bands.
- Each SBS has two strategy sets, i.e., $\left(\mathbb{S}_C^{\mathscr{B}},\mathbb{S}_S^{\mathscr{B}}\right)$. $\mathbb{S}_C^{\mathscr{B}}$ are the caching placement strategies to decide cache contents and $\mathbb{S}_S^{\mathscr{B}}$ is a set of bandwidths splitting strategies for licensed and unlicensed communications.
- In $\mathbb{S}_S^{\mathscr{B}} = \left\{\prod_{min\leq j\leq max} \mathfrak{P}_j^{\mathscr{B}} \middle| \mathfrak{P}_j^{\mathscr{B}} \in \left[\mathfrak{P}_{min}^{\mathscr{B}} \ldots \mathfrak{P}_j^{\mathscr{B}} \ldots \mathfrak{P}_{max}^{\mathscr{B}}\right]\right\}$, $\mathfrak{P}_j^{\mathscr{B}}$ means the jth bandwidth splitting ratio for unlicensed D2D communications. If the SBS selects the $\mathfrak{P}_j^{\mathscr{B}}$ strategy, the $\left[\mathfrak{P}_j^{\mathscr{B}} \times \mathbb{C}^{\mathscr{B}}\right]$

bandwidth amount is assigned to D2D communications and the remaining bandwidth amount $\left[\left(1-\mathfrak{P}_j^{\mathscr{B}}\right)\times\mathbb{C}^{\mathscr{B}}\right]$ is assigned to cellular communications.

- $\mathbb{S}^{\mathscr{E}}$ is the strategies of $\mathscr{E}\in\mathbb{E}$. The E decides his type, i.e., S-UE or D-UE, for his communications.
- $\mathbb{L}(\mathfrak{P},\mathscr{B})$ is the learning value for the B's strategy P; $\mathbb{L}(\mathfrak{P},\mathscr{B})$ is used to estimate the probability distribution ($\mathbb{P}^{\mathscr{B}}$) for the next bandwidth splitting strategy selection.
- In the $\{\mathbb{U}^{\mathscr{B}},\mathbb{U}^{\mathscr{E}}\}$, $\mathbb{U}^{\mathscr{B}}$ is the payoff received by the B and $\mathbb{U}^{\mathscr{E}}$ is the payoff received by the E during the D2D-enabled small cell network operation.
- $T=\left\{\mathscr{H}_1,\ldots,\mathscr{H}_t,\mathscr{H}_{t+1},\ldots\right\}$ denotes time, which is represented by a sequence of time steps with imperfect information for the dynamic Stackelberg game process.

The dynamic Stackelberg game (G) is a special case of traditional Stackelberg game. To solve the joint problem of cache placement and bandwidth splitting, it is natural that G is designed as a two-stage game approach. Regardless of whether they act as leaders or followers, all individual game participants independently and strategically choose actions that maximize their own payoffs. At the end of each game iteration, players periodically evaluate their outcomes and dynamically adjust their decisions in a fully distributed manner. This iterative feedback process continues step by step until an optimal solution is reached (Kim, 2017).

7.1.4 Cache Placement Algorithm in dynamic Stackelberg Game

Caching technology enables the local storage of popular content, allowing UEs to access frequently requested files without relying on the backhaul. Without caching, these files would need to be retrieved through the backhaul, increasing network congestion. By implementing caching, backhaul traffic and access latency can be significantly reduced, leading to enhanced system performance. However, due to the limited storage capacity of each SBS, it is not feasible to cache all files. Instead, only the most popular content is strategically stored to optimize content distribution. To facilitate an effective selection process for cached content, the *SCNC* scheme models the interaction between each SBS and its associated UEs using a novel dynamic Stackelberg game framework (Kim, 2017).

For the cache placement algorithm, the *SCNC* scheme considers a commercial small cell caching system comprising multiple SBSs and UEs. By leveraging multiple-leaders, single-follower Stackelberg game model, a cooperative SBS caching algorithm is developed. In practice, an efficient caching mechanism is closely integrated with the file placement strategy to ensure optimal content availability. In the small cell network architecture, the *SCNC* scheme assumes that a multimedia file set M = $\{\mathscr{M}_1,\ldots,\mathscr{M}_L\}$ consists of L popular files among total X multimedia files, and files in M can be possibly cached in each SBS. The popularity distribution among M is represented by a vector Q = $[g_1,\ldots,g_L]$, which is frequently requested by users. Generally, the vector Q can be modeled by a Zipf distribution, which is a discrete probability distribution commonly used in the modelling of rare events (Kim, 2017; Li, 2016).

The *SCNC* scheme considers the social relations of EUs and interactions among SBSs to adaptively obtained the g values in Q. In fact, social characteristics such as the external influence for users' relationships and ties have played a crucial role in information propagation over the Internet and will continue to shape the way information is accessed (Bastug, 2014). To exploit the correlation between users' social relations, centrality can be used to identify the most influential users, who may act as a conduit for information diffusion (Wu, 2016; Xia, 2013). In the *SCNC* scheme, centrality is considered weighty to estimate g values. At time $\mathcal{H}_t$, the $\mathcal{M}_l$ file's g value in the $\mathcal{B}_j$ $\left(g_{\mathcal{M}_l}(\mathcal{B}_j, \mathcal{H}_t)\right)$ is defined as follows (Kim, 2017);

$$g_{\mathcal{M}_l}(\mathcal{B}_j, \mathcal{H}_t) = \left(\frac{1}{(\Theta(\mathcal{M}_l))^{\tau(\mathcal{M}_l, \mathcal{B}_j, \mathcal{H}_t)}}\right) / \left(\sum_{f=1}^{L} \frac{1}{(\Theta(\mathcal{M}_f))^{\tau(\mathcal{M}_f, \mathcal{B}_j, \mathcal{H}_t)}}\right) \tag{7.1.1}$$

$$\text{s.t.,}\ \Theta(\mathcal{M}_l) = l + \left(\sum_{\mathcal{E}_i \in \mathfrak{X}(\mathcal{M}_l, \mathcal{B}_j)} \frac{\mathfrak{D}(\mathcal{E}_i)}{\mathcal{M}}\right) \text{ and } 1 \le l \le L$$

where $\mathfrak{X}(\mathcal{M}_l, \mathcal{B}_j)$ is the set of UEs, who request $\mathcal{M}_l$ file in the $\mathcal{B}_j$. $\mathfrak{D}(\mathcal{E}_i)$ is the number of first-degree social friends of $\mathcal{E}_i$. M is the maximum number of $\mathfrak{D}(\cdot)$ in the $\mathcal{B}_j$. In the equation (7.1.1), the $\tau(\mathcal{M}_l, \mathcal{B}_j, \mathcal{H}_t)$ function represents the skewness of the $\mathcal{M}_l$ distribution in the $\mathcal{B}_j$; a higher $\tau(\cdot)$ outcome corresponds to a higher file reuse. To adaptively obtain the $\tau(\cdot)$ outcome, we concentrate on the notion of global awareness networking, which attracts significantly the social and behavioral communities. By considering neighbor SBSs' file request situations, each SBS learns the global trends of UEs' propensity Finally, $\tau(\mathcal{M}_l, \mathcal{B}_j, \mathcal{H}_t)$ is given by (Kim, 2017);

$$\tau(\mathcal{M}_l, \mathcal{B}_j, \mathcal{H}_t) = \left(\max\left\{\min\left(\mathfrak{W}(\mathcal{M}_l, \mathcal{B}_j, \mathcal{H}_t), \beta\right), \xi\right\}\right) \tag{7.1.2}$$

$$\text{s.t.,}\ \mathfrak{W}(\mathcal{M}_l, \mathcal{B}_j, \mathcal{H}_t) = \frac{\left(\alpha \times \mathcal{H}(\mathcal{M}_l, \mathcal{B}_j, \mathcal{H}_{t-1})\right) + \left((1-\alpha) \times \frac{\sum_{k=1, k\neq j}^{L-1} \mathcal{H}(\mathcal{M}_l, \mathcal{B}_j, \mathcal{H}_{t-1})}{L-1}\right)}{\text{median}\left\{\mathcal{H}(\mathcal{M}_l, \mathfrak{B}, \mathcal{H}_{t-1})\right\}}$$

where $\mathcal{H}(\mathcal{M}_l, \mathcal{B}_j, \mathcal{H}_{t-1})$ function returns the g value of $\mathcal{M}_l$ file in the $\mathcal{B}_j$ at time $\mathcal{H}_{t-1}$. If the $\mathcal{M}_l$ file was not in M at time $\mathcal{H}_{t-1}$, $\mathcal{H}(\mathcal{M}_l, \mathcal{B}_j, \mathcal{H}_{t-1})$ function returns zero. α is a weighted average between local and global caching information, and β, ξ are the upper and lower bounds of τ function, respectively. median { · } returns the median value of all SBSs. Generally, the most popular files account for the majority of download requests. In the *SCNC* scheme, social and global properties of small cell networks can be used to design a cache placement protocol using È(·), $\tau(\cdot)$ and $\mathcal{H}(\cdot)$ functions. Finally, the file with a higher g value corresponds to be cached in each SBS from equations (7.1.1)-(7.1.2) (Kim, 2017).

7.1.5 Bandwidth Splitting Algorithm in dynamic Stackelberg Game

Recently, traffic offloading technology has been introduced as an effective solution to significantly enhance system capacity. By reducing the volume of data transmitted over cellular bands, this approach helps free up bandwidth for other UEs. However, given the constraints of limited spectrum resources, bandwidth allocation must be carefully optimized. The *SCNC* scheme examines a scenario where SBSs are allocated licensed bandwidth and have the flexibility to lease a portion of it to unlicensed UEs for

D2D communication. From various performance perspectives, unlicensed bandwidth can greatly enhance both capacity and coverage. As a result, the bandwidth splitting technique plays a crucial role in maximizing overall system capacity and ensuring QoS satisfaction for UEs (Hajmohammad, 2013; Kim, 2017; Lee, 2015).

For D2D-enabled small cell networks, we would face a two-tiered network structure where licensed bandwidth ($\mathbb{C}^{\mathcal{B}} \times (1 - \mathfrak{P}^{\mathcal{B}})$) is allocated for cellular communications and unlicensed bandwidth ($\mathbb{C}^{\mathcal{B}} \times \mathfrak{P}^{\mathcal{B}}$) is allocated for D2D communications. To support this mechanism, each SBS split the total bandwidth ($\mathbb{C}^{\mathcal{B}}$) for two kinds of communication paradigms (Hajmohammad, 2013; Lee, 2015). To cope with the design challenges of bandwidth splitting, the *SCNC* scheme designs a single-leader multiple-followers Stackelberg game model. In this model, traffic offloading through opportunistic communications exploits D2D communications in the unlicensed bandwidth bands. However, the bandwidth splitting problem for D2D communications is generally NP-hard. For this reason, the single-leader multiple-followers Stackelberg game model is reformulated based on the reinforcement learning algorithm with low computational complexity. To fine-tune the system performance, it is a suitable approach (Kim, 2017).

Under the co-existence situation of cellular and D2D communications, UEs should consider using the unlicensed bandwidth for D2D communications or the licensed bandwidth for cellular communications. From a standpoint of UE, two utility functions are defined for cellular and D2D communications. Both of them are formulated by considering the tradeoff between throughput and transmit power. For the $\mathscr{E}_m \in \mathbb{E}$, the utility function for cellular communications $\left(U_{\mathscr{E}_m}^{CC}\right)$ and for D2D communications $\left(U_{\mathscr{E}_m}^{D2D}\right)$ are given by (Kim, 2017);

$$\begin{cases} U_{\mathscr{E}_m}^{CC} = \left(\mathfrak{B}_{CC} \times \log_2\left(1 + \frac{\gamma^{\mathscr{E}_m}(\mathrm{P})}{\Omega}\right)\right) \Big/ \mathscr{P}_{\mathscr{E}_m}^{CC} \\ U_{\mathscr{E}_m}^{D2D} = \left(\mathfrak{B}_{D2D} \times \log_2\left(1 + \frac{\gamma^{\mathscr{E}_m}(\mathrm{P})}{\Omega}\right)\right) \Big/ \mathscr{P}_{\mathscr{E}_m}^{D2D} \end{cases} \tag{7.1.3}$$

where $\mathfrak{B}_{CC}, \mathfrak{B}_{D2D}$ are the assigned bandwidth channel for $\mathscr{E}_m$'s cellular and D2D communications, respectively. Ω ($\Omega \geq 1$) is the gap between uncoded M-QAM and the capacity, minus the coding gain. $\gamma^{\mathscr{E}_m}(\tilde{\mathrm{N}})$ is the $\mathscr{E}_m$'s signal-to-interference plus noise ratio (SINR) with power vector P for all UEs, and $\mathscr{P}_{\mathscr{E}_m}^{CC}, \mathscr{P}_{\mathscr{E}_m}^{D2D}$ are the $\mathscr{E}_m$'s power level for cellular and D2D communications, respectively. As a UE, the goal of each E is to maximize its own payoff by selecting a strategy in $\mathbb{S}^{\mathscr{E}}$; it decides his type, i.e., S-UE or D-UE (Kim, 2017).

$$\max_{\mathbb{S}^{\mathscr{E}}=\{\text{S-UE, D-UE}\}} \left\{U_{\mathscr{E}_m}^{CC}, U_{\mathscr{E}_m}^{D2D}\right\}, \text{ s.t., S-UE} \rightarrow U_{\mathscr{E}_m}^{CC} \text{ and D-UE} \rightarrow U_{\mathscr{E}_m}^{D2D} \tag{7.1.4}$$

From the standpoint of SBS, utility function is designed based on its the total system throughput; it is obtained from the sum of cellular and D2D communications in its coverage area. For the $\mathscr{B}_i$, the utility function $\left(U_{\mathscr{B}_i}\right)$ is defined as follows (Kim, 2017);

$$U_{\mathcal{B}_i} = \sum_{\mathcal{E}_g \in S^{S\text{-}UE}_{\mathcal{B}_i}} U^{CC}_{\mathcal{E}_g} + \sum_{\mathcal{E}_h \in S^{D\text{-}UE}_{\mathcal{B}_i}} U^{D2D}_{\mathcal{E}_h} \tag{7.1.5}$$

where $S^{S\text{-}UE}_{\mathcal{B}_i}, S^{D\text{-}UE}_{\mathcal{B}_i}$ are the set of S-UE type SBSs and D-UE type SBSs, respectively. To maximize the $U_{\mathcal{B}}$, each SBS selects his strategy, which decides the amounts of licensed and unlicensed bandwidth bands. To decide effectively the bandwidth splitting policy ($\mathfrak{P}^{\mathcal{B}}$), the *SCNC* scheme develops a new learning algorithm. In the *SCNC* scheme, learning is divided into two categories: local learning and global learning. Local learning refers to an insight temporal learning in its local SBS, and global learning refers to spatial leaning through neighboring SBSs. The main novelty of the bandwidth splitting algorithm is a joint-design manner concerning local and global learning approaches. To specify the global relationship of SBSs, the affinity indicator ($\rho^{\mathcal{B}_j}_{\mathcal{B}_i}$) between the $\mathcal{B}_i$ and $\mathcal{B}_j$ is defined as (Kim, 2017; Zhang, 2016a):

$$\rho^{\mathcal{B}_j}_{\mathcal{B}_i} = \frac{\mathbb{E}\left(\left(\mathcal{T}(\mathcal{B}_i) - \mathcal{A}(\mathcal{B}_i)\right) \times \left(\mathcal{T}(\mathcal{B}_j) - \mathcal{A}(\mathcal{B}_j)\right)\right)}{\sigma(\mathcal{B}_i) \times \sigma(\mathcal{B}_j)}, \text{ s.t., } -1 \leq \rho^{\mathcal{B}_j}_{\mathcal{B}_i} \leq 1 \tag{7.1.6}$$

where $\mathcal{T}(\mathcal{B}_i)$, $\mathcal{A}(\mathcal{B}_i)$, $\sigma(\mathcal{B}_i)$ are the current traffic amount, the mean and standard deviation of traffic history, respectively. E(·) is the expectation value. To learn the local traffic situation of each individual SBS, the current $U_{\mathcal{B}}$ value is periodically monitored. Based on the global (ρ) and local ($U_{\mathcal{B}}$) information, the *SCNC* scheme can estimate the learning values of SBS's bandwidth splitting strategies $\mathfrak{P}^{\mathcal{B}} \in \mathbb{S}^{\mathcal{B}}_{S}$. If the strategy $\mathfrak{P}^{\mathcal{B}_i}_j$ is selected at time $\mathcal{H}_{t-1}$ by the $\mathcal{B}_i$, the $\mathcal{B}_i$ updates the strategy $\mathfrak{P}^{\mathcal{B}_i}_j$'s learning value for the next time step $\left(\mathbb{L}_{\mathcal{H}_t}(\mathfrak{P}^{\mathcal{B}_i}_j, \mathcal{B}_i)\right)$ as follows (Kim, 2017);

$$\mathbb{L}_{\mathcal{H}_t}(\mathfrak{P}^{\mathcal{B}_i}_j, \mathcal{B}_i) = \left((1-\chi) \times \mathbb{U}^{\mathcal{B}_i}(\mathcal{H}_{t-1})\right) + \left(\chi \times [\mathcal{V} + \mathcal{W}]\right) \tag{7.1.7}$$

$$\text{s.t., } \mathcal{V} = \varrho \times \mathbb{L}_{\mathcal{H}_{t-1}}(\mathfrak{P}^{\mathcal{B}_i}_j, \mathcal{B}_i) \text{ and } \mathcal{W} = \left\{(1-\varrho) \times \mathbb{L}_{\mathcal{H}_{t-1}}\left(\mathfrak{P}^{\mathcal{B}}_j, \mathcal{B} = \left\{\mathcal{B}_k \mid \max_{\mathcal{B}_k \in \mathfrak{B}}\left(\rho^{\mathcal{B}_k}_{\mathcal{B}_i}\right)\right\}\right)\right\}$$

where χ is a learning rate that models how the L-values are updated. In the equation (7.1.7), V and W represent local and global learning values, respectively. Therefore, ϱ is a control factor for the weighted average between different learning approaches (Kim, 2017).

Based on the $\mathbb{L}(\cdot)$ values, a strategy selection distribution (P) for each SBS is defined. During the dynamic bandwidth splitting process, the *SCNC* scheme determines $\mathbb{P}^{\mathcal{B}_i}_{\mathcal{H}_t} = \left\{\mathcal{P}_{\mathcal{H}_t}(\mathfrak{P}^{\mathcal{B}_i}_{min}) \ldots \mathcal{P}_{\mathcal{H}_t}(\mathfrak{P}^{\mathcal{B}_i}_j) \ldots \mathcal{P}_{\mathcal{H}_t}(\mathfrak{P}^{\mathcal{B}_i}_{max})\right\}$ as the probability distribution of $\mathfrak{P}^{\mathcal{B}_i}$'s strategy selection at time $\mathcal{H}_t$; it is sequentially modified over time. In $\mathbb{P}^{\mathcal{B}_i}_{\mathcal{H}_t}$, the $\mathfrak{P}^{\mathcal{B}_i}_j$ strategy selection probability by the $\mathcal{B}_i$ at time $\mathcal{H}_t$ $\left(\mathcal{P}_{\mathcal{H}_t}(\mathfrak{P}^{\mathcal{B}_i}_j)\right)$ is defined as (Kim, 2017):

$$\mathcal{P}_{\mathcal{H}_t}(\mathfrak{P}^{\mathcal{B}_i}_j) = \frac{\text{EXP}\left(\mathbb{L}_{\mathcal{H}_t}(\mathfrak{P}^{\mathcal{B}_i}_j, \mathcal{B}_i)\right)}{\sum_{e=min}^{max} \text{EXP}\left(\mathbb{L}_{\mathcal{H}_t}(\mathfrak{P}^{\mathcal{B}_i}_e, \mathcal{B}_i)\right)} \tag{7.1.8}$$

According to the equation (7.1.8), the $\mathscr{B}_i$ stochastically selects the $\mathfrak{P}^{\mathscr{B}_i}$ strategy using his strategy selection distribution (P). Based on the selected $\mathfrak{P}^{\mathscr{B}_i}$ strategy, the $\mathscr{B}_i$ finally attempt to adjust the system performance. By using a simple bargaining process, the $\mathscr{B}_i$ carefully deliberates on his final decision. In particular, the outcome from the selected $\mathfrak{P}_i^{\mathscr{B}_i}$ strategy is considered as the status quo point of this bargaining process. At time $\mathscr{H}_{t-1}$, this point is the vector $\left(\mathbb{V}_{\mathscr{H}_{t-1}}\left(\overline{\mathfrak{P}_i^{\mathscr{B}_i}}\right)\right)$ of cellular and D2D communication payoffs achieved by the $\mathscr{B}_i$ with $\mathfrak{P}_i^{\mathscr{B}_i}$, that is (Kim, 2017);

$$\mathbb{V}_{\mathscr{H}_{t-1}}\left(\overline{\mathfrak{P}_i^{\mathscr{B}_i}}\right) = \left(U_{\mathscr{B}_i}^{CC}\left(\mathfrak{P}_i^{\mathscr{B}_i}, \mathscr{H}_{t-1}\right), U_{\mathscr{B}_i}^{D2D}\left(\mathfrak{P}_i^{\mathscr{B}_i}, \mathscr{H}_{t-1}\right)\right) \tag{7.1.9}$$

where $U_{\mathscr{B}_i}^{CC}\left(\mathfrak{P}_i^{\mathscr{B}_i}, \mathscr{H}_{t-1}\right)$, $U_{\mathscr{B}_i}^{D2D}\left(\mathfrak{P}_i^{\mathscr{B}_i}, \mathscr{H}_{t-1}\right)$ are $U_{\mathscr{B}_i}^{CC}$ and $U_{\mathscr{B}_i}^{D2D}$ values with the $\mathfrak{P}_i^{\mathscr{B}_i}$ strategy at time $\mathscr{H}_{t-1}$. Based on the selection strategy $\mathfrak{P}_i^{\mathscr{B}_i}$, which was obtained by the learning process, the $\mathscr{B}_i$ selects finally a strategy to obtain the best desirable solution as follows (Kim, 2017);

$$\mathfrak{P}_{min\leq k\leq max}^{\mathscr{B}_i} = \arg \max_{\mathfrak{P}_k^{\mathscr{B}_i} \in \mathbb{S}_S^{\mathscr{B}_i}} \prod_{\mathfrak{P}_k^{\mathscr{B}_i}} \left(\mathbb{V}_{\mathscr{H}_{t-1}}\left(\overline{\mathfrak{P}_k^{\mathscr{B}_i}}\right) - \mathbb{V}_{\mathscr{H}_{t-1}}\left(\overline{\mathfrak{P}_i^{\mathscr{B}_i}}\right)\right) \tag{7.1.10}$$

In each game round, the bargaining process is carried out sequentially, allowing each SBS to refine its decisions and mitigate unexpected outcomes. The fundamental concept of bargaining solutions has gained significant research interest due to its advantageous properties. However, conventional bargaining approaches are essentially equivalent to random optimization methods operating within a vast solution space, making convergence difficult (Han, 2007). As a result, their practical implementation in real-world network operations is challenging. To address this limitation, the *SCNC* scheme integrates a learning-based approach to enhance the bargaining model. This method offers a practical and efficient solution for network operations, delivering improved performance across diverse system environments (Kim, 2017).

7.1.6 Summary

In 5G cellular networks, small cell networks are expected to become a dominant and promising solution. The *SCNC* scheme introduces a novel D2D-enabled small cell network control framework that integrates caching strategies. This research is driven by the need to jointly address cache placement and bandwidth allocation challenges. By leveraging a learning-based game-theoretic approach, the *SCNC* scheme formulates a dynamic Stackelberg game model and designs optimized cache placement and bandwidth splitting algorithms to maximize system capacity. Game theory is increasingly recognized as a practical methodology for real-world network operations. In the multiple-leaders, single-follower Stackelberg game, the cache placement algorithm ensures that the most frequently requested files are stored, thereby alleviating fronthaul traffic congestion and reducing request latency. Meanwhile, in the single-leader, multiple-follower Stackelberg game, the bandwidth splitting algorithm optimally allocates resources between cellular and D2D communications, balancing overall system performance. Through feedback-driven self-monitoring and a combination of social, local, and global learning techniques,

SBSs and UEs dynamically adjust to evolving network conditions, effectively maximizing their expected benefits in D2D-enabled small cell networks (Kim, 2017).

7.2 Cooperative Dual-level Game based Control (*CDGC*) scheme

With advancements in wireless communications and the increasing intelligence of mobile devices, D2D communications are expected to become an integral component of future 5G networks. This emerging paradigm enhances user experiences while improving key system performance metrics such as network throughput, latency, fairness, and energy efficiency. Recently, S. Kim introduced the *Cooperative Dual-level Game-based Control* (*CDGC*) scheme, which involves multiple D2D operators and a network of mobile devices (Kim, 2019). To effectively model the interactions between these entities, the *CDGC* scheme employs two cooperative game-theoretic approaches based on incentive mechanism design and the *r-egalitarian Shapley* value. At the upper level, routing paths and incentive payments for multi-hop relay services are determined using an incentive mechanism. At the lower level, mobile devices distribute the received incentives based on the *r-egalitarian Shapley value*. These two control layers are interdependent, requiring proper coordination and collaboration to function effectively. By leveraging the key properties of both cooperative game models, the *CDGC* scheme achieves significant advantages in a fair and efficient manner (Kim, 2019).

7.2.1 Development motivation

The rapid proliferation of intelligent mobile devices equipped with smart applications has led to an unprecedented surge in mobile data traffic. This dramatic increase in network demand places a significant strain on existing wireless infrastructure. To support higher peak data rates and improved network capacity, the development of 5G mobile communication technology is essential, surpassing the capabilities of current 4G systems. By 2023, 5G networks are projected to handle more than 20% of global mobile data traffic, which is 1.5 times the total volume currently carried by 4G, 3G, and 2G combined. This remarkable growth in network traffic is expected to drive further research into novel cooperative communication techniques, primarily due to the increasing demand for seamless connectivity across mobile devices (Kim, 2019; Shakir, 2017).

Recently, D2D communication has emerged as a promising solution to address 5G network capacity and coverage challenges. Traditionally, D2D communication has been explored in scenarios involving natural disasters, where infrastructure-based networks are either partially or completely unavailable. This technology extends conventional wireless communication systems by enabling direct wireless links between mobile devices without relying on network infrastructure. However, single-hop D2D communication is typically constrained to a specific geographical area, limiting its applicability. To fully harness the benefits of D2D communication, multi-hop D2D communication is essential. This approach enables devices to relay data over multiple hops, creating an ad-hoc, decentralized communication model, which is becoming a key component of 5G network architecture (Desale, 2016; Kim, 2019; Shaikh, 2018).

In general, multi-hop D2D communication facilitates direct data exchange among multiple devices without requiring the involvement of wireless operators (WOs). However, conventional multi-hop D2D systems often struggle to guarantee QoS for data traffic. To address this limitation, a new paradigm has emerged: operator-controlled D2D communication. This model allows WOs to manage and optimize D2D communication within their networks, maximizing system efficiency and ensuring superior QoS for devices. Under this framework, WOs monitor devices within their coverage area and establish optimized routing paths in coordination with other operators. Since self-interested devices may not voluntarily participate in data relay transmissions, WOs must provide dynamic incentives to encourage their cooperation in multi-hop D2D communication (Kim, 2019; Lu, 2015; Park, 2007).

In an operator-controlled D2D system, both wireless operators and mobile devices must collaborate to enhance overall network performance. Additionally, mobile devices within a specific WO's coverage must reach an agreement on incentive distribution. This interdependent relationship between operators and devices can be modeled as a hierarchical cooperation problem. To effectively address this challenge, two critical questions must be answered: What is the appropriate incentive payment to encourage participation in multi-hop D2D communication? How should the incentive payment be distributed among relaying devices? Addressing these two questions requires distinct strategies, necessitating a cooperative game-theoretic approach (Kim, 2019; Lu, 2015; So, 2007).

Over the past decade, non-cooperative and cooperative game models have been widely used to analyze interactive decision-making in networked environments. However, traditional non-cooperative game models present several limitations that make them unsuitable for operator-controlled D2D communication. Key disadvantages of non-cooperative game models include i) high information acquisition overhead, ii) slow convergence to equilibrium, iii) inefficiency in terms of social welfare, and iv) complexity in characterizing equilibrium solutions. Given these challenges, cooperative game models are a more suitable approach for multi-hop D2D communication, as they inherently align with the collaborative nature of such systems. Consequently, a cooperative game-based framework is adopted to design an efficient operator-controlled D2D communication protocol (Karlin, 2017; Kim, 2014, Kim, 2019).

The *CDGC* scheme employs two distinct cooperative game-theoretic approaches to address the hierarchical cooperation problem in operator-controlled D2D communication. Recognizing the interdependent relationship between WOs and mobile devices, the *CDGC* scheme formulates a dual-level game model based on incentive mechanism design (IMD) and the *r-egalitarian Shapley value*. At the upper level, the IMD framework determines the incentive payments for WOs managing multi-hop D2D routes. At the lower level, each WO distributes the obtained incentives among its mobile devices using the *r-egalitarian Shapley value*. By leveraging cooperative game-theoretic principles, the *CDGC* scheme aims to achieve optimal network performance while addressing challenges related to asymmetric information and strategic interactions among operators and devices. This structured approach ensures fair and efficient incentive distribution, fostering collaboration in D2D-enabled 5G networks (Kim, 2019).

7.2.2 Contribution

The *CDGC* scheme models the interactions between WOs and mobile devices by formulating a dual-level hierarchical cooperative game. At the upper level, WOs act as game participants, with incentive payments determined using the IMD for relay WOs. At the lower level, data-relaying mobile devices serve as game players, distributing the received incentives from their respective WOs based on the *r-egalitarian Shapley value*. To fully exploit the advantages of this dual-level framework, the *CDGC* scheme systematically addresses key control challenges while incorporating all relevant practical considerations in operator-controlled multi-hop D2D communications (Kim, 2019).

7.2.3 Operator-controlled D2D communication infrastructure

The *CDGC* scheme operates within an operator-controlled D2D communication system, consisting of multiple mobile devices managed by different WOs. These mobile devices are randomly distributed across a large coverage area, which is subdivided into smaller regions, each served by a specific operator. In this operator-controlled D2D framework, WOs are interconnected via high-speed wired links, allowing for the exchange of system control information, while mobile devices establish wireless connections with their respective operators. Given the limited transmission power of individual devices, multi-hop relaying is employed to facilitate data transmission from source devices to destination devices, ensuring efficient and reliable communication (Lu, 2015; Zhang, 2018). Denote the set of operators as $\mathbb{O} = \{O_1, \ldots, O_n\}$ and $M^{O_{1\le i\le n}} = \{M_1^{O_i}, \ldots, M_m^{O_i}\}$ is the set of devices under the operator $O_i \in \mathbb{O}$. O_i is responsible to the devices in the set M^{O_i}, and $M_j^{O_i} \in M^{O_i}$ reports its status information to its own operator O_i, and each M is connected to neighboring devices for multi-hop D2D communications. For the multiple access at every hop, the *CDGC* scheme considers a OFDMA-based transmission, and communication capacity at each hop link is fixed (Byun, 2015; Lu, 2015; Zhang, 2018). The general infrastructure of operator-controlled D2D system is shown in Figure 1 (Kim, 2019).

Figure 1. Operator-controlled D2D communication infrastructure (Kim, 2019)

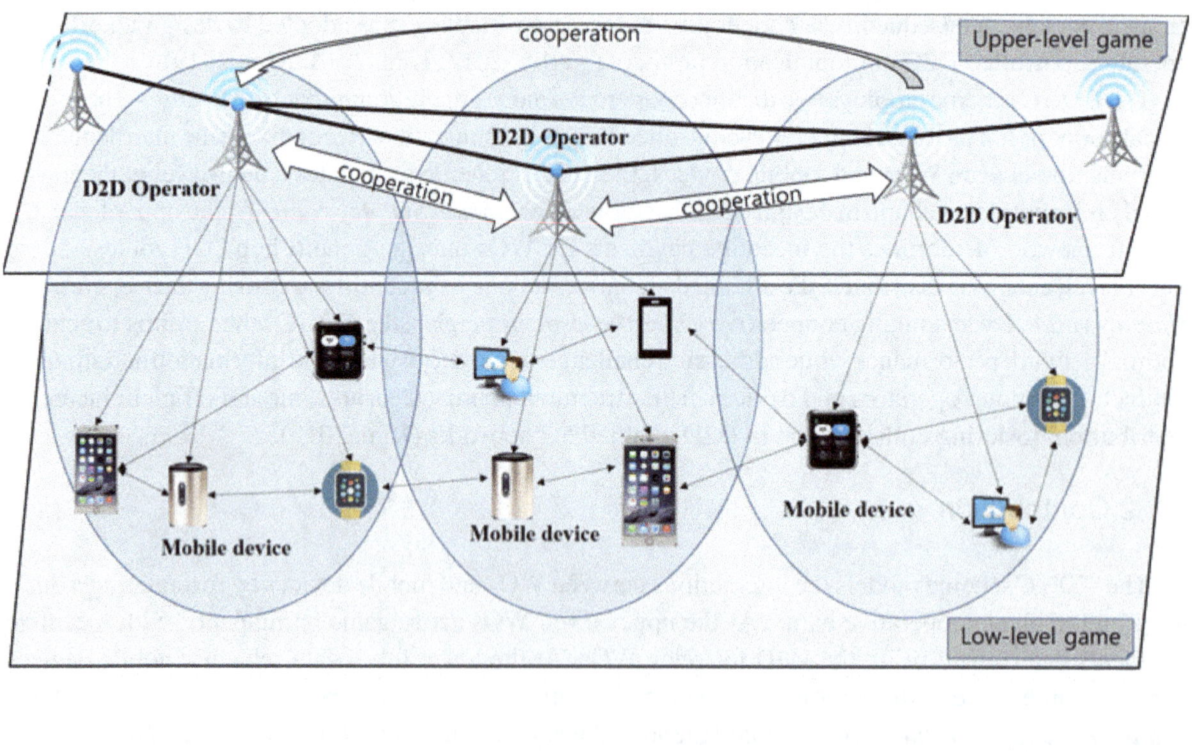

During D2D system operations, system agents, i.e., operators and mobile devices, make decisions individually. In this situation, the main issue for each agent is how to perform well by considering the mutual interaction relationship. To formulate this relationship, the *CDGC* scheme designs a new dual-level game model. At the upper level, operators $\{O_1, \ldots, O_n\}$ are game players and they establish routing paths while getting the relay service payment based on the IMD. At the lower level, mobile devices $\{M$

${}^{O_1}, \ldots, M^{O_n}\}$ are game players, and they share the service payment given by their corresponding operator according to the *r-egalitarian Shapley value*. For the implementation practicality, the lower-level games are carried out in an entirely distributed and parallel fashion. During the dual-level interaction, operators and mobile devices work together toward appropriate system performance. Let (M_k^O, M_l^O) denote the wireless link between two neighboring devices M_k^O and M_l^O, and $\mathcal{G}^{(M_k^O)}$ is denoted as the set of M_k^O's neighboring devices where $M_l^O \in \mathcal{G}^{(M_k^O)}$. $\mathcal{E}(M_k^O, M_l^O)$ is the communication cost degree of link (M_k^O, M_l^O) for D2D communications (Kim, 2019; Lu, 2015).

$$\mathcal{E}(M_k^O, M_l^O) = \beta_{M_k^O} \times \left(d(M_k^O, M_l^O)\right)^{\chi}, \text{ s.t., } \beta_{M_k^O} = \left(|\mathcal{G}^{(M_k^O)}|\right)^{\xi} \tag{7.2.1}$$

where $\beta_{M_k^O}$ is the interference parameter of M_k^O, and $|\mathcal{G}^{(M_k^O)}|$ is the cardinality of $\mathcal{G}^{(M_k^O)}$. χ and ξ are the path loss exponent and interference factors, respectively. $d_{(M_k^O, M_l^O)}$ is the distance between M_k^O and M_l^O. Let T is the multi-hop communication flow, and the source and destination device of flow T is represented as s(T) and d(T), respectively. T consists of multiple links, and mobile devices for these links can be covered by different operators (Lu, 2015). Mobile devices estimate the $\mathcal{E}(\cdot)$ values of all available connection links and report this information to their corresponding operators. Therefore, each operator can recognize the device topology of its covering area; operators interact with each other to configure the large area, which is partially covered by other operators (Kim, 2019).

If an operator includes a s(T) (*or* d(T)) device, it is called a source (*or* destination) operator. From the source operator to the destination operator, there can be some relay operators. Each relay operator reveals the total sum of $\mathcal{E}(\cdot)$ values about the relay links in its corresponding area; it can be interpreted as a relay cost of that operator. In the *CDGC* scheme, the source operator is responsible for gathering all necessary information and establishing a route through relay operators to reach the destination operator. Typically, multi-hop D2D communication algorithms operate under the assumption that devices under relay operators are willing to act as relay nodes along the multi-hop path. However, since relay devices must expend their own resources to forward data packets, incentivizing their participation is crucial to achieving a socially optimal outcome. To address this, the *CDGC* scheme employs an incentive payment mechanism during the upper-level game process to encourage cooperation among self-interested relay devices. Using the IMD framework, the source device compensates relay operators, who, in turn, redistribute the allocated incentives to their respective relay devices. In the lower-level game process, the problem of incentive distribution among relay devices is resolved using the *r-egalitarian Shapley value* approach (Kim, 2019).

7.2.4 Incentive mechanism design based upper-level game model

The primary challenge in the upper-level game process is determining the appropriate incentive payment for relay operators. In designing an incentive payment algorithm for D2D communications, a critical consideration is how much compensation each relay operator should receive for participating in relay services. The *CDGC* scheme addresses this issue by adopting the fundamental principles of IMD to compute incentive payments for relay operators. The IMD, also known as reverse game theory, is a field within economics and game theory that takes an engineering-oriented approach to achieving specific objectives, assuming that all players act rationally. The key feature of IMD is that a game designer, who is focused on achieving a socially optimal outcome, constructs the game's structure to guide players toward the desired equilibrium. IMD has been widely applied across various fields, including economics, politics, and network system management, particularly in private-information games. However, despite its successes, IMD is primarily effective in static environments and does not readily translate into an optimal mechanism for dynamic scenarios. Additionally, classical IMD models often overlook computational complexity in real-world applications (Dash, 2005; Kakade, 2011; Kim, 2019).

From the perspective of strategic players, a key objective in dynamic environments is to maximize long-term social welfare for all participants. While IMD-based mechanisms for dynamic settings have been explored, achieving an optimal incentive mechanism remains a complex challenge (Kakade, 2011). One particularly relevant extension of traditional IMD is the *Vickrey-Clarke-Groves* (VCG) mechanism, a generic, strategy-proof incentive model designed to achieve socially optimal outcomes in dynamic environments. The VCG mechanism ensures truthful reporting of players' preferences as a dominant strategy, making it resistant to strategic manipulation. This strategy-proof property serves as a normative guideline for optimal decision-making and provides better computational efficiency compared to classical IMD-based approaches (Dash, 2005; Kim, 2019).

In the upper-level game model, the VCG mechanism is used to define a strategic situation to make the D2D system exhibit better performance when independent operators pursue self-interested strategies. Let M be the payment mechanism for the upper-level game, and A is denoted as a set of possible outcomes based on inputs from relay operators. $\mathbb{R}^{\mathcal{T}}$ represents the set of relay operators for the T traffic relay service. Each relay operator $O_i \in \mathbb{R}^{\mathcal{T}}$ has his valuation function $\mathcal{V}_{O_i}(A,\mathcal{T})$, which quantifies the O_i's value to a specific outcome $A \in \mathbb{A}$. Usually, the $\mathcal{V}_{O_i}(A,\mathcal{T})$ maps A to a positive real number. In the *CDGC* scheme, this number represents the O_i's real contribution for the T relay service, and A is a set of relay operators to establish a routing path. Motivated by the basic idea of Dijkstra routing algorithm, A is given to establish the routing path while minimizing the total sum of $\mathcal{V}_{O_i}(A,\mathcal{T})$ values where $O_i \in A$. For the T relay service, the O_i's $\mathcal{V}_{O_i}(\bullet)$ function is defined as follows (Kim, 2019);

$$\mathcal{V}_{O_i}(A,\mathcal{T}) = \sum_{M_k^{O_i} \in \mathcal{R}_{\mathcal{T}}^{O_i}} \left(\left(\mathcal{E}\left(M_k^{O_i}, \mathcal{N}\left(M_k^{O_i}\right)\right)\right)^{\alpha} \times \mathbb{I}_{O_i} \right), \text{ s.t.}, \begin{cases} \mathbb{I}_{O_i} = 1, \text{If } O_i \in A \\ \mathbb{I}_{O_i} = 0, otherwise \end{cases} \tag{7.2.2}$$

where $\mathcal{R}_{\mathcal{T}}^{O_i}$ is the set of relay mobile devices under the O_i for the T service, and $\mathcal{N}\left(M_k^{O_i}\right)$ is the next relay device of $M_k^{O_i}$. $\left(\mathcal{E}\left(M_k^{O_i}, \mathcal{N}\left(M_k^{O_i}\right)\right)\right)^{\alpha}$ represents the $M_k^{O_i}$'s energy consumption to connect the $\mathcal{N}\left(M_k^{O_i}\right)$. With the control parameter α, the $\mathcal{E}(\cdot)$ value polynomially increases related to the communication cost degree between $M_k^{O_i}$ and $\mathcal{N}\left(M_k^{O_i}\right)$ (Kim, 2019).

From the M, each relay operator is asked to report his valuation function $\mathcal{V}(A,\mathcal{T})$ and M decides an outcome $A \in \mathbb{A}$ by selecting relay operators to set a multi-hop D2D path route. At this moment, each relay operator can submit his function, which may or may not equal $\mathcal{V}(A,\mathcal{T})$. Let $\widehat{\mathcal{V}}(A,\mathcal{T})$ represent that an operator misreports its true valuation function $\mathcal{V}(A,\mathcal{T})$, i.e., $\widehat{\mathcal{V}}(A,\mathcal{T}) \neq \mathcal{V}(A,\mathcal{T})$. If M is the VCG mechanism, the outcome $\left(A_{\widehat{\mathcal{V}},\mathcal{T}}^*\right)$ selected by M is (Kim, 2019);

$$A_{\widehat{\mathcal{V}},\mathcal{T}}^* := \arg\min_A \left(\sum_{O_k \in A} \widehat{\mathcal{V}}_{O_k}(A,\mathcal{T}) \middle| A \Rightarrow s(\mathcal{T}) | d(\mathcal{T}) \right), \text{ s.t.}, \widehat{\mathcal{V}} = \left[\widehat{\mathcal{V}}_{O_k \in A}(A,\mathcal{T}) \right] \tag{7.2.3}$$

where $A \Rightarrow s(\mathscr{T})|d(\mathscr{T})$ means that A consists of relay operators to relay the T service from $s(\mathscr{T})$ to $d(\mathscr{T})$. The incentive payment of O_i, i.e., $I_{O_i}(A^*_{\widehat{\mathscr{V}}}, \mathscr{T})$, is defined as the profit that his presence causes others with respect to the reported $\widehat{\mathscr{V}}_{O_k \in R^{\mathscr{T}}, O_k \neq O_i}(\mathscr{T})$ (Karlin, 2017; Kim, 2019); formally,

$$I_{O_i}(A^*_{\widehat{\mathscr{V}}}, \mathscr{T}) = \min_A \left(\sum_{O_j \neq O_i} \widehat{\mathscr{V}}_{O_j}(A) \right) - \sum_{O_j \neq O_i} \widehat{\mathscr{V}}_{O_j}(A^*_{\widehat{\mathscr{V}}}, \mathscr{T}) \tag{7.2.4}$$

In the equation (7.2.4), the first term is the total reported value the other operators would obtain when O_i was absent, and the second term is the total reported value the others obtain when O_i is present. If O_i's dominant strategy is to report his valuation truthfully, i.e., $\mathscr{V}_{O_i}(A, \mathscr{T})$, we say that M is truthful (Karlin, 2017). Formally, it can be expressed as follows (Kim, 2019);

$$U_{O_i}\left(\mathscr{V}_{O_i}(A, \mathscr{T}), \widehat{\mathscr{V}}_{-O_i}(A, \mathscr{T})\right) \geq U_{O_i}\left(\widehat{\mathscr{V}}_{O_i}(A, \mathscr{T}), \widehat{\mathscr{V}}_{-O_i}(A, \mathscr{T})\right) \tag{7.2.5}$$

$$\text{s.t.,} \begin{cases} U_{O_i}\left(\mathscr{V}_{O_i}(A, \mathscr{T}), \widehat{\mathscr{V}}_{-O_i}(A, \mathscr{T})\right) = I_{O_i}\left(\mathscr{V}_{O_i}(A, \mathscr{T}), \widehat{\mathscr{V}}_{-O_i}(A, \mathscr{T})\right) - C_{O_i}(A, \mathscr{T}) \\ U_{O_i}\left(\widehat{\mathscr{V}}_{O_i}(A, \mathscr{T}), \widehat{\mathscr{V}}_{-O_i}(A, \mathscr{T})\right) = I_{O_i}\left(\widehat{\mathscr{V}}_{O_i}(A, \mathscr{T}), \widehat{\mathscr{V}}_{-O_i}(A, \mathscr{T})\right) - C_{O_i}(A, \mathscr{T}) \\ \widehat{\mathscr{V}}_{-O_i}(A, \mathscr{T}) = \left[\widehat{\mathscr{V}}_{O_k \in R^{\mathscr{T}}, O_k \neq O_i}(A, \mathscr{T}),\right] \end{cases}$$

where $U_{O_i}(\cdot)$ is the O_i's utility function and $I_{O_i}\left(\mathscr{V}_{O_i}(A, \mathscr{T}), \widehat{\mathscr{V}}_{-O_i}(A, \mathscr{T})\right)$ is decided based on input $\mathscr{V}_{O_i}(A, \mathscr{T})$ and $\widehat{\mathscr{V}}_{-O_i}(A, \mathscr{T})$. $\widehat{\mathscr{V}}_{-O_i}(\cdot)$ is a vector of V except the O_i, and $C_{O_i}(A, \mathscr{T})$ is the real cost function of O_i. In the *CDGC* scheme, $C_{O_i}(A, \mathscr{T})$ is defined as the same manner as the communication cost degree. If the incentive payment for each relay operator is given according to (7.2.4), M is a truthful mechanism (Karlin, 2017; Kim, 2019).

Theorem 1. M *is a truthful mechanism for all relay operators* (Kim, 2019).

Proof. We can fix the reports $\widehat{\mathscr{V}}_{-O_i}$ of all relay operators except the O_i. Suppose that the O_i's true valuation function is $\mathscr{V}_{O_i}(\cdot)$, and can report his valuation function truthfully, i.e., $\mathscr{V}_{O_i}(\cdot)$, or untruthfully, i.e., $\widehat{\mathscr{V}}_{O_j}(\cdot)$.

Case I) If the O_i report his false valuation function $\widehat{\mathscr{V}}_{O_j}(\cdot)$, then the outcome of M is given by (7.2.3) and the O_i's incentive payment is given by (7.2.4). Based on this reason, the O_i's utility function can be defined as follows;

$$U_{O_i}\left(\widehat{\mathscr{V}}_{O_i}(A^*_{\widehat{\mathscr{V}}}, \mathscr{T}), \widehat{\mathscr{V}}_{-O_i}(A^*_{\widehat{\mathscr{V}}}, \mathscr{T})\right)$$

$$= I_{O_i}\left(\widehat{\mathscr{V}}_{O_i}(A^*_{\widehat{\mathscr{V}}}, \mathscr{T}), \widehat{\mathscr{V}}_{-O_i}(A^*_{\widehat{\mathscr{V}}}, \mathscr{T})\right) - C_{O_i}(A^*_{\widehat{\mathscr{V}}}, \mathscr{T}) = I_{O_i}(A^*_{\widehat{\mathscr{V}}}, \mathscr{T}) - C_{O_i}(A^*_{\widehat{\mathscr{V}}}, \mathscr{T})$$

$$= \arg\min_A \left(\sum_{O_j \neq O_i} \widehat{\mathscr{V}}_{O_j}(A) \middle| A \Rightarrow s(\mathscr{T})|d(\mathscr{T}) \right) - \sum_{O_j \neq O_i} \widehat{\mathscr{V}}_{O_j}(A^*_{\widehat{\mathscr{V}}}, \mathscr{T}) - C_{O_i}(A^*_{\widehat{\mathscr{V}}}, \mathscr{T})$$

(7.2.6)

Case II) If the O_i report truthfully his valuation function, i.e., $\mathscr{V}_i(\cdot)$, then the outcome of M is given by;

$$A'_{((\mathscr{V}_{O_i},\hat{\mathscr{V}}_{-O_i}),\mathscr{T})} = \arg\min_{A}\left(\left(\mathscr{V}_{O_i}(A,\mathscr{T}) + \sum_{O_j\in A, O_j\neq O_i}\hat{\mathscr{V}}_{O_j}(A)\right)\middle| A \Rightarrow s(\mathscr{T})|d(\mathscr{T})\right)$$

(7.2.7)

The O_i's incentive payment is

$$I_{O_i}\left(A'_{(\mathscr{V}_{O_i},\hat{\mathscr{V}}_{-O_i})},\mathscr{T}\right) =$$
$$\arg\min_{A}\left(\sum_{O_j\neq O_i}\hat{\mathscr{V}}_{O_j}\left(A'_{(\mathscr{V}_{O_i},\hat{\mathscr{V}}_{-O_i})},\mathscr{T}\right)\right) - \sum_{O_j\neq O_i}\hat{\mathscr{V}}_{O_j}\left(A'_{(\mathscr{V}_{O_i},\hat{\mathscr{V}}_{-O_i})},\mathscr{T}\right)$$

(7.2.8)

The O_i's utility function can be defined as follows;

$$U_{O_i}\left(\mathscr{V}_{O_i}\left(A'_{(\mathscr{V}_{O_i},\hat{\mathscr{V}}_{-O_i})},\mathscr{T}\right),\hat{\mathscr{V}}_{-O_i}\left(A'_{(\mathscr{V}_{O_i},\hat{\mathscr{V}}_{-O_i})},\mathscr{T}\right)\right)$$
$$= I_{O_i}\left(\mathscr{V}_{O_i}\left(A'_{(\mathscr{V}_{O_i},\hat{\mathscr{V}}_{-O_i})},\mathscr{T}\right),\hat{\mathscr{V}}_{-O_i}\left(A'_{(\mathscr{V}_{O_i},\hat{\mathscr{V}}_{-O_i})},\mathscr{T}\right)\right) - C_{O_i}\left(A'_{(\mathscr{V}_{O_i},\hat{\mathscr{V}}_{-O_i})},\mathscr{T}\right)$$
$$= I_{O_i}\left(A'_{(\mathscr{V}_{O_i},\hat{\mathscr{V}}_{-O_i})},\mathscr{T}\right) - C_{O_i}\left(A'_{(\mathscr{V}_{O_i},\hat{\mathscr{V}}_{-O_i})},\mathscr{T}\right)$$
$$= \min_{A}\left(\sum_{O_j\neq O_i}\hat{\mathscr{V}}_{O_j}\left(A'_{(\mathscr{V}_{O_i},\hat{\mathscr{V}}_{-O_i})},\mathscr{T}\right)\right) - \sum_{O_j\neq O_i}\hat{\mathscr{V}}_{O_j}\left(A'_{(\mathscr{V}_{O_i},\hat{\mathscr{V}}_{-O_i})},\mathscr{T}\right) - C_{O_i}\left(A'_{(\mathscr{V}_{O_i},\hat{\mathscr{V}}_{-O_i})},\mathscr{T}\right)$$

(7.2.9)

Finally,

$$U_{O_i}\left(\mathscr{V}_{O_i}\left(A'_{(\mathscr{V}_{O_i},\hat{\mathscr{V}}_{-O_i})},\mathscr{T}\right),\hat{\mathscr{V}}_{-O_i}\left(A'_{(\mathscr{V}_{O_i},\hat{\mathscr{V}}_{-O_i})},\mathscr{T}\right)\right) - U_{O_i}\left(\hat{\mathscr{V}}_{O_i}(A^*_{\hat{\mathscr{V}}},\mathscr{T}),\hat{\mathscr{V}}_{-O_i}(A^*_{\hat{\mathscr{V}}},\mathscr{T})\right)$$
$$= \left(I_{O_i}\left(A'_{(\mathscr{V}_{O_i},\hat{\mathscr{V}}_{-O_i})},\mathscr{T}\right) - \mathscr{V}_{O_i}\left(A'_{(\mathscr{V}_{O_i},\hat{\mathscr{V}}_{-O_i})},\mathscr{T}\right)\right) - \left(I_{O_i}(A^*_{\hat{\mathscr{V}}},\mathscr{T}) - \mathscr{V}_{O_i}(A^*_{\hat{\mathscr{V}}},\mathscr{T})\right)$$
$$= \min_{A}\left(\sum\nolimits_{O_j\neq O_i}\hat{\mathscr{V}}_{O_j}\left(A'_{(\mathscr{V}_{O_i},\hat{\mathscr{V}}_{-O_i})},\mathscr{T}\right)\right) - \sum\nolimits_{O_j\neq O_i}\mathscr{V}_{O_i}\left(A'_{(\mathscr{V}_{O_i},\hat{\mathscr{V}}_{-O_i})},\mathscr{T}\right) - C_{O_i}\left(A'_{(\mathscr{V}_{O_i},\hat{\mathscr{V}}_{-O_i})},\mathscr{T}\right)$$
$$-\min_{A}\left(\sum_{O_j\neq O_i}\hat{\mathscr{V}}_{O_j}(A)\right) + \sum_{O_j\neq O_i}\hat{\mathscr{V}}_{O_j}(A^*_{\hat{\mathscr{V}}},\mathscr{T}) + C_{O_i}(A^*_{\hat{\mathscr{V}}},\mathscr{T})$$

(7.2.10)

The O_i's report has no influence on $\min_A\left(\sum_{O_j \neq O_i} \widehat{\mathscr{V}}_{O_j}\left(A'_{(\mathscr{V}_{O_i}, \widehat{\mathscr{V}}_{-O_i})}, \mathscr{T}\right)\right)$ and $\min_A\left(\sum_{O_j \neq O_i} \widehat{\mathscr{V}}_{O_j}(A)\right)$. Therefore, they are the same constant from the viewpoint of player O_i. Therefore, the final equation (7.2.10) can be simplified as follows;

$$\left(C_{O_i}(A^*_{\widehat{\mathscr{V}}}, \mathscr{T}) + \sum_{O_j \neq O_i} \widehat{\mathscr{V}}_{O_j}(A^*_{\widehat{\mathscr{V}}}, \mathscr{T}) - C_{O_i}\left(A'_{(\mathscr{V}_{O_i}, \widehat{\mathscr{V}}_{-O_i})}, \mathscr{T}\right) - \sum_{O_j \neq O_i} \mathscr{V}_{O_i}\left(A'_{(\mathscr{V}_{O_i}, \widehat{\mathscr{V}}_{-O_i})}, \mathscr{T}\right)\right) \geq 0$$

$$\cong U_{O_i}\left(\mathscr{V}_{O_i}\left(A'_{(\mathscr{V}_{O_i}, \widehat{\mathscr{V}}_{-O_i})}, \mathscr{T}\right), \widehat{\mathscr{V}}_{-O_i}\left(A'_{(\mathscr{V}_{O_i}, \widehat{\mathscr{V}}_{-O_i})}, \mathscr{T}\right)\right) \geq U_{O_i}\left(\widehat{\mathscr{V}}_{O_i}(A^*_{\widehat{\mathscr{V}}}, \mathscr{T}), \widehat{\mathscr{V}}_{-O_i}(A^*_{\widehat{\mathscr{V}}}, \mathscr{T})\right)$$

(7.2.11)

In conclusion, the value of $U_{O_i}\left(\mathscr{V}_{O_i}\left(A'_{(\mathscr{V}_{O_i}, \widehat{\mathscr{V}}_{-O_i})}, \mathscr{T}\right), \widehat{\mathscr{V}}_{-O_i}\left(A'_{(\mathscr{V}_{O_i}, \widehat{\mathscr{V}}_{-O_i})}, \mathscr{T}\right)\right)$ is always higher than the value of $U_{O_i}\left(\widehat{\mathscr{V}}_{O_i}(A^*_{\widehat{\mathscr{V}}}, \mathscr{T}), \widehat{\mathscr{V}}_{-O_i}(A^*_{\widehat{\mathscr{V}}}, \mathscr{T})\right)$ for every $\widehat{\mathscr{V}}$, $\mathscr{V}_{O_i}$ and O_i. •

7.2.5 Lower-level game model based on the *r-egalitarian Shapley value*

In the upper-level game, the source device compensates relay operators for their participation in multi-hop D2D communication. In the lower-level game, each relay operator then redistributes the received incentive among its associated relay devices to offset their resource expenditure. Fundamentally, mobile devices under a relay operator are likely to form a coalition through a binding agreement if all participating relay devices benefit from improved individual payoffs. However, when certain relay devices contribute more significantly to the coalition than others, the incentive distribution must be both fair and optimal to ensure continued cooperation. Thus, the key challenge in the lower-level game is to sustain collaborative behavior among relay devices while fairly and efficiently distributing the allocated incentives. To address this issue, the *CDGC* scheme incorporates a novel cooperative game-theoretic solution designed to achieve an equitable and efficient incentive-sharing mechanism (Kim, 2019).

In 1953, L. Shapley characterized a solution concept that associates with canonical coalition games. This solution is known as the Shapley Value. In 2018, K. Yokote et al. modified the concept of Shapley Value and introduced the *r-egalitarian Shapley value* (ESh^r). To explain the ESh^r, we introduce some notations. Let (N,V) be a game with transferable utility, where $\mathbb{N} = \{a_1, a_2, \ldots, a_n\}$ is the set of players and V is the characteristic function, which assigns a real number V(C) to every coalition $\mathscr{C} \in \mathscr{N}$ where N is the set of non-empty subsets of N. A transferable utility game is a pair $(\mathscr{C}, \mathscr{V})$ consisting of a set of players C, and a coalition function $\mathscr{V} \in \mathbb{V}(\mathscr{C}) := \{f: 2^{\mathscr{C}} \rightarrow \mathbb{R}: f(\varnothing) = 0\}$. With the $(\mathscr{C}, \mathscr{V})$ and $\mathscr{S} \subseteq \mathscr{C}$, $\mathscr{S} \neq \varnothing$, let $(\mathscr{S}, \mathscr{V})$ denote the game in which the domain of V is restricted from $2^{\mathscr{C}}$ to $2^{\mathscr{S}}$, and the ESh^r of a_i is $ESh^r_{a_i}$ (Kim, 2019; Yokote, 2018).

Given the grand coalition form, the ESh^r investigates the problem of how to distribute the total payoff among players fairly. The solution of ESh^r assigns a payoff vector $ESh^r_{a_i \in \mathscr{C}}(\mathscr{C}, \mathscr{V}) \in \mathbb{R}$ to each game ($\mathscr{C}, \mathscr{V}$). To define the ESh^r, the idea of rescaling the worth of coalitions is necessary (Yokote, 2018). Let L denote the set of finite sequences of real numbers (Kim, 2019),

$$\mathbb{L} = \left\{(L_k)_{k=1}^{|\mathbb{N}|} \mid L_k \in \mathbb{R} \text{ for all } k \in \{1, \ldots, |\mathbb{N}|\}\right\}$$

(7.2.12)

where $|\mathbb{N}|$ represents the cardinality of N. According to (7.2.12) and $r \in \mathbb{L}$, $\mathcal{V}^r(\mathcal{S})$ is defined as (Kim, 2019);

$$\mathcal{V}^r(\mathcal{S}) = r_s \times \mathcal{V}(\mathcal{S}) \text{ s.t., } r \in \mathbb{L}, \text{ for all} \mathcal{S} \subseteq \mathcal{C}, \mathcal{S} \neq \emptyset \text{ and } \mathcal{V}^r \in \mathbb{V}(\mathcal{C}) \tag{7.2.13}$$

In the game $\mathcal{V}^r$, the worth of each coalition is rescaled by multiplying the sth entry of the sequence vector r, where s is the size of coalition S. This type of rescaling is often discussed in the context of the per-capita measure or discounting; $\mathcal{V}^r$can be interpreted as generalizing these ideas by allowing for any sequence of real numbers (Yokote, 2018). Finally, the ESh^r is defined as follows (Kim, 2019);

$$ESh^r_{a_i \in \mathcal{C}}(\mathcal{C}, \mathcal{V}) = \frac{\mathcal{V}(\mathcal{C}) - \mathcal{V}^r(\mathcal{C})}{|\mathcal{C}|} + \sum_{\mathcal{S} \subset \mathcal{C}, a_i \in \mathcal{S}} \left(X^r_{\mathcal{S},\mathcal{C}} \times Y^r_{\mathcal{S},a_i} \right) \tag{7.2.14}$$

$$\text{s.t.,} \begin{cases} X^r_{\mathcal{S},\mathcal{C}} = \dfrac{(|\mathcal{S}| - 1)! \times (|\mathcal{C}| - |\mathcal{S}|)!}{|\mathcal{C}|!} \\ Y^r_{\mathcal{S},a_i} = \left[\mathcal{V}^r(\mathcal{S}) - \mathcal{V}^r(\mathcal{S} - \{a_i\}) \right] \end{cases}$$

where $X^r_{\mathcal{S},\mathcal{C}}$ can be interpreted as the probability of a coalition containing the a_i with the size of $|\mathcal{S}|$ and $Y^r_{\mathcal{S},a_i}$ is the payoff difference between the coalitions with and without the a_i, which measures the contribution of the a_i to the coalition. According to $X^r_{\mathcal{S},\mathcal{C}}$ and $Y^r_{\mathcal{S},a_i}$, the original Shapley value is obtained based on the $\mathcal{V}^r$ game (Cai, 2004; Kim, 2014). Therefore, the *CDGC* scheme can interpret the $ESh^r_{a_i}$ like as; i) the worth of coalitions are rescaled based on the sequence vector r, ii) an imaginary $\mathcal{V}^r$ game is constructed, iii) the Shapley value idea is applied to the $\mathcal{V}^r$ game, and iv) the gap between the $\mathcal{V}(\mathcal{C})$ and $\mathcal{V}^r(\mathcal{C})$ is equally divided among players (Kim, 2019; Yokote, 2018).

In the lower-level game process, the given incentive of each individual relay operator is shared by corresponding relay devices. Therefore, relay devices are game players and form each coalition. In the *CDGC* scheme, the characteristic function V for the coalition C ($\mathcal{V}(\mathcal{C})$) is defined based on the bankruptcy problem in (Niyato, 2006). It is analogous to a distribution or entitlement problem by involving the allocation of a given amount of a perfectly divisible good. Therefore, the *CDGC* scheme can effectively estimate $\mathcal{V}(\mathcal{C})$ values for all possible coalitions. Based on the bankruptcy problem and ESh^r, relay devices can share their incentive payment using the equation (7.2.14); it is the most fair-efficient solution (Kim, 2019).

7.2.6 Summary

As the demand for traffic services continues to grow, the need for increased network capacity has driven the advancement of D2D communication in 5G networks. In traditional D2D communication systems, devices are typically restricted from communicating via multi-hop connections. To address this limitation, the *CDGC* scheme introduces an operator-assisted multi-hop D2D communication framework, where WOs play a crucial role in coordinating mobile devices in a distributed manner while receiving incentive payments from the source device. To encourage self-interested mobile devices to actively participate in multi-hop D2D communications, the *CDGC* scheme integrates two cooperative game-theoretic solutions: IMD and ESh^r. These two methodologies interact dynamically within the dual-level game model, allowing the *CDGC* scheme to create a win-win scenario for multi-hop D2D communication services. As a result, operators and mobile devices engage in mutual cooperation, working together to achieve optimized system performance (Kim, 2019).

7.3 Heterogeneous Network Spectrum Allocation (*HNSA*) scheme

To accommodate the rapid surge in mobile traffic, next-generation wireless networks integrate existing technologies such as dual-connectivity (DC) and network-assisted D2D communication. Recently, S. Kim introduced the *Heterogeneous Network Spectrum Allocation* (*HNSA*) scheme, designed for heterogeneous systems that incorporate both of these technologies (Kim, 2020b). To facilitate effective collaboration among individual network agents, the *HNSA* scheme leverages cooperative game theory, implementing its spectrum allocation algorithm as an innovative joint-bargaining process. This approach is built upon bargaining solutions derived from classical Nash bargaining, weighted Nash bargaining, and Nash bargaining with coalition structures. By employing a three-step interactive method, the *HNSA* scheme fully exploits the advantages of these different bargaining models, ensuring an efficient and balanced spectrum allocation. In dynamically changing HetNet environments, the *HNSA* scheme offers multiple advantages by systematically addressing DC-based D2D communication challenges while reaching a mutually beneficial agreement among network entities. Its key innovation lies in its ability to balance conflicting requirements, providing an optimized tradeoff during heterogeneous network operations (Kim, 2020b).

7.3.1 Development motivation

Over the past decade, the telecommunications industry has experienced an exponential increase in the number of connected mobile devices (MDs). This surge, driven by the widespread adoption of MDs and the growing demand for spectrum-intensive multimedia services such as mobile video streaming, gaming, and social networking, has led to steady growth in mobile data traffic. Reports from Cisco and Ericsson project that between 2016 and 2022, the compound annual growth rate (CAGR) of global mobile video traffic will rise by 11% annually, with total data traffic expanding nearly sixfold during this period, reaching 8.3 billion subscriptions worldwide by 2022. This trend is expected to accelerate further with the introduction of 6G networks, placing an even greater burden on traditional cellular infrastructure, which may struggle to support the required high data rates for next-generation applications (Kim, 2020b; Park, 2019; Shah, 2020).

To address these growing demands, HetNet architecture has emerged as a promising next-generation wireless network paradigm, gaining traction among both academic researchers and industry leaders. This approach transforms conventional single-tier cellular networks into multi-tier HetNets, where small base stations (SBSs) are deployed within the coverage area of macro base stations (MBSs). In HetNets, MBSs and SBSs are interconnected via backhaul links, forming an integrated network structure that enhances connectivity. The dense deployment of SBSs brings spectrum access points closer to MDs, improving traffic offloading from MBSs to SBSs and enhancing network efficiency. Leading global network operators, including AT&T, T-Mobile, Orange, and China Mobile, have actively deployed SBSs in high-density areas such as malls, markets, and cafés to boost local network capacity and alleviate congestion at overburdened MBSs (Jia, 2019; Kim, 2020b; Park, 2019).

Within the HetNet framework, D2D communication is considered a key enabling technology. Initially introduced to facilitate local peer-to-peer transmissions, D2D communication allows cellular MDs to exchange data without routing through traditional base stations. This technology offers several advantages, including; i) extended coverage, ii) reduced burden on cellular networks, iii) increased throughput, and iv) improved spectrum efficiency. In 5G networks, D2D communication has witnessed a surge in practical applications, ranging from peer-to-peer messaging and local multicasting to multi-player gaming, data flooding, and video dissemination. However, conventional D2D communication models have limitations. One of the key challenges is that establishing reliable direct links between MDs can be difficult, as potential D2D devices are often located far apart, making it challenging to meet QoS requirements (Hasan, 2014; Kim, 2020b).

To overcome the limitations of direct D2D links, network-assisted D2D communication has been introduced as a viable solution for real-world implementations. In this model, D2D-enabled devices do not need to be within the same cell area; instead, base stations (BSs) act as relay nodes to facilitate communication. The cellular spectrum resources at relay BSs are managed by network infrastructure, ensuring efficient spectrum sharing for D2D communication links. To fully leverage network-assisted D2D communication, the network system must i) manage the D2D pairing process, ii) coordinate time and spectrum allocation, and iii) guarantee QoS requirements. These improvements enhance the overall system throughput and spectrum efficiency. However, realizing these potential gains requires the development of advanced spectrum allocation algorithms that can effectively manage D2D device pairs within cellular environments. To fully exploit network-assisted D2D communication, critical control challenges must be addressed when designing efficient spectrum management algorithms (Fodor, 2012; Hasan, 2014; Kim, 2020b).

Among the key challenges in HetNet spectrum allocation, MD association is a crucial problem, as it directly influences both per-MD throughput and overall system capacity. Recently, dual connectivity (DC) has been introduced as a HetNet enhancement, allowing MDs to simultaneously utilize spectrum resources from both MBSs and SBSs. The DC technique introduces a new capability that enables MDs to maintain dual connections, thereby addressing increasing spectrum demand. By aggregating spectrum resources across multiple base stations, DC-based spectrum allocation improves both connectivity reliability and the user experience. This flexible association mechanism, enabled by interoperable BSs, represents a critical advancement in maximizing HetNet efficiency. Designing an optimized HetNet spectrum allocation algorithm requires an intelligent control paradigm that considers all critical spectrum management issues (Kim, 2020b; Lema, 2016).

Game theory has long been recognized as a powerful tool for analyzing decision-making scenarios in wireless networks. In particular, bargaining game theory has been widely used to achieve fair and efficient resource allocation. Due to its ability to balance efficiency and fairness, bargaining solutions have gained considerable attention in wireless resource management. The *Nash Bargaining Solution* (*NBS*), introduced by J. Nash in 1950, serves as a fundamental approach to fair and optimal resource sharing (Bergantiños, 2005; Chae, 2004). Building upon this foundation, the *Weighted Nash Bargaining Solution* (*WNBS*) was later developed, allowing for unequal bargaining power among heterogeneous players, based on their individual characteristics (Kim, 2014; Wang, 2014). In 2004, S. Chae and P. Heidhues introduced a more advanced framework: *the NBS with Coalition Structure* (*NBS-CS*). Unlike traditional *NBS*, the *NBS-CS* approach enables the simultaneous formation of multiple coalitions, where players cooperate within groups while negotiating resource allocation. This cooperative approach is particularly well-suited for HetNet spectrum allocation, as it allows network entities to form dynamic partnerships while optimizing resource utilization (Kim, 2020b).

7.3.2 Contribution

Building upon the principles of *NBS*, *WNBS*, and *NBS-CS*, the *HNSA* scheme introduces a novel spectrum allocation mechanism tailored for DC-based D2D communications. Within the HetNet framework, MBSs distribute spectrum resources to their respective SBSs based on *NBS* principles, ensuring a fair and optimal allocation. Each SBS, in turn, further allocates its assigned spectrum among different application services using the *WNBS* approach, which accounts for heterogeneous service demands and their varying bargaining power. Additionally, to enhance network performance, MBSs supplement running applications by providing additional spectrum resources as needed. Furthermore, to efficiently support network-assisted D2D communications, the *NBS-CS* model is integrated to optimize the DC technique, facilitating cooperative spectrum sharing. By synergizing these three bargaining models (*NBS*, *WNBS*, and *NBS-CS*), the *HNSA* scheme effectively addresses DC-based D2D communication challenges, ensuring a comprehensive and balanced spectrum allocation strategy within HetNet environments (Kim, 2020b).

7.3.3 DC based HetNet system infrastructure

The *HNSA* scheme considers a two-tier heterogeneous network platform, which comprises two types of BSs with orthogonal spectrum resource pools. At the upper tier, MBSs $\mathbb{M} = \{\mathcal{M}_1,\ldots,\mathcal{M}_n\}$ exist, and they can provide cellular network coverage for a large area. Within the coverage of each M, multiple SBSs break up a macro-cell site into much smaller areas. Therefore, SBSs $\mathbb{S} = \{\mathcal{S}_1,\ldots,\mathcal{S}_m\}$ coexist at the lower tier, and the cross-tier DC transmission is provided. One MBS (*or* one SBS) has a coverage area of radius $r_{\mathcal{M}}$ (*or* $r_{\mathcal{S}}$), and has a static portion of spectrum resource. It is more practical and suitable for the scenario where the different serving levels are ensured over different association spectrum bands. The *HNSA* scheme assumes that each MBS has its own spectrum resource $(\mathfrak{R}_{\mathcal{M}})$ and the spectrum resource $(\mathfrak{M}_{\mathcal{M}})$ for SBSs; $\mathfrak{M}_{\mathcal{M}}$ is adaptively distributed into corresponding SBSs. Afterward, individual SBSs also divide orthogonally their assigned spectrum pool into a set of different channels to support diverse applications (Kim, 2020b; Long, 2019).

In the HetNet cellular area, there are multiple MDs $\mathbb{D} = \{\mathcal{D}_1,\ldots,\mathcal{D}_k\}$; they are assumed randomly distributed, and equipped with two different radio interfaces to communicate simultaneously with the MBS and SBS. By dynamically monitoring both MBSs and SBSs, DC-enabled traffic scheduling is expected to further enhance data capacity and improve overall network efficiency. In addition to this, certain MDs are paired and utilize the D2D communication mode to transmit data directly to their corresponding MDs. In this mode, paired MDs periodically exchange status information, ensuring seamless connectivity. In emerging 5G networks, MBSs function as software-defined network (SDN) controllers, overseeing both cellular and D2D connections. Since BSs typically have comprehensive network awareness and high computational power, the *HNSA* scheme assumes that BSs are responsible for task scheduling of MDs in each time frame while simultaneously managing spectrum resources efficiently. As a result, during DC-based HetNet operations, the complex interactions between D2D pairs are modeled in coordination with their respective BSs, enabling a structured and optimized communication framework (Kim, 2020b; Pu, 2016).

In the *HNSA* scheme, a modeling situation for the HetNet spectrum allocation process is formulated as a joint bargaining game ($\mathbb{G}$) in a cooperative manner. The HetNet system operates in a slotted structure and each time frame has a time length t. Formally, the *HNSA* scheme defines game entities, i.e., $\mathbb{G} = \left\{\{\mathbb{M},\mathbb{S},\mathbb{D}\},\{\mathfrak{R}_{\mathscr{M}_i\in\mathbb{M}}\mathfrak{M}_{\mathscr{M}_i\in\mathbb{M}}\},\mathfrak{X}_{D2D}, \{U_{\mathscr{M}_i},U_{\mathscr{S}_j}^{\mathscr{M}_i},U_{\mathscr{D}_e}^{\mathscr{S}_j}\},T\right\}$ at each time period of gameplay (Kim, 2020b).

- $\{\mathbb{M},\mathbb{S},\mathbb{D}\}$ are the finite set of MBSs, SBSs and MDs; they are game entities for the joint bargaining games.
- $\mathfrak{R}_{\mathscr{M}_i}$ and $\mathfrak{M}_{\mathscr{M}_i}$ are the spectrum amount assigned for the $\mathscr{M}_i \in \mathbb{M}$; $\mathfrak{R}_{\mathscr{M}_i}$ is used by the $\mathscr{M}_i$ for the DC services, and $\mathfrak{M}_{\mathscr{M}_i}$ is distributed for the $\mathscr{S}_j \in \mathscr{W}_{\mathscr{M}_i} \subset \mathbb{S}$ where $\mathscr{W}_{\mathscr{M}_i}$ is the set of SBSs within the $\mathscr{M}_i$'s area.
- $\mathfrak{X}_{D2D}$ is the set of MDs, which are involved in the network-assisted D2D communications. Therefore, each element is paired with other element in the $\mathfrak{X}_{D2D}$.
- $U_{\mathscr{M}_i},U_{\mathscr{S}_j}^{\mathscr{M}_i},U_{\mathscr{D}_e}^{\mathscr{S}_j}$ are utility functions of $\mathscr{M}_i$, $\mathscr{S}_j$ and $\mathscr{D}_e$, respectively, where $\mathscr{S}_j \in \mathscr{W}_{\mathscr{M}_i}$ and $\mathscr{D}_e$ is in the $\mathscr{S}_j$'s covering area.
- $T = \{t_1,\ldots,t_c,t_{c+1},\ldots\}$ denotes time, which is represented by a sequence of time steps.

7.3.4 The basic concepts of cooperative bargaining solutions

In many HetNet environments, network agents can achieve greater benefits through cooperation. As a result, these agents often do not operate individually, but instead, form coalitions to optimize resource utilization. Given this collaborative structure, an important question arises: how should the collective benefits be distributed among different coalitions, as well as among individual agents within each coalition? To address this, various bargaining models, based on the generalization of the Nash solution, provide effective mechanisms for fair and efficient distribution benefit. In the standard cooperative game framework, a bargaining problem is defined as the process of dividing a surplus that players can collectively generate. Thus, bargaining solutions focus on determining equitable ways to distribute this shared surplus, ensuring desirable fairness and efficiency properties (Bergantiños, 2005; Chae, 2004; Kim, 2014, Kim, 2020b).

To define the basic idea of bargaining solution, we introduce the notation and preliminary definitions of bargaining solution. Let N be a finite set of players. We denote by $|N|$ the number of elements of N. We denote by $(x \cdot y)$ the vector $(x_i \cdot y_i)_{i\in N}$ and by $(x + y)$ the vector $(x_i + y_i)_{i\in N}$. Given $T \subsetneq N$, x_T is the restriction of x to $\mathbb{R}^T$. We denote by $\mathbb{R}^N_+$ the set $\{x \in \mathbb{R}^N{:}x_i \geq 0 \text{for every} i \in N\}$ and by $\mathbb{R}^N_{++}$ the set $\{x \in \mathbb{R}^N{:}x_i > 0 \text{ for every } i \in N\}$. Given $\gamma \in \mathbb{R}^N_{++}$, $(1/\gamma)$ is the vector $(1/\gamma_i)_{i\in N}$ (Bergantiños, 2005; Chae, 2004; Kim, 2020b).

A feasible set S is a subset of the payoff space, and a disagreement point d is an element of the payoff space. The points in S represent the feasible utility levels that the individuals can reach if they agree. Otherwise, if agreement is not reached, they obtain the utility levels given by the disagreement point. For every $S \subseteq \mathbb{R}^N$ and $\gamma,\beta \in \mathbb{R}^N$, we define $(\gamma \cdot S) + \beta = \{(\gamma \cdot x) + \beta | x \in S\}$. Given $\theta \in \mathbb{R}$ and $x \in \mathbb{R}^N$, we define $(\theta \cdot x)$ as the vector $(\theta \cdot x_i)_{i\in N}$. A coalition structure C over N is a partition of player set, i.e., $\mathscr{C} = \{C_1,\ldots,C_p\} \subsetneq 2^N$ where $\cup_{(C_q\in\mathscr{C})} C_q = N$ and $C_q \cap C_r = \varnothing$ whenever $q \neq r$. Each $C_q \in \mathscr{C}$ is called a coalition. We denote by $c \in \mathbb{R}^N$ the vector whose ith coordinate is given by $c_i = |C_q|$ if $i \in C_q$. Usually, bargaining game is a pair (N, υ) where õ is a characteristic function that assigns to each subset $T \subseteq N$ a number $\upsilon(T) \in \mathbb{R}$, with $\upsilon(\phi) = 0$. $\upsilon(T)$ represents the total utility payoff when players in T can cooperate. A bargaining problem over N is a pair (S, d) where $d \in S \subsetneq \mathbb{R}^N$. Let the boundary of S be ∂S, and there exists $x \in S$ such that $x > d$. The traditional Nash bargaining solution is the unique point $N(S, d) \in \partial S$ satisfying the below equation (Bergantiños, 2005; Chae, 2004; Kim, 2020b);

$$\prod_{i \in N} \left(N_i(S, d) - d_i\right) = \max_{x \in S, x \geq d} \prod_{i \in N} \left(x_i - d_i\right) \tag{7.3.1}$$

In the equation (7.3.1), all players involved in the bargaining game will be assigned the same bargaining powers. It may not be reasonable when players are significantly different. As a generalized Nash bargaining solution, *WNBS* is a variant of the *NBS* by assigning players with different bargaining powers. To get the *WNBS*, the objective function in (7.3.1) is modified as follows (Kim, 2020b; Wang, 2014);

$$\prod_{i \in N} \left(N_i(S, d) - d_i\right) = \max_{x \in S, x \geq d} \prod_{i \in N} \left(x_i - d_i\right)^{\omega_i}, \text{ s.t.}, \sum_{i \in N} \omega_i = 1 \tag{7.3.2}$$

where ω_i is the corresponding bargaining power of player $i \in N$. Bargaining power refers to the relative ability of game players to influence one another within a given strategic interaction. To enhance flexibility in solution selection while accounting for the diverse conditions of network environments, it is essential to incorporate differentiated bargaining powers among participants. This approach ensures that the variability in network conditions and the heterogeneous influence of different agents are effectively considered in the decision-making process (Kim, 2014; Kim, 2020b).

A bargaining game with coalition structure is a triple $(N, \upsilon, \mathscr{C})$ where C is a coalition structure over N, and the bargaining problem of $(N, \upsilon, \mathscr{C})$ is a triple $(S, d, \mathscr{C})$. A solution of $(S, d, \mathscr{C})$, which is characterized by the unique point $\Phi(S, d, \mathscr{C}) \in \partial S$, is the *NBS-CS*, and it is obtained as follows (Chae, 2004; Kim, 2020b);

$$\max_{U \in S, U \geq d} \prod_{C_j \in \mathscr{C}} \left(\prod_{i \in N} \left(x_i - d_i\right)^{f(C_j)} \right), \text{ s.t.}, \begin{cases} C = \{C_1, \ldots, C_p\} \\ f(C_j) > 0 \end{cases} \tag{7.3.3}$$

7.3.5 The bargaining control scheme for DC based HetNet system

The *HNSA* scheme introduces an innovative spectrum management mechanism tailored for the hierarchical HetNet platform. Its primary objective is to dynamically allocate limited spectrum resources to MDs in an adaptive manner. To minimize computational complexity, the spectrum allocation process is structured hierarchically, allowing the *HNSA* scheme to make step-by-step control decisions at each stage of the allocation process, ensuring efficient and scalable resource management. At the first phase, each individual MBS ($\mathscr{M}$) has the $\mathfrak{M}_{\mathscr{M}}$, and shares it with the set of SBSs $\left(\mathscr{W}_{\mathscr{M}}\right)$ within the $\mathscr{M}$'s area. Therefore, the $\mathfrak{M}_{\mathscr{M}}$ is partitioned into disjoint portions, and allocates them for corresponding SBSs. To implement this process for the $\mathscr{M}_i$ with $\mathscr{W}_{\mathscr{M}_i}$, the *HNSA* scheme focuses on the traditional *NBS* to get a fair-efficient solution. Let $\chi^{\mathcal{M}_i}_{\mathcal{S}_j}$ be the spectrum amount allocated for the $\mathcal{S}_j \in \mathcal{W}_{\mathcal{M}_i}$, and the $\mathcal{M}_i$'s spectrum distribution solution for $\mathcal{W}_{\mathcal{M}_i}$, i.e., $\check{C}^{\mathcal{M}_i}_{\mathcal{W}_{\mathcal{M}_i}}$, is a $\left|\mathcal{W}_{\mathcal{M}_i}\right|$-dimensional bandwidth allocation vector where $\check{C}^{\mathcal{M}_i}_{\mathcal{W}_{\mathcal{M}_i}} = \ldots\chi^{\mathcal{M}_i}_{\mathcal{S}_j \in \mathcal{W}_{\mathcal{M}_i}}\ldots$ (Kim, 2020b);

$$\check{C}^{\mathcal{M}_i}_{\mathcal{W}_{\mathcal{M}_i}} = \max_{\ldots\chi^{\mathcal{M}_i}_{\mathcal{S}_j \in \mathcal{W}_{\mathcal{M}_i}}\ldots} \left(\prod_{\mathcal{S}_j \in \mathcal{W}_{\mathcal{M}_i}} \left(U^{\mathcal{M}_i}_{\mathcal{S}_j}\left(\chi^{\mathcal{M}_i}_{\mathcal{S}_j}\right) - d_{\mathcal{S}_j} \right) \right)$$

(7.3.4)

$$\text{s.t.,}\begin{cases} U^{\mathcal{M}_i}_{\mathcal{S}_j}\left(\chi^{\mathcal{M}_i}_{\mathcal{S}_j}\right) = \left(\exp\left(\dfrac{\min\left(\chi^{\mathcal{M}_i}_{\mathcal{S}_j}, r^{t_c}\left(N^{\mathcal{M}_i}_{\mathcal{S}_j}\right)\right)}{r^{t_c}\left(N^{\mathcal{M}_i}_{\mathcal{S}_j}\right)} \times \Psi_{\mathcal{S}} \right) - \mu \right) \\ \left(\sum_{\mathcal{S}_j \in \mathcal{W}_{\mathcal{M}_i}} \chi^{\mathcal{M}_i}_{\mathcal{S}_j} \right) \leq \mathfrak{M}^{\mathcal{M}} \end{cases}$$

where $\Psi_{\mathcal{S}}$ is a profit coefficient factor and ì is a control parameter for the $U^{\mathcal{M}_i}_{\mathcal{S}_j}(\cdot)$, respectively. $N^{\mathcal{M}_i}_{\mathcal{S}_j}$ represents the set of MDs, which exist in the $\mathcal{S}_j$'s area covering by the $\mathcal{M}_i$. At time t_c, $r^{t_c}\left(N^{\mathcal{M}_i}_{\mathcal{S}_j}\right)$ is the total requested spectrum amounts of MDs in the $N^{\mathcal{M}_i}_{\mathcal{S}_j}$ (Kim, 2020b).

At the second phase, each individual S allocates its assigned spectrum pool $\left(\chi^{\mathcal{M}}_{\mathcal{S}}\right)$ for its corresponding MDs in $N^{\mathcal{M}}_{\mathcal{S}}$. In the *HNSA* scheme, the major challenge is to effectively support network-assisted D2D communications. Therefore, to implement the S's spectrum allocation process, the *HNSA* scheme prioritizes D2D services, and focuses on the *WNBS*. Let $\check{C}^{\mathcal{S}_j}\left[\mathcal{D}_e \in N^{\mathcal{M}_i}_{\mathcal{S}_j}\right] = \ldots,\chi^{\mathcal{S}_j}_{\mathcal{D}_e},\ldots$ be a $\left|N^{\mathcal{M}_i}_{\mathcal{S}_j}\right|$-dimensional spectrum allocation vector, and $\chi^{\mathcal{S}_j}_{\mathcal{D}_e}$ is the assigned spectrum amount for the $\mathcal{D}_e$. Finally, the $\check{C}^{\mathcal{S}_j}\left[\mathcal{D}_e \in N^{\mathcal{M}_i}_{\mathcal{S}_j}\right]$ is obtained as follows (Kim, 2020b);

$$\check{C}^{\mathcal{S}_j}\left[\mathcal{D}_e\right] = \max_{\ldots\chi^{\mathcal{S}_j}_{\mathcal{D}_e}\ldots} \left(\prod_{\mathcal{D}_e \in N^{\mathcal{M}_i}_{\mathcal{S}_j}} \left(U^{\mathcal{S}_j}_{\mathcal{D}_e}\left(\chi^{\mathcal{S}_j}_{\mathcal{D}_e}, r^{\mathcal{S}_j}_{\mathcal{D}_e}\right) - d_{\mathcal{D}_e} \right)^{\omega_{\mathcal{D}_e}} \right)$$

(7.3.5)

$$\text{s.t.,}\begin{cases} U^{\mathcal{S}_j}_{\mathcal{D}_e}\left(\chi^{\mathcal{S}_j}_{\mathcal{D}_e}, r^{\mathcal{S}_j}_{\mathcal{D}_e}\right) = \dfrac{\kappa}{\left(1 + exp\left(-\eta \times \dfrac{\min\left(\chi^{\mathcal{S}_j}_{\mathcal{D}_e}, r^{\mathcal{S}_j}_{\mathcal{D}_e}\right)}{r^{\mathcal{S}_j}_{\mathcal{D}_e}}\right)\right)} - \xi \\ \left(\sum_{\mathcal{D}_e \in N^{\mathcal{M}}_{\mathcal{S}_j}} \chi^{\mathcal{S}_j}_{\mathcal{D}_e} \right) \leq \chi^{\mathcal{M}_i}_{\mathcal{S}_j} and\, \omega_{\mathcal{D}_e} = \begin{cases} \varrho if \mathcal{D}_e \in N^{\mathcal{S}_j}_{C} \\ \beta if \mathcal{D}_e \in N^{\mathcal{S}_j}_{D} \end{cases} \end{cases}$$

where ê,η and î are coefficient factors for the $U^{\mathcal{S}_j}_{\mathcal{D}_e}(\cdot)$. $r_{\mathcal{D}_e}$ is the requested spectrum amount of $\mathcal{D}_e$. ϱ and â are bargaining for MDs. $N^{\mathcal{S}_j}_{C}$ is the set of simple cellular MDs and $N^{\mathcal{S}_j}_{D}$ is the set of network-assisted D2D MDs where $N^{\mathcal{M}_i}_{\mathcal{S}_j} = N^{\mathcal{S}_j}_{C} \cup N^{\mathcal{S}_j}_{D}$ (Kim, 2020b).

At the third phase, the MBS provides its own spectrum resource $\left(\mathfrak{R}_{\mathcal{M}}\right)$ for DC services. Let $N^{\mathcal{M}_i}_{C} = \bigcup_{\mathcal{S}_j \in \mathcal{W}_{\mathcal{M}_i}} N^{\mathcal{S}_j}_{C}$ and $N^{\mathcal{M}_i}_{D} = \bigcup_{\mathcal{S}_j \in \mathcal{W}_{\mathcal{M}_i}} N^{\mathcal{S}_j}_{D}$; $N^{\mathcal{M}_i}_{C}$ and $N^{\mathcal{M}_i}_{D}$ can be treated as a unit under diverse traffic environments. Therefore, bargaining with coalition structure is applied to provide spectrum resources for DC services. Let $\mathscr{E}_{\mathscr{D}}^{\mathscr{M}_i}$ be an added spectrum amount supplemented by the $\mathscr{M}_i$ for the $\mathscr{D}_e$. Finally, the solution $\left(\Theta\left[\mathscr{D}_e \in \left(N_C^{\mathscr{M}_i} \cup N_D^{\mathscr{M}_i}\right)\right]\right)$ is a $\left|N_C^{\mathscr{M}_i} \cup N_D^{\mathscr{M}_i}\right|$-dimensional spectrum allocation vector where $\Theta\left[\mathscr{D}_e \in \left(N_C^{\mathscr{M}_i} \cup N_D^{\mathscr{M}_i}\right)\right] = \ldots, \mathscr{E}_{\mathscr{D}_e}^{\mathscr{M}_i}, \ldots$; the $\Theta\left[\mathscr{D}_e \in \left(N_C^{\mathscr{M}_i} \cup N_D^{\mathscr{M}_i}\right)\right]$ is obtained as follows (Kim, 2020b);

$$\Theta\left[\mathscr{D}_e \in \left(N_C^{\mathscr{M}_i} \cup N_D^{\mathscr{M}_i}\right)\right] =$$

$$\max_{\ldots \mathscr{E}_{\mathscr{D}_e}^{\mathscr{M}_i} \ldots} \prod_{\mathscr{X} \in \left\{N_C^{\mathscr{M}_i}, N_D^{\mathscr{M}_i}\right\}} \left(\prod_{\mathscr{D}_e \in \mathscr{X}} \left(U_{\mathscr{D}_e}^{\mathscr{M}_i}\left(\mathscr{E}_{\mathscr{D}_e}^{\mathscr{M}_i}, \chi_{\mathscr{D}_e}^{\mathcal{S}_j}, r_{\mathscr{D}_e}^{\mathcal{S}_j}\right) - d_{\mathscr{D}_e}\right)^{F(\mathscr{D}_e, \mathscr{D}_k)}\right) \tag{7.3.6}$$

$$s.t. \begin{cases} U_{\mathscr{D}_e}^{\mathscr{M}_i}\left(\mathscr{E}_{\mathscr{D}_e}^{\mathscr{M}_i}, \chi_{\mathscr{D}_e}^{\mathcal{S}_j}, r_{\mathscr{D}_e}^{\mathcal{S}_j}\right) = \log\left(\dfrac{\min\left[\mathscr{E}_{\mathscr{D}_e}^{\mathscr{M}_i}, \left(r_{\mathscr{D}_e}^{\mathcal{S}_j} - \chi_{\mathscr{D}_e}^{\mathcal{S}_j}\right)\right]}{\left(r_{\mathscr{D}_e}^{\mathcal{S}_j} - \chi_{\mathscr{D}_e}^{\mathcal{S}_j}\right)} + \Gamma\right) \\ \left(\sum_{\mathscr{D}_e \in \left(N_C^{\mathscr{M}_i} \cup N_D^{\mathscr{M}_i}\right)} \mathscr{E}_{\mathscr{D}_e}^{\mathscr{M}_i}\right) \leq \mathfrak{R}_{\mathscr{M}_i} \text{ and } \left(\mathscr{D}_e, \mathscr{D}_k\right) \in \mathfrak{X}_{D2D} \\ F\left(\mathscr{D}_e, \mathscr{D}_k\right) = \begin{cases} \psi + \left(\phi \times \frac{X - Y}{X}\right), if \mathscr{D}_e \in N_D^{\mathscr{M}_i} \\ \alpha, if \mathscr{D}_e \notin N_C^{\mathscr{M}_i} \end{cases} \\ X = \left(\mathscr{E}_{\mathscr{D}_k}^{\mathscr{M}}\left[t_{c-1}\right] + \chi_{\mathscr{D}_k}^{\mathscr{M}}\left[t_{c-1}\right]\right) \text{ and } Y = \left(\mathscr{E}_{\mathscr{D}_e}^{\mathscr{M}_i}\left[t_{c-2}\right] + \chi_{\mathscr{D}_e}^{\mathcal{S}_j}\left[t_{c-2}\right]\right) \end{cases}$$

where Ã is a control parameter for the $U_{\mathscr{D}_e}^{\mathscr{M}_i}(\cdot)$. ø, á and ф are profit coefficient factors for the $F\left(\mathscr{D}_e, \mathscr{D}_k\right)$ when $\mathscr{D}_e, \mathscr{D}_k \in \mathfrak{X}_{D2D}$ and the $\mathscr{D}_k$ is paired with the $\mathscr{D}_e$ for D2D communications. $\mathscr{E}_{\mathscr{D}_k}^{\mathscr{M}}\left[t_{c-1}\right]$ and $\chi_{\mathscr{D}_k}^{\mathscr{M}}\left[t_{c-1}\right]$ are the $\mathscr{E}_{\mathscr{D}_k}^{\mathscr{M}}$ and $\chi_{\mathscr{D}_k}^{\mathscr{M}}$ values for the $\mathscr{D}_k$ at the time t_{c-1}. $\mathscr{E}_{\mathscr{D}_e}^{\mathscr{M}_i}\left[t_{c-2}\right]$ and $\chi_{\mathscr{D}_e}^{\mathcal{S}_j}\left[t_{c-2}\right]$ are the $\mathscr{E}_{\mathscr{D}_e}^{\mathscr{M}_i}$ and $\chi_{\mathscr{D}_e}^{\mathscr{M}_i}$ values for the $\mathscr{D}_e$ at the time t_{c-2}. By considering the traffic flow balance between $\mathscr{D}_e$ and $\mathscr{D}_k$, the function $F\left(\mathscr{D}_e, \mathscr{D}_k\right)$ is designed based on the repeated bargaining concept. Therefore, the outcome of $F\left(\mathscr{D}_e, \mathscr{D}_k\right)$ is dynamically adjusted based on the timed online manner for network-assisted D2D communications (Kim, 2020b).

7.3.6 Summary

Maintaining dual wireless connections is a promising strategy to enhance network capacity in 5G HetNets, enabling MDs to simultaneously utilize spectrum resources from two serving cells. The *HNSA* scheme addresses the challenges of network-assisted D2D communications within this emerging dual-connectivity paradigm, which serves as an efficient access method that significantly improves network performance and QoS. However, designing such an integrated spectrum management mechanism presents several challenges: i) spectrum distribution among different SBSs, ii) spectrum allocation for multiple MDs, and iii) spectrum sharing between D2D device pairs. To effectively tackle these issues, the *HNSA* scheme is built upon fundamental Nash principles, incorporating three bargaining solutions - *NBS*, *WNBS*, and *NBS-CS* - to develop an optimized spectrum control mechanism. Considering the hierarchical nature of the HetNet system, these bargaining models are seamlessly integrated within the *HNSA* scheme, operating cooperatively in a distributed manner. This ensures that limited spectrum resources are efficiently shared across BSs and multiple MDs, even in dynamically evolving HetNet environments (Kim, 2020b).

7.4 Cellular Network Spectrum Allocation (*CNSA*) scheme

In 5G cellular networks, D2D communication has emerged as a key technology to enhance spectral efficiency and reduce communication delays. Notably, D2D technology is increasingly supporting a wide range of services, enabling user-centric and personalized applications in future wireless networks. However, the deployment of D2D communications presents several technical challenges, particularly in the area of spectrum allocation. Recently, S. Kim addressed this issue by developing the *Cellular Network Spectrum Allocation* (*CNSA*) scheme for D2D-enabled cellular networks (Kim, 2020a). By leveraging key principles from cooperative game theory, the *CNSA* scheme integrates weighted utilitarian and meta-bargaining solutions to optimize spectrum allocation. The core objective of bargaining solutions is to determine the optimal way to distribute a jointly generated surplus among game players. Utilizing the advantages of a step-by-step interactive bargaining mechanism, the *CNSA* scheme ensures rational and efficient spectrum allocation, maximizing overall network performance.

7.4.1 Development motivation

Recent projections indicate that the number of wireless mobile devices is expected to exceed 50 billion by 2020, necessitating a 1,000-fold increase in data rates to accommodate the surge in wireless data traffic. Additionally, the continued expansion of emerging services such as virtual training, augmented reality, e-learning, and remote surgery is expected to pose significant technical challenges in supporting various multimedia services. As large-scale mobile traffic requirements and wireless applications continue to evolve, the traditional infrastructure-based cellular network may soon become insufficient. One of the key limitations of conventional cellular networks is that all network traffic is relayed through fixed infrastructure, even when the source and destination devices are physically close to one another (Chour, 2019; Kim, 2020a; Liu, 2019; Xiao, 2015).

With the advancement of 5G networks, innovative technical solutions have been introduced to enhance network capacity and improve QoS. Among these, D2D communication has emerged as a promising approach to increase network reliability and capacity. This technology allows mobile devices to communicate directly with one another over short distances, eliminating the need for data transmission through the core network infrastructure. The benefits of D2D communication include; i) reduced communication latency, ii) lower energy consumption, and iii) improved network reliability. Moreover, D2D-enabled devices can share spectrum resources assigned to mobile users, thereby enhancing spectral efficiency and optimizing energy utilization while improving the overall user experience (Anbiyaei, 2019; Kim, 2020a; Liu, 2019; Xiao, 2015).

With the increasing adoption of D2D communication, D2D multicast content sharing has been identified as an effective strategy to enhance local service quality in cellular networks. Since multiple users often request identical content, a significant portion of data traffic consists of repetitive transmissions. For example, applications such as live weather updates and video streaming frequently distribute the same content to geographically close users. By leveraging D2D multicast content sharing, redundant transmissions can be minimized, enabling large-scale parallel content distribution in a decentralized manner. As a result, this technology is gaining increasing attention as a highly efficient solution for data dissemination in modern networks (Bhardwaj, 2018; Kim, 2020a; Wang, 2019; Yang, 2018).

In a D2D multicast content-sharing scenario, one of the primary challenges lies in spectrum allocation, which determines how wireless spectrum resources are efficiently utilized. Since multiple nearby users often request the same service, they can be grouped into logical clusters or communities within a D2D multicast system. Within each content distribution cluster, a cluster head user is selected to relay content to multiple D2D requesters. The spectrum resources allocated to the cluster head are then shared within the cluster, maximizing overall network throughput. This spectrum-sharing mechanism creates new opportunities to enhance spectrum efficiency while optimizing network resource utilization (Kim, 2020a; Liu, 2019; Wang, 2019).

While traditional cellular networks offer broad coverage for mobile users, D2D multicast systems create a more diversified cluster structure. To mitigate co-channel interference, the *CNSA* scheme partitions the available spectrum into non-overlapping segments, assigning dedicated spectrum to each cluster. Given the irregular topology of D2D clusters, an efficient spectrum allocation strategy is essential to optimize spectrum utilization. Although various spectrum allocation techniques have been proposed for D2D multicasting networks in recent years, research in this area remains in its early stages. Several technical challenges continue to persist in both academic and industry sectors, emphasizing the need for a novel control paradigm that can provide a comprehensive and effective solution (Anbivaei, 2019; Bhardwaj, 2018; Kim, 2020a; Liu, 2019; Wang, 2019; Yang, 2018).

The *CNSA* scheme introduces an innovative spectrum allocation mechanism designed specifically for D2D multicast systems. To ensure effective spectrum distribution, the *CNSA* scheme integrates weighted utilitarian and meta-bargaining models into its decision-making process. The utilitarian bargaining model, extensively studied in social welfare theory, was formally axiomatized in 1975 by Harsanyi and has since been applied in bargaining contexts to ensure optimal and fair resource distribution. The meta-bargaining concept, originally introduced by van Damme, aims to resolve conflicts when cooperative game participants have different perceptions of the ideal solution concept. In this framework, the meta-bargaining game is a non-cooperative strategic game, derived from a cooperative bargaining game, where players select bargaining solutions as strategies to determine an optimal outcome. By incorporating these advanced bargaining models, the *CNSA* scheme ensures efficient and fair spectrum allocation while addressing the complex interactions in D2D multicast systems (Kim, 2020a; Trockel, 2002).

7.4.2 Contribution

Recognizing the diverse characteristics of cellular network users, the *CNSA* scheme assumes that each user has a different perspective on their ideal solution for spectrum allocation. To determine the most optimal and mutually beneficial solution, the *CNSA* scheme introduces an innovative resource allocation mechanism based on weighted utilitarian and meta-bargaining solutions. By employing a two-stage bargaining approach, the *CNSA* scheme fully utilizes the self-adaptive interactions among cellular users, ensuring a dynamic and cooperative decision-making process. As a result, the *CNSA* scheme efficiently distributes the limited spectrum resources in a fair and optimal manner, balancing network performance and user requirements (Kim, 2020a).

7.4.3 The D2D multicasting system infrastructure

The *CNSA* scheme tackles the spectrum allocation problem for an D2D multicasting enabled cellular network. The *CNSA* scheme considers a classic D2D cellular system which consists of a cellular base station (BS) and multiple cellular users (CUs) labeled as $\mathcal{Q}^{BS} = \{\mathcal{S}_1, \mathcal{S}_2, \ldots, \mathcal{S}_n\}$. CUs opportunistically access the spectrum resource of the cellular network. It is common that a specific hot content, e.g., the

World Cup or *Super Bowl*, is bound to attract the interest of multiple users. In many cases, multiple CUs may request the same content within a given time interval. To enhance spectral efficiency and reduce the burden on the BS, CUs collaborate to form content-sharing clusters. Within each cluster, a designated cluster head (CH) is responsible for caching the requested content from the BS and subsequently distributing it to the other content requesters (CRs) within the same cluster. The *CNSA* scheme assumes that CRs belong to distinct clusters with no overlaps, and they remain within their assigned cluster until the content-sharing process is completed. However, not all CUs may find a suitable cluster for content sharing. Those who remain outside clusters connect directly to the BS and operate in the traditional cellular mode; these users are referred to as direct cellular users (DCUs) (Kim, 2020a; Wu, 2019).

Without loss of generality, there exist clusters labeled as $\mathbb{C} = \{\mathcal{C}_1, \mathcal{C}_2, \ldots, \mathcal{C}_m\}$ where $m \ll n$. In the $\mathcal{C}_{1 \le i \le m}$, one CH and multiple CRs. Figure 2 gives a typical D2D multicast content sharing scenario, where three clusters, i.e., $\mathcal{C}_1, \mathcal{C}_2, \mathcal{C}_3$, are formed, and $\mathcal{S}_3$, $\mathcal{S}_5$ and $\mathcal{S}_8$ are the CHs, respectively. Taking $\mathcal{C}_1$ as an example, $\mathcal{S}_3$ firstly receives the content from the BS via cellular spectrum resource and then shares it with $\mathcal{S}_1$ and $\mathcal{S}_2$ simultaneously. Since the content sharing is also based on wireless transmissions, the spectrum resource should be assigned to avoid the mutual interference between the two communication types (Kim, 2020a; Wu, 2019).

Figure 2. D2D multicasting system infrastructure (Kim, 2020a)

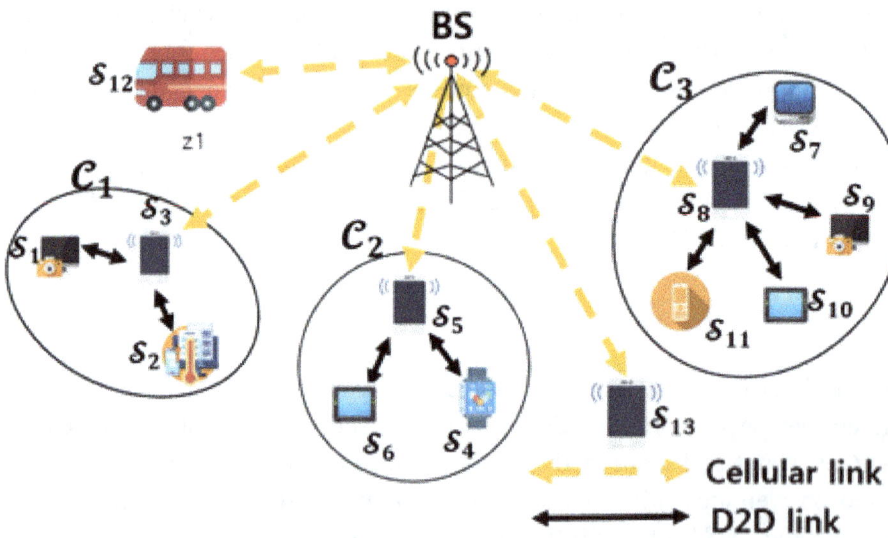

In an interference-prone environment, the *CNSA* scheme is designed to efficiently allocate cellular spectrum resources, ensuring maximum throughput for the D2D multicasting system. Achieving this goal presents two primary challenges; i) BS spectrum distribution - the BS must determine how to effectively allocate its spectrum resources between CHs and DCUs, and ii) CH resource sharing - each CH must decide how to distribute its allocated spectrum to support D2D multicasting services while maintaining direct communication with the BS. To address these challenges, the *CNSA* scheme leverages weighted utilitarian bargaining solutions to optimize BS spectrum allocation. For CH spectrum sharing, the meta-bargaining solution is applied to divide the assigned spectrum into two segments - one portion for direct communication with the BS, and the other portion for D2D multicasting services. By integrating these bargaining models, the *CNSA* scheme ensures an adaptive and efficient spectrum allocation strategy, optimizing both cellular and D2D communication performance (Kim, 2020a).

To fully leverage the benefits of D2D multicast content sharing, an effective user clustering method is essential. This clustering process comprises two key stages: CH selection and CR grouping. The clustering process begins with each CU broadcasting a *join-request message* to its neighboring CUs, indicating its interest in sharing specific content. Nearby CUs within the listening range receive this message and assess whether they wish to participate in the cluster. If a CU is interested in joining the cluster, it responds by sending a *join-acceptance message* back to the initiating CU. This dynamic and decentralized approach ensures that only relevant users are grouped together, optimizing content distribution efficiency and enhancing D2D multicast performance. In an entirely distributed online fashion, each individual CU can monitor the number of available CRs, i.e., $\mathcal{N}^{CR}$, for its own cluster, and reports this information to the BS. At this moment, the BS selects the CU with the highest $\mathcal{N}^{CR}$ number as the CH, and the first cluster is formed. With the remaining CUs, who are not included in the current generating cluster, this clustering process is repeated sequentially until all the possible clusters are configured (Kim, 2020a).

7.4.4 Bargaining solutions for cooperative game models

In 1950, J. Nash introduced the axiomatic approach to analyzing bargaining problems, laying the foundation for modern cooperative game theory. Over time, his framework has been expanded and refined, leading to the development of various bargaining solution concepts based on different rationality criteria. Some of the most widely recognized cooperative bargaining solutions include the *Nash solution*, *Kalai-Smorodinsky solution*, and *weighted utilitarian solution*. As a result of these theoretical advancements, a diverse range of bargaining models has emerged, each offering unique axiomatic justifications and applications in different strategic decision-making scenarios (Marco-Gil, 2012). The following sections introduce the notation and fundamental definitions of bargaining solutions, providing a structured framework for analyzing cooperative negotiations. Usually, bargaining games (S,d) consist of a set of feasible utility allocations over the set of feasible agreements, S, and the outcome which would result in case of disagreement point, d. $u(S,d)$ is defined by $u_1(S,d) := \max\left\{x_1 \geq d_1 \middle| \exists_{x_2} \geq d_2 : (x_1,x_2) \in S\right\}$ and $u_2(S,d) := \max\left\{x_2 \geq d_2 \middle| \exists_{x_1} \geq d_1 : (x_1,x_2) \in S\right\}$. Let Σ denote an arbitrary set of games, and a bargaining solution on Σ is a function $f : \Sigma \to \mathbb{R}^2$ such that $f(S,d) \in S$ holds for all $(S,d) \in \Sigma$. F denotes an arbitrary set of admissible solutions on Ó (Kim, 2020a; Naeve-Steinweg, 1999).

Some well-known bargaining solutions contained in F are *Nash bargaining solution* (*NBS*), *Kalai-Smorodinsky bargaining solution* (*KSBS*), and *weighted utilitarian bargaining solution* (*WUBS*) (Kalai, 1983; Kim, 2014, Kim, 2020a). For a simple two-person bargaining model, the *NBS*, *KSBS*, and *WUBS* are mathematically formulated as follows (Kim, 2020a);

- The *NBS* f^{NBS} is defined by (Naeve-Steinweg, 2004);

$$f^{NBS}(S,d) = \begin{cases} \arg\max\left\{\prod_{i=1}^{2}(x_i - d_i)\middle| x \in S_{\geq d}\right\}, \text{ if } d \notin \mathrm{WPO}(S) \\ \bar{x} \in \mathrm{PO}(S_{\geq d}), otherwise, i.e., d \in \mathrm{WPO}(S) \end{cases} \tag{7.4.1}$$

where $S_{\geq d} = \{x \in S | x \geq d\}$. To simplify notation, we define $S_{+} = S_{\geq 0}$. Let $WPO(S) = \{ x \in A | \{y \in A | y > x\}$

$= \varnothing\}$ denote the set of *weakly Pareto-optimal* allocation.

- The *KSBS* f^{KSBS} is defined by (Naeve-Steinweg, 2014);

$$f^{KSBS}(S,d) = \begin{cases} x \in \mathrm{PO}(S) \cap \mathrm{cvh}(\{d, u(S,d)\}), \mathrm{if} d \notin \mathrm{WPO}(S) \\ x \in \mathrm{PO}(S_{\geq d}), otherwise, i.e., d \in \mathrm{WPO}(S) \end{cases} \tag{7.4.2}$$

where $cvh(u(S,d)) := \{x \in \mathbb{R}^2 \mid x$

$= \lambda a + (1 - \lambda)b,\ \lambda \in [0,1],\ a, b \in u(S,d)\}$ denotes the convex hull of $u(S,d)$.

- The *WUBS* f^{WUBS} is defined by (Kalai, 1983)

$$f^{WUBS}(S,d) = x \in \max \sum_{i=1}^{2} (\omega_i \times x_i),\ \mathrm{s.t.},\ (x_1, x_2) \in S \text{ and } x \in PO(S) \tag{7.4.3}$$

where $(\omega_1,\ \omega_2) \in \mathbb{R}^2_+$ is a pair of weighting factors for two players. It is apparent that the *WUBS* is the one that maximizes the sum of the utility gains, and it is *Pareto-optimal* with respect to the reported utilities. However, a weight pair $(\omega_1,\ \omega_2)$ should be assigned to the utility scales of the two players, and then the *CNSA* scheme would be maximizing the weighted sum of the utilities (Kalai, 1983). The question of how to determine these weights will be addressed later (Kim, 2020a).

When game players agree on the same bargaining solution concept, they receive utility levels based on that solution. However, conflicts arise when players prefer different bargaining solutions, and their preferences do not align. To address such disagreements, a new bargaining framework, known as the *Meta Bargaining Solution* (*MBS*), was introduced. The core idea of *MBS* is to model a step-by-step bargaining process to reach a mutually acceptable resolution. When players are in disagreement, the *MBS* determines a compromise by selecting the midpoint of the line connecting their respective preferred bargaining allocations. If this midpoint lies within the feasible utility allocation set, it is adopted as the new disagreement point, allowing the bargaining process to continue towards a fair and rational outcome. In other words, the iterative procedure of *MBS* improves the d while satisfying the *Midpoint Domination* property, which requires that all players get at least as much as the average of their preferred allocations. The *MBS* procedure is iteratively executed until a *weakly Pareto-optimal* allocation is achieved. As this iterative process continues, the *MBS* outcome gradually converges to a fair and efficient allocation, balancing both individual and collective interests. The *MBS* framework can be analyzed from two different perspectives; i) cooperative perspective - this view focuses on formulating properties of fair compromises and developing functions that align with these fairness principles, and ii) non-cooperative perspective - here, game players treat bargaining solutions as strategic choices, and their payoff is determined through the step-by-step interactive process of *MBS*. In the *MBS* approach, the non-cooperative bargaining process is complemented by cooperative considerations, ensuring that both strategic decision-making and fairness principles contribute to the final allocation (Anbarci, 1992; Kalai, 1983; Kim, 2020a; Marco-Gil, 2012; Naeve-Steinweg, 1999, Naeve-Steinweg, 2004).

In the *MBS*, each player is assumed to have an intuitive idea about which point to select in every game. This assumption can be formalized as players always want to apply a certain bargaining solution. In a simple two-player *MBS* model, it is denoted as $[(S,d);f,g]$ where $(S \subseteq \mathbb{R}^2, d)$ is a game and f and g are bargaining solutions supported by player 1 and player 2, respectively (Naeve-Steinweg, 2004). Formally, $\Sigma_{\mathcal{F}} := \{[(S,d);f,g] | (S,d) \in \Sigma, f,g \in \mathcal{F}\}$denotes the class of meta games induced by Ó and F. Given a set $\Sigma_{\mathcal{F}}$ of meta games, a mechanism M is a function $\mathfrak{M}:\Sigma_{\mathcal{F}} \to \mathbb{R}^2, [(S,d);f,g] \mapsto \mathfrak{M}[(S,d);f,g] \in S$, which assigns every meta game $[(S,d);f,g]$to an allocation in $\Sigma_{\mathcal{F}}$. Mathematically, the *MBS* mechanism $\mathfrak{M}[(S,d);f,g]$, i.e., $\mathfrak{M}:\Sigma_{\mathcal{F}} \to \mathbb{R}^2$, is defined as follows, and Figure 3 shows the *MBS* graphically (Anbarci, 1992; Naeve-Steinweg, 2004; Kim, 2020a);

$$\mathfrak{M}[(S,d);f,g] := \lim_{t \to \infty} d^t \tag{7.4.4}$$

$$\text{s.t.,} \begin{cases} d^1 := d, \text{if } t = 1 \\ d^t := \frac{1}{2}(f(S,d^{t-1}) + g(S,d^{t-1})), \text{otherwise} (t > 1) \end{cases}$$

Figure 3. Meta bargaining solution for two-player cooperative game (Kim, 2020a)

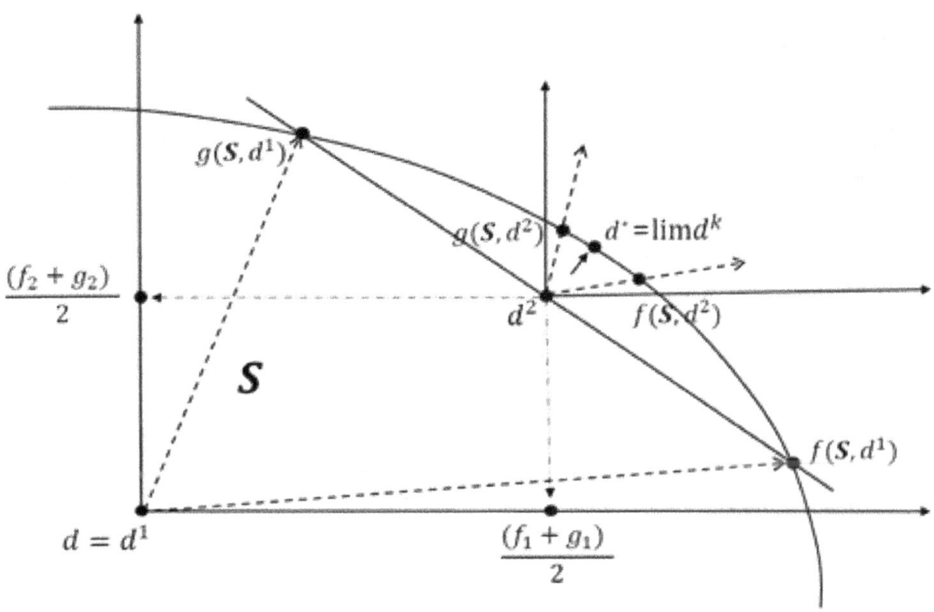

Analogously to bargaining theory, we can formulate some reasonable and desirable properties for the *MBS*. With some classical properties for bargaining solutions, MBS has new kinds of properties which capture a notion of meta-fairness and meta-equity to treat game players. Given an admissible family of bargaining solutions $\mathscr{F}$, a meta bargaining mechanism on F is a single valued function $\Phi(f,g) = \Phi_{fg} \in \mathscr{F}$ where $\Phi:\mathscr{F} \times \mathscr{F} \rightarrow \mathscr{F}$ and $f, g \in \mathscr{F}$ (Kim, 2020a; Marco-Gil, 2012; Naeve-Steinweg, 2004).

7.4.5 The D2D multicasting spectrum allocation scheme

In the D2D multicasting system, the BS needs to share its spectrum resource among CUs. The utility payoff obtained by the BS depends on the total traffic services for CUs. Therefore, the BS's payoff is directly proportional to a fraction of the allocated spectrum and the number of CUs. To quantify the BS service satisfaction, the BS's payoff ($\mathbb{U}^{BS}(\mathfrak{P})$) with the resource allocation strategy vector $\mathfrak{P} = \left[\ldots, \mathfrak{P}^{BS}_{\mathcal{S}_i}, \ldots\right]$ can be derived as follows (Kim, 2020a).

$$\mathbb{U}^{BS}(\mathfrak{P}) = \sum_{\mathcal{S}_i \in \mathcal{Q}^{BS}} \mathcal{U}^{\mathcal{S}_i}\left(\mathfrak{P}^{BS}_{\mathcal{S}_i}\right),$$

$$\text{s.t.,} \begin{cases} \mathcal{U}^{\mathcal{S}_i}\left(\mathfrak{P}^{BS}_{\mathcal{S}_i}\right) = \alpha \times \log\left(\dfrac{\min\left\{(\mathfrak{R}_M \times \gamma),\ \max\left\{0, \mathfrak{P}^{BS}_{\mathcal{S}_i} - (\mathscr{R}_m \times \mu)\right\}\right\}}{\mathscr{R}_m} + \chi\right) + \eta, \\ \quad if\, \mathcal{S}_i\, is CH \\ \mathcal{U}^{\mathcal{S}_i}\left(\mathfrak{P}^{BS}_{\mathcal{S}_i}\right) = \log\left(\dfrac{\min\left\{\mathfrak{R}_M,\ \max\left\{0, \mathfrak{P}^{BS}_{\mathcal{S}_i} - \mathscr{R}_m\right\}\right\}}{\mathscr{R}_m} + \chi\right) + \eta, if\, \mathcal{S}_i\, is DCU \end{cases} \tag{7.4.5}$$

where á is the number of CRs of its cluster, and ÷, ç are control parameter for the $\mathcal{U}^{\mathcal{S}_i}(\cdot)$. ã and ì are spectrum adjustment factors for the D2D multicasting service. $\mathfrak{R}_M$, $\mathscr{R}_m$ are the maximum and minimum spectrum requirements, respectively, for the requested service. For the CU spectrum assignment problem, the main concern of BS is the maximization of total system utilitarianism. Due to the characteristics of D2D multicasting, the *CNSA* scheme emphasizes the role of CHs. Therefore, the basic idea of *WUBS* is adopted to solve the BS's spectrum assignment problem, and the á value is considered as a weight factor for its corresponding CH. Finally, the *WUBS* for the BS, i.e., $WUBS_{BS}(\mathfrak{P})$, is defined as follows (Kim, 2020a);

$$WUBS_{BS}(\mathfrak{P}) = \max_{\mathfrak{P}=\left[\ldots,\mathfrak{P}^{BS}_{\mathcal{S}_i},\ldots\right]} \sum_{\mathcal{S}_i \in \mathcal{Q}^{BS}} \mathcal{U}^{\mathcal{S}_i}\left(\mathfrak{P}^{BS}_{\mathcal{S}_i}\right), \text{ s.t., } \mathbb{C} = \sum_{\mathcal{S}_i \in \mathcal{Q}^{BS}} \mathfrak{P}^{BS}_{\mathcal{S}_i} \tag{7.4.6}$$

where C is the total spectrum capacity of BS. In the viewpoint of CH, the main problem is how to share the assigned spectrum bandwidth $\left(\mathfrak{P}^{BS}_{\mathcal{S}}\right)$ between the direct communication to the BS, i.e., DC, and the multicasting among CRs, i.e., MC. In the *CNSA* scheme, the communication types - DC and MC - are modeled as game players that engage in negotiations to determine spectrum sharing for their respective services. From the DC perspective, the axiom of *Independence of Irrelevant Alternatives* is the preferred principle for spectrum allocation. This axiom suggests that a reasonable allocation remains valid and feasible even after the removal of certain payoff sets. Given this preference, NBS is a suitable approach for optimal resource allocation for DC services. On the other hand, MC favors *Individual Monotonicity*, which states that an expansion of the bargaining set (*S*) in a direction that benefits a specific player should always increase that player's payoff. As a result, the *KSBS* is the preferred allocation mechanism for MC services. To resolve the conflict arising from these different bargaining preferences, the *CNSA* scheme employs the *MBS*, ensuring a compromise-driven outcome that balances both communication types effectively (Kim, 2020a).

$$\mathfrak{M}[(S,d);f,g] := \lim_{\substack{t\to\infty \\ if|d^t - d^{t-1}| > \breve{A}}} d^t \tag{7.4.7}$$

$$\text{s.t.,}\begin{cases} d^t := \frac{1}{2}(f^{NBS}(S,d^{t-1}) + f^{KSBS}(S,d^{t-1})), if t > 1 \\ d^1 := (0,0), therwise(t = 1) \end{cases}$$

To obtain the $f^{NBS}(S,d^{t-1})$ and $f^{KSBS}(S,d^{t-1})$, we should define $x = (x_1,x_2)$ for game players. For the DC in the $\mathcal{S}_i$, who is the CH, the payoff value x_1 with the resource allocation strategy $\mathfrak{M}^{DC}_{\mathcal{S}_i}$ is defined through the following utility function (Kim, 2020a);

$$x_1 = U_{DC}\left(\mathfrak{M}^{DC}_{\mathcal{S}_i}\right) = \left(\frac{1}{1 + exp\left(-\left(\min\left\{\mathfrak{N}_M, \mathfrak{M}^{DC}_{\mathcal{S}_i}\right\} \times \frac{\beta}{\mathcal{R}_m}\right)\right)}\right) + \breve{A}, \tag{7.4.8}$$

$$\text{s.t.,}\ \mathfrak{M}^{DC}_{\mathcal{S}_i} \le \mathfrak{P}^{BS}_{\mathcal{S}_i}$$

where â, Ã are control parameters for the DC's payoff. For the MC in the $\mathcal{S}_i$, the payoff value x_2 with the resource allocation strategy $\mathfrak{M}^{MC}_{\mathcal{S}_i}$ is defined as follows (Kim, 2020a);

$$x_2 = U_{MC}\left(\mathfrak{M}^{MC}_{\mathcal{S}_i}\right) = \left(\frac{1}{1 + exp\left(-\left(\mathcal{T}^{MC} \times \frac{\beta}{\mathcal{R}_m} \times \check{C}\left(\mathfrak{B}^{C}_{\mathcal{S}_i}, \mathfrak{B}^{\mathcal{C}}_{\mathcal{S}_i}\right)\right)\right)}\right) + \breve{A}, \tag{7.4.9}$$

$$
\text{s.t.,}\begin{cases}
\mathcal{T}^{MC} = \min\left\{ \left(\mathfrak{N}_M \times \mu\right), \mathfrak{M}_{\mathcal{S}_i}^{MC} \right\} \\
\check{C}\left(\mathfrak{B}_{\mathcal{S}_i}^{C}, \mathfrak{B}_{\mathcal{S}_i}^{\mathscr{g}}\right) = \theta - \left(\dfrac{\mathfrak{B}_{\mathcal{S}_i}^{C} - \mathfrak{B}_{\mathcal{S}_i}^{\mathscr{g}}}{\mathfrak{B}_{\mathcal{S}_i}^{\mathscr{g}}} \right) \\
\mathfrak{M}_{\mathcal{S}_i}^{MC} = \mathfrak{P}_{\mathcal{S}_i}^{BS} - \mathfrak{M}_{\mathcal{S}_i}^{DC}
\end{cases}
$$

where $\mathfrak{B}_{\mathcal{S}_i}^{C}$, $\mathfrak{B}_{\mathcal{S}_i}^{\mathscr{g}}$ are current buffering, and desirable buffering amounts, respectively.θ is a calibrating constant for buffering (Kim, 2020a).

7.4.6 Summary

D2D communication has been recognized as a crucial technological component for future cellular networks; however, its implementation necessitates the development of efficient spectrum resource allocation protocols. The *CNSA* scheme introduces an innovative two-phase bargaining game model tailored for D2D multicasting systems. In the first phase, the *CNSA* scheme employs a weighted utilitarian bargaining solution to optimize spectrum resource allocation among multiple CUs, ensuring maximum system utilization, which is the most advantageous approach for the BS. In the second phase, the *CNSA* scheme utilizes a meta bargaining approach to address spectrum sharing conflicts between the CH and CRs. Since the CH and CRs often have differing perspectives on fairness in spectrum distribution, the meta bargaining method helps achieve an optimal balance between their conflicting interests. These two bargaining processes are conducted sequentially and independently, following a step-by-step approach to determine the most beneficial solution (Kim, 2020a).

7.5 Spectrum and Energy Efficiency Control (*SEEC*) scheme

D2D technology has been extensively studied by virtue of its ability to reduce traffic load and communication delay while improving network performance. However, it is a challenging issue to simultaneously coordinate energy efficiency (EE) and spectrum efficiency (SE) in D2D communications. Recently, S. Kim investigated a multi-object optimization problem with conflicting criteria and proposes a new solution paradigm to balance the EE and SE. Based on the ideas of multi-criteria decision-making method and bargaining game theory, he develops the *Spectrum and Energy Efficiency Control* (*SEEC*) scheme for D2D communications (Kim, 2025). The major novelty of the *SEEC* scheme is to strike a balance between the EE and SE through the developed multiple criteria bargaining process. The quantification of EE-SE tradeoff is appropriately negotiated to reach a mutually acceptable agreement. Therefore, the *SEEC* scheme selects the best power strategy that exploits the varying D2D communication conditions. To reduce complexity, each individual D2D communication pair executes the power control algorithm independently and iteratively in a distributed manner. Therefore, the *SEEC* scheme can be implemented with the acceptable polynomial complexity.

7.5.1 Development motivation

In recent years, the rapid proliferation of mobile devices and the surge in wireless data traffic have become increasingly evident. According to an Ericsson report, mobile data traffic is expected to grow exponentially in the coming years. This raises a critical question: how can the existing cellular network infrastructure accommodate the soaring demand for communication? In this context, D2D technology within cellular networks has garnered significant research interest as a promising solution to enhance network capacity. By enabling direct communication between two smart devices without routing through network infrastructure, D2D technology is anticipated to partially alleviate network congestion while allowing more mobile devices to connect in a faster and more reliable manner. Potential applications of D2D technology include proximity-based services such as peer-to-peer file sharing, content-aware services, and vehicle-to-everything (V2X) communications, among others (Gismalla, 2022; Huang, 2018; Kim, 2025; Wei, 2015).

Energy efficiency (EE) is a crucial factor in mobile wireless communications due to the limited battery capacity of mobile devices. Additionally, growing environmental concerns have further amplified the importance of energy-efficient communication, often referred to as 'green communication'. As a result, EE has emerged as a key enabler for future network development. However, while EE is a fundamental performance metric for environmentally friendly D2D communications, it often conflicts with spectrum efficiency (SE). SE, which refers to the amount of information transmitted per unit bandwidth, has attracted considerable interest due to the rapid growth in mobile data traffic and the inherent scarcity of spectrum resources. Unfortunately, EE and SE are inherently conflicting objectives - enhancing EE often results in a decrease in SE, and vice versa. Thus, investigating the tradeoff between EE and SE in D2D communications has become an essential research priority (Hao, 2018; Kim, 2025; Pham, 2017; Zhang, 2016b).

To address this challenge, there is a widely accepted view that balancing EE and SE should be approached as a multi-objective optimization problem, aiming to minimize total power consumption while maximizing spectrum utilization. EE is generally defined as the number of bits transmitted per unit of energy. Alternatively, it can be expressed as the ratio of SE to total power consumption. Based on this principle, the *SEEC* scheme examines the conflicting nature of power control in D2D communications (Hao, 2018; Pan, 2015). To analyze the tradeoff between EE and SE, the *SEEC* scheme formulates a novel multi-criteria control problem and proposes an integrated solution framework. The primary challenge in designing the *SEEC* scheme lies in leveraging the complementary and synergistic aspects of multiple decision criteria. By jointly optimizing EE and SE, the *SEEC* scheme aims to achieve higher data transmission rates while consuming less energy in D2D communications (Kim, 2025).

There are various approaches to solving multi-criteria decision-making problems, including lexicographic methods, weighted max-min strategies, no-preference techniques, and posteriori methods. These approaches seek to identify the optimal solution by considering multiple criteria simultaneously (Pham, 2017). However, conventional methods often fail to account for a consensus that benefits all criteria collectively, instead focusing on individual preferences. To overcome this limitation, the *SEEC* scheme introduces a novel solution framework that jointly maximizes EE and SE in D2D communications. By integrating multiple-criteria decision-making (MCDM) techniques with cooperative game theory, the *SEEC* scheme mathematically models multiple conflicting criteria to serve as a decision support tool. It effectively addresses the complex D2D control problem by optimizing the total data transfer rate per unit of energy consumption. Due to the inherent advantages of MCDM and cooperative game theory, this solution approach offers a fundamental framework applicable to structuring multi-objective problems in various engineering disciplines (Kim, 2025).

7.5.2 Contribution

To maximize the efficiency of the D2D communication system, a fundamental control challenge is to establish an optimal balance between EE and SE. The *SEEC* scheme addresses the D2D power control issue by incorporating a negotiation-based approach to resolve conflicts between these criteria. Following conventional MCDM methodologies, the *SEEC* scheme first normalizes the control factors related to EE and SE. It then applies the *asymmetric Nash bargaining solution* (*ANBS*) and the *Rubinstein bargaining solution* (*RBS*) to assess the interrelation between different criteria. The *ANBS*, named after mathematician John Nash, is a mathematical framework that determines the joint utility of all participating game players based on their respective disagreement points. Meanwhile, the *RBS*, developed by economist Ariel Rubinstein, facilitates benefit-sharing between two players. Essentially, it represents a non-cooperative extension of the *Nash bargaining solution*, enabling the division of surplus at a particular stage of the negotiation process (Anbarci, 2013; Forgó, 2003; Kim, 2014, Kim, 2015, Kim, 2025; Opricovic, 2004; Zolfani, 2020).

By integrating MCDM, *ANBS*, and *RBS*, a novel control approach - referred to as the *multiple-criteria bargaining solution* (*MCBS*) - is developed to determine the optimal solution. Through consensus among conflicting decision factors, the *MCBS* enables the achievement of a mutually beneficial resolution to the multi-objective problem. Utilizing the core principles of *MCBS*, the *SEEC* scheme effectively selects the most suitable power level for D2D communications (Kim, 2025).

7.5.3 D2D communication system description

As can be seen from Figure 4, the *SEEC* scheme considers a scenario of D2D communications under a single cellular network where n distinct D2D pairs. Let S denote the set of D2D pairs and i denote the pair index. The transmission power of ith pair in the S is set as P_i, which is consumed power in the transmitter device. For a computational simplicity, the values of P_i are discretized; P^{min} and P^{max} are the pre-defined minimum and maximum power thresholds where $P^{min} \leq P_i \leq P^{max}$ (Kim, 2025).

Figure 4. The infrastructure of cellular-assisted D2D communication system (Kim, 2025)

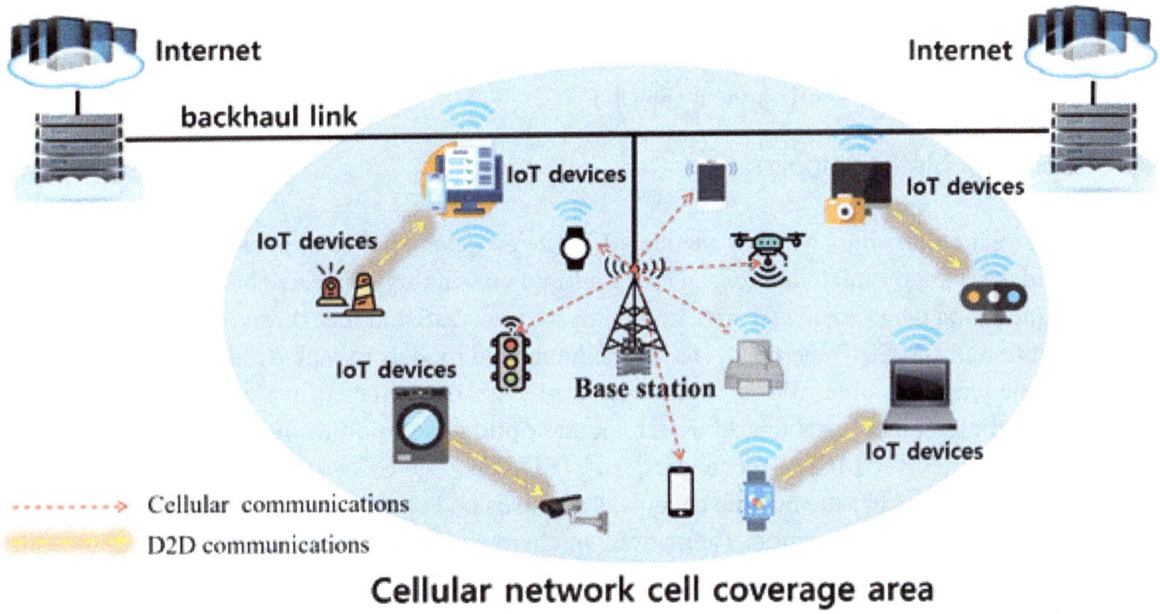

During D2D communications, one D2D pair exchanges information data through direct one-hop D2D link. The transmission capacity of each D2D pair is modeled based on the *signal-to-interference-plus-noise ratio* (*SINR*). Let the *SINR* of ith pair in the S, i.e., Γ_i, be given by (Kim, 2025; Wei, 2015);

$$\Gamma_i \cong \frac{1}{\mathscr{W}} \times \frac{P_i}{N_0 \times \vartheta_i} \tag{7.5.1}$$

where W is the channel bandwidth and N_0 is the power spectral density of additive white Gaussian noise (AWGN). ϑ_i is the path loss for the ith pair. With the P_i and W, the maximum transmission data rate in bits per second for the ith pair, i.e., χ_i, is defined as follows (Kim, 2025; Wei, 2015);

$$\chi_i(P_i, \mathscr{W}) = \mathscr{W} \times \log\left(1 + \left(\frac{1}{\mathscr{W}} \times \frac{P_i \times |h_0|^2}{N_0 \times \vartheta_i}\right)\right) = \mathscr{W} \times \log\left(1 + \left(\Gamma_i \times |h_0|^2\right)\right) \tag{7.5.2}$$

where h_0 is the channel coefficient, which follows a complex normal distribution $\mathscr{CN}(0, 1)$. According to (7.5.1) and (7.5.2), the overall EE and SE for the ith pair, i.e., EE_i and SE_i, can be denoted as follows (Kim, 2025; Wei, 2015);

$$\begin{cases} EE_i = \frac{\chi_i(P_i, \mathscr{W})}{P_i} = \frac{\mathscr{W} \times \log\left(1 + \left(\frac{1}{\mathscr{W}} \times \frac{P_i \times |h_0|^2}{N_0 \times \vartheta_i}\right)\right)}{P_i} \\ SE_i = \frac{\chi_i(P_i, \mathscr{W})}{\mathscr{W}} = \log\left(1 + \left(\frac{1}{\mathscr{W}} \times \frac{P_i \times |h_0|^2}{N_0 \times \vartheta_i}\right)\right) \end{cases} \quad (7.5.3)$$

7.5.4 Technical Methodologies

In recent years, numerous MCDM methods have been developed to support the decision-making process in identifying optimal solutions across multiple conflicting criteria. This research spans various fields, including operational research, computer science, decision theory, and management, among others. Traditionally, MCDM methods employ mathematical models to rank decision alternatives from the most to the least favorable. Within these frameworks, decision-makers evaluate scenarios based on their individual preferences, striving to determine the optimal solution despite competing objectives. Among the widely adopted MCDM techniques, the VIKOR and TOPSIS methods have gained significant popularity. Both approaches are utilized to rank alternatives and classify selected options by measuring their distance from the ideal solution. Their strong mathematical foundations and practical applicability have contributed to their widespread use in the MCDM domain (Hezer, 2021; Kim, 2025; Mateusz, 2018; Opricovic, 2004; Shekhovtsov, 2020).

The VIKOR and TOPSIS methods begin with a decision matrix that consists of alternatives and evaluation criteria, with data normalization serving as the initial step. Through the normalization process, the decision matrix is converted into a standardized form. In the VIKOR method, the best and worst values for each evaluation criterion are identified, and alternatives are assessed based on an aggregate function. Ultimately, the alternative closest to the ideal solution is selected. Conversely, in the TOPSIS method, both positive and negative ideal solutions are determined, followed by the calculation of separation measures for each alternative. Using the Euclidean distance metric, the most suitable alternative is identified. In both approaches, a compromise ranking is derived by comparing the proximity values of alternatives to the ideal solution, leading to the selection of the best option. However, these methods differ in terms of weighting, aggregation processes, and normalization techniques (Hezer, 2021; Kim, 2025; Mateusz, 2018; Opricovic, 2004; Shekhovtsov, 2020; Zolfani, 2020).

Although both methods can serve as a foundation for negotiation by incorporating criteria weights, they have certain limitations. The VIKOR method lacks a mechanism for weight elicitation and does not include a consistency check for judgment. Meanwhile, the TOPSIS method does not account for the relative importance of distance from the ideal solution (Hezer, 2021; Kim, 2025; Mateusz, 2018; Opricovic, 2004; Shekhovtsov, 2020).

The *ANBS*, first introduced in a foundational paper by J. F. Nash, is a conflict resolution method designed to model negotiation and group decision-making processes. It is one of the most widely recognized agreement concepts in cooperative game theory, enabling game participants to efficiently and fairly distribute their utilities in order to reach a mutually acceptable agreement. Based on the information about each player's preferences, the *ANBS* is also a distance-based method where the geometric distance from the disagreement vector (*d*) is maximized; d is the utilities that results if no agreement is reached by game players (Forgó, 2003). Simply, consider a two-player bargaining problem. It is defined as the set of utility possibilities (S) and d where $S \subset \mathbb{R}^2$ and $d \in S$. The *ANBS* with weight $\alpha \in (0,1)$, i.e., $ANBS^{\alpha}(S,d)$, is defined as follows (Anbarci, 2013; Kim, 2025);

$$ANBS^{\alpha}(S,d) = \arg\max_{(x_1,x_2)}\left\{(x_1 - d_1)^{\alpha} \times (x_2 - d_2)^{1-\alpha} | x \in S, x \geq d\right\} \tag{7.5.4}$$

In most real-world bargaining scenarios, multiple conflicting issues must be addressed. However, existing studies on the *ANBS* have not fully considered the impact of multiple interdependent criteria. As a result, the original *ANBS* framework cannot be directly applied to solve multi-objective decision-making problems (Kim, 2025).

A. Rubinstein initially defined a multi-stage, non-cooperative, two-player game with complete information. Within this extensive-form framework, the *RBS* was developed, utilizing an alternating-offers approach over an infinite time horizon. In the *RBS*, two players have their own bargaining powers, i.e., δ_1 and δ_2, and they negotiate with each other through several rounds of offers and counter offers to divide the benefit. According to the idea of *subgame perfect Nash equilibrium*, the *RBS* provides an agreement (x_1^*, x_2^*) where x_1^* and x_2^* are obtained profits for player 1 and 2, respectively (Kim, 2014, Kim, 2015, Kim, 2025).

$$(x_1^*, x_2^*) = \begin{cases} \left(\frac{1-\delta_2}{1-(\delta_1 \times \delta_2)}, \frac{\delta_2 \times (1-\delta_1)}{1-(\delta_1 \times \delta_2)}\right) ifplayer1offersfirst \\ \left(\frac{\delta_1 \times (1-\delta_2)}{1-(\delta_1 \times \delta_2)}, \frac{1-\delta_1}{1-(\delta_1 \times \delta_2)}\right) ifplayer2offersfirst \end{cases} \tag{7.5.5}$$

$$\text{s.t., } x_1^* + x_2^* = 1,\ x_1^* \geq 0, x_2^* \geq 0 \text{ and } 0 \leq \delta_1, \delta_s \leq 1$$

For a long time, finding a solution to non-cooperative bargaining games remained a challenge. As a *subgame-perfect equilibrium*, the *RBS* represents one of the most significant discoveries in cooperative bargaining theory. However, while the *RBS* offers an elegant uniqueness result for two-player negotiation scenarios, it becomes ineffective when extended to multiple participants (Kim, 2014, Kim, 2025).

7.5.5. The *MCBS* based power control algorithm

In the viewpoint of devices, the perceptions of EE and SE can be main factors for D2D communications, and they are defined directly related to the given power level. Therefore, the *SEEC* scheme assumes that power levels in $[P^{min}, \ldots, P^{max}]$ are available solution, and the EE and SE are decision criteria. For the MCDM process, a decision matrix $\mathbb{M} = \left[A_{ij}\right]_{n\times 2}$ is constructed where A_{ij} represents the value of ith (i = 1, 2, ..., n) available power levels under the jth (j =1, 2) evaluation criterion; j =1 is EE and j =2 is SE. And then, the M is transformed into a normalized matrix. The normalized value of A_{ij}, i.e., $\overline{A_{ij}}$, is calculated as follows (Kim, 2025);

$$\overline{A_{ij}} = \frac{A_{ij} - A_j^-}{A_j^* - A_j^-}\text{ , s.t., }\begin{cases} A_j^* = \max\limits_i \left\{ A_{ij} | j \in \{EE, SE\} \right\} \\ \\ A_j^- = \min\limits_i \left\{ A_{ij} | j \in \{EE, SE\} \right\} \end{cases}$$

(7.5.6)

Based on the normalized matrix, the *SEEC* scheme selects the best solution for the power decision problem. To satisfy this goal, each available solution should be evaluated and compared according to an agreed manner. The *SEEC* scheme adopts the bargaining game model to estimate each solution. Therefore, each criterion such as EE and SE is assumed as a game player and the power level is their strategy. In the **i**th solution, the utility functions of EE and SE, i.e., $U_{EE}^i(\cdot)$ and $U_{SE}^i(\cdot)$, are obtained based on the power levels $(P_i \in [P^{min}, \ldots, P^{max}])$ of transmitter device. According to the Weber-Fechner Law (Reichl, 2013), they are defined as follows (Kim, 2025);

$$\begin{cases} U_{EE}^i\left(\overline{A_{i,j=EE}}\right) = \exp\left(\eta \times \overline{A_{i,j=EE}}\right) - \gamma \\ \\ U_{SE}^i\left(\overline{A_{i,j=SE}}\right) = \log\left(\left(\theta \times \overline{A_{i,j=SE}}\right) + \gamma\right) \end{cases}$$

(7.5.7)

where ç, ã and è are adjustment parameters for the $U_{EE}^i(\cdot)$ and $U_{SE}^i(\cdot)$.

In the viewpoint of MCDM method, the EE and SE are considered as decision criteria. From the perspective of bargaining game theory, they are considered as game players. For the both cases, weights (Ψ) should be introduced to represent the relative importance of EE and SE. Usually, weights express the decision maker's preference for modelling the actual decision making process. Therefore, weights can significantly shift the outcome of negotiation process, leading to more advantageous positions for one who possesses a greater leverage (Anbarci, 2013; Forgó, 2003; Kim, 2014, Kim, 2025).

For the D2D power decision problem, the EE and SE have their weights. To adapt the highly dynamic D2D communication situation, the *SEEC* scheme has to decide carefully the weights of EE and SE, i.e., Ψ_{EE} and Ψ_{SE}, respectively. The *SEEC* scheme uses the *RBS* to decide the criteria weight values; it is the most adaptable negotiation solution for a two-player bargaining model while reflecting each player's current situation. For the suitable wireless communication, the energy status of data transmitter device is a major factor. Based on this consideration, the EE offers first, and its bargaining power (δ_{EE}) is defined as follows (Kim, 2025);

$$\delta_{EE} = \log\left(\left(\beta \times \frac{1 - \varepsilon_c^D}{\Delta^D}\right) + 1\right) + \eta$$

(7.5.8)

where â, ç are control factors for the δ_{EE}, and Δ^D, ε_c^D are the initial and current energy amount of the transmitter device, respectively. According to (7.5.5) and (7.5.8), the values of Ψ_{EE} and Ψ_{SE} are given by (Kim, 2025);

$$\begin{cases} \Psi_{EE} = \dfrac{1 - \delta_{SE}}{1 - (\delta_{EE} \times \delta_{SE})} \\ \qquad\qquad\qquad\qquad \text{s.t.,} \ \delta_{SE} = 1 - \delta_{EE} \ (7.5.9) \\ \Psi_{SE} = \dfrac{\delta_{SE} \times (1 - \delta_{EE})}{1 - (\delta_{EE} \times \delta_{SE})} \end{cases} \quad (7.5.9)$$

In the *SEEC* scheme, the payoff of ith solution, i.e., $\mathcal{Q}_i$, is estimated based on the idea of *ANBS*. According to (7.5.4) and (7.5.7)-(7.5.9), the $\mathcal{Q}_i$ is given by (Kim, 2025);

$$\mathcal{Q}_i = \prod_{j \in \{EE, SE\}} \left(U_j^i(\overline{A_{i,j}}) - d_j\right)^{\Psi_j}, \ \text{s.t.,} \ i \in \{1, 2, \ldots, n\}$$

(7.5.10)

Among the $\mathcal{Q}_{i \in \{1,2,\ldots,n\}}$ values of available solutions, the highest Q value, i.e., $k = \arg \max_{k \in \{1,2,\ldots,n\}} (\mathcal{Q}_k)$, is selected as the *MCBS*. Finally, the power of transmitter devices is decided as the kth power level for D2D communication. The diagram of the reciprocal combination of MCDM, *ANBS* and *RBS* is illustrated in Figure 5, (Kim, 2025).

Figure 5. The diagram of D2D power control model (Kim, 2025)

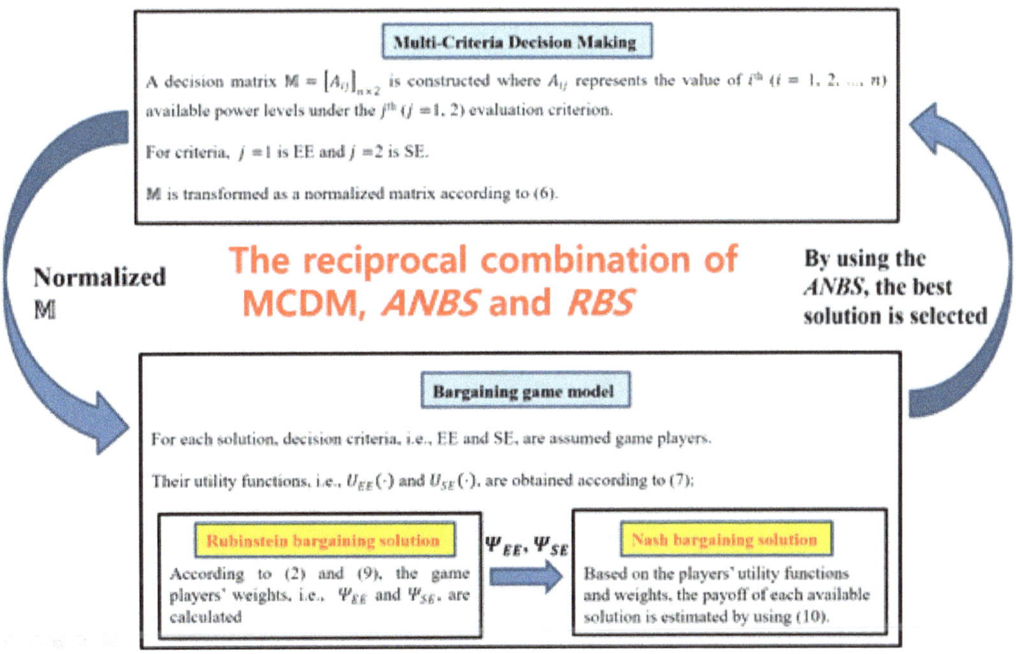

7.5.6 Summary

Maximizing both EE and SE in D2D communications is a key priority for future cellular networks. However, optimizing EE and SE does not always align and can sometimes lead to conflicting objectives. Therefore, finding an effective balance between EE and SE remains a critical research challenge. In the *SEEC* scheme, a robust multi-objective optimization approach has been explored for D2D communications, leading to the development of an innovative power control mechanism that balances EE and SE. To address the conflict between these criteria, the *SEEC* scheme incorporates a distance-based computation technique for multi-criteria decision problems and introduces a novel control framework known as the *MCBS*. By integrating the *ANBS*, *RBS*, and MCDM methods, the primary goal of *MCBS* is to enhance spectrum utilization while minimizing power consumption in devices. Leveraging this hybrid control mechanism, the *SEEC* scheme effectively resolves conflicting interests, ultimately improving system throughput, device performance, and energy efficiency within the D2D network platform (Kim, 2025).

REFERENCES

Afzal, A., Zaidi, S. A. R., McLernon, D., & Ghogho, M. (2016, May). On the analysis of cellular networks with caching and coordinated device-to-device communication. In *2016 IEEE International Conference on Communications (ICC)* (pp. 1-7). IEEE. DOI: 10.1109/ICC.2016.7510696

Anbarci, N., & Sun, C. J. (2013). Asymmetric Nash bargaining solutions: A simple Nash program. *Economics Letters, 120*(2), 211–214. DOI: 10.1016/j.econlet.2013.04.026

Anbarci, N., & Yi, G. (1992). A meta-allocation mechanism in cooperative bargaining. *Economics Letters, 38*(2), 175–179. DOI: 10.1016/0165-1765(92)90050-9

Anbiyaei, M., Faez, K., & Najimi, M. (2019). Energy-Efficient resource allocation for device-to-device underlay communications in cellular networks. *IET Signal Processing, 13*(6), 633–639. DOI: 10.1049/iet-spr.2018.5110

Atat, R., & Liu, L. (2016, December). On the achievable transmission capacity of secrecy-based D2D cellular networks. In *2016 IEEE Global Communications Conference (GLOBECOM)* (pp. 1-6). IEEE. DOI: 10.1109/GLOCOM.2016.7842047

Bastug, E., Bennis, M., & Debbah, M. (2014). Living on the edge: The role of proactive caching in 5G wireless networks. *IEEE Communications Magazine, 52*(8), 82–89. DOI: 10.1109/MCOM.2014.6871674

Bergantiños, G., Casas-Méndez, B., Fiestras-Janeiro, M. G., & Vidal-Puga, J. J. (2005). *A focal-point solution for bargaining problems with coalition structure*. University Library of Munich.

Bhardwaj, A., & Agnihotri, S. (2018). Energy-and spectral-efficiency trade-off for D2D-multicasts in underlay cellular networks. *IEEE Wireless Communications Letters, 7*(4), 546–549. DOI: 10.1109/LWC.2018.2794353

Bulti, D. A., & Raimond, K. (2015). Optimizing caching in a patch streaming multimedia-on-demand system. *Journal of Computing Science and Engineering : JCSE, 9*(3), 134–141. DOI: 10.5626/JCSE.2015.9.3.134

Byun, H., & So, J. (2015). Node scheduling control inspired by epidemic theory for data dissemination in wireless sensor-actuator networks with delay constraints. *IEEE Transactions on Wireless Communications, 15*(3), 1794–1807. DOI: 10.1109/TWC.2015.2496596

Cai, J., & Pooch, U. (2004, April). Allocate fair payoff for cooperation in wireless ad hoc networks using shapley value. In *18th International Parallel and Distributed Processing Symposium, 2004. Proceedings.* (p. 219). IEEE.

Chae, S., & Heidhues, P. (2004). A group bargaining solution. *Mathematical Social Sciences, 48*(1), 37–53. DOI: 10.1016/j.mathsocsci.2003.11.002

Chen, K. C. (2012). Machine-to-machine communications for healthcare. *Journal of Computing Science and Engineering : JCSE, 6*(2), 119–126. DOI: 10.5626/JCSE.2012.6.2.119

Cheng, F., Yu, Y., Zhao, Z., Zhao, N., Chen, Y., & Lin, H. (2017). Power allocation for cache-aided small-cell networks with limited backhaul. *IEEE Access : Practical Innovations, Open Solutions, 5*, 1272–1283. DOI: 10.1109/ACCESS.2017.2656384

Chour, H., Jorswieck, E. A., Bader, F., Nasser, Y., & Bazzi, O. (2019). Global optimal resource allocation for efficient FD-D2D enabled cellular network. *IEEE Access : Practical Innovations, Open Solutions*, *7*, 59690–59707. DOI: 10.1109/ACCESS.2019.2914973

Dash, R. K., Jennings, N. R., & Parkes, D. C. (2005). Computational-mechanism design: A call to arms. *IEEE Intelligent Systems*, *18*(6), 40–47. DOI: 10.1109/MIS.2003.1249168

Desale, Y., Bakade, K., & Wagh, S. (2016, November). Optimum protocol for route discovery in device to device communication. In *International Conference on Signal Processing (ICSP 2016)* (p. 19). Stevenage UK: IET. DOI: 10.1049/cp.2016.1458

Fodor, G., Dahlman, E., Mildh, G., Parkvall, S., Reider, N., Miklós, G., & Turányi, Z. (2012). Design aspects of network assisted device-to-device communications. *IEEE Communications Magazine*, *50*(3), 170–177. DOI: 10.1109/MCOM.2012.6163598

Forgó, F., & Szidarovszky, F. (2003). On the relation between the Nash bargaining solution and the weighting method. *European Journal of Operational Research*, *147*(1), 108–116. DOI: 10.1016/S0377-2217(02)00241-2

Gismalla, M. S. M., Azmi, A. I., Salim, M. R. B., Abdullah, M. F. L., Iqbal, F., Mabrouk, W. A., Othman, M. B., Ashyap, A. Y. I., & Supa'at, A. S. M. (2022). Survey on device to device (D2D) communication for 5GB/6G networks: Concept, applications, challenges, and future directions. *IEEE Access : Practical Innovations, Open Solutions*, *10*, 30792–30821. DOI: 10.1109/ACCESS.2022.3160215

Hajmohammad, S., & Elbiaze, H. (2013, June). Unlicensed spectrum splitting between femtocell and WiFi. In *2013 IEEE international conference on communications (ICC)* (pp. 1883-1888). IEEE. DOI: 10.1109/ICC.2013.6654796

Han, K., Li, J., Zhu, P., & Wang, X. (2007, September). The learning-bargaining algorithm used in unlicensed spectrum pre-allocation. In *2007 International Conference on Wireless Communications, Networking and Mobile Computing* (pp. 1264-1269). IEEE. DOI: 10.1109/WICOM.2007.322

Hao, Y., Ni, Q., Li, H., & Hou, S. (2018). Robust multi-objective optimization for EE-SE tradeoff in D2D communications underlaying heterogeneous networks. *IEEE Transactions on Communications*, *66*(10), 4936–4949. DOI: 10.1109/TCOMM.2018.2834920

Hasan, M., & Hossain, E. (2014). Distributed resource allocation for relay-aided device-to-device communication: A message passing approach. *IEEE Transactions on Wireless Communications*, *13*(11), 6326–6341. DOI: 10.1109/TWC.2014.2330831

Hezer, S., Gelmez, E., & Özceylan, E. (2021). Comparative analysis of TOPSIS, VIKOR and COPRAS methods for the COVID-19 Regional Safety Assessment. *Journal of Infection and Public Health*, *14*(6), 775–786. DOI: 10.1016/j.jiph.2021.03.003 PMID: 34022737

Huang, J., Huang, S., Xing, C. C., & Qian, Y. (2018). Game-theoretic power control mechanisms for device-to-device communications underlaying cellular system. *IEEE Transactions on Vehicular Technology*, *67*(6), 4890–4900. DOI: 10.1109/TVT.2018.2800051

Jia, X., Fan, Q., Xu, W., & Yang, L. (2019). Cross-tier dual-connectivity designs of three-tier hetnets with decoupled uplink/downlink and global coverage performance evaluation. *IEEE Access : Practical Innovations, Open Solutions*, *7*, 16816–16836. DOI: 10.1109/ACCESS.2019.2895389

Jiang, J., Zhang, S., Li, B., & Li, B. (2015). Maximized cellular traffic offloading via device-to-device content sharing. *IEEE Journal on Selected Areas in Communications*, *34*(1), 82–91. DOI: 10.1109/JSAC.2015.2452493

Kakade, S. M., Lobel, I., & Nazerzadeh, H. (2011). Optimal dynamic mechanism design via a virtual VCG mechanism. *ACM SIGecom Exchanges*, *10*(1), 27–30. DOI: 10.1145/1978721.1978728

Kalai, E. (1983). *Solutions to the bargaining problem* (Discussion Paper No. 556). Northwestern University, Center for Mathematical Studies in Economics and Management Science.

Karlin, A. R., & Peres, Y. (2017). *Game theory, alive* (Vol. 101). American Mathematical Soc. DOI: 10.1090/mbk/101

Kim, K. S., Uno, S., & Kim, M. W. (2010). Adaptive QoS mechanism for wireless mobile network. *Journal of Computing Science and Engineering : JCSE*, *4*(2), 153–172. DOI: 10.5626/JCSE.2010.4.2.153

Kim, S. (2012). Multi-leader multi-follower Stackelberg model for cognitive radio spectrum sharing scheme. *Computer Networks*, *56*(17), 3682–3692. DOI: 10.1016/j.comnet.2012.08.004

Kim, S. (Ed.). (2014). *Game theory applications in network design*. IGI Global. DOI: 10.4018/978-1-4666-6050-2

Kim, S. (2015). Nested game-based computation offloading scheme for mobile cloud IoT systems. *EURASIP Journal on Wireless Communications and Networking*, *2015*(1), 1–11. DOI: 10.1186/s13638-015-0456-5

Kim, S. (2017). D2D-Enabled Small Cell Network Control Scheme Based on the Dynamic Stackelberg Game. *Mobile Information Systems*, *2017*(1), 7965767. DOI: 10.1155/2017/7965767

Kim, S. (2019). A New Cooperative Dual-Level Game Approach for Operator-Controlled Multihop D2D Communications. *Mobile Information Systems*, *2019*(1), 6276872. DOI: 10.1155/2019/6276872

Kim, S. (2020a). D2D enabled cellular network spectrum allocation scheme based on the cooperative bargaining solution. *IEEE Access : Practical Innovations, Open Solutions*, *8*, 53710–53719. DOI: 10.1109/ACCESS.2020.2981290

Kim, S. (2020b). Heterogeneous Network Spectrum Allocation Scheme for Network-Assisted D2D Communications. *Mobile Information Systems*, *2020*(1), 8825119. DOI: 10.1155/2020/8825119

Kim, S. (2025). A New Spectrum and Energy Efficiency Trade-Off Control Paradigm for D2D Communications. *International Journal of Communication Systems*, *38*(5), e6030. DOI: 10.1002/dac.6030

Lee, K., & Shin, I. (2015). User mobility model based computation offloading decision for mobile cloud. *Journal of Computing Science and Engineering : JCSE*, *9*(3), 155–162. DOI: 10.5626/JCSE.2015.9.3.155

Lema, M. A., Pardo, E., Galinina, O., Andreev, S., & Dohler, M. (2016). Flexible dual-connectivity spectrum aggregation for decoupled uplink and downlink access in 5G heterogeneous systems. *IEEE Journal on Selected Areas in Communications*, *34*(11), 2851–2865. DOI: 10.1109/JSAC.2016.2615185

Li, J., Sun, J., Qian, Y., Shu, F., Xiao, M., & Xiang, W. (2016). A commercial video-caching system for small-cell cellular networks using game theory. *IEEE Access : Practical Innovations, Open Solutions*, *4*, 7519–7531. DOI: 10.1109/ACCESS.2016.2582836

Liu, Y., Wang, Y., Sun, R., & Miao, Z. (2019). Distributed resource allocation for D2D-assisted small cell networks with heterogeneous spectrum. *IEEE Access : Practical Innovations, Open Solutions*, *7*, 83900–83914. DOI: 10.1109/ACCESS.2019.2924245

Long, K., Wang, P., Li, W., & Chen, D. (2019). Spectrum resource and power allocation with adaptive proportional fair user pairing for NOMA systems. *IEEE Access : Practical Innovations, Open Solutions*, *7*, 80043–80057. DOI: 10.1109/ACCESS.2019.2908673

Lu, X., Wang, P., & Niyato, D. (2015, March). Hierarchical cooperation for operator-controlled device-to-device communications: A layered coalitional game approach. In *2015 IEEE Wireless Communications and Networking Conference (WCNC)* (pp. 2056-2061). IEEE.

Marco-Gil, M. D. C., Peris, J. E., & Subiza, B. (2012). *A concessions-based mechanism for meta-bargaining problems* (No. 12-13). University of Alicante, D. Quantitative Methods and Economic Theory.

Mateusz, P., Danuta, M., Małgorzata, Ł., Mariusz, B., & Kesra, N. (2018). TOPSIS and VIKOR methods in study of sustainable development in the EU countries. *Procedia Computer Science*, *126*, 1683–1692. DOI: 10.1016/j.procs.2018.08.109

Naeve-Steinweg, E. (1999). A note on van Damme's mechanism. *Review of Economic Design*, *4*(2), 179–187. DOI: 10.1007/s100580050032

Naeve-Steinweg, E. (2004). The averaging mechanism. *Games and Economic Behavior*, *46*(2), 410–424. DOI: 10.1016/S0899-8256(03)00123-4

Niyato, D., & Hossain, E. (2006, June). A cooperative game framework for bandwidth allocation in 4G heterogeneous wireless networks. In *2006* []. IEEE.]. *IEEE International Conference on Communications : [proceedings]. IEEE International Conference on Communications*, *9*, 4357–4362.

Opricovic, S., & Tzeng, G. H. (2004). Compromise solution by MCDM methods: A comparative analysis of VIKOR and TOPSIS. *European Journal of Operational Research*, *156*(2), 445–455. DOI: 10.1016/S0377-2217(03)00020-1

Pan, C., Xu, W., Zhang, W., Wang, J., Ren, H., & Chen, M. (2015). Weighted sum energy efficiency maximization in ad hoc networks. *IEEE Wireless Communications Letters*, *4*(3), 233–236. DOI: 10.1109/LWC.2015.2401580

Park, E. C., Kim, D. Y., Choi, C. H., & So, J. (2007). Improving quality of service and assuring fairness in WLAN access networks. *IEEE Transactions on Mobile Computing*, *6*(4), 337–350. DOI: 10.1109/TMC.2007.53

Park, G. S., & Song, H. (2019). Video quality-aware traffic offloading system for video streaming services over 5G networks with dual connectivity. *IEEE Transactions on Vehicular Technology, 68*(6), 5928–5943. DOI: 10.1109/TVT.2019.2909547

Pham, Q. V., & Hwang, W. J. (2017). Fairness-aware spectral and energy efficiency in spectrum-sharing wireless networks. *IEEE Transactions on Vehicular Technology, 66*(11), 10207–10219. DOI: 10.1109/TVT.2017.2717926

Pu, L., Chen, X., Xu, J., & Fu, X. (2016). D2D fogging: An energy-efficient and incentive-aware task offloading framework via network-assisted D2D collaboration. *IEEE Journal on Selected Areas in Communications, 34*(12), 3887–3901. DOI: 10.1109/JSAC.2016.2624118

Reichl, P., Tuffin, B., & Schatz, R. (2013). Logarithmic laws in service quality perception: Where microeconomics meets psychophysics and quality of experience. *Telecommunication Systems, 52*, 587–600.

Shah, S. W. H., Mian, A. N., & Crowcroft, J. (2020). Statistical QoS guarantees for licensed-unlicensed spectrum interoperable D2D communication. *IEEE Access : Practical Innovations, Open Solutions, 8*, 27277–27290. DOI: 10.1109/ACCESS.2020.2970829

Shaikh, F. S., & Wismüller, R. (2018). Routing in multi-hop cellular device-to-device (D2D) networks: A survey. *IEEE Communications Surveys and Tutorials, 20*(4), 2622–2657. DOI: 10.1109/COMST.2018.2848108

Shakir, M. Z., Ismail, M., Wang, X., Qaraqe, K. A., & Serpedin, E. (2017). From D2D to Ds2D: Prolonging the battery life of mobile devices via Ds2D communications. *IEEE Wireless Communications, 24*(4), 55–63. DOI: 10.1109/MWC.2017.1600348

Shekhovtsov, A., & Sałabun, W. (2020). A comparative case study of the VIKOR and TOPSIS rankings similarity. *Procedia Computer Science, 176*, 3730–3740. DOI: 10.1016/j.procs.2020.09.014

So, J., & Vaidya, N. H. (2007). Load-balancing routing in multichannel hybrid wireless networks with single network interface. *IEEE Transactions on Vehicular Technology, 56*(1), 342–348. DOI: 10.1109/TVT.2006.889569

Trockel, W. (2002). A universal meta bargaining implementation of the Nash solution. *Social Choice and Welfare, 19*(3), 581–586. DOI: 10.1007/s003550100133

Wang, J., Sun, Y., Wang, B., Wang, B., Wang, A., Li, S., & Sun, Z. (2019). Resource allocation for D2D video multicast using multi-leader multi-follower stackelberg game with personalized incentives. *IEEE Access : Practical Innovations, Open Solutions, 7*, 117019–117028. DOI: 10.1109/ACCESS.2019.2936283

Wang, X., Kwong, S., Xu, L., & Zhang, Y. (2014). Generalized Nash bargaining solution to rate control optimization for spatial scalable video coding. *IEEE Transactions on Image Processing, 23*(9), 4010–4021. DOI: 10.1109/TIP.2014.2341951 PMID: 25069116

Wei, L., Hu, R. Q., Qian, Y., & Wu, G. (2015). Energy efficiency and spectrum efficiency of multihop device-to-device communications underlaying cellular networks. *IEEE Transactions on Vehicular Technology, 65*(1), 367–380. DOI: 10.1109/TVT.2015.2389823

Wu, Y., Wu, D., Yang, L., Shi, X., Ao, L., & Fu, Q. (2019). Matching-coalition based cluster formation for D2D multicast content sharing. *IEEE Access : Practical Innovations, Open Solutions*, *7*, 73913–73928. DOI: 10.1109/ACCESS.2019.2918942

Wu, Y., Yao, S., Yang, Y., Zhou, T., Qian, H., Hu, H., & Hamalainen, M. (2016). Challenges of mobile social device caching. *IEEE Access : Practical Innovations, Open Solutions*, *4*, 8938–8947. DOI: 10.1109/ACCESS.2016.2633485

Xia, F., Liu, L., Li, J., Ma, J., & Vasilakos, A. V. (2013). Socially aware networking: A survey. *IEEE Systems Journal*, *9*(3), 904–921. DOI: 10.1109/JSYST.2013.2281262

Xiao, Y., Chen, K. C., Yuen, C., Han, Z., & DaSilva, L. A. (2015). A Bayesian overlapping coalition formation game for device-to-device spectrum sharing in cellular networks. *IEEE Transactions on Wireless Communications*, *14*(7), 4034–4051. DOI: 10.1109/TWC.2015.2416178

Xu, X., & Tao, M. (2017). Modeling, analysis, and optimization of coded caching in small-cell networks. *IEEE Transactions on Communications*, *65*(8), 3415–3428.

Yang, L., Wu, D., Xu, S., Zhang, G., & Cai, Y. (2018). Social-energy-aware user clustering for content sharing based on D2D multicast communications. *IEEE Access : Practical Innovations, Open Solutions*, *6*, 36092–36104. DOI: 10.1109/ACCESS.2018.2849204

Yokote, K., Kongo, T., & Funaki, Y. (2018). The balanced contributions property for equal contributors. *Games and Economic Behavior*, *108*, 113–124. DOI: 10.1016/j.geb.2017.08.007

Zhang, H., Song, L., & Zhang, Y. J. (2018). Load balancing for 5G ultra-dense networks using device-to-device communications. *IEEE Transactions on Wireless Communications*, *17*(6), 4039–4050. DOI: 10.1109/TWC.2018.2819648

Zhang, J., Zhang, X., Yan, Z., Li, Y., Wang, W., & Zhang, Y. (2016a, October). Social-aware cache information processing for 5G ultra-dense networks. In *2016 8th International Conference on Wireless Communications & Signal Processing (WCSP)* (pp. 1-5). IEEE.

Zhang, Y., Yang, Y., & Dai, L. (2016b). Energy efficiency maximization for device-to-device communication underlaying cellular networks on multiple bands. *IEEE Access : Practical Innovations, Open Solutions*, *4*, 7682–7691. DOI: 10.1109/ACCESS.2016.2623758

Zolfani, S., Yazdani, M., Pamucar, D., & Zarate, P. (2020). A VIKOR and TOPSIS focused reanalysis of the MADM methods based on logarithmic normalization. *arXiv preprint arXiv:2006.08150.*

Chapter 8
Vehicular Ad Hoc Networks

ABSTRACT

Vehicular ad hoc network consists of smart vehicles on the road and provides communication services among nearby vehicles or with roadside infrastructure. It is a kind of special wireless ad hoc network, which has the characteristics of high node mobility and fast topology changes. Vehicular networks can provide numerous interesting services, range from safety-related warning systems to improved navigation mechanisms as well as information and entertainment applications. In this chapter, the main features of vehicular networks are provided from novel emerging technologies to network control aspects used for vehicular network applications, as well as main issues and challenges. The objective of this chapter is to provide readers with a guideline to easily understand and objectively compare the different models, and eventually identify the one required for their needs.

8. OVERVIEW

Recent advancements in the automotive industry have led to the design of smart vehicles with high-capacity resources for communication, sensing, processing, and storage of data. Providing knowledge of smart vehicles or other entities involved with modern transportation is the key to proper traffic management, efficient use of resources, and to improving transportation safety. Vehicular ad hoc network (VANET) is a subclass of mobile ad hoc networks in which network nodes are road vehicles. In VANET, participating vehicles are equipped with wireless transceivers which allow them to exchange data with other neighboring vehicles. Currently, VANETs present a unique range of challenges and opportunities due to the semi-organized nature of vehicular movements. The deployment of VANETs in the real world is based on the message's correctness and timely delivery and assuredness of privacy protection and data security. In this regard, many researchers have conducted studies that present models and solutions related to the improvement of VANET from different aspects such as architectural design, networking, and data security.

Along with the benefits, there are a large number of challenges in VANET such as provisioning of service quality, high connectivity and bandwidth and security to vehicle and individual privacy. However, until now, there are many open issues and there are several unanswered questions for VANETs. One of the challenges posed by the study of VANETs is the definition of a vehicular mobility model providing an accurate and realistic vehicular mobility description. Another challenge is to be able to dynamically allocate the limited spectrum resources for each vehicle.

DOI: 10.4018/979-8-3373-2013-7.ch008

8.1 Joint Crowdsensing and Offloading Control (*JCOC*) scheme

Edge-assisted vehicular crowdsensing is an emerging paradigm in which intelligent vehicles gather and exchange sensing data at the edge of vehicular networks. With the growing adoption of the Internet of Intelligent Vehicles (IoIV), researchers have increasingly focused on optimizing service offloading. Recently, S. Kim introduced the *Joint Crowdsensing and Offloading Control* (*JCOC*) scheme, which incorporates social welfare considerations (Kim, 2023a). To encourage sensing participation, the *JCOC* scheme employs the principles of *cooperative multi-agent reinforcement learning* (*CMARL*) to design an innovative crowdsensing algorithm. Through cooperative learning, vehicles can autonomously determine the optimal sensing actions in a decentralized manner. Given the computational and communication constraints within the IoIV system, the *Lozano, Hinojosa, and Mármol solution* (*LHMS*) is utilized to efficiently allocate IoIV resources. Resource allocation for offloading services is dynamically adjusted based on each vehicle's sensing contribution. In the *JCOC* scheme, the crowdsensing and offloading mechanisms are tightly integrated, working collaboratively to achieve mutual benefits. The key advantages of this hybrid approach lie in its flexibility and adaptability to evolving IoIV system conditions (Kim, 2023a).

8.1.1 Development motivation

With advancements in IoT, communication technology, and data collection capabilities, smart objects are increasingly being embedded with processors and the ability to communicate over the public Internet. These smart objects can be assigned to IP addresses, allowing them to collect, process, and transmit crucial information across networks. In recent years, the IoT paradigm has gained widespread adoption, significantly transforming both business and consumer landscapes and driving a new industrial revolution. Within the IoT framework, vehicles play a crucial role in enabling safe and efficient travel while contributing to the development of intelligent transportation systems. Traditionally, vehicular ad hoc networks (VANETs) have been established based on mobile ad hoc network principles to enhance traffic flow and driving safety. However, in the era of IoT, VANETs are evolving further into the Internet of Intelligent Vehicles (IoIV) (Guan, 2018; Kim, 2023a).

As autonomous driving technology advances, IoIV is becoming an integral part of daily life. Modern intelligent vehicles are now equipped with multi-sensory platforms and onboard units (OBUs), enabling them to communicate with one another and with local network infrastructures, such as roadside units (RSUs). These RSUs, which serve as Internet access points along roadsides, primarily function to collect and analyze traffic data from intelligent vehicles while also providing drivers with Internet connectivity. Furthermore, the rapid expansion of IoIV has fueled the rise of computation-intensive and bandwidth-demanding applications. Industry forecasts predict that these applications will dominate IoIV traffic in the near future. As a result, the surge in computing-driven applications has led to the emergence of a new research domain known as mobile edge computing (MEC) (Chen, 2022; Kim, 2023a; Zhu, 2020).

IoIV applications are typically delay-sensitive and resource-intensive, requiring complex computations. However, individual vehicles often have limited computing power and storage capacity. To ensure the efficient and stable execution of onboard applications, MEC enhances real-time data processing by utilizing edge computing resources. By offloading computationally demanding tasks to MEC servers, smart vehicles can meet the diverse requirements of multiple onboard applications while extending the computing capabilities of existing IoIV infrastructure. This approach aligns with the vision of future networks, which emphasize high bandwidth, intensive computation, and real-time interaction. Therefore, RSUs equipped with MEC servers offer a viable solution to the computing offloading challenges in IoIV environments (Chen, 2022; Lang, 2022; Kim, 2023a).

In recent years, intelligent vehicles have been increasingly equipped with programmable and powerful mobile sensing components. Given their continuous mobility along roads, these vehicles have significant potential for collecting valuable data for traffic management and environmental monitoring. As a result, mobile crowdsensing has emerged as a novel sensing and computing paradigm, leveraging outsourced sensing tasks performed by connected vehicles. In a crowdsensing system, RSUs and sensor-equipped vehicles collaborate to execute large-scale sensing tasks, providing up-to-date and fine-grained information across wide geographical areas. However, transmitting large volumes of sensing data to remote cloud servers can result in significant bandwidth consumption and increased transmission delays. MEC addresses these challenges by bringing computational resources closer to vehicles, reducing access overheads and improving the efficiency of mobile crowdsensing applications (Gao, 2018; Kim, 2023a ; Liu, 2020).

Despite its technical advantages, the MEC-assisted crowdsensing (MECC) system faces several key challenges. First, smart vehicles operate autonomously and are not obligated to contribute their resources to crowdsensing tasks. Since most vehicles have limited computational and energy resources, they prioritize their own operational needs. Participating in crowdsensing requires time and energy for data collection and transmission, necessitating a well-designed incentive mechanism with fair compensation. Second, the interactions between independent decision-makers - including vehicles, RSUs, and MEC servers - are complex and difficult to model due to their self-interested nature. As a result, determining an optimal solution for resource allocation in such a dynamic environment remains a challenging task. Third, addressing the joint problem of crowdsensing and offloading services requires an effective control algorithm. While various algorithms have been proposed for crowdsensing, vehicular offloading, and MECC control, many existing studies fail to consider the interdependencies between these critical issues (Kim, 2023a; Li, 2022).

Decentralized approaches have gained traction in recent years for addressing complex network control problems. Many of these approaches fall within the domain of *cooperative multi-agent reinforcement learning* (*CMARL*), a methodology that has garnered significant interest in the machine learning community. CMARL enables agents to achieve their objectives through interactions with neighboring agents, enhancing both system robustness and efficiency. In this framework, agents actively sense their environment, evaluate possible actions, and adapt their future decisions accordingly. By mapping environmental states to specific actions, individual agents can achieve cooperative action selection, ensuring optimal collaboration and coordination. The *CMARL* focuses on decision-making strategies that enable agents to overcome cooperation dilemmas and reach joint agreements (Kim, 2023a; Wang, 2023).

The *JCOC* scheme addresses the challenges of crowdsensing and offloading control in the MECC environment. Building on the aforementioned concepts, the *JCOC* scheme integrates principles from dynamic game theory and *CMARL* to develop a unified crowdsensing and offloading control framework. To balance performance trade-offs across diverse service requirements, the *JCOC* scheme incorporates two key components: a crowdsensing algorithm and an offloading algorithm. The crowdsensing algorithm enables individual vehicles to determine their sensing contributions autonomously, leveraging the *CMARL* to optimize their sensing actions in a distributed manner. Meanwhile, the offloading algorithm facilitates collaboration between RSUs and MEC servers to efficiently handle offloading requests. Within the dynamic game model, RSUs allocate communication resources, while MEC servers provide computational resources in a centralized manner. These two tightly integrated control mechanisms enable seamless operation of crowdsensing and offloading services within the MECC infrastructure (Kim, 2023a)

In dynamic games, players repeatedly select and implement strategies from their available options, adjusting their choices based on successive interactions. A subfield of dynamic games, known as dynamic bargaining, games, specifically addresses decision-making problems involving complex and interactive strategies. The foundational concept of bargaining game theory was introduced by J. Nash, who formulated a bargaining model based on an expected utility function that defines feasible agreements and the corresponding outcomes in case of disagreement. The Nash bargaining game framework is designed to determine an equilibrium outcome that reflects the payoffs of all participating players. In recent years, multiple alternative approaches have been proposed to address classical bargaining problems (Kim 2023a; Lozano, 2019).

In 2019, Lozano, Hinojosa, and Mármol extended Nash's bargaining solution to the field of data envelopment analysis, introducing the *Lozano, Hinojosa, and Mármol solution* (*LHMS*). This approach retains several key properties of Nash's original solution while offering a new perspective on projecting inefficient decision-making units onto an efficient frontier. Specifically, the *LHMS* conceptualizes the projection process as a bargaining problem, where input and output dimensions serve as negotiating entities. Unlike conventional methods that globally determine target values from a system-wide perspective, the *LHMS* establishes input and output target values through a bargaining process that considers observed values as disagreement points. As a result, the final solution emerges from an impartial and rational arbitration between input and output variables (Kim 2023a; Lozano, 2019).

Reinforcement learning is a widely used technique for optimizing agent strategies in Markov decision processes. While reinforcement learning has primarily been applied to single-agent environments, many real-world problems require multiple agents to interact and collaborate. With the increasing adoption of game theory, multi-agent learning has become a prevalent approach for addressing cooperative decision-making challenges. CMARL enhances system robustness and efficiency by enabling multiple agents to coordinate their actions through interaction with neighboring agents. This method facilitates the identification of an optimal joint strategy that maximizes the expected cumulative reward for all participating agents. Game-theoretic analyses of *CMARL* have predominantly focused on repeated games involving a limited number of agents and action choices (Cui, 2021; Kim, 2023a).

8.1.2 Contribution

The *JCOC* scheme is designed to enhance the performance of the MECC system by integrating crowdsensing and offloading algorithms. Leveraging the principles of the *LHMS* and *CMARL*, the *JCOC* scheme intelligently incentivizes vehicles to engage in the crowdsensing process while dynamically managing offloading services. To ensure the efficient functioning of the MECC infrastructure, the crowdsensing and offloading algorithms are seamlessly integrated and executed iteratively in a step-by-step manner. By harnessing the synergies between cooperative learning and bargaining game theory, the *JCOC* scheme aims to achieve a socially optimal solution that balances system performance and resource allocation (Kim, 2023a)

8.1.3 IoIV system infrastructure and bargaining and learning models

We consider a typical edge assisted IoIV system platform, which consists of n RSUs, i.e., $\mathbb{R} = \{\mathcal{R}_1, \ldots, \mathcal{R}_n\}$, n MEC servers, i.e., $\mathbb{M} = \{\mathcal{M}_1, \ldots, \mathcal{M}_n\}$ and m intelligent vehicles, i.e., $\mathbb{V} = \{\mathcal{V}_1, \ldots, \mathcal{V}_m\}$. Each RSU is equipped with one MEC server where $\mathcal{R}_i \in \mathbb{R}$ and $\mathcal{M}_i \in \mathbb{M}$ are tightly coupled as a pair $[\mathcal{R}_i, \mathcal{M}_i]$. $\mathcal{R}_{1 \leq i \leq n}$ has a coverage area and a fixed spectrum band $\left(\mathfrak{M}_{\mathcal{R}_i}\right)$, and its equipped $\mathcal{M}_i$ has its own computation capacity $\left(\mathfrak{N}_{\mathcal{M}_i}\right)$. Each vehicle $\left(\mathcal{V}_{1 \leq j \leq m}\right)$ is covered by its corresponding RSU, and has the offload task $\mathcal{T}_{\mathcal{V}_j} = \left\{\mathcal{B}_{\mathcal{V}_j}, \mathcal{C}_{\mathcal{V}_j}\right\}$ where $\mathcal{B}_{\mathcal{V}_j}$ and $\mathcal{C}_{\mathcal{V}_j}$ are required communication and computation resources. When the $\mathcal{V}_j$ runs within the $\mathcal{R}_i$'s covering area, the $\mathcal{V}_j$ offloads his $\mathcal{T}_{\mathcal{V}_j}$ to the $\mathcal{M}_i$ through $\mathcal{R}_i$ (Liu, 2020, Kim, 2023a).

In the *JCOC* scheme, individual vehicles function as contributors to the mobile crowdsensing process, utilizing their onboard sensors to carry out sensing tasks and transmitting the collected data to their designated RSUs. However, executing these tasks consumes the vehicles' own computational resources and energy. As a result, most vehicles are hesitant to participate in crowdsensing activities unless they receive adequate compensation for their efforts. To address this, the *JCOC* scheme incorporates an incentive mechanism that rewards participating vehicles for their offloading services, ensuring fair compensation for the costs incurred during the crowdsensing process. For the participant recruitment in crowdsensing, the $\mathfrak{M}_{\mathcal{R}}$ and $\mathfrak{N}_{\mathcal{M}}$ resources are adaptively distributed in terms of compensation; they are used for vehicles' offloading services. From the viewpoint of R and M, their resources proportionally allocate based on the sensing contribution of each vehicle. From the viewpoint of vehicles, they act their crowdsensing work to maximize the profits for their offloading services (Kim, 2023a).

The *JCOC* scheme adopts the joint control paradigm using learning model ($\mathbb{L}$) and bargaining game ($\mathbb{G}$). Through the L and G, multiple RSUs, MEC servers and individual vehicles are sequentially interacted with each other to reach a desired outcome, which is called social optimum. It is noteworthy that the *JCOC* scheme formulates the$\mathcal{R}$-M-V association in an iterative manner. Formally, the *JCOC* scheme defines the tuple entities in the L and G, such as

$$\{\mathbb{L}, \mathbb{G}\} = \left\{\mathbb{R}, \mathbb{M}, \mathbb{V}, \left(\mathbb{L}_{\mathcal{V}_{1 \leq j \leq m}} = \left\{\mathbb{L}_{\mathcal{V}_j} | \mathcal{V}_j \in \mathbb{V}_{\mathcal{R}_i}, a^{\mathcal{V}_j}_{1 \leq k \leq l} \in \mathcal{L}_{\mathcal{V}_j}, \mathcal{R}^{\mathcal{V}_j}_a, \mathcal{H}_{\mathcal{V}_j}(a)\right\}\right), \left(\mathbb{G}^{\mathcal{M}_{1 \leq i \leq n}}_{\mathcal{R}_{1 \leq i \leq n}} = \left\{\mathbb{G}^{\mathcal{M}_i}_{\mathcal{R}_i} | \mathcal{V}_j \in \mathbb{V}_{\mathcal{R}_i}, \mathfrak{X}_{\mathcal{V}_j}, \mathfrak{M}_{\mathcal{R}_i}, \mathfrak{N}_{\mathcal{M}_i}, S_{\mathcal{V}_j}, \mathcal{U}_{\mathcal{V}_j}(\cdot)\right\}\right), T\right\}$$

(Kim, 2023a).

- R, M and V represent the sets of RSUs, MEC servers and vehicles, respectively.
- Each RSU has its fixed spectrum band $\left(\mathfrak{M}_{\mathcal{R}}\right)$ and is equipped with one MEC server. Each embedded MEC server has its computation capacity $\left(\mathfrak{N}_{\mathcal{M}}\right)$.
- Each vehicle in V participates in its learning model $\left(\mathbb{L}_{\mathcal{V}}\right)$. Let $\mathbb{V}_{\mathcal{R}_i}$ be the set of vehicles, which are covered by the $\mathcal{R}_i$.
- In the $\mathbb{L}_{\mathcal{V}}$, $\mathcal{V}_j \in \mathbb{V}_{\mathcal{R}_i}$ has his action set $\left(\mathcal{L}_{\mathcal{V}_j}\right)$, which consists of total l actions $\left(a^{\mathcal{V}_j}_{1 \leq k \leq l}\right)$. $\mathcal{R}^{\mathcal{V}_j}_a$ is the $\mathcal{V}_j$'s reward function with the joint action **a**, and $\mathcal{H}_{\mathcal{V}_j}(a)$ is a heuristic function to accelerate the learning process.
- Based on the *CMARL* model, the $\mathbb{L}_{\mathcal{V}_j}$ is operated in a distributed manner to learn the best action of $\mathcal{V}_j$.
- Each pair of RSU and MEC server operate their bargaining game $\left(\mathbb{G}^{\mathcal{M}}_{\mathcal{R}}\right)$. In the $\mathbb{G}^{\mathcal{M}_i}_{\mathcal{R}_i}$, $\mathfrak{X}_{\mathcal{V}_j}$ is the $\mathcal{V}_j$'s contribution to perform crowdsensing tasks where $\mathcal{V}_j \in \mathbb{V}_{\mathcal{R}_i}$.

- In the $\mathbb{G}^{\mathcal{M}_i}_{\mathcal{R}_i}$, the $\mathcal{V}_j$ is a game player, and $S_{\mathcal{V}_j}$ and $\mathcal{U}_{\mathcal{V}_j}(\cdot)$ are his strategy and utility function, respectively, to share the $\mathfrak{M}_{\mathcal{R}_i}$ and $\mathfrak{N}_{\mathcal{M}_i}$. According to the *LHMS*, the $\mathbb{G}^{\mathcal{M}_i}_{\mathcal{R}_i}$ is operated in a centralized manner.
- The $\mathbb{L}_{\mathcal{V}_j}$ and $\mathbb{G}^{\mathcal{M}_i}_{\mathcal{R}_i}$ are reciprocally interdependent each other based on $\mathfrak{X}_{\mathcal{V}_j}$, $\mathcal{H}_{\mathcal{V}_j}(a)$ and $\mathcal{U}_{\mathcal{V}_j}(\cdot)$, and work together, iteratively.
- Discrete time model $T \in \{t_1, \ldots, t_c, t_{c+1}, \ldots\}$ is represented by a sequence of time steps. The length of t_c matches the event timescale of $\mathbb{L}_{\mathcal{V}_j}$ and $\mathbb{G}^{\mathcal{M}_i}_{\mathcal{R}_i}$.

8.1.4 Technical Concepts and Ideas of *LHMS* and *CMARL*

Data Envelopment Analysis (DEA) is a non-parametric approach commonly employed to evaluate and enhance the relative efficiency of a group of homogeneous operational units. By analyzing observed input consumption and output production data, the DEA constructs a production possibility set for decision-making units. Building on the principles of bargaining theory, the *LHMS* has been introduced as a method to address the DEA model. To formally define the *LHMS* approach, the following notations will be utilized. J is the set of operating units, and I, O are the sets of inputs and outputs, respectively. For a given operating unit j, consider I_j^- and O_j^+ are sets of inputs and outputs, respectively. For the cases in which several inputs and/or outputs could improve their values, the classical Nash bargaining theory can be adapted. Assume that inputs in I_j^- are input players, and outputs in O_j^+ are output players. The bargaining set of the game, denoted by S_j, is a $|I_j^-| \times |O_j^+|$-dimensional feasible set in the utility space (Kim, 2023a; Lozano, 2019).

For the output players in O_j^+, the utilities coincide with the corresponding output variable. These utilities are represented by v, where for a given amount of output y_r, $r \in O_j^+$, $v_r(y_r)$ is its utility. However, for the input players in I_j^-, the utilities decrease with the amount of input consumed. Therefore, the *JCOC* scheme can assume the input utilities with respect to the worst case. These utilities are represented by u, where for a given amount of input x_k, $k \in I_j^-$, its utility is defined as $u_k(x_k) = \left(x_k^* = \max_{j\in J} x_{kj}\right) - x_k$, if the feature of linearity is adopted; x_{kj} is the amount of input k, consumed by operating unit j. Usually, the utilities coincide with the corresponding input and output variable. It means that there is a one-to-one correspondence between the operating points $(\hat{x}_k, \hat{y}_r)$ and their corresponding utility vectors $\left(\left(u_k(\hat{x}_k)\right)_{k\in I_j^-}, \left(v_r(\hat{y}_r)\right)_{r\in O_j^+}\right)$. The disagreement point corresponds to the utility of players is $d_j = \left(d_j^x, d_j^y\right) = \left(\left(d_{1j}^x, \ldots d_{kj}^x, \ldots, d_{|I_j^-|j}^x\right), \left(d_{1j}^y, \ldots d_{rj}^y, \ldots, d_{|O_j^+|j}^y\right)\right)$ (Kim, 2023a; Lozano, 2019).

Based on the Nash solution, the *LHMS* is defined by taking into account the one-to-one correspondence between the utility vectors $(u, v) \in S_d$ and their corresponding operating points $(\hat{x}, \hat{y})$. Mathematically, the *LHMS* for (u, v), i.e., *LHMS*(u, v), is given by (Kim, 2023a; Lozano, 2019);

$$LHMS(u,v) = \arg\max_{(\hat{x},\hat{y})} \left\{ \prod_{k\in I_j^-} \left(u_k(\hat{x}_k) - d_{kj}^x\right) \times \prod_{r\in O_j^+} \left(v_r(\hat{y}_r) - d_{rj}^y\right) \right\} \tag{8.1.1}$$

$$\text{s.t.,} \left(\left(u_k(\hat{x}_k)\right)_{k\in I_j^-}, \left(v_r(\hat{y}_r)\right)_{r\in O_j^+}\right) \in S_{d_j}$$

Traditionally, reinforcement learning is a model-free machine learning approach designed to determine the optimal action to take in any given state within an environment through an online learning framework. The desirability of an action is quantified by the expected cumulative reward, commonly referred to as the Q-value, which represents the long-term benefit of selecting a specific action in a particular state. These Q-values are typically stored in a two-dimensional array, mapping each action to every possible state. However, in certain scenarios, it is not necessary to represent the environment using explicit states. Instead, a one-dimensional array may be sufficient to identify the optimal action. In such cases, the primary objective of the stateless learning approach is to estimate the expected Q-value of a single reward for each action. Due to its simplicity and ease of implementation, the stateless learning method is particularly well-suited for CMARL. By eliminating the complexity associated with traditional multi-agent learning algorithms, this approach significantly enhances computational efficiency (Claus, 1998; Kim, 2023a; Morozs, 2015).

Under the multi-agent environment, each agent has to consider not only the effects of his own actions, but also the influence by the actions of the other agents. From this perspective, agents are assumed as joint action learners, and they are able to observe all actions taken by any agent. Therefore, multi-agent learning is intrinsically linked to the game theory, which analyzes the strategic interactions among multiple decision makers. In the stateless *CMARL* model, the *JCOC* scheme assumes the Q-value of agent i, i.e., $Q_i\,(a_i, a_{-i})$, that provides an estimate of the value of performing joint action $a\ =\ (a_i, a_{-i})$. The sample $(a_i, a_{-i}), \mathscr{R}$ is the experience obtained by the agent i where the joint action **a** was performed resulting in the reward R. At the beginning, all Q-values of all actions are initialized to zero, so all agents start learning with equal choice among all available actions. Each Q-value is updated by an agent each time. Based on the $(a_i, a_{-i}), \mathscr{R}$, the recursive update equation for $Q_i(a_i, a_{-i})$ is defined as follows (Claus, 1998; Kim, 2023a; Morozs, 2015).:

$$Q_i(a_i, a_{-i}) \ = \ \left[(1 - \alpha) \times Q_i(a_i, a_{-i})\right] + (\alpha \times \mathscr{R}) \tag{8.1.2}$$

where $\alpha\ \in\ [0, 1]$ is the learning rate parameter which weighs recent experience with respect to previous estimates of the Q-values. In addition, the *JCOC* scheme can take a heuristic acceleration technique for the reinforcement of the learning process. From external or internal additional knowledge, a heuristic function ($\mathscr{H}(a)$) is derived; it is not included in the learning process. Generally, the $\mathscr{H}(a)$ can influence the action selections of a learning agent to accelerate the learning process. The format and dimensions of $\mathscr{H}(a)$ should be compliant with the Q-table used by the given learning agent. By employing the $\mathscr{H}(a)$, the *JCOC* scheme can create the heuristically guided Q-value of agent i's action, i.e., $Q_i^{\mathscr{H}}(a_i, a_{-i})$, while a classical learning process takes place using the $Q_i(a_i, a_{-i})$, as defined in (8.1.2) (Kim, 2023a; Morozs, 2015).

$$Q_i^{\mathscr{H}}(a_i, a_{-i}) \ = \ Q_i(a_i, a_{-i}) + \mathscr{H}(a) \tag{8.1.3}$$

Let $\mathscr{L}_i$ be the action set of agent i, and $\mathbb{P}_i = \left(p_1^i, p_2^i, \ldots, p_{|\mathscr{L}_i|}^i\right)$ is the probability set for action selections where $p_{1\le j\le|\mathscr{L}_i|}^i$ is the probability of selecting the jth action of agent i. The p_j^i is updated as follows (Kim, 2023a):

$$p_j^i = exp\left(\frac{Q_i^{\mathscr{H}}(j)}{\tau}\right) / \sum_{k=1}^{|\mathscr{L}_i|} exp\left(\frac{Q_i^{\mathscr{H}}(k)}{\tau}\right) \tag{8.1.4}$$

where $Q_i^{\mathscr{H}}(j)$ is the $Q_i^{\mathscr{H}}(\cdot)$ value for the jth action, and τ is the temperature parameter that balances exploration and exploitation (Cui, 2021; Kim, 2023a).

8.1.5 The joint control scheme for the MECC based IoIV paradigm

To develop control algorithms for the MECC based IoIV paradigm, the *JCOC* scheme constructs the learning ($\mathbb{L}$) and game ($\mathbb{G}$) models for each vehicle. In the L, a stateless *CMARL* process is implemented in a distributed manner to learn the best sensing action. For the crowdsensing process, the incentive method plays an important role in ensuring the quantity and quality of sensed data. To recruit sensing participants, the IoIV platform provides offloading services to compensate vehicles' sensing costs based on the sensing contribution $\left(\mathfrak{X}_{\mathscr{V}}\right)$. Therefore, intelligent vehicles are motivated to participate in the crowdsensing process in the long term. From the viewpoint of vehicles, the major challenge is to maximize its reward while minimizing the crowdsensing cost. In the learning model of $\mathscr{V}_j \left(\mathbb{L}_{\mathscr{V}_j}\right)$, the crowdsensing activity of $\mathscr{V}_j$ is discretely varied, and each activity level is defined as the $\mathscr{V}_j$'s action. The *JCOC* scheme assumes that there are l different actions where $a_{1\le k\le l}^{\mathscr{V}_j} \in \mathscr{L}_{\mathscr{V}_j}$. After an action in the $\mathscr{L}_{\mathscr{V}_j}$ is selected, it is estimated by receiving a reward, and what is the best action is gradually learned (Kim, 2023a).

In the stateless *CMARL* setting, the *JCOC* scheme assumes that the $\mathscr{V}_j$ has his action set $\mathscr{L}_{\mathscr{V}_j} = \left\{a_1^{\mathscr{V}_j}, \ldots a_k^{\mathscr{V}_j} \ldots, a_l^{\mathscr{V}_j}\right\}$ where each action represents its crowdsensing participation level. For example, the *Q*-value of $\mathscr{V}_j$'s k^th^ action, i.e., $Q_{1\le k\le l}^{\mathscr{V}_j}\left(a_k^{\mathscr{V}_j}, a_{-\mathscr{V}_j}\right)$, provides an estimate value of performing the joint action $\mathbf{a} = \left(a_k^{\mathscr{V}_j}, a_{-\mathscr{V}_j}\right)$. Simply, sensing actions are mapped to real numbers, which represent vehicles' contributions. The $\mathscr{V}_j$ updates its estimate $Q_k^{\mathscr{V}_j}(\cdot)$ value based on the experience sample a, $\mathscr{R}_a^{\mathscr{V}_j}, \mathscr{H}_{\mathscr{V}_j}(a)$ where $\mathscr{R}_a^{\mathscr{V}_j}$ and $\mathscr{H}_{\mathscr{V}_j}(a)$ are the $\mathscr{V}_j$'s reward and heuristic functions of the joint action $\mathbf{a}$. The $\mathscr{R}_a^{\mathscr{V}_j}$ corresponds to the received $\mathscr{V}_j$'s benefit $\left(B_{\mathscr{V}_j}(\cdot)\right)$ minus the incurred sensing cost $\left(C_{\mathscr{V}_j}(\cdot)\right)$. The $B_{\mathscr{V}_j}(\cdot)$ is defined by considering the $\mathscr{V}_j$'s currently generating offload services and sensing action. The $C_{\mathscr{V}_j}(\cdot)$ increases monotonically with the sensing contribution. When the $\mathscr{V}_j$ selects the action $a_k^{\mathscr{V}_j}$, the $\mathscr{R}_a^{\mathscr{V}_j}$ is given by (Kim, 2023a);

$$\mathscr{R}_a^{\mathscr{V}_j}\left(a_k^{\mathscr{V}_j}\right) = B_{\mathscr{V}_j}\left(a_k^{\mathscr{V}_j}, \sigma_{\mathscr{V}_j}\right) - C_{\mathscr{V}_j}\left(a_k^{\mathscr{V}_j}\right) \tag{8.1.5}$$

$$\text{s.t.,}\begin{cases} B_{\mathscr{V}_j}\left(a_k^{\mathscr{V}_j},\sigma_{\mathscr{V}_j}\right) = \log\left(\lambda + \left(\frac{\sigma_{\mathscr{V}_j}}{\mathscr{X}_{\mathscr{V}_j}} \times a_k^{\mathscr{V}_j}\right)\right) \\ C_{\mathscr{V}_j}\left(a_k^{\mathscr{V}_j}\right) = \left(\beta \times \frac{\sigma_{\mathscr{V}_j}}{\mathscr{X}_{\mathscr{V}_j}} \times a_k^{\mathscr{V}_j}\right)^{\varepsilon} \end{cases}$$

where λ is an adjustment parameter for the $B_{\mathscr{V}_j}(\cdot)$, $\sigma_{\mathscr{V}_j}$ is the $\mathscr{V}_j$'s current offload task amount, $\mathscr{X}_{\mathscr{V}_j}$ is the $\mathscr{V}_j$'s local computation capacity, respectively. β and ε are cost control factors for the $C_{\mathscr{V}_j}(\cdot)$. The *JCOC* scheme adopts the heuristic function $\left(\mathscr{H}_{\mathscr{V}_j}(a)\right)$ to effectively estimate its incentive for the cooperative crowdsensing process. The $\mathscr{H}_{\mathscr{V}_j}(a)$ is strongly related to the $\mathscr{U}_{\mathscr{V}_j}(\cdot)$, which is defined in the game model G. Mathematically, the $\mathscr{H}_{\mathscr{V}_j}(a)$ is defined according to the rate of $\mathscr{V}_j$'s relative utility change (Kim, 2023a).

$$\mathscr{H}_{\mathscr{V}_j}(a) = \frac{\mathscr{U}_{\mathscr{V}_j}(a) - \mathscr{A}\_\mathscr{U}_{\mathscr{V}_j}(a)}{\mathscr{A}\_\mathscr{U}_{\mathscr{V}_j}(a)} \tag{8.1.6}$$

where $\mathscr{A}\_\mathscr{U}_{\mathscr{V}_j}(a)$ is the accumulated average utility of $\mathscr{V}_j$, which is obtained from its the offloading services. Finally, the stateless *CMARL Q*-function is defined according to (8.1.2)-(8.1.3),(8.1.5)-(8.1.6). From this *Q*-value, the $a_k^{\mathscr{V}_j}$'s selection probability by the $\mathscr{V}_j$, i.e., $p_{a_k}^{\mathscr{V}_j}$, is updated as follows (Kim, 2023a):

$$p_{a_k}^{\mathscr{V}_j} = exp\left(\frac{Q_{a_k}^{\mathscr{V}_j}\left(a_k^{\mathscr{V}_j},a_{-\mathscr{V}_j}\right)}{\tau}\right) / \sum_{l=1}^{|\mathscr{L}_{\mathscr{V}}|} exp\left(\frac{Q_{a_l}^{\mathscr{V}_j}\left(a_l^{\mathscr{V}_j},a_{-\mathscr{V}_j}\right)}{\tau}\right) \tag{8.1.7}$$

$$\text{s.t., } Q_{a_k}^{\mathscr{V}_j}\left(a_k^{\mathscr{V}_j},a_{-\mathscr{V}_j}\right) = \mathscr{R}_a^{\mathscr{V}_j}\left(a_k^{\mathscr{V}_j}\right) + \mathscr{H}_{\mathscr{V}_j}(a)$$

Based on its sensing action, the $\mathscr{V}_j$ can adaptively obtain the communication and computation resources for its offloading service. Usually, multiple vehicles generate different offloading tasks and request different amounts of resources. To effectively share the limited $\mathfrak{M}_{\mathscr{R}}$ and $\mathfrak{N}_{\mathscr{M}}$ resources, the *JCOC* scheme develops a cooperative game model $\mathbb{G}_{\mathscr{R}}^{\mathscr{M}}$. According to the resource type, the utility function for the $\mathscr{V}_j \in \mathbb{V}_{\mathscr{R}_i}$ $\left(\mathscr{U}_{\mathscr{V}_j}(\cdot)\right)$ consists of two sub-utility functions, i.e., $\mathscr{U}_{\mathscr{V}_j}^{\mathscr{R}_i}(\cdot)$ for $\mathfrak{M}_{\mathscr{R}_i}$ and $\mathscr{U}_{\mathscr{V}_j}^{\mathscr{M}_i}(\cdot)$ for $\mathfrak{N}_{\mathscr{M}_i}$. They are defined as follows (Kim, 2023a):

$$\mathscr{U}_{\mathscr{V}_j}\left(\mathfrak{M}_{\mathscr{R}_i},\mathfrak{N}_{\mathscr{M}_i},a,\mathfrak{X}_{\mathscr{V}_j},\mathcal{Q}_{\mathscr{V}_j}^{\mathscr{R}},S_{\mathscr{V}_j}^{\mathscr{R}}\right) = \mathscr{U}_{\mathscr{V}_j}^{\mathscr{R}_i}(\cdot) \times \mathscr{U}_{\mathscr{V}_j}^{\mathscr{M}_i}(\cdot) \tag{8.1.8}$$

$$
\text{s.t.,}\begin{cases}
\mathcal{U}_{\mathcal{V}_j}^{\mathcal{R}_i}\left(\mathfrak{M}_{\mathcal{R}_i}, a, \mathfrak{X}_{\mathcal{V}_j}, \mathcal{Q}_{\mathcal{V}_j}^{\mathcal{R}_i}, S_{\mathcal{V}_j}^{\mathcal{R}_i}\right) = \left(\Upsilon \times \log\left(\frac{\min\left(\mathcal{Q}_{\mathcal{V}_j}^{\mathcal{R}_i}, S_{\mathcal{V}_j}^{\mathcal{R}_i}\right)}{\mathcal{Q}_{\mathcal{V}_j}^{\mathcal{R}_i}} + \eta\right) + \psi\right)^{\mathcal{W}_{\mathcal{V}_j}} \\
\mathcal{U}_{\mathcal{V}_j}^{\mathcal{M}_i}\left(\mathfrak{N}_{\mathcal{M}_i}, a, \mathfrak{X}_{\mathcal{V}_j}, \mathcal{Q}_{\mathcal{V}_j}^{\mathcal{M}_i}, S_{\mathcal{V}_j}^{\mathcal{M}_i}\right) = \left(\frac{\phi}{\xi + \exp\left(\mu \times \frac{\min\left(\mathcal{Q}_{\mathcal{V}_j}^{\mathcal{M}_i}, S_{\mathcal{V}_j}^{\mathcal{M}_i}\right)}{\mathcal{Q}_{\mathcal{V}_j}^{\mathcal{M}_i}}\right)}\right)^{\mathcal{W}_{\mathcal{V}_j}} \\
\mathcal{W}_{\mathcal{V}_j} = \frac{\mathfrak{X}_{\mathcal{V}_j}}{\sum_{\mathcal{V}_s \in \mathbb{V}_{\mathcal{R}_i}} \mathfrak{X}_{\mathcal{V}_s}}
\end{cases}
$$

where Υ, η, ψ are adjustment parameters for the $\mathcal{U}_{\mathcal{V}_j}^{\mathcal{R}_i}(\cdot)$, and ϕ, ξ and μ are adjustment parameters for the $\mathcal{U}_{\mathcal{V}_j}^{\mathcal{M}_i}(\cdot)$. $\mathcal{Q}_{\mathcal{V}_j}^{\mathcal{R}_i}$ (*or* $\mathcal{Q}_{\mathcal{V}_j}^{\mathcal{M}_i}$) is the requested bandwidth (*or* computing power) amount from the $\mathcal{V}_j$, and $S_{\mathcal{V}_j}^{\mathcal{R}_i}$ (*or* $S_{\mathcal{V}_j}^{\mathcal{M}_i}$) is the allocated bandwidth (*or* computing power) amount from the $\mathcal{R}_i$ (*or* $\mathcal{M}_i$). To adaptively decide the $S_{\mathcal{V}_j}^{\mathcal{R}_i}$ and $S_{\mathcal{V}_j}^{\mathcal{M}_i}$, the *JCOC* scheme adopts the basic idea of *LHMS*. According to (1), the *LHMS* for $\left(\mathfrak{M}_{\mathcal{R}_i}, \mathfrak{N}_{\mathcal{M}_i}\right)$, i.e., $LHMS\left(\mathfrak{M}_{\mathcal{R}_i}, \mathfrak{N}_{\mathcal{M}_i}\right)$, is given by (Kim, 2023a);

$$
LHMS\left(\mathfrak{M}_{\mathcal{R}_i}, \mathfrak{N}_{\mathcal{M}_i}\right) = \arg\max_{\left(S_{\mathcal{V}}^{\mathcal{R}}, S_{\mathcal{V}}^{\mathcal{M}}\right)} \left\{ \prod_{\mathcal{V}_j \in \mathbb{V}_{\mathcal{R}_i}} \left(\mathcal{U}_{\mathcal{V}_j}^{\mathcal{R}_i}(\cdot) - d_{\mathcal{V}_j}^{\mathcal{R}}\right) \times \prod_{\mathcal{V}_j \in \mathbb{V}_{\mathcal{R}_i}} \left(\mathcal{U}_{\mathcal{V}_j}^{\mathcal{M}_i}(\cdot) - d_{\mathcal{V}_j}^{\mathcal{M}}\right) \right\} \tag{8.1.9}
$$

$$
\text{s.t., } \widehat{S_{\mathcal{V}}^{\mathcal{R}}} = \mathcal{V}_j \in \mathbb{V}_{\mathcal{R}_i} | \ldots, S_{\mathcal{V}_j}^{\mathcal{R}_i}, \ldots \text{ and } \widehat{S_{\mathcal{V}}^{\mathcal{M}}} = \mathcal{V}_j \in \mathbb{V}_{\mathcal{R}_i} | \ldots, S_{\mathcal{V}_j}^{\mathcal{M}_i}, \ldots
$$

where $d_{\mathcal{V}_j}^{\mathcal{R}}$ and $d_{\mathcal{V}_j}^{\mathcal{M}}$ are $\mathcal{V}_j$'s disagree points to share the $\mathfrak{M}_{\mathcal{R}_i}$ and $\mathfrak{N}_{\mathcal{M}_i}$, respectively.

8.1.6 Summary

The *JCOC* scheme addresses the joint crowdsensing and offloading challenge in the edge-assisted IoIV system. By integrating cooperative multi-agent learning with bargaining game theory, its primary objective is to achieve an optimal balance between vehicles' sensing contributions and the benefits of offloading. To accomplish this, the *JCOC* scheme first develops a multi-agent learning algorithm that enables vehicles to participate effectively in the crowdsensing process. Through this algorithm, each vehicle autonomously determines the optimal sensing action to maximize its individual payoff. Leveraging the *CMARL*, this learning process is executed in a decentralized manner. Additionally, from a practical system management perspective, the stateless learning approach is particularly well-suited due to its simplicity and efficiency. Second, the *JCOC* scheme develops a bargaining game model to share the limited $\mathfrak{M}_{\mathcal{R}}$ and $\mathfrak{N}_{\mathcal{M}}$ resources. To support vehicles' offload services, the idea of *LHMS* is adopted to distribute the $\mathfrak{M}_{\mathcal{R}}$ and $\mathfrak{N}_{\mathcal{M}}$ in a centralized manner. In the resource-sharing process, vehicles' sensing contributions serve as their bargaining power. Consequently, in the *JCOC* scheme, the crowdsensing and offloading control algorithms are closely integrated and operate in a coordinated manner. One of the key innovations of the *JCOC* scheme is its adaptability, which enables it to respond effectively to dynamic changes in the MECC-based IoIV platform. This adaptability ensures that the system remains responsive to varying conditions, optimizing performance in real-time (Kim, 2023a).

8.2 Cooperative Game based Traffic Control (*CGTC*) scheme

In recent years, automotive manufacturers have integrated multiple electronic control units (ECUs) into vehicles to enhance safety and driver convenience. Each ECU is responsible for managing specific vehicle functions and communicates with other units via the controller area network (CAN). The CAN bus is widely recognized for its low cost, efficient communication, and high flexibility. Currently, S. Kim is conducting research on the properties of the CAN bus and has developed the *Cooperative Game-based Traffic Control* (*CGTC*) scheme to optimize CAN system performance (Kim, 2021). By incorporating concepts from two distinct cooperative game solutions, the *CGTC* scheme examines the impact of ECU interactions and introduces an innovative two-step game model to maximize the benefits of CAN resource sharing. To dynamically accommodate varying data service requirements, control decisions in the two-step game are interdependent, ensuring that each ECU collaborates effectively. This interactive coordination process continues until a fair and efficient consensus is reached, making it a practical and applicable approach for real-world CAN system operations.

8.2.1 Development motivation

The complexity of automotive vehicle systems has increased significantly with the widespread adoption of electronic components such as engine management systems, ignition controls, infotainment systems, telematics, carputers, and electronic control units (ECUs). ECUs are compact devices embedded within a vehicle to regulate specific functions, with modern vehicles incorporating multiple ECUs for various tasks ranging from essential operational controls to comfort-enhancing features. Each ECU requires both power and data connections to function, receiving inputs from different vehicle components based on its designated role. In recent years, advancements in intelligent systems have facilitated enhanced data exchange between ECUs while minimizing wiring costs. Various ECU communication standards have been developed, including FlexRay, local interconnected networks, automotive Ethernet, in-vehicle networks, and the widely adopted controller area network (CAN) (Jeong, 2020; Kang, 2018; Kim, 2021).

The CAN protocol, originally developed by BOSCH, is a multi-master, event-driven serial communication bus protocol supporting speeds of up to 1 Mbit/s. Unlike traditional point-to-point communication networks like Ethernet, the CAN protocol does not transmit large data blocks but instead broadcasts numerous short messages across the network, ensuring data consistency across all ECUs within the vehicle. Among the various real-time network communication protocols for ECUs, the CAN protocol remains the most widely used due to its stability and reliability. In 2003, the International Organization for Standardization (ISO) formalized CAN as ISO 11898-1. Over the past three decades, the simplicity and dependability of the CAN protocol have solidified its status as the industry standard for ECU communication (Choi, 2018; Kim, 2021; Mubeen, 2014).

Recent studies have focused on improving the efficiency and safety of CAN-based networks. The CAN system supports data transmission rates ranging from 20 kb/s to 1 Mb/s while operating with multiple agents. Each CAN transmitter is connected to a common bus line with limited bandwidth. Despite its advantages, the CAN protocol was not originally designed to ensure fair resource allocation, making it vulnerable to bandwidth unfair-sharing issues. Typically, ECU-generated data in a CAN system are categorized into two classes: class I (real-time data) and class II (non-real-time data), based on their sensitivity to transmission delays. Since both types of data share the same communication medium, class I data transmissions may be delayed or interrupted by class II data, potentially exceeding predefined latency thresholds. This issue can result in performance inefficiencies and instability in real-time applications. Therefore, ensuring real-time requirements and increasing CAN system capacity are critical concerns (Hong, 2000; Kim, 2009; Kim, 2021; Zhang, 2018).

To enhance the performance and capacity of CAN systems, the dual-channel CAN bus platform has been introduced. Several modern microcontrollers now incorporate two communication bus channels, providing a universal solution for various CAN applications and enabling more efficient data exchange among ECUs. Effective operations in dual-channel CAN systems require autonomous, distributed, and rational ECUs to collaborate and coordinate effectively to optimize conflicting performance metrics. The primary objective in designing a new control protocol for dual-channel CAN networks is to ensure that ECUs make strategic, rational decisions that promote cooperative resource sharing while maximizing fair and efficient bandwidth utilization. However, achieving this goal presents significant technical challenges (Kim, 2009; Kim, 2021).

The control mechanism for a dual-channel CAN system can be analyzed through the lens of cooperative game theory. Accordingly, the *CGTC* scheme approaches the traffic management problem in dual-channel CAN networks using cooperative game-theoretic principles. In 2001, W. Bossert and H. Peters introduced a new cooperative game solution concept known as the *minimax regret and efficient bargaining solution* (*MREBS*). They further characterized the *MREBS* as a method for generating ex ante efficient outcome combinations under the assumption that players operate with minimax regret preferences. Monotone utopian bargaining solutions are typically characterized by efficiency based on the max-min criterion, which seeks to maximize the minimal gains relative to a disagreement point. The *MREBS* model incorporates the minimax regret principle, wherein regret is measured relative to utopian payoffs - i.e., the highest possible payoffs attainable. By integrating scale covariance into minimax regret preferences, the *MREBS* concept can be extended to general cooperative games, encompassing the *Raiffa-Kalai-Smorodinsky bargaining solution* (Bossert, 2001; Kim, 2021).

In 2017, M. Radzvilas introduced the *relative benefit equilibrating bargaining solution* (*RBEBS*), a model based on the comparison of ordinal relative individual benefit gains among game participants. Traditional cooperative game solutions assume that players are concerned not only with their absolute benefit gains from a particular agreement but also with how the agreement is achieved and how the benefits are distributed. In real-world cooperative games, agreements are typically reached only when at least one player makes a concession. Each feasible agreement is associated with a specific set of concessions that players must accept to reach a consensus. Given this dynamic, it is natural to expect players to seek agreements with the most equitable distribution of concessions. The *RBEBS* model is designed to mediate the relationship between the measurement of ordinal relative individual benefit gains and the measurement of ordinal relative concessions among game participants (Kim, 2021; Radzvilas, 2017).

8.2.2 Contribution

By incorporating the principles of *MREBS* and *RBEBS*, the *CGTC* scheme effectively manages traffic control in dual-channel CAN systems. The process unfolds in two key steps. First, the traffic volume allocated to each ECU is determined based on the *MREBS* framework. In dynamic and diverse CAN traffic environments, individual ECUs make decisions collaboratively to reach a mutually acceptable agreement. Second, the assigned traffic for each ECU is adaptively distributed across the dual communication channels in a coordinated manner. Since each channel exhibits distinct characteristics depending on the current traffic conditions, the algorithm employs a step-by-step interactive cooperative process. This process is structured as a repeated bargaining game model, ensuring the dynamic and efficient allocation of the limited CAN bandwidth. As a result, the two-step cooperative game-based approach is highly flexible and adaptable, allowing it to respond effectively to fluctuating CAN traffic conditions. This adaptability ultimately ensures a well-balanced and optimized system performance (Kim, 2021).

8.2.3 Dual channel CAN system infrastructure

The *CGTC* scheme is designed for a dual-channel CAN platform, which incorporates two communication channels to alleviate traffic congestion. These channels can be accessed by individual ECUs at any given time. Typically, the buses transmit two types of data: class I and class II. As a result, situations may arise where multiple ECUs attempt to access the bus simultaneously. To manage such conflicts, an adaptive arbitration process is employed. This arbitration mechanism prioritizes data based on predefined priority levels, ensuring that bandwidth allocation dynamically adjusts according to traffic demands and system preferences. Let $\mathbb{N} = \{\mathcal{N}_1, \ldots, \mathcal{N}_n\}$ be the set of ECUs and $\mathcal{L}_f$, $\mathcal{L}_s$ are the first and second bus lines, respectively. ECUs in N are interconnected together, and exchange information between themselves through $\mathcal{L}_f$ and $\mathcal{L}_s$. $\mathfrak{M}_f$ and $\mathfrak{M}_s$ are bandwidth capacities for the $\mathcal{L}_f$ and $\mathcal{L}_s$, respectively (Kim, 2021).

We consider a discrete time model $T \in \{t_1, \ldots, t_c, t_{c+1}, \ldots\}$, where the length of a time slot matches the event timescale at which the communication control decisions are updated. The *CGTC* scheme formulates the ECUs' interaction process as the upper-level cooperative game ($\mathbb{G}^U$), and each individual $\mathcal{N}_{1\leq i\leq n}$ gets its assigned bandwidth $\left(\grave{\mathrm{E}}_{\mathcal{N}_i}\right)$ through the concept of *MREBS*. Therefore, the $\mathcal{N}_i$ has a right to transfer his data within the $\grave{\mathrm{E}}_{\mathcal{N}_i}$'s acceptable limit. Since then, the $\mathcal{N}_i$ distributes its traffic amount, which is not larger than the $\grave{\mathrm{E}}_{\mathcal{N}_i}$, into the $\mathcal{L}_f$ and $\mathcal{L}_s$ by using the idea of *RBEBS*. This process is modeled as a lower-level cooperative game $\left(\mathbb{G}^L_{\mathcal{N}_i}\right)$. $\mathbb{G}^U$ and $\mathbb{G}^L_{\mathcal{N}_i}$ are repeated sequentially in a slotted time structure. Based on interactive feedback manner, the *CGTC* scheme is operated each time period during the step-by-step iteration. To reduce computation complexity, in the *CGTC* scheme, the amount of bandwidth allocation is specified in terms of basic bandwidth units (BBUs). Formally, the *CGTC* scheme defines game entities for the dual-channel CAN system infrastructure, i.e.,

$$\left\{\mathbb{G}^U, \mathbb{G}^L_{\mathcal{N}_{1\leq i\leq n}}\right\} = \left\{\mathbb{N}, \{\mathcal{L}_f, \mathcal{L}_s\}, \{\mathfrak{M}_T | \mathfrak{M}_f, \mathfrak{M}_s\}, \mathcal{B}_u, \left\{\mathcal{N}_i \in \mathbb{N} | U^U_{\mathcal{N}_i}, U^{\mathcal{L}_f}_{\mathcal{N}_i}, U^{\mathcal{L}_s}_{\mathcal{N}_i}, \mathbb{Q}_{\mathcal{N}_i}, \grave{\mathrm{E}}_{\mathcal{N}_i}\right\}, T\right\}$$

of gameplay (Kim, 2021).

- $\mathbb{G}^U$ and $\mathbb{G}^L_{\mathcal{N}_i}$ are upper and lower cooperative game models; they are related in a coordination manner of mutual and reciprocal interdependency.
- N is the set of the set of ECUs $\left(\mathcal{N}_{1\leq i\leq n}\right)$, and they are game players in the $\mathbb{G}^U$.
- $\mathcal{L}_f$ and $\mathcal{L}_s$ are the first and second bus lines, respectively, and they are game players in the $\mathbb{G}^L_{\mathcal{N}_i}$.
- $\mathfrak{M}_T$ is the total CAN system bandwidth capacity. $\mathfrak{M}_f$ and $\mathfrak{M}_s$ are bandwidth amounts for the $\mathcal{L}_f$ and $\mathcal{L}_s$, respectively, where $\mathfrak{M}_T = \mathfrak{M}_f + \mathfrak{M}_s$.
- $\mathcal{B}_u$ is a basic bandwidth unit to allocate the bus link bandwidth.
- $U^U_{\mathcal{N}_i}$ is the utility function of $\mathcal{N}_i$ at the $\mathbb{G}^U$.

- At the $\mathbb{G}^{L}_{\mathcal{N}_i}$, $U^{\mathcal{L}_f}_{\mathcal{N}_i}$ is the utility function of $\mathcal{L}_f$ and $U^{\mathcal{L}_s}_{\mathcal{N}_i}$ is the utility function of $\mathcal{L}_s$.
- $\mathcal{Q}_{\mathcal{N}_i}$ is the $\mathcal{N}_i$'s generated data amount, and $\grave{E}_{\mathcal{N}_i}$ is the assigned bandwidth to the $\mathcal{N}_i$. Therefore, $\mathbf{min}\left(\mathcal{Q}_{\mathcal{N}_i},\grave{E}_{\mathcal{N}_i}\right)$ is the traffic amount, which is distributed into the $\mathcal{L}_f$ and $\mathcal{L}_s$.
- $T = \{t_1, \ldots, t_c, t_{c+1}, \ldots\}$ denotes time, which is represented by a sequence of time steps.

8.2.4 The basic concepts of *MREBS* and *RBEBS*

Usually, a cooperative game solution assigns a feasible utility tuple to every bargaining problem. Therefore, it aggregates the individual preferences of game players into a collective outcome. In more technical terms, each player has a criterion to decide between several contingent contracts. To implement a new view on cooperative game solutions, the *CGTC* scheme adopts the ideas of *MREBS* and *RBEBS*. To explain these bargaining solutions, we introduce the notation and basic definitions. Denote $N = \{1, \ldots, n\}$ be a finite set of cooperative game players. Let $\mathbb{R}^n$ be the *n*-fold Cartesian product of real number set R, and the set of all nonnegative vectors in $\mathbb{R}^n$ is denoted by $\mathbb{R}^n_+$. An n-player cooperative game has a solution set $S \subset \mathbb{R}^n$ where elements of S are called feasible outcomes. If game players fail to reach some other outcome $x \in S$, a disagreement point $d = \{d_1 \ldots d_n\}$ results where $d \in S \subset \left(d + \mathbb{R}^n_+\right)$; S contains a vector $x > d$ for all $x \in S$. For any two vectors $x, y \in \mathbb{R}^n, x > (or \geq) y$ means $x_{1\leq i\leq n} > (or \geq) y_{1\leq i\leq n}$. The inequalities $<$ and $\leq$ are defined analogously (Bossert, 2001; Kim, 2021).

Denote C as the class of all n-player cooperative game, and a game solution is a mapping $\mathcal{F}:\mathbb{C} \rightarrow \mathbb{R}^n$ with $\mathcal{F}(S) \in S$ for all $S \in \mathbb{C}$. A solution F is called *Pareto optimal* if $\mathcal{F}(S) \in P(S)$ for all $S \in \mathbb{C}$ where $P(S) = \{x \in S | \forall y \in S[y \geq x \Rightarrow y = x]\}$. In S, each player i has a transitive and reflexive preference relation $\succeq_i$; $x \succ_i y$ means $x_i > y_i$, where $\succ_i$ denotes the asymmetric part of $\succeq_i$. Let F and $\succeq$ be a cooperative game solution and $(\succeq_1, \ldots, \succeq_n)$, respectively, and F is called to be *efficient with respect to* $\succeq$ if there is a player i with $\mathcal{F}(S) \succeq_i x$ for all $S \in \mathbb{C}$ and all $x \in S$. Finally, *efficiency with respect to* $\succeq$ implies *Pareto optimality*. A solution F is called *scale covariant* if $(a \cdot \mathcal{F}(S) + b) = \mathcal{F}(a \cdot S + b)$ for all $a, b \in \mathbb{R}^n$. With the *scale covariant* property, player i's minimax regret preference may be normalized to a preference $\succeq^u_i$ as follows (Bossert, 2001; Kim, 2021);

$$x \succeq^u_i y \equiv \mathbf{max}\left\{\frac{u_i(S) - x_i}{u_i(S) - d_i(S)}\right\} \leq \mathbf{max}\left\{\frac{u_i(S) - y_i}{u_i(S) - d_i(S)}\right\} \tag{8.2.1}$$

$$\text{s.t., } u_{1\leq i\leq n}(S) := \max_{x\in S} x_{1\leq i\leq n}, S \in \mathbb{C}, \text{ and } x, y \in S$$

where $u(S)$ is a utopia point of a problem $S \in \mathbb{C}$. Formally, the *MREBS* is defined as follows;

$$MREBS(S, d)_x = \left(P(S) \cap \{u(S) | x \succeq^u_i y\}\right) \geq d \tag{8.2.2}$$

The other bargaining solution, the *RBEBS*, recommends an implementation of a feasible agreement which minimizes the difference between players' relative concessions. To simply explain the concept of *RBEBS*, we assume that there are only two game players where $N = \{1,2\}$. Each agreement $x_i \in S$

is a pair of demands $x_i = (g_1, g_2)$ over some amount of divisible resource, the total amount of which is P. Each demand g_i of every $i \in N$ is selected from the interval $[d_i, \ldots, \mathfrak{P}]$. Let $\mathcal{S} \subseteq \mathcal{A}$ be a subset of feasible agreements of C. For any pair $x_i = (g_1, g_2) \in \mathcal{S}$ and $x_{j \neq i} = (h_1, h_2) \in \mathcal{S}$, such that $h_i > g_i$, the preferences of i are such that $h_i \succ_i g_i$, and so $x_j \succ_i x_i$. Since every $x_i \in \mathcal{S}$ is such that $g_1 + g_2 \in \mathcal{S}$, each player i has a strict preference relation $\succ_i$ over $\mathcal{S} \subseteq S$. For each i, the cardinality of the preferred set of agreements associated with some feasible agreement $x_i \in \mathcal{S}$ can be defined as follows (Kim, 2021; Radzvilas, 2017);

$$c_i(x_i, \{\mathcal{S}, d\}) = \left\{ |T|, \text{ where } x_{j \neq i} \in T \Rightarrow \left((x_j \in \{\mathcal{S}, d\}) \wedge (x_j \succ x_i) \right) \right\}$$

(8.2.3)

$$\text{s.t., } \mathcal{S} = \left\{ x_i = (g_1, g_2) \in \mathcal{S} : (g_1 + g_2 = \mathfrak{P}) \wedge (x_i \succcurlyeq_i d), \forall i \in N \right\}$$

From (8.2.3), it follows that $x_{j \neq i} \succ_i x_i$ if and only if $c_j(x_j, \{\mathcal{S}, d\}) < c_i(x_i, \{\mathcal{S}, d\})$. We can assume that each player's cardinality can be interpreted as a measure of the size of the player's concession. Commonsensically, each rational player seeks an implementation of an agreement which minimizes the cardinality of the preferred set of alternatives. Let $c_i^{max}(\{\mathcal{S}, d\})$ and $c_i^{min}(\{\mathcal{S}, d\})$ be player i's maximum and minimum possible concession associated with some outcome in the set $\{\mathcal{S}, d\}$, respectively. Since, $x_i \succ_i d_i\ \forall x_i \in \mathcal{S}$, it follows that $c_i^{max}(\{\mathcal{S}, d\}) = c_i(d)$ and $c_i^{min}(\{\mathcal{S}, d\}) = 0$. Therefore, player i's relative concession associated with some feasible outcome $y_i \in \{S, d\}$, i.e., $\mathcal{R}^{y_i}(y_i, \{S, d\}) \in [0, 1]$, can be defined as follows (Kim, 2021; Radzvilas, 2017):

$$\mathcal{R}^{y_i}(y_i, \{S, d\}) = \frac{c_i^{min}(\{S, d\}) - c_i(y_i, \{S, d\})}{c_i^{min}(\{S, d\}) - c_i^{max}(\{S, d\})}$$

(8.2.4)

$$\text{s.t.,} \begin{cases} c_i^{max}(\{\mathcal{S}, d\}) := \underset{x_i \in \{\mathcal{S}, d\}}{\textbf{argmax}} \left[c_i(x_i, \{\mathcal{S}, d\}) \right] \\ c_i^{min}(\{\mathcal{S}, d\}) := \underset{x_i \in \{\mathcal{S}, d\}}{\textbf{argmin}} \left[c_i(x_i, \{\mathcal{S}, d\}) \right] \end{cases}$$

Since $c_i^{min}(\{S, d\}) = 0$ and $c_i^{max}(\{S, d\}) = c_i(d)$, $\mathcal{R}^{y_i}(y_i, \{S, d\})$ can be simplified as follows (Kim, 2021):

$$\mathcal{R}^{y_i}(y_i, \{S, d\}) = \frac{-c_i(y_i, \{S, d\})}{-c_i(d)} = \frac{c_i(y_i, \{S, d\})}{c_i(d)}$$

(8.2.5)

Since player i's individual benefit gain represents the number of total possible concessions that he does not make if outcome $y \in \{S, d\}$ obtains, player i's relative individual benefit gain associated with $y_i \in \{S, d\}$, i.e., $\mathcal{G}^{y_i}(y_i, \{S, d\}) \in [0, 1]$, can be defined as follows (Kim, 2021; Radzvilas, 2017):

$$\mathcal{G}^{y_i}\left(y_i, \{S, d\}\right) = \frac{c_i^{max}(\{S, d\}) - c_i(y_i \{S, d\})}{c_i^{max}(\{S, d\}) - c_i^{min}(\{S, d\})}$$

(8.2.6)

Since $c_i^{min}(\{S, d\}) = 0$ and $c_i^{max}(\{S, d\}) = c_i(d)$ for player i, $\mathcal{G}^{y_i}\left(y_i, \{S, d\}\right)$ can be simplified as follows (Kim, 2021; Radzvilas, 2017):

$$\mathcal{G}^{y_i}\left(y_i, \{S, d\}\right) = \frac{c_i(d) - c_i(y_i, \{S, d\})}{c_i(d)}$$

(8.2.7)

Finally, the *RBEBS* can be defined in terms of relative concessions (Kim, 2021; Radzvilas, 2017):

$$RBEBS(S, d)_x = \arg\min_{x_i \in S}\left[\left|\left(\mathcal{G}^{x_i}\left(x_i, \{S, d\}\right) - \mathcal{G}^{x_{j \neq i}}\left(x_i, \{S, d\}\right)\right)\right|\right]$$

(8.2.8)

$$= \arg\min_{x_i \in S}\left[\left|\left(1 - \mathcal{R}^{x_i}\left(x_i, \{S, d\}\right)\right) - \left(1 - \mathcal{R}^{x_{j \neq i}}\left(x_i, \{S, d\}\right)\right)\right|\right]$$

$$= \arg\min_{x_i \in S}\left[\left|\mathcal{R}^{x_i}\left(x_i, \{S, d\}\right) - \mathcal{R}^{x_{j \neq i}}\left(x_j, \{S, d\}\right)\right|\right]$$

$$\text{s.t., } \mathcal{G}^{x_i}\left(x_i, \{S, d\}\right) = 1 - \frac{c_i\left(x_i, \{S, d\}\right)}{c_i(d)} = 1 - \mathcal{R}^{x_i}\left(x_i, \{S, d\}\right), \forall x_i \in S$$

8.2.5 The two-step cooperative game model for the CAN system

Each ECU $\left(\mathcal{N}_{1 \leq i \leq n}\right)$ is connected to the two CAN bus channels, and two types of data are generated probabilistically. In class I and class II data, service priority is differently decided. The *CGTC* scheme first assigns the bus bandwidth $\left(\mathfrak{M}_T\right)$ for individual ECUs at each time slot; it can effectively restrict the total traffic number of ECUs. And then, the ECUs' data is distributed into two channels. Theoretically, if two channels are used in the CAN platform, it should be possible to reduce the traffic on each channel to half (Kim, 2009). However, under a dynamically changing CAN traffic environment, it is difficult to create an ideal condition to provide fair-efficient traffic distribution. In the *CGTC* scheme, the bandwidth assignment problem is formulated in the $\mathbb{G}^U$, and each ECU's traffic distribution problem is modelled in the $\mathbb{G}^L_{\mathcal{N}}$. At the $\mathbb{G}^U$, $\mathcal{N}_{1 \leq i \leq n}$, are game players, and the utility function of $\mathcal{N}_i$ at time t_c, i.e., $U^U_{\mathcal{N}_i}(\cdot)$, is formally defined as follows (Kim, 2021);

$$U^U_{\mathcal{N}_i}\left(\mathcal{Q}^{t_c}_{\mathcal{N}_i}, \grave{\text{E}}^{t_c}_{\mathcal{N}_i}, \eta, \beta\right) =$$

$$\begin{cases} \gamma - exp\left(-\eta \times \frac{\min\left(\mathcal{Q}_{\mathcal{N}_i}^{t_c}, \grave{E}_{\mathcal{N}_i}^{t_c} \right)}{\grave{E}_{\mathcal{N}_i}^{t_c}} \right), if\, \mathcal{Q}_{\mathcal{N}_i}^{t_c}\, is\, class\, I \\ \frac{\sigma}{\alpha + exp\left(-\beta \times \frac{\min\left(\mathcal{Q}_{\mathcal{N}_i}^{t_c}, \grave{E}_{\mathcal{N}_i}^{t_c} \right)}{\grave{E}_{\mathcal{N}_i}^{t_c}} \right)}, otherwise \end{cases} \quad (8.2.9)$$

where $\mathcal{Q}_{\mathcal{N}_i}^{t_c}$ is the $\mathcal{N}_i$'s generated data amount, and $\grave{E}_{\mathcal{N}_i}^{t_c}$ is the assigned bandwidth to the $\mathcal{N}_i$ at time t_c. For the class I data service, γ is a coefficient factor and η is an adjustment parameter. For the class II data service, σ, α are coefficient factors, and β is an adjustment parameter. η and β values are dynamically decided based on the data preferences. In the $\mathbb{G}^U$, the $\mathfrak{M}_T$ is adaptively allocated to individual $\mathcal{N}_{1 \leq i \leq n}$ according to the *MREBS* (Kim, 2021).

$$MREBS_{\mathbb{X}=\left\{\grave{E}_{\mathcal{N}_1}^{t_c},\ldots,\grave{E}_{\mathcal{N}_n}^{t_c}\right\}}\left(\mathbb{U} = \left\{\ldots U_{\mathcal{N}_{1\leq i\leq n}}^{U}(\cdot)\ldots\right\}, \mathbb{D} = \left\{\ldots d_{\mathcal{N}_i}\ldots\right\}\right) = \quad (8.2.10)$$

$$\left(\max\left\{ \frac{\mathbb{U}_{\mathcal{N}_i}(\mathbb{X}) - U_{\mathcal{N}_i}^{U}\left(\mathcal{Q}_{\mathcal{N}_i}^{t_c}, \grave{E}_{\mathcal{N}_i}^{t_c}, \eta, \beta \right)}{\mathbb{U}_{\mathcal{N}_i}(\mathbb{X}) - \mathbb{D}_{\mathcal{N}_i}(\mathbb{X})} \right\} \leq \max\left\{ \frac{\mathbb{U}_{\mathcal{N}_i}(\mathbb{X}) - U_{\mathcal{N}_i}^{U}\left(\mathcal{Q}_{\mathcal{N}_i}^{t_c}, \grave{E}'^{t_c}_{\mathcal{N}_i}, \eta, \beta \right)}{\mathbb{U}_{\mathcal{N}_i}(\mathbb{X}) - \mathbb{D}_{\mathcal{N}_i}(\mathbb{X})} \right\} \right)$$

$$s.t., \mathbb{U}_{\mathcal{N}_{1\leq i\leq n}}(\mathbb{X}) := \max_{\grave{E}_{\mathcal{N}_i}^{t_c} \in \mathbb{X}} U_{\mathcal{N}_i}^{U}\left(\mathcal{Q}_{\mathcal{N}_i}^{t_c}, \grave{E}_{\mathcal{N}_i}^{t_c}, \eta, \beta \right), \Theta_{\mathcal{N}_i}^{t_c} \neq \Theta_{\mathcal{N}_i}^{t_c}\, and\, \mathfrak{M}_T \geq \sum\nolimits_{\mathcal{N}_i \in \mathbb{N}} \Theta_{\mathcal{N}_i}^{t_c}$$

The $\mathcal{N}_i$ gets the $\grave{E}_{\mathcal{N}_i}^{t_c}$ in the $\mathbb{G}^U$, and then the $\mathcal{N}_i$ distributes its data amount $\tilde{O}_{\mathcal{N}_i}^{t_c}$ into two channels, i.e., $\mathcal{L}_f$ and $\mathcal{L}_s$, at time t_c where $\tilde{O}_{\mathcal{N}_i}^{t_c} = \min\left(\mathcal{Q}_{\mathcal{N}_i}^{t_c}, \grave{E}_{\mathcal{N}_i}^{t_c} \right)$. In the $\mathbb{G}_{\mathcal{N}_i}^{L}$, the $\mathcal{N}_i$ attempts to smooth out the traffic overhead between $\mathcal{L}_f$ and $\mathcal{L}_s$; they have different traffic conditions based on the data service priorities. In the *CGTC* scheme, each N operates individually its lower-level cooperative game ($\mathbb{G}_{\mathcal{N}}^{L}$) to maximize the CAN system performance in a collaborative manner. In the $\mathbb{G}_{\mathcal{N}_i}^{L}$, $\mathcal{L}_f$ and $\mathcal{L}_s$ are game players, and the $\mathcal{L}_f$'s utility function, i.e., $U_{\mathcal{N}_i}^{\mathcal{L}_f}$, is defined as follows (Kim, 2021);

$$U_{\mathcal{N}_i}^{\mathcal{L}_f}\left(\chi_{\mathcal{N}_i}^{\mathcal{L}_f}, \tilde{O}_{\mathcal{N}_i}^{t_c}, \Gamma\_I_{\mathcal{L}_f}^{t_{c-1}}, \Gamma\_II_{\mathcal{L}_f}^{t_{c-1}}\right) =$$

$$\left[\left(\omega \times \log\left(\frac{\left(\Gamma\_I_{\mathcal{L}_f}^{t_{c-1}} + I_{\mathcal{N}_i}^{\mathcal{L}_f}\right)}{\mathfrak{M}_f} + \xi\right)\right) + \varepsilon\right] + \left[\tau \times exp\left(\frac{\left(\Gamma\_II_{\mathcal{L}_f}^{t_{c-1}} + II_{\mathcal{N}_i}^{\mathcal{L}_f}\right)}{\mathfrak{M}_f}\right) + \psi\right] \tag{8.2.11}$$

$$\begin{cases} I_{\mathcal{N}_i}^{\mathcal{L}_f} = \left(\chi_{\mathcal{N}_i}^{\mathcal{L}_f} \times \tilde{O}_{\mathcal{N}_i}^{t_c}\right) and II_{\mathcal{L}_f}^{t_c} = 0, if \tilde{O}_{\mathcal{N}_i}^{t_c} is class I \\ II_{\mathcal{N}_i}^{\mathcal{L}_f} = 0 and II_{\mathcal{L}_f}^{t_c} = \left(\chi_{\mathcal{N}_i}^{\mathcal{L}_f} \times \tilde{O}_{\mathcal{N}_i}^{t_c}\right), otherwise \end{cases}$$

where $\chi_{\mathcal{N}_i}^{\mathcal{L}_f}$ is the ratio of $\tilde{O}_{\mathcal{N}_i}^{t_c}$'s division for the $\mathcal{L}_f$ where $0 \leq \chi_{\mathcal{N}_i}^{\mathcal{L}_f} \leq 1$. $\Gamma\_I_{\mathcal{L}_f}^{t_{c-1}}$ and $\Gamma\_II_{\mathcal{L}_f}^{t_{c-1}}$ are the total amount of class I and class II data services in the $\mathcal{L}_f$ at time t_{c-1}. ω, ξ, ε are coefficient factors for the class I data service, and τ, ψ are coefficient factors for the class II data service. For the other bus channel $\mathcal{L}_s$, the $\mathcal{L}_s$'s utility function, i.e., $U_{\mathcal{N}_i}^{\mathcal{L}_s}(\cdot)$, is defined as the same manner as the $U_{\mathcal{N}_i}^{\mathcal{L}_f}(\cdot)$. In the $\mathbb{G}_{\mathcal{N}_i}^{L}$, the $\chi_{\mathcal{N}_i}^{\mathcal{L}_f}$ is dynamically decided for the $\mathcal{N}_i$ according to the *RBEBS* (Kim, 2021).

$$RBEBS_{(\chi_{\mathcal{N}_i}^{\mathcal{L}})}\left(\mathfrak{U} = \left\{U_{\mathcal{N}_i}^{\mathcal{L}_f}(\cdot), U_{\mathcal{N}_i}^{\mathcal{L}_s}(\cdot)\right\}, \mathcal{D} = \left\{d_{\mathcal{L}_f}, d_{\mathcal{L}_s}\right\}\right)$$

$$= \mathbf{arg} \min_{0 \leq \chi_{\mathcal{N}_i}^{\mathcal{L}_f}, \chi_{\mathcal{N}_i}^{\mathcal{L}_s} \leq 1} \left[\left\| \frac{\mathcal{C}_{\chi_{\mathcal{N}_i}^{\mathcal{L}_f}}\left(U_{\mathcal{N}_i}^{\mathcal{L}_f}(\cdot), \{\mathfrak{U}, \mathcal{D}\}\right)}{\mathcal{C}_{\chi_{\mathcal{N}_i}^{\mathcal{L}_f}}(\mathcal{D})} - \frac{\mathcal{C}_{\chi_{\mathcal{N}_i}^{\mathcal{L}_s}}\left(U_{\mathcal{N}_i}^{\mathcal{L}_s}(\cdot), \{\mathfrak{U}, \mathcal{D}\}\right)}{\mathcal{C}_{\chi_{\mathcal{N}_i}^{\mathcal{L}_s}}(\mathcal{D})} \right\|\right] \tag{8.2.12}$$

where $\mathcal{C}_{\chi_{\mathcal{N}_i}^{\mathcal{L}}}(\cdot)$ and $\mathcal{C}_{\chi_{\mathcal{N}_i}^{\mathcal{L}}}(\cdot)$ are the cardinality of $\chi_{\mathcal{N}_i}^{\mathcal{L}_f}$ and $\chi'^{\mathcal{L}_f}_{\mathcal{N}_i}$ where $\chi_{\mathcal{N}_i}^{\mathcal{L}_f} \neq \chi'^{\mathcal{L}_f}_{\mathcal{N}_i}$. The value of $\chi_{\mathcal{N}_i}^{\mathcal{L}_f}$ is multiples of $\Delta\chi_{\mathcal{N}_i}^{\mathcal{L}_f}$, and the CGTC scheme sets $\Delta\chi_{\mathcal{N}_i}^{\mathcal{L}_f} = 0.1$ for the computation simplicity. If the value of $\chi_{\mathcal{N}_i}^{\mathcal{L}_f}$ is obtained according to (8.2.12), the value of $\chi_{\mathcal{N}_i}^{\mathcal{L}_s}$ is spontaneously decided as $\chi_{\mathcal{N}_i}^{\mathcal{L}_s} = \left(1 - \chi_{\mathcal{N}_i}^{\mathcal{L}_f}\right)$ (Kim, 2021).

8.2.6 Summary

Since its introduction approximately 30 years ago, the CAN protocol has steadily gained popularity and has been widely adopted in various embedded and networked control systems. One of its distinguishing features is the bus-line mechanism, which provides an efficient communication framework for vehicular systems. The *CGTC* scheme addresses the CAN traffic control challenge by developing a novel two-phase cooperative game model, incorporating the principles of *MREBS* and *RBEBS*. In the first phase, the traffic volume allocated to each ECU is dynamically determined based on the *MREBS* framework. In the second phase, individual ECUs balance their service data between the two communication channels in a fair and efficient manner using the *RBEBS* approach. Through its iterative step-by-step process, the two-phase cooperative game model continuously optimizes data service distribution, ensuring the efficient utilization of CAN system resources (Kim, 2021).

8.3 Feedback Game based Service Control (*FGSC*) scheme

As a distinct category of mobile ad-hoc networks, VANETs have garnered significant interest among researchers. In VANETs, vehicles function both as routers and clients, establishing connections via inherently unreliable wireless links. The highly dynamic movement of vehicles presents a considerable challenge in designing effective communication protocols for VANETs. Recently, S. Kim addressed the complexities of real-world VANET operations and introduces the *Feedback Game-based Service Control* (*FGSC*) scheme, which integrates power and rate control algorithms (Kim, 2011b). Leveraging a game-theoretic framework, the *FGSC* scheme efficiently models the interactive dynamics among multiple vehicles.

8.3.1 Development motivation

In recent years, VANETs have introduced distinctive features and numerous potential applications, making them a subject of great interest. Consequently, inter-vehicle communication has attracted significant attention from both the wireless networking and automotive industries. Particularly, VANET-based inter-vehicle communication is regarded as a promising solution for enhancing traffic safety, optimizing driving efficiency, and improving traffic management at a relatively low cost. However, realizing these benefits requires overcoming substantial challenges. One of the primary obstacles is the necessity for reliable message transmission and frequent data exchange to disseminate real-time road and traffic information. However, the dynamic nature of network topology results in weak link connectivity and frequent fragmentation. To address these issues, ongoing research on VANETs has been focused on developing adaptive communication algorithms while mitigating network congestion (Alpcan, 2010; Hafeez, 2010; Kim, 2015; Morales, 2012).

Given the high mobility of vehicles, the communication link between two vehicles may only remain active for a short period. As a result, ensuring reliability is a critical factor for VANET applications. One possible approach to maintaining stable vehicle-to-vehicle connections is by increasing transmission power. However, higher transmission power leads to increased network overhead, whereas lower transmission power can cause excessive multi-hop communication, resulting in high latency. Consequently, fixed transmission power is not an effective solution for VANET management. Instead, dynamically adjusting transmission power is essential to maintain efficient network performance under varying conditions (Kim, 2015; Rawat, 2009).

In addition to adaptive power control, certain emergency messages in VANETs must be delivered promptly and accurately. For instance, accident-related alerts need to reach the intended area in time to prevent secondary collisions. Therefore, prioritizing emergency messages over other types of communication is highly desirable. To achieve this, VANET systems should implement message differentiation mechanisms. Similar to QoS differentiation in contention-based channel access, VANET messages should be assigned priority levels based on their urgency and delay sensitivity (Rawat, 2011). To ensure QoS across different types of traffic services, the *FGSC* scheme employs IEEE 802.11e, an extension of the IEEE 802.11 standard designed for MAC layer enhancements. By utilizing the *Contention Window* (*CW*), this protocol facilitates proportional service differentiation in VANETs by prioritizing messages based on delay requirements (Farrokhi, 2010; Kim, 2015; Saadia, 2014).

Due to the complexity of vehicle mobility, unpredictable link quality, dynamic topology changes, and message prioritization, conventional end-to-end communication methods have shown limited success in

VANET environments. Furthermore, many inter-vehicle communication techniques involve autonomous decision-makers - vehicles - with conflicting objectives. Each vehicle makes independent decisions based on its perception of the dynamically evolving environment shaped by the actions of others. To analyze the decision-making behavior of self-interested vehicles, game theory provides a valuable framework. As a branch of applied mathematics, game theory models interactions between independent agents, predicting their responses to the actions of others and evaluating scenarios involving both competition and cooperation. Currently, game theory has been widely adopted as a fundamental tool for modeling distributed systems and optimizing cross-layer interactions (Ghazvini, 2012; Kim, 2015).

8.3.2 Contribution

Building on the insights discussed earlier, the *FGSC* scheme introduces a novel game-theoretic communication mechanism for VANETs, designed to dynamically regulate both transmission power and packet rate. The *FGSC* scheme integrates adaptive power control and packet rate adjustment within a unified algorithmic framework. By considering the interdependent effects of power and packet rate, the *FGSC* scheme employs a cross-layer approach to approximate optimal network performance. Key characteristics of the *FGSC* scheme include i) a dynamically adaptable strategy that accounts for the real-time conditions of the VANET, ii) a fully distributed structure, facilitating practical deployment, iii) an interactive adaptation process grounded in a feedback-based model, and iv) a well-balanced tradeoff between competing objectives. Across a wide range of diverse and dynamic traffic scenarios, the *FGSC* scheme has demonstrated its ability to enhance VANET performance in real-world network environments (Kim, 2015). Usually, traditional game theory implicitly assumes that players are rational and have complete information about the game situation throughout the time period. However, this assumption is obviously not satisfied under the real-world environment; experiments have shown that players do not always act rationally. However, the game based interactive approach can adaptively approximate an efficient solution without unrealistic rationality requirement. It is a main contribution of the *FGSC* scheme for VANET management (Kim, 2015).

8.3.3 Power control algorithm for VANETs

In VANET communications, frequent message exchanges between vehicles are essential for executing tasks effectively. Consequently, the *Communication Range* (*CR*) is a crucial factor in enabling timely wireless communication. If all vehicles attempt to maximize their communication range for efficiency, messages can reach their destinations with fewer hops. However, this approach comes at the cost of increased co-channel interference, leading to a higher number of vehicles competing for access to the same channel simultaneously. As a result, excessive interference may lead to network congestion and increased packet collisions, ultimately degrading communication performance. Conversely, if vehicles minimize their transmission range to reduce interference, communication will require more hops, leading to greater packet delays, which could be unacceptable for many time-sensitive applications. Therefore, a fixed transmission range is inadequate for adapting to rapidly changing traffic conditions and may significantly degrade VANET performance. To balance power efficiency and network capacity, the *FGSC* scheme introduces a dynamic power control algorithm for VANETs, enabling the system to determine the optimal transmission range. This adaptive approach ensures communication connectivity while mitigating the negative impact of excessive transmission power, thereby enhancing overall network efficiency (Kim, 2015).

In the *FGSC* scheme, transmission power is dynamically adjusted to maximize the overall payoff. Since the transmitter's payoff serves as an indicator of the effectiveness of the chosen power level, its formulation is heavily influenced by network feedback. This feedback mechanism ensures that power adjustments are responsive to real-time network conditions, optimizing communication efficiency while maintaining stability (Rose, 2012). As a transmitter, the vehicle i's payoff with message m and k^{th} power level ($\mathfrak{R}_i^k(m)$) is defined as follows (Kim, 2015).

$$\mathfrak{R}_i^k(m) = \left(\gamma \times \left(1 - \frac{p_k}{p_{MAX}}\right)\right) + \frac{1}{1+|\mathcal{N}_i|} \times \left(\beta \times \sum_{x \in \mathcal{N}_i} \text{Feedback}_x^k(m)\right) \tag{8.3.1}$$

s.t., $\text{Feedback}_x^k(m) = 1_{\{\tilde{A}_x \geq \tilde{A}^{tar}\}}$ or $0_{\{\tilde{A}_x < \tilde{A}^{tar}\}}$

where γ and β are parameters to tune the interest in satisfying the power consumption and the network feedback, respectively. $p_k \in$ P is the selected power level among a finite possible power set $\mathcal{P} = \{p_1,.. p_k..,p_{MAX}\}$ where p_{MAX} is the maximum power level. $\mathcal{N}_i$ is the set of vehicles in the vehicle i's *CR* and the $\text{Feedback}_x^k(m)$ is a one-bit value, which depends on the nature of the feedback chosen in the VANET. $\tilde{A}_x$ is the *SINR* of $x \in \mathcal{N}_i$ and $\tilde{A}^{tar}$ is the target *SINR* for wireless communications. For each message (m) packet transmission, the outcome of Feedback(m) is obtained, periodically (Kim, 2015).

In an interactive feedback mechanism, there is typically a trade-off between feedback frequency and control overhead. The feedback frequency is determined based on the system's performance objectives. If the frequency is relatively low, the algorithm can respond more effectively and accurately to network fluctuations, bringing system performance closer to optimal. However, this comes at the cost of increased control overhead. Conversely, a higher feedback frequency reduces control overhead but compromises system efficiency, as it may respond too slowly to real-time traffic changes. To balance this trade-off, the *FGSC* scheme sets the feedback frequency at each packet transmission. According to (8.3.1), the payoff is designed to decrease monotonically with power consumption while increasing with the number of successful transmission feedback instances (Kim, 2015).

Due to the highly dynamic nature of VANET environments, the *FGSC* scheme cannot precisely determine the optimal transmission power level in advance. To address this challenge, the *FGSC* scheme must adaptively adjust power levels by continuously responding to real-time network conditions. Instead of relying on a fixed model, the *FGSC* scheme estimates the payoff of a selected power level by analyzing both historical payoff data and recent variations in payoff values. This adaptive approach enables the system to make informed power adjustments, improving network efficiency and maintaining stable communication. As a weighted average of two quantities, the accumulated payoff of kth power level ($EP_t[k]$) is modified as follows (Kim, 2015).

$$\begin{cases} EP_{t+1}[k] = (1 - \alpha) \times EP_t[k] + \alpha \times \mathfrak{R}_i^k(m) \\ EP_{t+1}[j] = EP_t[j], s.t. j \neq k \end{cases} \tag{8.3.2}$$

where $EP_{t+1}[k]$ is the accumulated payoff of kth power level for the next interval. Therefore, $EP[\cdot]$ is recursively updated during operation rounds. The parameter α controls the relative weights given to recent and past estimation histories in the decision. Under dynamic VANET environments, a fixed value of α cannot effectively adapt to the changing network conditions. Therefore, each vehicle dynamically modifies α at the end of each operation round to make it more responsive to the current VANET conditions. When the difference between $\mathfrak{R}_i^k(m)$ and $EP_t[k]$ is small, the value of α makes little difference on VANET performance. But under a dynamic network changing situation, the value of α can affect performance significantly. If the traffic is uniformly distributed over the entire VANETs and mobility is relatively uniform, the *FGSC* scheme can put more emphasis on the payoff history, i.e., on$EP_t[k]$. In this case, a lower value of α is more suitable. But if traffic distributions and mobility is non-uniform, due to temporal and spatial traffic fluctuations, the payoff estimation should strongly depend on the recent payoff, i.e., on $\mathfrak{R}_i^k(m)$. In this case, a higher value of α is more suitable. Therefore, the *FGSC* scheme must decide on the value of α by considering the current VANET traffic conditions and by treating it as an on-line decision problem. In the *FGSC* scheme, α is dynamically adjusted as follows (Kim, 2015).

$$\alpha = \frac{\left\|\mathfrak{R}_i^k(m) - EP_t[k]\right\|}{\max\left\{\mathfrak{R}_i^k(m),\ EP_t[k]\right\}} \tag{8.3.3}$$

To adaptively select the power level, the *FGSC* scheme defines the vector V[k] where $1 \leq k \leq |\mathcal{P}|$. Based on $EP_t[\cdot]$ values, V[k] = 1 where $k = \{k \mid \max_{1\leq k\leq|\mathcal{P}|} EP_t[k]\}$ and the remaining elements $V[l]_{1\leq l\leq|\mathcal{P}|, l\neq k}$ equal to 0. For the $(t+1)^{th}$ operation round in the vehicle i, the selection probability ($\mathbb{Q}_i^k(t+1)$) of kth power level (p_i^k) is defined as follows (Kim, 2015).

$$\mathbb{Q}_i^k(t+1) = \mathbb{Q}_i^k(t) + \frac{1}{t+1}\left(V[k] - \mathbb{Q}_i^k(t)\right) \tag{8.3.4}$$

In order to implement the time driven approach for the equation (8.3.4) operations, the *FGSC* scheme partitions the time-axis into equal intervals of length *unit_time*, which is a time period for operation round. $\mathbb{Q}_i^k(\cdot)$ value is periodically modified every *unit_time*. Finally, the vehicle *i*'s power level (p_i) for the next round operation is stochastically given according to the distribution by $\mathbb{Q}_i(\cdot)$. Therefore, based on the interactive feedback mechanism, each vehicle adaptively decides its power level to maximize the payoff (Kim, 2015).

8.3.4 The estimation for packet forwarding probability

In VANETs, the behavior of one vehicle can either positively or negatively influence another, particularly when deciding whether to forward a message. This phenomenon is known as interdependency. In such interdependent scenarios, vehicles must adopt strategic actions to optimize communication performance. According to strategic interaction theory, commonly referred to as game theory, a vehicle can determine the most effective strategy by anticipating the actions of other participating vehicles (Kim, 2015; Roberto, 2011).

Game theory enables distributed coordination of vehicle communication by modeling interactions among autonomous decision-makers. It helps predict possible responses to the actions of other vehicles and facilitates the analysis of both conflict and cooperation within the network. Typically, a formal game model consists of three key components: players (vehicles in this context), available strategies for each player, and utility functions that quantify the effectiveness of different strategies (Kim, 2011a). Therefore, to represent a traditional game G, the game model components are given by G = < *Players* (N), *Strategies* (S), *Utility_functions* (U) >, where N is a finite set of players $\{1,2,\ldots n\}$, S is a non-empty set of the strategies, i.e., $S = \{S_1,\ldots, S_n\}$, and U is the utility function set $(\{u_1,\ldots, u_n\})$ of players. Each player's satisfaction level can be translated into a quantifiable metric through a utility function. Therefore, a utility function is a measure of the satisfaction experienced by a player; it is a result of a selected strategy. Usually, the utility function maps the user-level satisfaction to a real number, which represents the resulting payoff. The goal of each player is to maximize his own payoff by selecting a specific strategy where $\max_{S_i}: u_i(S_i) \rightarrow R$, $S_i \in \mathbb{S}$ and $i \in \mathbb{N}$ (Kim, 2011a; Kim, 2015).

To formulate the packet broadcast mechanism as a game model, the vehicle i's strategy S_i represents $S_i = \{f, d\}$, where $S_i = f$ implies a message forwarding and $S_i = d$ implies a message dropping. The vehicle i that forwards a message will have utility. By taking into account the cost and the payoff of message propagation, it is calculated as $\mathfrak{R}_i^k(m) - \mathscr{C}_i^k(m)$ where $\mathfrak{R}_i^k(m)$ is the payoff of message forwarding and $\mathscr{C}_i^k(m)$ is the forwarding cost. According to (8.3.1), $\mathfrak{R}_i^k(m)$ is obtained. $\mathscr{C}_i^k(m)$ is defined by considering the cost/payoff relation to forward the message. In general, the cost function should be non-negative and convex in order to allow the existence of at least one non-negative minimum. In order to achieve this goal, it is sufficient to formulate the deviation from target *SINR* as a non-negative and convex function (Pasandshanjanai, 2011). In addition, it should be assumed that $\mathfrak{R}_i^k(m) > \mathscr{C}_i^k(m)$ to prevent no vehicle from participating the message forwarding. Finally, the vehicle i's cost function ($\mathscr{C}_i^k(m)$) with the message m is given by (Kim, 2015).

$$\mathscr{C}_i^k(m) = \min\left\{\mathfrak{R}_i^k(m), \left[\gamma \times \frac{p_k}{p_{MAX}} + \varepsilon \times \cosh\left(\tilde{A}_i - \tilde{A}^{tar}\right)\right]\right\} \tag{8.3.5}$$

where γ and ε are control parameters for the cost evaluation. They can be adjusted to put more emphasis on power level or *SINR* deviation. γ is the same parameter used in the equation (8.3.1). If the value of ε is increased, achieving target *SINR* ($\tilde{A}^{tar}$) is more emphasized, otherwise power consumption is prioritized (Kim, 2015).

In multi-hop communications, the broadcast mechanism can lead to excessive redundant packet forwarding in VANETs, increasing the likelihood of message collisions. This phenomenon, known as the *broadcast storm* problem, can significantly raise network overhead and introduce delays, particularly for time-sensitive emergency messages (Roberto, 2011). Therefore, determining the optimal number of vehicles that should forward a broadcast message is a critical factor in improving network efficiency. Since vehicles in VANETs operate in a self-organized and distributed manner, they must independently address this challenge while maintaining coordination. To mitigate the *broadcast storm* problem, the *FGSC* scheme models the message forwarding process as a broadcasting game. Each time a vehicle receives a broadcast message, it participates in this game to make a strategic decision on whether to forward the message, optimizing network performance while reducing redundancy and congestion (Kim, 2015).

Among vehicles that receive the message in a *CR*, a cost ($\mathscr{C}(m)$) is associated with the message forwarding vehicle. Because of that cost, some vehicles in the same *CR* expect that at least one vehicle forwards the message at the cost, and they can only benefit by not doing so. However, if no one vehicle forwards the message within the *CR*, the worst possible outcome is obtained for all vehicles. Based on this assumption, the *FGSC* scheme can develop the utility function for each vehicle. Let $P_F(i)$ be the probability of message being forwarded by vehicle i and $P_{N\_F}(i)$ be the probability of message not being forwarded by vehicle i, where $P_F(i) + P_{N\_F}(i) = 1$. According to $P_F(i)$ and $P_{N\_F}(i)$, the vehicle i's utilities with message forwarding ($U_F(i)$) and with non message forwarding ($U_{N\_F}(i)$) are defined as follows (Kim, 2015).

$$\begin{cases} U_F(i) = \left(1 - P_{N\_F}(i)\right) \times \left(\mathfrak{R}_i^k(m) - \mathscr{C}_i^k(m)\right) \\ U_{N\_F}(i) = P_{N\_F}(i) \times \left(1 - \prod_{j=1, j\neq i}^{\mathscr{N}_i} P_{N\_F}(j)\right) \times \mathfrak{R}_i^k(m) \end{cases} \tag{8.3.6}$$

Based on (8.3.6), the total utility of vehicle i ($U_T(i)$) is defined as follows (Kim, 2015).

$$U_T(i) = \left(\left(1 - P_{N\_F}(i)\right) \times \left(\mathfrak{R}_i^k(m) - \mathscr{C}_i^k(m)\right)\right) + \left(P_{N\_F}(i) \times \left(1 - \prod_{j=1, j\neq i}^{\mathscr{N}_i} P_{N\_F}(j)\right) \times \mathfrak{R}_i^k(m)\right) \tag{8.3.7}$$

To maximize $U_T(i)$, the best $P_F(i)$and $P_{N\_F}(i)$ values are obtained by differentiating the $U_T(i)$ function with respect to $P_{N\_F}(i)$ and equating it with zero (Kim, 2015).

$$\begin{aligned} \frac{\partial U_T(i)}{\partial P_{N\_F}(i)} &= \left(1 - \prod_{j=1, j\neq i}^{\mathscr{N}_i} P_{N\_F}(j)\right) \times \mathfrak{R}_i^k(m) - \mathfrak{R}_i^k(m) + \mathscr{C}_i^k(m) \\ &= q^{\mathscr{N}_i - 1} \times \mathfrak{R}_i^k(m) - \mathscr{C}_i^k(m) = 0 \end{aligned} \tag{8.3.8}$$

Finally, the FGSC scheme can obtain $P_{N\_F}(i) = \left(\frac{\mathscr{C}_i^k(m)}{\mathfrak{R}_i^k(m)}\right)^{\frac{1}{\mathscr{N}_i - 1}}$ and $P_F(i) = 1 - P_{N\_F}(i) = 1 - \left(\frac{\mathscr{C}_i^k(m)}{\mathfrak{R}_i^k(m)}\right)^{\frac{1}{\mathscr{N}_i - 1}}$. Based on the symmetric mixed strategy with probability $P_F(i)$, each vehicle may forward a message; Diekmann proved that this kind of game goes into Nash equilibrium (Diekmann, 1985; Kim, 2015).

8.3.5 Packet rate control algorithm for VANETs

IEEE 802.11e is an approved amendment to the IEEE 802.11 standard, introducing a set of QoS enhancements for wireless communications at the MAC (Medium Access Control) layer. The MAC layer plays a crucial role in efficient and fair channel allocation, ensuring optimal network performance while preventing collisions, which can occur when multiple network nodes attempt to transmit frames simultaneously (Ghazvini, 2012). One widely used MAC protocol for carrier transmission access is Carrier Sense Multiple Access (CSMA). In CSMA, any network node can initiate frame transmission at any time, leading to potential collisions when multiple nodes transmit concurrently. To mitigate simultaneous channel access conflicts, the CSMA protocol employs a backoff algorithm based on a *contention window* (*CW*). According to this algorithm, each node waits for a randomly determined period - bounded by its *CW* - before attempting transmission. If a collision occurs, the node retrieves transmission after another backoff period until successful (Kim, 2015; Rawat, 2011).

Initially, the *CW* size is assigned to the minimum *CW* value (CW_{min}) and the backoff process is started independently in every node; a node computes a random value called backoff time, in the range of 0 and *CW* size. Actually, when the network load increases, the backoff time should be increased to minimize the collision probability. Therefore, after each unsuccessful transmission, *CW* is multiplied by σ, which is called the persistence coefficient; σ multiplication decreases the collision probability through the *CW*'s increase. For each collision, the *CW* size increases up to its maximum value $CW_{max} = \sigma^n \times CW_{min}$ where n is the number of subsequent retransmissions. Since *CW* has increased till reaching the CW_{max}, this backoff time will be preserved until the packet is transmitted successfully. After successful transmission, the *CW* will be set back to CW_{min} (Kim, 2015; Rawat, 2011).

During real world VANET operations, each message competes with other messages to gain a transmission opportunity. However, emergency messages especially require limited end-to-end delay and a low packet loss rate; they should be processed preferentially. The *CW* size control approach can be an effective method to prioritize one message over the other messages. By shortening the *CW* size, some messages can have a higher probability of being transmitted with low latency. The *FGSC* scheme chooses σ as a QoS parameter. When a message is admitted, it will be attached with σ. However, different messages require different QoS. Therefore, the σ value is different for each message. In the *FGSC* scheme, messages are categorized into four priority classes: *class I* (delay-critical emergency messages), *class II* (accident-related messages), *class III* (warning messages) and *class IV* (general messages). According to the required QoS in vehicle *i*, σ value for the message *m* is assigned as follows (Kim, 2015):

$$\sigma_i(m) = \begin{cases} \left(\xi_I + \left[1 - P_F(i)\right]\right)^{\xi_I}, if m \in classI \\ \left(\xi_{II} + \left[1 - P_F(i)\right]\right)^{\xi_{II}}, if m \in classII \\ \left(\xi_{III} + \left[1 - P_F(i)\right]\right)^{\xi_{III}}, if m \in classIII \\ \left(\xi_{IV} + \left[1 - P_F(i)\right]\right)^{\xi_{IV}}, if mt \in classIV \end{cases} \quad (8.3.9)$$

$$\text{s.t., } P_F(i) = 1 - \left(\frac{\mathscr{C}_i^k(m)}{\mathfrak{R}_i^k(m)}\right)^{\frac{1}{\mathcal{N}_i - 1}}$$

where ξ_I, ξ_{II}, ξ_{III} and ξ_{IV} are persistence coefficients for each type messages. By modifying the σ parameter based on (8.3.9), each vehicle is able to adjust the backoff interval through *CW*. Therefore, the *FGSC* scheme can effectively support message differentiation according to the required QoS (Kim, 2015).

8.3.6 Summary

Due to their promising capabilities and broad range of potential applications, VANETs and their communication mechanisms have recently drawn significant attention from the research community. However, broadcasting over multi-hop VANETs presents several challenges, including link unreliability, hidden node issues, message redundancy, and the broadcast storm problem. These factors can significantly degrade overall network performance. To address these challenges, the *FGSC* scheme explores VANET communication algorithms by employing a dynamically adaptive approach that jointly adjusts transmission power and packet rate. Using an online feedback mechanism, the *FGSC* scheme continuously interacts with the real-time VANET environment, dynamically modifying power levels and QoS parameters to optimize network efficiency. The *FGSC* scheme's key contributions include calculating the optimal transmission power level, determining rebroadcast probability, and adjusting the *CW* size to enhance communication performance in VANETs (Kim, 2015).

8.4 Timed Bargaining-based Opportunistic Routing (*TBOR*) scheme

VANET is a type of mobile ad hoc network characterized by its highly dynamic nature, where vehicles exhibit high mobility and experience rapid topology changes. Due to these unique properties, VANETs have become a prominent area of research. Given their specific characteristics, specialized routing protocols are essential to enable short-range, high-speed communication among nearby vehicles. Recently, S. Kim introduced the *Timed Bargaining-based Opportunistic Routing* (*TBOR*) scheme, which is built on the timed bargaining game framework (Kim, 2016a). To dynamically adapt to changing VANET conditions, the *TBOR* scheme models the opportunistic routing process as an iterative bargaining model that incorporates a timed learning approach, allowing vehicles to make efficient routing decisions in real time.

8.4.1 Development motivation

In recent years, the rapid expansion of traffic and the widespread availability of information services have facilitated the seamless integration of communication networks with vehicular networks, leading to the emergence of VANETs as a prominent research area. VANETs are self-organizing, open-structured inter-vehicular communication networks characterized by dynamic, distributed, and multi-hop properties. Additionally, they possess unique features such as high vehicle speeds, frequent topology changes, predictable movement trajectories, and minimal hardware constraints. These factors have a direct impact on transmission performance within VANETs (Kim, 2016a; Morales, 2012; Wang, 2012).

Effective routing is crucial for ensuring optimal VANET performance, as network efficiency is closely tied to the effectiveness of routing algorithms. Over the past few decades, substantial research efforts have been dedicated to VANETs, resulting in the development of numerous specialized routing protocols. The primary objective of these protocols is to enhance network performance by ensuring the reliable delivery of messages to vehicles within the communication range and across the network, while also mitigating congestion, collisions, and traffic management issues. However, due to the dynamic nature of VANETs and their specific operational requirements, achieving widespread adaptation and standardization of these protocols remains a significant challenge (Kim, 2016; Morales, 2012; Wang, 2012).

Recently, game theory has gained widespread recognition as a practical approach for implementing real-world network operations. As a branch of applied mathematics, game theory serves as a powerful tool for modeling interactions among independent decision-makers. It provides a framework for predicting responses to the actions of other network participants while analyzing both conflict and cooperation scenarios. In recent years, game-theoretic approaches have been extensively applied to the design of routing algorithms, offering well-adapted models for describing interactions among vehicles within VANETs (Kim, 2011c; Kim, 2012; Kim, 2016a)

8.4.2 Contribution

Building on the insights discussed earlier, the *TBOR* scheme introduces a novel routing mechanism for VANETs. To accommodate the highly dynamic nature of VANET environments, the classical *NBS* model is modified by incorporating a sequential bargaining approach. By leveraging a timed game model, the *TBOR* scheme is structured as an iterative routing process, where each iteration consists of three essential steps: i) observing the current network environment, ii) estimating the expected payoff to determine the most suitable routing strategy, and iii) updating information to continuously adapt to network dynamics. One of the key advantages of the timed game approach is its ability to relax the traditional *NBS* assumption that assumes complete knowledge of network conditions. This flexibility allows the *TBOR* scheme to function effectively even under incomplete information scenarios. The *TBOR* scheme exhibits several important characteristics; i) adaptability, allowing it to dynamically adjust to changing VANET conditions, ii) an interactive process, driven by an iterative feedback mechanism, and iii) a practical distributed approach, enabling efficient and scalable routing decisions. Even in highly diverse and rapidly fluctuating network conditions, the *TBOR* scheme effectively approximates optimal VANET performance, making it a viable solution for real-world routing operations (Kim, 2016a).

8.4.3 Wireless link estimation

Generally, the formal game model consists of players, the possible strategies of players, and utility functions of the strategies (Kim, 2011; Kim, 2012). Therefore, to represent a traditional game *G*, the game model components are given by $G = \langle N, S, \{U_1,..,U_n\}\rangle$, where *N* is the number of players, S is a non-empty set of the strategies, $U_i \in \{U_1,..,U_n\}$ is the utility function of player *i*, and optimize$_{S_i}$: $U_i(S_i) \rightarrow \mathrm{R}$, $S_i \in \mathbb{S}$. The *TBOR* scheme changes the definition of the utility function. The *TBOR* scheme introduces the concept of time by considering a timed utility function that changes with time (Sarcia, 2013). Formally, the timed utility function is defined to map the actions and times to the player-level satisfaction (i.e., a real number) as follows (Kim, 2016a):

$$U_i(S_i(t), t) = \int_0^t U_i(S_i(t))dt \cong \sum_{n=0}^{n=t} U_i(S_i(n)), \text{ s.t., } S_i(t), S_i(n) \in \mathbb{S}_i \tag{8.4.1}$$

where $S_i(t)$ is the strategy of player i at time t. Therefore, the utility function is shaped based on the time variation (Kim, 2016a). Owing to the dynamic nature of VANET - high dynamic topology and communication environment - the wireless links among vehicles varies continually. Therefore, it is important to estimate the current link status by considering several control factors. For the estimation of the degree of communication adaptability, the *TBOR* scheme defines a link cost (*L_P*) for each link. In order to handle dynamic VANET conditions, the *L_P* value from the vehicle i to the vehicle j is obtained as (Kim, 2016a);

$$L\_P_{ij} = \left[\alpha \times \left(1 - {(D_M - d_{ij})}/{D_M}\right)\right] + \left[(1-\alpha) \times \left(1 - \Psi_{ij}(t)\right)\right] + \zeta_{ij}(t) \tag{8.4.2}$$

$$\text{s.t.}, \alpha = \frac{\mathbf{min}\left\{ \left|\vec{v}_i(t_i)\right|, \left|\vec{v}_j(t_j)\right| \right\}}{\mathbf{max}\left\{ \left|\vec{v}_i(t_i)\right|, \left|\vec{v}_j(t_j)\right| \right\}}$$

where D_M is the maximum coverage range of each vehicle and d_{ij} is the distance between the vehicles i and j. $\vec{v}_i(t_i), \vec{v}_j(t_j)$ are the velocity vectors of vehicle i and j, respectively, at time t. Therefore, the first term in equation (8.4.2) is the relative distance ratio between the vehicles i and j. A closer neighboring vehicle is more suitable for routing. Therefore, if the first term is close to 0, it is preferred for stable routing. $\Psi_{ij}(t)$is the entropy for vehicle j at time (t). In general, entropy is uncertainty and a measure of disorder in a system. It represents the topological change, which is a natural quantification of the effect of vehicle mobility on the connectivity service of VANET. In this sudy, the basic concept of entropy is adopted for supporting and evaluating stable routing routes. For the mobile vehicle j, the entropy $\Psi_{ij}(t)$is calculated as follows (Kim, 2016a; Shen, 2003):

$$\Psi_{ij}(t) = \frac{-\sum_{k \in F_j} P_k(t, \Delta_t) \times \log P_k(t, \Delta_t)}{\log C(F_j)}, \text{ s.t.}, P_k(t, \Delta_t) = \frac{a_{j,k}}{\sum_{i \in F_j} a_{j,i}} \tag{8.4.3}$$

where $\Delta_t$is a time interval, F_j denotes the set of the neighboring vehicles of vehicle j, and $C(F_j)$ is the cardinality (degree) of set F_j. In order to estimate the stability of a part of a specific route, $a_{j,i}$ represents a measure of the relative mobility between two vehicles j and i as follows (Kim, 2016a):

$$a_{j,i} = \frac{1}{I\_T} \times \sum_{l=1}^{I\_T} v(j, i, t_l), \text{ s.t.}, v(j, i, t) = \left|\vec{v}_j(t) - \vec{v}_i(t)\right| \tag{8.4.4}$$

where I_T is the number of discrete times t_l. The mobility information can be calculated and disseminated to other neighboring vehicles within time interval Δ_t. I_T is defined as integer multiples of Δ_t. $v(j,i,t)$is the relative velocity between vehicles j and i at time t. Any change can be described as a change of variable values $a_{j,i}$over time t, e.g., $a_{j,i}(t) \rightarrow a_{j,i}(t+\Delta_t)$. The entropy $\Psi_{ij}(t)$is normalized as

$0 \leq \Psi_{ij}(t) \leq 1$. If the value of $\Psi_{ij}(t)$ is close to 1, the part of the route that represents the links of the path associated with an intermediate vehicle *j* is stable. If the value of $\Psi_{ij}(t)$is close to 0, the local route is unstable (Shen, 2003). The third term, $\zeta_{ij}(t)$, is defined as a timed reinforcement function, which dynamically indicates the adaptability of the link *ij* from the source vehicle to the destination vehicle. As time elapses, the outcome of $\zeta_{ij}(t)$ is dynamically adjusted. This function is explained in detail in the next section II.B (Kim, 2016a).

In order to relatively estimate the current link situation, the control parameter α controls the relative weights assigned to the relative speed ratio and the entropy of the corresponding relay vehicle. In diverse network environments, a fixed value of α cannot effectively adapt to the changing conditions. The *TBOR* scheme treats this scenario as an on-line decision problem and adaptively modify the value of α. When the relative speed of two neighbor vehicles is high, the *TBOR* scheme can place more emphasis on the relative distance ratio. In this case, a higher value of α is more suitable. If the relative speed of two neighbor vehicles is low, the *TBOR* scheme can place more emphasis on $(1 - \Psi_{ij}(t))$. In this case, the path selection should strongly depend on a lower value of α. In the *TBOR* scheme, the value of α of the vehicle *i* is dynamically adjusted according to the relative speed ratio (i.e., $\frac{\mathbf{min}\left\{\left|\vec{v}_i(t_i)\right|, \left|\vec{v}_j(t_j)\right|\right\}}{\mathbf{max}\left\{\left|\vec{v}_i(t_i)\right|, \left|\vec{v}_j(t_j)\right|\right\}}$). Therefore, the system can be more responsive to current VANET conditions by using real-time network monitoring (Kim, 2016a).

8.4.4 Opportunistic packet forwarding algorithm

In the *TBOR* scheme, the *L_P* value can represent the normalized communication cost of each link. Using the *L_P* value, the *TBOR* scheme defines the path cost (*PC*) parameter to calculate the total routing path cost; *PC* is computed as the sum of all link costs from the source vehicle to the current vehicle. Based on the *PC* value, the routing algorithm opportunistically constructs adaptive multi-hop routing paths to reach the destination vehicle (Kim, 2016a).

At the start of routing operations, the source vehicle broadcasts its initial *PC* value (i.e., *PC* = 0). Within the communication coverage area, message-receiving relay vehicles individually estimate the link cost according to equation (8.4.2) and update the *PC* value to *PC* + log(*L_P*). Some relay vehicles can receive multiple *PC* values from different reachable neighbor vehicles. Each relay vehicle stores this information for self-organizing and independent-effective controlling. For example, the vehicle *i* may receive multiple *PC* values, i.e., $PC_1,,, PC_k,,, PC_{N_i}$, where the PC_k value is received from the message-sending neighbor vehicle k ($1 \leq k \leq N_i$) and N_i is the total number of reachable neighbor vehicles for the vehicle *i*. In this case, the vehicle *i* calculates its minimum PC_i value as follows (Kim, 2016a):

$$PC_i = \arg \mathbf{min}_{k \in N_i} \left(PC_k + \log(L\_P_{ik})\right) = \arg \mathbf{min}_{k \in N_i} \left(\sum_{L\_P_\ell \in \mathbb{I}_k} \log(L\_P_\ell) + \log(L\_P_{ik}) \right) \tag{8.4.5}$$

where $\mathbb{I}_k$ is the set of selected links between the source vehicle and the relay vehicle k. According to equation (8.4.5), the vehicle i adaptively selects one neighbor vehicle while minimizing the value of PC_i, which potentially incorporates more global network information (Kim, 2016a).

The *TBOR* scheme models the relay vehicle selections as an iterative bargaining game according to equation (8.4.5). Therefore, the opportunistic link selection and route establishment process is formulated in a sequential bargaining manner. During the step-by-step iteration, the estimated *PC* value is recursively forwarded while opportunistically selecting the most adaptable relay nodes. This route formation process is repeated until a multi-hop path from the source to the destination vehicle is established. Finally, multiple packets are received at the destination vehicle. Owing to the opportunistic link selection approach, multiple routing paths can be configured; packets from different routing paths have different *PC* values. Based on the packet forwarding history, the destination vehicle can select the most adaptable routing path (Γ) as follows (Kim, 2016a):

$$\Gamma = \arg\min_{\substack{\{\mathcal{P}_i,\mathcal{P}_i\in S\}\\ \{L\_P_k\in\mathcal{P}_i\}}} \int_{t=\mathcal{T}_s}^{t=\mathcal{T}_e} \log\left(L\_P_k\right)dt = \arg\min_{\{\mathcal{P}_i,\mathcal{P}_i\in S\}} \sum\nolimits_{L\_P_k\in\mathcal{P}_i} \log\left(L\_P_k\right)$$
$$= \arg\min_{\{\mathcal{P}_i,\mathcal{P}_i\in S\}} \log \prod_{L\_P_k\in\mathcal{P}_i} \left(L\_P_k\right) \cong \arg\min_{\{\mathcal{P}_i,\mathcal{P}_i\in S\}} \prod_{L\_P_k\in\mathcal{P}_i} \left(L\_P_k\right)$$

(8.4.6)

where S is the set of established routing paths and $\mathcal{P}_i$ is the ith routing path from the source vehicle to the destination vehicle. $\mathcal{T}_s$ and $\mathcal{T}_e$ are the packet forwarding start time and end time, respectively. Therefore, Γ is the path with the minimum product value of link costs. In the proposed game model, the payoff of players (i.e., routing source and destination vehicles) is assumed to be the sum of $(1/\log\left(L\_P_k\right))$, i.e., $U_i\left(S_i(t),t\right) = \int_{t=\mathcal{T}_s}^{t=\mathcal{T}_e}\left(1/\log(L\_P_k)\right)dt$. In order to maximize their payoff, players want to select the routing path while minimizing the sum of $\log(L\_P_k)$ by using equation (8.4.6) (Kim, 2016a).

In the *TBOR* scheme, each link selection decision is made in a myopic online manner. Therefore, as time elapses, the selected strategies are no longer the best ones. In order to induce relay vehicles to select globally desirable links, the destination vehicle should reinforce the most desirable path (Γ) for the next opportunistic routing decision. Periodically, the destination vehicle sends a backward packet to modify the result of the $\zeta(t)$ function of the relay vehicles in the desirable Γ path as follows (Kim, 2016a):

$$\zeta_{ij}(t) = \vartheta(t) = \begin{cases} \vartheta(t) = \vartheta(t-1) - \Delta_\vartheta, if\, l_{ij} \in \Gamma \\ \\ \vartheta(t) = \vartheta(t-1), otherwise \end{cases}$$

(8.4.7)

where l_{ij} is the link ij and Δ_ϑ is a discount factor. If l_{ij} is in the Γ path, the outcome of $\zeta_{ij}(t)$ monotonically decreases with Δ_ϑ over time. In addition, the *TBOR* scheme defines a particular period of time, $\mathcal{P}^{\mathcal{M}}$. After every period $\mathcal{P}^{\mathcal{M}}$, $\vartheta(\cdot)$ is reset to the initial value (η). Therefore, at the beginning of every period $\mathcal{P}^{\mathcal{M}}$, $\zeta(t)$ value is initialized. This periodic refreshment mechanism helps adaptively reflect the constantly changing network conditions (Kim, 2016a).

8.4.5 Summary

In recent years, VANETs have gained substantial research interest, leading to a growing number of studies in the field. One of the most challenging issues in VANETs is the establishment of stable and efficient routes. This challenge arises from the highly dynamic network topology and the inherent instability of routing links, making traditional routing approaches less effective. To address this issue, the *TBOR* scheme introduces a novel routing mechanism designed to enhance routing stability in vehicular networks. To effectively adapt to dynamic VANET environments, the *TBOR* scheme is formulated as a timed strategic game built on the sequential bargaining approach. By employing a step-by-step bargaining process, the *TBOR* scheme iteratively optimizes routing performance while maintaining a fully distributed framework, ensuring scalability and adaptability in real-world VANET scenarios (Kim, 2016a).

8.5 Vehicular Sensing, Clouding and Routing (*VSCR*) scheme

The transition from 4G to 5G in the wireless industry is expected to bring significant advancements in vehicular network environments, particularly in areas such as crowdsensing, cloud computing, and routing. Vehicular crowdsensing leverages vehicle mobility to facilitate location-based services, while vehicular cloud computing represents a hybrid approach that dynamically utilizes vehicular network resources - including computing power, storage, and internet connectivity - for real-time decision-making. Recently, S. Kim introduced innovative crowdsensing, cloud computing, and routing algorithms for next-generation vehicular networks, integrating them into a sophisticated framework known as the *Vehicular Sensing, Clouding, and Routing* (*VSCR*) scheme (Kim, 2019a). To establish an efficient and interactive vehicular network process, the *VSCR* scheme is built upon learning algorithms and game theory. Designed as a triple-plane game model, the *VSCR* scheme dynamically adapts to the evolving vehicular network system by promoting individual cooperative interactions. The game-based approach provides a key advantage by ensuring self-adaptability and real-time responsiveness to fluctuating network conditions, enhancing the overall efficiency and stability of vehicular networks.

8.5.1 Development motivation

With the rapid advancement of mobile communication and intelligent transportation technologies, VANETs have emerged as a crucial component of the Intelligent Transportation System (ITS). As a specialized form of Mobile Ad Hoc Networks (MANETs), VANETs share similarities with MANETs in terms of node mobility and self-organization. However, the key distinction lies in the fact that VANET nodes are moving vehicles, which operate at higher speeds and possess substantial power resources, providing an advantage over traditional MANETs. As an emerging technology, VANET research primarily focuses on advancing Vehicle-to-Vehicle (V2V) communication to facilitate fast and cost-efficient data exchange, enhancing both vehicle safety and driver comfort (Kim, 2016b, 2019a; Morales, 2012; Wang, 2016).

Typically, vehicles communicate by sending and receiving messages through an On-Board Unit (OBU), a portable data collection device installed in vehicles. By integrating information technology with transportation infrastructure, VANETs support a wide array of applications, including traffic management, road safety, route optimization, multimedia entertainment, and social networking. With recent advancements in vehicular technology, both automobile manufacturers and academic researchers are increasingly involved in shaping VANET systems. Currently, approximately 10% of vehicles on the road are wirelessly connected, and by 2020, reports predict that 90% of vehicles will have wireless connectivity (Chen, 2017; Kim, 2019a; Su, 2017; Wang, 2016b).

In VANET systems, each vehicle can transmit, receive, and route messages within a V2V communication framework. Additionally, vehicles engage in Vehicle-to-Roadside (V2R) communication by interacting with Roadside Units (RSUs) (Kim, 2019b). Acting as access points, RSUs enable seamless connectivity for moving vehicles, providing continuous network access and facilitating real-time services. Within VANET environments, RSUs and OBUs interact autonomously, making individual control decisions without relying on a centralized coordination mechanism (Chen, 2017; Kim, 2019a, 2019b; Oh, 2010; Su, 2017; Wang, 2016a, 2016b; Xiao, 2017).

VANET services can be categorized into three fundamental types: Vehicular Cloud Services (VCS), Vehicular Sensing Services (VSS), and Vehicular Routing Services (VRS). VCS leverages cloud computing to provide computational services to VANETs (Chaqfeh, 2016). As vehicular applications demand increasing computational power, the interaction between RSUs and vehicles within the VCS framework enables efficient allocation of computing, storage, and data resources. Vehicles requiring additional resources can access them through interconnected RSUs, creating valuable market opportunities while also presenting cost-related challenges (Ashok, 2016; Chaqfeh, 2016; Kim, 2019a). VSS plays a crucial role in public sensing applications. Vehicles equipped with OBUs function as mobile sensors, collecting localized information as requested by RSUs. In this model, RSUs act as service providers, compensating vehicles for their sensing contributions based on data quality. Examples of VSS applications include location-based road monitoring, real-time traffic condition reporting, air pollution measurement, and public information sharing. The mobility of vehicles extends sensing coverage, making VSS services highly valuable to a broad range of users (Ahn, 2017; Kim, 2019a; Wang, 2016a; Xiao, 2017). VRS facilitates efficient vehicle-to-vehicle communication, relying on effective routing strategies to address key VANET challenges. Although VANETs share some characteristics with MANETs, their high mobility, rapid topology changes, and abundant power supply introduce unique challenges for routing protocols. Developing adaptive vehicular routing mechanisms requires careful consideration of these factors to ensure robust VRS performance (Jang, 2013; Kim, 2016, 2011b, 2019a).

The *VSCR* scheme integrates VCS, VRS, and VSS into a unified VANET control framework. By combining these three components, the *VSCR* scheme maximizes synergistic benefits while maintaining an optimal balance between performance and resource utilization. However, integrating VCS, VRS, and VSS into a cohesive control scheme presents significant technical challenges, necessitating the development of a new control paradigm. Game theory has been widely applied in VANET performance analysis, where network agents are modeled as game players who interact according to strategic decision-making models (Kim, 2019b). The *VSCR* scheme employs distinct game models to develop innovative VCS, VRS, and VSS algorithms: In the VCS algorithm, vehicles share cloud resources provided by RSUs to enhance quality of service. The interaction between vehicles and RSUs is modeled as a non-cooperative game. In the VRS algorithm, vehicles aim to improve the reliability and efficiency of routing services through game-based decision-making mechanisms. In the VSS algorithm, vehicles function as mobile sensors, gathering necessary environmental data, while RSUs process and distribute the collected information. RSUs utilize a learning-based incentive mechanism to encourage participation, while individual vehicles adopt a cooperative bargaining strategy to optimize the balance among VCS, VSS, and VRS operations (Kim, 2019a).

The proposed VCS, VRS, and VSS algorithms must coexist and complement each other to accommodate diverse VANET requirements. To analyze the strategic interactions among these control mechanisms, a triple-plane negotiation model is introduced, linking VCS, VRS, and VSS decision-making processes. Within the VSS algorithm, vehicles select strategies based on bargaining solutions to achieve efficient VANET management. The classical Nash bargaining solution is implemented in a distributed manner, enabling triple-plane bargaining for each vehicle. By adopting a sequential game-based approach, the proposed triple-plane bargaining model effectively resolves conflicts, ensuring both practical applicability and system efficiency (Kim, 2019a).

8.5.2 Contribution

The key contributions of the *VSCR* scheme can be summarized as follows. The *VSCR* scheme introduces innovative VCS, VRS, and VSS algorithms, specifically designed to enhance VANET performance. It employs a game-theoretical framework to ensure efficient and adaptive service provisioning. A distributed triple-plane bargaining approach is utilized to balance conflicting network requirements, improving system stability and efficiency. Self-adaptability is seamlessly integrated into each control decision process, enabling dynamic real-time adjustments to network conditions. The *VSCR* scheme effectively models and captures dynamic interactions among game players, considering their diverse strategic perspectives. By achieving an optimal balance between theoretical efficiency and practical feasibility, the *VSCR* scheme offers a robust and scalable solution for real-world VANET applications. To the best of my knowledge, limited research has explored integrated bargaining-based algorithms for VANET control challenges, making the *VSCR* scheme a significant contribution to the field (Kim, 2019a).

8.5.3 Game models for the VCS, VRS and VSS algorithms

For VANET system operations, the *VSCR* scheme introduces three distinct game models tailored for the VCS, VRS, and VSS algorithms. In the *VSCR* scheme, vehicles and RSUs act as game players, making strategic decisions based on their interactions with other participants. Within the *VSCR* scheme, the VCS and VRS algorithms are modeled as non-cooperative games, where individual players optimize their strategies independently to maximize their own benefits. The VSS algorithm is formulated as a triple-plane bargaining model, integrating cooperation, coordination, and collaboration among the VCS, VRS, and VSS processes to achieve optimal performance balance. By incorporating these game-theoretic approaches, the *VSCR* scheme enhances adaptive decision-making and resource allocation, ensuring efficient and scalable VANET operations. For the VCS algorithm, the *VSCR* scheme formally defines the game model $\mathbb{G}^{VCS} = \{\mathrm{N}, \mathrm{C}, \left\{\mathbb{S}^{\mathcal{R}}, \mathbb{S}^{\mathcal{V}}_{\mathcal{V}_i \in \mathcal{V}}\right\}, \left\{\mathbb{U}^{\mathcal{R}}, \mathbb{U}^{\mathcal{V}}_{\mathcal{V}_i \in \mathcal{V}}\right\}, \mathbb{L}^{\mathfrak{P}^{\mathcal{R}}_j \in \mathbb{S}^{\mathcal{R}}}, \mathrm{T}\}$ at each time period of gameplay (Kim, 2019a).

- N is the finite set of VCS game players $\mathbb{N} = \{\mathcal{R}, \mathcal{V}\}$ where R represents a RSU and $\mathbf{V} = \left\{\mathcal{V}_1 \ldots \mathcal{V}_i, \ldots \mathcal{V}_k\right\}$ is a set of multiple vehicles in the R's coverage area.
- C is the cloud computation capacity, i.e., number of CPU cycles per second in the R.
- $\mathbb{S}^{\mathcal{R}} = \left\{\mathfrak{P}^{\mathcal{R}}_{min} \ldots \mathfrak{P}^{\mathcal{R}}_j \ldots \mathfrak{P}^{\mathcal{R}}_{max}\right\}$ is a set of R's strategies where $\mathfrak{P}^{\mathcal{R}}_j$ means the price level to execute one basic cloud service unit (BCSU). $\mathbb{S}^{\mathcal{V}}_{\mathcal{V}_i \in \mathcal{V}}$ is the $\mathcal{V}_i$'s available strategies $\mathbb{S}^{\mathcal{V}}_{\mathcal{V}_i} = \left[r^{\mathcal{V}_i}_{min} \ldots r^{\mathcal{V}_i}_k \ldots r^{\mathcal{V}_i}_{max}\right]$ where $r^{\mathcal{V}_i}$ represents the amount of cloud services of the $\mathcal{V}_i$; $r^{\mathcal{V}_i}$ is specified in terms of BCSUs.
- $\mathbb{U}^{\mathcal{R}}$ is the payoff received by the RSU, and $\mathbb{U}^{\mathcal{V}}_{\mathcal{V}_i \in \mathcal{V}}$ is the payoff received by the $\mathcal{V}_i$ during the VCS process.

- $\mathbb{L}^{\mathfrak{P}_j^{\mathscr{R}} \in S^{\mathscr{R}}}$ is the R's learning value for the strategy $\mathfrak{P}_j^{\mathscr{R}}$; L is used to estimate the probability distribution ($\mathbb{P}^{\mathscr{R}}$) for the next R's strategy selection.
- $T = \{\mathscr{H}_1, \ldots, \mathscr{H}_t, \mathscr{H}_{t+1}, \ldots\}$ denotes time, which is represented by a sequence of time steps with imperfect information for the VCS game process.

For the VRS algorithm, the *VSCR* scheme defines the game model $\mathbb{G}^{VRS} = \{$v, $\mathscr{N}_{\mathscr{V}_i}$, $\mathbb{A}_{\mathscr{V}_i \in \mathbb{V}}$, $U_{\mathscr{V}_i \in \mathbb{V}}$, $T\}$ at each time period of $\mathbb{G}^{VRS}$ gameplay. $\mathbb{G}^{VRS}$ can formulate the interactions of vehicles for VANET routing operations (Kim, 2019a).

- v is the finite set of game players $\mathbb{V} = \{\mathscr{V}_1 \ldots \mathscr{V}_n\}$, where n is the number of vehicles for the $\mathbb{G}^{VRS}$ game.
- $\mathscr{N}_{\mathscr{V}_i}$ is the set of $\mathscr{V}_i$'s one-hope neighboring vehicles.
- $\mathbb{A}_{\mathscr{V}_i \in \mathbb{V}} = \left\{ a_{\mathscr{V}_i}^{\mathscr{V}_m} | \mathscr{V}_m \in \mathscr{N}_{\mathscr{V}_i} \right\}$ is the finite set of $\mathscr{V}_i$'s available strategies where $a_{\mathscr{V}_i}^{\mathscr{V}_m}$ represents the selection of the $\mathscr{V}_m$ to relay a routing packet.
- $U_{\mathscr{V}_i \in \mathbb{V}}$ is the payoff received by the $\mathscr{V}_i$ during the VRS process.
- $T = \{\mathscr{H}_1, \ldots, \mathscr{H}_t, \mathscr{H}_{t+1}, \ldots\}$ is the same as the T in the $\mathbb{G}^{VCS}$.

For the VSS algorithm, the *VSCR* scheme formally defines the game model $\mathbb{G}^{VSS} = \{$N, **I**, $\left\{ S^{\mathscr{R}}, S_{\mathscr{V}_i \in \mathscr{V}}^{\mathscr{V}} \right\}$, $\left\{ u^{\mathscr{R}}, u_{\mathscr{V}_i \in \mathscr{V}}^{\mathscr{V}} \right\}$, $\mathfrak{Z}^{\mathfrak{P}_i^{\mathfrak{I}} \in \eta}$, T$\}$ at each time period of gameplay (Kim, 2019a).

- $\mathbb{N} = \{\mathscr{R}, \mathscr{V}\}$ is the finite set of game players for the $\mathbb{G}^{VSS}$ game; N is the same as the $\mathbb{N}$in the $\mathbb{G}^{VCS}$.
- **I** is the finite set of sensing tasks $\mathbf{I} = \{\mathfrak{I}_1 \ldots \mathfrak{I}_s\}$ in the R, where s is the number of total sensing tasks.
- $S^{\mathscr{R}} = [\eta_{\mathscr{R}}^{\mathfrak{I}_1} \ldots \eta_{\mathscr{R}}^{\mathfrak{I}_j} \ldots \eta_{\mathscr{R}}^{\mathfrak{I}_s}]$ is a vector of R's strategies where $\eta_{\mathscr{R}}^{\mathfrak{I}_j}$ represents the strategy set for the $\mathfrak{I}_j$. In the $\mathbb{G}^{VSS}$ game, $\eta_{\mathscr{R}}^{\mathfrak{I}_j}$ means the set of price levels for the crowdsensing work for the task $\mathfrak{I}_j$ during a time period ($\mathscr{H} \in T$). Like as the $\mathbb{G}^{VCS}$ game, $\eta_{\mathscr{R}}^{\mathfrak{I}_j}$ is defined as $\eta_{\mathscr{R}}^{\mathfrak{I}_j} = \left\{ \prod_{\mathfrak{I}_j \in N} \mathfrak{P}_i^{\mathfrak{I}_j} \middle| \mathfrak{P}_i^{\mathfrak{I}_j} \in [\mathfrak{P}_{min}^{\mathfrak{I}_j} \ldots \mathfrak{P}_l^{\mathfrak{I}_j} \ldots \mathfrak{P}_{max}^{\mathfrak{I}_j}] \right\}$.
- $S_{\mathscr{V}_i}^{\mathscr{V}} = [\mu_{\mathscr{V}_i}^{\mathfrak{I}_1} \ldots \mu_{\mathscr{V}_i}^{\mathfrak{I}_j} \ldots \mu_{\mathscr{V}_i}^{\mathfrak{I}_s}]$ is the vector of $\mathscr{V}_i$'s strategies where $\mu_{\mathscr{V}_i}^{\mathfrak{I}_j}$ represents the strategy for the $\mathfrak{I}_j$. In the $\mathbb{G}^{VSS}$ game, $\mu_{\mathscr{V}_i}^{\mathfrak{I}_j}$ means the $\mathscr{V}_i$'s active VSS participation, i.e., $\mu_{\mathscr{V}_i}^{\mathfrak{I}_j} = 1$, or not, i.e., $\mu_{\mathscr{V}_i}^{\mathfrak{I}_j} = 0$, for the $\mathfrak{I}_j$ during the H.
- $u^{\mathscr{R}}$ is the payoff received by the RSU, and $u_{\mathscr{V}_i \in \mathscr{V}}^{\mathscr{V}}$ is the payoff received by the $\mathscr{V}_i$ during the VSS process.
- $\mathfrak{Z}^{\mathfrak{P}_i^{\mathfrak{I}} \in \eta_{\mathscr{R}}^{\mathfrak{I}}}$ is the learning value for the strategy $\mathfrak{P}_i^{\mathfrak{I}_j}$; Z is used to estimate the probability distribution ($P^{\mathscr{R}}$) for the next R's strategy selection.
- $T = \{\mathscr{H}_1, \ldots, \mathscr{H}_t, \mathscr{H}_{t+1}, \ldots\}$ is the same as the T in the $\mathbb{G}^{VCS}$ and $\mathbb{G}^{VRS}$.

8.5.4 The VCS algorithm based on a non-cooperative game model

In the next-generation VANET paradigm, a wide range of applications demand substantial computational resources. However, most vehicles have limited processing capabilities, posing a significant challenge. To overcome this resource constraint, vehicular cloud technology has emerged as a promising approach to enhance VANET service performance. Within the VCS process, computational tasks can be offloaded to remote cloud servers, optimizing both resource utilization and computational efficiency (Kim, 2019a; Zhang, 2017). Vehicular cloud servers are deployed within RSUs, where they execute computational tasks received from vehicles. However, the service coverage of an RSU is restricted by its radio range, and given the high mobility of vehicles, a single vehicle may traverse multiple RSUs while offloading a task. Consequently, service results must be seamlessly migrated to another RSU to ensure successful delivery of the intended vehicle. From the perspectives of both RSUs and vehicles, the primary objective is payoff maximization. To achieve a mutually beneficial outcome, both entities strategically select their optimal strategies. Within the VCS process, RSUs act as *proposers*, while individual vehicles serve as *responders*. Through this interactive negotiation, RSUs and vehicles dynamically coordinate their actions to maximize efficiency and performance within the VCS framework (Kim, 2019a).

To design the VCS algorithm, the *VSCR* scheme considers a VCS platform consisting of one RSU (R), and a set of mobile vehicles (**V**); they formulate the game model $\mathbb{G}^{VCS}$. As a *proposer*, the R's strategy ($\mathfrak{P}_j^{\mathcal{R}} \in \mathbb{S}^{\mathcal{R}}$) indicates the offered price for one BCSU process. The R has its own utility function, which represents its net gain while providing the VCS process. The R's utility function with $\mathfrak{P}_j^{\mathcal{R}}$ strategy at the t'th time step $\mathcal{H}_t$ $\left(\mathbb{U}^{\mathcal{R}}(\mathcal{V},\mathfrak{P}_j^{\mathcal{R}}(\mathcal{H}_t))\right)$ is given by (Kim, 2019a);

$$\mathbb{U}^{\mathcal{R}}\left(\mathcal{V},\mathfrak{P}_j^{\mathcal{R}}(\mathcal{H}_t)\right) = \sum_{\mathcal{V}_i\in\mathcal{V}}\left(\xi_{\mathcal{V}_i}(\mathcal{H}_t)\times\left(\left\{\mathfrak{P}_j^{\mathcal{R}}(\mathcal{H}_t)\times r_k^{\mathcal{V}_i}(\mathcal{H}_t)\right\} - \mathcal{C}^{\mathcal{R}}\left(r_k^{\mathcal{V}_i}(\mathcal{H}_t)\right)\right)\right)$$

(8.5.1)

s.t., $\mathfrak{P}_j^{\mathcal{R}}(\mathcal{H}_t) \in S^{\mathcal{R}}, \mathcal{C}^{\mathcal{R}}\left(r_k^{\mathcal{V}_i}(\mathcal{H}_t)\right) = \Theta_{BCSU}^{\mathcal{R}}\times r_k^{\mathcal{V}_i}(\mathcal{H}_t), \Theta_{BCSU}^{\mathcal{R}} = {}^{(\theta\times\mathfrak{M})}/_{\mathbb{C}}$

and $\xi_{\mathcal{V}_i}(\mathcal{H}_t) = \begin{cases} 1, \mathcal{V}_i\, selects\, r_k^{\mathcal{V}_i} \in \mathbb{S}_{\mathcal{V}_i}^{\mathcal{V}}\, at\, \mathcal{H}_t \\ 0, otherwise\,(i.e., no VCS of \mathcal{V}_i) \end{cases}$

where $\mathcal{C}^{\mathcal{R}}\left(r_k^{\mathcal{V}_i}(\mathcal{H}_t)\right)$ is the cost function to execute the $\mathcal{V}_i$'s cloud service $\left(r_k^{\mathcal{V}_i}(\mathcal{H}_t)\right)$ at time $\mathcal{H}_t$. $\grave{E}_{BCSU}^{\mathcal{R}}$ is the processing cost to execute one BCSU and M is the currently using capacity of C. θ is the coefficient factor of cost calculation. In practice, the actual sensing cost $\mathcal{C}^{\mathcal{R}}(\cdot)$ is usually unknown by vehicles (Kim, 2019a).

As a *responder*, the vehicle $\mathcal{V}_i$'s strategy $\left(\mathbb{S}_{\mathcal{V}_i}^{\mathcal{V}}\right)$ represents the amount of cloud service. The payoff of $\mathcal{V}_i$ can be defined as a function of task offload level ($r^{\mathcal{V}_i}$) and service price ($\mathfrak{P}^{\mathcal{R}}$). Therefore, the $\mathcal{V}_i$'s utility function with $\mathfrak{P}_j^{\mathcal{R}}$ and $r_k^{\mathcal{V}_i}$ strategies at time $\mathcal{H}_t$ $\left(\mathbb{U}_{\mathcal{V}_i\in\mathcal{V}}^{\mathcal{V}}\left(\mathfrak{P}_j^{\mathcal{R}}(\mathcal{H}_t), r_k^{\mathcal{V}_i}(\mathcal{H}_t)\right)\right)$ is computed as follows (Kim, 2019a):

$$\mathbb{U}_{\mathcal{V}_i\in\mathcal{V}}^{\mathcal{V}}\left(\mathfrak{P}_j^{\mathcal{R}}(\mathcal{H}_t), r_k^{\mathcal{V}_i}(\mathcal{H}_t)\right) =$$

$$\varrho^{\mathcal{V}_i}(\mathcal{H}_t)\times\left(\left(\mathfrak{h}^{\mathcal{V}_i}\times r_k^{\mathcal{V}_i}(\mathcal{H}_t)\right) - \left(\mathfrak{P}_j^{\mathcal{R}}(\mathcal{H}_t)\times r_k^{\mathcal{V}_i}(\mathcal{H}_t)\right)\right)$$

(8.5.2)

$$\text{s.t., } \varrho^{\mathcal{V}_i}(\mathcal{H}_t) = \begin{cases} 1, and\ \Gamma_{\mathcal{H}_{t+1}}^{\mathcal{V}_i} = \Gamma_{\mathcal{H}_t}^{\mathcal{V}_i} - \left(\mathfrak{P}_j^{\mathcal{R}}(\mathcal{H}_t)\times r_k^{\mathcal{V}_i}(\mathcal{H}_t)\right) \\ if\left(\Gamma_{\mathcal{H}_t}^{\mathcal{V}_i} \geq \left(\mathfrak{P}_j^{\mathcal{R}}(\mathcal{H}_t)\times r_k^{\mathcal{V}_i}(\mathcal{H}_t)\right)\right) and\left(r_k^{\mathcal{V}_i} \in \mathbb{S}_{\mathcal{V}_i}^{\mathcal{V}} \text{ is selected}\right) \\ 0, otherwise\left(\text{i.e., } \Gamma_{\mathcal{H}_t}^{\mathcal{V}_i} < \left(\mathfrak{P}_j^{\mathcal{R}}(\mathcal{H}_t)\times r_k^{\mathcal{V}_i}(\mathcal{H}_t)\right)\right) \end{cases}$$

where $\mathfrak{h}^{\mathscr{V}_i}$ is the $\mathscr{V}_i$'s profit factor if one BCSU is processed, and $\Gamma_{\mathscr{H}_t}^{\mathscr{V}_i}$ is the amount of $\mathscr{V}_i$'s virtual money at time $\mathscr{H}_t$; if the $\mathscr{V}_i$ has enough money to pay the VCS price, the $\mathscr{V}_i$ can request its cloud service $\left(r_k^{\mathscr{V}_i}\right)$. As the $\mathbb{G}^{VCS}$ game players, RSU and vehicles attempt to maximize their utility functions. Interactions among game players continue repeatedly during the VCS process over time. In particular, the RSU should consider the reactions from vehicles to determine the price strategy ($\mathfrak{P}^{\mathscr{R}}$). The *VSCR* scheme develops a new learning method to decide an effective price policy for cloud services. If the strategy $\mathfrak{P}^{\mathscr{R}}$ is selected at time $\mathscr{H}_{t-1}$, the RSU updates the strategy $\mathfrak{P}_j^{\mathscr{R}}$'s learning value for the next time step $\left(\mathbb{L}_{\mathscr{H}_t}^{\mathfrak{P}_j^{\mathscr{R}}}\right)$ according to the following method (Kim, 2019a):

$$\mathbb{L}_{\mathscr{H}_t}^{\mathfrak{P}_j^{\mathscr{R}}} = \max\left\{\left[\left((1-\chi)\times \mathbb{L}_{\mathscr{H}_{t-1}}^{\mathfrak{P}_j^{\mathscr{R}}}\right) + \left(\chi \times \log_2\left(\mathbb{U}^{\mathscr{R}}\left(\mathscr{V}, \mathfrak{P}_j^{\mathscr{R}}(\mathscr{H}_{t-1})\right) / \sum_{\mathfrak{P}_k^{\mathscr{R}}\in\mathbb{S}^{\mathscr{R}}} \frac{\mathbb{L}_{\mathscr{H}_{t-1}}^{\mathfrak{P}_k^{\mathscr{R}}}}{\mathbb{S}^{\mathscr{R}}}\right)\right)\right], 0\right\} \tag{8.5.3}$$

where $\mathbb{S}^{\mathscr{R}}$ is the cardinality of $\mathbb{S}^{\mathscr{R}}$ and χ is a learning rate that models how the L-values are updated. To implement the price learning mechanism in the RSU, a strategy selection distribution ($\mathscr{P}^{\mathscr{R}} \in \mathbb{P}^{\mathscr{R}}$) for the VCS is defined based on the $\mathbb{L}(\cdot)$ values. During the $\mathbb{G}^{VCS}$ game process, the *VSCR* scheme sequentially determines $\mathbb{P}^{\mathscr{R}}(\mathscr{H}_t) = \left\{\mathscr{P}_{\mathfrak{P}_{min}^{\mathscr{R}}}^{\mathscr{R}}(\mathscr{H}_t)\ldots\mathscr{P}_{\mathfrak{P}_j^{\mathscr{R}}}^{\mathscr{R}}(\mathscr{H}_t)\ldots\mathscr{P}_{\mathfrak{P}_{max}^{\mathscr{R}}}^{\mathscr{R}}(\mathscr{H}_t)\right\}$ as the probability distribution of R's strategy selection at time $\mathscr{H}_t$. The $\mathfrak{P}_j^{\mathscr{R}}$ strategy selection probability at time $\mathscr{H}_t$ $\left(\mathscr{P}_{\mathfrak{P}_j^{\mathscr{R}}}^{\mathscr{R}}(\mathscr{H}_t)\right)$ is defined as (Kim, 2019a):

$$\mathscr{P}_{\mathfrak{P}_j^{\mathscr{R}}}^{\mathscr{R}}(\mathscr{H}_t) = \frac{\mathrm{EXP}\left(\mathbb{L}_{\mathscr{H}_{t-1}}^{\mathfrak{P}_j^{\mathscr{R}}}\right)}{\sum_{k=min}^{max}\left(\mathrm{EXP}\left(\mathbb{L}_{\mathscr{H}_{t-1}}^{\mathfrak{P}_k^{\mathscr{R}}}\right)\right)} \tag{8.5.4}$$

8.5.5 The VRS algorithm based on a non-cooperative game model

The primary objective of VRS is to efficiently transmit data from a source vehicle to a destination vehicle using wireless multi-hop transmission techniques. In this approach, intermediate vehicles along the routing path must relay data promptly, ensuring minimal delay in communication between the source and destination. For applications related to road safety, driver assistance, and emergency warnings, inter-vehicular communication plays a critical role in enabling real-time information exchange. Reliable VRS mechanisms are essential to enhance traffic safety, prevent accidents, and facilitate efficient vehicular coordination in dynamic VANET environments (Hieu, 2010; Kim, 2016, 2019b). The *VSCR* scheme develops a non-cooperative game model ($\mathbb{G}^{VRS}$) for vehicular routing services. As game players in $\mathbb{G}^{VRS}$, vehicles dynamically decide their routes. To configure the routing topology, a link cost ($\mathscr{LC}$) is defined to relatively handle dynamic VANET conditions. The *VSCR* scheme defines a wireless link status $\left(\mathscr{LC}_{\mathscr{V}_i}^{\mathscr{V}_j}(\mathscr{H}_t)\right)$ between the $\mathscr{V}_i$ and the $\mathscr{V}_j$ at time $\mathscr{H}_t$ like as (Kim, 2019a);

$$\mathscr{LC}_{\mathscr{V}_i}^{\mathscr{V}_j}(\mathscr{H}_t) = \left((1-\alpha)\times\left(\frac{d_{\mathscr{V}_i}^{\mathscr{V}_j}(\mathscr{H}_t)}{D_{\mathscr{V}_i}^{M}}\right)\right) + \left(\alpha\times\frac{\{\vec{v}_j(\mathscr{H}_t)\}}{\max_{\mathscr{V}_h\in\mathfrak{M}_{\mathscr{V}_i}^{\mathscr{H}_t}}\{\vec{v}_h(\mathscr{H}_t)\}}\right)$$

(8.5.5)

where $D^M_{\mathcal{V}_i}$ is the maximum coverage range of $\mathcal{V}_i$ and $d^{\mathcal{V}_j}_{\mathcal{V}_i}(\mathcal{H}_t)$ is the relative distance between the $\mathcal{V}_i$ and the $\mathcal{V}_j$ at time $\mathcal{H}_t$. Let $\mathfrak{M}^{\mathcal{H}_t}_{\mathcal{V}_i}$ be the set of the neighboring vehicles of $\mathcal{V}_i$ at time $\mathcal{H}_t$, and $\vec{v}_j(\mathcal{H}_t)$ is the relative velocity of the $\mathcal{V}_j$ at time $\mathcal{H}_t$. For the adaptive $\mathcal{LC}^{\mathcal{V}_j}_{\mathcal{V}_i}(\mathcal{H}_t)$ estimation, the parameter α controls the relative weights given to the distance and velocity of the corresponding vehicle (Kim, 2019a).

For the VRS, a source vehicle configures a multi-hop routing path by using the Bellman-Ford algorithm. As a routing game player, each source vehicle attempts to minimize total path cost ($\mathcal{PC}$). From the source vehicle $\mathcal{V}_s$ to the destination vehicle $\mathcal{V}_d$, the total path cost $\left(\mathcal{PC}^{\mathcal{V}_d}_{\mathcal{V}_s}(\mathcal{H}_t)\right)$ at time $\mathcal{H}_t$ is computed as the sum of all relay link costs as follows (Kim, 2019a);

$$\mathcal{PC}^{\mathcal{V}_d}_{\mathcal{V}_s}(\mathcal{H}_t) = \mathbf{min}\left\{ \sum_{\mathcal{LC}^{\mathcal{V}_d}_h(\mathcal{H}_t)\in Path^{\mathcal{V}_d}_{\mathcal{V}_s}} \mathcal{LC}^{\mathcal{V}_d}_h(\mathcal{H}_t) \right\} \tag{8.5.6}$$

where $Path^{\mathcal{V}_d}_{\mathcal{V}_s}$ is the routing path from the $\mathcal{V}_s$ to the $\mathcal{V}_d$. Usually, a vehicle acting as a relay node has to sacrifice its energy and bandwidth. Therefore, incentive payment algorithm should be developed to guide selfish vehicles toward the cooperative inter-vehicle routing paradigm (Kim, 2012, 2019b). By paying the incentive cost to relaying vehicles, the developed routing algorithm stimulates cooperative actions among selfish relay vehicles. As a $\mathbb{G}^{VRS}$ game player, a source vehicle in the proposed VRS algorithm pays its virtual money (Γ) to disseminate routing packets. If the source vehicle $\mathcal{V}_i$ selects the $Path^{\mathcal{V}_d}_{\mathcal{V}_i}$ according to (8.5.6), its utility function at time $\mathcal{H}_t$ $\left(U_{\mathcal{V}_i}\left(Path^{\mathcal{V}_d}_{\mathcal{V}_s}, \mathcal{H}_t\right)\right)$ can be defined as (Kim, 2019a):

$$U_{\mathcal{V}_i}\left(Path^{\mathcal{V}_d}_{\mathcal{V}_s}, \mathcal{H}_t\right) = \begin{cases} Q^{\mathcal{V}_i}_{\mathcal{H}_t} - \left(\mathcal{J} \times \left(\mathcal{PC}^{\mathcal{V}_d}_{\mathcal{V}_s}(\mathcal{H}_t)\right)\right) \\ and\ \Gamma^{\mathcal{V}_i}_{\mathcal{H}_{t+1}} = \Gamma^{\mathcal{V}_i}_{\mathcal{H}_t} - \left(\mathcal{J} \times \left(\mathcal{PC}^{\mathcal{V}_d}_{\mathcal{V}_s}(\mathcal{H}_t)\right)\right) \\ \\ if \Gamma^{\mathcal{V}_i}_{\mathcal{H}_t} \geq \left(\mathcal{J} \times \left(\mathcal{PC}^{\mathcal{V}_d}_{\mathcal{V}_s}(\mathcal{H}_t)\right)\right) \\ \\ 0, otherwise\left(\text{i.e.,}\ \Gamma^{\mathcal{V}_i}_{\mathcal{H}_t} < \left(\mathcal{J} \times \left(\mathcal{PC}^{\mathcal{V}_d}_{\mathcal{V}_s}(\mathcal{H}_t)\right)\right)\right) \end{cases} \tag{8.5.7}$$

where $Q^{\mathcal{V}_i}_{\mathcal{H}_t}$, $\Gamma^{\mathcal{V}_i}_{\mathcal{H}_t}$ are the outcome of $\mathcal{V}_i$'s routing, and the amount of $\mathcal{V}_i$'s virtual money at time $\mathcal{H}_t$, respectively. $\mathcal{J}$ is the coefficient factor to estimate the incentive payment for relaying vehicles (Kim, 2019a).

8.5.6 The VSS algorithm based on the triple-plane bargaining game

Recently, VSS has garnered significant interest, emerging as one of the most valuable features in VANET systems. Many VANET applications generate large volumes of sensed data, leveraging advancements in vehicular sensor technology. With OBUs, vehicles are increasingly capable of monitoring real-world conditions and contributing to intelligent transportation systems. Within VANET infrastructures, RSUs play a central role by requesting sensing tasks and gathering local information collected by vehicles. Upon receiving task requests, vehicles' OBUs detect and transmit the required sensing data to RSUs for further processing. Despite substantial research efforts in VSS, several critical challenges remain unresolved. A primary issue is that conducting sensing tasks and transmitting data consumes vehicular resources, such as energy, bandwidth, and computing power. Since self-interested vehicles may be reluctant to participate without adequate incentives, RSUs must compensate vehicles for their sensing contributions. To recruit an optimal set of vehicles capable of effectively covering the monitoring area while ensuring active participation, RSUs must provide appropriate incentives. By offering sufficient compensation, RSUs can motivate vehicles to fully engage in VSS tasks, thereby improving data accuracy and coverage in VANET environments (Han, 2015; Kim, 2019a, 2019b; Xiao, 2017).

The *VSCR* scheme designs the VSS algorithm to determine the incentive for vehicles to complete VSS tasks. Since vehicles may make different contributions to sensing work, the RSU may issue appropriate incentives in return for the collected sensing data (Kim, 2019b). The major goal of the algorithm is that the interactive trading between RSU and vehicles should benefit both of them. This game situation can be modeled effectively though the $\mathbb{G}^{VSS}$. In the VSS algorithm, the *VSCR* scheme considers a VSS platform consisting of one RSU (R), a set of mobile vehicles (**V**) and sensing tasks (**I**), and they formulate the sensing game model $\mathbb{G}^{VSS}$. To capture its own heterogeneity, a vehicle's strategy ($\mu \in S_{\mathscr{V}_i}^{\mathscr{V}}$) indicates the contribution for a specific task ($\mathfrak{I} \in \mathfrak{I}$). For example, the $\mathscr{V}_i$ can actually contribute for the task $\mathfrak{I}_s$ during the $\mathscr{H}_t$, i.e., $\mu_{\mathscr{V}_i}^{\mathfrak{I}_s} = 1$, or not, i.e., $\mu_{\mathscr{V}_i}^{\mathfrak{I}_s} = 0$, where $\mu_{\mathscr{V}_i}^{\mathfrak{I}_s} \in S_{\mathscr{V}_i}^{\mathscr{V}}$. Each vehicle has its own utility function, which represents its net gain. The $\mathscr{V}_i$'s utility function at time $\mathscr{H}_t$ $\left(u_{\mathscr{V}_i}^{\mathscr{V}}\left(S_{\mathscr{V}_i}^{\mathscr{V}}, \mathscr{H}_t\right)\right)$ is given by (Kim, 2019a);

$$u_{\mathscr{V}_i}^{\mathscr{V}}\left(S_{\mathscr{V}_i}^{\mathscr{V}} = \left[\mu_{\mathscr{V}_i}^{\mathfrak{I}_1} \ldots \mu_{\mathscr{V}_i}^{\mathfrak{I}_j} \ldots \mu_{\mathscr{V}_i}^{\mathfrak{I}_s}\right], \mathscr{H}_t\right) = \sum_{\mathfrak{I}_j = \mathfrak{I}_1}^{\mathfrak{I}_s} \left(\mu_{\mathscr{V}_i}^{\mathfrak{I}_j}(\mathscr{H}_t) \times \left(\eta_{\mathscr{R}}^{\mathfrak{I}_j}(\mathscr{H}_t) - \mathscr{C}_{\mathscr{V}_i}^{\mathfrak{I}_j}(\mathscr{H}_t)\right)\right) \tag{8.5.8}$$

$$\text{s.t.}, \eta_{\mathscr{R}}^{\mathfrak{I}_j}(\mathscr{H}_t) \in S^{\mathscr{R}} \text{ and } \mu_{\mathscr{V}_i}^{\mathfrak{I}_j}(\mathscr{H}_t) = \begin{cases} 1, if \mathscr{V}_i\, contributes for\, \mathfrak{I}_j\, at\, \mathscr{H}_t \\ 0, otherwise \end{cases}$$

where $\mathscr{C}_{\mathscr{V}_i}^{\mathfrak{I}_j}(\mathscr{H}_t)$ and $\eta_{\mathscr{R}}^{\mathfrak{I}_j}(\mathscr{H}_t)$ are the $\mathscr{V}_i$'s cost and the RSU's incentive payment for the $\mathfrak{I}_j$ sensing at time $\mathscr{H}_t$, respectively. Usually, game players simply attempt to maximize their utility functions. However, vehicles in the $\mathbb{G}^{VSS}$ game should consider not only VSS but also VCS and VRS. From the point of view of vehicles, this situation can be modeled effectively though a cooperative bargaining process. The *VSCR* scheme adopts a well-known bargaining solution concept, called the *Nash bargaining solution* (*NBS*) to effectively design the vehicle's VSS strategy decision mechanism; the *NBS* is an effective tool to achieve a mutually desirable solution among conflicting requirements (Kim, 2019a).

In the *VSCR* scheme, vehicles can earn the virtual money (Γ) from the VSS incentive payment and spend their Γ for their VCS and VRS. Due to this reason, the strategy decision in the VSS algorithm might directly affect the payoff of $\mathbb{G}^{VCS}$ and $\mathbb{G}^{VRS}$. During VANET operations, the individual vehicle $\mathscr{V}_i$ decides his VSS's strategy ($S_{\mathscr{V}_i}^{\mathscr{V}}$) to maximize the combined payoffs $\left(CP_{\mathscr{V}_i}\left(S_{\mathscr{V}_i}^{\mathscr{V}}, \mathscr{H}_t\right)\right)$ at time $\mathscr{H}_t$ (Kim, 2019a).

$$\mathbf{max} CP_{\mathscr{V}_i}\left(S_{\mathscr{V}_i}^{\mathscr{V}}, \mathscr{H}_t\right) = \max_{S_{\mathscr{V}_i}^{\mathscr{V}} = \left[\mu_{\mathscr{V}_i}^{\mathfrak{I}_1} \ldots \mu_{\mathscr{V}_i}^{\mathfrak{I}_j} \ldots \mu_{\mathscr{V}_i}^{\mathfrak{I}_s}\right]} \prod_{1 \le j \le 3} \left(\mathfrak{X}_j - d_j\right)^{\psi_j}, \text{ s.t.}, \sum_{j=1}^{3} \psi_j = 1$$

(8.5.9)

$$\text{s.t.,}\begin{cases} \mathfrak{X}_1 = \mathbb{U}^{\mathcal{V}}_{\mathcal{V}_i\in\mathcal{V}}\left(\mathfrak{P}^{\mathcal{R}}_j(\mathcal{H}_t), r^{\mathcal{V}_i}_k(\mathcal{H}_t)\right) \\ \mathfrak{X}_2 = U_{\mathcal{V}_i}\left(Path^{\mathcal{V}_d}_{\mathcal{V}_i}, \mathcal{H}_t\right) \\ \mathfrak{X}_3 = u^{\mathcal{V}}_{\mathcal{V}_i}\left(S^{\mathcal{V}}_{\mathcal{V}_i} = \left[\mu^{\mathfrak{I}_1}_{\mathcal{V}_i}\ldots\mu^{\mathfrak{I}_j}_{\mathcal{V}_i}\ldots\mu^{\mathfrak{I}_t}_{\mathcal{V}_i}\right], \mathcal{H}_t\right) \end{cases}$$

where d_j is a disagreement point; a disagreement point $d_{i,1\leq i\leq n}$ represents a minimum payoff of each game model. Therefore, d is the least guaranteed payoff for game players (i.e., zero in this model). $\psi_{j,1\leq j\leq 3}$ is a bargaining power, which is the relative ability to exert influence over bargaining process. Usually, the bargaining solution is strongly dependent on the bargaining powers. In the $\mathbb{G}^{VSS}$ game, $\psi_{j,1\leq j\leq 3}$ can be estimated as follows (Kim, 2019a);

$$\psi_{j,1\leq j\leq 3} = \frac{\mathfrak{Y}_j}{(\mathfrak{Y}_1 + \mathfrak{Y}_2 + \mathfrak{Y}_3)}$$

(8.5.10)

$$\text{s.t.,}\begin{cases} \mathfrak{Y}_1 = \max\left\{\left(R\Gamma^{VCS(\mathcal{V}_i)}_{\mathcal{H}_t} - U\Gamma^{VCS(\mathcal{V}_i)}_{\mathcal{H}_t}\right), \Gamma^{\mathcal{V}_i}_{\mathcal{H}_t}\right\} \\ \mathfrak{Y}_2 = \max\left\{\left(R\Gamma^{VRS(\mathcal{V}_i)}_{\mathcal{H}_t} - U\Gamma^{VRS(\mathcal{V}_i)}_{\mathcal{H}_t}\right), \Gamma^{\mathcal{V}_i}_{\mathcal{H}_t}\right\} \\ \mathfrak{Y}_3 = \Gamma^{\mathcal{V}_i}_{\mathcal{H}_t} \end{cases}$$

where $R\Gamma^{VCS(\mathcal{V}_i)}_{\mathcal{H}_t}$, $R\Gamma^{VRS(\mathcal{V}_i)}_{\mathcal{H}_t}$ are the requested virtual money to maximize the $\mathcal{V}_i$'s payoffs in $\mathbb{G}^{VCS}$ and $\mathbb{G}^{VRS}$ at time $\mathcal{H}_t$, respectively. $U\Gamma^{VCS(\mathcal{V}_i)}_{\mathcal{H}_t}$, $U\Gamma^{VRS(\mathcal{V}_i)}_{\mathcal{H}_t}$ are the used virtual money by the $\mathcal{V}_i$ in $\mathbb{G}^{VCS}$ and $\mathbb{G}^{VRS}$ at time $\mathcal{H}_t$, respectively. Finally, individual vehicles select their strategies $S^{\mathcal{V}}$ according to (8.5.9) and (8.5.10). In a distributed online fashion, each vehicle ensures a well-balanced performance among the VCS, VRS and VSS processes (Kim, 2019a).

In the VSS algorithm, the vehicles' actual sensing cost $\mathcal{C}^{\mathfrak{I}}_{\mathcal{V}}(\cdot)$ and each individual vehicle's situation is usually unknown by the RSU. Under this asymmetric information situation, the RSU should learn the vehicle's circumstance during the VSS process. Like the VCS process, each RSU is a *proposer* and individual vehicle is a *responder*, and they also interact with each other for their objectives based on the interactive feedback procedure. For the RSU, the payoff can be defined as a function of price levels for tasks and vehicles' responses. Therefore, the RSU's utility function $\left(u^{\mathcal{R}}(S^{\mathcal{R}}, \mathcal{H}_t)\right)$ at time $\mathcal{H}_t$ is computed as the sum of each task's outcome (Kim, 2019a):

$$u^{\mathcal{R}}\left(S^{\mathcal{R}} = \left[\eta^{\mathfrak{I}_1}_{\mathcal{R}}\ldots\eta^{\mathfrak{I}_j}_{\mathcal{R}}\ldots\eta^{\mathfrak{I}_t}_{\mathcal{R}}\right], \mathcal{H}_t\right) =$$

$$\sum_{\mathfrak{I}_h=\mathfrak{I}_1}^{\mathfrak{I}_t}\left(\left(\mathfrak{T}^{\mathfrak{I}_h}(\mathcal{H}_t) \times \Phi^{\mathfrak{I}_h}\right) - \sum_{\mathcal{V}_i\in\mathcal{V}}\left(\mu^{\mathfrak{I}_h}_{\mathcal{V}_i}(\mathcal{H}_t) \times \eta^{\mathfrak{I}_h}_{\mathcal{R}}(\mathcal{H}_t)\right)\right)$$

(8.5.11)

$$\text{s.t., } \mathfrak{T}^{\mathfrak{I}_h}(\mathscr{H}_t) = \begin{cases} 1, if \sum_{\mathscr{V}_i \in \mathscr{V}} \left(\mu_{\mathscr{V}_i}^{\mathfrak{I}_h}(\mathscr{H}_t)\right) \geq \mathscr{L}_{\mathscr{R}}^{\mathfrak{I}_h} \\ 0, otherwise \end{cases}$$

where $\Phi^{\mathfrak{I}_h}$ is the profit of $\mathfrak{I}_h$; when the VSS for the $\mathfrak{I}_h$ is successfully completed, i.e., $\sum_{\mathscr{V}_i \in \mathscr{V}} \left(\mu_{\mathscr{V}_i}^{\mathfrak{I}_h}(\mathscr{H}_t)\right) \geq \mathscr{L}_{\mathscr{R}}^{\mathfrak{I}_h}$, $\Phi^{\mathfrak{I}_h}$ is obtained. $\mathscr{L}_{\mathscr{R}}^{\mathfrak{I}_h}$ is the pre-defined VSS requirement to complete the task $\mathfrak{I}_h$ (Kim, 2019a).

As $\mathbb{G}^{VSS}$ game players, the RSU considers the inter-relationship among tasks and vehicles' reactions to determine its incentive payment strategy. To select the best strategy, a novel learning method is needed. The *VSCR* scheme develops a new learning method to determine an effective payment policy for each task. For the RSU, the strategy vector $S^{\mathscr{R}}(\mathscr{H}_t) = [\eta_{\mathscr{R}}^{\mathfrak{I}_1}(\mathscr{H}_t) \ldots \eta_{\mathscr{R}}^{\mathfrak{I}_s}(\mathscr{H}_t)]$ represents each price strategy for each task at time $\mathscr{H}_t$. Let $\mathfrak{Z}_{\mathscr{H}_t}^{\mathfrak{I}_s}(\mathfrak{P}_i^{\mathfrak{I}_s}, \eta_{\mathscr{R}}^{-\mathfrak{I}_s})$ be the learning value of taking strategy $\mathfrak{P}_i^{\mathfrak{I}_s} \in \eta_{\mathscr{R}}^{\mathfrak{I}_s}$ with a vector of payment strategies except the strategy $\mathfrak{P}_i^{\mathfrak{I}_s}$ $(\eta_{\mathscr{R}}^{-\mathfrak{I}_s})$. The RSU updates its $\mathfrak{Z}(\cdot)$ value over time according to the following method (Kim, 2019a):

$$\mathfrak{Z}_{\mathscr{H}_{t+1}}^{\mathfrak{P}_i^{\mathfrak{I}_s}}(\mathfrak{P}_i^{\mathfrak{I}_s} \in \eta_{\mathscr{R}}^{\mathfrak{I}_s}, \eta_{\mathscr{R}}^{-\mathfrak{I}_s}) = \left((1-\beta) \times \mathfrak{Z}_{\mathscr{H}_t}^{\mathfrak{I}_s}(\mathfrak{P}_i^{\mathfrak{I}_s}, \eta_{\mathscr{R}}^{-\mathfrak{I}_s})\right)$$

$$+\left(\beta \times \left(\left((1-\gamma^{\mathscr{R}}) \times \partial_{\mathscr{H}_t}^{\mathfrak{I}_s}\right) + \left\{\gamma^{\mathscr{R}} \times \sum_{\mathfrak{I}_h \in \mathfrak{I}^{-\mathfrak{I}_s}} \frac{\partial_{\mathscr{H}_t}^{\mathfrak{I}_h}}{(s-1)}\right\}\right)\right)$$

(8.5.12)

$$\text{s.t., } \partial_{\mathscr{H}_t}^{\mathfrak{I}_s} = \left(\left(\mathfrak{T}^{\mathfrak{I}_s}(\mathscr{H}_t) \times \Phi^{\mathfrak{I}_s}\right) - \sum_{\mathscr{V}_i \in \mathscr{V}} \left(\mu_{\mathscr{V}_i}^{\mathfrak{I}_s}(\mathscr{H}_t) \times \eta_{\mathscr{R}}^{\mathfrak{I}_s}(\mathscr{H}_t)\right)\right)$$

During the $\mathbb{G}^{VSS}$ game process, the RSU adaptively learns the current VSS state, and sequentially adjusts $\mathfrak{Z}(\cdot)$ values for each VSS task. Based on the $\mathfrak{Z}(\cdot)$ values, the *VSCR* scheme can determine $P_{\mathfrak{I}_s \in \mathfrak{I}}^{\mathscr{R}}(\mathscr{H}_t) = \left\{P_{\mathfrak{I}_s(\mathscr{H}_t)}^{\mathfrak{P}_{min}^{\mathfrak{I}_s}} \ldots P_{\mathfrak{I}_s(\mathscr{H}_t)}^{\mathfrak{P}_i^{\mathfrak{I}_s}} \ldots P_{\mathfrak{I}_s(\mathscr{H}_t)}^{\mathfrak{P}_{max}^{\mathfrak{I}_s}}\right\}$ as the probability distribution of R's strategy selection for the $\mathfrak{I}_s$ at time $\mathscr{H}_t$. The $\mathfrak{P}_i^{\mathfrak{I}_s}$ strategy selection probability for the $\mathfrak{I}_s$ at time $\mathscr{H}_{t+1}$ $\left(P_{\mathfrak{I}_s(\mathscr{H}_{t+1})}^{\mathfrak{P}_i^{\mathfrak{I}_s}}\right)$ is defined as (Kim, 2019a):

$$P_{\mathfrak{I}_s(\mathscr{H}_{t+1})}^{\mathfrak{P}_i^{\mathfrak{I}_s}} = \frac{\text{EXP}\left(\mathfrak{Z}_{\mathscr{H}_t}^{\mathfrak{P}_i^{\mathfrak{I}_s}}(\mathfrak{P}_i^{\mathfrak{I}_s} \in \eta_{\mathscr{R}}^{\mathfrak{I}_s}, \eta_{\mathscr{R}}^{-\mathfrak{I}_s})\right)}{\sum_{\mathfrak{P}_{j,j=min}^{\mathfrak{I}_s}}^{j=max} \text{EXP}\left(\mathfrak{Z}_{\mathscr{H}_t}^{\mathfrak{P}_i^{\mathfrak{I}_s}}(\mathfrak{P}_i^{\mathfrak{I}_s} \in \eta_{\mathscr{R}}^{\mathfrak{I}_s}, \eta_{\mathscr{R}}^{-\mathfrak{I}_s})\right)}$$

(8.5.13)

8.5.7 Summary

With the increasing adoption of vehicle-based sensory technology, VANETs are becoming more prominent, offering vehicular sensing, routing, and clouding services for 5G network applications. As a result, the development of next-generation VANET management protocols have become a crucial challenge to meet emerging demands. The *VSCR* scheme is designed based on learning algorithms and game theory, integrating VCS, VRS, and VSS algorithms into a unified framework. By leveraging collaborative interactions among VCS, VRS, and VSS, the *VSCR* scheme introduces a triple-plane bargaining game model to achieve an optimal performance balance. During VANET operations, RSUs continuously refine their strategies based on evolving network conditions, while vehicles assess the mutual interactions of their strategic choices. Acting as game players, both RSUs and vehicles utilize available information to dynamically adapt to VANET environments and make intelligent control decisions using a self-adaptive approach. Given the unique characteristics of VANETs, this integrated design approach proves to be highly effective in delivering satisfactory services even in incomplete information environments, ensuring efficient and adaptive network management (Kim, 2019a).

8.6 Radio Resource Management Control (*RRMC*) scheme

In vehicular networks, the integration of radar sensing and wireless communication has emerged as a promising approach to mitigate spectrum resource limitations. Recognizing this potential, S. Kim explores a Joint Radar Sensing and Communication (JSC) system, leveraging a cooperative spectrum-sharing strategy, and introduces the *Radio Resource Management Control* (*RRMC*) scheme (Kim, 2023b). By enabling full cooperation between radar sensing and wireless communication functionalities, the *RRMC* scheme effectively prevents underutilization of spectral resources while enhancing overall system efficiency. To implement JSC control, the *RRMC* scheme is built upon coalitional bargaining solutions, grounded in cooperative game theory. Specifically, the *RRMC* scheme employs i) *coalitional Nash bargaining solution* (*CNBS*) to distribute spectrum resources among roadside units (RSUs) and coordinate radar and communication functions within individual vehicles, and ii) *coalitional Kalai-Smorodinsky Bargaining solution* (*CKSBS*) to allocate spectrum resources among vehicles, ensuring payoff maximization for all participants. Within the *RRMC* scheme, the *CNBS* and *CKSBS* operate in tandem to share coalition surpluses while maintaining fairness and efficiency. The core contribution of the *RRMC* scheme lies in its reciprocal integration of *CNBS* and *CKSBS*, which generates a synergistic effect, optimizing radar-communication coexistence and enhancing vehicular network performance (Kim, 2023b).

8.6.1 Development motivation

Radio Detection and Ranging (RADAR) is a detection system that utilizes electromagnetic waves to determine the position, velocity, and characteristics of a target. By transmitting signals and analyzing the reflected waves, radar technology enables precise range and directional measurements. For high-resolution radar sensing, millimeter-wave (mmWave) communication has emerged as a critical technology, supporting applications such as autonomous driving, intelligent traffic systems, and virtual reality. A key component of Intelligent Transportation Systems (ITS) is Vehicle-to-Everything (V2X) communication, which enhances driving safety, efficiency, and comfort. To meet the increasing demands of next-generation V2X applications, the automotive industry has leveraged advancements in artificial intelligence and mobile communication technologies, leading to significant breakthroughs in vehicular networks (Bullock, 2000; Kim, 2023b; Zhang, 2021a)

Over the past decade, autonomous driving vehicles (ADVs) have been recognized as a major technological advancement in the transportation industry, with a projected market valuation of $81 billion by 2030 and $290 billion by 2040. A key advantage of ADVs is their potential to enhance global traffic safety, improve transport efficiency, reduce accidents, minimize transportation costs, and optimize lane capacity. To achieve these benefits, ADVs must integrate both radar sensing and communication capabilities. The radar function detects and analyzes surrounding environments, while the communication function enables the transmission of sensor data to roadside units (RSUs). These RSUs then aggregate and process the collected information, serving as a data hub for intelligent vehicles. By integrating wideband communication technologies, the seamless cooperation of radar and communication systems plays a fundamental role in establishing a fully functional intelligent traffic system (Kim, 2023b; Nguyen, 2021).

The integration of radar sensing and wireless communication has gained substantial research attention, particularly in developing efficient control techniques for spectrum sharing. Although these two functions can be co-located or physically integrated, their transmissions often overlap in time and frequency domains. Consequently, cooperative spectrum utilization is essential to optimize Joint Radar Sensing and Communication (JSC) systems. For 6G cellular networks, JSC techniques will play a pivotal role in joint-function system design, where a single waveform can simultaneously support both communication and radar sensing. Compared to conventional electronic devices, JSC-enabled devices demonstrate superior performance, offering enhanced efficiency and reduced hardware complexity. Furthermore, with the proliferation of electronic devices in future smart transportation systems, JSC technologies can significantly reduce hardware costs and lower power consumption (Kim, 2023b; Ni, 2021; Zhang, 2021b).

With the exponential rise of wireless devices and mobile services, wireless spectrum resources are becoming increasingly congested. To enhance spectrum efficiency, a promising solution is to enable cooperative spectrum sharing between radar sensing and communication systems. Recent advancements in coexistence strategies have led to the development of multi-platform spectrum sharing, allowing multiple applications to simultaneously utilize the same frequency bands. Within spectrum-sharing frameworks, both sensing and communication systems coordinate to achieve mutual objectives. Compared to traditional non-shared spectrum models, JSC spectrum sharing enhances spectrum utilization by introducing greater design flexibility and dynamic resource allocation (Kim, 2023b; Mu, 2022).

The millimeter-wave (mmWave) spectrum (30-300 GHz) has attracted significant research interest in 5G networks and is already widely used in radar sensing systems for collision avoidance in autonomous vehicles. With higher resolution and extended range, automotive mmWave radar enables safe driving in various conditions. The wireless industry is currently focusing on mmWave communication, which is set to be standardized within 5G NR based on 802.11ad/ay WLAN protocols. Additionally, the vast mmWave spectrum necessitates a re-evaluation of spectrum-sharing techniques. Early research indicates that spectrum-sharing gains can exceed 130% compared to traditional licensed spectrum models. Thus, in forthcoming 6G networks, a primary challenge will be to develop an effective spectrum-sharing protocol for radar sensing and communication services (Kim, 2023b; Liu, 2020a; Mu, 2022; Vrontos, 2020).

The *RRMC* scheme focuses on spectrum-sharing challenges among autonomous vehicles, which compete to maximize their utility. Each vehicle is equipped with radar sensing and communication functions, both of which must operate collaboratively to achieve optimal system performance. To ensure efficient resource allocation while meeting service requirements, inter- and intra-vehicle control mechanisms play a critical role. However, designing universal control mechanisms that effectively address spectrum-sharing challenges in dynamic JSC environments remains highly complex. In recent years, cooperative game theory has gained traction as an effective approach for resource-sharing algorithms, particularly in limited-spectrum environments. Game theory focuses on modeling rational decision-making, allowing game players to compete, collaborate, and negotiate for optimal outcomes. Within the JSC system, individual system agents act as intelligent game players, autonomously maximizing their payoffs without requiring human intervention (Kim, 2014, 2023b).

The *RRMC* scheme introduces a novel joint control mechanism for integrating radar sensing and communication within the JSC platform. To achieve an appropriate performance balance, the *RRMC* scheme employs the key principles of *coalitional Nash bargaining solution* (*CNBS*) and *coalitional Kalai-Smorodinsky bargaining solution* (*CKSBS*). To optimize sensing and communication services, the *RRMC* scheme incorporates three cooperative game models, which collaborate to ensure efficient resource allocation within JSC system infrastructures (Kim, 2023b).

In 2010, O. Compte and P. Jehiel introduced the *CNBS*, a bargaining model that maximizes the product of players' payoffs while ensuring fair resource allocation within coalition games. For coalition games with transferable utilities, the *CNBS* determines the payoff distribution that meets core constraints and maximizes group benefits. Building upon the *CNBS*, a new bargaining approach, known as the *CKSBS*, was developed (Pycia, 2012). The *CNBS* and *CKSBS* allow game players to fairly distribute coalition surpluses, ensuring both fairness and efficiency in JSC spectrum-sharing systems. Within the *RRMC* scheme, the *CNBS* and *CKSBS* are jointly utilized to enhance cooperation among autonomous vehicles, RSUs, and spectrum-sharing entities, fostering an optimized and adaptive vehicular communication environment (Kim, 2023b).

8.6.2 Contribution

As previously discussed, the JSC system was initially developed to address the coexistence challenge between radar sensing and commercial wireless communication networks. However, to meet the increasingly stringent spectral efficiency demands of next-generation wireless communications, further enhancements are necessary. The *RRMC* scheme focuses on solving the spectrum allocation problem by accounting for the mutual interactions between system components. To achieve this, the *RRMC* scheme integrates the fundamental principles of the *CNBS* and *CKSBS*. By leveraging these concepts, the *RRMC* scheme explores strategic interactions between RSUs and ADVs and introduces an innovative three-phase cooperative game model. Following the coalitional bargaining framework, individual system agents collaborate to efficiently share limited spectrum resources, while simultaneously negotiating conflicting interests to achieve an optimal balance between fairness and efficiency in the JSC system (Kim, 2023b).

8.6.3 Formulation for the joint radar and communication services

Consider an ADV-based JSC system operating in a coexistence scenario, where both radar sensing and communication functions are integrated into each vehicle. In this system, multicarrier waveforms are utilized for both functionalities, enabling efficient spectral sharing between radar detection and wireless communication. From the radar function's perspective, each ADV utilizes radar detection to acquire real-time road information, allowing users to identify and track targets. The collected radar data, which includes details about detected objects, must then be transmitted for further processing. From the communication function's perspective, vehicles relay their sensed data to Roadside Units (RSUs) through wireless links. Within the intelligent transportation system (ITS) and automated vehicular networks, RSUs primarily function as data aggregation and analysis units, gathering traffic-related information from connected vehicles. Additionally, RSUs distribute relevant information back to vehicles within their transmission range, enhancing situational awareness and decision-making. At this stage, the data transmitted by each vehicle includes not only target-specific information but also other contextual sensing data, such as weather conditions and environmental factors, further contributing to the efficiency of smart transportation systems (Kim, 2023b; Nguyen, 2022).

We consider a JSC system platform that consists of one base station ($\mathcal{B}$), a set of JSC-equipped vehicles, denoted by $\mathbb{V} = \{\mathcal{V}_1, \mathcal{V}_2, \ldots, \mathcal{V}_n\}$, and a set of RSUs, denoted by $\mathbb{R} = \{\mathcal{R}_1, \mathcal{R}_2, \ldots, \mathcal{R}_m\}$. The B has its spectrum resource, i.e., orthogonal subcarriers for sharing to the vehicles. Orthogonal subcarriers can be used by vehicles to modulate orthogonal frequency division multiplexing (OFDM) symbols. It makes use of a large number of closely spaced orthogonal subcarriers that are transmitted in parallel. For simplicity, each vehicle has a single antenna, and vehicles have to jointly process the received radar and communication signals. Therefore, how to handle the spectrum sharing problem is vital to improve the spectral resource efficiency in the JSC system (Kim, 2023b; Nguyen, 2021; Tian, 2019).

In the JSC platform, the B is connected by multiple $\mathcal{R}_{1 \le j \le m}$ through backhaul links, and the B's spectrum resource ($\mathcal{S}^{\mathcal{B}}$) is dynamically allocated to the Rs. In the viewpoint of each R, its assigned spectrum resource ($\mathcal{S}^{\mathcal{R}}$) is distributed for its corresponding ADVs, which are in the R's coverage area. From the R, each individual V gets its own spectrum resource ($\mathcal{S}^{\mathcal{V}}$), which should be shared by the V's radar and communication functions. Without loss of generality, assume that the total available $\mathcal{S}^{\mathcal{V}}$ is divided into two categories: radar-subcarrier and communication-subcarrier. In the *RRMC* scheme, the $\mathcal{S}^{\mathcal{B}}$, $\mathcal{S}^{\mathcal{R}}$, and $\mathcal{S}^{\mathcal{V}}$ resource sharing problems are formulated as bargaining games $\mathbb{G}_{\mathcal{B}}$, $\mathbb{G}_{\mathcal{R}}$, and $\mathbb{G}_{\mathcal{V}}$, respectively. As a consequence, these games are operated sequentially in a step-by-step interactive fashion based on the *CNBS* and *CKSBS*; it is noteworthy that the *RRMC* scheme formulates the B-R-V association in a cooperative manner. Formally, the *RRMC* scheme defines the three-phase game G entities, i.e.,

$$\mathbb{G} = \{\mathbb{G}_{\mathcal{B}}, \mathbb{G}_{\mathcal{R}}, \mathbb{G}_{\mathcal{V}}\} = \Big\{\mathcal{B}, \mathbb{R}, \mathbb{V}, \{\mathbb{G}_{\mathcal{B}} | \mathcal{S}^{\mathcal{B}}, \mathcal{R}_{1 \le j \le m}, \mathcal{S}^{\mathcal{R}_j}, U_{\mathcal{B}}(\cdot), U_{\mathcal{R}_j}(\cdot)\}, \{\mathbb{G}_{\mathcal{R}_{1 \le j \le m}} | \mathcal{S}^{\mathcal{R}_j}, \mathbb{V}^{\mathcal{R}_j}, \mathcal{S}^{\mathcal{V}_i}, \mathfrak{U}_{\mathcal{V}_i \in \mathbb{V}^{\mathcal{R}_j}}(\cdot)\}, \{\mathbb{G}_{\mathcal{V}_i} | \mathcal{S}^{\mathcal{V}_i}, F_R^{\mathcal{V}_i}, F_C^{\mathcal{V}_i}, \mathcal{S}_R^{\mathcal{V}_i}, \mathcal{S}_C^{\mathcal{V}_i}, \mathcal{U}_R^{\mathcal{V}_i}(\cdot), \mathcal{U}_C^{\mathcal{V}_i}(\cdot)\}, T\Big\}$$

of gameplay (Kim, 2023b).

- B, R and v represent the one base station, the set of RSUs, and the set of vehicles, respectively. They are mutually and reciprocally interdependent in a coordinated manner, and they work together to share the limited JSC spectrum resource.
- At the first phase, the $\mathbb{G}_{\mathcal{B}}$ is designed to share the $\mathcal{S}^{\mathcal{B}}$ among Rs. In the $\mathbb{G}_{\mathcal{B}}$, B and each $\mathcal{R}_{1 \le j \le m} \in \mathbb{R}$ are game players, and $\mathcal{S}^{\mathcal{R}_j}$ is the $\mathcal{R}_j$'s strategy; it is the allocated spectrum amount for the $\mathcal{R}_j$. The $U_{\mathcal{B}}(\cdot)$ and $U_{\mathcal{R}_j}(\cdot)$ are the utility functions of B and $\mathcal{R}_j$, respectively.
- At the second phase, the $\mathbb{G}_{\mathcal{R}_j}$ is designed to share the $\mathcal{S}^{\mathcal{R}_j}$ among $\mathcal{V}_i \in \mathbb{V}^{\mathcal{R}_j}$ where $\mathbb{V}^{\mathcal{R}_j}$ is the set of vehicles covered by the $\mathcal{R}_j$. In the $\mathbb{G}_{\mathcal{R}_j}$, each individual $\mathcal{V}_i$ is a game player, and $\mathcal{S}^{\mathcal{V}_i}$ is the allocated spectrum amount for the $\mathcal{V}_i$. The $\mathcal{S}^{\mathcal{V}_i}$ and $\mathfrak{U}_{\mathcal{V}_i \in \mathbb{V}^{\mathcal{R}_j}}(\cdot)$ are the $\mathcal{V}_i$'s strategy and utility function, respectively.
- At the third phase, the $\mathbb{G}_{\mathcal{V}_i}$ is designed to share the $\mathcal{S}^{\mathcal{V}_i}$ in each individual $\mathcal{V}_i$. $F_R^{\mathcal{V}_i}$ and $F_C^{\mathcal{V}_i}$ are the $\mathcal{V}_i$'s radar and communication functions. In the $\mathbb{G}_{\mathcal{V}_i}$, they are game players, and their obtained spectrum amounts, i.e., $\mathcal{S}_R^{\mathcal{V}_i}$ and $\mathcal{S}_C^{\mathcal{V}_i}$, are their strategies. The $\mathcal{U}_R^{\mathcal{V}_i}(\cdot)$ and $\mathcal{U}_C^{\mathcal{V}_i}(\cdot)$ are their utility functions.
- Discrete time model $T \in \{t_1, \ldots, t_c, t_{c+1}, \ldots\}$ is represented by a sequence of time steps. The length of t_c matches the event timescale of $\mathbb{G}_{\mathcal{B}}$, $\mathbb{G}_{\mathcal{R}}$, and $\mathbb{G}_{\mathcal{V}}$.

8.6.4 Main ideas of Coalitional Bargaining Solutions

In 2010, Compte and Jehiel defined a cooperative solution concept, called *CNBS*; it is the Nash bargaining solution for a coalition structure. They assume that the grand coalition is the one that generates the largest surplus. Therefore, the concept of *CNBS* is only defined for games with a non-empty core. To define the *CNBS*, a few preliminary definitions should be introduced. We consider n players labelled $i = 1,2,\ldots,n$. Any subset S of these players may form a team. The surplus that such a team generates is nonnegative and denoted $\nu(S)$. We denote by $N = \{1,2,\ldots,n\}$ the set that comprises all players note by S the set of subsets of N. In the *RRMC* scheme, it is assumed that the grand team is the only efficient coalition. That is, $\nu(N) > \nu(S)$ for all $S \neq N$. We define an allocation as a vector of payments $x = (x_1,\ldots,x_n)$ and we let $x(S) = \sum_{i\in S} x_i$. Given the assumption that $\nu(N) = \max_S \nu(S)$, an allocation x is *efficient* if $x(N) = \nu(N)$ (Burguet, 2020; Compte, 2010; Kim, 2023b).

We turn now to the notion of core stability. An allocation x is feasible if $x(N) \leq \nu(N)$. It cannot be blocked by a coalition S if $x(S) \geq \nu(S)$. The core consists of the allocations x that are feasible and that cannot be blocked by any coalition S. When the core is nonempty, the *CNBS* is the core allocation that maximizes the Nash product. This solution is independent of irrelevant alternatives. In other words, a reasonable outcome will be feasible after some payoffs have been removed. That is defined mathematically as follows (Compte, 2010; Kim, 2023b):

$$CNBS(N,x) = \arg\max_{x=(x_1,\ldots,x_n)} \prod_{i\in N} x_i \tag{8.6.1}$$

Based on the multi-object auctions, the role of the buyer-submodularity property has been considered between one seller and several buyers. Formally, we assume that player 1 is the seller and other players $i \neq 1$ are the buyers, so we have $\nu(S) > 0$ if and only if $1 \in S$. The function $\nu(\cdot)$ is buyer-submodular when $\nu(S) - \nu(S \setminus \{i\}) > \nu(N) - \nu(N \setminus \{i\})$ for any $S \subsetneq N$ and $i \in S$ with $i \neq 1$. In this situation, the *CNBS* with the buyer-submodularity property, i.e., $CNBS_{bs}(\cdot)$, is defined to find a profile x' that satisfies (Compte, 2010; Kim, 2023b):

$$CNBS_{bs}(N,x) = \arg\max_{x'=(x_2,\ldots,x_n)} \left(\left(\nu(N) - \sum_{i=2}^{n} x_i \right) \times \prod_{i\in N, i\neq 1} x_i \right) \tag{8.6.2}$$

$$\text{s.t., } x_i \leq \nu(N) - \nu(N \setminus \{i\}) \text{ for all } i \neq 1$$

CKSBS is another bargaining solution when all game players have preferences over the coalition they can join. The main idea of *CKSBS* is to select the maximum payoff that is proportional to the profile of ideal profit. Therefore, players achieve a compromise having the goal of decreasing the concessions to be made related to their ideal profits. Contrary to CNBS, CKSBS would satisfy the individual monotonicity property (Kim, 2014). To define the *CKSBS*, we denote by $\mathcal{Q}_\nu(S)$ the output coalition S produces and let $\mathcal{I}_i(\mathcal{Q}_\nu(S))$ be the i's ideal profit. The *CKSBS* selects the Pareto optimal profile of payoffs. Mathematically, the *CKSBS* is obtained for members of coalition S as follows (Pycia, 2012; Kim, 2023b):

$$\frac{x_i}{\mathcal{I}_i(\mathcal{Q}_\nu(S))} \text{ is constant across } i \in S \Leftrightarrow \frac{x_i}{\mathcal{I}_i(\mathcal{Q}_\nu(S))} = \frac{x_j}{\mathcal{I}_j(\mathcal{Q}_\nu(S))} \text{ with } i,j \in S$$

(8.6.3)

$$\text{s.t.,} \sum_{i \in S} x_i = \mathcal{Q}_\nu(S)$$

8.6.5 The spectrum allocation algorithm for the JSC system

In the JSC infrastructure, the *RRMC* scheme considers multiple RUSs within the B's coverage area, and they compete with each other to share the $\mathcal{S}^{\mathcal{B}}$. Usually, the utility payoff obtained by the $\mathcal{R}_{1 \le j \le m}$ depends on the traffic service load. Therefore, it is proportional to the allocated spectrum in the $\mathcal{R}_j$. At the first phase, the B and multiple $\mathcal{R}_{1 \le j \le m}$ are assumed as a seller and buyers of $\mathcal{S}^{\mathcal{B}}$, respectively, and they form a coalition $(\mathcal{C}_{\mathbb{R}}^{\mathcal{B}})$. To quantify the satisfactions of B and $\mathcal{R}_j$, the $U_{\mathcal{B}}(\cdot)$ and $U_{\mathcal{R}_j}(\cdot)$ can be derived with its strategy $\mathcal{S}^{\mathcal{R}_j}$. It is given by (Kim, 2023b):

$$\begin{cases} U_{\mathcal{B}}(\mathbb{R}, \mathcal{S}^{\mathcal{B}}, \mathcal{S}^{\mathcal{R}_{1 \le j \le m}}) = \beta - exp\left(-\frac{\alpha}{\mathcal{S}^{\mathcal{B}}} \times \sum_{\mathcal{R}_j \in \mathbb{R}} \mathcal{S}^{\mathcal{R}_j}\right) \\ \text{s.t.,} \sum_{\mathcal{R}_j \in \mathbb{R}} \mathcal{S}^{\mathcal{R}_j} \le \mathcal{S}^{\mathcal{B}} \\ U_{\mathcal{R}_j}(T^{\mathcal{R}_j}, \mathcal{S}^{\mathcal{B}}, \mathcal{S}^{\mathcal{R}_j}) = \log\left(\frac{exp\left(\frac{\min(T^{\mathcal{R}_j}, \mathcal{S}^{\mathcal{R}_j})}{T^{\mathcal{R}_j}}\right) - exp\left(-\frac{\min(T^{\mathcal{R}_j}, \mathcal{S}^{\mathcal{R}_j})}{T^{\mathcal{R}_j}}\right)}{exp\left(\frac{\min(T^{\mathcal{R}_j}, \mathcal{S}^{\mathcal{R}_j})}{T^{\mathcal{R}_j}}\right) + exp\left(-\frac{\min(T^{\mathcal{R}_j}, \mathcal{S}^{\mathcal{R}_j})}{T^{\mathcal{R}_j}}\right)} + \gamma\right) \end{cases}$$

(8.6.4)

where $T^{\mathcal{R}_j}$ is the total requested spectrum amount from the $\mathcal{R}_j$. β, α, and γ are control parameters for the $U_{\mathcal{B}}(\cdot)$ and $U_{\mathcal{R}_j}(\cdot)$, respectively. Based on the buyer-submodularity property, the $\mathcal{C}_{\mathbb{R}}^{\mathcal{B}}$ share the $\mathcal{S}^{\mathcal{B}}$ with the idea of $CNBS_{bs}$. According to (8.6.2), the amount of $\mathcal{S}^{\mathcal{R}_j}$ is decided (Kim, 2023b).

$$CNBS_{bs}\left(\mathcal{C}_{\mathbb{R}}^{\mathcal{B}}, (\mathcal{B}, \mathcal{R}_j \in \mathbb{R}), U_{\mathcal{B}}(\mathbb{R}, \mathcal{S}^{\mathcal{B}}, \mathcal{S}^{\mathcal{R}_j}), U_{\mathcal{R}_j}(T^{\mathcal{R}_j}, \mathcal{S}^{\mathcal{B}}, \mathcal{S}^{\mathcal{R}_j})\right) =$$

(8.6.5)

$$\underset{U'=\left(U_{\mathcal{R}_1}(\cdot),\ldots,U_{\mathcal{R}_m}(\cdot)\right)}{\arg\max} \left(\left(\nu(\mathcal{C}_{\mathbb{R}}^{\mathcal{B}}) - \sum_{\mathcal{R}_j \in \mathbb{R}} U_{\mathcal{R}_j}(\cdot)\right) \times \prod_{\mathcal{R}_j \in \mathbb{R}} U_{\mathcal{R}_j}(\cdot)\right)$$

$$\text{s.t.,} \ \nu(\mathcal{C}_{\mathbb{R}}^{\mathcal{B}}) = U_{\mathcal{B}}(\mathbb{R}, \mathcal{S}^{\mathcal{B}}, \mathcal{S}^{\mathcal{R}_j}) + \sum_{\mathcal{R}_j \in \mathbb{R}} U_{\mathcal{R}_j}(T^{\mathcal{R}_j}, \mathcal{S}^{\mathcal{B}}, \mathcal{S}^{\mathcal{R}_j})$$

At the second phase, each $\mathcal{R}_{1 \le j \le m}$ operates its $\mathbb{G}_{\mathcal{R}_j}$ game. Therefore, total m $\mathbb{G}_{\mathcal{R}}$ games are operated in a distributed manner. In this game, the multiple ADVs covered by the $\mathcal{R}_j$ are game players and they form a coalition $\mathcal{C}_{\mathcal{R}_j}$. Based on the $\mathcal{V}_i$'s strategy $\mathcal{S}^{\mathcal{V}_i}$, the utility function of $\mathcal{V}_i \in \mathbb{V}^{\mathcal{R}_j}$, i.e., $\mathfrak{U}_{\mathcal{V}_i}(\cdot)$, can be formulated as follows (Kim, 2023b):

$$\mathfrak{U}_{\mathcal{V}_i}(\mathcal{S}^{\mathcal{R}_j}, \mathbb{V}^{\mathcal{R}_j}, \mathcal{S}^{\mathcal{V}_i}) = \eta - exp\left(\phi \times \frac{\min(T^{\mathcal{V}_i}, \mathcal{S}^{\mathcal{V}_i})}{T^{\mathcal{V}_i}}\right)$$

(8.6.6)

$$\text{s.t.}, \sum_{\mathcal{V}_i \in \mathbb{V}^{\mathcal{R}_j}} \mathcal{S}^{\mathcal{V}_i} \le \mathcal{S}^{\mathcal{R}_j}$$

where $T^{\mathcal{V}_i}$ is the total requested spectrum amount from the $\mathcal{V}_i$, and η, ϕ are control parameters for the $\mathfrak{U}_{\mathcal{V}_i}(\cdot)$. Based on the individual monotonicity property, the *RRMC* scheme adopts the idea of *CKSBS* as the solution for the $\mathbb{G}_{\mathcal{R}_j}$. According to (8.6.3), the amount of $\mathcal{S}^{\mathcal{V}_i}$ is decided as follows (Kim, 2023b):

$$\frac{\mathfrak{U}_{\mathcal{V}_i}(\mathcal{S}^{\mathcal{R}_j}, \mathbb{V}^{\mathcal{R}_j}, \mathcal{S}^{\mathcal{V}_i})}{\mathcal{I}_{\mathcal{V}_i}\left(\mathcal{Q}_\nu(S_{\mathcal{R}_j})\right)} \text{is constant across } \mathcal{V}_i \in \left(\mathcal{C}_{\mathcal{R}_j} = \mathbb{V}^{\mathcal{R}_j}\right) :=$$

(8.6.7)

$$\frac{\mathfrak{U}_{\mathcal{V}_i}(\mathcal{S}^{\mathcal{R}_j}, \mathbb{V}^{\mathcal{R}_j}, \mathcal{S}^{\mathcal{V}_i})}{\mathcal{I}_{\mathcal{V}_i}\left(\mathcal{Q}_\nu\left(\mathcal{C}_{\mathcal{R}_j}\right)\right)} = \frac{\mathfrak{U}_{\mathcal{V}_k}(\mathcal{S}^{\mathcal{R}_j}, \mathbb{V}^{\mathcal{R}_j}, \mathcal{S}^{\mathcal{V}_k})}{\mathcal{I}_{\mathcal{V}_k}\left(\mathcal{Q}_\nu\left(\mathcal{C}_{\mathcal{R}_j}\right)\right)}, \text{ s.t., } \mathcal{V}_i, \mathcal{V}_k \in \left(\mathcal{C}_{\mathcal{R}_j} = \mathbb{V}^{\mathcal{R}_j}\right)$$

At the third phase, each $\mathcal{V}_i$ performs individually its $\mathbb{G}_{\mathcal{V}_i}$ game. Therefore, total n $\mathbb{G}_{\mathcal{V}}$ games are operated in a parallel fashion. In the $\mathbb{G}_{\mathcal{V}_i}$ game, the $F_R^{\mathcal{V}_i}$ and $F_C^{\mathcal{V}_i}$ are assumed as game players, and their utility functions, i.e., $\mathcal{U}_R^{\mathcal{V}_i}(\cdot)$ and $\mathcal{U}_C^{\mathcal{V}_i}(\cdot)$, are defined as follows (Kim, 2023b):

$$\begin{cases} \mathcal{U}_R^{\mathcal{V}_i}(\mathcal{S}^{\mathcal{V}_i}, \mathcal{S}_R^{\mathcal{V}_i}) = \psi - \dfrac{\psi}{exp\left(\theta \times \dfrac{\min(T_R^{\mathcal{V}_i}, \mathcal{S}_R^{\mathcal{V}_i})}{T_R^{\mathcal{V}_i}}\right)} \\ \text{s.t.}, \mathcal{S}^{\mathcal{V}_i} \ge \mathcal{S}_C^{\mathcal{V}_i} + \mathcal{S}_R^{\mathcal{V}_i} \\ \mathcal{U}_C^{\mathcal{V}_i}(\mathcal{S}^{\mathcal{V}_i}, \mathcal{S}_C^{\mathcal{V}_i}) = \log\left(\omega \times \dfrac{\min(T_C^{\mathcal{V}_i}, \mathcal{S}_C^{\mathcal{V}_i})}{T_C^{\mathcal{V}_i}} + \Gamma\right) \end{cases}$$

(8.6.8)

where $T_R^{\mathcal{V}_i}$ and $T_C^{\mathcal{V}_i}$ are the total requested spectrum amounts from the $F_R^{\mathcal{V}_i}$ and $F_C^{\mathcal{V}_i}$, respectively. ψ, θ are control parameters for the $\mathcal{U}_R^{\mathcal{V}_i}(\cdot)$, and ω, Γ are control parameters for the $\mathcal{U}_C^{\mathcal{V}_i}(\cdot)$. In the $\mathbb{G}_{\mathcal{V}_i}$ game, the $F_R^{\mathcal{V}_i}$ and $F_C^{\mathcal{V}_i}$ form a coalition $\mathcal{C}_{\mathcal{V}_i}$, and they attempt to share the $\mathcal{S}^{\mathcal{V}_i}$. However, in the viewpoint of $\mathcal{V}_i$, the $F_R^{\mathcal{V}_i}$ and $F_C^{\mathcal{V}_i}$ are their own components. Therefore, the original *CNBS*, which has the independent of irrelevant alternatives property, is used to provide the solution for the $\mathbb{G}_{\mathcal{V}_i}$. According to (8.6.1), the amounts of $\mathcal{S}_R^{\mathcal{V}_i}$ and $\mathcal{S}_C^{\mathcal{V}_i}$ are decided as follows (Kim, 2023b):

$$CNBS\left(S_{\mathcal{V}_i}, \left[\mathcal{U}_R^{\mathcal{V}_i}\left(\mathcal{S}^{\mathcal{V}_i}, \mathcal{S}_R^{\mathcal{V}_i}\right), \mathcal{U}_C^{\mathcal{V}_i}\left(\mathcal{S}^{\mathcal{V}_i}, \mathcal{S}_C^{\mathcal{V}_i}\right)\right]\right) = \arg\max_{\left[\mathcal{S}_R^{\mathcal{V}_i}, \mathcal{S}_C^{\mathcal{V}_i}\right]} \left(\mathcal{U}_R^{\mathcal{V}_i}(\cdot) \times \mathcal{U}_C^{\mathcal{V}_i}(\cdot)\right) \tag{8.6.9}$$

$$\text{s.t.}, \ \mathcal{S}^{\mathcal{V}_i} \geq \left(\mathcal{S}_R^{\mathcal{V}_i} + \mathcal{S}_C^{\mathcal{V}_i}\right)$$

8.6.6 Summary

For dual-function sensing and communication in the mmWave frequency bands, the *RRMC* scheme has been developed as a JSC platform, integrating radio sensing into existing mobile communication networks. As a novel spectrum allocation mechanism, the *RRMC* scheme leverages coalitional bargaining theory to optimize resource distribution and efficiency. Specifically, the *RRMC* scheme incorporates key principles from the *CNBS* and *CKSBS* to design a cooperative spectrum-sharing algorithm, which is structured into three interconnected game models; i) base station to RSU allocation - the spectrum resources from the base station are allocated to multiple RSUs, ii) RSU to ADV distribution - the spectrum assigned to RSUs is further distributed among individual ADVs, and iii) intra-ADV resource sharing - within each ADV, the allocated spectrum is dynamically shared between radar sensing and communication functions. Following a step-by-step approach, these game models enable reciprocal cooperation between RSUs and ADVs, ensuring efficient and adaptive spectrum utilization. Studies have demonstrated that the *RRMC* scheme effectively addresses the control challenges associated with radar-communication coexistence, making it a robust solution for next-generation JSC systems (Kim, 2023b).

8.7 Effective Crowdsensing and Routing Control (*ECRC*) scheme

VANETs have recently gained traction as a promising networking technology, drawing significant attention from both the automotive industry and the academic research community. As the demand for efficient vehicular communication continues to rise, the development of next-generation VANET management protocols have become a critical challenge. However, traditional control approaches - which have demonstrated reliability in conventional ad hoc wireless networks - are often difficult to directly apply to VANET environments due to their high mobility, dynamic topology, and varying traffic conditions. To address these challenges, S. Kim recently proposes the *Effective Crowdsensing and Routing Control* (*ECRC*) scheme for VANETs (Kim, 2019b). The *ECRC* scheme integrates reinforcement learning and game theory, introducing a dual-plane control mechanism structured as vertical and horizontal game models. In the vertical game, network agents and vehicles collaborate to optimize the crowdsensing process, ensuring efficient data collection and sharing. In the horizontal game, vehicles dynamically select their optimal routes for efficient VANET communication. Designed as a decentralized, distributed system, the dual-plane game paradigm effectively captures the real-time dynamics of VANET environments, allowing for adaptive decision-making and improved network performance (Kim, 2019b).

8.7.1 Development motivation

As a specialized form of MANETs, VANETs have become a crucial component of Intelligent Transportation Systems (ITS). In VANETs, vehicles are equipped with On-Board Units (OBUs), which integrate various functionalities, including event data recording, global positioning, radar sensing, and short-range wireless communication. With OBUs, each vehicle can function as a sender, receiver, and router, facilitating information exchange among other vehicles. Additionally, vehicles can communicate with fixed roadside infrastructures, known as Roadside Units (RSUs). Acting as Internet access points, RSUs play a role similar to base stations in cellular networks. A typical VANET consists of RSUs and vehicle OBUs, which continuously interact in response to dynamic network conditions (Wang, 2016a, 2016b; Morales, 2012; Chen, 2017; Su, 2017; Xiao, 2017; Kim, 2019b).

In VANET operations, two key data dissemination techniques support vehicular applications; Vehicle-to-Infrastructure (V2I) communication, where vehicles download or upload data via RSUs, and Vehicle-to-Vehicle (V2V) communication, where vehicles exchange data directly without RSU involvement. These two communication paradigms must coexist and complement each other to accommodate the diverse networking requirements of VANETs. Ensuring efficient data delivery via V2I and V2V communication is critical for maintaining seamless services in highly dynamic vehicular environments. Therefore, much research has been dedicated to managing control challenges related to V2I and V2V communications effectively (Chen, 2017; Kim, 2019b; Su, 2017).

With the rapid evolution of VANET technology, Vehicular Crowd Sensing (VCS) has emerged as a novel paradigm that utilizes the sensor-equipped OBUs in vehicles for real-time data collection. As a type of V2I communication, VCS enables location-based services, including; i) urban monitoring, ii) traffic and road condition analysis, iii) air pollution level measurements, and iv) public information sharing across multiple regions. Due to vehicle mobility, VCS enhances sensing coverage for location-based applications (Wang, 2016a; Xiao, 2017). During VCS operations, participating vehicles use OBUs to gather requested data and transmit it to RSUs. Acting as central servers, RSUs compensate vehicles based on the quality of their sensing contributions, encouraging accurate data reporting. Specifically, a vehicle is incentivized to submit a precise sensing report when its expected reward from the RSU exceeds its sensing costs (Kim, 2019b; Xiao, 2017).

Routing efficiency is a critical factor in network performance. In VANETs, V2V routing protocols are fundamental for reliable data transmission between source and destination vehicles. Since network performance is highly dependent on routing strategies, extensive research has been conducted to develop adaptive routing protocols. Although routing protocols designed for MANETs have been widely studied, VANETs possess distinct characteristics that necessitate tailored V2V routing algorithms. These unique factors include high mobility, frequent topology changes, and varying traffic conditions, which must be accounted for when designing efficient routing solutions (Jang, 2013; Kim, 2011b, 2016, 2019b).

Recent advancements in game theory have introduced the concept of learning in games, which focuses on how game players refine their strategies over time (Chalkiadakis, 2007). The *ECRC* scheme introduces innovative V2I and V2V communication algorithms to enhance VANET management. For V2I communication, the *ECRC* scheme employs a vertical game model to support an incentive-based VCS algorithm. In this vertical game, vehicles function as mobile sensors, collecting the required data, while RSUs act as central servers, processing and sharing collected information. For V2V communication, the *ECRC* scheme introduces a horizontal game model to implement a learning-based inter-vehicle communication protocol. Here, vehicles act as strategic players, optimizing routing quality and reliability. Designed as a dual-plane model, the *ECRC* scheme integrates both vertical and horizontal game interactions, capturing the dynamic nature of VANETs. By mimicking the interactive sequential game process, the *ECRC* approach provides a practical and adaptive solution for real-world VANET environments (Kim, 2019b).

To analyze strategic interactions among game players, the *ECRC* scheme integrates reinforcement learning techniques. Reinforcement learning focuses on optimizing decision-making processes, allowing system agents to select actions that maximize their cumulative rewards. Generally, reinforcement learning algorithms must balance exploration (discovering new strategies) and exploitation (leveraging known strategies) based on real-time performance. In game theory, reinforcement learning is used to model rational decision-making under bounded rationality, helping explain how effective solutions emerge over time. Given the increasing complexity of vehicular networks, research interest in combining game theory with reinforcement learning has grown significantly in recent years. This hybrid approach is now considered one of the most promising research directions for next-generation VANET control systems (Galindo, 2010; Hu, 2010; Kim, 2019b)

8.7.2 Contribution

The *ECRC* scheme introduces several innovative approaches to VANET management, contributing to both VCS and routing optimization. The primary contributions of the *ECRC* scheme are as follows; i) development of novel VCS and routing algorithms - the *ECRC* scheme introduces advanced VCS algorithms and adaptive routing protocols, specifically tailored for dynamic VANET environments, ii) implementation of a dual-plane game model - the *ECRC* scheme integrates a dual-plane game framework, incorporating both V2I and V2V communication models, ensuring efficient data dissemination and optimized vehicular interactions, iii) adoption of a distributed online learning approach - by utilizing a self-adaptive, real-time decision-making mechanism, the *ECRC* scheme enhances network responsiveness and efficiency under changing conditions, and iv) balancing optimality and practicality - the *ECRC* scheme dynamically manages control decisions, achieving a tradeoff between theoretical efficiency and practical applicability in real-world VANET scenarios. The *ECRC* scheme's significance lies in its ability to adapt to highly dynamic VANET environments, while effectively integrating game theory and learning algorithms. To the best of my knowledge, game-based learning algorithms for VANET control challenges have been minimally explored, making the *ECRC* scheme a pioneering contribution to the field (Kim, 2019b)

8.7.3 Dual-plane game model for V2I and V2V communications

In a VANET system, both RSUs and vehicles make individual control decisions while adapting to dynamic network conditions. Operating in a distributed online manner, these entities aim to maximize their own utility by leveraging learning-based decision-making processes. Building on this foundation, the *ECRC* scheme introduces a dual-plane game model, which structures V2I and V2V communications into two distinct interactive frameworks. The vertical game model for V2I communications captures the strategic interactions between RSUs and a set of vehicles, where RSUs function as network managers, optimizing data collection and service distribution. The horizontal game model for V2V communications defines the interactions among vehicles, focusing on routing decisions and cooperative packet forwarding to enhance communication efficiency. By formulating V2I interactions as a vertical game and V2V interactions as a horizontal game, the *ECRC* scheme ensures an adaptive and scalable network control framework, effectively addressing the challenges of real-world VANET environments. Formally, the *ECRC* scheme defines the vertical game model $\mathbb{G}^{V2I} = \{\mathrm{N}, \mathbf{I}, \left\{S^{\mathscr{R}}, S^{\mathscr{V}}_{\mathscr{V}_i \in \mathscr{V}}\right\}, \left\{u^{\mathscr{R}}, u^{\mathscr{V}}_{\mathscr{V}_i \in \mathscr{V}}\right\}, \mathfrak{Z}^{\mathfrak{P}_i^{\mathfrak{I}_j \in \mathfrak{I}}}, \mathrm{T}\}$ at each time period of gameplay (Kim, 2019b)

- N is the finite set of game players $\mathbb{N} = \{\mathscr{R}, \mathscr{V}\}$ where R represents a RSU and $\mathbf{V} = \left\{\mathscr{V}_1 \ldots \mathscr{V}_i, \ldots \mathscr{V}_k\right\}$ is a set of multiple vehicles where $\mathscr{V}_i$ is the ith vehicle and k is the number of vehicles for a $\mathbb{G}^{V2I}$ game; they are assumed as mobile nodes in the VCS process.
- **I** is the finite set of sensing tasks $\mathbf{I} = \{\mathfrak{I}_1 \ldots \mathfrak{I}_s\}$ in a RSU, where *s* is the number of total sensing tasks.
- $S^{\mathscr{R}} = \left[\eta_{\mathscr{R}}^{\mathfrak{I}_1} \ldots \eta_{\mathscr{R}}^{\mathfrak{I}_j} \ldots \eta_{\mathscr{R}}^{\mathfrak{I}_s}\right]$ is a vector of RSU's strategies where $\eta_{\mathscr{R}}^{\mathfrak{I}_j}$ represents the strategy for the $\mathfrak{I}_j$. In the $\mathbb{G}^{V2I}$ game, $\eta_{\mathscr{R}}$ means price levels for an unit sensing time like as $\eta_{\mathscr{R}}^{\mathfrak{I}_j} = \left\{\mathrm{Đ}_{\mathfrak{I}_j \in N} \mathfrak{P}_i^{\mathfrak{I}_j} \middle| \mathfrak{P}_i^{\mathfrak{I}_j} \in \left[\mathfrak{P}_{min}^{\mathfrak{I}_j} \ldots \mathfrak{P}_l^{\mathfrak{I}_j} \ldots \mathfrak{P}_{max}^{\mathfrak{I}_j}\right]\right\}$.
- $S^{\mathscr{V}}_{\mathscr{V}_i} = \left[\mu_{\mathscr{V}_i}^{\mathfrak{I}_1} \ldots \mu_{\mathscr{V}_i}^{\mathfrak{I}_j} \ldots \mu_{\mathscr{V}_i}^{\mathfrak{I}_s}\right]$ is the vector of $\mathscr{V}_i$'s strategies where $\mu_{\mathscr{V}_i}^{\mathfrak{I}_j}$ represents the strategy for the $\mathfrak{I}_j$. In the $\mathbb{G}^{V2I}$ game, $\mu_{\mathscr{V}_i}^{\mathfrak{I}_j}$ means the $\mathscr{V}_i$'s contribution indication for the $\mathfrak{I}_j$ at a unit sensing time.
- $u^{\mathscr{R}}$ is the payoff received by the RSU, and $u^{\mathscr{V}}_{\mathscr{V}_i \in \mathscr{V}}$ is the payoff received by the $\mathscr{V}_i$.
- $\mathfrak{Z}^{\mathfrak{P}_i^{\mathfrak{I}_j}}$ is the learning value for the $\mathfrak{P}_i^{\mathfrak{I}_j}$ in a RSU; Z is used to estimate the probability distribution for the next RSU's strategy selection.
- $T = \left\{\mathscr{H}_1, \ldots, \mathscr{H}_t, \mathscr{H}_{t+1}, \ldots\right\}$ denotes time, which is represented by a sequence of time steps with imperfect information for V2I communications.

For V2V communications, the *ECRC* scheme defines the horizontal game model $\mathbb{G}^{V2V} = \{\mathrm{v}, \mathscr{N}_{\mathscr{V}_i}, \mathbb{A}_{\mathscr{V}_i \in \mathbb{V}}, U_{\mathscr{V}_i \in \mathbb{V}}, \mathbb{L}_{\mathscr{V}_i}^{a_{\mathscr{V}_i}^{\mathscr{V}_m} \in \mathbb{A}_{\mathscr{V}_i}}, T\}$ at each time period of gameplay. $\mathbb{G}^{V2V}$ can formulate the interactions of vehicles for routing operations (Kim, 2019b).

- v is the finite set of game players $\mathbb{V} = \{\mathscr{V}_1 \ldots \mathscr{V}_n\}$, where *n* is the number of vehicles for the $\mathbb{G}^{V2V}$ game.
- $\mathscr{N}_{\mathscr{V}_i}$ is the set of $\mathscr{V}_i$'s neighboring vehicles.
- $\mathbb{A}_{\mathscr{V}_i \in \mathbb{V}} = \left\{a_{\mathscr{V}_i}^{\mathscr{V}_m} | \mathscr{V}_m \in \mathscr{N}_{\mathscr{V}_i}\right\}$ is the finite set of $\mathscr{V}_i$'s available strategies where $a_{\mathscr{V}_i}^{\mathscr{V}_m}$ represents the selection of the neighboring vehicle $\mathscr{V}_m$ to relay a routing packet.
- $U_{\mathscr{V}_i \in \mathbb{V}}$ is the payoff received by $\mathscr{V}_i$.
- $\mathbb{L}_{\mathscr{V}_i}^{a_{\mathscr{V}_i}^{\mathscr{V}_m} \in \mathbb{A}_{\mathscr{V}_i}}$ is the learning value for the $a_{\mathscr{V}_i}^{\mathscr{V}_m}$ in $\mathscr{V}_i$; L is used to estimate the probability distribution for the next $\mathscr{V}_i$'s strategy selection.
- $T = \left\{\mathscr{H}_1, \ldots, \mathscr{H}_t, \mathscr{H}_{t+1}, \ldots\right\}$ denotes time, which is represented by a sequence of time steps with imperfect information for V2V communications.

8.7.4 VCS algorithm based on the vertical game model

With the rapid advancement of sensor and vehicular technologies, modern OBUs are equipped with various integrated sensors, transforming them into powerful data collection platforms. Within the VCS process, the primary objective is to select an optimal set of vehicles to effectively cover the monitoring area, as the quality of gathered data largely depends on the sensing coverage ratio. However, achieving this optimal vehicle recruitment presents significant challenges (Han, 2015). In the VCS framework, RSUs serve as centralized data aggregators, collecting sensed information from participating vehicles.

Vehicles, in turn, utilize their OBUs to gather requested local information and transmit their sensing results to the RSU. Both data collection and transmission rely on wireless communication, which requires energy consumption. Since sensing tasks and data transmission deplete a vehicle's energy resources, and vehicles act as rational, self-interested entities, they require proper compensation for their participation. Vehicles are incentivized to contribute only when the expected payment from the RSU exceeds the cost of sensing and reporting. Therefore, to ensure sufficient vehicle participation, RSUs must design appropriate incentive mechanisms to encourage full sensing efforts from participating vehicles. Thus, determining optimal incentive strategies for VCS participants is a critical issue, as it directly impacts the effectiveness of data collection, the willingness of vehicles to participate, and the overall performance of the VANET sensing system (Kim, 2019b; Xiao, 2017).

The *ECRC* scheme focuses on designing incentive mechanisms for participatory VCS. In this framework, RSUs provide incentives to vehicles in exchange for collected sensing data, ensuring mutual benefit for both data providers (vehicles) and data aggregators (RSUs). To model this interactive exchange process, the *ECRC* scheme leverages game theory, which has been widely applied to analyze incentive structures in VCS environments. Specifically, game-based incentive mechanisms have emerged as effective solutions for addressing real-world engineering challenges, as they provide a strategic framework for decision-making in self-interested and decentralized vehicular networks (Sun, 2016). By adopting game-theoretic approaches, the *ECRC* scheme ensures that both RSUs and vehicles can optimize their respective utilities, creating a balanced, fair, and efficient incentive model for VCS participation (Kim, 2019b, Xiao, 2017).

To design the VCS algorithm, the *ECRC* scheme considers a VCS platform consisting of one RSU (R), a set of mobile vehicles (**V**) and sensing tasks (**I**), and they formulate the vertical game model $\mathbb{G}^{V2I}$. Due to their heterogeneity, mobile vehicles may make different contributions of sensing work. To capture this, a vehicle's strategy ($\mu \in S_{\mathscr{V}_i}^{\mathscr{V}}$) indicates the contribution of unit sensing time for a specific task ($\mathfrak{I} \in \mathfrak{I}$). During the unit sensing time, the $\mathscr{V}_i$ can actually contribute for the task $\mathfrak{I}_s$, i.e., $\mu_{\mathscr{V}_i}^{\mathfrak{I}_s}=1$, or not, i.e., $\mu_{\mathscr{V}_i}^{\mathfrak{I}_s} = 0$, where $\mu_{\mathscr{V}_i}^{\mathfrak{I}_s} \in S_{\mathscr{V}_i}^{\mathscr{V}}$. Each vehicle has its own utility function, which represents its net gain. For the $\mathscr{V}_i$'s utility function at the t'th unit time $\left(u_{\mathscr{V}_i}^{\mathscr{V}}\left(S_{\mathscr{V}_i}^{\mathscr{V}}, \mathscr{H}_t\right)\right)$ is given by (Kim, 2019b);

$$u_{\mathscr{V}_i}^{\mathscr{V}}\left(S_{\mathscr{V}_i}^{\mathscr{V}} = \left[\mu_{\mathscr{V}_i}^{\mathfrak{I}_1}\ldots\mu_{\mathscr{V}_i}^{\mathfrak{I}_j}\ldots\mu_{\mathscr{V}_i}^{\mathfrak{I}_s}\right], \mathscr{H}_t\right) = \sum_{\mathfrak{I}_j=\mathfrak{I}_1}^{\mathfrak{I}_s}\left(\mu_{\mathscr{V}_i}^{\mathfrak{I}_j}(\mathscr{H}_t) \times \left(\eta_{\mathscr{R}}^{\mathfrak{I}_j}(\mathscr{H}_t) - \mathscr{C}_{\mathscr{V}_i}^{\mathfrak{I}_j}(\mathscr{H}_t)\right)\right) \tag{8.7.1}$$

$$\text{s.t., } \eta_{\mathscr{R}}^{\mathfrak{I}_j}(\mathscr{H}_t) \in S^{\mathscr{R}} \text{ and } \mu_{\mathscr{V}_i}^{\mathfrak{I}_j}(\mathscr{H}_t) = \begin{cases} 1, if \mathscr{V}_i contributes for \mathfrak{I}_j at \mathscr{H}_t \\ 0, otherwise \end{cases}$$

where $\mathscr{C}_{\mathscr{V}_i}^{\mathfrak{I}_j}(\mathscr{H}_t)$ and $\eta_{\mathscr{R}}^{\mathfrak{I}_j}(\mathscr{H}_t)$ are the $\mathscr{V}_i$'s cost and the RSU's incentive for the $\mathfrak{I}_j$ at time $\mathscr{H}_t$, respectively. By using the reciprocal relationship between benefit and price, the $\mathscr{V}_i$'s payoff corresponds to the received benefit $(\eta_{\mathscr{R}}(\cdot))$ minus the incurred cost $\left(\mathscr{C}_{\mathscr{V}_i}(\cdot)\right)$. $\mathscr{V}_i$ can contribute multiple tasks. Therefore, $\mathscr{V}_i$'s utility is a summation of total payoffs for individual tasks. In practice, the actual sensing cost $\mathscr{C}_{\mathscr{V}}^{\mathfrak{I}}(\cdot)$ is usually unknown by the RSU. Under this asymmetric information situation, the RSU should learn the vehicles' circumstance through the interactive feedback process (Kim, 2019b).

For the RSU, the payoff can be defined as a function of price levels for tasks and vehicles' responses. Therefore, the RSU's utility function $\left(u^{\mathscr{R}}\left(S^{\mathscr{R}}, \mathscr{H}_t\right)\right)$ at time $\mathscr{H}_t$ is computed as the sum of each task's outcome (Kim, 2019b):

$$u^{\mathscr{R}}\left(S^{\mathscr{R}} = \left[\eta_{\mathscr{R}}^{\mathfrak{J}_1} \ldots \eta_{\mathscr{R}}^{\mathfrak{J}_j} \ldots \eta_{\mathscr{R}}^{\mathfrak{J}_s}\right], \mathscr{H}_t\right) =$$

$$\sum_{\mathfrak{J}_h=\mathfrak{J}_1}^{\mathfrak{J}_s}\left(\left(\mathfrak{T}^{\mathfrak{J}_h}\left(\mathscr{H}_t\right) \times \Phi^{\mathfrak{J}_h}\right) - \sum_{\mathscr{V}_i=\mathscr{V}_1}^{\mathscr{V}_k}\left(\mu_{\mathscr{V}_i}^{\mathfrak{J}_h}\left(\mathscr{H}_t\right) \times \eta_{\mathscr{R}}^{\mathfrak{J}_h}\left(\mathscr{H}_t\right)\right)\right) \tag{8.7.2}$$

$$\text{s.t., } \mathfrak{T}^{\mathfrak{J}_h}\left(\mathscr{H}_t\right) = \begin{cases} 1, if \sum_{\mathscr{V}_i=\mathscr{V}_1}^{\mathscr{V}_k}\left(\mu_{\mathscr{V}_i}^{\mathfrak{J}_h}\left(\mathscr{H}_t\right)\right) \geq \mathscr{L}_{\mathscr{R}}^{\mathfrak{J}_h} \\ 0, otherwise \end{cases}$$

where $\Phi^{\mathfrak{J}_h}$ is the profit of $\mathfrak{J}_h$ if the VCS for $\mathfrak{J}_h$ is successfully completed. $\mathscr{L}_{\mathscr{R}}^{\mathfrak{J}_h}$ is the pre-defined requirement to complete the task $\mathfrak{J}_h$. Like as the $\mathscr{V}_i$'s utility, the RSU's payoff corresponds to the received profit ($\Phi^{\mathfrak{J}_h}$) minus the payment $\left(\eta_{\mathscr{R}}\right)$. Therefore, the RSU's utility is a summation of all payoffs for successfully completed tasks. As $\mathbb{G}^{V2I}$ game players, the RSU and vehicles attempt to maximize their utility functions. Interactions among game players repeat for a given VCS task over time, yielding a dynamic vertical game. From the viewpoint of RSU, a novel learning method is needed to determine its payment strategy; the RSU should consider the inter-relationship among tasks and coordinated reactions from vehicles (Kim, 2019b).

The *ECRC* scheme develops a new reinforcement learning method to determine an effective payment policy for each task. For the RSU, the strategy vector $S^{\mathscr{R}}\left(\mathscr{H}_t\right) = \left[\eta_{\mathscr{R}}^{\mathfrak{J}_1}\left(\mathscr{H}_t\right) \ldots \eta_{\mathscr{R}}^{\mathfrak{J}_s}\left(\mathscr{H}_t\right)\right]$ represents each price strategy for each task at time $\mathscr{H}_t$. Let $\mathfrak{Z}_{\mathscr{H}_t}^{\mathfrak{J}_s}\left(\eta_{\mathscr{R}}^{\mathfrak{J}_s}, \eta_{\mathscr{R}}^{-\mathfrak{J}_s}\right)$ be the learning value of taking strategy $\eta_{\mathscr{R}}^{\mathfrak{J}_s}$ with a vector of payment strategies except the strategy $\eta_{\mathscr{R}}^{\mathfrak{J}_s}$ $\left(\eta_{\mathscr{R}}^{-\mathfrak{J}_s}\right)$. The RSU updates its $\mathfrak{Z}(\cdot)$ value over time according to the following method (Kim, 2019b):

$$\mathfrak{Z}_{\mathscr{H}_{t+1}}^{\mathfrak{P}_i^{\mathfrak{J}_s}}\left(\eta_{\mathscr{R}}^{\mathfrak{J}_s} = \mathfrak{P}_i^{\mathfrak{J}_s}, \eta_{\mathscr{R}}^{-\mathfrak{J}_s}\right) = \left(\left(1-\beta\right) \times \mathfrak{Z}_{\mathscr{H}_t}^{\mathfrak{J}_s}\left(\mathfrak{P}_i^{\mathfrak{J}_s}, \eta_{\mathscr{R}}^{-\mathfrak{J}_s}\right)\right)$$

$$+\left(\beta \times \left(\left(\left(1-\gamma^{\mathscr{R}}\right) \times \partial_{\mathscr{H}_t}^{\mathfrak{J}_s}\right) + \left\{\gamma^{\mathscr{R}} \times \sum_{\mathfrak{J}_h \in \mathfrak{J}^{-\mathfrak{J}_s}} \frac{\partial_{\mathscr{H}_t}^{\mathfrak{J}_h}}{(s-1)}\right\}\right)\right) \tag{8.7.3}$$

$$\text{s.t., } \partial_{\mathscr{H}_t}^{\mathfrak{J}_s} = \left(\left(\mathfrak{T}^{\mathfrak{J}_s}\left(\mathscr{H}_t\right) \times \Phi^{\mathfrak{J}_s}\right) - \sum_{\mathscr{V}_i=\mathscr{V}_1}^{\mathscr{V}_k}\left(\mu_{\mathscr{V}_i}^{\mathfrak{J}_s}\left(\mathscr{H}_t\right) \times \eta_{\mathscr{R}}^{\mathfrak{J}_s}\left(\mathscr{H}_t\right)\right)\right)$$

During the $\mathbb{G}^{V2I}$ game process, the RSU adaptively learns the current VCS state, and sequentially adjusts $\mathfrak{Z}(\cdot)$ values for each VCS task. In the formula (8.7.3), the first term captures the $\mathfrak{Z}(\cdot)$'s history, and the second term captures the relative impact of a strategy $\eta_{\mathscr{R}}^{\mathfrak{J}_s}$. The parameter β controls the relative weights given to two terms in the RSU's decision. To implement the task price learning mechanism in the RSU, the *ECRC* scheme defines a strategy selection distribution ($P^{\mathscr{R}}$) for VCS tasks. Based on the $\mathfrak{Z}(\cdot)$ values, the *ECRC* scheme can determine $P_{\mathfrak{J}_s}^{\mathscr{R}}\left(\mathscr{H}_t\right) = \left\{P_{\mathfrak{J}_s(\mathscr{H}_t)}^{\mathfrak{P}_{min}^{\mathfrak{J}_s}} \ldots P_{\mathfrak{J}_s(\mathscr{H}_t)}^{\mathfrak{P}_i^{\mathfrak{J}_s}} \ldots P_{\mathfrak{J}_s(\mathscr{H}_t)}^{\mathfrak{P}_{max}^{\mathfrak{J}_s}}\right\}$ as the probability distribution of strategies selected by the RSU. The $\mathfrak{P}_i$ strategy selection probability for the $\mathfrak{T}_s$ at time $\mathscr{H}_{t+1}$ $\left(P_{\mathfrak{T}_s(\mathscr{H}_{t+1})}^{\mathfrak{P}_i^{\mathfrak{T}_s}}\right)$ is defined as (Kim, 2019b):

$$P_{\mathfrak{T}_s(\mathscr{H}_{t+1})}^{\mathfrak{P}_i^{\mathfrak{T}_s}} = \frac{\mathrm{EXP}\left(\mathfrak{Z}_{\mathscr{H}_t}^{\mathfrak{P}_i^{\mathfrak{T}_s}}\left(\eta_{\mathscr{R}}^{\mathfrak{T}_s} = \mathfrak{P}_i^{\mathfrak{T}_s}, \eta_{\mathscr{R}}^{-\mathfrak{T}_s}\right)\right)}{\sum_{\mathfrak{P}_{j,j=min}^{\mathfrak{T}_s}}^{j=max} \mathrm{EXP}\left(\mathfrak{Z}_{\mathscr{H}_t}^{\mathfrak{P}_i^{\mathfrak{T}_s}}\left(\eta_{\mathscr{R}}^{\mathfrak{T}_s} = \mathfrak{P}_j^{\mathfrak{T}_s}, \eta_{\mathscr{R}}^{-\mathfrak{T}_s}\right)\right)}, \text{s.t.}, \mathrm{EXP}(x) = e^x \tag{8.7.4}$$

8.7.5 V2V routing algorithm based on the V2V game model

In modern vehicular networks, V2V communication has become a highly desirable technology, enabling context-aware applications through end-to-end multi-hop routing. These applications contribute to enhancing road safety, improving traffic efficiency, and increasing driver comfort, making inter-vehicular communication a key component of next-generation intelligent transportation systems. However, high vehicle mobility often leads to unstable network connectivity, resulting in inconsistent service quality. Due to the dynamic nature of vehicular movement, traditional MANET routing protocols have been found to perform inefficiently in VANET environments. Consequently, routing protocols designed for VANETs must be robust and adaptive, capable of addressing challenges related to frequent topology changes, rapid mobility, and intermittent connectivity (Kim, 2019; Nzouonta, 2009).

The *ECRC* scheme develops a new horizontal game model ($\mathbb{G}^{V2V}$) for V2V communications. As game players in $\mathbb{G}^{V2V}$, vehicles dynamically decide their routes. To configure the routing topology, each vehicle maintains two routing information: path costs ($\mathscr{PC}$) to reach the destination vehicle and learning values (L) for available paths. $\mathscr{PC}$ is computed as the sum of all relay link costs from the current vehicle to the source vehicle, and L is obtained for self-organizing and independent routing controls. Owing to the highly dynamic topology and communication environments, the wireless links in V2V communications vary continually. In order to handle dynamic VANET conditions, it is important to estimate the current link status by considering several factors. The *ECRC* scheme defines a wireless link status $\left(\mathscr{LC}_{\mathscr{V}_i}^{\mathscr{V}_j}(\mathscr{H}_t)\right)$ between the $\mathscr{V}_i$ and the $\mathscr{V}_j$ at time $\mathscr{H}_t$ like as (Kim, 2019b)

$$\mathscr{LC}_{\mathscr{V}_i}^{\mathscr{V}_j}(\mathscr{H}_t) = \left(\alpha \times \left(1 - \left(\left(D_{\mathscr{V}_i}^{M} - d_{\mathscr{V}_i}^{\mathscr{V}_j}\right) / D_{\mathscr{V}_i}^{M}\right)\right)\right) + \left((1-\alpha) \times \left(1 - \Psi_{\mathscr{V}_i}^{\mathscr{V}_j}(\mathscr{H}_t)\right)\right) \tag{8.7.5}$$

$$\text{s.t.}, \alpha = \left(\frac{\mathbb{F}\left(\mathfrak{M}_{\mathscr{V}_j}^{\mathscr{H}_t}\right)}{\max_{\mathscr{V}_r \in \mathfrak{M}_{\mathscr{V}_i}^{\mathscr{H}_t}} \left\{\mathbb{F}\left(\mathfrak{M}_{\mathscr{V}_r}^{\mathscr{H}_t}\right)\right\}}\right)$$

where $D_{\mathscr{V}_i}^{M}$ is the maximum coverage range of $\mathscr{V}_i$ and $d_{\mathscr{V}_i}^{\mathscr{V}_j}$ is the relative distance between the $\mathscr{V}_i$ and the $\mathscr{V}_j$. In wireless communications, energy consumption decreases proportional to the square of the propagation distance. Therefore, a closer neighboring vehicle is more suitable for routing. In the formula (8.7.5), the first term captures the relative distance, and the second term captures the vehicle's stability. The parameter α controls the relative weights given to two terms. $\mathfrak{M}_{\mathscr{V}_j}^{\mathscr{H}_t}$ denotes the set of the neighboring vehicles of $\mathscr{V}_j$ at time $\mathscr{H}_t$, and $\mathbb{F}\left(\mathfrak{M}_{\mathscr{V}_j}^{\mathscr{H}_t}\right)$ is the cardinality (degree) of set $\mathfrak{M}_{\mathscr{V}_j}$. Usually, $\mathbb{F}\left(\mathfrak{M}_{\mathscr{V}_j}^{\mathscr{H}_t}\right)$ represents the density level of $\mathscr{V}_j$. $\Psi_{\mathscr{V}_i}^{\mathscr{V}_j}(\mathscr{H}_t)$ is the entropy of the $\mathscr{V}_j$ at time $\mathscr{H}_t$ from the viewpoint of the $\mathscr{V}_i$. In general, entropy is uncertainty and a measure of the disorder in a system. It represents the topological change, which is a natural quantification of the effect of vehicle mobility on the connectivity service of VANET (Hieu, 2010; Jeon, 2014; Kim, 2016). In the *ECRC* scheme, the basic concept of entropy is adopted for supporting and evaluating stable routes. The $\Psi_{\mathscr{V}_i}^{\mathscr{V}_j}(\mathscr{H}_t)$ is calculated as follows (Kim, 2016a, 2019b):

$$\Psi_{\mathscr{V}_i}^{\mathscr{V}_j}(\mathscr{H}_t) = -\sum_{\mathscr{V}_k \in \mathfrak{M}_{\mathscr{V}_j}^{\mathscr{H}_t}} \left(\frac{\mathscr{F}_{\mathscr{V}_k}(\mathscr{H}_t, \Delta_y) \times \log \mathscr{F}_{\mathscr{V}_k}(\mathscr{H}_t, \Delta_y)}{\log\left(\mathbb{F}(\mathfrak{M}_{\mathscr{V}_j}^{\mathscr{H}_t})\right)} \right)$$

$$\text{s.t., } \mathscr{F}_{\mathscr{V}_k}(\mathscr{H}_t, \Delta_y) = \frac{1}{y} \times \left(\frac{\sum_{\mathscr{H}_x = \mathscr{H}_{t-y}}^{\mathscr{H}_t} \left(\left| \vec{v}_j(\mathscr{H}_x) - \vec{v}_i(\mathscr{H}_x) \right| \right)}{\sum_{\mathscr{H}_x = \mathscr{H}_{t-y}}^{\mathscr{H}_t} \sum_{\mathscr{V}_k \in \mathfrak{M}_{\mathscr{V}_j}^{\mathscr{H}_t}} \left(\left| \vec{v}_j(\mathscr{H}_x) - \vec{v}_k(\mathscr{H}_x) \right| \right)} \right) \tag{8.7.6}$$

where Δ_y is a time interval from $\mathscr{H}_{t-y}$ to $\mathscr{H}_t$; the mobility information can be calculated and disseminated to other neighboring vehicles within time interval Δ_y. $\vec{v}_i(\mathscr{H}_t), \vec{v}_j(\mathscr{H}_t)$ are the velocity vectors of $\mathscr{V}_i$ and $\mathscr{V}_j$, respectively, at time $\mathscr{H}_t$. The entropy $\Psi_{\mathscr{V}_i}^{\mathscr{V}_j}(\mathscr{H}_t)$ is normalized as $0 \leq \Psi_{\mathscr{V}_i}^{\mathscr{V}_j}(\mathscr{H}_t) \leq 1$. If the value of $\Psi_{\mathscr{V}_i}^{\mathscr{V}_j}(\mathscr{H}_t)$ is close to 1, the part of the route that represents the links of the path associated with an intermediate vehicle $\mathscr{V}_j$ is stable. If the value of $\Psi_{\mathscr{V}_i}^{\mathscr{V}_j}(\mathscr{H}_t)$ is close to 0, the local route is unstable (Kim, 2016a, 2019b).

For the adaptive $\mathscr{LC}_{\mathscr{V}_i}^{\mathscr{V}_j}(\mathscr{H}_t)$ estimation, the parameter α controls the relative weights given to the distance and the entropy of the corresponding vehicle. Under diverse VANET environments, the $\mathscr{V}_i$ dynamically modifies α at time $\mathscr{H}_t$. When the density level is light, the *ECRC* scheme can put more emphasis on vehicle stability. In this case, a lower value of α is more suitable. But if the density level is heavy, the link cost should strongly depend on the relative distance. In this case, a higher value of α is more suitable. Therefore, the *ECRC* scheme must decide on the value of α by considering the current density level of corresponding vehicle (Kim, 2019b).

During the routing path formation, each individual vehicle estimates the $\mathscr{LC}$ values of its neighboring vehicles. Vehicles within one-hop of the destination vehicle set their $\mathscr{PC}$ values to the one-hop $\mathscr{LC}$ value from themselves to the destination node. The remaining vehicles, i.e., $\mathscr{V}_j$, set their $\mathscr{LC}$ values as follows (Kim, 2019b):

$$\mathscr{PC}_{\mathscr{V}_j} = \min_{\mathscr{V}_e} \left\{ \mathscr{PC}_{\mathscr{V}_e} + \mathscr{LC}_{\mathscr{V}_j}^{\mathscr{V}_e} \right\}, \text{ s.t., } \mathscr{V}_e \in \mathfrak{M}_{\mathscr{V}_j} \tag{8.7.7}$$

Based on the $\mathscr{PC}$ values of the neighbor vehicles, the *ECRC* scheme can estimate a routing cost for V2V communications. Usually, most V2V routing algorithms are designed by relying on the assumption of vehicle collaboration assuming that vehicles are willing to act as relay nodes in a routing path. However, a vehicle acting as a relay node has to sacrifice its energy and bandwidth. Selfish vehicles have nothing to gain by conceding to forward a packet. Therefore, routing algorithms need to stimulate cooperative actions among VANET vehicles in order to be effective (Kim, 2010, 2012, 2019b; Raazi, 2010).

In the V2V routing algorithm, a source vehicle pays the cost to relaying vehicles. In the vertical game ($\mathbb{G}^{V2I}$), each vehicle earns its virtual money (Γ) through the VCS process; Γ is accumulated according to the RSU's incentive payment. In the horizontal game ($\mathbb{G}^{V2V}$), Γ is used to send routing packets. The source vehicle's incentive payment guides selfish relay vehicles toward a cooperative V2V routing. In the $\mathbb{G}^{V2V}$game model, the utility function of source vehicle can be interpreted as the received benefit minus the cost incurred. As a game player, the main goal of source vehicles is to maximize its payoff. If the $\mathcal{V}_i$ is source vehicle and selects the $\mathcal{V}_j$ as a next relay node (i.e., $a_{\mathcal{V}_i}^{\mathcal{V}_j}$), its utility function at time $\mathcal{H}_t\left(U_{\mathcal{V}_i}(\mathcal{V}_j,\mathcal{H}_t)\right)$ can be defined as (Kim, 2019b):

$$U_{\mathcal{V}_i}(\mathcal{V}_j,\mathcal{H}_t) = \begin{cases} \Psi_{\mathcal{V}_i} - J\times\left(Q_{\mathcal{V}_j}^{\mathcal{H}_{t-1}}\times\left(\mathcal{LC}_{\mathcal{V}_i}^{\mathcal{V}_j}+\mathcal{PC}_{\mathcal{V}_j}\right)\right)^{-1} and \\ \Gamma^{\mathcal{V}_i} = \Gamma^{\mathcal{V}_i} - J\times\left(\mathcal{LC}_{\mathcal{V}_i}^{\mathcal{V}_j}+\mathcal{PC}_{\mathcal{V}_j}\right) \\ if \Gamma^{\mathcal{V}_i} \geq J\times\left(\mathcal{LC}_{\mathcal{V}_i}^{\mathcal{V}_j}+\mathcal{PC}_{\mathcal{V}_j}\right) \\ 0, otherwise\left(\text{i.e., } \Gamma^{\mathcal{V}_i} < \mathcal{J}\times\left(\mathcal{LC}_{\mathcal{V}_i}^{\mathcal{V}_j}+\mathcal{PC}_{\mathcal{V}_j}\right)\right) \end{cases} \tag{8.7.8}$$

where $Q_{\mathcal{V}_j}^{\mathcal{H}_{t-1}}$ is the $\mathcal{V}_j$'s success ratio of packet transmission at time $\mathcal{H}_{t-1}$; if the $\mathcal{V}_j$ is not selected as a relay node at time $\mathcal{H}_{t-1}$, $Q_{\mathcal{V}_j}^{\mathcal{H}_{t-1}}$ is 1. $\Gamma^{\mathcal{V}_i}$ is the accumulated virtual money of $\mathcal{V}_i$ during the $\mathbb{G}^{V2I}$, and J is the coefficient factor of packet forwarding cost. The source vehicle $\mathcal{V}_i$ seeks to choose the routing path to maximize its payoff; $\mathbb{A}_{\mathcal{V}_i\in\mathbb{V}} = \left\{a_{\mathcal{V}_i}^{\mathcal{V}_j} \mid \mathcal{V}_j \in \mathcal{N}_{\mathcal{V}_i}\right\}$ is the $\mathcal{V}_i$'s available strategy set. In the $\mathbb{G}^{V2I}$, the *ECRC* scheme develops a new reinforcement learning method to select an efficient routing path. The learning value for the $a_{\mathcal{V}_i}^{\mathcal{V}_j}$ at time $\mathcal{H}_{t+1}\left(\mathbb{L}_{\mathcal{H}_{t+1}}^{\mathcal{V}_j}\right)$ is defined as follows (Kim, 2019b):

$$\mathbb{L}_{\mathcal{H}_{t+1}}^{\mathcal{V}_j} = \begin{cases} \left((1-\xi)\times\mathbb{L}_{\mathcal{H}_t}^{\mathcal{V}_j}\right) + \xi\times\left(\mathcal{U}_{\mathcal{H}_t}^{\mathcal{V}_i} + \left\{\gamma^{\mathcal{V}}\times\left(\mathcal{U}_{\mathcal{H}_t}^{\mathcal{V}_i} - \sum_{\substack{\mathcal{V}_x\in\mathcal{N}_{\mathcal{V}_i}\\ \mathcal{V}_x\neq\mathcal{V}_j}}\frac{\mathcal{W}_{\mathcal{H}_t}^{\mathcal{V}_x}}{\left|\mathcal{N}_{\mathcal{V}_i}\right|-1}\right)\right\}\right) \\ if \mathcal{V}_j is selected at time \mathcal{H}_t \\ \text{s.t., } \mathcal{U}_{\mathcal{H}_t}^{\mathcal{V}_i} = \left(Q_{\mathcal{V}_j}^{\mathcal{H}_t}\times U_{\mathcal{V}_i}(\mathcal{V}_j,\mathcal{H}_t)\right) \text{ and } \mathcal{W}_{\mathcal{H}_t}^{\mathcal{V}_x} = \left(Q_{\mathcal{V}_x}^{\mathcal{H}_{t-1}}\times\left(\mathcal{LC}_{\mathcal{V}_i}^{\mathcal{V}_x}+\mathcal{PC}_{\mathcal{V}_x}\right)\right) \\ \left((1-\xi)\times\mathbb{L}_{\mathcal{H}_t}^{\mathcal{V}_j}\right) + \xi\times\left(\mathcal{W}_{\mathcal{H}_t}^{\mathcal{V}_j} + \left\{\gamma^{\mathcal{V}}\times\left(\mathcal{W}_{\mathcal{H}_t}^{\mathcal{V}_j} - \sum_{\substack{\mathcal{V}_x\in\mathcal{N}_{\mathcal{V}_i}\\ \mathcal{V}_x\neq\mathcal{V}_j}}\frac{\mathcal{W}_{\mathcal{H}_t}^{\mathcal{V}_x}}{\left|\mathcal{N}_{\mathcal{V}_i}\right|-1}\right)\right\}\right) \\ otherwise \end{cases} \tag{8.7.9}$$

In the formula (8.7.9), the first term captures the L's history, and the second term captures the relative impact of selected strategy. The parameter ξ controls the relative weights given to two terms in the $\mathcal{V}_i$'s decision. During the $\mathbb{G}^{V2V}$ game process, a source vehicle learns the current VANET state and chooses the strategy with the highest L value, which leads to packets being relayed to the destination vehicle via the best route. To implement an effective path learning mechanism, the *ECRC* scheme defines a strategy selection distribution ($P^{\mathcal{V}}$) in the same manner as the vertical game model. Suppose there is a finite set of neighboring vehicles $\{\mathcal{V}_1 \ldots \mathcal{V}_m\}$ of $\mathcal{V}_i$ at time $\mathcal{H}_t$, and the $\mathcal{V}_i$ maintains individual L values for these vehicles. Based on this information, the *ECRC* scheme can obtain $P^{\mathcal{V}_i}(\mathcal{H}_{t+1}) = \left\{ P^{\mathcal{V}_1}_{\mathcal{H}_{t+1}} \ldots P^{\mathcal{V}_m}_{\mathcal{H}_{t+1}} \right\}$; it can be the probability distribution of strategies selected by the $\mathcal{V}_i$. The $a^{\mathcal{V}_j}_{\mathcal{V}_i}$ strategy selection probability at time $\mathcal{H}_{t+1}$ $\left(P^{\mathcal{V}_j}_{\mathcal{H}_{t+1}}\right)$ is defined as (Kim, 2019b):

$$P^{\mathcal{V}_j}_{\mathcal{H}_{t+1}} = \frac{\text{EXP}\left(\mathbb{L}^{\mathcal{V}_j}_{\mathcal{H}_{t+1}}\right)}{\sum_{\mathcal{V}_k \in \mathcal{N}_{\mathcal{V}_i}} \text{EXP}\left(\mathbb{L}^{\mathcal{V}_k}_{\mathcal{H}_{t+1}}\right)}, \text{ s.t.}, \text{EXP}(x) = e^x \tag{8.7.10}$$

8.7.6 Summary

The *ECRC* scheme introduces a novel dual-plane game model, integrating both V2I and V2V communications in VANETs. By leveraging the unique characteristics of VANETs, the *ECRC* scheme incorporates an interactive feedback mechanism, structuring the vertical and horizontal games based on reinforcement learning principles. The primary objective of the *ECRC* scheme is to develop efficient VCS algorithms and inter-vehicle routing protocols, particularly suited for incomplete information environments. Within the dual-plane game model, both RSUs and vehicles act as game players, continuously refining their strategies through dynamic learning mechanisms. By employing feedback-based self-monitoring and distributed learning techniques, game players (RSUs and vehicles) can adaptively respond to the current VANET conditions, ensuring opportunistic VCS participation and optimal routing decisions. Through this adaptive learning framework, the *ECRC* scheme effectively maximizes the expected benefits for all participants in highly dynamic VANET environments (Kim, 2019b).

REFERENCES

Ahn, J., Shin, D., Kim, K., & Yang, J. (2017). Indoor air quality analysis using deep learning with sensor data. *Sensors (Basel), 17*(11), 2476. DOI: 10.3390/s17112476 PMID: 29143797

Alpcan, T., & Buchegger, S. (2010). Security games for vehicular networks. *IEEE Transactions on Mobile Computing, 10*(2), 280–290. DOI: 10.1109/TMC.2010.146

Ashok, A., Steenkiste, P., & Bai, F. (2016, December). Adaptive cloud offloading for vehicular applications. In *2016 IEEE Vehicular Networking Conference (VNC)* (pp. 1-8). IEEE. DOI: 10.1109/VNC.2016.7835966

Bossert, W., & Peters, H. (2001). Minimax regret and efficient bargaining under uncertainty. *Games and Economic Behavior, 34*(1), 1–10. DOI: 10.1006/game.2000.0793

Bullock, S. R. (2000). *Transceiver and System Design for Digital Communications*. SciTech Publishing.

Burguet, R., & Caminal, R. (2020). Coalitional bargaining with consistent counterfactuals. *Journal of Economic Theory, 187*, 105013. DOI: 10.1016/j.jet.2020.105013

Chalkiadakis, G. (2007). *A Bayesian approach to multiagent reinforcement learning and coalition formation under uncertainty.*

Chaqfeh, M., Mohamed, N., Jawhar, I., & Wu, J. (2016, December). Vehicular cloud data collection for intelligent transportation systems. In *2016 3rd Smart Cloud Networks & Systems (SCNS)* (pp. 1-6). IEEE. DOI: 10.1109/SCNS.2016.7870555

Chen, C., Li, H., Li, H., Fu, R., Liu, Y., & Wan, S. (2022). Efficiency and fairness oriented dynamic task offloading in internet of vehicles. *IEEE Transactions on Green Communications and Networking, 6*(3), 1481–1493. DOI: 10.1109/TGCN.2022.3167643

Chen, J., Mao, G., Li, C., Zafar, A., & Zomaya, A. Y. (2017). Throughput of infrastructure-based cooperative vehicular networks. *IEEE Transactions on Intelligent Transportation Systems, 18*(11), 2964–2979. DOI: 10.1109/TITS.2017.2663434

Choi, W., Jo, H. J., Woo, S., Chun, J. Y., Park, J., & Lee, D. H. (2018). Identifying ecus using inimitable characteristics of signals in controller area networks. *IEEE Transactions on Vehicular Technology, 67*(6), 4757–4770. DOI: 10.1109/TVT.2018.2810232

Claus, C., & Boutilier, C. (1998). The dynamics of reinforcement learning in cooperative multiagent systems. *AAAI/IAAI, 1998*(746-752), 2.

Compte, O., & Jehiel, P. (2010). The coalitional Nash bargaining solution. *Econometrica, 78*(5), 1593–1623. DOI: 10.3982/ECTA7883

Cui, H., & Zhang, Z. (2021). A cooperative multi-agent reinforcement learning method based on coordination degree. *IEEE Access : Practical Innovations, Open Solutions, 9*, 123805–123814. DOI: 10.1109/ACCESS.2021.3110255

Diekmann, A. (1985). Volunteer's dilemma. *The Journal of Conflict Resolution*, *29*(4), 605–610. DOI: 10.1177/0022002785029004003

Farrokhi, G., Zokaei, S., Ghazisaeedi, E., & Khabazian, M. (2010, September). Improving safety messages dissemination in IEEE 802.11 E based VANETs using controlled repetition technique. In *2010 6th International Conference on Wireless and Mobile Communications* (pp. 395-399). IEEE.

Galindo-Serrano, A., & Giupponi, L. (2010). Distributed Q-learning for aggregated interference control in cognitive radio networks. *IEEE Transactions on Vehicular Technology*, *59*(4), 1823–1834. DOI: 10.1109/TVT.2010.2043124

Gao, G., Xiao, M., Wu, J., Huang, L., & Hu, C. (2018). Truthful incentive mechanism for nondeterministic crowdsensing with vehicles. *IEEE Transactions on Mobile Computing*, *17*(12), 2982–2997. DOI: 10.1109/TMC.2018.2829506

Ghazvini, M., Movahedinia, N., Jamshidi, K., & Moghim, N. (2012). Game theory applications in CSMA methods. *IEEE Communications Surveys and Tutorials*, *15*(3), 1062–1087. DOI: 10.1109/SURV.2012.111412.00167

Guan, K., He, D., Ai, B., Matolak, D. W., Wang, Q., Zhong, Z., & Kürner, T. (2018). 5-GHz obstructed vehicle-to-vehicle channel characterization for internet of intelligent vehicles. *IEEE Internet of Things Journal*, *6*(1), 100–110. DOI: 10.1109/JIOT.2018.2872437

Hafeez, K. A., Zhao, L., Liao, Z., & Ma, B. N. W. (2010, December). Impact of mobility on vanets' safety applications. In *2010 IEEE Global Telecommunications Conference GLOBECOM 2010* (pp. 1-5). IEEE. DOI: 10.1109/GLOCOM.2010.5684042

Han, K., Chen, C., Zhao, Q., & Guan, X. (2015, November). Trajectory-based node selection scheme in vehicular crowdsensing. In *2015 IEEE/CIC International Conference on Communications in China (ICCC)* (pp. 1-6). IEEE. DOI: 10.1109/ICCChina.2015.7448723

Hieu, C. T., & Hong, C. S. (2010). A Connection Entropy-based Multi-Rate Routing Protocol for Mobile Ad Hoc Networks. *Journal of Computing Science and Engineering : JCSE*, *4*(3), 225–239. DOI: 10.5626/JCSE.2010.4.3.225

Hong, S. H., & Kim, W. H. (2000). Bandwidth allocation scheme in CAN protocol. *IEE Proceedings. Control Theory and Applications*, *147*(1), 37–44. DOI: 10.1049/ip-cta:20000126

Hu, T., & Fei, Y. (2010). QELAR: A machine-learning-based adaptive routing protocol for energy-efficient and lifetime-extended underwater sensor networks. *IEEE Transactions on Mobile Computing*, *9*(6), 796–809. DOI: 10.1109/TMC.2010.28

Jang, I., Pyeon, D., Kim, S., & Yoon, H. (2013). A survey on communication protocols for wireless sensor networks. *Journal of Computing Science and Engineering : JCSE*, *7*(4), 231–241. DOI: 10.5626/JCSE.2013.7.4.231

Jeon, M., Kim, S. K., Yoon, J. H., Lee, J., & Yang, S. B. (2014). A direction entropy-based forwarding scheme in an opportunistic network. *Journal of Computing Science and Engineering : JCSE*, *8*(3), 173–179. DOI: 10.5626/JCSE.2014.8.3.173

Jeong, W., Han, S., Choi, E., Lee, S., & Choi, J. W. (2020). CNN-based adaptive source node identifier for controller area network (CAN). *IEEE Transactions on Vehicular Technology*, *69*(11), 13916–13920. DOI: 10.1109/TVT.2020.3016352

Kang, S., Seong, J., & Lee, M. (2018). Controller area network with flexible data rate transmitter design with low electromagnetic emission. *IEEE Transactions on Vehicular Technology*, *67*(8), 7290–7298. DOI: 10.1109/TVT.2018.2832659

Kim, H. H., Lee, S., Kim, M. H., & Lee, K. C. (2009). Development of traffic-balancing algorithm for CAN systems with dual communication channels. In *2009 6th International Conference on Electrical Engineering/Electronics, Computer, Telecommunications and Information Technology* (Vol. 1, pp. 344-347). IEEE. DOI: 10.1109/ECTICON.2009.5137023

Kim, J. H., Lee, K. J., Kim, T. H., & Yang, S. B. (2011b). Effective routing schemes for double-layered peer-to-peer systems in MANET. *Journal of Computing Science and Engineering : JCSE*, *5*(1), 19–31. DOI: 10.5626/JCSE.2011.5.1.019

Kim, K. S., Uno, S., & Kim, M. W. (2010). Adaptive QoS mechanism for wireless mobile network. *Journal of Computing Science and Engineering : JCSE*, *4*(2), 153–172. DOI: 10.5626/JCSE.2010.4.2.153

Kim, S. (2011a). Adaptive online power control scheme based on the evolutionary game theory. *IET Communications*, *5*(18), 2648–2655. DOI: 10.1049/iet-com.2011.0093

Kim, S. (2011c). Cellular network bandwidth management scheme by using nash bargaining solution. *IET Communications*, *5*(3), 371–380. DOI: 10.1049/iet-com.2010.0309

Kim, S. (2012). Adaptive Ad-hoc Network Routing Scheme by Using Incentive-based Model. *Ad-Hoc & Sensor Wireless Networks*, •••, 15.

Kim, S. (Ed.). (2014). *Game theory applications in network design*. IGI Global. DOI: 10.4018/978-1-4666-6050-2

Kim, S. (2015). QoS sensitive VANET control scheme based on feedback game model. [TIIS]. *KSII Transactions on Internet and Information Systems*, *9*(5), 1752–1767.

Kim, S. (2016a). Timed bargaining-based opportunistic routing model for dynamic vehicular ad hoc network. *EURASIP Journal on Wireless Communications and Networking*, *2016*(1), 1–9. DOI: 10.1186/s13638-016-0516-5

Kim, S. (2019a). New bargaining game model for collaborative vehicular network services. *Mobile Information Systems*, *2019*(1), 6269475. DOI: 10.1155/2019/6269475

Kim, S. (2019b). Effective crowdsensing and routing algorithms for next generation vehicular networks. *Wireless Networks*, *25*(4), 1815–1827. DOI: 10.1007/s11276-017-1632-9

Kim, S. (2021). Cooperative game based traffic control scheme for the dual-channel can platform. *IEEE Access : Practical Innovations, Open Solutions*, *9*, 34896–34905. DOI: 10.1109/ACCESS.2021.3062690

Kim, S. (2023a). Joint Crowdsensing and Offloading Algorithms for Edge-Assisted Internet of Intelligent Vehicles. *IEEE Access : Practical Innovations, Open Solutions*, *11*, 64897–64906. DOI: 10.1109/ACCESS.2023.3286851

Kim, S. (2023b). Radio resource management scheme in radar and communication spectral coexistence platform. *Computer Networks*, *230*, 109773. DOI: 10.1016/j.comnet.2023.109773

Kim, Y., Atchley, S., Vallee, G. R., Lee, S., & Shipman, G. M. (2016b). Optimizing end-to-end big data transfers over terabits network infrastructure. *IEEE Transactions on Parallel and Distributed Systems*, *28*(1), 188–201. DOI: 10.1109/TPDS.2016.2550439

Lang, P., Tian, D., Duan, X., Zhou, J., Sheng, Z., & Leung, V. C. (2022). Cooperative computation offloading in blockchain-based vehicular edge computing networks. *IEEE Transactions on Intelligent Vehicles*, *7*(3), 783–798. DOI: 10.1109/TIV.2022.3190308

Li, Y., Yang, B., Wu, H., Han, Q., Chen, C., & Guan, X. (2022). Joint offloading decision and resource allocation for vehicular fog-edge computing networks: A contract-stackelberg approach. *IEEE Internet of Things Journal*, *9*(17), 15969–15982. DOI: 10.1109/JIOT.2022.3150955

Liu, F., Masouros, C., Petropulu, A. P., Griffiths, H., & Hanzo, L. (2020a). Joint radar and communication design: Applications, state-of-the-art, and the road ahead. *IEEE Transactions on Communications*, *68*(6), 3834–3862. DOI: 10.1109/TCOMM.2020.2973976

Liu, L., Wen, X., Wang, L., Lu, Z., Jing, W., & Chen, Y. (2020b). Incentive-aware recruitment of intelligent vehicles for edge-assisted mobile crowdsensing. *IEEE Transactions on Vehicular Technology*, *69*(10), 12085–12097. DOI: 10.1109/TVT.2020.3011693

Lozano, S., Hinojosa, M. A., & Mármol, A. M. (2019). Extending the bargaining approach to DEA target setting. *Omega*, *85*, 94–102. DOI: 10.1016/j.omega.2018.05.015

Morales, M. M. C., Haw, R., Cho, E. J., Hong, C. S., & Lee, S. W. (2012). An adaptable destination-based dissemination algorithm using a publish/subscribe model in vehicular networks. *Journal of Computing Science and Engineering : JCSE*, *6*(3), 227–242. DOI: 10.5626/JCSE.2012.6.3.227

Morozs, N., Clarke, T., & Grace, D. (2015). Distributed heuristically accelerated Q-learning for robust cognitive spectrum management in LTE cellular systems. *IEEE Transactions on Mobile Computing*, *15*(4), 817–825. DOI: 10.1109/TMC.2015.2442529

Mu, X., Liu, Y., Guo, L., Lin, J., & Hanzo, L. (2022). NOMA-aided joint radar and multicast-unicast communication systems. *IEEE Journal on Selected Areas in Communications*, *40*(6), 1978–1992. DOI: 10.1109/JSAC.2022.3155524

Mubeen, S., Mäki-Turja, J., & Sjödin, M. (2014). Extending worst case response-time analysis for mixed messages in controller area network with priority and FIFO queues. *IEEE Access : Practical Innovations, Open Solutions*, *2*, 365–380. DOI: 10.1109/ACCESS.2014.2319255

Nguyen, H. T., Hoang, D. T., Luong, N. C., Niyato, D., & Kim, D. I. (2021). A hierarchical game model for OFDM integrated radar and communication systems. *IEEE Transactions on Vehicular Technology*, *70*(5), 5077–5082. DOI: 10.1109/TVT.2021.3069431

Nguyen, T. T., Elbassioni, K., Luong, N. C., Niyato, D., & Kim, D. I. (2022). Access management in joint sensing and communication systems: Efficiency versus fairness. *IEEE Transactions on Vehicular Technology*, *71*(5), 5128–5142. DOI: 10.1109/TVT.2022.3153612

Ni, Z., Zhang, J. A., Yang, K., Huang, X., & Tsiftsis, T. A. (2021). Multi-metric waveform optimization for multiple-input single-output joint communication and radar sensing. *IEEE Transactions on Communications*, *70*(2), 1276–1289. DOI: 10.1109/TCOMM.2021.3132368

Nzouonta, J., Rajgure, N., Wang, G., & Borcea, C. (2009). VANET routing on city roads using real-time vehicular traffic information. *IEEE Transactions on Vehicular Technology*, *58*(7), 3609–3626. DOI: 10.1109/TVT.2009.2014455

Oh, B., Na, Y., Yang, J., Park, S., Nang, J., & Kim, J. (2010). Genetic algorithm-based dynamic vehicle route search using car-to-car communication. *Advances in Electrical and Computer Engineering*, *10*(4), 81–86. DOI: 10.4316/aece.2010.04013

Pasandshanjani, E., Khalaj, B. H., & Moghaddam, M. S. (2011, July). A new cost function for game theoretic sir-based power control algorithms. In *2011 7th International Wireless Communications and Mobile Computing Conference* (pp. 1147-1151). IEEE. DOI: 10.1109/IWCMC.2011.5982702

Pycia, M. (2012). Stability and preference alignment in matching and coalition formation. *Econometrica*, *80*(1), 323–362. DOI: 10.3982/ECTA7143

Raazi, S. M. K. U. R., & Lee, S. (2010). A survey on key management strategies for different applications of wireless sensor networks. *Journal of Computing Science and Engineering : JCSE*, *4*(1), 23–51. DOI: 10.5626/JCSE.2010.4.1.023

Radzvilas, M. (2017). *Relative Benefit Equilibrating Bargaining Solution and the Ordinal Interpretation of Gauthier's Arbitration Scheme.*

Rawat, D. B., Popescu, D. C., Yan, G., & Olariu, S. (2011). Enhancing VANET performance by joint adaptation of transmission power and contention window size. *IEEE Transactions on Parallel and Distributed Systems*, *22*(9), 1528–1535. DOI: 10.1109/TPDS.2011.41

Rawat, D. B., Yan, G., Popescu, D. C., Weigle, M. C., & Olariu, S. (2009, September). Dynamic adaptation of joint transmission power and contention window in VANET. In *2009 IEEE 70th Vehicular Technology Conference Fall* (pp. 1-5). IEEE. DOI: 10.1109/VETECF.2009.5378793

Roberto, F. M., Júnior, J. C., & Schulzrinne, H. (2011, September). Using a symmetric game based in volunteer's dilemma to improve VANETs multihop broadcast communication. In *2011 IEEE 22nd International Symposium on Personal, Indoor and Mobile Radio Communications* (pp. 777-782). IEEE. DOI: 10.1109/PIMRC.2011.6140072

Saadia, R., & Qayum, A. (2014, January). Performance analysis of transport layer protocols for MPEG4 video traffic in IEEE 802.11 e ad-hoc wireless networks. In *Proceedings of 2014 11th International Bhurban Conference on Applied Sciences & Technology (IBCAST) Islamabad, Pakistan, 14th-18th January, 2014* (pp. 504-507). IEEE.

Sarcia, S. A. (2013, February). Timed strategic games A new game theory for managing strategic plans in the time dimension. In *2013 IEEE International Multi-Disciplinary Conference on Cognitive Methods in Situation Awareness and Decision Support (CogSIMA)* (pp. 187-194). IEEE. DOI: 10.1109/CogSIMA.2013.6523846

Su, Z., Hui, Y., & Yang, Q. (2017). The next generation vehicular networks: A content-centric framework. *IEEE Wireless Communications, 24*(1), 60–66. DOI: 10.1109/MWC.2017.1600195WC

Sun, J., Hou, F., Ma, S., & Shan, H. (2016, December). Social-aware incentive mechanism for participatory sensing. In *2016 IEEE Global Communications Conference (GLOBECOM)* (pp. 1-6). IEEE.

Tian, T., Zhang, T., Li, G., & Zhou, T. (2019). Mutual information-based power allocation and co-design for multicarrier radar and communication systems in coexistence. *IEEE Access : Practical Innovations, Open Solutions, 7*, 159300–159312. DOI: 10.1109/ACCESS.2019.2950890

Vrontos, C., Boccardi, F., Armour, S., Mellios, E., & Butler, J. (2020, May). Performance evaluation of spectrum sharing in mmWave cellular networks using ray-tracing. In *2020 IEEE Wireless Communications and Networking Conference (WCNC)* (pp. 1-6). IEEE. DOI: 10.1109/WCNC45663.2020.9120742

Wang, C., Zhang, Z., Shao, L., & Zhou, M. (2016a). Estimating travel speed via sparse vehicular crowdsensing data. In *2016 IEEE 3rd World Forum on Internet of Things (WF-IoT)* (pp. 643-648). IEEE. DOI: 10.1109/WF-IoT.2016.7845432

Wang, X., Zhao, C., Huang, T., Chakrabarti, P., & Kurths, J. (2023). Cooperative learning of multi-agent systems via reinforcement learning. *IEEE Transactions on Signal and Information Processing Over Networks, 9*, 13–23. DOI: 10.1109/TSIPN.2023.3239654

Wang, Y., Liu, Y., Zhang, J., Ye, H., & Tan, Z. (2016b). Cooperative store–carry–forward scheme for intermittently connected vehicular networks. *IEEE Transactions on Vehicular Technology, 66*(1), 777–784. DOI: 10.1109/TVT.2016.2536059

Wang, Z., Xie, X., & Zhao, D. (2012, February). Cross-layer design for TCP performance improvement in vehicular communication networks. *In 2012 14th International Conference on Advanced Communication Technology (ICACT)* (pp. 400-405). IEEE.

Xiao, L., Chen, T., Xie, C., Dai, H., & Poor, H. V. (2017). Mobile crowdsensing games in vehicular networks. *IEEE Transactions on Vehicular Technology, 67*(2), 1535–1545. DOI: 10.1109/TVT.2016.2647624

Zhang, H., Shi, Y., Wang, J., & Chen, H. (2018). A new delay-compensation scheme for networked control systems in controller area networks. *IEEE Transactions on Industrial Electronics, 65*(9), 7239–7247. DOI: 10.1109/TIE.2018.2795574

Zhang, J. A., Rahman, M. L., Wu, K., Huang, X., Guo, Y. J., Chen, S., & Yuan, J. (2021a). Enabling joint communication and radar sensing in mobile networks—A survey. *IEEE Communications Surveys and Tutorials, 24*(1), 306–345. DOI: 10.1109/COMST.2021.3122519

Zhang, K., Mao, Y., Leng, S., He, Y., & Zhang, Y. (2017). Predictive offloading in cloud-driven vehicles: Using mobile-edge computing for a promising network paradigm. *IEEE Vehicular Technology Magazine, 12*(2). Advance online publication. DOI: 10.1109/MVT.2017.2668838

Zhang, Q., Wang, X., Li, Z., & Wei, Z. (2021b). Design and performance evaluation of joint sensing and communication integrated system for 5G mmWave enabled CAVs. *IEEE Journal of Selected Topics in Signal Processing*, *15*(6), 1500–1514. DOI: 10.1109/JSTSP.2021.3109666

Zhu, L., He, D., Ai, B., Guan, K., Dang, S., Kim, J., Chung, H., & Zhong, Z. (2020). A ray tracing and joint spectrum based clustering and tracking algorithm for internet of intelligent vehicles. *Journal of Communications and Information Networks*, *5*(3), 265–281. DOI: 10.23919/JCIN.2020.9200890

Chapter 9
Non-Terrestrial Networks

ABSTRACT

Non-terrestrial network (NTN) is a wireless communication system that operates above the Earth's surface, using platforms in the air and in Earth's orbit. Currently, it has recently attracted elevated levels of interest in large-scale and ever-growing wireless communication networks through the utilization of flying objects, e.g., satellites, unmanned aerial vehicles, and drones. This chapter provides a comprehensive background on NTNs while highlighting the potential of game theoretic control approach in addressing various NTN challenges. In addition, the current development of NTN in 5G communications is discussed and the vision of the role of NTN in 6G networks is also presented.

9. OVERVIEW

Terrestrial networks form the fundamental infrastructure of modern communication systems, serving a huge number of users globally. A terrestrial network refers to a type of network that is used in various applications such as wireless sensor networks, vehicular networks, and emergency rescue and disaster relief operations. It is the backbone of modern communication systems, enabling the seamless exchange of information in our interconnected world while encompassing land-based infrastructures such as fiber optics, coaxial cables, and wireless transmission. However, terrestrial networks are facing a wide range of challenges, from coverage and reliability to interference and congestion. To address this problem, satellites and aerial platforms can be significant in the development of new telecommunications networks, allowing them to deploy networks with extremely high performance even in remote areas. Moreover, the decrease in launch costs and the use of new satellite technologies make the deployment of satellite constellations affordable.

Recently, non-terrestrial network (NTN) is recognized to play a crucial role and has emerged to be a promising solution. It is a communication network that leverages airborne, such as unmanned aerial vehicles, and space-borne vehicles, such as satellites, to facilitate ultra-reliable communications and connectivity with high data rates and low latency over expansive regions. Generally, we can understand that the terrestrial network alone cannot provide the flexibility, scalability, adaptability, and coverage required to meet the future network requirements. Nowadays, many organizations recognize the NTN as a key component to provide cost-effective and high-capacity connectivity in future 6G communications. As the demands of the 6G era are expected to be much higher, it is essential to address these challenges to ensure a robust and efficient communication infrastructure for the future.

DOI: 10.4018/979-8-3373-2013-7.ch009

9.1 Hierarchical Aerial Offload Computing (*HAOC*) scheme

Aerial access networks are anticipated to be a key technology in 6G, aimed at improving service availability in remote areas lacking terrestrial base stations. In these environments, a hierarchical structure comprising high-altitude platforms (HAPs) and unmanned aerial vehicles (UAVs) is utilized to offer aerial computing services for ground-based IoT devices. Recently, S. Kim introduced the *Hierarchical Aerial Offload Computing* (*HAOC*) scheme, which efficiently manages the constrained computational resources of both HAPs and UAVs (Kim, 2024a). The *HAOC* scheme formulates a joint resource allocation problem to optimize service quality for terrestrial IoT devices in offloading scenarios. To address this issue, it incorporates game theory principles, employing a centralized decision-making approach with decentralized execution. Within the Stackelberg-evolutionary game model, the HAP functions as the leader, determining its pricing strategy through an evolutionary learning process. Meanwhile, UAVs act as followers, making decisions about partial task offloading based on diverse objectives. By following this interactive control paradigm, the *HAOC* scheme ensures mutual benefits for HAPs, UAVs, and ground-based IoT devices while dynamically adapting to changing aerial network conditions (Kim, 2024a).

9.1.1 Development motivation

With the emergence and advancement of 5G networks, modern network systems have become an indispensable part of daily life. In recent years, a new technological surge has focused on the IoT paradigm due to the rapid proliferation of smart IoT devices. Typically, these devices are densely deployed in urban environments, including smart refrigerators, surveillance cameras, fire alarms, medical sensors, and fitness trackers. Additionally, IoT devices are also utilized in remote and challenging locations such as deserts, forests, and oceans for sensing and monitoring applications. For these remote IoT systems, fast and efficient communication is crucial for real-time data collection and assessment. In the evolution toward 6G, ubiquitous mobile technology services have garnered increasing attention (Dao, 2021; Kang, 2023; Kim, 2024a; Li, 2019).

Aerial access networks (AANs) are regarded as a complementary solution to conventional terrestrial networks, particularly for supporting smart IoT devices in remote regions where terrestrial infrastructure is either unavailable or insufficient to provide Internet services. AANs have been envisioned to enhance terrestrial communication by extending both coverage and capacity. Due to their high mobility and flexible deployment, high-altitude platforms (HAPs) and unmanned aerial vehicles (UAVs) play significant roles in AANs. HAPs are airships or aircraft positioned at an altitude of 20-25 km in the stratosphere, offering a broad range of wireless services. UAVs, commonly known as drones, are powered aerial systems that function as airborne base stations to assist in AAN operations. By leveraging their respective strengths, HAPs and UAVs can achieve synergy across various applications. Recently, industry leaders have deployed these technologies to deliver global broadband Internet, while academic research has explored the performance benefits of employing UAVs as mobile relay stations for AAN services (Joo, 2018; Kim, 2024a; Lakew, 2022; Zhang, 2022).

With the integration of HAPs and UAVs, 6G AANs have significantly expanded their coverage, effectively connecting IoT devices in remote locations. Generally, HAPs can remain stationary for extended periods, serving as stable airborne base stations with extensive coverage and strong computing capabilities. However, due to their high altitude, they cannot directly communicate with ground IoT devices over a single hop. In contrast, UAVs offer greater flexibility and cost-effectiveness in deployment, operating at lower altitudes where they can establish more reliable wireless connections with ground-based IoT devices. Nevertheless, UAVs have limited computational capacity, restricting their ability to process intensive computing tasks for IoT applications. Therefore, a collaborative approach between HAPs and UAVs is essential for the future development of AAN platforms (Jia, 2022; Kim, 2024a; Kang, 2023; Lakew, 2022).

In the IoT paradigm, vast amounts of data generated by connected devices require prompt processing. However, most IoT devices possess limited computing and storage capabilities, making it difficult to support complex applications with stringent service demands. To address this issue, mobile edge computing (MEC) has emerged as a promising solution, enabling computationally intensive IoT tasks to be offloaded to edge servers for real-time processing. By implementing MEC within hierarchical AANs, both HAPs and UAVs can function as edge computing providers, albeit with varying processing capacities. Equipped with substantial computing resources, HAPs have been introduced as effective MEC servers, facilitating efficient IoT task execution (Jia, 2022; Kang, 2023; Kim, 2024a; Tang, 2020).

As 6G AANs continue to evolve, hierarchical aerial computing systems have gained prominence, attracting research interest from both academia and industry. Within this aerial system, HAPs and UAVs serve as airborne base stations to assist terrestrial IoT devices. To optimize offloading service performance, these entities must collaborate, efficiently manage available resources, and accommodate diverse service requirements. Specifically, IoT devices offload their computational tasks to UAVs, which process them if they meet service constraints. Otherwise, the tasks are relayed to HAPs to leverage their superior computational capabilities. Effective offloading control requires balancing system efficiency and service quality, yet several open challenges remain. First, the cooperative framework between HAPs and UAVs within a hierarchical aerial platform is still underdeveloped. Second, resource allocation among HAPs and UAVs has yet to be optimized. Third, fairness in resource distribution is not adequately addressed in current frameworks. These challenges pose significant difficulties, and finding a universal solution remains elusive (Jia, 2022; Kim, 2024a; Zhang, 2022).

Recently, there has been increasing interest in applying Stackelberg game theory to model bi-level optimization problems. In this game-theoretic framework, players are asymmetrically positioned: a designated leader acts first, while subsequent followers respond to the leader's decisions. Within hierarchical systems with competing agents, Stackelberg game theory has become a standard analytical tool (Kulkarni, 2015). More recently, Stackelberg evolutionary game models have been introduced to capture interactions between an evolving leader and rational followers. In this framework, the leader drives the evolutionary process, while followers strategically adapt to achieve optimal outcomes. Through iterative interactions, both parties work toward a mutually beneficial equilibrium (Kim, 2024a; Stein, 2023).

The *HAOC* scheme introduces a novel aerial computing framework that integrates Stackelberg evolutionary game theory with a multi-objective bargaining approach. Within the hierarchical AAN structure, HAP serves as the leader, leveraging evolutionary learning to refine its pricing strategy over time. As followers, UAVs adjust their actions in response to the leader's decisions through multi-objective bargaining. By incorporating this leader-follower interaction, the *HAOC* scheme steers the AAN system toward an optimal state across diverse and dynamic network conditions (Kim, 2024a)

9.1.2 Contribution

The *HAOC* scheme introduces hierarchical AAN architecture designed to deliver edge computing services for diverse IoT applications. Building on the previous discussion and drawing inspiration from game theory, the *HAOC* scheme develops an innovative algorithm for computational resource allocation to address the hierarchical offloading decision challenge - an area that remains a complex research problem. To tackle this issue, the *HAOC* scheme proposes a novel computation offloading algorithm tailored for the hierarchical AAN platform. This framework formulates a bi-level optimization problem; a structure commonly found in Stackelberg game models. Within the Stackelberg evolutionary game framework, the *HAOC* scheme considers one HAP and multiple UAVs as asymmetric game participants. The HAP acts as an evolving leader, determining its pricing strategy through an evolutionary learning process. Meanwhile, UAVs function as followers, making autonomous decisions about the partial offloading of computing tasks received from their respective ground IoT devices. To dynamically respond to the leader's decisions, each UAV's offloading strategy is guided by a multi-objective bargaining solution. Through an interactive learning paradigm, the *HAOC* scheme facilitates reciprocal benefits for both the HAP and UAVs while achieving a well-balanced system performance, accommodating conflicting service demands (Kim, 2024a).

9.1.3 System description for the hierarchical aerial access network

The *HAOC* scheme considers a hierarchical aerial computing platform consisting of one HAP ($\mathcal{H}$), e relay UAVs i.e., $\mathbb{U} = \{\mathcal{U}_1, \ldots, \mathcal{U}_e\}$, and n terrestrial smart IoT devices, i.e., $\mathbb{D} = \{\mathcal{D}_1, \ldots, \mathcal{D}_n\}$. The H serves as a quasistatic base station in the air, and $\mathcal{U}_{1 \le j \le e} \in \mathbb{U}$ is distributed on the lower layer than the H, but within the coverage area of H. The ground IoT device $\mathcal{D}_{1 \le i \le n} \in \mathbb{D}$ generates various computing tasks that may not be completed locally due to the limited computing capability. The H and U are equipped with edge servers to provide offloading service for the D. Without loss of generality, the H is far away from the ground area, and has a stronger computation capacity than the U. Therefore, the H can assist UAVs to accomplish their offloaded computation tasks. To collaboratively complete the computation process, UAVs can serve as a relay node for the D's offloading data (Jia, 2022; Kang, 2023; Kim, 2024a).

On the ground, each terrestrial IoT device periodically produces computation-intensive tasks (τ) with distinct resource requirements. The τ is described as $\{c_\tau, d_\tau\}$ where c_τ and d_τ are the number of CPU cycles and the maximum tolerated delay, respectively (Kang, 2023). Due to the limited computing capability of IoT device, generated computing tasks are offloaded to UAV. The *HAOC* scheme assumes that each UAV flies autonomously based on a preprogrammed flight plan under the coverage of the HAP with no overlapping coverage area with other UAVs (Lakew, 2022). Therefore, each individual IoT device is associated with one UAV. Let $\mathbb{D}_{\mathcal{U}_j}$ denote the set IoT devices under the coverage of $\mathcal{U}_j$ where $\mathcal{U}_j \in \mathbb{U}$. When the $\mathcal{D}_i \in \mathbb{D}_{\mathcal{U}_j}$ generates an offloading task (τ), it first asks the resource to the $\mathcal{U}_j$, which hovers over its covering area (Kang, 2023; Kim, 2024a)

After collecting the offloading tasks, the $\mathcal{U}_j$ attempt to execute them within the delay requirement. Even though the $\mathcal{U}_j$ serves as an edge server, it has limited computing capability, i.e., $\mathfrak{M}_{\mathcal{U}_j}$, and the H has more powerful computing capability, i.e., $\mathcal{E}_{\mathcal{H}}$. According to the current task execution model, computing tasks can be partially offloaded to offer more flexibility; offloading task bits are bit-wise independent and can be arbitrarily partitioned to facilitate distributed parallel executions in the HAP and UAVs, simultaneously. Therefore, a part of offloading tasks can be offloaded to the H while minimizing the total processing delay. In the viewpoint of $\mathcal{U}_j$, the major challenge is to effectively decide the portion of the computation data bits to be offloaded to the H. From the H's point of view, the main goal is to optimize the usage of $\mathcal{E}_{\mathcal{H}}$ for its corresponding UAVs. In order to appropriately adjust the two aims, task offloading policy should be associated with the current conditions of $\mathscr{U}_j$ and H. The hierarchical aerial computing infrastructure is illustrated in Figure 1 (Jia, 2022; Kang, 2023; Kim 2024a; Lakew, 2022).

Figure 1. A hierarchical aerial computing infrastructure (Kim, 2024a)

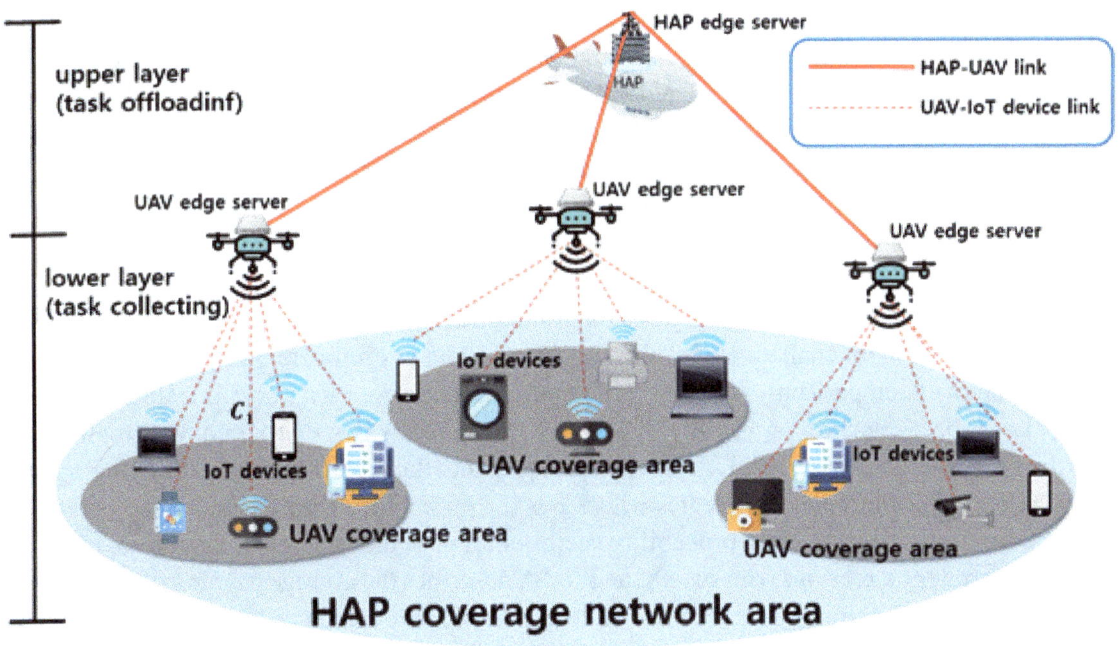

The relay offloading task amount of $\mathscr{U}_j$ is defined by two factors $\mathfrak{R}_{\mathscr{U}_j}$ and $\beta^{\mathscr{H}}_{\mathscr{U}_j}$; $\mathfrak{R}_{\mathscr{U}_j}$ is the total amount of collected computation tasks from devices in the $\mathbb{D}_{\mathscr{U}_j}$, and $\beta^{\mathscr{H}}_{\mathscr{U}_j}$ is the offload ratio from the $\mathscr{U}_j$ to the H. Therefore, the $\left(\beta^{\mathscr{H}}_{\mathscr{U}_j} \times \mathfrak{R}_{\mathscr{U}_j}\right)$ task amount is remotely executed by the H and the $\left(\left(1 - \beta^{\mathscr{H}}_{\mathscr{U}_j}\right) \times \mathfrak{R}_{\mathscr{U}_j}\right)$ task amount is locally executed by the $\mathscr{U}_j$. To proceed the remote computation process, a service price is charged by the H. Simply, the price is assumed as an access fee per unit of data bits for offloading services. Needless to say, setting an appropriate price is critical to the hierarchical AAN system. If the price is excessively low, it will cause excessive offloading requests from the UAVs. If a price is unduly high, it will inevitably result in reduced offloading requests. Therefore, the H should develop an effective pricing mechanism to perfectly strike a well-balanced AAN performance. Based on the H's price strategy, individual UAVs locally make their offloading decisions to maximize their profits by considering a tradeoff between service quality and cost (Kim, 2024a; Wang, 2022).

9.1.4 Stackelberg evolutionary game and bargaining solution

The game-theoretical control approach has been widely explored and is recognized as one of the most significant open challenges in scientific research. The *HAOC* scheme adopts the Stackelberg evolutionary game model, incorporating learning algorithms to analyze the interactions among intelligent system agents. This game-theoretic framework enables an understanding of dynamic control scenarios where agents' behaviors evolve over time through a bi-level interactive process, making it particularly suitable for hierarchical AAN systems (Kim, 2024a; Traulsen, 2023).

The Stackelberg evolutionary game model merges elements of classical Stackelberg game theory with evolutionary game dynamics to model interactions between an evolving leader and rational followers. In a traditional Stackelberg game, the leader takes the initial action, while multiple followers observe this decision and adjust their responses accordingly. Due to its hierarchical nature, the Stackelberg game serves as a fundamental example of a bi-level optimization problem (Stein, 2023). Within the Stackelberg evolutionary game framework, the HAP and UAVs are strategic players, where the HAP functions as the leader and UAVs act as followers. Each participant formulates its decisions while considering their impact on the actions and responses of the other players (Kim, 2024a).

Initially, evolutionary games centered around the stable evolutionary strategy as a likely outcome of evolution by natural selection. All developed concepts derived from the evolutionary stable strategy can drive changes in strategy frequencies, in contrast to the rational choice of the classical non-cooperative games (Stein, 2023). In the evolutionary game theory, replicator dynamics is a mathematical model to select a strategy. The main challenge of replicator dynamics is to eliminate dominated strategies over time based on payoff comparisons. A finite set of m pure strategies $\{s_1, s_2, \ldots, s_m\}$ is available for a player to play. Let x_i be a selection probability of strategy $s_{1\leq i\leq m}$, and A is the $m \times m$ payoff matrix, which has entries $s_{ij} = \mathscr{G}(s_i, s_j)$ for $i,j = 1,\ldots,m$. The s_{ij} represents the payoff to s_i when playing against s_j. $X = \{x_1, x_2, \ldots, x_m\}$ is the state of strategies where x_i is the probability with which strategy s_i is played. Therefore, X expresses the selection probability vector of all possible strategies. Therefore, $(\mathbb{A}X)_i$ is the payoff that the strategy s_i receives with state X, and $X \cdot \mathbb{A}X$ describes the average payoff in the evolutionary game. The growth rate of strategy s_i's selection strategy, i.e., $\hat{x}_i$, is equal to the difference between the s_i's current payoff and the average payoff; it is given as follows (Kim, 2024a; Sigmund, 2011; Tuyls, 2005);

$$\hat{x}_i = \left[(\mathbb{A}X)_i - X \cdot \mathbb{A}X\right] \times x_i \tag{9.1.1}$$

To measure the evolutionary success of a strategy, the *HAOC* scheme should consider the frequency of the strategies. Simply, the *HAOC* scheme can think that X is the m-dimensional simplex (Δ) of mixed strategies played by a fictitious player where $\Delta = \left\{X \in \mathbb{R}_+^m \mid \sum_{k=1}^{m} x_i = 1\right\}$ (Tuyls. Formally, the Stackelberg evolutionary game can provide a framework for modelling and managing a hierarchical evolving control system according to the interactive feedback mechanism. By taking into account the ecological dynamics, the leader selects his Stackelberg strategy to get a better outcome, and the corresponding followers adapt their rational strategies while converging to a stable system status based on the multi-dimensional simplex of mixed strategies (Kim, 2024a; Stein, 2023).

In recent years, cooperative control mechanisms have garnered significant attention. Traditional game theory operates under the assumption that players act as fully non-cooperative agents. However, this assumption may not be entirely realistic, especially in scenarios where players engage in long-term decision-making. In such cases, it is more natural for players to form coalitions, allowing them to take strategic actions that benefit all coalition members collectively. Currently, there is growing interest in cooperative game theory and bargaining solutions, which have become integral to coordinated decision-making processes. This approach is also applicable to hierarchical AAN control challenges, offering a structured framework for optimizing cooperative interactions among system agents (Avrachenkov, 2015; Kim, 2024a).

Multi-objective bargaining game theory extends classical cooperative game theory by addressing scenarios where players make decisions based on multiple, often conflicting, objectives. Unlike traditional cooperative games, multi-objective bargaining introduces challenges not only due to conflicts among players but also due to competing objectives within the decision-making process. The *HAOC* scheme incorporates the principles of cooperative game theory into the multi-objective optimization framework. Specifically, the multi-objective game model is decomposed into two single-objective subgames, allowing for the resolution of conflicts between different objectives. Using the same input data, each subgame independently determines its outcome, which is then integrated into a single value. This transformation effectively converts the multi-objective problem into a single-objective optimization task, simplifying decision-making and enhancing computational efficiency (Kim, 2024a; Wang, 2021).

Fortunately, several highly effective bargaining solutions exist to address conflicts of interest among game players. These solutions aim to determine a payoff allocation that all players unanimously agree upon while ensuring *Pareto optimality*. Depending on the desired properties of the final agreement, different bargaining frameworks have been proposed, including the *Nash Bargaining Solution* (*NBS*) and the *Kalai-Smorodinsky Bargaining Solution* (*KSBS*). The *NBS* is a widely recognized fair allocation rule that maximizes the product of players' payoffs. A key characteristic of *NBS* is its *independence of irrelevant alternatives* (*IIA*), meaning that if an *irrelevant alternative* is introduced, the relative preferences of players remain unchanged. In contrast, the defining feature of *KSBS* is *individual monotonicity* (*IM*), which ensures that if the bargaining set expands in a way that favors a particular player, that player will always benefit from the change. These bargaining solutions provide structured approaches for achieving fair and efficient outcomes in cooperative game scenarios. To define bargaining solutions, let S be the feasible solution set and r is an available solution. r^0 is the disagreement point that players can obtain in the absence of bargaining. Mathematically, the *NBS* for two players is given by (Ebrahimkhani, 2021; Kim, 2024a, Wang, 2021);

$$NBS = \max_{r\in S,\, r\geq r^0}\left(\sum_{i=1}^{2}\log\left(r_i - r_i^0\right)\right), \text{ s.t.}, r = \left(r_1, r_{2,}\right) \text{ and } r^0 = \left(r_1^0, r_2^0\right) \tag{9.1.2}$$

Formally, the *KSBS* for two players is given by (Ebrahimkhani, 2021);

$$KSBS = \max_{r\in S,\, r\geq r^0} \mathcal{Q}, \text{ s.t.}, \mathcal{Q} = \frac{r_i - r_i^0}{r_i^* - r_i^0}, i = 1,2 \tag{9.1.3}$$

where r_i^* is the ideal solution for player i; it is the upper bound for r_i.

9.1.5 Hierarchical offload computing model

In the *HAOC* scheme, they develop a game G, which is designed as a hierarchical Stackelberg model. In the upper-level, the H works as a leader and selects his price strategy ($p^{\mathcal{H}}$) from the S, which is represented by $\mathbb{S} = \{p_1^{\mathcal{H}},\ldots,p_k^{\mathcal{H}}\}$. To select the strategy, the H dynamically learns his best price through the evolutionary game approach. In the lower-level, multiple Us work as followers and make their offloading decisions based on the $p^{\mathcal{H}}$. Based on the different performance criteria, each individual U makes a partial offloading decision by using the multi-objective bargaining solution. As a consequence, H and Us is sequentially interacted with each other; it is noteworthy that the *HAOC* scheme formulates the H-Us association in a coordinated manner. Formally, the *HAOC* scheme defines the G game tuple entities, i.e., $\mathbb{G} = \left\{\mathcal{H},\mathbb{U},\mathbb{D},\mathbb{M}^{upper},\mathbb{M}^{lower}_{\mathcal{U}_{1\le j\le e}},T\right\}$ where $\mathbb{M}^{upper} = \{\mathcal{H},\mathbb{S},\mathfrak{U}_{\mathcal{H}}(\cdot)\}$ and $\mathbb{M}^{lower}_{\mathcal{U}_j} = \left\{\mathcal{U}_j \in \mathbb{U}|\mathbb{D}_{\mathcal{U}_j},\left(\mathfrak{U}^{QoS}_{\mathcal{U}_j}(\cdot),\mathfrak{U}^{QoE}_{\mathcal{U}_j}(\cdot)\right),\beta^{\mathcal{H}}_{\mathcal{U}_j}\right\}$ (Kim, 2024a).

- H, U and D represent one HAP, the set of UAVs and the set of IoT devices, respectively, and the H is a leader, and $\mathcal{U}_j \in \mathbb{U}$ are followers; they are asymmetric game players in the G.
- Discrete time model $T \in \{t_1,\ldots,t_c,t_{c+1},\ldots\}$ is represented by a sequence of time steps. The length of t_c matches the event timescale of G.
- $\mathbb{M}^{upper}$ and $\mathbb{M}^{lower}_{\mathcal{U}_j}$ are the upper and lower-level control processes.
- In the $\mathbb{M}^{upper}$, the H select his strategy from the strategy set $\mathbb{S} = \{p_1^{\mathcal{H}},\ldots,p_k^{\mathcal{H}}\}$ where $p^{\mathcal{H}}_{1\le i\le k}$ represents the predefined ith strategy. $\mathfrak{U}_{\mathcal{H}}(\cdot)$ represents the H's utility function.
- Each individual $\mathcal{U}_j \in \mathbb{U}$ independently operates its $\mathbb{M}^{lower}_{\mathcal{U}_j}$ in a distributed parallel fashion.
- In the $\mathbb{M}^{lower}_{\mathcal{U}_j}$, the $\mathcal{U}_j$ has two different utility functions where $\mathfrak{U}^{QoS}_{\mathcal{U}}(\cdot)$ is a function for QoS criterion and $\mathfrak{U}^{QoE}_{\mathcal{U}_j}(\cdot)$ is a function for QoE criterion. The $\beta^{\mathcal{H}}_{\mathcal{U}_j}$ is the $\mathcal{U}_j$'s strategy, which is the offload ratio from the $\mathcal{U}_j$ to the H.
- In each sequence of time steps, the H and $\mathcal{U}_{1\le j\le e}$ are reciprocally interdependent, and work together.

9.1.6 The resource allocation for the hierarchical AAN system

A bi-level control problem is a unique type of control problem in which one control process is embedded within another. The higher-level control problem is managed by an upper-level algorithm, while the lower-level algorithm addresses the inner control task. This concept was first introduced in game theory by H. von Stackelberg, who proposed a hierarchical structure as an effective approach for solving bi-level optimization problems. Within this framework, decision-making at each level occurs interactively, ensuring that the system converges toward a mutually beneficial solution (Kim, 2024a; Stein, 2023)

For the upper-level decision-making process, evolutionary dynamics can be applied. The *HAOC* scheme introduces an innovative evolutionary game model to address the offloading problem within the hierarchical AAN platform. It is important to highlight that while the Stackelberg evolutionary game approach is influenced by prior research (Stein, 2023), the *HAOC* scheme employs distinct formulations and solution strategies to enhance its effectiveness. At the upper-level, the H works as a leader and selects his price strategy from the S. For the H, its utility function, i.e., $\mathfrak{U}_{\mathcal{H}}(\cdot)$, is given by (Kim, 2024a);

$$\mathfrak{U}_{\mathcal{H}}\left(\mathbb{D},\left(\beta^{\mathcal{H}}_{\mathcal{U}},\mathfrak{R}_{\mathcal{U}}\right),p_l^{\mathcal{H}},\mathcal{E}_{\mathcal{H}}\right) = \max_{p_l^{\mathcal{H}}\in\mathbb{S}}\left\{\min\left[\sum_{\mathcal{U}_j\in\mathbb{D}}\left(\beta^{\mathcal{H}}_{\mathcal{U}_j}\times\mathfrak{R}_{\mathcal{U}_j}\right),\mathcal{E}_{\mathcal{H}}\right]\right\}\times p_l^{\mathcal{H}} \tag{9.1.4}$$

The $\mathfrak{U}_{\mathcal{H}}(\cdot)$ function can be interpreted as follows: i) providing total offload service cannot be larger than the H's computing capacity $\left(\mathcal{E}_{\mathcal{H}}\right)$, and ii) $p_l^{\mathcal{H}}$ is selected in the S to maximize the profit of H while inducing the UAVs's offloading. To obtain the selection probability of each H's strategy, the *HAOC* scheme should calculate the difference between the current strategy's payoff and the average payoff. Let $X_{\mathcal{H}} = \{x_1^{\mathcal{H}}, x_2^{\mathcal{H}}, \ldots, x_k^{\mathcal{H}}\}$ be the k-dimentional simplex of mixed strategies played by the $\mathcal{H}$. Let $\mathbb{A}_{\mathcal{H}}$ be the $k \times k$ utility matrix, which has entries $\Gamma_{ij} = \mathcal{G}\left(\mathfrak{U}_i^{\mathcal{H}}(\cdot), \mathfrak{U}_j^{\mathcal{H}}(\cdot)\right)$ for $i,j = 1,\ldots,k$ where $\Gamma_{ij} = \mathfrak{U}_i^{\mathcal{H}}(\cdot) - \mathfrak{U}_j^{\mathcal{H}}(\cdot)$. In these notations, $x_{1\le i\le k}^{\mathcal{H}}$ is the selection probability for the $p_i^{\mathcal{H}}$, and $\hat{x}_i^{\mathcal{H}}$ stands for the variation of x_i; it is the replicator dynamics for the strategy $p_i^{\mathcal{H}}$ (Kim, 2024a).

$$\hat{x}_i^{\mathcal{H}} = \left[(\mathbb{A}_{\mathcal{H}} X_{\mathcal{H}})_i - X_{\mathcal{H}} \cdot \mathbb{A}_{\mathcal{H}} X_{\mathcal{H}}\right] \times x_i^{\mathcal{H}} \tag{9.1.5}$$

$$= \left(J(p_i^{\mathcal{H}}, X_{\mathcal{H}}) - \sum_{j=1}^{k} \left(x_j \times J(p_j^{\mathcal{H}}, X_{\mathcal{H}})\right)\right) \times x_i^{\mathcal{H}}$$

$$= \left(\sum_{j=1}^{k} x_j^{\mathcal{H}} \times J(p_i^{\mathcal{H}}, p_j^{\mathcal{H}}) - \sum_{j=1}^{k} \sum_{l=1}^{k} x_j^{\mathcal{H}} \times J(p_j^{\mathcal{H}}, p_l^{\mathcal{H}}) \times x_l^{\mathcal{H}}\right)$$

where $J\left(p_i^{\mathcal{H}}, p_j^{\mathcal{H}}\right)$ is denoted by the relative payoff for the H using strategy $p_i^{\mathcal{H}}$ when comparing with the strategy $p_j^{\mathcal{H}}$, and $J(p_i^{\mathcal{H}}, X_{\mathcal{H}})$ is the relative payoff for the H using strategy $p_i^{\mathcal{H}}$ when comparing with the rest strategies distributed in $X_{\mathcal{H}}$, which can be expressed like as $\sum_{j=1}^{k} J\left(p_i^{\mathcal{H}}, p_j^{\mathcal{H}}\right) \times x_j^{\mathcal{H}}$ (Kim, 2011, 2024a).

At the lower-level, each individual $\mathcal{U}_{1\le j\le m}$ works as a follower and makes offloading decisions based on the leader's price decision ($p^{\mathcal{H}}$). At each time period, the $\mathcal{U}_j$ collects the offload tasks from its corresponding ground devices and proceeds with the offloading services. According to different service requests, the *HAOC* scheme can define two performance criteria; throughput based QoS, and delay associated with QoE. Therefore, the $\mathcal{U}_j$ has two different utility functions, i.e., $\mathfrak{U}_{\mathcal{U}_j}^{QoS}(\cdot)$ and $\mathfrak{U}_{\mathcal{U}_j}^{QoE}(\cdot)$, for QoS and QoE criteria, respectively. The $\mathfrak{U}_{\mathcal{U}_j}^{QoS}(\cdot)$ is defined as the sum of two QoS functions for the local and remote offloading services. Based on the $\mathfrak{R}_{\mathcal{U}_j}$ and $\beta_{\mathcal{U}_j}^{\mathcal{H}}$ values, the local QoS function of $\mathcal{U}_j$ $\left(\mathcal{P}_{\mathcal{U}_j}^{L}(\cdot)\right)$ is estimated as follows (Kim, 2024a);

$$\mathcal{P}_{\mathcal{U}_j}^{L}\left(\beta_{\mathcal{U}_j}^{\mathcal{H}}, \mathfrak{M}_{\mathcal{U}_j}, \mathfrak{R}_{\mathcal{U}_j}\right) = \left(exp\left(\omega \times \frac{\min\left(\left(1-\beta_{\mathcal{U}_j}^{\mathcal{H}}\right) \times \mathfrak{R}_{\mathcal{U}_j}, \mathfrak{M}_{\mathcal{U}_j}\right)}{\mathfrak{M}_{\mathcal{U}_j}}\right)\right)^{\varepsilon} - \theta \tag{9.1.6}$$

where ω, ε and θ are control parameters for the $\mathcal{P}_{\mathcal{U}_j}^{L}(\cdot)$. Simply, the $\mathcal{P}_{\mathcal{U}_j}^{L}(\cdot)$ function increases monotonically in proportion to the offloading task amount, i.e., $\left(1-\beta_{\mathcal{U}_j}^{\mathcal{H}}\right) \times \mathfrak{R}_{\mathcal{U}_j}$. Based on the relay task amount, the re-offloaded QoS function $\left(\mathcal{Q}_{\mathcal{U}_j}^{\mathcal{H}}(\cdot)\right)$ is estimated as follows (Kim, 2024a);

$$\mathcal{Q}_{\mathcal{U}_j}^{\mathcal{H}}\left(p^{\mathcal{H}}, \beta_{\mathcal{U}_j}^{\mathcal{H}}, \mathfrak{R}_{\mathcal{U}_j}, A_{\mathcal{U}_j}^{\mathcal{H}}\right) = \frac{\min\left(\beta_{\mathcal{U}_j}^{\mathcal{H}} \times \mathfrak{R}_{\mathcal{U}_j}, A_{\mathcal{U}_j}^{\mathcal{H}}\right)}{p^{\mathcal{H}} \times \left(\beta_{\mathcal{U}_j}^{\mathcal{H}} \times \mathfrak{R}_{\mathcal{U}_j}\right)} \times \log\left(\frac{\min\left(\beta_{\mathcal{U}_j}^{\mathcal{H}} \times \mathfrak{R}_{\mathcal{U}_j}, A_{\mathcal{U}_j}^{\mathcal{H}}\right)}{\beta_{\mathcal{U}_j}^{\mathcal{H}} \times \mathfrak{R}_{\mathcal{U}_j}} + \phi\right) \tag{9.1.7}$$

$$
\text{s.t.,}\begin{cases} A^{\mathcal{H}}_{\mathcal{U}_j} = \left(\beta^{\mathcal{H}}_{\mathcal{U}_j} \times \mathfrak{R}_{\mathcal{U}_j}\right), if \mathcal{E}_{\mathcal{H}} \geq \sum_{\mathcal{U}_j \in \mathrm{U}} \left(\beta^{\mathcal{H}}_{\mathcal{U}_j} \times \mathfrak{R}_{\mathcal{U}_j}\right) \\ A^{\mathcal{H}}_{\mathcal{U}_j} = \dfrac{\left(\beta^{\mathcal{H}}_{\mathcal{U}_j} \times \mathfrak{R}_{\mathcal{U}_j}\right)}{\sum_{\mathcal{U}_r \in \mathrm{U}} \left(\beta^{\mathcal{H}}_{\mathcal{U}_r} \times \mathfrak{R}_{\mathcal{U}_r}\right)} \times \mathcal{E}_{\mathcal{H}}, otherwise \end{cases}
$$

where ϕ is a control parameter for the $\mathcal{Q}^{\mathcal{H}}_{\mathcal{U}_j}(\cdot)$. The $\mathcal{Q}^{\mathcal{H}}_{\mathcal{U}_j}(\cdot)$ function can be interpreted as that if the total offloading requests from UAVs, i.e., $\sum_{\mathcal{U}_j \in \mathrm{U}} \left(\beta^{\mathcal{H}}_{\mathcal{U}_j} \times \mathfrak{R}_{\mathcal{U}_j}\right)$, is smaller than the $\mathcal{E}_{\mathcal{H}}$, the output is proportional to the actually processed offloading service. According to (9.1.6) and (9.1.7), the total QoS utility function $\mathfrak{U}^{QoS}_{\mathcal{U}_j}(\cdot)$ is defined as the sum of $\mathcal{P}^{L}_{\mathcal{U}_j}(\cdot)$ and $\mathcal{Q}^{\mathcal{H}}_{\mathcal{U}_j}(\cdot)$ functions (Kim, 2024a);

$$
\mathfrak{U}^{QoS}_{\mathcal{U}_j}\left(\beta^{\mathcal{H}}_{\mathcal{U}_j}, \mathcal{P}^{L}_{\mathcal{U}_j}(\cdot), \mathcal{Q}^{\mathcal{H}}_{\mathcal{U}_j}(\cdot)\right) = \mathcal{P}^{L}_{\mathcal{U}_j}\left(\beta^{\mathcal{H}}_{\mathcal{U}_j}, \mathfrak{M}_{\mathcal{U}_j}, \mathfrak{R}_{\mathcal{U}_j}\right) + \mathcal{Q}^{\mathcal{H}}_{\mathcal{U}_j}\left(p^{\mathcal{H}}, \beta^{\mathcal{H}}_{\mathcal{U}_j}, \mathfrak{R}_{\mathcal{U}_j}, A^{\mathcal{H}}_{\mathcal{U}_j}\right) \tag{9.1.8}
$$

According to the delay association, the $\mathfrak{U}^{QoE}_{\mathcal{U}_j}(\cdot)$ is also defined as the sum of two QoE functions for the local and remote offloading services. With the same function factors of $\mathfrak{U}^{QoS}_{\mathcal{U}_j}(\cdot)$, the local QoE satisfaction function of $\mathcal{U}_j$ $\left(\mathcal{F}^{L}_{\mathcal{U}_j}(\cdot)\right)$ is estimated as follows (Kim, 2024a);

$$
\mathcal{F}^{L}_{\mathcal{U}_j}\left(\beta^{\mathcal{H}}_{\mathcal{U}_j}, \mathfrak{M}_{\mathcal{U}_j}, \mathfrak{R}_{\mathcal{U}_j}\right) = \gamma \times \log\left(exp\left(\xi \times \frac{\min\left(\left(1-\beta^{\mathcal{H}}_{\mathcal{U}_j}\right) \times \mathfrak{R}_{\mathcal{U}_j}, \mathfrak{M}_{\mathcal{U}_j}\right)}{\mathfrak{M}_{\mathcal{U}_j}}\right)\right) \tag{9.1.9}
$$

where γ and ξ are control parameters for the $\mathcal{F}^{L}_{\mathcal{U}_j}(\cdot)$. The $\mathcal{F}^{L}_{\mathcal{U}_j}(\cdot)$ function is designed that the output increases in proportion to the local process ratio. Based on the H's price decision ($p^{\mathcal{H}}$), the remote QoE payoff function $\left(\mathcal{J}^{\mathcal{H}}_{\mathcal{U}_j}(\cdot)\right)$ is estimated as follows (Kim, 2024a);

$$
\mathcal{J}^{\mathcal{H}}_{\mathcal{U}_j}\left(p^{\mathcal{H}}, \beta^{\mathcal{H}}_{\mathcal{U}_j}, \mathfrak{R}_{\mathcal{U}_j}, A^{\mathcal{H}}_{\mathcal{U}_j}\right) = \left(\varrho^{\left(\frac{\min\left(\beta^{\mathcal{H}}_{\mathcal{U}_j} \times \mathfrak{R}_{\mathcal{U}_j}, A^{\mathcal{H}}_{\mathcal{U}_j}\right)}{\beta^{\mathcal{H}}_{\mathcal{U}_j} \times \mathfrak{R}_{\mathcal{U}_j} \times (\sigma + p^{\mathcal{H}})}\right)} - \eta\right) \tag{9.1.10}
$$

where ϱ, σ and η are control parameters for the $\mathcal{J}^{\mathcal{H}}_{\mathcal{U}_j}(\cdot)$. Simply, the output of $\mathcal{J}^{\mathcal{H}}_{\mathcal{U}_j}(\cdot)$ is strongly related to the H's price policy. According to (9.1.9) and (9.1.10), the total QoE utility function $\mathfrak{U}^{QoE}_{\mathcal{U}_j}(\cdot)$ is defined as the sum of $\mathcal{F}^{L}_{\mathcal{U}_j}(\cdot)$ and $\mathcal{J}^{\mathcal{H}}_{\mathcal{U}_j}(\cdot)$ functions (Kim, 2024a);

$$
\mathfrak{U}^{QoE}_{\mathcal{U}_j}\left(\beta^{\mathcal{H}}_{\mathcal{U}_j}, \mathcal{F}^{L}_{\mathcal{U}_j}(\cdot), \mathcal{J}^{\mathcal{H}}_{\mathcal{U}_j}(\cdot)\right) = \mathcal{F}^{L}_{\mathcal{U}_j}\left(\beta^{\mathcal{H}}_{\mathcal{U}_j}, \mathfrak{M}_{\mathcal{U}_j}, \mathfrak{R}_{\mathcal{U}_j}\right) + \mathcal{J}^{\mathcal{H}}_{\mathcal{U}_j}\left(p^{\mathcal{H}}, \beta^{\mathcal{H}}_{\mathcal{U}_j}, \mathfrak{R}_{\mathcal{U}_j}, A^{\mathcal{H}}_{\mathcal{U}_j}\right) \tag{9.1.11}
$$

For the computing resource allocation problem in the AAN platform, the *HAOC* scheme should decide the $\beta^{\mathcal{H}}_{\mathcal{U}_j}$ value to adaptively share the computation resources of H and U. For this decision process, the *HAOC* scheme develops a new multi-objective bargaining game model through the $\mathfrak{U}^{QoS}_{\mathcal{U}_j}(\cdot)$ and $\mathfrak{U}^{QoE}_{\mathcal{U}_j}(\cdot)$

functions. In the bargaining model, the H and U are game players, and the QoS and QoE criteria are jointly considered through two different bargaining problems. As a method to evaluate the QoS, the *HAOC* scheme adopts the *NBS* as the solution concept for the $\mathfrak{U}^{QoS}_{\mathcal{U}_j}(\cdot)$ function. To measure the QoE, the idea of *KSBS* is employed for the $\mathfrak{U}^{QoE}_{\mathcal{U}_j}(\cdot)$ function. Based on the preferences of players with respect to their own criteria, the *HAOC* scheme finally defines a new multi-objective bargaining solution, i.e., $MCBS^{\mathcal{H}}_{\mathcal{U}_j}$, as an average of *NBS*'s $\beta^{\mathcal{H}}_{\mathcal{U}_j}$ value, i.e., $\left(\beta^{\mathcal{H}}_{\mathcal{U}_j}\right)_{NBS}$, and *KSBS*'s $\beta^{\mathcal{H}}_{\mathcal{U}_j}$ value, i.e., $\left(\beta^{\mathcal{H}}_{\mathcal{U}_j}\right)_{KSBS}$ (Kim, 2024a);

$$MCBS^{\mathcal{H}}_{\mathcal{U}_j} = \beta^{\mathcal{H}}_{\mathcal{U}_j} = \frac{\left(\beta^{\mathcal{H}}_{\mathcal{U}_j}\right)_{NBS} + \left(\beta^{\mathcal{H}}_{\mathcal{U}_j}\right)_{KSBS}}{2} \tag{9.1.12}$$

$$\text{s.t.,}\begin{cases} NBS^{\mathcal{H}}_{\mathcal{U}_j} = \max\limits_{0 \le \left(\beta^{\mathcal{H}}_{\mathcal{U}_j}\right)_{NBS} \le 1} \left(\log\left(\mathcal{P}^{L}_{\mathcal{U}_j}(\cdot) - d^{L}_{\mathcal{U}_j}\right) + \log\left(\mathcal{Q}^{\mathcal{H}}_{\mathcal{U}_j}(\cdot) - d^{L}_{\mathcal{U}_j}\right)\right) \\ KSBS^{\mathcal{H}}_{\mathcal{U}_j} = \max\limits_{0 \le \left(\beta^{\mathcal{H}}_{\mathcal{U}_j}\right)_{KSBS} \le 1} \left(\mathcal{U}_{j,w} \in \mathbb{U} \left| \frac{\mathcal{F}^{L}_{\mathcal{U}_j}(\cdot) - d^{L}_{\mathcal{U}_j}}{\mathcal{F}^{L*}_{\mathcal{U}_j}(\cdot) - d^{L}_{\mathcal{U}_j}} = \frac{\mathcal{J}^{\mathcal{H}}_{\mathcal{U}_w}(\cdot) - d^{L}_{\mathcal{U}_w}}{\mathcal{J}^{\mathcal{H}*}_{\mathcal{U}_w}(\cdot) - d^{L}_{\mathcal{U}_w}}\right.\right) \end{cases}$$

The diagram of proposed hierarchical Stackelberg game model is illustrated in Figure 2. Based on the interactive control paradigm, the H works as a leader and selects the $p^{\mathcal{H}}$ at the upper level. At the lower level, multiple Us work followers and select their $\beta^{\mathcal{H}}_{\mathcal{U}}$ values (Kim, 2024a).

Figure 2. The diagram of proposed hierarchical Stackelberg game model (Kim, 2024a)

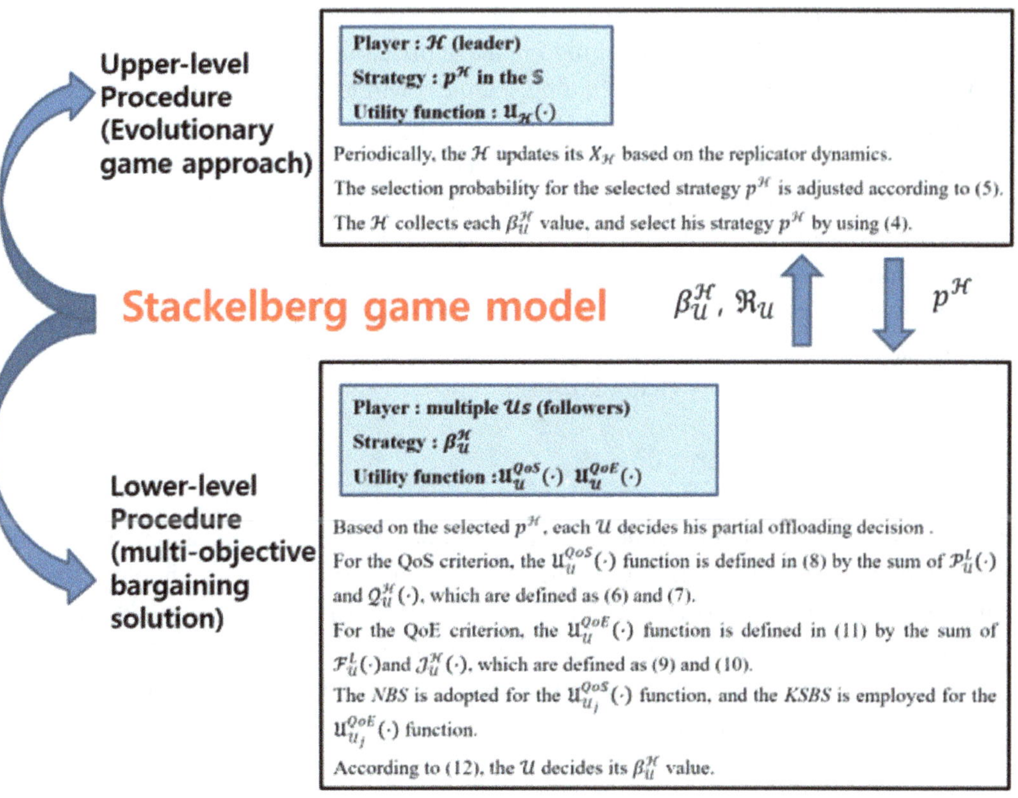

9.1.7 Summary

To enable broadband 6G wireless communications, HAPs and UAVs function as aerial base stations, forming a hybrid aerial-terrestrial network. Additionally, with the rapid increase in IoT devices' computational demands, the MEC paradigm has emerged as an effective solution. The *HAOC* scheme explores a hierarchical aerial computing system comprising HAPs, UAVs, and ground-based IoT devices. For offloading computation-intensive tasks, the *HAOC* scheme develops an innovative joint resource allocation algorithm designed to maximize the number of successfully processed tasks within the hierarchical AAN platform. Leveraging a hierarchical AAN structure, the *HAOC* scheme incorporates the Stackelberg evolutionary game model. In this framework, the HAP acts as the leader, determining its pricing strategy through an evolutionary learning process and progressively refining it to reach an optimal strategy. Based on the leader's decision, multiple UAVs, acting as followers, strive to maximize their individual profits. To balance QoS and QoE, each UAV adopts a novel multi-objective bargaining approach. Through a bi-level interactive control process, the evolving leader (HAP) and rational followers (UAVs) engage in dynamic negotiations, ensuring mutual benefits while adapting to fluctuating hierarchical AAN conditions (Kim, 2024a).

9.2 Collaborative Game based Task Offloading (*CGTO*) scheme

In 6G wireless networks, the deployment of unmanned aerial vehicles (UAVs) and tethered balloons (TBs) to support cellular networks has gained significant attention due to their rapid and flexible deployment at a relatively low cost. Recently, S. Kim introduced the *Collaborative Game-based Task Offloading* (*CGTO*) scheme, designed for smart devices operating in modern battlefield environments (Kim, 2024b). By leveraging an integrated platform consisting of TBs, UAVs, and smart devices, the *CGTO* scheme addresses two key challenges: (i) developing a task-splitting algorithm for partial offloading services and (ii) designing a TB resource-sharing algorithm to manage offloading requests efficiently. In the *CGTO* scheme, UAVs function as relay nodes, facilitating seamless wireless communication between TBs and individual smart devices. To achieve an optimal and mutually beneficial solution, the *CGTO* scheme is structured as a cooperative game model. First, the *sequential Raiffa bargaining solution* is employed to distribute the computational workload of each smart device in the battlefield. Second, the *average-surplus value* is utilized to allocate TB computing resources effectively. By integrating these two cooperative game-theoretic approaches, the *CGTO* scheme models the sequential interactions between TBs, UAVs, and battlefield devices, jointly optimizing the control mechanism for offloading services. Through this synergistic approach, the hybrid UAV-TB-assisted battlefield network infrastructure achieves a fair and efficient resource allocation strategy (Kim, 2024b).

9.2.1 Development motivation

The 6G wireless network system is anticipated to integrate terrestrial, aerial, and maritime communications to provide highly reliable, ultra-low latency services. In particular, 6G technology is expected to revolutionize customer services and applications through the IoT paradigm. As networks evolve towards full intelligence and autonomy, IoT advancements are being driven by cutting-edge technologies such as artificial intelligence, machine learning, quantum communication, blockchain, terahertz, and millimeter-wave communications. With IoT devices projected to reach 25 billion by 2025, they will enable advanced services with ubiquitous sensing and computing capabilities. Due to the immense potential of 6G in IoT applications, both academia and industry have shown significant interest in exploring its opportunities for the future Internet (Akhtar, 2020; Kim, 2024b; Nguyen, 2021).

As the IoT paradigm expands, the same principles apply to intelligent devices in military operations. Battlefield networks must be highly adaptable to ever-changing combat conditions. These networks must be designed to function in environments with limited communication infrastructure, heterogeneous devices, dynamic topologies, and the inherent unpredictability of combat scenarios. The advancements in IoT have significantly influenced modern military operations, which now incorporate thousands of interconnected 'things' to execute various tasks. Military applications generally involve environment sensing, communication, autonomous actions, and collaborative operations. Drawing from the IoT concept, the Internet of Battlefield Things (IoBT) has emerged as a specialized network designed to enhance military operations. As a subclass of IoT tailored for warfare, IoBT forms a complex interconnected system within the military domain, alleviating many burdens faced by soldiers on the battlefield (Fragkou, 2022; Kim, 2024b; Stocchero, 2022).

IoBT has the potential to transform modern warfare by leveraging data-driven intelligence to enhance combat effectiveness. In battlefield scenarios, IoBT facilitates computation and communication among strategic assets such as drones, armored vehicles, ground stations, and soldiers. However, these military devices are constrained by limited computational capacity and battery life. As a result, handling computationally intensive battlefield tasks is challenging for individual IoBT devices due to their restricted processing power and energy limitations. In conventional IoT frameworks, partial offloading is a viable approach to address this challenge. Through this strategy, computationally intensive tasks are divided into two subtasks: one executed locally on the IoT device, and the other offloaded to a remote cloud center for processing. This method allows for a more balanced workload distribution while optimizing resource utilization (An, 2022; Farooq, 2018; Kim, 2024b).

Most IoT applications, including smart homes, factories, and cities, rely on infrastructure-based networks where devices connect to the Internet through gateways or access points. However, in battlefield environments, conventional communication infrastructure such as cellular networks and base stations may be unavailable. To address this limitation, battlefield networks must adopt alternative communication frameworks. Due to their low deployment cost and extensive coverage, tethered balloons (TBs) have gained increasing attention for IoBT applications. Acting as airborne wireless base stations, TBs support a wide range of IoBT services. Each TB is connected to a ground charging station, ensuring a stable power supply and a wired backhaul link to the Internet. This allows TBs to function as communication hubs, bridging backhaul and access networks. Additionally, UAVs have been widely utilized in military operations for decades. Initially developed for military applications, UAVs have played a critical role in various defense missions, particularly as relays to extend battlefield communication coverage. By integrating TBs, UAVs, and IoBT devices, the battlefield network can enhance the capabilities of legacy wireless infrastructures, enabling seamless connectivity in combat environments (Farooq, 2018; Golam, 2020; Kim, 2024b; Wang, 2021b).

Traditional cloud offloading services introduce significant latency, which is particularly undesirable for time-sensitive IoBT applications. Due to this limitation, IoBT devices cannot rely on remote cloud centers for task offloading. To mitigate this issue, TB computing (TBC) presents a viable solution. As a mobile edge computing server, the *CGTO* scheme incorporates a cloud computing server directly into the TB, positioning it closer to IoBT devices. This setup enables IoBT devices to offload computational tasks to the TBC server, where they are processed locally, eliminating the need to transmit them to distant cloud centers. In this framework, UAVs play a crucial role in the TB-based IoBT platform. To optimize task offloading efficiency and cost-effectiveness, multiple UAVs are deployed between ground IoBT devices and TBC servers, serving as mobile relays. However, operating UAVs in conjunction with TBs within the IoBT system presents several challenges, including task offloading decision-making, computational task processing, and efficient allocation of limited TBC resources (An, 2022; Kim, 2024b; Wang, 2021b).

As research on UAV-TB-assisted IoBT systems remains in its early stages, the issue of controlling partial offloading services has received limited attention. The *CGTO* scheme focuses on cooperative game theory to coordinate multiple IoBT devices efficiently. Since the 1950s, various solution frameworks have been introduced to address resource allocation challenges. Among them, the *Shapley value* and *Nash bargaining solution* are the most commonly applied concepts in cooperative game theory. The *Shapley value*, based on mathematical formulation, provides a straightforward method for resource allocation. However, in certain scenarios, allocations based on the *Shapley value* may lack stability. Meanwhile, the *Nash bargaining solution* offers an optimal allocation by maximizing the product of utilities for cooperative players. Despite its efficiency, the Nash approach does not guarantee fair distribution among all participants (Kim, 2024b; Luo, 2022).

In 1951, H. Raiffa introduced the *Sequential Raiffa Bargaining Solution* (*SRBS*) for two-player bargaining problems. As an arbitration protocol, the *SRBS* is characterized by standard bargaining axioms. Additionally, a unique axiom emphasizes the iterative negotiation process, in which the procedure is repeatedly applied to a diminishing sequence of remaining games. Given a disagreement point, each player aims to secure the highest possible utility while ensuring that the other player retains at least their disagreement payoff. The interim agreement is then derived as the average of these two preferred outcomes. By iteratively using interim agreement as a new disagreement point, the *SRBS* continuously refines the bargaining process until it converges to a *Pareto-optimal* outcome. Through this mechanism, the *SRBS* bridges the gap between cooperative and non-cooperative bargaining models by incorporating elements of relative gains and concessions (Diskin, 2011; Kim, 2024b; Trockel, 2015).

In cooperative games, a playe's marginal contribution is a crucial metric for assessing their cooperative potential. Since the introduction of Shapley *value*, numerous value-based solutions have been proposed using marginal contribution as a core parameter. Most of these solutions allocate resources based on a weighted average of each player's marginal contributions across different coalitions. More recently, marginal surplus has emerged as an alternative metric for evaluating a player's contribution level. Defined as the difference between marginal contribution and individual worth, marginal surplus represents a player's net gain upon joining a coalition. Compared to marginal contribution, marginal surplus places greater emphasis on individual worth. Building on the concept of marginal surplus, *the average-surplus value* (*ASV*) has been introduced as a novel solution for cooperative games. Inspired by the *Shapley value*, the *ASV* is derived through a systematic process of distributing marginal surplus among players (Kim, 2024b; Li, 2022).

9.2.2 Contribution

As previously discussed, the integration of TBs, UAVs, and IoBT devices has garnered increasing interest, with the UAV-TB-assisted IoBT system recognized for its significant potential in both industry and academia. However, limited research has focused on developing an efficient offloading service mechanism within this platform. To address this gap, the *CGTO* scheme introduces a novel IoBT offloading mechanism by incorporating the fundamental principles of *SRBS* and *ASV*. In the *CGTO* scheme, each IoBT device utilizes TBs and UAVs to partially offload its computation-intensive tasks. These tasks are divided into two segments: one portion is processed locally on the IoBT device, while the other is offloaded to a TB for execution. The *CGTO* scheme employs the *SRBS* approach to determine the optimal task partitioning strategy. Each TB's TBC server receives offloaded tasks from corresponding IoBT devices and processes them in a coordinated manner. Additionally, leveraging the *ASV* concept, the computing resources of the TBC server are allocated efficiently, ensuring a balanced resolution of conflicting computational demands. For seamless communication, UAVs function as relay nodes between TBs and IoBT devices, facilitating efficient data transmission. The primary objective of the *CGTO* scheme is to enable effective negotiation among multiple system agents, ensuring an optimized performance for the IoBT network (Kim, 2024b).

9.2.3 UAV-TB assisted IoBT infrastructure and problem formulations

Consider a UAV-TB assisted IoBT network that consists of a set of TBs $\mathbb{B} = \{\mathcal{B}_1, \ldots, \mathcal{B}_k\}$, a set of UAVs $\mathbb{V} = \{\mathcal{V}_1, \ldots, \mathcal{V}_m\}$, and a set of IoBT devices $\mathbb{D} = \{\mathcal{D}_1, \ldots, \mathcal{D}_n\}$ in a given geographical area. As shown in Figure 3, there are k TBs placed in fixed locations and connected directly to ground charging stations, which are deployed in a particular battlefield zone while providing a stable power supply and wired backhaul links. It is assumed that there are several disjoint battlefield zones where each zone is covered by each TB. In a battlefield area, there are heterogeneous IoBT devices that are randomly distributed in the area and generate computation-intensive tasks regularly. The *CGTO* scheme also considers that a group of UAVs along with their corresponding TBs are deployed. The $\mathcal{B}_{1 \le i \le k}$ has its UAV subset $\mathbb{V}_{\mathcal{B}_i}$, and UAVs in $\mathbb{V}_{\mathcal{B}_i} \subset \mathbb{V}$ are associated with the $\mathcal{B}_i$ to supplement the TB-IoBT infrastructure. Specifically, the $\mathcal{V}_j \in \mathbb{V}_{\mathcal{B}_i}$ has its corresponding IoBT device subset $\left(\mathbb{D}^{\mathcal{B}_i}_{\mathcal{V}_j} \subset \mathbb{D}\right)$. Therefore, the $\mathcal{V}_j$ acts as flying relay node between the $\mathcal{B}_i$ and devices in $\mathbb{D}^{\mathcal{B}_i}_{\mathcal{V}_j}$. The *CGTO* scheme considers a TBC server, which is located at each TB, can exchange and manage the offloading services with the associated UAVs (Alzidaneen, 2019; Kim, 2024b; Zhang, 2021).

Figure 3. The infrastructure of UAV-TB assisted IoBT system (Kim, 2024b)

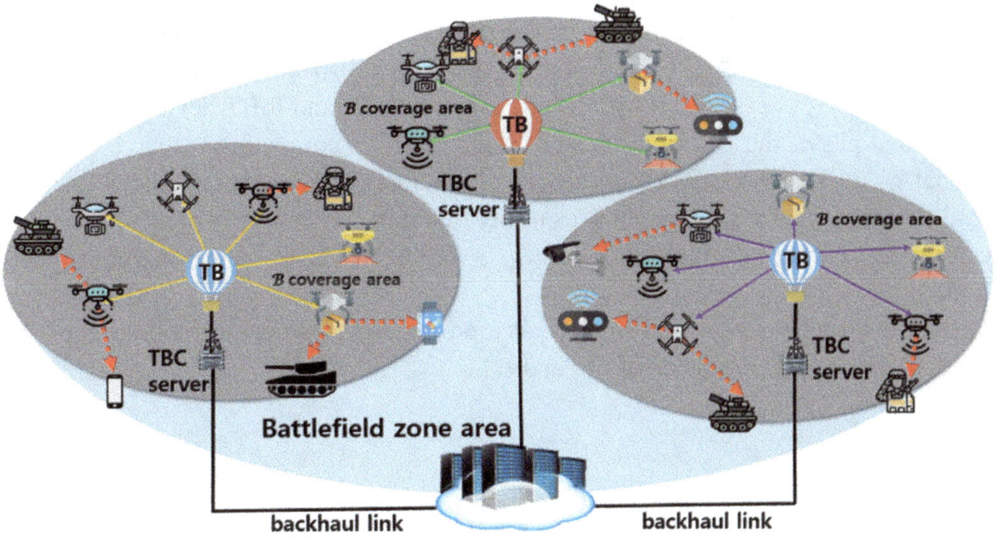

In the battlefield area, heterogeneous IoBT devices, such as humans, sensors, and vehicles, etc., carry out various tasks including environmental sensing, learning, and processing to meet multiple and diverse missions. Each $\mathscr{D}_{1 \le l \le n} \in \mathbb{D}$ has his computing power $\left(\mathfrak{M}_{\mathscr{D}_l}\right)$, and generates its computational workload task $\left(\mathscr{W}_{\mathscr{D}_l}\right)$; $\mathscr{W}_{\mathscr{D}_l}$ is assumed to be delay sensitive, and follows a Poisson process with an average generation rate at each time period. To reduce the computation load, the $\mathscr{D}_l$ can partially offload its task; $\mathscr{W}^L_{\mathscr{D}_l}$ is a subtask for the local processing and $\mathscr{W}^{\mathscr{B}}_{\mathscr{D}_l}$ is a subtask for the offloading service where $\mathscr{W}_{\mathscr{D}_l} = \mathscr{W}^L_{\mathscr{D}_l} + \mathscr{W}^{\mathscr{B}}_{\mathscr{D}_l}$. In the viewpoint of individual devices, their corresponding TBs exist a little farther away from them. Therefore, the *CGTO* scheme can make an access association between the TB and IoBT devices through the UAV. The $\mathscr{V}_j$ collects the total offloading tasks $\left(\mathscr{T}_{\mathscr{V}_j}\right)$ from its associated devices in the $\mathbb{D}^{\mathscr{B}_i}_{\mathscr{V}_j}$ where $\mathscr{T}_{\mathscr{V}_j} = \sum_{\mathscr{D}_l \in \mathbb{D}^{\mathscr{B}_i}_{\mathscr{V}_j}} \mathscr{W}^{\mathscr{B}}_{\mathscr{D}_l}$. For offloading services, the UAVs in $\mathbb{V}_{\mathscr{B}_i}$ offload their tasks to the $\mathscr{B}_i$. The $\mathscr{B}_i$ has a total computing power $\left(\mathfrak{M}_{\mathscr{B}_i}\right)$, and each $\mathscr{V}_j \in \mathbb{V}_{\mathscr{B}_i}$ shares a certain amount of computation capacity. Finally, the UAV-TB assisted IoBT system can process the offloaded tasks from the ground IoBT devices. However, this process faces two challenges: i) each device's task division issue, and ii) the $\mathfrak{M}_{\mathscr{B}_i}$ sharing issue. To maximize the UAV-TB assisted IoBT system performance, efficient task decision and the $\mathfrak{M}_{\mathscr{B}}$ sharing strategies may become key factors (Kim, 2024b).

In the *CGTO* scheme, the $\mathscr{D}_l$'s task division problem and the $\mathfrak{M}_{\mathscr{B}_i}$'s sharing problem are formulated as cooperative games $\mathbb{G}_{\mathscr{D}_l}$ and $\mathbb{G}_{\mathscr{B}_i}$, respectively. As a consequence, these two games are sequentially operated in an interactive fashion; it is noteworthy that the *CGTO* scheme formulates the TB-UAV-IoBT association in a cooperative manner. Formally, the *CGTO* scheme defines the two-phase game G entities, i.e., $\mathbb{G} = \left\{\mathbb{G}_{\mathscr{D}_l \in \mathbb{D}}, \mathbb{G}_{\mathscr{B}_i \in \mathbb{B}}\right\} = \left\{\mathbb{B}, \mathbb{V}, \mathbb{D}, \left\{\mathbb{G}_{\mathscr{D}_l} | \mathscr{D}_l \in \mathbb{D}, \mathfrak{M}_{\mathscr{D}_l}, \left(\mathscr{P}^L_{\mathscr{D}_l}, \mathscr{P}^{\mathscr{B}}_{\mathscr{D}_l}\right), S_{\mathscr{D}_l}, \left(U^L_{\mathscr{D}_l}(\cdot), U^{\mathscr{B}}_{\mathscr{D}_l}(\cdot)\right)\right\}, \mathbb{D}^{\mathscr{B}_i}_{\mathscr{V}_j}, \left\{\mathbb{G}_{\mathscr{B}_i} | \mathscr{B}_i \in \mathbb{B}, \mathfrak{M}_{\mathscr{B}_i}, \mathbb{V}_{\mathscr{B}_i}, \mathscr{V}_j \in \mathbb{V}_{\mathscr{B}_i}, v(\cdot), \mathscr{U}^{\mathscr{V}_j}_{\mathscr{B}_i}(\cdot)\right\}, T\right\}$

of gameplay (Kim, 2024b).

- B, V and D represent the sets of TBs, UAVs, and IoBT devices, respectively. They are mutually and reciprocally interdependent in a coordinated manner, and they work together in the UAV-TB assisted IoBT platform.
- At the first phase, the $\mathbb{G}_{\mathscr{D}_l}$ is designed to split the $\mathscr{D}_l$'s computation task $\left(\mathscr{W}_{\mathscr{D}_l}\right)$ for the offload service, and $\mathfrak{M}_{\mathscr{D}_l}$ is the $\mathscr{D}_l$'s local computing power. In the $\mathbb{G}_{\mathscr{D}_l}$, the subtasks for the local processing $\left(\mathscr{P}^L_{\mathscr{D}_l}\right)$ and for the offloading service $\left(\mathscr{P}^{\mathscr{B}}_{\mathscr{D}_l}\right)$ are game players.
- In the $\mathbb{G}_{\mathscr{D}_l}$, The $S_{\mathscr{D}_l}$ is the splitting ratio for the $\mathscr{W}_{\mathscr{D}_l}$. $S_{\mathscr{D}_l}$ and $U^L_{\mathscr{D}_l}(\cdot)$ are the strategy and utility function of $\mathscr{P}^L_{\mathscr{D}_l}$, and $\left(1 - S_{\mathscr{D}_l}\right)$ and $U^{\mathscr{B}}_{\mathscr{D}_l}(\cdot)$ are the strategy and utility function of $\mathscr{P}^{\mathscr{B}}_{\mathscr{D}_l}$, respectively.
- Individual device $\mathscr{D}_{1 \le l \le n} \in \mathbb{D}$ operates the $\mathbb{G}_{\mathscr{D}_l}$ game in a distributed manner.
- The UAV $\mathscr{V}_j \in \mathbb{V}_{\mathscr{B}_i}$ is associated with its corresponding $\mathscr{B}_i$ and IoBT devices in the $\mathbb{D}^{\mathscr{B}_i}_{\mathscr{V}_j}$.
- At the second phase, the $\mathbb{G}_{\mathscr{B}_i}$ is designed to share the $\mathscr{B}_i$'s computing resource $\left(\mathfrak{M}_{\mathscr{B}_i}\right)$ for each individual $\mathscr{V}_j \in \mathbb{V}_{\mathscr{B}_i}$. In the $\mathbb{G}_{\mathscr{B}_i}$, $\mathscr{V}_j$ is a game player, and $v(\cdot)$ is a characteristic function for each players' coalition. The $\mathscr{U}^{\mathscr{V}_j}_{\mathscr{B}_i}(\cdot)$ is the $\mathscr{V}_j$'s utility function.
- Like as the first phase, each individual TB $\mathscr{B}_i \in \mathbb{B}$ operates its $\mathbb{G}_{\mathscr{B}_i}$ game in a distributed parallel fashion.
- Discrete time model $T \in \{t_1, \ldots, t_c, t_{c+1}, \ldots\}$ is represented by a sequence of time steps. The length of t_c matches the event time-scale of $\mathbb{G}_{\mathscr{D}_l}$ and $\mathbb{G}_{\mathscr{B}_i}$.

9.2.4 Sequential Raiffa bargaining solution and average-surplus value

To characterize the fundamental idea of *SRBS*, we assume a two-player bargaining problem. R is a real number set, and $\mathbb{R}^2$ denotes the two-dimensional Euclidean space. Let $S \subset \mathbb{R}^2$ be a nonempty and finite set, and N is a set of natural numbers. For any $x, y \in \mathbb{R}^2$, we write $x \geq y$ ($x \gg y$, resp.) if for any $i \in S$, $x_i \geq y_i$ ($x_i > y_i$, resp.). A pair (S, d) is a bargaining game with $d \in S \subset \mathbb{R}^2$ where $d \leq y \leq x \in S \Rightarrow y \in S$. The set of all two-person bargaining games is denoted as B. For any non-empty subset C of B, a mapping $F{:}C \rightarrow \mathbb{R}^2 : (S,d) \mapsto F(S,d) \in S$ is called a bargaining solution on C. With $x \in \mathbb{R}^2$, let 1_S be indicator function of S where $1_S : \mathbb{R}^2 \rightarrow \mathbb{R} : x \mapsto (1, \text{if} x \in S|0, \text{if} x \notin S)$. For any game $(S,d) \in \mathbb{B}$, we consider the mapping $f^{(S,d)} : \mathbb{R}^2 \rightarrow \mathbb{R}^2$; it defined as follows (Kim, 2022a, 2024b; Trockel, 2015):

$$f^{(S,d)}\left(x_1, x_2\right) \; : = \left(f_1^{(S,d)}(x_2), f_2^{(S,d)}(x_1)\right) \tag{9.2.1}$$

$$\text{s.t.,}\begin{cases} f_1^{(S,d)}{:}R \rightarrow R \; : \; x_2 \mapsto f_1^{(S,d)}(x_2) \; : = \max\limits_{x_1 \in \mathbb{R}}\left(x_1 - d_1\right) \cdot \mathbf{1}_s(x_1, x_2) \\ f_2^{(S,d)}{:}R \rightarrow R \; : \; x_1 \mapsto f_2^{(S,d)}(x_1) \; : = \max\limits_{x_2 \in \mathbb{R}}\left(x_2 - d_2\right) \cdot \mathbf{1}_s(x_1, x_2) \end{cases}$$

Now consider the sequence $\left(m_k^{(S,d)}\right)_{k \in \mathbb{N}}$ defined by (Trockel, 2015):

$$m_{k \in \mathbb{N}}^{(S,d)} = \frac{1}{2} \times \left[\left(f_1^{(S,d)}\left(m_{k-1,2}^{(S,d)}\right), m_{k-1,2}^{(S,d)}\right) + \left(m_{k-1,1}^{(S,d)}, f_2^{(S,d)}\left(m_{k-1,1}^{(S,d)}\right)\right)\right] \tag{9.2.2}$$

$$\text{s.t.,}\; m_0^{(S,d)} \; : = \left(m_{0,1}^{(S,d)}, m_{0,2}^{(S,d)}\right) \; : = \left(d_1, d_2\right) \; = d$$

According to (9.2.1) and (9.2.2), the *SRBS*(S, d) is defined by (Trockel, 2015):

$$SRBS(S,\, d) \; : = \lim_{k \in (\mathbb{N} \cup \{0\})} m_k^{(S,d)}, \; \text{s.t.,}\; \forall (S,d) \in \mathbb{B} \tag{9.2.3}$$

To characterize the basic concept of *ASV*, let R be the sets of real numbers. A cooperative n-player game is a pair (N, v) where $N = \{1, \ldots, n\}$ is a finite set of n players, and $v : 2^n \rightarrow \mathbb{R}$ is a characteristic function assigning to each coalition $S \in 2^N$. The set of all non-empty coalitions of N is denoted by Ω, and a payoff vector for (N, v) is an n-dimensional vector $x \in \mathbb{R}^n$ assigning a payoff $x_{1 \leq i \leq n} \in \mathbb{R}^n$ to each player $i \in N$. Player i's marginal contribution to S $\left(MC_i(S)\right)$ and marginal surplus to S $\left(MS_i(S)\right)$ are defined as follows (Kim, 2024b; Li, 2022):

$$\begin{cases} MC_i(S) = v(S) - v(S \setminus i) \\ MS_i(S) = v(S) - v(S \setminus i) - v(i) \end{cases} \tag{9.2.4}$$

Obviously, marginal surplus is the difference between marginal contribution and the individual worth. It measures the contribution level of each player. Traditionally, well-known value solutions like Shapley value and Solidarity value, are developed based on the idea of marginal contribution. The Shapley value $(Sh\,V_i(N,v))$ and the Solidarity value $(So\,V_i(N,v))$ for player $i \in N$ in (N,v) are mathematically given by (Kim, 2024b; Li, 2022):

$$\begin{cases} Sh\,V_i(N,v) = \sum_{S \subseteq N, i \in S} \left(\left(\frac{(S-1)! \times (n-S)!}{n!} \right) \times MC_i(S) \right) \\ So\,V_i(N,v) = \sum_{S \subseteq N, i \in S} \left(\left(\frac{(S-1)! \times (n-S)!}{n!} \right) \times SMC_i(S) \right) \end{cases} \tag{9.2.5}$$

$$\text{s.t., } SMC_i(S) = \frac{1}{|S|} \times \sum_{j \in S} MC_j(S)$$

where $|S|$ is the cardinality of S. The Shapley value (*or* Solidarity value) assigns to every player a weighted average of all his marginal contributions (*or* the average marginal contributions) to all coalitions including himself. Based on the idea of marginal surplus, the *ASV* for player $i \in N$ in (N,v), i.e., $AS\,V_i(N,v)$, is mathematically given by (Kim, 2024b; Li, 2022):

$$AS\,V_i(N,v) = v(i) + \sum_{S \subseteq N, i \in S} \left(\left(\frac{(S-1)! \times (n-S)!}{n!} \right) \times MS^v(S) \right) \tag{9.2.6}$$

$$\text{s.t., } MS^v(S) = \frac{1}{|S|} \times \sum_{j \in S} \left(v(S) - v(S \setminus j) - v(j) \right)$$

Formally, the *ASV* is captured using the notion of a weighted average of the average marginal surpluses to all coalitions, and lays emphasis on taking into account the influence of the individual worth (Kim, 2024b; Li, 2022).

9.2.5 The task of offloading for the UAV-TB assisted IoBT platform

To develop the partial offloading algorithm for each IoBT device, the *CGTO* scheme constructs the $\mathbb{G}_{\mathcal{D}}$ games. They are operated independently during bargaining time steps. At each time period, the $\mathbb{G}_{\mathcal{D}_{1 \le l \le n}}$ is designed for the $\mathcal{D}_l$ to reach the collaborative strategy $S_{\mathcal{D}_l}$, which splits the $\mathscr{W}_{\mathcal{D}_l}$ for its partial offloading service. In the $\mathbb{G}_{\mathcal{D}_l}$ game, the $U^L_{\mathcal{D}_l}(\cdot)$ and $U^{\mathcal{B}}_{\mathcal{D}_l}(\cdot)$ are defined as follows (Kim, 2024b):

$$\begin{cases} U_{\mathcal{D}_i}^{L}\left(\mathcal{W}_{\mathcal{D}_i}, S_{\mathcal{D}_i}, \mathfrak{M}_{\mathcal{D}_i}\right) = \log(X)^{\alpha} - \left(\xi \times (X)^{\beta}\right) \\ U_{\mathcal{D}_i}^{\mathcal{B}}\left(\mathcal{W}_{\mathcal{D}_i}, S_{\mathcal{D}_i}, \mathfrak{M}_{\mathcal{D}_i}, \mathfrak{M}_{\mathcal{B}_i}\right) = \left(\frac{exp(Y) - exp(-Y)}{exp(Y) + exp(-Y)}\right) - \left(\mathcal{H}_{\mathfrak{B}}^{t_c} \times \log(Y + \varepsilon)\right) \end{cases} \tag{9.2.7}$$

$$\text{s.t., } \mathcal{H}_{\mathfrak{B}}^{t_c} = \frac{\mathfrak{R}_{\mathcal{B}_i}^{c}}{\mathfrak{M}_{\mathcal{B}_i}} \text{ and } \begin{cases} X = \dfrac{\min\left(\mathfrak{M}_{\mathcal{D}_i}, \mathcal{W}_{\mathcal{D}_i} \times S_{\mathcal{D}_i}\right)}{\mathfrak{M}_{\mathcal{D}_i}} + \eta \\ Y = \dfrac{\mathcal{W}_{\mathcal{D}_i} \times \left(1 - S_{\mathcal{D}_i}\right)}{\mathfrak{M}_{\mathcal{D}_i}} \end{cases}$$

where $\mathfrak{R}_{\mathcal{B}_i}^{c}$ is the currently using computing power in the $\mathcal{B}_i$, and $\mathfrak{M}_{\mathcal{B}_i}$, $\mathfrak{M}_{\mathcal{D}_i}$ are the $\mathcal{B}_i$ and $\mathcal{D}_i$'s computing powers, respectively. α, ξ, β, η are control parameters for the $U_{\mathcal{D}_i}^{L}(\cdot)$, and ε is a control parameter for the $U_{\mathcal{D}_i}^{\mathcal{B}}(\cdot)$. $\mathcal{H}_{\mathfrak{B}}^{t_c}$ is an adjustment factor for the $U_{\mathcal{D}_i}^{\mathcal{B}}(\cdot)$. In the $\mathbb{G}_{\mathcal{D}_i}$ game, the $\mathcal{W}_{\mathcal{D}_i}$ is split for the local service $\left(\mathcal{W}_{\mathcal{D}_i}^{L}\right)$ and the offload service $\left(\mathcal{W}_{\mathcal{D}_i}^{\mathcal{B}}\right)$. To decide the $\mathcal{W}_{\mathcal{D}_i}^{L}$ and $\mathcal{W}_{\mathcal{D}_i}^{\mathcal{B}}$ values, two game players in the $\mathbb{G}_{\mathcal{D}_i}$, i.e, $\mathcal{P}_{\mathcal{D}_i}^{L}$ and $\mathcal{P}_{\mathcal{D}_i}^{\mathcal{B}}$, should converge to a fair-efficient solution while maintaining their viewpoints. Therefore, the $\mathcal{P}_{\mathcal{D}_i}^{L}$ and $\mathcal{P}_{\mathcal{D}_i}^{\mathcal{B}}$ sequentially negotiate with each other to reach a mutual consensus. In the *CGTO* scheme, the *SRBS* is preferred for the solution concept of $\mathbb{G}_{\mathcal{D}_i}$. It is given by (Kim, 2024b):

$$SRBS\left(S_{\mathcal{D}_i}^{L,\mathcal{B}}, d_{\mathcal{D}_i}^{L,\mathcal{B}}\right) := \lim_{k \in (\mathbb{N} \cup \{0\})} m_k^{\left(S_{\mathcal{D}_i}^{L,\mathcal{B}}, d_{\mathcal{D}_i}^{L,\mathcal{B}}\right)} \tag{9.2.8}$$

$$:= \lim_{k \in (\mathbb{N} \cup \{0\})} \left(\frac{1}{2} \times \left[\begin{array}{l} \left(f_L^{\left(S_{\mathcal{D}_i}^{L,\mathcal{B}}, d_{\mathcal{D}_i}^{L,\mathcal{B}}\right)}\left(m_{k-1,\mathcal{B}}^{\left(S_{\mathcal{D}_i}^{L,\mathcal{B}}, d_{\mathcal{D}_i}^{L,\mathcal{B}}\right)}\right), m_{k-1,\mathcal{B}}^{\left(S_{\mathcal{D}_i}^{L,\mathcal{B}}, d_{\mathcal{D}_i}^{L,\mathcal{B}}\right)}\right) \\ + \left(m_{k-1,L}^{\left(S_{\mathcal{D}_i}^{L,\mathcal{B}}, d_{\mathcal{D}_i}^{L,\mathcal{B}}\right)}, f_{\mathcal{B}}^{\left(S_{\mathcal{D}_i}^{L,\mathcal{B}}, d_{\mathcal{D}_i}^{L,\mathcal{B}}\right)}\left(m_{k-1,L}^{\left(S_{\mathcal{D}_i}^{L,\mathcal{B}}, d_{\mathcal{D}_i}^{L,\mathcal{B}}\right)}\right)\right) \end{array} \right] \right)$$

$$s.t., \begin{cases} f_L^{\left(S_{\mathcal{D}_i}^{L,\mathcal{B}}, d_{\mathcal{D}_i}^{L,\mathcal{B}}\right)}\left(m_{k-1,\mathcal{B}}^{\left(S_{\mathcal{D}_i}^{L,\mathcal{B}}, d_{\mathcal{D}_i}^{L,\mathcal{B}}\right)}\right) = \max\limits_{m_{k-1,L}^{(S_{\mathcal{D}_i}^{L,\mathcal{B}}, d_{\mathcal{D}_i}^{L,\mathcal{B}})} \in \mathbb{R}} \left(m_{k-1,L}^{\left(S_{\mathcal{D}_i}^{L,\mathcal{B}}, d_{\mathcal{D}_i}^{L,\mathcal{B}}\right)} - d_L\right) \cdot \mathbf{1}_s\left(m_{k-1,L}^{\left(S_{\mathcal{D}_i}^{L,\mathcal{B}}, d_{\mathcal{D}_i}^{L,\mathcal{B}}\right)}, m_{k-1,\mathcal{B}}^{\left(S_{\mathcal{D}_i}^{L,\mathcal{B}}, d_{\mathcal{D}_i}^{L,\mathcal{B}}\right)}\right) \\ f_{\mathcal{B}}^{\left(S_{\mathcal{D}_i}^{L,\mathcal{B}}, d_{\mathcal{D}_i}^{L,\mathcal{B}}\right)}\left(m_{k-1,\mathcal{B}}^{\left(S_{\mathcal{D}_i}^{L,\mathcal{B}}, d_{\mathcal{D}_i}^{L,\mathcal{B}}\right)}\right) = \max\limits_{m_{k-1,\mathcal{B}}^{(S_{\mathcal{D}_i}^{L,\mathcal{B}}, d_{\mathcal{D}_i}^{L,\mathcal{B}})} \in \mathbb{R}} \left(m_{k-1,\mathcal{B}}^{\left(S_{\mathcal{D}_i}^{L,\mathcal{B}}, d_{\mathcal{D}_i}^{L,\mathcal{B}}\right)} - d_{\mathcal{B}}\right) \cdot \mathbf{1}_s\left(m_{k-1,\mathfrak{B}}^{\left(S_{\mathcal{V}}^{\mathfrak{B}}, d_{\mathcal{V}}^{\mathfrak{B}}\right)}, m_{k-1,\mathcal{B}}^{\left(S_{\mathcal{D}_i}^{L,\mathcal{B}}, d_{\mathcal{D}_i}^{L,\mathcal{B}}\right)}\right) \\ S_{\mathcal{D}_i}^{L,\mathcal{B}} = \left(U_{\mathcal{D}_i}^{L}(\cdot), U_{\mathcal{D}_i}^{\mathcal{B}}(\cdot)\right) \text{ and } d_{\mathcal{D}_i}^{L,\mathcal{B}} = \left(d_L, d_{\mathcal{B}}\right) \\ m_0^{\left(S_{\mathcal{D}_i}^{L,\mathcal{B}}, d_{\mathcal{D}_i}^{L,\mathcal{B}}\right)} := \left(m_{0,L}^{\left(S_{\mathcal{D}_i}^{L,\mathcal{B}}, d_{\mathcal{D}_i}^{L,\mathcal{B}}\right)}, m_{0,\mathcal{B}}^{\left(S_{\mathcal{D}_i}^{L,\mathcal{B}}, d_{\mathcal{D}_i}^{L,\mathcal{B}}\right)}\right) := \left(d_L, d_{\mathcal{B}}\right) = d_{\mathcal{D}_i}^{L,\mathcal{B}} \end{cases}$$

where d_L and $d_{\mathfrak{B}}$ are disagreement points for $\mathcal{P}_{\mathcal{D}_i}^{L}$ and $\mathcal{P}_{\mathcal{D}_i}^{\mathcal{B}}$, respectively. When the change between $m_{k-1,L}^{\left(S_{\mathcal{D}_i}^{L,\mathcal{B}}, d_{\mathcal{D}_i}^{L,\mathcal{B}}\right)}$ and $m_{k,L}^{\left(S_{\mathcal{D}_i}^{L,\mathcal{B}}, d_{\mathcal{D}_i}^{L,\mathcal{B}}\right)}$ is within a pre-defined minimum bound (Δ), this change can be negligible. At this time, we can think that the CGTO scheme converges with a fair-efficient bargaining solution, and the negotiation process is terminated. Finally, the $S_{\mathcal{D}_i}$ is decided according to the simple negotiative bargaining manner, and the CGTO scheme can get the $\mathcal{W}_{\mathcal{D}_i}^{L}$ and $\mathcal{W}_{\mathcal{D}_i}^{\mathcal{B}}$ values (Kim, 2024b).

In the *CGTO* scheme, the UAVs in $\mathbb{V}_{\mathcal{B}_i}$ are collecting the offloading tasks $\left(\mathcal{W}_{\mathcal{D}}^{\mathcal{B}}\right)$ from their corresponding devices in $\mathbb{D}_{\mathcal{V}_j}^{\mathcal{B}_i}$, and report them to the $\mathcal{B}_i$. For example, the $\mathcal{V}_j \in \mathbb{V}_{\mathcal{B}_i}$ gets his total offloading task $\left(\mathcal{T}_{\mathcal{V}_j}\right)$, and the $\mathcal{B}_i$ is asked to complete the all requested offloading services $\left(\mathcal{T}_{\mathcal{V}}^{\mathcal{B}_i}\right)$ from multiple UAVs in $\mathbb{V}_{\mathcal{B}_i}$ where $\mathcal{T}_{\mathcal{V}}^{\mathcal{B}_i} = \sum_{\mathcal{V} \in \mathbb{V}_{\mathcal{B}_i}} \mathcal{T}_{\mathcal{V}}$. However, the $\mathcal{B}_i$'s computing resource $\mathfrak{M}_{\mathcal{B}_i}$ is limited. Therefore, it should be shared effectively among the $\mathcal{B}_i$'s corresponding UAVs. Usually, the limited resource sharing problem is analogous to the bankruptcy problem. Formally, a bankruptcy problem is represented as (E, N, c, v) where a monetary estate $E \in \mathbb{R}_+$ has to be divided among a set of claimants $N = \{1, \ldots, n\}$, and $\mathbf{c} = (c_1, \ldots, c_n) \in \mathbb{R}_+^n$ is the vector of the claimants' claims where $0 < E < \sum_{i=1}^{n} c_i$. To estimate value solutions, the characteristic function for the coalition S, i.e., $v(S)$, denotes the minimal amount that the coalition $S \subset N$ will receive, once the claims of the creditors outside S have been fully compensated. In a game with n claimants, there are 2^n possible coalitions. In order to avoid secessions, the $v(S)$ should be guaranteed that the best coalition is the grand coalition grouping all claimants (Kim, 2021a, 2024b).

The *CGTO* scheme designs the $\mathfrak{M}_{\mathcal{B}_i}$ sharing algorithm as the second phase cooperative game $\left(\mathbb{G}_{\mathcal{B}_i}\right)$. In the $\mathbb{G}_{\mathcal{B}_i}$, the $\mathcal{V}_j \in \mathbb{V}_{\mathcal{B}_i}$ is a game player, and it's claim $\left(c_{\mathcal{V}_j}\right)$ is calculated as $c_{\mathcal{V}_j} = \sum_{\mathcal{D}_l \in \mathbb{D}_{\mathcal{V}_j}^{\mathcal{B}_i}} \mathcal{W}_{\mathcal{D}_l}^{\mathcal{B}_i}$. Therefore, the bankruptcy problem for the $\mathbb{G}_{\mathcal{B}_i}$ is formulated as $E = \mathfrak{M}_{\mathcal{B}_i}, N = \mathbb{D}_{\mathcal{V}_j}^{\mathcal{B}_i}$, and $c = \left\{\ldots c_{\mathcal{V}_j} \ldots\right\}$. Mathematically, the $v(S)$ is defined based on the bankruptcy model (Kim, 2021a, 2024b).

$$v\left(S | S \subset \mathbb{D}_{\mathcal{V}_j}^{\mathcal{B}_i}\right) = \max\left(0, \mathfrak{M}_{\mathcal{B}_i} - \sum_{\mathcal{V}_j \in \mathbb{D}_{\mathcal{V}_j}^{\mathcal{B}_i} \setminus S} c_{\mathcal{V}_j}\right) \tag{9.2.9}$$

In the *CGTO* scheme, the *ASV* is adopted as the solution concept of $\mathbb{G}_{\mathcal{B}_i}$. The *ASV* value for the $\mathcal{V}_j$, i.e., $ASV_{\mathcal{V}_j}(\cdot)$, is given by (Kim, 2024b);

$$ASV_{\mathcal{V}_j}\left(\mathbb{D}_{\mathcal{V}_j}^{\mathcal{B}_i}, v\right) = v\left(\mathcal{V}_j\right) + \sum_{S \subseteq \mathbb{D}_{\mathcal{V}_j}^{\mathcal{B}_i}, \mathcal{V}_j \in S}\left(\left(\frac{(S-1)! \times \left(\left|\mathbb{D}_{\mathcal{V}_j}^{\mathcal{B}_i}\right| - S\right)!}{\left|\mathbb{D}_{\mathcal{V}_j}^{\mathcal{B}_i}\right|!}\right) \times MS^{v}(S)\right) \tag{9.2.10}$$

$$\text{s.t., } MS^{v}(S) = \frac{1}{|S|} \times \sum_{\mathcal{V}_r \in S}\left(v(S) - v\left(S \setminus \mathcal{V}_r\right) - v\left(\mathcal{V}_r\right)\right)$$

Finally, the utility function for the $\mathcal{V}_j$, i.e., $\mathcal{U}_{\mathcal{B}_i}^{\mathcal{V}_j}(\cdot)$, is defined as follows (Kim, 2024b);

$$\mathcal{U}_{\mathcal{B}_i}^{\mathcal{V}_j}\left(ASV_{\mathcal{V}_j}\left(\mathbb{D}_{\mathcal{V}_j}^{\mathcal{B}_i}, v\right), \mathfrak{M}_{\mathcal{B}_i}\right) = \frac{\theta}{\varrho + exp\left(\psi \times \frac{\min\left(ASV_{\mathcal{V}_j}\left(\mathbb{D}_{\mathcal{V}_j}^{\mathcal{B}_i}, v\right), \mathfrak{M}_{\mathcal{B}_i}\right)}{\mathfrak{M}_{\mathcal{B}_i}}\right)} - \Gamma \tag{9.2.11}$$

where ϱ, θ, ψ and Γ are the adjustment parameters for the $\mathcal{U}_{\mathcal{B}_i}^{\mathcal{V}_j}(\cdot)$ (Kim, 2024b).

9.2.6 Summary

The *CGTO* scheme introduces an effective task offloading mechanism tailored for the UAV-TB-assisted battlefield network platform. By leveraging the interactive synergy among TBs, UAVs, and IoBT devices, the *CGTO* scheme models the sequential interactions of various system agents and formulates two cooperative game frameworks. First, individual IoBT devices partition their computational tasks for partial offloading. This task-splitting challenge is addressed in a distributed manner using the *SRBS* approach. Second, the computational resources of each TBC server are allocated based on the *ASV* concept, ensuring efficient resource sharing. To reach an optimal and mutually beneficial solution, system agents within the *CGTO* scheme engage in reciprocal negotiations, refining their interactions dynamically. By adopting a coordinated decision-making paradigm, the *CGTO* scheme enhances adaptability and flexibility, making it well-suited for the UAV-TB-assisted battlefield network infrastructure (Kim, 2024b).

9.3 Bargaining based Computation Offloading Control (*BCOC*) scheme

In recent years, the deployment of unmanned aerial vehicles (UAVs) has surged significantly across various civilian and commercial applications. UAVs are anticipated to play a crucial role in 5G and beyond-5G networks. However, developing UAV networks and applications presents numerous challenges. To address the issue of heavy computational tasks in UAV networks, S. Kim recently introduces the *Bargaining-based Computation Offloading Control* (*BCOC*) scheme (Kim, 2019). The *BCOC* scheme is designed to optimize the tradeoff between communication delay and computational cost. The *BCOC* scheme employs the fundamental principles of the (α,β)-bargaining solution and formulates a cooperative bargaining game model to tackle the UAV computation offloading challenge. By leveraging the characteristics of the (α,β)-bargaining solution, the *BCOC* scheme offers a strategic, axiom-based approach to solving the task offloading problem. Additionally, it enables UAVs to dynamically adjust their offloading decisions in response to real-time network conditions, ensuring efficient and adaptive computation management (Kim, 2019).

9.3.1 Development motivation

Unmanned aerial vehicles (UAVs) are a category of aircraft capable of operating without an onboard human pilot to control their movements. These small aerial systems can be managed either remotely or autonomously. Recent advancements in technology have facilitated the development of smaller, cost-effective UAVs, positioning them as key components in future wireless communication networks. As the number of UAVs is projected to grow exponentially in the coming years, it is crucial to develop innovative solutions not only to ensure secure and reliable remote command and control but also to support high-capacity mission-critical applications. Integrating UAVs into 5G networks is regarded as a promising approach for realizing next-generation communication systems (Kim, 2019; Messous, 2017a, 2019).

UAVs, which serve a wide range of missions, are closely linked to flying ad hoc networks (FANETs) and are expected to be deployed in numerous applications, including rescue operations, target identification, remote sensing, surveillance, service delivery, pollution monitoring, and precision agriculture. These applications span from basic environmental monitoring to complex, high-security military operations. Given their diverse functions, UAVs are often required to handle computationally intensive tasks such as video preprocessing, pattern recognition, and feature extraction - tasks that typically demand high processing power and strict latency requirements. However, despite ongoing improvements in UAV technology, significant challenges remain in their application development. Due to their compact size, one of the most critical challenges for many UAV applications is the limitation in computing capabilities. Computationally demanding applications necessitate substantial processing and energy resources, which can impact real-time operations, reduce UAV system longevity, and, in some cases, compromise mission success (Ateya, 2019; Kim, 2019; Messous, 2017b).

To mitigate the computational limitations of UAVs, mobile cloud computing (MCC) has emerged as a promising solution. Traditionally, cloud computing utilizes remote computational resources to enable scalable operations via the Internet. With the growing adoption of mobile devices, cloud computing has evolved into a mobile paradigm. MCC facilitates the development, execution, and hosting of mobile applications using cloud-based resources, allowing remote processing and storage to support mobile tasks. By leveraging MCC technology, UAVs can offload computationally intensive tasks to external cloud servers, significantly alleviating their processing burden and optimizing their operational efficiency (Ateya, 2019; Kim, 2019; Messous, 2017a, 2017b, 2019).

The *BCOC* scheme introduces an advanced computation offloading mechanism for UAVs. To harness MCC's computing power while minimizing UAV overhead, a key challenge lies in effectively partitioning UAV computational tasks for offloading services. To address this, each UAV's workload is divided into two components: one portion is executed locally, while the other is offloaded for remote processing. To determine an optimal task partitioning strategy, UAVs must dynamically assess the current FANET conditions and adjust their offloading decisions accordingly. The primary objective for each UAV is to maximize service efficiency by managing computational offloading effectively. However, due to the complexity of FANET environments, achieving this goal presents a significant challenge. Additionally, the *BCOC* scheme seeks to strike a balance between practical implementation and system-wide optimization (Kim, 2019)

The *BCOC* scheme is designed based on the (α,β)-bargaining solution. With a reference point, the (α,β)-bargaining solution introduces two parameters, i.e., α and β, in bargaining problems. The first parameter α can be interpreted as the salience of the reference point, whereas the second parameter β can be interpreted as its power in shaping players' aspirations. From a game-theoretical perspective, the (α,β)-bargaining solution exhibits a number of interesting axiomatic properties. Based on the infrastructure of FANETs, this solution can be adaptively applied to provide a fair-efficient solution for the task partitioning problem. Therefore, during the bargaining game process, the *BCOC* scheme makes control decisions to achieve a globally desirable system performance while maximizing UAVs' own profits (Kim, 2019).

9.3.2 Contribution

The key contributions of the *BCOC* scheme can be summarized as follows; i) it introduces a virtual task-splitting mechanism for computation-intensive tasks, enabling efficient offloading while addressing control challenges within the FANET system, ii) it leverages the (α,β)-bargaining solution to develop an innovative UAV computation offloading framework, iii) it formulates a distributed cooperative game model, allowing each UAV to effectively balance performance trade-offs between conflicting requirements, and iv) it validates the mutual benefits of the proposed approach, demonstrating its effectiveness in UAV computation offloading scenarios (Kim, 2019).

9.3.3 The FANET system infrastructure

The *BCOC* scheme considers a set of small UAVs $\mathbb{M} = \{\mathcal{M}_1,\mathcal{M}_2,\ldots,\mathcal{M}_n\}$ performing exploration missions that involve highly computation intensive tasks $\mathbb{T} = \left\{\mathcal{T}_{\mathcal{M}_1},\mathcal{T}_{\mathcal{M}_2},\ldots,\mathcal{T}_{\mathcal{M}_n}\right\}$ such as video-based surveillance applications. Without loss of generality, the *BCOC* scheme assumes that each individual $\mathcal{M}_{1\le i\le n}$ has at most one $\mathcal{T}_{\mathcal{M}_i}$ at a time, and it should be completed within a period of time. Most UAV applications require sufficient computing power to process and analyze collected data. However, small UAVs are constrained by limited computational capacity and minimal onboard battery resources. To address these limitations, computation-intensive tasks should be offloaded to a nearby base station (BS).

Within the FANET infrastructure, the BS integrates a cloudlet, which operates as a small-scale cloud datacenter. The primary function of the cloudlet is to support computation-intensive tasks in real-time by offering offloading services. As a novel architectural component, the cloudlet extends conventional cloud computing capabilities into the FANET system, enabling efficient task processing and resource management (Ateya, 2019; Kim, 2019; Lee, 2018).

Each task $\mathcal{T}_{\mathcal{M}_{1\le i\le n}}$ is defined through three numerical values $\mathcal{T}_{\mathcal{M}_i} = \left\{d_{\mathcal{M}_i}, \mathcal{X}_{\mathcal{M}_i}, t^{max}_{\mathcal{M}_i}\right\}$ where $d_{\mathcal{M}_i}$ denotes the input-bits for computation, which may involve programs codes, configuration files and so on. $\mathcal{X}_{\mathcal{M}_i}$ represents the number of CPU cycles required to process one-bit input, which may vary from different applications and can be obtained through task profilers. Therefore, it needs $d_{\mathcal{M}_i} \times \mathcal{X}_{\mathcal{M}_i}$ CPU cycles in order to successfully execute the $\mathcal{M}_i$. $t^{max}_{\mathcal{M}_i}$ is the maximum latency required by the $\mathcal{M}_i$. In other word, $d_{\mathcal{M}_i} \times \mathcal{X}_{\mathcal{M}_i}$ indicates the $\mathcal{M}_i$'s computation complexity and $t^{max}_{\mathcal{M}_i}$ is the maximum permissible latency to accomplish the $\mathcal{M}_i$. Let $\mathcal{L}_{\mathcal{T}_{\mathcal{M}_i}}$ be the rate of computation offloading for the $\mathcal{T}_{\mathcal{M}_i}$. The $\mathcal{L}_{\mathcal{T}_{\mathcal{M}_i}}$ is decided by the $\mathcal{M}_i$, and it can be viewed as a continuous variable between 0 and 1. Denote $\mathcal{T}^{O}_{\mathcal{M}_i} = \left(d_{\mathcal{M}_i} \times \mathcal{L}_{\mathcal{T}_{\mathcal{M}_i}}\right)$ be the number of task input-bits for the offloading computing, and $\mathcal{T}^{L}_{\mathcal{M}_i} = \left(d_{\mathcal{M}_i} \times \left(1 - \mathcal{L}_{\mathcal{T}_{\mathcal{M}_i}}\right)\right)$ represents the number of task input-bits for the local computing (Chen, 2015; Guo, 2018).

To complete the computation task, each UAV can dynamically adjust the $\mathcal{L}_{\mathcal{T}_{\mathcal{M}}}$ value; the $\left(d_{\mathcal{M}_i} \times \mathcal{X}_{\mathcal{M}_i} \times \left(1 - \mathcal{L}_{\mathcal{T}_{\mathcal{M}_i}}\right)\right)$ part of $\mathcal{T}_{\mathcal{M}_i}$ is computed locally, and the $\left(d_{\mathcal{M}_i} \times \mathcal{X}_{\mathcal{M}_i} \times \mathcal{L}_{\mathcal{T}_{\mathcal{M}_i}}\right)$ part of $\mathcal{T}_{\mathcal{M}_i}$ is offloaded to the cloudlet at the BS. Therefore, individual UAV can coordinate the offloading computation and local computation. The computation offloading process consists of three main phases: i) transmitting data to the cloudlet via wireless channels, ii) executing the offloaded task on the cloudlet, and iii) retrieving the processed results from the cloudlet to the respective UAV. In most cases, the time overhead associated with the third phase is negligible, as the size of the computed results is generally much smaller than the initial input data. The input typically consists of UAV configurations, program codes, and input parameters. Consequently, the *BCOC* scheme primarily focuses on the first two phases, which encompass communication and computational overheads (Chen, 2015; Guo, 2018; Kim, 2019).

9.3.4 The main concepts of (α, β)-bargaining solutions

Bargaining involves the coalitions of two or more players collaborating to achieve a shared objective, with the primary challenge being the equitable distribution of total profit among participants. Since rationality and intelligence are fundamental principles in bargaining, any cooperative engagement must consider the goal of maximizing individual payoffs. This field of study is commonly referred to as bargaining game theory (Herrero, 2010; Trockel, 2009). Most cooperative bargaining models rely on either a disagreement point or a reference point as an anchor that influences negotiated agreements. The key distinction between these two approaches lies in their bargaining power: the disagreement point can be enforced unilaterally, whereas the reference point requires mutual recognition. Consequently, bargaining power derived from a disagreement point is often categorized as 'hard power', whereas that derived from a reference point is considered 'soft power' (Karagözoğlu, 2019; Kim, 2019).

In a simplified framework, aspirations can be understood as players' expectations of the best-case scenario in negotiations. While anchor points define the worst-case scenario, aspiration points represent the most favorable outcome. Since multiple options exist for both anchor and aspiration points, different selections can lead to entirely different negotiation outcomes. The *Kalai-Smorodinsky bargaining solution* (*KSBS*) employs the disagreement point as its anchor and the ideal point as its aspiration point in determining a settlement. In contrast, the *Gupta-Livne bargaining solution* (*GLBS*) uses the reference point as its anchor while maintaining the ideal point as the aspiration. Recently, the *tempered aspirations bargaining solution* (*TABS*) introduced a novel approach by deriving aspirations from the reference point instead of the disagreement point. TABS incorporates the disagreement point as an anchor while utilizing the tempered aspirations point as an aspiration reference. When the reference point solely determines both the anchor and aspiration points, this framework is referred to as the *local Kalai-Smorodinsky bargaining solution* (*LKSBS*). Figure 4 illustrates these four bargaining solutions (Karagözoğlu, 2019; Kim, 2019).

Figure 4. Four bargaining solutions based on cooperative games (Kim, 2019)

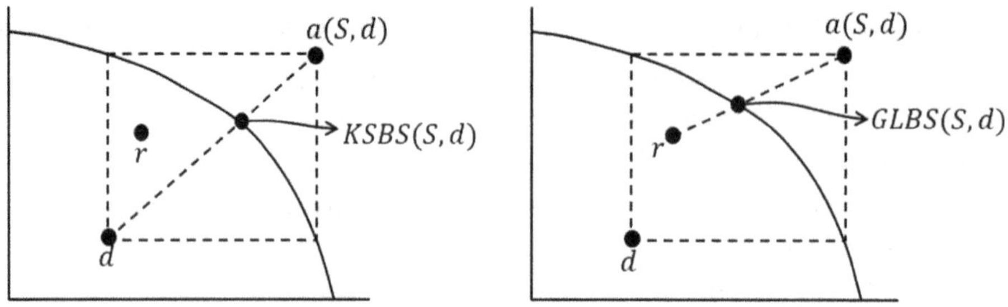

(a) *KSBS* (b) *GLBS*

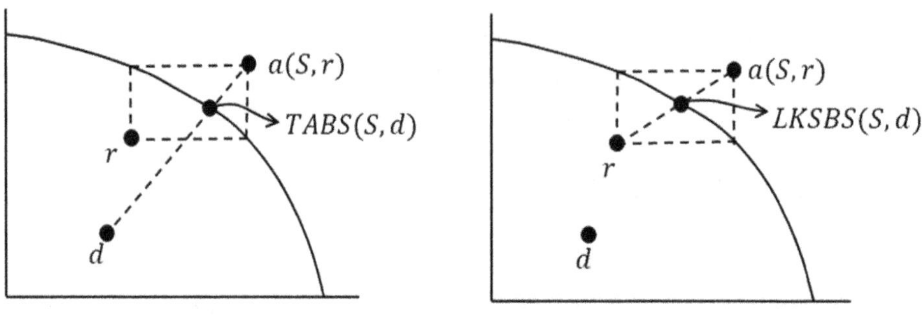

(c) *TABS* (d) *LKSBS*

A common feature of all four bargaining solutions mentioned above is that each establishes a feasible compromise between the worst-case and best-case scenarios. In essence, the specific context of a negotiation significantly impacts how the reference point shapes the final agreement. The role of contextual factors in bargaining outcomes is an intriguing area of research. In general, the prominence of the reference point is influenced by various contextual elements, which in turn affect players' aspirations. Consequently, two critical aspects in bargaining are the significance of the reference point and its impact on shaping players' expectations. To adaptively adjust these two factors, we introduce two parameters, $\alpha, \beta \in [0, 1]$, which capture the influence of the reference point on the anchor and its influence on players' aspirations, respectively. Higher values of α refer to higher influence on the anchor whereas higher values of β refer to higher influence on players' aspirations. This gives us a unifying framework for the study of bargaining problems with a reference point (Karagözoğlu, 2019; Kim, 2019).

Figure 5. The (α, β)-bargaining solution (Kim, 2019)

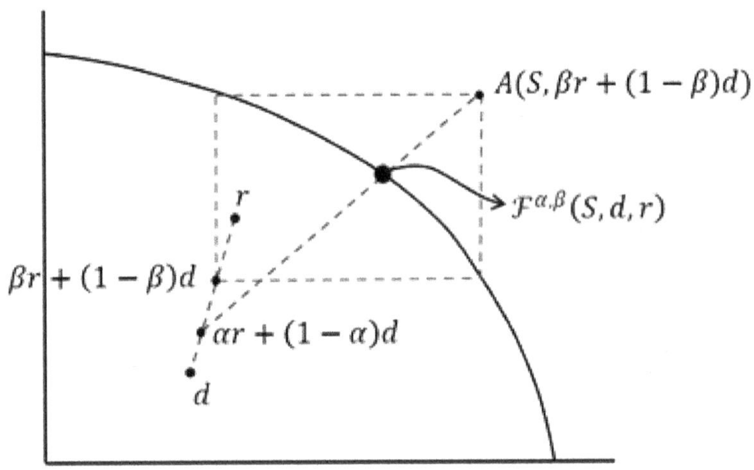

In bargaining problems, the two parameters α and β with a reference point allow us to obtain a family of bargaining solutions. This family encompasses some of the well-known solution concepts as special cases. For instance, If $\alpha = 0$ and $\beta = 0$, the reference point has no influence on the anchor point or the aspiration point. This bargaining solution is KSBS. If $\alpha = 1$ and $\beta = 0$, the reference point completely determines the anchor point but it has no influence on the aspiration point. This bargaining solution is GLBS. If $\alpha = 0$ and $\beta = 1$, the reference point has no influence on the anchor point but it completely determines the aspiration point. This bargaining solution is TABS. If $\alpha = 1$ and $\beta = 1$, the reference point completely determines both the anchor point and the aspiration point. This bargaining solution is LKSBS. Naturally, in between these four corner cases, there are infinitely many intermediate solution concepts that propose settlements by offering feasible compromises between the worst case and the best-case scenarios described by anchors and aspirations, respectively (Karagözoğlu, 2019; Kim, 2019).

To characterize the basic concepts of bargaining solutions, we preliminarily define some mathematical expressions. d is the disagreement point and r is the reference point where $r \geq d$ and $d, r \in$ S. $\mathbb{N} = \{1, 2 \ldots n\}$ is the set of game players, and game player i is an element of N where $i \in \mathbb{N}$. We assume that $\exists p \in \mathbb{R}^n_{++}, \exists q \in \mathbb{R}$ such that $\forall_x \in S: \sum_{i \in \mathbb{N}} (p_i \times x_i) \leq q$. For every $x \in S$ with $x > d$, let $\mathscr{A}(S,x)$ be the aspiration vector such that $\mathscr{A}_i(S,x) \equiv \max\{t \in \mathbb{R} | (t, x_{-i}) \in S\}$ where $x_{-i} = (x_1, \ldots, x_{i-1}, x_{i+1}, \ldots x_n)$. Accordingly, $\mathscr{A}(S,d)$ is the ideal point and $\mathscr{A}(S,r)$ is the tempered aspirations point. Let Σ^n be the class of all bargaining problems with a reference point. A solution concept for such problems is a function F: Σ^n $\mathbb{R}^n$ that associates each $(S, d, r) \in \Sigma^n$ with a unique point of S (Karagözoğlu, 2019; Kim, 2019).

By using the $\alpha, \beta \in [0,1]$ factors, we define the (α, β)-solution where α can be interpreted as the power of the reference point in determining the anchor point, and β can be interpreted as the influence of the reference point in shaping players' aspirations. For every $(S,d,r) \in \Sigma^n$, the bargaining solution with a given $(\alpha,\beta) \in [0,1]^2$ $(\mathscr{F}^{\alpha,\beta}(S,d,r))$ is defined as follows (Karagözoğlu, 2019; Kim, 2019);

$$\mathscr{F}^{\alpha,\beta}(S,d,r) = (\lambda^* \times F^{\mathscr{A}}) + [(1-\lambda^*) \times F^d] \tag{9.3.1}$$

$$\text{s.t.,} \begin{cases} \mathscr{F}^{\mathscr{A}} = A\big(S,((\beta \times r) + ((1-\beta) \times d))\big) \\ \mathscr{F}^{d} = ((\alpha \times r) + ((1-\alpha) \times d)) \\ \lambda^* = max\big\{\lambda \in [0,1] \,|\, ((\lambda \times \mathscr{F}^{\mathscr{A}}) + ((1-\lambda) \times \mathscr{F}^{d})) \in S\big\} \end{cases}$$

For every $\alpha,\beta \in [0,1]$, the (α,β)-solution proposes the maximum point of the bargaining set on the line segment connecting the aspiration point $\mathscr{F}^{\mathscr{A}}$ and the anchor point $\mathscr{F}^{d}$. As shown in Figure 9.3.1 and 9.3.2, the collection of all such possible solutions with $0 \leq \alpha,\beta \leq 1$ constitutes the (α,β)-family of bargaining solutions (Karagözoğlu, 2019; Kim, 2019).

9.3.5 The UAV computation task offloading algorithm

In a FANET system, energy consumption, time delay, and communication costs are critical control parameters. Given that most UAVs operate with constrained energy resources and limited computational capacity, optimizing these factors is essential for enhancing UAV performance. To address this challenge, the *BCOC* scheme introduces mathematical models to quantify the following overheads: i) computational overhead, ii) communication overhead, and iii) delay overhead. These formulations enable a systematic approach to managing UAV resources efficiently, ensuring improved operational effectiveness. First, to calculate the computing overhead, the *BCOC* scheme considers two different computing types; local computing and offloading computing. The local computing amount $\left(\mathscr{T}^{L}_{\mathscr{M}}\right)$ should be decided related to the current remaining energy of M $\left(\delta_{\mathscr{M}}\right)$, and the current computation amount in the cloudlet (I^t), which is the total sum of all UAVs offloading requests. According the $\delta_{\mathscr{M}}$ and I^t, the computing overhead for the $\mathscr{T}_{\mathscr{M}}$ $\left(\mathscr{Q}(\mathscr{T}_{\mathscr{M}}, \mathscr{L}_{\mathscr{M}}, \delta_{\mathscr{M}}, I^t)\right)$ is defined in an online manner (Kim, 2019);

$$\mathscr{Q}(\mathscr{T}_{\mathscr{M}}, \mathscr{L}_{\mathscr{M}}, \delta_{\mathscr{M}}, I^t) = \left(\mathscr{T}^{L}_{\mathscr{M}} \times \delta_{\mathscr{M}}\right) + \left(\mathscr{T}^{O}_{\mathscr{M}} \times \text{Ä}(I^t)\right) \tag{9.3.2}$$

$$\text{s.t.,} \begin{cases} \mathscr{T}^{L}_{\mathscr{M}} = \left(d_{\mathscr{M}} \times \mathscr{X}_{\mathscr{M}} \times \left(1 - \mathscr{L}_{\mathscr{T}_{\mathscr{M}}}\right)\right) and\, \mathscr{T}^{O}_{\mathscr{M}} = \left(d_{\mathscr{M}} \times \mathscr{X}_{\mathscr{M}} \times \mathscr{L}_{\mathscr{T}_{\mathscr{M}}}\right) \\ \delta_{\mathscr{M}} = \phi_{\mathscr{M}} \times \dfrac{(\mathscr{E}_{\mathscr{M}} - c_{\mathscr{M}})}{\mathscr{E}_{\mathscr{M}}} and\, \text{Ä}(I^t) = \log\left(\frac{I^t}{C}\right)^{\varepsilon} \end{cases}$$

where $\mathscr{L}_{\mathscr{T}_{\mathscr{M}}}$ is the task splitting ratio to partition the $\mathscr{T}_{\mathscr{M}}$ into two parts; $\mathscr{T}^{L}_{\mathscr{M}}$ and $\mathscr{T}^{O}_{\mathscr{M}}$. $c_{\mathscr{M}}$ and $\mathscr{E}_{\mathscr{M}}$ are the current remaining energy and the initial energy amount of M, respectuvely. C is the total computation capacity of cloudlet in the BS. $\phi_{\mathcal{M}}$ and ε are control parameters to estimate local and offload computing overheads, respectively. Second, to estimate the communication overhead, the *BCOC* scheme mainly concerns the extra energy consumption for data transmissions. Simply, the *BCOC* scheme assumes that ξ_C is the extra energy factor per bit for data offloading communications. According to the ξ_C, the communication overhead $\left(\mathcal{C}(\mathcal{T}_{\mathcal{M}}, \mathcal{L}_{\mathcal{M}})\right)$ is defined as $\mathcal{C}(\mathcal{T}_{\mathcal{M}}, \mathcal{L}_{\mathcal{M}}) = \mathcal{T}^{O}_{\mathcal{M}} \times \xi_C$ (Kim, 2019).

Third, to calculate the delay overhead, the *BCOC* scheme mainly focuses on the total time delay between $\mathcal{T}^{O}_{\mathcal{M}}$ and $\mathcal{T}^{L}_{\mathcal{M}}$. To estimate the time delay of $\mathcal{T}^{L}_{\mathcal{M}}$, the computing power of UAV is a critical issue. To estimate the time delay of $\mathcal{T}^{O}_{\mathcal{M}}$, the current computing power of cloudlet and data transmission rate (*DTR*) are major concerns. Usually, *DTR* is defined as the number of information bits that are transmitted without error. In the FANET's effective range, it can be achieved based on the *signal-to-interference-plus-noise* ratio. A general formula for the $\mathcal{M}_i$'s *DTR* ($DTR_{\mathcal{M}_i}(\mathbb{M})$) is defined as follows (Kim, 2019).

$$DTR_{\mathcal{M}_i}(\mathbb{M}) = \omega \times \log_2\left(1 + \frac{P_{\mathcal{M}_i} \times G_{\mathcal{M}_i BS}}{\sigma_0 + \sum_{\mathcal{M}_j \in \mathbb{M} \setminus \mathcal{M}_i}\left(P_{\mathcal{M}_j} \times G_{\mathcal{M}_j BS}\right)}\right) \tag{9.3.3}$$

where ω is the wireless channel bandwidth and $P_{\mathcal{M}_i}$ is the $\mathcal{M}_i$'s transmission power. $G_{\mathcal{M}_i BS}$ denotes the channel gain between the $\mathcal{M}_i$ and the BS. σ_0 is the background noise power. From the communication model in (9.3.3), we see that if too many UAVs choose to offload the computation task via the wireless channel simultaneously, they may incur severe interference, leading to low DTR; it would negatively affect the offloading performance of FANET system. According to the $DTR_{\mathcal{M}}(\mathbb{M})$, and computing powers of M and cloudlet, the delay overhead for the $\mathcal{T}_{\mathcal{M}}$ $\left(\mathcal{D}(\mathcal{T}_{\mathcal{M}}, \mathcal{L}_{\mathcal{M}})\right)$ is defined as follows (Kim, 2019);

$$\mathcal{D}(\mathcal{T}_{\mathcal{M}}, \mathcal{L}_{\mathcal{M}}) = \left(\left(\frac{\mathcal{T}^{O}_{\mathcal{M}}}{DTR_{\mathcal{M}}(\mathbb{M})} \times \frac{1}{\grave{E}_{\mathcal{M}}}\right) + \left(\mathcal{T}^{O}_{\mathcal{M}} \times \frac{1}{(\mathbb{C} - I^t)}\right)\right) + \left(\mathcal{T}^{L}_{\mathcal{M}} \times \frac{1}{\grave{E}_{\mathcal{M}}}\right) \tag{9.3.4}$$

where $\grave{E}_{\mathcal{M}}$ is the computing power of M. By jointly considering $\mathcal{Q}(\mathcal{T}_{\mathcal{M}}, \mathcal{L}_{\mathcal{M}}, \delta_{\mathcal{M}}, I^t)$, $\mathcal{C}(\mathcal{T}_{\mathcal{M}}, \mathcal{L}_{\mathcal{M}})$ and $\mathcal{D}(\mathcal{T}_{\mathcal{M}}, \mathcal{L}_{\mathcal{M}})$, the *BCOC* scheme can define the utility functions for the bargaining game. The *BCOC* scheme designs a new (α, β)-bargaining game model $\left(\mathbb{G}_{(\alpha,\beta)}\right)$ for the UAV's task offloading service. To implement the $\mathbb{G}_{(\alpha,\beta)}$ in each UAV, the *BCOC* scheme defines game players, their strategies, and utility functions, etc., where $\mathbb{G}_{(\alpha,\beta)} = \left\{\{\mathcal{T}^{L}_{\mathcal{M}}, \mathcal{T}^{O}_{\mathcal{M}}\}, \mathcal{L}_{\mathcal{T}_{\mathcal{M}}}, \{U^{L}_{\mathcal{M}}, U^{O}_{\mathcal{M}}\}, (\alpha,\beta), d_{(\mathcal{T}^{L}_{\mathcal{M}}, \mathcal{T}^{O}_{\mathcal{M}})}(S, x), r_{(\mathcal{T}^{L}_{\mathcal{M}}, \mathcal{T}^{O}_{\mathcal{M}})}(S, x), a_{(\mathcal{T}^{L}_{\mathcal{M}}, \mathcal{T}^{O}_{\mathcal{M}})}(S, x)\right\}$ (Kim, 2019):

- Players: in each UAV, local and global computing task parts. i.e., $\mathcal{T}^{L}_{\mathcal{M}}$ and $\mathcal{T}^{O}_{\mathcal{M}}$, are game players to process the UAV's task offloading service.
- Strategies: task splitting ratio, i.e., $0 \leq \mathcal{L}_{\mathcal{T}_{\mathcal{M}}} \leq 1$, is the strategy of the game $\mathbb{G}_{(\alpha,\beta)}$. To reduce computation complexity, strategies are specified in terms of basic control unit (Δ). Therefore, $\mathcal{L}_{\mathcal{T}_{\mathcal{M}}}$ is defined as discrete real-number levels.
- Utility functions: $\mathcal{T}^{L}_{\mathcal{M}}$ and $\mathcal{T}^{O}_{\mathcal{M}}$ players have their own utility functions $U^{L}_{\mathcal{M}}$ and $U^{O}_{\mathcal{M}}$, respectively. Each utility function maps player's satisfaction to a real number, which represents the resulting payoff in the game $\mathbb{G}_{(\alpha,\beta)}$. $U^{L}_{\mathcal{M}}$ and $U^{O}_{\mathcal{M}}$ are defined as follows (Kim, 2019);

$$
\begin{cases}
U^{L}_{\mathcal{M}}\left(\mathcal{T}_{\mathcal{M}}, \mathcal{L}_{\mathcal{T}_{\mathcal{M}}}\right) = \frac{\Psi}{\left(\mathcal{T}^{L}_{\mathcal{M}} \times \delta_{\mathcal{M}}\right)} + \frac{\Psi}{\left(\mathcal{T}^{L}_{\mathcal{M}} \times \frac{1}{\Theta_{\mathcal{M}}}\right)} \\
U^{O}_{\mathcal{M}}\left(\mathcal{T}_{\mathcal{M}}, \mathcal{L}_{\mathcal{T}_{\mathcal{M}}}\right) = \frac{\Psi}{\left(\mathcal{T}^{O}_{\mathcal{M}} \times \Delta(I^{t})\right)} + \frac{\Psi}{\mathcal{C}\left(\mathcal{T}_{\mathcal{M}}, \mathcal{L}_{\mathcal{M}}\right)} + \frac{\Psi}{\left(\Omega_{\mathcal{T}_{\mathcal{M}}} \times \left(\mathcal{T}^{O}_{\mathcal{M}} + \mathcal{T}^{L}_{\mathcal{M}}\right)\right)} \\
s.t., \begin{cases} \Psi = 1, if\, D\left(\mathcal{T}_{\mathcal{M}}, \mathcal{L}_{\mathcal{M}}\right) \leq \left(\left(\frac{\mathcal{T}^{L}_{\mathcal{M}}}{DTR_{\mathcal{M}}(\mathbb{M})} \times \frac{1}{\Theta_{\mathcal{M}}}\right) + \left(\mathcal{T}^{O}_{\mathcal{M}} \times \left(\frac{1}{(\mathbb{C} - I^{t})}\right)\right)\right) \\ \Psi = 0, otherwise \end{cases}
\end{cases}
\tag{9.3.5}
$$

where $\Omega_{\mathcal{T}_{\mathcal{M}}}$ is the maximum permissible latency of $\mathcal{T}_{\mathcal{M}}$.

- (α, β) values are used to determine the anchor and aspirations points;α can be interpreted as the power of the reference point in determining the anchor point, and β can be interpreted as the influence of the reference point in shaping players' aspirations.
- Disagreement point: the disagreement points of $\mathcal{T}^{L}_{\mathcal{M}}$ and $\mathcal{T}^{O}_{\mathcal{M}}$ $\left(d_{(\mathcal{T}^{L}_{\mathcal{M}}, \mathcal{T}^{O}_{\mathcal{M}})}(S, x)\right)$ is expected to be the result of non-cooperative actions; it means a failure of the bargaining process.
- Reference point: the reference point of $\mathcal{T}^{L}_{\mathcal{M}}$ and $\mathcal{T}^{O}_{\mathcal{M}}$ is denoted as $r_{(\mathcal{T}^{L}_{\mathcal{M}}, \mathcal{T}^{O}_{\mathcal{M}})}(S, x)$; it satisfies two features, i) $r_{(\mathcal{T}^{L}_{\mathcal{M}}, \mathcal{T}^{O}_{\mathcal{M}})} \in S \setminus WPO(S)$, and ii) $r_{(\mathcal{T}^{L}_{\mathcal{M}}, \mathcal{T}^{O}_{\mathcal{M}})} > d_{(\mathcal{T}^{L}_{\mathcal{M}}, \mathcal{T}^{O}_{\mathcal{M}})}$.
- Aspirations point: the aspirations point of $\mathcal{T}^{L}_{\mathcal{M}}$ and $\mathcal{T}^{O}_{\mathcal{M}}$ is denoted as $\mathcal{A}_{(\mathcal{T}^{L}_{\mathcal{M}}, \mathcal{T}^{O}_{\mathcal{M}})}(S, x)$; it is defined based on the $d_{(\mathcal{T}^{L}_{\mathcal{M}}, \mathcal{T}^{O}_{\mathcal{M}})}$ and $r_{(\mathcal{T}^{L}_{\mathcal{M}}, \mathcal{T}^{O}_{\mathcal{M}})}$ with the (α, β) values.

In the $\mathbb{G}_{(\alpha,\beta)}$, the *BCOC* scheme defines the $d_{(\mathcal{T}^{L}_{\mathcal{M}}, \mathcal{T}^{O}_{\mathcal{M}})} = \left(U^{L}_{\mathcal{M}}(\mathcal{T}_{\mathcal{M}}, 0), U^{O}_{\mathcal{M}}(\mathcal{T}_{\mathcal{M}}, 1)\right)$ and the $r_{(\mathcal{T}^{L}_{\mathcal{M}}, \mathcal{T}^{O}_{\mathcal{M}})} = \left(U^{L}_{\mathcal{M}}(\mathcal{T}_{\mathcal{M}}, ½), U^{O}_{\mathcal{M}}(\mathcal{T}_{\mathcal{M}}, ½)\right)$. Under diverse FANET environments, the α and β values are dynamically adjusted by considering the current UAV's energy status and the cloudlet's computing load. When the $c_{\mathcal{M}}$ value is low, the UAV should avoid the extra energy consumption for data transmissions current. If the I^{t} value is low, the cloudlet can have a more computing capacity to support offloading services. Therefore, in the *BCOC* scheme, each individual UAV adaptively adjusts the α and β values as follows (Kim, 2019);

$$
\alpha = \frac{c_{\mathcal{M}}}{\mathcal{E}_{\mathcal{M}}} \text{ and } \beta = \frac{I^{t}}{\mathbb{C}}
\tag{9.3.6}
$$

For every $(S, d, r) \in \Sigma^{n}$, the (α, β)-bargaining solution with a given $(\alpha, \beta) \in [0, 1]^{2}$ $\left(\mathcal{F}^{\alpha,\beta}\left(S, d_{(\mathcal{T}^{L}_{\mathcal{M}}, \mathcal{T}^{O}_{\mathcal{M}})}, r_{(\mathcal{T}^{L}_{\mathcal{M}}, \mathcal{T}^{O}_{\mathcal{M}})}\right)\right)$ is defined as follows (Kim, 2019);

$$
\mathcal{F}^{\alpha,\beta}\left(S, d_{(\mathcal{T}^{L}_{\mathcal{M}}, \mathcal{T}^{O}_{\mathcal{M}})}, r_{(\mathcal{T}^{L}_{\mathcal{M}}, \mathcal{T}^{O}_{\mathcal{M}})}\right) = (\lambda^{*} \times F^{\mathcal{A}}) + [(1 - \lambda^{*}) \times F^{d}]
\tag{9.3.7}
$$

$$
\text{s.t.,}\begin{cases}
\mathscr{F}^{\mathscr{A}} = \mathscr{A}_{(\mathscr{T}^{L}_{\mathscr{M}},\mathscr{T}^{O}_{\mathscr{M}})}\left(S,\left(\left(\frac{I^{t}}{\mathbb{C}}\times r_{(\mathscr{T}^{L}_{\mathscr{M}},\mathscr{T}^{O}_{\mathscr{M}})}\right)+\left(\left(1-\frac{I^{t}}{\mathbb{C}}\right)\times d_{(\mathscr{T}^{L}_{\mathscr{M}},\mathscr{T}^{O}_{\mathscr{M}})}\right)\right)\right) \\
\mathscr{F}^{d} = \left(\left(\frac{c_{\mathscr{M}}}{\mathscr{E}_{\mathscr{M}}}\times r_{(\mathscr{T}^{L}_{\mathscr{M}},\mathscr{T}^{O}_{\mathscr{M}})}\right)+\left(\left(1-\frac{c_{\mathscr{M}}}{\mathscr{E}_{\mathscr{M}}}\right)\times d_{(\mathscr{T}^{L}_{\mathscr{M}},\mathscr{T}^{O}_{\mathscr{M}})}\right)\right) \\
\lambda^{*} = max\left\{\lambda \in [0,1] \,\middle|\, ((\lambda\times\mathscr{F}^{\mathscr{A}})+((1-\lambda)\times\mathscr{F}^{d})) \in S\right\}
\end{cases}
$$

9.3.6 Summary

With recent technological advancements, FANETs are emerging as a promising solution for various UAV-based applications, enabling autonomous flight without human intervention. However, UAV applications often require complex and resource-intensive computations, while being constrained by limited onboard processing capabilities. This limitation makes computation offloading a viable approach to overcoming these challenges. To efficiently handle computation-intensive UAV applications while ensuring acceptable QoS and optimized energy consumption, the *BCOC* scheme addresses the task-splitting problem. To further enhance FANET performance, the *BCOC* scheme introduces an innovative UAV offloading mechanism based on a cooperative bargaining game model. According to the (α, β)-bargaining solution, individual UAVs can adaptively decide the ratio of the local computing and offload computing while maximizing their own utility payoffs under dynamic FANET environments (Kim, 2019).

9.4 High-speed Railway Communication Control (*HRCC*) scheme

The rapid expansion of data traffic in high-speed railway (HSR) communications has drawn significant research interest, driving technological advancements for next-generation mobile networks. However, conventional terrestrial communication networks struggle to meet the diverse service demands of HSR systems while maintaining seamless coverage across vast areas. To address this challenge, S. Kim recently introduces the *High-speed Railway Communication Control* (*HRCC*) scheme for HSR communications (Kim, 2021b). By integrating satellite communications, the *HRCC* scheme operates on an integrated terrestrial-satellite network (ITSN) platform. Utilizing weighted proportional gain and loss bargaining solutions, the *HRCC* scheme dynamically allocates the limited ITSN spectrum while effectively coordinating interactions between satellite and terrestrial networks. To maximize the benefits of collaborative control, the coordinated bargaining framework facilitates mutual advantages and ensures a reciprocal consensus in dynamically evolving HSR communication environments. The resulting solution offers globally optimal properties, providing strategic guidelines for ITSN operations (Kim, 2021b).

9.4.1 Development motivation

6G communication technology is poised to bring a transformative shift across various IoT applications, including smart traffic systems, environmental monitoring, virtual reality, and high-definition video streaming. As the anticipated successor to 5G, 6G will enable a new wave of intelligent applications by integrating cutting-edge technologies such as artificial intelligence, machine learning, quantum communication, tactile Internet, and big data analytics. Currently, the number of IoT users is increasing at an annual growth rate of approximately 25%, with projections estimating 80 billion connected devices by 2030. The integration of IoT with 6G communications is expected to redefine the future of hyper-connectivity, enabling a seamless digital experience for all users. To achieve the global vision of 6G, extensive coverage must be ensured through the deployment of highly heterogeneous networks, diverse communication scenarios, and innovative service models (Akhtar, 2020; Chettri, 2019; Kim, 2021b).

One of the key advancements in 6G communication is its ability to expand coverage and mobility through satellite communications. Traditional terrestrial networks predominantly serve urban users, often struggling with poor coverage in remote or challenging environments. Due to geographical constraints and fluctuating traffic demands in specific regions, terrestrial mobile networks may not always provide seamless service. In such cases, satellite-based mobile communications offer a viable solution for achieving universal connectivity. As a result, the 6G wireless communication network is expected to integrate terrestrial, aerial, and maritime communications, forming a comprehensive heterogeneous network. This hybrid system is designed to be highly reliable, capable of high-speed connectivity, and optimized to support a massive number of IoT devices. While extensive research has been conducted on integrating different network infrastructures, developing an efficient hybrid system that ensures seamless global coverage remains a significant challenge (Akhtar, 2020; Kim, 2021b; Yan, 2020).

Over the past decade, the rapid advancement of high-speed railway (HSR) communication has become a defining feature of intelligent transportation systems. HSR networks aim to deliver uninterrupted, high-quality connectivity, supporting a variety of smart applications such as train multimedia services, IoT-based railway monitoring, and onboard video surveillance. However, conventional HSR communication systems face several critical limitations, including inadequate network capacity, inefficient spectrum utilization, and insufficient support for high-speed data services. Furthermore, since HSR networks span vast geographical regions, establishing a dedicated terrestrial communication infrastructure is impractical and cost-prohibitive. To address these challenges, next-generation HSR communication systems must i) manage the growing volume of data traffic, ii) ensure train operational safety, and iii) deliver real-time multimedia entertainment services (Gao, 2020; Kim, 2021b; Xu, 2016).

With the evolution of wireless networks toward 6G, integrating satellite and terrestrial networks presents a promising solution for enhancing HSR services. In this framework, data traffic generated by HSR communications can be efficiently offloaded to satellite networks. Thus, an integrated terrestrial-satellite network (ITSN) is proposed to support multi-carrier wireless communications. By leveraging the unique advantages of satellite communications, the ITSN platform enhances the efficiency of conventional mobile networks, offering complementary functionalities. The convergence of satellite and terrestrial systems has the potential to significantly improve global information exchange, addressing the limitations of existing terrestrial networks. Given the rising demand for HSR connectivity, orchestrating satellite and terrestrial communications will become an inevitable trend in future network architectures (Kim, 2021b; Wang, 2019a, 2020).

As HSR service requirements continue to expand, spectrum resource management in the ITSN system has become a critical area of research. The increasing demand for HSR services, coupled with limited spectrum resources, necessitates the development of effective spectrum management strategies to optimize resource utilization while meeting diverse QoS requirements. However, due to the complexities of heterogeneous network integration and the dynamic nature of HSR services, traditional spectrum allocation methods may prove inadequate. Given the growing need for advanced QoS features and multi-carrier service support within the ITSN system, designing an optimized spectrum management algorithm has become a priority. This challenge has gained significant research attention, yet remains a complex and unresolved issue (Gao, 2020; Kim, 2021b; Xu, 2016).

The *HRCC* scheme introduces an innovative spectrum allocation mechanism tailored for ITSN-based HSR services. Recognizing the unique characteristics of ITSN infrastructure, the *HRCC* scheme optimizes system performance by refining spectrum allocation strategies for HSR communications. From the perspective of satellite and terrestrial network operators, efficiency and fairness are key performance indicators in ITSN systems. Given the conflicting requirements of different network elements, effective mediation of these control challenges is essential. To achieve an optimal balance, cooperative bargaining game theory serves as a suitable framework for negotiating trade-offs in ITSN spectrum allocation. The *HRCC* scheme applies bargaining game solutions, enabling rational network agents to make informed decisions and execute control actions that maximize overall system performance (Kim, 2021b).

In 1994, W. Thomson introduced the *weighted proportional gains solution* (*WPGS*), inspired by the fundamental principles of the *Kalai-Smorodinsky bargaining solution*. The *WPGS* approach accommodates asymmetries in bargaining power among players by determining the optimal solution along a trajectory extending from the disagreement point toward the utopia point. In the classical Kalai-Smorodinsky framework, the increments progress along a straight line between these two points. However, in the *WPGS*, these increments are adjusted proportionally to both bargaining power and ideal claims. In 2011, J. Segarra and M. Vilar proposed an alternative bargaining concept known as the *weighted proportional losses solution* (*WPLS*), which accounts for asymmetric bargaining power among participants. Unlike previous methods, the *WPLS* introduces an endogenous reference point, ensuring that mutual concessions between players are achieved even under uncertain probabilistic conditions. In asymmetric bargaining scenarios, the *WPLS* assigns losses to players in proportion to their ideal payoffs and bargaining powers. This solution maintains individual rationality while satisfying key bargaining principles. From a game-theoretic perspective, the *WPLS* can be interpreted as a normalized weighted equal losses solution, making it a valuable tool for spectrum allocation in ITSN-based HSR communications (Alós, 2018; Kim, 2021b; Segarra, 2011).

9.4.2 Contribution

Leveraging the principles of *WPGS* and *WPLS*, the *HRCC* scheme efficiently addresses the spectrum allocation challenge to enhance HSR services. Within the ITSN platform, HSR services can simultaneously connect to both satellite and terrestrial BSs. These services are classified into two categories - class I and class II -based on their QoS requirements. To manage the spectrum allocation effectively while accounting for mutual interactions, the *HRCC* scheme introduces a dual-level bargaining game model for resource distribution. The upper-level bargaining game is designed for class I data services where ITSN spectrum resources are dynamically allocated based on the *WPGS* approach. The lower-level bargaining game is developed for class II data services, in which ITSN spectrum is adaptively distributed using the *WPLS* framework. By implementing a step-by-step interactive cooperative process, this two-tier bargaining strategy ensures an optimal QoS balance in HSR communications, dynamically adapting to varying service demands (Kim, 2021b).

9.4.3 ITSN system infrastructure and bargaining game model

To ensure seamless wireless connectivity for HSR communications, the ITSN system integrates satellite and terrestrial cellular networks, forming a multi-bearer communication framework. The satellite network offers stable connections with shorter propagation delays, making it well-suited for HSR communications. In general, satellite communications experience significantly fewer fluctuations compared to terrestrial BS communications, particularly in high-mobility scenarios. As a result, satellite networks can complement terrestrial networks by providing not only extensive coverage but also enhanced adaptability to dynamic fluctuations in terrestrial network conditions. This capability enables high-mobility support for HSR communications. To maximize the flexibility and efficiency of the ITSN system, space-based satellites and ground BSs must operate in coordination, working cooperatively to optimize network performance. An overview of the ITSN platform is illustrated in Figure 6 (Kim, 2021b; Wang, 2019a; Yan, 2020).

Figure 6. The integrated terrestrial-satellite architecture for HSRs (Kim, 2021b)

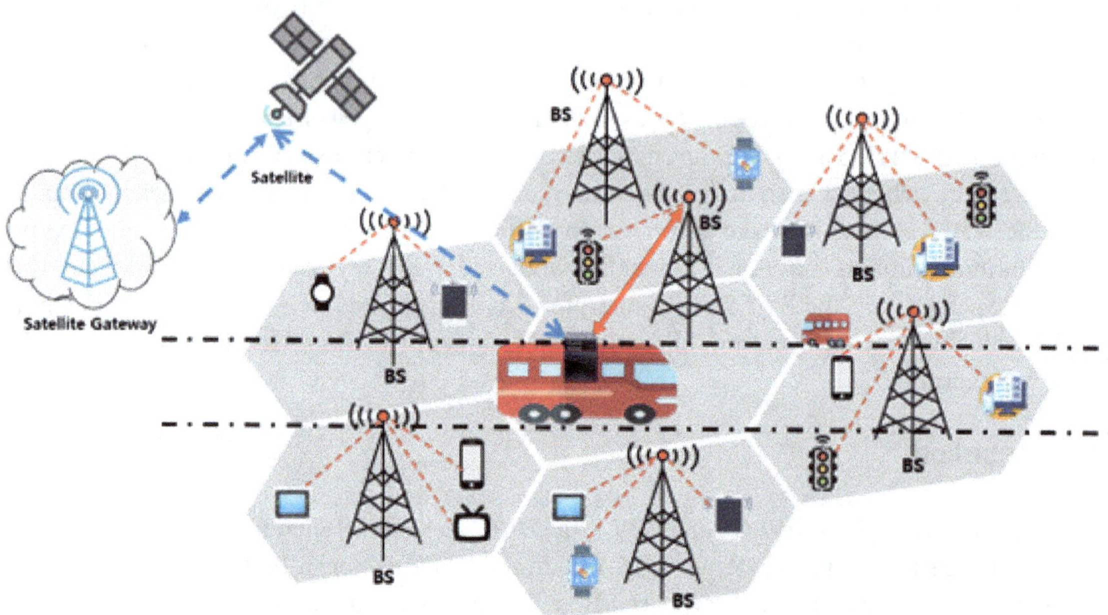

From a reliability perspective, HSR communication services can be categorized into two types: safety-critical (class I) data services and non-safety (class II) data services. To ensure train safety, class I services are given higher priority, as they demand exceptional performance in HSR communications. Meanwhile, class II services typically require higher data rates but are more tolerant to delays. By integrating satellite and terrestrial networks, the *HRCC* scheme aggregates diverse spectrum resources within a unified ITSN system and dynamically allocates them to optimize overall system performance. Since network resource capacities vary, different networks are capable of supporting different levels of data services. This leads to a critical challenge in such a complex environment: How can data services be integrated into a unified framework to efficiently process HSR communications? Addressing this question is essential for ensuring seamless and optimized connectivity in high-speed railway networks (Kim, 2021b; Yan, 2020).

In the ITSN infrastructure, the *HRCC* scheme assumes that there is one satellite ($\mathcal{S}$), one running train ($\mathcal{T}$) and the set ($\mathbb{B}$) of multiple BSs where $\mathbb{B} = \{\mathcal{B}_1, \mathcal{B}_2, \ldots, \mathcal{B}_n\}$. The T will generate class I and II data services, which should be sent simultaneously to the S and B. The *HRCC* scheme considers a discrete time model $T \in \{t_1, \ldots, t_c, t_{c+1}, \ldots\}$, where the length of a time slot matches the event timescale at which the running T can be contacting a specific B while moving along a railway. The *HRCC* scheme formulates the HSR's spectrum allocation process as a dual-level bargaining model ($\mathbb{G}$). The upper and lower bargaining games, i.e., $\mathbb{G}^U$ and $\mathbb{G}^L$, are modelled as two-player games, and the S and B are game players at each t. In the $\mathbb{G}^U$ (*or* $\mathbb{G}^L$), the spectrum resources of S and B are assigned for the T's class I (*or* class II) data services through the concept of *WPGS* (*or WPLS*). The $\mathbb{G}^U$ and $\mathbb{G}^L$ are sequentially repeated during the step-by-step iteration. Therefore, the spectrum allocation problem for HSR communications is solved based on the interactive and repeated manner at each time period. In the *HRCC* scheme, the amount of spectrum allocation is specified in terms of basic spectrum units to reduce computation complexity. Formally, the *HRCC* scheme defines the game G's entities for the HSR services in the ITSN system infrastructure, i.e,

$$\mathbb{G} = \{\mathbb{G}^U, \mathbb{G}^L\} = \left\{\mathcal{S}, \mathbb{B}, \mathcal{T}, \left\{\mathbb{G}^U | \left(\mathfrak{U}^U_{\mathcal{S}}, \mathfrak{U}^U_{\mathcal{B}}\right), \left(X^U_{\mathcal{S}}, X^U_{\mathcal{B}}\right)\right\}, \left\{\mathbb{G}^L | \left(U^L_{\mathcal{S}}, U^L_{\mathcal{B}}\right), \left(Y^L_{\mathcal{S}}, Y^L_{\mathcal{B}}\right)\right\}, \text{Ä}, T\right\}$$

of gameplay (Kim, 2021b).

- In the G, $\mathbb{G}^U$ and $\mathbb{G}^L$ are mutual and reciprocal interdependent in an interactive manner, and they work together to provide the T's class I and class II data traffic services.
- S is a satellite agent and B is the set of BSs $\left(\mathcal{B}_{1 \le i \le n}\right)$, which are located on the terrestrial ground and have good coverage along the rail tracks.
- S and B are game players in the $\mathbb{G}^U$ and $\mathbb{G}^L$ at each t for HSR services.
- In the $\mathbb{G}^U$, $\mathfrak{U}^U_{\mathcal{S}}$ and $\mathfrak{U}^U_{\mathcal{B}}$ are utility functions of S and B, respectively. During the bargaining process, $X^U_{\mathcal{S}}$ (*or* $X^U_{\mathcal{B}}$) is the allocated spectrum amount by S (*or* B).
- In the $\mathbb{G}^L$, $U^L_{\mathcal{S}}$ and $U^L_{\mathcal{B}}$ are utility functions of S and B, respectively. During the bargaining process, $Y^L_{\mathcal{S}}$(*or* $Y^L_{\mathcal{B}}$) is the allocated spectrum amount by S (*or* B).
- Δ is a basic spectrum unit to allocate the spectrum resource in the HSR service.
- $T = \{t_1, \ldots, t_c, t_{c+1}, \ldots\}$ denotes time, which is represented by a sequence of time steps.

9.4.4 The basic concepts of *WPGS* and *WPLS*

A bargaining problem is a pair (S, d), and Σ is defined as the set of bargaining problems. A solution is a function $f{:}\Sigma \rightarrow \mathbb{R}^2$ such that, for each $(S,d) \in \Sigma, f(S,d) \in S$. The point $f(S,d)$ is the solution point of (S,d). Without loss of generality, the domain Σ_0 is the problem whose disagreement point is zero. To reflect relative differences among each player $1 \le i \le n$, we define a parameter α; $\alpha \in \Delta^{n-1} = \left\{\beta \in \mathbb{R}^n_+ | \sum_{i=1}^{n} \beta_i = 1\right\}$ where $\mathbb{R}^n_+ = \left\{x \in \mathbb{R}^n | x \ge 0\right\}$ (Kim, 2021b; Segarra, 2011).

The utopia points of (S,d) is the highest possible utility payoffs for all players. For a given S, the point $m(S)$ is the utopia point where $m_i(S)$ defines the maximum outcome for player i on the set S. The *strong Pareto* frontier of S is the set $SPO(S) \equiv \{x \in S | ify \rangle x then y \notin S\}$ and the *weak Pareto* frontier is the set $WPO(S) \equiv \{x \in S | if y \gg x then y \notin S\}$. Formally, for a two-player bargaining game $S \in \Sigma_0$, the *WPGS* with weights $\alpha \in \text{int}\{\Delta^{n-1}\}$ is defined as follows (Kim, 2021b; Segarra, 2011);

$$WPGS^{\alpha}(S) = \left\{x \in \mathbb{R}^2_+ \middle| x_2 = \left(\frac{(1-\alpha) \times m_2(S)}{\alpha \times m_1(S)}\right) \times x_1 \right\} \cap WPO(S)$$

(9.4.1)

Originally, gain-based bargaining solutions are focused in the sense of proportional gains from disagreement points while loss-based bargaining solutions are focused in the sense of proportional concessions from the utopia points. In the line of mutual concessions philosophy, the mathematical definition of *WPLS* for a two-player bargaining game is defined as follows (Kim, 2021b; Segarra, 2011);

$$WPLS^{\alpha}(S) = \textbf{argmax}\left\{x \in S \middle| \frac{\alpha \times (m_1(S) - x_1)}{m_1(S)} = \frac{(1-\alpha) \times (m_2(S) - x_2)}{m_2(S)}\right\}$$

(9.4.2)

In Figure 7, the *WPGS* and *WPLS* are graphically represented (Kim, 2021b).

Figure 7. The WPGS and WPLS for a two-player bargaining game (Kim, 2021b)

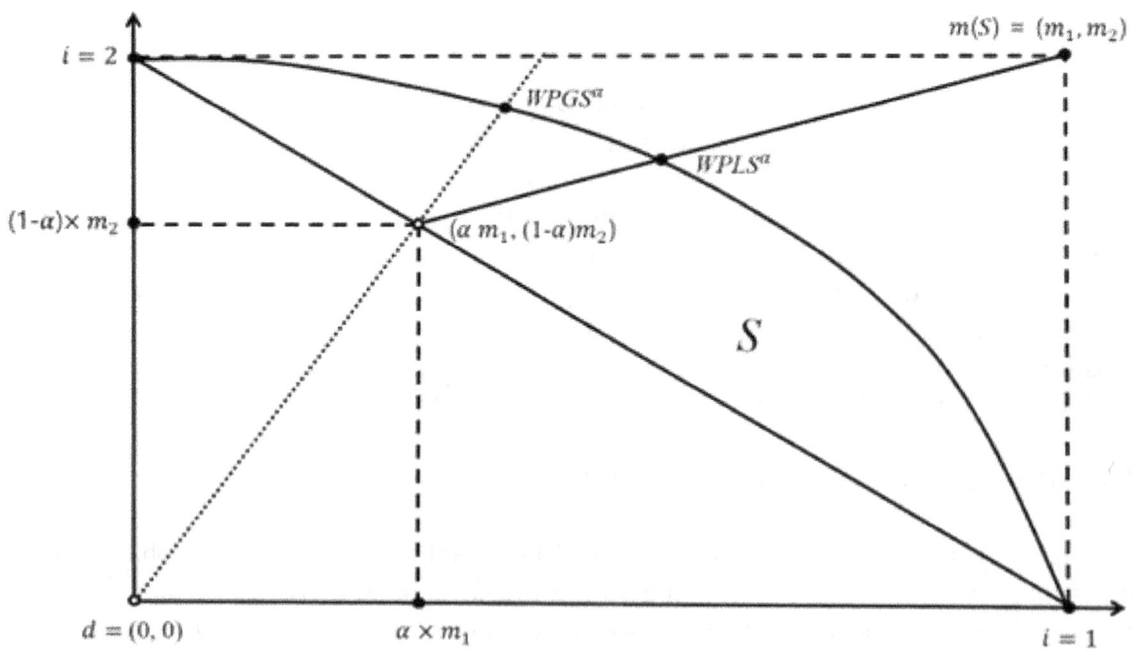

9.4.5 The dual-level bargaining game model for HSR services

At each time period, the T generates different types of HSR traffic data. According to the data type, the *HRCC* scheme is divided as a dual-level game model. At the upper-level, the T's class I data services are supported by the ITSN spectrum resource $(\mathscr{C}_I)$ where $\mathscr{C}_I = X_{\mathscr{S}}^U + X_{\mathscr{B}}^U$; $X_{\mathscr{S}}^U$ is assigned from the S, and $X_{\mathscr{B}}^U$ is assigned from the B. Through the $X_{\mathscr{S}}^U$ and $X_{\mathscr{B}}^U$ allocations, the resulting payoffs of S or B are defined as real numbers, which represent their satisfaction levels. In the $\mathbb{G}^U$, the utility functions of S and B at time t_c, i.e., $\mathfrak{U}_{\mathscr{S}}^U(\cdot)$ and $\mathfrak{U}_{\mathscr{B}}^U(\cdot)$, are formally defined as follows (Kim, 2021b);

$$\begin{cases} \mathfrak{U}_{\mathscr{S}}^U(\mathbb{T}_{\mathscr{S}}, \mathscr{A}_{\mathscr{S}}^{t_c}, X_{\mathscr{S}}^U, \mathscr{R}_I^{t_c}) = \dfrac{exp(\mathscr{Q}_I^{\mathscr{S}}) - exp(-\mathscr{Q}_I^{\mathscr{S}})}{exp(\mathscr{Q}_I^{\mathscr{S}}) + exp(-\mathscr{Q}_I^{\mathscr{S}})} - \mathscr{C}_{\mathscr{S}}^U(\mathscr{Q}_I^{\mathscr{S}}) \\ \mathfrak{U}_{\mathscr{B}}^U(\mathbb{T}_{\mathscr{B}}, \mathscr{A}_{\mathscr{B}}^{t_c}, X_{\mathscr{B}}^U, \mathscr{R}_I^{t_c}) = \log(\mathscr{W}_I^{\mathscr{B}} + \varepsilon) - \mathscr{C}_{\mathscr{B}}^U(\mathscr{W}_I^{\mathscr{B}}) \end{cases} \tag{9.4.3}$$

$$\text{s.t.,}\begin{cases} \mathscr{Q}_I^{\mathscr{S}} = \dfrac{\mathscr{A}_{\mathscr{S}}^{t_c} + \mathbf{min}(X_{\mathscr{S}}^U, \mathscr{R}_I^{t_c})}{\mathbb{T}_{\mathscr{S}}} and (X_{\mathscr{S}}^U + \mathscr{A}_{\mathscr{S}}^{t_c}) \le \mathbb{T}_{\mathscr{S}} \\ \mathscr{C}_{\mathscr{S}}^U(\mathscr{Q}_I^{\mathscr{S}}) = \left(exp(\mathscr{Q}_I^{\mathscr{S}}) + exp(-\mathscr{Q}_I^{\mathscr{S}}) - \mu\right) \\ \mathscr{W}_I^{\mathscr{B}} = \dfrac{\mathscr{A}_{\mathscr{B}}^{t_c} + \mathbf{min}(X_{\mathscr{B}}^U, \mathscr{R}_I^{t_c})}{\mathbb{T}_{\mathscr{B}}} and (X_{\mathscr{B}}^U + \mathscr{A}_{\mathscr{B}}^{t_c}) \le \mathbb{T}_{\mathscr{B}} \\ \mathscr{C}_{\mathscr{B}}^U(\mathscr{W}_I^{\mathscr{B}}) = \zeta \times \left(exp(\mathscr{W}_I^{\mathscr{B}}) + exp(-\mathscr{W}_I^{\mathscr{B}}) - \lambda\right) \end{cases}$$

where $\mathbb{T}_{\mathscr{S}}$, $\mathscr{A}_{\mathscr{S}}^{t_c}$ and $X_{\mathscr{S}}^U$ (*or* $\mathbb{T}_{\mathscr{B}}$, $\mathscr{A}_{\mathscr{B}}^{t_c}$ and $X_{\mathscr{B}}^U$) are total spectrum amount, available spectrum amount at time t_c, and assigned spectrum amount of S (or B), respectively. At time t_c, $\mathscr{R}_I^{t_c}$ is the requested class I data amount from the T. μ is a control parameter for the $\mathscr{C}_{\mathscr{S}}^U(\mathscr{Q}_I)$, and ε, ζ and λ are control parameters for the $\mathfrak{U}_{\mathscr{B}}^U(\cdot)$. In the formulas of $\mathfrak{U}_{\mathscr{S}}^U(\cdot)$ and $\mathfrak{U}_{\mathscr{B}}^U(\cdot)$, the first term captures the benefit of game players, and the second term represents the service cost. By using the reciprocal relationship between benefit and cost, player's payoff corresponds to the received benefit minus the incurred cost (Kim, 2014, 2021b).

From the viewpoint of fairness, class I data services should not be discriminated against between the identities of players but only depend on the service itself. Therefore, the *HRCC* scheme should guarantee that the value class I services does not change if scales of variables are multiplied by a common factor. Due to this reason, the *HRCC* scheme adopts the idea of *WPGS* to adjust the $X_{\mathscr{S}}^U$ and $X_{\mathscr{B}}^U$ values. According to (9.4.1), the fair-efficient solution for the $\mathbb{G}^U$, i.e., $WPGS(\mathbb{G}^U, X_{\mathscr{S}}^U, X_{\mathscr{B}}^U, \alpha_I)$ is given by (Kim, 2021b);

$$WPGS(\mathbb{G}^U, X_{\mathscr{S}}^U, X_{\mathscr{B}}^U, \alpha_I) = \max_{X_{\mathscr{S}}^U, X_{\mathscr{B}}^U}\left\{ \mathscr{R}_I^{t_c} \,\middle|\, \frac{\mathfrak{U}_{\mathscr{B}}^U(\mathbb{T}_{\mathscr{B}}, \mathscr{A}_{\mathscr{B}}^{t_c}, X_{\mathscr{B}}^U, \mathscr{R}_I^{t_c}) - d_{\mathscr{B}}^U}{\mathfrak{U}_{\mathscr{S}}^U(\mathbb{T}_{\mathscr{S}}, \mathscr{A}_{\mathscr{S}}^{t_c}, X_{\mathscr{S}}^U, \mathscr{R}_I^{t_c}) - d_{\mathscr{S}}^U} = \frac{(1 - \alpha_I) \times (\mathbb{M}_{\mathscr{B}}^U(\mathscr{R}_I^{t_c}) - d_{\mathscr{B}}^U)}{\alpha_I \times (\mathbb{M}_{\mathscr{S}}^U(\mathscr{R}_I^{t_c}) - d_{\mathscr{S}}^U)} \right\}$$

(9.4.4)

s.t., $\left(X^U_{\mathcal{S}} + X^U_{\mathcal{B}}\right) \leq \mathcal{R}^{t_c}_{I}$ and $0 \leq \alpha_I \leq 1$

where $\mathbb{M}^U_{\mathcal{S}}\left(\mathcal{R}^{t_c}_{I}\right), \mathbb{M}^U_{\mathcal{B}}\left(\mathcal{R}^{t_c}_{I}\right)$ are the S and B's ideal payoffs with $\mathcal{R}^{t_c}_{I}$, respectively. In the *WPGS* bargaining process, the parameter α_I controls the relative weights given to S and B. Under a dynamically changing ITSN system environment, the value of α_I is dynamically modified to ensure HSR data communications at each time period; it is adjusted adaptively based on the current traffic conditions of S and B (Kim, 2021b).

$$\alpha_I = \mathcal{A}^{t_c}_{\mathcal{S}} \Big/ \left(\mathcal{A}^{t_c}_{\mathcal{S}} + \mathcal{A}^{t_c}_{\mathcal{B}}\right)$$

(9.4.5)

After the $\mathbb{G}^U$ game process, the lower-level bargaining game is operated for the next round. At the $\mathbb{G}^L$, the T's class II data services are supported by the ITSN spectrum resource $\left(\mathcal{C}_{II}\right)$ where $\mathcal{C}_{II} = Y^L_{\mathcal{S}} + Y^L_{\mathcal{B}}$; $Y^L_{\mathcal{S}}$ is assigned from the S and $Y^L_{\mathcal{B}}$ is assigned from the B. As the same manner as the $\mathbb{G}^U$ game model, the utility functions of S and B in the $\mathbb{G}^L$, i.e., $U^L_{\mathcal{S}}(\cdot)$ and $U^L_{\mathcal{B}}(\cdot)$, are mathematically given as follows (Kim, 2021b);

$$\begin{cases} U^L_{\mathcal{S}}\left(\mathbb{T}_{\mathcal{S}}, \mathcal{A}^{t_c}_{\mathcal{S}}, Y^L_{\mathcal{S}}, \mathcal{R}^{t_c}_{II}\right) = \left(\beta - exp\left(\eta \times \mathcal{Q}^{\mathcal{S}}_{II}\right)\right) - \mathcal{C}^L_{\mathcal{S}}\left(\mathcal{Q}^{\mathcal{S}}_{II}\right) \\ U^L_{\mathcal{B}}\left(\mathbb{T}_{\mathcal{B}}, \mathcal{A}^{t_c}_{\mathcal{B}}, Y^L_{\mathcal{B}}, \mathcal{R}^{t_c}_{II}\right) = \phi \times \left(\dfrac{\gamma}{\kappa + exp\left(-\mathcal{W}^{\mathcal{B}}_{II}\right)} - \pi\right) - \mathcal{C}^L_{\mathcal{B}}\left(\mathcal{W}^{\mathcal{B}}_{II}\right) \end{cases} \tag{9.4.6}$$

$$\text{s.t.,}\begin{cases} \mathcal{Q}^{\mathcal{S}}_{II} = \dfrac{\mathcal{A}^{t_c}_{\mathcal{S}} + \mathbf{min}\left(Y^L_{\mathcal{S}}, \mathcal{R}^{t_c}_{II}\right)}{\mathbb{T}_{\mathcal{S}}} and \left(Y^L_{\mathcal{S}} + \mathcal{A}^{t_c}_{\mathcal{S}}\right) \leq \mathbb{T}_{\mathcal{S}} \\ \mathcal{C}^L_{\mathcal{S}}\left(\mathcal{Q}^{\mathcal{S}}_{II}\right) = \xi \times \left(exp\left(\mathcal{Q}^{\mathcal{S}}_{II}\right) + exp\left(-\mathcal{Q}^{\mathcal{S}}_{II}\right) - \rho\right) \\ \mathcal{W}^{\mathcal{B}}_{II} = \dfrac{\mathcal{A}^{t_c}_{\mathcal{B}} + \mathbf{min}\left(Y^L_{\mathcal{B}}, \mathcal{R}^{t_c}_{II}\right)}{\mathbb{T}_{\mathcal{B}}} and \left(Y^L_{\mathcal{B}} + \mathcal{A}^{t_c}_{\mathcal{B}}\right) \leq \mathbb{T}_{\mathcal{B}} \\ \mathcal{C}^L_{\mathcal{B}}\left(\mathcal{W}^{\mathcal{B}}_{II}\right) = \psi \times \left(exp\left(\mathcal{W}^{\mathcal{B}}_{II}\right) + exp\left(-\mathcal{W}^{\mathcal{B}}_{II}\right) - \tau\right) \end{cases}$$

where β, η are control parameters for the $U_{\mathcal{S}}^{L}(\cdot)$, and ϕ, γ, κ, π are adjustment factors for the $U_{\mathcal{B}}^{L}(\cdot)$. ξ, ρ are control factors for the $\mathscr{C}_{\mathcal{S}}^{L}(\cdot)$ and ψ, τ are control factors for the $\mathscr{C}_{\mathcal{B}}^{L}(\cdot)$. To ensure the efficiency for class II traffic services, the bargaining idea of *WPLS* is suitable to adjust the $Y_{\mathcal{S}}^{L}$ and $Y_{\mathcal{B}}^{L}$ values. According to (9.4.2), the effective solution for the $\mathbb{G}^{L}$, i.e., $WPLS(\mathbb{G}^{L}, Y_{\mathcal{S}}^{L}, Y_{\mathcal{B}}^{L}, \alpha_{II})$ is given by (Kim, 2021b);

$$WPLS(\mathbb{G}^{L}, Y_{\mathcal{S}}^{L}, Y_{\mathcal{B}}^{L}, \alpha_{II}) = \tag{9.4.7}$$

$$\max_{Y_{\mathcal{S}}^{L}, Y_{\mathcal{B}}^{L}} \left\{ \mathscr{R}_{II}^{t_c} \,\middle|\, \frac{\alpha_{II} \times \left(\mathbb{M}_{\mathcal{S}}^{L}(\mathscr{R}_{II}^{t_c}) - \mathcal{Q}_{II}^{\mathcal{S}}\right)}{\mathbb{M}_{\mathcal{S}}^{L}(\mathscr{R}_{II}^{t_c})} = \frac{(1-\alpha_{II}) \times \left(\mathbb{M}_{\mathcal{B}}^{L}(\mathscr{R}_{II}^{t_c}) - \mathcal{Q}_{II}^{\mathcal{B}}\right)}{\mathbb{M}_{\mathcal{B}}^{L}(\mathscr{R}_{II}^{t_c})} \right\}$$

$$\text{s.t.,} \begin{cases} \mathcal{Q}_{II}^{\mathcal{S}} = \left(Y_{\mathcal{S}}^{L} + \mathscr{A}_{\mathcal{S}}^{t_c}\right) and\, \mathcal{Q}_{II}^{\mathcal{B}} = \left(Y_{\mathcal{B}}^{L} + \mathscr{A}_{\mathcal{B}}^{t_c}\right) \\ \left(Y_{\mathcal{S}}^{L} + Y_{\mathcal{B}}^{L}\right) \le \mathscr{R}_{II}^{t_c} and 0 \le \alpha_{II} \le 1 \end{cases}$$

where α_{II} is a weight control parameter for class II traffic services while mediating the relative weights given to S and B. In the $\mathbb{G}^{L}$, the value of α_{II} is decided as the same manner as the α_{I} in $\mathbb{G}^{U}$ (Kim, 2021b).

According to (9.4.4) and (9.4.7), the *WPGS* and *WPLS* are obtained at each time t, and the *HRCC* scheme uses them to allocate the spectrum resources of S and B for the HSR data communications. By iteratively repeating the dual-level bargaining process, the *HRCC* scheme can adapt the current ITSN system condition while adjusting the different characteristics of satellite and terrestrial networks (Kim, 2021b).

9.4.6 Summary

The *HRCC* scheme addresses the spectrum allocation challenge within the ITSN infrastructure, ensuring efficient HSR communication. To achieve this, traffic services are categorized into two classes based on their QoS requirements, and the limited ITSN spectrum resources are conditionally shared among these different service types. A crucial aspect of this process is balancing the tradeoff between ensuring high-priority services meet their QoS demands and maximizing spectrum utilization efficiency. Utilizing the principles of *WPGS* and *WPLS*, the *HRCC* scheme is structured as a two-level bargaining game model through a dual decomposition approach. In the *HRCC* scheme, each control mechanism interacts dynamically, making real-time control decisions to determine the most optimal and profitable spectrum allocation strategy. By considering the distinct QoS requirements for class I (safety-critical) and class II (non-safety) services, the dual-level bargaining model aims to optimize ITSN spectrum usage while ensuring a mutually beneficial spectrum-sharing agreement. Given the diverse and dynamic nature of HSR traffic, the *HRCC* scheme provides significant and reciprocal advantages, offering an effective solution to the ITSN spectrum allocation problem (Kim, 2021b).

9.5 Cognitive Satellite Spectrum Management (*CSSM*) scheme

With the rapid expansion of satellite services, the need to enhance satellite spectrum efficiency has become increasingly important, drawing significant interest from researchers and industry professionals.

Recently, cognitive satellite communication mechanisms have emerged as a promising solution, enabling dynamic access to idle bands of licensed spectrum while facilitating spectrum sharing between satellite and terrestrial systems. In response to this challenge, S. Kim introduces the *Cognitive Satellite Spectrum Management* (*CSSM*) scheme for cognitive satellite communications (Kim, 2020). To optimize spectrum efficiency, the *CSSM* scheme integrates the principles of cooperative Solidarity values, striking a balance between productivity and solidarity in spectrum allocation. The *CSSM* scheme is built on two key Solidarity values - *Solidarity* and *Solidarity with size p values* - which guide a two-step interactive game approach to maximize the synergy between stakeholders, ensuring mutual benefits. The core innovation of the *CSSM* scheme lies in its ability to balance marginalism and egalitarianism, providing a rational framework for achieving a fair and efficient agreement under dynamically evolving satellite network conditions (Kim, 2020).

9.5.1 Development motivation

With the rapid expansion of high-data-rate multimedia services, 5G networks have gained significant attention as an enabler of innovative applications across multiple sectors. These include remote surgery, healthcare, tactile Internet, traffic safety, vehicular control systems, media, and entertainment. By fostering a world where anyone and anything can connect anytime and anywhere, advanced 5G communications are designed to support enhanced mobile broadband (eMBB), ultra-reliable low latency communications (URLLC), and massive machine-type communications (mMTC). While these features unlock new possibilities for applications and business models, meeting the demands of diverse, high-intensity traffic remains a challenge, prompting extensive research efforts in both academia and industry (Kim, 2020; Kodheli, 2019; Siddiqi, 2019).

While traditional 5G technologies have been extensively explored, their focus has primarily been on terrestrial networks. However, satellite communications offer distinct advantages in expanding 5G scalability, providing global coverage, efficient content distribution, network resilience, and energy savings. As a result, the integration of satellite and terrestrial networks is expected to become a defining feature of next-generation 5G architecture, reinforcing the reliability of 5G services. Recently, substantial research efforts have been dedicated to developing intelligent control algorithms that optimize resource utilization by leveraging commonalities between satellite and terrestrial network components (Kim, 2020; Wang, 2019; Zhang, 2017).

With the rapid growth of satellite communication services, the availability of licensed satellite spectrum - previously allocated for exclusive use - has become increasingly limited. This spectrum scarcity poses a challenge to improve the performance of integrated satellite-terrestrial network systems. To accommodate the growing demand for communication capacity, Cognitive Radio (CR) technology has emerged as a promising solution. CR enables dynamic spectrum access, allowing spectrum sharing between satellite and terrestrial networks. The primary advantage of CR technology lies in its ability to opportunistically and adaptively share spectrum resources, optimizing overall system efficiency. In recent years, research on spectrum sharing in satellite networks has explored the integration of CR technology, which is now considered an essential component in the development of effective satellite-terrestrial architectures (Chen, 2019; Kim, 2020; Lagunas, 2015).

For CR-based spectrum sharing to function effectively between satellite and terrestrial networks, spectrum sensing plays a crucial role in identifying available spectrum gaps for cognitive communications. Given this challenge, the *CSSM* scheme must incorporate an advanced spectrum-sharing mechanism.

Moreover, adaptive spectrum allocation must ensure fair resource distribution across multiple network agents. Since efficient spectrum management directly impacts total system capacity, it remains a central challenge - particularly within the dynamically evolving satellite environment. Addressing this issue requires the *CSSM* scheme to adopt a new intelligent control paradigm with a novel solution framework. To the best of my knowledge, spectrum allocation for cognitive satellite networks remains an underexplored research domain, necessitating further investigation (Kim, 2020).

A fundamental challenge in cooperative game theory is determining how to fairly distribute the gains derived from collaboration. A pioneering contribution in this field was made by L. Shapley, who introduced the concept of the *Shapley value*. This approach, based on a super-additive cooperative game model, ensures that coalition members receive rewards in proportion to their contributions. The core idea behind Shapley *value* is to allocate payoffs based on each player's productivity. A related concept, the *Solidarity value*, shares similarities with the *Shapley value*, but instead of relying solely on individual contributions, it averages the contributions across coalition members. This method provides a balanced approach between productivity and solidarity principles. Solidarity *value* integrates the productivity principle by incorporating a player's marginal contributions in the calculation. In simple terms, it represents a weighted sum of average contributions across different coalitions, ensuring a fair and equitable distribution of shared resources (Calvo, 2013; Hajir, 2016; Kim, 2014, 2020).

9.5.2 Contribution

Building upon the core principles of Solidarity value, the *CSSM* scheme introduces an innovative spectrum management mechanism for cognitive satellite communications. Within the satellite-terrestrial framework, each geostationary orbit (GEO) satellite distributes its spectrum resources to its associated low earth orbit (LEO) satellites, following the *Solidarity value* (*So*) principle. Once the spectrum resources have been allocated, each LEO satellite further distributes its assigned spectrum to its corresponding terrestrial relays (TRs) based on the *Solidarity value with size p* (So^p). This two-step interactive game approach maximizes synergistic benefits by integrating different value-based allocation strategies, ensuring a fair and efficient system performance for cognitive satellite-terrestrial communications (Kim, 2020).

9.5.3 Hierarchical cognitive satellite network infrastructure

The *CSSM* scheme assumes a multi-layer cognitive satellite network infrastructure with n GEO satellites $\mathbb{G} = \{\mathcal{G}_1,\ldots,\mathcal{G}_n\}$, m LEO satellites $\mathbb{L} = \{\mathcal{L}_1,\ldots,\mathcal{L}_m\}$ and k TRs $\mathbb{T} = \{\mathcal{T}_1,\ldots,\mathcal{T}_k\}$ constellation. A GEO satellite is a satellite in geosynchronous orbit, with an orbital period the same as the earth's rotation period. Such a satellite returns to the same position in the sky after each sidereal day. Instead of using GEO satellites, LEO satellites are smaller and lighter satellites while orbiting at lower altitude; they drastically reduce the cost of the solution because of satellite less expensive to manufacture and launch. TRs are terrestrial radio relay station designed for satellite communications; they may be located on the surface of the Earth, and are interconnected each other by backbone links (Chen, 2019; Kim, 2020; Kodheli, 2019; Lagunas, 2015; Siddiqi, 2019; Wang, 2019b; Zhang, 2017).

In the proposed architecture, end-user devices are not permitted to establish direct connections with satellites due to the excessive energy consumption, which exceeds the capabilities of mobile devices. Instead, terrestrial relays (TRs) serve as intermediaries, enabling end users to connect to the network while TRs communicate with low Earth orbit (LEO) satellites. Meanwhile, each geostationary orbit

(GEO) satellite maintains communication with its associated LEO satellites via inter-satellite links. Furthermore, it is assumed that the coverage areas of GEO satellites do not overlap to avoid interference (Wang, 2019c). To enhance coordination and cooperation among GEO satellites, LEO satellites, and TRs, the *CSSM* scheme introduces a two-tier cooperative game model for satellite spectrum management. This hierarchical structure enables each satellite entity to allocate its spectrum resources in a parallel and distributed manner, significantly reducing system complexity and simplifying spectrum management (Kim, 2020).

In the *CSSM* scheme, the available satellite spectrum is orthogonally divided among the respective network agents. However, as the demand for satellite broadband services continues to rise, the exclusively allocated spectrum is becoming increasingly scarce. The static frequency allocation policies currently in place limit the overall capacity of the integrated satellite-terrestrial network. To mitigate spectrum scarcity, existing research has primarily focused on CR-based spectrum reuse techniques, facilitating spectrum sharing between satellite and TR network segments (Chen, 2019). In cognitive satellite environments, spectrum sharing is structured as follows. At the upper layer, spectrum is dynamically shared among LEO satellites. At the lower layer, spectrum is shared among TRs. Within this two-tier satellite infrastructure, LEO satellites (*or* TRs) can be dynamically assigned as either primary or secondary users, depending on their layer position. This adaptive spectrum management enhances network efficiency and improves overall performance in the integrated satellite-terrestrial system (Kim, 2020).

In the *CSSM* scheme, a modeling situation for the satellite spectrum allocation process is formulated as a joint cooperative game ($\mathbb{G}$) in a distributed manner. The satellite system operates in a slotted time structure and spectrum allocation solution is obtained at each time period. Formally, the *CSSM* scheme defines game entities, i.e., $\mathbb{G} = \left\{ \{\mathbb{G}, \mathbb{L}, \mathbb{T}\}, \{\mathfrak{R}_{\mathscr{G}_i \in \mathbb{G}}, \mathfrak{M}_{\mathscr{L}_j \in \mathbb{L}}, \mathfrak{Y}_{\mathscr{T}_k \in \mathbb{T}}\}, \{U_{\mathscr{L}_j}^{\mathscr{G}_i}, U_{\mathscr{T}_k}^{\mathscr{L}_j}\}, p, T \right\}$ of gameplay (Kim, 2020).

- $\mathbb{G}, \mathbb{L}, \mathbb{T}$ are the finite sets of GEOs, LEOs and TRs; they are game entities for the joint cooperative games.
- $\mathfrak{R}_{\mathscr{G}_i}$ and $\mathfrak{M}_{\mathscr{L}_j}$ and $\mathfrak{Y}_{\mathscr{T}_k}$ are the spectrum amount assigned for the $\mathscr{G}_i \in \mathbb{G}$, for the $\mathscr{L}_j \in \mathbb{L}$, and $\mathscr{T}_k \in \mathbb{T}$, respectively.
- $U_{\mathscr{L}_j}^{\mathscr{G}_i}$ is a characteristic function of $\mathscr{L}_j$ where $\mathscr{L}_j \in \mathscr{W}_{\mathscr{G}_i}$; $\mathscr{W}_{\mathscr{G}_i}$ is the set of LEOs associated with the $\mathscr{G}_i$.
- $U_{\mathscr{T}_k}^{\mathscr{L}_j}$ is a characteristic function of $\mathscr{T}_k$ where $\mathscr{T}_k \in \mathbb{W}_{\mathscr{L}_j}$; $\mathbb{W}_{\mathscr{L}_j}$ is the set of TRs within the $\mathscr{L}_j$'s covering area.
- p is the size of coalition for the Solidarity value, which is used for the TR's spectrum allocation process.
- $T = \{t_1, \ldots, t_c, t_{c+1}, \ldots\}$ denotes time, which is represented by a sequence of time steps.

9.5.4 The basic concept of Solidarity value

One of the key challenges in economic allocation problems is finding a balance between marginalism and egalitarianism, particularly when utility is transferable. Two widely recognized allocation methods - the *Shapley value* and the *Egalitarian value* - each represent one of these opposing principles. The *Shapley value* is strictly based on the marginalistic principle, emphasizing individual contributions, whereas the *Egalitarian value* follows a highly solidaristic approach, distributing resources equally among participants (Béal, 2015).

In 1990, Y. Sprumont introduced the concept of Solidarity *value*, which was later axiomatically characterized by A. Nowak and T. Radzik. The *Solidarity value* provides a compromise between marginalism and egalitarianism, incorporating aspects of both principles. It respects the marginalism principle by considering each player's individual contributions to coalition outcomes. At the same time, it introduces a redistributive effect by incorporating not only a player's own contribution but also the average contribution of all coalition members. This approach ensures that a player's contribution to a coalition is not solely based on individual marginal impact, but rather on the collective contribution of all players involved. Thus, within the *CSSM* scheme, the *Shapley value* and the *Egalitarian value* can be considered extreme points, while the *Solidarity value* represents a balanced middle ground between the two (Béal, 2015; Calvo, 2013; Kim, 2020).

To characterize the basic concepts of Solidarity value, we preliminarily define some mathematical expressions. Let $N = \{1,2,\ldots,n\}$ be a fixed and finite set of n players. Each subset S of N is called a *coalition* while N is called the *grand coalition.* A cooperative game with transferable utility on a fixed player set N is a function $v{:}2^N \rightarrow \mathbb{R}$ such that $v(\emptyset) = 0$ where the collection of all subsets of S will be denoted by 2^S. For each coalition $S \subseteq N, v(S)$ describes the worth of the coalition S when its members cooperate. For any two games v and w in V_N and any $\alpha \in \mathbb{R}$, the game $(\alpha v + w) \in V_N$ is defined as follows: for each $S \subseteq N, (\alpha v + w)(S) = (\alpha v(S) + w(S))$. A game $v \in V_N$ is *positive* if, for each $S \subseteq N$, $v(S) \geq 0$. A game $v \in V_N$ is *monotone* if, for each $S \subseteq T \subseteq N, v(s) \leq v(T)$. A permutation σ on N assigns a position $\sigma(i)$ to each player $i \in N$. Let Σ_N be the set of $n!$ permutations on N. Given $v \in V_N$ and $\sigma \in \Sigma_N$, $\sigma v \in V_N$ is defined as: for each nonempty $S \subseteq N, \sigma v\left(\cup_{i \in S} \sigma(i)\right) = v(S)$ (Béal, 2015). An allocation rule Φ on V_N is a mapping $\Phi{:}V_N \rightarrow \mathbb{R}^n$ which uniquely determines, for each $v \in V_N$ and each $i \in N$, a payoff $\Phi_i(v) \in \mathbb{R}$ for participating to $v \in V_N$. The Egalitarian value, *Eg*, is defined on V_N as follows (Béal, 2015; Kim, 2020);

$$Eg_{i,\forall i \in N}(v) = \frac{v(N)}{n} \tag{9.5.1}$$

For any permutation σ and any player $i \in N$, the *CSSM* scheme can define the coalition containing player i and the set of its predecessors in σ as $P_i^\sigma = \{j \in N{:}\sigma(j) \leq \sigma(i)\}$. The Shapley value, *Sh*, is defined on V_N as follows (Béal, 2015; Kim, 2020);

$$Sh_{i,\forall i \in N}(v) = \frac{1}{n!} \times \left(\sum_{\sigma \in Ó_N} \left(v(P_i^\sigma) - v(P_i^\sigma \backslash i)\right)\right)$$
$$= \sum_{S \in 2^N : S \ni i} \left(\left(\frac{(n-s)! \times (s-1)!}{n!}\right) \times (v(S) - v(S \backslash i))\right) \tag{9.5.2}$$

The Solidarity value, *So*, attributes to any player $i \in N$ the average marginal contribution of all players in the same coalition $S \subseteq N$. Therefore, a poor player can achieve a more favorable status and be on a social level near to those players who participated in solidarity. *So* is defined on V_N as follows (Calvo, 2013; Kim, 2020):

$$So_{i,\forall i\in N}(v) = \sum_{S\subseteq N:i\in S}\left(\left(\frac{(n-s)!\times(s-1)!}{n!}\right)\times\triangle(v,S)\right)$$

(9.5.3)

$$\text{s.t., } \triangle(v,S) = \frac{1}{|S|}\times\left(\sum_{i\in S}(v(S)-v(S\backslash i))\right)$$

where $|S|$ is the cardinality of a finite set S, and $\triangle(v,S)$ is the average of the marginal contributions of players within coalition S. Based on the idea of *So*, a new value concept is introduced; the collective contribution to the grand coalition replaces the individual contributions as soon as the currently formed coalition has reached some size *p*. Therefore, there are two distinct steps. Before attaining the coalition size *p*, each entering player gets his individual contribution to the coalition he joins. When the coalition size *p* is reached, the remaining players keep on entering one by one, but instead of rewarding each of them when entering, they cumulate their contributions until the grand coalition is formed. Then, the collective contribution of the coalition of remaining players to the grand coalition is evenly distributed among them. This new value is called the Solidarity value with size *p*, So^p, and it is formally defined as follows (Béal, 2015; Kim, 2020);

$$So^p_{i,\forall i\in N}(v) = \frac{1}{n!}\times\sum_{\sigma\in\acute{O}_N} c_i^{\sigma,p}(v)$$

(9.5.4)

$$\text{s.t., } c_i^{\sigma,p}(v) = \begin{cases} v(P_i^\sigma) - v(P_i^\sigma\backslash i), if\sigma(i)\leq p \\ \dfrac{v(N)-v\left(P^\sigma_{\sigma^{-1}(p)}\right)}{n-p}, if\sigma(i) > p\end{cases} \text{ and } P^\sigma_{\sigma^{-1}(0)} = \varnothing$$

From (9.5.4), the $So^p_{i,\forall i\in N}(v)$ can be rewritten as follows (Béal, 2015; Kim, 2020);

$$So_i^p(v) = \frac{1}{n!}\times\left[\left(\sum_{\substack{\sigma\in\Sigma_N,\\ \sigma(i)\leq p}}\left(v(P_i^\sigma) - v(P_i^\sigma\backslash i)\right)\right) + \left(\sum_{\substack{\sigma\in\Sigma_N,\\ \sigma(i)>p}}\left(\frac{v(N)-v\left(P^\sigma_{\sigma^{-1}(p)}\right)}{n-p}\right)\right)\right]$$

$$=\frac{1}{n!}\times\left[\left(\sum_{\left(\substack{S\in 2^N,\\ S\ni i, s\leq p}\right)}\sum_{\left(\substack{\sigma\in\Sigma_N,\\ P_i^\sigma=S}\right)}(v(S)-v(S\backslash i))\right) + \left(\left(\sum_{\left(\substack{S\in 2^N,\\ S\notin i, s=p}\right)}\sum_{\left(\substack{\sigma\in\Sigma_N,\\ P^\sigma_{\sigma^{-1}(p)}=S}\right)}(v(N)-v(S))\right)\Big/\sum_{\left(\substack{\sigma\in\Sigma_N,\\ \sigma(i)>p}\right)}(n-p)\right)\right]$$

$$=\frac{1}{n!}\times\left[\left(\sum_{\substack{S\in 2^N,\\ S\ni i, s\leq p}}[(n-s)!\times(s-1)!\times(v(S)-v(S\backslash i))]\right) + \left(\left(\sum_{\left(\substack{S\in 2^N,\\ S\notin i, s=p}\right)}[(n-s)!\times s!\times(v(N)-v(S))]\right)\Big/\sum_{\left(\substack{\sigma\in\Sigma_N,\\ \sigma(i)>p}\right)}(n-p)\right)\right]$$

$$= \sum_{\substack{S\in 2^N,\\ S\ni i, s\le p}} \left[\frac{(n-s)!\times(s-1)!\times(v(S)-v(S\backslash i))}{n!}\right] + \sum_{\substack{S\in 2^N,\\ S\notin i, s=p}} \left[\frac{(n-s-1)!\times s!\times(v(N)-v(S))}{n!}\right]$$

$$= \sum_{\substack{S\in 2^N,\\ S\ni i, s\le p}} \left[\frac{(n-s)!\times(s-1)!\times(v(S)-v(S\backslash i))}{n!}\right] + \sum_{\substack{S\in 2^N,\\ S\ni i, s=p+1}} \left[\frac{(n-s)!\times(s-1)!\times(v(N)-v(S\backslash i))}{n!}\right] \quad (9.5.5)$$

9.5.5 The cooperative spectrum control scheme for satellite network

The *CSSM* scheme designs a new spectrum management mechanism for the two-tier satellite platform. The major goal of the *CSSM* scheme is to adaptively allocate the limited GEO spectrum resource to LEOs and TRs. To reduce computation complexity, the spectrum allocation process is divided hierarchically, and the *CSSM* scheme makes control decisions in a step-by-step manner at each stage. At the upper level, each individual $\mathscr{G}_i \in \mathbb{G}$ has the $\mathfrak{R}_{\mathscr{G}_i}$, and shares it with the set of LEOs $\left(\mathscr{W}_{\mathscr{G}_i}\right)$; they are entities for the game $\mathbb{G}_{\mathscr{G}_i}$. The $\mathfrak{R}_{\mathscr{G}_i}$ is partitioned into disjoint portions and allocates them for corresponding $\mathscr{L}_j \in \mathscr{W}_{\mathscr{G}_i} \subset \mathbb{L}$. To implement this satellite spectrum allocation process, the *CSSM* scheme focuses on the Solidarity value with the CR technique. For the CR implementation, users dynamically sense the radio environment to opportunistically utilize the unused spectrum bands. Therefore, the challenging function used in cognitive satellite communications is spectrum sensing (Kim, 2012, 2020).

The *CSSM* scheme presents the cooperative spectrum sensing technique as an effective sensing way. It occurs when a group of users voluntarily contributes to sensing and share their local sensing information to get a better picture of the spectrum usage. In the $\mathbb{G}_{\mathscr{G}_i}$ game model, all $\mathscr{L} \in \mathscr{W}_{\mathscr{G}_i}$ are game players; they work together in the spectrum sensing process. Therefore, LEOs sense other LEOs' spectrum, and send the sensing information to the GEO. By using appropriate data fusion rules, the GEO makes a judgment whether or not the spectrum is idle. With the decision of spectrum idleness, the GEO can determine the trust level of each individual LEO by considering the LEO's actions. It is an index of cooperation behavior and used to distribute the available spectrum more efficiently. Due to the dynamically changing CR network environment, the GEO needs to monitor the LEOs' behavior constantly. According to a sequence of LEO action observations, each LEO's trust value can be adaptively modified (Kim, 2012). Let $\Upsilon^t_{\mathscr{L}_j}$ be the trust value of $\mathscr{L}_j$ at the tth time period (Kim, 2020).

$$\Upsilon^t_{\mathscr{L}_j} = \begin{cases} (1-\alpha)\times\Upsilon^{t-1}_{\mathscr{L}_j} + \alpha\times(-1)^{\mathscr{D}^t_{\mathscr{L}_j}+\mathscr{R}^t_{\mathscr{L}_j}}, if \mathscr{R}^t_{\mathscr{L}_j} \neq 1 \\[2ex] (1-\beta)\times\Upsilon^{t-1}_{\mathscr{L}_j} + \beta\times(-1)^{\mathscr{R}^t_{\mathscr{L}_j}}, if \mathscr{R}^t_{\mathscr{L}_j} = 1 \end{cases} \quad (9.5.6)$$

$$\text{s.t.,} \begin{cases} \mathscr{D}^t_{\mathscr{L}_j} = \begin{cases} 0, if spectrum is idle \\ 1, else \end{cases} \\[3ex] \mathscr{R}^t_{\mathscr{L}_j} = \begin{cases} -1, if \mathscr{L}_j reports - spectrum is occupied \\ 0, if \mathscr{L}_j reports - spectrum is idle \\ 1, if \mathscr{L}_j reports - no report \end{cases} \end{cases}$$

where α and β are adjustment factors, which control the relative weights given to recent and past behavior history in trust value decision. $\mathscr{R}^t_{\mathscr{L}_j}$ and $\mathscr{D}^t_{\mathscr{L}_j}$ present the reporting result of the $\mathscr{L}_j$ and the final judgment by the GEO at the kth time period, respectively. Repeatedly, the GEO updates the $\Upsilon^t_{\mathscr{L}_j}$ value of each $\mathscr{L}_j \in \mathscr{W}_{\mathscr{G}_i}$ according to historical trust and the current trust evidence. Based on the $\Upsilon^t_{\mathscr{L}_j}$ information, the $\mathscr{L}_j$'s utility function $U^{\mathscr{G}_i}_{\mathscr{L}_j}$ is defined as follows (Kim, 2020);

$$U^{\mathscr{G}_i}_{\mathscr{L}_j}\left(\mathfrak{M}_{\mathscr{L}_j}, \chi_{\mathscr{L}_j}, \Upsilon^t_{\mathscr{L}_j}\right) = \left(\min\left(\mathfrak{M}_{\mathscr{L}_j}, \chi_{\mathscr{L}_j}\right) \times \log\left(\frac{\min\left(\mathfrak{M}_{\mathscr{L}_j}, \chi_{\mathscr{L}_j}\right)}{\chi_{\mathscr{L}_j}} + \xi\right)^{\Lambda}\right) - \mathscr{C}^{\mathscr{G}_i}_{\mathscr{L}_j} \tag{9.5.7}$$

$$\text{s.t.}, \mathscr{C}^{\mathscr{G}_i}_{\mathscr{L}_j} = c_{\mathscr{L}} \times \ell_{\mathscr{L}} \text{ and } \Lambda = \Upsilon^t_{\mathscr{L}_j} / \sum\nolimits_{\mathscr{L}_h \in \mathscr{W}_{\mathscr{G}_i}} \Upsilon^t_{\mathscr{L}_h}$$

where $\chi_{\mathscr{L}_j}$ is the requested spectrum amount from the $\mathscr{L}_j$ and ξ is a control parameter for the $U^{\mathscr{G}}_{\mathscr{L}}(\cdot)$. $c_{\mathscr{L}}$ and $\ell_{\mathscr{L}}$ are the cost factor for the CR band sensing and the number of bands by the $\mathscr{L}_j$, respectively. For the equations (9.5.3) and (9.5.4), the characteristic function ν for the coalition S, i.e., $\nu(S)$, is defined based on the bankruptcy game model (Dagan, 1993). Originally, the bankruptcy problem is a pair (E, c) where $c = \left(c_1, \ldots, c_n\right) \geq 0$, and $0 \leq E \leq \sum_{i=1}^{n} c_i$ where E represents the total value of the bankrupt company's estate and **c** is the vector of the creditors' claims. The sum of these claims is denoted by C. An allocation in such a problem is an n-tuple $x = \left(x_1, \ldots, x_n\right) \in \mathbb{R}^n$ with $\sum_{i=1}^{n} x_i = E$ and $0 \leq x_i \leq c_i, i = 1, \ldots, n$. In order to define the next rule, we need the following definition: For each bankruptcy problem (E, c) and creditor i, the amount of money conceded to creditor i by all other creditors, i.e., $\upsilon_i^{(E,c)}$, is given by (Dagan, 1993; Kim, 2020);

$$\upsilon_i^{(E,c)} = \max\left\{0;\ E - \sum_{\substack{1 \leq j \leq n, \\ j \neq i}} c_j\right\} \tag{9.5.8}$$

Based on the bankruptcy allocation rule, we simply replace E, c_j and $\nu(\cdot)$ by $\mathfrak{R}_{\mathscr{G}_i}$, $\chi_{\mathscr{L}_j}$ and $U^{\mathscr{G}_i}_{\mathscr{L}_j}(\cdot)$, respectively, in the $\mathbb{G}_{\mathscr{G}_i}$. And then, the *So* value for each LEO $\mathscr{L} \in \mathscr{W}_{\mathscr{G}_i}$, i.e., $So_{\mathscr{L} \in \mathscr{W}_{\mathscr{G}_i}}\left(U^{\mathscr{G}_i}_{\mathscr{L}_j}\right)$, is obtained according to (9.5.3). Finally, the spectrum distribution solution for the $\mathscr{W}_{\mathscr{G}_i}$, i.e., $\Theta_{\mathscr{W}_{\mathscr{G}_i}} = \mathscr{L}_j \in \mathscr{W}_{\mathscr{G}_i} | \ldots \mathfrak{M}_{\mathscr{L}_j} \ldots$, is calculated by using the *So* values; the $\grave{E}_{\mathscr{W}_{\mathscr{G}_i}}$ is a $\left|\mathscr{W}_{\mathscr{G}_i}\right|$-dimensional spectrum allocation vector (Kim, 2020).

$$\mathfrak{M}_{\mathscr{L}_j} = \mathfrak{R}_{\mathscr{G}_i} \times \left(So_{\mathscr{L}_j} / \sum_{\mathscr{L}_e \in \mathscr{W}_{\mathscr{G}_i}} So_{\mathscr{L}_e}\right), \text{ s.t., } \mathfrak{R}_{\mathscr{G}_i} = \sum_{\mathscr{L}_j \in \mathscr{W}_{\mathscr{G}_i}} \mathfrak{M}_{\mathscr{L}_j} \tag{9.5.9}$$

At the lower level, an individual $\mathscr{L}_j \in \mathbb{L}$ get the $\mathfrak{M}_{\mathscr{L}_j}$ according to (9.5.9), and shares it with the set of TRs within the $\mathscr{L}_j$'s covering area; they are entities for the game $\mathbb{G}_{\mathscr{L}_j}$. In the same manner as the game $\mathbb{G}_{\mathscr{G}_i}$, the $\mathfrak{M}_{\mathscr{L}_j}$ is partitioned into disjoint portions, and allocates them for corresponding $\mathscr{T}_k \in \mathbb{W}_{\mathscr{L}_j} \subset \mathbb{T}$. To implement this spectrum sharing process, the *CSSM* scheme focuses on the So^P value while emphasizing ***ETE*** feature, and defines the $\mathscr{T}_k$'s utility function $U^{\mathscr{L}_j}_{\mathscr{T}_k}$ as follows (Kim, 2020);

$$U_{\mathcal{T}_k}^{\mathcal{L}_j}\left(\mathfrak{Y}_{\mathcal{T}_k}, \chi_{\mathcal{T}_k}, \Upsilon_{\mathcal{T}_k}^t\right) = \left(\min\left(\mathfrak{Y}_{\mathcal{T}_k}, \chi_{\mathcal{T}_k}\right) \times \log\left(\frac{\min\left(\mathfrak{Y}_{\mathcal{T}_k}, \chi_{\mathcal{T}_k}\right)}{\chi_{\mathcal{T}_k}} + \varrho\right)^{\Theta}\right) - \mathscr{C}_{\mathcal{T}_k}^{\mathcal{L}_j} \tag{9.5.10}$$

$$\text{s.t.}, \mathscr{C}_{\mathcal{T}_k}^{\mathcal{L}_j} = c_{\mathcal{T}} \times \ell_{\mathcal{T}} \text{ and } \Theta = \Upsilon_{\mathcal{T}_k}^t / \sum_{\mathcal{T}_e \in \mathbb{W}_{\mathcal{L}_j}} \Upsilon_{\mathcal{T}_e}^t$$

where $\chi_{\mathcal{T}_k}$ is the requested spectrum amount of $\mathcal{T}_k$ and ϱ is a control parameter for the $U_{\mathcal{T}}^{\mathcal{L}}(\cdot)$. $c_{\mathcal{T}}$ and $\ell_{\mathcal{T}}$ are the cost factor for the CR band sensing and the number of bands by the $\mathcal{T}_k$, respectively. According to (9.5.4) or (9.5.5), the So^p value for the $\mathcal{T}_k \in \mathbb{W}_{\mathcal{L}_j}$, i.e., $So^p_{\mathcal{T}_k \in \mathbb{W}_{\mathcal{L}_j}}\left(U_{\mathcal{L}_j}^{\mathcal{G}_i}\right)$, is obtained. Finally, the spectrum distribution solution for the $\mathbb{W}_{\mathcal{L}_j}$, i.e., $\grave{\mathrm{E}}_{\mathbb{W}_{\mathcal{L}_j}} = \mathcal{T}_k \in \mathbb{W}_{\mathcal{L}_j} | \ldots \mathfrak{Y}_{\mathcal{T}_k} \ldots$, is calculated in the same way as the $\mathbb{G}_{\mathcal{G}_i}$ game model; $\grave{\mathrm{E}}_{\mathbb{W}_{\mathcal{L}_j}}$ is also a $\left|\mathbb{W}_{\mathcal{L}_j}\right|$-dimensional spectrum allocation vector (Kim, 2020).

9.5.6 Summary

With their ability to provide high-speed transmission and seamless global coverage, satellite communications have become increasingly essential. As 6G networks aim to deliver ubiquitous connectivity, satellites are expected to play a crucial role in complementing traditional terrestrial networks, ensuring continuous coverage anytime and anywhere. Within the *CSSM* scheme, the challenge of spectrum management in satellite networks is addressed using a joint cooperative game paradigm. To effectively tackle this issue, the *CSSM* scheme incorporates two cooperative value-based solutions: the *Solidarity value* and the *Solidarity value with size p*. By leveraging these principles, the *CSSM* scheme is designed to enhance multi-layer satellite system performance through a combined game-theoretic approach, making it an efficient spectrum access strategy. Considering the hierarchical structure of satellite networks, different value-based solutions are seamlessly integrated into the overall *CSSM* scheme, operating cooperatively and in a distributed manner. This ensures that limited spectrum resources are efficiently shared across various network entities, even under dynamically evolving satellite communication environments (Kim, 2020).

9.6 Cooperative Multi-path Routing Control (*CMRC*) scheme

Recent advancements in wireless communication technologies have enabled mobile users to access various data services anytime and anywhere. In response to the increasing demand for reliable maritime communication, S. Kim has investigated an Integrated Satellite-Maritime Network (ISMN) platform and develops the *Cooperative Multi-Path Routing Control* (*CMRC*) scheme (Kim, 2022b). Taking into account the geographical characteristics of maritime environments and the diverse nature of transmitted data, the *CMRC* scheme applies coalitional game theory to develop novel routing algorithms within the ISMN infrastructure. By utilizing different value-based solutions, routing packets are dynamically distributed across multiple routing paths, allowing them to self-adjust based on the highly dynamic conditions of the ISMN system. This approach facilitates cooperative interactions, ensuring that the mutual benefits derived from collaboration are maximized while maintaining reciprocal consensus across various routing control challenges (Kim, 2022b).

9.6.1 Development motivation

With the global standardization and deployment of 5G wireless communication networks, major advancements have been made, particularly in ultra-reliable low-latency communications (URLLC). Various key technologies have been introduced to meet these demands. However, despite its capabilities, 5G is still limited in providing connectivity to remote and underserved areas. To address these limitations, researchers in both academia and industry have shifted their focus to 6G wireless communication networks, which aim to achieve near 100% global coverage with massive connectivity. This level of worldwide accessibility will support an expansion of human activities across space, air, land, and sea. As a result, the successful deployment of 6G networks requires the integration of multi-dimensional space-air-ground-sea networks, particularly in remote maritime environments (Hassan, 2021; Kim, 2022b; You, 2021).

Oceans cover over 70% of the Earth's surface, and more than 80% of global cargo transportation relies on ocean shipping. Furthermore, maritime activities play a vital role in economic growth and societal development, with significant potential for future expansion. Given the increasing reliance on maritime operations, wireless broadband services must be established at sea to connect various maritime entities, including ships, buoys, oil drilling platforms, and beacons. Maritime ad-hoc networks (MANETs) are designed as multi-hop wireless networks, allowing maritime agents to communicate with one another and with shore-based backhaul stations positioned along coastlines. With technological advancements, maritime agents require seamless connectivity in any ocean region. MANETs facilitate data services that support maritime communication, navigation, and emergency response systems while maintaining satisfactory QoS (Kim, 2022b; Peng, 2020; Yau, 2019).

Developing a secure and efficient maritime communication system is particularly challenging due to the unique characteristics of MANETs and the presence of heterogeneous services. Several networking models can be implemented, including; i) self-assembling MANETs, utilizing ad hoc technology for maritime connectivity, and ii) satellite networks, providing wide-area coverage for global maritime communication. To enhance coverage and meet growing service demands, a hybrid network infrastructure has emerged, integrating satellite and maritime networks. However, satellite communication remains expensive, and the rising oceanic data traffic makes achieving global broadband coverage via satellite technology alone a costly challenge (Aslam, 2020; Li, 2020; Kim, 2022b).

The *CMRC* scheme is designed for the ISMN. It enables satellite and maritime networks to function collaboratively, providing seamless data services to maritime users. Given the vast ocean areas and variety of service demands, the ISMN infrastructure is considered an optimal compromise solution. Routing plays a critical role in the adaptive and efficient operation of ISMNs, as network performance is heavily influenced by routing algorithms. Administrators must determine how different types of network traffic should be handled. Maritime network traffic is typically categorized into the real-time (RT) traffic and non real-time (NRT) traffic. RT includes critical services that must be delivered with minimal delay and high quality. The accuracy of real-time traffic is determined by timeliness. NRT traffic consists of data that prioritizes average performance over immediate response time and can tolerate some delays. Motivated by these challenges, the *CMRC* scheme introduces a novel routing mechanism specifically for ISMN platforms. Given the unique characteristics of the ISMN topology, the *CMRC* scheme establishes multiple routing paths and dynamically distributes network packets across these routes. Although significant research has been conducted on multi-path routing, ISMN routing development remains in its early stages. Implementing an optimal data distribution algorithm is still a complex challenge in real-world applications. To improve efficiency, the *CMRC* scheme must consider service data types and prioritize RT traffic based on real-time network conditions (Karia, 2008; Kim, 2004, 2022b).

A fundamental approach in coalitional game theory is the *Shapley value*, which is based on two main assumptions: i) a player joining a coalition receives their full marginal contribution, and ii) all players have an equal probability of joining any coalition. The *Shapley value* is widely regarded as a normative solution in coalitional game theory. However, an alternative approach - the *Solidarity value* - shares similarities with the *Shapley value* but introduces a key distinction. While the *Shapley value* assigns a player's individual marginal contribution, the *Solidarity value* replaces it with the average contribution of all coalition members. This adjustment ensures that when a coalition forms, each player receives a fair share of the total coalition benefit, based on collective contributions rather than just individual input (Kim, 2014, 2022b; Oum, 2020).

When coalitional game solutions are analyzed from an axiomatic perspective, asymmetric value solutions emerge by removing the symmetric property from traditional axioms. These asymmetries arise due to differences in negotiation abilities and can be adjusted using weighted values to balance the utilitarian and egalitarian perspectives. Recently, several weighted value solutions have been proposed, including: *Weighted Shapley Value* (*WSV*), *Weighted Solidarity Value* (*WSoV*) and *Weighted Surplus Division Value* (*WSdV*). The *WSdV* closely aligns with the philosophical principles of the *Shapley value*, but it differs in how marginal contributions are allocated among coalition members. Instead of distributing benefits proportionally to individual contributions, it incorporates exogenous weight factors, ensuring that the surplus portion is distributed after each player has received their individual worth (Calvo, 2014; Kim, 2022b; Yang, 2019).

9.6.2 Contribution

Building on the *WSV*, *WSoV*, and *WSdV* solutions, the *CMRC* scheme effectively addresses data distribution challenges in the ISMN routing algorithm. To accommodate the differing preferences and requirements of RT and NRT data services, the *CMRC* scheme introduces a three-phase coalitional game model for routing optimization. In the first phase, RT data is partitioned into two streams - one designated for satellite communications and the other for MANETs. In the second phase, the RT data allocated to the MANET is further distributed across multiple routing paths. In the third phase, the NRT data is then routed through multiple paths within the MANET infrastructure. Each phase incorporates a specific value solution for control decision-making. The *WSdV* is applied in the first phase to optimize the initial RT data partitioning. The *WSV* is used in the second phase to ensure efficient multi-path RT data distribution. WSoV is utilized in the third phase to effectively allocate NRT data across multiple routes. Through an iterative and interactive process, the *CMRC* scheme dynamically manages data distribution, ensuring optimized system performance within the ISMN platform (Kim, 2022b).

9.6.3 ISMN infrastructure and three-phase coalitional game model

The *CMRC* scheme is based on a hybrid ISMN infrastructure, which comprises maritime ships, buoys, low Earth orbit (LEO) satellites, and terrestrial base stations (TBSs), as depicted in Figure 8. To facilitate high-rate backhaul communications, TBSs are strategically deployed along coastlines. For remote maritime access, LEO satellites serve as an optimal solution, effectively addressing maritime network challenges. Due to their line-of-sight (LOS) transmission capability, LEO satellites can reliably meet QoS requirements, making them well-suited for maritime communication scenarios. Additionally, the *CMRC* scheme assumes that LEO satellites maintain a backhaul link with either an earth gateway or TBSs, ensuring stable connectivity. Within this framework, maritime ships and buoys form a wireless multi-hop network, integrating satellite communication for extended coverage. As a result, maritime users can establish a connection to a TBS through one of two methods: multi-hop communication via other ships or buoys or direct communication through the LEO satellite. The multi-hop architecture of the ISMN enables a resource pooling platform, allowing network agents to share wireless services, satellite connectivity, and communication resources with maritime users (Hassan, 2021; Kim, 2022b; Yau, 2019).

The network topology has a set $\mathcal{S} = \{\mathcal{S}_1, \ldots \mathcal{S}_n\}$ of ships on the sea, and a set $\mathfrak{B} = \{\mathcal{B}_1, \ldots \mathcal{B}_m\}$ of floating buoys. Each $\mathcal{S}_{1 \le i \le n}$ has multiple marine user devices (MUDs) denoted by $\mathbb{D}^{\mathcal{S}_i} = \{\mathcal{D}_1^{\mathcal{S}_i}, \ldots \mathcal{D}_k^{\mathcal{S}_i}\}$. The *CMRC* scheme considers that one LEO satellite is optimally deployed at the desired location while fulfilling maritime network demands. Each ship and buoy have a transceiver with a high-gain antenna, which can collect and re-transmit data from/to neighboring network agents. Therefore, MANET provides maritime data communication services through multipath routing routes, which are established in an ad-hoc manner. Using agent-borne antennas, data can be transmitted or relayed in a multi-hop manner, which can be deployed flexibly while reducing the cost of communication infrastructure (Kim, 2022b; Xing, 2019).

Figure 8. The overview of ISMN system platform (Kim, 2022b)

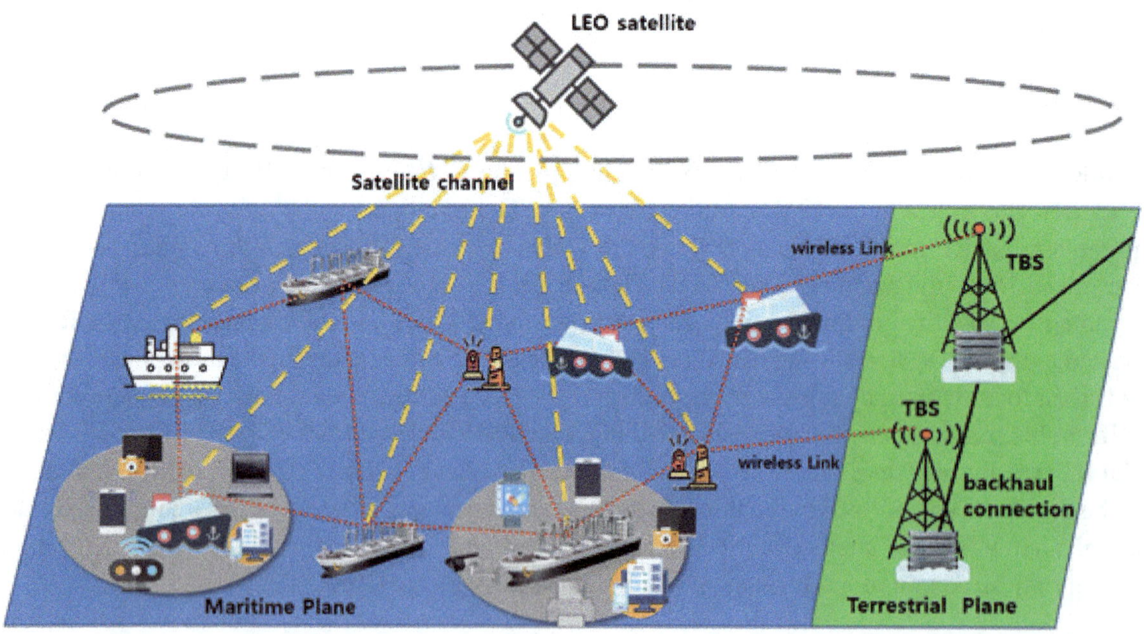

Ensuring QoS has become an increasingly critical and complex challenge in routing mechanisms, particularly in maritime networks. Among the various QoS requirements, prioritizing RT data services is essential, as the demand for RT communications in maritime environments is growing rapidly. Furthermore, the quality of RT services is highly sensitive to delays, making latency reduction a key performance factor. In the *CMRC* scheme, each MUD is assumed to generate a mix of RT and NRT data services. Consequently, the *CMRC* scheme must incorporate an effective differentiation strategy to manage these distinct data types while ensuring an optimized system performance balance between RT and NRT services. To solve this problem, the *CMRC* scheme develops a three-phase coalitional game model ($\mathbb{G}$), which consists of RT and NRT service processes to share the ISMN system communication resources. In the first and second phase games $\left(\mathbb{G}_{RT}^{F} \text{and} \, \mathbb{G}_{RT}^{S}\right)$, the RT data packets are divided and distributed. In the third-phase game $\left(\mathbb{G}_{NRT}^{T}\right)$, the NRT data packets are allocated to multiple paths of MANET. In an interactive and sequential manner, $\mathbb{G}_{RT}^{F}$, $\mathbb{G}_{RT}^{S}$ and $\mathbb{G}_{NRT}^{T}$ act independently and make decisions toward an appropriate ISMN system performance. Formally, the *CMRC* scheme defines the three-phase coalitional game entities, that is,

$$\mathbb{G} = \left\{\mathbb{G}_{RT}^{F}, \mathbb{G}_{RT}^{S}, \mathbb{G}_{NRT}^{T}\right\} = \left\{\left\{\mathbb{G}_{RT}^{F} | \mathfrak{M}_{RT}, \mathcal{W}, \mathcal{Z}, \mathcal{U}_{\mathcal{W}}^{RT}, \mathcal{U}_{\mathcal{Z}}^{RT}\right\}, \left\{\mathbb{G}_{RT}^{S} | \mathfrak{M}_{RT}^{\mathcal{Z}}, \mathbb{P}, \mathbb{U}_{P}^{RT}\right\}, \left\{\mathbb{G}_{NRT}^{T} | \mathfrak{M}_{NRT}^{\mathcal{Z}}, \mathbb{P}, \mathfrak{U}_{P}^{NRT}\right\}, T\right\}$$

(Kim, 2022b):

- The three-phase game G consists of $\mathbb{G}_{RT}^{F}$, $\mathbb{G}_{RT}^{S}$ and $\mathbb{G}_{NRT}^{T}$; which are mutually and reciprocally interdependent in an interactive manner.
- In the $\mathbb{G}_{RT}^{F}$, $\mathfrak{M}_{RT}$ is the total RT data packets generated from the source agent. W is the satellite connection path, and Z represents all possible MANET paths.
- The $\mathfrak{M}_{RT}$ is divided into the W and Z. $\mathfrak{M}_{RT}^{\mathcal{W}}$ is the allocated for the W, and $\mathfrak{M}_{RT}^{\mathcal{Z}}$ is the allocated to the Z where $\mathfrak{M}_{RT} = \mathfrak{M}_{RT}^{\mathcal{W}} + \mathfrak{M}_{RT}^{\mathcal{Z}}$.
- In the $\mathbb{G}_{RT}^{F}$, the W and Z are game players, and $\mathcal{U}_{\mathcal{W}}^{RT}$ and $\mathcal{U}_{\mathcal{Z}}^{RT}$ are their utility functions, respectively.
- In the $\mathbb{G}_{RT}^{S}$, $\mathbb{P} = \left\{P_1 \ldots P_q\right\}$ represents the set of selected paths in the MANET from the source to the destination agents.
- In the $\mathbb{G}_{RT}^{S}$, $P_{1 \le i \le q} \in \mathbb{P}$ is a game player, and $\mathbb{U}_{P_i}^{RT}$ is his utility function; $\mathfrak{M}_{RT}^{\mathcal{Z}}$ is distributed into multiple paths in P.
- In the $\mathbb{G}_{NRT}^{T}$, $\mathfrak{M}_{NRT}^{\mathcal{Z}}$ is the number of generated NRT data packets. $P_{1 \le i \le q} \in \mathbb{P}$ is a game player, and $\mathfrak{U}_{P_i}^{NRT}$ is his utility function.
- In the $\mathbb{G}_{NRT}^{T}$, $\mathfrak{M}_{NRT}^{\mathcal{Z}}$ is also distributed into multiple paths in P like to the $\mathbb{G}_{RT}^{S}$.
- $T = \left\{t_1, \ldots, t_c, t_{c+1}, \ldots\right\}$ denotes the time period, which is represented by a sequence of time steps.

9.6.4 The basic idea and implementation of value solutions

To characterize the basic concepts of value solutions, we begin with some definitions. A coalitional game on a finite player set $N = \{1 \ldots i \ldots n\}$ is an ordered pair (N, v), where utility function $v : 2^N \rightarrow \mathbb{R}$ assigns to each coalition $S \in 2^N$ as the worth $v(S)$, satisfying $v(\varnothing) = 0$. For each coalition S, $v(S)$ represents the reward that coalition S can be guaranteed by just itself without the collaboration of others. A value is a function γ which assigns to every game (N, v). For each player $i \in N$, a real number $\gamma_i(N, v)$ represents an assessment made by player i; it is his gain from participating in the game. For all $S \subseteq N$ and all $i \in S$, the marginal contribution of player i to coalition S is denoted as $\Delta^i(v, S)$, and let $\Gamma(v, S)$ be the average marginal contribution of all players in the coalition S; they are given by (Calvo, 2014; Kim, 2022b; Oum, 2020; Yang, 2019);

$$\begin{cases} \Delta^i(v,S) = v(S) - v(S\backslash i) \\ \Gamma(v,S) = \frac{1}{|S|} \times \left(\sum_{i\in S}(v(S) - v(S\backslash i))\right) \end{cases}$$

(9.6.1)

According to (9.6.1), player i's *Shapley value* $(Sh_i(\cdot))$ and *Solidarity value* $(So_i(\cdot))$ are defined as follows (Calvo, 2014; Kim, 2022b; Oum, 2020; Yang, 2019);

$$\begin{cases} Sh_i(N,v) = \sum_{S\subseteq N; i\in S}\left(\frac{(n-s)!(s-1)!}{n!} \times \Delta^i(v,S)\right) \\ So_i(N,v) = \sum_{S\subseteq N; i\in S}\left(\frac{(n-s)!(s-1)!}{n!} \times \Gamma(v,S)\right) \end{cases}$$

(9.6.2)

As a new viewpoint of Harsanyi, the *CMRC* scheme can define the *Shapley value* differently. The *Harsanyi dividend* identifies the surplus created by a coalition of players in a coalitional game. To specify this surplus, the worth of this coalition is corrected by the surplus created by sub-coalitions. To this end, the *Harsanyi dividend* of coalition N in game v $(d_v(N))$ is recursively determined by $d_v(N) = v(N) - \sum_{S\subset N} d_v(S)$. Using the *Harsanyi dividend*, the $Sh_i(\cdot)$ can be redefined as follows (Casajus, 2019; Haeringer, 2006; Kim, 2022b);

$$Sh_i(N,v) = \sum_{S\subseteq N; i\in S}\left(\frac{1}{|S|} \times d_v(s)\right)$$

(9.6.3)

According to (9.6.3), the *CMRC* scheme can think that the *Shapley value* allocates the payoff based on players' productivity and gives everyone his expected marginal contribution, presuming that all permutations of players' entrances are equal. In order to account for asymmetries among players beyond the game, the *weighted Shapley value* (*WSV*) was proposed, where these asymmetries are modeled by weights, which are explicitly given to players. With the weight vector $\omega \in \mathbb{R}^N_{++}$, player i's *WSV* $(WSV_i^{\omega}(\cdot))$ is defined as follows (Casajus, 2019; Haeringer, 2006; Kim, 2022b);

$$WSV_i^{\omega}(N,v) = \sum_{S\subseteq N; i\in S}\left(\frac{\omega_i}{\sum_{j\in S}(\omega_j)} \times d_v(s)\right)$$

(9.6.4)

$$\text{s.t.}, i \in N, \omega = [\omega_1, \ldots, \omega_n] \in \mathbb{R}^N_{++}$$

The *equal division* (*ED*) *solution* is the solution that distributes the worth $v(N)$ of the *grand coalition* equally among all players and thus assigns to every (N,v); the payoff is obtained as $ED_i(N,v) = v(N)/n$ for all $i \in N$; it averages the worth of the grand coalition to all players. Player i's equal *surplus division value* $(ES_i(\cdot))$ allocates the surplus evenly after giving every player his individual worth (Kim, 2022b; Yang, 2019);

$$ES_i(N,v) = v(\{i\}) + \left(\frac{1}{n} \times \left(v(N) - \sum_{j \in N} v(j)\right)\right), \text{ s.t.}, i \in N$$

(9.6.5)

Based on (9.6.5), player i's *WSdV* $\left(WSdV_i^{\omega}(\cdot)\right)$ is defined by employing the weight ω, which is the positive exogenous given to all players (Kim, 2022b; Yang, 2019);

$$WSdV_i^{\omega}(N,v) = v(\{i\}) + \left(\frac{\omega_i}{\sum_{j=1}^{n}(\omega_j)} \times \left(v(N) - \sum_{j \in N} v(j)\right)\right)$$

(9.6.6)

$$\text{s.t.}, \omega = [\omega_1, \ldots, \omega_n] \in \mathbb{R}_{++}^{N}$$

Based on the idea of *So*, the *weighted solidarity value* (*WSoV*) is defined by concerning the possibility of relative weight in the coalitional worth. It presents a new distribution rule for marginal contributions. With the weight ω, player i's *WSoV* $\left(WSoV_i^{\omega}(\cdot)\right)$ is the solution given by (Calvo, 2014, Kim, 2022b);

$$WSoV_i^{\omega}(N,v) = \sum_{S \subseteq N; i \in S} \left(\frac{(n-s)!(s-1)!}{n!} \times \Gamma_i^{\omega}(v,S)\right)$$

(9.6.7)

$$\Gamma_i^{\omega}(v,S) = \frac{\omega_i}{|\omega(S)|} \times \left(\sum_{i \in S} (v(S) - v(S \setminus i))\right)$$

9.6.5 The three-phase game for the ISMN routing algorithm

In MANETs, data may need to traverse long distances before reaching its final destination. Given the wireless communication characteristics and high mobility of MANET agents, the Ad hoc On-Demand Distance Vector (AODV) protocol has become one of the most widely used routing protocols for such networks. The AODV protocol operates by discovering potential routes before transmitting data from the source agent to the destination agent. By establishing multiple routes, AODV enhances the reliability of data transmission in dynamic MANET environments. To maintain network connectivity, mobile agents must periodically update and share local connection information with their neighboring agents. Additionally, for long-distance data delivery, agents use control messages to facilitate efficient data transmission across the network (Atto, 2020). To configure multi-hop AODV routing paths, each wireless link is estimated based on its distance and traffic load. The one-hop link cost between agents i and j, that is, $LC_{i,j}$, is obtained as (Kim, 2010, 2022b).

$$LC_{i,j} = \left((1-\alpha) \times \frac{d_{i,j}}{D_M}\right) + \left(\alpha \times \frac{l_i}{L_M}\right)$$

(9.6.8)

where D_M and L_M are the maximum coverage range of each agent and the maximum queue length, respectively. The $d_{i,j}$ is the distance from the agent i to the agent j, and l_i is the current queue length of agent i; l_i is defined as the amount of current traffic buffering, which is used as a threshold to detect traffic congestion. The α is a control parameter to estimate the relative current link cost. When the input data rate exceeds the output data rate in the agent i, the routing packets become congested, and l_i increases. In this case, a higher value of α is more suitable. By contrast, a lower value of α is more suitable. In the *CMRC* scheme, the α value is dynamically adjusted based on the current buffer space ratio where $\alpha = l_i/L_M$ (Kim, 2022b).

Based on the AODV protocol, multiple multi-hop routing paths in the MANET are constructed, and the q least cost paths are selected for the multipath routing algorithm, which adaptively distributes routing packets through these paths. To provide a fair and efficient solution to the packet distribution problem, the *CMRC* scheme adopts the concept of coalitional game, and the established paths are assumed as game players. Therefore, a coalition is formed by the combination of selected routing paths, and the coalitional game is a q-person game. Path $1 \leq h \leq q$ from the source agent s to the destination agent e has the set $\left(\mathbb{S}_h^{s,e}\right)$ of all one-hop links, and the path h's utility function $\left(U_h(s,e)\right)$ is defined as follows (Kim, 2022b);

$$U_h(s,e) = \frac{\psi}{\sum_{g \in \mathbb{S}_h^{s,e}} C(g)} \tag{9.6.9}$$

where ψ is a control factor and $C(g)$ is the cost function for the link g, which is defined using (9.6.8). Generally, the packet distribution problem for q routing paths leads to a conflict situation similar to the bankruptcy problem for q creditors. The bankruptcy problem assumes that a company becomes bankrupt and owes money to q creditors; this money must be divided among these creditors. The bankruptcy problem is a pair (E, c); E represents the company's remaining money and $\mathbf{c} = \left(c_1, \ldots, c_q\right)$ is the vector of creditors' claims where $0 < E < \sum_{i=1}^{q} c_i$. The utility function $\upsilon(S)^{E,c}$ for the coalition S is defined based on the bankruptcy model (Kim, 2010, 2022b);

$$\upsilon(S)^{E,c} = \max\left\{0;\ E - \sum_{i \in S} c_i\right\} \tag{9.6.10}$$

In the $\mathbb{G}_{RT}^F$, the *CMRC* scheme distributes the $\mathfrak{M}_{RT}$ for the W and Z. Simply, the *CMRC* scheme considers that one satellite is already deployed at the desired locations to ensure maritime network demands. Therefore, the *CMRC* scheme has two ways to connect from the source to the destination agents: one is established in the MANET (Z), and the other is connected by using the satellite network (W). The utility functions of W and Z from the source agent s to the destination agent e, that is, $\mathcal{U}_{\mathcal{W}}^{RT}(s,e)$ and $\mathcal{U}_{\mathcal{Z}}^{RT}(s,e)$, are defined as follows (Kim, 2022b);

$$\begin{cases} \mathcal{U}_{\mathcal{W}}^{RT}\left((s,e), \mathfrak{M}_{RT}^{\mathcal{W}}\right) = \left(\left(A_{\mathcal{W}} \times \frac{(\chi + f^{RT})}{\mathcal{C}_{\mathcal{W}}}\right) \times \mathfrak{M}_{RT}^{\mathcal{W}}\right) \\ \mathcal{U}_{\mathcal{Z}}^{RT}\left((s,e), \mathfrak{M}_{RT}^{\mathcal{Z}}\right) = \left(\sum_{P_k^{(s,e)} \in \mathbb{P}} \min_g\left\{\mathcal{A}_{\mathcal{Z}}(g) | g \in \mathbb{S}_{P_k}^{s,e}\right\}\right) \times \frac{\mathfrak{M}_{RT}^{\mathcal{Z}}}{\mathcal{C}_{\mathcal{Z}}} \end{cases}$$

(9.6.11)

where $A_{\mathscr{W}}$ is the currently available link capacity of W and f^{RT} is the current failure ratio of the RT data services. χ is a control parameter for $\mathscr{U}^{RT}(s,e)$, and $\mathscr{A}_{\mathscr{Z}}(g)$ is the currently available capacity of the one-hope link g. Therefore, $\mathscr{U}^{RT}_{\mathscr{Z}}(\cdot)$ is significantly affected by the sum of the bottleneck links of each path. $\mathscr{C}_{\mathscr{W}}$ and $\mathscr{C}_{\mathscr{Z}}$ are the cost factors of the W and Z, respectively. Based on the $\mathscr{U}^{RT}_{\mathscr{W}}(\cdot)$ and $\mathscr{U}^{RT}_{\mathscr{Z}}(\cdot)$, the *CMRC* scheme can get the $v(S)$ with (E,c). Especially, the $\mathbb{G}^{F}_{RT}$ is a two-player game with the W and Zwhere $v(\{\mathscr{W}\}) = c_{\mathscr{W}} = \mathscr{U}^{RT}_{\mathscr{W}}\left((s,e),\mathfrak{M}^{\mathscr{W}}_{RT}\right)$ with $\mathfrak{M}^{\mathscr{W}}_{RT} = \mathfrak{M}_{RT}$, and $v(\{\mathscr{Z}\}) = c_{\mathscr{Z}} = \mathscr{U}^{RT}_{\mathscr{Z}}\left((s,e),\mathfrak{M}^{\mathscr{Z}}_{RT}\right)$ with $\mathfrak{M}^{\mathscr{Z}}_{RT} = \mathfrak{M}_{RT}$, and $E = \vartheta \times \left(c_{\mathscr{W}} + c_{\mathscr{Z}}\right)$; ϑ is an adjustment factor to estimate the grand coalition value. The *CMRC* scheme adopts the *WSdV* as the solution of $\mathbb{G}^{F}_{RT}$; it is given by (Kim, 2022b);

$$WSdV^{\omega}_{\mathscr{P}_i \in \mathbb{N}}(\mathbb{N},v) = v\left(\{\mathscr{P}_i\}\right) + \left(\omega_{\mathscr{P}_i} \times \left(v(\mathbb{N}) - \sum_{\substack{\mathscr{P}_j \in N;\\ \mathscr{P}_i \neq \mathscr{P}_j}} v\left(\mathscr{P}_j\right)\right)\right) \tag{9.6.12}$$

$$\mathscr{P}_i \in \mathbb{N} = \{\mathscr{W},\mathscr{Z}\},\ \ \omega_{\mathscr{W}} = \max(\mu, f^{RT}) \text{ and } \omega_{\mathscr{Z}} = 1 - \omega_{\mathscr{W}}$$

where μ is an adjustment parameter for determining $\omega_{\mathscr{W}}$ value. According to (9.6.12), the W and Z obtain their *WSdVs* such as $WSdV^{\omega}_{\mathscr{W}}(\mathbb{N},v)$ and $WSdV^{\omega}_{\mathscr{Z}}(\mathbb{N},v)$. Finally, the $\mathfrak{M}_{RT}$ is divided into $\mathfrak{M}^{\mathscr{W}}_{RT}$ and $\mathfrak{M}^{\mathscr{Z}}_{RT}$ as follows (Kim, 2022b);

$$\begin{cases} \mathfrak{M}^{\mathscr{W}}_{RT} = \dfrac{WSdV^{\omega}_{\mathscr{W}}(\mathbb{N},v)}{WSdV^{\omega}_{\mathscr{W}}(\mathbb{N},v) + WSdV^{\omega}_{\mathscr{Z}}(\mathbb{N},v)} \times \mathfrak{M}_{RT} \\ \\ \mathfrak{M}^{\mathscr{Z}}_{RT} = \left(\mathfrak{M}_{RT} - \mathfrak{M}^{\mathscr{W}}_{RT}\right) = \dfrac{WSdV^{\omega}_{\mathscr{Z}}(\mathbb{N},v)}{WSdV^{\omega}_{\mathscr{W}}(\mathbb{N},v) + WSdV^{\omega}_{\mathscr{Z}}(\mathbb{N},v)} \times \mathfrak{M}_{RT} \end{cases} \tag{9.6.13}$$

In the $\mathbb{G}^{S}_{RT}$, the *CMRC* scheme distributes the $\mathfrak{M}^{\mathscr{Z}}_{RT}$ for multiple paths, that is, $P_{1\le k\le q}$, in the P. The $P_k \in \mathbb{P}$ is a game player and his utility function from the source agent s to the destination agent e, that is, $\mathbb{U}^{RT}_{P_k}(s,e)$, is defined as follows (Kim, 2022b);

$$\mathbb{U}^{RT}_{P_k}\left((s,e),\mathfrak{M}^{\mathscr{Z},P_k}_{RT}\right) = U_{P_k}(s,e) \times \frac{\mathfrak{M}^{\mathscr{Z},P_k}_{RT}}{\log\left(\delta + \mathbb{H}(P^{(s,e)}_k)\right)} \tag{9.6.14}$$

where $\mathfrak{M}^{\mathscr{Z},P_k}_{RT}$ is the allocated RT data packet amount for the P_k. δ is an adjustment parameter for $\mathbb{U}^{RT}_{P_k}(\cdot)$ and $\mathbb{H}\left(P^{(s,e)}_k\right)$ is the link hop number of P_k. In particular, the $\mathbb{G}^{S}_{RT}$ is a q-player game with $P_{1\le k\le q}$ where $v\left(P^{(s,e)}_k\right) = c_{P_k} = \mathbb{U}^{RT}_{P_k}\left((s,e),\mathfrak{M}^{\mathscr{Z},P_k}_{RT}\right)$ with $\mathfrak{M}^{\mathscr{Z},P_k}_{RT} = \mathfrak{M}^{\mathscr{Z}}_{RT}$, and $E = \vartheta \times \left(\sum_{1\le k\le q} v(P^{(s,e)}_k)\right)$. In the $\mathbb{G}^{S}_{RT}$, the *CMRC* scheme adopts the *WSV* as the solution concept; it is given by (Kim, 2022b);

$$WSV^{\omega}_{P_k}(\mathbb{P},v) = \sum_{S\subseteq\mathbb{P};P_k\in S}\left(\frac{\omega_{P_k}}{\sum_{P_i\in S}(\omega_{P_i})}\times d_v(S)\right)$$

(9.6.15)

$$\text{s.t., } P_k \in \mathbb{P}, \omega = \left[\omega_{P_1},\ldots,\omega_{P_q}\right] \in \mathbb{R}^q_{++} \text{ and } \omega_{P_k} = \frac{\mathfrak{P}^{\omega} - \mathbb{H}\left(P_k^{(s,e)}\right)}{\sum_{P_l\in\mathbb{P}}\left(\mathfrak{P}^{\omega} - \mathbb{H}\left(P_l^{(s,e)}\right)\right)}$$

where $\mathfrak{P}^{\omega}$ is the modification factor for the weight calculation. According to (9.6.15), the P_k obtains its *WSV*, such as $WSV^{\omega}_{P_k}(\mathbb{P},v)$, and the $\mathfrak{M}^{\mathcal{I}}_{RT}$ is divided for the P_k, that is, $\mathfrak{M}^{\mathcal{I},P_k}_{RT}$, as follows (Kim, 2022b);

$$\mathfrak{M}^{\mathcal{I},P_k}_{RT} = \frac{WSV^{\omega}_{P_k}(\mathbb{P},v)}{\sum_{P_i\in\mathbb{P}} WSV^{\omega}_{P_i}(\mathbb{P},v)} \times \mathfrak{M}^{\mathcal{I}}_{RT}$$

(9.6.16)

In the $\mathbb{G}^T_{NRT}$, the *CMRC* scheme distributes the $\mathfrak{M}^{\mathcal{I}}_{NRT}$ for the multiple paths in the P. Similar to the $\mathbb{G}^S_{RT}$, the P_k is a game player and his utility function, that is, $\mathfrak{U}^{NRT}_{P_k}(s,e)$ is defined as follows (Kim, 2022b);

$$\mathfrak{U}^{NRT}_{P_k}\left((s,e),\mathfrak{M}^{\mathcal{I},P_k}_{NRT}\right) = U_{P_k}(s,e) \times \left(1 - \left(\varepsilon \times \frac{l^M_{P_k}}{L_M}\right)\right) \times \mathfrak{M}^{\mathcal{I},P_k}_{NRT}$$

(9.6.17)

where $\mathfrak{M}^{\mathcal{I},P_k}_{NRT}$ is the allocated NRT data packet amount for the P_k. ε is a control factor for $\mathfrak{U}^{NRT}_{P}(\cdot)$ and $l^M_{P_k}$ is the queue length of the most congested link in the P_k. The $\mathbb{G}^T_{NRT}$ is also a q-player game with P_k where $v\left(P_k^{(s,e)}\right) = c_{P_k} = \mathfrak{U}^{NRT}_{P_k}\left((s,e),\mathfrak{M}^{\mathcal{I},P_k}_{NRT}\right)$ with $\mathfrak{M}^{\mathcal{I},P_k}_{NRT} = \mathfrak{M}^{\mathcal{I}}_{NRT}$, and $E = \vartheta \times \left(\sum_{1\le k\le q} v\left(P_k^{(s,e)}\right)\right)$. In the $\mathbb{G}^T_{NRT}$, the *CMRC* scheme adopts the *WSoV* as the solution concept; it is given by (Kim, 2022b);

$$WSoV^{\omega}_{P_k}(\mathbb{P},v) = \sum_{S\subseteq\mathbb{P};P_k\in S}\left(\frac{(q-|S|)!(|S|-1)!}{q!}\times \Gamma^{\omega}_{P_k}(v,S)\right)$$

(9.6.18)

$$\text{s.t., }\begin{cases} P_k \in P, \omega \in \mathbb{R}^q_{++}\ and\ \omega_k = \dfrac{L_M - l^M_{P_k}}{\sum_{P_l\in\mathbb{P}}\left(L_M - l^M_{P_l}\right)} \\[2ex] \Gamma^{\omega}_{P_k}(v,S) = \dfrac{\omega_k}{|\omega(S)|}\times\left(\sum_{P_l\in S}\left(v(S) - v(S\backslash P_l)\right)\right) \end{cases}$$

According to (9.6.18), the P_k obtains its *WSoV* similar to $WSoV^{\omega}_{P_k}(\mathbb{P},v)$, and the $\mathfrak{M}^{\mathcal{I}}_{NRT}$ is divided for the P_k, that is, $\mathfrak{M}^{\mathcal{I},P_k}_{NRT}$, as follows (Kim, 2022b);

$$\mathfrak{M}_{NRT}^{\mathcal{I},P_k} = \frac{WSoV_{P_k}^{\omega}(\mathbb{P},v)}{\sum_{P_i \in \mathbb{P}} WSoV_{P_i}^{\omega}(\mathbb{P},v)} \times \mathfrak{M}_{NRT}^{\mathcal{I}} \tag{9.6.19}$$

9.6.6 Summary

The emergence of 6G communication technologies in maritime networks has led to the development of ISMN infrastructure. To address packet distribution challenges across multiple AODV routing paths, the *CMRC* scheme introduces an innovative ISMN routing mechanism. By incorporating the *WSV*, *WSoV*, and *WSdV*, the *CMRC* scheme optimally distributes RT and NRT data packets. Taking into account data type characteristics and service preferences, the *CMRC* scheme formulates a three-phase coalitional game model to maximize efficiency and mutual benefits. In the *CMRC* scheme, the three-phase game model operates in a distributed online manner, where each phase interacts dynamically with the others, working collectively to achieve a mutually beneficial solution. Even under rapidly changing ISMN conditions, the *CMRC* scheme effectively navigates complex routing control challenges by leveraging reciprocal consensus across network entities (Kim, 2022b).

REFERENCES

Akhtar, M. W., Hassan, S. A., Ghaffar, R., Jung, H., Garg, S., & Hossain, M. S. (2020a). The shift to 6G communications: Vision and requirements. *Human-centric Computing and Information Sciences*, *10*(1), 1–27. DOI: 10.1186/s13673-020-00258-2

Alós-Ferrer, C., García-Segarra, J., & Ginés-Vilar, M. (2018). Anchoring on Utopia: A generalization of the Kalai–Smorodinsky solution. *Economic Theory Bulletin*, *6*(2), 141–155. DOI: 10.1007/s40505-017-0130-7

Alzidaneen, A., Alsharoa, A., & Alouini, M. S. (2019). Resource and placement optimization for multiple UAVs using backhaul tethered balloons. *IEEE Wireless Communications Letters*, *9*(4), 543–547. DOI: 10.1109/LWC.2019.2961906

An, X., Fan, R., Hu, H., Zhang, N., Atapattu, S., & Tsiftsis, T. A. (2022). Joint task offloading and resource allocation for IoT edge computing with sequential task dependency. *IEEE Internet of Things Journal*, *9*(17), 16546–16561. DOI: 10.1109/JIOT.2022.3150976

Aslam, S., Michaelides, M. P., & Herodotou, H. (2020). Internet of ships: A survey on architectures, emerging applications, and challenges. *IEEE Internet of Things Journal*, *7*(10), 9714–9727. DOI: 10.1109/JIOT.2020.2993411

Ateya, A. A. A., Muthanna, A., Kirichek, R., Hammoudeh, M., & Koucheryavy, A. (2019). Energy- and latency-aware hybrid offloading algorithm for UAVs. *IEEE Access : Practical Innovations, Open Solutions*, *7*, 37587–37600. DOI: 10.1109/ACCESS.2019.2905249

Atto, M., Mstafa, R. J., & Alkhayyat, A. (2020). Improving AODV routing protocol for image transmission over mobile video sensor networks. *IEEE Access : Practical Innovations, Open Solutions*, *8*, 169396–169407. DOI: 10.1109/ACCESS.2020.3024093

Avrachenkov, K., Elias, J., Martignon, F., Neglia, G., & Petrosyan, L. (2015). Cooperative network design: A Nash bargaining solution approach. *Computer Networks*, *83*, 265–279. DOI: 10.1016/j.comnet.2015.03.017

Béal, S., Rémila, E., & Solal, P. (2015). *A class of solidarity allocation rules for TU-games*. HAL.

Calvo, E., & Gutiérrez, E. (2013). The Shapley-solidarity value for games with a coalition structure. *International Game Theory Review*, *15*(01), 1350002. DOI: 10.1142/S0219198913500023

Calvo, E., & Gutiérrez-López, E. (2014). Axiomatic characterizations of the weighted solidarity values. *Mathematical Social Sciences*, *71*, 6–11. DOI: 10.1016/j.mathsocsci.2014.03.005

Casajus, A. (2019). Relaxations of symmetry and the weighted Shapley values. *Economics Letters*, *176*, 75–78. DOI: 10.1016/j.econlet.2018.12.031

Chen, X., Jiao, L., Li, W., & Fu, X. (2015). Efficient multi-user computation offloading for mobile-edge cloud computing. *IEEE/ACM Transactions on Networking*, *24*(5), 2795–2808. DOI: 10.1109/TNET.2015.2487344

Chen, Z., Guo, D., An, K., Zhang, B., Zhang, X., & Zhao, B. (2019). Efficient and fair resource allocation scheme for cognitive satellite-terrestrial networks. *IEEE Access : Practical Innovations, Open Solutions*, *7*, 145124–145133. DOI: 10.1109/ACCESS.2019.2943908

Chettri, L., & Bera, R. (2019). A comprehensive survey on Internet of Things (IoT) toward 5G wireless systems. *IEEE Internet of Things Journal*, *7*(1), 16–32. DOI: 10.1109/JIOT.2019.2948888

Dagan, N., & Volij, O. (1993). The bankruptcy problem: A cooperative bargaining approach. *Mathematical Social Sciences*, *26*(3), 287–297. DOI: 10.1016/0165-4896(93)90024-D

Dao, N. N., Pham, Q. V., Tu, N. H., Thanh, T. T., Bao, V. N. Q., Lakew, D. S., & Cho, S. (2021). Survey on aerial radio access networks: Toward a comprehensive 6G access infrastructure. *IEEE Communications Surveys and Tutorials*, *23*(2), 1193–1225. DOI: 10.1109/COMST.2021.3059644

Diskin, A., Koppel, M., & Samet, D. (2011). Generalized Raiffa solutions. *Games and Economic Behavior*, *73*(2), 452–458. DOI: 10.1016/j.geb.2011.04.002

Ebrahimkhani, A., & Akhbari, B. (2021). Bargaining solutions in heterogeneous networks: A reinforcement learning-based approach. *IET Communications*, *15*(18), 2315–2329. DOI: 10.1049/cmu2.12272

Farooq, M. J., & Zhu, Q. (2018). On the secure and reconfigurable multi-layer network design for critical information dissemination in the internet of battlefield things (IoBT). *IEEE Transactions on Wireless Communications*, *17*(4), 2618–2632. DOI: 10.1109/TWC.2018.2799860

Fragkou, E., Papakostas, D., Kasidakis, T., & Katsaros, D. (2022). Multilayer Backbones for Internet of Battlefield Things. *Future Internet*, *14*(6), 186. DOI: 10.3390/fi14060186

Gao, M., Ai, B., Niu, Y., Wu, W., Yang, P., Lyu, F., & Shen, X. (2020). Efficient hybrid beamforming with anti-blockage design for high-speed railway communications. *IEEE Transactions on Vehicular Technology*, *69*(9), 9643–9655. DOI: 10.1109/TVT.2020.3000757

Golam, M., Lee, J. M., & Kim, D. S. (2020, October). A uav-assisted blockchain based secure device-to-device communication in internet of military things. In *2020 International Conference on Information and Communication Technology Convergence (ICTC)* (pp. 1896-1898). IEEE. DOI: 10.1109/ICTC49870.2020.9289282

Guo, F., Zhang, H., Ji, H., Li, X., & Leung, V. C. (2018). An efficient computation offloading management scheme in the densely deployed small cell networks with mobile edge computing. *IEEE/ACM Transactions on Networking*, *26*(6), 2651–2664. DOI: 10.1109/TNET.2018.2873002

Haeringer, G. (2006). A new weight scheme for the Shapley value. *Mathematical Social Sciences*, *52*(1), 88–98. DOI: 10.1016/j.mathsocsci.2006.03.001

Hajir, M., Langar, R., & Gagnon, F. (2016). Coalitional games for joint co-tier and cross-tier cooperative spectrum sharing in dense heterogeneous networks. *IEEE Access : Practical Innovations, Open Solutions*, *4*, 2450–2464. DOI: 10.1109/ACCESS.2016.2562498

Hassan, S. S., Majeed, U., & Hong, C. S. (2021, January). Reliable integrated space-oceanic network profit maximization by Bender decomposition approach. In *2021 International Conference on Information Networking (ICOIN)* (pp. 565-570). IEEE. DOI: 10.1109/ICOIN50884.2021.9333882

Herrero, C., & Villar, A. (2010). The rights egalitarian solution for NTU sharing problems. *International Journal of Game Theory, 39*(1-2), 137–150. DOI: 10.1007/s00182-009-0177-2

Jia, Z., Wu, Q., Dong, C., Yuen, C., & Han, Z. (2022). Hierarchical aerial computing for Internet of Things via cooperation of HAPs and UAVs. *IEEE Internet of Things Journal, 10*(7), 5676–5688. DOI: 10.1109/JIOT.2022.3151639

Joo, C., & Choi, J. (2018). Low-delay broadband satellite communications with high-altitude unmanned aerial vehicles. *Journal of Communications and Networks (Seoul), 20*(1), 102–108. DOI: 10.1109/JCN.2018.000010

Kang, H., Chang, X., Mišić, J., Mišić, V. B., Fan, J., & Liu, Y. (2023). Cooperative UAV resource allocation and task offloading in hierarchical aerial computing systems: A MAPPO-based approach. *IEEE Internet of Things Journal, 10*(12), 10497–10509. DOI: 10.1109/JIOT.2023.3240173

Karagözoğlu, E., Keskin, K., & Özcan-Tok, E. (2019). Between anchors and aspirations: A new family of bargaining solutions. *Review of Economic Design, 23*(1-2), 53–73. DOI: 10.1007/s10058-019-00220-4

Karia, D. C., & Kolekar, U. D. (2008, September). Performance Analysis of Real and Non Real Time Traffic over WLAN Using Connection Admission Control Policy. In *2008 Second UKSIM European Symposium on Computer Modeling and Simulation* (pp. 548-554). IEEE. DOI: 10.1109/EMS.2008.25

Kim, S. (2010). Cooperative game theoretic online routing scheme for wireless network managements. *IET Communications, 4*(17), 2074–2083. DOI: 10.1049/iet-com.2009.0686

Kim, S. (2011). Adaptive online power control scheme based on the evolutionary game theory. *IET Communications, 5*(18), 2648–2655. DOI: 10.1049/iet-com.2011.0093

Kim, S. (2012). Trust-based bargaining game model for cognitive radio spectrum sharing scheme. *IEICE Transactions on Communications, 95*(12), 3925–3928. DOI: 10.1587/transcom.E95.B.3925

Kim, S. (Ed.). (2014). *Game theory applications in network design*. IGI Global. DOI: 10.4018/978-1-4666-6050-2

Kim, S. (2019). New bargaining game based computation offloading scheme for flying ad-hoc networks. *IEEE Access : Practical Innovations, Open Solutions, 7*, 147038–147047. DOI: 10.1109/ACCESS.2019.2946363

Kim, S. (2020). Cognitive Satellite Spectrum Management Scheme Based on the Cooperative Solidarity Values. *IEEE Access : Practical Innovations, Open Solutions, 8*, 113837–113846. DOI: 10.1109/ACCESS.2020.3003640

Kim, S. (2021a). A novel local and global cooperative approach for distributed mobile cloud computing. *IEEE Access : Practical Innovations, Open Solutions, 9*, 117813–117822. DOI: 10.1109/ACCESS.2021.3105060

Kim, S. (2021b). Bargaining-based spectrum allocation algorithm for high-speed railway communications. *IEEE Access : Practical Innovations, Open Solutions, 9*, 71651–71659. DOI: 10.1109/ACCESS.2021.3079134

Kim, S. (2022a). Hybrid RF/VLC network spectrum allocation scheme using bargaining solutions. *IEEE Access : Practical Innovations, Open Solutions*, *10*, 20019–20028. DOI: 10.1109/ACCESS.2022.3153327

Kim, S. (2022b). Cooperative Multi-Path Routing Algorithm for Integrated Satellite-Maritime Networks. *Mobile Information Systems*, *2022*(1), 8678007. DOI: 10.1155/2022/8678007

Kim, S. (2024a). Hierarchical aerial offload computing algorithm based on the Stackelberg-evolutionary game model. *Computer Networks*, *245*, 110348. DOI: 10.1016/j.comnet.2024.110348

Kim, S. (2024b). Collaborative game-based task offloading scheme in the UAV-TB-assisted battlefield network platform. *EURASIP Journal on Wireless Communications and Networking*, *2024*(1), 11. DOI: 10.1186/s13638-024-02337-9

Kim, S., & Varshney, P. K. (2004). An integrated adaptive bandwidth-management framework for QoS-sensitive multimedia cellular networks. *IEEE Transactions on Vehicular Technology*, *53*(3), 835–846. DOI: 10.1109/TVT.2004.825704

Kodheli, O., Andrenacci, S., Maturo, N., Chatzinotas, S., & Zimmer, F. (2019). An uplink UE group-based scheduling technique for 5G mMTC systems over LEO satellite. *IEEE Access : Practical Innovations, Open Solutions*, *7*, 67413–67427. DOI: 10.1109/ACCESS.2019.2918581

Kulkarni, A. A., & Shanbhag, U. V. (2015). An existence result for hierarchical Stackelberg v/s Stackelberg games. *IEEE Transactions on Automatic Control*, *60*(12), 3379–3384. DOI: 10.1109/TAC.2015.2423891

Lagunas, E., Sharma, S. K., Maleki, S., Chatzinotas, S., & Ottersten, B. (2015). Resource allocation for cognitive satellite communications with incumbent terrestrial networks. *IEEE Transactions on Cognitive Communications and Networking*, *1*(3), 305–317. DOI: 10.1109/TCCN.2015.2503286

Lakew, D. S., Tran, A. T., Dao, N. N., & Cho, S. (2022). Intelligent offloading and resource allocation in heterogeneous aerial access IoT networks. *IEEE Internet of Things Journal*, *10*(7), 5704–5718. DOI: 10.1109/JIOT.2022.3161571

Lee, H. S., & Lee, J. W. (2018). Task offloading in heterogeneous mobile cloud computing: Modeling, analysis, and cloudlet deployment. *IEEE Access : Practical Innovations, Open Solutions*, *6*, 14908–14925. DOI: 10.1109/ACCESS.2018.2812144

Li, W., Xu, G., Zou, R., & Hou, D. (2022). The allocation of marginal surplus for cooperative games with transferable utility. *International Journal of Game Theory*, *51*(2), 353–377. DOI: 10.1007/s00182-021-00795-9

Li, X., Feng, W., Chen, Y., Wang, C. X., & Ge, N. (2020). Maritime coverage enhancement using UAVs coordinated with hybrid satellite-terrestrial networks. *IEEE Transactions on Communications*, *68*(4), 2355–2369. DOI: 10.1109/TCOMM.2020.2966715

Li, Z., Wang, Y., Liu, M., Sun, R., Chen, Y., Yuan, J., & Li, J. (2019). Energy efficient resource allocation for UAV-assisted space-air-ground internet of remote things networks. *IEEE Access : Practical Innovations, Open Solutions*, *7*, 145348–145362. DOI: 10.1109/ACCESS.2019.2945478

Luo, C., Zhou, X., & Lev, B. (2022). Core, shapley value, nucleolus and nash bargaining solution: A Survey of recent developments and applications in operations management. *Omega, 110*, 102638. DOI: 10.1016/j.omega.2022.102638

Messous, M. A., Arfaoui, A., Alioua, A., & Senouci, S. M. (2017b). A sequential game approach for computation-offloading in an UAV network. In *GLOBECOM 2017-2017 IEEE Global Communications Conference* (pp. 1-7). IEEE. DOI: 10.1109/GLOCOM.2017.8253967

Messous, M. A., Sedjelmaci, H., Houari, N., & Senouci, S. M. (2017a). Computation offloading game for an UAV network in mobile edge computing. In *2017 IEEE International Conference on Communications (ICC)* (pp. 1-6). IEEE. DOI: 10.1109/ICC.2017.7996483

Messous, M. A., Senouci, S. M., Sedjelmaci, H., & Cherkaoui, S. (2019). A game theory based efficient computation offloading in an UAV network. *IEEE Transactions on Vehicular Technology, 68*(5), 4964–4974. DOI: 10.1109/TVT.2019.2902318

Nguyen, D. C., Ding, M., Pathirana, P. N., Seneviratne, A., Li, J., Niyato, D., Dobre, O., & Poor, H. V. (2021). 6G Internet of Things: A comprehensive survey. *IEEE Internet of Things Journal, 9*(1), 359–383. DOI: 10.1109/JIOT.2021.3103320

Oum, S. B. N., & Lambo, L. D. (2020). The Solidarity Value as a Probabilistic Solidarity Value. *Theoretical Economics Letters, 10*(06), 1397–1425. DOI: 10.4236/tel.2020.106085

Peng, S., Wang, Y., Xiao, H., & Lin, B. (2020). Implementation of an improved AODV routing protocol for maritime ad-hoc networks. In *2020 13th international congress on image and signal processing, biomedical engineering and informatics (CISP-BMEI)* (pp. 7-11). IEEE. DOI: 10.1109/CISP-BMEI51763.2020.9263519

Segarra, J. G., & Vilar, M. G. (2011). Weighted proportional losses solution. *Dept. Econ. Theory Econ. Hist. Univ. Granada, Granada, Spain, Working paper,* (10/21).

Siddiqi, M. A., Yu, H., & Joung, J. (2019). 5G ultra-reliable low-latency communication implementation challenges and operational issues with IoT devices. *Electronics (Basel), 8*(9), 981. DOI: 10.3390/electronics8090981

Sigmund, K. (2011). Introduction to evolutionary game theory. *Evolutionary game dynamics, 69*, 1-26.

Stein, A., Salvioli, M., Garjani, H., Dubbeldam, J., Viossat, Y., Brown, J. S., & Staňková, K. (2023). Stackelberg evolutionary game theory: how to manage evolving systems. *Philosophical Transactions of the Royal Society B, 378*(1876), 20210495.

Stocchero, J. M., Silva, C. A., de Souza Silva, L., Lawisch, M. A., dos Anjos, J. C. S., & de Freitas, E. P. (2022). Secure command and control for internet of battle things using novel network paradigms. *IEEE Communications Magazine, 61*(5), 166–172. DOI: 10.1109/MCOM.001.2101072

Tang, Q., Xie, R., Yu, F. R., Huang, T., & Liu, Y. (2020). Decentralized computation offloading in IoT fog computing system with energy harvesting: A Dec-POMDP approach. *IEEE Internet of Things Journal, 7*(6), 4898–4911. DOI: 10.1109/JIOT.2020.2971323

Traulsen, A., & Glynatsi, N. E. (2023). The future of theoretical evolutionary game theory. *Philosophical Transactions of the Royal Society B, 378*(1876), 20210508.

Trockel, W. (2009). *An axiomatization of the sequential Raiffa solution* (No. 425). Working Papers.

Trockel, W. (2015). Axiomatization of the discrete Raiffa solution. *Economic Theory Bulletin, 3*(1), 9–17. DOI: 10.1007/s40505-014-0046-4

Tuyls, K., & Nowé, A. (2005). Evolutionary game theory and multi-agent reinforcement learning. *The Knowledge Engineering Review, 20*(1), 63–90. DOI: 10.1017/S026988890500041X

Tuyls, K., Nowe, A., Lenaerts, T., & Manderick, B. (2004). An evolutionary game theoretic perspective on learning in multi-agent systems. *Synthese, 139*(2), 297–330. DOI: 10.1023/B:SYNT.0000024908.89191.f1

Wang, C., & Yang, H. (2021). Some kinds of bargaining equilibria of multi-objective games. *Acta Mathematicae Applicatae Sinica. English Series, 37*(2), 201–213.

Wang, E., Li, H., & Zhang, S. (2019). Load balancing based on cache resource allocation in satellite networks. *IEEE Access : Practical Innovations, Open Solutions, 7*, 56864–56879. DOI: 10.1109/ACCESS.2019.2914167

Wang, P., Zhang, J., Zhang, X., Yan, Z., Evans, B. G., & Wang, W. (2019). Convergence of satellite and terrestrial networks: A comprehensive survey. *IEEE Access : Practical Innovations, Open Solutions, 8*, 5550–5588. DOI: 10.1109/ACCESS.2019.2963223

Wang, S., Chen, M., Yin, C., Saad, W., Hong, C. S., Cui, S., & Poor, H. V. (2021). Federated learning for task and resource allocation in wireless high-altitude balloon networks. *IEEE Internet of Things Journal, 8*(24), 17460–17475. DOI: 10.1109/JIOT.2021.3080078

Wang, W., Tong, Y., Li, L., Lu, A. A., You, L., & Gao, X. (2019). Near optimal timing and frequency offset estimation for 5G integrated LEO satellite communication system. *IEEE Access : Practical Innovations, Open Solutions, 7*, 113298–113310. DOI: 10.1109/ACCESS.2019.2935038

Wang, X., Liy, H., Yao, W., Lany, T., & Wu, Q. (2020, May). Content delivery for high-speed railway via integrated terrestrial-satellite networks. In *2020 IEEE Wireless Communications and Networking Conference (WCNC)* (pp. 1-6). IEEE. DOI: 10.1109/WCNC45663.2020.9120643

Xing, J., Li, Q., Sun, H., Qi, J., Zhu, Z., & Wang, J. (2019). Multi-radio multi-channel assignment algorithm in maritime wireless mesh networks. *IEEE Access : Practical Innovations, Open Solutions, 7*, 85903–85912. DOI: 10.1109/ACCESS.2019.2924966

Xu, S., Zhu, G., Ai, B., & Zhong, Z. (2016). A survey on high-speed railway communications: A radio resource management perspective. *Computer Communications, 86*, 12–28. DOI: 10.1016/j.comcom.2016.04.003

Yan, L., Fang, X., Hao, L., & Fang, Y. (2020). Safety-oriented resource allocation for space-ground integrated cloud networks of high-speed railways. *IEEE Journal on Selected Areas in Communications, 38*(12), 2747–2759. DOI: 10.1109/JSAC.2020.3005487

Yang, H., Wang, W., & Ding, Z. (2019). The weighted surplus division value for cooperative games. *Symmetry*, *11*(9), 1169. DOI: 10.3390/sym11091169

Yau, K. L. A., Syed, A. R., Hashim, W., Qadir, J., Wu, C., & Hassan, N. (2019). Maritime networking: Bringing internet to the sea. *IEEE Access : Practical Innovations, Open Solutions*, *7*, 48236–48255. DOI: 10.1109/ACCESS.2019.2909921

You, X., Wang, C. X., Huang, J., Gao, X., Zhang, Z., Wang, M., Huang, Y., Zhang, C., Jiang, Y., Wang, J., Zhu, M., Sheng, B., Wang, D., Pan, Z., Zhu, P., Yang, Y., Liu, Z., Zhang, P., Tao, X., & Liang, Y. C. (2021). Towards 6G wireless communication networks: Vision, enabling technologies, and new paradigm shifts. *Science China. Information Sciences*, *64*(1), 1–74. DOI: 10.1007/s11432-020-2955-6

Zhang, G., Zhu, H. A. N., Hang, X. I. N., Xin, W. E. I., & Boya, D. I. (2022). Distortion minimization for multimedia transmission in NOMA HAP-UAV integrated aerial access networks. *Chinese Journal of Aeronautics*, *35*(9), 81–94. DOI: 10.1016/j.cja.2021.04.033

Zhang, J., Zhang, X., Imran, M. A., Evans, B., Zhang, Y., & Wang, W. (2017). Energy efficient hybrid satellite terrestrial 5G networks with software defined features. *Journal of Communications and Networks (Seoul)*, *19*(2), 147–161. DOI: 10.1109/JCN.2017.000024

Zhang, S., Liu, W., & Ansari, N. (2021). On tethered UAV-assisted heterogeneous network. *IEEE Transactions on Vehicular Technology*, *71*(1), 975–983. DOI: 10.1109/TVT.2021.3130130

Chapter 10
Wireless Body Area Networks

ABSTRACT

The increasing use of wireless networks and the constant miniaturization of electrical smart devices have empowered the development of wireless body area network (WBAN). It provides a continuous health monitoring of a patient without any constraint on his normal daily life activities. The successful implementation of WBAN will certainly improve quality of life whether it is medical or non-medical aspect. This chapter emphasizes on the implementation guidelines of WBANs based on the game theory, which will aid the future investigators, professionals and researchers. Furthermore, the current evolution of WBANs and their applications are discussed related to the resource allocation problem in WBAN communications.

10. OVERVIEW

The health of individuals is an important prerequisite to the health of communities, making health care a necessity for society that should be provided somehow. Wireless Body Area Network (WBAN) connects independent nodes, such as sensors, implantable biosensors and actuators that are situated in the clothes, on the body or under the skin of a person. The WBAN typically expands over the whole human body and the nodes are connected through a wireless communication channel. Current progress in WBANs has become one of the enabling technologies that provide many successful applications in the medical field; it has conquered the medical area with a wide range of possibilities. Especially, advances in wireless communication technologies along with recent developments in the embedded computing area are enabling the design, development, and implementation of WBANs.

With intelligent, low-power, micro and nano-technology devices, the WBAN is a communication standard optimized for low power devices and operation on, in or around the human body, to monitor the human health issues for further analysis. These devices provide continuous health monitoring and real-time feedback to the user or medical personnel. For what concerns the main issues to be accounted for in the design of a WBAN, the impact of wireless medium, the battery lifetime and the coexistence with other wireless networks are of fundamental importance. However, the wide applicability of WBANs makes their design challenging; formidable technical and social challenges must be dealt with to allow for their practical adoption.

DOI: 10.4018/979-8-3373-2013-7.ch010

10.1 Intra and Inter Network Control (*IINC*) Scheme

The Internet of Medical Things (IoMT) integrates smart devices to manage wireless body area networks (WBANs) using advanced networking technologies. To alleviate the computational load on WBANs, they are linked to mobile edge computing (MEC), which stores and analyzes medical data. Recently, S. Kim introduced the *Intra and Inter Network Control* (*IINC*) scheme to optimize resource sharing in MEC-supported WBANs by leveraging bargaining game theory (Kim, 2021). This approach fosters mutual benefits among WBANs, dynamically adjusting their goals to establish consensus for various applications. The *IINC* scheme utilizes two control factors to address bargaining issues, achieving a balance between efficiency and fairness.

10.1.1 Development Motivation

With the rapid advancement of IoT technology, numerous smart devices are now interconnected through the Internet, enabling ubiquitous services. This paradigm has been recognized as one of the most promising frameworks for bridging the gap between the cyber and physical realms. In recent years, the emergence of IoT has sparked a significant shift in how humans interact with machines. Projections by companies like Ericsson and Cisco Inc. indicate that by 2025, over 50 billion devices will be connected to the Internet, generating more than 500 Zettabytes (ZB) of data. These devices, widely distributed across various locations, collaboratively offer flexible and dynamic communication, computation, and storage capabilities. Advanced computational and communication technologies have empowered the IoT to transform both traditional industries and everyday life (Kim, 2021; Liu, 2017; Pan, 2017).

The healthcare sector has undergone a remarkable transformation due to IoT advancements, with traditional clinical practices being replaced by pervasive healthcare systems. As a result, the medical field must adapt to these changes by formulating strategies that ensure optimal patient care while maintaining profitability. One key innovation is the wireless body area network (WBAN), which facilitates real-time access to healthcare services. WBANs, which are wireless networks centered around the human body, collect and process physiological, behavioral, and other health-related data in real-time. By leveraging wearable or implanted medical sensors, WBANs, often referred to as the Internet of Medical Things (IoMT), provide cost-effective and reliable e-healthcare solutions, particularly for patients requiring immediate medical attention (Kim, 2021; Liu, 2017; Misra, 2014; Ning, 2020; Yuan, 2020).

Typically, WBANs consist of compact, lightweight IoMT devices placed on or around the human body. These devices utilize one-hop or two-hop star topologies within the physical and data link layers to monitor bodily functions and environmental conditions. The IEEE 802.15.6 standard, designed specifically for short-range communication among IoMT devices, ensures low-power, highly reliable wireless communication within the vicinity of the human body. This standard includes provisions for various data rates to meet the diverse application requirements of WBANs. Solutions for WBANs must align with the 802.15.6 standard to address their unique specifications and priorities. The IEEE 802.15.6 framework incorporates a coordinator node, known as the local processing unit, which manages multiple IoMT devices. By employing a beacon mode with a super-frame structure and duty cycle mechanism, this standard aims to optimize energy consumption while introducing new physical and medium access control layers for WBANs (Cicioğlu 2020, Cicioğlu 2021; Kim, 2021).

However, the advancement of IoMT presents certain challenges. Many IoMT devices have limited processing and storage capabilities, making it difficult to handle computation-intensive medical applications. To address this, mobile edge computing (MEC) has emerged as a promising solution. MEC enhances application performance by offloading computational tasks to nearby servers, thus reducing the processing burden on IoMT devices. This synergy between MEC and WBAN enables the efficient deployment of innovative IoMT applications and services (Kim, 2021; Ning, 2020).

To develop an MEC-assisted WBAN (M-W) system, an effective resource control algorithm is essential for fair and efficient resource allocation. The widespread use of WBANs leads to significant health-related data access and diverse service requirements, creating intra- and inter-WBAN resource allocation challenges. Implantable IoMT devices are particularly constrained by their limited computational power, which can impact the performance of resource-intensive applications. To mitigate these limitations, individual WBANs must decide whether to process tasks locally or offload them to MEC servers. Additionally, in urban M-W environments, the limited spectrum resources of MEC must be efficiently shared among multiple WBANs (Kim, 2021; Ning, 2020; Yuan, 2020; Zhang, 2018).

A key challenge in resource allocation is managing the self-interested behavior of IoMT devices while reconciling conflicting demands. The *IINC* scheme employs bargaining game theory to guide these devices toward achieving optimal system performance. Bargaining models often integrate disagreement and reference points, indirectly influencing negotiated outcomes by shaping players' aspirations. For example, the *Kalai-Smorodinsky bargaining solution* (*KSBS*) derives aspirations from disagreement points, representing worst-case scenarios. Alternative models, such as the *Gupta and Livne bargaining solution* (*GLBS*) and *Tempered Aspirations bargaining solution* (*TABS*), utilize reference points to define aspirations and anchors differently. A more recent approach, the *anchor and aspiration bargaining solution* (*AABS*), explicitly considers the influence of reference points on anchors and aspirations, offering a flexible framework for arbitrators to express their preferences in negotiation outcomes. Existing characterizations like *KSBS*, *GLBS*, *TABS*, and the *Local Kalai-Smorodinsky bargaining solution (LKSBS*) can be seen as specific instances of *AABS* (Karagözoğlu 2019; Kim, 2014, Kim, 2019a, Kim, 2021).

10.1.2 Contribution

The *IINC* scheme leverages bargaining solutions to effectively address resource allocation challenges in the M-W system. By considering the specific requirements of different IoMT applications, the *IINC* scheme develops an intra-WBAN bargaining game model to manage the offloading of computing tasks to the MEC. Additionally, it introduces an inter-WBAN bargaining game model to allocate the limited spectrum resources within the M-W system. To facilitate this dual bargaining process, concepts from the *AABS* and *KSBS* approaches are applied, tailored to the unique service characteristics. Through iterative bargaining steps, the *IINC* scheme efficiently manages the increasing demand for *WBAN* services, ensuring a balanced and optimized system performance (Kim, 2021).

10.1.3 M-W System Infrastructure and Dual Bargaining Game Model

In contemporary ubiquitous healthcare systems, individuals are equipped with IoMT-enabled WBANs to facilitate real-time monitoring of physiological parameters. The *IINC* scheme considers a scenario involving multiple WBANs connected to a single MEC server, which delivers health monitoring services. The MEC server ($\mathscr{M}$) is a mobility-enhanced small-scale cloud server that is located at the edge of the Internet. The M can provide powerful computing resources to WBANs with lower latency. Let there be n number of WBANs, which are represented as a set $\mathfrak{B} = \{\mathscr{B}_1, \ldots \mathscr{B}_n,\}$. In each individual WBAN, there are number of heterogeneous IoMT devices, which are wearable or implanted inside the human body, and one local processing unit (LPU), which serves the WBAN data aggregation and asks the M for both communication and computation resources. Therefore, multiple LPUs can offload several tasks to the M server based on the current M-W system condition. The overview of M-W system platform is shown in Figure 1 (Kim, 2020, Kim, 2021; Samanta, 2015).

Figure 1. The overview of M-W system platform (Kim, 2021)

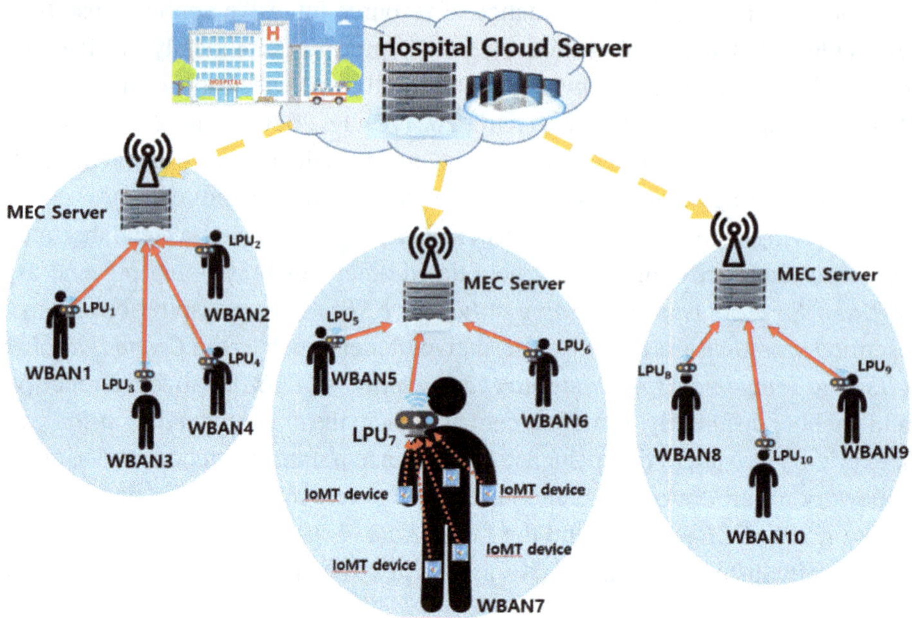

The two-stage bargaining game model includes intra- and inter-WBAN bargaining processes designed to allocate communication and computation resources within the M-W system. In the intra-WBAN bargaining process, each LPU periodically gathers health data from its associated IoMT devices. Based on its limited computational capacity, the LPU determines whether to execute the medical analysis task locally or offload it to the MEC server. In a parallel and distributed manner, n LPUs act independently and make their offload decisions for their corresponding WBANs. In the inter-WBAN bargaining process, the *IINC* scheme mainly concerns the spectrum allocation problem among WBANs, which contend for the M's spectrum resource. To share optimally the limited communication and computation resources of M-W system, the intra and inter-WBAN bargaining processes work together toward an appropriate system performance. Formally, the *IINC* scheme defines the two-stage bargaining game entities, i.e.,

$$\mathbb{G} = \left\{\mathbb{G}^{I}_{1\le i\le n}, \mathbb{G}^{II}\right\} = \left\{\left\{\mathbb{G}^{I}_{i} \mid \mathcal{L}_{i}, \mathfrak{P}_{\mathcal{L}}, \mathcal{S}_{i}, \left(\mathcal{S}^{L}_{i}, U^{L}_{i}\right), \left(\mathcal{S}^{O}_{i}, U^{O}_{i}\right)\right\}, \mathcal{M}, \left\{\mathbb{G}^{II} \mid \mathfrak{B}, \mathfrak{M}, \mathfrak{U}_{\mathcal{B}_i \in \mathfrak{B}}, N_{\mathcal{B}_i}\right\}, T\right\}$$

(Kim, 2021).

- In the two-stage game G, $\mathbb{G}_i^I$ is the intra-WBAN bargaining game for the $\mathscr{B}_i$, and $\mathbb{G}^{II}$ is the inter-WBAN bargaining game for the M-W system. $\mathbb{G}_i^I$ and $\mathbb{G}^{II}$ games are mutual and reciprocal inter-dependent in an interactive manner.
- In the $\mathbb{G}_i^I$, $\mathscr{L}_i$ is the $\mathscr{B}_i$'s LPU and $\mathfrak{P}_{\mathscr{L}}$ is the total computation capacity of $\mathscr{L}_i$.
- $\mathcal{S}_i$ is the current computation workload in the $\mathscr{B}_i$. The $\mathcal{S}_i$ is divided into two parts, i.e., $\mathcal{S}_i^L$ and $\mathcal{S}_i^O$, for the local and offload computing workloads where $\mathcal{S}_i = \mathcal{S}_i^L + \mathcal{S}_i^O$.
- In the $\mathbb{G}_i^I$, the $\mathcal{S}_i^L$ and $\mathcal{S}_i^O$ are game players. Therefore, the $\mathbb{G}_i^I$ is a two-player bargaining game.
- U_i^L is the utility function of local computing workload $\left(\mathcal{S}_i^L\right)$, and U_i^O is the utility function of off-load computing workload $\left(\mathcal{S}_i^O\right)$.
- M is a MEC server for the $\mathscr{B}_{1\le i\le n}$ and B is the set of WBANs where $\mathscr{B}_i \in \mathfrak{B}$. M is the total spectrum resource of M.
- In the $\mathbb{G}^{II}$, the $\mathscr{B}_{1\le i\le n}$ are game players, and $\mathfrak{U}_{\mathscr{B}_i}$ is the utility function of $\mathscr{B}_i$. Therefore, the $\mathbb{G}^{II}$ is a n-player bargaining game.
- Each application of IoMT devices has its own preference level where l_m, l_M are the minimum and maximum preference levels, respectively. The $N_{\mathscr{B}_i}$ is the set of $\mathscr{B}_i$' running applications' preferences.

10.1.4 The Basic Idea and Fundamental Concept of *AABS*

To characterize the basic concepts of bargaining solutions, let $A(S,x)$ be the aspiration vector such that $A_i(S,x) \equiv \max\{t \in \mathbb{R}|(t,x_{-i}) \in S\}$ where $x_{-i} = \left(x_1,\ldots,x_{i-1},x_{i+1},\ldots x_n\right)$. Accordingly, $A(S,d)$ is the ideal point and $A(S,r)$ is the tempered aspirations point. A bargaining solution concept for (S,d,r) is a function $f\colon \sum^n \to \mathbb{R}^n$ where $(S,d,r) \in \sum^n$. By using the two control factors, i.e., $0 \le \delta_I, \delta_{II} \le 1$, we can define the *AABS* where δ_I can be interpreted as the power of r in determining the anchor, and δ_{II} can be interpreted as the influence of r in shaping players' aspiration. For every $(S,d,r) \in \acute{O}^n$, the *AABS* with δ_I and δ_{II} values, i.e., $AABS^{\delta_I,\delta_{II}}(S,d,r)$, is defined as follows (Karagözoğlu 2019; Kim, 2021);

$$AABS^{\delta_I,\delta_{II}}(S,d,r) = \max_{\lambda\in[0,1]} \left\{ \lambda \mid \left(\left[\lambda \times \mathcal{Q}(S,\mathcal{X}^{\mathscr{A}})\right] + \left[(1-\lambda)\times \mathcal{X}^d\right] \right) \right\} \tag{10.1.1}$$

$$\text{s.t.,} \begin{cases} \mathcal{X}^{\mathscr{A}} = \left((\delta_{II}\times r) + \left((1-\delta_{II})\times d\right)\right) \\ \mathcal{X}^{d} = \left((\delta_{I}\times r) + \left((1-\delta_{I})\times d\right)\right) \\ AABS^{\delta_I,\delta_{II}}(S,d,r) \in S \end{cases}$$

Geometrically, the *AABS* proposes the maximum point of the bargaining set on the line segment connecting the aspiration, i.e., $\mathcal{X}^{\mathscr{A}}$, and the anchor, i.e., $\mathcal{X}^d$. With $0 \le \delta_I,\delta_{II} \le 1$, the collection of all such possible solutions constitutes the available set of *AABS*. Simply, the *KSBS* is obtained with $\delta_I = 0$ and $\delta_{II} = 0$. Therefore, we can think that the *KSBS* is the maximum point of the bargaining set on the line segment connecting the ideal point and the disagreement point. The ideas of different bargaining solutions are shown in Figure 2 (Karagözoğlu 2019; Kim, 2021).

Figure 2. The ideas of different bargaining solutions (Kim, 2021)

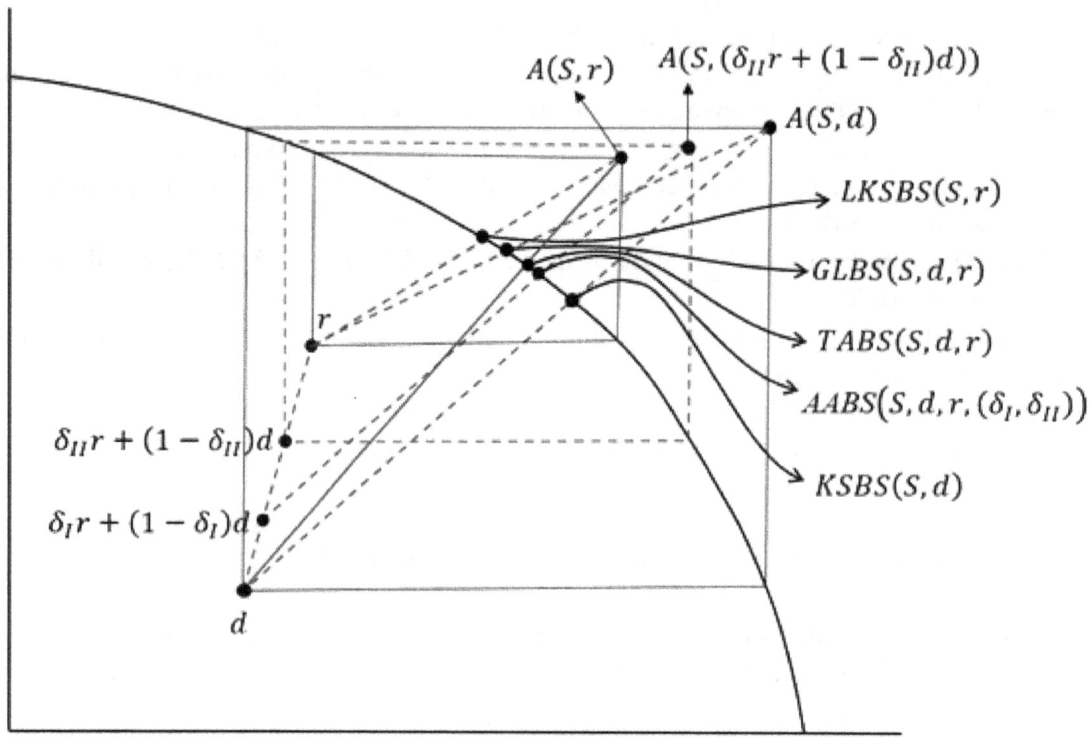

10.1.5 The Two-Stage Bargaining Game in the M-W Platform

At each time $t_c \in T$, IoMT devices in $\mathscr{B}_i$ generate independently their service data; $\mathcal{S}_i(t_c)$ is the $\mathcal{S}_i$ at time t_c. To process the $\mathcal{S}_i(t_c)$, the $\mathscr{L}_i$ needs an appropriate computing power capability to collect and analyze IoMT data. However, it has a limited computation capacity. To reduce the burden on the $\mathscr{L}_i$, the $\mathcal{S}_i(t_c)$ can be partially offloaded to its corresponding MEC. The *IINC* scheme designs the $\mathbb{G}_i^I$ for this computation offloading problem. To effectively exploit the computing capability in MEC while considering the current computation overhead of $\mathscr{L}_i$, one major challenge is how to split the $\mathcal{S}_i(t_c)$ to proceed the offload service; the $\mathcal{S}_i(t_c)$ is divided into two parts where the $\mathcal{S}_i^L(t_c)$ is executed locally by the $\mathscr{L}_i$, and $\mathcal{S}_i^O(t_c)$ is offloaded to the MEC. Therefore, the intra-WBAN data offloading problem is formulated as a two-player, i.e., $\mathcal{S}_i^L(t_c)$ and $\mathcal{S}_i^O(t_c)$, bargaining game. For the local computing service, the utility function of $\mathcal{S}_i^L(t)$, i.e., $U_i^L(\cdot)$, is defined according to the current M-W system conditions (Kim, 2021).

$$U_i^L\left(\tau_i, \mathcal{S}_i(t_c), \mathscr{W}_{\mathscr{M}}^{t_{c-1}}\right) = \left[\frac{\mathcal{Z}}{\eta + exp(\mathcal{Z})} \times \mathscr{H}_i^{t_c}\right] + \log\left(\left(\mathcal{Z} \times \left(1 - \mathscr{W}_{\mathscr{M}}^{t_{c-1}}\right)\right) + \theta\right) \tag{10.1.2}$$

$$\text{s.t.}, \mathscr{Z} = \frac{\left(\tau_i \times \mathcal{S}_i(t_c)\right)}{\mathcal{S}_i(t_c)} \text{ and } \mathscr{H}_i^t = \frac{\left(\tau_i \times \mathcal{S}_i(t_c)\right)}{\mathfrak{P}_{\mathscr{L}}}$$

where τ_i is the task offloading ratio where $0 \leq \tau_i \leq 1$. η, θ are the adjustment parameters for the $U_i^L(\cdot)$, and $\mathscr{W}_{\mathscr{M}}^{t_{c-1}}$ is the M's spectrum usage at time t_{c-1}. For the offload computing service, the utility function of $\mathcal{S}_i^O(t_c)$, i.e., $U_i^O(\cdot)$, is also defined according to the current information of $\mathscr{B}_i$ (Kim, 2021).

$$U_i^O\left(\tau_i, \mathcal{S}_i(t_c), \mathscr{W}_{\mathscr{M}}^{t_{c-1}}\right) = exp\left(\frac{\left(\tau_i \times \mathcal{S}_i(t_c)\right)}{\mathcal{S}_i(t_c)} \times \left(\left(1 - \mathscr{W}_{\mathscr{M}}^{t_{c-1}}\right) \times \beta\right)\right) - \sigma \tag{10.1.3}$$

where β and σ is the control parameter for the $U_i^O(\cdot)$. To adaptively adjust the τ_i value at time t_c, it is necessary to consider the current computation and communication situations of $\mathscr{B}_i$. These two factors are dynamically changeable, and equally important for the intra-WBAN bargaining process. Therefore, the *IINC* scheme adopts the idea of *AABS* for the $\mathbb{G}_i^I$ game, and the reference point at time t_c, i.e., $r_i(t_c) = \left(r_i^L(t_c), r_i^O(t_c)\right)$, and the disagreement point at time t_c, i.e., $d_i(t_c) = \left(d_i^L(t_c), d_i^O(t_c)\right)$, should be decided in an online fashion. To adapt the current intra-WBAN conditions, the *IINC* scheme assumes that the $r_i(t_c)$ decision problem is another bargaining problem. For this problem, the axiom of *individual monotonicity* is essential for local and offload computation tasks. Therefore, the concept of *KSBS* is suitable. The *KSBS* for the $r_i(t_c)$ decision, i.e., $KSBS_{r_i}(S^{r_i}, d)$, is obtained with the restricted service request domain, i.e., $S^{r_i} = \mathcal{S}_i(t_c)/\omega$ where ω is a diminishing factor. Therefore, the $r_i(t_c)$ is given by (Kim, 2021);

$$r_i(t_c) = KSBS_{r_i}(S^{r_i}, d) = \max_{\left(\mathcal{S}_i^L(t_c), \mathcal{S}_i^O(t_c)\right) \in (S^{r_i}, d)} \left(\min_{\mathscr{V} \in \left\{\mathcal{S}_i^L(t_c), \mathcal{S}_i^O(t_c)\right\}} \frac{\mathscr{V} - d_{\mathscr{V}}}{A_{\mathscr{V}}(S^{r_i}, d) - d_{\mathscr{V}}}\right) \tag{10.1.4}$$

The *IINC* scheme sets $d_i(t_c) = (0,0)$ as the result if the $\mathcal{S}_i^L(t_c)$ and $\mathcal{S}_i^O(t_c)$ cannot reach an agreement of offloading strategy. With the $r_i(t_c)$ and $d_i(t_c)$, the *IINC* scheme also decides the values of δ_I and δ_{II} to get the *AABS*. Under diverse M-W system environments, the values of δ_I and δ_{II} should be dynamically modified to improve the effectiveness of *AABS*. In the *IINC* scheme, they are adaptively decided based on the current information of each WBAN. At time t_c, the $\mathscr{B}_i$'s δ_I and δ_{II}, i.e., $\delta_I(t_c)$ and $\delta_{II}(t_c)$, are affected by the $\mathscr{B}_i$'s computation and M's communication conditions (Kim, 2021).

$$\delta_I(t_c) = \min\left(1, \frac{\mathcal{S}_i(t_c)}{\mathfrak{P}_{\mathscr{L}}}\right) \text{ and } \delta_{II}(t_c) = 1 - \mathscr{W}_{\mathscr{M}}^{t_{c-1}} \tag{10.1.5}$$

Finally, at time t_c, the *AABS* with a given $\left(\delta_I(t_c), \delta_{II}(t_c)\right) \in [0,1]^2$, i.e., $AABS_{(\delta_I(t_c), \delta_{II}(t_c))}(\cdot)$, is given as follows (Kim, 2021);

$$AABS_{(\delta_I(t_c), \delta_{II}(t_c))}\left(S^{\mathfrak{P}_{\mathscr{L}}}, d_i(t_c), r_i(t_c)\right) = \tag{10.1.6}$$

$$\max_{\lambda\in[0,1]} \left\{ \lambda \mid \left(\left[\lambda \times A(S^{\mathfrak{P}_{\mathcal{I}}}, \mathcal{X}_i^{\mathcal{A}}(t_c))\right] + \left[(1-\lambda)\times \mathcal{X}_i^{d}(t_c)\right] \right) \right\}$$

$$\text{s.t.,} \begin{cases} \mathcal{X}_i^{\mathcal{A}}(t_c) = \left((\delta_{II}(t_c)\times r_i(t_c)) + ((1-\delta_{II}(t_c))\times d_i(t_c)) \right) \\ \mathcal{X}_i^{d}(t_c) = \left((\delta_I(t_c)\times r_i(t_c)) + ((1-\delta_I(t_c))\times d_i(t_c)) \right) \\ AABS_{(\delta_I(t_c),\, \delta_{II}(t_c))}\left(S^{\mathfrak{P}_{\mathcal{I}}}, d_i(t_c), r_i(t_c)\right) \in S^{\mathfrak{P}_{\mathcal{I}}} \end{cases}$$

Each individual $\mathbb{G}^{I}_{1\le i\le n}$ games are operated independently in a parallel manner and decide each $\mathcal{B}_i$'s offloading strategy (τ_i). And then, as game players, each $\mathcal{B}_{1\le i\le n}$ participates in the $\mathbb{G}^{II}$ game to share the M for its task offloading service. Based on each WBAN's situation, the $\mathcal{B}_i$'s utility function in the $\mathbb{G}^{II}$, i.e., $\mathfrak{U}_{\mathcal{B}_i}(\cdot)$, is given by considering the allocated spectrum amount (Kim, 2021).

$$\mathfrak{U}_{\mathcal{B}_i}\left(\mathcal{R}^{t_c}_{\mathcal{B}_i}, \mathcal{C}^{t_c}_{\mathcal{B}_i}, N_{\mathcal{B}_i}, \mathfrak{M}\right) = \left(\gamma - \frac{\xi}{Y_{\mathcal{B}_i}}\right)^{\mathbb{C}_i \times Y_{\mathcal{B}_i}} - \varepsilon \tag{10.1.7}$$

$$\text{s.t., } Y_{\mathcal{B}_i} = \frac{\min\left(\mathcal{R}^{t_c}_{\mathcal{B}_i}, \mathcal{C}^{t_c}_{\mathcal{B}_i}\right)}{\mathcal{R}_{\mathcal{B}_i}}, \mathbb{C}_i = \sum_{\mathcal{T}\in N_{\mathcal{B}_i}} \frac{\mathcal{T}}{l_M} \text{ and } \sum_{\mathcal{B}_i\in\mathcal{B}} \mathcal{C}^{t_c}_{\mathcal{B}_i} \le \mathfrak{M}$$

where $\mathcal{R}^{t_c}_{\mathcal{B}_i}$ and $\mathcal{C}^{t_c}_{\mathcal{B}_i}$ are the $\mathcal{B}_i$'s requested and allocated spectrum resources, respectively, at time t_c. γ, ξ and ε is the adjustment parameter for the $\mathfrak{U}_{\mathcal{B}_i}(\cdot)$. According to (10.1.7), M distributes the M for the $\mathcal{B}_{1\le i\le n}$. In the $\mathbb{G}^{II}$ game, the *individual monotonicity* of each player is necessary to share the M. Therefore, the concept of *KSBS* is also adopted like as the reference point decision problem. The *KSBS* for the $\mathbb{G}^{II}$, i.e., $KSBS_{\mathcal{M}}(S^{\mathfrak{M}}, d^{\mathfrak{M}})$, is given by (Kim, 2021);

$$KSBS_{\mathcal{M}}(S^{\mathfrak{M}}, d^{\mathfrak{M}}) = \max_{\left(Y_{\mathcal{B}_1},\ldots,Y_{\mathcal{B}_n}\right)\in(S^{\mathfrak{M}},d^{\mathfrak{M}})} \left(\min_{F\in\left\{Y_{\mathcal{B}_1},\ldots,Y_{\mathcal{B}_n}\right\}} \frac{F - d_F^{\mathfrak{M}}}{A_F(S^{\mathfrak{M}}, d^{\mathfrak{M}}) - d_F^{\mathfrak{M}}} \right) \tag{10.1.8}$$

10.1.6 Summary

Edge computing offers a significant advantage in reducing overhead for IoMT-based WBAN healthcare services. The *IINC* scheme introduces a novel two-stage bargaining game model tailored to scenarios where multiple WBANs coexist. Utilizing an MEC server, each WBAN determines whether to handle medical analysis tasks locally or offload them to the MEC. In the first stage, the intra-WBAN bargaining process is established, leveraging the *AABS* to identify an optimal offloading strategy. In the second stage, the inter-WBAN bargaining process is implemented, using the *KSBS* to allocate the MEC's limited spectrum resources. These intra- and inter-WBAN bargaining games operate sequentially, enabling efficient sharing of computational and communication resources within the M-W platform. By dynamically adjusting to real-time bargaining factors, the *IINC* scheme demonstrates greater adaptability to changing system conditions, resulting in enhanced and mutually beneficial outcomes in the dynamic M-W environment (Kim, 2021).

10.2 Learning and Game Based Spectrum Allocation (*LGSA*) Scheme

The Internet of Medical Things (IoMT) paradigm enables pervasive healthcare services through in-home monitoring networks, becoming an integral part of modern human life. However, the increasing demand for healthcare services has led to challenges such as limited spectrum resources and service delays. To address these issues, S. Kim recently introduces the *Learning and Game-based Spectrum Allocation* (*LGSA*) scheme for the IoMT system platform (Kim, 2023). The primary challenge of the *LGSA* scheme is to efficiently share the constrained spectrum resources while dynamically accommodating diverse service requests. The *LGSA* scheme employs a bi-level control algorithm that integrates *multi-agent reinforcement learning* (*MARL*) and the *Balakrishnan-Gómez-Vohra* (*BGV*) solution to achieve mutually beneficial outcomes for multiple IoMT devices. At the first level, each IoMT device determines its salient point using the *MARL* model. At the second level, spectrum resources are allocated based on the *BGV* solution, which incorporates the salient points selected by the devices. By facilitating sequential interactions among intelligent devices, this bi-level control approach effectively encourages cooperation strategies while optimizing the overall spectrum allocation process (Kim, 2023).

10.2.1 Development Motivation

In recent years, wireless technology has emerged as one of the fastest-evolving fields in communications, becoming a primary carrier of digital data worldwide. Smart mobile devices leveraging seamless connectivity through wireless networks are now omnipresent, influencing diverse contexts. Among these advancements, the IoT has garnered significant attention, poised to play a central role in future networks. While initially gaining traction in industrial and business applications, the IoT's focus has shifted to integrating smart devices into homes and workplaces. As an integral component of future 6G networks, IoT holds immense potential to enhance various aspects of modern life (Gatouillat, 2018; Ghubaish, 2020; Kim, 2023).

The IoT paradigm connects the physical world to cyberspace, transforming everyday life through its diverse applications. This has driven significant advancements, particularly in healthcare, where aging populations and rising chronic diseases have placed enormous pressure on healthcare systems in developed nations. The Internet of Medical Things (IoMT), a specialized branch of IoT, merges the reliability of traditional medical devices with the scalability and adaptability of IoT technologies. By deploying smart medical devices, IoMT facilitates remote in-home health monitoring, addressing a wide array of healthcare needs. Pervasive monitoring systems further enable patients to move freely indoors without restrictions, fostering numerous innovative healthcare projects (Gatouillat, 2018; Kim, 2023; Ning, 2020).

Wireless IoMT devices generate vast amounts of complex, multidimensional data. Managing the efficient transmission of this data is challenging, given its volume and complexity. By 2020, IoMT devices account for 40% of the IoT market, and their expansion is projected to continue as they provide ubiquitous healthcare services. However, the rapid growth of IoMT devices and data generation imposes limitations, particularly the scarcity of spectrum resources and real-time processing demands. To meet the delay constraints of time-sensitive tasks, adaptive spectrum sharing and inter-device cooperation are critical requirements for IoMT's success. Intelligent management policies are essential to enable seamless interconnectivity among medical devices (Gatouillat, 2018; Ghubaish, 2020; Kim, 2023; Ning, 2020; Zhang, 2020).

In response to these challenges, the *LGSA* scheme was developed for IoMT networks, focusing on dynamic spectrum allocation. Spectrum sharing is crucial in network management to maintain communication quality, requiring innovative algorithms. Game theory and reinforcement learning have gained attention for their ability to model system dynamics intelligently. Game theory examines strategic interactions among players, while reinforcement learning enhances system adaptability through dynamic decision-making. Together, these methodologies form the foundation of the *LGSA* scheme, which integrates a learning-based game model for effective spectrum allocation in IoMT systems (Kim, 2023).

Although game theory is a powerful tool for analyzing decision-making, traditional game players often lack the intelligence to interact dynamically with their environment, limiting their ability to maximize payoffs. The advent of AI technologies, such as reinforcement learning, addresses this limitation by enabling dynamic game models to adapt and optimize decisions. Reinforcement learning has become a vital component in the AI domain, providing tools to analyze and improve outcomes in multi-agent environments. As a result, combining game theory and reinforcement learning has emerged as a promising approach to optimize IoMT spectrum allocation (Kim, 2023; Wang, 2019).

The *LGSA* scheme's goal is to create a hybrid spectrum allocation mechanism for IoMT systems, where each IoMT device operates as both a dynamic game player and a learning agent. By leveraging interactions between game models and learning algorithms, the *LGSA* scheme achieves balanced performance across IoMT devices, ensuring efficient spectrum usage for diverse medical services (Kim, 2023).

The *Balakrishnan-Gómez-Vohra* (*BGV*) solution, recently introduced as a game-theoretic concept, incorporates a salient point into bargaining problems. The *BGV* solution employs two reference points: the disagreement point, serving as an anchor, and the aspiration point, representing a tempered or modified utopia. It identifies the optimal utility profile by balancing individual rationality and proportional utility gains. This approach aligns well with the *LGSA* scheme's objective of fair and efficient resource allocation (Karagözoğlu, 2019; Kim, 2023; Thomson, 2022).

A multi-agent system consists of intelligent agents interacting with their environment to achieve predefined goals, offering a new paradigm for control applications. Reinforcement learning, especially in multi-agent domains, has advanced significantly, combining techniques such as policy search, dynamic programming, and game theory. *MARL* explores coordination and competition among agents, leveraging game-theoretic approaches to model and solve dynamic multi-agent problems. The *LGSA* scheme utilizes *MARL* to achieve fair and efficient solutions in the ever-changing IoMT environment (Kim, 2023; Hu, 2014).

10.2.2 Contribution

The *LGSA* scheme is designed to enhance IoMT system performance by integrating the *BGV* solution and the *MARL* model. The objective of optimization of the *LGSA* scheme is to maximize the normalized payoffs of individual devices and overall network throughput while maintaining fairness across IoMT devices. The *LGSA* scheme operates through two interconnected mechanisms; i) learning-level mechanism: at this level, each IoMT device determines its salient point using the *MARL* model, enabling dynamic and adaptive decision-making, ii) bargaining-level mechanism: here, the limited spectrum resource is allocated among IoMT devices in a fair and efficient manner, with the *BGV* solution applied to resolve the spectrum allocation challenge. By guiding self-interested IoMT devices toward socially optimal outcomes, the *LGSA* scheme ensures efficient utilization of IoMT system resources. The underlying methodology employs a coordinative learning game, fostering cooperation among devices while optimizing the overall system performance (Kim, 2023).

10.2.3 IoMT System Infrastructure and a Learning Game Model

Leveraging wearable technology, IoMT devices are widely utilized for continuous health monitoring in everyday life. The *LGSA* scheme presents the architectural framework of the IoMT platform and develops a bi-level learning game model tailored for health-focused application services. It consists of n access points (APs), i.e., $\mathbb{A} = \{\mathcal{A}_1,\ldots,\mathcal{A}_n\}$ and m IoMT devices, i.e., $\mathbb{D} = \{\mathcal{D}_1,\ldots,\mathcal{D}_m\}$. With low transmission power and cost, APs are envisioned to provide resilient communication services for multiple IoMT devices, which can gather, generate, fuse, analyze, and send medical data to their corresponding APs. Each $\mathcal{A}_{1\leq i\leq n} \in \mathbb{A}$ has its covering area, and $\mathbb{D}_{\mathcal{A}_i}$ is the set of $\mathcal{A}_i$'s corresponding IoMT devices where $\mathbb{D}_{\mathcal{A}_i} \subset \mathbb{D}$; they are randomly distributed in a given geographical area, and require a lot of spectrum resources for sending the medical data sensed from the patients every day. The $\mathcal{A}_i$ has its wireless spectrum capacity $\left(\mathfrak{M}_{\mathcal{A}_i}\right)$, and dynamically allocate the $\mathfrak{M}_{\mathcal{A}_i}$ for its corresponding IoMT devices. The IoMT device in the $\mathbb{D}_{\mathcal{A}_i}$, i.e., $\mathcal{D}_j \in \mathbb{D}_{\mathcal{A}_i}$, generates its health-aware data $\left(\mathcal{W}_{\mathcal{D}_j}\right)$. To transmit the $\mathcal{W}_{\mathcal{D}_j}$, it leads to the conflict situation to share the $\mathfrak{M}_{\mathcal{A}_i}$ among the devices in the $\mathbb{D}_{\mathcal{A}_i}$ (Kim, 2023).

In the *LGSA* scheme, the first-level *MARL* learning model M and the second-level cooperative game model G are formulated. Through the M and G, multiple IoMT devices are sequentially interacted with each other to reach a consensus. Formally, the *LGSA* scheme defines the tuple entities in the dual-phase control mechanism, such as

$$\{\mathbb{M},\mathbb{G}\} = \left\{\mathbb{A},\mathbb{D},\left\{\mathbb{M}_{\mathcal{D}_k}|\mathcal{D}_k \in \mathbb{D}, a_{1\leq k\leq l}^{\mathcal{D}_j} \in \mathcal{L}_{\mathcal{D}_j},\mathcal{R}_a^{\mathcal{D}_j}\right\},\left\{\mathbb{G}_{\mathcal{A}_i}|,\mathcal{A}_i \in \mathbb{A},\mathcal{D}_j \in \mathbb{D}_{\mathcal{A}_i},\mathfrak{M}_{\mathcal{A}_i},S_{\mathcal{D}_j},\mathcal{U}_{\mathcal{D}_j}(\cdot)\right\},T\right\}$$

(Kim, 2023).

- A and D represent the set of APs and the set of IoMT devices, respectively.
- At the first-level, the $\mathbb{M}_{\mathcal{D}_k}$ is developed as a *MARL* model for the $\mathcal{D}_k \in \mathbb{D}$. It is operated in a distributed manner to learn the best action of $\mathcal{D}_k$.
- In the $\mathbb{M}_{\mathcal{D}_k}$, $a_k^{\mathcal{D}_j}$ is the $\mathcal{D}_j$'s kth action, and $\mathcal{L}_{\mathcal{D}_j}$ is the $\mathcal{D}_j$'s action set, which consists of total l actions. $\mathcal{R}_a^{\mathcal{D}_j}$ is the $\mathcal{D}_j$'s reward function with the joint action **a**.
- At the second-level, each individual AP executes the $\mathbb{G}_{\mathcal{A}_{1\leq i\leq n}}$ in a centralized manner with its corresponding devices $\mathcal{D}_j \in \mathbb{D}_{\mathcal{A}_i}$.
- In the $\mathbb{G}_{\mathcal{A}_i}$, the $\mathcal{A}_i$'s spectrum resource $\left(\mathfrak{M}_{\mathcal{A}_i}\right)$ is shared by devices in $\mathbb{D}_{\mathcal{A}_i}$. The $\mathcal{D}_j \in \mathbb{D}_{\mathcal{A}_i}$ is a game player, and $S_{\mathcal{D}_j}$ and $\mathcal{U}_{\mathcal{D}_j}(\cdot)$ are his strategy and utility function, respectively.
- The $\mathbb{M}_{\mathcal{D}_k}$ and $\mathbb{G}_{\mathcal{A}_i}$ are reciprocally interdependent and work together. It is noteworthy that the *LGSA* scheme formulates the $\mathcal{A}_i$-$\mathbb{D}_{\mathcal{A}_i}$ association in an iterative manner.
- Discrete time model $T \in \{t_1,\ldots,t_c,t_{c+1},\ldots\}$ is represented by a sequence of time steps. The length of t_c matches the event's timescale of $\mathbb{M}_{\mathcal{D}_k}$ and $\mathbb{G}_{\mathcal{A}_i}$.

10.2.4 Fundamental Ideas of *BGV* Solution and *MARL* Model

To characterize the fundamental idea of *BGV* solution, let (l, x_{-i}) represent the vector $y \in \mathbb{R}^n$ such that $y_i = l$ and $y_j = x_j$ for any $j \neq i$. We denote $x \cdot y$ as the scalar product of the vectors, that is $x \cdot y = \sum_{i=1}^{n} (x_i \times y_i)$. We assume that $d, r \in S \subset \mathbb{R}^n$, and $\exists x \in S$ with $x > d$ and $r \geq d$. The assumption $d \in S$ means that players are able to agree to disagree, the assumption $r \in S$ means that the reference point is feasible, and the assumption $r \geq d$ means that the reference point is individually rational (Karagözoğlu 2019; Kim, 2023; Thomson, 2022).

By assuming that there exists $x \in S$ with $x > d$, we rule out degenerate problems where no agreement can make all agents better-off than the disagreement outcome. Let $\Upsilon(S,x)$ denote the aspiration vector such that for every $i \in \mathbb{N}$ and every $x \in S$: $\Upsilon_i(S,x) \equiv \max\{l \in \mathbb{R} | (l, x_{-i}) \in S\}$. Accordingly, $\Upsilon(S,d)$ is the ideal point and $\Upsilon(S,r)$ is the tempered aspirations point. Let Σ^n be the class of all bargaining problems with a reference point. A solution concept for such problems is a function F: $\Sigma^n \rightarrow \mathbb{R}^n$ that associates each $(S,d,r) \in \Sigma^n$ with a unique point of S. Finally, the mathematical definition of *BGV* solution, i.e., $BGV(S,d,r)$, is given by (Karagözoğlu, 2019; Kim, 2023; Thomson, 2022):

$$BGV(S,\ d,r) = (\lambda^* \cdot \Upsilon(S,r)) + ((1-\lambda^*) \cdot d) \tag{10.2.1}$$

$$\text{s.t.,}\begin{cases} \exists(S,\ d,r) \in \Sigma^n \\ \lambda^* = \max\{\lambda \in [0,1] | ((\lambda \cdot \Upsilon(S,r)) + ((1-\lambda) \cdot d)) \in S\} \end{cases}$$

Geometrically, the *BGV* solution is the maximum point of the bargaining set on the line segment connecting two points, which are $\Upsilon(S,r)$ and d (Kim, 2023).

Multi-agent learning focuses on the interactions among multiple agents operating within a shared dynamic environment. One of the most widely used approaches in *MARL* is Q-learning. The primary goal of Q-learning-based *MARL* is to reach an agreement that maximizes rewards, a cooperative learning challenge that has garnered significant attention over the past decade. *MARL* algorithms aim to estimate the Q-value for every possible action in each state. However, in some scenarios, the environment can be simplified by focusing solely on the action space rather than representing it with states. This approach simplifies the process by estimating the expected value of a single reward for each action available to the learning agent. To create a fully distributed learning system that effectively models intelligent agent behavior, single-state *MARL* has been introduced. In this model, the Q-values for actions are essentially evaluations of their expected utility in the subsequent step of the multi-agent learning process. (de Cote, 2006; Kim, 2023; Morozs, 2015).

The single state *MARL* model M is a tuple $N, \{A_i\}_{i \in N}, \{R_i\}_{i \in N}$, where N is a collection of agents, A_i is the set of actions available to the agent i, and R_i is its payoff. Multiple agents simultaneously choose their actions from their own action sets and receive their payoffs on the basis of the actions performed by all the agents. Let $a^t = [a_1^t, a_2^t, \ldots, a_n^t]$ be the joint action executed at iteration t, where a_{-i}^t is the joint action of all the agents except the agent i. At each learning stage, an agent's experience is characterized not only by its own action and payoff but also by all the actions actually executed by other agents in the multi-agent environment. Therefore, agents can learn to coordinate their actions through environmental feedback. The desired outcome of this process is to let the multiple agents collectively learn their best actions that maximize the total system profit. Usually, the Q-value is the expected cumulative reward for taking a particular action. For each pair <joint-action, action>, the agent i's Q-value, i.e., $Q_i(a^t, a_i^t)$, is represented as follows (de Cote, 2006; Kim, 2023; Morozs, 2015).

$$Q_i(a^t, a_i^t) = (1 - \alpha) \cdot Q_i(a^{t-1}, a_i^t) + \alpha \cdot \max_{a_i^{t+1}} \{ \mathcal{R}_t + \gamma \cdot Q_i(a^t, a_i^t) \} \tag{10.2.2}$$

where $\alpha \in [0, 1]$ is the learning rate, and $\gamma \in [0, 1)$ is the discount factor. To jointly reach a consensus, multiple intelligent agents should overcome the defect of selfishness through the collaboration (Kim, 2023).

10.2.5 The Spectrum Allocation Scheme for the IoMT Paradigm

The *LGSA* scheme introduces a bi-level control algorithm for each IoMT device, comprising a learning model M at the first level and a cooperative game model G at the second level. These two levels operate sequentially and interact dynamically over time to achieve consensus. While traditional Q-learning emphasizes state-action pairs, this approach cannot effectively model the complexities of real-world IoMT network systems. To address this, the *LGSA* scheme employs a single-state Q-learning mechanism, which simplifies the formulation by removing the dependence on state-action pairs. Initially developed for stateless games in computer science, this mechanism is well-suited for dynamically changing environments. Instead of relying on the successor state information, the Q-value for each action is updated based solely on the reward received. This reduces the model's complexity and enhances the applicability of *MARL* in distributed systems (Kim, 2023).

In the first control process, the *LGSA* scheme utilizes a single-state multi-agent Q-learning model M. Because the environment is stateless, the *LGSA* scheme reformulates equation (10.2.2). In this context, each agent maintains Q-values for its available actions, updating them after each learning step based on the reward obtained for the chosen action. In the stateless setting, the *LGSA* scheme assumes that each IoMT device has his action set $\mathcal{L}_{\mathcal{D}}$, which consists of device's available salient points. For example, the Q-value of $\mathcal{D}_j$'s k[th] action, i.e., $Q_k^{\mathcal{D}_j}\left(a_{-\mathcal{D}_j}, a_k^{\mathcal{D}_j}\right)$, provides an estimate of the value of performing the joint action $\mathbf{a} = \left(a_{-\mathcal{D}_j}, a_k^{\mathcal{D}_j}\right)$. The $\mathcal{D}_j$ updates its estimate $Q_k^{\mathcal{D}_j}(\cdot)$ value based on the experience sample *a*, $\mathcal{R}_a^{\mathcal{D}_j}$ where $\mathcal{R}_a^{\mathcal{D}_j}$ is the $\mathcal{D}_j$'s reward of **a**. Simply, $\mathcal{R}_a^{\mathcal{D}_j}$ is same as the $\mathcal{D}_j$'s utility function $\mathcal{U}_{\mathcal{D}_j}(\cdot)$ in the second-level cooperative game G; it is defined in the equation (10.2.4). Finally, the single-state *MARL* function is defined as follows (Claus, 1998; Kapetanakis, 2002; Kim, 2023).

$$Q_{1\le k\le l}^{\mathcal{D}_j}\left(a_{-\mathcal{D}_j}, a_k^{\mathcal{D}_j}\right) = (1 - \alpha) \cdot Q_k^{\mathcal{D}_j}\left(a_{-\mathcal{D}_j}, a_k^{\mathcal{D}_j}\right) + \alpha \cdot \mathcal{R}_a \tag{10.2.3}$$

$$= Q_k^{\mathcal{D}_j}\left(a_{-\mathcal{D}_j}, a_k^{\mathcal{D}_j}\right) + \alpha \cdot \left(\mathcal{R}_a - Q_k^{\mathcal{D}_j}\left(a_{-\mathcal{D}_j}, a_k^{\mathcal{D}_j}\right)\right)$$

A significant challenge in action selection is achieving a balance between exploration and exploitation. To address this, the *LGSA* scheme adopts the Boltzmann strategy, where each agent selects its next action probabilistically based on its current estimation of that action's usefulness. This approach ensures that actions with higher Q-values are more likely to be chosen, while still allowing less-explored actions to be tried occasionally, facilitating a more effective learning process (Claus, 1998; Kapetanakis, 2002; Kim, 2023).

For the second control process, the *LGSA* scheme develops a cooperative game model G. In the $\mathbb{G}_{\mathcal{A}_i}$, individual devices in the $\mathbb{D}_{\mathcal{A}_i}$ share the $\mathfrak{M}_{\mathcal{A}_i}$ in a centralized manner. In this game, the spectrum allocation process is operated based on the concept of *BGV* solution. In the viewpoint of $\mathcal{D}_j \in \mathbb{D}_{\mathcal{A}_i}$, the utility functions, i.e., $\mathcal{U}_{\mathcal{D}_j}(\cdot)$, is defined as follows (Kim, 2023):

$$\mathcal{U}_{\mathcal{D}_j}\left(\Gamma_{\mathcal{D}_j}, S_{\mathcal{D}_j}, \mathcal{I}_{\mathcal{D}_j}\right) = \exp\left(\mu \times \log\left(\frac{\min\left(\mathcal{W}_{\mathcal{D}_j}, S_{\mathcal{D}_j}\right)}{\mathcal{W}_{\mathcal{D}_j}}\right)\right) \times \frac{1}{\mathcal{I}_{\mathcal{D}_j}} \tag{10.2.4}$$

$$\text{s.t.}, \sum_{\mathcal{D}_j \in \mathbb{D}_{\mathcal{A}_i}} S_{\mathcal{D}_j} \leq \mathfrak{M}_{\mathcal{A}_i}$$

where μ is an adjustment parameter for the $\mathcal{U}_{\mathcal{D}_j}(\cdot)$, and $\mathcal{I}_{\mathcal{D}_j}$ is the service sensitivity of $\mathcal{D}_j$'s application. $\mathcal{W}_{\mathcal{D}_j}$ and $S_{\mathcal{D}_j}$ are the requested spectrum amount and allocated spectrum amount, respectively. According to (10.2.1) and (10.2.4), the spectrum resource $\mathfrak{M}_{\mathcal{A}_i}$ is shared based on the *BGV* solution idea. For the $\mathcal{D}_j$, the allocated spectrum amount, i.e., $BGV_{\mathcal{D}_j}(\cdot)$, is given by (Kim, 2023):

$$BGV_{\mathcal{D}_j}\left(\mathcal{D}_j \in \mathbb{D}_{\mathcal{A}_i} | \mathbb{U}, d_{\mathbb{D}_{\mathcal{A}_i}}, r_a\right) = \left(\lambda^* \cdot \Upsilon(\mathbb{U}, r_a)\right) + \left((1 - \lambda^*) \cdot d_{\mathbb{D}_{\mathcal{A}_i}}\right) \tag{10.2.5}$$

$$\text{s.t.}, \lambda^* = \begin{cases} U = \mathcal{D}_k \in \mathbb{D}_{\mathcal{A}_i} | \ldots, \mathcal{U}_{\mathcal{D}_k}, \ldots \\ \Upsilon(\mathbb{U}, r_a) = \Upsilon_{\mathcal{D}_j \in \mathbb{D}_{\mathcal{A}_i}} = \max\left\{r_{\mathcal{D}_j}, \max\left\{t \in \mathbb{R} | \left(t, r_{-\mathcal{D}_j}\right) \in \mathbb{U}\right\}\right\} \\ \max\left\{\lambda \in [0,1] | \left(\left(\lambda \cdot \Upsilon(\mathbb{U}, r_a)\right) + \left((1-\lambda) \cdot d_{\mathbb{D}_{\mathcal{A}_i}}\right)\right) \in \mathbb{U}\right\} \end{cases}$$

where r_a is a reference point to temper the IoMT devices' aspirations. In the *LGSA* scheme, the r_a is defined as devices' actions where $r_a = \mathcal{D}_k \in \mathbb{D}_{\mathcal{A}_i} | \ldots, r_{\mathcal{D}_j}, \ldots$ and $r_{\mathcal{D}_j}$ is the $\mathcal{D}_j$'s selected action, i.e., $a^{\mathcal{D}_j}_{1 \leq k \leq l}$. The $r_{-\mathcal{D}_j}$ is all devices' selected actions except the $\mathcal{D}_j$ (Kim, 2023).

10.2.6 Summary

The *LGSA* scheme explores an IoMT-assisted health monitoring system with limited spectrum resources and introduces a novel spectrum allocation mechanism based on a bi-level control approach. This approach integrates a *MARL* process for selecting each device's salient point and a cooperative game process for sharing the spectrum resources. In the *MARL* process, individual IoMT devices intelligently learn their optimal salient points within the dynamic IoMT network environment. This learning process operates in a parallel and distributed manner, enabling scalability and adaptability. For the cooperative game process, the *BGV* solution is employed to share spectrum resources among multiple IoMT devices in a centralized manner. These two control processes are meticulously combined, iteratively interacting to achieve a balanced system performance. By addressing contradictory service requirements, the bi-level approach aims to reach consensus while delivering reciprocal advantages. Additionally, the *LGSA* scheme adopts a single-state *MARL* model, which significantly reduces computational complexity compared to traditional *MARL* methods, enhancing efficiency without compromising performance (Kim, 2023).

10.3 Reverse Biform Based Resource Sharing (*RBRS*) Scheme

Recent advances in wireless sensor technologies have significantly contributed to the development of WBANs, which have been considered for applications in medical, healthcare, and sports fields. However, the specific features and reliability requirements of WBANs introduce a range of new challenges in designing novel WBAN protocols. To address these challenges, game-theoretic approaches offer a promising solution, enabling WBANs to enhance their performance while improving flexibility and adaptability. In this context, S. Kim has developed the *Reverse Biform-based Resource Sharing* (*RBRS*) scheme, utilizing a reverse-biform game model (Kim, 2019b). The *RBRS* scheme employs a dual-level framework to efficiently share limited WBAN resources through a coordinate-and-competitive game approach. Notably, the *RBRS* scheme accounts for the unique characteristics of WBAN applications, providing a generalized and robust solution to the resource-sharing problem.

10.3.1 Development Motivation

In modern times, the dramatic increase in average life expectancy ranks as one of society's greatest achievements. However, this situation brings two major challenges: the aging population and the rising cost of healthcare (Movassaghi, 2014). On a global level, millions of people die annually from chronic or fatal diseases. A common issue with these conditions is that many individuals experience symptoms and receive diagnoses too late. Research indicates that most diseases can be prevented if detected in their early stages. As a result, future healthcare systems should focus on proactive wellness management and prioritize early disease detection and prevention (Demner-Fushman, 2012; Kim, 2019b; Movassaghi, 2014; Salem, 2013; Son, 2016).

Over the last decade, an emerging short-range wireless communication technology, known as WBANs, has garnered remarkable attention from researchers. In WBANs, tiny physiological sensors are implanted in or on the human body to provide real-time health monitoring and ubiquitous healthcare services. These networks are essential in realizing the vision of a future healthcare paradigm, termed 'Pervasive Healthcare', which aims to deliver healthcare to anyone, anytime, and anywhere. WBANs are poised to revolutionize how individuals manage and perceive their health, akin to how the Internet transformed communication and information access (Jang, 2013; Kim, 2019b; Movassaghi, 2014; Qiu, 2016; Raazi, 2010).

WBANs interact with the Internet and other wireless technologies, offering new opportunities for intelligent and autonomous applications (Ahn, 2017; Movassaghi, 2014). Sensor nodes in WBANs periodically sample, monitor, process, and communicate data while providing greater mobility and flexibility to patients. Key objectives for WBAN sensor operations include minimum delay, maximum throughput, and maximum energy efficiency. However, existing wireless personal area network protocols often fall short in meeting WBANs' stringent requirements for high reliability, energy efficiency, and specific operational needs (Kim, 2019b; Zhou, 2014).

To enable decentralized and autonomous WBAN operations, game theory is an effective tool. Originally developed in economics, game theory is a mathematical framework for analyzing decision-making processes among self-interested players. It has been widely applied to microeconomics and, in recent years, has gained traction in telecommunications for solving distributed network control problems, including power control, resource sharing, routing, fault tolerance, and QoS management (Kim, 2019b).

In 1996, A. Brandenburger and H. Stuart introduced the concept of the biform game, a hybrid non-cooperative-cooperative game model formulated as a two-stage decision process (Brandenburger, 2007; Kim, 2012; Stuart Jr, 2005). In the biform game, the first stage represents non-cooperative decision-making in a competitive environment, while the second stage involves cooperative decision-making. During the second stage, players form coalitions to generate surplus, which is shared fairly and optimally, ensuring actual payoffs are realized. The biform game model has proven efficient when applied to business strategy scenarios (Brandenburger, 2007; Kim, 2012, Kim 2019b; Stuart Jr, 2005).

The *RBRS* scheme addresses the resource-sharing challenges in WBANs by developing a novel WBAN control mechanism that ensures QoS provisioning and network reliability. The WBAN control process is structured as a dual-level framework, comprising intra-WBAN and inter-WBAN operational algorithms. In the intra-WBAN algorithm, sensor nodes act as game players, forming coalitions within a WBAN to share limited resources fairly and optimally. This process is modeled as a cooperative game within each WBAN. In contrast, the inter-WBAN algorithm models the relationships between WBANs in a competitive environment as a non-cooperative game. To integrate these two algorithms, the *RBRS* scheme introduces the reverse-biform game, a novel paradigm that formalizes intra- and inter-WBAN interactions as a two-stage decision process. This reverse framework combines the characteristics of coordinated and competitive decision-making under different control scenarios, enabling self-adaptability and real-time effectiveness. To the best of my knowledge, the *RBRS* scheme represents the first attempt to jointly integrate intra- and inter-WBAN operations within a game-theoretic paradigm, offering a robust solution for resource-sharing challenges in WBANs (Kim 2019b).

10.3.2 Contribution

The *RBRS* scheme generalizes the relationship between intra- and inter-WBAN operational algorithms, comprehensively addressing all relevant practical factors to fully leverage the synergy of these control mechanisms. To model the interaction between the intra- and inter-WBAN algorithms, the *RBRS* scheme introduces a novel reverse-biform game model. This game-theoretic approach enables control decisions to strategically interact with one another, maximizing WBAN performance while ensuring that QoS requirements are met. The reverse-biform process operates recursively, integrating self-adaptability and real-time effectiveness into the system. As a result, the *RBRS* scheme achieves a fair and balanced solution under diverse WBAN network conditions, ensuring optimized resource allocation and robust network reliability (Kim 2019b).

10.3.3 Reverse-Biform Game Model for the WBAN System

In scenarios with multiple WBANs operating simultaneously, overlapping transmission ranges of nearby WBANs can lead to inter-WBAN interference, resulting in packet errors during their operations. Within each WBAN, there is one central coordinator, and several sensor nodes deployed at different positions on or in the human body, catering to diverse applications. During data transmissions among sensor nodes, data collisions may also occur within individual WBANs. The *RBRS* scheme simplifies and assumes this complex real-world situation as follows (Kim 2019b):

1. WBAN Structure: each individual WBAN consists of multiple sensors attached to or implanted in the body, along with a single central entity known as the Central Processing Coordinator (CPC).
2. Sensor Functionality: sensors in each WBAN continuously measure vital signs such as heart rate, blood pressure, body temperature, and oxygen saturation in blood. These measurements are then transmitted to the CPC.
3. CPC Role: the CPC collects data from the sensors and relays the aggregated information to an external network gateway. Through this gateway, all data is forwarded to a remote medical server over the Internet.
4. System Topology: the WBAN system forms a recursive star topology, rooted at the gateway. Each CPC acts as a sub-root for its respective WBAN. This topology is dynamically restructured as patients with WBANs move.
5. Intra-WBAN Interference: inside each WBAN, interference is effectively mitigated using a TDMA-based model. However, packet collisions may still occur at the CPC.
6. Inter-WBAN Interference: when multiple WBANs are deployed in proximity, serious inter-WBAN coexistence interference arises, affecting system performance.

In the WBAN system, the problems of intra-WBAN collision and inter-WBANs' interference seriously degrade the WBAN system performance. The *RBRS* scheme formulates a reverse-biform game model (G) to effectively mitigate the interference and collision problem. Mathematically, the G can be defined as$\mathbb{G} = \left\{N, \mathscr{E}_{i,i\in N}, \mathscr{C}, \mathscr{G}, S_{i,i\in N}, \mathscr{S}_{i,i\in N}^{1\leq j\leq m}, U_{i,i\in N}, \mathfrak{U}_{i,i\in N}^{1\leq j\leq m}, \mathscr{T}\right\}$ at each time period $t \in \mathscr{T}$ of gameplay (Kim 2019b).

- $\mathbf{N} = \{1,\ldots,n\}$ is the finite set of all WBANs.
- $\mathscr{E}_i = \left\{e_i^1, e_i^2, \ldots, e_i^m\right\}$ is the sensor nodes in the WBAN i $\in N$ where m is the total number of sensors in the ith WBAN.
- $\mathbf{C} = \left\{c_1, \ldots, c_n\right\}$is the set of CPCs of WBANs where $c_i \in \mathscr{C}$ is the CPC of ith WBAN.
- G is the gateway of WBANs' system; it is an access point connecting to the Internet.
- $S_{i,i\in N} = \left\{p_{min}^i, \ldots, p_{max}^i\right\}$ is the set of all possible strategies of c_i to connect the G; S_i represents the discrete power levels of c_i.
- $\mathscr{S}_{i,i\in N}^{1\leq j\leq m}$ is the set of all strategies of e_i^j; $\mathscr{S}_i^j$ represents the possible data tune levels of $c_i^j$for sensor communications.
- $U_{i,i\in N}$ is the payoff received by the c_i, and $\mathfrak{U}_{i,i\in N}^{1\leq j\leq m}$ is the payoff received by the e_i^j.
- $\mathrm{T} = \{0, 1, \ldots, t, t+1, \ldots\}$ is a time, which is represented by a sequence of time steps with imperfect information.

The reverse-biform game model is structured around two key algorithms: one for intra-WBAN operations and another for inter-WBAN interactions. In the inter-WBAN algorithm, individual WBANs function as game players, each striving to maximize their respective payoffs through interactions with other WBANs. Given the autonomous nature of WBANs, a non-cooperative strategy is considered more suitable for mitigating interference among them. Meanwhile, the intra-WBAN algorithm primarily focuses on resolving data collision issues within a single WBAN using the *RBRS* scheme. Since multiple sensors must efficiently share the limited resources available within each WBAN, independent resource management becomes essential. To effectively address the intra-WBAN collision problem, a cooperative bargaining approach is employed (Kim 2019b).

Building upon this framework, the *RBRS* scheme introduces a novel reverse-biform game ($\mathbb{G}$) to facilitate both intra-WBAN and inter-WBAN algorithms. This dual-level game model characterizes inter-WBAN interactions using a non-cooperative game framework, whereas intra-WBAN resource allocation is modeled as a cooperative game. Generally, conventional game models consist of players, their available strategies, and utility functions that define the outcomes of those strategies. Formally, the reverse-biform game ($\mathbb{G}$) is defined as $\text{G}=\left\{1 \le i \le n \middle| \mathbb{G}_i^{S\_S}, \mathbb{G}^{F\_S}\right\}$ where $\mathbb{G}_i^{S\_S} = \{\mathscr{E}_{i,i\in N}, \mathcal{S}_{i,i\in N}^{1\le j\le m}, \mathfrak{U}_{i,i\in N}^{1\le j\le m}\}$ is the first stage cooperative game to formulate interactions among the sensor nodes in the WBAN i $\in N$, and $\mathbb{G}^{F\_S}=\{\mathbf{C}, S_{k,k\in\mathscr{C}}, U_{k,k\in\mathscr{C}}\}$ is the second stage non-cooperative game to formulate the interactions among CPCs in $\mathbf{C}$. These two $\mathbb{G}_i^{S\_S}$ and $\mathbb{G}^{F\_S}$ sub-games are played sequentially based on the repeated approach. Under dynamically changing network environments, it is suitable for the ultimate practical WBAN system implementation (Kim 2019b).

10.3.4 Cooperative Game Based Intra-WBAN Management

For the intra-WBAN communications, star-topology based TDMA mechanism is introduced by arranging the time-slots for sensor nodes. In the G game model, the first phase game is the $\mathbb{G}_{i,1\le i\le n}^{S\_S}$ for the intra-WBAN management. In the $\mathbb{G}_{i,1\le i\le n}^{S\_S}$ game, game players are internal sensor nodes in each WBAN, i.e.,$e_i^j \in \mathscr{E}$, and they are rational to reach a win-win situation to maximize the payoff in their embedded WBAN. Therefore, they are more prone to coordinate and willing to play in a cooperative manner (Qiao, 2006; Zhao, 2015). To adapt to this network situation, the first phase $\mathbb{G}_{i,1\le i\le n}^{S\_S}$ game is designed as a cooperative game model (Kim 2019b).

Due to the specialty of intra-WBAN applications, the QoS of vital physiological information should be carefully considered. The *RBRS* scheme focuses on the three QoS factors such as i) the energy dissipation, ii) successful information transmission, and iii) data criticality. These three factors are concerned to effectively distribute the limited intra-WBAN resource. In general, sensor nodes are capacitated with limited amount of energy to survive on. Therefore, minimizing the rate of dissipation of energy is crucially important. Energy dissipation in sensor nodes may occur due to sensing energy (E_s) and transmission energy (E_{tr}). In each time period, E_s is the energy expanded due to one sensing the concerned health parameter and E_{tr} is the energy dissipated due to the one packet transmission. The jth sensor's total processing energy at time t ($\mathbb{E}_p^j(t)$) is defined as follows (Kim 2019b);

$$\mathbb{E}_p^j(t)=\left(E_s \times \mathfrak{S}_s^j(t)\right) + \left(E_{tr} \times N_{tr}^j(t)\right) \tag{10.3.1}$$

where $\mathfrak{S}_s^j(t)$ and $N_{tr}^j(t)$ are the number of sensings and transmitting packets, respectively. In the *RBRS* scheme, the utility function for each sensor node is defined as a ratio of the successfully transmitted number of bits versus the consumed processing energy. So, the utility function for the jth sensor in the ith WBAN at time t $\left(\mathfrak{U}_i^j(t)\right)$ is given by (Kim 2019b);

$$\mathfrak{U}_i^j(t) = \frac{\Gamma_t^j \times \left(\vartheta \times N_{tr}^j(t)\right)}{\mathbb{E}_p^j(t)}, \text{ s.t., } 1 \le j \le m \text{ and } i \in N \tag{10.3.2}$$

where Γ_t^j is the probability of successful packet transmission from the jth sensor node at time t, and ϑ is the total number of bits per each packet.

Information criticality index (ICI) represents the sensed data criticality. For a particular sensor, ICI is the ratio of deviation in sensed physiological data and the normal value of that physiological parameter (Misra, 2014, Misra, 2015). The jth sensor's ICI in the ith WBAN at time t ($\mathfrak{h}_i^j(t)$) is mathematically expressed as (Kim 2019b);

$$\mathfrak{h}_i^j(t) = \frac{\left(e^{\left(\psi_{max}^j - \psi^j(t)\right)} + e^{\left(\psi^j(t) - \psi_{min}^j\right)}\right)}{\phi \times e^{\left(\psi_{max}^j + \psi_{min}^j\right)}} \tag{10.3.3}$$

where ψ_{max}^j, ψ_{min}^j are the upper and lower limits of the jth sensor's reference range, respectively. $\psi^j(t)$ is the monitored value of the jth sensor at a given time t. ϕ is the control parameter for normalization. The numerator in the equation (10.3.3) indicates the difference between the sensed value and the normal value of a particular physiological parameter (Kim 2019b; Misra, 2014).

To effectively share intra-WBAN resources, each sensor node strives to improve its objectives without hindering the objectives of others. This necessitates a bargaining solution to ensure fair and optimal resource allocation among sensor nodes. Several solution concepts have been proposed for various cooperative games. Among these, the *Relative Utilitarian Bargaining Solution* (*RUBS*) emerges as a particularly compelling choice for modeling practical sensor node interactions. RUBS, like other well-known bargaining solutions, offers several attractive features that align with the cooperative nature of intra-WBAN resource sharing (Dhillon, 1998; Kim 2019b; Sobel, 2001).

In the *RBRS* scheme, the *RUBS* is obtained while ensuring each player's QoS provisioning. Based on socially cooperative sensors, the *RBRS* scheme can enhance the effectiveness of *RUBS*. At time t, the timed *RUBS* for the ith WBAN ($RUBS_t^i$) is mathematically defined as (Kim 2019b);

$$RUBS_t^i = \left\{\mathfrak{U}_i^1(\zeta)^*, \ldots \mathfrak{U}_i^m(\zeta')^*\right\} = \max_{\zeta \in \mathcal{S}_i^j} \sum_{j=1}^{m} \left(\frac{\mathfrak{X}_i^j(t) \times \mathfrak{U}_i^j(\zeta)}{\max_{\chi \in \mathcal{S}_i^j}\left(\mathfrak{U}_i^j(\chi)\right) - \min_{\chi' \in \mathcal{S}_i^j}\left(\mathfrak{U}_i^j(\chi')\right)}\right) \tag{10.3.4}$$

$$\text{s.t.}, \mathfrak{X}_i^j(t) = \mathfrak{h}_i^j(t) / \sum_{l=1}^{m} \mathfrak{h}_i^l(t), \; 0 \le \mathfrak{X}_i^j(t) \le 1 \text{ and } \sum_{l=1}^{m} \mathfrak{X}_i^l(t) = 1$$

where $\mathfrak{X}_i^j(t)$ is player e_i^j's bargaining power at time t, which is the relative ability to exert influence over other players. Based on various QoS requirements, $RUBS_t^i$ is a joint strategy, which is taken by the CPC at time t to satisfy the equation (10.3.4) (Kim 2019b).

10.3.5 Non-Cooperative game based Inter-WBAN management

In the G game model, the second phase game is the $\mathbb{G}^{F\_S}$ for the inter-WBANs' management. The goal of inter-WBANs' management is also to maximize the system performance while ensuring QoS. When multiple CPCs in individual WBANs are transmitted concurrently to the gateway G, it can cause interference to other WBANs. Therefore, the *RBRS* scheme considers the $\mathbb{G}^{F\_S}$ as a power control game among CPCs of WBANs, i.e., $c_i \in \mathscr{C}$. To effectively avoid the interference problem, each self-interested game player will decide independently its transmission power based on the evolutionary learning process. Taking this into consideration, the *RBRS* scheme applies to a non-cooperative game methodology to design the power control algorithm for inter-WBANs' management (Kim 2019b).

Let P be the transmit power vector ($\mathbb{P}$) for all CPCs where $p_{min\leq j\leq max}^{i,i\in N}$ is the power level of ith CPC. Therefore, P is assumed as below.

$$\mathbb{P} = \left\{ Đ_{i\in N} p_j^i \;\middle|\; p_j^i \in [\, p_{min}^i \cdots p_{max}^i] \right\}$$

(10.3.5)

where the p_{min}^i, p_{max}^i are the pre-defined minimum and maximum power levels are, respectively. Based on the interference, wireless communication performance is measured according to Shannon's capacity. With the channel bandwidth (*W*) and the *Signal-to-Interference plus Noise Ratio* (*SINR*), the ith CPC's transmission capacity ($\mathfrak{C}_i(\mathbb{P})$) is defined as follows (Kim, 2011, Kim 2019b).

$$\mathfrak{C}_i(\mathbb{P}) = W \times \log_2\left(1 + \left(\frac{p_l^i \times h_{ii}}{\sigma_i^2 + \sum_{j\neq i}(p_{l'}^j \times h_{ji})}\right)\right)$$

(10.3.6)

where σ_i^2 is the background noise is within the bandwidth, h_{ji} is the path gained from the transmitter of the jth CPC to the receiver i^{th} CPC. The *RBRS* scheme follows the assumption in (Kazemi, 2010; Kim, 2011, Kim 2019b).

The main issue of inter-WBANs' management is to choose CPC's transmission power to maximize its payoff from the viewpoint of individual WBAN. Based on the required QoS of transmitting information, each rational CPC will decide its transmission power in a non-cooperative manner. In the game $\mathbb{G}^{F\_S}$ $=\{\mathbf{C}, S_{i,i\in N}, U_{i,i\in N}\}$, CPCs in $\mathbf{C}$ are game players and S_i is the i th game player's strategy set, i.e., $p_j^i \in S_i$. The global strategy space P is given by the Cartesian product of all game players' strategies. $U_j^i(\mathbb{P})$ is the utility function for player i with the strategy p_j^i. To formulate the utility function, it is considered on how CPCs can increase or decrease their satisfaction. Usually, it can be assumed as a ratio of the successfully transmitted information with the QoS level versus the power consumed in transmission. Based on the tradeoff among the conflicting criteria, $U_j^i(\mathbb{P})$ is defined as follows (Kim 2019b).

$$U_j^i(\mathbb{P}) = \frac{\mathfrak{C}_i(\mathbb{P}) \times \varrho_i}{p_j^i}, \text{ s.t.}, \varrho_i = \sum_{l=1}^{m} \mathfrak{h}_i^l(t) \text{ and } p_j^i \in S_i \tag{10.3.7}$$

When the $\mathbb{G}^{F\_S}$ game process is repeatedly operated, the outcome of utility function would be varied over time. To represent the power control problem over the variable time t, let $\Theta_l^i(t+1)$ is player*i*'s propensity for the lth power level, i.e., p_l^i at time $t+1$. $\Theta_l^i(t+1)$ is obtained in a distributed manner for each individual player. Based on the *Q*-value idea, it is dynamically estimated as follows (Kim 2019b);

$$\grave{\mathrm{E}}_l^i(t+1) = \left(\eta \times \left(\frac{Q_i^l(t)}{\sum_{j=1}^{m} Q_i^j(t)} \times \left(1 + \frac{l}{m}\right)^{\varrho_i}\right)\right) + \left((1-\eta) \times \frac{Q_i^l(t)}{\sum_{j=1}^{m} Q_i^j(t)}\right) \tag{10.3.8}$$

$$\text{s.t., } Q_i^j(t) = Q_i^j(t-1) + \alpha \times \left(U_i(\mathbb{P}) + \left(\gamma \times \max_{1 \le \tau \le t-1} \left(Q_i^j(\tau)\right) - Q_i^j(t-1)\right)\right)$$

$$\text{and } \eta = \sum_{l=1}^{m} \mathfrak{h}_i^l(t),\ \alpha, \gamma \in (0,1]$$

where η is a weighted control parameter; the higher the ICI value of player i, the higher power level preferred. α and γ are the learning rate and the discount factor, respectively. Finally, the *RBRS* scheme can get the strategy*j* selection probability from player i for the time $t+1$ $\left(\mathscr{P}_j^i(t+1)\right)$ is defined based on the proportion to each power level's propensity ($\Theta^i(\cdot)$) (Kim 2019b).

$$\mathscr{P}_j^i(t+1) = \frac{\Theta_j^i(t+1)}{\sum_{l=1}^{m} \Theta_l^i(t+1)} \tag{10.3.9}$$

In a distributed online fashion, individual players dynamically estimate each $\mathscr{P}_{j \in S_i}^{i, 1 \le i \le n}(\cdot)$ values for each strategy. At the end of each time iteration, one player's strategy is used as the decision factor for the rest players, and the results of the other players' strategies are also used to make a player's decision for the next iteration. This multi-interactive feedback mechanism continues while approaching the stable system status (Kim 2019b).

10.3.6 Summary

The aging population and the growing prevalence of chronic diseases are placing an increasingly heavy burden on global healthcare systems. Recently, WBAN technology has emerged as a promising solution, enabling clinical improvements through continuous real-time remote monitoring of physiological parameters. However, the unique characteristics of WBANs - including their heterogeneity, distributed nature, and autonomy - present challenges not encountered in conventional wireless communication systems. In this context, game theory has shown great potential as a tool to address WBAN control issues effectively. The *RBRS* scheme introduces a novel reversed-biform game model to tackle WBAN control problems while ensuring QoS provisioning. This approach optimizes intra-WBAN resource allocation using the *RUBS* and manages wireless interference through an interactive learning mechanism. By accounting for the criticality of sensed data, the *RBRS* scheme adaptively allocates system resources to guarantee the QoS for vital data transmissions (Kim 2019b).

10.4 Bargaining Solution based Resource Allocation (*BSRA*) Scheme

WBANs are an emerging technology with immense potential for continuous health monitoring across various diseases. For diverse medical and healthcare applications, physiological data collected from multiple WBANs is transmitted to remote medical cloud servers for analysis and storage. However, integrating WBANs with cloud computing services introduces several technical challenges and issues. To address these challenges, S. Kim proposes the *Bargaining Solution-based Resource Allocation* (*BSRA*) scheme to optimize the utilization of limited system resources effectively (Kim, 2020). The *BSRA* scheme employs the principles of generalized *Gini* and *Choquet* bargaining solutions, implementing a dual stage bargaining process to enhance flexibility and adaptability. Notably, the approach considers the distinctive features of cloud-assisted WBAN applications, offering a generalized fair and efficient solution for resource allocation problems (Kim, 2020).

10.4.1 Development Motivation

With the rapid advancements in smart devices and IoT technologies, WBANs offer a viable alternative to traditional, complex, and offline-based healthcare processes. By continuously monitoring a patient's vital signs, WBANs enable real-time feedback, thereby enhancing patient care. The key benefits of WBANs include: (i) accelerating patient recovery through online monitoring, (ii) reducing healthcare costs, and (iii) improving the accuracy of critical medical decisions. Typically, WBANs consist of lightweight, wearable, or implanted smart edge devices that track physiological parameters. To facilitate medical diagnosis and healthcare management, WBANs must gather real-time patient data - such as blood pressure, electrocardiogram signals, body temperature, and heart rate - without disrupting daily activities (Kim, 2020; Liu, 2018; Sodagari, 2018).

However, as the number of WBANs in a specific region increases, network performance may degrade due to end-to-end packet delivery delays and reduced throughput. For timely medical diagnoses and prompt responses, seamless real-time access to a patient's vital signs is essential. Furthermore, WBANs must not only process collected data efficiently but also provide timely feedback. Most WBAN applications involve multiple computational tasks, which are becoming increasingly resource-intensive. Given the compact and portable nature of WBANs, their computational capabilities are inherently limited, making it challenging to meet these demands using conventional WBAN infrastructure alone. To address this challenge, various methods have been proposed, with a primary focus on optimizing the utilization of limited WBAN resources (He, 2015; Kim, 2020; Misra, 2018; Quwaider, 2014).

In recent years, cloud computing has emerged as a rapidly expanding innovation, gaining significant attention from both academia and industry. As a novel IT paradigm, cloud computing is a service-oriented approach that leverages distributed networks to provide computing resources and efficiently handle resource-intensive tasks. This paradigm addresses many of the inherent limitations associated with conventional WBANs (He, 2015). Importantly, WBANs and cloud computing are not mutually exclusive; rather, they complement each other's weaknesses. A promising solution to enhance WBAN capabilities is to offload computational tasks to remote cloud servers. This realization has led to the development of integrated approaches that exploit the interdependent features of WBANs and cloud computing (Kim, 2020; Quwaider, 2014).

By merging WBANs with cloud computing, cloud-assisted WBAN (CWBAN) systems have been developed to create cost-effective, scalable, and real-time healthcare solutions. These systems facilitate patient data analysis across various environments by offloading computational tasks from individual WBANs to cloud servers. Consequently, CWBANs address critical WBAN control issues while enhancing computational capacity. Additionally, they reduce deployment costs and enable real-time sharing of critical medical data to improve decision-making processes (Hassan, 2017; Kim, 2020; Quwaider, 2014; Wan, 2013).

The seamless integration of WBANs with cloud computing brings several advantages. Most notably, cloud computing extends WBAN capabilities by providing a broader range of functional services, thereby meeting more diverse application requirements. This integration enhances performance efficiency while mitigating resource constraints in WBAN edge devices. Several real-world implementations of CWBANs exist. For instance, Harvard University's CodeBlue Project focuses on hospital environments by deploying multiple router nodes within healthcare facilities. These nodes, operating with ZigBee radios, do not rely on a centralized or distributed server for data control and storage. Instead, the system utilizes MoteTrack localization, achieving an accuracy of approximately 1 meter and supporting a packet transmission rate of up to 40 kb/s (Cao, 2009). Similarly, Johns Hopkins University's Advanced Health and Disaster Aid Network is being developed for mass casualty incidents. This system establishes communication between personal servers and central servers, enabling various users—such as emergency department personnel, incident commanders, and medical specialists—to access real-time patient data through a web-based portal (Gao, 2007). Meanwhile, the Wearable Health Monitoring System at the University of Alabama focuses on large-scale telemedicine applications. By employing time-division multiple access mechanisms, this system coordinates data collection from sensor nodes and transfers the information to remote central servers. Physicians can access patient data via the internet, and the system can generate automated alerts through server-based agents (Kim, 2020; Milenković, 2006).

Despite these advancements, research on CWBAN systems is still in its early stages, with several technical challenges remaining unresolved. Addressing these challenges is a top priority for developing practical, real-world CWBAN solutions (Wan, 2013). Game theory, a framework rooted in the '*rational choice theory*', provides a strategic decision-making approach that mathematically models interactions between rational players. Game theory is particularly concerned with evaluating the consequences of different strategic choices, which are defined through players' utility functions. Recently, game-theoretical methods have been widely adopted for resource allocation in network management. In particular, non-cooperative game theory examines scenarios where players' strategic choices are not externally enforceable, while cooperative game theory allows for binding agreements among players through enforceable contracts (Kim, 2014, Kim, 2020).

Given the characteristics of CWBAN resource allocation, cooperative game models offer a more suitable approach. These models provide an effective framework for managing both communication and computational resources in CWBANs. Specifically, cooperative bargaining game models have been adopted to design an optimized resource allocation strategy. The *BSRA* scheme incorporates concepts from the *Generalized Gini Bargaining Solution* (*GGBS*) and the *Choquet Bargaining Solution* (*CBS*) to efficiently allocate resources in CWBANs. Under the *GGBS* approach, communication resources are distributed among WBANs, while the *CBS* ensures optimal allocation of cloud computing resources.

Both solutions are grounded in rationality principles, aligning with utility space maximization strategies (Hinojosa, 2008; Kim, 2020).

10.4.2 Contribution

To realize the full potential of CWBAN platforms, several technical challenges and issues must be addressed. Among these, the allocation of communication and computation resources is particularly critical. The *BSRA* scheme tackles this problem by employing a novel dual-phase cooperative game process to model the interaction between each WBAN and the cloud server. The *BSRA* scheme integrates two distinct bargaining solutions to achieve fair and efficient resource allocation; i) the *GGBS*: optimally allocates communication resources among WBANs, and ii) the *CBS*: ensures efficient distribution of computation resources in the cloud. By combining the strengths of these solutions, the *BSRA* scheme leverages their synergy to provide a well-balanced approach to resource management. Furthermore, its design emphasizes self-adaptability and real-time effectiveness, ensuring optimal performance in dynamic and diverse operational scenarios. This innovative approach enables CWBAN systems to deliver fair, efficient, and reliable healthcare services (Kim, 2020).

10.4.3 Multi-Utilitarian Bargaining Solutions

A cooperative bargaining problem consists of a set of game players and a feasible set of payoffs. Guided by a common principle of rationality, most solutions to bargaining problems align with the maximization of some ordering of the utility space. Over time, various solutions have emerged, each rooted in slightly different assumptions about the properties desired for the final agreement point. Among these solutions, multi-utilitarian bargaining solutions have gained prominence. These solutions are rationalized by piecewise linear increasing social welfare functions defined on players' utility gains. They depend on a set of weighted utilitarian criteria and the decision rules used to combine them. Each multi-utilitarian bargaining solution is linked to a probability measure over the set of weighted utilitarian rules, derived as the expectation of this probability distribution (Hinojosa, 2007, Hinojosa, 2008). The *BSRA* scheme introduces and characterizes two bargaining solutions from this class: the *generalized Gini Bargaining Solution* (*GGBS*) and the *Choquet Bargaining Solution* (*CBS*) (Kim, 2020).

Introduced in 1994 by C. Blackorby et al., the *GGBS* is rationalized by generalized Gini social welfare functions. It represents a compromise between the two polar bargaining solutions - utilitarian and egalitarian. As a social choice rule, the *GGBS* balances equity and efficiency, making it a compelling approach to resource allocation (Blackorby, 1994). Proposed in 2000 by E. Ok and L. Zhou, the *CBS* is derived from social welfare functions represented by Choquet integrals with respect to a monotonic capacity. Unlike the *GGBS*, the *CBS* does not require anonymity, and its representing social welfare function is not necessarily quasi-concave. The *CBS* generalizes the *GGBS* by allowing for broader flexibility in decision-making. All weighted utilitarian bargaining solutions fall under the *CBS* framework, as they can be rationalized by a linear combination of the utilitarian and max-max social welfare functions (Oka, 2000). By incorporating these two solutions, the *BSRA* scheme effectively addresses the resource allocation challenges in cloud-assisted WBAN systems, ensuring both fairness and adaptability in dynamic healthcare environments (Kim, 2020).

An n-player bargaining problem can be described by a set of feasible utility vectors, $S \subseteq \mathbb{R}^n_+$, where we assume that S is convex, compact, and comprehensive and that there exists $x \in S \cap \mathbb{R}^n_{++}$. The set of

n-player bargaining problems is denoted by $\sum$. A bargaining solution is a correspondence $F:\sum \rightarrow \mathbb{R}^n_+$ such that $\emptyset \neq F(S) \subseteq S$ for all $S \in \sum$. Notice that F is allowed to yield multiple outcomes to a problem. Some simple bargaining solutions are the following (Hinojosa, 2008; Kim, 2020).

- The egalitarian bargaining solution $E(S)$: for each $S \in \sum$, $E(S) = \{x \in S | x_i = x_j$ or all $i,j = 1,2,\ldots,n$, and there is no $y \in S$ such that $y \gg x\}$.
- The utilitarian bargaining solution $U(S)$: for each $S \in \sum$, $U(S) = \{x \in S | \sum_{i=1}^{n} x_i \geq \sum_{i=1}^{n} y_i$ for all $y \in S\}$ with $\in \mathbb{R}^n_{++}$.
- The multi-utilitarian bargaining solution $M(S)$: for each $S \in \sum$, $M(S) = \{x \in S | \max U^l(x)$ where $U^l(x) = \psi^l \cdot x\}$ with $\psi^l \in \mathbb{R}^n_{++}$ for $l \in \{1, 2, \ldots, k\}$.
- The weighted utilitarian (λ-utilitarian) bargaining solution U_λ: for each $S \in \sum$, $U_\lambda(S) = \{x \in S | \sum_{i=1}^{n} (\lambda_i \times x_i) \geq \sum_{i=1}^{n} (\lambda_i \times y_i)$ for all $y \in S\}$ with $\lambda \in \mathbb{R}^n_{++}$.

$U^l(x)$ is a social welfare function that is increasing and linear in every rank-ordered subset of the set of n-dimensional vectors with non-negative components. Note that the weighted utilitarian solution is a particular case of multi-utilitarian solution in $\mathbb{R}^n_+$ (Hinojosa, 2008; Kim, 2020).

Now, the idea of *GGBS* is explained. Let $\pi:N \rightarrow N$ denote a permutation function, in which $\pi(i) = j$ means that the ith component of the n-dimensional permuted vector is j. Let x_π be the π-permutation of vector $x \in \mathbb{R}^n$, $x_\eth = (x_{\pi(1)}, x_{\pi(2)}, \ldots, x_{\pi(n)})$. The set of all possible permutation functions is denoted by $\prod$. For $\Upsilon = \{\Upsilon_1, \Upsilon_2, \ldots, \Upsilon_n\} \in \mathbb{R}^n_+$, such that $\Upsilon_1 \geq \Upsilon_2 \geq \ldots \geq \Upsilon_n > 0, G_\Upsilon:\mathbb{R}^n \rightarrow \mathbb{R}$ represents the *generalized Gini ordering* (Blackorby, 1994; Hinojosa, 2007, Hinojosa, 2008; Kim, 2020);

$$G_\Upsilon(x) = \sum_{i=1}^{n} (\Upsilon_i \times x_{(i)}) \tag{10.4.1}$$

where $(x_{(1)}, x_{(2)}, \ldots, x_{(n)})$ denotes a rank-ordered permutation vector of $x \in \mathbb{R}^n_+$, that is $x_{(1)} \leq x_{(2)} \leq, \ldots, \leq x_{(n)}$. Let Δ^n be the simplex in $\mathbb{R}^n_+$ and $\Delta^n = \{\Upsilon \in \mathbb{R}^n_+ | \Upsilon \cdot 1^n = 1\}$. For each $S \in \sum$, the *GGBS* with $G_\Upsilon(x)$ is defined as follows (Blackorby, 1994; Hinojosa, 2007, Hinojosa, 2008; Kim, 2020);

$$GGBS_\Upsilon(S) = \arg\max_{x \in S} G_\Upsilon(x) = \arg\max_{x \in S} \sum_{i=1}^{n} (\Upsilon_i \times x_{(i)}), \text{ s.t.}, \Upsilon \in \Delta^n \tag{10.4.2}$$

Now, the idea of *CBS* is explained. Let V be a monotonic real-valued set function on 2^N with $\mathscr{V}(\emptyset) = 0$. It is supposed that V is 1-normalized, that is, $\mathscr{V}(N) = 1$, and it is monotonic if $\mathscr{V}(K) < \mathscr{V}(L)$ for all $K \subset L$. Associated to the monotonic feature, the *Choquet* social welfare function is defined as (Hinojosa, 2007, Hinojosa, 2008; Kim, 2020; Oka, 2000);

$$\mathscr{W}_{\mathscr{V}}(x) = \sum_{i=1}^{n} \left([\mathscr{V}(\{(i),(i+1),\ldots,(n)\}) - \mathscr{V}(\{(i+1),\ldots,(n)\})] \times x_{(i)} \right) \tag{10.4.3}$$

where $(x_{(1)}, x_{(2)}, \ldots, x_{(n)})$ denotes a rank-ordered permutation of $\mathbf{x} \in \mathbb{R}^n_+$, that is $x_{(1)} \leq x_{(2)} \leq, \ldots, \leq x_{(n)}$. For each $S \in \sum$, the *CBS*,$\mathscr{C}_{\mathscr{V}}(S)$,is defined as follows (Hinojosa, 2007, Hinojosa, 2008; Kim, 2020; Oka, 2000);

$$\mathscr{C}_{\mathscr{V}}(S) = \arg \max_{x \in S} \mathscr{W}_{\mathscr{V}}(x) \tag{10.4.4}$$

10.4.4 Dual-phase bargaining model for the CWBAN system

In modern ubiquitous healthcare environments, a new architectural element of cloud computing, known as the cloudlet, extends traditional cloud infrastructure. A cloudlet is a mobility-enhanced, small-scale cloud server located at the edge of the Internet. By being closer to the end devices, the cloudlet provides powerful computing resources to WBANs while significantly reducing latency. In the *BSRA* scheme, the cloud-assisted WBAN (CWBAN) system is designed to utilize cloudlets for enhanced performance. Each individual WBAN is equipped with a Central Processing Coordinator (CPC), which acts as the interface between the WBAN and the cloudlet C. The CPC collects vital physiological data from the sensors within its WBAN and coordinates with the C for further processing and resource allocation (Kim, 2020).

The dual-phase bargaining game model incorporates two distinct resource allocation algorithms. In the initial phase, the *BSRA* scheme considers individual CPCs as game players, each striving to maximize their respective payoffs through the allocation of communication resources. The second phase of the *BSRA* scheme primarily focuses on optimizing the distribution of computational resources among offloaded tasks. To ensure the efficient utilization of the limited resources available in the CWBAN system, the communication and computation resource allocation processes are structured based on the principles of *GGBS* and *CBS*. Given the dynamic nature of CWBAN system environments, this approach is well-suited for practical and adaptive WBAN system operations. To obtain a fair-efficient solution, the allocation decision process is formulated as a cooperative bargaining game

$$\mathbb{G} = \left\{\mathbb{T} = \{\Gamma_1, \ldots, \Gamma_n\}, \mathbb{P}_\Gamma = \left\{\mathscr{P}^\Gamma_1, \mathscr{P}^\Gamma_2, \ldots, \mathscr{P}^\Gamma_m\right\}, \left\{U_\Gamma, \mathfrak{U}_{\mathscr{V}}\right\}, \mathbb{A}_\Gamma = \left\{\ldots \mathscr{A}^\Gamma_k \ldots\right\}, \left\{\mathscr{C}, \mathfrak{P}_{\mathscr{C}}, \mathfrak{C}_{\mathscr{C}}\right\}, \mathbb{S} = \{x, y\}\right\}$$

(Kim, 2020):

- T is the set of CPCs where $\Gamma_{1 \leq i \leq n}$ is the CPC of ith WBAN.
- $\mathbb{P}_\Gamma$ is the set of Γ's possible price policies for CWBAN services where $\mathscr{P}^\Gamma_{1 \leq j \leq m}$ represents the jth price policy of Γ to get the communication and computation resources.
- U_Γ is the Γ's utility function, and $\mathfrak{U}_{\mathscr{V}}$ is the social welfare function. Each function maps player's satisfaction to a real number, which represents the resulting payoff in the game $\mathbb{G}$.
- $\mathbb{A}_\Gamma$ is the set of Γ's application tasks, which are generated from the Γ's WBAN, and ask for offloading services.
- The cloudlet C has its communication resources, i.e., $\mathfrak{P}_{\mathscr{C}}$, and computation capacity, i.e., $\mathfrak{C}_{\mathscr{C}}$, to provide offload services.
- Bargaining solution set S consists of two n-dimensional vectors x and y, which correspond to the communication and computation resource allocations for each Γ.

At the first phase, the $\mathfrak{P}_{\mathscr{C}}$ of CWBAN system is adaptively allocated to each CPC according to the idea of *GGBS*. To quantify the communication service satisfaction, the payoff of player Γ_i with the $\mathbb{P}_{\Gamma_i}$ for the $\mathbb{A}_{\Gamma_i}$ $\left(U_{\Gamma_i}(\mathbb{P}_{\Gamma_i}, \mathbb{A}_{\Gamma_i})\right)$ can be derived as follows (Kim, 2020).

$$U_{\Gamma_i}\left(\mathbb{P}_{\Gamma_i}, \mathbb{A}_{\Gamma_i}\right) = x_{\Gamma_i} \times \left(\tilde{\mathbb{A}} - \frac{1}{exp\left(\mu \times \frac{x_{\Gamma_i}}{\mathfrak{M}\left(\mathbb{A}_{\Gamma_i}\right)}\right)}\right)^{\mathscr{X}} \tag{10.4.5}$$

$$\text{s.t.,} \begin{cases} M\left(\mathbb{A}_{\Gamma_i}\right) = \sum_{\mathscr{A}_k^{\Gamma_i} \in \mathbb{A}_{\Gamma_i}} \mathscr{R}_{\mathfrak{P}_{\mathscr{C}}}\left(\mathscr{A}_k^{\Gamma_i}\right) \\ X = \beta + \log\left(1 + \left(\frac{\sum_{\mathscr{A}_k^{\Gamma_i} \in \mathbb{A}_{\Gamma_i}} \mathscr{S}_{\mathbb{P}_{\Gamma_i}}\left(\mathscr{A}_k^{\Gamma_i}\right)}{\left|\mathbb{A}_{\Gamma_i}\right|}\right)\right) \end{cases}$$

where μ, β are the control parameters for the $U_{\Gamma_i}\left(\mathbb{P}_{\Gamma_i}, \mathbb{A}_{\Gamma_i}\right)$ and x_{Γ_i} is the total amount of allocated communication resource for the Γ_i. $\mathscr{R}_{\mathfrak{P}_{\mathscr{C}}}\left(\mathscr{A}_k^{\Gamma_i}\right)$ is the requested amount of communication resources for the $\mathscr{A}_k^{\Gamma_i}$, and $\mathscr{S}_{\mathbb{P}_{\Gamma_i}}\left(\mathscr{A}_k^{\Gamma_i}\right)$ is the price policy of the $\mathscr{A}_k^{\Gamma_i}$. $\left|\mathbb{A}_{\Gamma_i}\right|$ is the cardinality of the set $\mathbb{A}_{\Gamma_i}$ (Kim, 2020).

Each CPC periodically reports its $\mathfrak{M}\left(\mathbb{A}_{\Gamma}\right)$ and X information to the cloudlet server C. At this time, the communication resource, i.e., wireless bandwidth, should be dynamically allocated to the corresponding CPCs. The *BSRA* scheme adopts the method of *GGBS* to solve the bandwidth allocation problem in the CWBAN system. To obtain the *GGBS*, the *BSRA* scheme should define the permutation vector $x_{\eth}$ and the simplex A in $\mathbb{R}_+^n$. In the *BSRA* scheme, the order of $x_{(i)}$ in $x_{\eth}$ and the order of $\mathscr{A}_i$ in A are decided based on each Γ's quantitative concern and qualitative attention. The rationale behind this decision mechanism is a mix of the egalitarian and utilitarian principles. Therefore, the combination of $x_{\eth}$ and A are inequality averse but include weighted utilitarian criteria to achieve compromise bargaining solution. Based on the amount of requested communication resources and the price policy, the relative importance of the Γ_i $\left(I_{\Gamma_i}\right)$ can be defined as follows (Kim, 2020);

$$I_{\Gamma_i} = \log\left(1 + \frac{m_{\Gamma_i}}{M_{\Gamma}}\right), \text{ s.t.,} \begin{cases} m_{\Gamma_i} = \sum_{\mathscr{A}_k^{\Gamma_i} \in \mathbb{A}_{\Gamma_i}} \left(\mathscr{S}_{\mathbb{P}_{\Gamma_i}}\left(\mathscr{A}_k^{\Gamma_i}\right) \times \mathscr{R}_{\mathfrak{P}_{\mathscr{C}}}\left(\mathscr{A}_k^{\Gamma_i}\right)\right) \\ M_{\Gamma} = \max_{\Gamma_i \in \mathbb{T}} m_{\Gamma_i} \end{cases} \tag{10.4.6}$$

According to (10.4.2), the *BSRA* scheme can get the simplex vector V_{Γ} as follows;

$$V_{\Gamma} = \left(v_{\Gamma_1}, v_{\Gamma_2}, \ldots, v_{\Gamma_n}\right), \text{ s.t., } v_{\Gamma_i} = \frac{I_{\Gamma_i}}{\sum_{\Gamma_i \in \mathbb{T}} I_{\Gamma_i}} \tag{10.4.7}$$

Using the vector V_{Γ}, the *BSRA* scheme defines the vector $\Upsilon_{\Gamma} = \left\{\Upsilon_1, \Upsilon_2, \ldots, \Upsilon_n\right\}$ based on the one-to-one mapping with the V_{Γ} while satisfying the order $\Upsilon_1 \geq \Upsilon_2 \geq \ldots \geq \Upsilon_n > 0$. The π-permutation of vector $x_{\eth(\Gamma)}$ is obtained based on the m_{Γ_i} value of each Γ. Therefore, $x_{\pi(\Gamma)} = \left(x_{(\Gamma_1)}, x_{(\Gamma_2)}, \ldots, x_{(\Gamma_n)}\right)$ denotes a rank-ordered permutation vector of $x_{\pi(\Gamma)} \in \mathbb{R}_+^n$ where $x_{(\Gamma_1)} \leq x_{(\Gamma_2)} \leq, \ldots, \leq x_{(\Gamma_n)}$. Based on the I_Γ and $x_{\eth(\Gamma)}$, the *GGBS* for each $S \in \sum$ is finally given by (Kim, 2020).

$$GGBS_{I_\Gamma}(S) = \arg\max_{x \in S} \sum_{i=1}^{n} \left(\Upsilon_i \times x_{(\Gamma_i)}\right), \text{ s.t., } I_\Gamma \in \ddot{A}^n$$

(10.4.8)

Like communication resources, the limited computation resources of C can negatively impact the CWBAN performance. Hence, the C needs a viable solution to effectively distribute the cloud computing power to the offloaded tasks. The *BSRA* scheme adopts the method of *CBS* to solve this computation resource allocation problem in the CWBAN system. The *BSRA* scheme assumes that each Γ's computation tasks can be offloaded to the C. At this time, the C should distribute its computation power $\mathfrak{C}_{\mathscr{C}}$ to each Γ while considering the system efficiency and fairness. Based on the amount of requested computation resource and the price policy, the *BSRA* scheme defines the 1-normalized monotonic set function V as follows (Kim, 2020).

$$\mathscr{V}\left(\{\Gamma_i, \Gamma_{i+1}, \ldots, \Gamma_n\}\right) = \ln\left(\eta + \left(\mathfrak{X}(\Gamma_i)/\mathfrak{Z}(\mathbb{T})\right)\right)^{\psi}$$

(10.4.9)

$$\text{s.t.,}\begin{cases} X(\Gamma_i) = \sum_{j=i}^{n} \sum_{\mathscr{A}_k^{\Gamma} \in \mathbb{A}_{\Gamma_j}} r_{\Gamma_j}\left(\mathbb{A}_{\Gamma_j}\right) and Z(\mathbb{T}) = \sum_{j=1}^{n} \sum_{\mathscr{A}_k^{\Gamma} \in \mathbb{A}_{\Gamma_j}} r_{\Gamma_j}\left(\mathbb{A}_{\Gamma_j}\right) \\ \mathfrak{r}_{\Gamma_j}\left(\mathbb{A}_{\Gamma_j}\right) = \mathscr{S}_{\mathrm{P}_{\Gamma_j}}\left(\mathscr{A}_k^{\Gamma_j}\right) \times \mathscr{R}_{\mathfrak{C}_{\mathscr{C}}}\left(\mathscr{A}_k^{j}\right) \end{cases}$$

where η, ψ are the control parameters for the $\mathscr{V}\left(\{\Gamma_i, \Gamma_{i+1}, \ldots, \Gamma_n\}\right)$ and $\mathscr{R}_{\mathfrak{C}_{\mathscr{C}}}\left(\mathscr{A}_k^{j}\right)$ is the requested amount of computing power for the $\mathscr{A}_k^{\Gamma_i}$. It is assumed that $\mathscr{V}(\varnothing)=0$ and $\mathscr{V}(N) = 1$. Based on the set function V, the *Choquet* social welfare function $\mathfrak{U}_{\mathscr{V}}(y)$ can be defined as follows (Kim, 2020);

$$\mathfrak{U}_{\mathscr{V}}(y) = \sum_{i=1}^{n} \left(\left[\mathscr{V}\left(\{(\Gamma_i), (\Gamma_{i+1}), \ldots, \Gamma_n\}\right) - \left(\{(\Gamma_{i+1}), \ldots, \Gamma_n\}\right)\right] \times y_{(i)}\right)$$

(10.4.10)

The π-permutation of vector $y_{\eth(\Gamma)}$ is obtained based on the $r_{\Gamma_j}\left(\mathbb{A}_{\Gamma_j}\right)$ value of each Γ. Therefore, The π-permutation of vector $y_{\eth(\Gamma)} = \left(y_{(\Gamma_1)}, y_{(\Gamma_2)}, \ldots, y_{(\Gamma_n)}\right)$ also denotes the rank-ordered permutation vector of $y_{\eth(\Gamma)} \in \mathbb{R}_+^n$ where $y_{(\Gamma_1)} \leq y_{(\Gamma_2)} \leq, \ldots, \leq y_{(\Gamma_n)}$. Finally, the *CBS*, $\mathscr{C}_{\mathscr{V}}(S)$, is given by (Kim, 2020);

$$\mathscr{C}_{\mathscr{V}}(S) = \arg\max_{y \in S} \mathfrak{U}_{\mathscr{V}}(x)$$

(10.4.11)

10.4.5 Summary

Cloudlet technology plays a crucial role in enhancing the performance of many mobile applications. Its seamless integration with WBANs opens up significant opportunities for the advancement of future network systems. However, limited research has been conducted on cloudlet placement in WBANs, and studies on CWBANs remain sparse. To address the CWBAN resource allocation problem, the *BSRA* scheme introduces a dual-phase bargaining game model. This model employs; i) the *GGBS* for communication resource allocation, leveraging its mix of egalitarian and utilitarian principles, ii) the *CBS* for computation resource allocation, ensuring adaptability to computation-intensive tasks. These two bargaining processes are executed sequentially, enabling the tuning of CWBAN resources while achieving a balanced trade-off between fairness and efficiency. This dual-phase approach ensures an effective allocation of resources to meet the demands of real-time, critical healthcare applications (Kim, 2020).

REFERENCES

Ahn, J., Shin, D., Kim, K., & Yang, J. (2017). Indoor Air Quality Analysis Using Deep Learning with Sensor Data. *Sensors (Basel), 17*(11), 2476. DOI: 10.3390/s17112476 PMID: 29143797

Blackorby, C., Bossert, W., & Donaldson, D. (1994). Generalized Ginis and cooperative bargaining solutions. *Econometrica, 62*(5), 1161–1178. DOI: 10.2307/2951511

Brandenburger, A., & Stuart, H. (2007). Biform games. *Management Science, 53*(4), 537–549. DOI: 10.1287/mnsc.1060.0591

Cao, H., Leung, V., Chow, C., & Chan, H. (2009). Enabling technologies for wireless body area networks: A survey and outlook. *IEEE Communications Magazine, 47*(12), 84–93. DOI: 10.1109/MCOM.2009.5350373

Cicioğlu, M., & Çalhan, A. (2020). Energy-efficient and SDN-enabled routing algorithm for wireless body area networks. *Computer Communications, 160*, 228–239. DOI: 10.1016/j.comcom.2020.06.003

Cicioğlu, M., & Çalhan, A. (2021). Energy efficiency solutions for IEEE 802.15. 6 based wireless body sensor networks. *Wireless Personal Communications, 119*(2), 1499–1513. DOI: 10.1007/s11277-021-08292-8

Claus, C., & Boutilier, C. (1998). The dynamics of reinforcement learning in cooperative multiagent systems. *AAAI/IAAI, 1998*(746-752), 2.

de Cote, E. M., Lazaric, A., & Restelli, M. (2006, May). Learning to cooperate in multi-agent social dilemmas. In *Proceedings of the fifth international joint conference on Autonomous agents and multi-agent systems* (pp. 783-785). DOI: 10.1145/1160633.1160770

Demner-Fushman, D., Antani, S., Simpson, M., & Thoma, G. R. (2012). Design and development of a multimodal biomedical information retrieval system. *Journal of Computing Science and Engineering : JCSE, 6*(2), 168–177. DOI: 10.5626/JCSE.2012.6.2.168

Dhillon, A. (1998). Extended Pareto rules and relative utilitarianism. *Social Choice and Welfare, 15*(4), 521–542. DOI: 10.1007/s003550050121

Gao, T., Massey, T., Selavo, L., Crawford, D., Chen, B. R., Lorincz, K., Shnayder, V., Hauenstein, L., Dabiri, F., Jeng, J., Chanmugam, A., White, D., Sarrafzadeh, M., & Welsh, M. (2007). The advanced health and disaster aid network: A light-weight wireless medical system for triage. *IEEE Transactions on Biomedical Circuits and Systems, 1*(3), 203–216. DOI: 10.1109/TBCAS.2007.910901 PMID: 23852414

Gatouillat, A., Badr, Y., Massot, B., & Sejdić, E. (2018). Internet of medical things: A review of recent contributions dealing with cyber-physical systems in medicine. *IEEE Internet of Things Journal, 5*(5), 3810–3822. DOI: 10.1109/JIOT.2018.2849014

Ghubaish, A., Salman, T., Zolanvari, M., Unal, D., Al-Ali, A., & Jain, R. (2020). Recent advances in the internet-of-medical-things (IoMT) systems security. *IEEE Internet of Things Journal, 8*(11), 8707–8718. DOI: 10.1109/JIOT.2020.3045653

Hassan, M. M., Albakr, H., Al-Dossari, H., & Mohamed, A. (2017). Resource provisioning for cloud-assisted body area network in a smart home environment. *IEEE Access : Practical Innovations, Open Solutions*, *5*, 13213–13224. DOI: 10.1109/ACCESS.2017.2726012

He, D., Zeadally, S., & Wu, L. (2015). Certificateless public auditing scheme for cloud-assisted wireless body area networks. *IEEE Systems Journal*, *12*(1), 64–73. DOI: 10.1109/JSYST.2015.2428620

Hinojosa, M. Á., Mármol, A. M., & Zarzuelo, J. M. (2007). *Multi-utilitarian bargaining solutions* (No. 07.13).

Hinojosa, M. A., Mármol, A. M., & Zarzuelo, J. M. (2008). Inequality averse multi-utilitarian bargaining solutions. *International Journal of Game Theory*, *37*(4), 597–618. DOI: 10.1007/s00182-008-0137-2

Hu, Y., Gao, Y., & An, B. (2014). Multiagent reinforcement learning with unshared value functions. *IEEE Transactions on Cybernetics*, *45*(4), 647–662. DOI: 10.1109/TCYB.2014.2332042 PMID: 25014990

Jang, I., Pyeon, D., Kim, S., & Yoon, H. (2013). A survey on communication protocols for wireless sensor networks. *Journal of Computing Science and Engineering : JCSE*, *7*(4), 231–241. DOI: 10.5626/JCSE.2013.7.4.231

Jiang, T., Zhao, Q., Grace, D., Burr, A. G., & Clarke, T. (2016). Single-state Q-learning for self-organised radio resource management in dual-hop 5G high capacity density networks. *Transactions on Emerging Telecommunications Technologies*, *27*(12), 1628–1640. DOI: 10.1002/ett.3019

Kapetanakis, S., & Kudenko, D. (2002). Reinforcement learning of coordination in cooperative multi-agent systems. In *AAAI'2002* (pp. 326-331).

Karagözoğlu, E., Keskin, K., & Özcan-Tok, E. (2019). Between anchors and aspirations: A new family of bargaining solutions. *Review of Economic Design*, *23*(1-2), 53–73. DOI: 10.1007/s10058-019-00220-4

Kazemi, R., Vesilo, R., Dutkiewicz, E., & Fang, G. (2010, October). Inter-network interference mitigation in wireless body area networks using power control games. In *2010 10th International Symposium on Communications and Information Technologies* (pp. 81-86). IEEE. DOI: 10.1109/ISCIT.2010.5664908

Kim, S. (2011). Adaptive online power control scheme based on the evolutionary game theory. *IET Communications*, *5*(18), 2648–2655. DOI: 10.1049/iet-com.2011.0093

Kim, S. (2012). Biform game based cognitive radio scheme for smart grid communications. *Journal of Communications and Networks (Seoul)*, *14*(6), 614–618. DOI: 10.1109/JCN.2012.00027

Kim, S. (Ed.). (2014). *Game theory applications in network design*. IGI Global. DOI: 10.4018/978-1-4666-6050-2

Kim, S. (2019a). New bargaining game based computation offloading scheme for flying ad-hoc networks. *IEEE Access : Practical Innovations, Open Solutions*, *7*, 147038–147047. DOI: 10.1109/ACCESS.2019.2946363

Kim, S. (2019b). Reverse-biform game based resource sharing scheme for wireless body area networks. *International Journal of Ad Hoc and Ubiquitous Computing*, *31*(4), 219–229. DOI: 10.1504/IJAHUC.2019.10022741

Kim, S. (2020). Bargaining Solution-Based Resource Allocation Scheme for Cloud-Assisted Wireless Body Area Networks. *Wireless Communications and Mobile Computing*, *2020*(1), 5470428. DOI: 10.1155/2020/5470428

Kim, S. (2021). A New Two-Stage Bargaining Game Approach for Intra-and Inter-WBAN Management. *Mobile Information Systems*, *2021*(1), 5798741. DOI: 10.1155/2021/5798741

Kim, S. (2023). Learning and game based spectrum allocation model for internet of medical things (IoMT) platform. *IEEE Access : Practical Innovations, Open Solutions*, *11*, 48059–48068. DOI: 10.1109/ACCESS.2023.3266331

Liu, Z., Liu, B., & Chen, C. W. (2017). Transmission-rate-adaption assisted energy-efficient resource allocation with QoS support in WBANs. *IEEE Sensors Journal*, *17*(17), 5767–5780. DOI: 10.1109/JSEN.2017.2725906

Liu, Z., Liu, B., & Chen, C. W. (2018). Joint power-rate-slot resource allocation in energy harvesting-powered wireless body area networks. *IEEE Transactions on Vehicular Technology*, *67*(12), 12152–12164. DOI: 10.1109/TVT.2018.2872960

Milenković, A., Otto, C., & Jovanov, E. (2006). Wireless sensor networks for personal health monitoring: Issues and an implementation. *Computer Communications*, *29*(13-14), 2521–2533. DOI: 10.1016/j.comcom.2006.02.011

Misra, S., Moulik, S., & Chao, H. C. (2015). A cooperative bargaining solution for priority-based data-rate tuning in a wireless body area network. *IEEE Transactions on Wireless Communications*, *14*(5), 2769–2777. DOI: 10.1109/TWC.2015.2393303

Misra, S., & Samanta, A. (2018). Traffic-aware efficient mapping of wireless body area networks to health cloud service providers in critical emergency situations. *IEEE Transactions on Mobile Computing*, *17*(12), 2968–2981. DOI: 10.1109/TMC.2018.2822279

Misra, S., & Sarkar, S. (2014). Priority-based time-slot allocation in wireless body area networks during medical emergency situations: An evolutionary game-theoretic perspective. *IEEE Journal of Biomedical and Health Informatics*, *19*(2), 541–548. DOI: 10.1109/JBHI.2014.2313374 PMID: 24686307

Morozs, N., Clarke, T., & Grace, D. (2015). Distributed heuristically accelerated Q-learning for robust cognitive spectrum management in LTE cellular systems. *IEEE Transactions on Mobile Computing*, *15*(4), 817–825. DOI: 10.1109/TMC.2015.2442529

Movassaghi, S., Abolhasan, M., Lipman, J., Smith, D., & Jamalipour, A. (2014). Wireless body area networks: A survey. *IEEE Communications Surveys and Tutorials*, *16*(3), 1658–1686. DOI: 10.1109/SURV.2013.121313.00064

Ning, Z., Dong, P., Wang, X., Hu, X., Guo, L., Hu, B., Guo, Y., Qiu, T., & Kwok, R. Y. (2020). Mobile edge computing enabled 5G health monitoring for Internet of medical things: A decentralized game theoretic approach. *IEEE Journal on Selected Areas in Communications*, *39*(2), 463–478. DOI: 10.1109/JSAC.2020.3020645

Oka, E. A., & Zhou, L. (2000). The Choquet bargaining solutions. *Games and Economic Behavior*, *33*(2), 249–264. DOI: 10.1006/game.1999.0778

Pan, J., & McElhannon, J. (2017). Future edge cloud and edge computing for internet of things applications. *IEEE Internet of Things Journal*, *5*(1), 439–449. DOI: 10.1109/JIOT.2017.2767608

Qiao, H., Rozenblit, J., Szidarovszky, F., & Yang, L. (2006, December). Multi-agent learning model with bargaining. In *Proceedings of the 2006 winter simulation conference* (pp. 934-940). IEEE. DOI: 10.1109/WSC.2006.323178

Qiu, Y., Haley, D., Chan, T., & Davis, L. (2016, January). Game theoretic framework for studying WBAN coexistence: 2-Player game analysis and n-player game estimation. In *2016 Australian Communications Theory Workshop (AusCTW)* (pp. 53-58). IEEE. DOI: 10.1109/AusCTW.2016.7433609

Quwaider, M., & Jararweh, Y. (2014, April). An efficient big data collection in body area networks. In *2014 5th international conference on information and communication systems (ICICS)* (pp. 1-6). IEEE. DOI: 10.1109/IACS.2014.6841986

Raazi, S. M. K. U. R., & Lee, S. (2010). A survey on key management strategies for different applications of wireless sensor networks. *Journal of Computing Science and Engineering : JCSE*, *4*(1), 23–51. DOI: 10.5626/JCSE.2010.4.1.023

Salem, O., Liu, Y., & Mehaoua, A. (2013). Anomaly detection in medical wireless sensor networks. *Journal of Computing Science and Engineering : JCSE*, *7*(4), 272–284. DOI: 10.5626/JCSE.2013.7.4.272

Samanta, A., Bera, S., & Misra, S. (2015). Link-quality-aware resource allocation with load balance in wireless body area networks. *IEEE Systems Journal*, *12*(1), 74–81. DOI: 10.1109/JSYST.2015.2458586

Sobel, J. (2001). Manipulation of preferences and relative utilitarianism. *Games and Economic Behavior*, *37*(1), 196–215. DOI: 10.1006/game.2000.0839

Sodagari, S., Bozorgchami, B., & Aghvami, H. (2018). Technologies and challenges for cognitive radio enabled medical wireless body area networks. *IEEE Access : Practical Innovations, Open Solutions*, *6*, 29567–29586. DOI: 10.1109/ACCESS.2018.2843259

Son, S., So, H., Kim, J., Choi, D., & Lee, H. J. (2016). Energy-efficient adaptive optical character recognition for wearable devices. *Electronics Letters*, *52*(2), 113–115. DOI: 10.1049/el.2015.2959

Stuart, H. W.Jr. (2005). Biform analysis of inventory competition. *Manufacturing & Service Operations Management*, *7*(4), 347–359. DOI: 10.1287/msom.1050.0090

Thomson, W. (2022). On the axiomatic theory of bargaining: A survey of recent results. *Review of Economic Design*, *26*(4), 491–542. DOI: 10.1007/s10058-022-00319-1

Wan, J., Zou, C., Ullah, S., Lai, C. F., Zhou, M., & Wang, X. (2013). Cloud-enabled wireless body area networks for pervasive healthcare. *IEEE Network*, *27*(5), 56–61. DOI: 10.1109/MNET.2013.6616116

Wang, D., Zhang, W., Song, B., Du, X., & Guizani, M. (2019). Market-based model in CR-IoT: A Q-probabilistic multi-agent reinforcement learning approach. *IEEE Transactions on Cognitive Communications and Networking*, *6*(1), 179–188. DOI: 10.1109/TCCN.2019.2950242

Yuan, X., Tian, H., Wang, H., Su, H., Liu, J., & Taherkordi, A. (2020). Edge-enabled wbans for efficient qos provisioning healthcare monitoring: A two-stage potential game-based computation offloading strategy. *IEEE Access : Practical Innovations, Open Solutions*, *8*, 92718–92730. DOI: 10.1109/ACCESS.2020.2992639

Zhang, B., & Zhang, Y. (2018). An individual differentiated coexisting mechanism for multiple wireless body area networks based on game theory. *IEEE Access : Practical Innovations, Open Solutions*, *6*, 54564–54581. DOI: 10.1109/ACCESS.2018.2872746

Zhang, T., Liu, M., Yuan, T., & Al-Nabhan, N. (2020). Emotion-aware and intelligent internet of medical things toward emotion recognition during COVID-19 pandemic. *IEEE Internet of Things Journal*, *8*(21), 16002–16013. DOI: 10.1109/JIOT.2020.3038631 PMID: 35782178

Zhao, X., Liu, B., Chen, C., & Chen, C. W. (2015, December). QoS-driven power control for inter-WBAN interference mitigation. In *2015 IEEE global communications conference (GLOBECOM)* (pp. 1-6). IEEE. DOI: 10.1109/GLOCOM.2015.7417058

Zhou, J., Guo, A., Xu, J., Nguyen, H., & Su, S. (2014, September). A game theory control scheme in medium access for wireless body area network. In *10th International Conference on Wireless Communications, Networking and Mobile Computing (WiCOM 2014)* (pp. 404-409). IET.

Chapter 11
New Technologies for Future Communications

ABSTRACT

As new communications technologies are developed and brought to market, it is vital for communications professionals and engineers to keep abreast of these issues. Since the technologies and applications also affect our daily lives, it is important to understand how they will affect the 6G networks and communication infrastructures. This chapter examines new technologies for 6G communications, and highlights their fundamentals, current mathematical models and system designs to make these technologies more practical in future networks. Especially, this chapter covers various topics ranging from energy transferring to multiple access techniques.

11. OVERVIEW

Due to the limitation of WiFi coverage, WiFi tethering is a useful and economic means of providing on-the-go mobile devices' Internet connection. With WiFi tethering technology, mobile devices can go online from their WiFi-enabled tablets or laptops even when no public hotspots are available. By enabling revenue sharing between the network operators and the sponsors, the sponsored data has been proven to be a promising solution. This technique is becoming a ubiquitous trend in the 5G networks for improving data connectivity while increasing mobile engagement for the sponsors and ensuring revenue for the network operators. Wireless energy transferring technology is emerging as a viable option for recharging the sensing nodes wirelessly. Therefore, Simultaneous wireless information and power transfer (SWIPT) method is to integrate the wireless power transfer along with near-field magnetic communication. It is a promising technology to provide cost-effective information and energy access to wireless devices in various applications in future.

With the rapid growth of big data traffic, smart home is facing challenges from security, home network and data usage. Generally, the smart home network refers to the modern home automation system while enabling the interconnectivity and control of different smart devices. Rate splitting multiple access (RSMA) technology has been taken as one of the promising solutions to improve the transmission efficacy due to its capability of interference management. It has a variety of advantages, including higher throughput, enhanced capacity, and network efficiency, which makes it suitable for beyond 6G networks. User-centric cell-free network architecture defines a cooperative serving cluster of spatially distributed

DOI: 10.4018/979-8-3373-2013-7.ch011

transmitters specifically for smart devices. To meet the unprecedented mobile traffic demands of future wireless networks, user-centric cell-free network represents a practical and scalable embodiment of distributed network systems. Visible light communication (VLC) is a wireless method that enables high-speed transmission of data with visible light. It has attracted significant attention due to its dual functionalities including illumination and high transmission rate.

11.1 The Wireless Energy Harvesting Control (*WEHC*) Scheme

Given the nature of radio frequency-based energy harvesting technology, numerous smart devices rely on energy-constrained relays, which must obtain energy from the hybrid access point (HAP). Recently, S. Kim introduced the *Wireless Energy Harvesting Control* (*WEHC*) scheme (Kim, 2021a). The *WEHC* scheme aims to develop an innovative energy harvesting mechanism by integrating simultaneous wireless information and power transfer (SWIPT) with a wireless powered communication network (WPCN) to enhance dual-hop wireless communication. The *WEHC* scheme employs a twofold bargaining game model to dynamically regulate control decisions in response to competing demands. By utilizing two distinct bargaining solutions, the *WEHC* scheme identifies the inherent conflict between energy harvesting and information transmission, leading to the development of two time-sharing strategies. These strategies operate iteratively in a compact manner to optimize the performance of the dual-hop communication system (Kim, 2021a).

11.1.1 Development Motivation

The IoT represents a fundamental paradigm in next-generation wireless communication systems. With the emergence of the IoT concept, innovative applications such as smart mirrors, robotic vacuum cleaners, and wireless kitchen appliances have been introduced, fundamentally transforming daily life through Internet-enabled smart devices. In the forthcoming era of 6G communication, advancements in smart homes, pollution monitoring, energy conservation, intelligent transportation, and industrial automation are reshaping traditional lifestyles into a high-tech, interconnected environment. Presently, there are millions of IoT devices, with an estimated annual growth rate of approximately 25%, and projections indicate that this number will rise to 80 billion by 2030. The evolution of smart IoT technologies is expected to redefine the way we perceive the Internet and technological advancements within 6G networks (Chettri, 2019; Kim, 2021a; Kumar, 2019; Yang, 2018b).

Ongoing progress in IoT technology has garnered significant interest from researchers and developers worldwide. However, various challenges must be overcome to fully realize the potential impact of IoT. Notably, certain wireless devices operate on batteries with limited lifespans, making their longevity dependent on battery performance. Although replacing or recharging batteries extends device functionality, this process can be inconvenient, costly, hazardous in toxic environments, or even unfeasible. These challenges have stimulated interest in alternative solutions, particularly wireless energy transfer technology, which has recently gained widespread attention due to its potential to provide continuous power to energy-limited devices. This technology enables smart devices to harvest energy from wireless signals, thereby extending their operational lifetime for designated applications (Ju, 2013; Kim, 2021a; Kumar, 2019).

Among the various energy transfer techniques, radio frequency (RF)-based technology offers a viable long-range energy transfer alternative compared to other methods such as inductive coupling and magnetic resonance coupling. One key advantage of RF signals is their dual function of carrying both information and energy, facilitating RF-enabled wireless-powered communications. Additionally, recent advancements in high-efficiency rectifying antennas are expected to improve the effectiveness of wireless energy harvesting from RF signals. As a result, RF-based energy harvesting presents significant potential for replacing small batteries in low-power electronic devices and systems (Ju, 2013; Kim, 2021a; Ma, 2015).

Research in RF energy harvesting is generally categorized into two main areas: Simultaneous Wireless Information and Power Transfer (SWIPT) and Wireless Powered Communication Networks (WPCN). SWIPT enables IoT devices to receive partition signals, allocating one portion for data transmission and the other for energy harvesting. This approach allows devices to extract energy while simultaneously processing information from the same signal. In contrast, WPCN involves energy-limited IoT devices initially harvesting energy from RF signal transmissions and subsequently utilizing this energy for communication purposes. In this model, IoT devices operate based on a 'harvest-then-transmit' protocol. Recently, SWIPT and WPCN have been integrated to enhance the efficiency of RF energy harvesting technologies (Kim, 2021a; Luy, 2019; Ramezani, 2017; Yu, 2019).

Despite considerable research efforts in wireless energy technologies, many existing studies have not adequately addressed cooperative control strategies within a coordinated framework. The *WEHC* scheme integrates SWIPT and WPCN within a unified platform (S-W joint platform) to optimize wireless communication and energy harvesting for point-to-point systems. By refining the conventional harvest-then-transmit protocol, different control mechanisms are applied for uplink and downlink communication. Specifically, the WPCN is utilized for downlink transmission, whereas the SWIPT is employed for uplink communication. Consequently, the *WEHC* scheme necessitates a time adjustment mechanism to balance varying demands. Addressing this issue requires the development of an effective control algorithm capable of dynamically managing competing requirements, a task that presents notable complexity and challenges (Kim, 2014).

A cooperative bargaining game is a strategic process in which participants determine how to distribute a collectively generated surplus (Kim, 2014). One of the earliest models incorporating reference points in bargaining theory is the *Reference Aspiration Bargaining Situation* (*RABS*). The *RABS* is formally defined as the maximum feasible point along the line connecting the aspiration point and disagreement point, where the aspiration point originates from the reference point rather than the disagreement point. Essentially, the *RABS* selects the maximum individual payoff while ensuring that each player's payoff gain from the disagreement point maintains the same proportion relative to the difference between their aspiration and disagreement points. Within the *RABS*, the disagreement point serves as a strategic threat in negotiations while also determining proportional payoffs (Karagözoğlu, 2015; Kim, 2021a)

The *Negotiated Aspiration Bargaining Solution* (*NABS*) is closely related to proportional bargaining models, which identify the optimal feasible point along a trajectory starting from the disagreement point in a direction influenced by an externally determined bargaining weight vector. Specifically, the *NABS* determines the optimal resource allocation along the utopia direction, originating from an endogenous reference point that incorporates both the utopia point and bargaining power. The characterization of *NABS* is derived by relaxing Pareto optimality while incorporating the standard axiom of restricted contraction independence, along with the minimal requirement that utopia should be chosen when feasible (Alós-Ferrer, 2018; Balakrishnan, 2011; Kim, 2021a).

11.1.2 Contribution

By applying the principles of *RABS* and *NABS*, the *WEHC* scheme efficiently manages energy harvesting and dual-hop wireless communications within the S-W joint platform. Taking into account the mutual interactions among multiple devices, the *WEHC* scheme introduces a novel two-stage bargaining game model to regulate the uplink time allocation for each service. In the initial bargaining stage, the uplink time duration assigned to each IoT device is determined based on the *RABS* framework. In the subsequent bargaining stage, the proportion of each relay's energy harvesting and uplink data transmission is dynamically adjusted according to the *NABS* approach. Within this twofold bargaining model, control decisions are strategically made to establish a mutually beneficial agreement among multiple IoT devices. Through an iterative, step-by-step negotiation process, the *WEHC* scheme achieves an optimal balance in dual-hop communications while efficiently harvesting energy, ultimately enhancing overall system throughput (Kim, 2021a).

11.1.3 RF-Enabled Energy Transfer and Twofold Bargaining Model

In traditional wireless communication systems, a key observation is that devices located farther from the hybrid access point (HAP) must be transmitted at higher power levels compared to those positioned closer. This phenomenon occurs due to the significant signal attenuation experienced by distant devices. The resulting fairness issue, known as the doubly near-far problem, arises from variations in the distances between IoT devices and HAP. To address this challenge, energy-constrained relays are introduced, enabling these relays to recharge their batteries by harvesting energy from the RF signals received from HAP. Subsequently, these relays facilitate wireless communication between the HAP and their associated devices through an amplify-and-forward strategy, utilizing the harvested energy. This approach, based on the harvest-then-cooperation concept, is categorized as a dual-hop wireless communication mechanism. To enhance the efficiency of this system, the S-W joint platform has been developed, integrating WPCN for downlink communication and SWIPT for uplink transmission, allowing relays to harvest energy from both HAP and IoT device signals (Kim, 2021a; Yu, 2019).

The *WEHC* scheme considers a dual-hop wireless communication platform consisting of one HAP, n IoT devices $\mathbb{N} = \{\mathcal{D}_1, \ldots, \mathcal{D}_n\}$, and n energy-constrained relays $\mathbb{R} = \{\mathcal{R}_1, \ldots, \mathcal{R}_n\}$, where each agent is equipped with a single antenna to transfer energy and information. The *WEHC* scheme assumes that there are no direct links between the HAP and IoT devices, and relays are responsible for assisting the data transmission of IoT devices and the HAP. To be specific, $\mathcal{R}_i$ is exclusively used for forwarding the data between $\mathcal{D}_i$ and the HAP. The HAP and IoT devices have stable energy supply, but energy-constrained relays need to harvest the energy from RF signals broadcast by the HAP and IoT devices. That is to say, relays harvest their energy through the downlink signal from the HAP and uplink signal from its corresponding IoT device and then retransmit the device's data to the HAP by consuming the harvested energy. The overview of dual-hop wireless communication platforms is shown in Figure 1. (Kim, 2021a; Kyu, 2019; Ramezani, 2017; Yu, 2019).

Figure 1. The dual-hop wireless communication platform (Kim. 2021a)

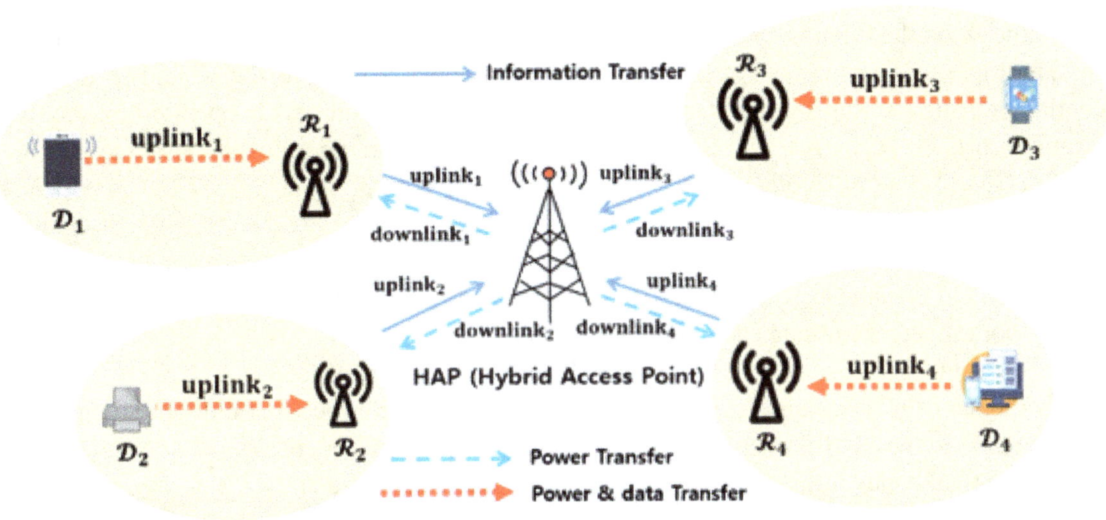

To maximize the performance of dual-hop wireless communication system, the *WEHC* scheme proposes a new RF energy harvesting mechanism based on the S-W joint platform. In the *WEHC* scheme, the control issues of downlink and uplink operations are related to the time-sharing strategy. At time t_c, the allocated time duration in the WPCN downlink (τ_0) is used only for the energy transfer; the HAP broadcasts RF energy signal and the relays harvest their energy through the received signal. On the other hand, the time duration of SWIPT uplink $\left(L_t^{UP}\right)$ is shared by n relays, and the time duration τ_i is allocated to the $\mathcal{R}_i$ where $L_t^{UP} = \sum_{1 \le i \le n} \tau_i$. The τ_i is split by two time sub-slots, i.e., $\left[(1-\rho_i)\cdot\tau_i\right]$ and $\left[\rho_i\cdot\tau_i\right]$; $\left[(1-\rho_i)\cdot\tau_i\right]$ is used to transfer the data and energy from $\mathcal{D}_{1 \le i \le n}$ to $\mathcal{R}_{1 \le i \le n}$, and $\left[\rho_i\cdot\tau_i\right]$ is used to retransfer the data from $\mathcal{R}_{1 \le i \le n}$ to the HAP. To effectively adapt the current dual-hop communication environments, the ρ_i value should be dynamically adjusted to maximize the system throughput. M is the total bandwidth amount in the S-W joint platform. The downlink and uplink frame structure in the S-W joint platform is illustrated in Figure 2 (Kim, 2021a; Kyu, 2019; Ramezani, 2017; Yu, 2019).

Figure 2. The downlink and uplink frame structure in the S-W joint platform (Kim, 2021a)

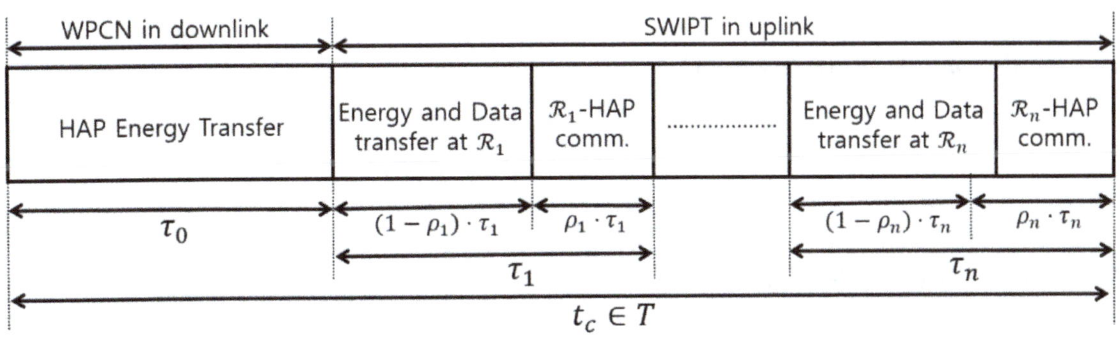

At each time t, the *WEHC* scheme formulates the time-sharing problem as a twofold bargaining game ($\mathbb{G}$) where the G consists of the first ($\mathbb{G}^F$) and second $\left(\mathbb{G}^S_{1\le i\le n}\right)$bargaining games. In the $\mathbb{G}^F$, IoT devices in the N are game players, and the uplink time duration $\left(L^{UP}_t\right)$ is shared through the concept of *RABS*. In the $\mathbb{G}^S_i$, the $\mathcal{R}_i$ in the R adjusts its own ρ_i value based on the idea of *NABS*. For the second bargaining process, multiple $\mathbb{G}^S_{1\le i\le n}$ games are operated in a parallel manner. Therefore, the $\mathbb{G}^F$ and $\mathbb{G}^S_i$ are repeated sequentially during the step-by-step iteration, and the time-sharing process is operated interactively at each time step $t_c \in \{t_1,\dots,t_c,t_{c+1},\dots\}$. Formally, the *WEHC* scheme defines the G game entities, i.e, $\mathbb{G} = \left\{\mathbb{G}^F,\mathbb{G}^S_{1\le i\le n}\right\} = \left\{\mathbb{N},\mathbb{R},\mathfrak{M},\left\{\mathbb{G}^F\middle|\mathbb{N},\tau_{1\le i\le n},U_{\mathcal{D}_{1\le i\le n}}\right\},\left\{\mathbb{G}^S_{1\le i\le n}\middle|\mathbb{R},\rho_{1\le i\le n},\mathfrak{U}_{\mathcal{R}_{1\le i\le n}}\right\},T\right\}$ of gameplay (Kim, 2021a).

- In the G, the $\mathbb{G}^F$ and $\mathbb{G}^S_{1\le i\le n}$ are mutual and reciprocal interdependent in an interactive manner, and they work together to share each limited time step.
- N is the set of IoT devices, i.e., $\mathcal{D}_{1\le i\le n}$, which are game players of $\mathbb{G}^F$.
- R is the set of relays, i.e., $\mathcal{R}_{1\le i\le n}$, and the second bargaining process is individually executed in each relay. In the $\mathcal{R}_i$, the energy and data entities are game players of $\mathbb{G}^S_i$.
- M is the total bandwidth amount in the S-W joint platform where $\mathfrak{M} = \sum_{0\le i\le n}\tau_i$.
- In the $\mathbb{G}^F$, the strategy profile is $\{\tau_1,\dots\tau_i,\dots,\tau_n\}$ where $L^{UP}_t = \sum_{1\le i\le n}\tau_i$. The τ_i is the size of the allocated time duration for the $\mathcal{D}_i$, and $U_{\mathcal{D}_i}$ is the utility function of $\mathcal{D}_i$.
- In the $\mathbb{G}^S_i$, the strategy is ρ_i where $0\le \rho_i \le 1$; it is the ratio for the $\mathcal{R}_i$'s time splitting. The $\mathfrak{U}^D_{\mathcal{R}_i}$ and $\mathfrak{U}^E_{\mathcal{R}_i}$ are utility functions of $\mathcal{R}_i$'s data and energy entities, respectively.
- Discrete time model $T \in \{t_1,\dots,t_c,t_{c+1},\dots\}$ is represented by a sequence of time steps. The length of a time step matches the G's event timescale.

11.1.4 The basic concepts of *RABS* and *NABS*

Let $\sum^n$ be the family of all n-person bargaining problems of the form (S,d). A bargaining solution is a function $f : \sum^n \mapsto \mathbb{R}^n_+$ satisfying $f(S,d) \in S$ for every $(S,d) \in \sum^n$; that is, given a bargaining problem (S,d), the solution f prescribes $f(S,d)$. For a given $S \in \sum^n$, the utopia point $m(S) \in \mathbb{R}^n_+$ is the point where each coordinate i contains the maximum conceivable outcome for player i among individually rational alternatives. Usually, the bargaining solution is strongly dependent on the bargaining powers. If different bargaining powers are used, players with a higher bargaining power obtain a higher resource than the other players. To measure asymmetries in bargaining power, let $\alpha = (\alpha_1,\dots,\alpha_n)$ be a vector of bargaining weights. The endogenous reference point ($m^\alpha(S)$) is always feasible for every $S \in \Sigma_0$, i.e. $m^\alpha(S) \in S$. Formally, it is defined as follows (Alós-Ferrer, 2018; Karagözoğlu, 2015; Kim, 2021a)

$$m^\alpha(S) = \left(\left(\alpha_1\times m_1(S)\right),\dots,\left(\alpha_n\times m_n(S)\right)\right) \tag{11.1.1}$$

$$\text{s.t.,}\begin{cases}\alpha \in \Delta^{n-1} = \left\{\beta \in \mathbb{R}^n_+\middle|\sum^n_{i=1}\beta_i = 1\right\}\\ m_i(S) = max\left\{x_i\middle|x \ge d \text{ and } x \in S\right\}\end{cases}$$

where $m_i(S)$ is player i's utopia point. At the endogenous reference point, each player achieves a fraction of his utopian payoff which is proportional to players' bargaining power. In general, $m^{\alpha}(S)$ might be in the interior of S, and it captures a starting point for negotiated aspirations. A bargaining problem with a reference point is a triple (S, d, r) where the reference point $r \in S \setminus WPO(S)$ can satisfy $r \geq d$. With weights $\alpha \in$ Ä$^{n-1}$, the $NABS(S, d, r)$ is mathematically given as follows (Alós-Ferrer, 2018; Karagözoğlu, 2015; Kim, 2021a);

$$NABS(S, d, r = m^{\alpha}(S)) = ((1 - \lambda^{*}) \times m^{\alpha}(S)) + (\lambda^{*} \times m(S)) \tag{11.1.2}$$

$$\text{s.t., } \lambda^{*} = \mathbf{max}\,\{\lambda \in [0, 1] | [((1 - \lambda) \times m^{\alpha}(S)) + (\lambda \times m(S))] \in S\}$$

Another bargaining solution, RABS, also incorporates the reference point (r) into the definition of bargaining problems. Especially, the r is interpreted as an intermediate agreement which facilitates the conflict resolution process, and it has a strong potential to influence bargaining agreements. In addition, RABS employs the aspiration point $\mathcal{A}(S, r)$, which is a new salient point into bargaining problems. For every $x \in S$, an aspiration vector$\mathcal{A}(S, r)$ is defined in such a way that for every $i \in \{1, \ldots, n\}: \mathcal{A}_i(S, x) = \mathbf{max}\,\{k \in \mathbb{R} | (k, x_{-i}) \in S\}$. Simply, the ideal point in the $NABS$ is assumed as $\mathcal{A}(S, d)$. Since $r \geq d$, the aspiration point in the $RABS$ will be lower compared to the ideal point in the $NABS$ where $\mathcal{A}(S, r) \leq \mathcal{A}(S, d)$. With the $\mathcal{A}(S, r)$, the $RABS$ (S, d, r) is formally defined as follows (Karagözoğlu, 2015; Kim, 2021a);

$$RABS(S, d, r) = (\Gamma^{*} \times \mathcal{A}(S, r)) + ((1 - \Gamma^{*}) \times d) \tag{11.1.3}$$

$$\text{s.t., } \Gamma^{*} = \mathbf{max}\,\{\Gamma \in [0, 1] | [(\Gamma \times \mathcal{A}(S, r)) + ((1 - \Gamma) \times d)] \in S\}$$

Conceptually, for each game player, the ideal point, i.e., $\mathcal{A}(S, d)$, gives the maximum payoff level that can be reached in an individually rational agreement, and the aspiration point, i.e., $\mathcal{A}(S, r)$, is muted since $r \geq d$. Figure 3 provides a graphical illustration of the $RABS$ and $NABS$ for a two-player bargaining game (Kim, 2021a).

Figure 3. The RABS and NABS for a two-player bargaining game (Kim, 2021a)

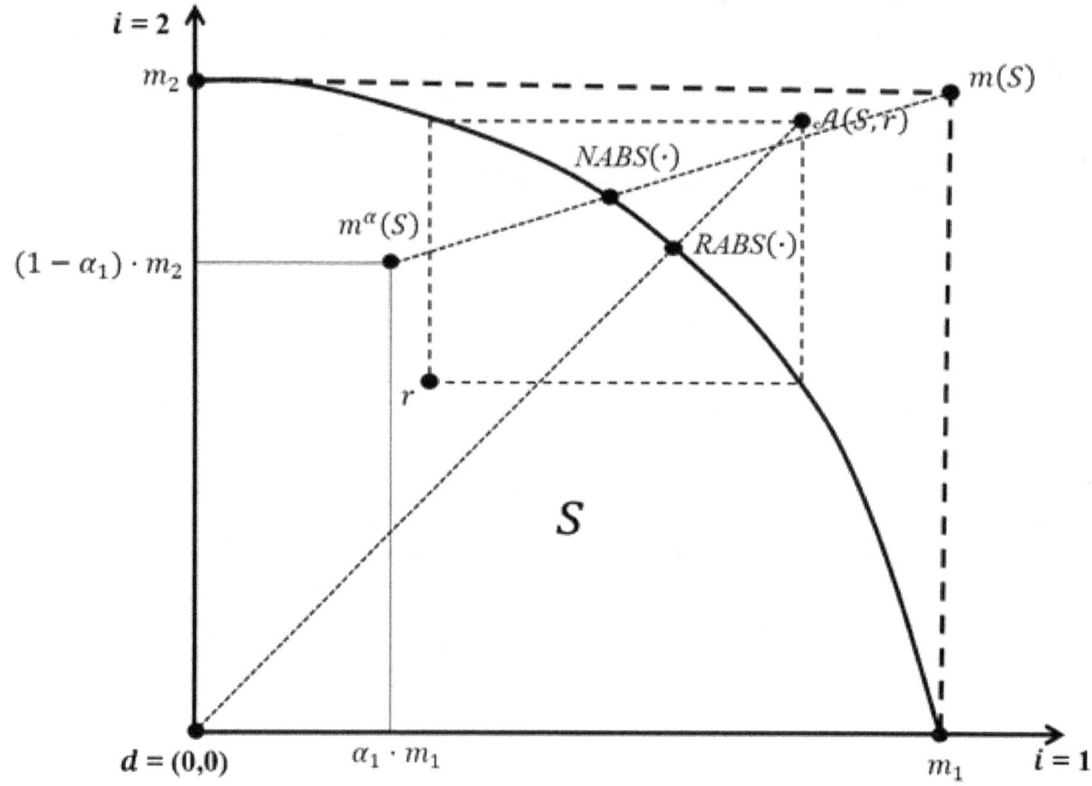

11.1.5 The twofold bargaining game in the S-W joint platform

At each time period $t \in T$, IoT devices have their data information and attempt to transfer them to the HAP through their corresponding relays. In the frame structure, the first τ_0 duration is dedicated for the WPCN downlink. During the time period τ_0, the HAP broadcasts the RF energy signal, and the relays harvest the energy for their forwarding tasks. At time t_c, the $\mathscr{R}_i$'s harvested energy function $\left(Ç_{\mathscr{R}_i}(\cdot)\right)$ from the HAP is expressed as follows (Kim, 2021a);

$$Ç_{\mathscr{R}_i}\left(\tau_0, \mathfrak{M}, d_{\mathscr{R}_i}, t_c\right) = \Delta_e \times \tau_0 \times \mathfrak{M} \times \left(\frac{D_M^{HAP} - d_{\mathscr{R}_i}}{D_M^{HAP}}\right)^{\phi} \times t_c \tag{11.1.4}$$

where Δ_e is the energy harvesting efficiency and ϕ is an adjustment factor for distance. $d_{\mathscr{R}_i}$ is the distance from the HAP to $\mathscr{R}_i$ and D_M^{HAP} is the maximum coverage range of HAP. The remaining time $\left(L_{t_c}^{UP}\right)$ in the frame structure is assigned, i) to transmit the data and energy from devices to relays, and ii) to retransmit the data from relays to HAP. It is done by taking two sub-slots for each device in the

SWIPT uplink. First, the $L_{t_c}^{UP}$ is split into n parts for each device, e.g., the time duration τ_i is dynamically allocated for the $\mathscr{D}_i$. Second, the $\tau_{1\le i\le n}$ is divided by the ρ_i factor. These processes are formulated as $\mathbb{G}^F$ and $\mathbb{G}_i^S$ based on the twofold bargaining model. For the $\mathbb{G}^F$, the utility function of $\mathscr{D}_i$ at time t_c $\left(U_{\mathscr{D}_i}(\cdot)\right)$ is formally defined as follows (Kim, 2021a);

$$U_{\mathscr{D}_i}\left(\tau_i, t_c, \mathcal{Q}_{\mathscr{D}_i}^{t_c}, \mathscr{E}_{\mathscr{R}_i}^{t_c}\right) = \frac{\mathscr{X}}{\eta + exp(\mathscr{X})} + \left(\left(\frac{E_{\mathscr{R}_i}^T - \mathscr{E}_{\mathscr{R}_i}^{t_c}}{E_{\mathscr{R}_i}^T}\right) \times \log(\mathscr{X} + \theta)\right) \tag{11.1.5}$$

$$\text{s.t., } \mathscr{X} = \frac{\min\left(\left(\tau_i \times \mathfrak{M}\right), \mathcal{Q}_{\mathscr{D}_i}^{t_c}\right)}{\mathcal{Q}_{\mathscr{D}_i}^{t_c}}$$

where $\mathcal{Q}_{\mathscr{D}_i}^{t_c}$ is the $\mathscr{D}_i$'s data amount at time t_c. $\mathscr{E}_{\mathscr{R}_i}^{t_c}$ is the $\mathscr{R}_i$'s remaining energy at time t_c, and $E_{\mathscr{R}_i}^T$ is the maximum capacity of $\mathscr{R}_i$'s battery. η and θ are control parameters for the $U_{\mathscr{D}_i}(\cdot)$. Usually, the utility function estimates the satisfaction of service quality. Therefore, it is defined as related to the amount of actually allocated time duration. To adaptively adjust the $\tau_{1\le i\le n}$ values at each time period, the *WEHC* scheme should consider the relative fairness among different IoT devices. Therefore, it is reasonable that the aspiration point is preferred to the utopian point to get the time splitting solution. Due to this reason, the *WEHC* scheme adopts the idea of *RABS* for the first bargaining process. According to (11.1.3), the *RABS* solution for the $\mathbb{G}^F$, i.e., $RABS_{\mathbb{G}^F}(\cdot)$ is given by (Kim, 2021a);

$$RABS_{\mathbb{G}^F}\left(U_{\mathscr{D}_{1\le i\le n}}(\cdot), S_{\mathbb{G}^F}, r_{\mathbb{G}^F}, d_{\mathbb{G}^F}\right) = \tag{11.1.6}$$

$$\max_{\Gamma\in[0,1]}\left\{\mathscr{D}_i \in \mathbb{N} \middle| \left(\Gamma \times \mathscr{A}_{\mathscr{D}_i}(S_{\mathbb{G}^F}, r_{\mathbb{G}^F})\right) + \left((1-\Gamma)\times d_{\mathbb{G}^F}\right)\right\}$$

$$\text{s.t., } \mathscr{A}_{\mathscr{D}_i}(S_{\mathbb{G}^F}, r_{\mathbb{G}^F}) = \mathbf{max}\left\{U_{\mathscr{D}_i}(\cdot) \in \mathbb{R} \middle| \left(U_{\mathscr{D}_i}(\cdot), (r_{\mathbb{G}^F})_{-i}\right) \in S_{\mathbb{G}^F}\right\}$$

where $S_{\mathbb{G}^F}$, $r_{\mathbb{G}^F}$ and $d_{\mathbb{G}^F}$ are the feasible payoff sets, reference and disagree points in the $\mathbb{G}^F$, respectively. In the game model, the $d_{\mathbb{G}^F}$ is set to zeros, and the $r_{\mathbb{G}^F}$ is defined based on the relative traffic weight of each device. It is given by (Kim, 2021a);

$$r_{\mathbb{G}^F} = \left\{r_{\mathbb{G}^F}^{\mathscr{D}_{1\le i\le n}} \middle| r_{\mathbb{G}^F}^{\mathscr{D}_i} = \Upsilon \times \left(\frac{\mathcal{Q}_{\mathscr{D}_i}^{t_c}}{\eta + exp\left(\mathcal{Q}_{\mathscr{D}_i}^{t_c}\right)} + \left(\left(\frac{E_{\mathscr{R}_i}^T - \mathscr{E}_{\mathscr{R}_i}^{t_c}}{E_{\mathscr{R}_i}^T}\right) \times \log\left(\mathcal{Q}_{\mathscr{D}_i}^{t_c} + \theta\right)\right)\right)\right\} \tag{11.1.7}$$

$$\text{s.t., } \Upsilon = \mathcal{Q}_{\mathscr{D}_i}^{t_c} / \sum_{\mathscr{D}_j \in \mathbb{N}} \mathcal{Q}_{\mathscr{D}_j}^{t_c}$$

From the $\mathbb{G}^F$, IoT devices obtain their time durations $\left(\tau_{1\le i\le n}\right)$. And then, for the next round, the n individual second bargaining processes are operated simultaneously and independently. At the $\mathbb{G}_{1\le i\le n}^S$, each individual relay $\mathscr{R}_{1\leq i\leq n}$ decides its ρ_i value to transfer the data and energy. The utility functions of data and energy entities in the $\mathscr{R}_i$, i.e., $\mathfrak{U}^D_{\mathscr{R}_i}(\cdot)$ and $\mathfrak{U}^E_{\mathscr{R}_i}(\cdot)$, are mathematically given as follows (Kim, 2021a);

$$\begin{cases} \mathfrak{U}^D_{\mathscr{R}_i}\left(\rho_i, \tau_i, t_c, \mathcal{Q}^{t_c}_{\mathscr{D}_i}\right) & = \dfrac{\mu \times \mathscr{Y}}{\sigma + exp(\mathscr{Y})} + \left(\dfrac{\mathcal{Q}^{t_c}_{\mathscr{D}_i}}{\tau_i \times \mathfrak{M}} \times \log(\mathscr{Y} + \xi)\right) \\ \mathfrak{U}^E_{\mathscr{R}_i}\left(\rho_i, \tau_i, t_c, \mathscr{E}^{t_c}_{\mathscr{R}_i}\right) & = \dfrac{exp(\mathscr{Z}) - exp(-\mathscr{Z})}{exp(\mathscr{Z}) + exp(-\mathscr{Z})} - \left(\left(\dfrac{d_{\mathscr{D}_i}}{D^{\mathscr{D}_i}_M}\right)^{\phi} \times \log(\mathscr{Z} + \beta)\right) \end{cases} \tag{11.1.8}$$

$$\text{s.t.,}\begin{cases} Y = \dfrac{\min\left(\left(\rho_i \times \tau_i \times \mathfrak{M}\right), \mathcal{Q}^{t_c}_{\mathscr{D}_i}\right)}{\mathcal{Q}^{t_c}_{\mathscr{D}_i}} \\ Z = \dfrac{\min\left(E^T_{\mathscr{R}_i}, \left(\mathscr{H}^{\mathscr{D}_i}_{\mathscr{R}_i}\left(\rho_i, \tau_i, d_{\mathscr{D}_i}, t_c\right) + \mathscr{E}^{t_c}_{\mathscr{R}_i}\right)\right)}{E^T_{\mathscr{R}_i}} \end{cases}$$

where μ, σ, ξ are control parameters for the $\mathfrak{U}^D_{\mathscr{R}_i}(\cdot)$, and ϕ, β are adjustment factors for the $\mathfrak{U}^E_{\mathscr{R}_i}(\cdot)$. $d_{\mathscr{D}_i}$ is the distance from $\mathscr{D}_i$ to $\mathscr{R}_i$, and $D^{\mathscr{D}_i}_M$ is the maximum coverage range of $\mathscr{D}_i$. $\mathscr{H}^{\mathscr{D}_i}_{\mathscr{R}_i}(\cdot)$ is the $\mathscr{R}_i$'s energy harvesting function from the $\mathscr{D}_i$. From the viewpoint of each relay, its energy can be harvested according to the HAP and its corresponding device. Therefore, the $\mathscr{R}_i$'s harvesting energy depends on the $\mathscr{H}^{\mathscr{D}_i}_{\mathscr{R}_i}(\cdot)$ and $\text{Ç}_{\mathscr{R}_i}$. At time t_c, $\mathscr{H}^{\mathscr{D}_i}_{\mathscr{R}_i}(\cdot)$ and the energy consumption function $\left(E_{\mathscr{R}_i}(\cdot)\right)$ for wireless communications between the HAP and $\mathscr{R}_i$ are given by (Kim, 2021a);

$$\begin{cases} \mathscr{H}^{\mathscr{D}_i}_{\mathscr{R}_i}\left(\rho_i, \tau_i, d_{\mathscr{D}_i}, t_c\right) & = \Delta_e \times (1 - \rho_i) \times \tau_i \times M \times \left(\dfrac{D^{\mathscr{D}_i}_M - d_{\mathscr{D}_i}}{D^{\mathscr{D}_i}_M}\right)^{\phi} \times t_c \\ E_{\mathscr{R}_i}\left(\mathcal{Q}_{\mathscr{R}_i}, d_{\mathscr{R}_i}, t_c\right) & = \Delta_c \times \mathcal{Q}^{t_c}_{\mathscr{D}_i} \times \left(\dfrac{d_{\mathscr{R}_i}}{D^{HAP}_M}\right)^{\phi} \end{cases} \tag{11.1.9}$$

where Δ_c is the energy consuming parameter for each data bit, and $\mathcal{Q}^{t_c}_{\mathscr{D}_i}$, which is the $\mathscr{R}_i$'s data amount at time t_c, is transferred to the HAP. In the $\mathbb{G}^S_i$, it is worthless if the $\mathscr{R}_i$'s energy is exhausted. Therefore, it is appropriate that disagreement should be decided according to energy preference. Because of this, the *WEHC* scheme uses the concept of *NABS* to adjust the ρ_i value. According to (11.1.2), the *NABS* solution for the $\mathbb{G}^S_i$, i.e., $NABS_{\mathbb{G}^S_i}(\cdot)$ is given by (Kim, 2021a);

$$NABS_{\mathbb{G}^S_i}\left(\mathfrak{U}^D_{\mathscr{R}_i}(\cdot), \mathfrak{U}^E_{\mathscr{R}_i}(\cdot), \rho_i, S_{\mathscr{R}_i}, r_{\mathscr{R}_i}, \mathscr{I}_{\mathscr{R}_i}, d_{\mathscr{R}_i}\right) \tag{11.1.10}$$

$$= \max_{\rho_i \in [0,1]}\left\{\mathscr{R}_i \in \mathbb{R} \middle| \left((1 - \rho_i) \times r_{\mathscr{R}_i}\right) + \left(\rho_i \times \mathscr{I}_{\mathscr{R}_i}\right)\right\}$$

$$
s.t.\begin{cases} m_E\left(S_{\mathcal{R}_i}\right) = \mathbf{max}\left\{\mathfrak{U}^E_{\mathcal{R}_i}(\cdot) \in \mathbb{R} \middle| x \geq d_{\mathcal{R}_i} \text{ and } x \in S_{\mathcal{R}_i}\right\} \\ m_D\left(S_{\mathcal{R}_i}\right) = \mathbf{max}\left\{\mathfrak{U}^D_{\mathcal{R}_i}(\cdot) \in \mathbb{R} \middle| x \geq d_{\mathcal{R}_i} \text{ and } x \in S_{\mathcal{R}_i}\right\} \\ \omega_E = \frac{E^T_{\mathcal{R}_i} - \mathcal{E}^{t_c}_{\mathcal{R}_i}}{E^T_{\mathcal{R}_i}} \text{ and } \omega_D = 1 - \omega_E \\ r_{\mathcal{R}_i} = \left(\left(\frac{\omega_E}{\alpha} \times m_E\left(S_{\mathcal{R}_i}\right)\right), \left(\frac{\omega_D}{\alpha} \times m_D\left(S_{\mathcal{R}_i}\right)\right)\right) \\ \mathcal{I}_{\mathcal{R}_i} = \left(m_E\left(S_{\mathcal{R}_i}\right), m_D\left(S_{\mathcal{R}_i}\right)\right) \end{cases}
$$

$$
m_D\left(S_{\mathcal{R}_i}\right) = \mathbf{max}\left\{\mathfrak{U}^D_{\mathcal{R}_i}(\cdot) \in \mathbb{R} \middle| x \geq d_{\mathcal{R}_i} \text{ and } x \in S_{\mathcal{R}_i}\right\}
$$

$$
\omega_E = \frac{E^T_{\mathcal{R}_i} - \mathcal{E}^{t_c}_{\mathcal{R}_i}}{E^T_{\mathcal{R}_i}} \text{ and } \omega_D = 1 - \omega_E
$$

$$
r_{\mathcal{R}_i} = \left(\left(\frac{\omega_E}{\alpha} \times m_E\left(S_{\mathcal{R}_i}\right)\right), \left(\frac{\omega_D}{\alpha} \times m_D\left(S_{\mathcal{R}_i}\right)\right)\right)
$$

$$
\mathcal{I}_{\mathcal{R}_i} = \left(m_E\left(S_{\mathcal{R}_i}\right), m_D\left(S_{\mathcal{R}_i}\right)\right)
$$

where $d_{\mathcal{R}_i}, \mathcal{I}_{\mathcal{R}_i}, r_{\mathcal{R}_i}$ are the disagree, utopian, and reference points in the $\mathbb{G}^S_i$, respectively. ω_E (*or* ω_D) is a weight control parameter for the energy (*or* data) entity, and α is a ratio correction factor for the $r_{\mathcal{R}_i}$; ω_E and ω_D are used to decide the $r_{\mathcal{R}_i}$. According to (11.1.6) and (11.1.10), the *RABS* and *NABS* are obtained, and the *WEHC* scheme uses them to share the frame structure at time t_c for the dual-hop communication. By iteratively repeating the twofold bargaining model, the *WEHC* scheme can adapt the current S-W joint platform conditions while maximizing the system throughput (Kim, 2021a).

11.1.6 Summary

The *WEHC* scheme introduces an innovative two-stage bargaining game model for allocating uplink time within the S-W joint infrastructure. Leveraging the dual-hop communication framework, relays obtain the necessary energy from the RF-enabled WPCN downlink and the energy-carrying SWIPT uplink. In the latter scenario, the limited uplink resource is distributed based on the two-stage bargaining approach. In the initial bargaining stage, the *RABS* principle is employed to divide uplink capacity among multiple IoT devices. In the subsequent stage, the *NABS* concept is applied to allocate the assigned time more effectively, ensuring mutual benefits. The primary objective of the *WEHC* scheme is to strike a balance between energy harvesting and data relaying, which is crucial for enabling dual-hop communication while addressing the doubly near-far problem. Throughout the iterative negotiation process, the first and second bargaining stages are executed sequentially, optimizing the overall performance of the S-W joint system (Kim, 2021a).

11.2 Meta-Bargaining based Tethering Control (*MBTC*) scheme

The exchange of resources and services within sharing economy networks has garnered growing interest. In line with this trend, the convergence of IoT and the sharing economy represents two emerging paradigms that promote the sharing of personal assets, including mobile devices, among users. Recently, S. Kim introduced the *Meta-Bargaining based Tethering Control* (*MBTC*) scheme, designed to optimize resource allocation for multiple mobile devices (Kim, 2020). Within the dynamically fluctuating WiFi environment, a key challenge is the efficient distribution of limited spectrum resources. To address this, the *MBTC* scheme incorporates four distinct bargaining solutions and employs a meta-bargaining approach to comprehensively manage spectrum allocation. By modeling competitive interactions among network agents as a cooperative game, the *MBTC* scheme enables participants to negotiate and reach mutually beneficial agreements while enhancing overall system performance. The core innovation of the meta-bargaining approach lies in balancing tradeoffs across different bargaining solutions, thereby ensuring a fair and efficient WiFi tethering mechanism (Kim, 2020).

11.2.1 Development motivation

The fundamental concept of IoT is envisioned to enhance the quality of modern life. In recent years, IoT has been widely implemented across various intelligent systems, including connected industries, smart cities, smart homes, autonomous vehicles, intelligent buildings, and smart campuses, among other domains. This widespread adoption has led to a remarkable surge in the number of IoT devices deployed and operational worldwide. According to the Ericsson Mobility Report 2018, the number of IoT devices is projected to surpass 4.1 billion by 2024, exhibiting a compound annual growth rate of 33% since 2013. However, many IoT devices face significant constraints in computational power and communication capabilities, necessitating the offloading of intensive computing tasks. Given these limitations, a critical challenge in IoT operations is the efficient utilization of available resources while mitigating traffic congestion caused by the simultaneous access demands of a vast number of devices (Baek, 2020; Kim, 2020; Yao, 2018).

As a rising economic and technological paradigm, the sharing economy represents an innovative economic model that fosters resource and service sharing. Over the past few years, the sharing economy has witnessed exponential growth and has drawn global attention. This model primarily involves peer-to-peer sharing of access to goods and services, facilitated by community-driven online platforms that leverage underutilized resources to enhance productivity. Unlike conventional ownership-centric models, the sharing economy emphasizes collaborative consumption rather than individual ownership. The success of this paradigm is evident in its diverse implementation models, with one of the most promising being IoT-based WiFi tethering. This model enables smart devices and digital services to facilitate the sharing of physical resources among users (Georgiadis, 2019; Guo, 2018; Kim, 2020).

WiFi tethering technology enables mobile IoT devices to share their Internet connectivity with nearby devices. Typically, a mobile IoT device gains Internet access via wireless cellular networks or existing WiFi connections. When tethering is established over WiFi, the Internet-connected IoT device functions as a portable WiFi access point (AP), commonly referred to as a mobile AP, relaying data between the Internet and nearby IoT devices - known as client devices. These client devices, lacking direct Internet access, connect to the mobile AP using their WiFi interfaces. Through WiFi tethering, users can maintain Internet connectivity on their WiFi-enabled devices even in the absence of public access points. Due to its flexibility and cost-effectiveness, WiFi tethering technology serves as a viable tool for implementing the sharing economy while ensuring seamless Internet access for mobile users on the move (Kim, 2020; Lim, 2016; Tabrizi, 2015;).

To enhance system performance, the *MBTC* scheme introduces an advanced WiFi tethering control algorithm. By incorporating load balancing among mobile APs, the *MBTC* scheme establishes a novel approach for WiFi spectrum resource allocation. However, achieving this objective presents considerable challenges, necessitating the adoption of an intelligent control framework. To address this, the *MBTC* scheme leverages the foundational principles of cooperative game theory. Initially, it integrates four distinct bargaining solutions: *Nash Bargaining Solution* (*NBS*), *Kalai-Smorodinsky Solution* (*KSS*), *Egalitarian Bargaining Solution* (*EBS*), and *Asymmetric Kalai-Smorodinsky Solution* (*AKSS*). By applying the principles of these bargaining solutions, the *MBTC* scheme develops a meta-cooperative game model to derive optimal outcomes. These methodologies are subfields of game theory, a discipline that has gained widespread recognition as a powerful analytical tool across various domains, particularly in the advancement of next-generation networks and data communication systems (Dubra, 2001; Karos, 2018; Kim, 2020).

When game participants encounter bargaining challenges, they may adopt different solution approaches based on varying interpretations of fairness and equity. To achieve a unanimous consensus, E. Damme introduced the concept of meta-bargaining, which aims to determine a reasonable bargaining solution when players hold divergent perspectives on the optimal solution model. Typically, a meta-bargaining game is structured as a non-cooperative strategic game derived from an underlying cooperative bargaining process. Consequently, players' strategies consist of selecting from a set of bargaining solutions. The *MBTC* scheme innovates by formulating a novel meta-bargaining model. Unlike Damme's meta-game framework, the *MBTC* meta-bargaining approach is more comprehensive, allowing for the integration of multiple bargaining solutions as strategic choices. As a result, the game outcome inherently incorporates characteristics from different bargaining solutions, making it adaptable to a wide range of strategic interactions (Kim, 2020; Trockel, 2002).

11.2.2 Contribution

Utilizing the principles of *NBS*, *AKSS*, *KSS*, and *EBS*, the *MBTC* scheme introduces an innovative meta-bargaining game model to optimize the allocation of limited spectrum resources. Within the WiFi tethering framework, mobile APs distribute their spectrum resources in a manner that balances both cooperation and competition. Given the dynamic nature of traffic conditions, each mobile AP holds a distinct perspective on the optimal allocation of WiFi spectrum. To facilitate a mutually agreeable distribution, the meta-bargaining approach is employed. By integrating the *NBS*, *AKSS*, *KSS*, and *EBS*, the *MBTC* scheme harnesses the collective advantages of multiple bargaining solutions, ensuring a comprehensive and adaptive strategy for managing WiFi tethering decisions in real time (Kim, 2020).

11.2.3 WiFi tethering system architecture

In conventional wireless communication settings, mobile devices typically access the Internet independently via separate 4G/5G connections, without engaging in cooperative networking. However, modern mobile devices are generally equipped with at least two network interfaces: one for cellular connectivity and another for WiFi access. Given that cellular network communications consume significantly more power than WiFi connections, WiFi is the preferred mode of communication whenever available. Nevertheless, some mobile devices may be positioned too far from public APs to establish a direct WiFi connection. In such scenarios, overall power consumption across the networked devices can be minimized by designating a subset of devices as mobile APs, enabling the remaining devices to access the Internet via WiFi through these selected mobile APs. Furthermore, this collaborative approach fosters energy efficiency, as mobile APs that provide service to others can, in turn, be supported by different devices when their battery levels decline. Consequently, this cooperative networking system offers long-term benefits to all participating mobile devices by optimizing power consumption and connectivity management (Dong, 2013; Kim 2020).

The *MBTC* scheme considers a small WiFi tethering platform based on the WiFi cooperation topology, which comprises two types of mobile devices with the WiFi spectrum resource pool (χ); χ can be assigned with an approximately interference-free spectrum band. There are a set $\mathbb{A} = \{\mathcal{A}_1, \ldots, \mathcal{A}_n\}$ of mobile APs and a set $\mathbb{M} = \{\mathcal{M}_1, \ldots, \mathcal{M}_m\}$ of client mobile devices. Simply, we assume each individual $\mathcal{M}_{1 \le j \le m}$ subscribes to only one $\mathcal{A}_{1 \le i \le n}$, which provides a mobile Internet service for each of its subscribers. Let $\mathfrak{M}_{\mathcal{A}_i}$ and $\mathcal{R}_{\mathcal{A}_i}$ denote the set of subscribers of $\mathcal{A}_i$ and the allocated spectrum resource of $\mathcal{A}_i$, respectively, where $\bigcup_{\mathcal{A}_i \in \mathbb{A}} \mathcal{R}_{\mathcal{A}_i} = \chi$. Each M generates application tasks. Usually, these tasks can be categorized into two classes: class I (real-time data) tasks and class II (non-real-time data) tasks. $\mathfrak{R}_{\mathcal{M}_j}$ is the spectrum request of $\mathcal{M}_j$'s tasks. All A and M remain relatively still, e.g., as students in a lecture hall. χ is the total WiFi spectrum resource pool, and it is orthogonally assigned to multiple As in a cooperative manner. Traditionally, the WiFi tethering system is considered in a centralized way; a centralized controller needs to collect the control information by methods like extending IEEE 802.11 protocol (Dong, 2013; Kim, 2020; Zhang, 2019).

The *MBTC* scheme considers the constant spectrum resource constraint of each M, since this is the main restriction in practical cellular traffic. To effectively distribute the χ among As in A, a modeling situation is formulated as a meta-bargaining game ($\mathbb{G}$). In the G, game players are $\mathcal{A}_{1 \le i \le n} \in$ A and the assigned WiFi spectrum resources, i.e., $\mathcal{R}_{\mathcal{A}_i}$, are their strategies where $\chi \ge \sum_{\mathcal{A}_i \in \mathbb{A}} \mathcal{R}_{\mathcal{A}_i}$. Based on the $\mathcal{R}_{\mathcal{A}_i}$, the $\mathcal{A}_i$'s utility function $\left(U_{\mathcal{A}_i}(\cdot)\right)$ is defined as follows (Kim, 2020);

$$U_{\mathcal{A}_i}\left(\mathcal{R}_{\mathcal{A}_i}, \mathcal{S}_{\mathcal{A}_i}, r_{\mathcal{A}_i}\right) = \mathbb{I}_{\mathcal{A}_i} - \left(\mathcal{R}_{\mathcal{A}_i} \times r_{\mathcal{A}_i}\right) \tag{11.2.1}$$

$$s.t.\begin{cases} \mathbb{I}_{\mathcal{A}_i} = \left(\mathcal{R}_{\mathcal{A}_i} \times \left[\xi + \dfrac{\kappa}{\left(1 + exp\left(-\eta \times \dfrac{\min\left(\mathcal{R}_{\mathcal{A}_i}, \mathcal{S}_{\mathcal{A}_i}\right)}{\mathcal{R}_{\mathcal{A}_i}}\right)\right)}\right] \times r_{\mathcal{A}_i}\right) \\ \mathcal{S}_{\mathcal{A}_i} = \sum_{\mathcal{M}_j \in \mathfrak{M}_{\mathcal{A}_i}} \mathfrak{R}_{\mathcal{M}_j} \end{cases}$$

where κ, η and ξ are coefficient factors for the $U_{\mathcal{A}_i}(\cdot)$. $r_{\mathcal{A}_i}$ is the price per bit for the WiFi tethering service. According to $U_{\mathcal{A}}(\cdot)$, how to decide the amount of $\mathcal{R}_{\mathcal{A}}$ for each A is the main challenge in the *MBTC* scheme (Kim, 2020).

11.2.4 The basic concepts of different bargaining solutions

Nearly seventy years ago, J. Nash introduced his groundbreaking work on what is now recognized as the axiomatic bargaining solution. In this framework, game participants are presented with a feasible set of alternatives from which they must reach a mutually agreeable compromise. Once an agreement is established, the chosen alternative becomes the final outcome. The primary objective of bargaining solutions is to predict how game participants will resolve conflicts or, alternatively, how a neutral arbitrator would determine and recommend a fair compromise (Thomson, 2009). In 1975, Kalai and Smorodinsky identified certain limitations in Nash's bargaining solution, noting that it failed to meet specific desirable properties. To address these shortcomings, they introduced a new class of bargaining solutions that focus on equalizing the ratios between each player's actual payoff and their ideal payoff - the maximum achievable payoff under the condition that all other players receive their minimum acceptable payoff. Based on these relative ratios, the *AKSS*, *KSS*, and *EBS* were subsequently developed, forming a family of proportional bargaining solutions (Kim, 2009, 2020).

Player i's ideal payoff is $a_i(S,d) = \max\left\{s_i : s \in S, s_j \geq d_j \text{for all} j \neq i\right\}$. Typically, this point is not feasible where $a(S,d) = \left(a_1(S,d),\ldots,a_n(S,d)\right) \notin S$. A bargaining solution is a map P, that assigns to every problem (S,d) a unique feasible point $\mathfrak{P}(S,d) \in S$. Different bargaining solutions are characterized by certain properties. A crucial issue of bargaining solutions is how the individual payoffs compare to each other. The *NBS* maps any problem S to the unique point in S; it is mathematically defined as follows (Karos, 2018; Kim 2020);

$$\mathfrak{P}^{NBS}(S) = \mathbf{max}\prod_{i \in N} (x_i - d_i) \tag{11.2.2}$$

where $x = \left[x_1 \ldots x_i \ldots x_n\right]$ is a vector and $x \in S$. Usually, the *NBS* is a prime example of bargaining problems. Since the early 1970s, from another angle, the focus has shifted to the dependence of proportional solutions with ideal points. With a non-negative number ρ, the *EBS*, *AKSS* and *KSS* and is defined as follows (Karos, 2018; Kim 2020);

$$\begin{cases} \left(\mathfrak{P}_j^{EBS}(S) \middle/ \mathfrak{P}_i^{EBS}(S)\right) = \left(a_j(S) \middle/ a_i(S)\right)^{\rho=0} \\ \left(\mathfrak{P}_j^{AKSS}(S) \middle/ \mathfrak{P}_i^{AKSS}(S)\right) = \left(a_j(S) \middle/ a_i(S)\right)^{0<\rho<1}, s.t., i,j \in N \\ \left(\mathfrak{P}_j^{KSS}(S) \middle/ \mathfrak{P}_i^{KSS}(S)\right) = \left(a_j(S) \middle/ a_i(S)\right)^{\rho=1} \end{cases} \tag{11.2.3}$$

For instance, the *EBS* equalizes the payoffs and the *KSS* equalizes the fractions of the ideal payoffs that players achieve; that is, $\left(\mathfrak{P}_j^{EBS}(S)/\mathfrak{P}_i^{EBS}(S)\right) = 1$ and $\left(\mathfrak{P}_j^{KSS}(S)/\mathfrak{P}_i^{KSS}(S)\right) = \left(a_j(S)/a_i(S)\right)$, respectively, in S. More generally, *EBS* and *KSS* belong to the following class of bargaining solutions given by $0 \leq \rho \leq 1$. The *AKSS* for any problem S can be expressed as $\lambda \cdot \left(a_1(S)^\rho \ldots a_n(S)^\rho\right)$ where λ is the maximum possible. In general, the parameter ρ measures the advantage of having a large ideal payoff. There is no advantage for the *EBS*, and as ρ increases this advantage increases as well (Karos, 2018; Kim, 2020).

11.2.5 The meta-bargaining control scheme for WiFi tethering

The *MBTC* scheme designs a new spectrum allocation mechanism for the WiFi tethering platform. The major goal of the *MBTC* scheme is to adaptively allocate the limited spectrum resource to multiple As. To reduce computation complexity, the spectrum allocation process is specified in terms of basic spectrum units (BSUs), where one BSU is the minimum amount (e.g., 512 Kb in the *MBTC* scheme) of spectrum allocation. Each individual A has its $\mathcal{S}_{\mathcal{A}}$, and shares the χ with other As. To implement the spectrum allocation process, the *MBTC* scheme consists of two phases. At the first phase, the *MBTC* scheme obtains the *NBS*, *EBS*, *AKSS* and *KSS* for the As, independently. To get them, let $U_{\mathcal{A}} = \left[U_{\mathcal{A}_1}\ldots.U_{\mathcal{A}_n}\right] \subseteq \mathbb{R}^{\mathbb{A}}$ be a set of feasible utility payoffs, and $d_{\mathcal{A}} = \left[d_{\mathcal{A}_1}\ldots.d_{\mathcal{A}_n}\right]$ is the disagree point vector for each A. Mathematically, the *NBS* for the bargaining problem $(U_{\mathcal{A}}, d_{\mathcal{A}})$, i.e., $\mathfrak{P}_{\mathcal{A}}^{NBS}(U_{\mathcal{A}}, d_{\mathcal{A}})$, is given by (Kim, 2020);

$$\mathfrak{P}_{\mathcal{A}}^{NBS}(U_{\mathcal{A}}, d_{\mathcal{A}}) = \mathbf{max}\prod\nolimits_{\mathcal{A}_i\in\mathbb{A}}\left(U_{\mathcal{A}_i}(\mathcal{R}_{\mathcal{A}_i}, \mathcal{S}_{\mathcal{A}_i}, r_{\mathcal{A}_i}) - d_{\mathcal{A}_i}\right) \tag{11.2.4}$$

$$\text{s.t.,}\ \chi = \sum\nolimits_{\mathcal{A}_i\in\mathbb{A}} \mathcal{S}_{\mathcal{A}_i}$$

The EBS for $(U_{\mathcal{A}}, d_{\mathcal{A}})$, i.e., $\mathfrak{P}_{\mathcal{A}}^{EBS}(U_{\mathcal{A}}, d_{\mathcal{A}})$, is given by;

$$\mathfrak{P}_{\mathcal{A}}^{EBS}(U_{\mathcal{A}}, d_{\mathcal{A}}) = \left[\frac{U_{\mathcal{A}_j}(\mathcal{R}_{\mathcal{A}_j}, \mathcal{S}_{\mathcal{A}_j}, r_{\mathcal{A}_j}) - d_{\mathcal{A}_j}}{U_{\mathcal{A}_i}(\mathcal{R}_{\mathcal{A}_i}, \mathcal{S}_{\mathcal{A}_i}, r_{\mathcal{A}_i}) - d_{\mathcal{A}_i}} = 1\right] \tag{11.2.5}$$

$$\cong \left(\left(U_{\mathcal{A}_j}(\mathcal{R}_{\mathcal{A}_j}, \mathcal{S}_{\mathcal{A}_j}, r_{\mathcal{A}_j}) - d_{\mathcal{A}_j}\right) = \left(U_{\mathcal{A}_i}(\mathcal{R}_{\mathcal{A}_i}, \mathcal{S}_{\mathcal{A}_i}, r_{\mathcal{A}_i}) - d_{\mathcal{A}_i}\right)\right)$$

$$\text{s.t.,}\ \mathcal{A}_i, \mathcal{A}_j \in \mathbb{A}, \mathcal{A}_i \neq \mathcal{A}_j\ \text{and}\ \chi = \sum\nolimits_{\mathcal{A}_i\in\mathbb{A}} \mathcal{S}_{\mathcal{A}_i}$$

To calculate the *AKSS* for $(U_{\mathcal{A}}, d_{\mathcal{A}})$, i.e., $\mathfrak{P}_{\mathcal{A}}^{AKSS}(U_{\mathcal{A}}, d_{\mathcal{A}})$, the *MBTC* scheme should defined the power parameters, i.e., $\rho = [\rho_1\ldots\rho_n]$. Traditionally, class I task services have a higher priority than class II task services. By considering the user's QoE, the $\rho_{1\leq i\leq n}$ values are decided according to the ratio of class I data tasks. For the $\mathcal{A}_i$, the power parameter ρ_i is given by (Kim, 2020);

$$\rho_i = {r_{\mathcal{A}_i}}\Big/\sum\nolimits_{\mathcal{A}_k\in\mathbb{A}}(r_{\mathcal{A}_k})\ \text{s.t.,}\ r_{\mathcal{A}_i} = \frac{\Sigma^{I}_{\mathcal{A}_i}}{\Sigma^{I}_{\mathcal{A}_i} + \Sigma^{II}_{\mathcal{A}_i}} \tag{11.2.6}$$

where $\grave{E}^{I}_{\mathcal{A}_i}$ and $\grave{E}^{II}_{\mathcal{A}_i}$ are the total requested spectrum sums of class I and class II tasks, respectively, from the $\mathcal{A}_i$. Using (11.2.6), the $\mathfrak{P}_{\mathcal{A}}^{AKSS}(U_{\mathcal{A}}, d_{\mathcal{A}})$ is given by (Kim, 2020);

$$\mathfrak{P}_{\mathcal{A}}^{AKSS}(U_{\mathcal{A}}, d_{\mathcal{A}}) =$$

$$\left[\left(\frac{U_{\mathcal{A}_j}(\mathcal{R}_{\mathcal{A}_j}, \mathcal{S}_{\mathcal{A}_j}, r_{\mathcal{A}_j}) - d_{\mathcal{A}_j}}{U_{\mathcal{A}_i}(\mathcal{R}_{\mathcal{A}_i}, \mathcal{S}_{\mathcal{A}_i}, r_{\mathcal{A}_i}) - d_{\mathcal{A}_i}}\right) = \left(\frac{a_{\mathcal{A}_j}(\mathcal{R}_{\mathcal{A}_j}, \mathcal{S}_{\mathcal{A}_j}, r_{\mathcal{A}_j}) - d_{\mathcal{A}_j}}{a_{\mathcal{A}_i}(\mathcal{R}_{\mathcal{A}_i}, \mathcal{S}_{\mathcal{A}_i}, r_{\mathcal{A}_i}) - d_{\mathcal{A}_i}}\right)^{\rho}\right]$$
$$\cong \left(\left(\frac{U_{\mathcal{A}_j}(\mathcal{R}_{\mathcal{A}_j}, \mathcal{S}_{\mathcal{A}_j}, r_{\mathcal{A}_j}) - d_{\mathcal{A}_j}}{\left(a_{\mathcal{A}_j}(\mathcal{R}_{\mathcal{A}_j}, \mathcal{S}_{\mathcal{A}_j}, r_{\mathcal{A}_j}) - d_{\mathcal{A}_j}\right)^{\rho_j}}\right) = \left(\frac{U_{\mathcal{A}_i}(\mathcal{R}_{\mathcal{A}_i}, \mathcal{S}_{\mathcal{A}_i}, r_{\mathcal{A}_i}) - d_{\mathcal{A}_i}}{\left(a_{\mathcal{A}_i}(\mathcal{R}_{\mathcal{A}_i}, \mathcal{S}_{\mathcal{A}_i}, r_{\mathcal{A}_i}) - d_{\mathcal{A}_i}\right)^{\rho_i}}\right)\right)$$
$$\text{s.t., } \mathcal{A}_i, \mathcal{A}_j \in \mathbb{A}, \mathcal{A}_i \neq \mathcal{A}_j,\ \rho_i, \rho_j \in \rho,\ \rho_i \neq \rho_j \text{ and } \chi = \sum\nolimits_{\mathcal{A}_i \in \mathbb{A}} \mathcal{S}_{\mathcal{A}_i} \tag{11.2.7}$$

Simply, the KSS for $(U_{\mathcal{A}}, d_{\mathcal{A}})$, i.e., $\mathfrak{P}_{\mathcal{A}}^{KSS}(U_{\mathcal{A}}, d_{\mathcal{A}})$, is obtained based on the $\mathfrak{P}_{\mathcal{A}}^{AKSS}(U_{\mathcal{A}}, d_{\mathcal{A}})$ with $\rho = 1$, i.e., $\rho_{1 \le i \le n} = 1$ (Kim, 2020).

At the second phase, the *MBTC* scheme formulates the meta-bargaining model to get the final solution for the WiFi spectrum allocation. At the first phase, the *MBTC* scheme can get the four different bargaining solutions. In the viewpoint of $\mathcal{A}_i$, the *NBS*, *EBS*, *AKSS* and *KSS* are $\mathfrak{P}_{\mathcal{A}_i}^{NBS}$, $\mathfrak{P}_{\mathcal{A}_i}^{EBS}$, $\mathfrak{P}_{\mathcal{A}_i}^{AKSS}$, and $\mathfrak{P}_{\mathcal{A}_i}^{KSS}$, respectively. According to this information, the *MBTC* scheme re-negotiates the WiFi spectrum sharing problem, and determines the weighting factor for each bargaining solution. The *MBTC* scheme can decide the weighting factor of $\mathfrak{P}_{\mathcal{A}}^{EBS}(U_{\mathcal{A}}, d_{\mathcal{A}})$, i.e., $\tilde{A}_{\mathcal{A}}^{EBS}$, based on the load balancing idea (Kim, 2020).

$$\Gamma_{\mathcal{A}}^{EBS} = \beta - \frac{\max_{\mathcal{A}_i \in \mathbb{A}}(\mathcal{S}_{\mathcal{A}_i}) - \min_{\mathcal{A}_i \in \mathbb{A}}(\mathcal{S}_{\mathcal{A}_i})}{\max_{\mathcal{A}_i \in \mathbb{A}}(\mathcal{S}_{\mathcal{A}_i})} \tag{11.2.8}$$

where β is an adjustment constant for the $\Gamma_{\mathcal{A}}^{EBS}$. By considering the end user's QoE, the *MBTC* scheme may emphasize the class I traffic services. Therefore, the *MBTC* scheme can decide the weighting factor of $\mathfrak{P}_{\mathcal{A}}^{AKSS}(U_{\mathcal{A}}, d_{\mathcal{A}})$, i.e., $\Gamma_{\mathcal{A}}^{AKSS}$, based on the total class I task ratio (Kim, 2020).

$$\Gamma_{\mathcal{A}}^{AKSS} = \Theta_{\mathcal{T}}^{I} \Big/ (\Theta_{\mathcal{T}}^{I} + \Theta_{\mathcal{T}}^{II}),\ \text{s.t., } \Theta_{\mathcal{T}}^{I} = \sum_{\mathcal{A}_i \in \mathbb{A}} \Theta_{\mathcal{A}_i}^{I} \text{ and } \Theta_{\mathcal{T}}^{II} = \sum_{\mathcal{A}_i \in \mathbb{A}} \Theta_{\mathcal{A}_i}^{II} \tag{11.2.9}$$

According to the reciprocal complement relationship between $\mathfrak{P}_{\mathcal{A}}^{AKSS}(U_{\mathcal{A}}, d_{\mathcal{A}})$and $\mathfrak{P}_{\mathcal{A}}^{KSS}(U_{\mathcal{A}}, d_{\mathcal{A}})$, the weighting factor of $\mathfrak{P}_{\mathcal{A}}^{KSS}(U_{\mathcal{A}}, d_{\mathcal{A}})$, i.e., $\Gamma_{\mathcal{A}}^{KSS}$, is simply given by (Kim, 2020);

$$\Gamma_{\mathcal{A}}^{KSS} = \Theta_{\mathcal{T}}^{II} \Big/ (\Theta_{\mathcal{T}}^{I} + \Theta_{\mathcal{T}}^{II}) = (1 - \Gamma_{\mathcal{A}}^{AKSS}) \tag{11.2.10}$$

The prominent feature associated with *NBS* is that if the bargaining solution of the larger set is found on a smaller domain, then the solution is not affected by expanding the domain. This feature provides a powerful property for the WiFi spectrum allocation algorithm when there are utility limits for each user (Park, 2007). For example, if the class II task services are strongly concerned with their service reliability, it might be possible that in class II tasks, a higher QoE level is not required. For this reason, the weighting factor of $\mathfrak{P}_{\mathscr{A}}^{NBS}(U_{\mathscr{A}}, d_{\mathscr{A}})$, i.e., $\Gamma_{\mathscr{A}}^{NBS}$, is decided as follows (Kim, 2020);

$$\Gamma_{\mathscr{A}}^{NBS} = \left(\Theta_{\mathcal{T}}^{I} + \Theta_{m}^{II}\right) \Big/ \left(\Theta_{\mathcal{T}}^{I} + \Theta_{\mathcal{T}}^{II}\right), \text{ s.t., } \Theta_{m}^{II} = \sum_{\mathscr{A}_i \in \mathbb{A}} \sum_{\mathscr{M}_j \in \mathfrak{M}_{\mathscr{A}_i}} m_{\mathscr{M}_j}^{II} \tag{11.2.11}$$

where $m_{\mathscr{M}_j}^{II}$ is the minimum spectrum requirement for class II services from the $\mathscr{M}_j$. According to (11.2.8)-(11.2.11), the *MBTC* scheme can get the weighting factors for the *NBS*, *EBS*, *AKSS* and *KSS*, respectively, and formulate the meta game model to get the final solution for the WiFi spectrum sharing problem (Kim, 2020).

The initial meta-bargaining mechanism was originally developed to support a strategic non-cooperative framework. However, unlike this traditional approach, the meta-bargaining model is more comprehensive, as it allows various bargaining solutions to be incorporated as strategic options, each with different weighting factors. Consequently, the *MBTC* scheme is capable of encompassing all bargaining solutions as viable strategies, ensuring greater flexibility in decision-making. Given an admissible family of bargaining solutions $U_{\mathscr{A}}$, the meta-bargaining solution on $U_{\mathscr{A}}$ is a single valued function $\Phi_{\mathscr{A}}(\cdot)$ (Kim, 2020);

$$\Phi_{\mathscr{A}}(\varrho) : \left[\mathfrak{P}_{\mathscr{A}}^{NBS}(\varrho) \times \mathfrak{P}_{\mathscr{A}}^{EBS}(\varrho) \times \mathfrak{P}_{\mathscr{A}}^{AKSS}(\varrho) \times \mathfrak{P}_{\mathscr{A}}^{KSS}(\varrho)\right]_{\tilde{\mathbb{A}}} \rightarrow \varrho \tag{11.2.12}$$

$$\text{s.t., } \Gamma = \left(\Gamma_{\mathscr{A}}^{NBS}, \Gamma_{\mathscr{A}}^{EBS}, \Gamma_{\mathscr{A}}^{AKSS}, \Gamma_{\mathscr{A}}^{KSS}\right) \text{ and } \varrho = \left(U_{\mathscr{A}}, d_{\mathscr{A}}\right)$$

Based on the $\Phi_{\mathscr{A}}(\varrho)$, the *MBTC* scheme calculates the final spectrum allocation amount for the $\mathscr{A}_i$, i.e., $\psi_{\mathscr{A}_i}$; it is estimated according to a social choice rule, which can be interpreted as a means to give a relative justification for the *NBS*, *EBS*, *AKSS* and *KSS* based on the current WiFi system situation. Finally, the $\psi_{\mathscr{A}_i}$ is given by (Kim, 2020);

$$\psi_{\mathscr{A}_i} = \left(\frac{\sum_{\varphi \in \{NBS,\, EBS,\, AKSS,\, KSS\}} \left(\Gamma_{\mathscr{A}}^{\varphi} \times \mathfrak{P}_{\mathscr{A}_i}^{\varphi}\left(U_{\mathscr{A}_i}, d_{\mathscr{A}_i}\right)\right)}{\sum_{\mathscr{A}_j \in \mathbb{A}} \left(\sum_{\varphi \in \{NBS,\, EBS,\, AKSS,\, KSS\}} \left(\Gamma_{\mathscr{A}}^{\varphi} \times \mathfrak{P}_{\mathscr{A}_j}^{\varphi}\left(U_{\mathscr{A}_j}, d_{\mathscr{A}_j}\right)\right)\right)} \times \chi\right) \tag{11.2.13}$$

$$\text{s.t., } \chi = \sum_{\mathscr{A}_i \in \mathbb{A}} \psi_{\mathscr{A}_i}$$

11.2.6 Summary

Recently, coordinated tethering within small-scale WiFi platforms has been explored as a means to enhance mobile spectrum efficiency in densely populated areas. The *MBTC* scheme introduces an innovative spectrum allocation mechanism for mobile AP-controlled WiFi tethering, enabling heterogeneous users to gain specific benefits upon participation. However, designing an effective WiFi spectrum allocation strategy is particularly challenging due to the dynamic fluctuations in network traffic. To address this complexity, the *MBTC* scheme incorporates four distinct bargaining solutions - *NBS*, *EBS*, *AKSS*, and *KSS* - while utilizing the meta-bargaining approach to optimize spectrum allocation. By considering the diverse perspectives of individual mobile APs, these four bargaining solutions are strategically integrated within the meta-bargaining framework, fostering cooperation and collaboration among them. As a result, the limited WiFi spectrum is efficiently distributed among multiple mobile APs, ensuring improved WiFi tethering performance (Kim, 2020).

11.3 Price and Service Distribution Control (*PSDC*) scheme

In recent years, millimeter-wave (mmWave)-based user-centric cell-free (UCCF) networks have garnered significant attention within the wireless industry. In the UCCF network architecture, multiple distributed access points (APs) concurrently serve numerous smart devices within a given geographical region. Recently, S. Kim explored a resource allocation challenge and proposes the *Price and Service Distribution Control* (*PSDC*) scheme for multi-connective mobile devices (Kim, 2023). By leveraging key characteristics of mmWave communications, the *PSDC* scheme organizes APs into clusters and groups devices accordingly. To determine the optimal service price for each AP, the *PSDC* scheme employs the principles of *cooperative multi-agent learning* (*CMAL*). For managing multi-connective devices, the *generalized Gini bargaining solution* (*GGBS*) is utilized to address the distribution of communication tasks. Through cooperative learning, individual APs iteratively refine their pricing strategies in an interactive manner. Meanwhile, leveraging GGBS, each device dynamically allocates its service requests to the most appropriate APs. The learning and bargaining algorithms are intricately integrated, working in tandem to achieve a socially optimal outcome. A key advantage of this integrated approach is its adaptability in reaching mutually beneficial agreements within the dynamic UCCF network environment (Kim, 2023).

11.3.1 Development motivation

Over the past decade, both academic researchers and industry experts have widely recognized the substantial growth in mobile multimedia data. This surge is primarily attributed to the rapid proliferation of smart wireless devices and the increasing demand for bandwidth-intensive applications. A major enabler of this trend is network densification, which involves deploying additional base stations (BSs) and access points (APs) to enhance the spatial reuse of wireless spectrum. Generally, network densification is considered a fundamental approach to achieving the anticipated 1000-fold capacity increase required for the next-generation 6G networks. In its early stages, densification was primarily achieved by installing additional BSs to accommodate more users and extend coverage. As network architecture evolved, this strategy expanded to include the deployment of multiple APs with smaller coverage areas. Despite certain technical challenges, network densification is expected to remain a key component of future wireless networks, continuously improving their efficiency and performance (Kim, 2023; Romanous, 2015; Zaidi, 2019).

Next-generation multimedia cellular networks are expected to be highly heterogeneous and densely populated, primarily through the integration of small cells managed by APs. This assumption is based on the linear scaling of network capacity with the number of deployed small cells. Additionally, recent advancements in millimeter-wave (mmWave) communications offer the potential for reliable, high-capacity wireless links. The mmWave spectrum, which ranges from 24 GHz to 300 GHz, is a crucial technology for achieving multi-gigabit per second (Gbps) data rates. Unlike conventional sub-6 GHz wireless communications, mmWave technology presents several advantages, including abundant spectral resources, reduced component costs, and the ability to employ highly directional antennas. However, despite these benefits, mmWave communication faces a major limitation: restricted coverage and susceptibility to physical obstructions. High-frequency electromagnetic waves have a shorter propagation range and are more vulnerable to environmental obstacles. Consequently, mmWave networks require line-of-sight (LoS) communication, where physical obstructions can significantly degrade signal strength and reduce transmission range (Kim, 2023; Kountouris, 2017; Yang, 2018a).

To address the blockage problem in mmWave communications, two primary solutions have been proposed: multi-connectivity and network densification. Multi-connectivity allows individual mobile devices to establish simultaneous connections with multiple BSs and APs, thereby improving the likelihood of LoS communication through macro diversity. Meanwhile, network densification enhances communication quality by ensuring that mobile devices remain within close proximity to their serving APs. These APs are interconnected via fiber or wireless fronthaul links, enabling efficient data exchange within the network. The combined implementation of multi-connectivity and network densification is a key driver in advancing IoT-driven communications. Recent research has focused on optimizing these techniques, leading to the development of the *PSDC* scheme, which aims to jointly optimize multi-connectivity and fair connection association to address the dynamic nature of mmWave networks (Choi, 2021; Kim, 2023).

With the growing adoption of network densification and multi-connectivity, the user-centric cell-free (UCCF) paradigm has emerged as a promising approach for mmWave communication networks. Historically, cellular network operations have been structured around cell-based architectures, where mobile devices establish direct connections with designated BSs to access communication services. However, the UCCF paradigm challenges this traditional approach by shifting towards a cell-free design, where mobile devices exchange multiple data flows through a network of heterogeneous APs. In this model, multiple network edge points provide connectivity to each mobile device, ensuring that QoS requirements are met, regardless of location. Achieving this level of adaptability and efficiency in UCCF networks requires the implementation of innovative techniques (Boccardi, 2014; Chen, 2016; Kim, 2023).

The shift from a cell-oriented network design to a user-centric framework is becoming increasingly significant in the development of 5G and beyond. The UCCF paradigm not only enhances ubiquitous coverage and device-specific QoS in dense small-cell deployments but also improves spectral efficiency through optimized resource allocation. From an AP's perspective, one of the primary challenges is determining the optimal pricing strategy for mmWave communication services. Conversely, from a device's standpoint, the key concern is effectively distributing service requests among available APs. While various algorithms have been proposed to address these issues, the complex interactions between APs and mobile devices make it difficult to formulate a universal solution. As a result, optimizing control mechanisms within UCCF networks remains a challenging problem (Kim, 2023).

Recent advancements in multi-agent learning have led to significant breakthroughs in solving complex sequential decision-making problems in machine learning. A multi-agent environment involves multiple agents interacting with one another to achieve a common goal. In this context, *cooperative multi-agent learning* (*CMAL*) has gained considerable attention in the machine learning community. Within the *CMAL* framework, agents actively sense their environment, evaluate possible actions, and adjust their strategies accordingly. This interactive learning process enhances cooperative decision-making and enables all agents to reach a mutually beneficial agreement. Due to its ability to facilitate cooperative resource allocation, the *CMAL* aligns with the principles of cooperative bargaining game theory (Kim, 2023; Vidhate, 2016; Wang, 2023).

The *PSDC* scheme introduces an innovative UCCF network control mechanism aimed at optimizing price control and service distribution. Inspired by the principles of cooperative bargaining game theory and *CMAL*, the *PSDC* scheme integrates a joint control process to guide both APs and IoT devices toward a socially optimal outcome. The *PSDC* scheme consists of two primary algorithms: price control algorithm and service distribution algorithm. In the price control algorithm, APs determine their pricing strategies for mmWave-based UCCF communication services. Utilizing the *CMAL*, individual APs collaborate to refine their pricing decisions through coordinated learning. In the service distribution algorithm, each IoT device dynamically distributes its service requests among corresponding APs. This process is governed by a bargaining game model, ensuring repeated cooperative decision-making for optimal service allocation. These two control mechanisms are tightly integrated, enabling the system to balance performance trade-offs and resolve conflicting demands within the UCCF network infrastructure (Kim, 2023).

A cooperative bargaining problem involves a set of players and a feasible range of utility payoffs. Most bargaining solutions in cooperative game theory adhere to the rationality principle, ensuring that outcomes are consistent with maximizing specific utility functions. As a sub-class of bargaining solutions, various multi-utilitarian solutions have been explored in recent studies. These solutions rely on weighted utilitarian criteria to prioritize different objectives, often incorporating an inequality-averse perspective. Following this principle, C. Blackorby, W. Bossert, and D. Donaldson introduced the *generalized Gini bargaining solution* (*GGBS*), which is based on generalized Gini orderings of utility vectors. The *GGBS*, as an inequality-averse multi-utilitarian solution, is derived by analyzing polyhedral structures generated by permutations of positive utility vectors. As a weighted utilitarian approach, the *GGBS* possesses desirable properties for ensuring fair and efficient resource allocation in bargaining-based decision-making scenarios (Hinojosa, 2008, 2007; Kim, 2023).

Cooperative multi-agent systems are designed to enable multiple agents to collaboratively solve complex tasks or maximize collective utility. Many real-world control problems can naturally be modeled using multi-agent frameworks. Consequently, recent research has focused on the application of machine learning techniques in multi-agent environments. Machine learning, through iterative trials, facilitates adaptive decision-making processes. The *PSDC* scheme specifically addresses problem domains where multiple agents cooperate to optimize outcomes rather than compete against one another. This cooperative learning approach, referred to as the *CMAL*, has its foundation in cooperative game theory. As a general analytical framework, cooperative game theory provides both a structured methodology for defining the cooperative learning environment, and the tools required to evaluate its outcomes. The integration of cooperative bargaining models and *CMAL* techniques is poised to have far-reaching applications in the field of wireless communications (Kim, 2023; Panait, 2005).

11.3.2 Contribution

The *PSDC* scheme is designed to optimize the performance of UCCF networks. By integrating the *CMAL* and *GGBS*, the *PSDC* scheme employs a sophisticated combination of learning and bargaining algorithms, which operate iteratively in a step-by-step manner. Within the UCCF system infrastructure, this joint control approach leverages the cooperative interactions between APs and IoT devices to attain a socially optimal outcome. To maximize system efficiency, the *PSDC* scheme introduces a joint price control and service distribution mechanism for mmWave-based UCCF networks. To the best of my knowledge, this joint control paradigm is the first of its kind in the literature specifically designed for UCCF network infrastructure. In the price control algorithm, individual APs learn and refine their pricing strategies through the *CMAL*, ensuring a coordinated multi-agent learning process. Meanwhile, in the service distribution algorithm, each IoT device dynamically allocates its service requests based on the principles of *GGBS*. Using a cooperative bargaining game model, service requests are effectively distributed among the corresponding APs. Given the dynamic nature of UCCF environments, the bargaining process is repeatedly executed, adapting to network conditions in real-time. By seamlessly integrating cooperative learning and bargaining algorithms, the system ensures mutual benefits for both APs and IoT devices. This hybrid approach captures the sequential interactions between APs and IoT devices, facilitating a mutually beneficial agreement that ultimately enhances UCCF system performance (Kim, 2023).

11.3.3 UCCF system bargaining and learning models

We consider a radio access network architecture in which the BS is functionally divided into n APs, i.e., $\mathbb{A} = \{\mathcal{A}_1,\ldots,\mathcal{A}_n\}$. The n APs are connected by means of a backhaul network to the BS. $\mathcal{A}_{1\leq i\leq n}$ has a coverage area, and a fixed mmWave spectrum band $\left(\mathfrak{M}_{\mathcal{A}_i}\right)$. The *PSDC* scheme assumes one BS supervises all APs in the considered area, and individual APs provide access links to m mobile IoT devices, i.e., $\mathbb{M} = \{\mathcal{M}_1,\ldots,\mathcal{M}_m\}$. They are randomly deployed in the area. In the UCCF system platform, each device can select several APs to form a cooperation cluster where $\mathcal{M}_j$ generates its own cluster $\mathcal{C}_{\mathcal{M}_j} \subseteq \mathbb{A}$; the clusters are formed from the perspective of each device to ensure QoS. To form a cluster, the *PSDC* scheme should consider the dynamic blockage of mmWave communications. Due to the moving blockers, the LoS blocked intervals are demonstrated. To mitigate the problem of blockage-induced outages, mmWave-specific resource allocation solutions should include the ratified multi-connectivity functionality based on the dynamic re-clustering technique (Chen, 2015; Kim, 2023).

From the viewpoint of $\mathcal{A}_i$, a set of devices, i.e., $\mathbb{S}_{\mathcal{A}_i}$, is dynamically maintained where $\mathcal{M}_j \in \mathbb{S}_{\mathcal{A}_i}$ is the $\mathcal{A}_i$'s corresponding device. The performance of APs largely depends on the effective sharing of $\mathfrak{M}_{\mathcal{A}_i}$. From the point of view of the $\mathcal{M}_j$, the cluster $\mathcal{C}_{\mathcal{M}_j}$ should be kept adaptively in the UCCF platform to proceed its service request $\left(r_{\mathcal{M}_j}\right)$. To maximize the $\mathcal{M}_j$'s QoS satisfaction, the main issue is how to distribute the $\mathcal{R}_{\mathcal{M}_j}$ among APs in the $\mathcal{C}_{\mathcal{M}}$. This situation inevitably leads to overlapped $\mathcal{C}_{\mathcal{M}}$ and $\mathbb{S}_{\mathcal{A}}$ where $\mathcal{C}_{\mathcal{M}_j} \cap \mathcal{C}_{\mathcal{M}_k} \neq \emptyset, \mathbb{S}_{\mathcal{A}_i} \cap \mathbb{S}_{\mathcal{A}_c} \neq \emptyset, \mathcal{M}_j,\mathcal{M}_k \in \mathbb{M}, j \neq k, \mathcal{A}_i,\mathcal{A}_c \in \mathbb{A}, i \neq c$. In other words, each AP and device may belong to different $\mathcal{C}_{\mathcal{M}}$ and $\mathbb{S}_{\mathcal{A}}$, respectively, and they are dynamically changeable. Control decisions to share the $\mathfrak{M}_{\mathcal{A}}$ are mutually dependent on the $\mathcal{C}_{\mathcal{M}}$ and $\mathbb{S}_{\mathcal{A}}$. Therefore, the proper coordination and collaboration of between APs and IoT devices are necessary. A general UCCF system infrastructure is shown in Figure 4 (Chen, 2015; Kim, 2023).

Figure 4. A general user centric cell free platform infrastructure (Kim, 2023)

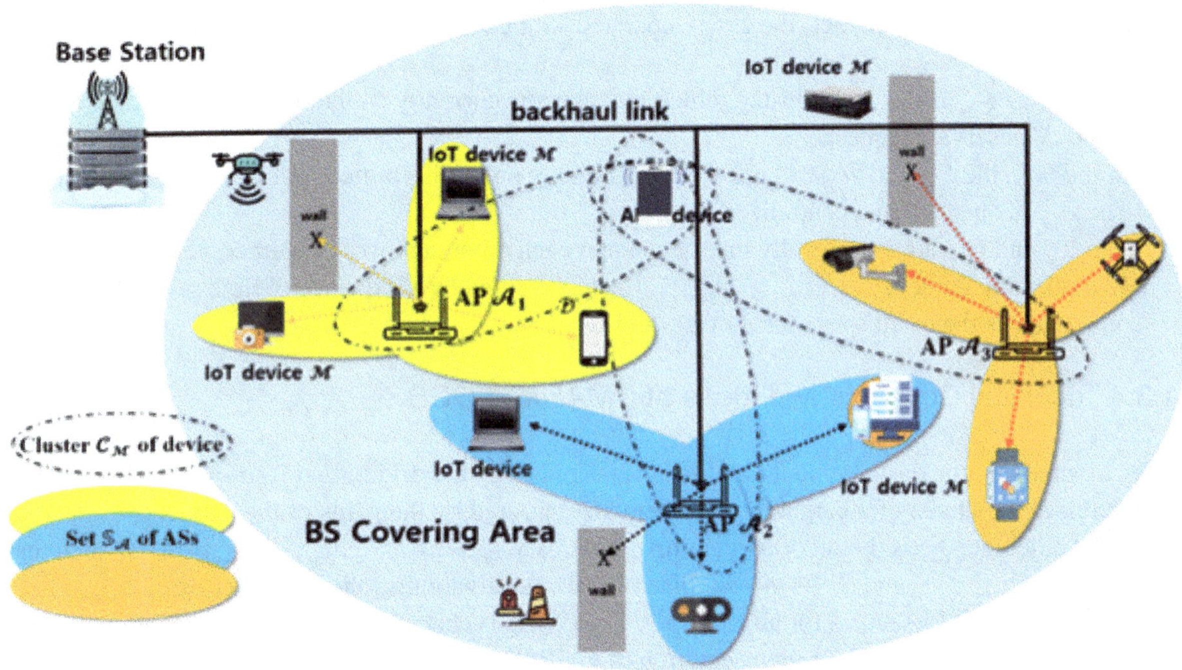

The *PSDC* scheme needs learners in the multi-agent learning process ($\mathbb{L}$) and game players in the bargaining game ($\mathbb{G}$). In the UCCF platform, APs learn what is the best price policy to provide cell free mmWave communication services. Through the learning process, APs interact with their corresponding devices in the $\mathbb{S}_{\mathcal{A}}$ and neighboring APs to adaptively adjust their price ($\mathcal{P}$). In the point view of devices, the major challenge is how to distribute their service requests $\left(r_{\mathcal{M}}\right)$ into their contacting APs. To optimize their payoffs, the $r_{\mathcal{M}}$ is dynamically assigned based on the bargaining process. The *PSDC* scheme adopts the joint control paradigm based on the dynamic learning game. Through the L and G, multiple APs and individual devices are sequentially interacted with each other to reach a mutually acceptable solution, which is called social optimal consensus. It is noteworthy that the *PSDC* scheme formulates the A-M association in an iterative manner based on the overlapped $\mathcal{C}_{\mathcal{M}}$ and $\mathbb{S}_{\mathcal{A}}$. Formally, the *PSDC* scheme define the tuple entities in the L and G, such as

$$\{\mathbb{L},\mathbb{G}\} = \left\{\mathbb{A},\mathbb{M},\mathbb{L}_{\mathcal{A}_{1\leq i\leq n}} = \left\{\mathbb{L}_{\mathcal{A}_i}\middle|\mathbb{S}_{\mathcal{A}_i},\mathcal{P}^{\mathcal{A}_i}_{1\leq k\leq l} \in \mathbb{P}_{\mathcal{A}_i},\mathcal{R}^{\mathcal{A}_i}_{p}\right\},\mathbb{G}_{\mathcal{M}_{1\leq j\leq m}} = \left\{\mathbb{G}_{\mathcal{M}_j}\middle|\mathcal{A}_i \in \mathcal{C}_{\mathcal{M}_j},r_{\mathcal{M}_j},\mathfrak{M}_{\mathcal{A}_i},S^{\mathcal{M}_j}_{\mathcal{A}_i},\mathcal{U}^{\mathcal{M}_j}_{\mathcal{A}_i}(\cdot)\right\},T\right\}$$

(Kim, 2023).

- A and M represent the sets of APs and IoT devices, respectively.
- Each AP A has its fixed mmWave spectrum band $\left(\mathfrak{M}_{\mathcal{A}}\right)$ and each IoT device M generates its service request $\left(r_{\mathcal{M}}\right)$.
- In the $\mathbb{L}_{\mathcal{A}_i}$, $\mathbb{S}_{\mathcal{A}_i}$ is the set of $\mathcal{A}_i$'s corresponding devices, and $\mathcal{A}_i$ has his action set $\left(\mathbb{P}_{\mathcal{A}_i}\right)$, which consists of total l actions $\left(\mathcal{P}^{\mathcal{A}_i}_{1\leq k\leq l}\right)$.

- In the $\mathbb{L}_{\mathcal{A}_i}$, $\mathcal{R}_p^{\mathcal{A}_i}$ is the $\mathcal{A}_i$'s reward function with the joint action $p = \left(\mathcal{P}^{\mathcal{A}_i}, p_{-\mathcal{A}_i}\right)$ in the multi-agent learning process. $\mathcal{R}_p^{\mathcal{A}_i}$ depends on the behavior of $\mathcal{M}_j \in \mathbb{S}_{\mathcal{A}_i}$.
- Based on the *CMAL* model, the $\mathbb{L}_{\mathcal{A}_i}$ is operated in a coordinated manner to learn the best price policy of $\mathcal{A}_i$.
- In the $\mathbb{G}_{\mathcal{M}_j}$, $\mathcal{C}_{\mathcal{M}_j}$ is the $\mathcal{M}_j$'s cluster, which includes the currently contacting APs with $\mathcal{M}_j$. The $r_{\mathcal{M}_j}$ is the $\mathcal{M}_j$'s service request.
- In the $\mathbb{G}_{\mathcal{M}_j}$, the $S_{\mathcal{A}_i}^{\mathcal{M}_j}$ and $\mathcal{U}_{\mathcal{A}_i}^{\mathcal{M}_j}(\cdot)$ are the $\mathcal{A}_i$'s strategy and utility function for the $\mathcal{M}_j$, respectively. The $\mathbb{G}_{\mathcal{M}_j}$ is operated in a centralized manner.
- The $\mathbb{L}_{\mathcal{A}_i}$ and $\mathbb{G}_{\mathcal{M}_j}$ are reciprocally interdependent each other, and work together, iteratively.
- Discrete time model $T \in \{t_1, \ldots, t_c, t_{c+1}, \ldots\}$ is represented by a sequence of time steps. The length of t_c matches the event's timescale of $\mathbb{L}_{\mathcal{A}_i}$ and $\mathbb{G}_{\mathcal{M}_j}$.

11.3.4 Technical Concepts and Ideas of *CMAL* and *GGBS*

We assume that the disagreement point is normalized to be the origin, that is, $d = 0^n$. This implies that the relevant features of a bargaining problem are captured by the gains of the agents over the disagreement outcome. $\sum^n$ is the set of bargaining problems $S \subseteq \mathbb{R}^n$ where $\exists c \in \mathbb{R}^n_{++}$, $m \in \mathbb{R}$ such that $c \cdot x \leq m$ for all $x \in S$, and $\exists x \in S$ such that $x \geq 0^n$. A bargaining solution is a correspondence F^n: $\sum^n \rightarrow \mathbb{R}^n_+$ such that $F^n(S) \subseteq S$ for all $S \in \sum^n$. The *GGBS* is characterized based on generalized Gini orderings. To define the standard representation of *GGBS*, let x^r denote a rank-ordered permutation of $x \in \mathbb{R}^n$ such that $x_1^r \geq x_2^r \geq \cdots \geq x_n^r$. To get a rank-ordered permutation of x, a permutation function, $\pi: N \rightarrow N$, is introduced where $\pi(i) = j$ means that the ith component of the n-dimensional permuted vector is j. Let x_π be the π-permutation of vector $x \in \mathbb{R}^n$, $x_\eth = \left(x_{\pi(1)}, x_{\pi(2)} \cdots x_{\pi(n)}\right)$, and Δ^n is the simplex in $\mathbb{R}^n_+$ where $\Delta^n = \{\omega \in \mathbb{R}^n_+ | \omega \cdot 1^n = 1\}$. With $\omega = \{\omega_1, \omega_2 \cdots \omega_n\}$, the generalized Gini ordering is represented by a function $G_\omega: \mathbb{R}^n \rightarrow \mathbb{R}$ (Blackorby, 1994; Hinojosa, 2008, 2007; Kim, 2023);

$$G_\omega(x) = \sum_{i=1}^{n} (\omega_i \times x_i^r), \text{s.t.}, \begin{cases} x_i^r = x_{(\pi(j)=i)} = x_{(i)} \\ 0 \leq \omega_1 \leq \omega_2 \leq \ldots \leq \omega_n \, and \, \omega_n > 0 \\ \omega \in \mathbb{R}^n_+ \, and \, \omega \cdot 1^n = 1 \end{cases} \tag{11.3.1}$$

Therefore, $(x_1^r, x_2^r \cdots x_n^r)$ is same as $\left(x_{(1)}, x_{(2)} \cdots x_{(n)}\right)$, which denotes a rank-ordered permutation vector of $x \in \mathbb{R}^n_+$, that is $x_{(1)} \geq x_{(2)} \geq \ldots \geq x_{(n)}$. The *GGBS* is given by the following definition (Blackorby, 1994; Kim, 2023);

$$GGBS_\omega(S) = \{x \in S | x \geq 0^n \text{and } G_\omega(x) \geq G_\omega(y)\} \tag{11.3.2}$$

$$= \arg\max_{x \in S} G_\omega(x) = \arg\max_{x \in S} \sum_{i=1}^{n} (\omega_i \times x_i^r)$$

$$\text{s.t., all} \, y \in S \, \text{and} \, y \geq 0^n$$

The $GGBS$ is rationalized by the corresponding generalized Gini ordering in the sense that it selects, for each $S \in \sum^n$, the best elements in $S \cap \mathbb{R}^n_+$ according to this ordering. Therefore, the utilitarian bargaining solution is the special case of $GGBS_\omega(S)$ with $\omega_i = \omega_j$ for all $i,j = 1,\ldots,n$, and the egalitarian bargaining solution is obtained by setting $\omega_1 = \ldots = \omega_{n-1} = 0$. As is the case for the utilitarian and egalitarian bargaining solutions, the $GGBS$ requires an informational environment that accommodates cardinal unit comparability (Blackorby, 1994; Hinojosa, 2008, 2007; Kim, 2023).

In a multi-agent environment, multiple agents interact with one another to achieve consensus-based decision-making. The process of multi-agent consensus involves the integration of knowledge, objectives, skills, and strategic planning to enable agents to take coordinated actions and collaboratively solve complex problems. The primary objective of multi-agent consensus is to establish an optimal control policy by independently optimizing each agent's performance while ensuring collective efficiency. To achieve this, multi-agent consensus requires multiple agent learners capable of understanding the context and behavior of other agents to refine their collaboration strategies. Additionally, in partially observed environments, agent learners must determine coordinated actions that effectively align behaviors to solve the given task. One of the prominent techniques for multi-agent consensus is the value-based method, which focuses on evaluating the desirability of different actions through an online learning framework. This approach enables agents to continuously update their decision-making strategies based on real-time interactions. Consequently, the dynamic game model serves as a fundamental framework for facilitating multi-agent consensus, providing a structured methodology for optimizing cooperative behaviors in complex environments (Jiang, 2020; Kim, 2023; Vidhate, 2016).

In traditional value-based learning, the desirability of an action is represented by the expected cumulative reward, as known as Q-value, for taking a particular action. Under the multi-agent environment, each agent has to consider not only the effects of his own actions, but also the influence by the actions of the other agents. From this perspective, agents are able to observe the effects of all actions taken by any agent. Based on the idea of stateless value-based learning paradigm, the $PSDC$ scheme assumes the Q-value of agent j's action i (a_i), i.e., $Q^j_{a_i}(\cdot)$, that provides an estimate of the value of performing the a_i. At the beginning, all Q-values of all actions are initialized to zero, so all agents start learning with equal choice among all available actions. The sample $\langle j, a_i, \mathscr{R} \rangle$ is the experience obtained by the agent j where the a_i was performed resulting in the reward R. Recursively, each Q-value is updated by an agent each time; it is defined as follows (Claus, 1998; Kim, 2023; Morozs, 2015):

$$Q^j_{a_i}(a_i) = \left[(1 - \alpha) \times Q^j_{a_i}(a_i)\right] + (\alpha \times \mathscr{R}) \tag{11.3.3}$$

where $\alpha \in [0, 1]$ is the learning rate parameter which weighs recent experience with respect to previous estimates of the Q-values. In multi-agent settings, one of the major concerns is to analyze the strategic interactions among multiple decision makers. However, each agent has incomplete and fixed knowledge. Therefore, the Q-value according to (11.3.3) cannot be adequately efficient in a dynamic, complex or changing multi-agent environment. In the $CMAL$, the $PSDC$ scheme adopts the strategy averaging method for the multi-agent knowledge sharing. It means that agent team ($\mathbb{T}$) learn to manage the fresh, hidden and dynamic situations with its team APs, which is celled as teammates. By taking the weighted average of Q-values, each agent learns from all of the team agents. From the viewpoint of agent j, the weighted average Q-value of the a_i with T, i.e., $WA\_Q^j_{a_i}(\mathbb{T})$, is given by (Kim, 2023; Vidhate, 2016);

$$WA\_Q^{j}_{a_i}(\mathbb{T}) = \left(\beta \times \sum_{k \in \mathbb{T}/j} \frac{Q^{k}_{a_i}(a_i)}{|\mathbb{T}| - 1}\right) + \left((1-\beta) \times Q^{j}_{a_i}(a_i)\right)$$

(11.3.4)

Let $\mathbb{P}_j = \left\{p^j_{a_1}, p^j_{a_2} \cdots p^j_{a_{|\mathscr{A}|}}\right\}$ be the probability set for the agent j's action selections where $p^j_{a_k}$ is the probability for selecting the a_k. The $p^j_{a_k}$ is updated as follows (Kim, 2023):

$$p^j_{a_k} = exp\left(\frac{WA\_Q^{j}_{a_k}(\mathbb{T})}{\tau}\right) / \sum_{l=1}^{|\mathbb{P}_j|} exp\left(\frac{WA\_Q^{j}_{a_l}(\mathbb{T})}{\tau}\right)$$

(11.3.5)

where τ is the temperature parameter that balances exploration and exploitation (Cui, 2021).

11.3.5 The joint control scheme for the UCCF paradigm

To develop the joint control mechanism for the mmWave based UCCF paradigm, the *PSDC* scheme constructs the learning model ($\mathbb{L}$) and bargaining game model ($\mathbb{G}$). In the L, an online *CMAL* process is implemented in a coordinated manner to learn the best price strategy. To maximize the profit, the $\mathscr{A}_i$ attempts to adaptively select his price strategy ($\mathscr{P}^{\mathscr{A}_i}$) in the $\mathbb{P}_{\mathscr{A}_i}$. For this selection, the $\mathscr{A}_i$ should consider not only its own price selection, but also its team APs' selections. If the $\mathscr{P}^{\mathscr{A}_i}$ is too high, devices in the $\mathbb{S}_{\mathscr{A}_i}$ try to rely on other APs in their $\mathscr{C}_{\mathscr{M}}$. If the $\mathscr{P}^{\mathscr{A}_i}$ is too low, its profit $\left(\mathscr{R}^{\mathscr{A}_i}_p\right)$ is significantly reduced. Therefore, price strategy selection is coupled with other APs' selections; the result of each AP's selection is the input back to the other AP's selection process. The dynamics of the interactive feedback mechanism can cause cascade interactions of APs and the $\mathscr{A}_i$ can make his selections to quickly find the most profitable solution. Finally, it can lead all team APs to an efficient consensus (Kim, 2023).

The *PSDC* scheme adopts the idea of *CMAL* to decide the price strategy of AP. In the learning model of $\mathscr{A}_i$ $\left(\mathbb{L}_{\mathscr{A}_i}\right)$, the price strategies of $\mathscr{A}_i$ is discretely varied, and each strategy in the $\mathbb{P}_{\mathscr{A}_i}$ is defined as the $\mathscr{A}_i$'s action. After an action in the $\mathscr{A}_i$ is selected, it is estimated by receiving a reward and *Q*-values of other APs. Based on the current reward and collected information from team APs, the *CMAL* is processed, and the $\mathscr{A}_i$ can gradually learn what is the best action. In the stateless *CMAL* setting, the *Q*-value of $\mathscr{A}_i$'s k[th] action $\left(\mathscr{P}^{\mathscr{A}_i}_k\right)$, i.e., $Q^{\mathscr{A}_i}_k\left(\mathscr{P}^{\mathscr{A}_i}_k, p_{-\mathscr{A}_i}\right)$, provides an estimate value of performing the joint action $p = \left(\mathscr{P}^{\mathscr{A}_i}_k, p_{-\mathscr{A}_i}\right)$. The $\mathscr{A}_i$ updates its estimate $Q^{\mathscr{A}_i}_k(\cdot)$ value based on the experience sample $\langle p, \mathscr{R}^{\mathscr{A}_i}_p\left(\mathscr{P}^{\mathscr{A}_i}_k, \mathbb{S}_{\mathscr{A}_i}\right), \mathbb{Q}_p(\mathbb{T})\rangle$ where $\mathscr{R}^{\mathscr{A}_i}_p\left(\mathscr{P}^{\mathscr{A}_i}_k, \mathbb{S}_{\mathscr{A}_i}\right)$ is the $\mathscr{A}_i$'s reward, and $\mathbb{Q}_p(\mathbb{T})$ is the average *Q*-value table of team APs ($\mathbb{T}$) with the joint action **p**. When the $\mathscr{A}_i$ selects the action $\mathscr{P}^{\mathscr{A}_i}_k$, the reward $\mathscr{R}^{\mathscr{A}_i}_p\left(\mathscr{P}^{\mathscr{A}_i}_k, \mathbb{S}_{\mathscr{A}_i}\right)$ is given by (Kim, 2023);

$$\mathscr{R}^{\mathscr{A}_i}_p\left(\mathscr{P}^{\mathscr{A}_i}_k, \mathbb{S}_{\mathscr{A}_i}\right) = \sum_{\mathscr{M}_j \in \mathbb{S}_{\mathscr{A}_i}} \left(\mathscr{P}^{\mathscr{A}_i}_k \times \mathfrak{M}^{\mathscr{M}_j}_{\mathscr{A}_i}\right)$$

(11.3.6)

where $\mathbb{S}_{\mathscr{A}_i}$ is the set of $\mathscr{A}_i$'s corresponding devices, and $\mathfrak{M}^{\mathscr{M}_j}_{\mathscr{A}_i}$ is the allocated spectrum amount for the $\mathscr{M}_j$. The $\mathbb{Q}_p$ table, i.e., $\mathbb{Q}_p(\mathbb{T}) = \left[WA\_Q^{\mathscr{A}_i}_{p_1}(\mathbb{T}), WA\_Q^{\mathscr{A}_i}_{p_2}(\mathbb{T}) \cdots WA\_Q^{\mathscr{A}_i}_{p_l}(\mathbb{T})\right]$ is updated based on the collected Q-values of APs in the team T. Finally, the stateless CMAL Q-function is defined according to (11.3.3),(11.3.4) and (11.3.6) (Kim, 2023).

$$WA\_Q_{p_{1\leq k\leq l}}^{\mathscr{A}_i}(\mathbb{T}) = \left(\beta \times \sum_{\mathscr{A}_j \in \mathbb{T}/\mathscr{A}_i} \frac{Q_{p_k}^{\mathscr{A}_j}}{|\mathbb{T}| - 1}\right) + \left((1-\beta) \times Q_{p_k}^{\mathscr{A}_i}\right)$$

(11.3.7)

$$\text{s.t., } Q_{p_k}^{\mathscr{A}_i} = \mathscr{R}_p^{\mathscr{A}_i}\left(\mathscr{P}_k^{\mathscr{A}_i}, \mathbb{S}_{\mathscr{A}_i}\right)$$

where β is a weighting factor for the *CMAL*. From the $WA\_Q_{p_k}^{\mathscr{A}_i}(\mathbb{T})$ value, the $\mathscr{P}_k^{\mathscr{A}_i}$'s selection probability by the $\mathscr{A}_i$, i.e., $\Gamma_k^{\mathscr{A}_i}$, is updated as follows (Kim, 2023):

$$\Gamma_k^{\mathscr{A}_i} = exp\left(\frac{WA\_Q_{p_k}^{\mathscr{A}_i}(\mathbb{T})}{\tau}\right) / \sum_{c=1}^{l} exp\left(\frac{WA\_Q_{p_c}^{\mathscr{A}_i}(\mathbb{T})}{\tau}\right)$$

(11.3.8)

In the $\mathbb{G}_{\mathscr{M}_j}$, a main concern of $\mathscr{M}_j$ is to effectively distribute its $r_{\mathscr{M}_j}$ into APs in the $\mathscr{C}_{\mathscr{M}_j}$. To proceed with this service distribution, the $\mathscr{M}_j$ divides its $r_{\mathscr{M}_j}$ based on the *GGBS*. By considering the current workload of APs in the $\mathscr{C}_{\mathscr{M}_j}$, the *PSDC* scheme can get the simplex vector $V\left[\mathscr{C}_{\mathscr{M}_j}\right]$ as follows (Kim, 2023);

$$V\left[\mathscr{C}_{\mathscr{M}_j}\right] = \left(\cdots, v_{\mathscr{A}_i}, \cdots\right), \text{ s.t., } v_{\mathscr{A}_i} = \frac{\mathfrak{T}_{\mathscr{A}_i}}{\sum_{\mathscr{A}_k \in \mathscr{C}_{\mathscr{M}_j}} \mathfrak{T}_{\mathscr{A}_k}}$$

(11.3.9)

where $\mathfrak{T}_{\mathscr{A}_i}$ is the allocated spectrum amount of $\mathscr{A}_i$. Therefore, the $\mathscr{A}_i$'s currently available spectrum amount is $\mathfrak{M}_{\mathscr{A}_i} - \mathfrak{T}_{\mathscr{A}_i}$. Using the vector $V\left[\mathscr{C}_{\mathscr{M}_j}\right]$, the *PSDC* scheme defines the vector $\omega = \left[\omega_1, \omega_2 \cdots \omega_{|\mathscr{C}_{\mathscr{M}}|}\right]$ based on the one-to-one mapping with the $V\left[\mathscr{C}_{\mathscr{M}_j}\right]$ while satisfying the order $\omega_1 \leq \omega_2 \leq \ldots \leq \omega_{|\mathscr{C}_{\mathscr{M}}|}$ (Kim, 2023).

In wireless communications, the *Signal to Interference plus Noise Ratio* (*SINR*) is a quantity used to give theoretical upper bounds on channel capacity. It is defined as the power of a certain signal of interest divided by the sum of the interference power and the power of some background noise. In each wireless connection of AP-device pair, the *PSDC* scheme can estimate its *SINR*. Based on the *SINR* between $\mathscr{M}_j$ and $\mathscr{A}_i$, i.e., $SINR_{\mathscr{M}_j}^{\mathscr{A}_i}$, the $\mathscr{A}_i$'s spectrum efficiency for the $\mathscr{M}_j$ $\left(EE_{\mathscr{M}_j}^{\mathscr{A}_i}\right)$ can be expressed as (Kim, 2011, 2023).

$$EE_{\mathscr{M}_j}^{\mathscr{A}_i} = W \times \log_2\left(1 + SINR_{\mathscr{M}_j}^{\mathscr{A}_i}\right)$$

(11.3.10)

where W is the channel bandwidth. The *PSDC* scheme orders the APs by considering the *SINR* information. Let $\left[\mathcal{C}_{\mathcal{M}_j}\right]_\pi$ be the π-permutation of vector of APs in the $\mathcal{C}_{\mathcal{M}_j}$. To get a rank-ordered permutation of $\mathcal{C}_{\mathcal{M}_j}$, a permutation function π is given by (Kim, 2023);

$$\pi(\mathcal{C}_{\mathcal{M}_j} \Big| \mathcal{A}_e, \mathcal{A}_i, \mathcal{A}_n \in \mathcal{C}_{\mathcal{M}_j}) = \left(x_{\pi\left(SINR_{\mathcal{M}_j}^{\mathcal{A}_e}\right)} \cdots x_{\pi\left(SINR_{\mathcal{M}_j}^{\mathcal{A}_i}\right)} \cdots x_{\pi\left(SINR_{\mathcal{M}_j}^{\mathcal{A}_n}\right)}\right)$$

$$\text{s.t.,} \begin{cases} \pi\left(SINR_{\mathcal{M}_j}^{\mathcal{A}_e}\right) = 1 and \pi\left(SINR_{\mathcal{M}_j}^{\mathcal{A}_n}\right) = \left|\mathcal{C}_{\mathcal{M}_j}\right| \\ SINR_{\mathcal{M}_j}^{\mathcal{A}_e} \geq \cdots \geq SINR_{\mathcal{M}_j}^{\mathcal{A}_i} \geq \cdots \geq SINR_{\mathcal{M}_j}^{\mathcal{A}_n} \end{cases} \tag{11.3.11}$$

In the viewpoint of $\mathcal{M}_j$, the generalized Gini ordering for APs is represented according to (11.3.1) (Kim, 2023);

$$G_\omega\left(SINR_{\mathcal{M}_j}, \mathcal{C}_{\mathcal{M}_j}\right) = \sum_{i=1}^{\left|\mathcal{C}_{\mathcal{M}_j}\right|} \left(\omega_i \times x_i^r\right)$$

$$\text{s.t.,} \begin{cases} x_i^r = x_{\pi\left(SINR_{\mathcal{M}_j}^{\mathcal{A}_k}\right)=i} and \mathcal{A}_k \in \mathcal{C}_{\mathcal{M}_j} \\ 0 \leq \omega_1 \leq \omega_2 \leq \ldots \leq \omega_n and \omega_n > 0 \\ \omega \in \mathbb{R}_+^n and \omega \cdot 1^n = 1 \end{cases} \tag{11.3.12}$$

where $\left(x_1^r, x_2^r, \cdots, x_{\left(\left|\mathcal{C}_{\mathcal{M}_j}\right|\right)}^r\right)$ denotes a rank-ordered permutation vector of *SINR* values between $\mathcal{M}_j$ and each AP in the $\mathcal{C}_{\mathcal{M}_j}$. To get the *GGBS*, the *PSDC* scheme replaces the x_i^r with the $\mathcal{A}_i$'s utility function for the $\mathcal{M}_j$, i.e., $\mathcal{U}_{\mathcal{M}_j}^{\mathcal{A}_i}(\cdot)$, where $\pi\left(SINR_{\mathcal{M}_j}^{\mathcal{A}_k}\right) = i$ and $\mathcal{A}_i \in \mathcal{C}_{\mathcal{M}_j}$. According to the service assignment, the $\mathcal{U}_{\mathcal{M}_j}^{\mathcal{A}_i}(\cdot)$ is defined as follows (Kim, 2023):

$$\mathcal{U}_{\mathcal{M}_j}^{\mathcal{A}_i}\left(p_{(\cdot)}^{\mathcal{A}_i}, r_{\mathcal{M}_j}, S_{\mathcal{A}_i}^{\mathcal{M}_j}\right) = \Upsilon \times \left(\frac{\log\left(\frac{\min\left(S_{\mathcal{A}_i}^{\mathcal{M}_j}, r_{\mathcal{M}_j}\right)}{r_{\mathcal{M}_j}} + \eta\right)}{p_{(\cdot)}^{\mathcal{A}_i}} \times Y\right)$$

$$\text{s.t.,} \begin{cases} Y = \left(\frac{\min\left(S_{\mathcal{A}_i}^{\mathcal{M}_j}, r_{\mathcal{M}_j}\right)}{r_{\mathcal{M}_j}} \times \log_2\left(1 + SINR_{\mathcal{M}_j}^{\mathcal{A}_i}\right)\right) \\ \sum_{\mathcal{A}_v \in \mathcal{C}_{\mathcal{M}_j}} S_{\mathcal{A}_v}^{\mathcal{M}_j} \leq r_{\mathcal{M}_j} \end{cases} \tag{11.3.13}$$

where ϒ and η are adjustment parameters for the $\mathcal{U}_{\mathcal{M}_j}^{\mathcal{A}_i}(\cdot)$, and $r_{\mathcal{M}_j}$ is the $\mathcal{M}_j$'s service request. The $S_{\mathcal{A}_i}^{\mathcal{M}_j}$ means the $\mathcal{A}_i$'s allocated spectrum amount to process the $r_{\mathcal{M}_j}$, and the $p_{(\cdot)}^{\mathcal{A}_i}$ is the $\mathcal{A}_i$'s price strategy. Finally, the *GGBS* for the $\mathcal{M}_j$, i.e., $GGBS_{\mathcal{M}_j}(\cdot)$, is given as follows (Kim, 2023);

$$GGBS_{\mathcal{M}_j}\left(G_\omega\left(SINR_{\mathcal{M}_j},\mathcal{C}_{\mathcal{M}_j}\right)\right) = \arg\max_{\langle\ldots,S^{\mathcal{M}_j}_{\mathcal{A}_i},\ldots\rangle} \sum_{i=1}^{\left|\mathcal{C}_{\mathcal{M}_j}\right|}\left(\omega_i \times x_i^r\right) \tag{11.3.14}$$

$$\text{s.t., } x_i^r := \mathcal{U}^{\mathcal{A}_i}_{\mathcal{M}_j}\left(p^{\mathcal{A}_i}_{(\cdot)}, r_{\mathcal{M}_j}, S^{\mathcal{M}_j}_{\mathcal{A}_i}\right) \text{ with } \pi\left(SINR^{\mathcal{A}_k}_{\mathcal{M}_j}\right) = i$$

From the AP's viewpoint, service requests are received from corresponding IoT devices. If the total sum of requests is larger than the $\mathfrak{M}_{\mathcal{A}}$, the *PSDC* scheme adopts the utilitarian idea to distribute the $\mathfrak{M}_{\mathcal{A}}$. Therefore, the $\mathfrak{M}_{\mathcal{A}}$ is shared based on the special case of *GGBS*. Especially, to share the $\mathfrak{M}_{\mathcal{A}_i}$, the *GGBS* for the $\mathcal{A}_i$, i.e., $GGBS_{\mathcal{A}_i}(\cdot)$, is given as follows (Kim, 2023);

$$GGBS_{\mathcal{A}_i}\left(G_\omega\left(SINR_{\mathcal{M}_j},\mathbb{S}_{\mathcal{A}_i}\right)\right) = \arg\max_{\langle\ldots,S^{\mathcal{M}_j}_{\mathcal{A}_i},\ldots\rangle} \sum_{i=1}^{\left|\mathbb{S}_{\mathcal{A}_i}\right|}\left(\omega_i \times x_i^r\right) \tag{11.3.15}$$

$$\text{s.t., }\begin{cases}\omega_1 = \omega_2 = \ldots = \omega_g\, with \left|\mathbb{S}_{\mathcal{A}_i}\right| = g\\ x_i^r := \mathcal{U}^{\mathcal{A}_i}_{\mathcal{M}_j}\left(p^{\mathcal{A}_i}_{(\cdot)}, r_{\mathcal{M}_j}, S^{\mathcal{M}_j}_{\mathcal{A}_i}\right) with\, \pi\left(SINR^{\mathcal{A}_k}_{\mathcal{M}_j}\right) = i\end{cases}$$

11.3.6 Summary

The *PSDC* scheme addresses the challenges of price control and service distribution in mmWave-based UCCF networks. By integrating the *CMAL* and bargaining game theory, the *PSDC* scheme introduces a novel learning game paradigm to enhance system performance. First, the *PSDC* scheme develops a price control algorithm for APs, leveraging the principles of *CMAL* to optimize pricing strategies. Second, a service distribution algorithm is designed for each device, utilizing the *GGBS* to allocate service requests efficiently. Finally, these two control mechanisms are seamlessly integrated to maximize UCCF system performance while achieving a socially optimal outcome. These algorithms operate in a step-by-step iterative manner, allowing the system to dynamically adapt to changing network conditions. A key innovation of this joint approach is its ability to leverage reciprocal advantages through sequential interactions between APs and IoT devices. In dynamic mmWave-based UCCF network environments, the *PSDC* scheme effectively reaches a cooperative consensus, ensuring efficient conflict resolution and improved network resource utilization (Kim, 2023).

11.4 Smart Home Traffic Control (*SHTC*) scheme

A smart home network (SHN) is a type of communication network that enables seamless interaction among smart devices within a home environment. When connected to the Internet, SHNs gain enhanced functionalities, including home security, automation, and entertainment. However, fluctuations in SHN data distributions often lead to traffic congestion, posing challenges in maintaining efficient network performance. To address this issue, S. Kim recently introduces the *Smart Home Traffic Control* (*SHTC*) scheme, which is designed to adaptively manage different QoS traffic services (Kim, 2021b). The *SHTC* scheme classifies SHN traffic services into three distinct categories based on their characteristics, allowing for optimized allocation of the limited SHN bandwidth resources. By leveraging bargaining theory, the *SHTC* scheme formulates a two-level game model, where different bargaining solutions are applied at each level to maximize the benefits of a collaborative operational framework. This interactive coordination paradigm ensures mutual benefits in dynamically evolving SHN environments. A key innovation of the two-level bargaining approach is its ability to balance diverse QoS requirements while fostering a reciprocal consensus among different traffic services. Through a coordinated decision-making process, the *SHTC* scheme effectively achieves a globally desirable solution, ensuring a fair yet efficient trade-off in resource allocation and traffic management (Kim, 2021b).

11.4.1 Development motivation

The evolution of new technologies has been largely driven by innovation and expansion, both of which play a crucial role in enhancing social well-being and reshaping human interactions with their environment. In the modern era, a new paradigm has emerged, enabling individuals to effectively manage advanced information and communication technologies while optimizing their behavior in domestic settings. A smart home refers to a technologically enhanced living space where appliances and devices can be remotely controlled via an Internet connection using mobile or networked devices. Within a smart home environment, IoT devices are interconnected through a home area network (HAN), allowing users to remotely control functions such as home security, temperature regulation, lighting, and entertainment systems. The expansion of the smart home concept has been significantly driven by the widespread adoption of emerging technologies, which facilitate seamless interaction between services and functionalities (Attia, 2019; Kim, 2021b; Mendes, 2015).

A smart home network (SHN) interconnects various devices, including smart TVs, sensors, electronic appliances, and intelligent IoT systems, enabling them to interact either through user commands or automated system controls. The rapid growth of SHNs has resulted in heterogeneous service communications and an increasing number of IoT devices generating diverse traffic patterns. Smart home technologies are particularly prominent in areas such as home security, energy management, and remote healthcare monitoring. However, due to the coexistence of numerous IoT devices transmitting data simultaneously, traffic congestion issues arise in SHNs, requiring efficient traffic load management. This necessitates an adaptive traffic scheduling mechanism capable of addressing constraints such as network congestion and data transmission delays. From the perspective of smart home service providers, ensuring QoS differentiation while optimizing SHN performance is a critical control challenge, for which multiple approaches have been explored (Attia, 2019, 2020; Kim, 2021b; Kua, 2017; Lu, 2018).

The smart home gateway (SHG) plays a central role in processing traffic from multiple IoT devices, ensuring that each traffic type meets its specific QoS requirements. Typically, real-time services such as medical monitoring, fire detection, and online video streaming require immediate processing, whereas best-effort services may be temporarily queued before being processed. Current QoS-aware scheduling mechanisms categorize SHN traffic into three distinct classes: class I: real-time random data (e.g., emergency alerts and medical data), class II: real-time periodic data (e.g., sensor updates and smart appliance control), and class III: best-effort data (e.g., non-critical background traffic). These classifications correspond to critical, delay-sensitive, and delay-tolerant services, respectively. To enhance SHN performance, the SHG must implement an optimal strategy that efficiently balances traffic processing while maintaining fairness (Attia, 2020; Kim, 2021b).

Traditionally, SHGs employ multiple queues to classify and prioritize incoming data packets. The SHG scheduling process relies on predefined priority-based strategies to handle classified traffic efficiently. However, due to the dynamic nature of SHNs, the increasing number of connected IoT devices places a heavy burden on network resources, exacerbating payload congestion. Moreover, SHN bandwidth is a highly valuable and limited resource. Existing SHG scheduling methods often fail to account for simultaneous traffic demands, resulting in inefficient bandwidth allocation. The most critical challenge in SHG operation lies in determining an optimal bandwidth allocation strategy that accommodates the heterogeneous traffic demands while adapting to dynamic network conditions. Addressing this issue necessitates the development of an intelligent control paradigm and innovative bandwidth allocation model (Attia, 2019, 2020; Kim, 2021b; Kua, 2017).

In 2014, S. Rachmilevitch introduced a novel bargaining solution concept, referred to as the *Bridging Bargaining Solution* (*BBS*). This model integrates formal axioms that merge two well-established bargaining frameworks: i) *Nash Bargaining Solution* (*NBS*) - recognized as the unique independent standard solution, and ii) *Kalai-Smorodinsky Bargaining Solution* (*KSBS*) - known as the unique monotonic standard solution. The *BBS* was developed to reconcile the principles of independence and monotonicity in the *NBS* and *KSBS*. It ensures that each game participant receives at least the minimum payoff guaranteed under the *NBS* and *KSBS*, thereby bridging the gap between the two bargaining approaches. Within game theory, the *BBS* framework provides a method for harmonizing conflicting bargaining solutions while ensuring fairness in resource allocation (Kim, 2021b; Shiran, 2014).

In 2018, the *Iterated Egalitarian Bargaining Solution* (*IEBS*) was introduced, building upon the principles of *Egalitarian Bargaining Solution* (*EBS*) and *Equal Loss Bargaining Solution* (*ELBS*). Both *EBS* and *ELBS* incorporate egalitarian principles when proposing bargaining outcomes. EBS maximizes the payoff profile by ensuring that each player receives an equal gain over their disagreement outcome. The *ELBS* maximizes the payoff profile by guaranteeing that each player incurs an equal loss relative to their best possible outcome. Despite their advantages, EBS and *ELBS* fail to fully satisfy the normative requirement that players should secure at least half of their best outcome in all bargaining scenarios. To address this limitation, the *IEBS* approach introduces a progressive, iterative compromise mechanism, where bargaining outcomes are refined step-by-step based on successive iterations of *EBS* and *ELBS*. With the increasing complexity of smart home traffic management, intelligent bargaining-based control frameworks are gaining prominence as viable solutions. The integration of multi-agent learning with bargaining solutions such as *BBS* and *IEBS* presents a promising direction for optimizing QoS-aware traffic scheduling in SHNs. Future research should explore adaptive, dynamic, and fair-efficient scheduling models that leverage bargaining game theory to ensure optimal performance in smart home networks (Karagözoğlu, 2018; Kim, 2021b).

11.4.2 Contribution

Leveraging the principles of *BBS* and *IEBS*, the *SHTC* scheme efficiently addresses traffic management challenges within SHNs. The *SHTC* scheme introduces a dual-level bargaining game model to optimize the allocation of limited bandwidth resources within the SHG. First, the *SHTC* scheme implements an 'over' bargaining game involving two players. Player 1 represents critical and delay-sensitive traffic services, which include class I (real-time random data) and class II (real-time periodic data) services. Player 2 represents best-effort services, corresponding to class III (delay-tolerant traffic). Within a dynamic and heterogeneous SHN environment, both players negotiate bandwidth allocation to establish a mutually acceptable resource-sharing agreement through the *BBS*. Second, the *SHTC* scheme formulates an 'under' bargaining game, which further refines bandwidth distribution. Player I corresponds to class I traffic (critical real-time data) and player II corresponds to class II traffic (delay-sensitive periodic data). Building upon the *IEBS*, both players adaptively negotiate their bandwidth allocation based on the outcome determined in the over bargaining game. By employing a step-by-step interactive cooperative process, this two-tier bargaining approach is highly flexible and adaptive, allowing it to respond effectively to dynamic fluctuations in SHN traffic patterns. This structured negotiation process ensures an equitable and efficient QoS balance, optimizing the overall performance and stability of SHN operations (Kim, 2021b).

11.4.3 SHN system infrastructure and bargaining game model

Each SHN consists of a diverse range of smart IoT devices, delivering services across various applications, including monitoring, health assistance, safety, and energy efficiency. To support these functionalities, multiple devices generate heterogeneous traffic data, each with distinct QoS requirements, which are categorized into different traffic classes. In scenarios where SHN traffic congestion occurs, the implementation of a strategic and rational control mechanism becomes essential. The primary challenge lies in simultaneously ensuring user satisfaction in terms of QoS while maximizing the overall performance of the SHN. Addressing this issue requires intelligent traffic management strategies that optimize resource allocation and system efficiency (Attia, 2019; Kim, 2021b).

In the SHN, the *SHTC* scheme assumes that there is a set ($\mathbb{D}$) of multiple smart devices where $\mathbb{D} = \{\mathscr{D}_1, \mathscr{D}_2, \dots, \mathscr{D}_n\}$, and these devices will generate data with different types and distributions. When these devices are activated, they have to send traffic simultaneously to the SHG. The *SHTC* scheme has a single SHG in each SHN, and three different data queues, i.e., $\mathscr{Q}_I$, $\mathscr{Q}_{II}$ and $\mathscr{Q}_{III}$, are implemented in the SHG to handle the class I, II and III traffic data, respectively. The SHG contains two modules: classifier and allocator. The classifier classifies the incoming traffic depending on their data types, and traffic data are buffered into their corresponding queues. And then, the operator can process the traffic services based on the allocated bandwidth for each queue. The *SHTC* scheme considers a discrete time model $T \in \{t_1, \dots, t_c, t_{c+1}, \dots\}$, where the length of a time slot matches the event timescale at which control decisions are updated (Kim, 2021b).

The *SHTC* scheme formulates SHG's bandwidth allocation process as a two-level bargaining model ($\mathbb{G}$). The over bargaining game ($\mathbb{G}^O$) is modelled as a two-player game; class I and II traffic services are represented as one game player (P_I^O) and the class III traffic service is defined as the other game player (P_{II}^O). They get their assigned bandwidth amounts, i.e., $\mathfrak{B}_I^O$ and $\mathfrak{B}_{II}^O$, through the concept of *BBS*. Since then, the $\mathfrak{B}_I^O$ amount should be distributed into class I and II traffic services. The under bargaining game ($\mathbb{G}^U$) is also modelled as a two-player game; the class I traffic service is one game player (P_I^U) and the class II traffic service is another game player (P_{II}^U). They get their assigned bandwidth amounts, i.e., $\mathfrak{B}_I^U$ and $\mathfrak{B}_{II}^U$, through the idea of *IEBS*. The $\mathbb{G}^O$ and $\mathbb{G}^U$ are repeated sequentially in a slotted time structure. Based on the interactive feedback manner, the *SHTC* scheme is operated each time period during the step-by-step iteration. In the *SHTC* scheme, the amount of bandwidth allocation is specified in terms of basic bandwidth units to reduce computation complexity. Formally, the *SHTC* scheme defines G game entities for the SHN system infrastructure, i.e,

$$\mathbb{G} = \{\mathbb{G}^O, \mathbb{G}^U\} = \left\{\mathbb{D}, \{\mathscr{Q}_I, \mathscr{Q}_{II}, \mathscr{Q}_{III}\}, \mathfrak{B}, \left\{\mathbb{G}^O \middle| (P_I^O, P_{II}^O), (\mathfrak{U}_I^O, \mathfrak{U}_{II}^O), (\mathfrak{B}_I^O, \mathfrak{B}_{II}^O)\right\}, \left\{\mathbb{G}^U \middle| (P_I^U, P_{II}^U), (\mathfrak{U}_I^U, \mathfrak{U}_{II}^U), (\mathfrak{B}_I^U, \mathfrak{B}_{II}^U)\right\}, \mathscr{B}_u, T\right\}$$

of gameplay (Kim, 2021b).

- $\mathbb{G}^O$ and $\mathbb{G}^U$ are over and under bargaining game models; they are related in an interactive manner of mutual and reciprocal interdependency.
- D is the set of the set of smart devices $\left(\mathscr{D}_{1\leq i\leq n}\right)$, which generate different types of traffic data.
- $\mathcal{Q}_I$, $\mathcal{Q}_{II}$ and $\mathcal{Q}_{III}$ are the queues in the SHG for class I, II and III data, respectively.
- B is the total SHG bandwidth for SHN traffic services.
- In the $\mathbb{G}^O$, P_I^O and P_{II}^O are game players, and $\mathfrak{U}_I^O$ and $\mathfrak{U}_{II}^O$ are their utility functions, respectively. $\mathfrak{B}_I^O$ and $\mathfrak{B}_{II}^O$ are the allocated bandwidth for the P_I^O and P_{II}^O during the bargaining process.
- In the $\mathbb{G}^U$, P_I^U and P_{II}^U are game players, and $\mathfrak{U}_I^U$ and $\mathfrak{U}_{II}^U$ are their utility functions, respectively. $\mathfrak{B}_I^U$ and $\mathfrak{B}_{II}^U$ are the allocated bandwidth for the P_I^U and P_{II}^U during the bargaining process.
- $\mathscr{B}_u$ is a basic unit to allocate the bandwidth resource in the SHG.
- $T=\left\{t_1,\ldots,t_c,t_{c+1},\ldots\right\}$ denotes time, which is represented by a sequence of time steps.

11.4.4 The basic concepts of *BBS* and *IEBS*

A solution is any function $\mathscr{F}:\mathbb{C}\rightarrow\mathbb{R}^n$ that satisfies $\mathscr{F}(S,d)\in S$ for all $(S,d)\in\mathbb{C}$. Given a feasible set S, the weak *Pareto* frontier of S is defined as $WP(S)\equiv\{x\in S:y>x\Rightarrow y\notin S\}$. The best that player $1\leq i\leq n$ can hope for in the problem (S,d) is $a_i(S,d)\equiv\max\left\{x_i:x\in S_d\right\}$ where $S_d\equiv\{x\in S:x\geq d\}$. Therefore, for all game players, the point $a(S,d)=\left(a_1(S,d).\ .\ .\ ,a_n(S,d)\right)$ is the ideal point of the problem (S,d). The *KSBS* is defined by $KSBS(S,d)\equiv WP(S)\cap[d;a(S,d)]$. On the other hand, the *NBS* is defined to be the unique maximizer of $Đ_{1\leq i\leq n}\left(x_i-d_i\right)$ over S_d (Kim, 2021b; Rachmilevitch, 2013).

To bridge the gap between NBS and *KSBS*, a new bargaining solution, i.e., *BBS*, has been introduced. For each$(S,d)\in\mathbb{C}$, this solution assumes that player i temporarily gets his payoff as follows (Kim, 2021b; Shiran, 2014);

$$m_i(S,d)=\mathbf{min}\left\{NBS_i(S,d),KSBS_i(S,d)\right\}\tag{11.4.1}$$

According to (11.4.1), the *BBS* for player i, i.e., $BBS_i(S,d)$, is given by (Shiran, 2014);

$$BBS_{1\leq i\leq n}(S,\ d)\equiv\begin{cases}m_i(S,d), if NBS(S,\ d)=KSBS(S,\ d)\\ KSBS_{1\leq i\leq n}\left(\{x\in S:x\geq m(S,d)\},m(S,d)\right), otherwise\end{cases}\tag{11.4.2}$$

It is easy to see that the *BBS* is a standard solution which is at least as independent as the *KSBS*. Simply, we assume a two-player bargaining game where $S\subset\mathbb{R}^2$. Given $x,y\in\mathbb{R}^2$, $x\preccurlyeq y$ is a partial order on the plane if $x_1\leq y_1$ and $x_2\geq y_2$. That is, $x\preccurlyeq y$ means that x is weakly to the north-west of y. If the following proposition (11.4.3) is true for every $(\mathcal{Q},e)\in\mathbb{C}$, we can say that the *BBS* is at least as independent as the *KSBS* given (S,d) (Kim, 2021b; Shiran, 2014).

$$\begin{cases} (KSBS(\mathcal{Q},e) \preccurlyeq KSBS(S,\ d)) \Rightarrow ((KSBS(\mathcal{Q},e) \preccurlyeq BBS(\mathcal{Q},e) \preccurlyeq KSBS(S,\ d))) \\ (KSBS(S,\ d) \preccurlyeq KSBS(\mathcal{Q},e)) \Rightarrow ((KSBS(S,\ d) \preccurlyeq BBS(\mathcal{Q},e) \preccurlyeq KSBS(\mathcal{Q},e))) \end{cases}$$

(11.4.3)

s.t., $e = d, \mathcal{Q} \subset S$ and $x \in \mathcal{Q}$

Another new bargaining solution, the *IEBS*, has been introduced based on the *EBS* and *ELBS*. The *EBS* equalizes players' gains over their disagreement outcomes. If we simply assume a two-person bargaining problem, this corresponds to selecting the intersection point of the Pareto frontier and the 45-degree line drawn from the disagreement point. ELBS equalizes players' losses from their ideal point outcomes. Therefore, the *ELBS* assigns to each $S \subset \mathbb{R}^2$, the point $ELBS(S) = a(S) - (\mathbb{M}, \mathbb{M})$, where m is the minimum possible. This corresponds to selecting the point at the intersection of the Pareto frontier and the 45-degree line drawn from the ideal point. Note that if $a_1(S) > a_2(S)$, then $ELBS_1(S) > EBS_1$ and $EBS_2(S) > ELBS_2(S)$, and vice versa. Let N be the set of positive integer number. The *IEBS* assigns to each S, the point x, if $EBS(S) = ELBS(S) = x$ and assigns the point $y \equiv \cap_{\mathcal{T} \in \mathrm{N}} PO(S_{\mathcal{T}})$, where $S_0 \equiv S$ and the bargaining problem in iteration step T, i.e., $S_{\mathcal{T}}$, for $\mathcal{T} \geq 1$ is derived by applying *EBS* and *ELBS* to $S_{\mathcal{T}-1}$ in a way that, the disagreement point and ideal point of $S_{\mathcal{T}}$ denoted by $d(S_t)$ and $a(S_t)$, respectively (Karagözoğlu, 2018; Kim, 2021b).

$$d(S_t) = (\mathcal{I}, \mathcal{H}) \text{ and } a(S_t) = (\mathcal{K}, \mathcal{L})$$

(11.4.4)

$$\text{s.t.,} \begin{cases} I = \left(\mathbf{min}\{EBS_1(S_{t-1}),\ ELBS_1(S_{t-1})\}\right) \\ H = \left(\mathbf{min}\{EBS_2(S_{t-1}), ELBS_2(S_{t-1})\}\right) \\ K = \left(\mathbf{max}\{EBS_1(S_{t-1}), ELBS_1(S_{t-1})\}\right) \\ L = \left(\mathbf{max}\{EBS_2(S_{t-1}), ELBS_2(S_{t-1})\}\right) \end{cases}$$

The *IEBS* can be understood as a conflict resolution mechanism that systematically reconciles the differences between the *EBS* and the *ELBS*. This resolution process follows a step-by-step iterative approach, where the minimal outcomes from each iteration serve as starting points, while the maximal outcomes are treated as ideal reference points for the subsequent bargaining step. Figure 5 illustrates the operational dynamics of *IEBS*, demonstrating how it functions in scenarios where the *EBS* and *ELBS* propose divergent outcomes (Karagözoğlu, 2018; Kim, 2021b).

Figure 5. Iterated egalitarian bargaining solution (IEBS) (Kim 2021b)

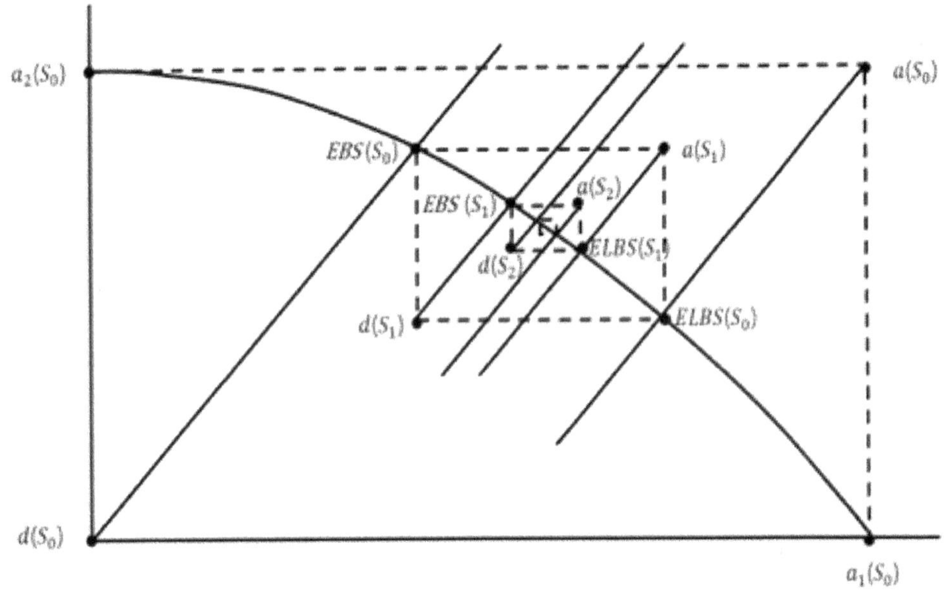

The *IEBS* satisfies the axiom of *Midpoint domination* (*MD*), which requires the solution to provide players' payoffs that are at least as large as the average of their best and worst payoffs where $ELBS(S) \geq (1/2 \cdot (a(S) + d(S)))$; the *MD* axiom is violated by the *EBS* and the *ELBS*. If we want to utilize an egalitarian justice issue in problems where the *EBS* and the *ELBS* disagree, the *IEBS* is a reasonable alternative in a domain that allows inter-players' utility comparisons (Karagözoğlu, 2018; Kim, 2021b).

11.4.5 The two-level bargaining game model for the SHG

Each $\mathcal{D}_{1 \leq i \leq n}$ generates independently different types of traffic data. According to the data type, service priority is differently decided. The *SHTC* scheme is developed as a two-level bargaining game mode. At the over-level, the $\mathbb{G}^{O}$ is designed as a two-player game model, and it divides the total SHG bandwidth ($\mathfrak{B}$) into two parts; $\mathfrak{B}_{I}^{O}$ is assigned for the P_{I}^{O} and $\mathfrak{B}_{II}^{O}$ is assigned for the P_{II}^{O} where $\mathfrak{B} = \mathfrak{B}_{I}^{O} + \mathfrak{B}_{II}^{O}$. Under a dynamically changing SHN traffic environment, an utility function maps player-level satisfaction to a real number, which represents the resulting payoff. The utility functions of P_{I}^{O} and P_{II}^{O} at time t_c, i.e., $\mathfrak{U}_{I}^{O}(\cdot)$ and $\mathfrak{U}_{II}^{O}(\cdot)$, are formally defined as follows (Kim, 2021b);

$$\begin{cases} \mathfrak{U}_{I}^{O}\left(\mathcal{S}I_{\mathcal{D}_i \in \mathrm{D}}^{t_c}, \mathfrak{B}_{I}^{O}\right) = \phi \times \left(\dfrac{\lambda}{\kappa + exp\left(-\mu \times \dfrac{\min\left(\sum_{\mathcal{D}_i \in \mathrm{D}} \left(\mathcal{S}I_{\mathcal{D}_i}^{t_c}\right), \mathfrak{B}_{I}^{O}\right)}{\sum_{\mathcal{D}_i \in \mathrm{D}} \left(\mathcal{S}I_{\mathcal{D}_i}^{t_c}\right)} \right)} - \pi \right) \\ \mathfrak{U}_{II}^{O}\left(\mathcal{S}II_{\mathcal{D}_i \in \mathrm{D}}^{t_c}, \mathfrak{B}_{II}^{O}\right) = \zeta \times \log\left(\dfrac{\min\left(\sum_{\mathcal{D}_i \in \mathrm{D}} \left(\mathcal{S}II_{\mathcal{D}_i}^{t_c}\right), \mathfrak{B}_{II}^{O}\right)}{\sum_{\mathcal{D}_i \in \mathrm{D}} \left(\mathcal{S}II_{\mathcal{D}_i}^{t_c}\right)} + \varepsilon \right) \end{cases} \quad (11.4.5)$$

where $\mathcal{S}I^{t_c}_{\mathcal{D}_i}$ is the generated class I and class II traffic data, and $\mathcal{S}II^{t_c}_{\mathcal{D}_i}$ is the generated class III traffic data from the $\mathcal{D}_i$ at time t_c. ϕ, λ, κ, μ, π are coefficient factors for the $\mathfrak{U}^O_I(\cdot)$, and ζ, ε are coefficient factors for the $\mathfrak{U}^O_{II}(\cdot)$. The *SHTC* scheme adopts the idea of *BBS* to adjust the $\mathfrak{B}^O_I$ and $\mathfrak{B}^O_{II}$ values. Therefore, beforehand, the *SHTC* scheme should know the *NBS* and *KSBS* for the $\mathbb{G}^O$ game. The $NBS_{P^O_I,P^O_{II}}$ is the *NBS* and the $KSBS_{P^O_I,P^O_{II}}$ is the *KSBS* for the P^O_I and P^O_{II}, respectively (Kim, 2021b).

$$\begin{cases} NBS_{P^O_I,P^O_{II}} = \max_{\mathfrak{B}^O_I,\mathfrak{B}^O_{II}}\left(\left(\mathfrak{U}^O_I\left(\mathcal{Q}I^{t_c}_{\mathcal{D}_i},\mathfrak{B}^O_I\right)-d_{P^O_I}\right)\times\left(\mathfrak{U}^O_{II}\left(\mathcal{Q}II^{t_c}_{\mathcal{D}_i},\mathfrak{B}^O_{II}\right)-d_{P^O_{II}}\right)\right) \\ KSBS_{P^O_I,P^O_{II}} = \max_{\mathfrak{B}^O_I,\mathfrak{B}^O_{II}}\left(\frac{\left(\mathfrak{U}^O_I\left(\mathcal{Q}I^{t_c}_{\mathcal{D}_i},\mathfrak{B}^O_I\right)-d_{P^O_I}\right)}{\left(\mathfrak{U}^O_I\left(\mathcal{Q}I^{t_c}_{\mathcal{D}_i},\mathfrak{B}\right)-d_{P^O_I}\right)} = \frac{\left(\mathfrak{U}^O_{II}\left(\mathcal{Q}II^{t_c}_{\mathcal{D}_i},\mathfrak{B}^O_{II}\right)-d_{P^O_{II}}\right)}{\left(\mathfrak{U}^O_{II}\left(\mathcal{Q}II^{t_c}_{\mathcal{D}_i},\mathfrak{B}\right)-d_{P^O_{II}}\right)}\right) \end{cases} \tag{11.4.6}$$

where $d_{P^O_I}$ and $d_{P^O_{II}}$ are the disagree points of P^O_I and P^O_I, respectively. $\mathfrak{U}^O_I\left(\mathcal{Q}I^{t_c}_{\mathcal{D}_i\in D},\mathfrak{B}\right)$ and $\mathfrak{U}^O_{II}\left(\mathcal{Q}II^{t_c}_{\mathcal{D}_i\in D},\mathfrak{B}\right)$ are the ideal points of P^O_I and P^O_I according to (11.4.5). Based on the values of $NBS_{P^O_I,P^O_{II}}$ and the $KSBS_{P^O_I,P^O_{II}}$, the BBS for the P^O_I and P^O_{II} is given by (Kim, 2021b);

$$BBS_{P^O_I,P^O_{II}} \equiv \begin{cases} \mathfrak{N}_{P^O_I,P^O_{II}}, \mathit{if} NBS_{P^O_I,P^O_{II}} = KSBS_{P^O_I,P^O_{II}} \\ KSBS_{P^O_I,P^O_{II}} \mathit{with}\, d_{P^O_I,P^O_{II}} = \mathfrak{N}_{P^O_I,P^O_{II}}, \mathit{otherwise} \end{cases} \tag{11.4.7}$$

$$\text{s.t., } \mathfrak{N}_{P^O_I,P^O_{II}} = \mathbf{min}\left\{NBS_{P^O_I,P^O_{II}}, KSBS_{P^O_I,P^O_{II}}\right\}$$

where $d_{P^O_I,P^O_{II}}$ is a new disagreement point in the *BBS*. By using (11.4.7), the results of $\mathbb{G}^O$, i.e., the $\mathfrak{B}^O_I$ and $\mathfrak{B}^O_{II}$ amounts, are obtained (Kim, 2021b).

At the under-level, the $\mathbb{G}^U$ is also developed as a two-player game model, and it divides the allocated $\mathfrak{B}^O_I$ into two parts; $\mathfrak{B}^U_I$ is assigned for the P^U_I and $\mathfrak{B}^U_{II}$ is assigned for the P^U_{II} where $\mathfrak{B}^O_I = \mathfrak{B}^U_I + \mathfrak{B}^U_{II}$. To effectively share the $\mathfrak{B}^O_I$ between P^U_I and P^U_{II}, the utility functions of P^U_I and P^U_{II} at time t_c, i.e., $\mathfrak{U}^U_I(\cdot)$ and $\mathfrak{U}^U_{II}(\cdot)$, are mathematically defined as follows (Kim, 2021b);

$$\begin{cases} \mathfrak{U}^U_I\left(\mathcal{W}I^{t_c}_{\mathcal{D}_j\in D},\mathfrak{B}^U_I\right) = \sigma\times exp\left(-\alpha\times\frac{\min\left(\sum_{\mathcal{D}_j\in D}\left(\mathcal{W}I^{t_c}_{\mathcal{D}_j}\right),\mathfrak{B}^U_I\right)}{\sum_{\mathcal{D}_j\in D}\left(\mathcal{W}I^{t_c}_{\mathcal{D}_j}\right)}\right) \\ \mathfrak{U}^U_{II}\left(\mathcal{W}II^{t_c}_{\mathcal{D}_j\in D},\mathfrak{B}^U_{II}\right) = \beta - exp\left(-\eta\times\frac{\min\left(\sum_{\mathcal{D}_j\in D}\left(\mathcal{W}II^{t_c}_{\mathcal{D}_j}\right),\mathfrak{B}^U_{II}\right)}{\sum_{\mathcal{D}_j\in D}\left(\mathcal{W}II^{t_c}_{\mathcal{D}_j}\right)}\right) \end{cases} \tag{11.4.8}$$

where $\mathcal{W}I^{t_c}_{\mathcal{D}_j}$ and $\mathcal{W}II^{t_c}_{\mathcal{D}_j}$ are the class I and class II traffic data generated from the $\mathcal{D}_j$ at time t_c, respectively. σ, α are coefficient factors for class I data services, and β, η are coefficient factors for class II data services. The *SHTC* scheme uses the concept of *IEBS* to adjust the $\mathfrak{B}^U_I$ and $\mathfrak{B}^U_{II}$ values. Therefore, in advance, the *SHTC* scheme should know the *EBS* and *ELBS* for the $\mathbb{G}^U$ game. The $EBS_{P^U_I,P^U_{II}}$ is the *EBS* and the $ELBS_{P^U_I,P^U_{II}}$ is the *ELBS* for the P^U_I and P^U_{II}, respectively (Kim, 2021b).

$$\begin{cases} EBS_{P_I^U,P_{II}^U} = \max_{\mathfrak{B}_I^U,\mathfrak{B}_{II}^U}\left(\left(\mathfrak{U}_I^U\left(\mathscr{W}I_{\mathscr{D}_j\in\mathbb{D}}^{t_c},\mathfrak{B}_I^U\right) - d_{P_I^U}\right) = \left(\mathfrak{U}_{II}^U\left(\mathscr{W}II_{\mathscr{D}_j\in\mathbb{D}}^{t_c},\mathfrak{B}_{II}^U\right) - d_{P_{II}^U}\right)\right) \\ ELBS_{P_I^U,P_{II}^U} = \max_{\mathfrak{B}_I^U,\mathfrak{B}_{II}^U}(X_I = X_{II}) \end{cases} \tag{11.4.9}$$

$$\text{s.t.,} \begin{cases} X_I = \left|\mathfrak{U}_I^U\left(\mathscr{W}I_{\mathscr{D}_j}^{t_c},\mathfrak{B}\right) - \mathfrak{U}_I^U\left(\mathscr{W}I_{\mathscr{D}_j\in\mathbb{D}}^{t_c},\mathfrak{B}_I^U\right)\right| \\ X_{II} = \left|\mathfrak{U}_{II}^u\left(\mathscr{W}II_{\mathscr{D}_j}^{t_c},\mathfrak{B}\right) - \mathfrak{U}_{II}^U\left(\mathscr{W}II_{\mathscr{D}_j\in\mathbb{D}}^{t_c},\mathfrak{B}_{II}^U\right)\right| \end{cases}$$

According to (11.4.9), the $EBS_{P_I^U,P_{II}^U}$ and the $ELBS_{P_I^U,P_{II}^U}$ are obtained, and the *SHTC* scheme sets them as the first round temporal solutions. And then, for the next round, the *SHTC* scheme can get a new ideal point, i.e., a_{i+1}, and a new disagree point, i.e., d_{i+1}, as follows (Kim, 2021b);

$$\begin{cases} a_{i+1} = \left(a_{i+1}^I, a_{i+1}^{II}\right) = \left(\max\left(EBS_{P_I^U}, EBS_{P_I^U}\right), \max\left(EBS_{P_{II}^U}, EBS_{P_{II}^U}\right)\right) \\ d_{i+1} = \left(d_{i+1}^I, d_{i+1}^{II}\right) = \left(\min\left(EBS_{P_I^U}, EBS_{P_I^U}\right), \min\left(EBS_{P_{II}^U}, EBS_{P_{II}^U}\right)\right) \end{cases} \tag{11.4.10}$$

Based on the values of a_{i+1} and d_{i+1}, the second round's temporal solutions are obtained through the negotiation of *EBS* and *ELBS*. By iteratively repeating this process, the *SHTC* scheme can approach the $IEBS_{P_I^U,P_{II}^U}$ while adjusting the difference between the *EBS* and *ELBS* (Kim, 2021b).

11.4.6 Summary

With the increasing integration of smart appliances into modern households, SHNs have become an essential component of home automation and connectivity. In a dynamic SHN environment, implementing an efficient bandwidth-sharing strategy is crucial for maximizing network performance. The *SHTC* scheme introduces an optimized SHN traffic management framework based on cooperative game theory. To address the bandwidth allocation challenge in the SHG, the *SHTC* scheme incorporates the principles of the *BBS* and the *IEBS* within a two-tier bargaining game model. At the over-level, the *BBS* framework bridges the gap between *NBS* and *KSBS* to distribute SHG bandwidth among class I (critical), class II (delay-sensitive), and class III (best-effort) traffic. At the under-level, the *IEBS* model resolves the differences between the *EBS* and ELBS to further allocate the assigned bandwidth between class I and class II traffic. Through a step-by-step interactive process, the two-tier bargaining mechanism ensures that diverse QoS requirements are met while optimizing SHG bandwidth utilization. Under varied SHN traffic conditions, the *SHTC* scheme achieves globally desirable system-wide efficiency, providing mutually beneficial advantages across different QoS service demands (Kim, 2021b).

11.5 Data Sponsoring Model Control (*DSMC*) scheme

As the transition towards future 5G networks progresses, the demand for mobile data usage continues to increase significantly, with high data costs emerging as a critical concern. In response to this challenge, *data sponsoring* has been introduced as a transformative paradigm aimed at benefiting the telecommunications industry. The fundamental concept of *data sponsoring* involves subsidizing service payments to attract more users, ultimately generating greater financial gains for both users and network operators. Recently, S. Kim develops the *Data Sponsoring Model Control* (*DSMC*) scheme, which employs a novel dual-leader Stackelberg game framework (Kim, 2019). Within the *DSMC* scheme, the strategic interactions between three key network entities - users, service providers, and content providers - are analyzed using game-theoretic principles. As dual leaders, service providers and content providers cooperatively adjust service pricing, while mobile users, acting as followers, independently respond to pricing decisions in a competitive manner. By integrating various game-theoretic methodologies, the *DSMC* scheme explores the mutual dependencies between these network participants, striving to achieve an optimal balance between user benefits and overall system profitability.

11.5.1 Development motivation

The global mobile data traffic has been experiencing an exponential surge, reaching 2.5 exabytes per month in 2014 and projected to surpass 24.3 exabytes per month by 2020. Additionally, the total number of mobile devices worldwide is expected to reach 75 billion, with these devices becoming increasingly intelligent. According to Cisco, mobile video traffic will constitute nearly 75% of total mobile data usage. As more users stream videos on their smart devices, content providers (CPs) - such as YouTube and Twitter - must adopt flexible data access strategies to accommodate this shift. While the rapid growth in mobile video consumption presents significant opportunities, it also introduces new operational challenges for CPs (Pang, 2016; Chen, 2018; Jang, 2013; Rizvi, 2015; Kim, 2019) (Chen, 2018; Jang, 2013; Kim, 2019; Rizvi, 2015; Pang, 2016).

Since mobile users (MUs) are highly price-sensitive when consuming video content, attracting more users while ensuring profitability has become a critical challenge for CPs (Kim, 2010; Pande, 2013). To address this, AT&T introduced '*data sponsoring*' in 2014, a service model that enables content providers to subsidize mobile users' data costs. This approach allows CPs to attract more users while increasing their revenue through in-app advertising. By covering users' video data expenses, CPs encourage more engagement, leading to greater advertisement-driven revenue growth. A real-world example is AT&T's Sponsored Data program, which enables advertisers to sponsor video streaming for users. This strategy increases ad exposure by encouraging more users to watch advertisements that they might have otherwise skipped. Thus, *data sponsoring* creates a mutually beneficial arrangement for both CPs and MUs (Kim, 2019; Pang, 2016; Xiong, 2018).

Between CPs and MUs, service providers (SPs) serve as intermediaries, offering network access, bandwidth provisioning, and data relay services. As business entities, SPs charge MUs for service deliveries, meaning that *data sponsoring* also holds potential advantages for SPs. Since MUs are more inclined to use data-intensive services when costs are subsidized by CPs, SPs benefit from increased network usage and revenue generation (Joe-Wong, 2018). Due to the economic incentives for CPs, SPs, and MUs, researchers have taken an interest in exploring the interdependencies between these three network entities. This has led to the development of *data sponsoring* control models aimed at optimizing revenue distribution and fostering a 'triple-win' outcome. The service-oriented architecture for *data sponsoring* typically includes; Policy and Charging Rules Function (PCRF), Online Charging System, Deep Packet Inspection, Captive Portal, Ad Server, and Partner Settlement module. These components collectively support operators in managing sponsored data plans effectively (Joe-Wong, 2018; Kim, 2019; Xiong, 2018).

Conventional network management methods are inadequate for handling the complexities of *data sponsoring*, necessitating the adoption of a new control framework. Recently, game theory has gained widespread application in computer networks, including resource allocation and competitive analysis among network entities. In environments where multiple intelligent agents interact, game theory provides a robust framework for identifying optimal decision-making strategies. Game models can be categorized into; i) cooperative games - where rational decision-makers can reach binding agreements, and ii) non-cooperative games - where binding agreements are not possible, requiring each entity to act independently. Today, game theory is applied across a broad range of strategic decision-making problems, making it a key analytical tool in network management and optimization (Kim, 2014, 2015, 2019; Raazi, 2010).

Among various game models, the Stackelberg game has attracted significant attention for its hierarchical decision-making framework. In a traditional Stackelberg game, two types of players exist; leaders who make strategic decisions while anticipating the responses of followers, and followers who react rationally to optimize their own objectives. This model is well-suited for static bi-level optimization problems, where leaders influence the actions of multiple followers. The *DSMC* scheme incorporates the Stackelberg game framework but extends it to address *data sponsoring* challenges. Unlike the conventional model, the *DSMC* scheme introduces a dual-leader Stackelberg game, where the CP and SP act as dual game leaders, and multiple MUs are modeled as followers. Within this framework, CPs and SPs dynamically interact to optimize system performance. While some studies have explored multi-leader Stackelberg games, few have effectively addressed the coexistence challenges of multiple leaders (Kim, 2019).

Building upon the multi-leader Stackelberg model, the *DSMC* scheme structures decision-making as a systematic feedback process: leaders (CP & SP) independently set service prices using an iterative learning algorithm, and followers (MUs) dynamically respond to maximize their own payoffs. This two-stage decision-making framework is based on a non-cooperative game approach. As a result of user interactions, CPs and SPs secure their profits, while a cooperative game approach ensures fair revenue sharing between the two leaders. Using the Rubinstein bargaining model, CPs and SPs negotiate their mutual benefits while balancing competition and cooperation (Kim, 2019).

In a sponsored data framework, users can access websites, stream videos, and use applications without consuming their personal data plans. In return, users are exposed to advertisements, which are funded by advertising vendors and paid directly to network operators. This model effectively aligns the interests of advertisers/promoters, content service operators, and mobile users. A Sponsored Data Solution incorporates key components, including PCRF, Online Charging System, Deep Packet Inspection, Captive Portal, Ad Server and Partner Settlement module. These elements enable telecom operators to efficiently implement sponsored data plans (Kim, 2019; Sterlite Technologies,2003).

- **Communications Service Provider (CDR) Reconciliation:** Sponsored data generates event-based Call Detail Records (CDRs), which log user interactions with advertisements. These records facilitate billing and revenue calculations for advertising partners.
- **Integrated Policy and Charging:** The PCRF-integrated online charging system enables QoS provisioning, partner pricing, revenue sharing, and settlement while allowing personalized data plans.
- **Multi-Network Compatibility:** Sponsored data solutions can be deployed across LTE, 3G, Wi-Fi, and Fixed Line networks.
- **Seamless Integration with Existing Systems:** Components can be integrated into existing third-party operator platforms.

- **Location-Based Services:** Operators can utilize location-specific partnerships, targeted promotions, and real-time localized offerings to enhance user engagement.

11.5.2 Contribution

The *DSMC* scheme is designed by incorporating smart data pricing strategies, leveraging two distinct game-theoretic methodologies to derive a mutually beneficial solution. To facilitate efficient *data sponsoring*, the *DSMC* scheme emphasizes the cooperation between leaders (content providers and service providers) and the competition between leaders and followers (mobile users). The key contributions of the *DSMC* scheme can be summarized as follows. The *DSMC* scheme introduces a dynamic price adjustment algorithm for the dual leaders (CPs and SPs). Using the iterative firefly learning mechanism, the prices of content and services are adaptively modified at each time interval to respond to real-time market conditions. The *DSMC* scheme proposes a novel profit-sharing algorithm for leaders, ensuring equitable revenue distribution. Based on the Rubinstein bargaining model, both CPs and SPs can reach a fair and efficient agreement to equitably split profits. The *DSMC* scheme examines the sequential interactions between price adjustment strategies, mobile users' reactions, and profit-sharing negotiations. By jointly integrating these processes, the *DSMC* scheme develops a comprehensive dual-leader Stackelberg game model, ensuring mutual benefits for all entities involved. The synergistic advantage of this approach stems from the strategic combination of multiple game-theoretic models. The *DSMC* scheme accounts for incomplete information environments, making it more practical for real-world implementation. Existing approaches have predominantly focused on idealized scenarios with complete system information, which rarely reflect real-world conditions. By addressing uncertainties in *data sponsoring* control, the *DSMC* scheme enhances robustness and adaptability in real-world telecommunications environments (Kim, 2019).

11.5.3 Dual-leader Stackelberg game model for CP, SPs, and MUs

The *DSMC* scheme is based on the premise that multimedia data comprises both real content and advertisements, with the CP subsidizing MUs' data consumption costs. Meanwhile, the SP determines the pricing structure for data transmission. Under this *data sponsoring* framework, the *DSMC* scheme introduces a dual-leader Stackelberg game model, where CPs, SPs, and MUs act as strategic players. The CP and SP serve as dual leaders, making pricing decisions based on MUs' interactive behavior. The MUs act as followers, deciding whether to consume multimedia data based on the pricing strategies set by the leaders. Each participant aims to maximize its individual payoff, while considering the influence of other players' decisions. This game-theoretic approach allows dynamic pricing adjustments and strategic interactions between content providers, service providers, and mobile users, optimizing the *data sponsoring* process and ensuring a mutually beneficial outcome for all entities involved. The *DSMC* scheme formulates the dual-leader Stackelberg game as a three-stage sequential decision process: i) the CP and the SP individually decide their price policies based on the online learning mechanism, ii) the MUs react to the leaders' decisions in a non-cooperative manner, and iii) according to the *Rubinstein* bargaining solution, the CP and the SP share the profit, which comes from the MUs' payments. These processes are dynamically repeated until the best solution is obtained. Formally, the *DSMC* scheme defines the dual-leader Stackelberg game model

$$G = \left\{ \mathbb{N} = \left\{ \left\{ CP, SP_{1 \le i \le n} \right\}, \left\{ MU^{SP_i}_{1 \le j \le m} \right\} \right\}, \mathbb{S} = \left\{ \mathbb{S}^{CP}, \mathbb{S}^{SP}, \mathbb{S}^{MU} \right\}, \mathbb{L} = \left\{ \mathbb{L}^{CP}, \mathbb{L}^{SP} \right\}, \mathbb{U} = \left\{ \mathbb{U}^{CP}, \mathbb{U}^{SP}, \mathbb{U}^{MU} \right\}, \eta, T \right\}$$

at each time period of the gameplay (Kim, 2019).

- N is the finite set of game players. The *DSMC* scheme models a monopoly CP scenario to analyze the impact of *data sponsoring* on multiple SPs and MUs; there are n SPs and m MUs where $MU_{1 \le j \le m}^{P_i}$ represents the jth MU in the covering area of SP_i.
- S is the finite set of each player's strategies, which are defined discretely. $\mathbb{S}^{CP} = \left\{ \mathfrak{P}^{CP}_{min} \ldots \mathfrak{P}^{CP}_{w} \ldots \mathfrak{P}^{CP}_{max} \right\}$ is the strategy set of the CP where $\mathfrak{P}^{CP}_{w}$ represents the w^{th} *data sponsoring* level for the MUs. $\mathbb{S}^{SP} = \left\{ \mathfrak{Z}^{SP}_{min} \ldots \mathfrak{Z}^{SP}_{k} \ldots \mathfrak{Z}^{SP}_{max} \right\}$ is the strategy set of the SP where $\mathfrak{Z}^{SP}_{k}$ represents the kth price level for the MUs.$\mathbb{S}^{MU} = \{on, off\}$ is the strategy set of the MUs where *on, off* represent whether the MU's is consuming data or not.
- U is the utility set of each player's payoffs where $\mathbb{U}^{CP}$, $\mathbb{U}^{SP}$, and $\mathbb{U}^{MU}$ represent the payoffs of the CP, the SP, and the MU, respectively.
- L is the finite set of leaning vectors. $\mathbb{L}^{CP} = [\psi^{\mathfrak{P}^{CP}_{min}} \ldots \psi^{\mathfrak{P}^{CP}_{w}} \ldots \psi^{\mathfrak{P}^{CP}_{max}}]$ is the CP's leaning vector where $\psi^{\mathfrak{P}^{CP}_{w}}$ represents the strategy $\mathfrak{P}^{CP}_{w}$'s propensity. $\mathbb{L}^{SP} = [\xi^{\mathfrak{Z}^{SP}_{min}} \ldots \xi^{\mathfrak{Z}^{SP}_{k}} \ldots \xi^{\mathfrak{Z}^{SP}_{max}}]$ is the SP's leaning vector where $\xi^{\mathfrak{Z}^{SP}_{k}}$ represents the strategy $\mathfrak{Z}^{SP}_{k}$'s propensity; in the next time period, L is used to estimate the strategy selection probability for the CP and the SP's strategy decisions.
- η is the CP's fixed unit price per bit for the provision of the data content.
- $T = \left\{ \mathcal{H}_1, \ldots, \mathcal{H}_t, \mathcal{H}_{t+1}, \ldots \right\}$ denotes time, which is represented by a sequence of time steps with imperfect information for the dual-leader Stackelberg game process.

The first process in the dual-leader Stackelberg game is the leader's price decision algorithm. From the viewpoint of the CP and the SP, they select their strategies in the $\mathbb{S}^{CP}$and the $\mathbb{S}^{SP}$to maximize their payoffs. Simply, the utility functions of the CP and the SP_i ($\mathbb{U}^{CP}_{\mathcal{H}_t}$ and $\mathbb{U}^{SP_i}_{\mathcal{H}_t}$) at time $\mathcal{H}_t$ are be defined as follows (Kim, 2019);

$$\begin{cases} \mathbb{U}^{CP}_{\mathcal{H}_t}(\mathfrak{P}^{CP}_{w}) = \left(\mathcal{S}^{CP}_{\mathcal{H}_t}(\mathfrak{P}^{CP}_{w}, N^{SP_i}) - \left(\mathcal{C}^{CP}_{\mathcal{H}_t}(N^{SP_i}) + \Theta^{CP}_{\mathcal{H}_t}(\mathfrak{P}^{CP}_{w}, N^{SP_i}) \right) + \mathcal{V}^{CP}_{\mathcal{H}_t}(\mathfrak{P}^{CP}_{w}, N^{SP_i}) \right) \\ \mathbb{U}^{SP_i}_{\mathcal{H}_t}(\mathfrak{Z}^{SP_i}_{k}) = \left(\mathcal{S}^{SP}_{\mathcal{H}_t}(\mathfrak{Z}^{SP_i}_{k}, N^{SP_i}) - \mathcal{C}^{SP_i}_{\mathcal{H}_t}(N^{SP_i}) \right) \end{cases} \tag{11.5.1}$$

$$s.t., \begin{cases} \mathcal{C}^{CP}_{\mathcal{H}_t}(N^{SP_i}) = \mathcal{F}^{data}_{\mathcal{H}_t}\left(N^{SP_i}, \mathcal{H}_t\right) \times \eta \\ \Theta^{CP}_{\mathcal{H}_t}\left(\mathfrak{P}^{CP}_{w}, N^{SP_i}\right) = \left(\mathcal{F}^{data}_{\mathcal{H}_t}\left(N^{SP_i}, \mathcal{H}_t\right) \times \eta \right) \times \mathfrak{P}^{CP}_{w} \\ \mathcal{V}^{CP}_{\mathcal{H}_t}\left(\mathfrak{P}^{CP}_{w}, N^{SP_i}\right) = \mathcal{F}^{data}_{\mathcal{H}_t}\left(N^{SP_i}, \mathcal{H}_t\right) \times \mathcal{R}^{CP} \\ \mathcal{C}^{SP_i}_{\mathcal{H}_t}(N^{SP_i}) = \mathcal{F}^{data}_{\mathcal{H}_t}\left(N^{SP_i}, \mathcal{H}_t\right) \times r^{SP}_{c} \end{cases}$$

where N^{SP_i} is the set of MUs who actively consume the data through the SP_i. $\mathcal{S}_{\mathcal{H}_t}(\cdot)$ is a function that represents the profit-sharing between the CP and the SP at time $\mathcal{H}_t$; it is defined in the leader's profit-sharing algorithm based on the *Rubinstein* bargaining approach. $\mathcal{C}^{CP}_{\mathcal{H}_t}(N^{SP_i})$ and $\mathcal{C}^{SP_i}_{\mathcal{H}_t}(N^{SP_i})$ are the CP and the SP's cost functions to provide data services for the MUs in N^{SP_i}, respectively. $\mathcal{F}^{data}_{\mathcal{H}_t}\left(N^{SP_i}, \mathcal{H}_t\right)$ is a function that represents the total amount of data consumed from the N^{SP_i} at time $\mathcal{H}_t$. r^{SP}_{c} is a cost factor for the data service of the SP. $\Theta_{\mathscr{H}}^{CP}(\mathfrak{P}_w^{CP}, N^{SP_i})$ is the *data sponsoring* subsidy by the CP's strategy $\mathfrak{P}_w^{CP}$ for the MUs in N^{SP_i}. $\mathscr{V}_{\mathscr{H}_t}^{CP}(\mathfrak{P}_w^{CP}, N^{SP_i})$ is the extra revenue function from the advertising selling, and $\mathscr{R}^{CP}$ is the revenue rate per amount of data consumed (Kim, 2019).

As game leaders, the main goal of the CP and the SP is to maximize their utility functions by rationally selecting their strategies, i.e., $\mathfrak{P}^{CP}$ and $\mathfrak{Z}^{SP}$. In a situation where information is incomplete, it is necessary for the CP and the SP to learn the condition of the current system while self-adapting their actions to respond quickly to real-time network dynamics. Therefore, the CP and the SP should adopt a learning technique to decide their next strategies by using their accumulated knowledge. So far, several learning methods have been developed to help game players who have incomplete information about their dynamic network environments (Kim, 2019).

In 2008, Xin-She Yang proposed a novel swarm intelligence method, called the firefly mechanism, which was inspired by the flashing lights of fireflies in nature. It is a kind of stochastic, nature-inspired, meta-heuristic algorithm that can be applied for solving the hardest optimization problems. In the firefly mechanism, a solution is discovered by trial and error (Fister, 2013). The *DSMC* scheme adopts the basic idea of the firefly mechanism to design the CP and SP's strategy decision algorithm. At time $\mathscr{H}_{t-1}$, the strategy $\mathfrak{P}_w^{CP}$ in $\mathbb{S}^{CP}$ is selected by the CP, the selection propensity of $\mathfrak{P}^{CP}$ for the time $\mathscr{H}_t$ $\left(\psi_{\mathscr{H}_t}^{\mathfrak{P}^{CP}}\right)$ in $\mathbb{L}^{CP}$ is modified as follows (Kim, 2019):

$$\begin{cases} \psi_{\mathscr{H}_t}^{\mathfrak{P}_w^{CP}} = \psi_{\mathscr{H}_{t-1}}^{\mathfrak{P}_w^{CP}} + (\alpha^{CP} \times \mathscr{R}) \\ +\left(\frac{1}{\| \mathbb{S}^{CP} \| - 1} \times \sum_{\substack{\mathfrak{P}_g^{CP} \in \mathbb{S}^{CP}, \\ \mathfrak{P}_g^{CP} \neq \mathfrak{P}_w^{CP}}} \left([1 - (\beta^{CP})] \times \mathfrak{M}^{CP} \right)\right) \\ if \mathfrak{P}_w^{CP} is selected at time \mathscr{H}_{t-1} \\ \psi_{\mathscr{H}_t}^{\mathfrak{P}_f^{CP}} = \psi_{\mathscr{H}_{t-1}}^{\mathfrak{P}_f^{CP}}, if \mathfrak{P}_f^{CP} \neq \mathfrak{P}_w^{CP} \end{cases} \tag{11.5.2}$$

$$\text{s.t.}, \begin{cases} F\left(\psi_{\mathscr{H}_{t-1}}^{\mathfrak{P}_g^{CP}}, \psi_{\mathscr{H}_{t-1}}^{\mathfrak{P}_w^{CP}}\right) = \sqrt{\left(\psi_{\mathscr{H}_{t-1}}^{\mathfrak{P}_g^{CP}} - \psi_{\mathscr{H}_{t-1}}^{\mathfrak{P}_w^{CP}}\right)^2} \\ \mathfrak{M}^{CP} = \frac{\mathbb{U}_{\mathscr{H}_{t-1}}^{CP}\left(\mathfrak{P}_w^{CP}\right) - \mathbb{U}_{\mathscr{H}_{t-1}}^{CP}\left(\mathfrak{P}_g^{CP}\right)}{\mathbb{U}_{\mathscr{H}_{t-1}}^{CP}\left(\mathfrak{P}_w^{CP}\right)} \\ R = \left(rand_0^1 - \epsilon\right) \end{cases}$$

where β^{CP} is the attractiveness when the difference between $\psi_{\mathscr{H}_{t-1}}^{\mathfrak{P}_g^{CP}}$ and $\psi_{\mathscr{H}_{t-1}}^{\mathfrak{P}_w^{CP}}$ is the highest. Therefore, the $\psi^{\mathfrak{P}_w^{CP}}$ value is adjusted based on the strategy $\mathfrak{P}_w^{CP}$'s relative preference to the strategy $\mathfrak{P}_g^{CP}$; γ is a coefficient, which controls the convergence rate. In addition, the randomized searching process is used to explore, with a regulating parameter α^{CP} and the function $rand_0^1$, which generates a random number within [0,1]. According to $\psi_{\mathscr{H}_t}^{\mathfrak{P}^{CP}}$ values in $\mathbb{L}^{CP}$, the $\mathfrak{P}_w^{CP}$ strategy selection probability at time $\mathscr{H}_t$ $\left(P_{\mathscr{H}_t}^{\mathfrak{P}_w^{CP}}\right)$ is finally defined as follows (Kim, 2019):

$$P_{\mathscr{H}_t}^{\mathfrak{P}_w^{CP}} = \frac{\text{EXP}\left(\psi_{\mathscr{H}_t}^{\mathfrak{P}_w^{CP}}\right)}{\sum_{\mathfrak{P}_g^{CP} \in \mathbb{S}^{CP}} \left(\text{EXP}\left(\psi_{\mathscr{H}_t}^{\mathfrak{P}_g^{CP}}\right)\right)} \tag{11.5.3}$$

At each time period, the CP dynamically selects his strategies based on the probability distribution $P_{\mathcal{H}}^{\mathfrak{P}^{CP}} = \left[P_{\mathcal{H}}^{\mathfrak{P}_{min}^{CP}} \ldots P_{\mathcal{H}}^{\mathfrak{P}_{max}^{CP}}\right]$; this approach can make the CP more responsive to the current system conditions. With the CP, the SP also periodically modifies the selection propensity ($\xi\mathfrak{Z}^{SP}$) of strategy $\mathfrak{Z}^{SP}$ in $\mathbb{S}^{SP}$ and his selection probability $P_{\mathcal{H}}^{\mathfrak{Z}^{SP}}$ in the same manner as the CP does. In the *DSMC* scheme, a single CP carries out the dual-leader Stackelberg game with multiple SPs in parallel. Therefore, the CP's strategy is selected separately to the MUs, who access data contents through different SPs, and each SP also interacts with the CP independently to decide its strategy $\mathfrak{Z}^{SP}$ based on the probability distribution $P_{\mathcal{H}}^{\mathfrak{Z}^{SP}} = \left[P_{\mathcal{H}}^{\mathfrak{Z}_{min}^{SP}} \ldots P_{\mathcal{H}}^{\mathfrak{Z}_{max}^{SP}}\right]$ (Kim, 2019).

The second process in the dual-leader Stackelberg game is the followers' reaction mechanism, which deals with the problem of maximizing the MU's utility function and decides the leaders' profit at time $\mathcal{H}_t\left(\mathfrak{X}_{\mathcal{H}_t}^{CP,SP}\right)$. Traditionally, the MU's utility function is defined by considering the reciprocal relationship between the satisfaction and the paying cost. As followers, the major goal of the jth MU $\left(MU_j\right)$ is to decide his strategies to satisfy the following equation (Kim, 2019).

$$\max_{\phi_{\mathcal{H}_t}^{MU_j} \in \mathbb{S}^{MU}} \mathbb{U}_{\mathcal{H}_t}^{MU_j}\left(\phi_{\mathcal{H}_t}^{MU_j}, \chi_{MU_j}, \left(\mathfrak{P}_w^{CP}, \mathfrak{Z}_k^{SP_i}\right)\right) \tag{11.5.4}$$

$$= \max_{\phi_{\mathcal{H}_t}^{MU_j} \in \mathbb{S}^{MU}} \left(\left[\mathfrak{A}\left(\chi_{MU_j}\right) + \Theta_{\mathcal{H}_t}^{CP}\left(\mathfrak{P}_w^{CP}, \chi_{MU_j}\right) - \mathfrak{C}\left(\left(\mathfrak{P}_w^{CP}, \mathfrak{Z}_k^{SP_i}\right), \chi_{MU_j}\right)\right] \times \phi_{\mathcal{H}_t}^{MU_j}\right)$$

$$\text{s.t.,}\begin{cases} A\left(\chi_{MU_j}\right) = \left[\left(\vartheta \times \log\left(1 + \frac{\chi_{MU_j}}{\text{Gbps}}\right)\right) \times \chi_{MU_j}\right] \\ C\left(\left(\mathfrak{P}_w^{CP}, \mathfrak{Z}_k^{SP_i}\right), \chi_{MU_j}\right) = \left(\chi_{MU_j} \times \left(\eta + \mathfrak{Z}_k^{SP_i}\right)\right) \end{cases} \text{and } \phi_{\mathcal{H}_t}^{MU_j} = \begin{cases} 1, if\, \phi_{\mathcal{H}_t}^{MU_j} is on \\ 0, if\, \phi_{\mathcal{H}_t}^{MU_j} is off \end{cases}$$

where χ_{MU_j} is the amount of data consumed by MU_j, and SP_i is MU_j's connecting SP. $\mathfrak{A}(\cdot)$ and $\mathfrak{C}(\cdot)$ are MU_j's satisfaction and cost functions. In a non-cooperative game manner, MU_j rationally selects his strategy $\phi_{\mathcal{H}_t}^{MU_j}$ to maximize the $\mathbb{U}_{\mathcal{H}_t}^{MU_j}(\cdot)$ (Kim, 2019).

11.5.4 The leaders' profit-sharing algorithm

In cooperative games, bargaining solutions serve as a fundamental approach to predict how profits should be fairly distributed when players reach a joint agreement. Various bargaining models have been developed, each based on different assumptions regarding the desired fairness criteria. To establish a unique solution in bargaining scenarios, Israeli economist Ariel Rubinstein introduced the alternating-offer bargaining model, known as the *Rubinstein Bargaining Solution* (*RBS*). This model, built upon Selten's concept of subgame-perfect equilibrium, provides a structured solution for bargaining games that involve alternating offers extending over an infinite time horizon. For a considerable period, solving this type of bargaining problem was regarded as a complex challenge. Thus, Rubinstein's work remains one of the most influential contributions to cooperative game theory, offering a fair and efficient approach to profit-sharing among players (Park, 2016; Zhao, 2002; Zhou, 2011). In the *DSMC* scheme, the profit-sharing negotiation between the CP and SP is formulated according to the *RBS* framework. A key insight from the basic *RBS* model is that frictionless bargaining leads to indeterminate outcomes. If bargaining players are indifferent to the timing of an agreement, they may engage in prolonged negotiations without reaching a resolution. Conversely, the presence of frictions in the bargaining process creates incentives for both parties to finalize an agreement within a reasonable timeframe (Kim, 2019).

In the scenario of the *RBS*, two players are bargaining about the division of the profit while having their own bargaining powers (δ). Based on their bargaining power, players negotiate with each other by proposing offers alternately. A player with more bargaining power benefits more from bargaining. After several rounds of negotiation, an agreement between the CP and the SP $\left(RBS\left(x_{CP}^{*},x_{SP}^{*}\right)\right)$ is given by (Kim, 2019; Park, 2016; Zhao, 2002; Zhou, 2011).

$$RBS\left(x_{CP}^{*},x_{SP}^{*}\right) =$$

$$\begin{cases} \left(\frac{1-\delta_{SP}}{1-(\delta_{CP}\times\delta_{SP})},\frac{\delta_{SP}\times(1-\delta_{CP})}{1-(\delta_{CP}\times\delta_{SP})}\right), ifplayerCPoffersfirst \\ \left(\frac{\delta_{CP}\times(1-\delta_{SP})}{1-(\delta_{CP}\times\delta_{SP})},\frac{1-\delta_{CP}}{1-(\delta_{CP}\times\delta_{SP})}\right), ifplayerSPoffersfirst \end{cases} \tag{11.5.5}$$

$$\text{s.t.}, \left(x_{CP}^{*},x_{SP}^{*}\right) \in \mathbf{R}^2 : x_{CP}^{*}+x_{SP}^{*} = 1,\ x_{CP}^{*} \geq 0,\ x_{SP}^{*} \geq 0 \text{ and } 0 \leq \delta_{CP},\delta_{SP} \leq 1$$

where δ_{CP} and δ_{SP} are the bargaining powers of the CP and the SP, respectively. x_{CP}^{*} and x_{SP}^{*} are the profit division ratios of the CP and the SP, respectively. Therefore, the leaders' profit X, which is the sum of the data consuming MU's $\mathfrak{C}(\cdot)$, is divided by the rate of x_{CP}^{*}: x_{SP}^{*}. It means that the $\mathcal{S}_{\mathcal{H}_t}^{CP}(\mathfrak{P}^{CP},N^{SP})$ and $\mathcal{S}_{\mathcal{H}_t}^{SP}(\mathfrak{Z}^{SP},N^{SP})$ functions in the equation (11.5.1) are defined as follows (Kim, 2019);

$$\begin{cases} \mathcal{S}_{\mathcal{H}_t}^{CP}(\mathfrak{P}^{CP},N^{SP}) = \mathfrak{X}_{\mathcal{H}_t}^{CP,SP} \times \frac{x_{CP}^{*}}{(x_{CP}^{*}+x_{SP}^{*})} \\ \mathcal{S}_{\mathcal{H}_t}^{SP}(\mathfrak{Z}^{SP},N^{SP}) = \mathfrak{X}_{\mathcal{H}_t}^{CP,SP} \times \frac{x_{SP}^{*}}{(x_{CP}^{*}+x_{SP}^{*})} \end{cases} \tag{11.5.6}$$

It is obvious that there is a first-proposer advantage in the bargaining process. In the *DSMC* scheme, the leader, who has a larger value of its cost function $\mathscr{C}(\cdot)$, becomes a first-proposer to compensate for its control overhead. Traditionally, the bargaining power in the *Rubinstein-Stahl's* model is defined as follows (Kim, 2019).

$$\delta_{CP} = \frac{exp(\mathcal{J}_{CP}\times\Delta)}{exp(\mathcal{J}_{CP}\times\Delta)+exp(\mathcal{J}_{SP}\times\Delta)} \text{ and } \delta_{SP} = \frac{exp(\mathcal{J}_{SP}\times\Delta)}{exp(\mathcal{J}_{CP}\times\Delta)+exp(\mathcal{J}_{SP}\times\Delta)} \tag{11.5.7}$$

where Δ is the fixed time period of a negotiation round, and J is a weight factor to adaptively adjust the bargaining power δ, which increases monotonically with J. During the *data sponsoring* operations, each leader has a different weight factor. To provide more fair-efficient control over the collaboration between the CP and the SP, this factor value should be adjusted dynamically under various system situations. In the dual-leader profit-sharing algorithm, the leader's operating cost is used to set the value of J. If one leader has a relatively higher cost to provide data services, the *DSMC* scheme gives preference to that leader with a higher value of the weight factor. At time $\mathscr{H}_t$, the $\mathcal{J}_{CP}$and $\mathcal{J}_{SP}$ values are given by (Kim, 2019):

$$\begin{cases} \mathcal{J}_{CP} = \dfrac{\left(\mathscr{C}_{\mathscr{H}_t}^{CP}(N^{SP}) + \grave{\mathrm{E}}_{\mathscr{H}_t}^{CP}(\mathfrak{P}_w^{CP}, N^{SP})\right)}{\left(\left(\mathscr{C}_{\mathscr{H}_t}^{CP}(N^{SP}) + \grave{\mathrm{E}}_{\mathscr{H}_t}^{CP}(\mathfrak{P}_w^{CP}, N^{SP})\right) + \mathscr{C}_{\mathscr{H}_t}^{SP}(N^{SP})\right)} \\ \mathcal{J}_{SP} = \dfrac{\mathscr{C}_{\mathscr{H}_t}^{SP}(N^{SP})}{\left(\left(\mathscr{C}_{\mathscr{H}_t}^{CP}(N^{SP}) + \grave{\mathrm{E}}_{\mathscr{H}_t}^{CP}(\mathfrak{P}_w^{CP}, N^{SP})\right) + \mathscr{C}_{\mathscr{H}_t}^{SP}(N^{SP})\right)} \end{cases} \tag{11.5.8}$$

During the *data sponsoring* system operations, the discount factors of leaders, i.e., $\mathcal{J}_{CP}$and $\mathcal{J}_{SP}$, are adaptively adjusted to find the best solution. Therefore, the obtained profit can be dynamically shared by the CP and the SP while adaptively responding to the current *data sponsoring* system conditions (Kim, 2019).

11.5.5 Summary

Recently, *data sponsoring* has emerged as a transformative strategy in the mobile data market, enabling CPs to subsidize MUs' data costs. As sponsored data gains traction in the industry, understanding its impact on CPs, SPs, and MUs becomes increasingly important. To address the *data sponsoring* challenge in 5G networks, the *DSMC* scheme introduces a novel control mechanism based on the dual-leader Stackelberg game model. The *DSMC* scheme operates through the following key steps. First, the CP and SP determine their pricing strategies using the firefly learning method, which enables dynamic price adjustments based on real-time market conditions. Second, MUs individually optimize their payoffs in a distributed and non-cooperative game framework. Their responses are influenced by the pricing strategies set by the CP and SP. Third, the profit-sharing agreement between the CP and SP is formulated using the *RBS*, ensuring fair and efficient revenue distribution. In the *DSMC* scheme, the decisions of CPs, SPs, and MUs are interdependent, with each entity's decision-making process being influenced by the actions of the others. This creates a feedback-driven iterative game, where each decision feeds into the next, continuously refining the system until the optimal solution is reached (Kim, 2019).

11.6 Hybrid Network Spectrum Allocation (*HNSA*) scheme

Visible Light Communication (VLC) is emerging as a promising wireless communication technology, and its co-integration with radio frequency (RF) networks has been extensively explored in indoor environments to support high data rate communications. To fully harness the advantages of both RF and VLC technologies, it is crucial to develop intelligent resource management algorithms that optimize their coexistence. Recently, S. Kim proposed the *Hybrid Network Spectrum Allocation (HNSA)* scheme, which is designed based on cooperative bargaining theory (Kim, 2022a). The *HNSA* scheme leverages the fundamental principles of *asymmetric Nash*, *Kalai-Smorodinsky*, and *egalitarian bargaining solutions* to establish a fair and efficient spectrum allocation strategy. To achieve optimal resource distribution, the *HNSA* scheme incorporates four distinct allocation rules to assess the asymmetry among network agents. By employing a step-by-step interactive bargaining approach, the *HNSA* scheme enables disproportionate agents to coordinate their decisions and effectively share RF and VLC spectrum resources. This adaptive resource management strategy is essential for balancing conflicting requirements while dynamically responding to variations in hybrid RF/VLC network environments. By integrating cooperative decision-making with real-time system adjustments, the *HNSA* scheme ensures an efficient trade-off between competing network demands, ultimately enhancing spectrum utilization and communication performance (Kim, 2022a).

11.6.1 Development motivation

The past century witnessed transformative shifts through the Industrial Revolution and the Internet Revolution. The coming era, however, is set to be dominated by Industry 4.0, characterized by advancements in interconnectivity, automation, machine learning, and real-time data communications. The fundamental goal of Industry 4.0 is to integrate the physical components of the manufacturing process with cyber-based networks, enabling intelligent industrial operations. Unlike previous industrial advancements, Industry 4.0 is distinct in that it operates on highly decentralized and widely distributed industrial devices, influenced by geographical dispersion. As the adoption of Industry 4.0 accelerates, it is imperative to develop innovative solutions that optimize operational costs and system reliability. However, traditional network infrastructures and communication platforms are constrained by high transmission costs and communication delays, making them inadequate for the requirements of Industry 4.0. The mainstream approach to addressing these challenges is to develop self-organizing and interactive communication mechanisms that leverage vast amounts of real-time industrial data (Bajic, 2020; Kim, 2022a; Zhang, 2021).

A cornerstone technology for realizing Industry 4.0 is the Industrial Internet of Things (IIoT). IIoT represents a new paradigm in information and communication technologies, facilitating the interconnection of industrial devices with intelligent capabilities. These smart devices are capable of sensing, processing, and exchanging data within interconnected networks. To achieve efficient data processing, wireless communication technologies have become indispensable in Industry 4.0, primarily due to their flexibility and mobility. However, despite the promise of IIoT, several technical challenges persist in ensuring reliable IIoT services. A vast number of IIoT devices connect to the Internet, generating enormous volumes of data within short time intervals. Operators struggle to accommodate diverse service preferences among industrial users. With limited wireless communication resources, managing efficient spectrum allocation has become a key research priority. The lack of adequate telecommunication infrastructure raises concerns about the reliability of IIoT services, especially in geographically dispersed industrial environments (Kim, 2022a; Liao, 2020; Zhang, 2021).

The exponential growth in IIoT device deployment has led to an unprecedented increase in data rate demands in recent years. This has necessitated the development of high-speed transmission strategies, wider network coverage, and more efficient communication protocols. However, conventional radio frequency (RF)-based systems face inherent limitations: escalating spectrum costs, and severe network congestion and interference. As a result, Industry 4.0 requires alternative communication technologies capable of operating at higher frequency ranges while addressing spectrum scarcity (Kim, 2022a).

To mitigate the limitations of RF-based systems, Visible Light Communication (VLC) has emerged as a promising alternative. VLC technology leverages light-emitting diode (LED) sources to transmit data through light intensity modulation, a process imperceptible to the human eye. VLC technology offers several advantages: significantly higher transmission capacity, enhanced security due to localized transmission, extensive frequency reuse, lower cost compared to RF-based wireless communications, reduced latency, and ensuring real-time data exchange. These attributes position VLC as an effective solution to complement existing RF communication frameworks in Industry 4.0 applications (Kim, 2022a; Shrivastava, 2020; Tabassum, 2018).

Despite its advantages, VLC technology faces several limitations that hinder its independent deployment. VLC networks rely heavily on LOS propagation, resulting in performance degradation in non-LOS environments. Each VLC AP covers only small, localized areas, unlike RF-based APs. Data transmission rates are highly susceptible to spatial variations. To overcome these challenges, recent studies have focused on hybrid RF/VLC network architectures, which allow both RF and VLC systems to coexist. The complementary nature of these technologies has gained significant interest: RF networks provide ubiquitous coverage with moderate data rates, and VLC networks support high-speed transmission in localized areas. In indoor industrial environments, this hybrid approach enables seamless network performance: RF systems ensure continuous connectivity, preventing coverage gaps. VLC systems enhance data rates in areas with high user density. RF and VLC signals do not interfere, maintaining electromagnetic immunity. This RF/VLC hybrid model delivers high-speed, low-latency data transmission, ensuring QoS guarantees for IIoT applications (Hammouda, 2018; Kim, 2022a; Kong, 2019; Li, 2015).

Hybrid RF/VLC networks fall under the category of heterogeneous networks (HetNets). HetNets integrate multiple wireless access technologies to enhance spectral efficiency. Traffic offloading, a widely adopted technique in HetNets, helps manage network congestion by diverting excess traffic to alternative network infrastructures. In the context of hybrid RF/VLC networks, traffic offloading enables i) relief of RF spectrum congestion by offloading data-intensive transmissions to VLC networks, and ii) efficient spectrum utilization, improving overall network performance. However, effective spectrum resource allocation remains a key challenge in cooperative RF/VLC network deployments. The *HNSA* scheme investigates spectrum management strategies in RF/VLC hybrid networks and develops an optimal cooperative spectrum allocation mechanism. This approach ensures maximum system performance under diverse network conditions (Kim, 2022a; Shao, 2016; Shrivastava, 2020).

The *HNSA* scheme introduces a unified framework to compare different bargaining solutions, aiming to i) bridge the gap between axiomatic and strategic bargaining models, ii) harmonize multiple resource allocation strategies, and iii) leverage cooperative game theory to achieve fair and efficient spectrum distribution. By integrating interactive network cooperation, the *HNSA* scheme provides an adaptive spectrum allocation framework, facilitating optimized RF/VLC spectrum sharing while ensuring high system performance in Industry 4.0 applications (Kim, 2022a).

11.6.2 Contribution

The *HNSA* scheme leverages a hybrid RF/VLC framework in conjunction with dynamic spectrum allocation techniques. Within the integrated VLC and RF network environment, diverse IIoT devices can concurrently connect to the respective RF and VLC access points to enhance their overall satisfaction. The primary goal of the proposed spectrum allocation strategy is to efficiently distribute constrained spectrum resources while maximizing social welfare. To achieve this, multiple IIoT devices collaborate and operate cooperatively, adapting to dynamic system conditions. The *HNSA* scheme incorporates key principles from asymmetric *Nash*, *Kalai-Smorodinsky*, and *egalitarian bargaining solutions*, implementing a unified bargaining mechanism that unfolds through sequential negotiations between device-AP pairs. Each participant in the bargaining process pursues a shared objective, making binding commitments based on real-time traffic information exchange. Consequently, the game strategies become interdependent, directly influencing system performance. To the best of my knowledge, this study is the first to selectively integrate different bargaining approaches into the design of a spectrum allocation algorithm for hybrid RF/VLC networks (Kim, 2022a).

11.6.3 Hybrid RF/VLC platform and unified bargaining game

The *HNSA* scheme incorporates a unified network model to establish a hybrid RF/VLC platform. It takes into account heterogeneous APs, including RF and VLC APs, which facilitate wireless connectivity for multiple IIoT devices. These APs are strategically positioned at various locations within an indoor setting. To ensure efficient resource management, RF and VLC APs interact frequently and maintain connectivity to the Internet via backhaul links. Although both RF and VLC transmissions rely on electromagnetic waves for data transfer, their characteristics vary significantly. As a result, they can coexist while supporting different data access rates. The *HNSA* scheme assumes that the hybrid RF/VLC network infrastructure composed of a single RF AP ($\mathscr{R}$) and multiple VLC APs, i.e., $\mathbb{V} = \{\mathscr{V}_1,\ldots,\mathscr{V}_n\}$. A number of IIoT devices, i.e., $\mathbb{D} = \{\mathscr{D}_1,\ldots,\mathscr{D}_m\}$, are located in the coverage region of both APs, which provide data transmission services for IIoT devices. This framework can easily be extended to a multiple RF APs scenario. Figure 6 illustrates the hybrid RF/VLC platform architecture (Hammouda, 2018; Kim, 2022a; Yang, 2020).

Figure 6. The hybrid RF/VLC platform architecture (Kim, 2022a)

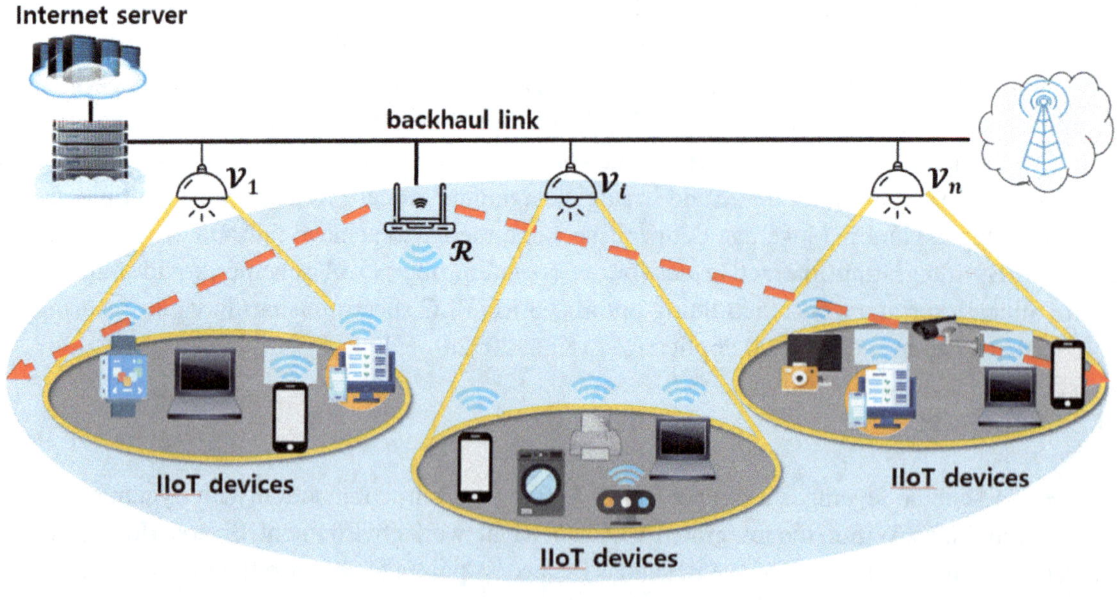

Each individual device is considered to be equipped with an RF front-end and a photodiode. Therefore, the collaboration between RF and VLC is realized by the dual connectivity capability, which enables IIoT devices to aggregate different spectrum resources from both VLC and RF networks. Therefore, at a given time, IIoT devices can connect simultaneously both networks through two parallel communication links. The R and each $\mathscr{V}_{1\leq i\leq n}$ have their spectrum resources, i.e., $\mathfrak{M}_{\mathscr{R}}$ and $\mathfrak{M}_{\mathscr{V}_i}$, respectively. They have multiple signal reception interfaces that allow simultaneous connections to corresponding IIoT devices. Each IIoT device $\left(\mathscr{D}_{1\leq j\leq m}\right)$ generate communication tasks according to a Poisson process and accesses its corresponding APs via wireless communications. The operational timeline is discretized into time slots to make spectrum allocation; it is the same time scale of communication task arrivals, which is a common assumption in the hybrid RF/VLC system (Al Hammadi, 2020; Kim, 2022a).

For the $\mathscr{D}_j$, the totally generated communication workload is $\Upsilon_{\mathscr{D}_j} = \Upsilon^{RF}_{\mathscr{D}_j} + \Upsilon^{VLC}_{\mathscr{D}_j}$ where $\Upsilon^{RF}_{\mathscr{D}_j}$ and $\Upsilon^{VLC}_{\mathscr{D}_j}$ are the $\mathscr{D}_j$'s RF and VLC communication amounts, respectively. However, the RF and VLC spectrum resources are limited. Therefore, effective spectrum allocation strategies should be considered to improve resource efficiency. To address the RF/VLC spectrum sharing problem, the *HNSA* scheme formulates three bargaining games. At a time period, each individual $\mathscr{D}_j$ processes its bargaining games with its corresponding $\mathscr{V}_i$ and the R. In the $\mathscr{D}_j$, the bargaining game $\left(\mathbb{G}_{\mathscr{D}_j}\right)$ is designed to decide the amounts of $\Upsilon^{RF}_{\mathscr{D}_j}$ and $\Upsilon^{VLC}_{\mathscr{D}_j}$. In the $\mathscr{V}_i$ and R, two bargaining games, i.e., $\mathbb{G}_{\mathscr{V}_i}$ and $\mathbb{G}_{\mathscr{R}}$, are formulated to share the $\mathfrak{M}_{\mathscr{V}_i}$ and $\mathfrak{M}_{\mathscr{R}}$, respectively. It is noteworthy that the unified bargaining game formulates the R-V-D association, and the $\mathbb{G}_{\mathscr{D}_j}$, $\mathbb{G}_{\mathscr{V}_i}$ and $\mathbb{G}_{\mathscr{R}}$ games are designed based on the concepts of different bargaining solutions. These games are operated in a cooperative manner, and they are repeated in a step-by-step interactive fashion. Formally, the *HNSA* scheme defines the $\mathbb{G}_{\mathscr{D}_j}$, $\mathbb{G}_{\mathscr{V}_i}$ and $\mathbb{G}_{\mathscr{R}}$ game entities, i.e.,

$$\mathbb{G} = \left\{\mathbb{G}_{\mathscr{D}_{1\leq j\leq m}}, \mathbb{G}_{\mathscr{V}_{1\leq i\leq n}}, \mathbb{G}_{\mathscr{R}}\right\} = \left\{\mathbb{D}, \mathbb{V}, \left\{\mathbb{G}_{\mathscr{D}_j} \middle| \left(\Upsilon^{RF}_{\mathscr{D}_j}, \Upsilon^{VLC}_{\mathscr{D}_j}\right), U^{RF}_{\mathscr{D}_j}(\cdot), U^{VLC}_{\mathscr{D}_j}(\cdot)\right\}, \left\{\mathbb{G}_{\mathscr{V}_i} \middle| \mathfrak{M}_{\mathscr{V}_i}, \mathbb{D}_{\mathscr{V}_i}, \mathfrak{M}^{\mathscr{V}_i}_{\mathscr{D}_j\in\mathbb{D}_{\mathscr{V}_i}}, \mathscr{U}^{\mathscr{V}_i}_{\mathscr{D}_j\in\mathbb{D}_{\mathscr{V}_i}}(\cdot)\right\}, \left\{\mathbb{G}_{\mathscr{R}} \middle| \mathfrak{M}_{\mathscr{R}}, \mathbb{D}, \mathfrak{M}^{\mathscr{R}}_{\mathscr{D}_j\in\mathbb{D}}, \mathfrak{U}^{\mathscr{R}}_{\mathscr{D}_j\in\mathbb{D}}(\cdot)\right\}, T\right\}$$

of gameplay (Kim, 2022a).

- The $\mathbb{G}_{\mathscr{D}_j}$, $\mathbb{G}_{\mathscr{V}_i}$ and $\mathbb{G}_{\mathscr{R}}$ are mutually and reciprocally interdependent in an interactive manner, and they work together to share the $\mathfrak{M}_{\mathscr{V}_i}$ and $\mathfrak{M}_{\mathscr{R}}$.
- D is the set of IIoT devices and V is the set of VLC APs in the heterogeneous RF/VLC platform.
- In the $\mathbb{G}_{\mathscr{D}_j}$, RF and VLC communication services are game players. $\Upsilon^{RF}_{\mathscr{D}_j}$ and $\Upsilon^{VLC}_{\mathscr{D}_j}$ are their strategies, and $U^{RF}_{\mathscr{D}_j}(\cdot)$ and $U^{VLC}_{\mathscr{D}_j}(\cdot)$ are their utility functions.
- In the $\mathbb{G}_{\mathscr{V}_i}$, $\mathfrak{M}_{\mathscr{V}_i}$ is the $\mathscr{V}_i$'s total VLC spectrum capacity, and $\mathbb{D}_{\mathscr{V}_i}$ is the set of IIoT devices covered by the $\mathscr{V}_i$. $\mathscr{D}_j \in \mathbb{D}_{\mathscr{V}_i}$ is a game player, and $\mathfrak{M}^{\mathscr{V}_i}_{\mathscr{D}_j}$ is the allocated VLC spectrum amount for the $\mathscr{D}_j$; it is the $\mathscr{D}_j$'s strategy. The $\mathscr{U}^{\mathscr{V}_i}_{\mathscr{D}_j}(\cdot)$ is the $\mathscr{D}_j$'s utility function.
- In the $\mathbb{G}_{\mathscr{R}}$, $\mathfrak{M}_{\mathscr{R}}$ is the R's total RF spectrum capacity, and the $\mathscr{D}_j \in \mathbb{D}$ is a game player, and $\mathfrak{M}^{\mathscr{R}}_{\mathscr{D}_j}$ is the allocated RF spectrum amount for the $\mathscr{D}_j$; it is the $\mathscr{D}_j$'s strategy. The $\mathfrak{U}^{\mathscr{R}}_{\mathscr{D}_j}(\cdot)$ is the $\mathscr{D}_j$'s utility function.
- Discrete time model $T \in \left\{t_1, \ldots, t_c, t_{c+1}, \ldots\right\}$ is represented by a sequence of time steps. The length of t_c matches the event time-scale of $\mathbb{G}_{\mathscr{D}_j}$, $\mathbb{G}_{\mathscr{V}_i}$ and $\mathbb{G}_{\mathscr{R}}$.

11.6.4 The fundamental ideas of three bargaining solutions

Let (S,N,f) be a bargaining problem: $S \subset \mathbb{R}^n$ is a nonempty and finite set, $N = \{1 \ldots i \ldots n\}$ be the set of players, and f is a function from $\mathbb{R}^n$ to R. The relationship between f and U is analogous to that between a transformation function and a production possibility set in producer theory. As the Euclidean metric on $\mathbb{R}^n$, $d \in S$ is the disagreement point. $\mathscr{W} \in \mathbb{R}^n_{++}$ is a vector of players' asymmetry where $\mathbb{R}^n_{++} = \left\{x \in \mathbb{R}^n \middle| x > 0\right\}$. Weighted by W, the *asymmetric Nash bargaining solution* (*ANBS*) is defined as the following maximization problem (Kawamori, 2014; Kim, 2022a):

$$ANBS(S,d,\mathscr{W}) = \arg\max_{x \in U} \prod_{i \in N} (x_i - d_i)^{\mathscr{W}_i}, \text{s.t.}, x \in f(S,d) \geq d \tag{11.6.1}$$

To define the *asymmetric Kalai-Smorodinsky bargaining solution* (*AKSBS*), we define the utopia point $\mathscr{A}(S)$ where $\mathscr{A}(S) \equiv \max\{x | x_{1 \leq i \leq n} \in S\}$ and the weak Pareto set $WP(S) \equiv \{x \in S | y \gg x \Rightarrow y \notin S\}$. The main idea of *AKSBS* is to equalize the fractions of their ideal payoffs that players achieve. Therefore, the *AKSBS* is defined as follows (Dubra, 2001; Kim 2022a):

$$AKSBS(S,d,\mathscr{W}) = \left\{ i,j \in N \middle| \frac{(x_j - d_j)}{(x_i - d_i)} = \frac{\mathscr{W}_j \times \mathscr{A}_j(S)}{\mathscr{W}_i \times \mathscr{A}_i(S)} \right\} \cap WP(S) \tag{11.6.2}$$

As a proportional bargaining solution based on the weight, the *asymmetric egalitarian bargaining solution* (*AEBS*) is defined as follows (Bossert, 1995; Kim, 2022a):

$$AEBS(S,d,\mathscr{W}) = \left\{ i,j \in N \middle| \left(\mathscr{W}_i \times (x_i - d_i)\right) = \left(\mathscr{W}_j \times (x_j - d_j)\right) \right\} \tag{11.6.3}$$

To define the W, the *HNSA* scheme considers the bankruptcy problem under the assumption that claimants have reference-dependent preferences. In a bankruptcy problem, an arbitrator must allocate a finite and perfectly divisible resource among several claimants whose claims sum to a greater amount than is available. To estimate different specifications for claimants' reference, four most prominent rules, such as *Proportional* (*P*), *Constrained Equal Awards* (*CEA*), *Constrained Equal Losses* (*CEL*), and *Small Claims First* (*SCF*), are proposed on the basis of the level of utilitarian and maximin welfare that they generate. The *HNSA* scheme assumes that claimants in the bankruptcy problem are players in the bargaining games, and players' references are given as their asymmetric bargain weights (Gallice, 2019; Kim, 2022a).

Let $E \in \mathbb{R}_+$ denote the endowment of the resource to be allocated and each claimant $i \in N$ has a claim $c_i \in \mathbb{R}_+$ on E. The vector $c = (c_1, \ldots, c_n)$ collects individual claims where $\mathscr{C} = \sum_{i \in N} c_i$. A bankruptcy problem is a pair $(c,E) \in \mathbb{R}^n_+ \times \mathbb{R}_+$ where c is such that $\mathscr{C} \geq E$. Each rule is a function that associates to any problem (c,E), and provides a unique awards vector. The *P* rule ($P(c,E)$) allocates endowment proportional to claims. For the c_i, the *P* award is given as follows (Gallice, 2019; Kim, 2022a):

$$P_{i \in N}(c,E) = \Gamma \times c_i, \text{ s.t.}, \Gamma = {}^{E}\!/_{\mathscr{C}} \tag{11.6.4}$$

The *CEA* rule ($CEA(c,E)$) assigns equal awards to all claimants subject to the requirement that no one receives more than his claim. For the c_i, the *CEA* award is given by (Gallice, 2019; Kim, 2022a):

$$CEA_{i\in N}(c,E) = \min\{c_i,\Gamma\},\ \text{s.t.},\ \sum_{i\in N}\min\{c_i,\Gamma\} = E \tag{11.6.5}$$

The CEL rule ($CEL(c,E)$) assigns an equal number of losses to all claimants subject to the requirement that no one receives a negative amount. For the c_i, the CEL award is given by (Gallice, 2019; Kim, 2022a):

$$CEL_{i\in N}(c,E) = \max\{0,c_i-\Gamma\},\ \text{s.t.},\ \sum_{i\in N}\max\{0,c_i-\Gamma\} = E \tag{11.6.6}$$

The SCF rule ($SCF(c,E)$) assigns to each agent the minimum amount between his claim and what remains of the endowment, starting from the first. To define the $SCF_i(c,E)$, let $\succ_c$ denote an order on the set of claimants according to their claims and starting from the lowest; ties are broken randomly. The $SCF_i(c,E)$ is given by (Gallice, 2019; Kim, 2022a):

$$SCF_{i\in N}(c,E) = \min\left\{c_i,\max\left\{\left(E-\sum_{j\succ_c i}c_j\right),0\right\}\right\} \tag{11.6.7}$$

11.6.5 The interactive bargaining game in the hybrid RF/VLC platform

To develop the interactive bargaining game, the *HNSA* scheme constructs the $\mathbb{G}_{\mathcal{D}_{1\le j\le m}}$, $\mathbb{G}_{\mathcal{V}_{1\le i\le n}}$ and $\mathbb{G}_{\mathcal{R}}$ games. They are interacting with one another during a sequence of time steps. At each time period, the $\mathbb{G}_{\mathcal{D}_j}$ is designed for the $\mathcal{D}_j$'s RF and VLC communication services. This game decides the $\Upsilon^{RF}_{\mathcal{D}_j}$ and $\Upsilon^{VLC}_{\mathcal{D}_j}$ values to maximize the $\mathcal{D}_j$'s payoff. At the time t_c, the $U^{RF}_{\mathcal{D}_j}(\cdot)$ and $U^{VLC}_{\mathcal{D}_j}(\cdot)$ are defined as follows (Kim, 2022a):

$$\begin{cases} U^{RF}_{\mathcal{D}_j}\left(\Upsilon_{\mathcal{D}_j},\Upsilon^{RF}_{\mathcal{D}_j},\rho^{RF}_{t_c}\right) = \left(\dfrac{exp\left(\frac{\Upsilon^{RF}_{\mathcal{D}_j}}{\Upsilon_{\mathcal{D}_j}}\right)-exp\left(-\frac{\Upsilon^{RF}_{\mathcal{D}_j}}{\Upsilon_{\mathcal{D}_j}}\right)}{exp\left(\frac{\Upsilon^{RF}_{\mathcal{D}_j}}{\Upsilon_{\mathcal{D}_j}}\right)+exp\left(-\frac{\Upsilon^{RF}_{\mathcal{D}_j}}{\Upsilon_{\mathcal{D}_j}}\right)}\right)\times\left[\alpha-\dfrac{\rho^{RF}_{t_c}}{\mathfrak{M}_{\mathcal{R}}}\right] \\ U^{VLC}_{\mathcal{D}_j}\left(\Upsilon_{\mathcal{D}_j},\Upsilon^{VLC}_{\mathcal{D}_j},\rho^{VLC}_{t_c}\right) = \log\left(\dfrac{\Upsilon^{VLC}_{\mathcal{D}_j}}{\Upsilon_{\mathcal{D}_j}}+\eta\right)\times\left[\beta-\dfrac{\rho^{VLC}_{t_c}}{\mathfrak{M}_{\mathcal{V}_i}}\right] \end{cases} \tag{11.6.8}$$

$$\text{s.t.},\ \Upsilon_{\mathcal{D}_j}\ge\Upsilon^{RF}_{\mathcal{D}_j},\ \Upsilon_{\mathcal{D}_j}\ge\Upsilon^{VLC}_{\mathcal{D}_j}\ \text{and}\ \Upsilon_{\mathcal{D}_j} = \Upsilon^{RF}_{\mathcal{D}_j}+\Upsilon^{VLC}_{\mathcal{D}_j}$$

where $\rho^{RF}_{t_c}$ and $\rho^{VLC}_{t_c}$ are the using RF and VLC spectrum amounts at time t_c, respectively. α is a control parameter for the $U^{RF}_{\mathcal{D}_j}(\cdot)$, and β, η are control parameters for the $U^{VLC}_{\mathcal{D}_j}(\cdot)$. Since all game players in the $\mathbb{G}_{\mathcal{D}_j}$ are the $\mathcal{D}_i$'s communication services, the solution concept should be strongly concerned the egalitarianism for game players. In this case, the *AEBS* is preferred for the solution concept of $\mathbb{G}_{\mathcal{D}_j}$. It is given by (Kim, 2022a):

$$AEBS\left(\left(U_{\mathscr{D}_j}^{RF}(\cdot),U_{\mathscr{D}_j}^{VLC}(\cdot)\right),\left(\Upsilon_{\mathscr{D}_j}^{RF},\Upsilon_{\mathscr{D}_j}^{VLC}\right),d_{\mathscr{D}_j},\mathscr{W}_{\mathscr{D}_j}\right) = \tag{11.6.9}$$

$$\left\{\begin{array}{l}\left(\mathscr{W}_{\mathscr{D}_j}^{RF}\times\left(U_{\mathscr{D}_j}^{RF}\left(\Upsilon_{\mathscr{D}_j},\Upsilon_{\mathscr{D}_j}^{RF},\rho_{t_c}^{RF}\right)-d_{\mathscr{D}_j}^{RF}\right)\right)\\ =\left(\mathscr{W}_{\mathscr{D}_j}^{VLC}\times\left(U_{\mathscr{D}_j}^{VLC}\left(\Upsilon_{\mathscr{D}_j},\Upsilon_{\mathscr{D}_j}^{VLC},\rho_{t_c}^{VLC}\right)-d_{\mathscr{D}_j}^{VLC}\right)\right)\end{array}\right\}$$

$$\text{s.t., } \mathscr{W}_{\mathscr{D}_j}^{RF} = \frac{\rho_i^{VLC}}{\mathfrak{M}_{\mathscr{V}_i}}\Big/\left(\frac{\rho_i^{RF}}{\mathfrak{M}_{\mathscr{R}}}+\frac{\rho_i^{VLC}}{\mathfrak{M}_{\mathscr{V}_i}}\right) \text{ and } \mathscr{W}_{\mathscr{D}_j}^{VLC} = \frac{\rho_i^{RF}}{\mathfrak{M}_{\mathscr{R}}}\Big/\left(\frac{\rho_i^{RF}}{\mathfrak{M}_{\mathscr{R}}}+\frac{\rho_i^{VLC}}{\mathfrak{M}_{\mathscr{V}_i}}\right)$$

where $d_{\mathscr{D}_j}^{RF}$ and $d_{\mathscr{D}_j}^{VLC}$ (*or* $\mathscr{W}_{\mathscr{D}_j}^{RF}$ and $\mathscr{W}_{\mathscr{D}_j}^{VLC}$) are disagreement points (*or* weight factors) for RF and VLC communication services, respectively. Weight factors are decided by considering the comparative traffic congestion ratio. For example, if the RF network is heavily congested than the VLC network, the weight factor of VLC network increases, relatively (Kim, 2022a).

Contrast to the $\mathbb{G}_{\mathscr{D}_j}$, the $\mathbb{G}_{\mathscr{V}_i}$ and $\mathbb{G}_{\mathscr{R}}$ games are operated in the $\mathscr{V}_i$ and R, respectively. First, the $\mathbb{G}_{\mathscr{V}_i}$ game is developed to share the $\mathfrak{M}_{\mathscr{V}_i}$, and all game players are devices in the $\mathbb{D}_{\mathscr{V}_i}$. As a game player, the utility function of $\mathscr{D}_j \in \mathbb{D}_{\mathscr{V}_i}$, i.e., $\mathscr{U}_{\mathscr{D}_j}^{\mathscr{V}_i}(\cdot)$, is defined as follows (Kim, 2022a):

$$\mathscr{U}_{\mathscr{D}_j}^{\mathscr{V}_i}\left(\mathfrak{M}_{\mathscr{V}_i},\mathfrak{M}_{\mathscr{D}_j}^{\mathscr{V}_i},\mathbb{D}_{\mathscr{V}_i}\right) = exp\left(\frac{exp\left(\frac{\mathfrak{M}_{\mathscr{D}_j}^{\mathscr{V}_i}}{\mathfrak{M}_{\mathscr{V}_i}}\right)-exp\left(-\frac{\mathfrak{M}_{\mathscr{D}_j}^{\mathscr{V}_i}}{\mathfrak{M}_{\mathscr{V}_i}}\right)}{exp\left(\frac{\mathfrak{M}_{\mathscr{D}_j}^{\mathscr{V}_i}}{\mathfrak{M}_{\mathscr{V}_i}}\right)+exp\left(-\frac{\mathfrak{M}_{\mathscr{D}_j}^{\mathscr{V}_i}}{\mathfrak{M}_{\mathscr{V}_i}}\right)}\right)-\zeta \tag{11.6.10}$$

$$\text{s.t., } \mathfrak{M}_{\mathscr{V}_i} \geq \sum\nolimits_{\mathscr{D}_j\in\mathbb{D}_{\mathscr{V}_i}}\mathfrak{M}_{\mathscr{D}_j}^{\mathscr{V}_i}$$

where ζ is a control parameter for the $\mathscr{U}_{\mathscr{D}_j}^{\mathscr{V}_i}(\cdot)$. In the $\mathbb{G}_{\mathscr{V}_i}$, relatively small game players share the $\mathfrak{M}_{\mathscr{V}_i}$ for the higher data rate service. In this case, the difference of players' requirements for the $\mathfrak{M}_{\mathscr{V}_i}$ needs the independence of alternatives other than the disagreement point. Therefore, the *ANBS* is preferred for the solution concept of $\mathbb{G}_{\mathscr{V}_i}$. It is given by (Kim, 2022a):

$$ANBS\left(\mathscr{U}_{\mathscr{D}_j\in\mathbb{D}_{\mathscr{V}_i}}^{\mathscr{V}_i}(\cdot),d^{\mathscr{V}_i},\mathscr{W}^{\mathscr{V}_i}\right) = \tag{11.6.11}$$

$$\arg\max_{(\ldots\mathfrak{M}_{\mathscr{D}_j}^{\mathscr{V}_i}\ldots)}\prod\nolimits_{\mathscr{D}_j\in\mathbb{D}_{\mathscr{V}_i}}\left(\mathscr{U}_{\mathscr{D}_j}^{\mathscr{V}_i}\left(\mathfrak{M}_{\mathscr{V}_i},\mathfrak{M}_{\mathscr{D}_j}^{\mathscr{V}_i},\mathbb{D}_{\mathscr{V}_i}\right)-d_{\mathscr{D}_j}^{\mathscr{V}_i}\right)^{\mathscr{W}_{\mathscr{D}_j}^{\mathscr{V}_i}}$$

where $d_{\mathscr{D}_j}^{\mathscr{V}_i}$ and $\mathscr{W}_{\mathscr{D}_j}^{\mathscr{V}_i}$ are the disagreement point and weight factor of $\mathscr{D}_j$, respectively, in the $\mathbb{G}_{\mathscr{V}_i}$. To get the $\mathscr{W}_{\mathscr{D}_j}^{\mathscr{V}_i}$ value, the *HNSA* scheme adopts the *P*, *CEA*, *CEL* and *SCF* allocation rules using (11.6.4)-(11.6.7), and estimates a mathematical average of them. Finally, it is given by considering the minimum weight and relative normalization (Kim, 2022a):

$$\mathscr{W}_{\mathscr{D}_j}^{\mathscr{V}_i} = {}^{\min(\mathcal{I}_{\mathscr{D}_j}^{\mathscr{V}_i}/\Upsilon_{\mathscr{D}_j}^{VLC},\Theta)}\Big/\left(\sum_{\mathscr{D}_k \in \mathbb{D}_{\mathscr{V}_i}} \min(\mathcal{I}_{\mathscr{D}_k}^{\mathscr{V}_i}/\Upsilon_{\mathscr{D}_k}^{VLC},\Theta)\right)$$

$$\text{s.t.}, \mathcal{I}_{\mathscr{D}_j}^{\mathscr{V}_i} = \left(\begin{array}{c} P_{\mathscr{D}_j}(R^{\mathscr{V}_i}, \mathfrak{M}_{\mathscr{V}_i}) + CEA_{\mathscr{D}_j}(R^{\mathscr{V}_i}, \mathfrak{M}_{\mathscr{V}_i}) \\ + CEL_{\mathscr{D}_j}(R^{\mathscr{V}_i}, \mathfrak{M}_{\mathscr{V}_i}) + SCF_{\mathscr{D}_j}(R^{\mathscr{V}_i}, \mathfrak{M}_{\mathscr{V}_i}) \end{array}\right)\Big/4 \quad (11.6.12)$$

where $\Upsilon_{\mathscr{D}_j}^{VLC}$ is the requested $\mathscr{V}_i$ spectrum amount from the $\mathscr{D}_j$, and $R^{\mathscr{V}_i}$ is the vector of $\Upsilon_{\mathscr{D}_k \in \mathbb{D}_{\mathscr{V}_i}}^{VLC}$. Θ is the upper bound for the $\mathscr{W}_{\mathscr{D}_j}^{\mathscr{V}_i}$. Second, the $\mathbb{G}_{\mathscr{R}}$ game determines the $\mathfrak{M}_{\mathscr{D}_j}^{\mathscr{R}}$ value to distribute the limited $\mathfrak{M}_{\mathscr{R}}$ resource. In this game, the $\mathscr{D}_j \in \mathbb{D}$ is a game player, and its utility function $\left(\mathfrak{U}_{\mathscr{D}_j}^{\mathscr{R}}(\cdot)\right)$ is defined as follows (Kim, 2022a):

$$\mathfrak{U}_{\mathscr{D}_j}^{\mathscr{R}}\left(\mathfrak{M}_{\mathscr{R}}, \mathbb{D}, \mathfrak{M}_{\mathscr{D}_j}^{\mathscr{R}}\right) = \frac{\log\left(\psi + \frac{\mathfrak{M}_{\mathscr{D}_j}^{\mathscr{R}}}{\mathfrak{M}_{\mathscr{R}}}\right)}{\log\left(\varrho + \frac{\mathfrak{M}_{\mathscr{D}_j}^{\mathscr{R}}}{\mathfrak{M}_{\mathscr{R}}}\right)}, \text{ s.t., } \mathfrak{M}_{\mathscr{R}} \geq \sum_{\mathscr{D}_j \in \mathbb{D}} \mathfrak{M}_{\mathscr{D}_j}^{\mathscr{R}} \quad (11.6.13)$$

where ψ and ϱ are adjustment factors for the $\mathfrak{U}_{\mathscr{D}_j}^{\mathscr{R}}(\cdot)$. In the $\mathbb{G}_{\mathscr{R}}$, relatively large game players share the $\mathfrak{M}_{\mathscr{R}}$ for the lower data rate service. In this case, it is clear that one player's gain in a negotiation relative to the other players must increase or decrease monotonically with respect to that player's weight factor relative to that of the other players. Therefore, the *AKSBS* is preferred for the solution concept of $\mathbb{G}_{\mathscr{R}}$. It is obtained as follows (Kim, 2022a):

$$AKSBS\left(\mathfrak{U}_{\mathscr{D}_j \in \mathbb{D}}^{\mathscr{R}}(\cdot), d^{\mathscr{R}}, \mathscr{W}^{\mathscr{R}}\right) = \quad (11.6.14)$$

$$\left\{ \frac{\left(\mathfrak{U}_{\mathscr{D}_j}^{\mathscr{R}}\left(\mathfrak{M}_{\mathscr{R}}, \mathbb{D}, \mathfrak{M}_{\mathscr{D}_j}^{\mathscr{R}}\right) - d_{\mathscr{D}_j}^{\mathscr{R}}\right)}{\left(\mathfrak{U}_{\mathscr{D}_i}^{\mathscr{R}}\left(\mathfrak{M}_{\mathscr{R}}, \mathbb{D}, \mathfrak{M}_{\mathscr{D}_i}^{\mathscr{R}}\right) - d_{\mathscr{D}_i}^{\mathscr{R}}\right)} = \frac{\mathscr{W}_{\mathscr{D}_j}^{\mathscr{R}} \times \mathscr{A}_{\mathscr{D}_j}\left(\mathfrak{U}_{\mathscr{D}_j \in \mathbb{D}}^{\mathscr{R}}(\cdot)\right)}{\mathscr{W}_{\mathscr{D}_i}^{\mathscr{R}} \times \mathscr{A}_{\mathscr{D}_i}\left(\mathfrak{U}_{\mathscr{D}_i \in \mathbb{D}}^{\mathscr{R}}(\cdot)\right)} \right\} \cap WP\left(\mathfrak{U}_{\mathbb{D}}^{\mathscr{R}}(\cdot)\right)$$

where $d_{\mathscr{D}_i}^{\mathscr{R}}$ and $\mathscr{W}_{\mathscr{D}_i}^{\mathscr{R}}$ (*or* $d_{\mathscr{D}_j}^{\mathscr{R}}$ and $\mathscr{W}_{\mathscr{D}_j}^{\mathscr{R}}$) are the disagreement point and weight factor of $\mathscr{D}_i$ (*or* $\mathscr{D}_j$), respectively. In the $\mathbb{G}_{\mathscr{R}}$, $\mathscr{W}_{\mathscr{D}_i}^{\mathscr{R}}$ is estimated as the same manner as the $\mathbb{G}_{\mathscr{V}_i}$ game (Kim, 2022a).

11.6.6 Summary

The *HNSA* scheme addresses the spectrum allocation challenge within a hybrid RF/VLC system. To efficiently manage this issue, bargaining principles derived from *ANBS*, *AKSBS*, and *AEBS* are incorporated, while the asymmetrical factors of network entities are dynamically assessed based on the *P*, *CEA*, *CEL*, and *SCF* rules. Consequently, the unified bargaining framework facilitates mutual consensus, ensuring a fair and efficient allocation strategy. Through an iterative interactive process, the *HNSA* scheme distributes the constrained RF and VLC spectrum resources, accommodating the competing demands of multiple IIoT devices. A key innovation of the *HNSA* scheme lies in its adoption of best response dynamics, particularly in handling diverse and fluctuating RF/VLC system conditions (Kim, 2022a).

11.7 Rate Splitting Multiple Access (*RSMA*) scheme

With the rapid advancement of communication technologies and the increasing demand for data traffic, traditional dynamic spectrum allocation methods are becoming inadequate to fully satisfy key wireless communication requirements such as spectrum and energy efficiency. To address these challenges, rate splitting multiple access (RSMA) has recently emerged as a generalized multiple access technique. Notably, S. Kim has developed the *Rate Splitting Multiple Access* (*RSMA*) scheme grounded in cooperative bargaining game theory (Kim, 2022b). Drawing on the core principles of pessimistic utilitarian and optimistic max-min bargaining solutions, the *RSMA* scheme facilitates mutual consensus among diverse network users to enhance overall system performance. Utilizing an interactive bargaining mechanism, the *RSMA* scheme dynamically manages spectrum and power resources within the RSMA system. To achieve an optimal balance between fairness and efficiency, the bargaining strategy enables coordinated decision-making while effectively handling trade-offs among competing objectives.

11.7.1 Development motivation

The rapid proliferation of smart mobile devices and the emergence of the IoT paradigm have been significant driving forces behind advancements in wireless communication technologies. The ultimate goal is to support a vast array of IoT services, including extended reality, smart homes, healthcare applications, autonomous driving, and tactile internet. To accommodate the increasing demand for future data services, research and development efforts in academia, industry, and regulatory bodies have focused on novel network architectures. By leveraging full-duplex wireless transmission, these systems aim to deliver data rates of approximately 1,000 Mbps, surpassing user expectations. The vision for future networks revolves around constructing a large-scale, interconnected system capable of dynamically adapting to user service requests. To date, numerous studies have yielded valuable insights through innovative communication techniques and novel approaches (Kim, 2022b; Li, 2020; Naser, 2020).

From the first to the fourth generation of mobile networks, multiple access protocols have played a crucial role in differentiating the capacity of wireless communications. Orthogonal multiple access (OMA) techniques regulate user access to the network, preventing transmission collisions. Various methods have maintained transmission orthogonality, including frequency division multiple access (FDMA) in 1G, time division multiple access (TDMA) in 2G, and code division multiple access (CDMA) in 3G. For 4G networks, orthogonal frequency division multiple access (OFDMA) was introduced, allocating specific subcarriers to users, thereby mapping signals into orthogonal time-frequency resources. This approach enhances wireless network performance by independently modulating subcarriers within frequency bands. The fundamental principle of OMA is the separation of user data streams to minimize inter-user interference, thereby achieving multiplexing gains with manageable complexity (Gamal, 2021; Kim, 2022b; Zeng, 2018).

As mobile communications progress into the 5G era, conventional OMA techniques struggle to meet the stringent demands of ultra-high data rates, ultra-low latency, ultra-reliability, and massive connectivity required by smart mobile devices. Non-orthogonal multiple access (NOMA) has emerged as a promising alternative to address these challenges in 5G networks. The core concept of NOMA involves serving multiple users by utilizing the same resources across time, frequency, and space. By enabling simultaneous transmission of multiple user signals within a single time-frequency resource element (one subcarrier in an OFDMA symbol), NOMA enhances spectrum efficiency beyond conventional OMA methods. However, since NOMA requires individual users to decode interference from overlapping signals, it imposes a significant computational burden due to the complexity of signal processing (Kim, 2022b; Yang, 2021; Zeng, 2018).

To accommodate a larger number of users in 5G networks, space division multiple access (SDMA) exploits spatial separation to enable full-time and frequency resource allocation. As a variant of OMA, SDMA employs linear precoding to distinguish users in the spatial domain, treating residual multi-user interference as noise. By superimposing users on the same time-frequency resources while differentiating them spatially, SDMA enhances spectrum efficiency. However, due to rate saturation, SDMA struggles to support high data rates or a large number of simultaneous users. To overcome these limitations, a novel multiple access technique known as rate splitting multiple access (RSMA) has been developed, addressing the drawbacks of both NOMA and SDMA. In next-generation networks, RSMA serves as a bridging solution between SDMA, which treats interference as noise, and NOMA, which relies on interference decoding (Gamal¸ 2021; Kim, 2021b; Lin, 2021; Mao, 2021; Yang, 2021; Zhou, 2021b).

Rate splitting technology has recently gained attention in multi-antenna broadcast channels as an advanced and resilient non-orthogonal transmission method for interference management in wireless communications. This approach involves splitting each user's message into two components: a common part and a private part. The transmitter jointly encodes both parts, broadcasting all superimposed data streams to users. Each user sequentially decodes and removes the common part before processing the private part using successive interference cancellation (SIC). RSMA enables users to partially decode interference while treating residual interference as noise. By strategically allocating power between the common and private parts, RSMA is particularly suitable for integration with multicast communication scenarios in 6G wireless networks (Kim, 2022b; Zhou, 2021a, 2021b).

Leveraging rate splitting technology, RSMA can efficiently support heterogeneous users in diverse environments. As one of the key enablers of 6G networks, RSMA effectively mitigates the challenges associated with NOMA and SDMA. Spectrum and energy efficiency are critical considerations for future communication system design - spectrum efficiency quantifies data transmission per unit time, while energy efficiency measures data transmission per unit energy. Recent studies demonstrate that RSMA surpasses conventional multiple access techniques, such as NOMA and SDMA, in both spectrum and energy efficiency. However, implementing RSMA in future wireless networks presents several challenges, including the optimal partitioning of common and private messages and efficient resource management for private message transmission (Kim, 2022b; Naser, 2020; Yang, 2021; Zhou, 2021a).

The primary objective of the *RSMA* scheme is to develop a robust control mechanism that optimizes resource management. A key challenge in the *RSMA* scheme is coordinating individual users while ensuring an optimal balance between common and private message transmissions. Achieving this is particularly complex in dynamic wireless environments, necessitating an intelligent control framework and novel solution methodologies. The *RSMA* scheme incorporates cooperative game theory to establish an effective control paradigm for RSMA-based communications. By applying cooperative game theory principles, RSMA systems intelligently and adaptively manage resources to balance competing requirements. To the best of my knowledge, cooperative approaches to RSMA control protocols have not yet been thoroughly explored (Kim, 2022b).

Most cooperative bargaining solutions in the literature are influenced by the principles of egalitarianism and utilitarianism. Egalitarianism advocates for equal treatment of all participants, while utilitarianism seeks to maximize the aggregate utility of all players. These principles often represent opposing axiomatic foundations for developing bargaining solutions to multi-criteria decision-making problems. When an additive value function is used for decision selection, a utilitarian approach is applied, whereas egalitarianism motivates the adoption of max-min solutions. Recent research has explored various bargaining solutions that incorporate both egalitarian and utilitarian principles to address multi-criteria bargaining challenges (Hinojosa, 2011; Kim, 2022b).

Multi-criteria bargaining solutions are concerned with scenarios in which multiple players must consider several decision-making criteria, each influenced by all players' choices. When a set of weighted criteria is available, egalitarianism leads to weighted max-min solutions, while utilitarianism results in weighted utilitarian solutions. However, when only partial information on criterion importance is available, alternative bargaining solutions extending the traditional utilitarian and max-min approaches have been developed. From a utilitarian perspective, the *Pessimistic Utilitarian Bargaining Solution* (*PUBS*) was formulated based on a pessimistic rule, favoring inequality aversion. Conversely, the *Optimistic Max-min Bargaining Solution* (*OMBS*) is rooted in egalitarian principles and applies an optimistic rule for selecting criterion weights (Hinojosa, 2011; Kim, 2022b).

11.7.2 Contribution

The *RSMA* scheme utilizes the RSMA system infrastructure in conjunction with cooperative game theory to develop an advanced RSMA control mechanism. In an environment where common and private data streams coexist, heterogeneous IoT devices collaboratively share limited system resources, aiming to maximize their individual payoffs while maintaining energy and spectrum efficiency. The primary objective in designing a novel RSMA resource management protocol is to steer self-interested IoT devices toward a socially optimal outcome while improving overall system performance. To achieve this, multiple IoT devices coordinate and operate cooperatively, adapting to dynamically fluctuating network traffic conditions. Consequently, the *RSMA* scheme is structured as two cooperative games, incorporating key principles from the *PUBS* and *OMBS*. Within the *RSMA* scheme, control decisions are interdependent and dynamically adjusted through an iterative, step-by-step interactive process. To the best of my knowledge, this study is the first to integrate bargaining solutions into the design of an RSMA system control algorithm (Kim, 2022b).

11.7.3 RSMA system platform and control problem formulation

The *RSMA* scheme considers a downlink multi-antenna multi-user wireless communication system. The network system consists of a set of multiple-antenna base stations (BSs), i.e., $\mathbb{B} = \{\mathscr{B}_1, \ldots, \mathscr{B}_n\}$; they are positioned at different locations in a cellular network environment. Each BS is equipped with $\mathscr{L} > 1$ antenna and serves a group of single-antenna IoT devices, i.e., $\mathbb{D}_{\mathscr{B}} = \{\mathscr{D}_1^{\mathscr{B}}, \ldots, \mathscr{D}_m^{\mathscr{B}}\}$; they are randomly located in the B's coverage area. In the RSMA system, the message $\mathfrak{M}_{\mathscr{B},\mathscr{D}_j}$ intended for the $\mathscr{D}_{1 \le j \le m}^{\mathscr{B}}$ is split into two parts where a common part $\mathfrak{M}^{C,\mathscr{B}}$ and a private part $\mathfrak{M}^{p}_{\mathscr{B},\mathscr{D}_j}$. The all devices' common part is encoded into the common stream $\mathscr{S}_{C_{\mathscr{B}}}$ using a codebook shared by all devices, i.e., $\mathscr{D}_{1 \le j \le m}^{\mathscr{B}} \in \mathbb{D}_{\mathscr{B}}$; the $\mathscr{S}_{C_{\mathscr{B}}}$ is required to be decoded by all devices. The private part for the $\mathscr{D}_j^{\mathscr{B}}$, i.e., $\mathfrak{M}^{p}_{\mathscr{B},\mathscr{D}_j}$, is the remaining part of the message $\mathfrak{M}_{\mathscr{B},\mathscr{D}_j}$; it is encoded into the private stream $\mathscr{S}_{\mathscr{D}_j^{\mathscr{B}}}$ for the $\mathscr{D}_j^{\mathscr{B}}$. Each

BS sends the transmitted signal, i.e., $\div_{\mathscr{B}}$, and the received signal at the $\mathscr{D}_j^{\mathscr{B}}$, i.e, $\mathscr{R}_{\mathscr{D}_j}^{\mathscr{B}}$, are expressed as follows (Kim, 2022b; Yang¸ 2021).

$$\begin{cases} \div_{\mathscr{B}} = \left(\sqrt{p_C} \times \mathscr{S}_{C_{\mathscr{B}}}\right) + \sum_{\mathscr{D}_j^{\mathscr{B}} \in \mathbb{D}_{\mathscr{B}}} \left(\sqrt{p_{\mathscr{D}_j^{\mathscr{B}}}} \times \mathscr{S}_{\mathscr{D}_j^{\mathscr{B}}}\right) \\ \mathscr{R}_{\mathscr{D}_j}^{\mathscr{B}} = \left(\sqrt{Ç_{\mathscr{D}_j^{\mathscr{B}}}} \times \div_{\mathscr{B}}\right) + \mathscr{N}_{\mathscr{D}_j^{\mathscr{B}}} \\ = \left(\sqrt{\left(Ç_{\mathscr{D}_j^{\mathscr{B}}} \times p_C\right)} \times \mathscr{S}_{C_{\mathscr{B}}}\right) + \sum_{\mathscr{D}_k^{\mathscr{B}} \in \mathbb{D}_{\mathscr{B}}} \left(\sqrt{\left(Ç_{\mathscr{D}_j^{\mathscr{B}}} \times p_{\mathscr{D}_k^{\mathscr{B}}}\right)} \times \mathscr{S}_{\mathscr{D}_j^{\mathscr{B}}}\right) + \mathscr{N}_{\mathscr{D}_j^{\mathscr{B}}} \end{cases} \quad (11.7.1)$$

where p_C and $p_{\mathscr{D}_j^{\mathscr{B}}}$ are the transmit powers for the $\mathscr{S}_{C_{\mathscr{B}}}$ and $\mathscr{S}_{\mathscr{D}_j^{\mathscr{B}}}$, respectively. $Ç_{\mathscr{D}_j^{\mathscr{B}}}$ represents the channel gain between the $\mathscr{D}_j^{\mathscr{B}}$ and B, and $\mathscr{N}_{\mathscr{D}_j^{\mathscr{B}}}$ is the additive white Gaussian noise with variance σ^2. The achievable rate of the $\mathscr{D}_j^{\mathscr{B}}$ decoding common stream $\mathscr{S}_{C_{\mathscr{B}}}$, i.e., $C_{\mathscr{D}_j^{\mathscr{B}}}$, can be expressed as (Kim, 2022b; Yang¸ 2021).:

$$C_{\mathscr{D}_j^{\mathscr{B}}} = A_{C_{\mathscr{B}}} \times \log_2\left(1 + \frac{\mathscr{R}_{\mathscr{D}_j}^{\mathscr{B}} \times p_C}{\left(Ç_{\mathscr{D}_j^{\mathscr{B}}} \times \sum_{\mathscr{D}_k^{\mathscr{B}} \in \mathbb{D}_{\mathscr{B}}} \left(p_{\mathscr{D}_k^{\mathscr{B}}}\right)\right) + \sigma^2}\right) \quad (11.7.2)$$

where $A_{C_{\mathscr{B}}}$ is the assigned spectrum for the common stream $\mathscr{S}_{C_{\mathscr{B}}}$. After having decoded the $\mathscr{S}_{C_{\mathscr{B}}}$, each device can decode its private stream, the achievable private stream rate of $\mathscr{D}_j^{\mathscr{B}}$, i.e., $\gamma_{\mathscr{D}_j^{\mathscr{B}}}$, is given by (Kim, 2022b; Yang¸ 2021).:

$$\gamma_{\mathscr{D}_j^{\mathscr{B}}} = A_{\mathscr{D}_j^{\mathscr{B}}} \times \log_2\left(1 + \frac{\mathscr{R}_{\mathscr{D}_j}^{\mathscr{B}} \times p_{\mathscr{D}_j^{\mathscr{B}}}}{\left(\mathscr{R}_{\mathscr{D}_j}^{\mathscr{B}} \times \sum_{\substack{\mathscr{D}_k^{\mathscr{B}} \in \mathbb{D}_{\mathscr{B}} \\ \mathscr{D}_k^{\mathscr{B}} \neq \mathscr{D}_j^{\mathscr{B}}}} \left(p_{\mathscr{D}_k^{\mathscr{B}}}\right)\right) + \sigma^2}\right) \quad (11.7.3)$$

where $A_{\mathscr{D}_j^{\mathscr{B}}}$ is the assigned spectrum for the private stream, i.e., $\mathscr{S}_{\mathscr{D}_j^{\mathscr{B}}}$. Finally, the total transmission rate of $\mathscr{D}_j^{\mathscr{B}}$, i.e., $T_{\mathscr{D}_j}^{\mathscr{B}}$, is estimated based on the common stream rate and achievable private stream rate where $T_{\mathscr{D}_j}^{\mathscr{B}} = C_{\mathscr{D}_j^{\mathscr{B}}} + \gamma_{\mathscr{D}_j^{\mathscr{B}}}$. In the viewpoint of $\mathscr{D}_j^{\mathscr{B}}$, the major goal is to maximize the $T_{\mathscr{D}_j}^{\mathscr{B}}$. However, communication resources in the RSMA system are limited. Therefore, effective resource control strategies should be considered to improve the spectrum and power efficiency. To address this control problem, the *RSMA* scheme formulates two bargaining games, i.e., $\mathbb{G}_A^{\mathscr{B}}$ and $\mathbb{G}_P^{\mathscr{B}}$ for each B. It is noteworthy that these games are operated in a cooperative manner, and they are repeated in a step-by-step interactive fashion. The $\mathbb{G}_A^{\mathscr{B}}$ game decides the spectrum amounts for the $\mathscr{S}_{C_{\mathscr{B}}}$ and $\mathscr{S}_{\mathscr{D}_j^{\mathscr{B}} \in \mathbb{D}_{\mathscr{B}}}$, i.e., $A_{C_{\mathscr{B}}}$ and $A_{\mathscr{D}_j^{\mathscr{B}}}$. The $\mathbb{G}_P^{\mathscr{B}}$ game adjusts the power levels for the $\mathscr{S}_{C_{\mathscr{B}}}$ and $\mathscr{S}_{\mathscr{D}_j^{\mathscr{B}} \in \mathbb{D}_{\mathscr{B}}}$, i.e., p_C and $p_{\mathscr{D}_j^{\mathscr{B}}}$. Formally, the *RSMA* scheme defines the $\mathbb{G}_A^{\mathscr{B}}$ and $\mathbb{G}_P^{\mathscr{B}}$ game entities for each B, i.e.,

$$\mathbb{G} = \left\{\mathbb{G}_A^{\mathscr{B}}, \mathbb{G}_P^{\mathscr{B}}\right\} = \left\{\mathscr{B} \in \mathbb{B}, \mathbb{D}_{\mathscr{B}} \subset \mathbb{D}, \left\{\mathbb{G}_A^{\mathscr{B}} \middle| \mathscr{S}_{C_{\mathscr{B}}}, \mathscr{S}_{\mathscr{D}_j^{\mathscr{B}} \in \mathbb{D}_{\mathscr{B}}}, \left(A_{C_{\mathscr{B}}}, A_{\mathscr{D}_j^{\mathscr{B}}}\right), U_{C_{\mathscr{B}}}(\cdot), U_{\mathscr{D}_j^{\mathscr{B}}}(\cdot)\right\}, \left\{\mathbb{G}_P^{\mathscr{B}} \middle| \mathscr{S}_{C_{\mathscr{B}}}, \mathscr{S}_{\mathscr{D}_j^{\mathscr{B}} \in \mathbb{D}_{\mathscr{B}}}, \left(p_{C_{\mathscr{B}}}, p_{\mathscr{D}_j^{\mathscr{B}}}\right), \mathfrak{U}_{C_{\mathscr{B}}}(\cdot), \mathfrak{U}_{\mathscr{D}_j^{\mathscr{B}}}(\cdot)\right\}, T\right\}$$

of gameplay (Kim, 2022b).

- The $\mathbb{G}_A^{\mathscr{B}}$ and $\mathbb{G}_P^{\mathscr{B}}$ are mutually and reciprocally interdependent in an interactive manner, and they work together to allocate spectrum amounts and adjust power levels for common and private streams.
- B is the set of BSs and D is the set of total IoT devices in the RSMA system platform. B is one BS in B, and $\mathbb{D}_{\mathscr{B}}$ is the subset of D; devices in $\mathbb{D}_{\mathscr{B}}$ are covered by the B.
- In the $\mathbb{G}_A^{\mathscr{B}}$, $\mathcal{S}_{C_{\mathscr{B}}}$ and $\mathcal{S}_{\mathscr{D}_j^{\mathscr{B}}}$ are game players where $\mathscr{D}_j^{\mathscr{B}} \in \mathbb{D}_{\mathscr{B}}$, and $A_{C_{\mathscr{B}}}$ and $A_{\mathscr{D}_j^{\mathscr{B}}}$ are their strategies, which are the allocated spectrum amounts for the $\mathcal{S}_{C_{\mathscr{B}}}$ and $\mathcal{S}_{\mathscr{D}_j^{\mathscr{B}}}$, respectively. $U_{C_{\mathscr{B}}}(\cdot)$ and $U_{\mathscr{D}_j^{\mathscr{B}}}(\cdot)$ are their utility functions.
- In the $\mathbb{G}_P^{\mathscr{B}}$, $\mathcal{S}_{C_{\mathscr{B}}}$ and $\mathcal{S}_{\mathscr{D}_j^{\mathscr{B}}}$ are also game players like as the $\mathbb{G}_A^{\mathscr{B}}$. $p_{C_{\mathscr{B}}}$ and $p_{\mathscr{D}_j^{\mathscr{B}}}$ are their strategies, which are the power levels for the $\mathcal{S}_{C_{\mathscr{B}}}$ and $\mathcal{S}_{\mathscr{D}_j^{\mathscr{B}}}$, respectively. $\mathfrak{U}_{C_{\mathscr{B}}}(\cdot)$ and $\mathfrak{U}_{\mathscr{D}_j^{\mathscr{B}}}(\cdot)$ are their utility functions.
- Discrete time model $T \in \{t_1, \ldots, t_c, t_{c+1}, \ldots\}$ is represented by a sequence of time steps. The length of t_c matches the event time-scale of $\mathbb{G}_A^{\mathscr{B}}$ and $\mathbb{G}_P^{\mathscr{B}}$.

11.7.4 The fundamental ideas of *PUBS* and *OMBS*

To characterize the fundamental ideas of *PUBS* and *OMBS*, let $\triangle^k = \{\lambda \in \mathbb{R}_+^k | \lambda \cdot \mathbb{I}^k = 1\}$ denote the simplex in $\mathbb{R}_+^k$ where $\mathbb{I}^k$ is an *k*-dimensional vector with components equal to one. For a polyhedric subset $\Lambda \subseteq \triangle^k$, let $ext(\Lambda)$ denote the set of its extreme points, i.e.,$ext(\Lambda) = \{\lambda \in \Lambda | \nexists \lambda_1, \lambda_2 \in \Lambda,$ $\lambda_1 \neq \lambda_2$ and $\alpha \in (0,1)$such that $\lambda = ((\alpha \times \lambda_1) + ((1-\alpha) \times \lambda_2))\}$. A multi-criteria problem with partial information is represented by a triplet (X, z, Λ), where i) $X \subseteq \mathbb{R}^n$ is the set of feasible points in the decision space, each $x \in X$ is as n-dimensional vector of decision variables; ii) z: $\mathbb{R}^n \rightarrow \mathbb{R}_+^k$ is a vector function, whose k components, $z_1(x), \ldots, z_k(x)$, are k scalar functions of the decision vector x which represent the criteria that have to be taken into account to evaluate each feasible decision; iii) $\Lambda \subseteq \Delta^k$ is a polyhedron of weights, which represent the information available with respect to the importance of the criteria. A solution for the class of multi-criteria problems is a map, F, which associates a subset of X to each problem $(X, z, \Lambda) \in \mathscr{P}$ where P is the class of these multi-criteria problems, and $F(X, z, \Lambda) \subseteq X$. Therefore, a solution is single valued in P if for all $(X, z, \Lambda) \in \mathscr{P}$ (Hinojosa, 2011; Kim, 2022b).

The *PUBS* is developed based on the idea that the evaluation of a decision may depend on the minimum weighted value among all the feasible weights in Λ. In the *PUBS*, the minimum feasible value assigned to a decision across the whole set of weights is maximized. Therefore, the *PUBS* is characterized as a solution for single-criterion optimization problems which depend on the extreme points of the polyhedron of weights. Note that this problem is linear provided that X is a polyhedron and z are linear criteria functions. Mathematically, the *PUBS* for each $(X, z, \Lambda) \in \mathscr{P}$, i.e., $F_{PUBS}(X, z, \Lambda)$, is defined as follows (Hinojosa, 2011; Kim, 2022b):

$$F_{PUBS}(X, z, \Lambda) = \arg\max_{x \in X} m_\Lambda(x) = \arg\max_{x \in X}\left[\min_{\lambda \in \Lambda}\left\{\sum\nolimits_{i=1}^{k} (\lambda_i \times z_i(x))\right\}\right] \tag{11.7.4}$$

The *OMBS* is designed according to the concept that the evaluation of a decision depends upon its worst performance across all the criteria but selecting the best of this minimum across all feasible weighting vectors. Therefore, the *OMBS* is the weighted max-min solution, which corresponds to the vectors in Λ that give the best max-min values. Formally, the *OMBS* for each $(X,z,\Lambda) \in \mathcal{P}$, i.e., F_{OMBS} (X,z,Λ), is defined as follows (Hinojosa, 2011; Kim, 2022b):

$$F_{OMBS}(X,z,\Lambda) = \arg\max_{x\in X} N_{\Lambda}(x) = \arg\max_{x\in X}\left[\min_{i=1\ldots k}\left\{\frac{z_i(x)}{\lambda_i}\right\}\right] \tag{11.7.5}$$

11.7.5 Two bargaining game models for the RSMA system

To develop the *RSMA* scheme, the $\mathbb{G}_A^{\mathcal{B}}$ and $\mathbb{G}_P^{\mathcal{B}}$ games are constructed. They interact with each other during the spectrum and power control process. To maximize the RSMA system performance, the $\mathbb{G}_A^{\mathcal{B}}$ decides the values of $A_{C_{\mathcal{B}}}$ and $A_{\mathcal{D}_j^{\mathcal{B}}\in \mathbb{D}_{\mathcal{B}}}$, which are the allocated spectrum amounts for the $\mathcal{S}_{C_{\mathcal{B}}}$ and $\mathcal{S}_{\mathcal{D}_j^{\mathcal{B}}}$, respectively. At each time period, $U_{C_{\mathcal{B}}}(\cdot)$ and $U_{\mathcal{D}_j^{\mathcal{B}}}(\cdot)$ in the $\mathbb{G}_A^{\mathcal{B}}$ are defined as follows (Kim, 2022b):

$$\begin{cases} U_{C_{\mathcal{B}}}\left(A_{C_{\mathcal{B}}}, M_{C_{\mathcal{B}}}\right) = exp\left(\alpha + \log\left(\dfrac{\min\left(A_{C_{\mathcal{B}}}, M_{C_{\mathcal{B}}}\right)}{M_{C_{\mathcal{B}}}}\right)\right) \\ U_{\mathcal{D}_j^{\mathcal{B}}}\left(A_{\mathcal{D}_j^{\mathcal{B}}}, M_{\mathcal{D}_j^{\mathcal{B}}}\right) = \log\left(exp\left(\dfrac{\min\left(A_{\mathcal{D}_j^{\mathcal{B}}}, M_{\mathcal{D}_j^{\mathcal{B}}}\right)}{M_{\mathcal{D}_j^{\mathcal{B}}}}\right)^{\beta}\right) \end{cases} \tag{11.7.6}$$

$$\text{s.t.}, \ \mathfrak{R}_{\mathcal{B}} = A_{C_{\mathcal{B}}} + \sum\nolimits_{\mathcal{D}_j^{\mathcal{B}}\in \mathbb{D}_{\mathcal{B}}} A_{\mathcal{D}_j^{\mathcal{B}}}$$

where $M_{C_{\mathcal{B}}}$ and $M_{\mathcal{D}_j^{\mathcal{B}}}$ are the requested spectrum amounts for the $\mathcal{S}_{C_{\mathcal{B}}}$ and $\mathcal{S}_{\mathcal{D}_j^{\mathcal{B}}}$, respectively. α is a control parameter for the $U_{C_{\mathcal{B}}}(\cdot)$, and β is a control parameter for the $U_{\mathcal{D}_j^{\mathcal{B}}}(\cdot)$. $\mathfrak{R}_{\mathcal{B}}$ is the total spectrum amount for the B. To adaptively allocate the spectrum for each stream, the *RSMA* scheme can consider two criteria; delay sensitivity and packet loss intolerance, which will be set up for each individually services. Therefore, the criteria vector Λ for the $\mathbb{G}_A^{\mathcal{B}}$ consists of this information. To maximize the spectrum efficiency, the solution concept should be strongly concerned the utilitarianism. In this case, the *PUBS* is preferred for the solution concept of $\mathbb{G}_A^{\mathcal{B}}$. It is given by (Kim, 2022b):

$$\left[\mathbb{G}_A^{\mathcal{B}}\right]_{PUBS}\left(\left(U_{C_{\mathcal{B}}}(\cdot), U_{\mathcal{D}_j^{\mathcal{B}}}(\cdot)\right), \left(A_{C_{\mathcal{B}}}, A_{\mathcal{D}_j^{\mathcal{B}}}\right), \Lambda_{\mathbb{G}_A^{\mathcal{B}}}\right) = \tag{11.7.7}$$

$$\arg\max_{A_{C_{\mathcal{B}}}, A_{\mathcal{D}_j^{\mathcal{B}}\in \mathbb{D}_{\mathcal{B}}}}\left[\min_{\lambda_{C_{\mathcal{B}}}, \lambda_{\mathcal{D}_j^{\mathcal{B}}}\in \Lambda_{\mathbb{G}_A^{\mathcal{B}}}}\left\{\left(\lambda_{C_{\mathcal{B}}} \times U_{C_{\mathcal{B}}}(\cdot)\right) + \sum\nolimits_{\mathcal{D}_j^{\mathcal{B}}\in \mathbb{D}_{\mathcal{B}}}\left(\lambda_{\mathcal{D}_j^{\mathcal{B}}} \times U_{\mathcal{D}_j^{\mathcal{B}}}(\cdot)\right)\right\}\right]$$

where $\lambda_{C_{\mathcal{B}}}$ and $\lambda_{\mathcal{D}_j^{\mathcal{B}}}$ are weight factors in the $\Lambda_{\mathbb{G}_A^{\mathcal{B}}}$ for $C_{\mathcal{B}}$ and $\mathcal{D}_j^{\mathcal{B}}$ communication services, respectively. Contrast to the $\mathbb{G}_A^{\mathcal{B}}$, the $\mathbb{G}_P^{\mathcal{B}}$ game is operated to set the power levels for game players. As game players, the utility function of $C_{\mathcal{B}}$ and $\mathcal{D}_j^{\mathcal{B}}$, i.e., $\mathfrak{U}_{C_{\mathcal{B}}}(\cdot)$ and $\mathfrak{U}_{\mathcal{D}_j^{\mathcal{B}}}(\cdot)$, are defined as follows (Kim, 2022b):

$$\mathfrak{U}_{C_{\mathcal{B}}}\left(p_{C_{\mathcal{B}}},\Gamma_{C_{\mathcal{B}}}\right) = \eta \times \left(exp\left(\frac{exp\left(\frac{\min\left(p_{C_{\mathcal{B}}},\Gamma_{C_{\mathcal{B}}}\right)}{\Gamma_{C_{\mathcal{B}}}}\right) - exp\left(-\frac{\min\left(p_{C_{\mathcal{B}}},\Gamma_{C_{\mathcal{B}}}\right)}{\Gamma_{C_{\mathcal{B}}}}\right)}{exp\left(\frac{\min\left(p_{C_{\mathcal{B}}},\Gamma_{C_{\mathcal{B}}}\right)}{\Gamma_{C_{\mathcal{B}}}}\right) + exp\left(-\frac{\min\left(p_{C_{\mathcal{B}}},\Gamma_{C_{\mathcal{B}}}\right)}{\Gamma_{C_{\mathcal{B}}}}\right)} \right) - \mu \right)$$

$$\mathfrak{U}_{\mathcal{D}_j^{\mathcal{B}}}\left(p_{\mathcal{D}_j^{\mathcal{B}}},\Gamma_{\mathcal{D}_j^{\mathcal{B}}}\right) = \tau \times \log\left(\frac{exp\left(\frac{\min\left(p_{\mathcal{D}_j^{\mathcal{B}}},\Gamma_{\mathcal{D}_j^{\mathcal{B}}}\right)}{\Gamma_{C_{\mathcal{B}}}}\right) - exp\left(-\frac{\min\left(p_{\mathcal{D}_j^{\mathcal{B}}},\Gamma_{\mathcal{D}_j^{\mathcal{B}}}\right)}{\Gamma_{C_{\mathcal{B}}}}\right)}{exp\left(\frac{\min\left(p_{\mathcal{D}_j^{\mathcal{B}}},\Gamma_{\mathcal{D}_j^{\mathcal{B}}}\right)}{\Gamma_{C_{\mathcal{B}}}}\right) + exp\left(-\frac{\min\left(p_{\mathcal{D}_j^{\mathcal{B}}},\Gamma_{\mathcal{D}_j^{\mathcal{B}}}\right)}{\Gamma_{C_{\mathcal{B}}}}\right)} + \varphi \right)$$

(11.7.8)

$$\text{s.t., } \mathfrak{P}_{\mathcal{B}} = p_{C_{\mathcal{B}}} + \sum\nolimits_{\mathcal{D}_j^{\mathcal{B}} \in \mathbb{D}_{\mathcal{B}}} p_{\mathcal{D}_j^{\mathcal{B}}}$$

where η and μ are control parameters for the $\mathfrak{U}_{C_{\mathcal{B}}}(\cdot)$, and τ and φ are control parameters for the $\mathfrak{U}_{\mathcal{D}_j^{\mathcal{B}}}(\cdot)$. $\mathfrak{P}_{\mathcal{B}}$ is the total power energy of B. In the *RSMA* scheme, it is defined as a conceptual power unit. To adjust the power levels for each stream, the *RSMA* scheme also makes decisions according to the same performance criteria. Therefore, the Λ for the $\mathbb{G}_P^{\mathcal{B}}$, i.e., $\Lambda_{\mathbb{G}_P^{\mathcal{B}}}$, is defined as the same manner as the $\Lambda_{\mathbb{G}_\lambda^{\mathcal{B}}}$. In the $\mathbb{G}_P^{\mathcal{B}}$, the power levels of game players should be decided by considering the relative fairness. In this case, the *OMBS* is preferred for the solution concept of $\mathbb{G}_P^{\mathcal{B}}$. It is given by (Kim, 2022b):

$$\left[\mathbb{G}_P^{\mathcal{B}}\right]_{OMBS}\left(\left(\mathfrak{U}_{C_{\mathcal{B}}}(\cdot),\mathfrak{U}_{\mathcal{D}_j^{\mathcal{B}}}(\cdot)\right),\left(p_{C_{\mathcal{B}}},p_{\mathcal{D}_j^{\mathcal{B}}}\right),\Lambda_{\mathbb{G}_P^{\mathcal{B}}}\right) =$$

(11.7.9)

$$= \arg\max_{p_{C_{\mathcal{B}}},p_{\mathcal{D}_j^{\mathcal{B}}}}\left[\min_{\lambda_{C_{\mathcal{B}}},\lambda_{\mathcal{D}_j^{\mathcal{B}}}\in\Lambda_{\mathbb{G}_P^{\mathcal{B}}}}\left\{\mathcal{D}_j^{\mathcal{B}} \in \mathbb{D}_{\mathcal{B}}\middle|\frac{\mathfrak{U}_{C_{\mathcal{B}}}(\cdot)}{\lambda_{C_{\mathcal{B}}}},\frac{\mathfrak{U}_{\mathcal{D}_j^{\mathcal{B}}}(\cdot)}{\lambda_{\mathcal{D}_j^{\mathcal{B}}}}\right\}\right]$$

11.7.6 Summary

In next-generation networks, the RSMA technology serves as a versatile and efficient multiple access approach. Effective resource management is essential to achieving optimal RSMA system performance. The *RSMA* scheme specifically addresses the challenge of multi-criteria resource management within RSMA systems. To efficiently tackle this issue, it incorporates bargaining principles derived from the *PUBS* and the *OMBS*, structuring the problem into two cooperative game models. This approach facilitates mutual consensus, enabling a balanced trade-off between spectrum and power efficiency. Through an interactive bargaining process, the *RSMA* scheme efficiently allocates limited system resources while accommodating varying service task intensities (Kim, 2022b).

REFERENCES

Al Hammadi, A., Sofotasios, P. C., Muhaidat, S., Al-Qutayri, M., & Elgala, H. (2020). Non-orthogonal multiple access for hybrid VLC-RF networks with imperfect channel state information. *IEEE Transactions on Vehicular Technology*, *70*(1), 398–411. DOI: 10.1109/TVT.2020.3044837

Alós-Ferrer, C., García-Segarra, J., & Ginés-Vilar, M. (2018). Anchoring on Utopia: A generalization of the Kalai–Smorodinsky solution. *Economic Theory Bulletin*, *6*(2), 141–155. DOI: 10.1007/s40505-017-0130-7

Attia, M. B., Nguyen, K. K., & Cheriet, M. (2019). Dynamic QoE/QoS-aware queuing for heterogeneous traffic in smart home. *IEEE Access : Practical Innovations, Open Solutions*, *7*, 58990–59001. DOI: 10.1109/ACCESS.2019.2914658

Attia, M. B., Nguyen, K. K., & Cheriet, M. (2020). Dynamic QoS-aware scheduling for concurrent traffic in smart home. *IEEE Internet of Things Journal*, *7*(6), 5412–5425. DOI: 10.1109/JIOT.2020.2978160

Baek, B., Lee, J., Peng, Y., & Park, S. (2020). Three dynamic pricing schemes for resource allocation of edge computing for IoT environment. *IEEE Internet of Things Journal*, *7*(5), 4292–4303. DOI: 10.1109/JIOT.2020.2966627

Bajic, B., Rikalovic, A., Suzic, N., & Piuri, V. (2020). Industry 4.0 implementation challenges and opportunities: A managerial perspective. *IEEE Systems Journal*, *15*(1), 546–559. DOI: 10.1109/JSYST.2020.3023041

Balakrishnan, P. S., Gómez, J. C., & Vohra, R. V. (2011). The tempered aspirations solution for bargaining problems with a reference point. *Mathematical Social Sciences*, *62*(3), 144–150. DOI: 10.1016/j.mathsocsci.2011.09.003

Blackorby, C., Bossert, W., & Donaldson, D. (1994). Generalized Ginis and cooperative bargaining solutions. *Econometrica*, *62*(5), 1161–1178. DOI: 10.2307/2951511

Boccardi, F., Heath, R. W., Lozano, A., Marzetta, T. L., & Popovski, P. (2014). Five disruptive technology directions for 5G. *IEEE Communications Magazine*, *52*(2), 74–80. DOI: 10.1109/MCOM.2014.6736746

Bossert, W., & Tan, G. (1995). An arbitration game and the egalitarian bargaining solution. *Social Choice and Welfare*, *12*(1), 29–41. DOI: 10.1007/BF00182191

Chen, M., & Hao, Y. (2018). Task offloading for mobile edge computing in software defined ultra-dense network. *IEEE Journal on Selected Areas in Communications*, *36*(3), 587–597. DOI: 10.1109/JSAC.2018.2815360

Chen, Z., Hou, X., & Yang, C. (2015). Training resource allocation for user-centric base station cooperation networks. *IEEE Transactions on Vehicular Technology*, *65*(4), 2729–2735. DOI: 10.1109/TVT.2015.2420114

Chettri, L., & Bera, R. (2019). A comprehensive survey on Internet of Things (IoT) toward 5G wireless systems. *IEEE Internet of Things Journal*, *7*(1), 16–32. DOI: 10.1109/JIOT.2019.2948888

Choi, S., Choi, J. G., & Bahk, S. (2021). Association control for user centric millimeter wave communication systems. *IEEE Transactions on Vehicular Technology, 70*(11), 12001–12012. DOI: 10.1109/TVT.2021.3113470

Claus, C., & Boutilier, C. (1998). The dynamics of reinforcement learning in cooperative multiagent systems. *AAAI/IAAI,* 1998(746-752), 2.

Cui, H., & Zhang, Z. (2021). A cooperative multi-agent reinforcement learning method based on coordination degree. *IEEE Access : Practical Innovations, Open Solutions, 9,* 123805–123814. DOI: 10.1109/ACCESS.2021.3110255

Dong, J., Ou, Z., Ylä-Jääski, A., & Cui, Y. (2013, December). Mobile hotspots cooperation towards better energy efficiency. In *2013 IEEE Globecom Workshops (GC Wkshps)* (pp. 760-765). IEEE.

Dubra, J. (2001). An asymmetric Kalai–Smorodinsky solution. *Economics Letters, 73*(2), 131–136. DOI: 10.1016/S0165-1765(01)00426-8

Fister, I., Fister, I.Jr, Yang, X. S., & Brest, J. (2013). A comprehensive review of firefly algorithms. *Swarm and Evolutionary Computation, 13,* 34–46. DOI: 10.1016/j.swevo.2013.06.001

Gallice, A. (2019). Bankruptcy problems with reference-dependent preferences. *International Journal of Game Theory, 48*(1), 311–336. DOI: 10.1007/s00182-018-0647-5

Gamal, S., Rihan, M., Hussin, S., Zaghloul, A., & Salem, A. A. (2021). Multiple access in cognitive radio networks: From orthogonal and non-orthogonal to rate-splitting. *IEEE Access : Practical Innovations, Open Solutions, 9,* 95569–95584. DOI: 10.1109/ACCESS.2021.3095142

Georgiadis, L., Iosifidis, G., & Tassiulas, L. (2019). On the efficiency of sharing economy networks. *IEEE Transactions on Network Science and Engineering, 7*(3), 1094–1110. DOI: 10.1109/TNSE.2019.2904431

Guo, L., Ning, Z., Hou, W., Hu, B., & Guo, P. (2018). Quick answer for big data in sharing economy: Innovative computer architecture design facilitating optimal service-demand matching. *IEEE Transactions on Automation Science and Engineering, 15*(4), 1494–1506. DOI: 10.1109/TASE.2018.2838340

Hammouda, M., Akın, S., Vegni, A. M., Haas, H., & Peissig, J. (2018). Link selection in hybrid RF/VLC systems under statistical queueing constraints. *IEEE Transactions on Wireless Communications, 17*(4), 2738–2754. DOI: 10.1109/TWC.2018.2802937

Hinojosa, M. A., & Mármol, A. M. (2011). Egalitarianism and utilitarianism in multiple criteria decision problems with partial information. *Group Decision and Negotiation, 20*(6), 707–724. DOI: 10.1007/s10726-009-9184-8

Hinojosa, M. Á., Mármol, A. M., & Zarzuelo, J. M. (2007). *Multi-utilitarian bargaining solutions* (No. 07.13).

Hinojosa, M. A., Mármol, A. M., & Zarzuelo, J. M. (2008). Inequality averse multi-utilitarian bargaining solutions. *International Journal of Game Theory, 37*(4), 597–618. DOI: 10.1007/s00182-008-0137-2

Jang, I., Pyeon, D., Kim, S., & Yoon, H. (2013). A survey on communication protocols for wireless sensor networks. *Journal of Computing Science and Engineering : JCSE, 7*(4), 231–241. DOI: 10.5626/JCSE.2013.7.4.231

Jiang, Y., & Tan, F. (2020). An enhanced model-free reinforcement learning algorithm to solve Nash equilibrium for multi-agent cooperative game systems. *IEEE Access : Practical Innovations, Open Solutions, 8*, 223743–223755. DOI: 10.1109/ACCESS.2020.3043806

Joe-Wong, C., Sen, S., & Ha, S. (2018). Sponsoring mobile data: Analyzing the impact on internet stakeholders. *IEEE/ACM Transactions on Networking, 26*(3), 1179–1192. DOI: 10.1109/TNET.2018.2826531

Ju, H., & Zhang, R. (2013). Throughput maximization in wireless powered communication networks. *IEEE Transactions on Wireless Communications, 13*(1), 418–428. DOI: 10.1109/TWC.2013.112513.130760

Karagözoğlu, E., & Keskin, K. (2015). A tale of two bargaining solutions. *Games, 6*(2), 161–174. DOI: 10.3390/g6020161

Karagözoğlu, E., & Tok, E. Ö. (2018). Iterated egalitarian compromise solution to bargaining problems and midpoint domination. *Operations Research Letters, 46*(3), 282–285. DOI: 10.1016/j.orl.2018.02.005

Karos, D., Muto, N., & Rachmilevitch, S. (2018). A generalization of the Egalitarian and the Kalai–Smorodinsky bargaining solutions. *International Journal of Game Theory, 47*(4), 1169–1182. DOI: 10.1007/s00182-018-0611-4

Kawamori, T. (2014). A noncooperative foundation of the asymmetric Nash bargaining solution. *Journal of Mathematical Economics, 52*, 12–15. DOI: 10.1016/j.jmateco.2014.03.004

Kim, K. S., Uno, S., & Kim, M. W. (2010). Adaptive QoS mechanism for wireless mobile network. *Journal of Computing Science and Engineering : JCSE, 4*(2), 153–172. DOI: 10.5626/JCSE.2010.4.2.153

Kim, S. (2011). Adaptive online power control scheme based on the evolutionary game theory. *IET Communications, 5*(18), 2648–2655. DOI: 10.1049/iet-com.2011.0093

Kim, S. (Ed.). (2014). *Game theory applications in network design*. IGI Global. DOI: 10.4018/978-1-4666-6050-2

Kim, S. (2015). Nested game-based computation offloading scheme for mobile cloud IoT systems. *EURASIP Journal on Wireless Communications and Networking, 2015*(1), 1–11. DOI: 10.1186/s13638-015-0456-5

Kim, S. (2019). Novel Data Sponsoring Control Scheme Based on the Dual-Leader Stackelberg Game Model. *Mobile Information Systems, 2019*(1), 1231574. DOI: 10.1155/2019/1231574

Kim, S. (2020). Cooperative WiFi Tethering Control Algorithm Based on the Meta-Bargaining Approach. *IEEE Access : Practical Innovations, Open Solutions, 8*, 213370–213377. DOI: 10.1109/ACCESS.2020.3040301

Kim, S. (2021a). A New Twofold Bargaining Game Paradigm for Wireless Energy Harvesting Algorithm. *Mobile Information Systems, 2021*(1), 8565467. DOI: 10.1155/2021/8565467

Kim, S. (2021b). Home network traffic control scheme based on two-level bargaining game model. *IEEE Access : Practical Innovations, Open Solutions*, *9*, 59665–59674. DOI: 10.1109/ACCESS.2021.3073485

Kim, S. (2022a). Hybrid RF/VLC network spectrum allocation scheme using bargaining solutions. *IEEE Access : Practical Innovations, Open Solutions*, *10*, 20019–20028. DOI: 10.1109/ACCESS.2022.3153327

Kim, S. (2022b). Efficient rate splitting multiple access scheme based cooperative bargaining solutions. *IEEE Access : Practical Innovations, Open Solutions*, *10*, 55081–55089. DOI: 10.1109/ACCESS.2022.3177654

Kim, S. (2023). Joint Price Control and Service Distribution Scheme for User-Centric Cell-Free Network Systems. *IEEE Access : Practical Innovations, Open Solutions*, *11*, 72884–72894. DOI: 10.1109/ACCESS.2023.3291407

Kong, J., Ismail, M., Serpedin, E., & Qaraqe, K. A. (2019). Energy efficient optimization of base station intensities for hybrid RF/VLC networks. *IEEE Transactions on Wireless Communications*, *18*(8), 4171–4183. DOI: 10.1109/TWC.2019.2922611

Kountouris, M. (2017). Performance limits of network densification. *IEEE Journal on Selected Areas in Communications*, *35*(6), 1294–1308. DOI: 10.1109/JSAC.2017.2687638

Kua, J., Nguyen, S. H., Armitage, G., & Branch, P. (2017). Using active queue management to assist IoT application flows in home broadband networks. *IEEE Internet of Things Journal*, *4*(5), 1399–1407. DOI: 10.1109/JIOT.2017.2722683

Kumar, S., Tiwari, P., & Zymbler, M. (2019). Internet of Things is a revolutionary approach for future technology enhancement: A review. *Journal of Big Data*, *6*(1), 1–21. DOI: 10.1186/s40537-019-0268-2

Li, S., Sun, W., Zhang, H., & Zhang, Y. (2020, December). Physical layer security for edge caching in 6G networks. In *GLOBECOM 2020-2020 IEEE Global Communications Conference* (pp. 1-6). IEEE.

Li, X., Zhang, R., & Hanzo, L. (2015). Cooperative load balancing in hybrid visible light communications and WiFi. *IEEE Transactions on Communications*, *63*(4), 1319–1329. DOI: 10.1109/TCOMM.2015.2409172

Liao, Y., Shou, L., Yu, Q., Ai, Q., & Liu, Q. (2020). An intelligent computation demand response framework for IIoT-MEC interactive networks. *IEEE Networking Letters*, *2*(3), 154–158. DOI: 10.1109/LNET.2020.3001178

Lim, W. S., & Shin, K. G. (2016). POEM: Minimizing energy consumption for WiFi tethering service. *IEEE/ACM Transactions on Networking*, *24*(6), 3785–3797. DOI: 10.1109/TNET.2016.2556689

Lin, Z., Lin, M., De Cola, T., Wang, J. B., Zhu, W. P., & Cheng, J. (2021). Supporting IoT with rate-splitting multiple access in satellite and aerial-integrated networks. *IEEE Internet of Things Journal*, *8*(14), 11123–11134. DOI: 10.1109/JIOT.2021.3051603

Lu, C. H. (2018). Context-aware service provisioning via agentized and reconfigurable multimodel cooperation for real-life IoT-enabled smart home systems. *IEEE Transactions on Systems, Man, and Cybernetics. Systems*, *50*(8), 2914–2925. DOI: 10.1109/TSMC.2018.2831711

Lyu, B., Qi, T., Guo, H., & Yang, Z. (2019). Throughput maximization in full-duplex dual-hop wireless powered communication networks. *IEEE Access : Practical Innovations, Open Solutions*, *7*, 158584–158593. DOI: 10.1109/ACCESS.2019.2950519

Ma, Y., Chen, H., Lin, Z., Li, Y., & Vucetic, B. (2015). Distributed and optimal resource allocation for power beacon-assisted wireless-powered communications. *IEEE Transactions on Communications*, *63*(10), 3569–3583. DOI: 10.1109/TCOMM.2015.2468215

Mao, Y., Piovano, E., & Clerckx, B. (2021). Rate-splitting multiple access for overloaded cellular internet of things. *IEEE Transactions on Communications*, *69*(7), 4504–4519. DOI: 10.1109/TCOMM.2021.3067642

Mendes, T. D., Godina, R., Rodrigues, E. M., Matias, J. C., & Catalão, J. P. (2015). Smart home communication technologies and applications: Wireless protocol assessment for home area network resources. *Energies*, *8*(7), 7279–7311. DOI: 10.3390/en8077279

Morozs, N., Clarke, T., & Grace, D. (2015). Distributed heuristically accelerated Q-learning for robust cognitive spectrum management in LTE cellular systems. *IEEE Transactions on Mobile Computing*, *15*(4), 817–825. DOI: 10.1109/TMC.2015.2442529

Naser, S., Sofotasios, P. C., Bariah, L., Jaafar, W., Muhaidat, S., Al-Qutayri, M., & Dobre, O. A. (2020). Rate-splitting multiple access: Unifying NOMA and SDMA in MISO VLC channels. *IEEE Open Journal of Vehicular Technology*, *1*, 393–413. DOI: 10.1109/OJVT.2020.3031656

Panait, L., & Luke, S. (2005). Cooperative multi-agent learning: The state of the art. *Autonomous Agents and Multi-Agent Systems*, *11*(3), 387–434. DOI: 10.1007/s10458-005-2631-2

Pande, A., Ramamurthi, V., & Mohapatra, P. (2013). Quality-oriented video delivery over LTE. *Journal of Computing Science and Engineering : JCSE*, *7*(3), 168–176. DOI: 10.5626/JCSE.2013.7.3.168

Pang, H., Gao, L., & Sun, L. (2016, December). Joint optimization of data sponsoring and edge caching for mobile video delivery. In *2016 IEEE Global Communications Conference (GLOBECOM)* (pp. 1-7). IEEE. DOI: 10.1109/GLOCOM.2016.7841928

Park, H., & van der Schaar, M. (2007). Bargaining strategies for networked multimedia resource management. *IEEE Transactions on Signal Processing*, *55*(7), 3496–3511. DOI: 10.1109/TSP.2007.893755

Park, Y., & Kim, S. (2016). Game theory-based bi-level pricing scheme for smart grid scheduling control algorithm. *Journal of Communications and Networks (Seoul)*, *18*(3), 484–492. DOI: 10.1109/JCN.2016.000063

Raazi, S. M. K. U. R., & Lee, S. (2010). A survey on key management strategies for different applications of wireless sensor networks. *Journal of Computing Science and Engineering : JCSE*, *4*(1), 23–51. DOI: 10.5626/JCSE.2010.4.1.023

Rachmilevitch, S. (2013). Cooperative bargaining: Independence and monotonicity imply disagreement. *Economics Letters*, *118*(1), 240–242. DOI: 10.1016/j.econlet.2012.10.024

Ramezani, P., & Jamalipour, A. (2017). Throughput maximization in dual-hop wireless powered communication networks. *IEEE Transactions on Vehicular Technology*, *66*(10), 9304–9312. DOI: 10.1109/TVT.2017.2697499

Rizvi, S., Karpinski, K., & Razaque, A. (2015). Novel architecture of self-organized mobile wireless sensor networks. *Journal of Computing Science and Engineering : JCSE, 9*(4), 163–176. DOI: 10.5626/JCSE.2015.9.4.163

Romanous, B., Bitar, N., Imran, A., & Refai, H. (2015, September). Network densification: Challenges and opportunities in enabling 5G. *In 2015 IEEE 20th international workshop on computer aided modelling and design of communication links and networks (CAMAD)* (pp. 129-134). IEEE.

Shao, S., & Khreishah, A. (2016). Delay analysis of unsaturated heterogeneous omnidirectional-directional small cell wireless networks: The case of RF-VLC coexistence. *IEEE Transactions on Wireless Communications, 15*(12), 8406–8421. DOI: 10.1109/TWC.2016.2614822

Shiran, R. (2014). Bridging the gap between the Nash and Kalai-Smorodinsky bargaining solutions. *Contributions to Game Theory and Management, 7*, 300–312.

Shrivastava, S., Chen, B., Chen, C., Wang, H., & Dai, M. (2020). Deep Q-network learning based downlink resource allocation for hybrid RF/VLC systems. *IEEE Access : Practical Innovations, Open Solutions, 8*, 149412–149434. DOI: 10.1109/ACCESS.2020.3014427

Sterlite Technologies. (2003, December 3). Sponsored data. *Sterlite Technologies.* https://www.sterlitetech.com/sponsored-data.html

Tabassum, H., & Hossain, E. (2018). Coverage and rate analysis for co-existing RF/VLC downlink cellular networks. *IEEE Transactions on Wireless Communications, 17*(4), 2588–2601. DOI: 10.1109/TWC.2018.2799204

Tabrizi, H., Farhadi, G., Cioffi, J. M., & Aldabbagh, G. (2014). Coordinated tethering over white spaces. *IEEE Transactions on Vehicular Technology, 64*(9), 4170–4179. DOI: 10.1109/TVT.2014.2361349

Thomson, W. (2009). Bargaining and the theory of cooperative games: John Nash and beyond.

Trockel, W. (2002). A universal meta bargaining implementation of the Nash solution. *Social Choice and Welfare, 19*(3), 581–586. DOI: 10.1007/s003550100133

Vidhate, D. A., & Kulkarni, P. (2016, December). Enhanced cooperative multi-agent learning algorithms (ECMLA) using reinforcement learning. *In 2016 International Conference on Computing, Analytics and Security Trends (CAST)* (pp. 556-561). IEEE. DOI: 10.1109/CAST.2016.7915030

Wang, X., Zhao, C., Huang, T., Chakrabarti, P., & Kurths, J. (2023). Cooperative learning of multi-agent systems via reinforcement learning. *IEEE Transactions on Signal and Information Processing Over Networks, 9*, 13–23. DOI: 10.1109/TSIPN.2023.3239654

Xiong, Z., Feng, S., Niyato, D., Wang, P., & Zhang, Y. (2018, April). Competition and cooperation analysis for data sponsored market: A network effects model. *In 2018 IEEE Wireless Communications and Networking Conference (WCNC)* (pp. 1-6). IEEE. DOI: 10.1109/WCNC.2018.8376985

Yang, G., Xiao, M., & Poor, H. V. (2018a). Low-latency millimeter-wave communications: Traffic dispersion or network densification. *IEEE Transactions on Communications, 66*(8), 3526–3539. DOI: 10.1109/TCOMM.2018.2817199

Yang, J., Hu, J., Lv, K., Yu, Q., & Yang, K. (2018b). Multi-dimensional resource allocation for uplink throughput maximisation in integrated data and energy communication networks. *IEEE Access : Practical Innovations, Open Solutions*, *6*, 47163–47180. DOI: 10.1109/ACCESS.2018.2865402

Yang, L., Zhang, Q., Zhang, W., Deng, L., & Yang, H. (2020). Resource allocation for hybrid visible light communications (VLC)-WiFi networks. *IEEE Access : Practical Innovations, Open Solutions*, *8*, 176588–176597. DOI: 10.1109/ACCESS.2020.3026388

Yang, Z., Chen, M., Saad, W., & Shikh-Bahaei, M. (2021). Optimization of rate allocation and power control for rate splitting multiple access (RSMA). *IEEE Transactions on Communications*, *69*(9), 5988–6002. DOI: 10.1109/TCOMM.2021.3091133

Yao, J., & Ansari, N. (2018). QoS-aware fog resource provisioning and mobile device power control in IoT networks. *IEEE Transactions on Network and Service Management*, *16*(1), 167–175. DOI: 10.1109/TNSM.2018.2888481

Yu, H., Ju, M., & Jeong, C. (2019). Sum-throughput maximization in dual-hop wireless powered DF-based relay networks. *IEEE Access : Practical Innovations, Open Solutions*, *7*, 162268–162281. DOI: 10.1109/ACCESS.2019.2952162

Zaidi, S., Smida, O. B., Affes, S., Vilaipornsawai, U., Zhang, L., & Zhu, P. (2019). User-centric base-station wireless access virtualization for future 5G networks. *IEEE Transactions on Communications*, *67*(7), 5190–5202. DOI: 10.1109/TCOMM.2019.2910258

Zeng, J., Lv, T., Liu, R. P., Su, X., Peng, M., Wang, C., & Mei, J. (2018). Investigation on evolving single-carrier NOMA into multi-carrier NOMA in 5G. *IEEE Access : Practical Innovations, Open Solutions*, *6*, 48268–48288. DOI: 10.1109/ACCESS.2018.2868093

Zhang, M., Gao, L., Huang, J., & Honig, M. L. (2019). Hybrid pricing for mobile collaborative Internet access. *IEEE/ACM Transactions on Networking*, *27*(3), 986–999. DOI: 10.1109/TNET.2019.2911123

Zhang, P., Wang, C., Jiang, C., & Han, Z. (2021). Deep reinforcement learning assisted federated learning algorithm for data management of IIoT. *IEEE Transactions on Industrial Informatics*, *17*(12), 8475–8484. DOI: 10.1109/TII.2021.3064351

Zhao, Y., & Zhao, H. (2002, October). Study on negotiation strategy [power markets]. In *Proceedings. International Conference on Power System Technology* (Vol. 3, pp. 1335-1338). IEEE. DOI: 10.1109/ICPST.2002.1067747

Zhou, G., Mao, Y., & Clerckx, B. (2021a). Rate-splitting multiple access for multi-antenna downlink communication systems: Spectral and energy efficiency tradeoff. *IEEE Transactions on Wireless Communications*, *21*(7), 4816–4828. DOI: 10.1109/TWC.2021.3133433

Zhou, J., Sun, Y., Chen, R., & Tellambura, C. (2021b). Rate splitting multiple access for multigroup multicast beamforming in cache-enabled C-RAN. *IEEE Transactions on Vehicular Technology*, *70*(12), 12758–12770. DOI: 10.1109/TVT.2021.3120093

Zhou, W., Xie, B., & Hao, C. (2011, February). A novel bargaining based power allocation for coordinated multiple point transmission/reception. *In 13th International Conference on Advanced Communication Technology (ICACT2011)* (pp. 458-462). IEEE.

Chapter 12
Emerging Network Control Frameworks

ABSTRACT

Nowadays, 5G is in its initial phase of commercialization. The 5G network will revolutionize the existing wireless network with its enhanced capabilities and novel features. The development of 6G networks represents a groundbreaking leap in mobile communication, targeting the creation of an intelligent network of everything that seamlessly integrates advanced technologies such as AI, learning algorithms, and the IoT. This chapter has an extensive discussion of emerging network control frameworks. Furthermore, some novel technologies for 6G networks are presented to offer several challenges and some potential research directions.

12. OVERVIEW

The concept of using balloons to provide internet connectivity is known as balloon network. It involves deploying large balloons, equipped with transmitters and receivers, into the stratosphere to provide Internet access to devices on the ground. In balloon networks, balloons are acting as flying wireless base stations, and provide their powerful computational capabilities to process the computational tasks offloaded from their associated ground devices. Therefore, balloons dynamically allocate their resource to meet their corresponding devices' needs. Unmanned aerial vehicle (UAV) network is widely applied in the traffic monitoring, remote sensing, search and rescue and military reconnaissance, and so on. Among applications and enabling technologies for UAVs, wireless communication is absolutely essential. In contrast to conventional terrestrial communications, UAV-enabled communications and networking offer new opportunities in 5G and beyond. The railway network is undergoing profound transformations through digitalization and emerging wireless technologies to provide new services and improve passenger comfort while increasing the security and safety of train operation. Studies on the railway network have looked at high-speed railway specific research problems, such as, demand estimation, and competition and cooperation with network operators.

Space-air-ground integrated network, as an integration of satellite systems, aerial networks, and terrestrial communications, has been becoming an emerging architecture and attracted intensive research interest. Leveraging the intrinsic merits of high altitude and the ability of multicasting or broadcasting, satellite communication systems provide an opportunity for novel mobile communication networks with

DOI: 10.4018/979-8-3373-2013-7.ch012

its tight interaction and complementary characteristics to traditional terrestrial networks. Currently, the maritime industry is increasingly moving toward the adaptation of the 5G communication paradigm, which will enable ships to provide ubiquitous access of data and information to maritime stakeholders, including shipping agents and passengers while improving safety, efficiency, and environmental sustainability. Nevertheless, as the future communication technologies are growing, the novel requirement of ship networks is to transfer scalable, robust, seamless, and secure information between ships and ports.

12.1 Cooperative Network Traffic Control (*CNTC*) scheme

The emergence of new applications in 5G networks has significantly increased their popularity, providing internet service providers (ISPs) with opportunities to introduce high-quality services to their customers. One of the major challenges ISPs face today is balancing profitability with optimal network performance. Recently, S. Kim investigates the fundamental dynamics of interactions among ISPs and develops the *Cooperative Network Traffic Control* (*CNTC*) scheme to optimize network resource sharing (Kim, 2021). The *CNTC* scheme examines the effects of ISP peering relationships and bandwidth allocation to enhance overall system revenue. Specifically, the *CNTC* scheme identifies key characteristics of ISP interactions and applies the concept of Yu bargaining solutions to address traffic management challenges. By utilizing this bargaining approach, the *CNTC* scheme minimizes the Euclidean distance between an ISP's ideal payoff and the feasible solution set, ensuring a fair and efficient outcome. The primary innovation of the *CNTC* scheme lies in its ability to reconcile conflicting requirements while optimizing network performance.

12.1.1 Development motivation

In recent years, the Internet has transitioned from a flat structure to a more hierarchical network model due to the rapid surge in end-user demands. This transformation has led to the emergence of a highly interconnected ecosystem of content providers within a multi-tier framework. Internet service providers (ISPs), which enable users to access and utilize Internet services, play a crucial role in this infrastructure. ISPs are typically categorized within a three-tier model based on the nature of services they offer. Tier-1 ISPs form the backbone of the global Internet, ensuring interconnectivity across networks. Tier-2 ISPs act as intermediaries, linking tier-1 and tier-3 ISPs by leveraging a combination of paid transit from tier-1 ISPs and peering arrangements with other tier-2 ISPs to facilitate end-user access through tier-3 ISPs. Tier-3 ISPs, in turn, provide Internet access to residential and business customers. The intricate interactions between ISPs at various tiers enable the seamless operation of the global Internet platform (Kim,2018; Shao, 2018).

A fundamental aspect of ISP operations is managing how traffic flows in and out of an ISP's network domain. This capability is essential for ensuring high-quality service and maintaining system efficiency, particularly when dealing with unbalanced traffic loads. Given the hierarchical structure of the Internet, two primary types of ISP relationships govern traffic control: transit and peering. Transit services facilitate the exchange of network traffic across different tiers, typically requiring lower-tier ISPs to pay for access to broader Internet connectivity. Conversely, peering relationships enable ISPs to exchange traffic directly, providing mutual benefits without additional transit costs. As the significance of both transit and peering relationships continues to grow, there is an increasing need for dynamic inter-domain traffic control protocols that emphasize explicit coordination among ISPs to maintain stability, efficiency, and predictability (Kim,2018; Shao, 2018; Shrimali, 2009).

Caching has long been a strategy used by ISPs to reduce transit costs by storing frequently accessed content within local networks. By minimizing the volume of traffic that must traverse external networks, caching helps lower transmission delays and operational expenses. With the rapid expansion of ISP-driven networks, caching has regained attention as an effective traffic management tool. However, the storage capacity limitations of local ISP caches present a challenge. To address this issue, content-peering has been introduced as a collaborative strategy among ISPs. By enabling lower-tier ISPs to share cached content, this approach can significantly reduce transit costs. Nevertheless, implementing content-peering in the current Internet architecture remains difficult due to its reliance on peering links, which directly influence inter-ISP connectivity. To fully realize the benefits of inter-domain cache sharing, efficient management of peering links is essential (Kim,2018; Shao, 2018).

One of the primary challenges associated with ISP link management is finding ways to maximize profitability while sustaining optimal network performance as Internet traffic continues to grow. In the multi-tier ISP model, ISPs must negotiate the purchase and sale of transit and peering services, determining prices based on network provisioning and data transmission volumes. Currently, ISP pricing models are primarily based on bandwidth access and usage. However, as Internet applications become increasingly diverse, there is growing interest in developing a multi-level ISP pricing framework that reflects the value ISPs place on network services. Service-dependent pricing has been proposed as an efficient traffic management strategy that allows ISPs to allocate bandwidth according to the value assigned to their network demands. This approach ensures that ISPs are incentivized to regulate congestion and optimize network resources according to their pricing preferences (Kim, 2021; Lee, 2008a; Reiter, 2011).

Determining optimal pricing and bandwidth allocation is a complex process, as ISPs seek to maximize their profits while minimizing operational costs. An effective pricing strategy must address multiple objectives, including i) maximizing ISP profitability, ii) optimizing bandwidth distribution, and iii) enhancing overall network capacity. However, since an ISP's decisions are influenced by the strategies of other ISPs, a more strategic approach to control is required. To tackle this challenge, the *CNTC* scheme employs a game-theoretic framework. Based on cooperative game theory principles, the *CNTC* scheme models pricing and bandwidth allocation as a two-stage interactive bargaining process. In the first stage, each ISP dynamically determines its service pricing. In the second stage, available bandwidth resources are adaptively allocated between transit and peering services. These two stages work in tandem to enhance system-wide network performance through a coordinated decision-making process (Kim, 2021).

For multi-objective optimization problems, P. L. Yu introduced the *Yu bargaining solution* (*YBS*), a method that provides a fair compromise by minimizing the Euclidean distance between a feasible solution set and the idealized utopia point. This approach aligns closely with bargaining models that consider competing claims among participants. Recently, the *YBS* framework has been further analyzed to highlight its efficiency and fairness properties. Notably, its theoretical structure is dual to that of the *Nash bargaining solution*. If network participants insist on achieving a minimum guaranteed payoff while remaining open to cooperative resource-sharing, the *YBS* emerges as a suitable solution framework (Conley, 2008; Kim, 2021).

12.1.2 Contribution

The two-step bargaining game model enables the *CNTC* scheme to efficiently manage inter-ISP traffic control. Through a cooperative and coordinated approach, individual ISPs engage in decision-making processes to arrive at mutually beneficial agreements. Given their varying network conditions, ISPs hold different perspectives on service pricing and resource allocation strategies. By employing a step-by-step interactive feedback mechanism, the outcome of the pricing decision directly influences the bandwidth allocation process. The dynamic nature of this two-stage bargaining model facilitates cascading interactions among multiple ISPs, allowing them to identify the most profitable solution while addressing conflicting demands. Consequently, the key innovation of the *CNTC* scheme lies in its adaptability, flexibility, and responsiveness to real-time ISP traffic conditions (Kim, 2021).

12.1.3 Inter-ISP network system infrastructure

This study focuses on a regional network composed of tier-2 and tier-3 ISPs, both of which utilize local caches to store frequently accessed content. However, the *CNTC* scheme does not account for caching mechanisms at the tier-1 ISP level. From an inter-ISP platform perspective, the *CNTC* scheme assumes that each ISP periodically broadcasts information regarding the content available in its local cache. When an end user requests content that is not stored locally, the tier-3 ISP prioritizes retrieving it from neighboring ISPs or its associated tier-2 ISP, as these options are more cost-effective compared to acquiring the content from a tier-1 ISP. Let $\mathbb{I} = \mathfrak{P}_1 \cup \mathbb{N} \cup \mathbb{M}$ denote the set of ISPs where $\mathfrak{P}_1$ is the tier-1 ISP, and$\mathbb{N} = \{\mathscr{N}_1, \ldots, \mathscr{N}_n\}$, $\mathbb{M} = \{\mathscr{M}_1, \ldots, \mathscr{M}_m\}$ are the sets of tier-2 and tier-3 ISPs, respectively. Tier-2 and tier-3 ISPs are geographically dispersed, and $M_{\mathscr{N}_i}$ is the set of tier-3 ISPs, which are connected to the $\mathscr{N}_i \in \mathbb{N}$ where $\mathscr{M}_{1 \le j \le m} \in M_{\mathscr{N}_i \subseteq \mathbb{N}} \subseteq \mathbb{M}$. Tier-3 ISPs in the $M_{\mathscr{N}_i}$ are geographically close to each other and interconnected together by signing private peering agreements. Therefore, there is an opportunity to exchange information between themselves to effectively operate the inter-ISP traffic (Kim, 2021; Lee, 2008b; Shao, 2018).

We consider a discrete time model $T \in \{t_1, \ldots, t_c, t_{c+1}, \ldots\}$, where the length of a time slot matches the timescale at which the control decisions are updated. In the *CNTC* scheme, a modeling situation for the ISPs' interaction process is formulated as the game $(\mathbb{G})$; G is subdivided into $\mathbb{G}^F_{t_c}$ and $\mathbb{G}^S_{t_c}$, and they are repeated sequentially in a slotted time structure. Based on the interactive feedback manner, the *CNTC* scheme is operated each time period during the step-by-step iteration. Formally, the *CNTC* scheme defines game entities for the multi-tier network system infrastructure, i.e,

$$\mathbb{G} = \left\{\mathbb{G}^F_{t_c}, \mathbb{G}^S_{t_c}\right\} = \left\{\{\mathbb{N}, \mathbb{M}\}, \mathfrak{M}_{BBU}, \{\mathbb{P}^{\mathscr{N}}, \mathbb{P}^{\mathscr{M}}\}, \left\{\mathscr{N}_i \in \mathbb{N} \middle| \mathbb{A}^{\mathscr{N}_i}, \mathscr{I}^{\mathscr{N}_i}, U^F_{\mathscr{N}_i}, U^{\mathscr{N}_i}_{SC}, U^S_{\mathscr{N}_i}, \mathcal{Q}^{\mathscr{N}_i}\right\}, \left\{\mathscr{M}_j \in \mathbb{M} \middle| \mathbb{A}^{\mathscr{M}_j}, \mathscr{I}^{\mathscr{M}_j}, U^F_{\mathscr{M}_j}, U^{\mathscr{M}_j}_{SC}, U^S_{\mathscr{M}_j}, \mathcal{Q}^{\mathscr{M}_j}\right\}, T\right\}$$

of gameplay (Kim, 2021).

- G is a two-step bargaining game consisting of $\mathbb{G}^F_{t_c}$ and $\mathbb{G}^S_{t_c}$ subgames; they are related in a coordination manner of mutual and reciprocal interdependency.
- $\mathbb{G}^F_{t_c}$ is a first-phase bargaining game at time t_c to decide the service price.
- $\mathbb{G}^S_{t_c}$ is a second-phase bargaining game at time t_c to allocate the bandwidth resource.
- N is the set of tier-2 ISPs $(\mathscr{N}_{1 \le i \le n})$, and M the set of tier-3 ISPs $(\mathscr{M}_{1 \le j \le m})$.
- $\mathfrak{M}_{BBU}$ is a basic bandwidth unit to allocate the ISP traffic service.

- $\mathbb{P}^{\mathcal{N}} = \{\Theta_{min}^{\mathcal{N}} \ldots \Theta_{h}^{\mathcal{N}} \ldots \Theta_{max}^{\mathcal{N}}\}$ is a set of N's price strategies where $\Theta_{h}^{\mathcal{N}}$ means the hth price level to service one $\mathfrak{M}_{BBU}$. $\mathbb{P}^{\mathcal{M}} = \{\Theta_{min}^{\mathcal{M}} \ldots \Theta_{h}^{\mathcal{M}} \ldots \Theta_{max}^{\mathcal{M}}\}$ is a set of M's price strategies; $\mathbb{P}^{\mathcal{N}}$ and $\mathbb{P}^{\mathcal{M}}$ are the $\mathbb{G}_{t_c}^{F}$ game's strategies.
- $\mathbb{A}^{\mathcal{N}_i}$ is the bandwidth allocation vector of $\mathcal{N}_i \in \mathbb{N}$, and $\mathbb{A}^{\mathcal{M}_j}$ is the bandwidth allocation vector of $\mathcal{M}_j \in \mathbb{M}$; $\mathbb{A}^{\mathcal{N}_i}$ and $\mathbb{A}^{\mathcal{M}_j}$ are the $\mathbb{G}_{t_c}^{S}$ game's strategies.
- $\mathcal{J}^{\mathcal{N}_i}$ and $\mathcal{J}^{\mathcal{M}_j}$ are the total link bandwidth amount of $\mathcal{N}_i$ and $\mathcal{M}_j$, respectively.
- At the $\mathbb{G}_{t_c}^{F}$, $U_{\mathcal{N}_i}^{F}$ and $U_{\mathcal{M}_j}^{F}$ are the utility functions of $\mathcal{N}_i$ and $\mathcal{M}_j$, respectively. The $U_{SC}^{\mathcal{N}_i}$ is the utility function of $\mathcal{N}_i$'s service customers, and $U_{SC}^{\mathcal{M}_j}$ is the utility function of $\mathcal{M}_j$'s service customers.
- At the $\mathbb{G}_{t_c}^{S}$, $U_{\mathcal{N}_i}^{S}$ is the utility function of $\mathcal{N}_i$, and $U_{\mathcal{M}_j}^{S}$ is the utility function of $\mathcal{M}_j$.
- $\mathcal{Q}^{\mathcal{N}_i}$ and $\mathcal{Q}^{\mathcal{M}_j}$ are the local cache of $\mathcal{N}_i$ and $\mathcal{M}_j$, respectively.
- $T = \{t_1, \ldots, t_c, t_{c+1}, \ldots\}$ denotes time, which is represented by a sequence of time steps.

12.1.4 The basic concepts of *YBS* and *WYBS*

Cooperative bargaining theory is commonly used to address social choice problems, fundamentally serving as a framework for selecting payoff distributions. In bargaining theory, solutions can be classified into three main categories. The first approach focuses on distributing the gains among participants relative to an origin point, adhering to a set of axioms that enhance its theoretical appeal. The second approach involves bargaining based on predetermined claims to the social surplus, where participants negotiate a compromise and agree on a feasible payoff allocation. The third alternative requires players to negotiate based on an endogenously determined yet typically unattainable utopia point, where each coordinate represents the highest feasible payoff achievable by the participants (Conley, 2008; Kim, 2021).

As a solution in the third category bargaining solutions, the *YBS* has been introduced by P. Yu. The *CNTC* scheme adapts the concept of *YBS* to design the inter-ISP traffic control mechanism. Let $\Sigma^n \equiv \{S \in \mathbb{R}^n | S \text{is nonempty and compact}\}$. The utopia point of a feasible set $S \in \Sigma^n$ is defined as $u(S) = (u_1(S), \ldots, u_n(S)) \in \mathbb{R}^n$ where $u_i(S) = \max_{x \in S} x_i$, and the nadir point is defined as $m(S) = (m_1(S), \ldots, m_n(S)) \in \mathbb{R}^n$ where $m_i(S) = \min_{x \in S} x_i$. Let define several classes of feasible sets such as $\Sigma^{con} \equiv \{S \in \Sigma \mid S \text{is convex}\}$ and $\Sigma_0^{con} \equiv \{S \in \Sigma^{con} | x \gg m(S) \text{for some} x \in S\}$. Let e denote the vector (1, .., 1), and for each i, e^i is the standard basis vector in $\mathbb{R}^n$. The convex hull of a set $S \subseteq \mathbb{R}^n$ will be denoted as *con(S)*, and we define $\Delta = con\{e^1, \ldots, e^n\}$. If $p \in \mathbb{R}_{++}^n$ and a is a number, let $H(p,a) = \{x \in \mathbb{R}^n | p \cdot x = a\}$ denote the associated hyperplane. Let $\mathcal{L}^+ = \{S \in \Sigma_0^{con} | P(S) = H(p,a) \cap [n(S) + \mathbb{R}_+^n]$ for some $p \in \mathbb{R}_{++}^n\}$ and let $\mathcal{L}^- = \{S \in \Sigma_0^{con} | P(S) = H(p,a) \cap [u(S) - \mathbb{R}_+^n]$ for some $p \in \mathbb{R}_{++}^n\}$. Formally, the *YBS* for all $S \in \Sigma_0^{con}$ is defined by using an endogenously determined aspiration point; it is given by (Conley, 2008; Kim, 2021);

$$YBS(S) = \arg\min_{x \in S} \left(\sum_{i \in N} (u_i(S) - x_i)^{\psi} \right)^{\frac{1}{\psi}}, \text{ s.t., } 1 \leq \psi < \infty \tag{12.1.1}$$

Based on the value of ψ, we can define the concept *YBS*. If $\psi = 1$, the *YBS* becomes the utilitarian solution. If $\psi = 2$, it may minimize the Euclidean distance between the utopia point and the feasible set. If $\psi = \infty$, it corresponds to minimizing the Chebychev distance from the utopia point to the feasible set. Therefore, we can think that the *YBS* is a compromise bargaining solution. The idea of *YBS* can be generalized to a weighted version, i.e., *WYBS*, based on the asymmetric compromise; it is given by (Conley, 2008; Kim, 2021);

$$WYBS(S) \equiv \arg\min_{x \in S} \left(\sum_{j \in N} \left(\mathscr{W}_j \times \left[u_j(S) - x_j \right]^{\psi} \right) \right)^{\frac{1}{\psi}}, \tag{12.1.2}$$

s.t., $\mathscr{W} \in \mathbb{R}^n_{++}$, $S \in \Sigma^{con}$

The *WYBS* minimizes the distance between the utopia point and the feasible set where distance is defined in terms of a weighted metric on $\mathbb{R}^n$. Simply, the *WYBS* is the same as the *YBS* if W is a positive scalar multiple of e (Conley, 2008; Kim, 2021).

12.1.5 The two-step bargaining game model for the inter-ISP system

To decide the ISP's service price, each individual ISP designs the $\mathbb{G}^F_{t_c}$ as a two-player game model; one player is service provider, i.e., each ISP, and the other player is all service customers, who are neighboring ISPs and represented as a single player. To provide fairness for different game players in the $\mathbb{G}^F_{t_c}$, the bargaining idea of *YBS* is suitable to implement the $\mathbb{G}^F_{t_c}$. As a service provider, the $\mathcal{M}_j$'s utility function with price strategy $\Theta^{\mathcal{M}}_h \in \mathbb{P}^{\mathcal{M}}$, i.e., $U^F_{\mathcal{M}_j}(\Theta^{\mathcal{M}}_h)$, is defined with purely selfish and altruistic subjects. Formally, $U^F_{\mathcal{M}_j}(\Theta^{\mathcal{M}}_h)$ is given by (Kim, 2021);

$$U^F_{\mathcal{M}_j}(\Theta^{\mathcal{M}}_h) = \mathfrak{T}_{\mathcal{M}_j} \times \left(\left(\frac{\sigma}{exp(-\beta \times \Theta^{\mathcal{M}}_h)} \right) - \log(\Theta^{\mathcal{M}}_h + \alpha) \right) \tag{12.1.3}$$

where $\mathfrak{T}_{\mathcal{M}_j}$ is the total traffic service amount of $\mathcal{M}_j$. σ, β, α are adjustment parameters for the $U^F_{\mathcal{M}_j}(\Theta^{\mathcal{M}}_h)$. As a service customer, the utility function with price strategy $\Theta^{\mathcal{M}}_h$, i.e., $U^{\mathcal{M}_j}_{SC}(\Theta^{\mathcal{M}}_h)$, is formally derived as follows (Kim, 2021).

$$U^{\mathcal{M}_j}_{SC}(\Theta^{\mathcal{M}}_h) = \mathfrak{T}_{\mathcal{M}_j} \times \left(\frac{\zeta}{(\gamma + exp(-\Theta^{\mathcal{M}}_h))} + \mu \right) \tag{12.1.4}$$

where ζ, γ and μ are control parameters for the $U^{\mathcal{M}_j}_{SC}(\Theta^{\mathcal{M}}_h)$. According to (12.1.3) and (12.1.4), the tier-2 ISP $\mathcal{N}_i$'s utility function, i.e., $U^F_{\mathcal{N}_i}(\cdot)$, and its customer's utility function, i.e., $U^{\mathcal{N}_i}_{SC}(\cdot)$ are defined as the same manner as the $U^F_{\mathcal{M}_j}(\cdot)$ and $U^{\mathcal{M}_j}_{SC}(\cdot)$, respectively. By using the solution concept of *YBS*, the $\mathcal{M}_j$'s price strategy at time t_c, i.e., $\Theta^{\mathcal{M}_j}_{min \le k \le max}(t_c) \in \mathbb{P}^{\mathcal{M}}$, is decided as follows (Kim, 2021);

$$YBS = \arg \min_{\Theta^{\mathcal{M}_j}_{min \le k \le max}(t_c)} \left(\sum_{g \in \{P,C\}} \left(U^*_g - \mathbb{I}_g(\Theta^{\mathcal{M}_j}_k(t_c)) \right)^{\chi} \right)^{\frac{1}{\chi}}$$

(12.1.5)

$$\text{s.t.,}\ \mathbb{I}_P\left(\Theta_k^{\mathcal{M}_j}(t_c)\right) = U_{\mathcal{M}_j}\left(\Theta_h^{\mathcal{M}}\right) \text{ and } \mathbb{I}_C\left(\Theta_k^{\mathcal{M}_j}(t_c)\right) = U_{SC}^{\mathcal{M}_j}\left(\Theta_h^{\mathcal{M}}\right)$$

where U_P^* and U_C^* are the utopia payoffs of service providers and service customer, respectively. According to (12.1.5), the service price for each individual ISPs in the tier-3 area is decided, and this information is announced to its neighboring ISPs. The service price for ISPs in the tier-2 area is also decided as the same manner as the tier-3 ISP (Kim, 2021).

In the second-step game $\left(\mathbb{G}_{t_c}^S\right)$, service request ISPs are symmetric game players, and the limited bandwidth resource is dynamically allocated to maximize the system performance. When the $\mathcal{M}_j$ selects its price strategy $\Theta_k^{\mathcal{M}_j}$ at time t_c, the $\mathcal{M}_l$'s utility function for the peering service, i.e., $U_{\mathcal{M}_l}^S\left(\Theta_k^{\mathcal{M}_j}(t_c)\right)$, is formally defined as follows (Kim, 2021);

$$U_{\mathcal{M}_l}^S\left(\mathcal{R}^{\mathcal{M}_l}, \Theta_k^{\mathcal{M}_j}(t_c), \mathfrak{N}_{\mathcal{M}_j}\right) = \left(Z_{\mathcal{M}_l}\left(\mathcal{R}_I^{\mathcal{M}_l}, \mathfrak{N}_{\mathcal{M}_j}\right) + H_{\mathcal{M}_l}\left(\mathcal{R}_{II}^{\mathcal{M}_l}, \mathfrak{N}_{\mathcal{M}_j}\right)\right)$$

(12.1.6)

$$\text{s.t.,}\begin{cases} Z_{\mathcal{M}_l}\left(\mathcal{R}^{\mathcal{M}_l}, \mathfrak{N}_{\mathcal{M}_j}\right) = \dfrac{\mathcal{R}_I^{\mathcal{M}_l}}{\Theta_k^{\mathcal{M}_j}(t_c)} \times \left(\left(\xi + \exp(-\omega \times (\mathcal{R}_I^{\mathcal{M}_l}/\mathfrak{N}_{\mathcal{M}_j})\right)\right)^{-1} - \varepsilon) \\ H_{\mathcal{M}_l}\left(\mathcal{R}^{\mathcal{M}_l}, \mathfrak{N}_{\mathcal{M}_j}\right) = \dfrac{\mathcal{R}_{II}^{\mathcal{M}_l}}{\Theta_k^{\mathcal{M}_j}(t_c)} \times \log\left(\mathcal{R}_{II}^{\mathcal{M}_l}/\mathfrak{N}_{\mathcal{M}_j} + \eta\right)^{\varphi} \\ \mathcal{M}_l \in \mathbb{V}^{\mathcal{M}_j}, \mathcal{R}^{\mathcal{M}_l} = \mathcal{R}_I^{\mathcal{M}_l} + \mathcal{R}_{II}^{\mathcal{M}_l} \text{and}\, \mathfrak{N}_{\mathcal{M}_j} = \sum_{\mathcal{M}_d \in \mathbb{V}^{\mathcal{M}_j}} \mathcal{R}^{\mathcal{M}_d} \end{cases}$$

where ω, ε and ξ are control factors for the $Z_{\mathcal{M}}(\cdot)$, and η, φ are control factors for the $H_{\mathcal{M}}(\cdot)$. $\mathbb{V}^{\mathcal{M}_j}$ is the set of ISPs, which request peer service to the $\mathcal{M}_j$. $\mathcal{R}_I^{\mathcal{M}_l}$ and $\mathcal{R}_{II}^{\mathcal{M}_l}$ are the real-time traffic amount and non real-time traffic amount in the $\mathcal{R}^{\mathcal{M}_l}$, respectively. Usually, real-time traffic has higher priority than non real-time traffic during network operations based on different tolerance characteristics (Kim, 2021).

In the *CNTC* scheme, ISPs are assumed to work together in a cooperative manner, and they negotiate with each other to fairly share the overloaded communication link. From the viewpoint of fairness, payoff should not discriminate between the identities of players but only depend on utility functions. Therefore, the *CNTC* scheme should guarantee that if players' utilities are exactly the same, they should get symmetric payoffs. Due to this reason, the *CNTC* scheme chooses the concept of *WYBS* to solve the bandwidth allocation problem. By using the solution concept of *WYBS*, the $\mathcal{M}_j$'s bandwidth allocation strategy at time t_c, i.e., $\mathbb{A}^{\mathcal{M}_j} = [\ldots, \mathcal{R}^{\mathcal{M}_l}, \ldots]$, is decided as follows (Kim, 2021);

$$WYBS = \arg \min_{\mathbb{A}^{\mathcal{M}_j} = [\ldots, \mathcal{R}^{\mathcal{M}_l}, \ldots]} \left(\sum_{\mathcal{M}_l \in \mathbb{V}^{\mathcal{M}_j}} \left(\mathcal{S}_{\mathcal{M}_l} \times \left[\mathbb{U}_{\mathcal{M}_l}^* - I_{\mathcal{M}_l}^S(\mathbb{A}^{\mathcal{M}_j})\right]\right)^{\chi}\right)^{\frac{1}{\chi}}$$

(12.1.7)

$$\text{s.t.,}\begin{cases} I_{\mathcal{M}_l}^S(\mathbb{A}^{\mathcal{M}_j}) = U_{\mathcal{M}_l}^S\left(\mathcal{R}^{\mathcal{M}_l}, \Theta_k^{\mathcal{M}_j}(t_c), \mathfrak{N}_{\mathcal{M}_j}\right) \\ \mathcal{S}_{\mathcal{M}_l} = \mathcal{R}_I^{\mathcal{M}_l} \Big/ \sum_{\mathcal{M}_d \in \mathbb{V}^{\mathcal{M}_j}} \mathcal{R}_I^{\mathcal{M}_d} \text{and}\, \mathfrak{N}_{\mathcal{M}_j} = \sum_{\mathcal{M}_l \in \mathbb{V}^{\mathcal{M}_j}} \mathcal{R}^{\mathcal{M}_l} \le \mathcal{J}^{\mathcal{M}_j} \end{cases}$$

where $\mathbb{U}^{*}_{\mathcal{M}_l}$ is the utopia payoff of $\mathcal{M}_l$. The value of $\mathcal{R}^{\mathcal{M}_l}$ in $\mathbb{V}^{\mathcal{M}_j}$ is adaptively adjusted according to (12.1.7), and the bandwidth resource of $\mathcal{M}_j$ is distributed to the $\mathcal{M}_l \in \mathbb{V}^{\mathcal{M}_j}$. By using (12.1.6), the tier-2 ISP $\mathcal{N}_i$'s utility function, i.e., $U^{S}_{\mathcal{N}_i}(\cdot)$ is defined as the same manner as the $U^{S}_{\mathcal{M}_l}(\cdot)$, and the bandwidth resource of $\mathcal{N}_i$ is also distributed based on the same idea of the equation (12.1.7) (Kim, 2021).

12.1.6 Summary

The Internet is composed of economically self-interested network entities that manage access, peering, and transit services. These competitive behaviors often result in techno-economic inefficiencies, such as unstable peering arrangements and traffic imbalances. Recent studies suggest that cooperative game theory can provide a fair and efficient resource-sharing solution within the multi-tier ISP infrastructure. The *CNTC* scheme explores the complex interactions among ISPs and proposes a structured methodology to address inter-ISP traffic control challenges. By incorporating the principles of the *YBS* and the *WYBS*, the *CNTC* scheme introduces a two-phase bargaining game model designed to efficiently allocate hierarchical Internet resources. In the first phase, each ISP dynamically determines its service pricing based on the *YBS*, where pricing decisions can influence the strategies of other ISPs through a feedback mechanism. In the second phase, ISP link bandwidth is adaptively allocated using the *WYBS* to manage both intra- and inter-domain traffic flows. This approach enables ISPs to regulate traffic routing within their networks, optimizing resource utilization while maintaining fair and efficient traffic control (Kim, 2021).

12.2 Data Trading and Power Control (*DTPC*) scheme

Edge computing (EC) has emerged as an efficient and cost-effective solution to enhance the computational capabilities of IoT devices with hardware constraints. Recently, the EC-assisted Metaverse system has been recognized as a next-generation Internet paradigm, enabling users to engage in various activities such as gaming, work, and social interactions within a virtual environment. Leveraging widespread wireless connectivity and advanced EC technologies, the Metaverse system effectively coordinates interactions among its system agents. In recent research, S. Kim introduces the *Data Trading and Power Control* (*DTPC*) scheme, which integrates auction theory, machine learning, and bargaining approaches to optimize system performance (Kim, 2024). Specifically, the McAfee double auction mechanism is employed to facilitate data transactions between service providers and IoT devices. Additionally, learning algorithms and bargaining solutions are utilized to optimize resource allocation for wireless communication among these devices. To model the sequential interactions between system agents, the *DTPC* scheme incorporates a unified control mechanism aimed at achieving an optimal balance in Metaverse system performance. Through its synergistic integration of multiple methodologies, this hybrid protocol represents an innovative approach to EC-assisted Metaverse infrastructure.

12.2.1 Development motivation

The COVID-19 pandemic has significantly altered lifestyles, accelerating the adoption of online virtual environments as a practical alternative for various human activities. The rapid advancements in technology have fueled both business and scientific efforts to develop virtual spaces, creating substantial value across multiple domains. The increasing availability of live and stored data sources has further facilitated the convergence of physical and digital realities. Today, these virtual environments are commonly referred to as the 'Metaverse'. The term Metaverse originates from the 1992 science fiction novel Snow Crash, where it combines 'meta', signifying virtuality and transcendence, with 'universe', denoting an all-encompassing digital space. Conceptually, Metaverse integrates virtual and mixed-reality environments, providing a multi-user, computer-generated 3D interface that enables real-time interactions across distances (Almarzouqi, 2022; Chang, 2022; Han, 2022; Kim, 2024).

Both academic and industrial sectors have actively explored the development of an IoT-based Metaverse, transitioning from physical prototypes to immersive virtual interactions aimed at enhancing productivity and efficiency. The proliferation of information technology and smart IoT devices has led to an exponential increase in data generation. Cisco estimated that by 2020, IoT devices would produce over 500 Zettabytes (ZB) of data. However, the majority of IoT devices have limited processing power and storage capacity, requiring cloud-based assistance to augment their capabilities (Kim, 2010; Van, 2022; Wang, 2010;).

Edge computing (EC) has emerged as a next-generation solution that integrates cloud systems and IoT networks in a dynamic and flexible manner. By leveraging a decentralized architecture, EC enables data processing at the edge of network nodes, allowing autonomous decision-making without relying solely on centralized cloud servers. To reduce the burden of transmitting large volumes of data, EC-based services execute tasks locally before connecting to the cloud, thereby mitigating network congestion, improving processing efficiency, and enhancing response times for offloaded services. This decentralized approach is particularly crucial for reducing latency in virtual environments, allowing developers to create immersive experiences capable of supporting millions of concurrent users. The primary goal of EC is to optimize network operations and service distribution while improving user experience. Additionally, EC operates within a distributed framework, deploying edge servers across different locations to enhance performance and reliability (El-Sayed, 2017; Kim, 2024; Pan, 2017; Singh, 2020).

Metaverse can be understood as a digitally immersive representation of an environment, which may either replicate real-world settings or create entirely fictional spaces. To realize the Metaverse, it is essential to digitize the physical world, a process facilitated by digital twin technology. A digital twin creates a virtual replica of a real-world entity by leveraging information technologies. In essence, a digital twin mirrors a physical object within a digital environment, ensuring that the virtual and physical counterparts remain nearly identical. The adoption of digital twin technology within the Metaverse provides several benefits, including continuous monitoring, enhanced understanding, and real-time feedback collection from the physical world. As a result, digital twins offer valuable insights into Metaverse operations and play a crucial role in optimizing decision-making processes (Chang, 2022; Han, 2022; Kim, 2024).

A key challenge in developing an effective digital twin within the EC-assisted Metaverse is ensuring seamless data synchronization between the physical and virtual worlds. Maintaining real-time updates requires continuously acquiring fresh data from the physical environment. To achieve this, smart IoT devices are employed to gather data from their respective regions. For example, virtual driving service providers need real-world data on road conditions, traffic, driver behavior, and weather to enhance the realism of their services. Similarly, virtual drone service providers require information on environmental factors such as temperature, wind speed, chemical signals, light, and sound. To encourage IoT devices to participate in data collection, an economic incentive system is necessary within the Metaverse platform. The Metaverse service mechanism can be conceptualized as a bilateral market involving service providers (SPs) and IoT devices, which function as buyers and sellers, respectively. This economic structure ensures that IoT devices are motivated to contribute valuable data while maintaining an efficient data-sharing ecosystem (Han, 2022; Kim, 2024; Xu, 2022).

As research on EC-assisted Metaverse systems remains in its early stages, limited attention has been given to the control mechanisms required to synchronize IoT-collected data with the Metaverse. Additionally, Metaverse system agents must engage in real-time, interactive experiences that align with their mutual interests. Although existing studies have made significant contributions to specific aspects of the EC-assisted Metaverse, they often address individual components rather than an integrated system-wide approach. The *DTPC* scheme introduces a novel interactive mechanism between SPs and IoT devices to establish a new control paradigm. First, the *DTPC* scheme incorporates McAfee's double auction theory to manage bilateral interactions in the data trading market. Second, IoT devices utilize a Q-learning model to dynamically adjust their power levels based on system conditions. Third, the limited spectrum resources within the EC system are allocated to IoT devices using the *status quo proportional bargaining* (*SQPB*) solution. By integrating auction mechanisms, learning algorithms, and bargaining strategies, this hybrid control framework enhances Metaverse system throughput while optimizing IoT device payoffs and ensuring fairness across the system (Kim, 2024).

12.2.2 Contribution

As previously discussed, the advancement of EC-assisted Metaverse platforms has garnered increasing interest from both industry and academia, particularly due to their immense potential to seamlessly integrate the physical and digital worlds. However, despite the promising prospects of the Metaverse paradigm, several challenges remain. It is crucial to recognize that EC-assisted Metaverse platforms differ fundamentally from traditional edge-based network systems, making conventional control methods unsuitable for their deployment and operation. To address these challenges, the McAfee double auction, reinforcement learning, and *SQPB* solution are integrated into a novel hybrid control framework for the Metaverse. By combining these three technologies, an intelligent EC-assisted Metaverse system can be designed to support the future digital society. The double auction mechanism facilitates interactive data trading between SPs and IoT devices, ensuring efficient market-driven exchanges. Meanwhile, the learning-based *SQPB* solution enables individual IoT devices to dynamically allocate limited communication resources. The primary objective of the *DTPC* scheme is to achieve an optimal balance in Metaverse system performance while effectively managing conflicting interests among various system agents (Kim, 2024).

12.2.3 Technical Methodologies

In the McAfee double auction, there are n sellers whose set is $S = \{1,2,\dots,n\}$, and m buyers whose set is $B = \{1,2,\dots,m\}$. Sellers apply to the auction platform to sell the same type of items, and buyers apply to the platform to purchase the same type of product. The valuation of seller $i \in S$ means that the seller i is willing to sell the item at a price no lower than s_i. The buyer $j \in B$ has a valuation, which indicates that the buyer j is willing to buy an item at a price no higher than his valuation. Specifically, the McAfee double auction model can be used to motivate sellers and buyers in an online fashion. In this case, sellers and buyers are randomly joining the platform anytime, and the platform has to make irrevocable decisions based on the current situation. Usually, the McAfee double auction consists of three processes; bid sorting, winners' selection and pricing (Liu, 1992; McAfee, 1992; Sergei, 2007; Xu, 2020).

- Bids sorting process: Sort the buyers' bids $B^b_{1 \le h \le m}$ in non-increasing order and sort the sellers' bids $S^b_{1 \le h \le n}$ in nondecreasing order. The bids of buyers and prices of sellers are their valuations.

$$\begin{cases} B^b_1 \ge B^b_2 \ge \ldots \ge B^b_h \ge \ldots \ge B^b_m \\ S^b_1 \le S^b_2 \le \ldots \le S^b_h \le \ldots \le S^b_n \end{cases} \tag{12.2.1}$$

- Winners' selection process: Find the index of the least profitable transaction, $h = \arg\max\left(B^b_h \ge S^b_h\right)$. The first h buyers and sellers in (12.2.1) can be the auction winners. Specifically, we should ensure that the bid of the ith winning buyer is higher than that of the ith winning seller where h is the maximum value of i.
- Pricing process: the auction operator defines the auction price (p) where $p = \left(B^b_{h+1} + S^b_{h+1}\right)/2$. If $p \in \left[S^b_h, B^b_h\right]$, all auction winners trade at price p. If $p \notin \left[S^b_h, B^b_h\right]$, only first $h - 1$ buyers and sellers in (12.2.1) are the auction winners; winning buyers pay B^b_h, and the payment of all winning sellers is S^b_h. It is called the direct implementation of the dominant strategy auction.

The primary objective of the McAfee double auction model is to develop an economically robust and market-truthful online auction mechanism. Among its various properties, *truthfulness* is the most critical aspect in auction theory (Liu, 2022). In a two-sided market, multiple sellers offer items for sale while multiple buyers evaluate the possibility of purchasing them. Each trader assigns a unique valuation to different bundles of items, and these valuations remain private information. Consequently, buyers and sellers may engage in strategic behavior to maximize their gains. In 1992, R. McAfee introduced a truthful double auction mechanism, which enables the efficient operation of a two-sided market without requiring prior knowledge of traders' valuations. To achieve an asymptotically optimal trade gain between buyers and sellers, McAfee's double auction process treats participants as single-parametric agents, where each seller can offer a single unit, and each buyer can purchase one unit. The key characteristics of the McAfee mechanism include i) *Truthfulness*: participants are incentivized to disclose their true valuations, ii) *Individual Rationality*: each agent benefits from auction participation and does not incur a loss, iii) *Budget Balance*: the auction operator does not need to subsidize market transactions, iv) *Asymptotic Efficiency*: as the market size increases, auction efficiency approaches its theoretical maximum, and v) *Prior-Free Design*: no statistical information about traders' valuations is required for auction execution. By incorporating these properties, the McAfee double auction provides a robust framework for fair and efficient market transactions (Kim 2024; Segal-Halevi, 2018).

Reinforcement learning (RL) focuses on enabling system agents to make decisions that maximize their cumulative rewards. It involves maintaining a balance between exploration, where the agent seeks new strategies, and exploitation, where it leverages existing knowledge to optimize performance. In an RL framework, agents learn through trial-and-error interactions with their dynamic environment. At each step, an agent selects an action that causes the environment to transition into a new state, receiving a corresponding scalar reward as feedback. One widely used RL algorithm is Q-Learning, which provides a straightforward method for agents to determine optimal actions in controlled Markovian environments. As a model-free, off-policy learning approach, Q-Learning enables agents to independently discover the best course of action based on their current state. By continuously evaluating its position within the environment, the agent determines the most appropriate action to take next (Busoniu, 2008; Kim, 2018, 2024).

The environment of RL is described by a Markov decision process, which is a mathematical framework for modeling the decision-making process. It is useful for optimization problems that are solved via learning algorithms. A finite Markov decision process is a tuple $\langle S, \mathbb{A}, P, \mathscr{R} \rangle$ where **S** is the finite set of environment states, A is the finite set of agent actions, **P** is the state transition probability function, and R is the reward. The state signal $s_k \in S$ describes the environment at each discrete time step k. At each step, the agent can alter the state by taking actions $a_k \in \mathbb{A}$. As a result of action a_k, the environment changes state from s_k to some $s_{k+1} \in S$ according to the state transition probability $P^{a_k}_{s_k \to s_{k+1}} \in \mathbf{P}$ where $P^{a_k}_{s_k \to s_{k+1}}$ is the probability of moving to state s_{k+1} from state s_k with a given action a_k. As the effect of action a_k, the agent receives a scalar reward $r_k(s_k, a_k) \in \mathscr{R}$ where $r_k(s_k, a_k)$ is a reward function. In the reinforcement learning algorithm, a policy ψ_k is mapped from state s_k and action a_k to the probability $\psi_k(s_k, a_k)$ of taking action a_k when in state s_k. After taking an action according to policy ψ_k, the expected reward $\left(r^{\psi_k}(s_k)\right)$ in state s_k under policy ψ_k is received; it is defined as follows (Busoniu, 2018; Kim, 2018, 2024):

$$r^{\psi_k}(s_k) = E_{\psi_k}\left\{\sum_{k=0}^{\infty}\left(\gamma^k \times \mathscr{R}_{t+k}(s_{t+k}, a_{t+k})\right)\right\}, \text{ s.t.}, \gamma \in [0,1) \tag{12.2.2}$$

where γ is the discount factor is that discounts are the rewards in the future; it is typically time-varying, decreasing with time. The expectation is taken over the probabilistic state transitions. The agent's goal is to maximize its long-term reward while only receiving feedback about its immediate, one-step performance. One way it can achieve this is by using reinforcement to learn algorithms. The *DTPC* scheme adopts a well-known reinforcement learning technique called Q-learning. In the Q-learning algorithm, the value of state-action pairs is given by Q-function, i.e., $Q^{\psi_k}(s_k, a_k)$, which is the value under policy ψ_k. This is defined as (Busoniu, 2018; Kim, 2018, 2024):

$$Q^{\psi_k}(s_k, a_k) = E_{\psi_k}\left\{\sum_{k=0}^{\infty}\left(\gamma^k \times \mathscr{R}_{t+k}(s_{t+k}, a_{t+k})\right)\right\} \tag{12.2.3}$$

where γ^k is a separate learning rate, which is used for each state-action pair. The optimal Q-function is defined as $Q^*(s,a) = \max_{\psi} Q^{\psi}(s,a)$. It satisfies the Bellman optimality equation (Busoniu, 2018; Kim, 2018, 2024):

$$Q^*(s,a) = \sum_{s' \in \mathbb{S}}\left(P^a_{s \to s'} \times \left(r(s,a) + \left(\gamma \times \max_{a' \in \mathbb{A}} Q^*(s', a')\right)\right)\right) \tag{12.2.4}$$

This equation states that the optimal value of taking the action a in the state s is the expected immediate reward plus the discounted optimal value attainable from the next state. Especially, the samples of the right-hand side in (12.2.4) are estimated using actual experience with the task, in the form of rewards and pairs of subsequent states. Using the Equation (12.2.4), the agent can achieve his goal by first computing Q^* values for all states and actions, and then choosing actions by the greedy policy, which can maximize the expected reward. The Q-learning algorithm turns the equation (12.2.4) into an iterative approximation procedure (Busoniu, 2018; Kim, 2018, 2024).

$$Q_{k+1}(s_k, a_k) = Q_k(s_k, a_k) + (\beta_k \times \Gamma_k)$$

$$\text{s.t.}, \Gamma_k = \left(r_{k+1} + \left[\gamma \times \max_{a'} Q_k(s_{k+1}, a') - Q_k(s_k, a_k)\right]\right) \text{ and } \beta_k \in (0, 1] \tag{12.2.5}$$

where β_k is the learning rate is, and it specifies how far the current estimate $Q_k(s_k, a_k)$ is adjusted towards the update target Γ_k. The Γ_k is the temporal difference between estimates of $Q(s_k, a_k)$ at two successive time steps, $k + 1$ and k. The equation (12.2.5) does not require knowledge about the transition and reward functions. Therefore, we can say that the Q-learning algorithm is model-free (Busoniu, 2018; Kim, 2018, 2024).

In 2009, S. Pechersky developed the *status quo proportional bargaining* (*SQPB*) solution with the corresponding axiomatic characterization (Kim, 2014b; Pecherskii, 2009; Zhang, 2010). To characterize the fundamental ideas of *SQPB* solution, the following notations will be used. Let N be the set of natural numbers and let P be the family of nonempty, finite subsets of N. The members of P may be thought of as the set of all potential players, and P is the family of all subsets of those players that may conceivably become involved in some bargaining problem. Let $\mathcal{N} \in \mathbb{N}$ be a non-empty finite set of players, and R, $\mathbb{R}_+$ are the sets of all and non-negative real numbers, respectively. $\mathbb{R}^n$ is denoted as the n-fold Cartesian product of R (Kim 2024; Pecherskii, 2009).

A bargaining game with player's set $\mathcal{N} \in \mathcal{P}$ is a pair $(q, \mathcal{Q})$, where $q \in \mathbb{R}^{|\mathcal{N}|}$ is the status quo point, and $\mathcal{Q} \subset \mathbb{R}^{|\mathcal{N}|}$. When players cannot cooperate, the only outcome for players is q. If all players cooperate, they can get the outcome $x \in \mathcal{Q}$. For player $i \in \mathcal{N}$, q_i in q and x_i in Q are possible giving payoffs. Denote by $\grave{U}^{\mathcal{N}}$ the family of all bargaining games with N, and a bargaining solution is a map F as follows (Kim, 2024; Pecherskii, 2009; Sergei, 2007):

$$\mathcal{F}: \bigcup_{\mathcal{N} \in \mathcal{P}} \Omega^{\mathcal{N}} \rightarrow \bigcup_{\mathcal{N} \in \mathcal{P}} \mathbb{R}_+^{|\mathcal{N}|}, \text{s.t.}, \mathbb{R}_+^{|\mathcal{N}|} = \left\{x \in \mathbb{R}^{|\mathcal{N}|} \middle| x \geq 0\right\} \tag{12.2.6}$$

For each $(q, \mathcal{Q}) \in \Omega^{\mathcal{N}}$, a unique point of Q, i.e., $\mathcal{F}(q, \mathcal{Q})$, is called the solution of bargaining game ($q, \mathcal{Q}$). The *SQPB* solution considers bargaining games $(q, \mathcal{Q})$ with positive status quo points q. For a bargaining game $(q, \mathcal{Q})$, there is such $x \in \mathcal{Q}$ that $x \geq q$, and the *SQPB* solution is mathematically defined as follows (Kim, 2024; Pecherskii, 2009; Sergei, 2007):

$$SQPB(q, \mathcal{Q}) = \mu(q, \mathcal{Q}) \cdot q, \text{ s.t.}, \mu(q, \mathcal{Q}) = \max\left\{t \in \mathbb{R}_+ \middle| (t \cdot q \in \mathcal{Q})\right\} \tag{12.2.7}$$

Briefly, the *DTPC* scheme considers some relations between well-known bargaining solutions: *Nash bargaining solution* (*NBS*), the *egalitarian bargaining solution* (*EBS*), and *utilitarian bargaining solution* (*UBS*). First it is necessary to consider what happens with solutions when the *DTPC* scheme applies the logarithmic transformation (*LN*) of utilities. $LN(x \in \mathbb{R}_{+}^{|\mathcal{N}|})$ is given by (Kim, 2024; Pecherskii, 2009);

$$LN(x) = \left(\ln(x_1), \ln(x_2), \ldots, \ln(x_n)\right) \tag{12.2.8}$$

According to (12.2.8), the *DTPC* scheme can define the bargaining game $LN(q, \mathcal{Q}) = (LN(q), LN(\mathcal{Q}))$ where $LN(\mathcal{Q}) = \left\{(z_1, z_2, \ldots, z_n) \middle| \exists x \in \mathcal{Q}, z = LN(x)\right\}$. It is noteworthy that the *UBS* does not depend on the status quo point (q). Therefore, it is not difficult to check that $LN(NBS(\mathcal{Q})) = UBS(LN(\mathcal{Q}))$ for every $\mathcal{Q} \in \grave{\mathrm{U}}^{\mathcal{N}}$. And then, it is clear that $LN(SQPB(q, \mathcal{Q})) = EBS(LN(q, \mathcal{Q}))$ (Kim, 2024; Pecherskii, 2009).

12.2.4 EC-assisted Metaverse problem formulation

The *DTPC* scheme considers a EC-assisted Metaverse infrastructure, which consists of the physical world and the virtual world. In the physical world, there are one edge server ($\mathscr{E}$), a set of virtual SPs $\mathbb{S} = \{\mathscr{P}_1^{\mathscr{E}}, \ldots, \mathscr{P}_m^{\mathscr{E}}\}$ and a set of smart IoT devices $\mathbb{D} = \{\mathscr{D}_1^{\mathscr{E}}, \ldots, \mathscr{D}_n^{\mathscr{E}}\}$. Each SP is employed by the Metaverse platform to collect geo-spatial data from a target region in the physical world; SPs maintain the digital twin in the Metaverse system for their virtual services. Individual IoT devices in the D are randomly distributed in the area covered by the E, and they are connected with the E through wireless connections. As a control centre, the E makes control decisions to efficiently manage the EC-assisted Metaverse system. As shown in Figure 1, the *DTPC* scheme formulates the double auction model where SPs are buyers, IoT devices are sellers, and the E is an auctioneer. For the collected data trading, each auction is independent, and the auction frequency can be dynamically adjusted according to the inter-auction interval (Kim, 2024; Van, 2022; Xu, 2022).

Figure 1. The infrastructure of the edge assisted Metaverse system (Kim, 2024)

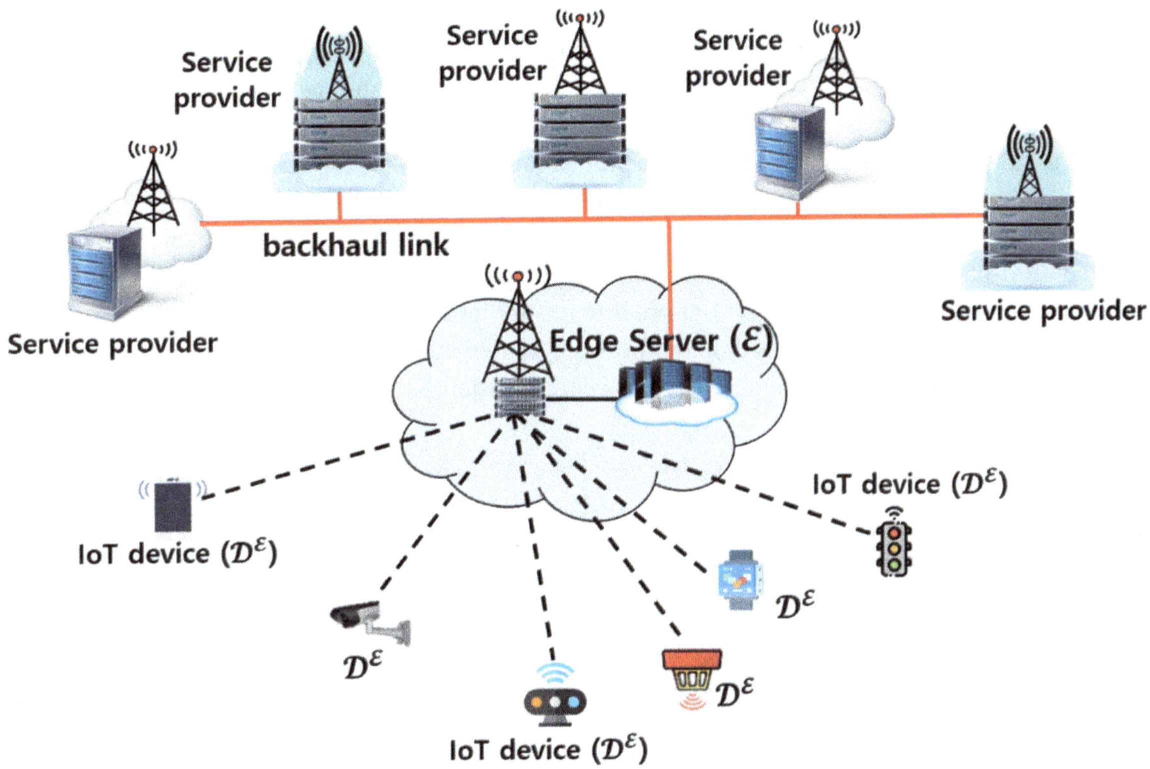

After the auction process, individual IoT devices report their collected data to the E; the $\mathscr{D}^{\mathscr{E}}_{1\leq i\leq n}$ generates different data units $\mathbb{I}^{\mathscr{D}_i} = \{I_1^{\mathscr{D}_i}, I_2^{\mathscr{D}_i}, I_3^{\mathscr{D}_i}, \dots\}$, which are assumed to be delay sensitive and mutually independent, and they follow a Poisson process with an average generation rate at each time period. For data trading, there are limitations when buyers and sellers give their bids and prices. Buyers offer buyers' bids within the range of $[0, \mathscr{Q}_{max}]$, and sellers ask sellers' prices within the range of $[0, \mathscr{S}_{max}]$ where bids and prices are randomly generated following a uniform distribution. The E has a total spectrum amount $(\mathfrak{M}_{\mathscr{E}})$, and each IoT device needs a certain amount of spectrum resource for their wireless communications. Whereas in reality, the $\mathfrak{M}_{\mathscr{E}}$ is limited and raced. When a lot of communications are executed, the $\mathfrak{M}_{\mathscr{E}}$ will become exhausted rapidly, and the Metaverse system suffers to the communication resource scarcity. Therefore, to maximize the EC-assisted Metaverse performance, efficient spectrum management strategy becomes a key factor. To tackle the $\mathfrak{M}_{\mathscr{E}}$ allocation problem, the *DTPC* scheme develops a cooperative game model ($\mathbb{G}$) based on the idea of *SQPB* solution. In the G, $\mathscr{D}^{\mathscr{E}}_{1\leq i\leq n}$ in the D is a game player, and $U^{\mathscr{E}}_{\mathscr{D}_i}$ is the $\mathscr{D}_i^{\mathscr{E}}$' s utility function. To obtain the *SQPB* solution, each individual device estimates the current communication environment and decides its power level based on the reinforcement learning algorithm. The information about the devices' power levels is used as the status quo point for the *SQPB* solution (Kim, 2024).

12.2.5 Hybrid EC-assisted Metaverse control scheme

As mentioned above, numerous $\mathscr{D}^{\mathscr{E}}_{1\leq i\leq n}$ and $\mathscr{P}^{\mathscr{E}}_{1\leq j\leq m}$ connect to the E and the double auction model is proposed to depict the two-sided relationship between devices in D and SPs in S. To trade the collected data, let S and D denote the sets of buyers and sellers, respectively, and the E is a auctioneer. For simplicity, the *DTPC* scheme assumes that the sizes of all trading data are the same. To the auctioneer, the seller $\mathscr{D}^{\mathscr{E}}_i$ asks the price $p^{\mathscr{E}}_i$ for his collected data provisioning, and $\mathbb{P} = \{p^{\mathscr{E}}_1,\ldots,p^{\mathscr{E}}_i\ldots,p^{\mathscr{E}}_n\}$ represents the set of prices submitted by all sellers. To the auctioneer, the buyer $\mathscr{P}^{\mathscr{E}}_j$ submits his bid to get the data, and $\mathbb{B} = \{b^{\mathscr{E}}_1,\ldots,b^{\mathscr{E}}_j\ldots,b^{\mathscr{E}}_m\}$ represents the set of bids submitted by all buyers. The auctioneer E plays the role of trusted third-party, and collects the prices and bids from the sellers and buyers, respectively. Besides, the$\mathscr{E}$ determine the trade payment $(\tilde{N}^{\mathscr{E}})$ in order to ensure fairness and transparency of the transaction between the buyers and sellers. Based on the McAfee double auction, the E sorts of all buyers' bids in descending order, and sorts all sellers' prices in ascending order (Kim, 2024; McAfee, 1992; Sun, 2022):

$$\begin{cases} B^{\mathscr{E}}_1 \geq B^{\mathscr{E}}_2 \geq \ldots\geq B^{\mathscr{E}}_h \geq B^{\mathscr{E}}_{h+1} \geq \ldots\geq B^{\mathscr{E}}_m \\ S^{\mathscr{E}}_1 \leq S^{\mathscr{E}}_2 \leq \ldots\leq S^{\mathscr{E}}_h \leq S^{\mathscr{E}}_{h+1} \leq \ldots\leq S^{\mathscr{E}}_n \end{cases} \text{s.t.,} \begin{cases} B^{\mathscr{E}}_h = b^{\mathscr{E}}_{1\leq j\leq m} \in B \\ S^{\mathscr{E}}_h = p^{\mathscr{E}}_{1\leq i\leq n} \in P \end{cases} \tag{12.2.9}$$

According to (12.2.9), the *DTPC* scheme can find a breakeven index h, i.e., $B^{\mathscr{E}}_h \geq S^{\mathscr{E}}_h$ and $B^{\mathscr{E}}_{h+1} < S^{\mathscr{E}}_{h+1}$, and the index h is used as a threshold to determine the $\tilde{N}^{\mathscr{E}}$ and auction winners where $\tilde{N}^{\mathscr{E}} = (B^{\mathscr{E}}_{h+1} + S^{\mathscr{E}}_{h+1})/2$. Buyers who submit higher bids than the $\tilde{N}^{\mathscr{E}}$ and sellers who ask lower prices than the $\mathscr{P}^{\mathscr{E}}$ are auction winners. The primary goal of the auction model between IoT devices and SPs is to ensure *truthfulness*, also known as *incentive computability. Truthfulness* is a fundamental property that guarantees auction efficiency while preventing market manipulation. A truthful auction ensures that buyers maximize their utility by bidding their true valuation. If buyers were allowed to manipulate their bids strategically, they could gain an unfair advantage while negatively impacting other participants. To maintain *truthfulness* in the EC-assisted Metaverse system, the *DTPC* scheme must ensure that neither IoT devices nor SPs can increase their utility by misrepresenting their bid or price, regardless of how other participants behave. This property is a top priority in auction model design. The *DTPC* scheme rigorously validates that its auction mechanism adheres to *truthfulness*, ensuring a fair and efficient market environment (Jiang, 2019; Kim, 2024; McAfee, 1992; Sun, 2022).

Theorem 1: *Truthfulness* in the bidding and asking is a dominant strategy for IoT devices and SPs (Kim, 2024).

Proof. First, we consider that a buyer ($\mathscr{T}$) with valuation b, and let r denote the buyer's bidding report. Order the remaining buyers' reported bids is defined as $b_1 \geq b_2 \geq \cdots \geq b_{m-1}$, where we are not assuming the T has the highest value. The order of sellers' prices is defined as $s_1 \leq s_2 \leq \cdots \leq s_n$. Let k denote the efficient number of trades given that the T trades, i.e., $b_k \geq s_k$ and $b_{k+1} < s_{k+1}$. In the case that the T trades, he pays

$$\begin{cases} b_k, if p \notin [s_k, b_k] and r \geq b_k \\ p, if p \in [s_k, b_k] and r \geq p \end{cases}, \text{s.t.}, p = \frac{(b_{k+1} + s_{k+1})}{2}$$

(12.2.10)

If $p \notin [s_k, b_k]$, the T's utility is either $b - b_k$ or 0, depending on whether $r \geq b_k$. Similarly, if $p \in [s_k, b_k]$, the T trades at p unless the reports a value less than p. Otherwise, the T is excluded from trading. Now if there is a tie for being the kth rank and the T reports $r = b_k$, the T buys with certainty when $p \in [s_k, b_k]$; so the preceding analysis applies. However, if there is a tie involving the T and $p \notin [s_k, b_k]$, the mechanism randomizes over whether the T is excluded from trading. This includes the case when $b_k = b_{k+1}$, for this implies $p = \frac{(b_{k+1} + s_{k+1})}{2} > \frac{(b_{k+1} + b_{k+1})}{2} = b_{k+1} = b_k$; that is $p \notin [s_k, b_k]$. In this case, the T receives $b - b_k$ if $r > b_k$. It follows immediately that the T with $b \neq b_k$ will never report $r = b_k$, and that the T with $b = b_k$ cannot do better than reporting $r = b_k$ (McAfee, 1992).

It is useful to note that the only way the T can affect the price is by eliminating himself from trading. Thus, by reporting a value $r \in [s_k, b_k)$, when $p \notin [s_k, b_k]$, the T can reduce the price, but it eliminates himself from purchasing by doing so, for the T is now the kth rank. Similarly, if $p \in [s_k, b_k]$, a report of $r < p$ can lower the price, but also eliminates the T from trade, because there are k efficient trades if $r \geq s_k$, and the report $r < p$ places the buyer at k or higher. The case of the seller is similar (McAfee, 1992). Based on this fact, we can proof that *Truthfulness* in the bidding and asking is a dominant strategy for IoT devices and SPs.

After the double auction process, the IoT devices, which are winning sellers, report their collected data information to the E. In IoT-based systems, specific communication protocols are designed to guide self-interested IoT devices toward achieving a socially optimal outcome. These protocols play a crucial role in ensuring the effectiveness of EC-assisted Metaverse platforms. Given the inherent interference interactions in wireless communication, the power control decisions of one IoT device directly impact the performance of others. From an individual device's perspective, the primary objective of power control is to maximize its utility by adjusting transmission power levels. Increasing power levels results in higher *Signal-to-Interference plus Noise Ratio* (*SINR*), which is defined as the ratio of the power of the desired signal to the sum of interference and background noise power. SINR serves as a key metric in wireless communication systems, as it determines the theoretical upper bounds on channel capacity. A higher *SINR* leads to lower bit error rates and higher throughput, enhancing communication efficiency. However, achieving a higher *SINR* comes at the cost of increased wireless interference and higher energy consumption, which can negatively impact overall network performance (Kim, 2011, 2024).

In the *DTPC* scheme, each individual IoT device selects its power level by employing Q-learning algorithm. Let $p_{\mathcal{D}_i^g} \in [p_{min}, p_{max}]$ be the set of available power levels of $\mathcal{D}_i^g$ where p_{min} and p_{max} are the pre-defined minimum and maximum power levels, respectively. The $\mathcal{D}_i^g$ selects his action from the P, and estimates the expected reward through a reward function $\left(\mathcal{R}_{\mathcal{D}_i^g}(\cdot)\right)$ (Kim, 2024).

$$\mathcal{R}_{\mathcal{D}_i^g}\left(\mathbb{P}, p_{\mathcal{D}_i^g}\right) = \left(W \times \log_2\left(1 + \gamma_{\mathcal{D}_i^g}(\mathbb{P})\right)\right) / p_{\mathcal{D}_i^g}$$

(12.2.11)

$$\text{s.t.}\begin{cases} v = \left\{ \prod_{\mathscr{D}_i^{\mathscr{E}} \in \mathrm{D}} p_{\mathscr{D}_i^{\mathscr{E}}} \middle| p_{\mathscr{D}_i^{\mathscr{E}}} \in \left[p_{min}, p_{max}\right] \right\} \\ \gamma_{\mathscr{D}_i^{\mathscr{E}}}(\mathbb{P}) = \dfrac{p_{\mathscr{D}_i^{\mathscr{E}}} \times h_{(\mathscr{D}_i^{\mathscr{E}}, \mathscr{D}_i^{\mathscr{E}})}}{\sigma_{\mathscr{D}_i^{\mathscr{E}}} + \sum_{\mathscr{D}_j^{\mathscr{E}} \in \mathrm{D}, \mathscr{D}_j^{\mathscr{E}} \neq \mathscr{D}_i^{\mathscr{E}}} \left(p_{\mathscr{D}_j^{\mathscr{E}}} \times h_{(\mathscr{D}_j^{\mathscr{E}}, \mathscr{D}_i^{\mathscr{E}})} \right)} \end{cases}$$

where v is the transmit power vector $\langle p_{\mathscr{D}_1^{\mathscr{E}}}, \ldots, p_{\mathscr{D}_n^{\mathscr{E}}} \rangle$ for all IoT devices and W is the channel bandwidth. $\gamma_{\mathscr{D}_i^{\mathscr{E}}}(\mathbb{P})$ is the general formula for the *SINR* of $\mathscr{D}_i^{\mathscr{E}}$. In the $\gamma_{\mathscr{D}_i^{\mathscr{E}}}(\mathbb{P})$, $\sigma_{\mathscr{D}_i^{\mathscr{E}}}$ is the background noise within the $\mathscr{D}_i^{\mathscr{E}}$'s bandwidth, and $h_{(\mathscr{D}_j^{\mathscr{E}}, \mathscr{D}_i^{\mathscr{E}})}$ is the path gain from the transmitter of player $\mathscr{D}_j^{\mathscr{E}}$ to the receiver of player $\mathscr{D}_i^{\mathscr{E}}$. The *DTPC* scheme follows the assumption in the previous paper (Kim, 2011) for $\sigma_{\mathscr{D}_i^{\mathscr{E}}}$ and $h_{(\mathscr{D}_j^{\mathscr{E}}, \mathscr{D}_i^{\mathscr{E}})}$ values; device transmitters use variable-rate M-QAM, with a bounded probability of symbol error and trellis coding with a nominal coding gain. Simply, the numerator and denominator of (12.2.11) are the $\mathscr{D}_j^{\mathscr{E}}$'s throughput, which is defined as the number of information bits that are successfully transmitted, and the $\mathscr{D}_j^{\mathscr{E}}$'s energy consumption. From the viewpoint of $\mathscr{D}_i^{\mathscr{E}}$, the main interest is to maximize his reward; it is obtained by increasing the amount of transmitted data with the lower energy consumption. However, there is a fundamental tradeoff (Kim, 2011, 2024).

The *DTPC* scheme assumes that individual IoT devices work together under with bounded rationality to strike a balance between the exploration of uncharted territory and the exploitation of current knowledge based on online performance. Therefore, the *DTPC* scheme adopts the idea of Q-learning algorithm to design each device's power control algorithm. Let $s_{\mathscr{D}_i^{\mathscr{E}}}(t), a_{\mathscr{D}_i^{\mathscr{E}}}(t)$ be the state and action of $\mathscr{D}_i^{\mathscr{E}}$ at time step t, respectively, where L is the total number of power levels, $s_{\mathscr{D}_i^{\mathscr{E}}}(t) \in \mathbb{S}^p = \left\{ s_{p_{min}=1}, \ldots, s_{p_{max}=L} \right\}$ and $a_{\mathscr{D}_i^{\mathscr{E}}}(t) \in \left[p_{min=1}, p_{max=L}\right]$. In $Q^{\psi}(s,a)$, $\mathscr{R}_{\mathscr{D}_i^{\mathscr{E}}}\left(v, a_{\mathscr{D}_i^{\mathscr{E}}}(t)\right)$ is the reward to $\mathscr{D}_i^{\mathscr{E}}$ at time t when action $a_{\mathscr{D}_i^{\mathscr{E}}}(t)$ occurs in state $s_{\mathscr{D}_i^{\mathscr{E}}}(t)$, and $\mathscr{R}_{\mathscr{D}_i^{\mathscr{E}}}\left(v, a_{\mathscr{D}_i^{\mathscr{E}}}(t)\right) \to \mathbb{R}$. According to (5), $Q_{t+1}\left(s_{\mathscr{D}_i^{\mathscr{E}}}(t), a_{\mathscr{D}_i^{\mathscr{E}}}(t)\right)$ is defined as follows (Busoniu, 2008; Kim, 2011, 2024):

$$Q_{t+1}\left(s_{\mathscr{D}_i^{\mathscr{E}}}(t),\ a_{\mathscr{D}_i^{\mathscr{E}}}(t)\right) = Q_t\left(s_{\mathscr{D}_i^{\mathscr{E}}}(t),\ a_{\mathscr{D}_i^{\mathscr{E}}}(t)\right) + \left(\beta_t \times \Gamma_t\right) \tag{12.2.12}$$

$$\text{s.t.}, \Gamma_t = \left(\mathscr{R}_{\mathscr{D}_i^{\mathscr{E}}}\left(v, a_{\mathscr{D}_i^{\mathscr{E}}}(t)\right) + \left[\gamma \times \max_{a'_{\mathscr{D}_i^{\mathscr{E}}} \in [p_1, p_L]} Q_t\left(s_{\mathscr{D}_i^{\mathscr{E}}}(t+1), a'_{\mathscr{D}_i^{\mathscr{E}}}\right) - Q_t\left(s_{\mathscr{D}_i^{\mathscr{E}}}(t), a_{\mathscr{D}_i^{\mathscr{E}}}(t)\right) \right] \right)$$

A greedy Q-learning algorithm always chooses the action with the highest Q value, which leads to maximizing the $\mathscr{D}_i^{\mathscr{E}}$'s reward. To develop an effective Q-learning algorithm, we must define $P(a_{\mathscr{D}_i^{\mathscr{E}}}(t) | s_{\mathscr{D}_i^{\mathscr{E}}}(t) \to s_{\mathscr{D}_i^{\mathscr{E}}}(t+1))$, which is the probability of moving to state $s_{\mathscr{D}_i^{\mathscr{E}}}(t+1)$ from state $s_{\mathscr{D}_i^{\mathscr{E}}}(t)$ with a given action $a_{\mathscr{D}_i^{\mathscr{E}}}(t)$; it is given by (Busoniu, 2008; Kim, 2011, 2024):

$$P(a_{\mathscr{D}_i^{\mathscr{E}}}(t) \,|\, s_{\mathscr{D}_i^{\mathscr{E}}}(t) \to s_{\mathscr{D}_i^{\mathscr{E}}}(t+1)) = \frac{exp\left(Q_t\left(s_{\mathscr{D}_i^{\mathscr{E}}}(t), a_{\mathscr{D}_i^{\mathscr{E}}}(t)\right)\right)}{\sum_{a'_{\mathscr{D}_i^{\mathscr{E}}} \in [p_1, p_L]} exp\left(Q_t\left(s_{\mathscr{D}_i^{\mathscr{E}}}(t), a'_{\mathscr{D}_i^{\mathscr{E}}}\right)\right)} \tag{12.2.13}$$

According to (12.2.13), the $\mathcal{D}_i^{\mathcal{g}}$ selects his power level $p_{\mathcal{D}_i^{\mathcal{g}}}$. Therefore, the E can get the information of $v = \langle p_{\mathcal{D}_1^{\mathcal{g}}},\ldots,p_{\mathcal{D}_n^{\mathcal{g}}}\rangle$.

Nowadays, the game theory uses reinforcement of learning algorithms to explain how an effective solution may arise under bounded rationality. Therefore, there has been growing interest in research incorporating game theory and learning algorithms. In recent years, learning algorithmic game theory has become an increasingly important area of resource allocation research fields (Kim, 2011). The *DTPC* scheme develops Q-learning algorithmic game theory to handle the spectrum allocation problem for IoT devices. In the *DTPC* scheme, each IoT device is assumed to be a game player, and the *SQPB* solution is applied to share the $\mathfrak{M}_{\mathcal{g}}$. Under the dynamic EC-assisted Metaverse communication environment, the power level of each device can be interpreted as a default point, which gives the payoff for player in the event that the bargain fails. Therefore, the *DTPC* scheme sets the status quo point ($\mathbb{Q}$) based on the players' power vector v. For the time t, the $\mathcal{D}_i^{\mathcal{g}}$'s status quo point $\left(q_{\mathcal{D}_i^{\mathcal{g}}}\right)$ is defined as follows (Kim, 2024):

$$q_{\mathcal{D}_i^{\mathcal{g}}} = u_{\mathfrak{M}} \times \left(\vartheta + \frac{a_{\mathcal{D}_i^{\mathcal{g}}}(t)}{P_{max}}\right) \text{ s.t., } q_{\mathcal{D}_i^{\mathcal{g}}} \in \mathbb{Q} = \langle q_{\mathcal{D}_1^{\mathcal{g}}},\ldots,q_{\mathcal{D}_n^{\mathcal{g}}}\rangle \tag{12.2.14}$$

where $u_{\mathfrak{M}}$ is the spectrum unit for allocation process, and ϑ is an adjustment factor to decide the $q_{\mathcal{D}_i^{\mathcal{g}}}$. According to (12.2.7) and (12.2.14), the *SQPB* solution $\left(SQPB_{\mathrm{G}}(\mathbb{Q},\mathfrak{M}_{\mathcal{g}})\right)$ for the $\mathfrak{M}_{\mathcal{g}}$ allocation problem is given by (Kim, 2024):

$$SQPB_{\mathrm{G}}(\mathbb{Q},\mathfrak{M}_{\mathcal{g}}) = \max_{\mathcal{K}}\left\{\mathcal{K} \in \mathbb{R}_+ \middle| (\mathcal{K}\cdot\mathbb{Q} \in \mathbb{R}^n)\right\}\cdot\mathbb{Q} \tag{12.2.15}$$

$$\text{s.t., } \sum\nolimits_{q_{\mathcal{D}_i^{\mathcal{g}}}\in\mathbb{Q}}\left(\mathcal{K}\times q_{\mathcal{D}_i^{\mathcal{g}}}\right) \le \mathfrak{M}_{\mathcal{g}}$$

As a game player, the utility function of $\mathcal{D}_i^{\mathcal{g}}$ $\left(U_{\mathcal{D}_i}^{\mathcal{g}}(\cdot)\right)$ is defined based on the allocated spectrum amount, i.e., $A_{\mathcal{D}_i}^{\mathcal{g}} = \left(\mathcal{K}\times q_{\mathcal{D}_i^{\mathcal{g}}}\right)$, which is obtained from the equation (12.2.15) (Kim, 2024):

$$U_{\mathcal{D}_i}^{\mathcal{g}}\left(\mathcal{S}_{\mathcal{D}_i}^{\mathcal{g}},A_{\mathcal{D}_i}^{\mathcal{g}}\right) = \left[\mathcal{G}_{\mathcal{D}_i^{\mathcal{g}}}\times\log\left(\frac{\min\left(\mathcal{S}_{\mathcal{D}_i}^{\mathcal{g}},A_{\mathcal{D}_i}^{\mathcal{g}}\right)}{\mathcal{S}_{\mathcal{D}_i}^{\mathcal{g}}}+\xi\right)\right] - exp\left(-\frac{\min\left(\mathcal{S}_{\mathcal{D}_i}^{\mathcal{g}},A_{\mathcal{D}_i}^{\mathcal{g}}\right)}{\mathcal{S}_{\mathcal{D}_i}^{\mathcal{g}}}\right) \tag{12.2.16}$$

$$\text{s.t., } \mathcal{G}_{\mathcal{D}_i^{\mathcal{g}}} = \left(\eta + \frac{\mathcal{R}_{\mathcal{D}_i^{\mathcal{g}}}\left(\mathbb{P},p_{\mathcal{D}_i^{\mathcal{g}}}\right)}{\max_{\mathcal{D}_j^{\mathcal{g}}\in\mathrm{D}}\mathcal{R}_{\mathcal{D}_j^{\mathcal{g}}}\left(\mathbb{P},p_{\mathcal{D}_j^{\mathcal{g}}}\right)}\right)$$

where $\mathcal{G}_{\mathcal{D}_i^{\mathcal{g}}}$ is a gaining weight and ξ, η are control parameters for the $U_{\mathcal{D}_i}^{\mathcal{g}}(\cdot)$.

As a key paradigm for 6G networks, the EC-based Metaverse system operates as close to end users as possible, significantly reducing network latency and enhancing the performance of IoT devices. While this architecture offers numerous advantages, it is essential to develop an economic model that incentivizes both SPs and IoT devices to actively engage in Metaverse interactions. Furthermore, individual IoT devices require communication resources to efficiently transmit their collected data. To address these challenges, the *DTPC* scheme introduces a novel Metaverse control mechanism that integrates auction theory, reinforcement learning, and bargaining algorithms. First, the McAfee double auction model is employed to regulate the interactive relationship between SPs and IoT devices, ensuring a fair and efficient data trading process. Second, each IoT device autonomously determines its power level using the Q-learning algorithm, allowing adaptive optimization of communication parameters. Third, the limited spectrum resources are allocated based on the *SQPB* solution, ensuring equitable distribution among network participants. The reciprocal integration of auction, learning, and bargaining models forms the foundation of the *DTPC* scheme, as depicted in Figure 2 (Kim, 2024).

Figure 2. The diagram of edge assisted Metaverse control model (Kim, 2024)

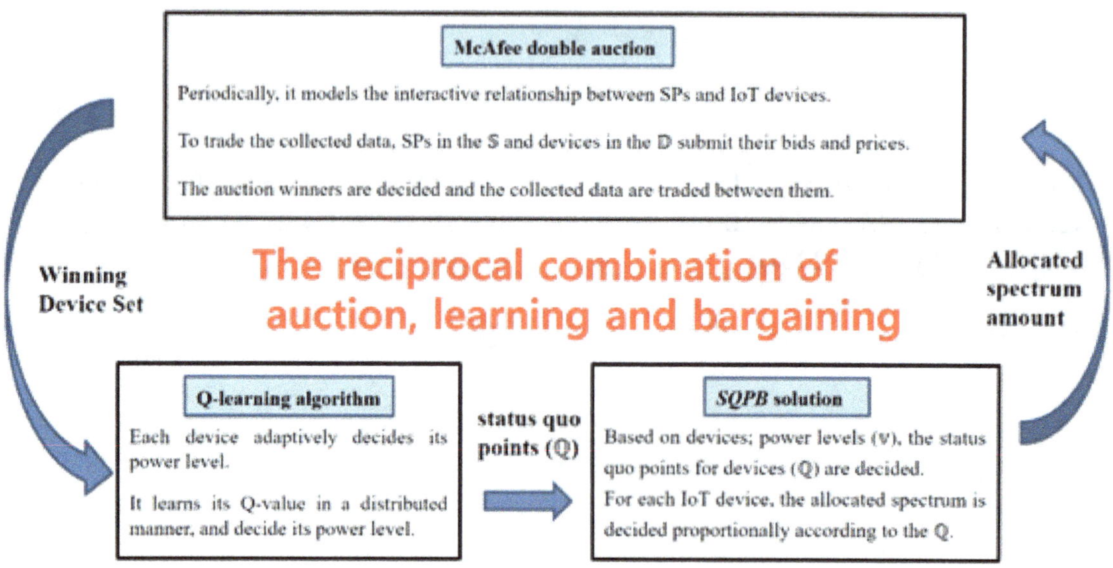

12.2.6 Summary

With the rise of emerging technologies, auction, learning, and bargaining algorithms have become fundamental components of Metaverse system operations. While these technologies hold great potential for enhancing the efficiency of Metaverse platforms, they are still in their early stages of development. The *DTPC* scheme addresses the Metaverse control challenge by integrating McAfee's double auction, Q-learning, and the *SQPB* solution. To manage the mutual interaction between SPs and IoT devices, the *DTPC* scheme incorporates a double auction model for data trading. Using Q-learning, individual IoT devices dynamically adjust their power levels, with this information serving as the status quo point ($\mathbb{Q}$). The EC spectrum resource is then allocated among IoT devices based on the *SQPB* solution, ensuring efficient and fair resource distribution. The auction, learning, and bargaining algorithms are intricately combined, each mutually dependent on the others. By jointly leveraging their advantages, these three algorithms work cooperatively and interactively, allowing the *DTPC* scheme to achieve a 'win-win' outcome even in the highly dynamic EC-assisted Metaverse environment. Through its real-time online approach, the *DTPC* scheme exhibits remarkable adaptability and flexibility, catering to diverse Metaverse performance requirements. This model offers mutual benefits by optimizing system throughput, IoT device payoffs, and overall device fairness (Kim, 2024).

12.3 User Provided Network Control (*UPNC*) scheme

In recent years, wireless mobile services have undergone a significant paradigm shift driven by three key factors; i) the increasing demand for ubiquitous connectivity, ii) the unprecedented surge in mobile data traffic, and iii) advancements in mobile devices equipped with enhanced capabilities. To address these challenges, User Provided Network (UPN) has emerged as a promising technology, offering substantial improvements in cellular coverage and capacity. Recently, S. Kim introduced the *User Provided Network Control* (*UPNC*) scheme, which leverages game theory to optimize UPN performance (Sergei, 2007). Inspired by the Stackelberg game model, the *UPNC* scheme enables mobile devices to dynamically switch roles between hosts and clients, thereby enhancing the efficiency of UPN systems. Given the highly dynamic nature of UPN environments, this approach proves to be particularly effective. By adopting a decentralized, non-cooperative framework, the *UPNC* scheme successfully models complex interactions within UPN systems. This mechanism not only captures the evolving dynamics of user-provided networks but also marks a significant step forward in the evolution of next-generation mobile networks.

12.3.1 Development motivation

Technological advances are ushering in a new era in telecommunication networks, driven by the widespread adoption of wireless technologies and an increasing number of user-friendly mobile devices. However, these developments also bring new challenges for cellular networks, particularly as Machine-to-Machine (M2M) communications generate an unprecedented volume of data traffic with diverse service requirements (Chen, 2012; Dighriri, 2016). With the emergence of novel wireless network paradigms, a shift toward personalized services has become evident, placing greater emphasis on user experience (QoE). Among these advancements, 5G network standards have emerged as a promising solution to enhance QoE, making it a fundamental component of future networks (Kim, 2017; Wang, 2016a).

QoE is a measure of how users perceive and experience network services, offering a holistic evaluation that extends beyond conventional QoS metrics. Unlike QoS, which focuses primarily on technical performance, QoE incorporates subjective user perceptions and evaluates how well a service meets users' expectations, including aesthetic and even hedonic preferences (Kim, 2010; Wang, 2015). While QoE is a crucial aspect of next-generation networks, supporting it within 5G environments - particularly for device-to-device communication, ultra-reliable services, and large-scale machine-based systems - remains a complex challenge (Kim, 2017).

The increasing demand for mobile data and the proliferation of advanced handheld devices have led to the development of a novel networking paradigm known as the User Provided Network (UPN). This concept, first introduced by R. Sofia, defines UPN as a model where end users function as both consumers and providers of network access (Sofia, 2018). With advances in social and mobile technologies, modern smart devices possess surplus computational and networking resources, allowing them to act as micro-providers that extend connectivity to other nearby users without relying on additional network infrastructure. In particular, the rise of social networking platforms has motivated users to share their connectivity, fostering community-driven networking models that promote shared interests and collective benefits (Wang, 2015; Kim, 2017).

Two primary UPN models have been developed using different approaches. One method enables mobile devices to create mesh networks, allowing users to share Internet connections without intervention from network operators. The other approach involves a virtual mobile operator, where users function as mobile Wi-Fi hotspots and provide network access to non-subscribers in exchange for free data quotas. Well-known examples of these UPN models include Open Garden and Karma, which have implemented these mechanisms through startup initiatives (Lorenzo, 2015). Although these models employ different control strategies, they share a common challenge - UPN operations depend on self-organizing User Equipments (UEs) that must voluntarily participate. However, user participation presents a challenge due to conflicting interests, as existing UPN models lack control algorithms that effectively coordinate these conflicts (Iosifidis, 2014; Kim, 2017).

To address these challenges, the *UPNC* scheme introduces a new control mechanism that enables users to share mobile connections and function as hosts for nearby clients. Within the *UPNC* scheme, users benefit from resource sharing, allowing them to utilize surplus connectivity from other users. However, ensuring fair and efficient sharing requires an advanced control paradigm. To model the complex interactions between multiple hosts and clients, the Stackelberg game is employed as a suitable decision-making framework. Originally introduced by H. von Stackelberg in 1934, this model describes leader-follower dynamics in monopoly markets. In a classical Stackelberg game, one player acts as a leader, setting strategic decisions, while others respond as followers, optimizing their objectives based on the leader's actions (Kim, 2012, 2014b).

To adapt the Stackelberg model to the UPN context, the *UPNC* scheme introduces the Position Changeable Stackelberg (PCS) game. In this model, users dynamically switch roles between leaders (hosts) and followers (clients) based on real-time network conditions, optimizing their payoffs accordingly. Each user learns from system conditions and adjusts their strategy dynamically, ensuring an efficient and flexible UPN operation (Kim, 2017). The success of UPN systems depends on users' willingness to share their connectivity and network resources (Khalili, 2015). To encourage participation, effective incentive mechanisms must be designed to motivate hosts. Existing UPN models provide some basic incentives - for example, the Karma UPN system rewards host with free data quotas, while Open Garden employs a virtual currency model based on bargaining mechanisms for fair resource allocation. However, previous studies have largely overlooked critical interactions between hosts and clients, as well as the dynamic nature of UPN topologies (Khalili, 2015; Kim, 2017).

In the *UPNC* scheme, hosts charge UPN service fees to clients, providing them with a monetary incentive for resource sharing. To determine the optimal pricing strategy, hosts utilize a reinforcement learning algorithm in a distributed manner. By continuously analyzing the temporal and spatial conditions of the UPN system, individual hosts make intelligent pricing decisions that adapt to dynamic network environments. Building on the PCS game model and reinforcement learning, the *UPNC* scheme focuses on three key design and operational questions: i) optimal positioning - how should users determine whether to act as a host or client in a given UPN topology?, ii) incentive mechanisms - what is the appropriate incentive structure to ensure active host participation?, and iii) strategic decision-making - how should users select their strategies under imperfect information conditions? To effectively address these issues, the *UPNC* scheme prioritizes feasibility, self-adaptability, and efficiency, ensuring that UPN services are practical, responsive, and scalable. Although various UPN control models have been proposed, no comprehensive study has systematically examined a realistic UPN scenario within an adaptive and dynamic framework (Kim, 2017).

12.3.2 Contribution

The *UPNC* scheme extends the UPN control algorithm by incorporating several key aspects; i) users within the UPN can adaptively determine their roles as either hosts or clients. To effectively model interactions among position-flexible users, the UPNC scheme utilizes the PCS game, ii) the *UPNC* scheme introduces a dynamic pricing policy for UPN services, designed to motivate hosts to participate in resource sharing, iii) a novel price decision process is formulated under practical assumptions, leveraging reinforcement learning. Through both individual and social learning, hosts optimize their pricing strategies based on network conditions, iv) by incorporating a fair-balanced solution, the *UPNC* scheme ensures efficient service allocation across varied UPN environments, addressing the diverse needs of users. Through these enhancements, the *UPNC* scheme provides a flexible, adaptive, and efficient control framework for next-generation UPN systems (Kim, 2017).

12.3.3 PCS game model and players' utility functions

Several UPN models exist today, each differing in architecture and service offerings. However, a fundamental characteristic shared by these models is that UPNs consist of multiple network agents arranged in a mesh topology, where all agents collaborate to enhance network performance. Generally, UPNs feature two primary types of network agents: eNBs and UEs. The function of eNBs is to support Internet connectivity by acting as gateways to connect backbone networks. UEs participate in UPN services to advance traffic management and reduce energy consumption. The core objective of the UPN concept is to leverage the diverse capabilities of UEs, including their resource availability and Internet access, to establish an autonomous, self-regulating intervention mechanism that maximizes network efficiency (Kim, 2017).

We consider the operation of a multicellular network that consists of a set B = $\{B_1 \ldots B_n\}$ of eNBs, and a set G = $\{G_1 \ldots G_y\}$ of UEs. Some UEs may be in range with a B $\in$ B while some others may not. Each eNB has a set of bandwidth resource blocks that can be allocated to UEs in each time period. The *UPNC* scheme assumes that UEs may change their locations and hence have different neighbors in different time periods. Such a mobility assumption allows us to get closed form expressions and gain valuable insights into the design of UPN (Kim, 2017).

UEs operate under varying conditions. Within the model, two distinct types of UEs are identified: hosts and clients within the UPN. Hosts have the capability to establish a direct connection with the eNB and can share their internet access with clients, effectively acting as temporary gateways for them. Clients, on the other hand, may either connect directly to the eNB for their own services or access the network through a nearby host. The primary objective of a host is to maximize their profit, which is determined by the revenue earned from serving clients. Similarly, clients aim to optimize their own benefits, which are influenced by the outcomes of UPN services. Depending on their real-time status, UEs dynamically adjust their positioning, and their decisions evolve adaptively over time. The *UPNC* scheme is represented as a novel PCS game model, where UEs, as game participants, determine their positions and strategies based on their interactions with other UEs. At each time period of gameplay, the *UPNC* scheme formally defines the PCS game model $\mathbb{G}^{PCS} = \{\mathrm{B}, \mathrm{G}, \{S_h^G, S_c^G\}, \{U_h^G, U_c^G\}, L_G^{\mathfrak{P}}, \mathrm{T}\}$ as follows (Kim, 2017);

- $\mathrm{B} = \{B_1 \ldots B_n\}$ represents a set of eNBs, and each B has orthogonal bandwidth blocks.
- $\mathrm{G} = \{G_1 \ldots G_y\}$ is the finite set of PCS game players, i.e., UEs. The G can be divided into three subsets G^h, G^c and G^n; G^h is the subset of hosts $G_k^h \in \mathrm{G}^h \subset \mathrm{G}$, G^c is the subset of clients $G_i^c \in \mathrm{G}^c \subset \mathrm{G}$ and G^n is the subset of UEs, which temporarily give up service requests $G_l^n \in \mathrm{G}^n \subset \mathrm{G}$.
- In the $\{S_h^G, S_c^G\}$, $S_h^G = \{s_{min}^h \ldots s_i^h \ldots s_{max}^h\}$ is the set of host's strategies where $s_i^h \in S_h^G$ means the ith price level for client's communications. S_c^G is the set of client's available strategies; each client has three strategy options; i) selects one host among his neighboring hosts for relay communications, ii) directly connects to his corresponding eNB, or iii) to be a member of G^n.
- U_h^G is the payoff received by the host, and U_c^G is the payoff received by the client during the UPN operation.
- L_G^s is the learning value for the host's strategy s; L_G^s is used to estimate the probability distribution (P^G) for the next strategy selection.
- $T = \{H_1, \ldots, H_t, H_{t+1}, \ldots\}$ denotes time, which is represented by a sequence of time steps with imperfect information for the PCS game process.

In the UPN system, eNBs charge their subscribers, i.e., hosts and clients, based on the usage-based price mechanism. If unexpected growth of traffic may develop in a specific eNB, it may create local traffic congestion. To alleviate this kind of traffic overload, the current price (P_B) should increase. If few subscribers access in a eNB, the current price should decrease to attract subscribers to participate in UPN services. As the same manner, hosts charge their clients based on the current traffic load. According to the current price, individual clients act independently and selfishly; all game players select their strategies to maximize their payoffs (Kim, 2017).

Each host can connect to the eNB anytime via his device. At the same time, this device can provide Internet connections for nearby clients as a gateway. Therefore, the assigned communication resource for a host can be shared with clients, and clients need to pay the price for their UPN services. Depending on the payment, cost and expense functions of UPN services, the utility function of the host G_k^h at time H_t $\left(U_h^{G_k^h}(H_t)\right)$ can be defined as follows (Kim, 2017);

$$U_h^{G_k^h}(H_t) = \sum_{G_i^c \in N_{G_k^h}} \left(P_h^{G_k^h}\left(r_c^{G_i^c}(H_t)\right) - C_h^{G_k^h}\left(\grave{\mathrm{E}}_{G_k^h}^{B}(H_t), r_c^{G_i^c}(H_t)\right) - D_h^{G_k^h}\left(r_c^{G_i^c}(H_t)\right) \right) \tag{12.3.1}$$

$$\text{s.t.,}\begin{cases} P_h^{G_k^h}\left(r_c^{G_i^c}(H_t)\right) = F^s\left(G_k^h, H_t\right) \times r_c^{G_i^c}\left(H_t\right) \\ C_h^{G_k^h}\left(\grave{E}_{G_k^h}^{B}(H_t), r_c^{G_i^c}(H_t)\right) = \left[\alpha + \grave{E}^{B}(H_t)\right] \times r_c^{G_i^c}\left(H_t\right) \\ D_h^{G_k^h}\left(r_c^{G_i^c}(H_t)\right) = m_h^{G_k^h} \times r_c^{G_i^c}\left(H_t\right) \end{cases}$$

where $P_h^{G_k^h}(\,)$ is the price function to clients at time H_t, and $C_h^{G_k^h}(\,)$ and $D_h^{G_k^h}(\,)$ are the cost and expense functions of G_k^h, respectively, for the UPN services. $N_{G_k^h}$ is the set of G_k^h's neighboring hosts, and $r_c^{G_i^c}$ is the bandwidth amount of G_i^c's UPN service. $F^s\left(G_k^h, H_t\right)$ is the G_k^h's UPN price strategy per bit at time H_t where $F^s\left(G_k^h, H_t\right) \in S_h^{G_k^h}$. Let α and $\grave{E}^{B}(H_t)$ denote the connected eNB's minimum charge and bandwidth utilization at time H_t, respectively. $m_h^{G_k^h}$ is the G_k^h's the expense factor per bit of UPN services (Kim, 2017).

Each client can connect to the backbone Internet directly via the eNB or indirectly via a nearby host. If there are multiple hosts in his coverage area, the client selects one of them to maximize his payoff. If G_k^h is selected as a host, the utility function of the client G_i^c at time H_t $\left(U_c^{G_i^c}\left(G_k^h, H_t\right)\right)$ can be defined as follows (Kim, 2017);

$$U_c^{G_i^c}\left(G_k^h, H_t\right) = Q_c^{G_i^c}\left(r_c^{G_i^c}(H_t)\right) - I_c^{G_i^c}\left(G_k^h, r_c^{G_i^c}(H_t)\right) - K_c^{G_i^c}\left(r_c^{G_i^c}(H_t)\right) \tag{12.3.2}$$

$$\text{s.t.,}\begin{cases} Q_c^{G_i^c}\left(r_c^{G_i^c}(H_t)\right) = \psi_c^{G_i^c} \times r_c^{G_i^c}\left(H_t\right) \\ I_c^{G_i^c}\left(G_k^h,\ r_c^{G_i^c}(H_t)\right) \\ = \min\left\{\min\left\{\ G_k^h \in N_{G_i^c} \middle| F^s\left(G_k^h, H_t\right)\right\}, PS_B(H_t)\right\} \times r_c^{G_i^c}\left(H_t\right) \\ K_c^{G_i^c}\left(r_c^{G_i^c}(H_t)\right) = \zeta_c^{G_i^c} \times r_c^{G_i^c}\left(H_t\right) \end{cases}$$

where $Q_c^{G_i^c}(\,)$, $I_c^{G_i^c}(\,)$ and $K_c^{G_i^c}(\,)$ are outcome, charge and expense functions of G_i^c, respectively. Let $\psi_c^{G_i^c}$ and $\zeta_c^{G_i^c}$ denote the G_i^c's satisfaction and expense factors per bit of UPN service, respectively. $PS_B(H_t)$ represents the corresponding eNB's price strategy at time H_t, e.g., $\left[\alpha + \Theta^B(H_t)\right]$ (Kim, 2017).

As discussed previously, there are three positions of UEs; host (G^h), client (G^c), or a member of $\mathbf{G}^n$. At each time, UEs try to decide their best positions. Hence, each UE selects his position X to maximize his payoff (U) where $\times \in \mathrm{G}^h \cup \mathrm{G}^c \cup \mathrm{G}^n$ (Kim, 2017).

$$\times = \max_{\times \in \mathbf{G}^h \cup \mathbf{G}^c \cup \mathbf{G}^n}\left\{\ U_h^{\times \in \mathbf{G}^h},\ U_c^{\times \in \mathbf{G}^c},\ U_c^{\times \in \mathbf{G}^n}\ \right\}$$

(12.3.3)

$$\text{s.t.,}\ U_h^{\times \in \mathbf{G}^h} = U_h^{G^h}(H_t),\ U_c^{\times \in \mathbf{G}^c} = U_c^{G^c}(G^h, H_t) \text{ and } U_c^{\times \in \mathbf{G}^n} = 0$$

12.3.4 Strategy selection mechanisms in the PCS game

During UPN operations, eNBs are responsible for managing radio resources to optimize communication for all UEs. Consequently, the performance of the UPN is largely influenced by the efficiency of its radio resource management algorithm and its implementation. Each eNB oversees resource allocation through the Radio Resource Management (RRM) entity, which dynamically assigns radio resources to active UEs within its coverage area. The allocation is quantified in terms of Resource Units (RUs), with one RU representing the minimum allocation unit, such as 128 Mbps. The RRM must effectively satisfy the requirements of UEs while maximizing resource utilization within the flat radio network architecture (Kim, 2017; Mustika, 2015).

The *UPNC* scheme adopts the fundamental principle of the *Nash Bargaining Solution* (*NBS*) for RU allocation. The *NBS* serves as an effective mechanism to achieve a mutually beneficial outcome, striking a balance between efficiency and fairness. However, in many instances, the assumptions underlying traditional *NBS* do not align with real-world conditions. Specifically, in the context of dynamic UPN scenarios, the conventional *NBS* framework cannot be directly applied to the RU allocation process. To mitigate the inherent challenges posed by uncertainty, a modified *NBS* approach is required to accommodate ongoing network changes. (Kim, 2014b, 2017).

Taking into account the mobility of UEs, the *UPNC* scheme introduces an iterative resource allocation algorithm structured in a sequential Nash bargaining framework. The core idea of sequential Nash bargaining is to continuously monitor the current system environment and dynamically update the *NBS* to accommodate network fluctuations. This method alleviates the traditional *NBS* assumption that complete information is always available. Consequently, through a step-by-step iterative process, the eNB adaptively allocates RUs to UEs requesting resources. In the B_j, the allocated RU amount for the G_k at time H_t $\left(\phi_{H_t}^{G_k}(B_j)\right)$ is defined as follows (Kim, 2017);

$$\phi_{H_t}^{G_k}(B_j) =$$

$$\begin{cases} \vartheta_{H_t}^{G_k}, \text{ if } R_{H_t}^{B_j} \geq \sum_{G_l \in A(B_j)} \vartheta_{H_t}^{G_l} \\ \arg \max_{[\dots \vartheta_{H_t}^{G_l \in A(B_j)} \dots]} \left\{ \sum_{G_l \in A(B_j)} \left(\gamma_{H_t}^{G_l} \times \log \left(\vartheta_{H_t}^{G_l} \right) \right) \right\}, \text{ otherwise} \\ = \arg \max_{[\dots \vartheta_{H_t}^{G_l \in A(B_j)} \dots]} \left\{ \log \prod_{G_l \in A(B_j)} \left(\vartheta_{H_t}^{G_l} \right)^{\gamma_{H_t}^{G_l}} \right\}, \text{ s.t.}, \sum_{G_l \in A(B_j)} \vartheta_{H_t}^{G_l} = R_{H_t}^{B_j} \end{cases} \quad (12.3.4)$$

where $\vartheta_{H_t}^{G_k}$ is the amount of RU requested by the G_k at time H_t, and $R_{H_t}^{B_j}$ is the available RU amount of B_j. $A(B_j)$ is the set of UEs, which now request new RUs from the B_j. $\gamma_{H_t}^{G_l}$ is the G_l's bargaining power at time H_t, which is the relative ability to exert influence over other UEs. Usually, the bargaining solution is strongly dependent on the bargaining powers. If different bargaining powers are used, the G with a higher bargaining power obtains a higher RU. In the *UPNC* scheme, determining the bargaining powers depends on the G's mobility. Let $\mathscr{E}_{H_t}^{G_l}$ is the entropy for the G_l at the time H_t. Usually, the basic concept of entropy is uncertainty and a measure of the disorder in a system (Cao, 2010; Jeon, 2014; Kim, 2014a). To evaluate G_l's stability in the UPN, $\mathscr{E}^{G_l}_{H_t}$ represents the topological change, which is a natural quantification of the effect of G_l mobility in the UPN; it is calculated as follows (Kim, 2017);

$$\mathscr{E}^{G_l}_{H_t} = \left(-\sum_{G_k \in F_{G_l}} \left(P_{G_k}(H_t, \Delta_H) \times \log P_{G_k}(H_t, \Delta_H)\right)\right) / \left(\log C\left(F_{G_l}\right)\right) \tag{12.3.5}$$

$$\text{s.t., } P_{G_k}(H_t, \Delta_H) = O(G_l, G_k) / \sum_{G_f \in F_{G_l}} O(G_l, G_f)$$

where Δ_H is a time interval, and F_{G_l} denotes the set of the neighboring UEs of G_l, and $C\left(F_{G_l}\right)$ is the cardinality (degree) of set F_{G_l}. $O\left(G_l, G_f\right)$ represent a measure of the relative mobility among two G_l and G_f. Recently, new sensor technology has been developing rapidly. Nowadays, On-Board Units (OBUs) are embedded in most sensors of UEs (Wang, 2016b). OBUs are entities capable of event data recorder, global positioning system, forward and backward radar, sensing facility, and short-range wireless interface. Using OBUs, each UEs can easily estimate the relative mobility of other UEs (Kim, 2017).

The entropy $\mathscr{E}^{G_l}_{H_t}$ is normalized as $0 \leq \mathscr{E}^{G_l}_{H_t} \leq 1$. If the value of $\mathscr{E}^{G_l}_{H_t}$ is close to 1 (*or* close to 0), the G_l is stable (*or* unstable) (Kim, 2014a). Finally, the G_l's bargaining power at time H_t ($\gamma^{G_l}_{H_t}$) is calculated as follows (Kim, 2017);

$$\gamma^{G_l}_{H_t} = \frac{\mathscr{E}^{G_l}_{H_t}}{\sum_{G_k \in A(B_j)} \mathscr{E}^{G_k}_{H_t}}, \tag{12.3.6}$$

As a host, the G^h's strategy (S^h) represents the price level of client's communications. Therefore, the payoff of G^h $\left(U^{G^h}_h\right)$ can be defined as a function of S^h and requested client services $\left(r^{G^c}_c\right)$. Interactions between G^h and G^c continue repeatedly during the UPN process over time. Therefore, the G^h should consider the reactions from G^c to determine his strategy S^h. Under incomplete and asymmetric information situation, G^h should learn from the current environment, builds knowledge, and ultimately makes strategy decisions to maximize his payoff. Until now, several learning algorithms have been developed to learn from dynamic environments. However, traditional learning algorithms are not sufficient to follow rapid UPN changes. In particular, slow reaction to system fluctuations is a main drawback (Kim, 2017).

The *UPNC* scheme develops a new learning algorithm to decide effectively the host's price policy for UPN services. Usually, learning is divided into two categories: individual learning and social learning. Individual learning refers to trial and error, or insight temporal learning, and social learning refers to spatial leaning through interactions with other individuals. The main novelty of the learning method is joint-design manner concerning individual and social learning approaches. If the strategy S^h_j is selected at time H_{t-1} by the G^h_k in the B_g's coverage area, the G^h_k updates the strategy S^h_j's learning value for the next time step $\left(L^{s^h_j}_{G^h_k}(H_t)\right)$ according to the following method (Kim, 2017):

$$L^{s^h_j}_{G^h_k}(H_t) = \left((1 - x) \times L^{s^h_j}_{G^h_k}(H_{t-1})\right) + (x \times [v + w])$$

(12.3.7)

$$\text{s.t.}, v = (1-\beta) \times U_h^{G_k^h}(H_{t-1}) \text{ and } w = \left\{ \beta \times \sum_{G_i^h \in \Lambda(B_g)} \frac{L_{G_i^h}^{s_j^h}(H_{t-1})}{\|\Lambda(B_g)\|} \right\}$$

where x is a learning rate that models how the L-values are updated, and $\Lambda(B_g)$ is the set of hosts, which are connected to the B_g. $\|B_g\|$ is the cardinality of the set $\Lambda(B_g)$. In the equation (12.3.7), v and w represent individual and social learning values, respectively. Therefore, β is a control factor for the weighted average between different learning approaches. When a UE's mobility in the G_k^h's coverage area is high, the *UPNC* scheme places more emphasis on the social learning, i.e. on w. In this case, a higher value of β is more suitable. When a UE's mobility in the G_k^h's coverage area is low, the $L(\cdot)$ value should be strongly dependent on other UEs' responses in the G_k^h's area. In this case, a lower value of β is preferable to emphasize individual learning, i.e., on v. In the *UPNC* scheme, the value of β is dynamically adjusted based on the current entropy average of UEs, which are covered by the G_k^h. That is to say, β is the arithmetic mean of each UE's entropy, i.e., $\mathscr{E}_{H_{t-1}}^{G}$, according to the equation (12.3.5). Therefore, the *UPNC* scheme can learn the finest strategy both the individual and the social level with the incomplete information of UPN conditions (Kim, 2017).

Based on the $L(\cdot)$ values, a strategy selection distribution (**P**) for each host can be defined. During the UPN process, the *UPNC* scheme determines $\mathbf{P}^{G_k^h}(H_t) = \left\{ p_{s_{min}^h}^{G_k^h}(H_t) \ldots p_{s_j^h}^{G_k^h}(H_t) \ldots p_{s_{max}^h}^{G_k^h}(H_t) \right\}$ as the probability distribution of G_k^h's strategy selection at time H_t; it is sequentially modified over time. In the *UPNC* scheme, $p_s^G(\cdot)$ is defined based on the concept of Boltzmann distribution method, which has been used expensively in various machine learning algorithms. Finally, the s_j^h strategy selection probability by the G_k^h at time H_t $\left(\mathscr{P}_{s_j^h}^{G_k^h}(H_t)\right)$ is given by (Kim, 2017):

$$p_{s_j^h}^{G_k^h}(H_t) = \frac{\text{EXP}\left(L_{G_k^h}^{s_j^h}(H_{t-1})\right) \times \left(\min\left\{\max\left(U_h^{G_k^h}(H_{t-1}, s_j^h), 0\right), 1\right\}\right)}{\sum_{i=min}^{max}\left(\text{EXP}\left(L_{G_k^h}^{s_i^h}(H_{t-1})\right) \times \left(\min\left\{\max\left(U_h^{G_k^h}(H_{t-1}, s_j^h), 0\right), 1\right\}\right)\right)}$$

(12.3.8)

where $U_h^{G_k^h}(H_{t-1}, s_j^h)$ is the G_k^h's payoff with the strategy s_j^h at time H_{t-1}. Therefore, the $\min\{\max_{()}\}$ term is used to exclude non-profitable strategies from selection at time H_t.

12.3.5 Summary

Modern technology enables extensive communication without relying on conventional network infrastructures. From a global standpoint, UPN holds significant potential, driving a paradigm shift in network services by allowing end users to function as either hosts or clients. The *UPNC* scheme is designed based on the PCS game model, where UEs, as game participants, determine their positions while considering mutual interactions and refining their strategies in dynamic UPN environments. By leveraging feedback-based self-monitoring and distributed learning techniques, UEs continuously adapt to the evolving UPN landscape, optimizing their expected benefits. Given the distinctive characteristics of UPNs, the *UPNC* scheme ensures satisfactory service delivery even in scenarios with incomplete information (Kim, 2017).

12.4 Triple Bargaining based Energy Management (*TBEM*) scheme

The emergence of hierarchical smart-grid infrastructure has garnered significant research interest, leading to the development of intelligent electrical power distribution networks. These networks are composed of multiple micro-grids, which play a crucial role in smart-grid systems. The micro-grid concept has demonstrated great potential in delivering key attributes such as reliability, security, stability, and energy sustainability. Recently, S. Kim introduced the *Triple Bargaining-based Energy Management* (*TBEM*) scheme (Kim, 2019a). The *TBEM* scheme leverages three bargaining solutions - *Nash with claim*, *midpoint-constrained egalitarian*, and *proportional bargaining* - to dynamically allocate the limited energy resources within the smart-grid system. To enhance computational efficiency, these bargaining solutions are hierarchically applied within a structured triple-game model. Due to the unique advantages of each bargaining approach, the control decisions in the *TBEM* scheme are interdependent, triggering cascading interactions to achieve optimal system performance.

12.4.1 Development motivation

Since the conclusion of the First World War, the global urban population has increased nearly ninefold. By 2050, the world's population is projected to reach approximately 9 billion, with an estimated 70% residing in urban areas. Currently, cities contribute over 80% of the global GDP, making it increasingly critical to enhance the management of urban populations and the services they rely on. Given this rapid urban expansion, densely populated cities have emerged as major energy consumers, accounting for two-thirds of the world's total energy demand. As urbanization continues, cities are integrating innovative technologies to enhance infrastructure efficiency, enabling a more advanced and intelligent urban lifestyle. Consequently, there is an urgent need to shift towards urban sustainability, particularly in the planning and management of energy consumption (Carli, 2016; Kim, 2019a; Navidi, 2017).

In response to this trend, cities are increasingly adopting smart technologies to drive continuous innovation and development. The concept of a smart city is relatively new and has been recently defined. In essence, a smart city utilizes information and communication technologies (ICT) to enhance urban living by making it more efficient, comfortable, and sustainable. The implementation of smart city management varies significantly depending on the adopted approach. Various attempts have been made to establish a clear definition of a smart city, considering multiple perspectives. Governments worldwide have already implemented smart city initiatives as key programs aimed at: (i) optimizing energy efficiency, (ii) strengthening network infrastructures, and (iii) providing innovative services. As a result, urban planners, policymakers, and city administrators are able to make optimal or near-optimal decisions regarding resource allocation, ultimately reducing energy consumption while improving QoS (Carli, 2016; Kim, 2019a; Mahapatra, 2017).

For the future development of smart cities, the introduction of smart grids is essential to ensure a sustainable and high-quality lifestyle for urban residents. Designed as an advanced electricity network, the smart-grid system overcomes the limitations of traditional electrical grids by incorporating two-way digital communication. Since achieving smart city goals is closely linked to the deployment of smart grids, these two components are deeply interconnected - smart grids are central to smart cities and cannot fully function without them. Consequently, smart grids play a crucial role in enhancing the reliability and efficiency of power distribution systems while modernizing energy infrastructure (Kim, 2019a; Morello, 2017).

With the adoption of smart-grid technology, both the energy business model and the relationship between energy service providers and consumers are undergoing a transformation. The growing integration of distributed energy generation from renewable sources, along with the expansion of energy storage systems, is paving the way for new market structures that promote localized energy consumption. Small-scale power generation units, located closer to points of use, are becoming increasingly popular as they facilitate more efficient energy utilization at the local level. This shift has led to the development of the micro-grid concept - a small-scale, self-sustaining energy unit. Functioning as both a producer and consumer of energy (prosumer), a micro-grid consists of interconnected loads and distributed energy resources (DERs) within defined electrical boundaries, operating as a single controllable entity within the smart grid. Thus, micro-grids can be considered integral components of the broader smart-grid system, offering decentralized and consumer-centric energy solutions (Cui, 2014; Kim, 2019a).

Despite the numerous advantages of micro-grids, integrating them into the smart grid presents several challenges. One of the most significant obstacles is the need for adaptive control mechanisms that can intelligently coordinate distributed energy resources, ensuring that individual micro-grids function as manageable entities. Effective coordination would allow distributed resources to operate more efficiently and consistently. However, the lack of adequate control over the increasing number of distributed energy sources in micro-grids can lead to operational difficulties in managing the distribution network. To address this issue, hierarchical smart-grid architecture has been proposed. This approach enhances scalability within the grid model while enabling efficient resource sharing among micro-grid units. Additionally, the hierarchical structure is designed to be resilient against cascading failures by incorporating self-contained mechanisms at multiple levels of granularity. As a result, this architecture significantly improves the reliability of the smart grid by facilitating power distribution within and across various micro-grid units while mitigating uncertainties in a distributed manner (Kim, 2019a; Li, 2014).

The *TBEM* scheme introduces an advanced energy management mechanism based on hierarchical smart-grid architecture. Implementing efficient smart-grid energy management requires a well-structured strategy for coordinating power exchange among interconnected micro-grid units. This coordination strategy is primarily inspired by bargaining game theory, which analyzes negotiation scenarios where multiple parties seek to divide resources fairly. Bargaining theory, a key branch of game theory, provides solutions to resource allocation disputes by determining equitable divisions. Applications of bargaining theory extend beyond economics to disciplines such as biology, politics, and social sciences, where both conflict and cooperation play crucial roles. Since the early 2000s, ICT and network management have also been incorporated into this framework, further expanding its applicability (Kim, 2014b, 2019a).

12.4.2 Contribution

Building upon the aforementioned discussion, the *TBEM* scheme is developed based on fundamental principles of bargaining solutions, including *Nash with claim*, *midpoint-constrained egalitarian*, and *proportional bargaining solutions*. These approaches exhibit several notable axiomatic properties and can be justified from a game-theoretic standpoint. Utilizing the hierarchical smart-grid infrastructure, these three bargaining strategies are integrated into a triple-game model, operating in a coordinated manner. Throughout the triple bargaining game process, the hierarchical cooperation strategy ensures mutual benefits while optimizing overall grid system performance. To the best of my knowledge, the *TBEM* scheme represents the first attempt to explore hierarchical smart-grid control through the application of triple bargaining solutions (Kim, 2019a).

12.4.3 The hierarchical smart-grid system infrastructure

A defining characteristic of hierarchical smart-grid architecture is the distribution of multiple micro-grids across a broad geographical area, interconnected through distribution power networks. These decentralized micro-grids may also be linked to the main smart-grid at specific points. Within the *TBEM* scheme, the hierarchical smart-grid infrastructure is structured into three control levels. Based on these levels, grid controllers are classified into micro-grid local controllers (MLCs), micro-grid global controllers (MGCs), and the smart-grid central controller (SCC). At the primary level, the SCC continuously monitors multiple MGCs and regulates power flow, ensuring overall power balance within the smart-grid system. At the secondary level, the MGCs are strategically positioned within designated coverage areas, managing their associated micro-grids. These controllers serve as gateways, facilitating the bidirectional exchange of power and information between the smart-grid system and micro-grids. At the tertiary level, the MLCs function as the core operational units of individual micro-grids, ensuring a stable and reliable electricity supply. They also address technical challenges related to micro-grid operations while considering cost factors and environmental conditions affecting energy production. The hierarchical smart-grid system follows a cascading implementation approach across these three levels (Kim, 2019a).

Each micro-grid operates as an autonomous intelligent entity within the distribution system, serving as a reliable localized power source that efficiently integrates various forms of distributed generation (DG), particularly Renewable Energy Sources (RES). RESs have recently emerged as viable alternatives to traditional power sources, driven by advancements such as increased efficiency, lower generator costs, and reduced uncertainty in energy availability. Each MLC possesses self-sustaining capabilities, allowing it to dynamically manage the integrity, reliability, and stability of its micro-grid. Additionally, multiple MLCs located in the same geographic region are grouped together and linked to their respective MGC (Hatziargyriou, 2014; Kim, 2019a).

The implementation of hierarchical architecture introduces a multi-resolution control mechanism. At the primary control level, the main objective is to determine the optimal energy distribution strategy for self-interested and rational MGCs. The SCC is primarily responsible for overseeing the operational status of MGCs across different districts, working towards achieving this goal. This level is also where overall power grid planning takes shape, incorporating broader energy policy decisions. At the secondary level, the allocated energy for each MGC is further distributed among its corresponding MLCs. In a decentralized and parallel manner, each MGC independently manages its connected MLCs. Functioning as an intermediary gateway, the MGC is responsible for making high-level managerial control decisions, which in turn influence the operations of lower-level controllers. At the tertiary level, the energy assigned to each MLC is further allocated to individual micro-grid devices. Positioned at the lowest tier of the hierarchy, MLCs operate on the smallest planning scale to ensure the efficient functioning of their respective micro-grids. Their primary role is to facilitate local micro-grid operations while contributing to the overall controllability of the smart-grid system (Kim, 2019a).

Based on the combined design of three different phases, the integrated approach attempts to ensure self-adaptability and real-time effectiveness. To leverage the full synergy of hierarchical interdependence among phases, the *TBEM* scheme develops a novel triple bargaining game model and provides a new solution concept. The *TBEM* scheme assumes the hierarchical smart-grid system comprised of one SCC, multiple MGCs $\mathbb{G} = \{\mathcal{G}_1, \mathcal{G}_2, \ldots, \mathcal{G}_n\}$, various MLCs $\mathbb{L} = \{\mathcal{L}_1, \mathcal{L}_2, \ldots, \mathcal{L}_m\}$ and numerous micro-grid devices $\mathbb{D} = \{\mathcal{D}_1, \mathcal{D}_2, \ldots, \mathcal{D}_k\}$ where $n > m \geqslant k$. This model is flexible to be applied expansively into a system with multiple SCCs. Each device D is associated with a specific MLC $\mathcal{L}_{1 \leq j \leq m}$, which is also associated with a specific MGC. Finally, each MGC is connected to the SCC; expressively $\mathcal{L}_j^{\mathcal{G}_i}\left(or\mathcal{D}_l^{\mathcal{L}_j}\right)$ represents the $\mathcal{L}_j\left(or\mathcal{D}_l\right)$ associated with the $\mathcal{G}_i\left(or\mathcal{L}_j\right)$. Let $\mathfrak{N} = \left\{\mathfrak{N}_{min}\ldots\mathfrak{N}_b\ldots\mathfrak{N}_{max}\right\}$ be a set of energy allocation strategies where $\mathfrak{N}_b$ means the allocation amount level of power resource. For computation simplicity, strategies $\mathfrak{N}_{min\le b\le max}$ are discretely specified in terms of basic power units (BPUs) (Kim, 2019a).

At the primary level, the SCC is a game planner and associated $\mathcal{G}_{1\le i\le n} \in \mathbb{G}$ are bargaining game players. In this game, the total energy resource (E) in the smart-grid system is allocated to each MGC in a fair-efficient manner. At the secondary level, each individual MGC G is a game planner and corresponding $\mathcal{L}^{\mathcal{G}}_{1\le j\le m} \in \mathbb{L}$ are bargaining game players. In a parallel manner, the second-tier games are operated independently by multiple MGCs. At the tertiary level, each individual MLC is a game planner and associated $\mathcal{D}^{\mathcal{L}}_{1\le l\le k} \in \mathbb{D}$ are bargaining game players. In the same manner as the second-tier games, third-tier games are operated independently and dispersively by various MLCs (Kim, 2019a).

Let Σ^n be the class of all n-person problems, and a n-person bargaining problem with claims is a triple (S, d, c) where S is a subset of $\mathbb{R}^n$ and d, c are points in $\mathbb{R}^n$. There exists $x \in S$ with $x > d, c \notin S$ and $c \ge d$ where the coordinates of x are the utility levels, attained by the n players through the choice of some joint action. A solution is a function, $\mathcal{F}:\Sigma^n \to \mathbb{R}^n$ that associates with each $(S, d, c) \in \Sigma^n$ a unique point of S, $\mathcal{F}(S, d, c)$, called the solution outcome of (S, d, c). Consider a reference function given by a mapping $g:\Sigma^c \to \mathbb{R}^n$. The $g(S, d, c)$ function will serve as an origin from which relative utility gains or losses are measured; it is not necessarily to be a feasible utility allocation (Chun, 1992; Gerber, 2005; Herrero, 1997; Karos, 2019; Kim, 2019a).

The best that player $i \in N$ can hope for in (S, d, c), i.e., her ideal payoff in (S, d, c), is $a_i(S, d, c) = \max\left\{s_i | s \in S \text{ and } s \ge d\right\}$. The point $a(S, d, c) = \left(a_1(S, d, c), \ldots, a_n(S, d, c)\right)$ is called the ideal point of (S, d, c); it is typically not a feasible solution. The midpoint of (S, d, c) is $m(S, d, c) = \frac{1}{n}\cdot a(S, d, c)$ and always lies in S. Let $\overline{B}$ be the collections of pairs (S, d, c), and the set of all bargaining problems is denoted by $B \subseteq \overline{B}$. A bargaining solution with domain $B' \subseteq \overline{B}$ is a map F that assigns to each pair ($S, d, c) \in B'$ a point $\mathcal{F}(S, d, c) \in S$. Normally, a bargaining solution is defined on a subset of B rather than $\overline{B}$. The TBEM scheme assumes that bargaining solutions are translation invariant, that is for any $t \in \mathbb{R}^n$ it holds that $F(S + t, d + t, c + t) = F(S, d, c) + t$. A point $s \in S$ is weakly Pareto optimal if there is no $t \in S$ with $t \gg s$, and strongly Pareto optimal if there is no $t \in S$ with $t > s$ (Karos, 2019; Kim, 2019a).

Let $\mathcal{D}_H$ denote the Hausdorff distance. The Hausdorff distance between two subsets X,Y of a metric space with metric D is defined as $\mathcal{D}_H(X, Y) = \max\left\{\sup_{x\in X}\inf_{y\in Y}\mathcal{D}(x, y), \sup_{y\in Y}\inf_{x\in X}\mathcal{D}(x, y)\right\}$. The sequence ($S^p)_{p\in N}$ is a Cauchy sequence if for any $\varepsilon > 0$ there is p such that $\mathcal{D}_H(S^{p'}, S^{p''}) < \varepsilon$ for all $p', p'' \ge p$. For any two accumulation points S, S' one has $\overline{S} = \overline{S'}$, that is S and S' have identical closures. Hence $\overline{S}$ is the unique accumulation point of $(S^p)_{p\in N}$ in $\overline{B}$. In particular, $\left(\overline{B}, \mathcal{D}_H\right)$ is a complete metric space and $\overline{B}$ is the closure of B. Let $e = (1, \ldots, 1) \in \mathbb{R}^n$ and define the diagonal Δ in $\mathbb{R}^n$ by $\Ddot{A} = \left\{x | x = \lambda\cdot e \text{for-} \text{some} \lambda \in \mathbb{R}\right\}$. A mapping $L:\mathbb{R}^n \to \mathbb{R}^n$ is called a *positive affine transformation* if there exist $a \in \mathbb{R}_{++}$ and $b \in \mathbb{R}^n$ such that for all $x \in \mathbb{R}^n$ and all i, $(L(x))_i = a_i x_i + b_i$. Let L be the class of all *positive affine transformations* on $\mathbb{R}^n$. For a set $A \subset \mathbb{R}^n$ and $L \in \mathcal{L}$, let $L(A) = \{y | \exists x \in A \text{with} y = L(x)\}$. Note that if $(S, d, c) \in \sum^n$, then $(L(S), L(d), L(c)) \in \sum^n$ as well (Chun, 1992; Gerber, 2005; Herrero, 1997; Karos, 2019; Kim, 2019a).

The proportional solution, P: for all $(S, d, c) \in \Sigma^n$, $\mathcal{P}(S, d, c)$ is the maximal point of S on the segment connecting the point d and the point c (Chun, 1992; Herrero, 1997). It can be easily calculated by using the bisection or iterative water-filling algorithms. The Nash bargaining solution with respect to g, $\mathcal{N}^g$: for all $(S,d,c) \in \Sigma_g^c \rightarrow \mathbb{R}^n$, $\mathcal{N}^g(S,d,c)$ is defined based on the reference function $g{:}\Sigma^c \rightarrow \mathbb{R}^n$ (Gerber, 2005; Kim, 2019a);

$$\mathcal{N}^g(S,d,c) = \arg\mathbf{max} \prod_{i=1}^{n} \left(x_i - g_i(S,d,c)\right) \tag{12.4.1}$$

s.t., $x \in S, i \in N, x \geq g(S,d,c)$ and $(S,d,c) \in \Sigma_g^c$

The egalitarian solution $E(S,d,c)$ of a bargaining problem may select a payoff configuration which is not midpoint dominant. The $E(S,d,c)$ is the most prominent monotonic solution that maps each problem S to the weakly Pareto optimal point $E(S,d,c) \in S$ with $E_i(S,d,c) = E_j(S,d,c)$ for all $i,j \in N$. Hence, the arbitrator assigns the same and the highest possible payoff to all players. Based on the idea of $E(S,d,c)$, the midpoint-constrained egalitarian bargaining solution $\mathcal{ME}(S,d,c)$ is formally defined as follows (Karos, 2019; Kim, 2019a);

$$\mathcal{ME}(S,d,c) = \min_{s \in S} \|s - E(S,d,c)\|, \text{ s.t., } s \geq m(S,d,c) \tag{12.4.2}$$

where " · " denotes the Euclidean norm. If $s \neq t$ are such that $\|s - E(S,d,c)\| = \|t - E(S,d,c)\|$, then $\|\alpha s + (1-\alpha)t - E(S,d,c)\| < \|s - E(S,d,c)\|$ for all á $\in (0,1)$. Therefore, the solution of (2) is unique, which means that $\mathcal{ME}(S,d,c)$ is well-defined (Karos, 2019).

12.4.4 The triple bargaining game-based energy management

To implement the smart-grid management paradigm, the time axis is partitioned into equal intervals of length *unit_time* (Δ_t). At each Δ_t, energy requirements are received in phases, and each level controllers distribute their energy resources. In the triple game model, SCC, MGCs, MLCs and micro-devices are assumed as game players, and they have their own utility functions. Usually, utility functions are functions describing the level of satisfaction or happiness of game players while consuming resources; they can take one or more arguments that affect the overall satisfaction of players. Simply, the utility function of SCC is defined as the total sum of MGCs' utilities. Therefore, the SCC utility function $\left(U_{SCC}(\cdot)\right)$ is defined based on the allocated energy resources of associated MGCs, and it maps service quality of the target to a benefit value (Kim, 2019a).

$$U_{SCC}\left(\mathbb{E}_{SCC}^{\Delta_t}, \mathbb{G}\right) = \sum_{\mathcal{G}_i \in \mathbb{G}} \left(exp\left(\alpha \times \frac{1}{exp\left(\beta \times \frac{A_{\mathcal{G}_i}^{\Delta_t}}{\max_{\mathcal{G}_j \in \mathbb{G}}\left(\mathcal{C}_{\mathcal{G}_j}^{\Delta_t}\right)}\right)} \right) - \gamma \right) \tag{12.4.3}$$

$$\text{s.t.,} \mathbb{E}_{SCC}^{\Delta_t} = \sum\nolimits_{\mathcal{G}_i \in \mathbb{G}} A_{\mathcal{G}_i}^{\Delta_t} \text{ and } \mathcal{C}_{\mathcal{G}_j}^{\Delta_t} = \sum\nolimits_{\mathcal{L}_i \in \mathbb{L}_{\mathcal{G}_j}} \mathcal{C}_{\mathcal{L}_i}^{\Delta_t}$$

where $\mathbb{E}_{SCC}^{\Delta_t}$ is the total available energy amount of SCC at Δ_t. α, β and γ are control parameters to calculate the $U_{SCC}(\cdot)$ function. $A_{\mathcal{G}_i}^{\Delta_t}$ and $\mathcal{C}_{\mathcal{G}_i}^{\Delta_t}$ are the allocated and idealistic energy amounts for the $\mathcal{G}_i$ at time Δ_t, respectively. $\mathbb{L}_{\mathcal{G}_i}$ is the set of MLCs covered by the $\mathcal{G}_i$. The *TBEM* scheme adopts the midpoint-constrained egalitarian bargaining solution to distribute the $\mathbb{E}_{SCC}^{\Delta_t}$ at the primary level. Based on the allocated resources of MLCs, the MGC $\mathcal{G}_i$'s utility function $\left(U_{\mathcal{G}_i}(\cdot)\right)$ is defined in the same manner as the $U_{SCC}(\cdot)$ (Kim, 2019a).

$$U_{\mathcal{G}_i}\left(\mathbb{E}_{\mathcal{G}_i}^{\Delta_t},\mathbb{L}_{\mathcal{G}_i}\right) = \sum_{\mathcal{L}_k \in \mathbb{L}_{\mathcal{G}_i}} \left(exp\left(\delta \times \frac{1}{exp\left(\zeta \times \frac{A_{\mathcal{L}_k}^{\Delta_t}}{\max_{\mathcal{L}_j \in \mathbb{L}_{\mathcal{G}_i}}\left(\mathcal{C}_{\mathcal{L}_j}^{\Delta_t}\right)}\right)}\right) - \mu \right) \tag{12.4.4}$$

$$\text{s.t.,} \mathbb{E}_{\mathcal{G}_i}^{\Delta_t} = \sum\nolimits_{\mathcal{L}_k \in \mathbb{L}_{\mathcal{G}_i}} A_{\mathcal{L}_k}^{\Delta_t} \text{ and } \mathcal{C}_{\mathcal{L}_l}^{\Delta_t} = \sum\nolimits_{\mathcal{D}_e \in \mathbb{D}_{\mathcal{L}_l}} \mathcal{C}_{\mathcal{D}_e}^{\Delta_t}$$

where $\mathbb{E}_{\mathcal{G}_i}^{\Delta_t}$ is the assigned energy amount for the $\mathcal{G}_i$ at Δ_t. $A_{\mathcal{L}_k}^{\Delta_t}$ and $\mathcal{C}_{\mathcal{L}_k}^{\Delta_t}$ are the allocated and idealistic energy amounts of $\mathcal{L}_k$ at time Δ_t, respectively. δ, ζ and μ are control parameters to calculate the $U_{\mathcal{G}}(\cdot)$ function. $\mathbb{D}_{\mathcal{L}_l}$ is the set of micro-devices connected to the $\mathcal{L}_l$. The *TBEM* scheme adopts the Nash bargaining solution with claim to assign the $\mathbb{E}_{\mathcal{G}}$ to corresponding MLCs. In the *TBEM* scheme, the $g(S,d,c)$ for the $\mathcal{G}_i$ $\left(\mathcal{G}_{\mathcal{G}_i}(S,d,c)\right)$ is defined as follows (Kim, 2019a);

$$\mathcal{G}_{\mathcal{G}_i}(S,d,c) = \mathcal{C}_{\mathcal{G}_i}^{\Delta_t} \times \left[\Gamma \times \left(\sigma - \log\left(\min\left(\frac{\sum_{\mathcal{L}_l \in \mathbb{L}_{\mathcal{G}_i}}\left(\text{RES}_{\mathcal{L}_l}\right)}{\mathcal{C}_{\mathcal{G}_i}^{\Delta_t}}, \theta\right) + \eta\right)\right)\right] \tag{12.4.5}$$

where Γ, σ, θ and η are control parameters to adjust the function $\mathcal{G}_{\mathcal{G}_i}(S,d,c)$. Based on the allocated resources of associated micro-grid devices, the MLC $\mathcal{L}_k$'s utility function $\left(U_{\mathcal{L}_k}(\cdot)\right)$ is defined in the same manner as the functions (12.4.3) and (12.4.4) (Kim, 2019a).

$$U_{\mathcal{L}_k}\left(\mathbb{E}_{\mathcal{L}_k}^{\Delta_t},\mathbb{D}_{\mathcal{L}_k}\right) = \sum_{\mathcal{D}_c \in \mathbb{D}_{\mathcal{L}_k}} \left(exp\left(\omega \times \frac{1}{exp\left(\psi \times \frac{A_{\mathcal{D}_c}^{\Delta_t}}{\max_{\mathcal{D}_e \in \mathbb{D}_{\mathcal{L}_k}}\left(\mathcal{C}_{\mathcal{D}_e}^{\Delta_t}\right)}\right)}\right) - \chi \right) \tag{12.4.6}$$

$$\text{s.t.,} \mathbb{E}_{\mathcal{L}_k}^{\Delta_t} = \sum\nolimits_{\mathcal{D}_c \in \mathbb{D}_{\mathcal{L}_k}} A_{\mathcal{D}_c}^{\Delta_t}$$

where $\mathbb{E}_{\mathcal{L}_k}^{\Delta_t}$ is the assigned energy amount for the MLC $\mathcal{L}_k$ at Δ_t. ω, ψ and χ are control parameters to calculate the $U_{\mathcal{L}}(\cdot)$ function. $A_{\mathcal{D}_c}^{\Delta_t}$ and $\mathcal{C}_{\mathcal{D}_c}^{\Delta_t}$ are the allocated and idealistic energy amounts of $\mathcal{D}_c$ at time Δ_t, respectively. From the viewpoint of MLC, the major concern for the energy allocation is to ensure the proportionality of idealistic allocations. Therefore, the *TBEM* scheme uses the proportional bargaining solution to distribute the $\mathbb{E}_{\mathcal{L}}$ to corresponding micro-devices. Finally, the micro-grid device $\mathcal{D}_c$'s utility function $\left(U_{\mathscr{D}_c}(\cdot)\right)$ is defined in the same manner as the functions (12.4.3),(12.4.4) and (12.4.6) (Kim, 2019a).

$$U_{\mathscr{D}_c}\left(A_{\mathscr{D}_c}^{\Delta_i}, \mathscr{C}_{\mathscr{D}_c}^{\Delta_i}\right) = exp\left(\xi \times \frac{1}{exp\left(\varphi \times \frac{A_{\mathscr{D}_c}^{\Delta_i}}{\mathscr{C}_{\mathscr{D}_c}^{\Delta_i}}\right)}\right) - \tau \qquad (12.4.7)$$

where ξ, φ and τ are control parameters to calculate the $U_{\mathscr{D}_c}(\cdot)$ function.

12.4.5 Summary

The emergence of smart-grid technology has sparked significant research interest in the development of intelligent power distribution networks. Ensuring an efficient and equitable system operation is crucial for the successful implementation of future smart-grid systems. The *TBEM* scheme introduces an innovative energy management mechanism tailored for hierarchical smart-grid structures. Built upon *Nash, extended-egalitarian*, and *proportional bargaining solutions*, the *TBEM* scheme is structured as a triple-game model. Within the *TBEM* scheme, controllers at each hierarchical level act as game players, making strategic control decisions that trigger cascading interactions to achieve a balanced and efficient outcome. Through coordination algorithms implemented across three control tiers, the overall energy management of the smart-grid system is systematically regulated, ensuring adaptive utility maximization while enhancing system reliability (Kim, 2019a).

12.5 Underwater Network Distributed Routing (*UNDR*) scheme

Underwater sensor networks (USNs) have recently gained attention as a promising networking technology for various underwater applications. However, acoustic routing in USNs faces several challenges due to the dynamic nature of the aquatic environment, high energy consumption, long propagation delays, and limited bandwidth. As a result, conventional routing protocols, which are highly effective in terrestrial wireless networks, struggle to adapt to underwater conditions. To address these challenges, S. Kim introduces the *Underwater Network Distributed Routing* (*UNDR*) scheme, designed specifically to overcome acoustic transmission issues in USNs (Kim, 2018). The *UNDR* scheme leverages reinforcement learning and game theory, structuring the routing process as a game-theoretic model to enhance packet forwarding efficiency. In particular, the Q-learning game framework enables decentralized and distributed decision-making, allowing the USN system to dynamically adapt to environmental changes.

12.5.1 Development motivation

With over 70% of the Earth's surface covered by bodies of water, including rivers, canals, seas, and oceans, underwater acoustic transmission has gained increasing attention due to its crucial role in applications such as distributed tactical surveillance, disaster prevention, mine reconnaissance, and environmental monitoring. Recent advancements in communication technology have facilitated the development of underwater sensor networks (USNs). Utilizing cutting-edge underwater exploration technologies, USNs consist of sink nodes and autonomous micromechanical sensors. These networks rely on spatially dispersed sensors that collect water-related data and communicate wirelessly through acoustic signals within the underwater environment. Sink nodes located at the surface aggregate data from these underwater sensors and relay the information to a monitoring center via satellite for further analysis (Ghoreyshi, 2016; Kim, 2018; Shah, 2016; Wahid, 2012).

Routing is a critical aspect of any network infrastructure. In wireless sensor networks, routing protocols play a key role in ensuring efficient data transmission from source to destination nodes. Since network performance is closely linked to the efficiency of the routing algorithm, significant research efforts have been devoted to developing adaptive network routing protocols. While terrestrial sensor networks have been extensively studied, leading to numerous proposed communication protocols (Jang, 2013; Raazi, 2010; Rizvi, 2015), the direct application of these solutions to USNs is not feasible due to the distinct characteristics of underwater environments. The development of an effective routing algorithm for USNs must account for unique underwater challenges, including high signal absorption, long propagation delays, limited bandwidth, and a high bit-error rate (Kim, 2018; Shah, 2016).

Existing routing algorithms are categorized based on their design goals and operational scenarios. Some operate in a static offline manner, while others function dynamically in real time. Additionally, routing strategies can be classified as either centralized or distributed. However, centralized control is impractical for large-scale sensor networks, and static algorithms fail to significantly enhance USN performance in dynamic underwater environments. Since underwater conditions are constantly changing, routing decisions must be made in real time without prior knowledge of future system states. Given these constraints, distributed and dynamic routing methodologies are the most suitable approaches for large-scale USN implementations (Kim 2014b, 2018; Raazi, 2010).

The *UNDR* scheme introduces a novel routing mechanism designed to address the unique challenges of USNs, including energy efficiency, long propagation delays, dynamic topology shifts, and node localization in underwater environments. To effectively tackle these issues, the *UNDR* scheme seeks to balance overall network performance while ensuring practical applicability for real-world USN operations. The design principles of the *UNDR* scheme emphasize feasibility, self-adaptability, and efficiency in providing an optimal routing solution. Despite numerous proposed USN routing protocols, no comprehensive routing framework based on an integrated approach has been developed. Existing research highlights that this remains a complex and unresolved challenge (Kim, 2018).

The *UNDR* scheme leverages a game-theoretic learning algorithm to facilitate point-to-point routing decisions. In this model, each sensor node is considered a game player that autonomously makes routing decisions in a distributed online manner, adapting to the dynamic changes within the USN. The core concept of the proposed game framework is an extension of the widely recognized Markov decision process into a multi-player routing game model. To determine the optimal routing strategy, each node learns from its environment using a reinforcement learning algorithm. Specifically, the *UNDR* scheme demonstrates that game-theoretic insights can be applied to derive a distributed Q-learning algorithm. In essence, learning is defined as the ability to make intelligent routing decisions based on both past and present system states (Kim 2014b, 2018).

12.5.2 Contribution

The *UNDR* scheme offers several key contributions: i) it introduces an innovative routing mechanism specifically designed for USNs, ii) it integrates game theory with the Q-learning algorithm to optimize routing decisions, iii) it employs a distributed online approach, enabling self-adaptability and real-time decision-making, iv) it develops a routing algorithm that effectively balances conflicting requirements, and v) it determines routing decision probabilities based on real-time USN conditions. The significance of the *UNDR* scheme lies in its ability to dynamically respond to the current state of the USN system while accounting for the ever-changing underwater environment. To the best of my knowledge, research on game-theoretic distributed learning algorithms for USN routing remains scarce (Kim, 2018).

12.5.3 Routing game model

In the operation of a USN system, sensor nodes independently make routing decisions by evaluating factors such as end-to-end delay, hop count, packet error rate, and energy efficiency. Given the dynamic nature of acoustic conditions, these nodes aim to optimize their own performance in a distributed online manner through a learning-based approach. The *UNDR* scheme models sensor nodes as rational game players that strategically determine their routing choices. Based on this premise, the *UNDR* scheme introduces a novel game-theoretic framework, referred to as the routing game, specifically designed for the USN system. Formally, the *UNDR* scheme defines the routing game model G=$\{\mathbb{N}, \mathbb{A}_{i,1\leq i\leq n}, U_{i,1\leq i\leq n}, \mathbb{U}_{i,1\leq i\leq n}[m], \mathbb{S}_{i,1\leq i\leq n}, T\}$ at each time period t of gameplay (Kim, 2018).

- N is the finite set of game players $\mathbb{N} = \{\delta_1 \dots \delta_i, \dots \delta_n\}$, where δ_i is the ith sensor node and n is the number of sensors.
- $\mathbb{A}_i = \{a_i^1, a_i^2 \dots a_i^m\}$ is the finite set of δ_i's available actions, where a_i^m represents the selection of the mth neighboring node to relay the routing packet; m is the total number of neighbors.
- $U_{i,1\leq i\leq n}$ is the payoff received by δ_i.
- The array $\mathbb{U}_i[m] = [\mathcal{U}_i^1, \mathcal{U}_i^2 \dots \mathcal{U}_i^m]$ is the payoff history of δ_i; $\mathcal{U}_i^m$ represents the accumulated payoff of strategy a_i^m. $\mathbb{U}_i[\cdot]$ is used to estimate the strategy selection probabilities.
- $\mathbb{S}_i = \{s_i^1, s_i^2 \dots s_i^m\}$ is the set of states in δ_i. The definition of system states is related to routing strategies. If δ_i takes action a_i^m, then δ_i's next state is s_i^m.
- $T = \{0, 1, \dots, t, t+1, \dots\}$ denotes time, which is represented by a sequence of time steps with imperfect information.

12.5.4 Reinforcement learning mechanism

Reinforcement learning focuses on enabling system agents to make optimal decisions that maximize their rewards. These algorithms strike a balance between exploring new possibilities and leveraging existing knowledge, adapting their strategies based on real-time performance. In the context of game theory, reinforcement learning provides insights into how effective solutions can emerge under conditions of bounded rationality. As a result, there has been a growing research interest in integrating game theory with learning algorithms. In recent years, game-theoretic learning algorithms have gained significant attention as a crucial area of study (Galindo-Serrano, 2010; Hu, 2010; Kim, 2014b, 2018).

The environment is typically formulated as a Markov decision process (MDP), which is therefore the basic concept of reinforcement learning. The MDP is a mathematical framework for modeling the decision-making process and is useful for optimization problems that are solved via learning algorithms (Galindo-Serrano, 2010; Hu, 2010). In the current scenario, the MDP consists of the set of states (S), set of actions (A), state transition probabilities (**P**), and rewards (R). **P** is composed of elements $P^a_{s\to s'}$, which are the probabilities of moving to state s' from state s (i.e. $s',s \in \mathbb{S}$) with a given action $a \in \mathbb{A}$. $\mathcal{R}(s,a)$ is the reward for taking action a from state s. Mathematically, at time t, $P^{a}_{s \to s'}$ and $\mathcal{R}(s,a)$ are defined as follows (Hu, 2010; Kim, 2018):

$$\begin{cases} P^{a}_{s \to s'} = Pr\{S_{t+1} = S' |_{s=s_t,\, a=a_t}\}, s.t., \sum_{s' \in S} P^{a}_{s \to s'} = 1 \\ R(s,a) = \{\mathcal{R}_t(s_t,a_t)|_{s=s_t,\, a=a_t}\} = \sum_{s_{t+1} \in S} \left(P^{a_t}_{s_t \to s_{t+1}} \times \mathcal{R}^{a_t}_{s_t \to s_{t+1}}\right) \end{cases}$$

(12.5.1)

where s_t and a_t denote the state and action at time t, respectively. $\mathcal{R}^{a}_{s \to s'}$ is the reward expected for taking action a at state s and entering the next state s'. In the reinforcement learning algorithm, a policy ψ is mapped from each state $s \in \mathbb{S}$ and action $a \in \mathbb{A}(s)$ to the probability $\psi(s,a)$ of taking action a when in state s. The expected reward ($V^{\psi}(s)$) in state s under policy ψ is received after taking an action according to policy ψ at time t, and for following policy ψ thereafter. $V^{\psi}(s)$ is defined as (Hu, 2010; Kim, 2018):

$$V^{\psi}(s) = E_{\psi}\{\mathcal{R}_t(s_t,a_t)\} = E_{\psi}\left\{\sum_{k=0}^{\infty}\left(\gamma^k \times \mathcal{R}_{t+k}(s_{t+k},a_{t+k})\right)\right\},\ s.t., \gamma \in [0,1)$$

(12.5.2)

where γ is a control factor that discounts the rewards in the future. Based on Bellman's principle of optimality, equation (12.5.2) can be rewritten as a recursive equation (Kim, 2018):

$$V^{*}(s) = \max_{\psi} V^{\psi}(s) = \max_{\psi} E_{\psi}\left\{\sum_{k=0}^{\infty}\left(\gamma^k \times \mathcal{R}_{t+k}(s_{t+k},a_{t+k})\right)\right\}$$

$$= \max_{\psi} E_{\psi}\left\{\mathcal{R}_t(s_t,a_t) + \sum_{k=1}^{\infty}\left(\gamma^k \times \mathcal{R}_{t+k}(s_{t+k},a_{t+k})\right)\right\}$$

$$= \max_{a}\left[\mathcal{R}_t(s_t,a_t) + \left(\gamma \times \left(\sum_{s_{t+1} \in S}\left(\left(P^{a_t}_{s_t \to s_{t+1}} \times V^{*}(S_{t+1})\right)\right)\right)\right)\right]$$

(12.5.3)

To solve equations (12.5.3), reinforcement learning algorithms can be used. The *UNDR* scheme adopts a well-known reinforcement learning technique called Q-learning. Q-learning works by successively improving its evaluations of the quality of particular actions at particular states. In addition, Q-learning provides agents with the ability to learn to act optimally in Markovian domains by experiencing the consequences of their actions but does not require the agents to build maps of the domains (Galindo-Serrano, 2010; Hu, 2010). In the Q-learning algorithm, the value of state–action pairs is given by $Q^{\psi}(s,a)$, which is the value of taking action a in state s under policy ψ. This is defined as (Kim, 2018):

$$Q^{\psi}(s, a) = E_{\psi}\left\{ \sum_{k=0}^{\infty} \left(\gamma^k \times \mathscr{R}_{t+k}(s_{t+k}, a_{t+k})\right) \right\}$$

$$= \left(\mathscr{R}_t(s_t, a_t) + \gamma \times \left(\sum_{s_{t+1} \in S} \left(\left(P^{a_t}_{s_t \to s_{t+1}} \times V^*(S_{t+1}) \right) \right) \right) \right)$$

(12.5.4)

According to equations (12.5.3) and (12.5.4), the *UNDR* scheme can compute $V^*(s)$ as (Galindo-Serrano, 2010; Hu, 2010):

$$V^*(s) = \max_a Q^*(s, a) = \max_a Q^*(s_t, a_t)$$

$$= \max_{a_t} \left[\mathscr{R}_t(s_t, a_t) + \gamma \times \left(\sum_{s_{t+1} \in S} \left(\left(P^{a_t}_{s_t \to s_{t+1}} \times \max_a Q^*(S_{t+1}, a) \right) \right) \right) \right]$$

(12.5.5)

In equation (12.5.5), $Q^*(s_t, a_t)$ can be approximated as (Hu, 2010; Kim, 2018):

$$Q^*(s_t, a_t) \cong$$

$$\lim_{t \to \infty} \left[\{(1 - \beta) \times Q(s_t, a_t)\} + \left(\beta \times \left(\mathscr{R}_t(s_t, a_t) + \{\gamma \times \max_a Q(s_{t+1}, a)\} \right) \right) \right]$$

(12.5.6)

$$\text{s.t.}, \beta \in (0, 1]$$

where β is a learning rate that models how the Q-values are updated. The advantage of Q-learning is that it simply calculates the Q values of all actions based on the system model to update V, which is less computationally demanding (Galindo-Serrano, 2010; Hu, 2010). However, conventional Q-learning has the limitation that the reinforcement learning process is very slow (Kim, 2018).

12.5.5 USN configuration and routing algorithms

The key concept of the configuration algorithm is to configure the sensor network topology. Each sensor node maintains routing information in a *control_record* for self-organizing and independent control. The *control_record* consists of two system parameters: path cost ($\mathscr{PC}$) and the payoff history array ($\mathbb{U}[\cdot]$). $\mathscr{PC}$ represents the degree to which communication can be adapted to reach the nearest sink node, whereas $\mathbb{U}[\cdot]$ contains the payoff history for each action. The information in the *control_record* is used to determine adaptive routing decisions (Kim, 2018).

$\mathscr{PC}$ is estimated based on the link costs in the route to the sink node. To estimate each wireless link between two sensor nodes, the *UNDR* scheme defines a link cost (L_c). In each node, $\mathscr{PC}$ is estimated as the sum of all L_c values from the current node to the sink node. Using real-time monitoring of the remaining energy level of the node and the energy consumption of wireless communications, the value of L_c from δ_i to δ_j ($L_c(i,j)$) is obtained as follows (Kim, 2018):

$$L_c(i,j) = \left[(1-\alpha_i) \times \frac{d_{i,j}}{D_M}\right] + \left[\alpha_i \times \left(1 - \frac{\mathscr{E}_j}{E_M}\right)\right] \tag{12.5.7}$$

where $d_{i,j}$ is the distance from δ_i to δ_j, and $\mathscr{E}_j$ is the remaining energy of δ_j. E_M and D_M denote the initial energy and the maximum coverage range of each sensor node, respectively. Therefore, $d_{i,j}$ and $\mathscr{E}_j$ are normalized by D_M and E_M (Kim, 2018).

For the adaptive $L_c(i,j)$ estimation, the parameter α_i controls the relative weights given to the distance and the remaining energy of the corresponding node. When the remaining energy of the current node δ_i is high, the *UNDR* scheme places more emphasis on the energy status of the next node δ_j, i.e. on $\left(1 - (\mathscr{E}_j/E_M)\right)$. In this case, a higher value of α_i is more suitable. When the remaining energy of node i is insufficient, the path selection should be strongly dependent on the energy dissipation for data transmission. In this case, a lower value of α_i is preferable for the energy consumption rate, i.e., on $\left(d_{i,j}/D_M\right)$, as the distance between two neighbor nodes directly affects the energy consumption rate. In the proposed configuration algorithm, the value of α_i for the corresponding δ_i is dynamically adjusted based on the rate of remaining energy with respect to the initially assigned energy ($\mathscr{E}_i/E_M$) (Kim, 2018).

During network topology formation, each individual node estimates the L_c values of its neighboring nodes. Nodes within one-hop of the sink node set their $\mathscr{PC}$ values to the one-hop L_c value from themselves to the sink node. The remaining sensor nodes δ_i set their $\mathscr{PC}$ values as follows (Kim, 2018):

$$\mathscr{PC}_i = \min_j\left\{\mathscr{PC}_j + L_c(i,j)\right\}, \text{ s.t.}, j \in \mathscr{N}_i \tag{12.5.8}$$

where $\mathscr{N}_i$ is the set of neighboring nodes of δ_i. Based on the $\mathscr{PC}$ values of the neighbor nodes, $\mathbb{U}[\cdot]$ is initialized as (Kim, 2018):

$$\mathbb{U}_i[j]_{j\in\mathscr{N}_i} = \mathscr{PC}_j + L_c(i,j) \tag{12.5.9}$$

When the proposed configuration algorithm has terminated, the virtual topology of the USN can be considered as a spanning tree structure that is rooted at the sink node while minimizing the $\mathscr{PC}$ values. However, this is a temporary topology. During the routing operation, each node's $\mathscr{PC}$ and $\mathbb{U}[\cdot]$ values are dynamically re-estimated, and the network topology is also dynamically restructured (Kim, 2018).

In designing the USN routing algorithm, the *UNDR* scheme considers the multi-player MDP. Each sensor node is assumed to be a player, and the objective of player i (δ_i) is to find the best routing policy $\psi_i^* \in \mathbb{A}_i$, where $\mathbb{A}_i$ is the action set of δ_i; $|\mathbb{A}_i| = |\mathscr{N}_i|$ and ψ_i represents a one-hop routing decision involving one of the neighboring nodes. Based on the routing decisions, each player statically determines its subsequent state. Let $s_i(t), a_i(t)$ be the state and action of δ_i at time step t, respectively, where $s_i(t) \in \mathbb{S}_i$

and $a_i(t) \in \mathbb{A}_i$. $U_i(s_i(t), a_i(t))$ is the payoff to δ_i at time t when action $a_i(t)$ occurs in state $s_i(t)$, and $U_i(s_i(t), a_i(t)) \rightarrow \mathbb{R}$. $U_i(s, a)$ can be interpreted as the reward (R) in $Q^{\psi}(s, a)$, i.e. the expected discounted reinforcement of taking action a in state s. The function $U_i(s, a)$ is critical to Q-learning, as it determines the behavior and performance of δ_i (Kim, 2018).

By employing Q-learning, the main goal of the USN routing algorithm is to deliver the packet to the sink node with the maximum payoff. If the mth neighboring node is selected as a relay node at time period t, i.e. $a_i(t) = a_i^m$, $U_i(\cdot)$ can be defined as (Kim, 2018):

$$U_i\left(s_i(t), a_i(t) = a_i^m\right) = \mathcal{R}_t\left(s_i(t), a_i^m\right) = \left(\frac{1}{\mathcal{PC}_m^t + L_c(i,j)}\right) \times (-\vartheta)^{\mathcal{Q}(a_i^m)}$$

(12.5.10)

$$\text{s.t.}, \mathcal{Q}\left(a_i^m\right) = \begin{cases} 0, \mathit{ifthispacketissuccessfullydeliveredbyaction}\, a_i^m \\ 1, \mathit{otherwise} \end{cases}$$

where $\mathcal{PC}_m^t$ is the $\mathcal{PC}$ value of the mth neighboring node at time t, and ϑ is a penalty factor for packet delivery failure. According to $U_i\left(s_i(t), a_i(t) = a_i^m\right)$, the corresponding $\mathcal{U}_i^m$ value in $\mathbb{U}_i[m]$ is updated; $\mathbb{U}_i[1 \ldots m] = \left\{\mathcal{U}_i^1(\cdot), \mathcal{U}_i^2(\cdot), \ldots \mathcal{U}_i^m(\cdot)\right\}$ is a vector specifying the payoffs for δ_i. $\mathbb{U}_i[m]$ is updated as follows (Kim, 2018):

$$\mathbb{U}_i[m] = \mathcal{U}_i^m(t+1) = \mathfrak{J} + \left(\frac{1}{\eta} \times \left\{U_i\left(s_i(t), a_i(t) = a_i^m\right) - \mathfrak{J}\right\}\right)$$

(12.5.11)

$$\text{s.t.}, \mathfrak{J} = \frac{1}{t} \times \left(\sum_{e=1}^{t} U_i\left(s_i(e), a_i(e)\right)\right)$$

where η is a weight parameter for adjusting the current impact. At each time period, δ_i chooses action a_i to maximize $U_i(\cdot)$, and thus maximize V, i.e., $V^*(s) = \max_a Q^*(s, a)$ according to equation (12.5.5). A greedy Q-learning algorithm always chooses the action with the highest Q value, which leads to packets being relayed to the sink node via the best route. To develop an effective Q-learning algorithm, the *UNDR* scheme must define $P^{a_i}_{s_t \rightarrow s_{t+1}}$ in equations (12.5.4) and (12.5.5). Defining $P^{a_i}_{s_t \rightarrow s_{t+1}}$ is a probability decision problem. Through the multi-sensor routing game model, each sensor node adaptively learns the current USN state to dynamically determine $P^{a_i}_{s_t \rightarrow s_{t+1}}$. Suppose there is a finite set of neighboring sensors $\{\delta_1 \ldots \delta_m\}$ chosen by δ_i at game iteration t. According to equations (12.5.10) and (12.5.11), each individual node obtains the U value, and updates $\mathbb{U}[\cdot]$. Based on this information, the *UNDR* scheme can determine $P^{a_i}_{s_t \rightarrow s_{t+1}}$ in equations (12.5.4) and (12.5.5). Let $P_i(t) = \{P^{a_i^1}_{s_t \rightarrow s_{t+1}}, \ldots, P^{a_i^m}_{s_t \rightarrow s_{t+1}}\}$ be the probability distribution of the actions selected by δ_i. The values of $P_i(t)$ are defined as (Kim, 2018):

$$P^{a_i^m}_{s_t \rightarrow s_{t+1}} = P_i^m\left(S' \middle| S, a_i^m(t)\right) = \frac{\text{EXP}\left(\mathcal{U}_i^m(t)\right)}{\sum_{l \in \mathcal{N}_i} \text{EXP}\left(\mathcal{U}_i^l(t)\right)}, \text{s.t.}, \text{EXP}(x) = e^x$$

(12.5.12)

12.5.6 Summary

USNs have recently garnered significant attention due to their advanced capabilities in acoustic monitoring and resource discovery. These networks support a diverse range of applications, from long-term environmental observations to addressing immediate threats. The *UNDR* scheme introduces a learning-based game approach to tackle routing challenges in USNs, with the primary objective of developing an efficient routing algorithm suited for the complex underwater communication environment. Considering the distinct characteristics of USNs, the *UNDR* scheme incorporates an interactive feedback mechanism and formulates a routing game model based on the well-established Q-learning framework. Within the *UNDR* scheme, sensor nodes continuously refine their routing strategies and dynamically make routing decisions to enable adaptive and opportunistic data transmission. By leveraging online self-monitoring and distributed learning techniques, the *UNDR* scheme adjusts in real time to the evolving USN environment, effectively maximizing overall network performance and expected benefits (Kim, 2018).

12.6 Ultra-Dense Network Control (*UDNC*) scheme

Future 5G cellular networks are being developed to accommodate the rapid surge in mobile traffic. Among emerging wireless system paradigms, the ultra-dense network (UDN) is recognized as a key technology for achieving the next-generation network capacity. However, one of the primary challenges in UDN operations is designing efficient bandwidth resource allocation algorithms while ensuring QoS provisioning. Recently, S. Kim introduced the *Ultra-Dense Network Control* (*UDNC*) scheme, which leverages efficient and innovative methodologies to address these challenges (Kim, 2019b). Rooted in cooperative game theory, the *UDNC* scheme dynamically allocates limited bandwidth resources among small cell operators. The proposed approach is formulated as a dual-game model, employing two distinct game-theoretic solutions. Initially, each cell operator estimates its bandwidth requirements using an *iterative Nash bargaining solution*. Subsequently, the limited bandwidth is allocated to each operator based on the *τ-value*. To enhance computational efficiency, the *UDNC* scheme is implemented hierarchically, incorporating cascade interactions to streamline the resource allocation process.

12.6.1 Development motivation

By 2020, the number of mobile devices was projected to reach 75 billion units, with wireless traffic surpassing 24.3 exabytes per month. The increasing prevalence of data-intensive applications such as mobile social media, 3D virtual reality streaming, wearable devices, and vehicular video is shaping a new era in mobile communications (Chen, 2018; Cisco, 2017). As mobile data traffic continues to grow exponentially, wireless networks are facing unprecedented challenges. To accommodate this surge in demand, next-generation wireless systems, including 5G networks, must be designed to deliver high-quality mobile services. Achieving this objective requires a fundamental transformation in wireless cellular architecture. Notably, network densification and cell splitting have emerged as two of the most effective strategies for increasing capacity and enhancing user experience (Chen, 2018; Kamel, 2016; Kim, 2019b; Shi, 2018).

The concept of ultra-dense networks (UDNs) offers a new vision for the evolution of cellular networks. As a fundamental paradigm shift, UDNs aim to position network access points as close as possible to end users. This is primarily accomplished through the dense deployment of small cells in high-traffic areas. When integrated with macrocells, low-power small cells, including microcells and picocells, significantly enhance network coverage and capacity. In addition, dense small-cell deployment alleviates congestion on macrocells by offloading a substantial portion of wireless data traffic, particularly in indoor environments, where more than 80% of total data consumption occurs. Based on the principle of partitioning space into smaller cell units, network capacity increases sub-linearly with growing cell density (Chen, 2018; Kamel, 2016; Kim, 2019b; Shi, 2018).

Despite the growing interest in UDNs, several control challenges remain for real-world implementation. Among them, efficient bandwidth management across multiple cells is a critical factor in optimizing UDN performance. To maximize system capacity, UDNs employ a multi-layer heterogeneous architecture, where macrocells coexist with a dense layer of small cells that can be deployed in various locations. This structural design creates opportunities for enhanced inter-cell cooperation, improving bandwidth utilization. Given the constraints of limited bandwidth resources, both academic and industry researchers have devoted considerable efforts to developing bandwidth management algorithms that also uphold high QoS standards (Jang, 2013; Kim, 2010, 2019b; Marabissi, 2018; Pande, 2013).

The primary objective of the *UDNC* scheme is to establish an innovative bandwidth control mechanism tailored for diverse cell types within UDNs. To achieve this goal, advanced bandwidth estimation and allocation algorithms are developed to enhance UDN efficiency. However, the inherently complex and dynamic nature of UDN communication environments results in significant spatial and temporal fluctuations in traffic load. Notably, as cell sizes decrease, traffic distribution becomes increasingly unbalanced. Consequently, managing load balancing across different-sized cells is a crucial aspect of UDN bandwidth management. Traditional bandwidth allocation methods prove ineffective in addressing these unique challenges. Therefore, a novel control paradigm is essential for the adaptive development of bandwidth control algorithms in UDN environments (Kim 2019b; Rizvi, 2015; Yu, 2018).

To address these challenges, the *UDNC* scheme is formulated as a dual-game model based on cooperative game theory principles. For bandwidth estimation within each cell, the *Nash Bargaining Solution* (*NBS*) is iteratively applied to predict bandwidth requirements for the subsequent time period. Meanwhile, for bandwidth distribution across different cells, the *τ-value* mechanism is utilized to allocate the limited bandwidth resources effectively. These two cooperative game approaches are integrated into a dual-game framework, ensuring interactive feedback mechanisms that provide mutual benefits. Throughout the two-phase game process, individual cells leverage the Nash bargaining strategy to make autonomous control decisions that not only maximize their own performance but also contribute to achieving an optimal global UDN system outcome (Kim, 2019b).

12.6.2 Contribution

As service demands continue to grow, developing flexible and adaptive resource control mechanisms for UDNs has become a critical challenge. The primary goal of the *UDNC* scheme is to optimize UDN system performance while maintaining QoS. The *UDNC* scheme primarily focuses on designing bandwidth estimation and allocation algorithms tailored for heterogeneous UDN cell environments. By integrating two cooperative game-theoretic solutions, the *UDNC* scheme constructs a fair and efficient dual-game model, ensuring a well-balanced performance across various network cells. While multiple UDN bandwidth management protocols have been introduced, limited research has explored the intricate interactions between individual cells while maintaining globally optimal system properties (Kim, 2019b).

12.6.3 Bandwidth estimation algorithm in a UDN

The *UDNC* scheme is designed for a multitier ultra-dense cellular network, where a dense deployment of various small cells operates within the coverage of a larger macrocell. In this UDN system infrastructure, the geographic region is divided into three distinct zones: global, local, and hotspot areas. Each of these areas is managed by different types of access point operators - macrocells for global coverage, microcells for local coverage, and picocells for hotspot areas. These access point operators are interconnected via high-speed communication links, facilitating the seamless exchange of control information. The network allocates an orthogonal bandwidth spectrum to each cell operator, ensuring efficient resource distribution. Within this hierarchical UDN system, heterogeneous cells coexist while overlapping coverage areas. Mobile users are randomly distributed across different cell regions and dynamically select the most suitable cell operator based on adaptability and network conditions (Liu, 2016; Shi, 2018). The overall architecture of the UDN system is illustrated in Figure 3. (Kim, 2019b).

Figure 3. Multitier ultra-dense cellular network structure (Kim, 2019b)

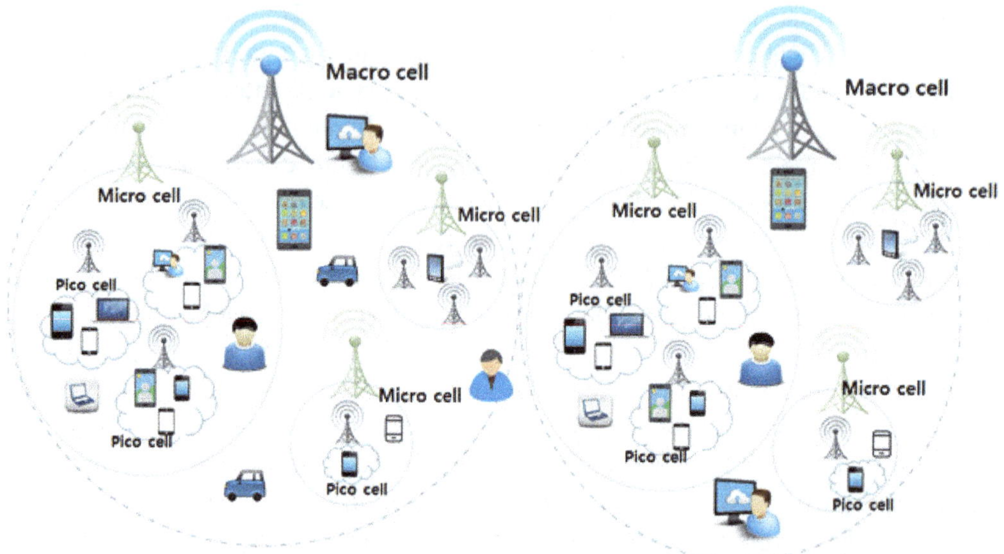

In cooperative game theory, players form binding agreements, relying on the possibility of external enforcement to ensure cooperative behavior. The primary focus in cooperative games is determining how payoffs are distributed among players, which contrasts with non-cooperative game theory. While non-cooperative game theory aims to predict individual players' strategies, payoffs, and Nash equilibria, cooperative game theory examines coalitions, or groups of players, that collaborate to achieve shared benefits. Each coalition has a set of feasible payoffs that can be distributed among its members, and the final allocation determines the rewards received by individual players. Depending on the method used to distribute payoffs, different solution concepts emerge within cooperative game theory (Kim, 2014b, 2019b).

Effective bandwidth control is a critical factor in maximizing UDN system performance. Typically, the bandwidth control process is divided into two key phases: bandwidth estimation and bandwidth allocation. In the *UDNC* scheme, the *NBS* and the *τ-value* are utilized to address these challenges. These cooperative game-theoretic approaches ensure a fair distribution of total surplus among participating players (cell operators) while adhering to fundamental rationality principles. Specifically, in the *UDNC* scheme, the *NBS* is applied to develop the bandwidth estimation algorithm, while the *τ-value* is employed to design the bandwidth allocation algorithm, enabling efficient and adaptive resource distribution (Kim, 2019b).

Efficient distribution of limited wireless bandwidth among overlapping cell operators requires precise bandwidth estimation for the upcoming time interval. Various probing-based bandwidth estimation techniques have been developed; however, temporal and spatial traffic fluctuations in UDN cell areas pose significant challenges to design an effective estimation algorithm. To address this issue, the *UDNC* scheme employs an *iterative Nash bargaining* model to develop a novel bandwidth estimation algorithm. The *NBS*, a widely recognized method in cooperative game theory, is utilized to ensure a mutually beneficial and optimal bandwidth allocation among network operators (Kim, 2019b).

Each small cell operator, i.e., microcell access point (MAP) or picocell access point (PAP), has two factors, i.e., cost and QoS, to define its own payoff. Traditionally, a sigmoid function has been used to represent the QoS with respect to the allocated bandwidth. For the ith PAP's QoS function under the jth MAP covering area at time t, $\left(\mathscr{F}Q_{PAP_i}^{MAP_j}\right)$ is defined as follows (Kim, 2019b):

$$\mathscr{F}Q_{PAP_i}^{MAP_j}\left(\mathscr{R}^t_{PAP_i}\right) = \frac{\varrho}{\eta + \left(\psi \times e^{-\gamma \times \left(\left(\mathscr{R}^t_{PAP_i} - \mathfrak{A}^t_{PAP_i}\right)/\mathfrak{W}\right)}\right)} \tag{12.6.1}$$

$$\text{s.t.}, \mathfrak{W} = \min\left(\mathscr{R}^t_{PAP_i}, \mathfrak{A}^t_{PAP_i}\right)$$

where η, γ, ψ and ϱ are the pre-determined parameters to shape the QoS function; γ affects the slope of the curve and η, ψ and ϱ affect the range of the QoS function. $\mathscr{R}^t_{PAP_i}$ and $\mathfrak{A}^t_{PAP_i}$ denote the amount of actively allocated bandwidth and the summation of the users' bandwidth requirements for the PAP_i at time t, respectively. When $\mathscr{R}^t_{PAP_i}$ is less than $\mathfrak{A}^t_{PAP_i}$, the QoS function is concave, which indicates that the PAP_i's demand for the bandwidth resource becomes stronger. Conversely, when $\mathscr{R}^t_{PAP_i}$ is greater than $\mathfrak{A}^t_{PAP_i}$, the QoS function is convex, which indicates that the PAP_i's demand for the bandwidth resource is not as strong (Kim, 2015,2019b).

The cost function of PAP_i $\left(\mathscr{F}C_{PAP_i}^{MAP_j}\left(\mathscr{R}^t_{PAP_i}\right)\right)$ defines the instantaneous expense for $\mathscr{R}^t_{PAP_i}$. This is a proportional function of the amount of bandwidth allocation and is given by $K \times \left(\mathscr{R}^t_{PAP_i}\right)^q$, where K and q are estimation parameters. Because the utilities of the cost and QoS are clearly conflicting, the main goal is to maximize the $\mathscr{F}Q$ function while minimizing the $\mathscr{F}C$ function. From the viewpoint of PAP_i, it is necessary to determine the most suitable amount of bandwidth. The *UDNC* scheme adopts the *NBS* methodology to estimate a desired bandwidth amount for each small cell. Based on Nash theory, this leads to a unique optimal solution (Benkacem, 2018). Therefore, the expected bandwidth amount (χ) for PAP_i is obtained through the following *NBS* that balances the utilities of cost and QoS objectives to determine a trade-off solution (Kim, 2019b).

$$\max_{\chi} \prod_{i=1}^{2} \mathbb{F}_i(\chi), \text{ s.t.}, \begin{cases} \mathbb{F}_1(\chi) = \mathscr{F}\, Q_{PAP_i}^{MAP_j}(\chi) - \mathscr{F}\, Q_{PAP_i}^{MAP_j}(\Omega) \\ \mathbb{F}_2(\chi) = \dfrac{\mathscr{F}\, C_{PAP_i}^{MAP_j}(\chi) - \mathscr{F}\, C_{PAP_i}^{MAP_j}(\Omega)}{\chi} \end{cases}$$

(12.6.2)

where $\mathscr{F}\, Q_{PAP_i}^{MAP_j}(\Omega)$ and $\mathscr{F}\, C_{PAP_i}^{MAP_j}(\Omega)$ are the elements of threat point, which represents the pair of payoffs whereby the two functions fail to achieve a cooperative agreement (Benkacem, 2018). At each time period, the χ value is obtained according to (12.6.2). Finally, PAP_i's bandwidth requirement for the t time interval $\left(\mathscr{R}_Q^t\right)$ is estimated as follows (Kim, 2019b):

$$\mathscr{R}_Q^t = \left(\omega \times \mathscr{R}_Q^{t-1}\right) + ((1-\omega) \times \chi)$$

(12.6.3)

where χ is obtained from the *NBS* using Equation (12.6.2) and the parameter ω controls the relative weights given to recent and past bandwidth estimation histories in the PAP_i decision. Therefore, the bandwidth estimation algorithm is designed using the iterative *NBS* model with a timely learning approach. In the UDNs, all PAPs can estimate their bandwidth requirements according to (12.6.1)-(12.6.3). From the viewpoint of MAPs, each PAP can be assumed to be an individual user. Therefore, MAPs also estimate their bandwidth requirements in the same manner as the PAP (Kim, 2019b).

12.6.4 τ-value based bandwidth allocation algorithm

In 1981, S.H. Tijs introduced the concept of the *τ-value*, a unique solution framework characterized by effective axiomatic properties (Yanovskaya, 2010). The fundamental principle behind the *τ-value* is that each player's dividend is allocated proportionally to their marginal contribution within the grand coalition (Tijs, 1987). The study of the *τ-value* initially emerged as an extension of classical cooperative game theory into interval games - a model of cooperation in which only the upper and lower bounds of coalition payoffs are precisely known. Interval games are particularly relevant in real-world scenarios where players must collaborate while having certainty only about their minimum and maximum potential gains. Consequently, the *τ-value* serves as an effective mechanism for determining how payoffs are equitably distributed among players (Branzei, 2008; Kim, 2019b; Tijs, 1987; Yanovskaya, 2010).

To define the *τ-value*, we consider *n*-person cooperative games. Let $\mathbb{N} = \{1,2,\ldots,n\}$ be the player set and a real-valued function $v:2^{\mathbb{N}} \rightarrow \mathbb{R}$ with $v(\varnothing) = 0$. For each coalition $S \in 2^{\mathbb{N}}$, $v(S)$ is the worth of S; it can be interpreted as the reward, which players in S can obtain by working together. The *τ-value* characterizes unique mapping $\mathbf{T} = [T_1 \ldots T_i \ldots T_n]$, where T_i is the payoff given to player i by the *τ-value* **T**. Therefore, the *τ-value* provides a solution on how to divide $v(\mathbb{N})$ among players when the grand coalition forms. Based on the $v(\cdot)$ function, the marginal vector $\mathscr{M}[\mathbb{N}|v]$, which is the vector with player $i \in \mathbb{N}$, is defined as follows (Branzei, 2008; Kim, 2019b; Núñez, 2003; Tijs, 1987; Yanovskaya, 2010):

$$\mathscr{M}[\mathbb{N}|v] = (\mathscr{M}(1|v) \ldots \mathscr{M}(i|v) \ldots \mathscr{M}(n|v)),$$

(12.6.4)

$$\text{s.t.}, \mathscr{M}(i|v) = v(\mathbb{N}) - v(\mathbb{N} - \{i\})$$

In (4), $\mathscr{M}(i|v)$ measures the payoff increase when player i joins the coalition of all the other players. From the viewpoint of player i, $\mathscr{M}(i|v)$ is an idealistic payoff for player i; it can be interpreted as an upper bound dividend. If player i is willing to pay the idealistic payoffs of other players in a coalition, that coalition is attractive for those players. If S is such a coalition, and player i pays the other players' idealistic payoffs, then player i can obtain the remaining payoff (Branzei, 2008; Kim, 2019b; Núñez, 2003; Tijs, 1987; Yanovskaya, 2010).

$$\mathscr{R}(i|S;v) = v(S) - \sum_{j\in S-\{i\}} \mathscr{M}[j|v] \tag{12.6.5}$$

where $\mathscr{R}(i|S;v)$ is the remainder for player i in coalition S. As opposed to the concept of idealistic payoff, $\mathscr{R}(i|S;v)$ is a lower bound for player i's dividend in the grand coalition. As a rational game player, player i can select a specific coalition with the maximal remainder $(m(i|v))$ for him among all possible coalitions. The payoff for player i in such a coalition is called the minimal right and is given by (Branzei, 2008; Kim, 2019b; Núñez, 2003; Tijs, 1987; Yanovskaya, 2010);

$$m(i|v) = \max_{i\in S} \mathscr{R}(i|S;v) \tag{12.6.6}$$

According to (12.6.6), $m[i \in \mathbb{N}|v] = (m(1|v)\ldots m(i|v)\ldots m(n|v))$ is defined as the minimal right vector. Based on the upper and lower payoff vectors, the *τ-value* **T** vector can be defined as the unique efficient payoff vector on the line segment between the $\mathscr{M}[\mathbb{N}|v]$ and $m[\mathbb{N}|v]$ vectors (Núñez, 2003; Tijs, 1987; Yanovskaya, 2010). Formally,

$$T = (\delta \times m[\mathbb{N}|v]) + ((1-\delta) \times \mathscr{M}[\mathbb{N}|v]) \tag{12.6.7}$$

$$\text{s.t.}, \begin{cases} m(i|v) \le M(i|v) \\ \sum_{i\in\mathbb{N}} m(i|v) \le v(\mathbb{N}) \le \sum_{i\in\mathbb{N}} \mathscr{M}(i|v) \\ \delta isuniquelydeterminedby \sum_{i\in\mathbb{N}} T_i = v(\mathbb{N}) \end{cases}$$

Based on the definition, the *τ-value* is essentially a compromise value between $\mathscr{M}[\mathbb{N}|v]$ and $m[\mathbb{N}|v]$ for the cooperative assignment (Branzei, 2008; Kim, 2019b; Núñez, 2003; Tijs, 1987; Yanovskaya, 2010).

In general, the resource distribution problem for small cells leads to the similar conflict situation as in the money division problem for n claimants. Therefore, the *UDNC* scheme adopts the bankruptcy game model to estimate the payoff of each small cell's coalition. The bankruptcy game assumes that a company becomes bankrupt, and the total amount of debt owed by the bankrupt company is greater than the company-owned assets. The remaining assets must be divided among the debtors, and debtors choose to form independent coalitions to capture more benefits. The bankruptcy game is defined as $G(\mathcal{N}, v)$, where $\mathcal{N} = \{\mathcal{D}_1 \ldots \mathcal{D}_i \ldots \mathcal{D}_n\}$ represents the company debtors and a coalition S is defined as a subset of N where $S \subset \mathcal{N}$. Therefore, there are 2^n possible coalitions with n debtors. v is the characteristic function that associates to each coalition (S) of claimants: $2^n \rightarrow \mathbb{R}$ where $v(\varnothing) = 0$ (Jia, 2018; Kim, 2014b, 2019b; Niyato, 2006):

$$v(S) = \max(0, E - \sum_{\mathcal{D}_i \in \mathcal{N} \backslash S} \mathcal{R}_{\mathcal{D}_i}) \tag{12.6.8}$$

$$\text{s.t.}, S \subset \mathcal{N} \backslash \{\varnothing\}, E < \sum_{\mathcal{D}_i \in \mathcal{N}} \mathcal{R}_{\mathcal{D}_i} \text{ and } \mathcal{R}_{\mathcal{D}_i} \in R$$

where $E \geq 0$ is an asset that must be divided among the debtor set N and $\mathcal{R}_{\mathcal{D}_i}$ is the $\mathcal{D}_i$'s demand. $R \in \mathbb{R}_+^{|\mathcal{N}|}$ is the claim vector of all debtors. In the bankruptcy game, the solution $\mathcal{S} = \{\mathcal{S}_{\mathcal{D}_1} \ldots \mathcal{S}_{\mathcal{D}_i} \ldots \mathcal{S}_{\mathcal{D}_n}\}$ is stable only when each debtor satisfies the current distribution condition. The following are the constraints of the solution (Kim, 2019b):

$$(i) \sum_{\mathcal{D}_i \in \mathcal{N}} \mathcal{S}_{\mathcal{D}_i} = E, \ (ii)\ v(\{\mathcal{D}_i\}) \leq \mathcal{S}_{\mathcal{D}_i}, \ (iii) \sum_{\mathcal{D}_i \in S} \mathcal{S}_{\mathcal{D}_i} \geq v(S) \tag{12.6.9}$$

Constraint (i) ensures that the total allocated assets do not exceed the resources available to the debtors. Constraint (ii) states that the assigned assets must be at least equal to the amount the debtor would receive if they did not participate in any coalition. Constraint (iii) implies that a debtor cannot form a new coalition to secure a higher payoff than what is already allocated (Jia, 2018; Kim, 2014b, 2019b; Niyato, 2006).

By incorporating the *τ-value* and the bankruptcy game model, the *UDNC* scheme redefines the small cell bandwidth allocation problem. In the *UDNC* scheme, multiple microcells are overlaid by a macrocell, while multiple picocells are overlaid by a microcell. Due to this multilayer UDN architecture, the bandwidth allocation process follows a hierarchical structure. At regular intervals, each PAP estimates its bandwidth requirements using the bandwidth estimation algorithm and reports this data to its corresponding MAP. The MAP, in turn, gathers the bandwidth requests from all PAPs within its coverage area and allocates bandwidth accordingly. However, if the available bandwidth is insufficient to meet all PAPs' demands, the allocation follows the *τ-value* principle. To determine the *τ-value* for each small cell, the payoffs of all possible coalitions (S) are estimated based on the bankruptcy game framework. This process is applied recursively - each MAP estimates its own bandwidth requirement and submits the request to its corresponding macrocell, treating each PAP as an individual user in the process. The macrocell then distributes the available bandwidth among its MAPs following the same allocation methodology. Throughout each time period, these cascading interactions among heterogeneous cell operators are cooperatively managed using a dual-game approach. This enables cell operators to efficiently determine the most profitable bandwidth allocation strategy in real time (Kim, 2019b).

12.6.5 Summary

The *UDNC* scheme provides a solution to the bandwidth management challenge in UDN operations by introducing a hierarchical interactive control mechanism based on a dual-game approach. To accommodate diverse service demands, the *UDNC* scheme effectively integrates bandwidth estimation and allocation algorithms, which are formulated using *iterative Nash bargaining* and the τ-*value*, respectively. These two control mechanisms dynamically interact within the dual-game model, enabling a mutually beneficial 'win-win' solution in the ever-changing UDN environment (Kim, 2019b).

REFERENCES

Almarzouqi, A., Aburayya, A., & Salloum, S. A. (2022). Prediction of user's intention to use metaverse system in medical education: A hybrid SEM-ML learning approach. *IEEE Access : Practical Innovations, Open Solutions, 10*, 43421–43434. DOI: 10.1109/ACCESS.2022.3169285

Benkacem, I., Taleb, T., Bagaa, M., & Flinck, H. (2018). Optimal VNFs placement in CDN slicing over multi-cloud environment. *IEEE Journal on Selected Areas in Communications, 36*(3), 616–627. DOI: 10.1109/JSAC.2018.2815441

Branzei, R., Dimitrov, D., & Tijs, S. (2008). *Models in cooperative game theory* (Vol. 556). Springer Science & Business Media.

Busoniu, L., Babuska, R., & De Schutter, B. (2008). A comprehensive survey of multiagent reinforcement learning. *IEEE Transactions on Systems, Man, and Cybernetics. Part C, Applications and Reviews, 38*(2), 156–172. DOI: 10.1109/TSMCC.2007.913919

Carli, R., Dotoli, M., & Pellegrino, R. (2016). A hierarchical decision-making strategy for the energy management of smart cities. *IEEE Transactions on Automation Science and Engineering, 14*(2), 505–523. DOI: 10.1109/TASE.2016.2593101

Chang, L., Zhang, Z., Li, P., Xi, S., Guo, W., Shen, Y., Xiong, Z., Kang, J., Niyato, D., Qiao, X., & Wu, Y. (2022). 6G-enabled edge AI for metaverse: Challenges, methods, and future research directions. *Journal of Communications and Information Networks, 7*(2), 107–121. DOI: 10.23919/JCIN.2022.9815195

Chen, K. C. (2012). Machine-to-machine communications for healthcare. *Journal of Computing Science and Engineering : JCSE, 6*(2), 119–126. DOI: 10.5626/JCSE.2012.6.2.119

Chen, M., & Hao, Y. (2018). Task offloading for mobile edge computing in software defined ultra-dense network. *IEEE Journal on Selected Areas in Communications, 36*(3), 587–597. DOI: 10.1109/JSAC.2018.2815360

Chun, Y., & Thomson, W. (1992). Bargaining problems with claims. *Mathematical Social Sciences, 24*(1), 19–33. DOI: 10.1016/0165-4896(92)90003-N

Cisco. (2017, February). Cisco visual networking index: Global mobile data traffic forecast update 2016–2021 white paper. Cisco. http://www.cisco.com/c/en/us/solutions/collateral/ serviceprovider/visual-networking-index-vni/mobile-white-paper-c11-520862.pdf

Conley, J. P., McLean, R., & Wilkie, S. (2008). Axiomatic Foundations for Compromise Theory. Working Paper.: Available at: https://www.researchgate.net/publication/241537525

Cui, T., Wang, Y., Nazarian, S., & Pedram, M. (2014, February). An electricity trade model for microgrid communities in smart grid. In *ISGT 2014* (pp. 1–5). IEEE.

Dighriri, M., Alfoudi, A. S. D., Lee, G. M., & Baker, T. (2016, August). Data traffic model in machine to machine communications over 5G network slicing. In *2016 9th International Conference on Developments in eSystems Engineering (DeSE)* (pp. 239-244). IEEE.

El-Sayed, H., Sankar, S., Prasad, M., Puthal, D., Gupta, A., Mohanty, M., & Lin, C. T. (2017). Edge of things: The big picture on the integration of edge, IoT and the cloud in a distributed computing environment. *IEEE Access : Practical Innovations, Open Solutions, 6*, 1706–1717. DOI: 10.1109/ACCESS.2017.2780087

Galindo-Serrano, A., & Giupponi, L. (2010). Distributed Q-learning for aggregated interference control in cognitive radio networks. *IEEE Transactions on Vehicular Technology, 59*(4), 1823–1834. DOI: 10.1109/TVT.2010.2043124

Gerber, A. (2005). Reference functions and solutions to bargaining problems with and without claims. *Social Choice and Welfare, 24*(3), 527–541. DOI: 10.1007/s00355-004-0315-5

Ghoreyshi, S. M., Shahrabi, A., & Boutaleb, T. (2016, March). An opportunistic void avoidance routing protocol for underwater sensor networks. In *2016 IEEE 30th international conference on advanced information networking and applications (AINA)* (pp. 316-323). IEEE. DOI: 10.1109/AINA.2016.96

Han, Y., Niyato, D., Leung, C., Miao, C., & Kim, D. I. (2022, May). A dynamic resource allocation framework for synchronizing metaverse with iot service and data. In *ICC 2022-IEEE International conference on Communications* (pp. 1196–1201). IEEE. DOI: 10.1109/ICC45855.2022.9838422

Hatziargyriou, N. (Ed.). (2014). *Microgrids: architectures and control.* John Wiley & Sons.

Herrero, C. (1997). Endogenous reference points and the adjusted proportional solution for bargaining problems with claims. *Social Choice and Welfare, 15*(1), 113–119. DOI: 10.1007/s003550050095

Hieu, C. T., & Hong, C. S. (2010). A Connection Entropy-based Multi-Rate Routing Protocol for Mobile Ad Hoc Networks. *Journal of Computing Science and Engineering : JCSE, 4*(3), 225–239. DOI: 10.5626/JCSE.2010.4.3.225

Hu, T., & Fei, Y. (2010). QELAR: A machine-learning-based adaptive routing protocol for energy-efficient and lifetime-extended underwater sensor networks. *IEEE Transactions on Mobile Computing, 9*(6), 796–809. DOI: 10.1109/TMC.2010.28

Iosifidis, G., Gao, L., Huang, J., & Tassiulas, L. (2014). Incentive mechanisms for user-provided networks. *IEEE Communications Magazine, 52*(9), 20–27. DOI: 10.1109/MCOM.2014.6894448

Jang, I., Pyeon, D., Kim, S., & Yoon, H. (2013). A survey on communication protocols for wireless sensor networks. *Journal of Computing Science and Engineering : JCSE, 7*(4), 231–241. DOI: 10.5626/JCSE.2013.7.4.231

Jeon, M., Kim, S. K., Yoon, J. H., Lee, J., & Yang, S. B. (2014). A direction entropy-based forwarding scheme in an opportunistic network. *Journal of Computing Science and Engineering : JCSE, 8*(3), 173–179. DOI: 10.5626/JCSE.2014.8.3.173

Jia, Y., Tian, H., Fan, S., Zhao, P., & Zhao, K. (2018, April). Bankruptcy game based resource allocation algorithm for 5G Cloud-RAN slicing. In *2018 IEEE Wireless Communications and Networking Conference (WCNC)* (pp. 1-6). IEEE.

Jiang, L., Wang, Q., Song, R., Ye, B., & Dai, J. (2019). A truthful double auction framework for promoting femtocell access. *IEEE Access : Practical Innovations, Open Solutions*, *7*, 34991–35000. DOI: 10.1109/ACCESS.2019.2904548

Kamel, M., Hamouda, W., & Youssef, A. (2016). Ultra-dense networks: A survey. *IEEE Communications Surveys and Tutorials*, *18*(4), 2522–2545. DOI: 10.1109/COMST.2016.2571730

Karos, D., & Rachmilevitch, S. (2019). The midpoint-constrained egalitarian bargaining solution. *Mathematical Social Sciences*, *101*, 107–112. DOI: 10.1016/j.mathsocsci.2019.07.006

Khalili, M. M., Gao, L., Huang, J., & Khalaj, B. H. (2015, April). Incentive design and market evolution of mobile user-provided networks. In *2015 IEEE Conference on Computer Communications Workshops (INFOCOM WKSHPS)* (pp. 498-503). IEEE. DOI: 10.1109/INFCOMW.2015.7179434

Kim, J., Bang, I., Sung, D. K., Yi, Y., & Kim, B. H. (2015, August). Design of a multi-variable QoE function based on the remaining battery energy. In *2015 IEEE 26th Annual International Symposium on Personal, Indoor, and Mobile Radio Communications (PIMRC)* (pp. 976-980). IEEE. DOI: 10.1109/PIMRC.2015.7343439

Kim, K. S., Uno, S., & Kim, M. W. (2010). Adaptive QoS mechanism for wireless mobile network. *Journal of Computing Science and Engineering : JCSE*, *4*(2), 153–172. DOI: 10.5626/JCSE.2010.4.2.153

Kim, S. (2011). Adaptive online power control scheme based on the evolutionary game theory. *IET Communications*, *5*(18), 2648–2655. DOI: 10.1049/iet-com.2011.0093

Kim, S. (2012). Multi-leader multi-follower Stackelberg model for cognitive radio spectrum sharing scheme. *Computer Networks*, *56*(17), 3682–3692. DOI: 10.1016/j.comnet.2012.08.004

Kim, S. (2014a). Adaptive MANET multipath routing algorithm based on the simulated annealing approach. *TheScientificWorldJournal*, *2014*(1), 872526. DOI: 10.1155/2014/872526 PMID: 25032241

Kim, S. (Ed.). (2014b). *Game theory applications in network design*. IGI Global. DOI: 10.4018/978-1-4666-6050-2

Kim, S. (2017). Dynamic position changeable Stackelberg game for user-provided network control algorithms. *EURASIP Journal on Wireless Communications and Networking*, *2017*(1), 1–13. DOI: 10.1186/s13638-017-1004-2

Kim, S. (2018). A better-performing Q-learning game-theoretic distributed routing for underwater wireless sensor networks. *International Journal of Distributed Sensor Networks*, *14*(1), 1550147718754728. DOI: 10.1177/1550147718754728

Kim, S. (2019a). A new triple bargaining game-based energy management scheme for hierarchical smart grids. *IEEE Access : Practical Innovations, Open Solutions*, *7*, 161131–161140. DOI: 10.1109/ACCESS.2019.2951499

Kim, S. (2019b). New dual-game-based cooperative bandwidth control scheme for ultra-dense networks. *Wireless Networks*, *25*(6), 3665–3674. DOI: 10.1007/s11276-019-01961-4

Kim, S. (2021). Cooperative inter-ISP traffic control scheme based on bargaining game approach. *IEEE Access : Practical Innovations, Open Solutions*, *9*, 31782–31791. DOI: 10.1109/ACCESS.2021.3058445

Kim, S. (2024). Data trading, power control and resource allocation algorithms for metaverse platform. *International Journal of Ad Hoc and Ubiquitous Computing*, *46*(3), 181–194. DOI: 10.1504/IJAHUC.2024.140038

Lee, S. C., Jiang, J. W., Chiu, D. M. C., & Lui, J. C. (2008b). Interaction of ISPs: Distributed resource allocation and revenue maximization. *IEEE Transactions on Parallel and Distributed Systems*, *19*(2), 204–218. DOI: 10.1109/TPDS.2007.70714

Lee, S. C., & Lui, J. C. (2008a). On the interaction and competition among Internet service providers. *IEEE Journal on Selected Areas in Communications*, *26*(7), 1277–1283. DOI: 10.1109/JSAC.2008.080924

Li, D., & Jayaweera, S. K. (2014). Distributed smart-home decision-making in a hierarchical interactive smart grid architecture. *IEEE Transactions on Parallel and Distributed Systems*, *26*(1), 75–84. DOI: 10.1109/TPDS.2014.2308204

Liu, S., Yu, Y., Guo, L., Yeoh, P. L., Ni, Q., Vucetic, B., & Li, Y. (2022). Truthful online double auctions for mobile crowdsourcing: An on-demand service strategy. *IEEE Internet of Things Journal*, *9*(17), 16096–16112. DOI: 10.1109/JIOT.2022.3151924

Liu, Y., Peng, M., Shou, G., Chen, Y., & Chen, S. (2020). Toward edge intelligence: Multiaccess edge computing for 5G and Internet of Things. *IEEE Internet of Things Journal*, *7*(8), 6722–6747. DOI: 10.1109/JIOT.2020.3004500

Liu, Y., Wang, Y., Zhang, Y., Sun, R., & Jiang, L. (2016, September). Game-theoretic hierarchical resource allocation in ultra-dense networks. In *2016 IEEE 27th Annual International Symposium on Personal, Indoor, and Mobile Radio Communications (PIMRC)* (pp. 1-6). IEEE. DOI: 10.1109/PIMRC.2016.7794819

Lorenzo, B., Gómez-Cuba, F., García-Rois, J., Gonzalez-Castano, F. J., & Burguillo, J. C. (2015, December). A microeconomic approach to data trading in user provided networks. In *2015 IEEE Globecom Workshops (GC Wkshps)* (pp. 1-7). IEEE. DOI: 10.1109/GLOCOMW.2015.7413997

Mahapatra, C., Moharana, A. K., & Leung, V. C. (2017). Energy management in smart cities based on internet of things: Peak demand reduction and energy savings. *Sensors (Basel)*, *17*(12), 2812. DOI: 10.3390/s17122812 PMID: 29206159

Marabissi, D., Bartoli, G., Fantacci, R., & Micciullo, L. (2018). Energy efficient cooperative multicast beamforming in ultra dense networks. *IET Communications*, *12*(5), 573–578. DOI: 10.1049/iet-com.2017.0618

McAfee, R. P. (1992). A dominant strategy double auction. *Journal of Economic Theory*, *56*(2), 434–450. DOI: 10.1016/0022-0531(92)90091-U

Morello, R., Mukhopadhyay, S. C., Liu, Z., Slomovitz, D., & Samantaray, S. R. (2017). Advances on sensing technologies for smart cities and power grids: A review. *IEEE Sensors Journal*, *17*(23), 7596–7610. DOI: 10.1109/JSEN.2017.2735539

Mustika, I. W., & Nurcahyani, I. (2015, December). Proportional fairness with adaptive bandwidth allocation for video service in downlink LTE. In *2015 IEEE International Conference on Communication, Networks and Satellite (COMNESTAT)* (pp. 54-59). IEEE. DOI: 10.1109/COMNETSAT.2015.7434297

Navidi, A., & Khatami, F. A. S. (2017). Energy management and planning in smart cities. *CIRED 24, 2017*(1), 2723-2725.

Niyato, D., & Hossain, E. (2006, June). A cooperative game framework for bandwidth allocation in 4G heterogeneous wireless networks. In *2006* [). IEEE.]. *IEEE International Conference on Communications : [proceedings]. IEEE International Conference on Communications*, *9*, 4357–4362.

Núñez, M., & Rafels, C. (2003). The assignment game: The τ-value. *International Journal of Game Theory*, *31*, 411–422. DOI: 10.1007/s001820300127

Pan, J., & McElhannon, J. (2017). Future edge cloud and edge computing for internet of things applications. *IEEE Internet of Things Journal*, *5*(1), 439–449. DOI: 10.1109/JIOT.2017.2767608

Pande, A., Ramamurthi, V., & Mohapatra, P. (2013). Quality-oriented video delivery over LTE. *Journal of Computing Science and Engineering : JCSE*, *7*(3), 168–176. DOI: 10.5626/JCSE.2013.7.3.168

Pecherskii, S. L. (2009). Proportionality in bargaining games: Status quo-proportional solution and consistency. *Contributions to Game Theory and Management*, *2*(0), 334–343.

Raazi, S. M. K. U. R., & Lee, S. (2010). A survey on key management strategies for different applications of wireless sensor networks. *Journal of Computing Science and Engineering : JCSE*, *4*(1), 23–51. DOI: 10.5626/JCSE.2010.4.1.023

Reiter, M., & Steinberg, R. (2011). Congestion-dependent pricing and forward contracts for complementary segments of a communication network. *IEEE/ACM Transactions on Networking*, *20*(2), 436–449. DOI: 10.1109/TNET.2011.2160997

Rizvi, S., Karpinski, K., & Razaque, A. (2015). Novel architecture of self-organized mobile wireless sensor networks. *Journal of Computing Science and Engineering : JCSE*, *9*(4), 163–176. DOI: 10.5626/JCSE.2015.9.4.163

Segal-Halevi, E., Hassidim, A., & Aumann, Y. (2018, April). Muda: A truthful multi-unit double-auction mechanism. *Proceedings of the AAAI Conference on Artificial Intelligence*, *32*(1), •••. DOI: 10.1609/aaai.v32i1.11450

Sergei, P. (2007). Proportionality in NTU games: Proportional excess, nucleolus and prenucleolus. *Contributions to Game Theory and Management*, *1*, 394–412.

Shah, P. M., Ullah, I., Khan, T., Hussain, M. S., Khan, Z. A., Qasim, U., & Javaid, N. (2016, March). MobiSink: Cooperative routing protocol for underwater sensor networks with sink mobility. In *2016 IEEE 30th international conference on advanced information networking and applications (AINA)* (pp. 189-197). IEEE.

Shao, X., Asaeda, H., Dong, M., & Ma, Z. (2018). Cooperative inter-domain cache sharing for information-centric networking via a bargaining game approach. *IEEE Transactions on Network Science and Engineering*, *6*(4), 698–710. DOI: 10.1109/TNSE.2018.2871021

Shi, Y., Zhang, J., Chen, W., & Letaief, K. B. (2018). Generalized sparse and low-rank optimization for ultra-dense networks. *IEEE Communications Magazine*, *56*(6), 42–48. DOI: 10.1109/MCOM.2018.1700472

Shrimali, G., Akella, A., & Mutapcic, A. (2009). Cooperative interdomain traffic engineering using nash bargaining and decomposition. *IEEE/ACM Transactions on Networking*, *18*(2), 341–352. DOI: 10.1109/TNET.2009.2026748

Singh, P., Kaur, A., Aujla, G. S., Batth, R. S., & Kanhere, S. (2020). Daas: Dew computing as a service for intelligent intrusion detection in edge-of-things ecosystem. *IEEE Internet of Things Journal*, *8*(16), 12569–12577. DOI: 10.1109/JIOT.2020.3029248

Sofia, R., & Mendes, P. (2008). User-provided networks: Consumer as provider. *IEEE Communications Magazine*, *46*(12), 86–91. DOI: 10.1109/MCOM.2008.4689212

Sun, L., Xue, G., & Yu, R. (2022). TAFS: A truthful auction for IoT application offloading in fog computing networks. *IEEE Internet of Things Journal*, *10*(4), 3252–3263. DOI: 10.1109/JIOT.2022.3143101

Tijs, S. H. (1987). An axiomatization of the τ-value. *Mathematical Social Sciences*, *13*(2), 177–181. DOI: 10.1016/0165-4896(87)90054-0

Van Huynh, D., Khosravirad, S. R., Masaracchia, A., Dobre, O. A., & Duong, T. Q. (2022). Edge intelligence-based ultra-reliable and low-latency communications for digital twin-enabled metaverse. *IEEE Wireless Communications Letters*, *11*(8), 1733–1737. DOI: 10.1109/LWC.2022.3179207

Wahid, A., & Kim, D. (2012). An energy efficient localization-free routing protocol for underwater wireless sensor networks. *International Journal of Distributed Sensor Networks*, *8*(4), 307246. DOI: 10.1155/2012/307246

Wang, Y., Li, P., Jiao, L., Su, Z., Cheng, N., Shen, X. S., & Zhang, P. (2016a). A data-driven architecture for personalized QoE management in 5G wireless networks. *IEEE Wireless Communications*, *24*(1), 102–110. DOI: 10.1109/MWC.2016.1500184WC

Wang, Y., & Lin, X. (2015). User-provided networking for QoE provisioning in mobile networks. *IEEE Wireless Communications*, *22*(4), 26–33. DOI: 10.1109/MWC.2015.7224724

Wang, Y., Liu, Y., Zhang, J., Ye, H., & Tan, Z. (2016b). Cooperative store–carry–forward scheme for intermittently connected vehicular networks. *IEEE Transactions on Vehicular Technology*, *66*(1), 777–784. DOI: 10.1109/TVT.2016.2536059

Xu, J., & He, X. (2020). Design of double auction mechanism based on social network. *IEEE Access : Practical Innovations, Open Solutions*, *8*, 8324–8335. DOI: 10.1109/ACCESS.2020.2964654

Xu, M., Niyato, D., Kang, J., Xiong, Z., Miao, C., & Kim, D. I. (2022, May). Wireless edge-empowered metaverse: A learning-based incentive mechanism for virtual reality. In *ICC 2022-IEEE International conference on Communications* (pp. 5220–5225). IEEE. DOI: 10.1109/ICC45855.2022.9838736

Yanovskaya, E. B. (2010). The nucleolus and the τ-value of interval games. *Contributions to Game Theory and Management*, *3*(0), 421–430.

Yu, G., Liu, R., Chen, Q., & Tang, Z. (2018). A hierarchical SDN architecture for ultra-dense millimeter-wave cellular networks. *IEEE Communications Magazine*, *56*(6), 79–85. DOI: 10.1109/MCOM.2018.1700490

Zhang, N., & Zheng, Y. (2010, August). Application of Management Game Theory to the Practice of Human Resource Management. In *2010 International Conference on Management and Service Science* (pp. 1-3). IEEE. DOI: 10.1109/ICMSS.2010.5577233

About The Author

Sungwook Kim received BS and MS degrees in computer science from Sogang University, Seoul, Rep. of Korea in 1993 and 1995, respectively. In 2003, he received a PhD degree in computer science from Syracuse University, Syracuse, New York, supervised by Prof. Pramod K. Varshney. He has held faculty positions in the Department of Computer Science of ChoongAng University, Seoul, Rep. of Korea. In 2006, he returned to Sogang University, where he is currently a professor in the Department of Computer Science and Engineering, and he is a research director of the Network Research Laboratory. His research interests include resource management, online algorithms, adaptive quality-of-service control, and game theory for network design.

Index

Symbols

A

B

C

D

E

F

G

S

T

U

V

W

www.ingramcontent.com/pod-product-compliance
Lightning Source LLC
LaVergne TN
LVHW060251011026
849161LV00011B/2692